D0083245

Morphology and Evolution of Vascular Plants

A Series of Books in Biology

EDITORS
Donald Kennedy
Roderic B. Park

Third Edition

MORPHOLOGY AND EVOLUTION OF VASCULAR PLANTS

Ernest M. Gifford

University of California, Davis

Adriance S. Foster

University of California, Berkeley

W. H. FREEMAN AND COMPANY

New York

Dedicated to the late Professor Adriance S. Foster—
outstanding teacher, dedicated researcher, and true friend

Cover Image:
from G. C. Oeder's *Flora Danica,* vol. 4, Copenhagen, 1777.
Provided by the library of the New York Botanical Garden,
Bronx, New York.

Library of Congress Cataloging-in-Publication Data

Gifford, Ernest M.
Morphology and evolution of vascular plants/Ernest M. Gifford,
Adriance S. Foster.—3d ed.
p. cm.—(A Series of books in biology)
Rev. ed. of: Comparative morphology of vascular plants/by
Adriance S. Foster, Ernest M. Gifford. 2d ed. 1974.
Bibliography: p.
Includes index.
ISBN 0-7167-1946-0
1. Botany—Morphology. 2. Plants—Evolution. I. Foster,
Adriance S. (Adriance Sherwood), 1901–1973. II. Foster, Adriance S.
(Adriance Sherwood), 1901–1973. Comparative morphology of vascular
plants. III. Title. IV. Series: Series of books in biology (W. H.
Freeman and Company)
QK641.G46 1988
582′.04—dc19 88-28153
CIP

Printed in the United States of America

1 2 3 4 5 6 7 8 9 0 HL 7 6 5 4 3 2 1 0 8 9

Contents

Preface

THIS edition represents a thorough revision of the subject matter of *Comparative Morphology of Vascular Plants,* Second Edition, published in 1974.

During the past 14 years, some of the advances in vascular plant morphology have not only extended our general knowledge, but have either provided continuing support for certain theories or become the basis for the modification of others. For one example, the cytological and physiological evidence now favors the origin of land plants from some member(s) of the alga class Charophyceae rather than the Chlorophyceae. For another, the unique development of an antheridium of certain leptosporangiate ferns, described many years ago but contested in the intervening years, has been confirmed. And more and more evidence points to the importance of insect pollination in cycads and possibly in the fossil cycadeoids. Additional advances in our knowledge and techniques have resulted in

1 The establishment of a general phylogenetic classification for ferns based upon the cladistic approach
2 Research leading to new information on polyploidy and sexuality in lycopods and ferns
3 Confirmation of the phylogenetic origin of the ovuliferous scale in conifers
4 Additional information on the physiology of the bizarre African plant *Welwitschia mirabilis*
5 Proper recognition of similarities of embryo development in dicotyledons and monocotyledons
6 Paleobotanical discoveries of a variety of flower types from the Lower and mid-Cretaceous Period

Anyone familiar with the Second Edition will note a change in the system of classification of vascular plants. The composition of the basic groups remains the same; only the suffixes of the major taxa have been changed. In the Second Edition vascular plants were placed in the Tracheophyta, as a major division of the plant kingdom. In the present edition, the subgroups (classes) of the Tracheophyta are raised to divisional rank. The rationale for this change is discussed in Chapter 2. Some teachers may wish to continue using the taxon Tracheophyta. If so, the divisions recognized in this edition would then become classes.

As in the Second Edition, Chapters 1 through 6 have been updated and retained as useful information for orienting the reader on subjects such as classification, the appearance of plant groups in geologic time, general morphological characteristics of vascular plants and their life cycles, anatomy of the vegetative body, and development of reproductive structures.

Although many of the illustrations from the Second Edition have been retained, this revision has many new photographs and line drawings that are essential for the understanding of structures and morphological concepts. Many colleagues have graciously donated illustrations; they are mentioned in the figure captions or in the acknowledgment section.

As in almost every field of plant biology, advances in our knowledge have resulted in the appearance of numerous research publications on plant morphology. Not all the new discoveries are discussed at great length in this book. Literature citations at the end of each chapter provide the reader with a list of pertinent publications on topics discussed in the text. Chapter 21 of the second edition has been eliminated because much of the information it contained is found in other books.

I am pleased to pay tribute to my wife, Jean D. Gifford, for typing all newly added material to the revision, for her assistance in proofreading, and for her enduring patience during the entire project.

Acknowledgments

I assume responsibility for the content and viewpoints expressed in this book, but I would like to express my appreciation to many individuals for their assistance in numerous and diverse ways. The following individuals performed an invaluable service in reviewing all or portions of the original revised manuscript: Professors David E. Bilderback, David D. Cass, Dennis W. Stevenson, and Wilson N. Stewart. Many of their suggestions were incorporated in this edition. I appreciate the assistance of Professor Harlan P. Banks for his assistance with the geologic time-scale table.

I express my gratitude to a colleague, Professor James A. Doyle, for valuable discussions on pollen morphology and evolution. Special thanks are due to Professor Warren H. Wagner for the construction of a "phylogenetic tree" of ferns based on the cladistic method. I am grateful to Professor Richard L. Hauke for his critique of the morphological characteristics I have used to separate the subgenera in *Equisetum*.

The following individuals also have provided new illustrations or made corrections on existing ones upon my request and I am grateful for their cooperation: Charles B. Beck, David W. Bierhorst, David E. Bilderback, James Bruce, Elizabeth G. Cutter, Ted Delevoryas, William A. DiMichele, Donald A. Eggert, William E. Friedman, Alan L. Koller, Tom L. Phillips, Phillip M. Rury, Rudolf Schmid, Elizabeth Sheffield, Leslie Sunell, Thomas N. Taylor, Jerome Ward, Dean P. Whittier.

Finally, I would like to express my appreciation to the artist, Linda A. Vorobik, for the excellent preparation of new illustrations.

DAVIS, CALIFORNIA
SEPTEMBER 1988

Ernest M. Gifford

Morphology and Evolution of Vascular Plants

CHAPTER 1

The Science of Plant Morphology

EVERYONE, even the scientifically untrained observer, recognizes from experience the extraordinary diversity in the form, stature, and habit of plants. The "seaweeds" of the ocean, the lowly "mosses" and graceful "ferns" of the woodlands, the towering cone-bearing trees, and the infinitely varied flowering plants of orchard and garden all are recognized as different kinds of plants.

Casual inspection of the *surface aspects* of plants, however, is a highly unreliable method either for separating plants into natural groups or for gaining a proper understanding of the nature and relationships of their parts. In marked contrast with such undisciplined regard of form and structure, the science of plant morphology attempts, by rigorous techniques and meticulous observations, to probe beneath the surface aspects of plants—in short, to explore and to compare those *hidden aspects* of form, structure, and reproduction that constitute the basis for the interpretation of similarities and differences among plants. One of the most fruitful results of early morphologic studies was the recognition that a relatively few fundamental types of organs underly the construction of the plant body. Thus, the leaf, stem, and root were regarded as the principal types of vegetative organs, the size, form, proportions, and arrangement of which are subject to the most varied development or modification. As knowledge of the reproductive cycles of plants increased, sporangia and gametangia were added to this short list of major organ categories, and the importance of a broad comparative study of the resemblances, or homologies, of plant organs thus became established.

The Concept of Homology

Introduced by R. Owen in 1843, the term *homology* is derived from the Greek *homologia* which means "agreement" and is applied to corresponding organs and structures in plants and in animals. The homology of organs is based upon their structural similarities regardless of their present-day function. An example of homologous structures would be the wings of a bird and the forelegs of a reptile. On the other hand, *analogous* structures and organs may perform the same function but are not derived from the same prototype, e.g., wings of a bee and wings of a bird. The identification of homologous structures remains, even today, unresolved, although research continues to provide insights (de Beer, 1971; Stebbins, 1974).

The essence of the idea of homology was expressed by the great poet and philosopher Johann Wolfgang von Goethe, to whom we owe the word "morphology" (literally, the science of form). Goethe sought for the nature of the morphological relationships among the various kinds of leafy appendages in higher plants. In his celebrated essay, *Metamorphosis in Plants,* published in 1790, he concluded that no real boundary exists between such organs as cotyledons, foliage leaves, bracts, and the organs of the flower—all are expressions of the same type of organ, i.e., the leaf. To quote from R. H. Eyde (1975, p. 431): "Goethe explained that a term is needed to cover all manifestations of the metamorphosed organ: hence he has adopted the word *leaf.*" He believed that "all plant appendages are variants of an intuitively perceived ideal appendage, the primal leaf, which somehow contains all its own transformations." Although Goethe's theory has been criticized as an example of idealistic, even metaphysical morphology, it has proved an extremely astute viewpoint and indeed constitutes the theoretical basis for the current view that the flower is a determinate axis with modified foliar appendages (see Chapter 19).

With the rapid expansion of botanical knowledge in the nineteenth century came an emphasis on the importance of the concept of homology and the need for interpreting homologies in the broadest possible light. Goethe's ideas, and the earlier observations of C. F. Wolff (1774) on the origin of leaves at the growing point of the shoot, paved the way to a better understanding of *serial homology* in plants. With reference to a shoot, this term designates the equivalence in *method of origin* and *positional relationships* of the successive foliar appendages of a shoot. Thus, a bud scale, or a floral bract, is considered serially homologous with a foliage leaf because, like the latter, it arises as a lateral outgrowth from the shoot apex. Ontogenetic studies have shown the very close resemblances in detail of origin and early histogenesis among the varied types of foliar organs of both vegetative and flowering shoots. Moreover, the different types of foliar appendages in the same plant are often intercon-

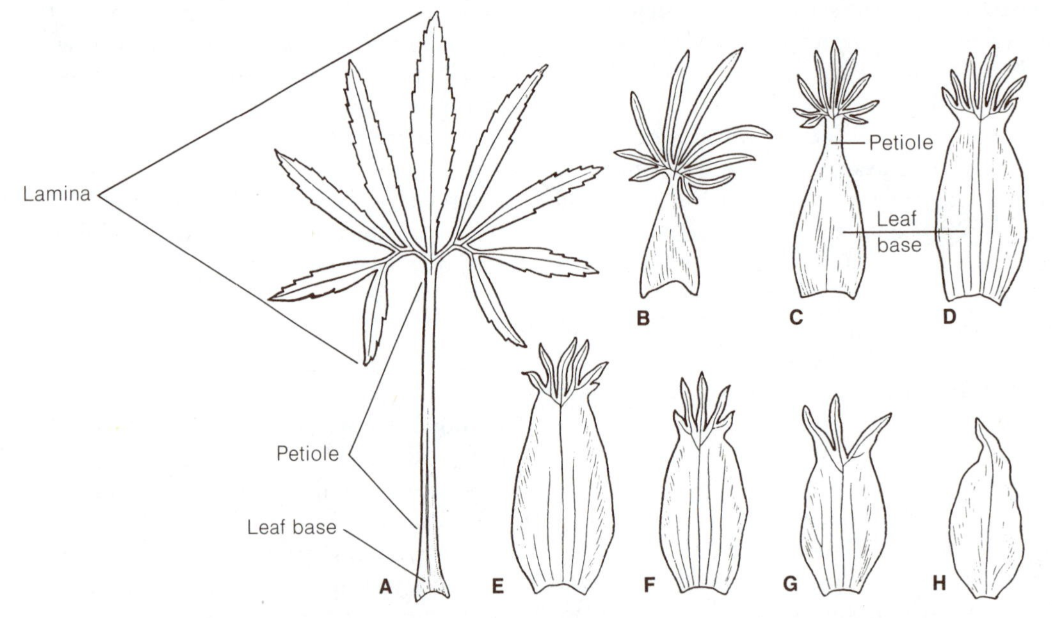

FIGURE 1-1 Serial homology between the foliage leaf (**A**) and the succession of floral bracts (**B–H**) in *Helleborus foetidus.* Note that the gradual suppression of a petiolar zone (bracts, **B–D**) and the progressive reduction of the lamina (bracts, **E–H**) are accompanied by a corresponding increase in the prominence of the leaf base. [Redrawn from *Vergleichende Morphologie der höheren Pflanzen* by W. Troll. Gebrüder Borntraeger, Berlin. 1935.]

nected by intermediate forms or transitional organs (Fig. 1-1).

The concept of *general homology* in plants is much more difficult to demonstrate ontogenetically (see Mason, 1957, for a critical discussion). This is so because, unlike higher animals, plants are characterized by an open system of growth — a plant embryo is not a miniature of the adult, and hence homologies based on the resemblance in position, development, and form of two organs in different kinds of plants may be open to serious question.

The concept of homologies in plants was placed in an entirely new position as the result of the publication in 1859 of Charles Darwin's classic, *The Origin of Species.* His theory of the role of natural selection in producing the gradual adaptive changes in the form and organography of both plants and animals exerted a profound effect on all questions of homologies. The goal of morphology now became very clear: the interpretation of form and structure from a historic (i.e., phylogenetic) point of view. Resemblances or homologies between organs were to be viewed as the result of descent from a common ancestral "type." Thus, the strong trend toward the phylogenetic interpretation of form and structure that arose during the latter part of the past century has continued to this day. In addition to its effect on all concepts of homology, the phylogenetic approach to morphology has provided the basis for a more realistic and natural classification of the plant kingdom.

Although many of the widely recognized similarities in basic structure between the organs and tissues of closely *related plants* clearly seem to be "homogenetic," i.e., because of the plants' origin from a common ancestor, there are remarkable structural resemblances between systematically *unrelated species* or groups of plants. In the latter case, the morphological correspondence is "homoplastic" and the result of convergent evolution. A possible example of convergent evolution in vascular plants is provided by the presence of seeds in such widely divergent groups as the extinct seed ferns and the Cycadeoidales (fossil cycadlike plants of the Mesozoic Era), the modern conifers and flowering plants.

Additional interesting examples of homoplastic developments in vascular plants are described and analyzed by Wardlaw (1952, 1955, 1965, 1968a, 1968b), who regards them as *homologies of organization* and worthy of intensive study (Fig. 1-2). It must be emphasized, however, that in actual practice it proves very difficult to separate homogenetic from homoplastic resemblances. In this connection it must be constantly borne in mind that structural resemblances — whether they are interpreted as homogenetic or homoplastic — are the result of the interaction of genetic, physiological, and environmental factors which have been operating in *different ways* and to *different degrees* during the long evolutionary history of vascular plants. Wardlaw (1965) has admirably summarized the task as follows: "To obtain a balanced view, it is therefore necessary to enquire to what extent prevalent homologies of organization can be accounted for in terms of genetical factors, on the one hand, and of what, for convenience, may be described as com-

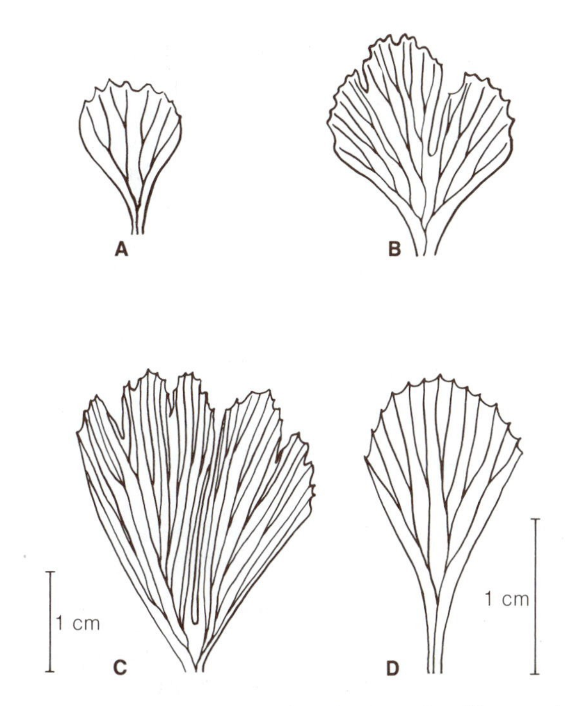

FIGURE 1-2 "Homology of organization" as illustrated by similar venation patterns in the leaves of unrelated plants. **A, B,** juvenile, dichotomously veined leaves of *Anemia adiantifolia,* a species of fern; **C,** dichotomous venation of a lamina segment of *Kingdonia uniflora,* a dicotyledon; **D,** open dichotomous venation in the lamina of *Circaeaster agrestis,* a dicotyledon. [Adapted from *Organization and Evolution in Plants* by C. W. Wardlaw. Longmans, Green, London. 1965.]

mon, intrinsic, or not specifically genetical, factors, on the other."

Clearly, reliable interpretations require consideration of evidence that is derived from a wide variety of sources. Morphological theories increase in plausibility in relation to the extent to which collateral lines of evidence can be harmonized with one another. The rest of this chapter, therefore, contains a brief, critical review of the sources of evidence that should be considered and evaluated in interpreting any problems of form and structure in plants.

Sources of Evidence in Morphological Interpretation

Adult Form and Structure

By far, the most voluminous data of comparative morphology have resulted from the study of the form of the adult plant with fully developed tissues and organs. Information derived from such study has contributed significantly to our knowledge of the wide variations in: (1) the form, venation, and phyllotaxy (arrangement on the axis) of foliar organs, (2) the patterns of branching of root and shoot systems, and (3) the morphological construction of such spore-producing structures as sporophylls, strobili, and flowers. During the second half of the nineteenth century, increasing emphasis was placed upon the study of the primary vascular system of the plant as the key to the interpretation of the morphological nature or homology of plant organs. The wide and continued use today of vascular patterns in morphology is based upon the fundamental assumption that the vascular system is more stable, or conservative, in a phylogenetic sense, than other tissue systems. (See Schmid, 1972, for a critique of the concept of vascular conservatism.) Considerable support for this assumption is provided not only from comparative study of living plants but also by the beautifully preserved patterns of vasculation in the vegetative and reproductive structures of extinct plants. In addition to the emphasis on primary vascular systems, much attention has also been given to extensive surveys of the minute structure or histology of secondary xylem, or wood (see Chapter 19). The results of such surveys have been applied in the appraisal of the taxonomic aspects of genera and families in the seed plants, and particularly in the effort to determine the origin and trends of evolutionary specialization of tracheids and vessels (Metcalfe and Chalk, 1950; Bailey, 1954).

The Fossil Record

A salient problem common to all phylogenetic interpretations is the need for determining the sequence in the evolutionary development of organs, tissues, and cells. Unfortunately, however, the known fossil record, as revealed by paleobotanical studies, is extremely fragmentary. Consequently, phylogenetic theories are still based largely on circumstantial or indirect evidence derived from the comparative study of living plants. The history of plant morphology is replete with examples of how the same series of morphological types has been interpreted by some investigators as a sequence of *advancing complexity,* and by others as a series in *progressive reduction.* Therefore it is clear that inferences regarding the phylogeny of an organ must be based upon the wise evaluation of the evidence from extinct as well as living types of plants. New paleobotanical discoveries will continue to force morphologists to reconsider and revise many of the so-called classic viewpoints that were based solely on living plants. The reader is referred to the books by Taylor (1981) and Stewart (1983) for up-to-date discussions of advances in paleobotany.

To determine the age of rocks containing fossils it is necessary to establish geologic chronology by converting the relative time scale of geologic events (and the presence of fossils) into a quantitative scale having standard units of time. The procedures used have been termed the "hourglass" methods. In the middle of the last century the hourglass methods were based on the salinity of the oceans, sedimentation rates, and thickness of sediments, and these measurements provided a framework for a rough time scale, measured in years.

The historian and astronomer describe events in absolute units of time, and it is now possible for geologists and paleobotanists to develop a satisfactory quantitative geologic time scale (Table 1-1). This is an achievement of the present century and is based upon the radioactivity of isotopes of certain elements. The age of many minerals and rocks can now be determined within fairly narrow limits by radiometric methods. The dates of extremely ancient rocks and strata can be established by analyzing the decay of uranium and thorium, for example, to lead and helium, or potassium (^{40}K) to argon (^{40}Ar) and rubidium (^{87}Rb) to strontium (^{87}Sr). Uranium-238 (^{238}U) disintegrates radioactively in a series of steps to a lead isotope (^{206}Pb), and half of a given amount of ^{238}U will disintegrate in 4.5 billion years. If, in a sample of a uranium ore, all ^{206}Pb has been formed by the process of radioactive disintegration since the original uranium-containing mineral was formed, the ratio $^{206}Pb/^{238}U$ related to the half-life of ^{238}U is a measure of this time. By use of radiometric methods with appropriate refinements, the age of the earth has been estimated to be 4.5 to 4.6 billion years.

Although these methods, using the isotopes mentioned, have proved to be essential in dating past eras and periods, they are of only limited value in the dating of recent strata because of the extremely long half-life of, for example, ^{238}U.

The carbon-14 (^{14}C) method has proved to be extremely useful in dating biological specimens back to about 60,000 years. Cosmic rays slam into the earth's atmosphere at high velocities and produce nuclear particles, including some neutrons. Most of the neutrons are absorbed by nitrogen, changing it into the radioactive isotope carbon-14 (^{14}C). The ^{14}C then becomes part of the world's reservoir of carbon. The ^{14}C isotope decays with a half-life of about 5,700 years back to nitrogen-14 (^{14}N). Once an organism dies there is no longer an exchange with atmospheric radiocarbon, and the ^{14}C "clock" starts. The radiocarbon (^{14}C) begins decaying (reverting) to ^{14}N. The ratio of the amount of carbon-14 in a fossil sample to the amount of ^{14}C in modern wood or tissue indicates how long ago the organism died. There are several sources of error in this method but, when due precautions are taken, much useful dating information can be obtained for fossil remains and artifacts of prehistoric cultures.

Fossils themselves have been used as indicators of the age of rocks in which they occur. This is based on the observation that certain strata around the world contain characteristic types of plants or animals *(index fossils)*. Organisms that had wide geographic distribution, that are easily identifiable, and lived for only a short period in geologic time before becoming extinct, constitute the best index fossils. The presence of index fossils in rocks of unknown age, together with certain geologic criteria, makes it possible to determine the age of the rocks by comparing them with rocks from regions where the age of the specific fossil-containing rocks are well known.

In concluding this brief discussion of the use and limitations of the *known* fossil record in evaluating phylogenetic theories, it seems desirable to outline some of the principal modes of preservation of the organs and tissues of extinct plants.

One of the most common types of preservation of fossil material is termed a *compression.* This results when an organ such as a leaf is flattened by the weight of the sediment deposited upon it. Water is squeezed out, and often only a thin carbonaceous film is left, but the outline of the leaf remains as an imprint on the sedimentary rock. In some instances, however, enough preserved tissue remains so that even ultrastructural cellular details can be observed with the transmission electron microscope (Niklas et al., 1978). Although little or no internal cellular structure generally remains in a typical compression, the cuticle (i.e., the external layer of waxy material which covered the epidermis) is extremely resistant to decay and commonly displays, *in relief,* the pattern of arrangement and the form of the original epidermal cells and stomata. The organization of stomata varies according to family, genus, or even species of plants, and hence it proves of great value in the identification of the leaves and other organs of plants of the remote past (Florin, 1931, 1951).

Imprints of plant parts, devoid of organic material, which are formed when the sediment separates from the surface of a fossil, are called *impressions.* As Delevoryas (1962) has indicated, an impression actually is the "negative" of a compression and at best may only show the contour and general venation pattern of such a structure as a leaf. The formation of an impression is analogous to the creation of a leaf imprint in a concrete sidewalk when a leaf falls onto or is pressed into the surface of wet concrete. The concrete hardens and the imprint remains, revealing the contour and often the venation pattern of the leaf blade.

Another type of fossil impression, useful in reconstructing the form and three-dimensional aspects of the parts (e.g., the stems) of extinct plants is termed a *cast.* In some cases, a cast is the result of

Table 1-1 Geologic time scale.

Era		Period and Epoch (Beginning of time period in millions of years ago)	Characteristic Plants and Events	Representative Geologic Events and Animals
CENOZOIC	QUATERNARY	Recent 0.01		
		Pleistocene 1.9	Extinction of many temperate tree genera in Europe and North America.	Retreats and advances of vegetation coinciding with several advances and retreats of major continental ice sheets; rise of modern man; wooly mammoths.
	TERTIARY	Pleiocene 5.1	Spread of grasslands; local extinction or restriction of range of many species because of climatic change in temperate latitudes.	Elevation of Andes and continued continental uplift in many regions of the world, e.g., Sierra Nevada and Rocky Mountains in the United States; camels, horses; hominids.
		Miocene 25	Establishment of present-day forests; climatic cooling and restriction of broad-leaved evergreens to lower latitudes; prairie grasses.	Alps commence to rise; continental uplifts elsewhere.
		Oligocene 38	Appearance of temperate forests at middle latitudes.	Dry climates in southwestern North America; rodents, cats, dogs, rhinoceroses.
		Eocene 55	Widespread occurrence of now relic taxa: *Metasequoia, Cercidiphyllum;* many now extinct woody angiosperms, though generally assignable to modern families.	Continued inundation of Gulf and Atlantic states in the United States; primitive horses.
		Paleocene 65		
MESOZOIC		Cretaceous 144	Angiosperms rise to dominance in Upper Cretaceous, but present in Lower Cretaceous; a few modern genera by end of Cretaceous and an increase in angiosperm families.	Uniformity of climate and widespread inundation of the continents; elevation of Rocky Mountains; last of dinosaurs.
		Jurassic 213	Conifers worldwide in distribution; "age of cycads and cycadeoids"; greatest cosmopolitanism of floras.	Uniform, mild climates; first birds; dinosaurs abundant.
		Triassic 248	Rise of cycadophytes and Ginkgoales; diversification of conifers and ferns.	Beginning of the breakup of Pangea; continued widespread semiarid-type climates; rise of dinosaurs; first mammals.

PALEOZOIC		Permian 286	Rise of Voltziales and Coniferales; extinction of most Carboniferous groups except for derivative forms: herbaceous lycopods, horsetails; glossopterids in Southern Hemisphere.	Greatest consolidation of continents into one super land mass (Pangea); semiarid climate; glaciation in Southern Hemisphere; diversification of reptiles.
PALEOZOIC	CARBONIFEROUS	Pennsylvanian 320	Widespread forest trees and coal swamps; seed ferns, sphenopsids, calamites, arborescent lycopods, ferns, cordaites, mosses.	Glaciation in Southern Hemisphere; collision of Laurasia and Gondwanaland land masses; widespread epicontinental seas; wet, tropical climate; formation of great coal beds; origin of reptiles, insects abundant.
PALEOZOIC	CARBONIFEROUS	Mississippian 360	Expansion of seed ferns, arborescent lycopods, sphenopsids, calamites, ferns.	Widespread epicontinental seas and limestone deposition in the interior of the United States; reefs rich in crinoids; spread of amphibians and sharks; insects evolved wings.
PALEOZOIC		Devonian 408	Early vascular plants: *Rhynia, Horneophyton, Zosterophyllum, Asteroxylon, Psilophyton;* primitive sphenopsids and ferns; liverworts; progymnosperms; first seed plants by end of Devonian.	Collision of North America and Europe; abundance of fish in seas and fresh water; gastropods (ammonoids); decline of trilobites; first amphibians.
PALEOZOIC		Silurian 438	Simple vascular land plants *(Cooksonia)* in mid-Silurian.	Brachiopods, bryozoans, echinoderms, arthropods (eurypterids), corals; jawed fish; first insects (wingless).
PALEOZOIC		Ordovician 505	Algae abundant; spores of possible first land plants.	Early jawless and armored fish; brachiopods and bryozoans abundant; corals; graptolites.
PALEOZOIC		Cambrian 590	Multicellular marine algae abundant; evidence of fungi.	Trilobites abundant; other marine invertebrates representing most of the major groups continued to appear as the period progressed.
		Precambrian ~4,600	Definite evidence of procaryotic algal and bacterial life by 3.5 billion years ago; unicellular, eucaryotic algae by the end of era.	Various marine protozoa; worms, jellyfish by end of era.

Source: Time divisions based on *A Geologic Time Scale* by W. B. Harland et al. Cambridge University Press, Cambridge, 1982.

the deposition of sediment in the cavity formed by the decay of some or all of the central tissues of a trunk or stem of a plant. Even though no organic material remains, casts of this type may show markings which are the counter-parts of the inner surface of the wood.

The most useful type of preservation, termed a *petrifaction or permineralization,* reveals, often in exquisite detail, cellular organization. Petrifactions are formed by the infiltration and subsequent crystallization, in the lumina of the cells and in the intercellular spaces in plant tissues, of calcium carbonate, magnesium carbonate, or iron carbonate, and of silica or other mineral substances. Petrifactions can be prepared for examination by cutting out and mounting a small portion of the fossil on a slide and then grinding and polishing the surface until the specimen is thin enough to be studied under the microscope with transmitted light. A less wasteful technique, which also makes it possible to obtain serial sections, has been used very successfully for petrifactions in which the carbonized plant material is embedded in a matrix of calcium or magnesium carbonate. The polished surface of the specimen is first treated with dilute hydrochloric acid, which removes the mineral matrix and leaves the cell walls of the tissue projecting above the rock surface; if the matrix is siliceous, hydrofluoric rather than hydrochloric acid must be employed. After the acid has been washed away, the surface of the specimen is flooded with acetone. Then a thin sheet of cellulose acetate is laid on the surface, allowed to dry, and then peeled off. The "peel" contains a thin section of the organic remains of the fossil. This process may be repeated as many times as the specimen allows and in accordance with the number of serial sections desired. Each peel can be examined directly or mounted permanently on a slide for detailed microscopic study. (For more detailed information on types of plant fossils and the techniques for studying them, see Delevoryas, 1962; Schopf, 1975; Taylor, 1981; and Stewart, 1983.)

Ontogeny

A highly important source of evidence for morphological interpretation is derived from the study of ontogeny — the actual development of a plant or of one of its component organs, tissues, or cells from the primordial stage to maturity. Histogenesis is a phase of ontogenetic study concerned with the origin of cells and tissues, and embryogenesis and organogenesis are concerned with the history of development of embryos and organs, respectively. It should be emphasized, however, that the boundaries between these lines of ontogenetic study are drawn largely as a matter of convenience in dealing with varied aspects of development characteristic of the plant as a whole.

The importance of detailed ontogenetic information is clearly shown in the interpretation of the life cycles of vascular plants. Since the classic studies of Hofmeister (1862) on *Alternation of Generations,* it has been repeatedly verified that each of the two generations (sporophyte and gametophyte) begins ontogenetically as a single cell. The spore, which results from meiosis, is the primordial cell which develops into the gametophyte, whereas the zygote, resulting from gametic union or fertilization, is the starting point of the sporophyte (see Fig. 2-1).

Ontogenetic studies have also proved essential to the solution of many special morphological problems. For example, the distinction between the two main types of sporangia in vascular plants is based primarily upon differences in their method of origin and early development (Chapter 4).

Despite the demonstrated value of ontogenetic evidence, there are certain limitations to the ontogenetic interpretation of morphological problems which must be clearly understood. One of the most important of these limitations concerns the assumed occurrence of recapitulation in plants. According to the theory of recapitulation, the ontogeny of an organism tends — in abbreviated fashion — to repeat or recapitulate its evolutionary history. A frequently cited example of so-called recapitulation in plants is the development in seedlings of juvenile leaves that differ conspicuously in size, form, and venation from the foliage characteristic of the adult phase of the same plant. However, whether these juvenile leaves provide a reliable clue to the morphology of the ancestral leaf type in any given case is open to serious question. In general, ontogenetic sequences, whether in the succession of foliar types in the young plant or the stages in the ontogeny of an organ or tissue, do not fully or accurately depict the complex path of evolutionary history, and recapitulatory hypotheses should always be made with great caution. This is true because ontogenetic sequences vary so widely in their extent and character. In some instances the ontogeny of a structure may be relatively protracted, thus permitting the phylogenetic evaluation of a well-defined series of stages. Vessel elements, for example, acquire their characteristic perforations late in ontogeny, and their early development may closely resemble that of tracheids from which they undoubtedly have phylogenetically evolved. Very commonly, however, the ontogenetic history of a structure may be conspicuously abbreviated or telescoped and therefore may be of little or no value for phylogenetic questions.

Thus, it is well to realize the important interrelationship between the processes of ontogeny and phylogeny (Mason, 1957). Evolution or phylogeny involves historic change, but from our present point of view "change" is effected by factors that cause the gradual or abrupt modification of ontogenetic processes. For example, a mature tracheid does not give rise to a vessel element, nor does a simple leaf give rise to a complex leaf. As Bailey (1944) has so clearly said: "Both comparative and developmental morphology will be more productive of valid generalizations when problems of mutual interest are attacked from a broadened viewpoint of the phylogeny of modified ontogenies in the vascular plants as a whole."

Physiology and Morphogenesis

If plant morphology is regarded as a phase of botanical science exclusively concerned with the description and phylogenetic interpretation of form and structure , it might seem to have little or nothing to do with those dynamic activities of plants that fall within the designation of plant physiology.

We have attempted, in this chapter, to show that comparative morphology is fundamentally concerned with solving the complex problems of the *evolutionary relationships* among plants. In contrast, the field of *morphogenesis* is preoccupied with the *totality of factors,* genetic, biochemical, physiological, and environmental, that *together* are responsible for the inception and development of form in plants. One of the many goals of morphogenesis is a better understanding of the "homologies of organization"—i.e., those structural resemblances, between unrelated organisms, that presumably have been produced by parallel and independent paths of evolution. In this connection, morphogenesis is more than simply the precise observation and description of the successive developmental stages of homogenetic or homoplastic structures, but is more comprehensive and clearly indicated by the expression "causal morphology" —i.e., the *experimental study* of the genesis of form.

In the future, morphogenesis may begin to give answers to some of the following fundamental types of questions. (1) Why is the process of the initiation and differentiation of organs and tissues characterized by such an orderly and *integrated series* of developmental stages? Each step in the development of an embryo, for example, is *epigenetic,* i.e., dependent upon the preceding stage, and yet a growing embryo is an integrated "organism" at each and every phase of its development. (2) How and why do structurally similar primordia, which arise successively at the shoot apex, develop into such diverse appendages as bud scales, foliage leaves, and floral organs? (3) And what factors control or "determine" the divergent paths of development of cells that arise from a common meristem?

There is a great need, in the present age of excessive specialization in botany, for a broad multidisciplinary approach to the science of plant morphology. The danger of a limited approach to morphological problems was recognized by Wardlaw (1968a); he explains the dilemma in the following passage from his treatise on plant morphogenesis.

> All too often, the morphologist stops just at the point where the more searching study of the underlying physiological-genetical factors really begins; nevertheless, he performs a vital service: he indicates what is there to be investigated. But one can also think of excellently conceived physiological studies that have obviously been undertaken without an adequate knowledge of the *observable* morphogenetic developments. Somewhere between these extremes we ought to do better!

REFERENCES

Bailey, I. W.
1944. The development of vessels in angiosperms and its significance in morphological research. *Amer. Jour. Bot.* 31:421–428.
1954. *Contributions to Plant Anatomy.* Chronica Botanica Co., Waltham, Mass.

Cutter, E. G.
1965. Recent experimental studies of the shoot apex and shoot morphogenesis. *Bot. Rev.* 31:7–113.
1966. *Trends in Plant Morphogenesis.* Wiley, New York.

Darwin, C.
1859. *The Origin of Species.* J. Murray, London.

De Beer, Gavin.
1971. Homology, an Unsolved Problem. *Oxford Biology Readers,* 11. Oxford University Press, London.

Delevoryas, T.
1962. *Morphology and Evolution of Fossil Plants.* Holt, Rinehart and Winston, New York.

Eyde, R. H.
1975. The foliar theory of the flower. *Amer. Scientist* 63:430–437.

Florin, R.
1931. Untersuchungen zer Stammesgeschichte der Coniferales und Cordaitales. *Svenska Vetensk. Akad. Handl.* Ser. 5. 10:1–588
1951. Evolution in Cordaites and Conifers. *Acta Horti Bergiani.* Bd. 15. No. 11.

Goethe, J. W., Von
1790. *Versuch die Metamorphose der Pflanzen zu erklären.* Gotha.

Hofmeister, W.
1862. *On the Germination, Development and Fructification of the Higher Cryptogamia and on the Fructification of the Coniferae.* Published for the Ray Society by Robert Hardwicke, London.

Mason, H. L.
1957. The concept of the flower and the theory of homology. *Madroño* 14:81–95.

Metcalfe, C. R., and L. Chalk
1950. *Anatomy of the Dicotyledons.* 2 v. Clarendon Press, Oxford.

Niklas, K. J., R. M. Brown, Jr., R. Santos, and B. Vian.
1978. Ultrastructure and cytochemistry of Miocene angiosperm leaf tissues. Proc. Nat. Acad. Sci. 75:3263–3267.

Schmid, R.
1972. Floral bundle fusion and vascular conservatism. *Taxon* 21:429–446.

Schopf, J. M.
1975. Modes of fossil preservation. *Rev. Palaeobot. Palynol.* 20:27–53.

Stebbins, G. L.
1974. *Flowering Plants. Evolution above the Species Level.* Harvard University Press, Cambridge, Mass.

Stewart, W. N.
1983. *Paleobotany and The Evolution of Plants.* Cambridge University Press. Cambridge.

Taylor, T. N.
1981. *Paleobotany. An Introduction to Fossil Plant Biology.* McGraw-Hill, New York.

Wardlaw, C. W.
1952. *Phylogeny and Morphogenesis.* Macmillan, London.
1955. *Embryogenesis in Plants.* Methuen, London.
1965. *Organization and Evolution in Plants.* Longmans, Green, London.
1968a. *Morphogenesis in Plants. A Contemporary Study.* Methuen, London.
1968b. *Essays on Form in Plants.* University Press, Manchester.

Wolff, C. F.
1774. *Theoria Generationis,* Editio nova. Halle.

CHAPTER 2

Early Land Plants and the Salient Features of Vascular Plants

ONE of the most significant events in the long evolutionary development of the plant kingdom was the origin of the first land plants. These organisms, which in all probability arose from some ancient group of green algae, were extremely simple in organography. For example, the ancient plant *Rhynia gwynne-vaughanii* was a rootless, leafless plant that was low in stature, with a simple and primitive vascular system and simple reproductive structures located at the tips of the aerial branches.

A land flora was possible only after the detrimental effects of solar radiation, the ultraviolet (UV) radiation, became shielded by an ozone layer in the atmosphere. During the exceedingly long Precambrian and Cambrian Eras (Chapter 1, Table 1-1) little oxygen was evolved from photosynthesis, and hence little ozone was produced. Aquatic algae would have been shielded from the UV effects by immersion. With an increasing density of algae, more oxygen would have been formed and an ozone layer created.

In addition to formation of the protective ozone layer, there was the evolution of biochemical conversion, by plants living in the littoral zone, of certain primary metabolites (amino acids and sugars) to secondary products (alkaloids, anthocyanins, tannin, lignin, and other phenolic substances) which could provide some protection from UV, parasitic fungi, protozoa, and other predators.

For vascular plant evolution the formation of the polymer lignin is of paramount importance. This substance provides the strength to keep a plant upright so that it can present a maximum photosynthetic surface. It made possible the dominance of vascular plants on earth, thereby permitting the evolution of animals. Lignin is a tough, hard substance that fills the microcapillaries between cellulose microfibrils in the cell walls of water-conducting cells (tracheary elements) and fibers; it is not present in bryophytes (Niklas and Pratt, 1980).

Whether the bryophytes (liverworts and mosses) and vascular plants were derived from an aquatic multicellular ancestral algal group(s) which then migrated to land, or originated from algae already present in terrestrial habitats, may never be resolved. This subject will be discussed more fully later in this chapter in the section on Alternation of Generations. However, there are certain features that land plants developed during millions of years for successful existence on land. Some of these modifications are shared by bryophytes and vascular plants, whereas others are exclusive to vascular plants and have made possible their eminent success. The following features pertain mainly to vascular plants:

1 Development of an anchorage and water- (and dissolved minerals) absorbing system, such as underground stems (rhizomes) with rhizoids (extensions of epidermal cells), performing the function of roots. True roots apparently developed later.
2 Development of a water-and-mineral–conducting system (xylem), and a system (phloem) for the conduction of photosynthates. The xylem would also provide support by the incorporation of lignin in walls of cells.
3 Prevention of desiccation by the formation of a waxy layer (cuticle) on aerial branch systems.
4 Development of structures for aerial gas exchange (stomata).
5 Development of specialized photosynthetic tissue in aerial stems; leaves, as we know them, were undoubtedly formed later in evolution.
6 Production of spores with the cell wall impregnated with sporopollenin, a substance that prevents desiccation and is virtually indestructible by microorganisms.

Except for the previously existent photosynthetic system in algae, all of the above adaptive modifications developed over an exceedingly long time. (For expanded discussions of this subject, the reader is referred to Niklas, 1976, 1981; Lowry, Lee, and Hébant, 1980; Stebbins and Hill, 1980; Niklas and Pratt, 1980.)

In this chapter an effort will be made to select and to describe briefly the salient morphological features that are common to most living vascular plants. Such an orientation is necessary for the more detailed treatment of comparative morphology presented in subsequent chapters. For convenience in exposition, the definitive features of sporophyte and gametophyte generations will be described separately.

Sporophyte Generation

The normal origin of this generation is from the zygote, a diploid cell which results from the fertilization of the egg by a male gamete (a motile sperm in ferns, for example) (Fig. 2-1). There is considerable variation in vascular plants with respect to the length of the period of attachment and physiological dependency of the young multicellular sporophyte upon the gametophyte generation. Ultimately, the sporophyte gains physiological independence (except for certain parasitic plants) and develops into the dominant, typically photosynthetic, phase of the life cycle. *This physiological independence and dominance of the sporophyte constitutes one of the most definitive characters of vascular plants.*

From an organographic viewpoint, the sporophyte typically consists of a shoot system (usually aerial) made up of stems, various types of foliar organs, and roots. The latter may vary widely in their origin during embryogeny and later phases of growth. Shoots and roots are theoretically capable of unlimited apical growth, branching, and organ formation because of the maintenance at their tips of apical meristems. This capacity for continued indefinite apical growth is not found in the small dependent sporophytes of mosses and liverworts and hence represents a fundamentally important distinction between these organisms and vascular plants.

From an anatomical viewpoint, one of the most definitive features of the sporophyte is the presence of a vascular system. The form and arrangement of the vascular system vary not only among different groups of vascular plants but among different organs of the same plant as well. Because of its remarkably well-preserved condition in fossils, the vascular system often provides important clues to relationships and to trends of phylogenetic development. Except for the vein endings, which in leaves or floral organs consist largely of tracheary

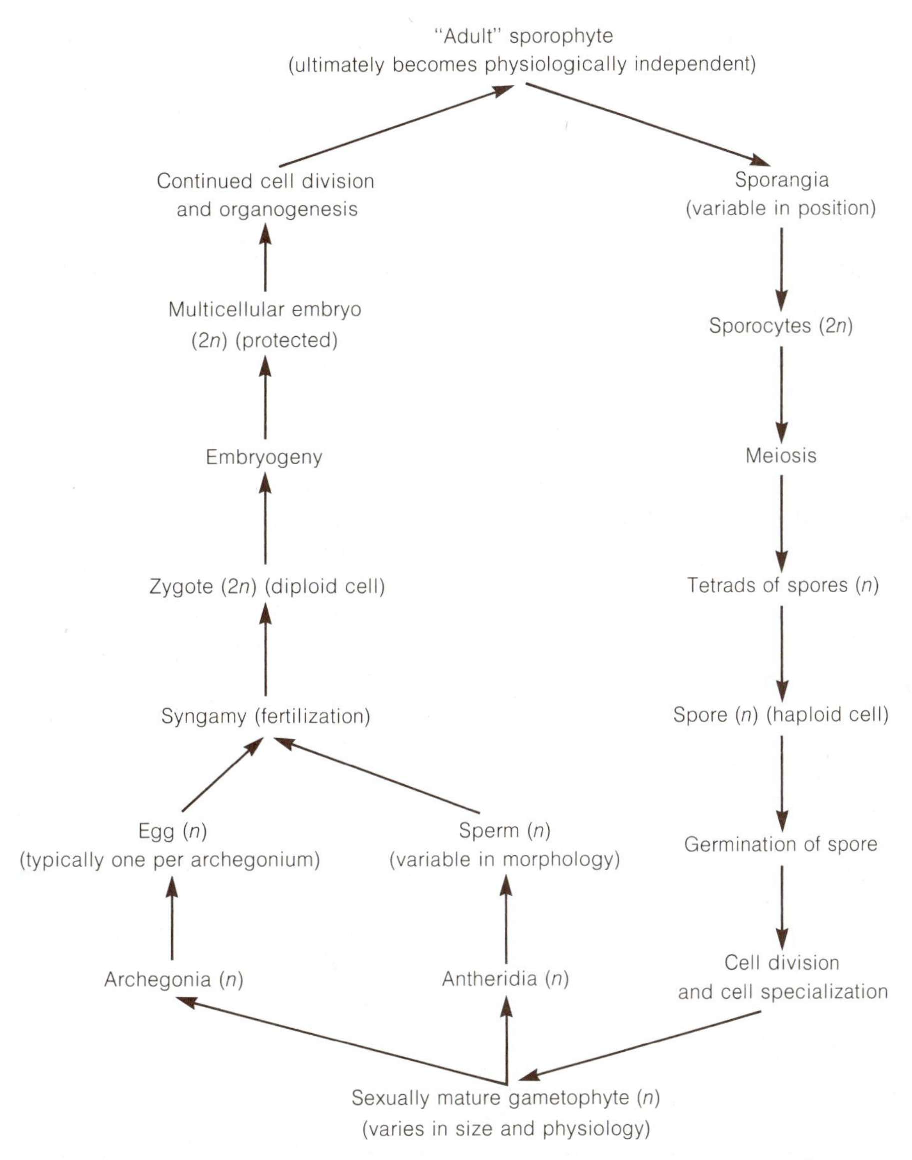

FIGURE 2-1 Alternation of sporophyte and gametophyte generations in the complete life cycle of lower vascular plants. The processes and structures represented are discussed in detail in the text.

cells, the vascular system consists typically of two distinctive tissues: *phloem,* with conducting sieve elements, and *xylem,* with tracheary elements. Vascular tissues are essential to the conduction of water, mineral solutes, and soluble organic compounds — upon their coordinated functioning depends the normal physiology and growth of the sporophyte. Evidently, in the early occupation of the land by plants, the rapid evolution of an effective vascular system must have been of primary importance.

The characteristic reproductive structure of the sporophyte is the sporangium. As will be shown in detail in Chapter 4, sporangia differ significantly in their position, methods of origin and development, and mature structure including sporangium wall and spore type. These differences provide reliable criteria, in many instances, for classifying vascular plants.

Gametophyte Generation

The normal origin of the gametophyte generation is from a spore, which, in turn, is a product of the meiotic divisions of sporocytes in the sporangium (see Fig. 2-1). The gametophyte may be a free-living photosynthetic plant or subterranean and apparently dependent upon the presence of an endophytic fungus for its existence. In marked contrast, the gametophytes of angiosperms are much smaller and are physiologically dependent upon the sporophyte generation. Aside from these morphological and physiological variations, the chief importance of the gametophyte is the production of male gametes (or sperms) and female gametes (or eggs). These gametes are developed in distinctive multicellular gametangia in all lower vascular plants, the sperms arising in antheridia and the solitary egg developing within the archegonium. In angiosperms, the male and female gametes are produced directly by the greatly reduced and modified gametophytes, and morphologically definable sex organs are not developed. The position, ontogeny, and structure of gametangia and gametes furnish important characters used in the comparison and classification of the groups of lower vascular plants. A full discussion of the comparative morphology of gametangia is presented in Chapter 5.

Although typical vascular tissue is usually restricted to the sporophyte generation, tracheary tissue has been observed in the gametophytes of certain plants. The best known example is the subterranean gametophyte of *Psilotum,* a vascular plant of uncertain phylogenetic origin and systematic relationships (see Chapter 8). Gametophytes, 1 millimeter or more in diameter, may have a central conducting strand, demarcated by an endodermis. A well-developed strand consists (in transection) of one to three tracheids with annular or scalariform thickenings, surrounded by elongated cells appearing to be the exact counterparts of the sporophytic sieve elements (Hébant, 1976). The vascular strand arises at the apical meristem of the gametophyte but commonly is discontinuous, fading out and reappearing usually several times in the length of a few millimeters.

Several "explanations" for the occurrence of a vascular strand in *Psilotum* gametophytes were made by Holloway (1939) and deserve brief examination especially in the light of additional studies made since the publication of his paper.

First of all, it might be argued that the vascularized gametophytes of *Psilotum* are "abnormal" in chromosome number which "conditions" the development of a conducting strand. Manton (1950) discovered that *all sizes* of gametophytes were diploid (2*n*) and that the sporophytes, to which they were related, were tetraploid (4*n*), and Bierhorst (1968) only found vascular tissue in the larger diploid gametophytes of *P. nudum.* Haploid gametophytes of the same species were entirely devoid of a vascular system.

A second possible interpretation of vascularized gametophytes in *Psilotum* is that some type of physiological change occurs after a given gametophyte has reached a certain size in its growth. Holloway (1939) correlated "the presence of the conducting strand in *Psilotum* . . . with special robustness of growth." Bierhorst (1953) reached a similar conclusion and stated that "all the gametophytes are potential producers of vascular strands, and that a given gametophyte apex can, if the bulk of it meristem reaches a given threshhold, produce a strand."

A third interpretation of the vascular strand in the gametophytes is that it represents the persistence of a type of vasculature that was characteristic of the gametophytes of at least *some* of the ancient and primitive vascular plants of the Devonian, such as *Rhynia.* (For a more complete description of the morphology of rhyniophytes, see Chapter 7, p. 76. During the past 20 years suggestions have been made that the axes of *Rhynia gwynne-vaughanii,*which had previously been interpreted by Kidston and Lang (1921) as portions of the sporophyte, might actually represent vascularized gametophytes, comparable with those of *Psilotum.* The axes of *R. gwynne-vaughanii* have hemispherical projections on their surfaces, and Pant (1962) suggested that these might be young attached sporophytes. Lemoigne (1968, 1969, 1970) described what appear to be archegonia on certain vascularized axes of *R. gwynne-vaughanii,* although no antheridia were observed as is characteristic of *Psi-*

lotum. However, Pant (1962) cautiously suggested that the presumptive gametophytes of *Rhynia gwynne-vaughanii* might eventually prove to be the sexual generation of *Rhynia major*. Recently this idea has been questioned by Edwards (1980, 1986) who was able to establish that sporangia were attached to some axes of *R. gwynne-vaughanii*—thereby identifying it as a sporophyte. Also, Edwards (1986) believes that *R. major* is not a vascular plant because the conducting strand in the stems resembles that of certain nonvascular plants such as mosses. It was renamed *Aglaophyton major*. (See Chapter 7 for a more complete discussion of the basis for this change in name.)

There is, however, evidence of gametophytic plants from the Devonian Rhynie Chert in Scotland (Remy and Remy, 1980; Remy, 1982). These gametophytes consist of radially symmetrical axes that terminate in concave, disklike pads (gametophores) that bear either antheridia or archegonia, or both, that resemble those of certain bryophytes. The internal anatomy of the axes and the presence of stomata, however, resemble that of *Rhynia*. The investigators are careful to point out that it is difficult to decide where the taxonomic affinities of these fossils lie. For the present, they suggest that one should perhaps regard them as gametophytes of plants "representing an evolutionary stage previous to the separation of land plants into bryophytes and higher land plants." However, the results of their investigations offer the hope that future paleobotanical discoveries may reveal whether there is a relationship between these gametophytic plants and the well-known sporophytes (such as *Rhynia* and *Horneophyton*) of the Rhynie Chert (Chapter 7).

Alternation of Generations

Alternation of generations, or phases in the life cycle is a consistent feature of all groups of vascular plants and hence represents the basic pattern of reproduction in these dominant plants of the modern world. Figure 2-1 represents schematically the structures and processes common to the complete life cycle of *lower vascular plants*. The reader must understand this generalized life cycle clearly before becoming involved in the infinite variations of detail which occur in the cycle of reproduction of specific genera or groups of plants.

Because the sporophyte generation is the obvious and dominant phase of the life cycle, we may properly begin our analysis of Fig. 2-1 with the zygote. This diploid cell results from the union of a male gamete with the egg. The next event is the process of embryogeny, which involves the production from the zygote of a multicellular embryo, the early development, form, and organography of which are often specific in a given group. By means of further growth and differentiation from the shoot and root apices of the embryo, the adult and independent sporophyte is developed. Ultimately the sporophyte plant forms sporangia in which spore-mother cells, or sporocytes, are produced. These cells, like all normal cells of the vegetative sporophyte, are diploid. However, each sporocyte can by meiotic division give rise to a tetrad or group of four haploid spores. When circumstances are favorable a spore germinates and by cell division and cell specialization produces the gametophyte generation. The salient function of this phase is the production of male gametes and eggs. The union—called fertilization or syngamy—of a male gamete and an egg restores the diploid chromosome number and produces a zygote from which a new sporophytic plant may develop.

It should be clear from this description that the alternation of sporophyte and gametophyte phases in the life cycle is normally coordinated with a periodic doubling followed by a halving of the chromosome number. The diploid zygote ($2n$), from which the sporophyte arises contains twice the number of chromosomes typical of the spore (n) that produces the gametophyte. Is this difference in chromosome number of spore and zygote a clue to the remarkable morphological and functional differences between the generations which arise from these cells? Or, to ask the question in another way: Are syngamy and meiosis *always essential processes* in the production, respectively, of sporophyte and gametophyte? These questions apparently deserve an answer in the negative because of certain deviations from a "normal" reproductive cycle, namely, *parthenogenesis, apospory,* and *apogamy*.

In certain angiosperms and a few ferns, the embryo arises from an unfertilized egg, a phenomenon designated as parthenogenesis (i.e., "virgin birth").

The phenomenon termed *apospory* is the development of gametophytes, without a haploid spore stage, from vegetative cells of the sporophyte. Very commonly, aposporous gametophytes bear functional antheridia and archegonia, and chromosome studies have shown that their gametes are diploid and that a sporophyte resulting from their union is tetraploid (see Steil, 1939, 1951 for further information on apospory).

A third deviation from the usual reproductive cycle of a vascular plant is the phenomenon of *apogamy*, which is the formation of a sporophyte, without the act of fertilization, from vegetative cells of the gametophyte. Depending upon the chromosome number of the gametophyte (i.e., whether normal n or aposporous $2n$), the apogamous sporophyte may be either haploid or diploid in chromosome number.

Quite apart from the interesting questions raised by the phenomena of parthenogenesis, apospory, and apogamy, there remains the broader and still unsolved problem of the phylogenetic origin of the type of alternation of generations which is "normal" and prevalent throughout all living vascular plants. If, as is commonly believed, the remote ancestors of terrestrial plants were green algae, the central question at issue is the evolutionary origin of the diploid, vascularized, independent sporophyte typical of the vast majority of vascular plants.

A century ago, Celakovsky (1874) recognized two principal forms of alternation which he distinguished by the terms "antithetic" and "homologous." In his opinion, antithetic alternation, which is characteristic of archegoniate plants (i.e., bryophytes and vascular plants with archegonia), arose during evolution by the interpolation of a new phase, i.e., the sporophyte, between successive gametophytes. In contrast, he limited his concept of homologous alternation to the succession of morphologically similar phases which occurs in the reproductive cycles of certain fungi and algae.

Today the terms "antithetic" and "homologous" are usually applied to two divergent theories which both seek to explain phylogenetically the characteristic type of life cycle in vascular plants. Bower (1935) has urged that in place of "antithetic" and "homologous" theories it would be more appropriate to speak, respectively, of the "interpolation theory" and the "transformation theory." In his view, these substitute terms are explicit in that they convey the alternative methods of phylogenetic origin of the sporophyte generation. This point may be clarified by the following brief contrast between the two theories. For a more detailed analysis, the reader is referred to Bower (1935), Fritsch (1945), and Wardlaw (1952).

According to the *interpolation theory*, which was strongly championed by Bower, the origin of the sporophyte generation was fundamentally the result of the postponement of meiosis and the development from the zygote of a "new" diploid vegetative phase. During subsequent evolution, this vegetative phase or sporophyte became increasingly complex, both organographically and anatomically but "meiosis and spore formation, though delayed, would still be the final result" (Fig. 2-2). Bower also maintained that the gradual specialization of an independent spore-producing diploid phase was closely connected, in a biological sense, with the transition from aquatic to terrestrial life by the algal-like "progenitors" of the first land plants.

The *transformation theory* of alternation, in contrast, postulates that the spore-producing and gamete-producing *individuals* of the *original* parental green algae were morphologically similar and hence that the life cycle was *isomorphic* in type (Fig. 2-3). This idea was developed in great detail by Fritsch (1945) and some of his interesting ideas deserve brief attention at this point. According to Fritsch, the first step in the long evolutionary development of archegoniate plants from aquatic green algae was the attainment of a morphological distinction between *erect* branches and a creeping prostrate system. This resulted in what is termed a *heterotrichous habit* of growth. Further elaboration of the upright system by means of longitudinal divisions in some of the cells then produced a more-or-less "parenchymatous" type of organization. Fritsch regarded this second step a "most significant evolutionary advance, since in it lay the germ for the development of a plant body of almost unlimited size." Evolutionary modifications then occurred that would have led to morphological divergence between sporophytic and gametophytic individuals and the adoption of a heteromorphic type of life cycle (Fig. 2-3). In Fritsch's view, the gametophyte, at this stage in evolution, corresponded in form and dichotomous branching to the

A

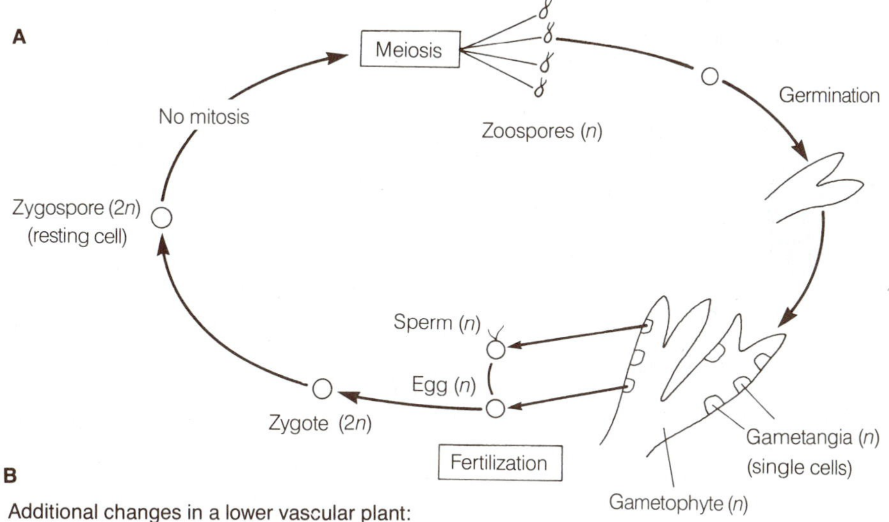

B

Additional changes in a lower vascular plant:

1. Organographic elaboration of sporophyte
2. Anatomical elaboration, e.g., vascular tissue and epidermis with stomata

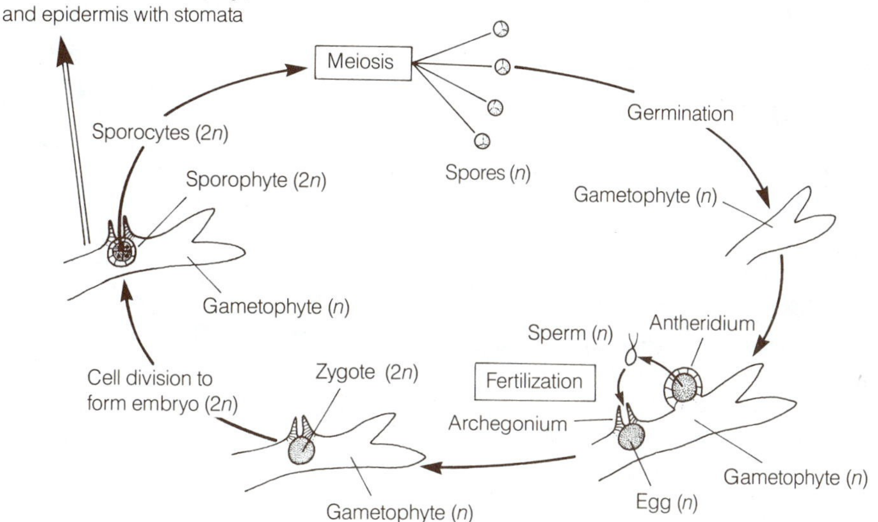

Evolutionary changes:

* 1. Postponement of meiosis
* 2. Development of a "new" 2*n* phase
3. Cutinized spore wall
4. Multicellular gametangia
5. Retention of egg
6. Embryo development within archegonium

*Changes unique to this theory

FIGURE 2-2 Schematic representation of the origin of alternation of generations or phases in the life cycle according to the interpolation (or antithetic) theory. **A**, Hypothetical haploid green alga with oogamy. **B**, Development of a simple multicellular sporophyte; note additional features required for the evolution of a simple vascular plant. [Modified from class handout by Dr. R. Schmid.]

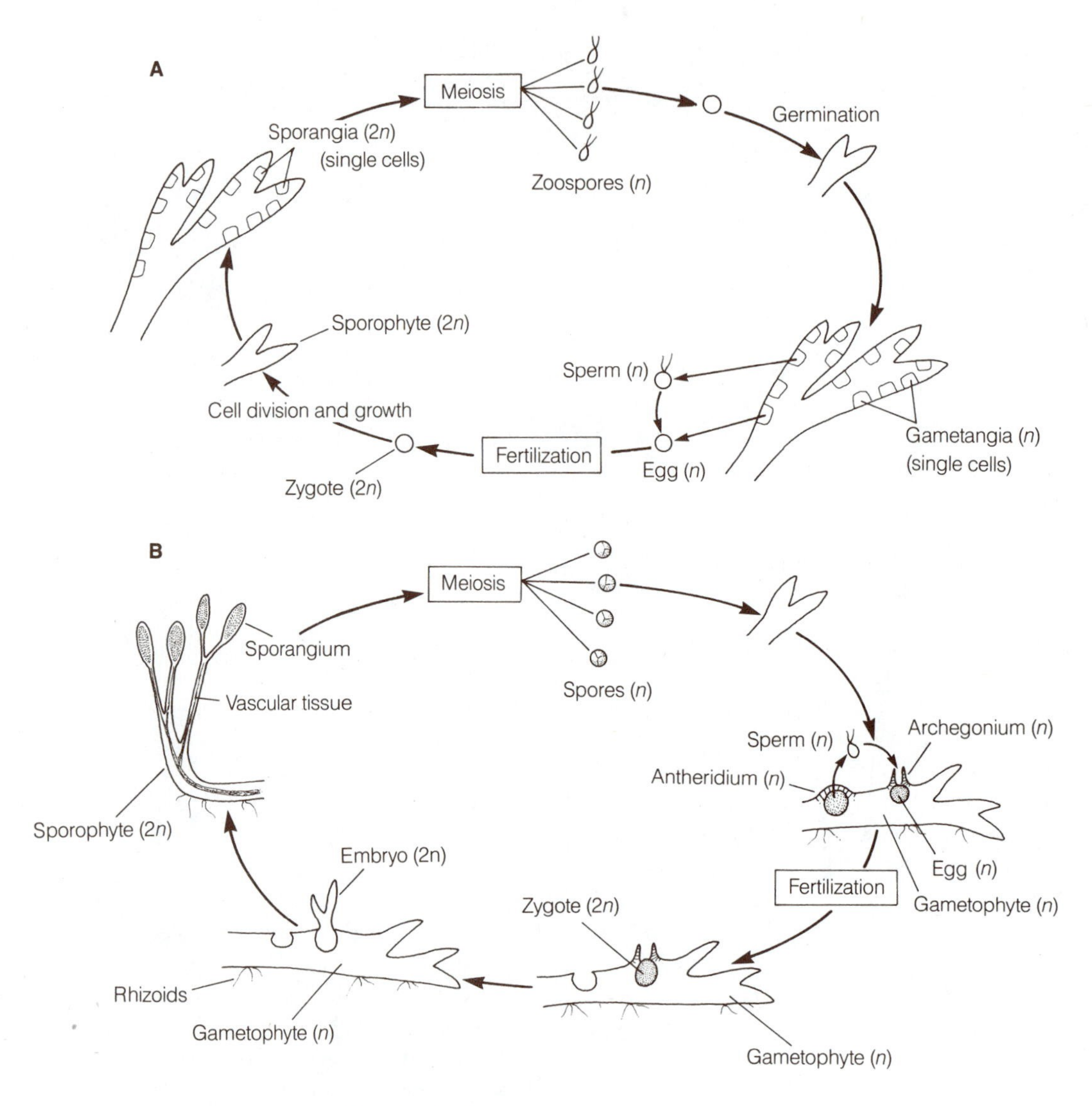

Evolutionary changes:

1. Cutinized spore wall
2. Vascular tissue in sporophyte
3. Multicellular sporangia and gametangia
4. Cuticle on epidermis; stomata
5. Retention of egg
6. Embryo development within archegonium

FIGURE 2-3 Schematic representation of the origin of alternation of generations or phases in the life cycle according to the transformation (or homologous) theory. **A**, Hypothetical green alga with isomorphic alternation of generations and oogamy. **B**, Homosporous lower vascular plant (generalized). [Modified from class handout by Dr. R. Schmid.]

primary underground system of the sporophyte and had "lost the capacity for emergence over-ground."

This conclusion is particularly interesting in light of discoveries of possible "rhizomelike" gametophytes (Lemoigne, 1968, 1969; Remy,1982) from the Devonian.

For many years it has been assumed that some group in the class Chlorophyceae of the green algae (Chlorophyta) was the progenitor of land plants, from which bryophytes and vascular plants later evolved. Indeed, an argument in support of this assumption is that vascular plants, bryophytes, and the Chlorophyceae share certain basic features: chlorophylls *a* and *b*, a cellulose wall, and starch as a reserve product. Some of the living Chlorophyceae have an isomorphic alternation of generations, so essential to Fritsch's concept. While the idea of shared attributes is appealing, recent research has cast doubt on the concept that the Chlorophyceae is the ancestral group from which bryophytes and vascular plants originated. It is probably impossible to select a specific living algal progenitor, but recent studies have pointed toward the green-algal class Charophyceae (stoneworts), including *Coleochaete* (Stewart and Mattox, 1975; Pickett-Heaps, 1976), as being more likely candidates. These algae possess the attributes enumerated previously for the Chlorophyceae, but ultrastructural and biochemical studies have revealed basic differences between the two classes. Charophytes, bryophytes, and vascular plants display a persistent mitotic spindle, followed by formation of the new cell wall. In the Chlorophyceae, however, the spindle collapses at telophase, the daughter nuclei often approach each other, and microtubules are formed at right angles to the original spindle. The new cell wall is then formed parallel to the new microtubules by cell plate formation or cleavage. In addition, the Charophyceae possess glycolate oxidase similar to that in higher plants but which has not been identified in Chlorophyceae. Furthermore, planes of cell division and branching patterns in *Coleochaete* have been cited as examples of how land-plant meristems and parenchyma may have evolved from branched, filamentous charophycean green algae (Graham, 1982; Hagemann, 1978). Also, details of sexual reproduction in *Coleochaete* provide support for the interpolation theory of the origin of alternation of generations (Graham, 1985).

Classification of Vascular Plants

The dominant land plants of the earth, despite their extraordinary diversity in habit, organography, and method of reproduction, share one highly important character: the presence of a vascular system. This anatomical character has been widely accepted as the basis for designating all plants with *tracheary tissue,* i.e., xylem, as "tracheophytes" (Sinnott, 1935), or more technically, "Tracheophyta" (Eames, 1936).

Certainly from a broad evolutionary viewpoint, the origin of a vascular system must have been highly significant in the development of the sporophyte generation and in its continued adaptation to life on land. In the second edition of this book the term Tracheophyta (a division) was used to designate all vascular plants. Twelve classes were recognized as constituting the Tracheophyta. These classes consisted of groups of plants that are generally recognizable based on consistent morphological features. Acceptance of the Tracheophyta as a major taxon does imply a monophyletic origin of vascular plants (Stewart, 1983). In recent years there has been a tendency to recognize several divisions of vascular plants, instead of only one. In these systems, what were formerly classes, for example, have been raised to divisional status. Embodied in these systems is the concept of polyphyletic origin of vascular plants, whether stated or implied. Paleobotanical research in recent years has resulted in the recognition of what appear to be separate evolutionary lines of vascular plants as far back as the Lower Devonian and even into the mid-Silurian period. Thus far no megafossils of vascular plants from earlier periods have been discovered, although microfossils (e.g., isolated tracheidlike cells, spores) that could be from vascular plants or bryophytes have been described. Thus there are many uncertainties as to whether all vascular plants can be traced back to one origin from an algal group, or whether the acquisition of vascular tissue is the expression of an adaptive tendency appearing in seemingly unrelated groups, as is the case with the development of leaves, vessels, and the seed habit. Only with additional paleobotanical evidence may we ultimately be able to answer this question.

All phylogenetic schemes of classification are, at best, tentative. No one scheme satisfies contemporary morphologists and paleobotanists with respect to the number of divisions, orders, and families. One of the difficulties, just mentioned, is the absence of convincing phyletic interconnections between the various groups of vascular plants that have been proposed. Moreover, there is considerable confusion at the level of formal schemes of classification because of the shifting of lower taxa (e.g., classes) to higher categories (e.g., divisions) or vice versa, merely by changing the suffix of certain names.

In view of these difficulties and nomenclature problems, no attempt will be made to contrast the various schemes of classifying vascular plants. However, instead of using "Tracheophyta" as the one division for vascular plants, in this edition we recognize fifteen divisions. Each of these formerly represented a class or even an order. This system is more in line with those of some contemporary botanists. The plants in each division have characters that are recognizable and can be used to differentiate them from those in other divisions, as well as to provide the bases for making comparisons and for establishing tentative phylogenetic relationships.

The conspectus shown in Table 2-1 lists the divisions of vascular plants recognized in the present text and relates them to the categories used in the previous edition. We introduce the conspectus at this point simply to orient the reader to the names and representative examples of the divisions of vascular plants. More detailed taxonomic treatment of each of the divisions will be found in subsequent chapters.

For additional reading on systems of classification and phylogeny, the reader is referred to Newman, 1947; Stewart, 1960, 1961, 1983; Banks, 1970; Bold et al., 1980; Taylor, 1981.

Table 2-1 Classification of vascular plants

Classification from *Comparative Morphology of Vascular Plants,* 2d edition	Classification used in present edition (Names in parentheses are used in some other systems for the same group)
Tracheophyta (division)	Divisions
Rhyniopsida* →	1. Rhyniophyta (extinct plants, e.g., *Rhynia, Cooksonia*)
Zosterophyllopsida →	2. Zosterophyllophyta (extinct plants, e.g., *Zosterophyllum*)
Trimerophytopsida →	3. Trimerophytophyta (extinct plants, e.g., *Psilophyton, Trimerophyton*)
Psilopsida →	4. Psilophyta (Psilotophyta) (two living genera, *Psilotum* and *Tmesipteris*)
Lycopsida →	5. Lycophyta (Microphyllophyta) (extinct forms such as *Protolepidodendron* and *Lepidodendron,* and the living genera *Lycopodium, Selaginella, Phylloglossum, Isoetes,* and *Stylites*)
Sphenopsida →	6. Sphenophyta (Arthrophyta) (mostly extinct plants, e.g., *Calamites, Sphenophyllum;* the one living genus is *Equisetum*)
Filicopsida →	7. Filicophyta (Pteridophyta) (living and extinct ferns)
Progymnospermopsida →	8. Progymnospermophyta (extinct plants, e.g., *Archaeopteris*)
Cycadopsida	
Pteridospermales (order) →	9. Pteridospermophyta (extinct seed ferns, e.g., *Lyginopteris, Medullosa*)
Cycadales (order) →	10. Cycadophyta (extinct and living cycads, e.g., *Cycas, Zamia*)
Cycadeoidales (order) →	11. Cycadeoidophyta (extinct cycadeoids, e.g., *Cycadeoidea*)
Coniferopsida	
Ginkgoales (order) →	12. Ginkgophyta (extinct members and living *Ginkgo biloba*)
Coniferales (order) →	13. Coniferophyta (extinct and living conifers, e.g., pine, fir)
Gnetopsida →	14. Gnetophyta (*Ephedra, Gnetum, Welwitschia*)
Angiospermopsida →	15. Magnoliophyta (Anthophyta) (angiosperms or flowering plants)

* The taxonomic level in the left column is class unless otherwise indicated.

REFERENCES

Banks, H. P.
1970. *Evolution and Plants of the Past.* Wadsworth, Belmont, Calif.

Bierhorst, D. W.
1953. Structure and development of the gametophyte of *Psilotum nudum. Amer. Jour. Bot.* 40:649–658
1968. On the Stromatopteridaceae (Fam. Nov.) and on the Psilotaceae. *Phytomorphology* 18:232–268.

Bold, H. C., C. J. Alexopoulos, and T. Delevoryas
1980. *Morphology of Plants and Fungi,* 4th edition. Harper and Row, New York.

Bower, F. O.
1935. *Primitive Land Plants.* Macmillan, London.

Celakovsky, L.
1874. *Bedeutung des Generationswechsels der Pflanzen.* Prag.

Eames, A. J.
1936. *Morphology of Vascular Plants. Lower Groups.* McGraw-Hill, New York.

Edwards, D. S.
1980. Evidence for the sporophytic status of the Lower Devonian plant *Rhynia gwynne-vaughanii* Kidston and Lang. *Rev. Palaeobot. Palynol.* 29:177–188.
1986. *Aglaophyton major,* a non-vascular land-plant from the Devonian Rhynie Chert. *Bot. J. Linnean Soc.* 93:173–204.

Fritsch, F. E.
1945. Studies in the comparative morphology of the algae. IV. Algae and archegoniate plants. *Ann. Bot.* n. s. 9:1–29.

Graham, L. E.
1982. The occurrence, evolution, and phylogenetic significance of parenchyma in *Coleochaete* Bréb. (Chlorophyta). *Amer. Jour. Bot.* 69:447–454.
1985. The origin of the life cycle of land plants. *Am. Scientist* 73:178–186.

Hagemann, W.
1978. Zur Phylogenese der terminalen Sprossmeristeme. *Ber. Deut. Bot. Ges.* 91:699–716.

Hébant, C.
1976. Evidence for the presence of sieve elements in the vascularised gametophytes of *Psilotum* from Holloway's collections. *New Zealand J. Bot.* 14:187–191.

Holloway, J. E.
1939. The gametophyte, embryo, and young rhizome of *Psilotum triquetrum* Swartz. *Ann. Bot.* n.s. 3:313–336.

Kidston, R., and W. H. Lang
1921. Old Red Sandstone plants showing structure, from the Rhynie Chert Bed, Aberdeenshire. IV. Restorations of the vascular cryptogams and discussion of their bearing on the general morphology of the Pteridophyta and the origin of the organization of land plants. *Trans. Roy. Soc. Edinb.* 52:831–854.

Lemoigne, Y.
1968. Observation d'archégones portés par des axes du type *Rhynia gwynne-vaughanii* Kidston et Lang. Existence de gamétophytes vascularisés au Dévonien. *Compt. Rend. Acad. Sci.* (Paris) 266:1655–1657.
1969. Contribution à la connaissance du gamétophyte *Rhynia gwynne-vaughanii* Kidston et Lang. Problème des protuberances et processus de ramification. *Bull. Mens. Soc. Linn. Lyon* 38(4):94–102.
1970. Nouvelles diagnoses du genre *Rhynia* et de l'espèce *Rhynia gwynne-vaughanii. Bull. Soc. Bot. France* 117:307–320.

Lowry, B., D. Lee, and C. Hébant
1980. The origin of land plants: a new look at an old problem. *Taxon* 29:183–197.

Manton, I.
1950. *Problems of Cytology and Evolution in the Pteridophyta.* Cambridge University Press, London.

Newman, I. V.
1947. The place of ferns and seed plants in classification. *Trans. Roy. Soc. New Zealand* 77:154–160.

Niklas, K. J.
1976. The role of morphological biochemical reciprocity in early plant evolution. *Ann. Bot.* 40:1239–1254.
1981. The chemistry of fossil plants. *BioScience* 31:820–825.

Niklas, K. J., and L. M. Pratt
1980. Evidence for lignin-like constituents in Early Silurian (Llandoverian) plant fossils. *Science* 209:396–397.

Pant, D. D.
1962. The gametophyte of the Psilophytales. *Proc. Summer School Bot. Darjeeling, India* (Held June 2–15, 1960) Pp. 276–301.

Pickett-Heaps, J.
1976. Cell division in eucaryotic algae. *BioScience* 26:445–450.

Remy, W.
1982. Lower Devonian gametophytes: relation to the phylogeny of land plants. *Science* 215:1625–1627.

Remy, W., and R. Remy
1980. Devonian gametophytes with anatomically preserved gametangia. *Science* 208:295–308.

Sinnott, E. W.
1935. *Botany. Principles and Problems,* 3d edition. McGraw-Hill, New York.

Stebbins, G. L., and G. J. C. Hill
1980. Did multicellular plants invade the land? *Am. Nat.* 115:342–353.

Steil, W. N.
1939. Apogamy, apospory, and parthenogenesis in the pteridophytes. *Bot. Rev.* 5:433–453.
1951. Apogamy, apospory, and parthenogenesis, II. *Bot. Rev.* 17:90–104.

Stewart, K. D., and K. R. Mattox
1975. Comparative cytology, evolution and classification of the green algae with some consideration of the origin of other organisms with chlorophylls *a* and *b*. *Bot. Rev.* 41:104–135.

Stewart, W. N.
1960. More about the origin of vascular plants. *Plant Sci. Bull.* 6:1–5.
1961. The origin of vascular plants: monophyletic or polyphyletic? *Rec. Adv. Bot.* 2:960–963.
1983. *Paleobotany and the Evolution of Plants.* Cambridge University Press, Cambridge.

Taylor, T. N.
1981. *Paleobotany: An Introduction to Fossil Plant Biology.* McGraw-Hill, New York.

Wardlaw, C. W.
1952. *Phylogeny and Morphogenesis.* Macmillan, London.

CHAPTER 3

The Vegetative Sporophyte

THIS chapter reviews those features of the vegetative sporophyte of vascular plants that figure most commonly in morphological comparisons and interpretations. For the convenience of the reader, the material is presented in five categories: (1) the contrasts between shoot and root, (2) the methods of branching or ramification of shoots, (3) the concept of microphylls and megaphylls, (4) the comparative anatomy of the sporophyte, and (5) the stelar theory.

Shoots and Roots

In the great majority of living vascular plants the developing embryo gives rise to a leafy stem, or shoot, and a primary root. Further development of the young sporophyte results, through the activity of apical meristems, in the formation of additions to the original shoot and root components.

From the standpoints of organography, function, and anatomy, shoots and roots are very different types of systems. Roots develop no superficial extensions other than the absorbing root hairs and are to be regarded as naked axes; their chief functions are anchorage and the absorption of water and solutes. In contrast, shoots have a jointed or segmental organography because the axis or stem bears conspicuous lateral appendages, or leaves. The chief functions of shoots are photosynthesis, storage, and reproduction.

In regard to apical growth and branching, roots and shoots differ in several important respects. The root apex consists of a root cap that functions as a protective buffer to the delicate meristem, which lies beneath it. This subterminal meristem of roots is the point of origin of two different patterns of cell formation. One adds new cells outwardly to the root cap, the other contributes the cells which become a part of the root body (Fig. 3-1, A). Except for hairs, the only lateral extensions that may occur are lateral roots, and these structures, unlike the usually superficially developed branches of the shoot, originate deep within the tissue of the parent root. In marked contrast to the root apex, the apex of the shoot consists of the terminal meristem itself, and no cap of tissue comparable to a root cap is developed. Aside from giving rise to the primary stem tissues, a very important function of the shoot apex is the formation of new leaves (Fig. 3-1, B).

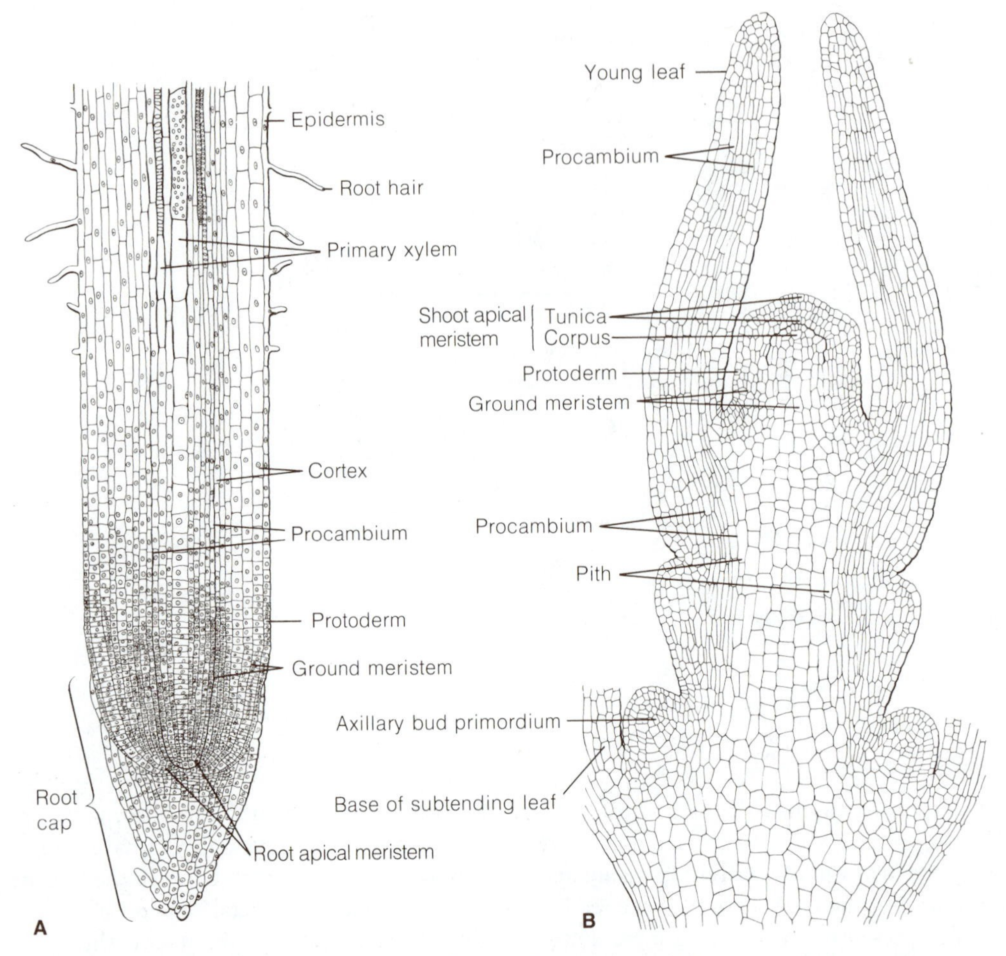

FIGURE 3-1 Longisections showing the apical meristems and the origin of the primary meristematic tissues in the root and shoot of angiosperms. **A,** root tip of *Hordeum sativum;* **B,** shoot tip of *Hypericum uralum.* [**A** redrawn from *A Textbook of General Botany,* 4th edition, by R. M. Holman and W. W. Robbins. Wiley, New York. 1951; **B** redrawn from Zimmermann, *Jahrb. Wiss. Bot.* 68:289, 1928.]

Leaves originate as primordia (singular, primordium) by means of localized cell division and cell extension at discrete loci or nodes on the flanks of the shoot apex. Leaf primordia typically arise in acropetal sequence; that is, the youngest stages in leaf development are found nearest the summit of the meristem. In addition to their acropetal order of development, the primordia of leaves are laid down usually in an orderly and often distinctive arrangement or *phyllotaxis* with reference to the stem. In the most common type of phyllotaxis (or phyllotaxy) a single leaf is initiated at each node, and the leaves form a spiral (or helix) around the stem—an arrangement designated *alternate spiral* or *alternate helical* (Fig. 3-2, A). In other plants the leaves are formed in pairs (*opposite phyllotaxis,* Fig. 3-2, B). If the successive pairs are at right angles to one another, the arrangement is termed *decussate phyllotaxis.* Leaves may also occur in groups of three or more *(whorled phyllotaxis).* The term *distichous* refers to a phyllotactic pattern in which leaves are in two rows. The causal factors responsible for phyllotactic patterns are poorly understood. (See Wardlaw, 1952, 1965, and Cutter, 1965, for résumés of experimental studies, and Young, 1978, for computer models.)

The differentiation between distinct root and shoot systems in most living vascular plants is of

FIGURE 3-2 Two types of leaf arrangement (phyllotaxis). **A,** alternate spiral or alternate helical *(Elaeagnus)*. **B,** opposite *(Lonicera)*. Note buds in the axils of leaves (flowers and floral buds associated with the lower two leaf pairs in **B**).

considerable interest from an evolutionary viewpoint. It is now rather generally agreed that this differentiation did not exist in such Devonian land plants as *Horneophyton* and *Rhynia* (Chapter 7). In these archaic organisms portions of the underground system of stems apparently served physiologically as roots. It seems reasonable to postulate that roots were acquired later in the evolution of vascular plants. However, the steps in the evolutionary divergence of roots from primitive shoots, which led to the acquisition of a root cap, a prevalent internal or endogenous origin of roots, and the retention of a primitive type of vascular system, are unfortunately obscure today. (See Chapter 19 for a more detailed account of root morphology and anatomy.)

Types of Branching in Shoots

The types of branching of the shoot system, in both extinct and living vascular plants, are highly varied and often definitive for the larger taxa, i.e., divisions or classes. At this point in our discussion of the vegetative sporophyte, we will briefly consider the major patterns of branching in vascular plants. Additional details on the types of ramification in the various groups of vascular plants will be found in later chapters of this book.

From the broadest possible morphological viewpoint, two principal types of ramification may be distinguished: dichotomous and lateral.

Dichotomous Branching

Two of the widely accepted morphological characteristics of a dichotomously branched shoot system are (1) the absence of a dominant, or "major," axis and (2) the occurrence of a *sequence* of paired branches which are *not* associated with subtending leaves. These characteristics are the result of a particular type of apical ontogeny in which the terminal meristem itself divides into two more-or-less equal and divergent apices (Figs. 3-3, A–C;

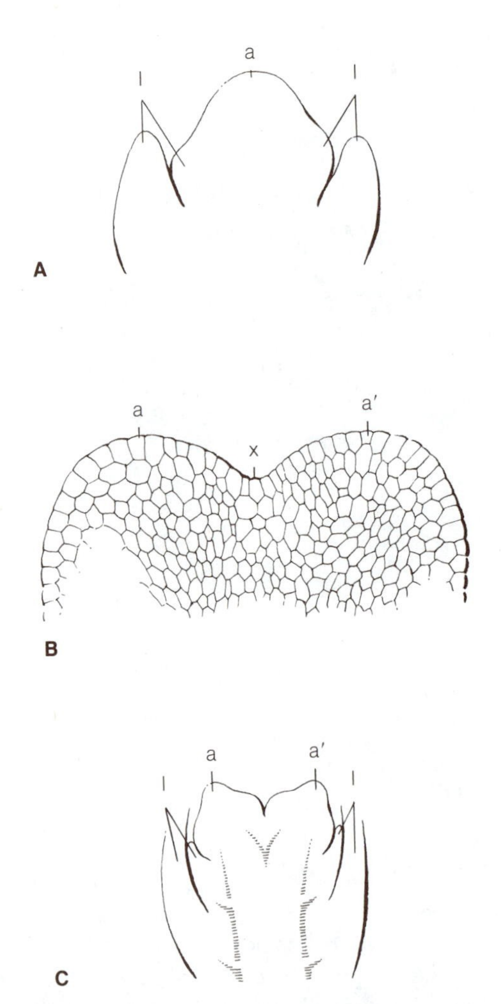

FIGURE 3-3 Dichotomous branching in *Lycopodium alpinum*. **A,** vegetative shoot apex (a) with leaf primordia (1); **B,** dichotomous division of a reproductive shoot apex (x) into two new apices, (a and a′); **C,** later stage in dichotomy of vegetative shoot apex. [From *Vergleichende Morphologie der höheren Pflanzen* by W. Troll. Gebrüder Borntraeger, Berlin. 1937.]

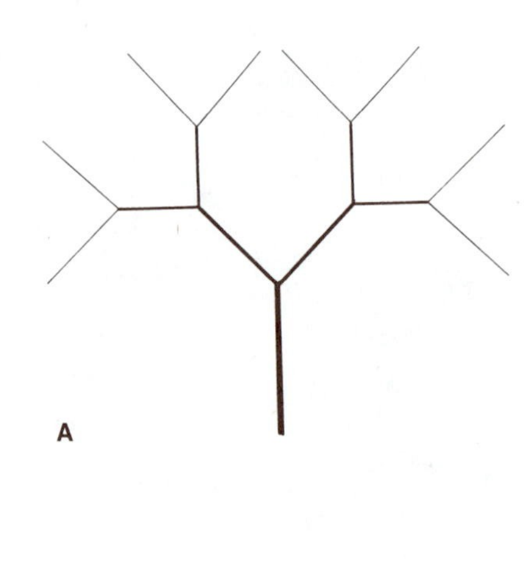

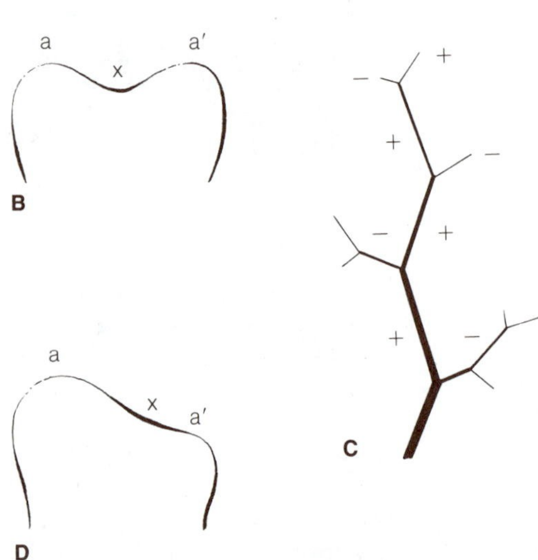

FIGURE 3-4 Isotomous and anisotomous types of dichotomous branching. **A,** isotomous branching, showing equal development following each dichotomy; **B,** origin of isotomous branch primordia (a, a′) by equal division of shoot apex (x); **C,** anisotomous branching, showing unequal development of stronger (+) and weaker (−) branches following each dichotomy; **D,** origin of anisotomous branch primordia (a, a′) by unequal division of shoot apex (x). [From *Vergleichende Morphologie der höheren Pflanzen* by W. Troll. Gebrüder Borntraeger, Berlin. 1937.]

3-4, B). This process of apical forking, or dichotomy, may be repeated on an indefinite scheme, leading in many cases to a very regular and distinctive kind of shoot system (Fig. 3-4, A).

If the successive bifurcations occur in one plane, a flattened dorsiventral system is formed. Troll (1937) terms this type of branching "flabellate dichotomy." In other types of plants, each dichotomy of the axis is at approximately right angles to the preceding bifurcation and the resulting shrublike shoot system is termed "cruciate dichotomy" by Troll.

Many of the nineteenth century botanists limited their concept of dichotomy to those plants in which the initiation of branching resulted from the longitudinal division of the apical cell in the shoot apex. Each of the two daughter cells produced by such a division was then observed to function as the new apical cell of a branch primordium.

However, if "dichotomy" is defined in the above restrictive sense, this type of branching would have to be regarded as extremely rare because the shoot apex of the majority of vascular plants lacks a definable apical cell.

The dichotomy of the shoot apex may produce successive pairs of branches approximately equal in size and degree of development. Troll (1937) terms this kind of symmetrical branching *isotomous dichotomy* (Fig. 3-4, A, B). The two members of a bifurcation also may develop unequally. One member—or "shank"—of each pair develops more strongly and soon overtops its weaker sister branch (Fig. 3-4, C, D). This derivative form of dichotomy is termed *dichopodial branching* by Bock (1962) who characterizes it, in its simplest expression, as the alternate promotion of the right and left shanks of successive dichotomies. The repetition of this *anisotomous dichotomy,* as Troll designates it, results in the formation of a zig-zag "axis" and shorter, more-or-less determinate "lateral branches." Bock regards the origin of dichopodial branching as a highly significant event in evolution that led to lateral branching, pinnate leaves, and the gradual formation of a midvein in the early phylogeny of venation patterns in leaves. (For further details, consult Bock, 1962, 1969).

Lateral Branching

In contrast to dichotomy, lateral branching of a shoot system originates by the expansion of buds more-or-less distal to the shoot apex of the main or dominant axis (Fig. 3-1, B). This kind of ramification is commonly termed "monopodial branching" to distinguish it from typical dichotomous or dichopodial branching.

Lateral buds, as shown by Troll (1937), most frequently arise in some type of relationship to the leaves of a shoot. In some fern species that have dorsiventral rhizomes buds arise without reference to leaves ("acrogenous branching," according to Troll's terminology), but it is more common in ferns for buds to originate near or from the abaxial side of the leaf bases or from the petiole (Troop and Mickel, 1968). *Equisetum* appears unique among vascular plants with monopodial branching in that the buds *alternate in position* with the fused leaves at each node (Chapter 10). In seed plants, however, the dominant type of branching is *axillary* (Fig. 3-1, B).

One of the most remarkable details of axillary bud formation in the angiosperms is the development, in a single leaf axil, of a series of *accessory buds* in addition to the main bud. Accessory buds, arranged in one or two vertical rows in the leaf axil, are known as "serial buds." If the accessory buds are arranged in a crescentic pattern in the leaf axil, they are termed "collateral buds." (Detailed treatments of accessory buds are given by Sandt, 1925, and Troll, 1937).

In many dicotyledons with normal axillary branching, it is common for the development of a terminal inflorescence or the abortion and subsequent abscission of the entire tip of a vegetative shoot to result in the continuation of the growth of such shoots from one or more of the uppermost axillary buds (Garrison and Wetmore, 1961; Millington, 1963). If shoot-tip abortion occurs in a plant with decussate phyllotaxy, the uppermost pair of axillary buds may subsequently expand into shoots and give the false impression that branching is dichotomous. Troll (1935) designates this type of pseudodichotomy as a *dichasial sympodium,* in contrast to the *monochasial sympodium,* which is produced when only a single axillary bud continues the development of the shoot.

Microphylls and Megaphylls

The problem of interpretation of the morphology and evolutionary history of the leaves of vascular plants has attracted much attention. Leaves, regardless of their size, form, or structure, arise as lateral protuberances from a shoot apex and at maturity represent the typical lateral appendages of the axis or stem. From an ontogenetic viewpoint, a leaf is a determinate organ. In contrast to the theoretically unlimited or open type of apical growth characteristic of the stem, the apical growth of the leaf primordium in most plants ceases early in ontogeny and is followed by a phase of tissue specialization and enlargement which culminates in the production of the final shape and structure of the adult foliar organ. The question is: How did the distinction between an indeterminate axis and its lateral foliar appendages arise during the evolutionary history of vascular plants? The answer to this question is particularly important in the light of paleobotanical studies on certain Devonian and Silurian plants, some of which represent the simplest vascular plants known to science. The sporophytes of *Rhynia* and *Horneophyton* are of exceptional interest because their aerial dichotomously branched axes were entirely devoid of foliar appendages (see Chapter 7).

Morphological Contrasts

MICROPHYLLS. Unfortunately the term "microphyll"—literally "small leaf"—which botanists generally apply to a certain leaf type unduly, and often inaccurately, emphasizes the small dimension of an appendage rather than its distinctive vascular anatomy and phylogenetic mode of origin. From a more precise morphological standpoint, one of the salient characters of a microphyll is its extremely simple and presumably primitive vascular system. As is shown in Fig. 3-5, D, the vascular supply of a typical microphyll consists of a single strand—the leaf trace—which diverges from the periphery of the *stele* (i.e., the vascular cylinder) of the stem and extends as an unbranched midvein through the leaf. It must be emphasized that the divergence of the single trace at the node of a microphyll is *not* associated with a corresponding break, or *leaf gap*, in the stele. This is also true for certain taxa that have microphylls and are siphonostelic (their stem has a central pith).

There are two exceptions to the presence of an unbranched vein in the leaves generally considered to be microphylls. For example, in some extinct arborescent lycopods there were two veins in the leaf blade which presumably arose from a dichotomy of the leaf trace. Also, there is a *Selaginella* species with a branched venation system (Chapter 9).

MEGAPHYLLS. The megaphyll (called "macrophyll" by some authors) is illustrated by the comparatively large pinnate leaves of ferns, although here too the salient feature is not size but rather morphological and anatomical organization. In contrast to a microphyll, the divergence of the leaf trace (or traces) in most of the living ferns is associated with the formation of parenchymatous areas, or *leaf gaps*, in the vascular cylinder of the stem (Figs. 3-6, 3-19, B). There are, however, notable exceptions to this anatomical correlation between

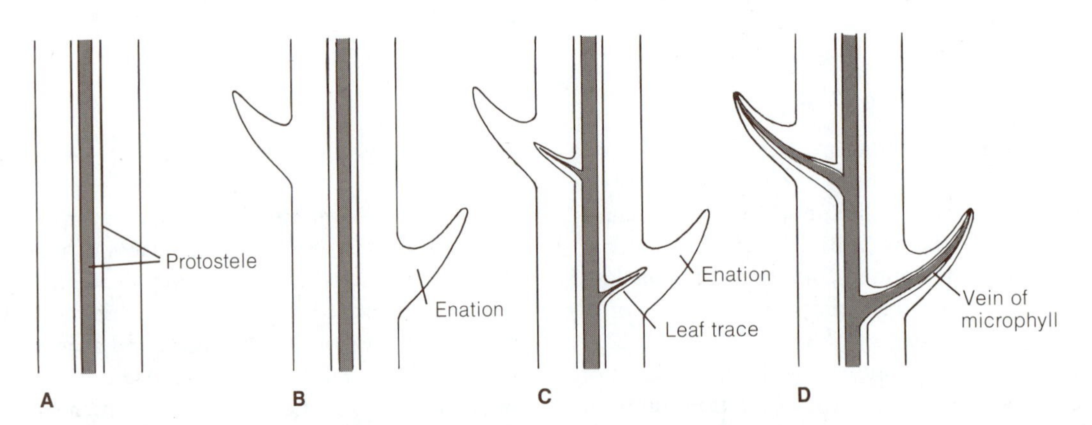

FIGURE 3-5 Longisectional diagrams showing the origin and evolutionary development of microphylls according to the enation theory. **A,** the leafless, protostelic axis of *Rhynia;* **B,** primitive enations, devoid of leaf traces, as illustrated by the shoot of *Psilotum;* **C,** portion of the "shoot" of the extinct lycopod *Asteroxylon,* showing the termination of leaf traces at the bases of the veinless enations; **D,** typical microphylls, in which a leaf trace extends as an unbranched midvein into each of the foliar appendages. The condition in **D** is the prevailing pattern in *Lycopodium, Selaginella,* and other members of the Lycophyta. [Redrawn from Lemoigne, *Bull. Mens. Soc. Linn. Lyon,* 37(9):367, 1968.]

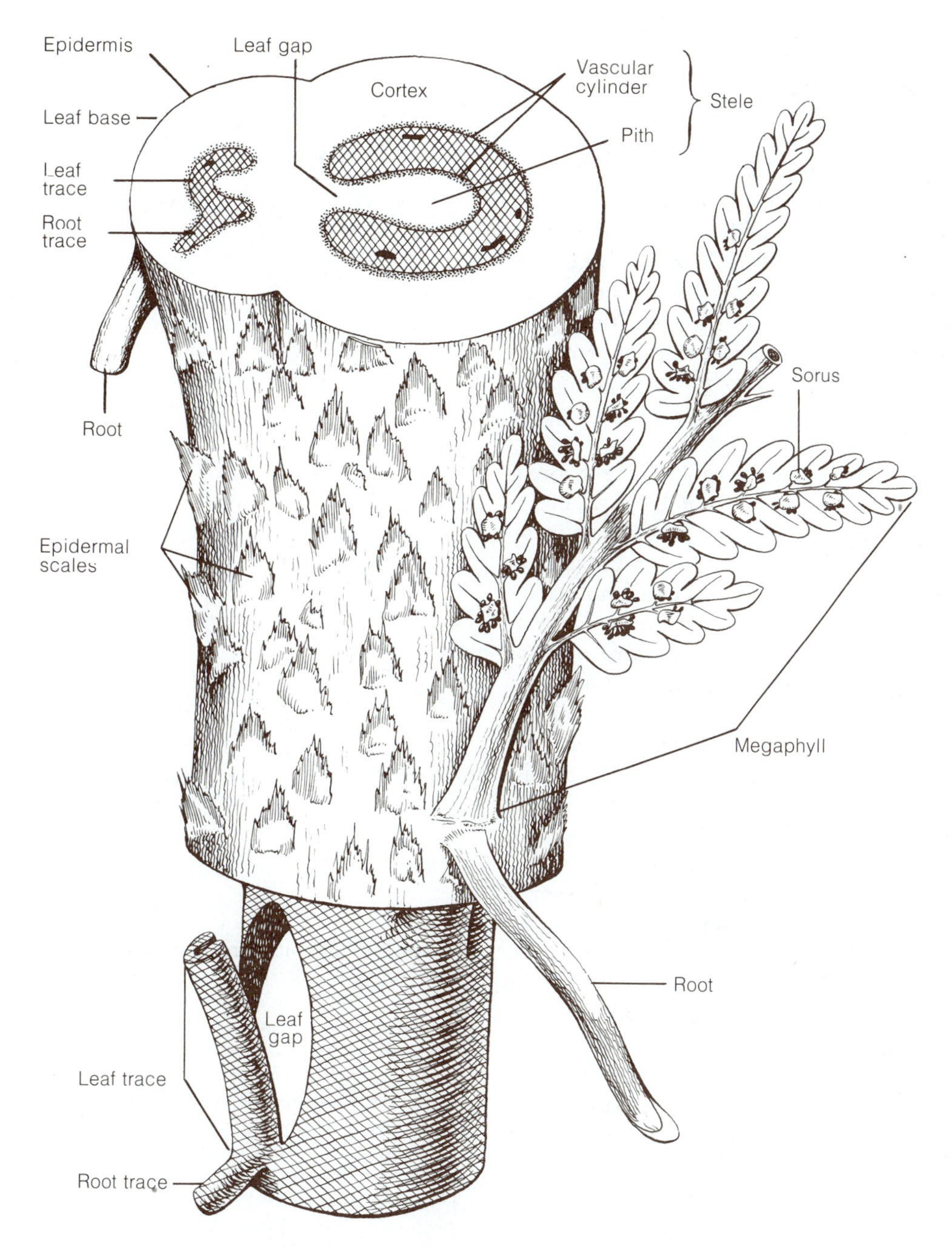

FIGURE 3-6 Organography and general vascular anatomy of a small portion of a fern shoot. A pinnatifid megaphyll with its abaxial sori is seen in surface view at right. Note that the divergence of a leaf trace into a megaphyll (shown at the top and bottom of the figure) is associated with a leaf gap in the stele of the stem. [From *The Anatomy of Woody Plants* by E. C. Jeffrey. University of Chicago Press, Chicago. 1917.]

leaf traces and leaf gaps where the vascular cylinder of the axis is a protostele and devoid of leaf gaps. Fern megaphylls differ from microphylls by their more complex patterns of venation. In contrast to the univeined microphyll, the lamina of a fern leaf, whether simple or pinnate in organization, develops a relatively complex system of branched vascular strands.

Phylogenetic Origin

As Bower (1935) has clearly pointed out, the evolutionary significance of the morphological differences between microphylls and megaphylls "can only be solved by comparison between various types living and fossil, aided where possible by reference to individual development." From this broad outlook, Bower postulated that the microphylls and megaphylls of lower vascular plants are the results of separate paths of foliar evolution, i.e., the two leaf types are not homologous from a phyletic point of view.

Figure 3-5 shows diagrammatically the theoretical steps in the evolution of the microphyll. Beginning with the leafless type of axis found in an ancient plant such as *Rhynia,* the earliest stage in microphyll evolution may have been a simple emergence or "enation" devoid of a leaf trace (Fig. 3-5, A). The further elaboration of primitive enations would be indicated by the initiation of "leaf traces" which diverged from the periphery of the protostele and terminated at the bases of the appendages. This theoretical stage in phylogeny is illustrated by the extinct lycopod *Asteroxylon* and possibly by the enation "leaves" of the living *Psilotum* (Fig. 3-5, B, C). In the final stage of mircophyll evolution, a vascular strand, or leaf trace, continued as a single midvein into the microphyll (Fig. 3-5, D). This type of univeined microphyll is typical of the living members of the Lycophyta. (For further details on the microphylls of the Lycophyta, see Chapter 9.)

According to another theory, microphylls may have evolved phylogenetically through progressive reduction of dichotomously branched, lateral, leaflike appendages. This concept will be discussed at the end of the next section (The Telome Theory).

In contrast to the progressive evolutionary specialization of microphylls from enations, megaphylls are considered by Bower to have evolved by the specialization of the distal regions of dichotomized branch systems. His theory, based on the earlier ideas of Lignier (1903), postulates that the fern megaphyll is a "cladode leaf," and that the first stage in its origin was a gradual change from equal dichotomous branching to a dichopodial type of growth. This consisted in the unequal development of the sister branches of dichotomizing axes, one continuing as the main axis or "stem," the other becoming laterally "overtopped" and representing the precursor of a megaphyll (Fig. 3-7, A, B). Subsequent flattening or "planation" occurred, and finally the ultimate divisions of each of the overtopped branch systems became united by development of laminar tissue, forming a simply di-

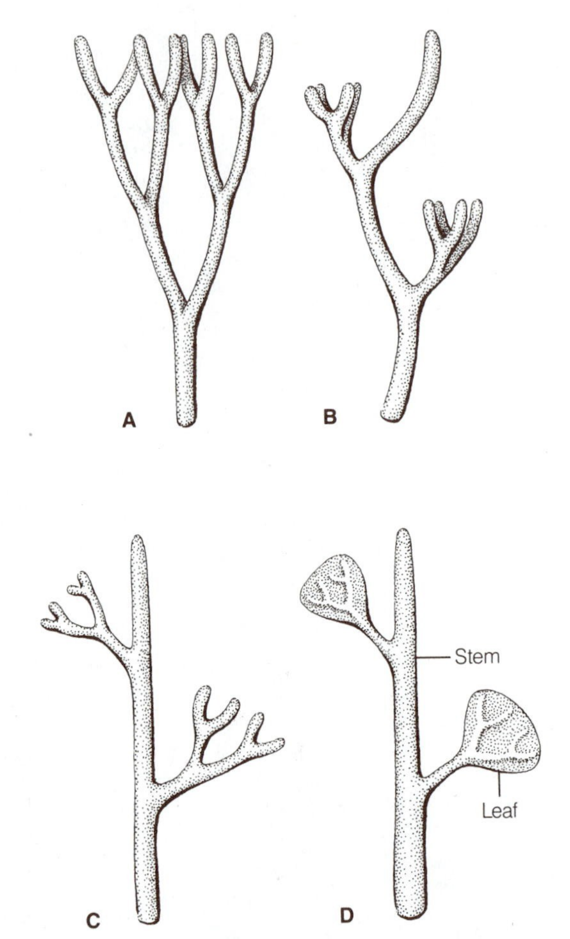

FIGURE 3-7 Schematic representation of the phylogenetic origin and development of the megaphyll according to the telome theory. **A,** isotomous branch system, without distinction between axis and megaphylls; **B,** unequal dichotomy or "overtopping," the weaker branches representing initial stages in megaphyll evolution; **C,** dichotomous branching of primitive megaphylls in a single plane ("planation"); **D,** union between forked divisions of megaphylls ("webbing") has produced a flat, dichotomously veined lamina. [Adapted from *Cryptogamic Botany, Vol. II. Bryophytes and Pteridophytes* by G. M. Smith. McGraw-Hill, New York. 1955.]

chotomously veined megaphyll (Fig. 3-7, C, D). Possible examples of overtopping can be found in the extinct division Trimerophytophyta from the Devonian (Chapter 7).

The Telome Theory

During the early development of plant morphology, the study of homologies was based on the belief that the organography of the sporophyte in vascular plants *as a whole* could be interpreted with reference to such angiospermic organs as leaf, stem, and root. Paleobotanical studies made during the early part of the present century suggested the need for a "new outlook" in phyletic morphology, especially a reappraisal of the evolutionary history of the sporophyte. The most comprehensive synthesis of the major steps in the evolution of vascular plants was made by Walter Zimmermann, who, in 1930, originated the telome theory (Zimmermann, 1953, 1959, 1965).

We will outline only the most salient features of the telome theory. Readers interested in the historical aspects of this theory and a more detailed analysis of it should consult Zimmermann (1965), Wilson (1953), and Stewart (1964).

Zimmermann selected *Rhynia* as a type of very ancient "primordial" land plant that provides a relatively simple example of the application of his telome theory. According to his interpretation, the dichotomously branched sporophyte of *Rhynia* was composed of morphological "units," which he designated as "telomes" and "mesomes." Reference to Fig. 3-8 will serve to explain his use of these terms. A telome, in the broadest sense of the term, is one of the distal branches of a dichotomized axis. Each telome ends at the point of forking of the axis whereas a mesome represents the "internodal" region between two successive dichotomies of an axis. In ontogeny, a given telome becomes "converted" to a mesome if dichotomous branching continues (see Fig. 3-8, upper right). From a functional standpoint, telomes are classified as "fertile" when they terminate in sporangia, and as "vegetative" when they constitute "phylloides." A system of united telomes and mesomes, as illustrated in Fig. 3-8, is termed a "primordial syntelome." Anatomically, the entire telome system is vascularized by a continuous protostele and is further morpho-

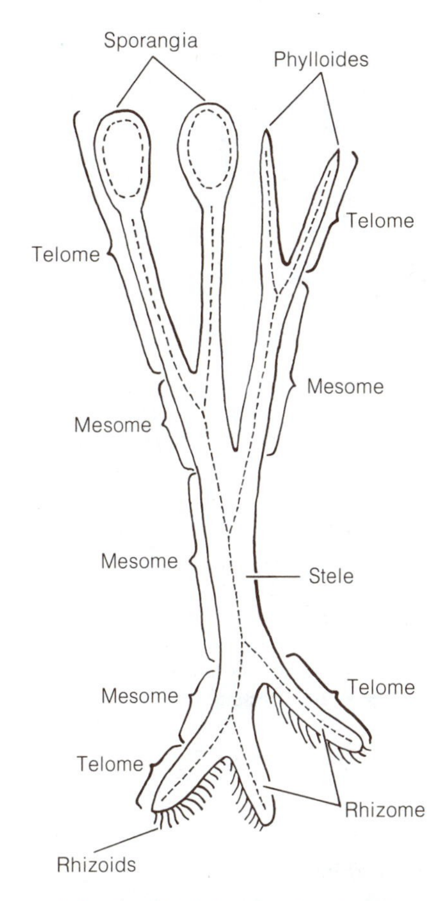

FIGURE 3-8 Simplified diagram showing the general organography of a primordial vascular land plant of the type of *Rhynia*. The morphological "units" of this elemental sporophyte (or "primordial syntelome"), according to Zimmermann's concept, are telomes (i.e., sporangia and vegetative phylloides) and mesomes. The telomes of the underground creeping rhizome bore rhizoids. See text for further explanation. [Adapted from *Die Telomtheorie* by W. Zimmermann. Gustav Fischer Verlag, Stuttgart. 1965.]

logically differentiated into a creeping subterranean portion, composed of rhizoid-bearing telomes and an upright aerial portion terminating in sporangia and phylloides.

According to Zimmermann, the further phylogenetic development of the *Rhynia* type of telome system resulted from the operation of five "elementary processes" (Fig. 3-9). Independently — or more commonly in various combinations — these processes were responsible for the gradual evolution of the diverse types of leaves and sporophylls (sporan-

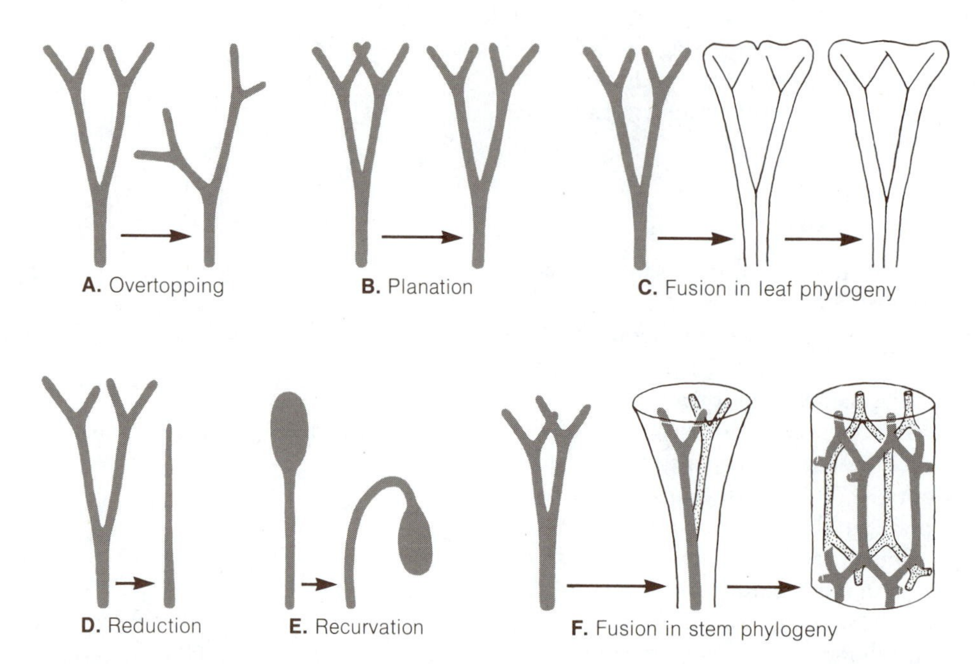

FIGURE 3-9 Diagrams illustrating the five elementary processes that, beginning with primordial system of telomes (see Fig. 3-8), were operative in the phylogenetic specialization of vascular plants. See text for further explanation. [Adapted from *Die Telomtheorie* by W. Zimmermann. Gustav Fischer Verlag, Stuttgart. 1965.]

gium-bearing structures) characteristic of the main taxa of vascular plants.

One of the most important of the elementary processes was "overtopping," i.e., the unequal development of certain parts of a dichotomously branched system, the subordinated overtopped lateral members representing the beginning of "leaves," and the overtopping portions, stemlike axes (Fig. 3-9, A). The additional processes of planation and "fusion" (syngenesis) are considered by Zimmermann to have been particularly significant with reference to leaf and sporophyll evolution. Planation resulted in the arrangement of groups of telomes and mesomes in a single plane (Fig. 3-9, B) while fusion, which at first only entailed parenchyma formation, connected these units into a flat dichotomously veined lamina. Subsequent fusion of the vascular strands in the leaf led to the formation of reticulate venation and fusion in the stem resulted in the anastomosis of the originally separate steles (Fig. 3-9, C, F).

According to the telome theory the elementary process of "reduction" accounts for the origin of the microphyllous type of leaf (Figs. 3-9, D; 3-10). This interpretation rejects the "enation theory" of microphyll evolution (see pp. 30–31) and holds that *both* microphylls and megaphylls originated from subordinated and dichotomously branched portions of a primordial syntelome (see Zimmermann, 1965). Following planation and webbing, a group of telomes became progressively reduced to a single univeined "needle-leaf" or microphyll (Figs. 3-9, D; 3-10). It is of interest to note that Zimmermann's phylogenetic concept of the origin of microphylls includes not only the foliar organs of the Lycophyta but also the univeined leaves of such members of the Sphenophyta as *Equisetum* and the needle-leaves of the Coniferales.

Some support for the reduction concept is derived from the study of certain Devonian lycopods (e.g., *Protolepidodendron, Leclercqia*) in which leaves were bifurcated or had more than two distal segments (Chapter 9). It is possible, of course, that microphylls may have originated phylogenetically in various groups either by the progressive method (enation concept) or by the processes of reduction. The final solution will depend upon future paleobotanical discoveries.

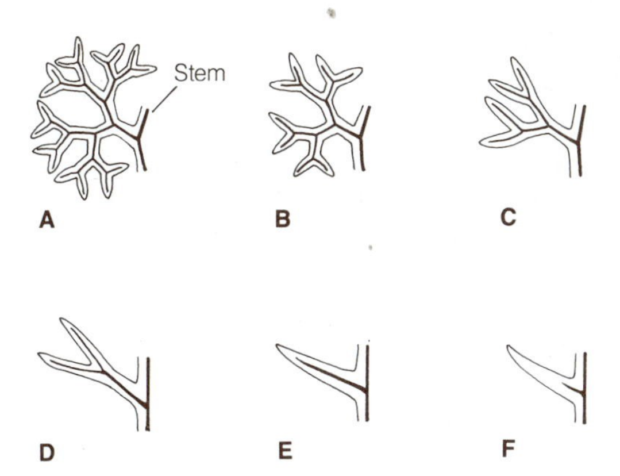

FIGURE 3-10 Origin of the microphyll by processes of reduction from a dichotomizing lateral branchlike system, according to the telome theory (**A–F**). [Redrawn from *Die Telomtheorie* by W. Zimmermann. Gustav Fischer Verlag, Stuttgart. 1965.]

The elementary process of "recurvation" was responsible for the phylogenetic origin of the sporangium-bearing organs of the Sphenophyta. Recurvation, as conceived by Zimmermann, is a process resulting in the bending or "anatropous curvature" of the stalks of a group of sporangia, and is illustrated by the single curved fertile telome represented in Fig. 3-9, E. Subsequent fusion of the bases of the curved sporangium stalks led to the distinctive peltate sporangiophores of the strobilus of the modern genus *Equisetum* (see Zimmermann, 1965; Chapter 10, Fig. 10-12, A).

It was, of course, inevitable that the telome theory, because of its comprehensiveness, would elicit decidedly mixed reactions from paleobotanists and morphologists. In his most recent monograph, Zimmermann (1965) vigorously defended his theory and rejected the numerous objections that have been raised against it. He concludes his book with the statement that the "universality" of his concept has dealt with all the "test cases" and that, so far, no alternative theory has been proposed that so satisfactorily accounts for the enormous diversity of vascular plants as the telome theory.

It should be strongly emphasized that Zimmermann's concept of phylogeny is "hologenetic"—i.e., he regards evolution as the result of the *progressive modification* of ontogenetic processes which have been in operation over a time span of many millions of years. In more explicit terms, an ontogenetically *mature* structure, such as a telome or a system of telomes, does not give rise to a more advanced type of mature organ, such as a microphyll or a megaphyll. On the contrary, from the standpoint of hologeny, new types of fertile and sterile organs are the result of *changes in the genotype* that in turn are expressed ontogenetically by the gradual modification in the organs and structures formed by the embryo and the apical meristems of the developing plant. (For a more detailed discussion of hologeny, see Zimmermann, 1966.)

Recent paleobotanical discoveries have detracted somewhat from the telome theory. Zimmermann selected *Rhynia* (see Chapter 7) from the Devonian as the prototype. One can also cite another genus, *Cooksonia* (mid-Silurian), unknown to Zimmermann when he formulated his concept, in support of his theory. *Cooksonia* was basically dichotomously branched and had terminal sporangia. However, several genera from the Lower Devonian had lateral sporangia on their stems, although the rhyniophyte *Renalia* (see Chapter 7) could be cited as showing how overtopping and reduction functioned to produce short lateral branches with terminal sporangia (see Chapter 7). Paleobotanists will continue to test the validity of the telome concept as they discover other plants of the Silurian and Lower Devonian.

Comparative Anatomy of the Sporophyte

The organographic evolution of the sporophyte has reached various levels in the different groups of living vascular plants. Paleobotanical evidence reveals that the evolution of cell types and tissues was similarly complex and variable. Indeed, one of the most important achievements in paleobotany has been the discovery and description of the well-preserved anatomy of several ancient groups of vascular plants. This has made possible at least the beginning of a true phylogenetic interpretation and classification of cell types and tissues (Foster, 1972). It seems reasonable to conclude this chapter with a brief review of the outstanding aspects of sporophyte anatomy. Of course, this resume should in no sense be regarded as a satisfactory condensation of the vast subject matter of plant anatomy (Esau,

1965), but we hope that it will provide the indispensable orientation needed by readers in an introductory approach to the comparative anatomy of stem, leaf, and root.

Sachs' Classification of Tissue Systems

One of the most useful schemes for understanding the general topographical anatomy of the adult sporophyte was devised by the celebrated German botanist, Julius von Sachs (1875). The great merit of his classification is its simplicity and its wide applicability to the primary structure of the stem, leaf, and root. According to Sachs, the early phylogenetic development of vascular plants resulted in the differentiation of three principal systems of tissues: the external epidermal and cork layers collectively termed the *dermal system;* the strands of conducting phloem and xylem tissue which compose the *fascicular system;* and the remaining internal tissue or tissues designated the *fundamental* or *ground tissue system.* Sachs emphasized that each of these tissue systems may comprise the most varied cell types and that his scheme of classification was concerned with the broadest possible contrast between systems of tissues.

Figure 3-11 shows diagrammatically the application of Sachs' classification and terminology to the gross anatomy of the stem, leaf, and root of flowering plants. This figure indicates plainly that the fascicular tissue system is the most variable of the three, from the standpoint of its pattern of development within plant organs. In stems the fascicular system appears either as a central cylinder of phloem and xylem or in the form of vascular bundles arranged in a cylinder or scattered throughout the ground tissue system (see Fig. 3-11, A; also Fig. 3-20). The form of the fascicular system of the leaf ranges from one to many bundles or a cylinder in the petiole, to a complex system of veins, usually arranged in a single plane, in the lamina (Fig. 3-11, B). In the root the pattern of the fascicular system is very distinctive, consisting of a radial and alternate series of phloem and xylem strands; commonly the latter are joined at their inner edge to form a solid core of xylem as shown in Fig. 3-11, C.

In contrast with these diverse patterns exhibited by the fascicular system, the form and arrangement of Sachs' other tissue systems are comparatively simple. The dermal system, in all foliar organs and in young stems and roots, is represented by the epidermis (Figs. 3-11, 3-12). Typically this is a single layer of superficial cells which are tightly joined except for the stomatal openings; in some groups of dicotyledons the leaf may develop on one or both surfaces a multiple epidermis consisting of two or

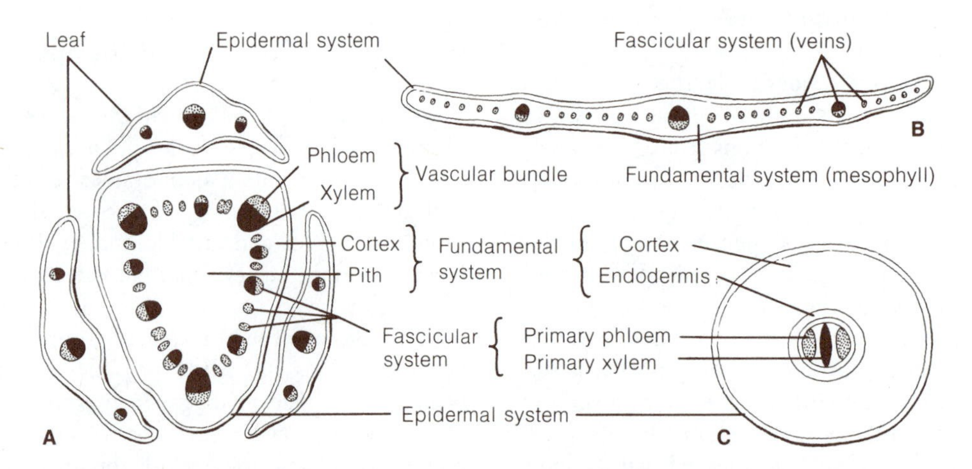

FIGURE 3-11 Diagrams (based on *Linum usitatissimum*) illustrating the positions and patterns of the epidermal, fascicular, and fundamental tissue systems in the vegetative organs of a dicotyledon. **A,** transection of stem and three leaf bases; **B,** transection of lamina of leaf; **C,** transection of root. [Redrawn from *Plant Anatomy* by K. Esau. Wiley, New York. 1953.]

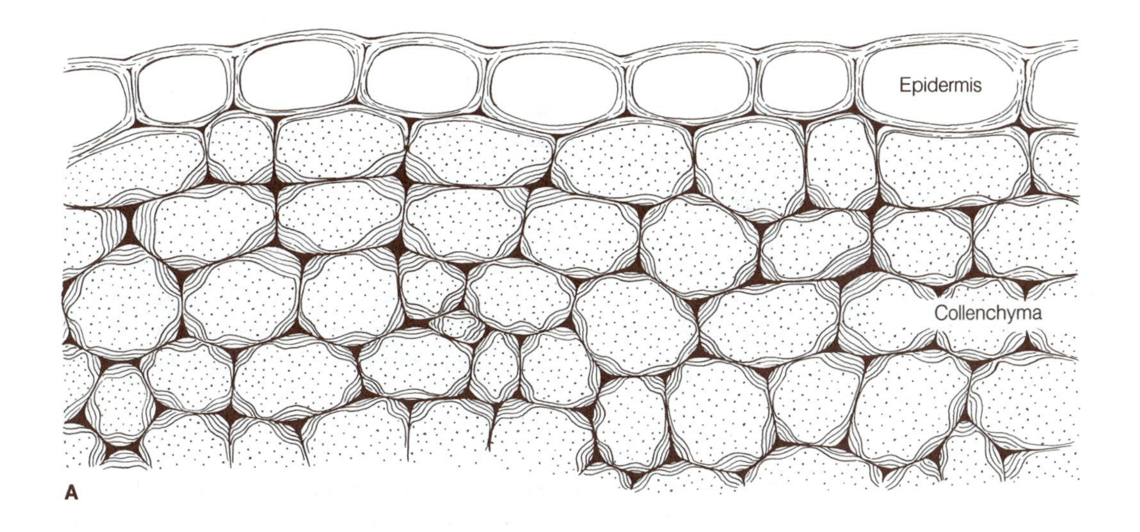

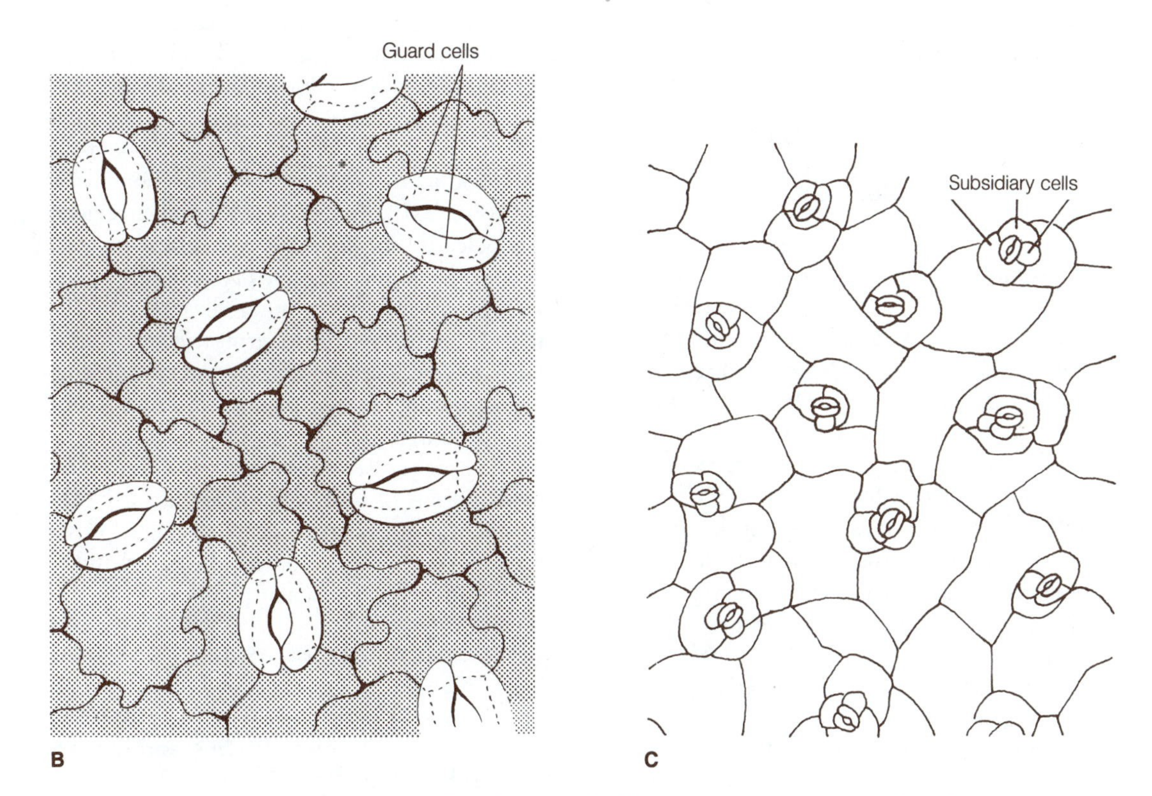

FIGURE 3-12 **A,** transection of epidermis and collenchyma tissue in stem of *Cucurbita;* **B,** surface view of stomata in lower epidermis of leaf of *Capsicum;* **C,** surface view of stomata with subsidiary cells in lower epidermis of leaf of *Sedum.* [**A,** redrawn from *Plant Anatomy* by K. Esau. Wiley, New York. 1953; **B,** courtesy of E. F. Artschwager; **C** redrawn from *Anatomy of Seed Plants,* 2d edition, by K. Esau. Wiley, New York. 1977.]

many layers of cells, all of which have originated from the subdivision of the original surface layer of cells (Fig. 3-13, C). In the stems and roots of many vascular plants the epidermis is eventually sloughed away by the development beneath it of *cork*. Like the cells of the epidermis, cork cells are compactly arranged, and it is only at certain areas known as lenticels that well-developed intercellular air-space systems are found. Situated below the dermal system and external to or surrounding the fascicular system is the fundamental tissue system (Fig. 3-11). In dicotyledonous stems, for example, this tissue system is represented by the cortex—a cylinder of tissues between epidermis and phloem, and the pith—a central column of parenchyma (see Fig. 3-11, A). The stems of the majority of the Lycophyta and the roots of most plants develop the cortical portion of the fundamental system but lack a pith (Fig. 3-11, C). In the lamina of leaves the photosynthetic mesophyll represents the fundamental tissue system (Fig. 3-11, B).

Structure and Development of Tissue Systems

In addition to its value in the topographical description and comparison of primary tissue systems, Sachs' scheme is very helpful in the morphological interpretation of *tissue development* at the apices of shoots and roots. Apical meristems, as shown by numerous studies, are extraordinarily variable in their histology. (See Esau, 1965; Gifford, 1954; Gifford and Corson, 1971, for literature on the subject.) In some plants, for example, *Equisetum* and the leptosporangiate ferns, a single well-defined apical cell occupies the tip of the axis and represents the ultimate point of origin of all the meristematic tissue of the apex of root and shoot (see Fig. 19-28). However, the shoot apices of *Lycopodium,* of certain eusporangiate ferns, and of a great many gymnosperms possess several superficial apical initials. And finally, the shoot apices of the angiosperms have a typical stratified arrangement of cells, the outer layer or layers being the *tunica,* which surrounds a central mass of meristem designated the *corpus;* in this type of apex, the number and position of individualized apical initials is very uncertain in most instances (Figs. 3-1, B; 19-18). Despite the histological differences among the types of apices in vascular plants, an essentially similar pattern of early histogenesis or tissue formation is common to all of them. This consists in the ultimate segregation, behind the apex of shoot and root, of three primary meristematic tissues: the *protoderm,* the *procambium,* and the *ground meristem* (Fig. 3-1). These tissues are the precursors of the epidermal, fascicular, and fundamental tissue systems, respectively; we will now briefly examine their salient features.

THE EPIDERMIS. The protoderm, derived from the cells of the shoot or root apex, is a uniseriate layer of dividing cells that ultimately forms the epidermis of mature organs (Figs. 3-1, 3-12, 3-13). Certain protoderm cells enlarge, acquire cutinized outer walls, and differentiate as the typical epidermal cells of leaves and stems. In many plants large numbers of protoderm cells develop into the various types of epidermal appendages or trichomes. The stomata are characteristic of the epidermal system of foliage leaves, and many types of stems, floral organs, and fruits (Fig. 3-12, B, C). Paleobotanical evidence shows clearly that stomata were present in such ancient and simple vascular plants as the Rhyniophyta. Stomata develop by the division and differentiation of certain protoderm cells into pairs of guard cells between which a stomatal opening or pore is formed. In many plants two or more of the cells bordering upon the stoma are distinctive in form and are termed subsidiary cells (Fig. 3-12, C). The epidermis of fossil plants, particularly the arrangement and structure of its stomata, provides very important clues about the phylogeny and relationships of extinct vascular plants.

THE FUNDAMENTAL TISSUE SYSTEM. This system of tissues originates from the ground meristem and is represented by the tissues found in the cortex of stems and roots, the pith of stems, and the mesophyll of foliar organs (Figs. 3-1; 3-11). In contrast to the elongated and often spindle-shaped procambial cells, ground meristem cells, prior to differentiation, are polyhedral cells which closely approximate tetrakaidecahedra (fourteen-sided bodies) in form. Cells of this type very commonly enlarge,

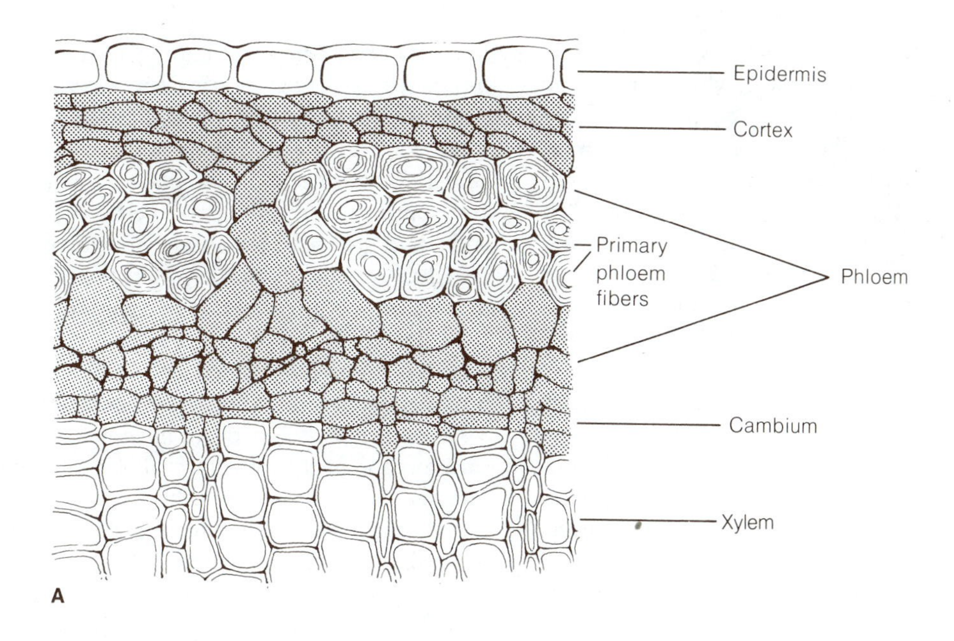

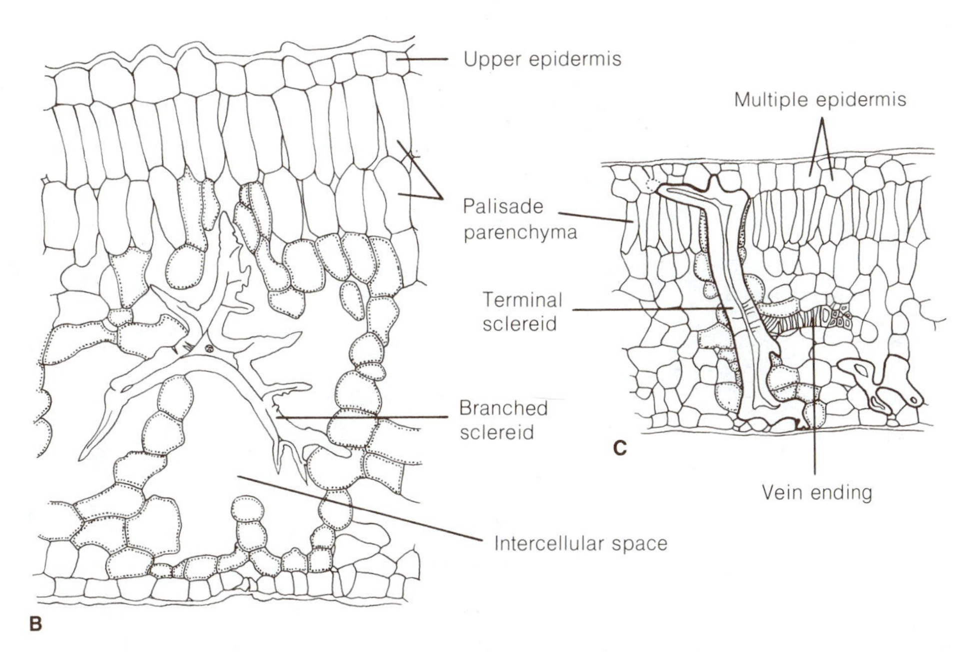

FIGURE 3-13 **A,** transection, stem of *Linum usitatissimum* illustrating position of strands of primary phloem fibers; **B,** transection, portion of lamina of *Trochodendron aralioides* showing a branched sclereid; **C,** transection, portion of lamina of *Mouriria huberi,* showing a columnar, ramified terminal sclereid. [**A,** redrawn from *Plant Anatomy* Ed. 2, by K. Esau, Wiley, New York, 1965; **B,** redrawn from A. S. Foster, *Amer. J. Bot.* 32:456, 1945; **C,** redrawn from A. S. Foster, *Amer. J. Bot.* 34:501, 1947.]

become separated by intercellular spaces, and mature into the parenchyma tissue, which often is the principal component of the fundamental tissue system. But additional cell types and tissues may originate from unspecialized ground meristem and become part of the fundamental tissue system of plant organs. A common example of this is *collenchyma,* which is very commonly developed in the outer region of the cortex of stems and in the subepidermal region of petioles (Fig. 3-12, A). In the ontogeny of collenchyma the ground meristem cells divide and elongate, ultimately producing compact strands or a cylinder of living cells with unevenly thickened primary walls; very commonly the thickest portions of the wall are laid down at the angles or corners where several collenchyma cells meet. Regarded functionally, collenchyma tissue provides support and flexibility for growing organs and at maturity is characterized by considerable tensile strength.

Another extremely common type of tissue in the fundamental tissue system is *sclerenchyma,* which is composed of cells with thick, lignified secondary walls. Two fairly well-demarcated cell types are included under sclerenchyma: *fibers,* which typically are conspicuously elongated cells with pointed ends, and *sclereids,* which are polygonal (the so-called stone cells), columnar, or profusely branched in form. At maturity fibers very commonly are dead cells, devoid of cytoplasm, and occur as strands or cylinders of tightly joined cells which evidently provide mechanical strength to plant organs (Fig. 3-13, A). Sclereids may occur in compact masses in various parts of the fundamental tissue system, but they also occur as isolated cells or *idioblasts.* In the leaves of many dicotyledons, branched idioblastic sclereids are frequently diffuse in their distribution in the mesophyll (Fig. 3-13, B), but in certain genera the sclereids are terminal, or restricted to the vein endings (Fig. 3-13, C).

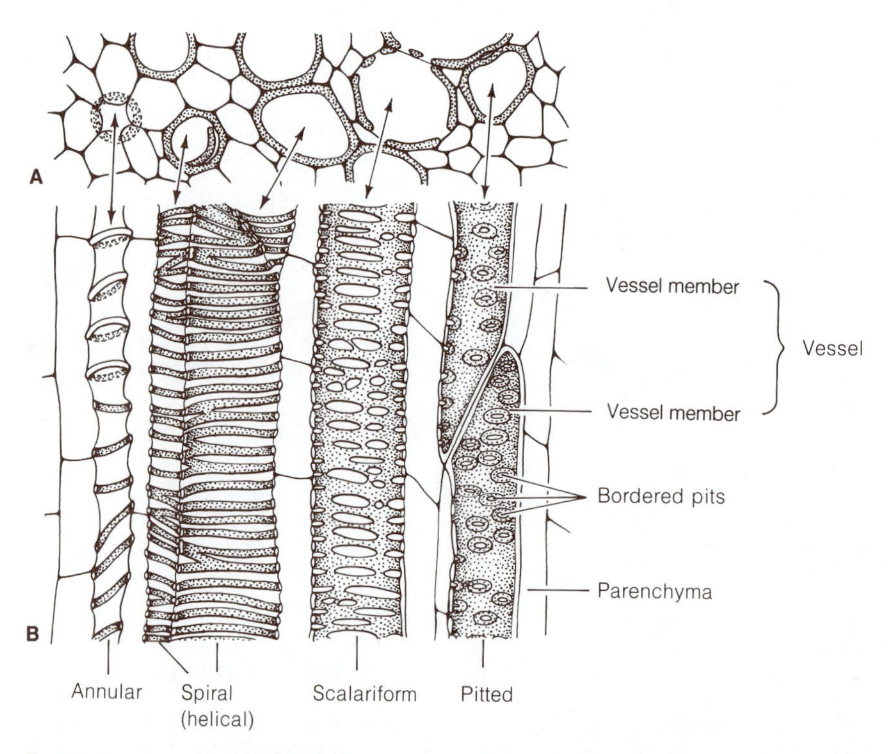

FIGURE 3-14 A portion of the primary xylem of the stem of *Aristolochia* in transverse (**A**) and longisectional (**B**) view. Note diversity in types of secondary wall patterns of tracheary elements in progressing from protoxylem at left to metaxylem at right. [From *Plant Anatomy* by K. Esau. Wiley, New York. 1953.]

THE FASCICULAR TISSUE SYSTEM. From what we have said regarding the variable patterns of the fascicular system, a corresponding variability is to be expected in the patterns of procambium formation in young organs. In the young, terminal regions of many stems (particularly the stems of gymnosperms and angiosperms) and in differentiating leaves the procambium consists of discrete cellular strands composed of elongated cells (Fig. 3-1, B); each strand matures as a vascular bundle composed of primary phloem and primary xylem tissue. In the stems of certain lower vascular plants (*Lycopodium,* and certain ferns) and in many roots, the procambium is a central core or column of tissue from which the vascular cylinder, devoid of pith, originates.

It is important first to understand the general structure and organization of primary xylem and the meaning of the terms *protoxylem* and *metaxylem.* Protoxylem designates the pole of earliest developed primary xylem and includes all tracheary tissue that differentiates (i.e., completes its growth and secondary wall development) during the period of organ elongation. Tracheary elements of the protoxylem, as shown in Fig. 3-14, often develop their secondary walls as a series of rings (annular elements) or as one or more spiral bands (spiral elements). The metaxylem is the remaining portion of the primary xylem, which completes its differentiation after the organ in which it occurs has ceased to elongate. Metaxylem cells, also shown in Fig. 3-14, usually have more extensively developed secondary walls which commonly appear as a series of connected bars (scalariform elements) or a network (reticulate elements), or else the wall is pitted.

The common and basic type of tracheary element in the xylem of vascular plants is the *tracheid.* A tracheid is a water-conducting cell, dead at maturity, generally elongate, with a lignified secondary wall. The dimensions of tracheids vary considerably among different vascular plant groups. In ferns, cycads, and conifers the length varies from about 1 millimeter to 5 to 7 millimeters, whereas in most flowering plants (angiosperms) a tracheid may be 1 to 2 millimeters or less in length. A conspicuous morphological feature of tracheids with extensive secondary walls is the occurrence of pits (Fig. 3-15). Pits are not complete holes, but are small areas lacking secondary wall deposition. The pits occur

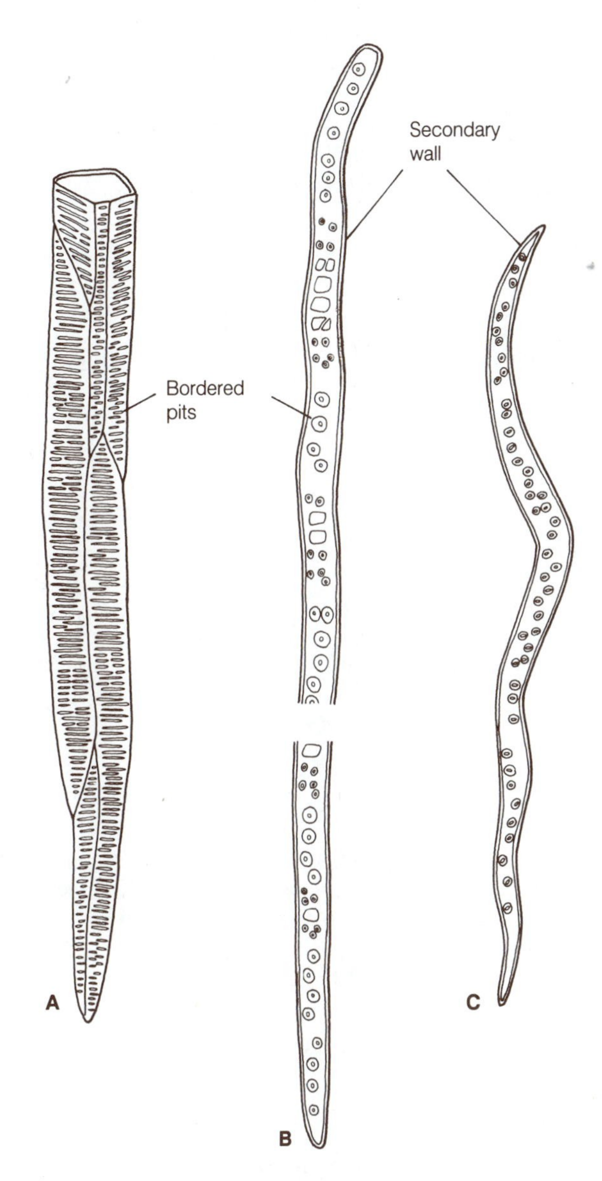

FIGURE 3-15 Representative tracheids from three major groups of vascular plants. **A,** *Woodwardia,* a fern (one-sixth of cell shown). **B,** *Pinus,* a conifer (one-third of cell shown). **C,** *Quercus,* oak. [Redrawn from *An Introduction to Plant Anatomy* by A. J. Eames and L. H. MacDaniels, McGraw-Hill, New York. 1947.]

opposite each other (pit pairs) in adjacent cells. This alignment more readily permits the movement of water through the xylem.

The second type of tracheary element is the *vessel member,* which may resemble a tracheid but has at maturity several actual holes (perforations) or

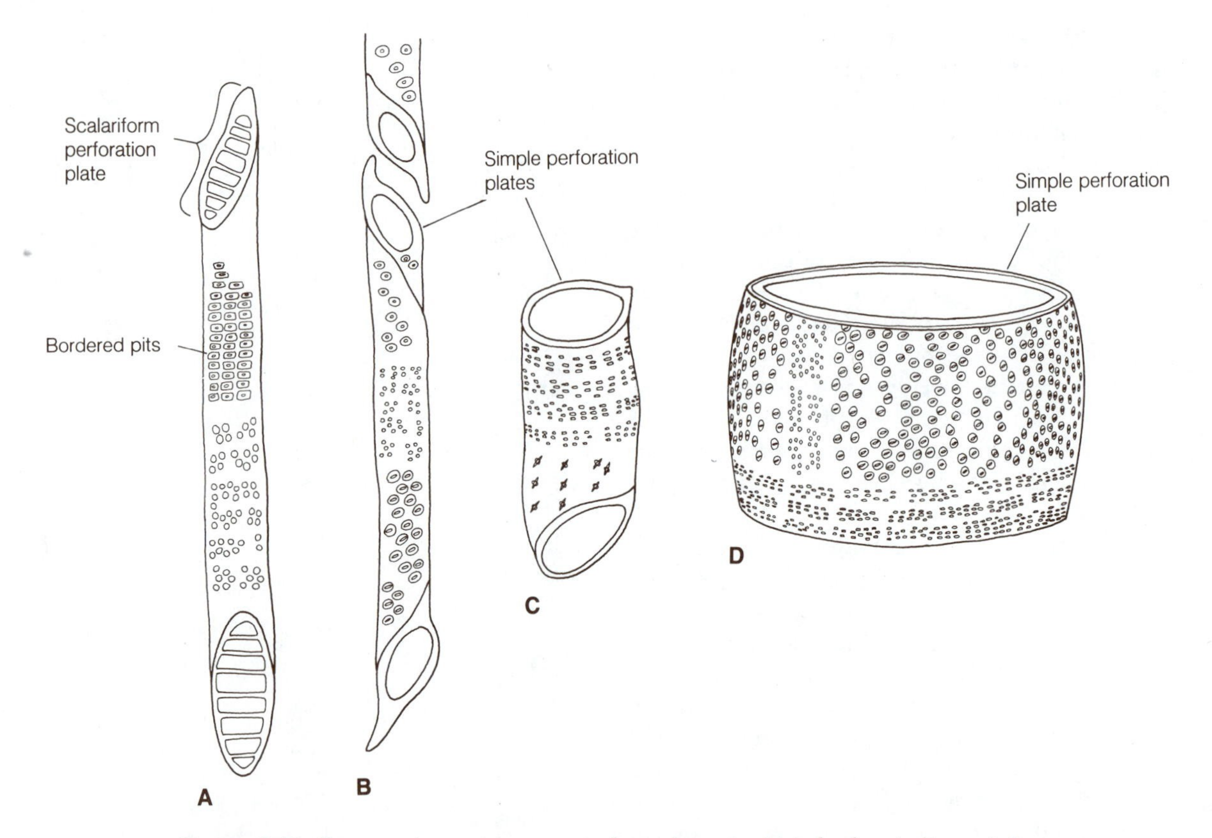

FIGURE 3-16 Representive angiosperm vessel members. **A,** *Liriodendron* (tulip tree); **B,** *Quercus* (oak); **C,** *Acer* (maple); **D,** *Quercus.* [Redrawn from *An Introduction to Plant Anatomy* by A. J. Eames and L. H. MacDaniels. McGraw-Hill, New York. 1947.]

one large hole at each end of the cell (Fig. 3-16). Several to many vessel members, joined end to end, constitute a *vessel.* The possession of vessels is the prevailing condition in angiosperms (Magnoliophyta), and they also occur in some other groups of vascular plants. Pits occur in the lateral walls where each vessel member is in contact with the other tracheary elements and parenchyma (Fig. 3-16). Perforations permit the conduction of a larger volume of water per unit of time by reducing the impedance to water.

It must be emphatically stated that the primary xylem may consist wholly of tracheids or of both tracheids and vessel members. The type of secondary wall pattern is therefore not necessarily correlated with the presence or absence of vessel members.

Because the xylem of extinct plants is often well preserved in fossils, the recognition of the position of protoxylem in relation to metaxylem is a matter of considerable significance in phylogenetic interpretations. In the roots of all vascular plants, in the stem of *Psilotum,* and in the majority of all investigated Lycophyta, the first protoxylem cells to acquire secondary walls occur at the outermost edge (in the direction of the surface of stem and root) of the procambial cylinder; these cells establish the future pattern of xylem differentiation which

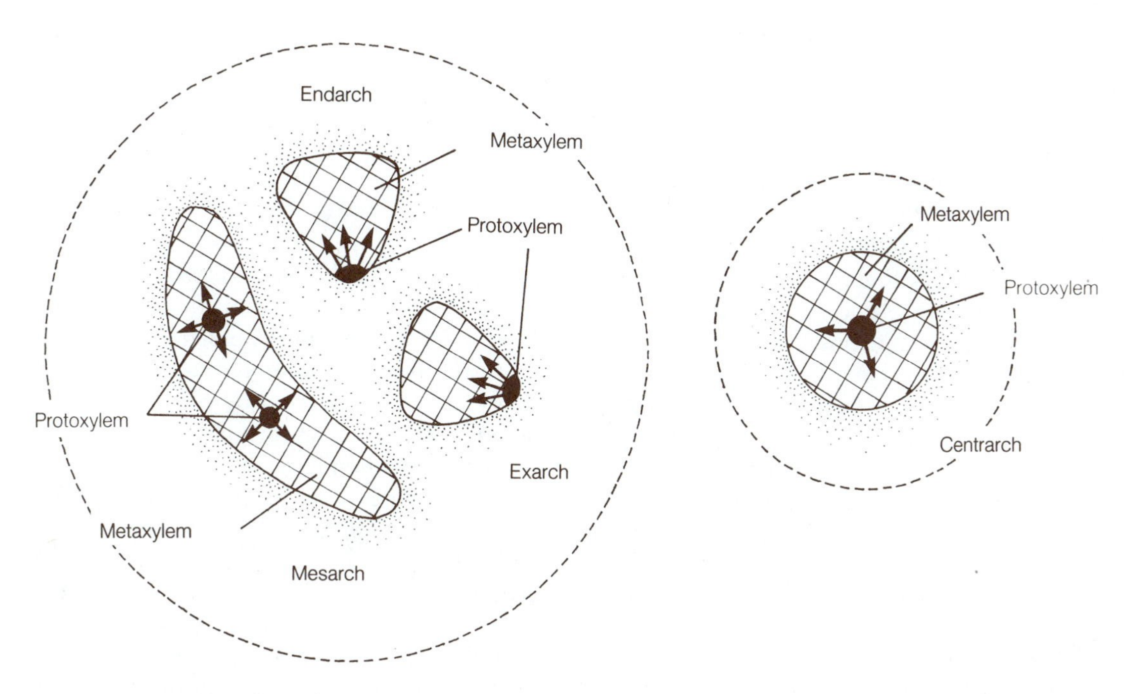

FIGURE 3-17 Schematic representation of the directions of radial maturation of tracheary elements in the primary xylem of vascular plants. Broken lines indicate the periphery of the stems.

occurs *centripetally*—toward the center of the axis. Primary xylem of this type is termed *exarch* and is regarded as a primitive condition in vascular plants (Fig. 3-17). In the stems of modern seed plants, however, the protoxylem begins its development from the innermost procambial cells—those situated next to the pith—and the remainder of the process of primary xylem differentiation occurs *centrifugally* or toward the periphery of the stem (Fig. 3-17). This type of primary xylem is termed *endarch xylem* and is believed to be the most highly advanced type. In the leaf and stem bundles of many ferns the primary xylem is *mesarch*. This means that the protoxylem begins development within a procambial strand and that further xylem formation occurs centripetally as well as centrifugally; consequently, at maturity the protoxylem cells are surrounded by the metaxylem, as can be seen in Fig. 3-17. Another type, *centrarch* xylem, is now recognized in certain extinct plants of the Devonian. In these plants (e.g., *Rhynia*) there is but one vascular strand in the stem and the protoxylem is in the center (hence, centrarch) surrounded by metaxylem (Fig. 3-17; see also Chapter 7). The distinction between these types of xylem maturation is of considerable importance, not only among different groups of vascular plants but even between root and stem of the same plant.

The specialized types of cells involved in translocation of photosynthetic products are termed *sieve elements*. A definitive feature, at least in angiosperms, is the absence at maturity of a nucleus and the presence of more or less specialized *sieve areas* and *sieve plates* in the cell walls. A sieve area is a modified portion of the primary wall traversed by connecting strands of protoplasm. Each of these, at an early ontogenetic stage, is surrounded by a cylinder of *callose* (1 → 3(-β-d)-glucan), a carbohydrate differing from cellulose in the type of linkage between molecules. One type of sieve element is the *sieve cell* that occurs in lower vascular plants and gymnosperms. The sieve cells are elongate and often have numerous sieve areas (Fig. 3-18, A, B). Each sieve area is matched by another one in an adjacent sieve cell, thus providing an interconnected system throughout the phloem. In many angiosperms the sieve elements (termed *sieve-tube members*) occur in superposed series collectively termed *sieve tubes*. The end walls of sieve-tube members may have several large sieve areas (Fig. 3-18, C, D) or one large *sieve plate* in which the pores are larger than in sieve areas (Fig. 3-18, E–G). In addition to the specialized sieve elements (sieve cells and sieve-tube members), primary as well as secondary phloem may contain parenchyma, sclereids, and fibers. For detailed treatments of ontogeny and structure of both primary and secondary phloem consult Esau, Cheadle, and Gifford, 1953, and Esau, 1965, 1969, 1977.

Last we need to consider the concept of secondary vascular tissues and their demarcation from the primary vascular system. Secondary vascular tissues are produced from the *vascular cambium,* and since the cells of the secondary xylem are often arranged in radial rows this system of tissues often is clearly demarcated from the more irregular pattern of cells of the primary vascular system. But the criterion of orderly versus irregular cell arrangement is not always valid; in some angiosperms and gymnosperms the tracheary cells of the primary as well as the secondary xylem are in regular radial alignment. As a consequence, the boundary between the primary and secondary vascular systems can be only approximately determined even when the entire ontogenetic development has been studied. Secondary growth by means of a vascular cambium has repeatedly arisen during the evolution of vascular plants. Many extinct groups (for example, *Lepidodendron* and *Calamites;* see Figs. 10-27, B; 9-44) showed conspicuous secondary growth, and secondary growth is a prominent feature of all living gymnosperms and of a large number of the angiosperms. (See Chapters 17 and 19 for detailed descriptions of the secondary xylem of conifers and dicotyledons.) However, most of the lower vascular plants of today are devoid of cambial activity. In these organisms, as in most of the monocotyledons, the fascicular system is entirely primary and derived ontogenetically from the procambium.

The Stelar Theory

During the latter half of the nineteenth century the increasing emphasis placed on the importance of the vascular system in morphological interpretation led to the formulation of the stelar theory. This theory, which was developed by Van Tieghem and his students (Van Tieghem and Douliot, 1886) deserves our attention because of its far-reaching effects on modern concepts of the morphology and evolution of the primary vascular system. According to Van Tieghem, the primary structure of the stem and root are fundamentally similar in that each organ consists of a central *stele* enveloped by the cortex, the outer layer of which is the epidermis. The term stele was used in a collective sense by Van Tieghem to designate not only the primary vascular tissues but also the so-called conjunctive tissues associated with them: pericycle, vascular rays, and, when it occurs, the pith tissue.

One of the critical—and, in the light of modern studies, controversial—aspects of the stelar theory is the nature of the anatomical boundaries which separate the cortex from the stele. Van Tieghem considered that the inner boundary of the cortex is the endodermis, a cylinder of living cells which, from a strict histological viewpoint, are characterized by the presence of casparian strips. These strips or bands are chemically modified portions of the radial and end walls of the endodermal cells and are thought to contain both lignin and suberin.

In roots and in the stems of many of the lower vascular plants an endodermis is present and represents a tangible boundary between cortex and stele (Fig. 8-4, A). But an endodermis, in the sense just defined, is absent from the stems of a large proportion of the seed plants, especially woody types, and in plants lacking an endodermis the limits between cortex and stele are more difficult to establish. The pericycle likewise is not present in all vascular plants. Although the pericycle is a recognizable cylinder of cells at the outer edge of the stele of roots and the stems of lower vascular plants, ontogenetic

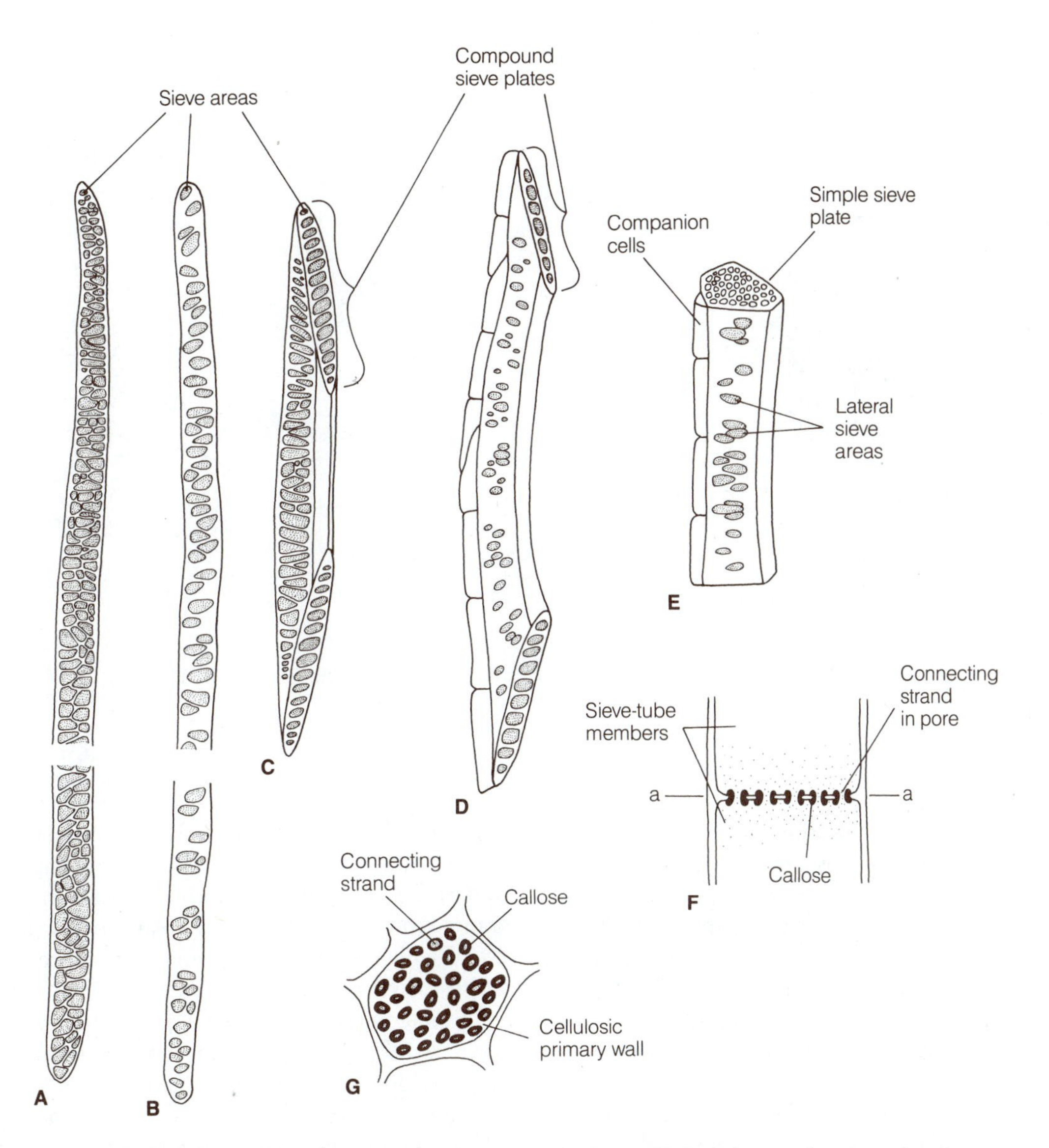

FIGURE 3-18 Sieve elements of various taxa. **A,** sieve cell, *Pteridium,* a fern (one-fourth of cell shown); **B,** sieve cell, *Tsuga,* a conifer (one-third of cell shown); **C,** sieve-tube member, Juglans, an angiosperm; **D,** sieve-tube member, *Liriodendron,* an angiosperm; **E,** sieve-tube member, *Robinia,* an angiosperm; **F,** schematic representation of a longitudinal section of a sieve plate; **G,** sieve plate in transverse section (plane a-a in **F**). [**A–E** redrawn from *An Introduction to Plant Anatomy* by A. J. Eames and L. H. MacDaniels. McGraw-Hill, New York. 1947.]

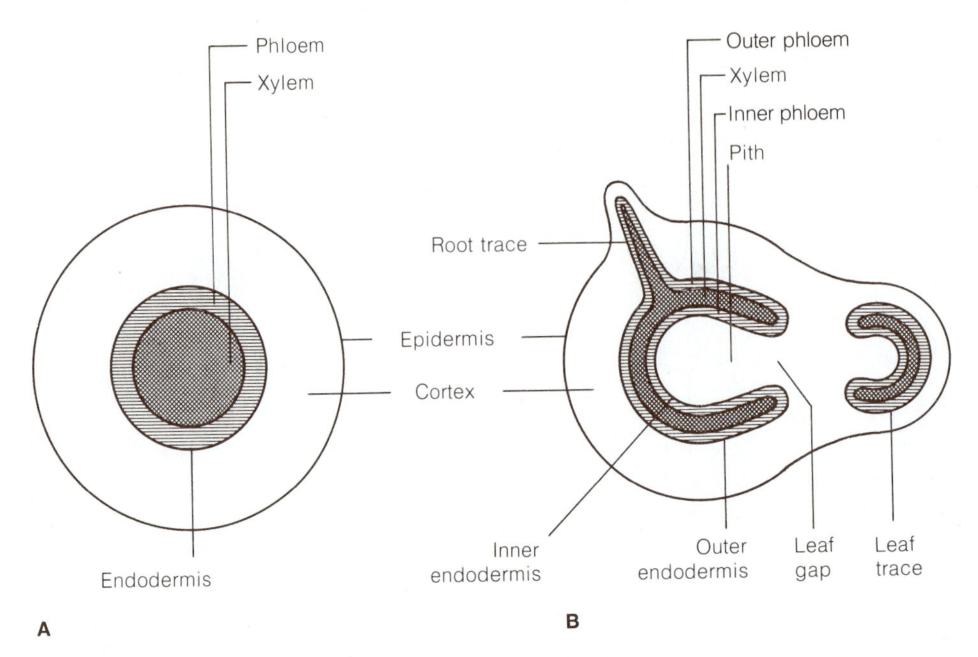

FIGURE 3-19 Types of steles in the stems of vascular plants. **A,** the protostele; **B,** the siphonostele. [Redrawn from *The Anatomy of Woody Plants* by E. C. Jeffrey. University of Chicago Press, Chicago. 1917.]

studies have shown that the so-called pericycle in the stems of many angiosperms is actually the outermost portion of the primary phloem (Blyth, 1958). In such instances there is no independent tissue zone separating the cortex from the stele.

Despite the absence of consistent histological boundaries between the cortex and stele, the value of the stelar theory as a unified concept has been widely recognized and has led to efforts to classify and interpret phylogenetically the varied types of vascular cylinders that occur in stems and roots.

Stelar Types

It is now rather generally agreed that from a phylogenetic as well as an ontogenetic standpoint the most primitive type of stele is the *protostele,* which is characterized by the absence of a central column of pith. In its simplest form the protostele is merely a central strand of primary xylem sheathed by a cylinder of phloem (see Fig. 3-19, A). This particular form of protostele is often termed a *haplostele.* In certain plants the contour of the core of xylem is lobed or star-shaped in transectional view — this form is designated an *actinostele* (Chapters 8 and 9). Interconnected strands of xylem, which in transection appear as separate plates of tissue between which occurs the phloem is a specialized type of protostele termed a *plectostele.*

A stele with a central column of pith, called a *siphonostele,* is regarded as an advance in anatomical development. There is considerable difference of opinion among morphologists concerning the phylogenetic origin of the pith which will be discussed in later chapters.

The histological structure of the siphonostele varies widely. In many fern genera the vascular cylinder consists of xylem surrounded on both sides by phloem and endodermal layers (outer endodermis, outer phloem and inner endodermis, inner phloem; Fig. 3-19). This condition is termed *amphiphloic,* and a stele with this construction is specifically termed a *solenostele.* Solenosteles commonly have *leaf gaps* (Fig. 3-19, B). A leaf gap occurs where parenchyma interrupts the vascular cylinder of the stem above the departure of a leaf trace and connects the cortex with the pith. The cylinder is complete again in the internode before another leaf trace departs at a higher level of the stem. In this case the leaf gaps are said to be nonoverlapping. In the region of the leaf gap, the inner and outer endodermal layers are continuous around the margins of the gap (Fig. 3-19, B).

Evolutionary modification of the solenostele in ferns is believed to have resulted in a type of stele with two or more overlapping leaf gaps at any given level of the stem. This type of vascular cylinder is termed a *dictyostele*. When visualized in three dimensions, a stele of this type is a network of interconnected vascular strands (Chapter 13). In some dictyosteles some of the parenchymatous regions are not truly leaf gaps, but interfascicular strips of parenchyma without a consistent relationship to the leaf trace system. This type of stele is termed a "dissected" dictyostele. A dictyostele, as viewed in transverse section of the stem, appears as a ring of separate bundles each of which is *amphicribral,* i.e., it is composed of a central strand of xylem surrounded by a sheath of phloem, followed by a pericyclic and an endodermal layer. A concentric bundle of this type is commonly called a "meristele"—a portion of the interconnected vascular system (Chapter 13). In some ferns only the outer phloem is present; this type of siphonostele is termed *ectophloic* (Chapter 13, Fig. 13-20, B).

In dicotyledons, the primary vascular cylinder also appears, in transections of the stem, as a ring of more-or-less discrete vascular bundles separated by areas of parenchyma (see Fig 3-11, A; and Chapter 19, Fig. 19-20, A). This stelar type, termed the *eustele* by Brebner (1902), is distinguished from the fern dictyostele by the collateral, or bicollateral, organization of the vascular strands. Most commonly, each vascular bundle is collateral, consisting of a strand of xylem flanked externally by a strip of phloem tissue (Fig. 3-11, A). In certain taxa of the dicotyledons (e.g., *Cucurbita*) the vascular bundles are bicollateral with phloem on both sides of xylem strands. It should be emphasized at this point that the divergence of leaf traces from a eustele is associated with parenchymatous areas that are usually regarded as leaf gaps. However, it proves very difficult to demarcate the limits of each leaf gap and, as we will show in Chapter 14, the concept of leaf gap in the eusteles of both dicotyledons and conifers has been rejected by several investigators.

In our brief account of stelar types, we have attempted to show the considerable value of the concepts of protostele, siphonostele, solenostele, dictyostele, and eustele in the interpretation of the diverse patterns of the primary vascular system of the axis. Monocotyledons, however, pose a very difficult problem because the primary vascular system of the stems in this group of angiosperms usually does not conform to any of the stelar types found in other groups of vascular plants. In place of a concentric cylinder of primary xylem and phloem, or a tube of anastomosed strands, the vascular bundles in the stems of monocotyledons appear dispersed or "scattered" in arrangement; no clear boundaries exist between cortex and pith, and the divergence of the numerous leaf traces at each node is obviously not associated with definable leaf gaps. Brebner (1902, p. 520), to whom we owe many terms used to designate stelar types, proposed the term *atactostele* (from the Greek root word "atactos" meaning "without order") for the stele in monocotyledons. He defined it as consisting of "a number of more or less irregularly arranged vascular bundles together with the ground tissue in which they are imbedded." However, in a recent review, Schmid (1982) has recommended that the term atactostele be abandoned and the term eustele be applied to monocotyledons also. Although the term "atactostele" has been widely adopted for descriptive purposes (Esau, 1965, 1977; Fahn, 1967; Zimmermann, 1959), determining the significance —ontogenetic or phylogenetic—of the scattered arrangement of bundles in monocotyledonous stems has proved a most difficult and elusive problem. The magnitude of the problem is well illustrated by reference to bundle number in certain palms. According to Zimmermann and Tomlinson (1965, p. 165), about 1,000 bundles occur in the small stem (2 to 3 centimeters in diameter) of *Rhapis excelsa,* and in stems of *Cocos,* one-half meter in diameter, there may be more than 20,000 vascular bundles "at any level"! In grasses, commonly used as examples of "typical" monocotyledons, the number of bundles is considerably less. Transections of the stem of *Secale,* for example, reveal a primary vascular system consisting of two rings of vascular bundles located in the outer peripheral region of the internode (Fig. 3-20, A). A contrasting pattern of bundle arrangement is illus-

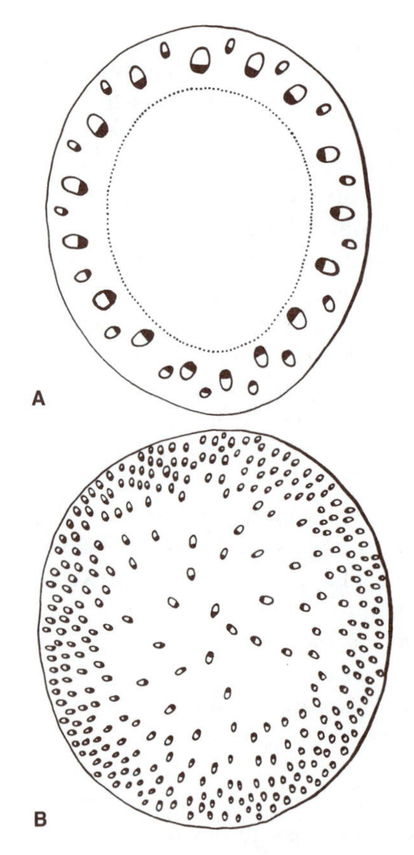

FIGURE 3-20 Transectional diagrams showing two patterns of vascular bundle arrangement in the stems of grasses. **A,** *Secale* (rye)—two "circles" of bundles at peripheral region of hollow stem; **B,** *Zea* (corn)—numerous bundles scattered throughout the transection. [Redrawn from *Allgemeine Botanik* by W. Troll. Ferdinand Enke Verlag, Stuttgart. 1959.]

trated by the stem of corn *(Zea)* in which numerous small bundles are closely spaced at the periphery, and larger, more widely spaced strands are distributed in the central region (Fig. 3-20, B).

To follow the course of vascular bundles in monocotyledonous stems is a difficult task. It is only in recent years that a series of studies was initiated by Zimmermann and Tomlinson designed to explain the course of vascular bundles in stems and their complex behavior with reference to the vascularization of leaves and axillary inflorescences. These investigators devised an "optical shuttle method" for photographing, by means of a motion picture camera, successive transections of the stems of *Rhapis* (Palmaceae) and *Prionium* (Juncaceae), enabling them to follow the course of vascular bundles. This method has resulted in considerable progress in reconstructing the extremely complex vascular system characteristic of these monocotyledons. Readers wishing further information should consult the original papers by Zimmermann and Tomlinson (1965, 1966, 1967, 1968, 1972) and the critical review by Beck et al. (1982).

REFERENCES

Beck, C. B., R. Schmid, and G. W. Rothwell
1982. Stelar morphology and the primary vascular system of seed plants. *Bot. Rev.* 48:691–815.

Blyth, A.
1958. Origin of primary extraxylary stem fibers in dicotyledons. *Univ. Calif. Publ. Bot.* 30:145–232.

Bock, W.
1962. *Systematics of Dichotomy and Evolution.* (Geological Center Research Series. Vol. 2.) Geological Center, North Wales, Pa.
1969. *The American Triassic Flora and Global Distribution.* (Geological Center Research Series. Vols. 3 and 4.) Geological Center, North Wales, Pa.

Bower, F. O.
1935. *Primitive Land Plants.* Macmillan, London.

Brebner, G.
1902. On the anatomy of *Danaea* and other Marattiaceae. *Ann. Bot.* 16:517–552.

Cutter, E. G.
1965. Recent experimental studies of the shoot apex and shoot morphogenesis. *Bot. Rev.* 31:7–113.

Esau, K.
1965. *Plant Anatomy,* 2d edition. Wiley, New York.
1969. *The Phloem.* (Handbuch d. Pflanzenanatomie. Band V, Teil 5.) Gebrüder Borntraeger, Berlin-Stuttgart.
1977. *Anatomy of Seed Plants,* 2d edition. Wiley, New York.

Esau, K., V. I. Cheadle, and E. M. Gifford, Jr.
1953. Comparative structure and possible trends of specialization of the phloem. *Amer. Jour. Bot.* 40:9-19.

Fahn, A.
1967. *Plant Anatomy.* Pergamon, Oxford.

Foster, A. S.
1972. Cell types: Spermatophytes. Table 12, Part I, pp. 132–135 in *Biology Data Book,* 2d

edition, Vol. I. Federation of Amer. Soc. for Exper. Biology, Bethesda, Maryland.

Garrison, R., and R. H. Wetmore
1961. Studies in shoot-tip abortion: *Syringa vulgaris*. *Amer. Jour. Bot.* 48:789–795.

Gifford, E. M., Jr.
1954. The shoot apex in angiosperms. *Bot. Rev.* 20:477–529.

Gifford, E. M., Jr., and G. E. Corson, Jr.
1971. The shoot apex in seed plants. *Bot. Rev.* 37:143–229

Lignier, O.
1903. Equisétales et Sphenophyllales. Leur origine filicinéenne commune. *Bull. Soc. Linn. Normandie* Ser. 5. 7:9–137.

Millington, W. F.
1963. Shoot-tip abortion in *Ulmus americana*. *Amer. Jour. Bot.* 50:371–378.

Sachs, J., von
1875. *Textbook of Botany*. Clarendon Press, Oxford.

Sandt, W.
1925. *Zur Kenntnis der Beiknospen*. Bot. Abh., herausg. von K. Goebel. Heft 7. Gustav Fischer, Jena.

Schmid, R.
1982. The terminology and classification of steles: historical perspective and the outlines of a system.*Bot. Rev.* 48:817–931.

Stewart, W. N.
1964. An upward outlook in plant morphology. *Phytomorphology* 14:120–134.

Troll, W.
1935. *Vergleichende Morphologie der höheren Pflanzen*. Bd. 1, Lieferung 1. Pp. 101–107. Gebrüder Borntraeger, Berlin.
1937. *Vergleichende Morphologie der höheren Pflanzen*. Bd. 1, Lieferung 2. Pp. 465–660. Gebrüder Borntraeger, Berlin.

Troop, J. E., and J. T. Mickel
1968. Petiolar shoots in the Dennstaedtioid and related ferns. *Amer. Fern Jour.* 58:64–70.

Van Tieghem, P., and H. Douliot
1886. Sur la polystélie. *Ann. Sci. Nat. Bot.* Sér. 7. 3:275–322.

Wardlaw, C. W.
1952. *Phylogeny and Morphogenesis*. St. Martin's Press, New York.
1957. Experimental and analytical studies of pteridophytes. XXXVII. A note on the inception of microphylls and macrophylls. *Ann. Bot.* n.s. 21:427–437.
1965. *Organization and Evolution in Plants*. Longmans, Green, London.

Wilson, C. I.
1953. The telome theory. *Bot. Rev.* 19:417–437.

Young, D. A.
1978. On the diffusion theory of phyllotaxis. *J. Theor. Biol.* 71:421–432.

Zimmermann, M. H., and P. B. Tomlinson
1965. Anatomy of the palm *Rhapis excelsa*. I. Mature vegetative axis. *Jour. Arnold Arboretum* 46:160–178.
1966. Analysis of complex vascular systems in plants: optical shuttle method. *Science* 152:72–73.
1967. Anatomy of the palm *Rhapis excelsa*. IV. Vascular development in apex of vegetative aerial axis and rhizome. *Jour. Arnold Arboretum* 48:122–142.
1968. Vascular construction and development in the aerial stem of *Prionium* (Juncaceae). *Amer. Jour. Bot.* 55:1100–1109.
1972. The vascular system of monocotyledonous stems. *Bot. Gaz.* 133:141–155.

Zimmermann, W.
1953. Main results of the "telome theory." *The Palaeobotanist* 1:456–470.
1959. *Die Phylogenie der Pflanzen,* 2d edition. Gustav Fischer Verlag, Stuttgart.
1965. *Die Telomtheorie*. Gustav Fischer Verlag, Stuttgart.
1966. Kritische Beiträge zu einigen biologischen Problemen. VII. Die Hologenie. *Zeit. Pflanzenphysiol.* 54:125–144.

CHAPTER 4

Sporangia

A salient and definitive feature of the sporophyte generation in vascular plants is the production of spore sacs or sporangia. In marked contrast with a moss or liverwort, wherein the entire sporophyte normally forms only a single, nonseptate sporangium, vascular plants are polysporangiate, and the number of sporangia and spores developed by a single individual may be enormous. For example, a single, well-developed male shield fern *(Dryopteris filix-mas)* may produce approximately 50,000,000 spores in a single season. A larger plant, such as a full-grown pine or fir tree, can produce an astronomical number of pollen grains. The so-called sulfur showers which are familiar phenomena each year in regions of coniferous forests are countless yellow pollen grains that have been released from the sporangia of cones and are buoyed by wind currents. From a broad biological point of view, the apparently wasteful overproduction of spores, especially by lower vascular plants and wind-pollinated seed plants, actually tends to compensate for the high proportion that do not survive after dispersal from the parent plant. Thus, as is true also of the prodigious development of gametes in animals and in many kinds of plants, it seems that overproduction of spores and pollen grains is a mechanism for insuring the perpetuation of the race.

All normal sporangia of vascular plants share one important feature: they are *the specific* structures of the sporophyte where meiosis occurs. This process can be localized further to the sporocytes, or spore mother cells. Each sporocyte is normally a diploid cell which, as a result of meiotic divisions, yields a group of four spores known as a spore tetrad (Fig. 4-1). Aside from this functional identity, however, the sporangia in the various vascular plant groups differ widely in position, form, size, structure, and method of development and provide consistent and useful criteria for morphological comparison and taxonomic utilization. Indeed, such major groups of vascular plants as the Psilophyta, Lycophyta, Sphenophyta, and Filicophyta are sharply distinguished from one another on the basis of sporangium morphology alone.

It should now be clear that sporangia are structures which, although conservative in their functional aspects, vary considerably in other ways. This points to a long, complex evolutionary history, the details of which may never become entirely clear. However, as paleobotanical research continues, fossilized sporangia in an excellent state of preservation continually become available for study and comparison. This chapter emphasizes the important topographical, ontogenetic, and structural features of the sporangia of the *lower groups* of vascular plants. Later chapters will discuss sporangium morphology in the gymnosperms and angiosperms.

Position of Sporangia

The evidence provided by such extinct members of the Rhyniophyta as *Horneophyton* and *Rhynia* indicates that sporangia antedated leaves in evolution (Chapter 7). In other words, a primitive and elemental type of sporangium appears to have been a *cauline* (i.e., belonging to the stem) structure and was merely a sporogenous tip of a main axis, or occurred laterally on the stem *(Zosterophyllum)*.

In other groups of lower vascular plants the sporangia are related to leaves. A foliar structure which subtends or bears one or more sporangia is termed a *sporophyll*. The sporophylls may be photosynthetic organs like the sterile foliage leaves. In some plants, however, special areas or even complete leaves, are devoted exclusively to spore production and are usually nonphotosynthetic.

Certain groups of vascular plants are distinguished by the fact that the sporophylls, instead of being intermingled with ordinary foliage leaves, are aggregated into a compact conelike structure termed a *strobilus*. The Lycophyta display the various degrees of development and distinctness of the strobilus. In lycopsid plants a solitary sporangium is associated with each sporophyll and is either situated in its axil or attached adaxially to the basal region of the sporophyll. Most species of *Lycopodium* develop well-defined strobili at the tips of main or lateral shoots. But in *Lycopodium lucidulum* and *Lycopodium selago*, for example, the spor-

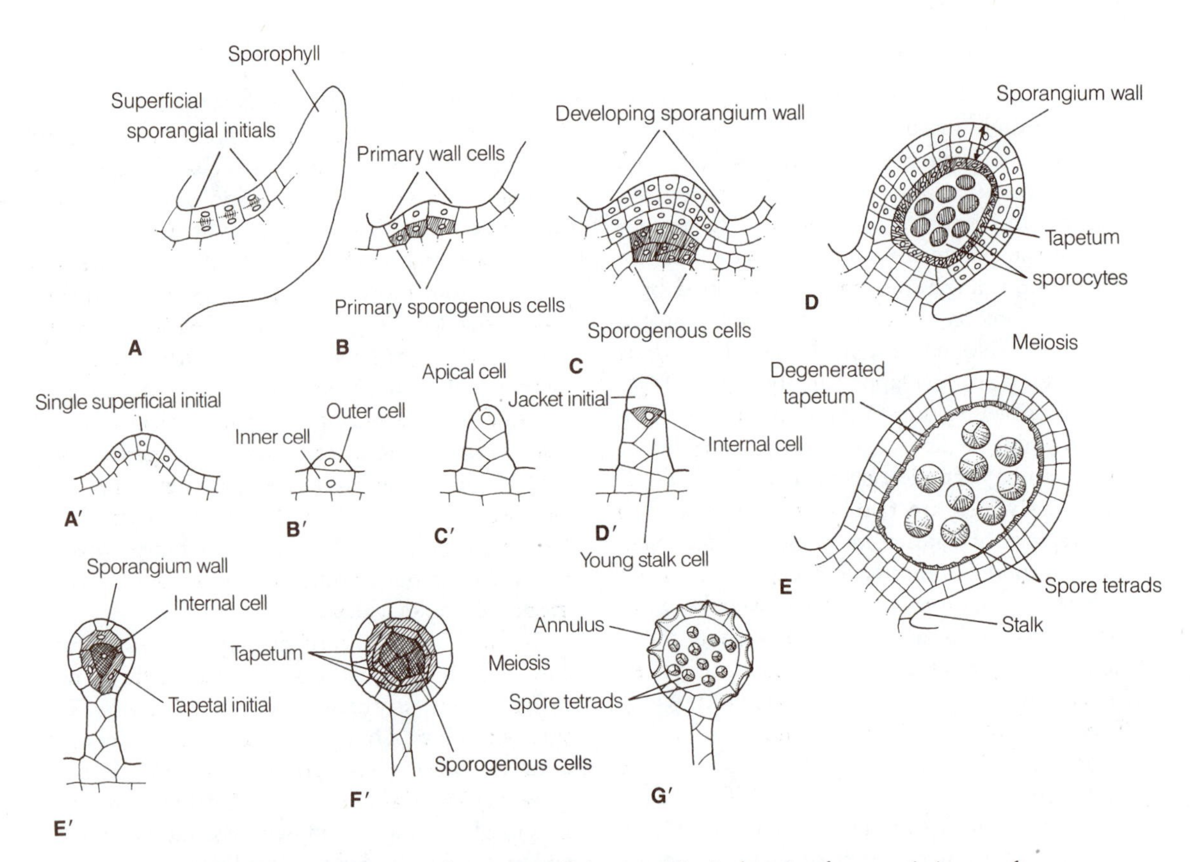

FIGURE 4-1 Ontogeny and structure of the two principal types of sporangia in vascular plants. **A–E,** the eusporangium; **A′–G′,** the leptosporangium. See text for detailed discussion of this diagram.

angia occur in poorly defined patches or zones which alternate with purely vegetative regions of the shoot system (see Chapter 9). The strobiloid forms of *Lycopodium* represent an evolutionary development from the morphological condition represented in such species as *L. selago*.

Structure of Sporangia

From a broadly defined structural standpoint, the mature sporangium in lower vascular plants consists of one or of many spores enclosed by a protective wall (Fig. 4-1). The term "wall" is the layer or layers distinct in origin from the sporocytes that constitute the sterile protective jacket of the sporangium. When a sporangium is sunken within the stem or the sporophyll the wall is merely an external multilayered cover which is not sharply demarcated from the adjacent sterile tissue. But in emergent sporangia the wall is much more clearly defined. In addition to its protective role, certain of the surface cells of the sporangia of many lower vascular plants are unevenly thickened and collectively form a distal plate, a ridge, or an incomplete ring; these varied cell patterns constitute dehiscence mechanisms. Besides the essential spores and the variously constructed wall, most sporangia develop a *tapetum* during the early or middle stages in their ontogeny. From a functional standpoint, the tapetum consists of cells which probably provide nourishment for the developing sporocytes and spores. According to Goebel (1905), there are two principal types of tapeta between which transition forms occur. The tapetum may be a *plasmodial* type characterized by the breaking down of the cell walls and the intrusion of the protoplasts between the sporocytes and spores. By contrast, with a *secretion type* of tapetum, the cells do not separate but remain in position and apparently secrete nutritive substances which are used by the developing spores.

The tapetum, whatever its physiological type, is structurally a continuous jacket of cells that completely surrounds the central mass of sporogenous tissue (Fig. 4-1, D; Chapter 9). From a histogenetic point of view, however, the *outer part* of the tapetum (the portion toward the outer surface of the sporangium) and the *inner part* (the portion adjacent to the tissue of the sporophyll or the sporangial stalk) differ in their origin. In most eusporangia, the inner part of the tapetum develops from "sterile" tissue, i.e., from vegetative cells which have not arisen from the inner derivatives of the original sporangial initials. In contrast, a large portion of the outer tapetum most frequently is the innermost layer of the sporangium wall or is derived from the outer layer of potentially sporogenous cells. These two alternative methods of tapetum origin in eusporangia are represented diagrammatically in Fig. 4-2. Possibly, as has been suggested in the literature, the different methods of origin of the tapetum may somehow be correlated with the size of the sporangium. For example, in the small, delicate sporangia of advanced ferns, the wall is a single cell layer in thickness, and the tapetal initials originate in a most precise manner from the divisions of a single large tetrahedral internal cell (see Figs. 4-3, 13-13). But in *Isoetes*, notable for the enormous size of its sporan-

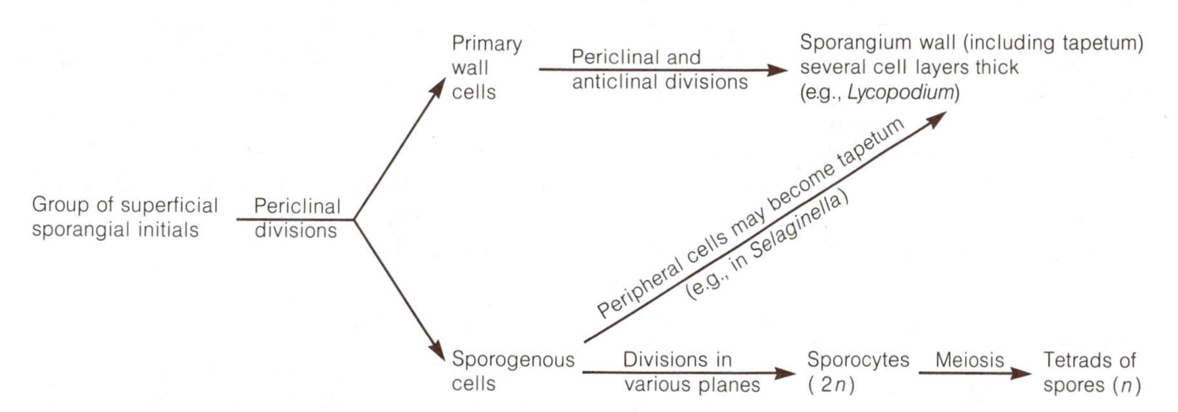

FIGURE 4-2 Summary of the ontogeny of the eusporangium in lower vascular plants.

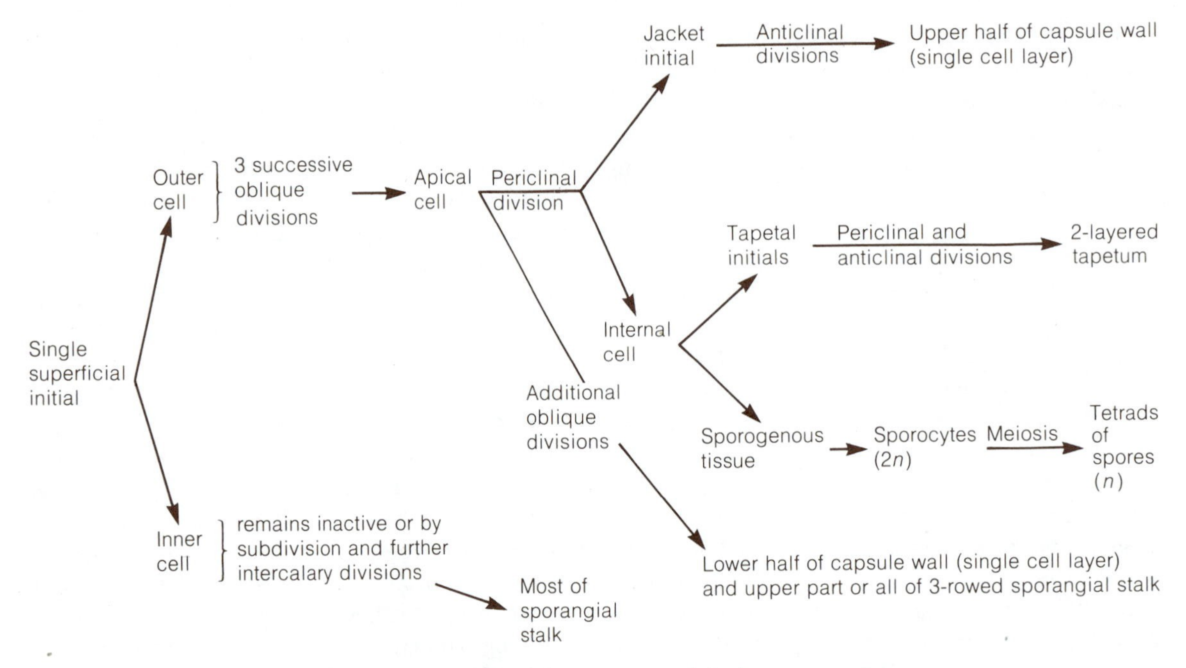

FIGURE 4-3 Summary of the ontogeny of the leptosporangium

gia, the extensive tapetum originates in part from the innermost layer of the wall and in part from the potentially sporogenous cells.

Ontogeny and Classification of Sporangia

Our present terminologies and classification of sporangia in vascular plants are based upon the important research of the great German morphologist Karl Goebel. In a paper published in 1881 he suggested that there are two principal types of sporangia: the eusporangium and the leptosporangium. The former type arises ontogenetically from several initial cells, and, at least before final maturation, develops a wall of more than one layer of cells. The eusporangium, as Goebel discovered, is the prevailing type throughout vascular plants. In contrast, the leptosporangium arises from a single parent cell or initial, and its wall is composed of but a single layer of cells. This more delicate type of sporangium, as Goebel correctly supposed, is restricted to the most highly advanced families of ferns which are now commonly called the leptosporangiate ferns.

Goebel's (1880, 1881) emphasis on ontogeny as a basis for the interpretation and classification of sporangia has had far-reaching effects, and the numerous studies since his original work have tended generally to support the distinction that he made between eusporangia and leptosporangia. Although recognizing the distinctions that can usually be made between the eusporangiate and leptosporangiate methods of sporangial development, Bower (1889, 1891, 1935), however, has drawn attention repeatedly to the existence of intermediate patterns of sporangial ontogeny, as illustrated in the ferns by the family Osmundaceae. Such intermediate patterns justify, in Bower's opinion, the belief that the eusporangiate and leptosporangiate modes of development represent the end points of a continuous morphological series and that a rigid morphological distinction should not be made between them. This series, he believes, also represents the probable phylogenetic development of sporangia which, at least in the ferns, began with the eusporangiate type and terminated in the highly evolved leptosporangium.

The Eusporangium

In all lower vascular plants and in the microsporangia of certain conifers (Fagerlind, 1961) the parent cells, or initials, of the eusporangium are superficial (i.e., at the surface) in position (Chapter 17). Sometimes these initials are referred to as epidermal cells, but this seems an inappropriate term for the highly meristematic cells from which the sporangium arises. The first step in the development of the sporangium is the division of the initials by walls parallel to the surface (Fig. 4-1, A). Divisions in this plane are termed *periclinal,* and they result in the formation of an outer and an inner series of cells. In a very general sense, these two series of cells represent the starting points for the sporangium wall and the sporogenous tissue, respectively. Because of this, the outer series is commonly called jacket cells, primary wall cells, or parietal cells, and the inner series primary sporogenous cells (Fig. 4-1, B). But the intensive studies of Bower and other investigators show that the first periclinal divisions in the sporangial initials do not always sharply define the future sporogenous tissue. On the contrary, further additions to the potential spore-forming cells may be made by periclinal divisions in the original outer cell series; moreover, the number of surface cells that function as parent cells for the sporogenous tissue may be somewhat variable even within a single genus.

The genus *Lycopodium* will serve to illustrate the general eusporangiate pattern; see Fig. 4-2 for the general scheme of ontogeny.

In this plant, the number of *rows of surface cells* that collectively function as sporangial initials varies from one in *Lycopodium selago* to as many as three in *Lycopodium alpinum. Periclinal* divisions in the initials result in a rather clear-cut distinction between (1) an outer cell layer which by further periclinal and *anticlinal* divisions (i.e., divisions resulting in new walls oriented perpendicular to the surface) builds up the several-layered wall, and (2) an inner fertile layer of potentially sporogenous cells from which by irregularly oriented planes of division the sporocytes ultimately arise (Fig. 4-1, A-E). Occasionally, according to Bower, subsequent periclinal divisions in the superficial cells resulting from the first periclinal divisions may contribute additional cells to the sporogenous tissue.

Although all typical eusporangia have a wall two or more layers thick at a middle stage in ontogeny, the *inner wall layers* (including of course the tapetum when this is of parietal origin) commonly become stretched, compressed, and ultimately destroyed. Thus, at maturity the walls of many eusporangia may appear to consist of only a single layer of cells.

The output of spores from eusporangia is variable but is frequently much greater than that in the leptosporangium. Estimates of spore number made by counting (in serial sections) the number of spore-mother cells and multiplying by four reveal numbers that are in the hundreds, or in the thousands in various eusporangiate ferns. Apparently, the greatest spore output occurs in *Isoetes,* where it is estimated that from 150,000 to 1,000,000 spores may develop in a single microsporangium.

The form of spores is quite variable and often is of considerable diagnostic value. Tetrahedral spores, with conspicuous triradiate ridges, are characteristic of most species in the Lycophyta, whereas bilateral spores are characteristic of *Psilotum* and *Tmesipteris.*

The Leptosporangium

In contrast to the multicellular origin of the eusporangium, the ontogeny of a typical leptosporangium begins with the transverse or oblique division of a *single superficial initial* (Fig. 4-1, A′, B′). As Fig. 4-3 shows, the *inner* of the two cells produced by the first division of the sporangial initial may (1) contribute cells which by further intercalary divisions produce a large part of the sporangial stalk or (2) remain inactive and play no role in the ontogeny of the sporangium. The latter pattern of development appears to be characteristic of leptosporangia with three-rowed stalks, an organization that Bower (1935) regarded as "the commonest of all in Leptosporangiate Ferns." We will first describe the main steps in development of a sporangium with a three-rowed stalk; the reader should refer frequently to Figs. 4-1, A′-G′, and 4-3 in order to follow the precise sequence of events. Following this description, there are a few comments regarding the interesting divergence in the method of early development of the stalk in certain genera of leptosporangiate ferns.

As shown in Figs. 4-1, A′-G′, and 4-3, the *outer cell*—formed by the division of the sporangial initial—may function as the parent cell from which both the stalk and the spore-containing capsule of the sporangium are derived. This outer cell, by means of three successive obliquely oriented divisions, forms a distal *apical cell* that is tetrahedral in shape and which by means of further oblique divisions, parallel to its three lateral faces, produces additional segments. The lower segments give rise to stalk cells, and the three uppermost segments, by subsequent anticlinal divisions, form the lower portion of the capsule wall (Fig. 4-3).

A very critical stage in the development of the leptosporangium now occurs as the result of a distal periclinal division in the apical cell. This division yields an outer *jacket initial* and a pyramidal *internal cell* (Figs. 4-1, D′; 4-3). The jacket initial, by means of anticlinal divisions, completes the formation of the one-layered wall of the capsule (Fig. 4-3). Concomitantly, the internal cell, by divisions parallel to its sides, produces four *tapetal* initials; from these cells, as a result of both periclinal and anticlinal divisions, there arises the two-layered *tapetum* characteristic of leptosporangia (Figs. 4-1, E′, F′; 4-3). Following the formation of the four peripheral tapetal initials, the tetrahedral inner cell divides in various planes and forms a mass of sporogenous tissue from which in turn the *sporocytes* arise (Figs. 4-3; 13-13, F, G, H.).

Maturation of the leptosporangium includes the meiotic division of the sporocytes to form spore tetrads, the disintegration of the tapetum, the conspicuous elongation of the stalk cells, and the ultimate dehiscence of the capsule. In all the advanced families among the leptosporangiate ferns the *annulus* is vertical (longitudinal) and the dehiscence of the capsule is transverse or brevicidal (Fig. 13-13, I). But in the phylogenetically less evolved groups, the annulus varies from apical to oblique, and the plane of dehiscence of the capsule is longitudinal or oblique (Fig. 13-14).

As stated previously, the three-rowed type of sporangial stalk is very common in leptosporangiate ferns, and its component cells are all apparently derived from lateral segments of the apical cell. In the genus *Diellia*, however, Wagner (1952) found that the first few divisions of the sporangial initial all take place in the transverse plane. After this three- to four-celled "filamentous stage" of development, the distal cell begins to function as an apical cell while the cells below it ultimately give rise, by intercalary divisions, to the characteristic one-rowed stalk; in *Diellia*, the stalk is composed of three cell rows only immediately below the capsule.

Although there is considerable variation in the average spore output of the leptosporangia in the various families, the number is, in general, less than for eusporangia. In none of the genera for which Bower made spore counts does the number of spores per sporangium equal the huge numbers produced by the sporangia in the Lycophyta or in eusporangiate ferns. In typical leptosporangia the number of spores is a power of two: 16, 32, 64, 128, 256, or 512. According to Bower (1935) "in the vast majority of the leptosporangiate ferns each capsule contains 64 spores or less." Such numbers as 24 and 48, though not powers of two, do occur and are the result of reductions in the number of cell divisions which lead to the formation of sporocytes. Mature fern spores exhibit various types of wall sculpturing which may be of value in taxonomic differentiation (Wagner, 1952; Tryon and Tryon, 1982). In the majority of ferns an outer layer *(perispore)* is formed from substances derived from the tapetum.

As stated earlier in this chapter, development and structure of the stalk of the sporangium in the Osmundaceae do not easily fit the scheme of either a typical leptosporangium or a typical eusporangium (Willliams, 1928). In the first place, as shown in Fig. 4-4, A, B, the form of the initial cell from which the sporogenous cell and the tapetum originate is variable. Sometimes the cell is truncated at the base, thus resembling the form of comparable cells in a eusporangium; at other times, in the same sorus, the cell is pointed at the base as in a typical leptosporangiate fern. Second, Fig. 4-4 reveals the fact that the entire sporangium cannot be traced in origin to a single cell, as can a typical leptosporangium. On the contrary, neighboring cells contribute, as they do in a eusporangium, to the development of the massive stalk which contrasts strikingly with its more slender counterpart in a typical leptosporangium (compare Fig. 4-1, F′ with Fig. 4-4, C, and Fig. 13-13, I). Finally, the spore output in the transitional sporangia of the Osmundaceae is, in general, higher than in ordinary leptosporangiate ferns; in *Osmunda regalis*, for example, it ranges from 256 to 512.

Before concluding this chapter and as an introduction to the next one, the terms homospory and heterospory will be defined. A plant is homo-

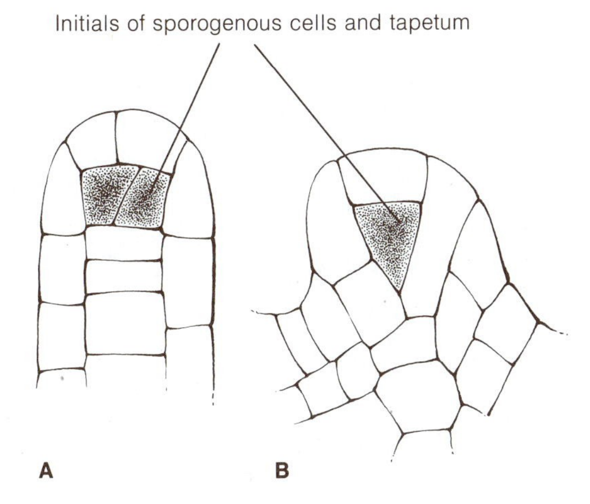

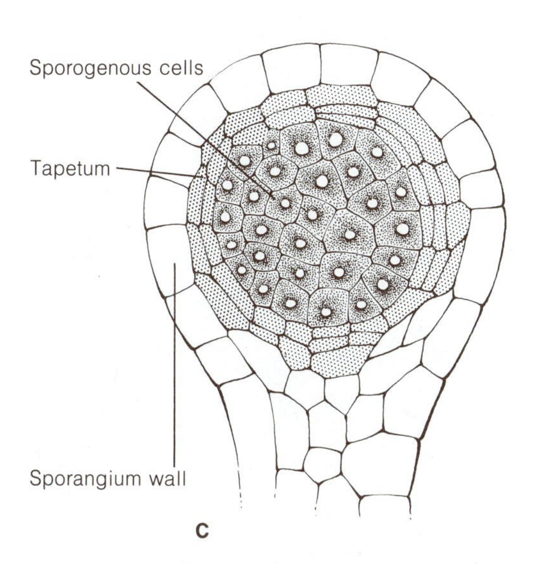

FIGURE 4-4 Development of sporangia in the Osmundaceae. **A, B,** *Todea barbara,* illustrating variations in early stages of differentiation; **C,** *Osmunda regalis,* longisection of sporangium showing wall, tapetum, sporogenous tissue, and massive stalk. [From *Primitive Land Plants* by F. O. Bower. Macmillan, London. 1935.]

sporous if all of the spores produced are similar in size and give rise to gametophytes that usually develop both the male (antheridium) and female (archegonium) gametangia. In a heterosporous plant, two types of spores are formed. Microspores are produced in microsporangia and develop into male gametophytes that give rise to sperms. Larger spores, the megaspores, are produced in megasporangia and they develop into female gametophytes (megagametophytes). There is variation among the major groups of vascular plants as to whether the megaspores are shed from or are retained within the megasporangia. These variations will be discussed in the next chapter and in chapters relating to specific plant groups.

REFERENCES

Bower, F. O

1889. The comparative examination of the meristems of ferns as a phylogenetic study. *Ann. Bot.* 3:305–392.

1891. Is the Eusporangiate or the Leptosporangiate the more primitive type in the ferns? *Ann. Bot.* 5:109–134.

1935. *Primitive Land Plants.* Macmillan, London.

Fagerlind, F.

1961. The initiation and early development of the sporangium in vascular plants. *Svensk Bot. Tidskr.* 55:299–312.

Goebel, K.

1880. Beiträge zur vergleichenden Entwickelungsgeschichte der Sporangien. *Bot. Zeit.* 38:545–552.

1881. Beiträge zur vergleichenden Entwickelungsgeschichte der Sporangien. *Bot. Zeit.* 39:681–694, 697–706, 713–720.

1905. *Organography of Plants,* Part II, English edition. By I. B. Balfour. Clarendon Press, Oxford.

Tryon, R. M., and A. F. Tryon

1982. *Ferns and Allied Plants.* Springer-Verlag. New York.

Wagner, W. H., Jr.

1952. The fern genus *Diellia. Univ. Calif. Publ. Bot.* 26:1–212.

Williams, S.

1928. Sporangial variation in the Osmundaceae. *Trans. Roy. Soc. Edinb.* 55:795–805.

CHAPTER 5

Gametangia

ALL vascular plants achieve sexual reproduction by the pairing of morphologically unlike gametes. In all lower vascular plants the male gamete, usually termed the sperm or spermatozoid, is a motile, flagellated cell that requires liquid water to reach the passive, nonmotile egg. From an evolutionary point of view such motile sperms, like their counterparts in liverworts and mosses, exemplify the persistence of the kind of gamete typical of many aquatic algae. In contrast with the somewhat casual dispersal of flagellated sperms in lower vascular plants, modern-day seed plants, because of the development of a pollen tube by the male gametophyte, do not depend on water for transportation of sperms—the pollen tube conveys the male gametes directly to the immediate vicinity of the egg. This is true of *Ginkgo* and the cycads, which have retained flagellated sperms, but it is particularly significant in all the other living groups of gymnosperms and angiosperms, for in these the male gametes are devoid of flagella.

In all the lower groups of vascular plants the sex cells are normally produced in separate organs or *gametangia*. The male gametangium, or *antheridium,* produces the sperms, which vary in number from four in *Isoetes* to several thousand in certain eusporangiate ferns. The female gametangium, or *archegonium,* is quite different in that typically it produces only a single, nonmotile egg which at the time of fertilization is situated at the base of the archegonial canal; the canal provides the channel through which the sperm must pass to reach the female gamete.

Antheridia and archegonia, as is true of the functionally equivalent sex organs of the bryophytes, are complex organs consisting of a jacket (usually one cell thick) of sterile cells which encloses and shelters the gametes. This is illustrated in Fig. 5-1. Such construction is quite unlike the simple unicellular sex organs that are typical of the algae, and it has led to speculation regarding the evolutionary history of gametangia and the possible homology between antheridia and archegonia.

As is true of sporangia, an exceptional amount of attention has been paid to the ontogeny and comparative structure of gametangia. One of the most interesting results of comparative research is the discovery that with respect to gamete production the archegonium is a remarkably uniform sex organ. Isolated cases have been reported of archegonia with several eggs, but generally only a single functional gamete is formed (Fig. 5-1, D′, E′). On the basis of this standardized structure, plants with archegonia are designated collectively as Arche-

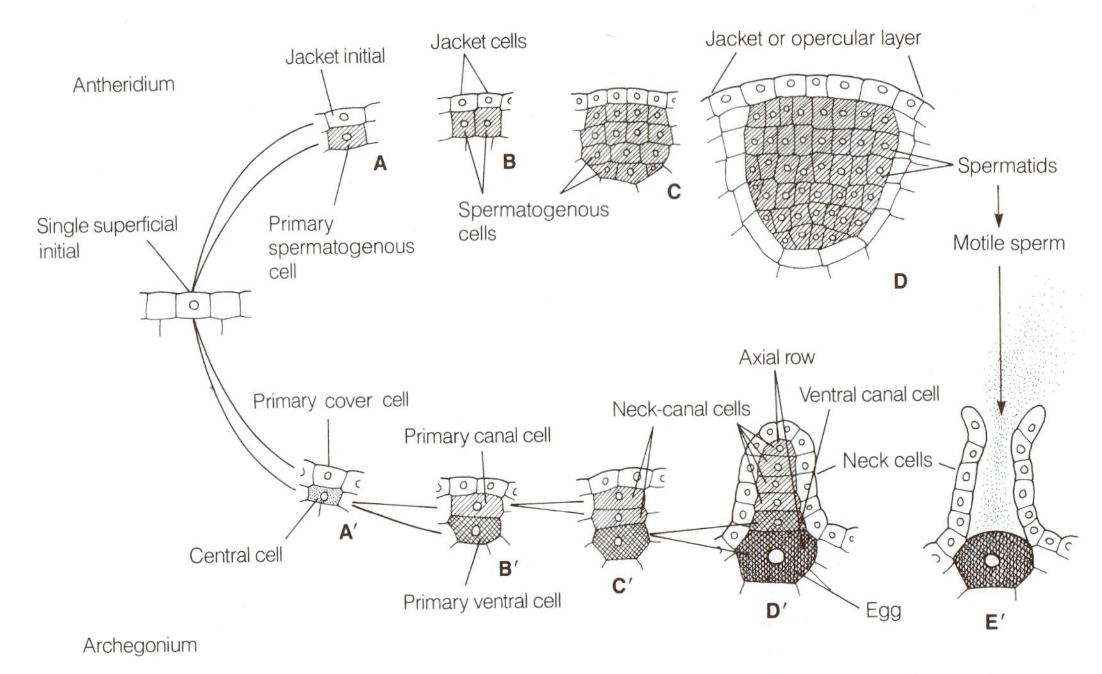

FIGURE 5-1 Ontogeny and structure of gametangia in lower vascular plants (exclusive of the leptosporangiate ferns): **A–D,** the antheridium; **A′–E′,** the archegonium. Note the similarity in the mode of origin and the first stage in development of the two types of sex organs.

goniatae. Included in this group are the bryophytes, all lower vascular plants including the ferns, and the majority of the gymnosperms. In the gymnosperms archegonia are modified in structure, but ontogenetically and structurally they are strictly comparable with the archegonia of the lower groups. In angiosperms the egg is produced in a highly specialized embryo sac, and archegonia as definable structures are absent.

With reference to the evolutionary origin and modification of sex organs in vascular plants, antheridia, as definable multicellular structures, are restricted to the lower groups and, unlike archegonia, do not occur in living gymnosperms. This contrast between antheridium and archegonium seems to be correlated with the marked reduction in size of the male gametophyte generation in some lower vascular plants and in the seed plants. In the male gametophytes of *Selaginella* and *Isoetes*, for example, only the equivalent of a single antheridium is produced, whereas in gymnosperms the equivalent structure is further reduced to a few vegetative cells and usually two male gametes. This trend toward the elimination of sterile cells culminates in the male gametophyte of angiosperms, which normally consists of a single vegetative nucleus and two male gametes. Doubtless, the evolution of the pollen tube was a very significant factor in the drastic reduction and final elimination of the antheridium as a definable structure. On the other hand, the persistence of the archegonium as a distinct structures among the relatively advanced gymnosperms is related to the larger size of the female gametophyte and to the prime biological importance of the archegonium as a structure which produces the egg and shelters the young embryo (Chapter 6).

This chapter is largely a general account of the position, structure, and ontogeny of the antheridium and the archegonium. Details of the morphology of gametophytes and gametangia will be given in later chapters that deal with specific groups of plants.

Position of Gametangia

It will be desirable first to contrast the gametophytes and sex organs of homosporous and heterosporous lower vascular plants. In the former the gametophyte is *exosporic*—i.e., free-living and not enclosed by the spore wall — whereas in all heterosporous groups the male and female gametophytes are *endosporic* — i.e., entirely or for the most part enclosed by the wall of the *microspore* or *megaspore,* respectively. These differences are important from a biological as well as a morphological standpoint, and there is little doubt that the relatively large exosporic type of gametophyte represents the original and primitive condition. We therefore begin with this type.

Exosporic gametophytes are typically bisexual, which means they produce or have the capacity to produce both types of gametangia. In some instances the environmental conditions surrounding the developing gametophyte may influence the type or types of sex organs that are produced.

The gametangia of exosporic gametophytes are usually embedded in vegetative tissue. This is always the case with the venter (egg-containing portion) of the archegonium, but the antheridium is less consistent and may protrude slightly above the surface of the gametophyte or be conspicuously emergent organs.

The distribution pattern of antheridia and archegonia varies considerably and is correlated to some degree with the form and level of evolutionary development of the gametophyte. In the fleshy, radial, nonphotosynthetic gametophytes of the Psilophyta the two kinds of gametangia are intermingled and tend to occur over the entire surface of the gametophyte (Fig. 8-10, A). An intermingling of gametangia also occurs in certain presumably primitive species of *Lycopodium* but in other members of this genus the antheridia and archegonia are formed in distinct patches or clusters on the upper surface of the gametophyte (Chapter 9). With reference to the thalloid, dorsiventral, photosynthetic type of gametophyte characteristic of the leptosporangiate ferns, both kinds of gametangia are commonly restricted to the lower surface with the archegonia limited to the cushion of tissue situated behind the notch of the heart-shaped prothallus; the more numerous antheridia generally occur near the basal end of the gametophyte as well as on the wings (Chapter 13).

Endosporic gametophytes, being restricted to heterosporous plants, are usually strictly unisexual. As mentioned before, the male gametophyte is extremely reduced in structure, consisting in *Selaginella* and *Isoetes* of a few sterile cells that enclose the spermatogenous tissue. The female gametophyte, by contrast, is more robust, consisting of a mass of food-storing tissue which usually fills the cavity of the megaspore and which, in *Selaginella* and *Isoetes,* is exposed by the cracking of the spore wall along the triradiate ridge (Fig. 9-38, A). The small archegonia are restricted in occurrence to the surface of the protuberant cushion of gametophyte tissue.

Structure of Gametangia

Antheridium

In the embedded type of antheridium a well-defined sterile jacket is always present. Because this layer is instrumental in the mechanism of sperm discharge, Goebel has proposed that it be designated the "opercular layer" (Fig. 5-1, D). This term is appropriate because, generally, one or more centrally located opercular or cap cells of this layer separate or become broken, thus creating an opening through which the sperm can escape. Most commonly, the jacket layer of embedded antheridia is only a single cell in thickness. But in *Botrychium dissectum,* for example, a member of the Ophioglossaceae, the jacket is two cells thick, except where one to four scattered opercular cells occur in the exposed jacket layer (Bierhorst, 1971).

In the emergent type of antheridium the jacket is always a single layer of cells thick. Some variation occurs in the total number of cells which constitute this layer in the various families of the leptosporangiate ferns, although there is usually developed but a single cap or opercular cell. At one end of the series are the apparently highly specialized antheridia of the majority of the Filicales, the jacket here consisting of two girdling cells and a terminal opercular (cap) cell. At the other extreme are the massive complex antheridia of the Gleicheniaceae, wherein the jacket may consist of as many as ten or twelve cells, one of which functions as an opercular cell (Stokey, 1950, 1951).

In lower vascular plants there are two principal types of sperm, based upon the number of flagella developed. In *Lycopodium* and *Selaginella* the sperm are biflagellate and in this respect are quite unlike the prevailingly multiflagellate sperm developed in all other lower vascular plants, and in

Ginkgo and the cycads. Since the sperm of many algae are biflagellate, considerable interest, from the standpoint of evolution, is attached to the similar condition in *Lycopodium* and *Selaginella*. The taxonomic (and possibly phyletic) importance of the number of flagella developed by sperm is illustrated by the genus *Isoetes*. This plant is commonly regarded as a member of the Lycophyta, but its sperm are multiflagellate, and in this respect they are quite unlike the male gametes of either *Lycopodium* or *Selaginella*.

As mentioned earlier in the chapter, the number of sperm formed by a single antheridium varies widely according to the genus or group of plants. In Bower's (1935) opinion, there is a close parallel between the output of spores and the number of sperm, as revealed by comparison of the sporangia and antheridia of eusporangiate and leptosporangiate ferns. In the first group there tends to be a relatively high number of spores and sperm, but in the advanced leptosporangiates there is a marked reduction in number of both spores and male gametes. These differences, Bower believed, show that the progressive refinement in structure of spore-bearing structures has been extended in a comparable manner to the sex organs.

Archegonium

Just prior to its complete maturation the archegonium typical of many ferns consists of the *neck*, which projects conspicuously above the surface of the gametophyte (Fig. 5-1 D′, E′), and the so-called *axial row* of cells. The upper members of this row are enclosed by the neck, and the lowermost cells, including the egg, are sunk in the tissue of the gametophyte, as shown by Fig. 5-1 D′. The venter, or basal embedded portion of the archegonium, contains the *ventral canal cell* and the *egg*. In members of both the eusporangiate and leptosporangiate ferns the venter is demarcated from adjacent prothallial cells by a rather discrete cellular jacket, but in many vascular plants the cells bounding the lower portion of the axial row of the archegonium are not morphologically distinguishable from other cells of the gametophyte. The archegonial neck, which is the morphological and functional equivalent of the jacket of the antheridium, is structurally uniform, usually comprising four vertically arranged rows of cells (Fig, 5-4, B). The number of cell rows is remarkably consistent, despite the wide variation in the total number of cells which comprise the neck.

From a morphological viewpoint, the most important and significantly variable part of the immature archegonium is the axial row (Fig. 5-1, D′). In certain species of *Lycopodium* the so-called axial row may actually consist of a partially or completely double series of cells (Spessard, 1922; Lyon, 1904). More commonly, however, the axial row is composed of a single series of cells. The lowest cell in the axial row is the egg itself and normally is the only component of the row which survives, all the others eventually disintegrating as the archegonium reaches full maturity (Fig. 5-1, E′). The cell located directly above the egg is known as the ventral canal cell; this cell ontogenetically is a sister cell of the egg and in exceptional cases apparently functions as a gamete. All the remaining cells of the axial row, situated above the ventral canal cell, are termed *neck-canal cells* (Fig. 5-1, D′) and on theoretical grounds may be regarded as potential gametes also.

From a comparative viewpoint it is possible to arrange the archegonia of lower vascular plants in a *reductional series*, beginning with those in which there are many neck-canal cells (e.g., *Lycopodium*) and culminating in plants that have only a single binucleate neck-canal cell (e.g., many leptosporangiate ferns). Whether this series actually corresponds to the general phylogenetic development of the archegonium is difficult to determine, since in reputedly primitive plants there may be only two or possibly only a single binucleate neck-canal cell. The behavior of the ventral canal cell also is variable and hence of considerable comparative interest. In many lower vascular plants it is a well-defined cell, differing from the egg cell in its smaller size and more flattened form. But in the Ophioglossaceae, which also apparently represents an ancient group of vascular plants, a distinct ventral canal cell (Campbell, 1911) is difficult to demonstrate. This is true in many gymnosperms where the ventral canal cell may be represented by only a short-lived nucleus. Furthermore, in these plants neck-canal cells have been entirely eliminated, and the mature archegonium consists of a small neck of one or several cell tiers and a huge egg (Maheshwari and Sanwal, 1963).

Very little systematic investigation has been devoted to the morphology of neck cells and the mechanism of the opening of the archegonial neck. In ferns the neck cells may be more or less the same size. In certain species of *Equisetum* the terminal cells of the four rows of neck cells may be elongate, whereas in others all of the neck cells may become exceedingly long and twisted (Duckett, 1979; Hauke, 1968). At maturity the neck cells spread apart, or just the terminal four separate and become strongly reflexed. An interesting account of the opening of archegonia has been supplied by Ward (1954) for the fern *Phlebodium*. In the presence of water the four terminal neck cells separate and cells of the axial row, except the egg, are forcibly ejected from the mouth of the archegonium within a few minutes to an hour. Initially, a tiny stream of mucilaginous material flows out, followed by the ejection of more mucilage and the contents of the neck canal, thus producing a channel through which the motile sperm can pass (Fig. 5-1, E′). In genera of four groups of lower vascular plants, it has been shown that salts of malic acid exert a chemotactic stimulus upon the sperm. Sperm initially may swim at random, but they swim toward and down the archegonial neck when they come under the influence of an attractant (Doyle, 1970). However, specific attractants have not been identified from nature.

Ontogeny of Gametangia

Despite the striking divergence between fully developed antheridia and archegonia, both kinds of sex organs are remarkably alike in their method of initiation and in the earliest stages of their ontogeny (Figs. 5-2, 5-3). In all investigated cases each type of gametangium originates from a *single surface cell* of the gametophyte (Fig. 5-1). Whether there are definable cytological differences between the parent cells of antheridia and those of archegonia is a question that has never been intensively studied. Spessard (1922) in his description of the gametophyte of *Lycopodium lucidulum*, where the sex organs are intermingled, states that "it is impossible to distinguish an antheridium from an archegonium in the very earliest stages of their development. There are some indications that the archegonial initial is slightly larger and longer than the antheridial initial, but this is so uncertain that it is useless as a criterion."

In addition to their identical mode of origin, antheridia and archegonia begin their development by the periclinal division of the parent cells (Fig. 5-1, A, A′, 5-2, and 5-3). With reference to the embedded type of antheridium, this results in an outer *jacket cell,* the anticlinal subdivision of which produces the jacket layer, and an inner *primary sper-*

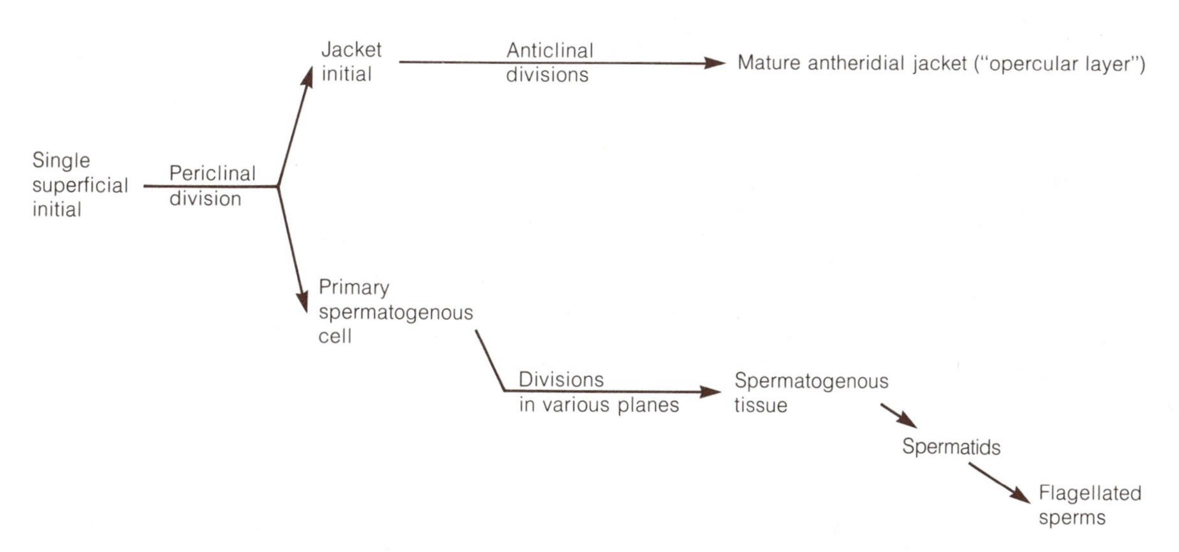

FIGURE 5-2 Ontogeny of antheridium in lower vascular plants (exclusive of leptosporangiate ferns).

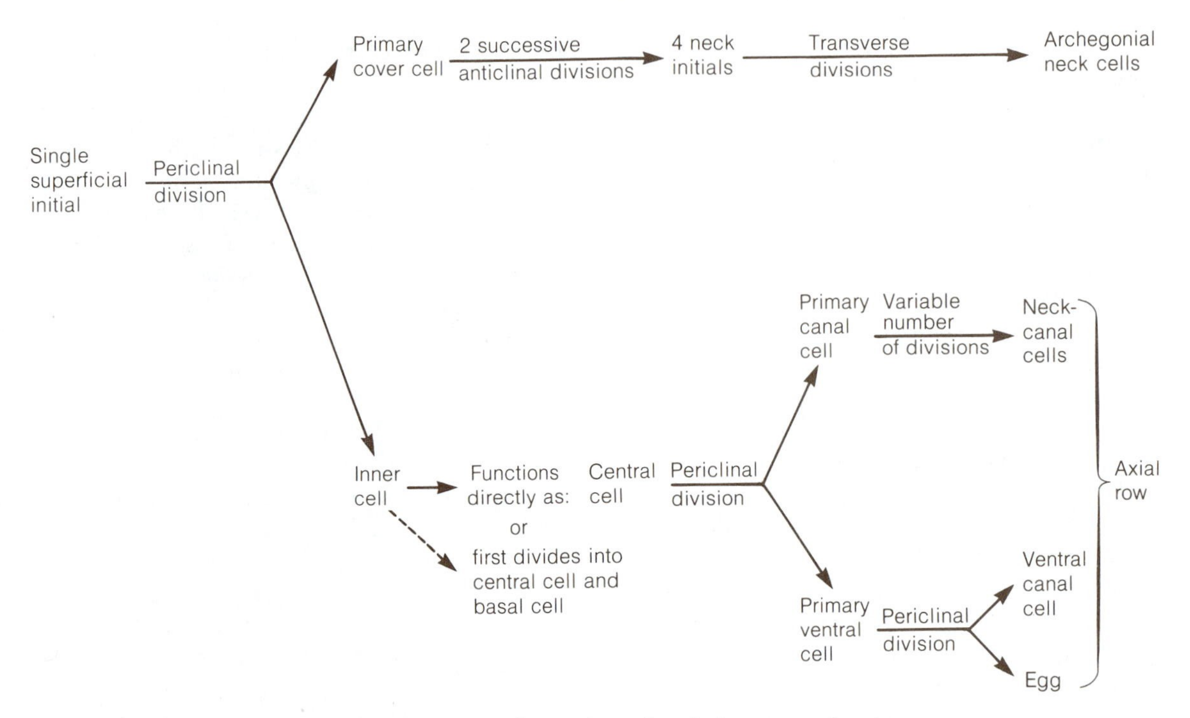

FIGURE 5-3 Ontogeny of an archegonium in lower vascular plants.

matogenous cell (sometimes referred to as the primary androgonial cell) from which the usually numerous spermatids and ultimately the sperm originate (Figs. 5-1, A–D, and 5-2). There is a comparable distinction between outer sterile cells and inner potentially fertile cells, due to the first periclinal division of an archegonial initial (Figs. 5-1, A′, and 5-3). In this case, the outer cell or *primary cover cell,* by means of two successive anticlinal divisions at right angles to one another, produces four neck initials from which, by transverse divisions, the four vertical rows of the archegonial neck are developed. The inner of the two derivatives of the archegonial initial may first produce a *basal cell,* as in many ferns, and then serve as the parent cell of the axial row; or, as is more commonly the case, it may function directly without first forming a basal cell. In either instance, the parent cell of the axial row is usually designated as the *central cell* (Figs. 5-1, A′, and 5-3).

The presence or absence of a basal cell is not a consistent feature of archegonium development within a family or even a genus.

The central cell in the archegonia of lower vascular plants divides periclinally into an outer *primary canal cell* and an inner *primary ventral cell* (Figs. 5-1, B′, and 5-3). From the former, one or as many as sixteen neck-canal cells originate. The primary ventral cell is believed to function is some cases as an egg. Perhaps more typically, however, it divides periclinally, forming the evanescent *ventral canal cell* and the basally situated *egg* (Figs. 5-1, D′, and the 5-3). In the gymnosperms, as mentioned earlier, neck-canal cells are not developed. On the contrary, the central cell functions as the initial from which the large egg and the transitory ventral canal cell (or its nucleus) originate.

There remains for final consideration the ontogeny of the antheridium in leptosporangiate ferns, which differs in several interesting respects from that characteristic of embedded antheridia. Some of the differences parallel to a remarkable degree the ontogenetic differences between leptosporangia and eusporangia that have been described in Chapter 4. Perhaps one of the most definitive ontogenetic features of the emergent type of antheridium in leptosporangiate ferns is that the first division wall of the initial does not set apart an outer sterile and an inner fertile cell. On the contrary, the formation of the primary spermatogenous cell is delayed until two or more sterile cells have been cut off by the antheridial initial. In such primitive fern groups as the Gleicheniaceae and Osmundaceae the antheridial initial, by means of alternating oblique divisions, first produces a series of basal cells which constitute the short stalk and the lower portion of the antheridial jacket. Ultimately the initial becomes divided by a curved periclinal wall into an outer *jacket cell* and an inner *primary spermatogenous cell*. From the former, by anticlinal divisions, the upper part of the antheridial jacket including an opercular cell is formed, and the spermatids develop from the subdivision of the primary spermatogenous cell. The periclinal wall, which sets apart the primary spermatogenous cell from the sterile jacket initial, is a remarkable parallel with the similarly oriented wall delimiting the primary wall and primary sporogenous cells of a young leptosporangium.

The early ontogeny of the delicate antheridium (Fig. 5-4, A) of more specialized homosporous leptosporangiate ferns likewise results first in the production of sterile jacket cells. These cells, however, are fewer in number and of unusual shape. There has been considerable controversy over the years on the exact ontogeny of an antheridium of this type, but the classic model of Kny (1869) and Goebel (1905) was confirmed by Schraudolf (1968), and more recently by Kotenko (1985, 1986) who described antheridial development for *Onoclea sensibilis*.

In *Onoclea*, the antheridial initial is formed by an unequal division of a gametophyte cell (basal wall, 1-1, Fig. 5-5, C, D). The initial enlarges and then undergoes an unusual division. The mitotic spindle is oriented at 45 degrees so that the new cell plate will intercept wall 1-1 at an angle, resulting in the formation of a basal cell ("funnel cell") and an upper cell (Fig. 5-5, E–G). In median longitudinal section the new wall (2-2) is evident at both sides of the antheridium, because it is curvilinear, completely separating the upper cell from the basal cell (Fig. 5-5, F–H). The upper cell then divides in the formation of a bell-shaped cell and the primary spermatogenous cell (Fig. 5-5, H–J). The bell-shaped cell undergoes an unequal division (similar to the process described for the origin of the basal cell) resulting in the formation of a doughnut-shaped ring cell and a disc-shaped cap cell (Fig. 5-5, K–M). The antheridial jacket at maturity consists

A

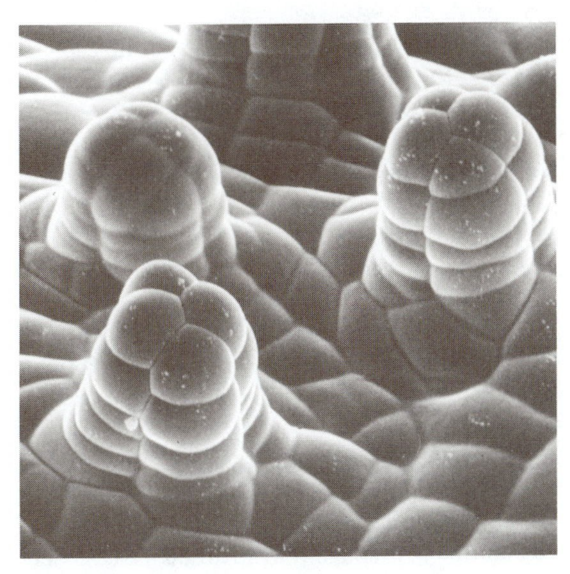

B

FIGURE 5-4 Scanning electron micrographs of fern gametangia. **A,** antheridia of *Onoclea;* **B,** archegonia of *Todea;* note the four tiers of neck cells. [Courtesy of E. G. Cutter.]

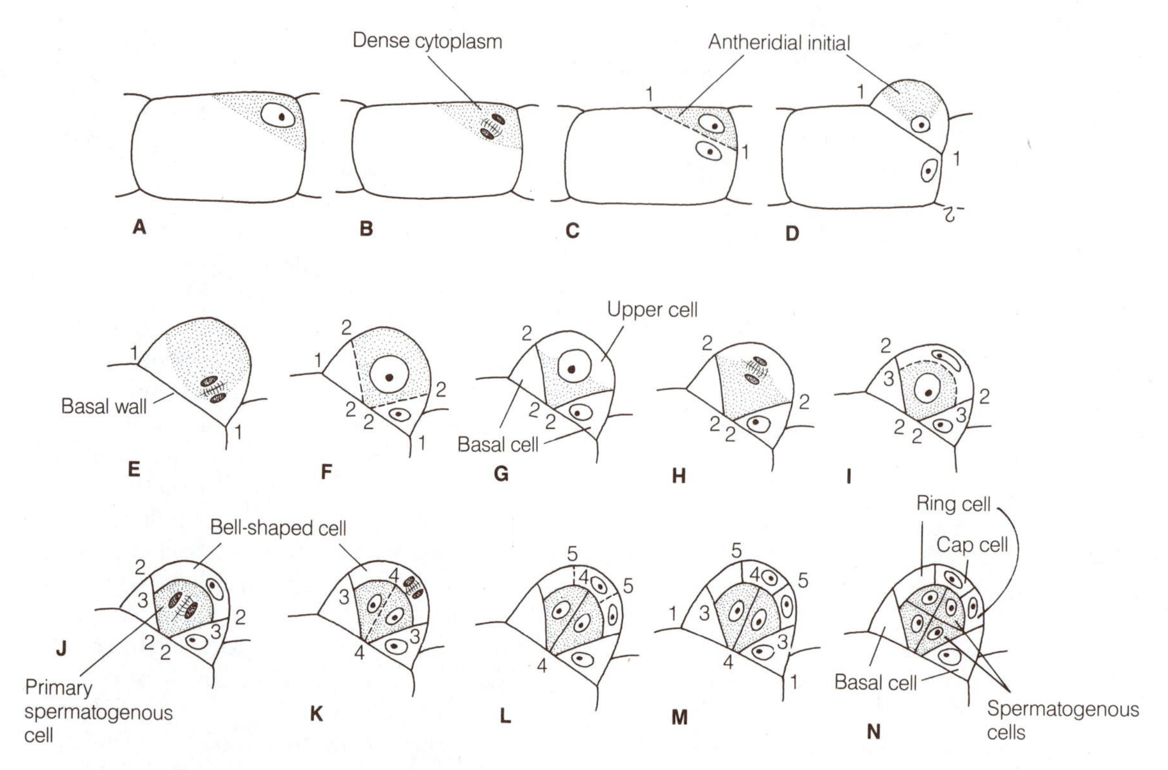

FIGURE 5-5 Ontogeny of the antheridium of *Onoclea sensibilis*. See text for details. [Based on J. L. Kotenko, *Bot. Gaz.* 147:28–39, 1986.]

of three cells: basal cell, ring cell, and the cap cell (Fig. 5-5, M, N). Spermatogenous cells continue to divide, resulting in the formation of spermatids. Loosening of the cap cell permits the escape of the spermatids, which quickly become motile sperm (Fig. 5-6).

One developmental feature is worthy of note: the basal cell and ring cell are formed in a manner quite unique in vascular plants. Leung and Näf (1979) agreed that a funnel or basal cell is formed in *Onoclea*, but that the new cell plate (2-2) is formed from base to rim, rather than proceeding in circular fashion.

Earlier, Davie (1951) reported a different ontogeny for *Pityrogramma calomelanos* and other ferns, including *Onoclea sensibilis*. The first division wall (2-2) in the antheridial initial is transverse or only slightly concave and not funnel shaped. With increasing turgor in the upper cell, the transverse wall becomes depressed in the middle portion and comes into contact with the original basal wall (1-1). The second ring cell and cap cell are formed in a somewhat similar fashion. Stone (1962) took exception to Davie's conclusions, saying that in some ferns the first division in the antheridial initial was transverse, but the wall never came into contact with the basal wall. A ring cell and cap cell were produced later in the jacket initial by an unequal division as shown in Fig. 5-5, K, L.

In *Polypodium polycarpon* (Schraudolf, 1968) antheridial development is entirely comparable to that described by Kotenko (1985, 1986). In addition, Schraudolf and Richter (1978) demonstrated that in two species of *Polypodium* plasmodesmata are present in that portion of the wall between the primary spermatogenous cell and the underlying cell, as well as between the cap cell and the spermatogenous cell. It is doubtful that plasmodesmata would form secondarily between these cells if an antheridium developed as described by Davie.

Clearly, from these studies that we have just briefly reviewed, differences exist in *details* of an-

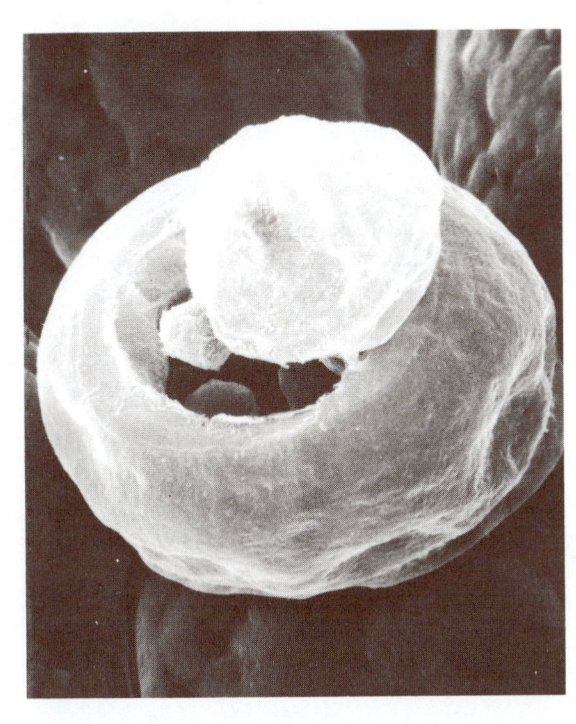

FIGURE 5-6 Scanning electron micrograph of mature antheridium of *Onoclea sensibilis*, showing displacement of cap cell and the underlying spermatids. ×1350. [Courtesy of J. L. Kotenko.]

theridial development, not only among fern genera (Schraudolf, 1963) but even among antheridia of the same species (Verma and Khullar, 1966), including teratological (abnormal) forms (Schraudolf, 1968). Continued research is obviously needed and should help eventually to resolve some of the present controversial aspects of antheridial ontogeny in homosporous, leptosporangiate ferns, and to supply information of systematic value.

REFERENCES

Bierhorst, D. W.
1971. *Morphology of Vascular Plants.* Macmillan, New York.

Bower, F. O.
1935. *Primitive Land Plants.* Macmillan, London.

Campbell, D. H.
1911. *The Eusproangiatae. The Comparative Morphology of the Ophioglossaceae and Marattiaceae.* Carnegie Institution of Washington, Washington, D.C. (Publication No. 140.)

Davie, J. H.
1951. The development of the antheridium in the Polypodiaceae. *Amer. Jour. Bot.* 38:621–628.

Doyle, W. T.
1970. *The Biology of Higher Cryptogams.* Macmillan, London.

Duckett, J. G.
1979. Comparative morphology of the gametophytes of *Equisetum* subgenus *Hippochaete* and the sexual behaviour or *E. ramosissimum* subsp. *debile,* (Roxb.) Hauke, *E. hyemale* var. *affine* (Engelm.) A. A., and *E. laevigatum* A. Br. *Bot. J. Linn. Soc.* 79:179–203.

Goebel, K.
1905. *Organography of Plants,* Part II, English edition by I. B. Balfour. Clarendon Press, Oxford.

Hauke, R. L.
1968. Gametangia of *Equisetum bogotense. Bull. Torrey Bot. Club* 95:341–345.

Kny, L.
1869. Über den Bau und die Entwicklung des Farn-Antheridiums. *Monatsber. Kgl. Preuss. Akad. Wiss. zu Berlin,* pp. 416–431.

Kotenko, J. L.
1985. Antheridial formation in *Onoclea sensibilis* L.: genesis of the jacket cell walls. *Amer. J. Bot.* 72:596–605.
1986. Antheridium formation in *Onoclea sensibilis* L.: cytoplasmic polarity and determination of wall positions. *Bot. Gaz.* 147:28–39.

Leung, C., and U. Näf
1979. On the cytology of antheridium formation in the fern species *Onoclea sensibilis.* I. Cytology of cell plate and cell wall formation. *Amer. J. Bot.* 66:765–775.

Lyon, F.
1904. The evolution of the sex organs of plants. *Bot. Gaz.* 37:280–293.

Maheshwari, P., and M. Sanwal
1963. The archegonium in gymnosperms. A review. *Mem. Indian Bot. Soc.* No. 4, 103–199.

Schraudolf, H.
1963. Einige Beobachtungen zur Entwicklung der Antheridien von *Anemia phyllitidis. Flora* 153:282–290.
1968. Einige Beobachtungen zur Ausbildung des Antheridiums von Polypodiaceen. *Flora* Abt. B. 157:379–385.

Schraudolf, H., and U. Richter
1978. Elektronenmikroskopische Analyse der Musterbildung im Antheridium der Polypodiaceae. *Plant Syst. Evol.* 129:291–297.

Spessard, E. A.
1922. Prothallia of *Lycopodium* in America. II. *L. lucidulum* and L. *obscurum* var. *dendroideum*. *Bot. Gaz.* 74:392–413.

Stokey, A. G.
1950. The gametophyte of the Gleicheniaceae. *Bull. Torrey Bot. Club* 77:323–339.
1951. The contribution by the gametophyte to classification of the homosporous ferns. *Phytomorphology* 1:39–58.

Stone, I. G.
1962. The ontogeny of the antheridium in some leptosporangiate ferns with particular reference to the funnel-shaped wall. *Aust. Jour. Bot.* 10:76–92.

Verma, S. C., and S. P. Khullar
1966. Ontogeny of the polypodiaceous fern antheridium with particular reference to some Adiantaceae *Phytomorphology* 16:302–314.

Ward, M.
1954. Fertilization in *Phlebodium aureum* J. Sm. *Phytomorphology* 4:1–17.

CHAPTER 6

Embryogeny

THE term embryogeny designates the successive steps in the growth and differentiation of a zygote into a young sporophyte. Although one cannot establish sharp boundaries, it is convenient to distinguish between the early definitive stages in embryogeny and the later phases of growth during which the organs of the embryo arise and become functionally significant.

In all the Archegoniatae (including the majority of gymnosperms) fertilization and the first critical phases of embryogeny occur within the shelter provided by the venter of the archegonium and the adjacent gametophytic tissue. At this first period the polarity (i.e., the distinction between the apex and the base) of the embryo is determined. The subsequent period of enlargement and organ development usually results, in lower vascular plants, in the formation of a shoot apex together with one or more leaves, a root, a well-defined foot, and, in many genera, a suspensor. The foot serves as a haustorial organ which attaches the embryo to the nutritive tissue of the gametophyte. In the majority of archegoniate plants the apex of the young embryo faces inwardly toward the gametophytic tissue and away from the neck of the archegonium, as can be seen in Fig. 6-1, B, C. An embryo of this type illustrates strikingly the intimate nutritive relationship between it and the gametophyte since, during enlargement and organogenesis, it must digest its way through a considerable volume of gametophytic tissue before emerging at the surface of the gametophyte. Ultimately, in contrast with the permanently dependent embryos of all the bryophytes, the embryo of lower vascular plants becomes free from the nursing gametophyte and grows into an independent sporophyte. In the majority of the gymnosperms and in angiosperms the embryo is shed from the parent plant in a somewhat dormant condition, is enveloped by the seed coat (s), and is often provided with a special nutritive tissue; here the embryo constitutes the essential part of that remarkable structure that we term a seed.

In this chapter the discussion will be restricted mainly to the principles of comparative embryogeny as illustrated by *lower vascular plants*. For an account of the embryogeny of seed plants, the reader should consult Chapters 15 to 18 and 20.

General Organography of Embryos

Before discussing the origin and early development of embryos we must have a clear idea of the main categories of organs found in the embryos of vascu-

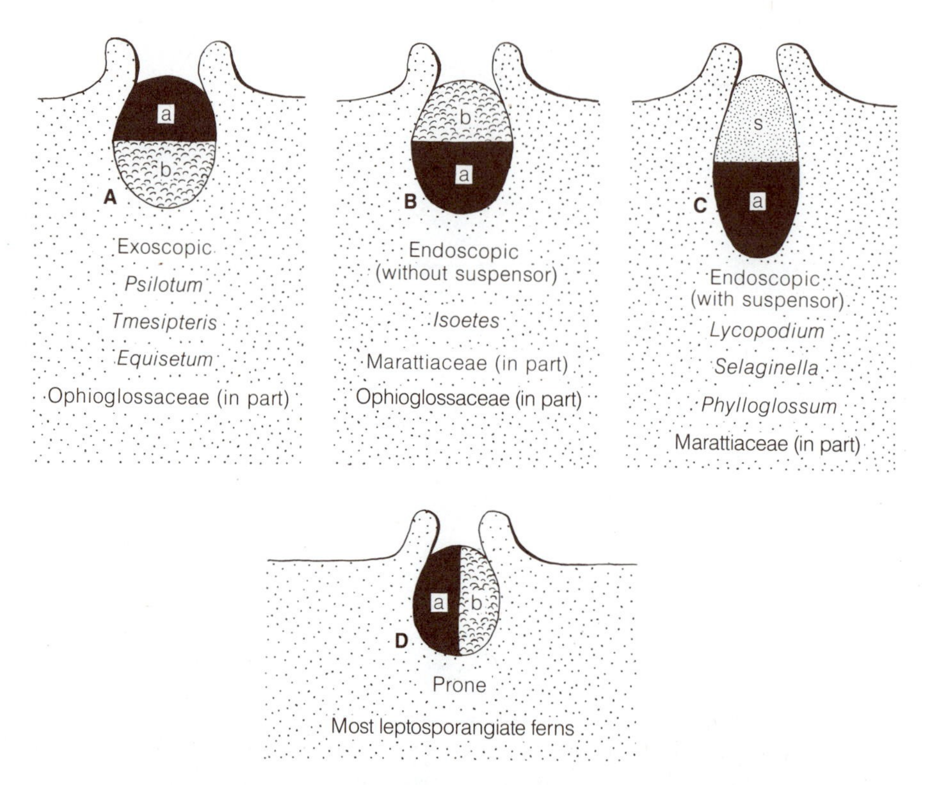

FIGURE 6-1 Main types of polarity in the development of the embryo in lower vascular plants. In all figures, the apical pole (cell a) is shown in black and the basal pole (cell b) is shaded. **A,** exoscopic polarity; **B,** endoscopic polarity without a suspensor; **C,** endoscopic polarity with a suspensor; **D,** orientation of two-celled embryo in leptosporangiate ferns.

lar plants. From a broad morphological viewpoint, as was repeatedly emphasized by Bower (1922, 1935), a very young embryo is somewhat filamentous or spindle shaped, and consists of two definable poles: the *apical pole,* which gives rise to the shoot apex and one or more leaves, and the *basal pole,* which in many groups of the lower vascular plants is represented by the *suspensor.* Let us now examine these components of the embryo.

The apical pole is the portion from which the first or primary shoot takes its origin; it is a very consistent feature of the embryo of the large majority of vascular plants. In much of the literature dealing with the embryos of lower vascular plants the first leaf or leaves of the embryo are designated as "cotyledons," a term which probably should be reserved for the first foliar organs of the embryos of seed plants (Wagner, 1952). The size of the first leaf or leaves of the embryo in lower vascular plants fluctuates widely; in some genera—for example, *Lycopodium*—they are very small and rudimentary, whereas in others—*Isoetes, Botrychium virginianum,* and many leptosporangiate ferns—the first leaf is large, precociously developed, and may even be significantly photosynthetic (see Chapter 13). In seed plants, the number of cotyledons ranges from one to many.

As stated above, the base of the embryo in certain genera is represented by the suspensor; this structure, highly developed in most gymnosperms and found also in the embryos of angiosperms, is variable in its occurrence in lower vascular plants (Fig. 6-2). The suspensor is a temporary organ, the growth and enlargement of which serve to orient and keep the embryo in intimate contact with nutritive tissue of the gametophyte.

In addition to the shoot apex, leaf primordia, and suspensor, the embryo of most vascular plants

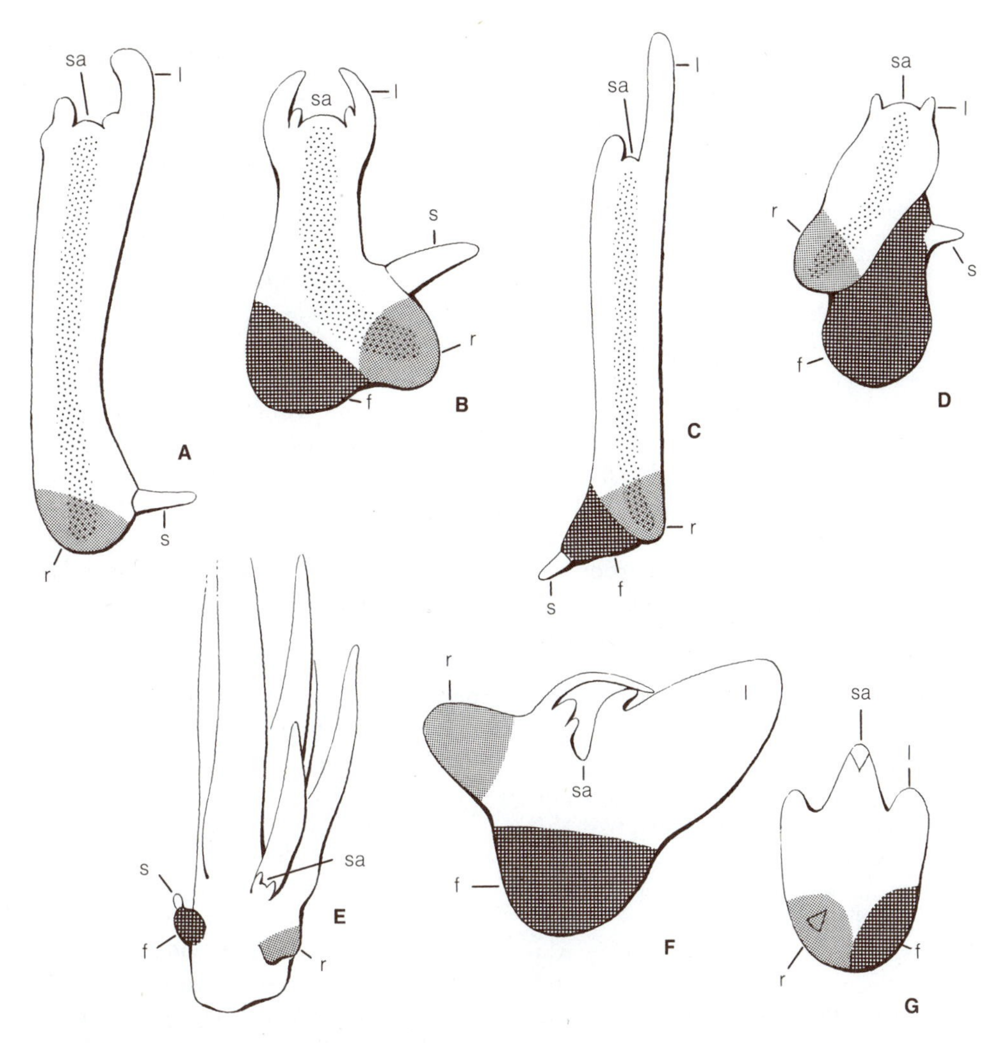

FIGURE 6-2 Variations in form and position of the organs of the embryos in lower vascular plants. **A,** *Selaginella spinulosa;* **B,** *Selaginella martensii;***C,** *Lycopodium selago;* **D,** *Lycopodium clavatum;* **E,** *Lycopodium cernuum;* **F,** *Isoetes;* **G,** *Equisetum.* Abbreviations: f, foot; l, leaf; r, root; s, suspensor; sa, shoot apex. [Redrawn from *Primitive Land Plants* by F. O. Bower. Macmillan, London. 1935.]

develops a root. Conspicuous exceptions are found in the Psilophyta in which the sporophyte, at all stages in its growth, is devoid of roots (Chapter 8). Because there is no evidence in these genera of even a "vestigial" root during embryogenesis, the absence of roots at later phases in the ontogeny of their sporophytes is regarded as a primitive rather than a derived condition.

As Bower (1935) strongly emphasized, in lower vascular plants the root is consistently *lateral* with reference to the longitudinal axis of the embryo (Fig. 6-2). Plants with this type of orientation of the first root of the embryo were termed *homorhizic* by Goebel (1930) in contrast to the *allorhizic* seed plants, in which shoot and root poles lie at opposite ends of the axis of the embryo (Chapter 19).

We must also briefly consider the so-called *foot* of the embryo, whose occurrence and degree of development vary more than that of the suspensor, even within species of the same genus (Fig. 6-2). Bower (1935) regarded the foot as an "opportunist growth" which arises in a position convenient for performing its function as a suctorial or nursing structure. In some embryos, the foot is conspicuous, and its role in anchoring and conveying nutrients from the gametophyte is evident.

Polarity and Early Embryogeny

During the middle and latter part of the nineteenth century morphologists traced in minute detail the origin and development of the organs of many types of embryos. One of the results was the concept that there is a single or basic plan of segmentation throughout plant embryogeny. In particular, the embryos of *Equisetum* and leptosporangiate ferns were considered fundamental and typical of all embryos, since the major organs, leaf, stem, root, and the foot were traced back to specific quadrants (i.e., quarter sections) formed at the very beginning of embryonic development. In the light of our extended knowledge of embryogeny, it is now evident that great variation exists in the definition of these so-called quadrants and their role in organogenesis, and that the study of embryogeny should be pursued in as flexible a manner as possible, with due regard to the effect of such external factors as gravity, light, and nutrition on the embryonic pattern (Ward, 1954; Wardlaw, 1955).

According to Bower (1908, 1922, 1935), who paved the way toward a more dynamic understanding of embryogeny, the first step in the development of the embryo is the definition of its axial polarity. Except for the leptosporangiate ferns, the first division wall in the zygote is approximately transverse to the long axis of the archegonium and results in a definite distinction between the future shoot apex and base of the embryo. However, two types of polarity arise from this first division of the zygote. In the more common polarity type, termed *endoscopic,* the apical pole—cell a in Fig. 6-1, B, C—is directed toward the base of the archegonium; in the opposed, or *exoscopic,* type the apex faces outward toward the neck of the archegonium—cell a in Fig. 6-1, A.

The occurrence of these two main types of embryo polarity in vascular plants poses some interesting morphological and phylogenetic questions. Embryos of such lower vascular plants as *Lycopodium, Selaginella,* and *Isoetes,* as well as seed plants, are of the endoscopic type. It is particularly significant to note that seed plants are of endoscopic polarity even though the mode of early development of their proembryos may be remarkably different (see Chapters 15 to 18 and 20).

The less common exoscopic polarity characterizes the embryogeny of *Psilotum, Tmesipteris, Equisetum,* some members of the Ophioglossaceae, and the leptosporangiate fern genus *Actinostachys* (Bierhorst, 1968). Generally speaking, either endoscopic or exoscopic polarity characterizes the major groups within lower vascular plants, although both patterns may be encountered within a single genus, as in *Botrychium.* In a strict sense, the embryos of most of the investigated members of the leptosporangiate ferns do not fall into either the endoscopic or exoscopic categories because the first division wall in the zygote tends to be parallel to the long axis of the archegonium (Fig. 6-1, D). Consequently, the apical and basal poles of the embryo tend to be oriented laterally with respect to the archegonial axis, and the embryo is prone rather than vertical or curved with respect to the gametophyte (Fig. 6-1, D).

The presence of a suspensor is invariably correlated with an endoscopic type of polarity. Figure 6-3 illustrates schematically the method of origin of the suspensor in the endoscopic type of embryogeny characteristic of *Lycopodium* and *Selaginella.* The zygote divides by the formation of a transverse wall I–I into two cells (Fig. 6-3, A). The upper cell (labeled s) is directed toward the neck of the archegonium and represents the parent cell of the future suspensor, which may enlarge without further division or give rise to a short multicellular filamentous suspensor. The lower cell (labeled em), termed the *embryonic cell,* subsequently gives rise to the main body of the embryo. The plane of the first division of the embryonic cell may be transverse, yielding a *hypobasal cell* (in contact with the suspensor) and an *epibasal cell* (Fig. 6-3, B). Two successive longitudinal divisions in each of these cells results in an eight-celled embryo, consisting of an epibasal tier and a hypobasal tier of four cells each (Fig. 6-3, C). Although this tiered arrangement of cells apparently is characteristic of the young embryos of many vascular plants, the sequence of wall formation in the embryonic cell appears to vary. The plane of the first division wall in the embryonic cell

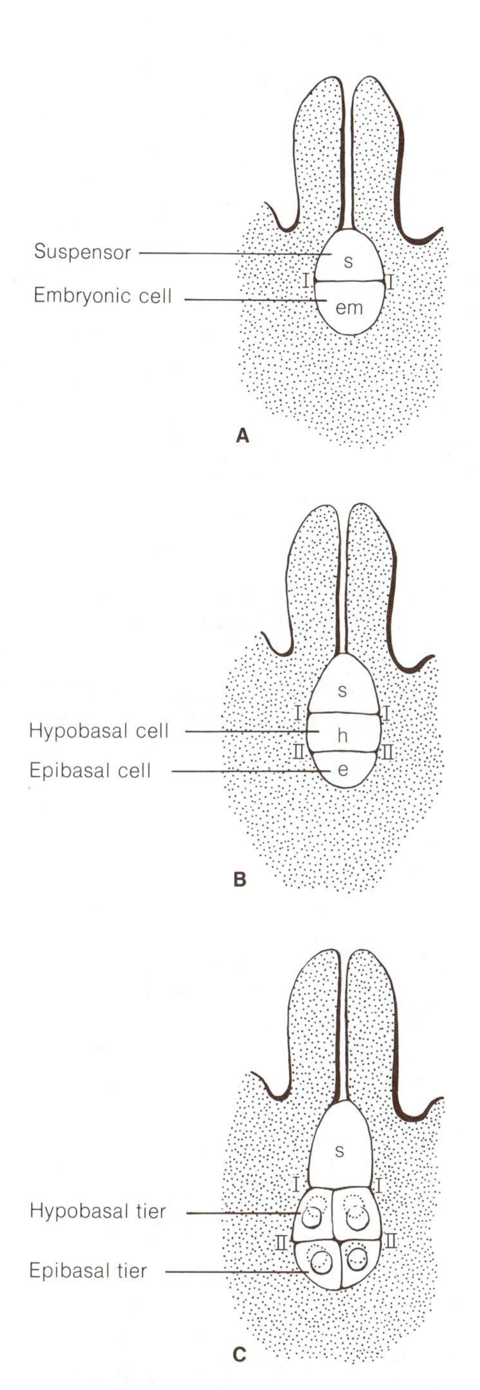

Figure 6-3 Successive stages in early development of an endoscopic type of embryo with a suspensor. **A,** the first division of a zygote is transverse (wall I–I) and yields a suspensor cell (s) directed toward the archegonial neck and an embryonic cell (em) directed inwardly; **B,** transverse division of embryonic cell (wall II–II) has produced an epibasal (e) and a hypobasal (h) cell; **C,** the octant stage, resulting from two successive longitudinal divisions of the epibasal and hypobasal cells.

may be longitudinal rather than transverse; each of the two cells then divides in a similar plane yielding a group of four cells—a quadrant stage. Transverse divisions of each of these quadrant cells then results in the formation of an eight-celled embryo comprising an epibasal and a hypobasal tier.

In embryos devoid of a suspensor the zygote itself functions directly as the embryonic cell, and its first division results, as before, in an epibasal and a hypobasal cell (Wardlaw, 1955). But in contrast with the invariably endoscopic polarity of embryos with a suspensor, the suspensorless types vary widely in orientation of the epibasal cell or apical pole. *Isoetes* and certain genera in the Marattiaceae lack suspensors, yet are characterized by endoscopic polarity. On the other hand, all exoscopic embryos are devoid of a suspensor, and in them the epibasal region faces outward toward the neck of the archegonium.

There appears to be no entirely satisfactory explanation for the "prone" orientation of the suspensorless embryo in leptosporangiate ferns. Bower (1935) has pointed out that gravity may play a role in determining the plane of the first division of the zygote. In any event, the first leaf and the first root emerge from beneath the lower surface of the gametophyte rather than penetrating the gametophyte as in endoscopic embryos (Chapter 13).

Origin and Development of Organs

Following the early definition of polarity and the establishment of a multicellular embryo, a young sporophyte enters a phase of enlargement and organ formation. Classical accounts of the embryogeny of leptosporangiate ferns and *Equisetum* reported the shoot apex, first leaf, first root, and foot originating from specific quadrants of the embryo. However, this rigid interpretation has been challenged in recent years (see Chapter 13).

Organogenesis in the embryos of other lower vascular plants follows no clear-cut uniform or standardized pattern (Fig. 6-2). The shoot apex and the first foliar appendages show the greatest constancy of all the parts of the embryo because they invariably arise from the epibasal tier. According to the analysis of various embryonic patterns made by Bower, the shoot apex originates at or near the geometrical center of the epibasal half of the embryo. In terms of cells, this center corresponds closely to the point of intersection of the octant walls and is plainly defined in those plants in which a single definitive apical cell is produced by obliquely oriented divisions. The first foliar organs, whatever their size, number, or arrangement, likewise originate from the epibasal half of the embryo. In many plants, these primary leaves are arranged in a pair or a whorl around the embryo apex, thus resembling the relation of later leaf primordia of the sporophyte to the shoot apex. But in some embryos, (*Isoetes* and in members of the Ophioglossaceae), the first leaf is precociously large and tends to displace the shoot apex to the side or even to cause a delay in its initiation.

The root is, without question, the most variable organ of the embryo. Sometimes it arises jointly with the first leaf and shoot apex from the epibasal half of the embryo. In other embryos, the primary root develops from the hypobasal portion. In addition to its variable point of origin, the position of the first root with reference to other organs of the embryo also is inconstant (Fig. 6-4).

In contrast to many lower vascular plants, the root apex of seed-plant embryos often becomes organized as early as, or earlier than, the shoot apical meristem. This developmental feature may well be correlated with the early establishment of the root during seed germination.

With the foot, it is evident that although this portion of the embryo arises from all or a part of the cells of the hypobasal tier, its form and degree of development fluctuate even within members of the same genus (see Fig. 6-2).

Embryogeny from a Phylogenetic Standpoint

As a conclusion to this chapter it is appropriate to examine briefly the phylogenetic implications of those various aspects of embryogeny.

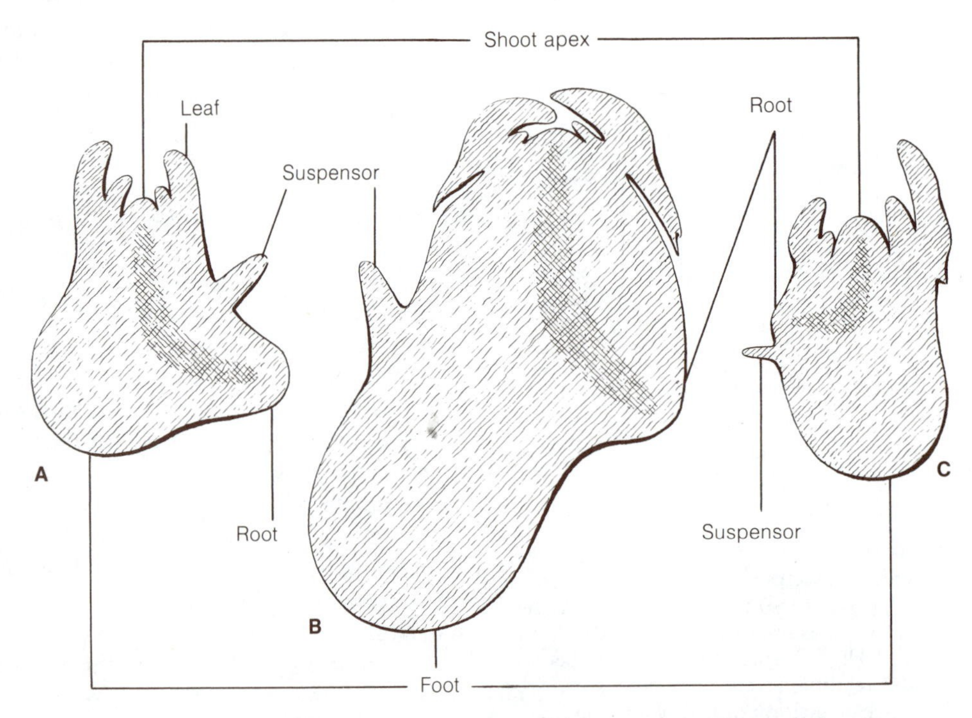

FIGURE 6-4 Embryos of three species of *Selaginella* showing variability in the position of the root with reference to other organs. **A**, *S. denticulata;* **B**, *S. poulteri;* **C**, *S. galeottii* [Redrawn from *Primitive Land Plants* by F. O. Bower, Macmillan, London, 1935.]

First, we may raise the question of the historical aspects and evolutionary significance of the two principal types of embryo polarity. Which is the more primitive form, the exoscopic or the endoscopic? It might be assumed that an endoscopic embryo with a suspensor represents the "primitive spindle" form, from which the suspensorless types have taken origin. According to Bower, the abolition of the suspensor has taken place repeatedly and independently, as evidenced by its variable occurrence. Once the embryo became free of a suspensor its former obligatory endoscopic polarity might be retained, if this were selectively advantageous from the standpoint of nutrition, or a complete inversion to the exoscopic type of polarity might take place. Goebel (1928) also held that a rotation in polarity from endoscopic to exoscopic is admissable in some cases. On the other hand, the occurrence of exoscopic polarity in such presumably primitive vascular plants of the Psilophyta, as well as throughout the Bryophyta, might equally well suggest that endoscopic polarity, which is typical of the majority of higher vascular plants, is the derived condition. Very likely, these phyletic questions cannot be answered with our present knowledge of embryogeny and plant phylogeny.

Paleobotanical investigations, however, do offer some hope of eventually answering these questions. Very small undifferentiated embryos have been described for arborescent lycopods from the Lower Carboniferous — the earliest records of vascular plant embryos. Suspensors were not evident, although they may have not been preserved (Schopf, 1941; Galtier, 1964). More recently, much larger (older) vascularized embryos of *Lepidocarpon* were discovered (Phillips, 1979). This lycopod had a footlike structure, and the embryo shoot tip dichotomized to form a stem and a *stigmarian*-like axis (see Chapter 9, p. 000). In another arborescent lycopod *(Bothrodendrostrobus)* the well-preserved embryo consisted of a foot and the first root and leaf. The embryo probably developed much as it does in the living genus *Isoetes*. There was no suspensor (Stubblefield and Rothwell, 1981). Embryos of fossil gymnosperms have been described from the Pennsylvanian (Carboniferous) and from the Lower Permian, but no suspensor appeared to be present (Miller and Brown, 1973; Stidd and Cosentino, 1976). Embryos from the Jurassic Period of the Mesozoic, in an excellent state of preservation, have been reported. These embryos were endoscopic and possessed a massive suspensor, much like modern-day conifers (Stockey, 1975). These and future paleobotanical discoveries may lead to some important phylogenetic conclusions regarding early embryogeny in lower vascular plants and seed plants.

There is great need for experimental investigation of the factors influencing the polarity of embryos. In certain of the liverworts it seems clear that the orientation of the archegonium is not a determining factor because, regardless of whether it is upright, horizontal, or pendent, the embryo is strictly exoscopic. Probably, as Bower suggested, the basal nutrition of the young embryo in these cases is an important factor in maintaining the exoscopic polarity. Of particular interest, from a morphogenetic viewpoint, would be a study of genera such as *Botrychium,* in which the polarity varies with the species. There is no real basis for a rigid and highly deterministic interpretation of either histogenesis or organogenesis, and phylogenetic interpretations of plant embryos should be made with due regard to the possible effects of gravity, light, and the source and type of nutrition on the polarity and pattern of development of the young encapsulated sporophyte (Wardlaw, 1955).

Experimental Embryogenesis

Thus far we have discussed the development of embryos in their normal environment. However, very young isolated embryos also can be grown to maturity by in vitro techniques. Outgrowths can develop from the epidermis of cultured stems, leaves, and floral parts of *Ranunculus sceleratus,* which display rather faithfully the stages of normal embryogeny (Konar and Nataraja, 1965; also see Raghavan, 1976, for other examples). These structures are termed *embryoids*. The discovery that embryoids can develop from single isolated cells obtained from loosely organized callus (proliferative parenchymatous cells) of carrot demonstrates the totipotency of somatic plant cells. These single cells exhibit polarity and pass through embryonic stages comparable to normal zygotic embryos, and can be grown into mature plants (Steward, 1970).

Even more remarkably, embryoids can be obtained from haploid pollen grains when grown on a suitable nutrient medium. These embryoids also pass through developmental stages quite similar to those of normal embryos and grow to maturity. (See Raghavan, 1976, for specific examples.)

What do these interesting and remarkable discoveries tell us? They emphasize the plasticity and totipotency of living plant cells. Even those cells that have undergone a considerable degree of differentiation can return to a meristematic state and produce an organized embryo. Under normal conditions gene expression for embryo development occurs in the environment (physical and biochemical) of the archegonium (in lower vascular plants, gymnosperms) or in the embryo sac of angiosperms. However, by simulating the normal biochemical environment, and by adding certain growth substances, diploid or haploid cells can be induced to return to the embryonic state without the process of fertilization. The ability to grow plants from single cells, coupled with recombinant-DNA techniques, constitute the basis for genetic engineering in plants.

REFERENCES

Bierhorst, D. W.
1968. Observations on *Schizaea* and *Actinostachys* spp., including *A. oligostachys*, sp. nov. *Amer. Jour. Bot.* 55:87–108.

Bower, F. O.
1908. *The Origin of a Land Flora.* Macmillan, London.
1922. The primitive spindle as a fundamental feature in the embryology of plants. *Proc. Roy. Soc. Edinb.* 43:1–36.
1935. *Primitive Land Plants.* Macmillan, London.

Galtier, J.
1964. Sur le gamétophyte femelle des Lépidodendracées. *C.R. Acad. Sci. (Paris).* 258:2625–2628.

Goebel, K.
1928. *Organographie der Pflanzen,* 3d edition. Erster Teil. G. Fischer, Jena.
1930. *Organographie der Pflanzen.* Dritte Auf. Zweiter Teil. G. Fischer, Jena.

Konar, R. N., and K. Nataraja
1965. Experimental studies in *Ranuculus sceleratus* L. Development of embryos from the stem epidermis. *Phytomorphology* 15:132–137.

Miller, C. N., and J. T. Brown
1973. Paleozoic seeds with embryos. *Science* 179:184–185.

Phillips, T. L.
1979. Reproduction of heterosporous arborescent lycopods in the Mississippian-Pennsylvanian of Euramerica. *Rev. Palaeobot. Palynol.* 27:239–289.

Raghavan, V.
1976. *Experimental Embryogenesis in Vascular Plants.* Academic, London.

Schopf, J. M.
1941. Contribution to Pennsylvanian paleobotany; *Mazocarpon oedipterum* sp. nov., and sigillarian relationships. *Ill. Geol. Surv. Rep. Inv.* 75. 53 pp.

Steward, F. C.
1970. Totipotency, variation and clonal development of cultured cells. *Endeavour* 29:117–124.

Stidd, B. M., and K. Cosentino
1976. *Nucellangium*: gametophytic structure and relationship to *Cordaites. Bot. Gaz.* 137:242–249.

Stockey, R. A.
1975. Seeds and embryos of *Araucaria mirabilis. Amer. J. Bot.* 62:856–868.

Stubblefield, S. P., and G. W. Rothwell
1981. Embryogeny and reproductive biology of *Bothrodendrostrobus mundus* (Lycopsida). *Amer. J. Bot.* 68:625–634.

Wagner, W. H., Jr.
1952. The fern genus *Diellia. Univ. Calif. Publ. Bot.* 26:1–212.

Ward, M.
1954. The development of the embryo of *Phlebodium aureum* J. Sm. *Phytomorphology* 4:18–26.

Wardlaw, C. W.
1955. *Embryogenesis in Plants.* Wiley, New York.

CHAPTER 7

Early Vascular Plants

RHYNIOPHYTA, ZOSTEROPHYLLOPHYTA, AND TRIMEROPHYTOPHYTA

BETWEEN the first occurrence of prokaryotic algal-like plants (cyanobacteria) in the Precambrian and the first well-documented vascular plants there is a gap of about 3 billion years (see Chapter 1, Table 1-1). In considering the history of early terrestrial plants, it is essential to distinguish between the first authentic vascular plants and the more primitive land plants that undoubtedly preceded them.

In the search for the most primitive vascular plant, some paleobotanists have called attention to certain enigmatic organisms that they believe might represent early land plants or even progenitors of the earliest vascular plants — earlier than late Silurian. One form, represented by sheets of cells, had tubes with annular or helical thickenings that resemble tracheids. Also, thick-walled resistant spores with triradiate ridges have been described, but byrophytes or some extinct algae could have produced this type of spore. Some of these organisms might even be animals. (See Banks, 1975a, 1975b, and Gensel and Andrews, 1984, for reviews of this subject.)

Rocks containing undisputed remains of early vascular plant "megafossils" (e.g., stems, connected plant parts) occur frequently in what is now the Appalachians of the United States, eastern Canada, Great Britain, Scandanavia, and central Europe. Past geologic events may explain this distribution. It is postulated that during Silurian-Devonian times the continents were joined and the northern United States, eastern Canada, and central Europe were on the equator. The uplifting of mountains (Caledonian orogeny) and the movement apart of the continents by plate tectonics could account for the present geographical location of the remains of many of the early vascular plants. They were deposited and fossilized in sediments washed into marine waters or deposited in brackish lagoons, or became fossilized *in situ* on mud flats or in bogs.

Knowledge about ancient plants dates back to 1859 when Sir J. William Dawson described a vascular plant (*Psilophyton princeps*) of Devonian age from the Gaspé Peninsula, Quebec, Canada. This fossil was dichotomously branched, with lateral masses of sporangia. In 1870 he modified his description, stating that the sporangia occurred in pairs at the tips of slender branches. It is now apparent that Dawson was examining the remains of several different taxa. In 1967, Hueber and Banks, after a careful reexamination of Dawson's collections and notes, concluded that *Psilophyton princeps* was a plant that branched pseudomonopodially (as a result of anisotomous branching) and isotomously, had naked or spinous stems, possessed a central xylem strand, and had sporangia that were borne terminally on lateral branches (Fig. 7-7).

Little attention was given to Dawson's work because the descriptions were based upon rather fragmentary evidence, and doubt existed about the morphological interpretation of the remains. Also, the extreme age of the plants contributed to the skepticism of morphologists.

It was not until almost sixty years later that the real significance of Dawson's discoveries was appreciated. From 1917 to 1921 Kidston and Lang described well-preserved vascular plants from deposits of Devonian age (now known to be Lower Devonian) in the Rhynie Chert beds, Aberdeenshire, Scotland. It is believed that these plants grew in an intermontane bog and became fossilized *in situ*. Kidston and Lang recognized similarities between *Psilophyton* described by Dawson and their discoveries, and the order "Psilophytales" was established to include all of these presumably primitive extinct plants. Following these early studies, additional genera were discovered in other parts of the world and the "Psilophytales" became a catch-all group for extinct Devonian plants that could not easily be placed in any other lower vascular plant group.

With continuing research on fossil plants, however, it has become increasingly clear that there were at least three main groups within the Psilophytales. In one line the plants had terminal sporangia that were predominantly fusiform. The stems were naked (lacked appendages). Examples are *Rhynia* and *Horneophyton* (Fig. 7-2, A; 7-3, B). To this group, Banks (1968b) gave the name Rhyniophytina (Rhyniophyta, as used in this text). In the second group, sporangia were lateral on axes, predominantly globose or reniform, and each sporangium dehisced along the distal edge. The stems were either devoid of appendages or had spinelike outgrowths. This second group is the Zosterophyllophyta (e.g., *Sawdonia*, *Gosslingia*, *Zosterophyllum*, Figs. 7-5; 7-6). Banks (1968b) also proposed a third division, the Trimerophytina (Trimerophytophyta) to accommodate genera (e.g., *Psilophyton*, *Pertica*, Figs. 7-7; 7-8; 7-9) that do not fit into the other two groups. In this third group the plants had a main axis that branched pseudomonopodially. Each lateral branched a number of times and finally terminated in several or a mass of sporangia. The trimerophytes seem to some paleobotanists to be a natural outgrowth of the Rhyniophyta. Within the Trimerophytophyta the distinction between a central axis (stem) and lateral branches may represent the precursors of fronds. Through planation and development of laminar tissue, a dorsiventral leaf could be evolved (see Chapter 3 for expanded discussion).

Establishment of the three divisions of early vascular plants is based upon morphological-anatomical features. Analysis of the chemical constituents of fossil plants, e.g., sterols, fatty acids, lignin, and aromatic acids corroborates the naturalness of the classification (Niklas, 1979). Undoubtedly, the chemical approach will become increasingly important in providing valuable information when structural features of fossils are poorly preserved.

Rhyniophyta

Some of the simplest vascular plants are placed in the division Rhyniophyta. The genera *Rhynia* and *Horneophyton* first described by Kidston and Lang (1917–1921) are now considered to be of Lower Devonian age. The plants became fossilized as petrifactions (permineralizations) in sediments that formed quartzlike rocks (chert). Preservation was remarkable in that structural details, including stem anatomy and the details of spore morphology are evident.

Originally Kidston and Lang described *Rhynia gwynne-vaughanii* as having a prostrate dichotomously branched rhizome system with rhizoids and an upright dichotomously branched aerial system with smaller lateral branches. The sporangia were fusiform and terminated some of the upright main branches, although Kidston and Lang were unable to document conclusively the connection between the stems and homosporous sporangia. Stomata and cuticle were present on the stems. The stem vascular cylinder was a protostele with centrally placed tracheids that had annular thickenings, surrounded by a thin-walled phloemlike region (Fig. 7-1).

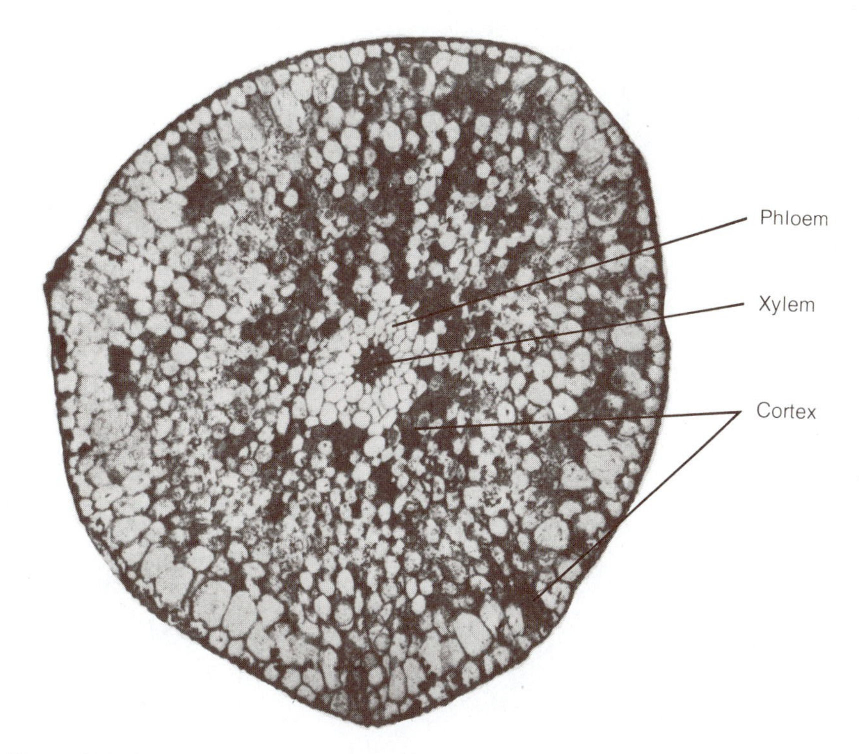

FIGURE 7-1 Transverse section, stem of *Rhynia*. [From *Plant Life through the Ages* by A. C. Seward. Cambridge University Press, New York. 1931.]

A recent study (Edwards, 1980) has shown that the aerial branch system was much more branched than originally suggested. There was a limited number of dichotomies near the base of the aerial system with lateral ("adventitious") branches arising on the tapering main axes. An aerial branch system was about 20 centimeters in height. Sporangia were terminal on the main axes and probably abscised after spore dispersal and were usually overtopped by the development of lateral branches (Fig. 7-2, A). No specialized dehiscence mechanism was observed. Edwards (1980) confirmed the presence of a central strand of annular or helically thickened tracheids surrounded by phloem.

Kidston and Lang described another species, *Rhynia major*, from the Rhynie Chert. This species was more robust than *R. gwynne-vaughanii* and about 60 centimeters in height. The aerial stems branched dichotomously, and homosporous sporangia terminated the ultimate stem dichotomies. In a detailed study of numerous specimens, Edwards (1986) has produced a reconstruction that differs from that of Kidston and Lang. According to Edwards there was an extensive dichotomously branched rhizome system with a limited number of upright stems that branched dichotomously at wide angles. All axes terminated in paired, homosporous sporangia above a dichotomy (Fig. 7-2, B). The upright branch system is estimated to have been about 18 centimeters in height—much less than that estimated by Kidston and Lang. The sporangia were less than 12 millimeters long and about 4 millimeters wide with stomata in the outer layer; dehiscence was longitudinal. Of special interest, the central conducting strand of stems consisted of elongate thin-walled cells of a phloemlike region surrounding numerous elongate cells with *uniformly* thickened cell walls, the innermost cells of which were narrower and had thinner walls than the peripheral cells. The important feature is that those cells that would constitute the xylem of a vascular plant did not have annular or helical thick-

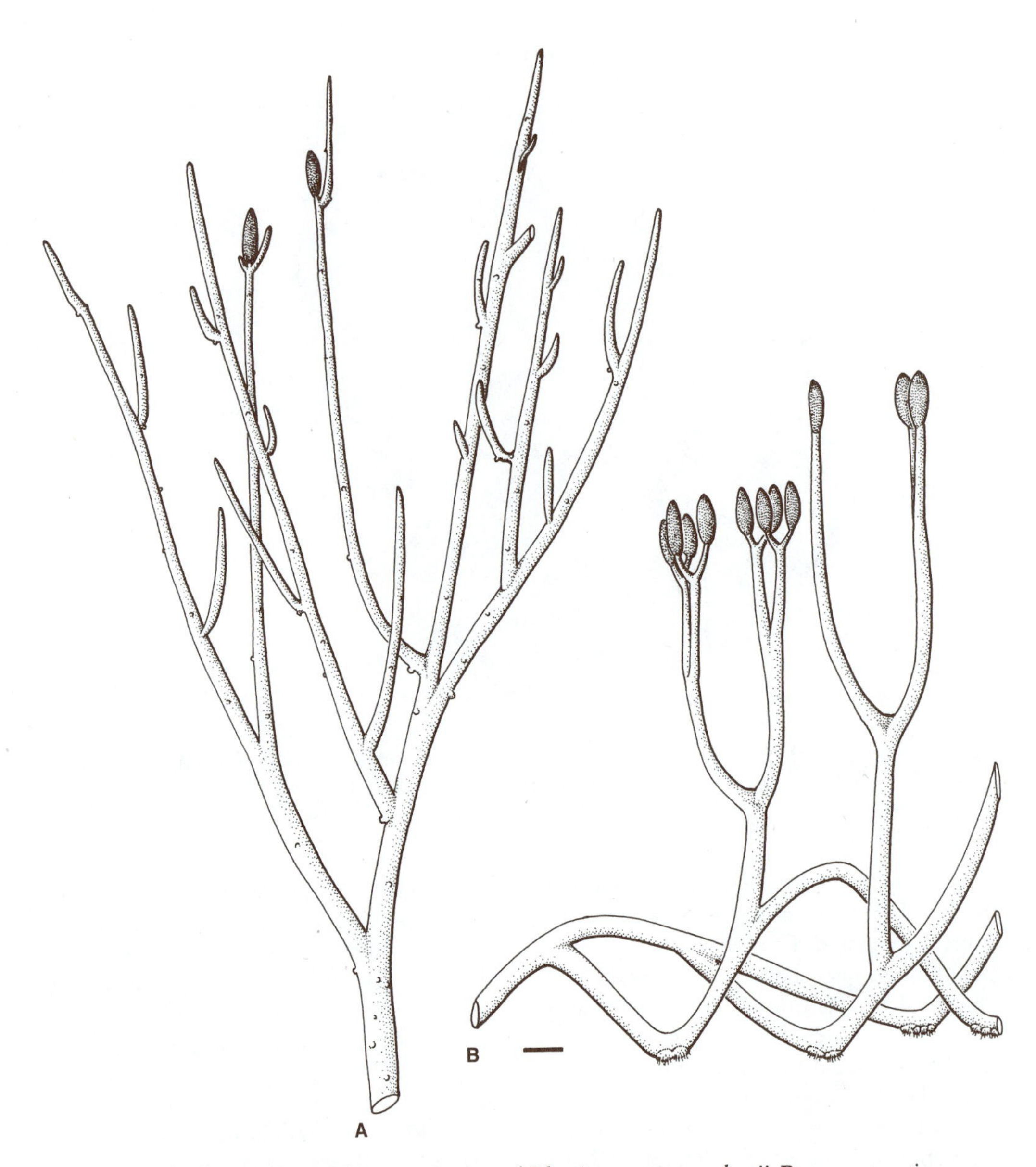

FIGURE 7-2 **A,** suggested reconstruction of *Rhynia gwynne-vaughanii.* **B,** reconstruction of *Aglaophyton* (*Rhynia*) *major;* scale bar = 10 mm. [**A** redrawn from Edwards, *Rev. Palaeobot. Palynol.* 29:177–188, 1980; **B** redrawn from Edwards, *Bot. J. Linnean Soc. 93:173–204, 1986.]*

enings and should not be considered tracheids. Therefore, Edwards (1986) likens the conducting strand to that of certain mosses in which central water-conducting cells (*hydrom*) is surrounded by a region of thin-walled food-conducting cells *(lepton)*. For this reason, Edwards (1986) believes that *Rhynia major* is not a vascular plant and has renamed it *Aglaophyton major. Aglaophyton* is not comparable to any living bryophyte nor is it a vascular plant and appears to be an intermediate form between the two. Edwards (1986) concludes that it is not surprising to find plants from the Late Silurian and Devonian that illustrate different "evolutionary experiments."

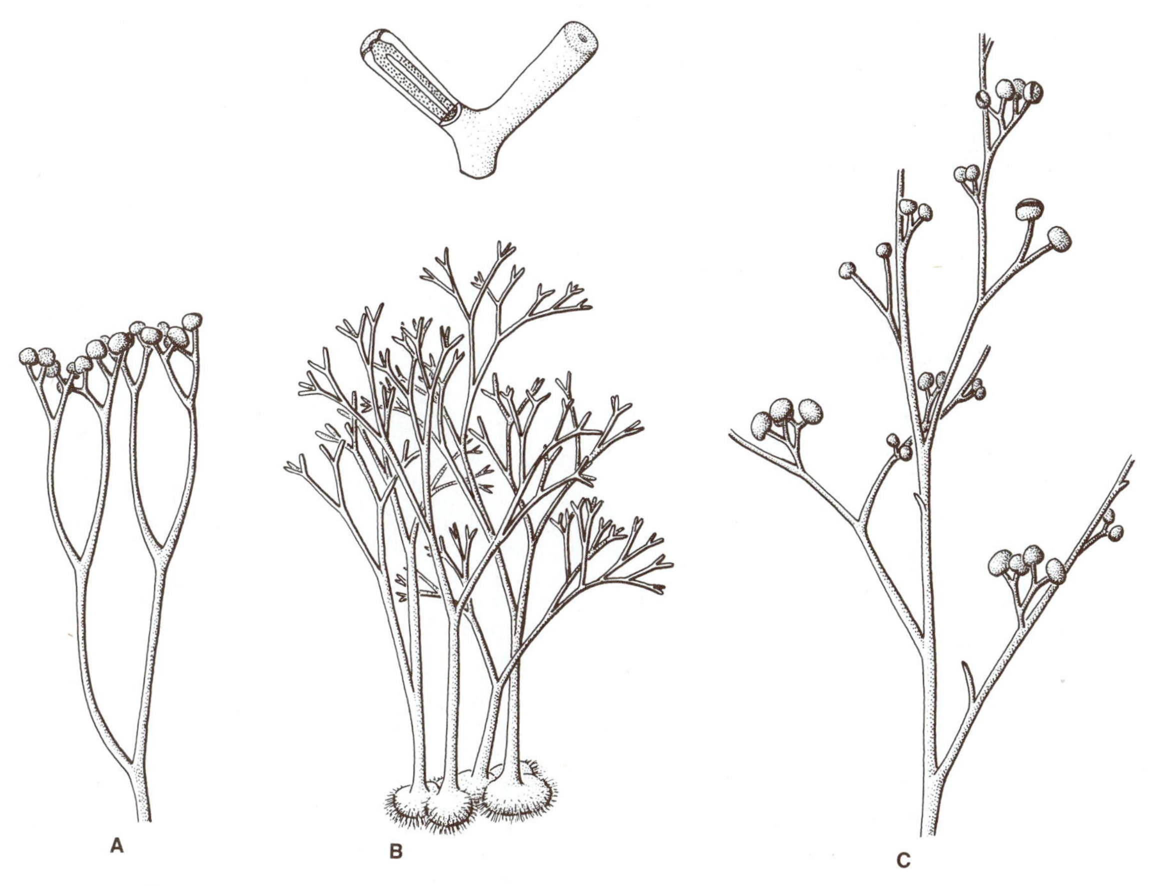

FIGURE 7-3 Reconstructions of members of the Rhyniophyta. **A**, *Cooksonia caledonica;* **B**, *Horneophyton lignieri;* **C**, *Renalia hueberi.* [**A** redrawn from Edwards, *Palaeontol.* 13:451–461, 1970; **B** from Eggert, *Amer. J. Bot* 61:405–413, 1974; **C** redrawn from Gensel, *Rev. Palaeobot. Palynol.* 22:19:–37, 1976.]

Horneophyton lignieri

The taxon *Horneophyton lignieri* (Fig. 7-3, B) is found with *Rhynia*—often in dense aggregations—in the Rhynie Chert. The basal part consisted of a tuberous, cormlike structure with numerous rhizoids. The aerial axes were smooth and branched dichotomously. A vascular cylinder was present in the aerial branches, but disappeared in the basal corm. Based upon recent studies, the sporangia have been shown to be unique (Eggert, 1974). The tips of the branches end in two to four cylindrical lobes, each having a sterile columella surrounded by spores. Although it has not been conclusively demonstrated, the sporangia may have opened by means of a small apical pore at the tips of the lobed sporangium. Each spore had a triradiate ridge (trilete type of spore), evidence of its contact with the other three spores of a tetrad. As in *Rhynia*, *Horneophyton* was homosporous.

Cooksonia, Renalia

At the present time, *Cooksonia* is generally accepted as the oldest known vascular plant. Specimens are from the United States, Canada, Wales, Scotland, and Czechoslovakia. Those from Wales are from deposits as old as mid-Silurian; others are from the Lower Devonian. Being the smallest and simplest vascular plants known from the fossil record, species of *Cooksonia* were slender, leafless, dichotomously branched plants 5 to 7 centimeters in length and from 1 millimeter or less to about 2 millimeters in diameter. Some specimens had vascular strands with annular tracheids (Lang, 1937; Edwards and Davies, 1976). The terminal sporangia were globose to reniform (Fig. 7-3, A) and contained trilete spores (with triradiate ridges). (For further descriptions, see Obrhel, 1962; Banks, 1971, 1980; Edwards, 1970a; Edwards and Feehan, 1980.)

Gensel (1976) described another rhyniophyte, *Renalia hueberi*, from the coastline of Gaspé Bay, Canada, that had main axes at least 11 centimeters tall and exhibited dichotomous to anisotomous (pseudomonopodial) branching. The laterals branched dichotomously and terminated in round to reniform sporangia that dehisced along the distal margin (Fig. 7-3, C). Spores were homosporous and trilete. Gensel believes that *Renalia* may have evolved from the Late Silurian *Cooksonia*-like plants, but the mode of dehiscence is more like that of zosterophylls, strengthening the idea that an interrelationship exists between the Rhyniophyta and the Zosterophyllophyta (Banks, 1968a).

Taeniocrada

Most members of the Rhyniophyta probably lived in marshes, bogs, or on mud flats, although their remains may have been washed to other sites where they became fossilized. *Taeniocrada*, however, may have been more of an aquatic, because stomata have not been found on the lower part of the plants, although the epidermis is fairly well preserved. The stems were flat or ribbonlike, exhibited dichotomous branching, and had clusters of terminal sporangia (Fig. 7-4). The conducting strand may have consisted of helically strengthened tubes rather than typical tracheids (Hueber, 1982).

Thus, members of the Rhyniophyta exhibit certain common morphological characteristics: (1) dichotomous or anisotomous (pseudomonopodial) branching; (2) a protostelic xylem cylinder with centrarch xylem; and (3) terminal sporangia, although showing some variation in form. They are of mid-Silurian (*Cooksonia*)-Lower Devonian age (Banks, 1980).

FIGURE 7-4 Reconstruction of *Taeniocrada decheniana*. [Redrawn from Kräusel and Weyland, *Abh. Preuss. Geol. Landesanstalt* 131:1–92, 1930.]

ZOSTEROPHYLLOPHYTA. Members of this division are often found in the same fossil beds with rhyniophytes and led to a complication for Dawson in his study of the Gaspé flora. One of the main organographic differences between the two divisions is the placement of sporangia. In the Zosterophyllophyta, the sporangia were *lateral* on the stems.

Sawdonia, Crenaticaulis

Upon reexamination of Dawson's collections, Hueber (1971) proposed the binomial *Sawdonia ornata* for what Dawson named as *Psilophyton princeps* var. *ornatum*. The genus is an anagram of the surname Dawson, so named in honor of the nineteenth-century geologist. The plant branched dichotomously from a rhizome system, had multicellular emergences or spines, and had reniform sporangia borne laterally and singly on short stalks that were located toward the curled (circinate) tips of branches (Fig. 7-5, A). The protostele was exarch with tracheids that had helical, scalariform, or reticulate secondary wall thickenings.

Another species, *Sawdonia acanthotheca* (Fig. 7-5, B), of Middle Devonian age is quite similar to *S. ornata* but differs in the nature of the emergences on the axes and their presence on sporangia (Gensel et *al.*, 1975). Also, *Crenaticaulis verruculosus* (Lower Devonian, Gaspé Bay) is similar to *Sawdonia* but had one or two vertical series of emergences. Preservation of the xylem is sufficiently good to reveal that the cylinder was elliptical in outline and that xylem maturation was exarch, one of the unifying characteristics of the Zosterophyllophyta (Banks and Davis, 1969).

Gosslingia

Another zosterophyllophyte from the Lower Devonian is *Gosslingia breconensis* (Fig. 7-6, A). This plant was at least 50 centimeters high and branched predominantly by unequal dichotomies. As in *Sawdonia* the axis tips were circinately coiled, but the stems were smooth except for curious vascularized tubercles at points of branching. The oval

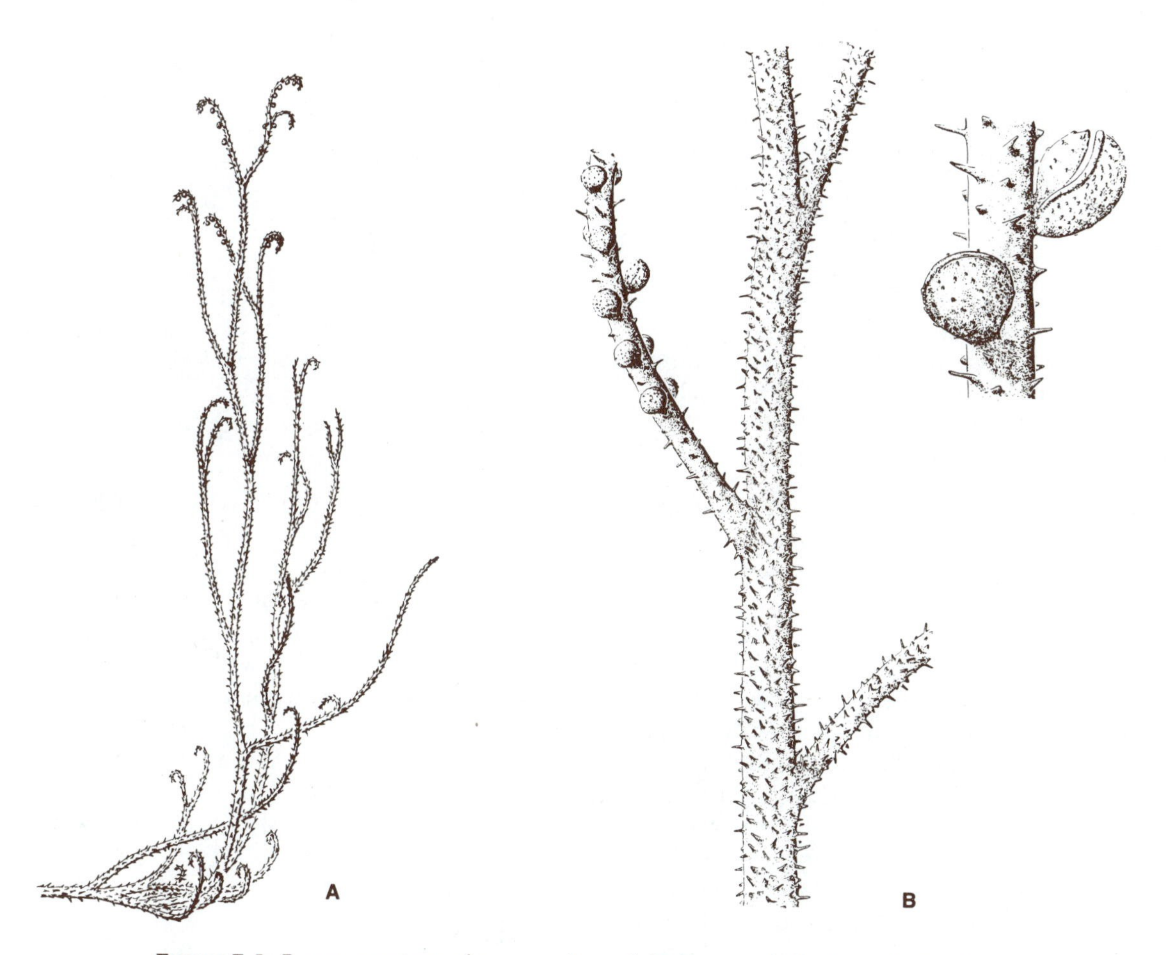

FIGURE 7-5 Reconstructions of two members of the Zosterophyllophyta. **A,** *Sawdonia ornata;* note lateral sporangia on terminal branches; **B,** *Sawdonia acanthotheca;* note line of dehiscence on sporangium at right. [**A** from Andrews, *Ann. Missouri Bot. Garden* 61:179–202, 1974; **B** from Gensel et al., *Bot. Gaz.* 136:50–62, 1975.]

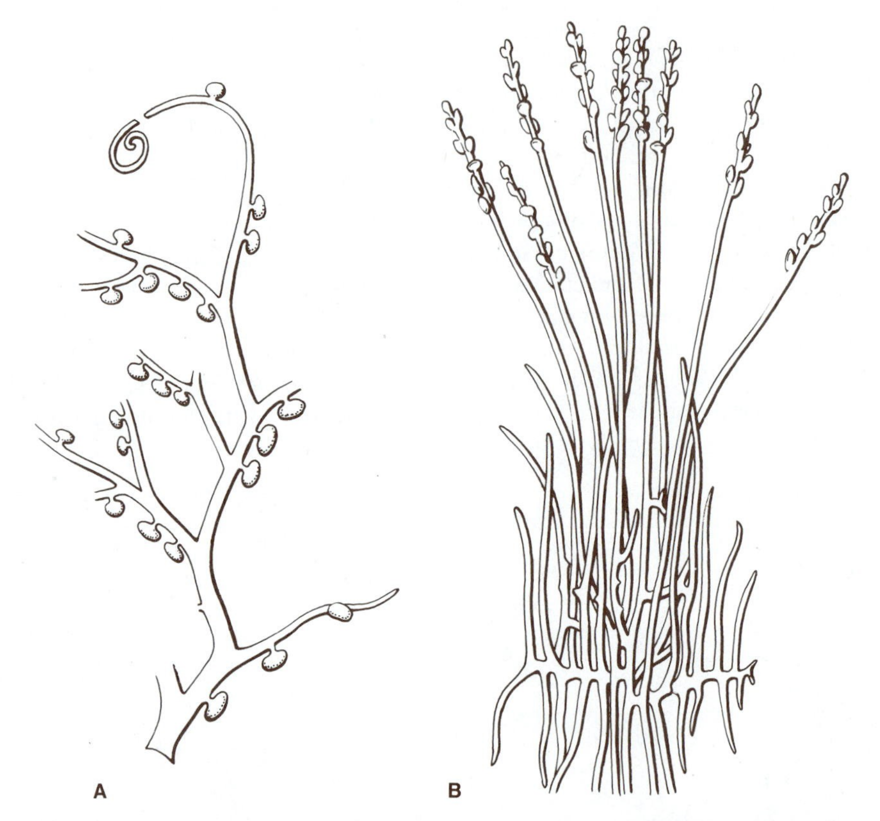

FIGURE 7-6 Reconstructions of two members of the Zosterophyllophyta. **A,** *Gosslingia breconensis;* **B,** *Zosterophyllum myretonianum.* [**A** redrawn from *Studies in Paleobotany* by H. N. Andrews. Wiley, New York. 1961: **B** redrawn from Walton, *Phytomorphology* 14:155–160, 1964.]

to reniform sporangia were scattered along the main and lateral branches. The species was apparently homosporous (Edwards, 1970b).

Zosterophyllum

The genus *Zosterophyllum* is known from Lower Devonian rocks of Scotland, South Wales, Belgium, France, Russia, and Australia. It differs from *Gosslingia* in that the former has sporangia that were aggregated into spikes. Branching was basically dichotomous, but a rather unique type, especially in *Z. myretonianum* (Walton, 1964), with an upward directed system and a downward directed rhizome system (Fig. 7-6, B). The sporangia dehisced by means of a slit along the distal edge. The xylem strand in *Z. llanoveranum* was elliptical to terete, and the xylem was exarch (Edwards, 1969).

Characteristically, then, the Zosterophyllophyta exhibited dichotomous branching, and the stems were either smooth or had multicellular spines. Sporangia occurred laterally on the stems and in some were aggregated into spikes. The presence of distal dehiscence has been well established in the group. The xylem strand of stems was exarch.

TRIMEROPHYTOPHYTA. This major group was established by Banks (1968b) for early vascular plants that were not members of the Rhyniophyta or Zosterophyllophyta.

Psilophyton

In 1967, Hueber and Banks (1967) reexamined Dawson's collections and additional specimens and selected a type, *Psilophyton princeps*. Morphologically, the plants had stout main axes and lateral branches that branched dichotomously, the ultimate branches ending in pendulous paired sporangia. The axes had blunt spines and interrupted longitudinal ridges. The sporangia were elongate and elliptical, about 7.5 to 8 millimeters long and 1.2 millimeters wide, and were homosporous (Fig. 7-7).

Several other species have been described. *Psilophyton dawsonii*, from the Lower Devonian of the Gaspé, and Ontario, Canada, had a system of stout main axes and smaller lateral branches were undoubtedly produced by unequal dichotomous (anisotomous) branching (Banks et al., 1975). Lateral branches were arranged alternately in two rows. Fertile lateral branches usually dichotomized six times before terminating in clusters of approximately 32 sporangia (Fig. 7-8, A). The axes were smooth, unlike *P. princeps* (Fig. 7-7). In main axes the xylem cylinder occupied about one-fourth of the stem and was a terete, centrarch protostele in which the protoxylem tracheids had helical and scalariform thickenings. The metaxylem had scalariform tracheids and some tracheids with circular bordered pits.

FIGURE 7-7 Reconstruction of *Psilophyton princeps*. [From Andrews, *Ann. Missouri Bot. Garden* 61:179–202, 1974.]

Another species, *Psilophyton charientos*, from New Brunswick Province, Canada, probably grew to a maximum height of about 50 centimeters and exhibited the common pseudomonopodial habit (Gensel, 1979). All but the most distal segments were densely covered with slender, tapered spines. The fertile lateral branches dichotomized, often terminating in gracefully recurved tips bearing sporangia (Fig. 7-8, B) that dehisced longitudinally. The xylem was centrarch, in which the protoxylem tracheids had annular thickenings and the metaxylem tracheids were scalariformly pitted.

Pertica

Pertica quadrifaria (Fig. 7-9) is a very interesting plant from the Lower Devonian found in northern Maine (Kasper and Andrews, 1972). It was erect, perhaps a meter tall, with a distinct pseudomonopodial main axis, and dichotomous lateral branches arranged in four rows, although there is evidence that they were initiated in helical fashion at the tip of the main axis (Fig. 7-9). The laterals dichotomized numerous times and some of the ultimate branchlets bore numerous sporangia in dense clusters. Although vascular tissue was not preserved in the compressions and impressions, there is little doubt about its assignment to the Trimerophytophyta. This plant, probably more than others, illustrates how precursors of megaphylls might have originated by overtopping (see Chapter 3, pp. 30–31). Niklas and Kerchner (1984) have presented computer simulations of various branching patterns that are correlated with photosynthetic efficiency. Repeated three-dimensional, isotomous branching in a plant with limited growth would produce considerable shading. By the processes of overtopping (leading to pseudomonopodial branching) and planation of side branches in the course of evolution would produce a minimum of

FIGURE 7-8 Trimerophytophyta. Reconstructions of *Psilophyton dawsonii* (**A**) and *Psilophyton charientos* (**B**). [**A** from Banks et al., *Palaeontographica Amer.* 8:77–127, 1975; **B** from Gensel, *Palaeontographica* 168 B:81–99, 1979.]

self-shading. The trimerophytes, especially *Pertica,* illustrate the beginnings of these trends found in Middle and Upper Devonian vascular plants. (For additional information on Devonian plants, refer to Gensel and Andrews, 1984, 1987.)

Interrelationships of the Early Vascular Plants

The divisions Rhyniophyta, Zosterophyllophyta, and Trimerophytophyta are particularly interesting and significant from an evolutionary standpoint because: (1) they include some of the most ancient, least specialized, and most primitive known vascular plants; (2) some of the forms may well represent the type of early land plants from which the more highly developed sporophytes of ferns and other lower vascular plant groups have originated; and (3) the arrangement and histology of vascular systems of these ancient plants gives us a basis for making comparisons between them and their possible living derivatives. Well-preserved fossil remains have been found in rocks from the mid-Silurian into the Devonian, and the reconstructions and descriptions of these plants are among the most significant paleobotanical achievements in the twentieth century. The earliest known undisputed vascular plant is *Cooksonia* from the mid-Silurian. For many years *Baragwanathia* (a lycopod) from Australia was considered to be the oldest known vascular plant, but the age of the rocks that contain it was determined to be Devonian rather then Silu-

FIGURE 7-9 Reconstruction of *Pertica quadrifaria*. [From Kasper and Andrews, *Amer. J. Bot.* 59:897–911, 1972.]

rian (Jaeger, 1962), although the possibility that it was actually of Silurian age has been raised again (Garratt, 1981). Another example of changes resulting from increased paleobotanical research is the shifting of the genus *Asteroxylon* (originally described by Kidston and Lang from the Lower Devonian Rhynie Chert) from the "psilophytes" to the Lycophyta (see Lyon, 1964; Banks, 1968b; Chapter 9).

Are the three divisions interrelated, and how do they relate to derivative groups? The oldest vascular plant, *Cooksonia*, which bore terminal, homosporous sporangia containing resistant spores with triradiate marks, is found in rocks from the mid-Silurian into the Lower Devonian. *Rhynia* and *Horneophyton*, which are more perfectly fossilized, lived in the lower Devonian, but at a later time than *Cooksonia*. In addition to displaying dichotomous branching and possessing smooth axes and terminal sporangia, they were protostelic with centrarch xylem. Coexisting with the rhyniophytes was *Psilophyton*, and at a slightly later date, *Pertica*. It is quite possible that these latter two genera were derivatives of *Rhynia*-like plants by a modification of growth habit: pseudomonopodial branching (result of anisotomous branching) which resulted in the overtopping of determinant lateral branches by main axes. The presence of a centrarch protostele in *Rhynia gwynne-vaughanii*, for example, and in the Trimerophytophyta also supports this concept. The Trimerophytophyta may well have been the stock from which ferns, extinct seed ferns, and progymnosperms were derived.

The third group, Zosterophyllophyta, was present in the Lower Devonian. *Zosterophyllum* exhibited dichotomous branching, had no leaves, and the sporangia were lateral, occurring in terminal groups. Others, *Sawdonia* and *Crenaticaulis*, had multicellular spines or emergences and lateral sporangia. However, Stewart (1983) considers the sporangia to be terminal on greatly shortened axes, even though the presence of vascular tissue has not been demonstrated. The occurrence of vascular tissue would support the view that the Zosterophyllophyta and the Rhyniophyta share a common ancestry, and would also be supportive of the telome theory (Chapter 3). The vegetative stems, where known, are protostelic with exarch xylem. Extinct members of the Lycophyta, as well as living members, share certain characteristics, e.g., exarch xylem, with the zosterophylls, and it is reasonable to assume that they were derived from some member(s) of the Zosterophyllophyta (Chapter 9).

An obvious omission from the discussion is an account of the gametophyte generation. It has long been assumed that none of the known remains of early vascular plants are gametophytes. Upon reexamination of the original material of *Rhynia gwynne-vaughanii* from the Rhynie Chert, Pant (1962) and Lemoigne (1968, 1969, 1970) have concluded that the vascularized axes of this species are probably the gametophytes of *R. (Aglaophyton) major*. This assumption is based upon several morphological features and upon Kidston and Lang's inability to ever prove actual connections between axes and sporangia for *R. gwynne-vaughnii*. However, the recent discovery of sporangia attached to axes is a challenge to the theory (Edwards, 1980). Other reputed gametophytic axes have been reported from the Rhynie Chert in which antheridia and archegonia are described as occurring on pads at the ends of axes (Remy and Remy, 1980; Remy, 1982). Future paleobotanical discoveries may reveal whether there is a definite relationship between these presumed gametophytes and the well-known sporophytes of the Rhynie Chert. (See Chapter 2, p. 15 for a more complete discussion.

When Did Vascular Plants Originate?

Over the past fifteen years the time of the origin of vascular land plants has been the subject of controversy. Paleobotanists agree that land plants were present in the Lower Silurian, but whether all of them were vascular plants (or potentially vascular plants) is another matter. The presence of dispersed resistant spores with the triradiate mark, fragments of epidermal tissue with cuticle, and tubes that simulate tracheids have been reported from the Lower Silurian. The spores and cuticle could be from bryophytelike plants, or from some algal-like land plants that were not the progenitors of vascular plants, but only the remnants of evolutionary "experiments." Some paleobotanists contend that the stratigraphic occurrence of early vascular plants should be reported very precisely and that it should be clearly shown that the fossils possess tracheids and show evidence of other features such as cuticle, stomata, and trilete spores. They argue that errors are likely in studies of dispersed microfossils (e.g., spores)—for example, from the "contamination" of older rocks by spores from younger rocks (Banks, 1968b, 1975a, b, 1980; Chaloner, 1964, 1970). *Cooksonia* is the oldest demonstrable vascular plant, occurring in the latter part of the Silurian, with one report of *Cooksonia*-type sporangia and stems in the middle of the Silurian (Edwards and Feehan, 1980), although tracheids were not observed. Those who hold to a Silurian origin of vascular plants believe that there was sufficient time for the appearance of simple, naked, dichotomously branched plants such as *Cooksonia* and the further elaboration and diversity of vascular plants that occurred by the end of the Devonian (Chaloner and Sheerin, 1979).

A different point of view has been expressed by other paleobotanists (Gray and Boucot, 1978, 1980; Boucot and Gray, 1982). They contend that land plants, potentially including vascular plants, arose in the Late Ordovician or earlier and they assume that some type of plant(s) preceded the appearance of *Cooksonia*-like plants. Spore tetrads and cuticular sheets of cells have been reported, for example, from the Late Ordovician rocks of Libya. To them, the production of resistant spores was initially more important for existence of potential vascular plants on land than was the development of tracheids. The advent of tracheids came with the deposition of lignin in cell walls permitting an increase in plant height. They acknowledge that much of the evidence on dispersed spores comes from marine rocks near shore or off shore, but cite examples of the remains of plants that were growing on land in the earliest part of the Silurian consisting of sheets of cells some of which had banded tubular elements. Dispersed trilete spores have been found, but without organic connection to any other fossil material (Pratt et al., 1978).

Those who champion the concept of a pre-Silurian origin view megafossil evidence of pre-Devonian vascular plants as a poor source of information about the *time* and *place* of origin of vascular plants. For example, the number of discovered megafossils from the Silurian is meager compared to the numerous dispersed spore types described. This would indicate to them that a large land flora existed, some of which might have been vascular plants. The Silurian was of relatively short duration in geologic time, but evolutionary "experiments" leading to the origin of land plants in the Ordovician, followed by the appearance of simple *Cooksonia*-like plants in the Silurian could be a reasonable scenario.

REFERENCES

Banks, H. P.

1968a. The stratigraphic occurrence of early land plants and its bearing on their origin. In D. H. Oswald (ed.), *Proceedings of the International Symposium on the Devonian System*, Vol. I. pp. 721–730. Calgary, Canada.

1968b. The early history of land plants. In E. T. Drake (ed.), *Evolution and Environment.* (Symposium volume of papers at Centennial Celebration of Peabody Museum, pp. 73–107). Yale University Press, New Haven.

1971. Occurrence of *Cooksonia*, the oldest vascular land plant macrofossil, in the Upper Silurian of New York state. *J. Indian Bot. Soc.* 50A:227–235.

1975a. Early vascular land plants: proof and conjecture. *BioScience* 25:730–737.

1975b. The oldest vascular land plants: a note of caution. *Rev. Palaeobot. Palynol.* 20:13–25.

1980. Floral assemblages in the Siluro-Devonian. In D. L. Dilcher and T. N. Taylor (eds.), *Biostratigraphy of Fossil Plants*, pp. 1–24. Dowden, Hutchinson and Ross, Pennsylvania.

Banks, H. P., and M. R. Davis

1969. *Crenaticaulis*, a new genus of Devonian plants allied to *Zosterophyllum*, and its bearing on the classification of early land plants. *Amer. J. Bot.* 56:436–449.

Banks, H. P., S. Leclercq, and F. M. Hueber

1975. Anatomy and morphology of *Psilophyton dawsonii*, sp. n. from the late Lower Devonian of Quebec (Gaspé) and Ontario, Canada. *Palaeontogr. Amer.* 8:73–127.

Boucot, A. J., and J. Gray

1982. Geologic correlates of early land plant evolution. *Third North American Paleontological Convention, Proceedings.* Vol. 1:61–66.

Chaloner, W. G.

1964. An outline of Pre-Cambrian and Pre-Devonian microfossil records: evidence of early land plants from microfossils. In *Abstracts, Tenth International Bot. Congress, Edinburgh.* pp. 16–17.

1970. The rise of the first land plants. *Biol. Rev.* 45:353–377.

Chaloner, W. G., and A. Sheerin

1979. Devonian macrofloras. *Special Papers in Palaeontology* No. 23:145–161.

Edwards, D.

1969. Further observations on *Zosterophyllum llanoveranum* from the Lower Devonian of South Wales. *Amer J. Bot.* 56:201–210.

1970a. Fertile Rhyniophytina from the Lower Devonian of Britain. *Palaeontol.* 13:451–461.

1970b. Further observations on the Lower Devonian plant, *Gosslingia breconensis* Heard. *Phil. Trans. Roy. Soc.* London 258:225–243.

Edwards, D. S.

1980. Evidence for the sporophytic status of the Lower Devonian plant *Rhynia gwynne-vaughanii* Kidston and Lang. *Rev. Palaeobot. Palynol.* 29:177–188.

1986. *Aglaophyton major*, a non-vascular land-plant from the Devonian Rhynie Chert. *Bot. J. Linnean Soc.* 93:173–204.

Edwards, D., and E. C. W. Davies

1976. Oldest recorded *in situ* tracheids. *Nature* 263:494–495.

Edwards, D., and J. Feehan

1980. Records of *Cooksonia*-type sporangia from late Wenlock strata in Ireland. *Nature* 287:41–42.

Eggert, D. A.

1974. The sporangium of *Horneophyton lignieri* (Rhyniophytina). *Amer. J. Bot.* 61:405–413.

Garratt, M. J.

1981. The earliest vascular land plants: Comment on the age of the oldest *Baragwanathia* flora. *Lethaia* 14:8.

Gensel, P. G.

1976. *Renalia hueberi*, a new plant from the Lower Devonian of Gaspé. *Rev. Palaeobot. Palynol.* 22:19–37.

1979. Two *Psilophyton* species from the Lower Devonian of eastern Canada with a discussion of morphological variation within the genus. *Palaeontographica* Abt. B 168:81–99.

Gensel, P. G., and H. N. Andrews

1984. *Plant Life in the Devonian.* Praeger, New York.

Gensel, P. G., and H. N. Andrews.

1987. The evolution of early land plants. *Am. Scientist* 75:478–489.

Gensel, P. G., H. N. Andrews, and W. H. Forbes

1975. A new species of *Sawdonia* with notes on the

origin of microphylls and lateral sporangia. *Bot. Gaz.* 136:50–62.

Gray, J., and A. J. Boucot
1978. The advent of land plant life. *Geology* 6:489–492.
1980. Microfossils and evidence of land plant evolution. *Lethaia* 13:174.

Hueber, F. M.
1971. *Sawdonia ornata*: a new name for *Psilophyton princeps* var. *ornatum*. *Taxon* 20:641–642.
1982. *Taeniocrada dubia* Kr. and W.: its conducting strand of helically strengthened tubes. Bot. Soc. Am. Misc. Ser. 162:58–59.

Hueber, F. M., and H. P. Banks.
1967. *Psilophyton princeps*: the search for organic connection. *Taxon* 16:81–85.

Jaeger, H.
1962. Das Alter der ältesten bekannten Landpflanzen *(Baragwanathia-flora)* in Australien auf Grund der begleitenden Graptolithen. *Palaeontol. Zeit.* 36:7.

Kasper, A. E., Jr., and H. N. Andrews, Jr.
1972. *Pertica*, a new genus of Devonian plants from northern Maine. *Amer. J. Bot.* 59:897–911.

Kidston, R., and W. H. Lang
1917–1921. On Old Red Sandstone plants showing structure, from the Rhynie Chert Bed, Aberdeenshire. Parts I-V. *Trans. Roy. Soc. Edinb.* 51–52.

Lang, W. H.
1937. On the plant-remains from the Downtonian of England and Wales. *Phil. Trans. Royal Soc.* Ser. B, 227:245–291.

Lemoigne, Y.
1968. Observation d'archégones portés par des axes du type *Rhynia gwynne-vaughanii* Kidston et Lang. Existence de gamétophytes vascularisés au Dévonien. *Compt. Rend. Acad. Sci.* (Paris), Ser. D, 266:1655–1657.
1969. Contribution à la connaissance du gamétophyte *Rhynia gsynne-vaughanii* Kidston et Lang. Problème des protuberances et processus de ramification. *Bull. Mens. Soc. Linn. Lyon* 38(4):94–102.
1970. Nouvelles diagnoses du genre *Rhynia* et de l'espèce *Rhynia gwynne-vaughanii. Bull. Soc. Bot. France* 117:307–320.

Niklas, K. J.
1979. An assessment of chemical features for the classification of plant fossils. *Taxon* 28:505–516.

Niklas, K. J., and V. Kerchner
1984. Mechanical and photosynthetic constraints on the evolution of plant shape. *Paleobiology* 10:79–101.

Obrhel, J.
1962. Die Flora der Pridolí-Schichten (Budňany-Stufe) des mittelböhmischen Silurs. *Geologie* 11:83–97.

Pant, D. D.
1962. The gametophyte of the Psilophytales. In P. Maheshwari, B. M. Johri, and I. K. Vasil (eds.), *Proc. Summer School of Botany at Darjeeling* (1960) pp. 276–301. Ministry of Scientific Research and Cultural Affairs, India.

Pratt, L. M., T. L. Phillips, and J. M. Dennison
1978. Evidence of non-vascular land plants from the Early Silurian (Llandoverian) of Virginia, U.S.A. *Rev. Palaeobot. Palynol.* 25:121–149.

Remy, W.
1982. Lower Devonian gametophytes: relation to the phylogeny of land plants. *Science* 215:1625–1627.

Remy, W., and R. Remy
1980. Devonian gametophytes with anatomically preserved gametangia. *Science* 208:295–296.

Stewart, W. N.
1983. *Paleobotany and the evolution of plants.* Cambridge University Press, Cambridge.

Walton, J.
1964. On the morphology of *Zosterophyllum* and some other early Devonian plants. *Phytomorphology* 14:155–160.

CHAPTER 8

Psilophyta

THE division Psilophyta is made up of living plants comprising one order, one family, and two genera (*Psilotum,* Fig. 8-1; and *Tmesipteris,* Fig. 8-8). These plants are rather simple in organization, and traditionally they have been aligned with the extinct "psilophytes" (Chapter 7). Indeed, there are some similarities between some of the extinct early vascular plants (Chapter 7) and the Psilophyta: (1) the sporophytes are dichotomously branched with an underground rhizome system and an upright system of branches; (2) there are no roots; (3) the stems have a relatively simple vascular cylinder; (4) the sporangia are eusporangiate and homosporous; and (5) the sporangia may be interpreted as occurring at the ends of shortened axes. If this interpretation is correct, the Psilophyta resemble the ancient vascular plants in the Rhyniophyta and Trimerophytophyta (Chapter 7).

Paleobotany contributes little to proving the ancestry of *Psilotum* and *Tmesipteris* because of the lack of fossils of the two genera. If either genus is a relict form of some ancient or primitive vascular plant, we have no direct knowledge of it today. Approximately 400,000,000 years separate the Psilophyta and the earliest vascular plants and naturally, this has led to skepticism regarding the closeness of relationship. The two genera undoubtedly will be shifted from one major taxon to another in the years to come. For example, one botanist (Bierhorst, 1968a, b; 1969; 1971) would abolish a separate division or class and remove the two genera, as a family, to the leptosporangiate ferns. However, *Psilotum* and *Tmesipteris* are rootless and eusporangiate, whereas the fern (*Stromatopteris*) that is used for comparison has roots and is leptosporangiate.

Psilotin, a specific phenolic substance, has been found in *Psilotum* and *Tmesipteris* (Tse and Towers, 1967). This phenolic has not been found in the Lycophyta, which would support the conclusion that the two genera constitute a natural group. Also, the distribution of flavonoid compounds in *Psilotum*, *Tmesipteris*, and the primitive filicalean ferns shows that it is unlikely that the two genera are closely related to the ferns (Cooper-Driver, 1977).

Psilotales

Psilotaceae — Sporophyte Generation of *Psilotum*

DISTRIBUTION AND ORGANOGRAPHY. *Psilotum,* consisting of at least two species, *Psilotum nudum* (Fig. 8-1) and *Psilotum complanatum* (Reed, 1966),

FIGURE 8-1 Growth habit of *Psilotum nudum* under greenhouse culture. Note dichotomies of branches at left.

is pantropical and subtropical in distribution, reaching north to Okinawa, Japan, Florida, Bermuda, and Hawaii. *P. nudum* has been reported in Nigeria (Savory, 1949), Basutoland (Morgan, 1962), Texas and Arizona (Mason, 1968), and in Spain (Allen, 1966). Its presence in some of these localities may be the result of introduction by man, intentionally or inadvertently. Plants occur as epiphytes on tree ferns, on coconut palm trunks, or at the base of trees, or they may be terrestrial, growing in soil or among exposed rocks. *P. nudum* grows remarkably well under greenhouse conditions and is cultivated in most botanical gardens in temperate regions. Depending upon its location and environment, the sporophyte of *P. nudum* may be pendent or erect and dwarfed (8 centimeters high) or as tall as 75 to 100 centimeters. The plant body consists of a basal branched rhizome system generally hidden beneath the soil or humus, and slender, upright, green aerial portions that are dichotomously branched and bear small appendages and synangia (Fig. 8-2, A, B). The branched rhizome system, which bears numerous rhizoids, grows by means of apical meristems located at the tips of ultimate branches. According to Bierhorst (1954b), the degree of branching of the rhizome is related to obstacles which the apical meristem encounters in its growth through the soil. No roots are present, although the underground rhizome system anchors the plant and rhizoids serve as absorptive structures. A mycorrhizal intracellular fungus, gaining entrance through rhizoids, is present in cells of the outer cortex (Verdoorn, 1938; Bierhorst, 1954b). This fungus may be related intimately to the physiology of the plant. Any one of the rhizome tips may turn upward and undergo several to many dichotomies that establish the basic plan of aerial branch organization. The basal part of the shoot may be

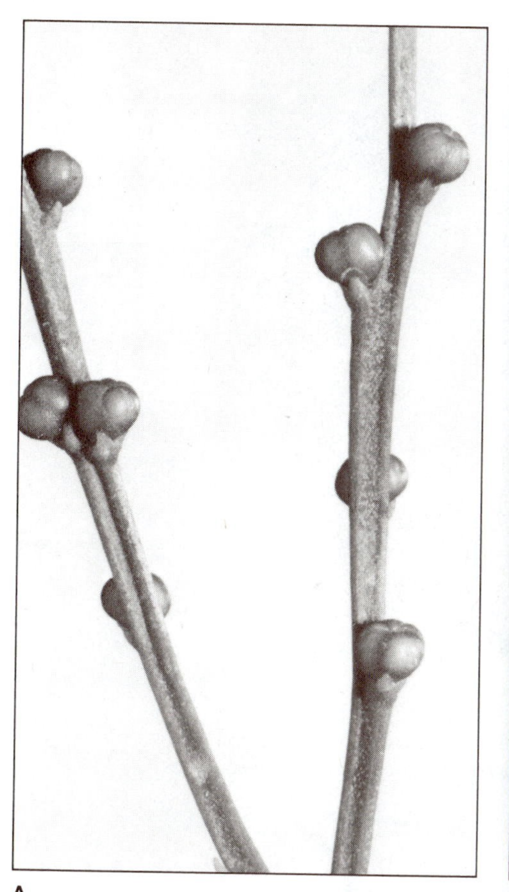

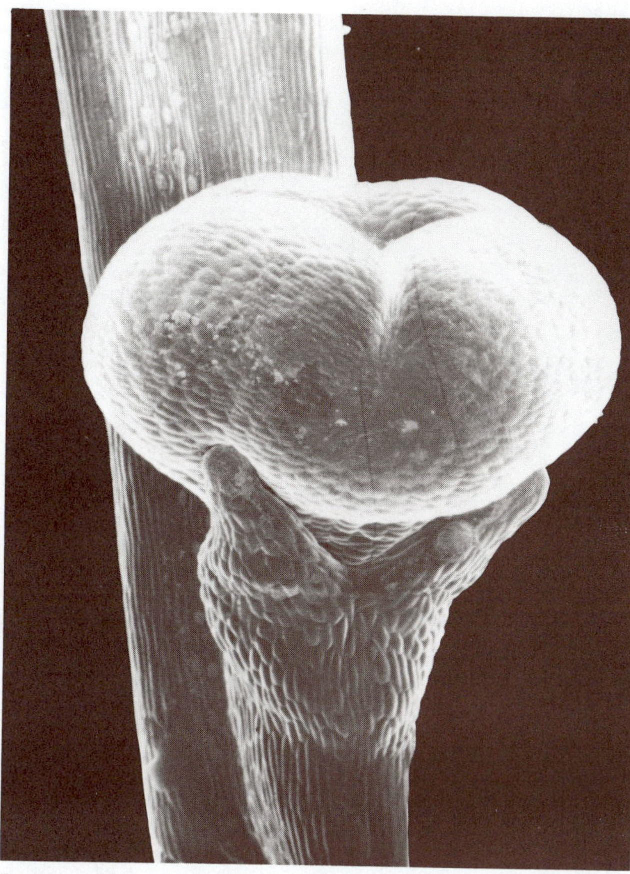

FIGURE 8-2 Organography of *Psilotum nudum*. **A,** portions of two branches showing three-lobed synangia and their associated forked appendages; **B,** scanning electron micrograph of a synangium and the forked foliar appendage.

cylindrical with longitudinal ribs, whereas the more distal aerial stems have three longitudinal ridges.

STEM ANATOMY. The apical meristem of rhizomes and aerial branches has a single, large apical cell (Bierhorst, 1954b; Marsden and Wetmore, 1954; Roth, 1963; Siegert, 1964) which divides repeatedly, giving rise to additional meristematic cells that differentiate eventually into tissues constituting the three primary tissue systems.

The aerial system is covered by an epidermis, as shown in Fig. 8-3, in which the outer tangential cell walls are heavily cutinized and covered by a cuticle. Stomata are present mainly in areas between the longitudinal ribs and are without special subsidiary cells much like the type in certain gymnosperms (Pant and Mehra, 1963). Internal to the epidermis there is a rather broad cortex which can be resolved into three regions (Fig. 8-3). The outer portion, directly beneath the epidermis, consists of elongated, lobed parenchyma cells with intercellular spaces between the vertical rows. Starch grains are present in great numbers. Internal to this zone there is a cylinder of vertically elongated and thick-walled cells, with small intercellular air spaces and few or no starch grains. In the lower portions of the aerial stems the walls of these cells apparently become lignified. In progressing from this zone to the vascular cylinder, the cell walls become thinner and thinner and less lignified with an increase in the number of starch grains per cell.

The boundary between the fundamental tissue

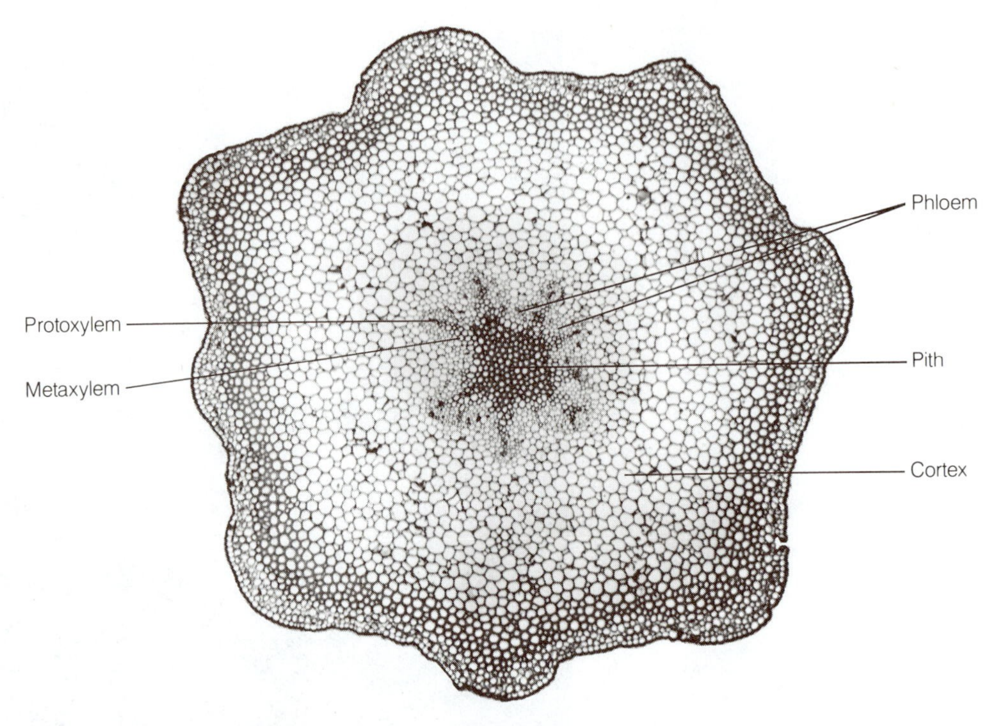

FIGURE 8-3 Transection of stem of *Psilotum nudum* near base of an aerial branch. Note differences in thickness of cell walls in outer and inner cortex, and xylem lobes and sclerenchymatous pith.

(cortex) system and the vascular cylinder is marked by the endodermis (Fig. 8-4, A), whose vertically elongated cells have a conspicuous casparian strip in the radial and end walls. Occupying the center of the rhizome in *P. nudum* is a slender cylinder of primary xylem which may be greatly reduced or even interrupted in small axes (Bierhorst, 1954b) and is a ridged or fluted cylinder in the aerial branches. Near the transition region from rhizome to aerial stem this cylinder may have as many as ten lobes (Bower, 1935; Pitot, 1950), whereas fewer lobes are present in the more distal parts of the aerial branch system (Figs. 8-3; 8-4, A). At levels where several xylem lobes are present, the center of the stem in *P. nudum* is generally occupied by elongate sclerenchymatous cells. Partially disorganized protoxylem tracheids, with helical or annular thickenings, occupy the extreme tips of the xylem lobes in aerial branches while the remainder is composed of metaxylem tracheids with predominantly scalariform or circular bordered pits. In summary, the rhizome is protostelic, becoming an exarch siphonostele throughout a considerable portion of the aerial branch system; the uppermost branches are, however, strictly actinostelic.

Internal to the endodermis is a cylinder of parenchyma-like cells, generally one layer thick, which is designated as the pericycle. The phloem is internal to the pericycle and occupies the regions between the lobes or flanges of the xylem. The smaller, somewhat angular cells are the sieve cells (Fig. 8-4, A). At maturity these cells are elongate, the walls are relatively thick and often lignified, and they possess many spherical bodies (refractive spherules). The sieve cells of *Psilotum* lack callose lining the pores of the sieve areas of the end walls and pores in the lateral walls (Lamoureux, 1961; Perry and Evert, 1975). However, most of the tissue in the bays between the xylem arms is composed of elongate parenchyma cells.

FOLIAR APPENDAGES. The "foliar" appendages in *Psilotum* are small scalelike structures that are helically arranged on the upper part of the aerial stem

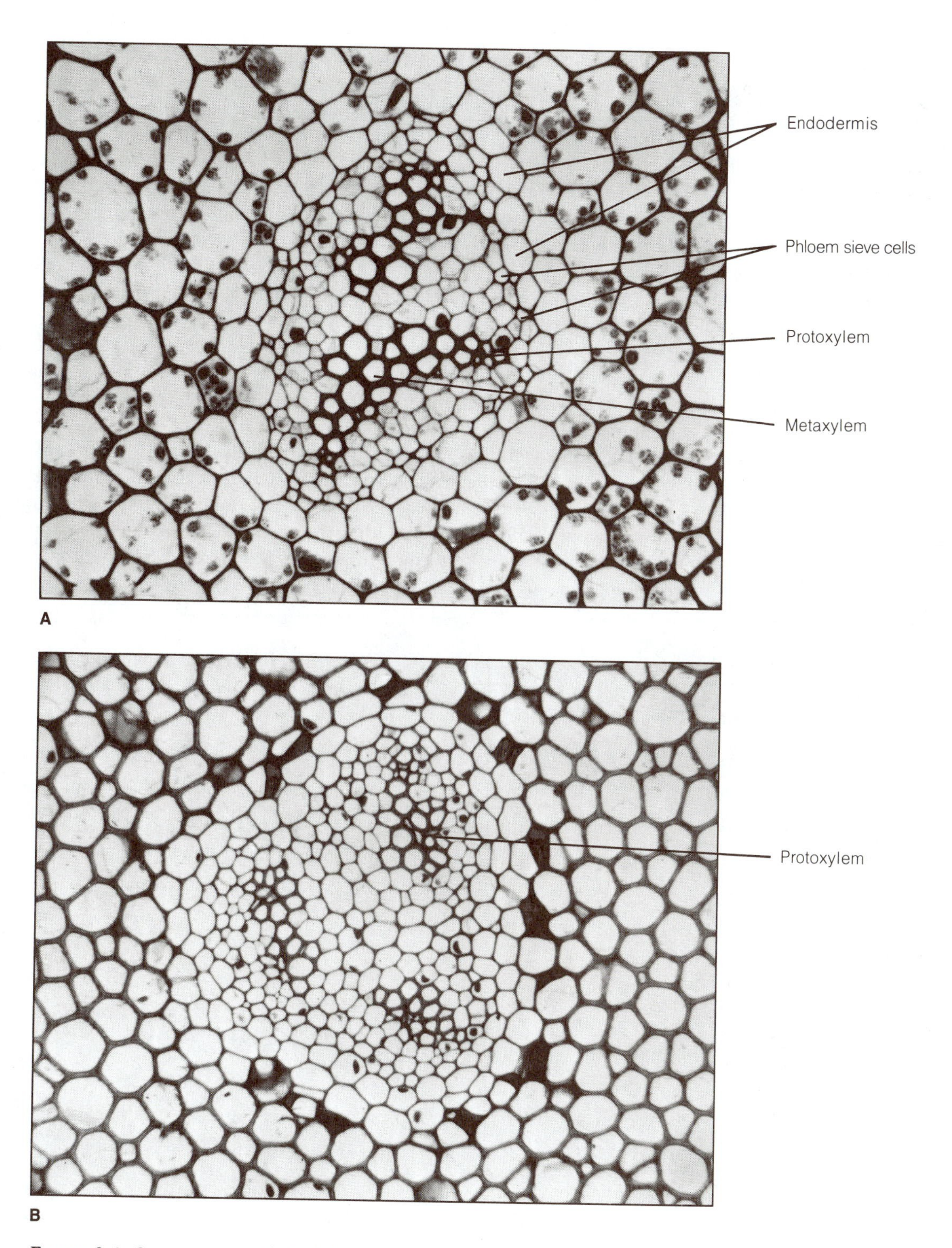

FIGURE 8-4 Stem anatomy of the Psilotaceae. **A,** transection of the vascular cylinder and adjacent cortex of a small aerial stem of *Psilotum nudum;* **B,** transection near the base of an aerial branch of *Tmesipteris* sp.

(Fig. 8-2, A). Internally the appendage consists of photosynthetic parenchyma cells that are continuous, lower down, with similar tissue of the stem. There is no vascular bundle in the appendage of *P. nudum*, although in *P. complanatum* a "leaf" trace ends at the base of the foliar structure. Grouped generally on the upper portions of the stems are bilobed appendages, each of which is associated with a three-lobed synangium, the fusion product of two or more sporangia. A morphological interpretation of the foliar appendages is presented in Chapter 3.

STRUCTURE AND DEVELOPMENT OF THE SYNANGIUM. Interpretations of the spore-producing structure in the Psilotales are varied and controversial; see, for example, Solms-Laubach, 1884; Bower, 1894, 1935; Eames, 1936; Campbell, 1940; Smith, 1955; Bierhorst, 1956; Zimmermann, 1959; Roth, 1963. To comment at length on all of the various theories is beyond the scope of this book. Thus, only a descriptive account will be presented with selected interpretive theories.

The spore-producing structure of *Psilotum* has been described as a trilocular sporangium and as a trisporangiate structure, i.e., a *synangium*. Results of recent investigations support the latter interpretation.

The mature synangium of *Psilotum* is generally a three-lobed structure (Fig. 8-2, A, B; and Fig. 8-5, B), 1 to 2 millimeters wide, located at the tip of a very short axis, and closely asssociated with a forked, foliar appendage. Each lobe of the synangium, corresponding to a sporangium, exhibits loculicidal dehiscence at maturity.

Some investigators believe that the short-stalked synangium arises from the adaxial side of the forked appendage or in its axil (Bower, 1894, 1935; Roth, 1963). Others have reported that the original primordium is, in fact, the "fertile axis" and that the

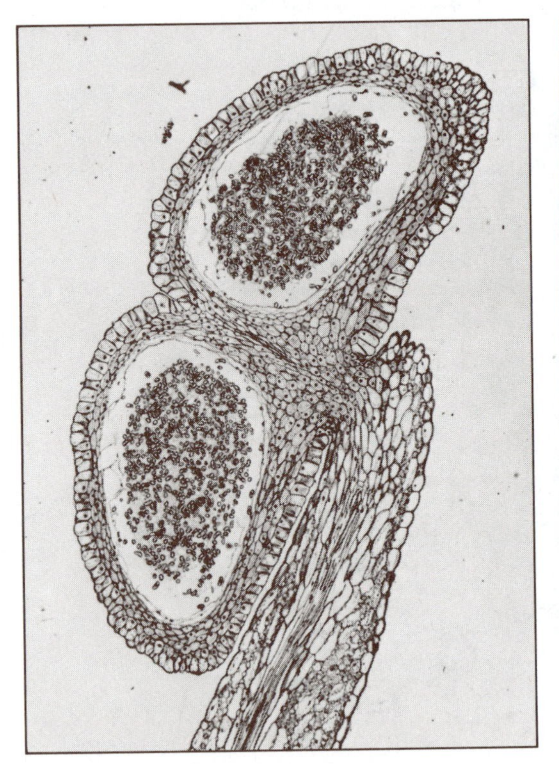

A

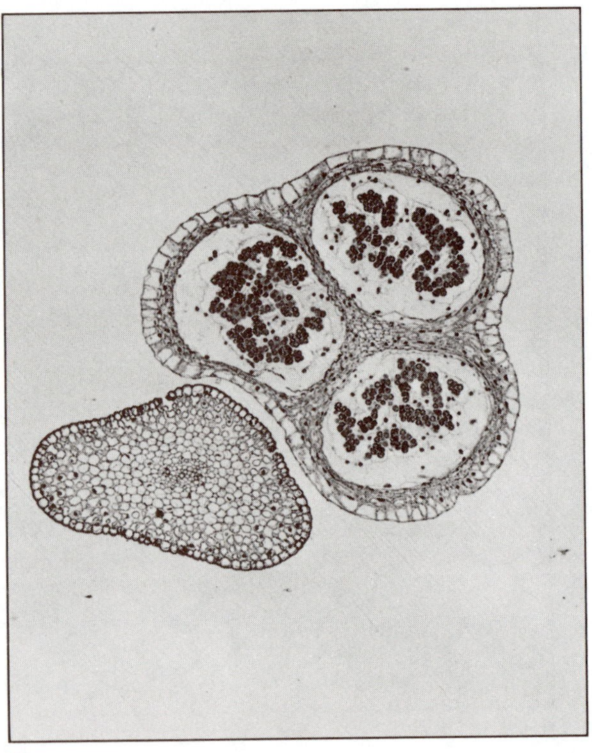

B

FIGURE 8-5 Synangia of the Psilotaceae. **A,** longisection, bilobed synangium of *Tmesipteris* sp.; **B,** transection, trilobed synangium of *Psilotum nudum*: note that the sporocytes are surrounded by an irregular fluid-like tapetum.

forked appendage is a lateral outgrowth on it (Bierhorst, 1956; Rouffa, 1978). Support for the latter concept is based upon experimental work. With *P. nudum* maintained on a long daily photoperiod (16 hours at 200 to 400 footcandles), the typically short fertile axes proliferated into definite branchlike structures (Rouffa, 1967). Comparable growth forms can be found occasionally under natural conditions. A variety of *Psilotum* from Japan, historically known as *Bunryu-zan*, has no sterile foliar appendages, and the synangia are borne at the tips of branches (Rouffa, 1971, 1978; Fig. 8-6).

As a result of his intensive studies on typical clones of *Psilotum* and the many varieties of *Psilotum*, such as *Bunryu-zan*, Rouffa (1978) has postulated that the fertile axes of ancestral forms may have ended in several short-stalked unfused sporangia. In the course of evolution the stalks became shortened and the sporangia became fused into synangia, culminating in the highly reduced fertile axis of the typical and more commonly known *Psilotum* clones. Ancestral forms that might provide support for this conclusion are found in the extinct group Rhyniophyta, especially *Renalia* (Chapter 7).

The development of each sporangium of the synangium in *P. nudum* is eusporangiate, i.e., separate groups of surface initials divide periclinally, setting apart primary-wall initials and primary sporogenous cells. By repeated periclinal and anticlinal divisions of the primary-wall initials, a sporangial wall of four or five layers is produced. Derivatives of the primary sporogenous cells divide in various planes to form the sporogenous tissue (Fig. 8-7). The sporogenous mass becomes irregular in outline and surrounded by a massive tapetum (Fig. 8-5, B). As is true of other plants with eusporangiate development, numerous spores are produced as a result of meiosis. Individual spores are bilaterally symmetrical and the wall has an irregular pattern of rounded, raised ridges (rugose) in some clones. Ultrastructurally, the spore-wall layers resemble more specifically those of the fern family Gleicheniaceae,

A

B

FIGURE 8-6 Two views of an appendageless form of *Psilotum* from Japan known as *Bunryu-zan*: note that the terminal synangia generally consist of more than three fused sporangia. [Courtesy Dr. A. S. Rouffa.]

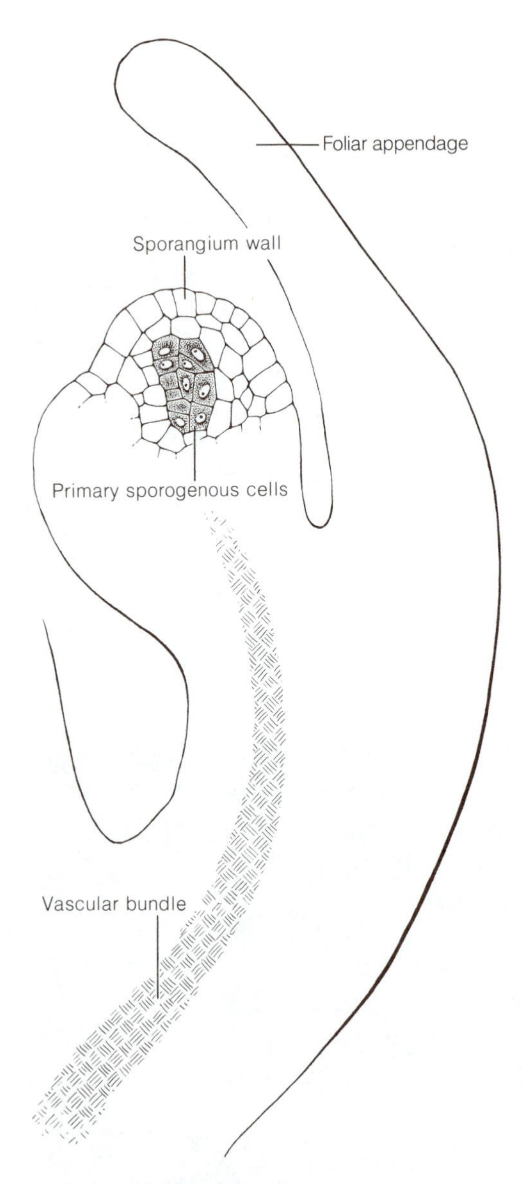

FIGURE 8-7 A young developing synangium and its associated appendage in *Psilotum nudum*. The details of only one lobe (sporangium) are shown. The vegetative stem is to the left.

but not those of the Lycophyta or *Equisetum* (Lugardon, 1979).

Unlike the foliar appendages, the fertile axis is vasculated (Fig. 8-7). A vascular bundle extends into the synangium and may become divided into three parts, corresponding to the three sporangia (Bierhorst, 1971).

Psilotaceae — Sporophyte Generation of *Tmesipteris*

DISTRIBUTION AND ORGANOGRAPHY. In contrast to *Psilotum*, which is widespread in its distribution, *Tmesipteris* is confined to Australia, New Caledonia, New Zealand, and other islands of the South Pacific Ocean. *Tmesipteris* (Fig. 8-8) generally grows as a pendulous epiphyte, 5 to 20 centimeters long, on the trunks of tree ferns or other trees, but often it may be found on mounds of humus. For many years only two species of *Tmesipteris* were recognized. *Tmesipteris tannensis* was the more widespread species. However, several forms are sufficiently different to be considered as separate species, and the five recognizable populations in southeast Australia and at least one other in New Zealand (Barber, 1954, 1957) appear to be reproductively isolated from one another.

FIGURE 8-8 The pendant aerial branches of *Tmesipteris* sp. Synangia can be seen near the tip of the middle branch. In some of the appendages, the single unbranched midvein is evident.

The *Tmesipteris* sporophyte is very similar to that of *Psilotum,* with a branching rhizome system and aerial shoots. However, significant morphological differences do exist between the genera. Each aerial shoot of *Tmesipteris* may exhibit only one dichotomy. The "foliar" appendages are scalelike at the base, gradually increasing in size toward the tip. The majority of appendages are larger than those of *Psilotum,* and are flat and broadly lanceolate with a mucronate tip. The larger leaves are supplied with a single, unbranched vascular bundle. The bases of the leaves are strongly decurrent, and the distinction between stem and leaf is difficult to determine, particularly near the shoot tip, because a foliar appendage often terminates the axis. Roots are absent – a feature characteristic of ancient vascular plants (Chapter 7) and *Psilotum.*

STEM ANATOMY. The rhizome of *Tmesipteris* is protostelic, gradually becoming siphonostelic in the aerial system (Fig. 8-4, B) with three or more protoxylem poles surrounded by metaxylem. The primary xylem of aerial branches is mesarch in development, in contrast with *Psilotum* in which the primary xylem is exarch. The center of the stem consists of parenchymalike cells which may have relatively thick walls. External to the strands of xylem is a cylinder of phloem in which the sieve cells have spherical inclusions and lignified cell walls. These cells are elongate and have numerous sieve areas on their tapering end walls and lateral walls (Sykes, 1908; Lamoureux, 1961). An endodermis is present in the rhizome but lacking in the aerial portions. Between the phloem cylinder and an inner layer of cortex, cells containing brown tanniferous or phenolic substances may physiologically represent the endodermis (Fig. 8-4, B).

The cortex consists of a compact tissue of parenchymalike cells with evenly thickened walls. There may be small groups of photosynthetic parenchyma directly beneath the epidermis; the outer tangential walls of the epidermis are cutinized and covered by a definite cuticle.

FOLIAR APPENDAGES. The foliar appendages of *Tmesipteris* are larger than those of *Psilotum* and, moreover, exhibit a more diversified anatomy. The flattened appendage is covered by a uniseriate epidermis with cutinized outer tangential cell walls in which some of the thickening is laid down in the form of striations; stomata may occur on both surfaces. The internal ground tissue is uniformly arranged, consisting of lobed parenchyma cells. The single concentric vascular bundle, located centrally, is composed of several protoxylem elements surrounded incompletely by metaxylem, which, in turn, is enclosed by phloem. As in the aerial stem, no definable endodermis is present, although a compact zone of parenchyma cells occupies the expected position of such a layer.

SYNANGIUM. The mature two-lobed synangium of *Tmesipteris* is interpreted as occupying the terminus of a short lateral branch, although the axis tip is recurved and the synangium appears to be adaxial (Figs. 8-5, A; 8-9). The two foliar appendages that are attached to the fertile axis just below the synangium extend some distance beyond the synangium.

FIGURE 8-9 Small portion, branch of *Tmesipteris* sp. showing three bilobed synangia and their associated leaves.

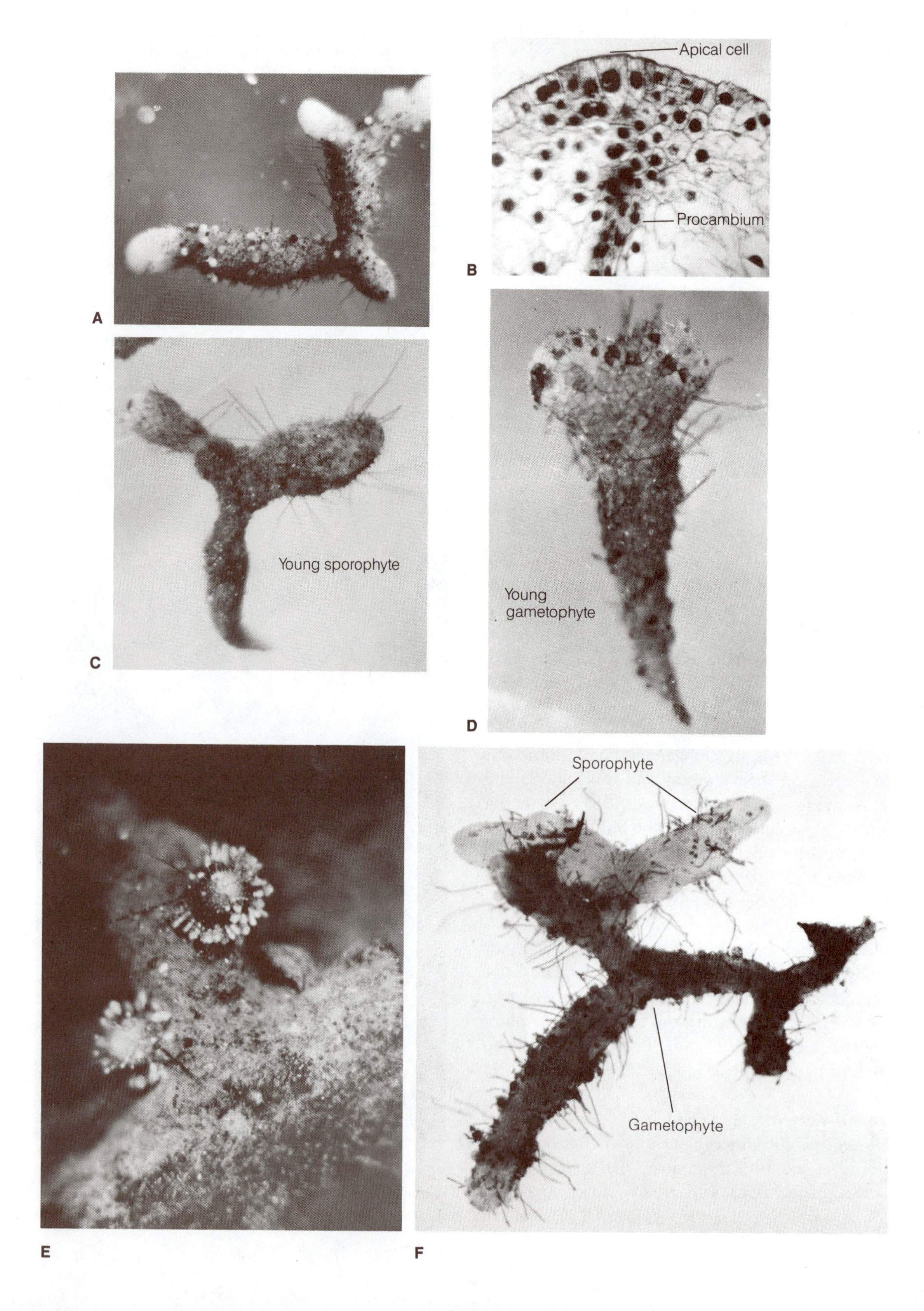
Apical cell
Procambium
Young sporophyte
Young gametophyte
Sporophyte
Gametophyte
A
B
C
D
E
F

The single vascular bundle of the axis divides into three strands at the level of the foliar appendages. The lateral bundles traverse the appendages, whereas the median strand continues up the axis, and in some species ends in a trichotomy. The two lateral traces traverse the septum between the sporangia, and the central strand ends medianly in the septum (Sykes, 1908; Bierhorst, 1971).

According to Bierhorst (1956), the development of the fertile axis is similar to that in *Psilotum*: appearance of a primordium near the vegetative shoot tip, apical growth of the primordium, appearance of separate groups of sporangial initials, and the formation of the two foliar appendages from a common outgrowth on the original fertile axis. Subsequent development also is similar to that in *Psilotum,* with the ultimate development of a synangium with two thick-walled sporangia devoid of a well-defined tapetum and containing a large number of spores. Dehiscence of each sporangium is effected through the formation of a longitudinal cleft.

In the discussion thus far we have identified the aerial unit of *Psilotum* and *Tmesipteris* as a shoot or branch. However, Bierhorst (1968b, 1969, 1971) has interpreted each aerial entity as the morphological equivalent of a frond similar to leaves of ferns. The frond consists of sterile pinnae and fertile pinnae. The sterile pinnae are the foliar appendages not associated with synangia. The fertile pinnae are the forked appendages with synangia. On the basis of his morphological analysis, as well as upon other features of the life cycle, Bierhorst places the family Psilotaceae in the fern order Filicales (Bierhorst, 1971). The reader is referred to publications presenting arguments, pro and con, for such a concept (Bierhorst, 1977; Kaplan, 1977; Wagner, 1977).

Gametophyte Generation

The nature of the gametophyte generation in the Psilophyta long remained a gap in our knowledge of vascular plants; only in the twentieth century has it been discovered and described *(Tmesipteris,* Lawson, 1917a; *Psilotum,* Darnell-Smith, 1917, and Bierhorst, 1953). Original descriptions were based upon collections from natural habitats. The gametophytes were subterranean or grew in the crevices of rocks. In 1949 sexually mature gametophytes of *P. nudum* were discovered growing in undisturbed pots of greenhouse plants (Moseley and Zimmerly, 1949; Zimmerly and Banks, 1950; Bierhorst, 1953). Mature plants resemble pieces of the sporophyte rhizome in that they are brown, radially symmetrical, often dichotomously branched but frequently irregularly branched, and invested with rhizoids (Fig. 8-10, A, D, F). As is true of underground sporophytic axes, the branching of gametophytic plants is correlated with apical injury (Bierhorst, 1953).

There have been three successful attempts to germinate *Psilotum* spores. Darnell-Smith (1917) successfully germinated spores under simulated natural conditions. Bierhorst (1955) mixed spores in soil, and recovered young gametophytes after fifteen months. Whittier (1973, 1975) obtained germination on nutrient agar, but only after the cultures had been in the dark for six or more months. These experiments indicate that lodgment in rock crevices or burial of spores in humus or soil may be necessary for germination under natural conditions.

ANATOMY AND CYTOLOGY. Growth of a gametophyte is initiated by apical cells located at the tips of the ultimate dichotomies (Fig. 8-10, B). The gametophyte is devoid of chlorophyll, living a saprophytic existence, and is presumably aided by the presence of an endophytic fungus, which gains entrance through the rhizoids and invades nearly all cells of the plant except the apical meristems and young gametangia. The aseptate fungal hyphae may form large masses in the cells. The hyphae store quantities of lipid which appear to be released into the host cytoplasm upon fungal degeneration. The lipid may be used as an energy source by the game-

FIGURE 8-10 Gametophytes and rhizomes of *Psilotum nudum.* **A,** gametophyte showing meristematic apices (white) and prominent globular antheridia; **B,** longisection, apex of gametophyte (note large apical cell); **C,** young sporophyte of gemmaceous origin (gemmae are vegetative propagules formed both on rhizomes and gametophytes); **D,** Young gametophyte of gemmaceous origin (dark areas are presumably archegonia); **E,** clusters of gemmae on a sporophytic rhizome; **F,** gametophyte with attached sporophyte. [**A-E** courtesy Dr. D. W. Bierhorst.]

tophytes. The identity of the fungus has not been determined conclusively (Peterson, Howarth, and Whittier, 1981; Fig. 8-ll, B).

Cells of the gametophyte are parenchymatous; however, there are instances in certain races of *Psilotum* where annular and scalariform or scalariform-reticulate tracheids, surrounded by phloem and an endodermis, have been shown to occupy the center of the gametophyte (Holloway, 1938, 1939; Bierhorst, 1953). The presence of vascular tissue in the gametophytic plant is somewhat unusual among vascular plants (see Chapter 2). The external similarity of the gametophyte and the sporophytic rhizome, coupled with the presence of vascular tissue in the gametophyte, has been cited as evidence for the transformation theory as it relates to the origin of alternation of generations or phases in the life cycle (Chapter 2).

Chromosome numbers are relatively high in *Psilotum* and *Tmesipteris*. A wild "diploid" sporophyte of *Psilotum* occurs in Ceylon (Manton, 1950), and has a chromosome number $n = 52$ to 54. Bierhorst (1968b) has described similar races from Fiji and New Caledonia. Populations from widely separated regions such as India, Jamaica, Australia, New Zealand, and Japan are largely tetraploid, $n = 104$. The highest chromosome number reported, n = approximately 210 (octoploid), is from New Caledonia (see Tryon and Tryon, 1982). The form commonly grown in greenhouses is tetraploid and it may have vascular tissue in the gametophyte. There is no vascular tissue in gametophytes of diploid races. That tetraploidy has been operative in the genus is supported by the presence of quadrivalents during meiosis (Ninan, 1956). A similar type of polyploid series appears to exist for *Tmesipteris* (Barber, 1957) by replications of entire chromosome sets, and is possibly based upon the theoretical basic number, $x = 13$, for both genera.

GAMETANGIA. Sex organs (antheridia and archegonia) are scattered over the surface of the gametophyte and are intermingled (Fig. 8-10, A). Young sex organs generally begin development very close to the apices of the gametophyte. The first indication of antheridial development is the presence of a periclinal division in a single surface cell, which sets aside an outer jacket initial and an inner primary spermatogenous cell (Holloway, 1918, 1939; Lawson, 1917b). By anticlinal divisions a single-layered jacket of several cells is produced enclosing a developing spermatogenous mass in which cell divisions occur in many planes. Ultimately the antheridium projects above the surface (see Fig. 8-11, A). Each spermatid eventually becomes a multiflagellate sperm and escapes through an opercular cell on the side of the antheridium (Bierhorst, 1954a).

The archegonium likewise is initiated from a single superficial cell. The initial periclinal division sets aside an outer cover cell and an inner central cell. Subsequent development is, in general, similar to that of archegonia of other lower vascular plants (Chapter 5, Fig. 5-1). At maturity there are four rows of four to six tiers of neck cells and generally one binucleate neck-canal cell (Bierhorst, 1954a). The basal two tiers of neck cells have brown, thick walls and remain in place, but the distal tiers break off with the slightest disturbance. Archegonia with only one or two tiers of neck cells are generally seen in gametophytes that have been cleaned from soil. *Psilotum nudum* gametophytes grown in vitro, however, have provided views of archegonial opening. At maturity, in the presence of water, cells of the apical tier separate and the mucilaginous contents of the neck cell are released. Within a few minutes the four rows of neck cells separate and become greatly reflexed and the apical tiers almost touch the gametophyte at the base of the archegonial neck. If undisturbed, the neck cells remain in this position for several days (Whittier and Peterson, 1980). Fertilization is accomplished by the union of a multiflagellate sperm and egg.

The Embryo

In the method of early segmentation of the zygote, and in the structure and subsequent development of the embryo, there is close similarity between *Psilotum* and *Tmesipteris* (Holloway, 1921, 1939). The first division of the fertilized egg (zygote) results in a wall formed at right angles to the long axis of the archegonium. The cell directed toward the neck of the archegonium is designated as the epibasal cell, the lower is the hypobasal cell. Members of the Psilophyta are illustrative of exoscopic polarity (Chapter 6). The apical epibasal cell will ultimately give rise to the sporophytic branch system (aerial and underground), while the hypoba-

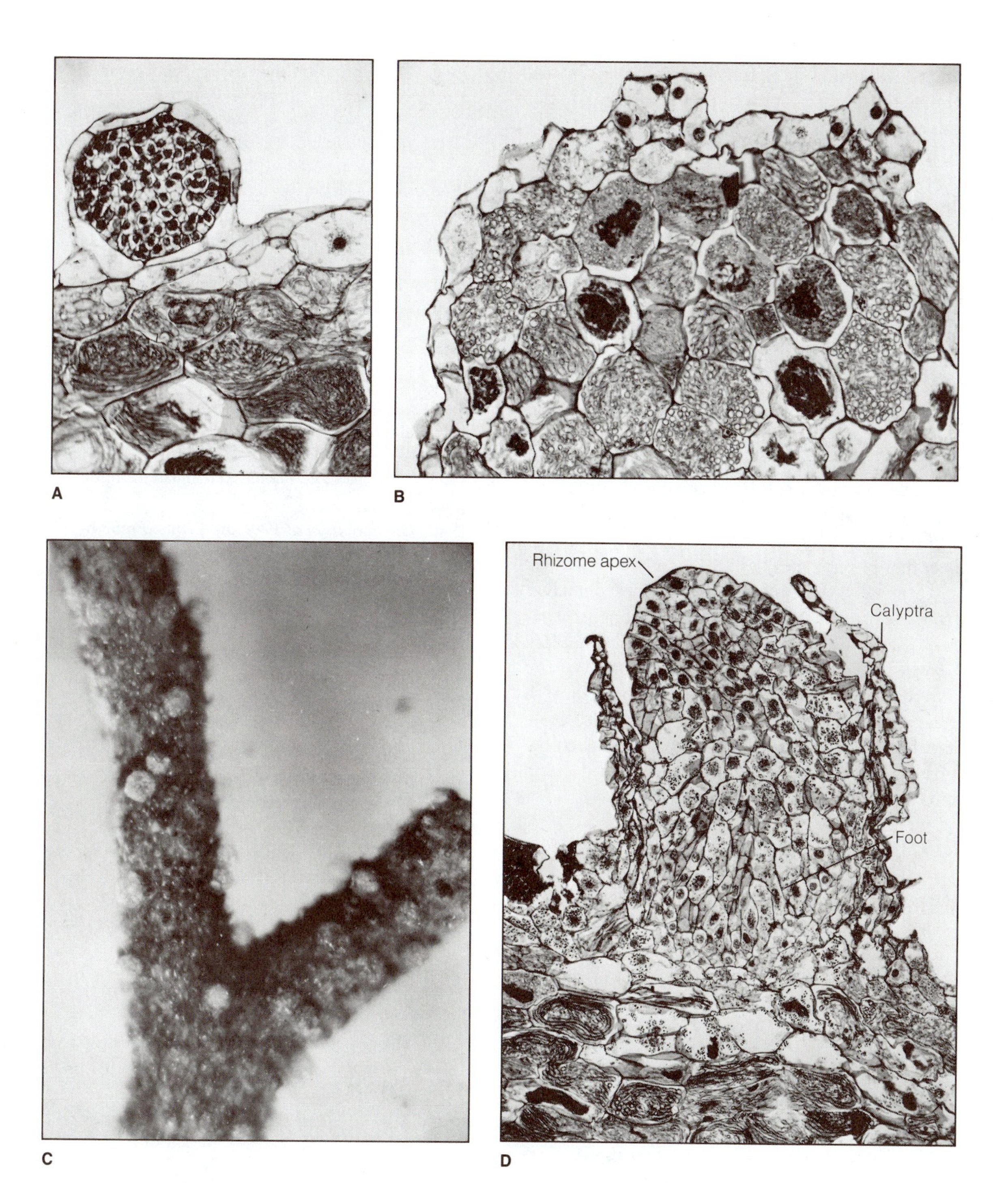

FIGURE 8-11 Gametangia and embryo of *Psilotum nudum*. **A,** section of nearly mature antheridium, showing jacket layer and spermatids; **B,** transection of gametophyte showing two mature archegonia (consult text for details of development; note that most cells of the gametophyte contain hyphae of an endophytic fungus); **C,** portion of a gametophyte with emergent antheridia; **D,** developing embryo attached to gametophyte by enlarged foot. [From slides prepared by Dr. D. W. Bierhorst.]

sal cell will produce the foot—a structure that anchors the young sporophyte securely to the gametophyte.

By repeated cell divisions the apical portion increases in size, and an apical cell is established at the distal end. Continued growth of the shoot is due in large measure to the activity of this apical cell; frequently in *Tmesipteris* two apical cells are present on the flanks of the embryo, resulting in two precociously formed horizontal branches. Concomitant with embryonic development the gametophyte forms a calyptralike outgrowth through which the young sporophyte eventually emerges (Fig. 8-11, D). While the rhizome portion is assuming form, the foot enlarges by repeated cell divisions, sending haustorial outgrowths into the gametophytic tissue. The foot, by virtue of its position and organization, is well suited for the functions of anchorage and absorption of nutrients until the sporophyte becomes physiologically independent.

Ultimately the sporophyte becomes detached from the foot and the gametophyte through a separation layer in the vicinity of the original boundary between rhizome and foot. Throughout all of the differentiation process this original boundary is clearly discernible. The rhizome continues to grow and branch, and eventually some of the rhizome tips emerge above the soil or humus and develop into photosynthetic aerial branches.

REFERENCES

Allen B. M.
1966. *Psilotum nudum* in Europe. *Taxon* 15: 82–83.

Barber, H. N.
1954. New species of *Tmesipteris*. *Victorian Natur.* 71:97–99.
1957. Polyploidy in the Psilotales. *Proc. Linn. Soc. New South Wales* 82:201–208.

Bierhorst, D. W.
1953. Structure and development of the gametophyte of *Psilotum nudum*. *Amer. Jour. Bot.* 40:649–658.
1954a. The gametangia and embryo of *Psilotum nudum*. *Amer. Jour. Bot.* 41:274–281.
1954b. The subterranean sporophytic axes of *Psilotum nudum*. *Amer. Jour. Bot.* 41:732–739.
1955. A note on spore germination in *Psilotum nudum*. *Virginia J. Sci* 6:96.
1956. Observations on the aerial appendages in the Psilotaceae. *Phytomorphology* 6:176–184.
1968a. Observations on *Schizaea* and *Actinostachys* spp., including *A. oligostachys*, sp. nov. *Amer. Jour. Bot.* 55:87–108.
1968b. On the Stromatopteridaceae (fam. nov.) and the Psilotaceae. *Phytomorphology* 18:232–268.
1969. On *Stromatopteris* and its ill-defined organs. *Amer. Jour. Bot.* 56:160–174.
1971. *Morphology of Vascular Plants*. Macmillan, New York.
1977. The systematic position of *Psilotum* and *Tmesipteris*. *Brittonia* 29:3–13.

Bower, F. O.
1894. Studies in the morphology of spore-producing members: Equisetinae and Lycopodineae. *Phil. Trans. Roy. Soc.* London 185B:473–572.
1935. *Primitive Land Plants*. Macmillan, London.

Campbell, D. H.
1940. *The Evolution of the Land Plants (Embryophyta)*. Stanford University Press, Stanford, California.

Cooper-Driver, G.
1977. Chemical evidence for separating the Psilotaceae from the Filicales. *Science* 198:1260–1262.

Darnell-Smith, G. P.
1917. The gametophyte of *Psilotum*. *Trans. Roy. Soc. Edinb.* 52:79–91.

Eames, A. J.
1936. *Morphology of Vascular Plants. Lower Groups*. McGraw-Hill, New York.

Holloway, J. E.
1918. The prothallus and young plant of *Tmesipteris*. *Trans. Proc. N. Z. Inst.* 50:1–44.
1921. Further studies on the prothallus, embryo, and young sporophyte of *Tmesipteris*. *Trans. Proc. N. Z. Inst.* 53:386–422.
1938. The embryo and gametophyte of *Psilotum triquetrum*. A preliminary note. *Ann. Bot.* n.s. 2:807–809.
1939. The gametophyte, embryo, and young rhizome of *Psilotum triquetrum* Swartz. *Ann. Bot.* n.s. 3:313–336.

Kaplan, D. R.
1977. Morphological status of the shoot systems of Psilotaceae. *Brittonia* 29:30–53

Lamoureux, C. H.
1961. Comparative studies on phloem of vascular cryptogams. Ph.D. dissertation, University of California, Davis, Calif.

Lawson, A. A.
1917a. The prothallus of *Tmesipteris tannensis*. *Trans. Roy. Soc. Edinb.* 51:785–794.

1917b. The gametophyte generation of the Psilotaceae. *Trans. Roy. Soc. Edinb.* 52:93–113.

Lugardon, B.
1979. Sur la formation du sporoderme chez *Psilotum triquetrum* Sw. (Psilotaceae). *Grana* 18:145–165.

Manton, I.
1950. *Problems of Cytology and Evolution in the Pteridophyta.* Cambridge University Press, London.

Marsden, M. P. F., and R. H. Wetmore
1954. In vitro culture of the shoot tips of *Psilotum nudum.* Amer. *Jour. Bot. 41:640–645.*

Mason, C. T., Jr.
1968. A new family of vascular plants (Psilotaceae) for Arizona. *Madroño.* 19:224.

Morgan, D.
1962. *Psilotum triquetrum,* Swartz in Basutoland. *Nature* 195:1121.

Moseley, M. F., Jr., and B. C. Zimmerly
1949. *Psilotum* gametophytes matured under greenhouse conditions from self-sown spores. *Science* 110:482.

Ninan, C. A.
1956. Cytology of *Psilotum nudum* (L.) Beauv. *(P. triquetrum* Sw.*). Cellule* 57:307–318.

Pant, D. D., and B. Mehra
1963. Development of stomata in *Psilotum nudum* (L.) Beauv. *Curr. Sci.* 32:420–422.

Perry, J. W., and R. F. Evert
1975. Structure and development of the sieve elements in *Psilotum nudum. Amer. Jour. Bot.* 62:1038–1052.

Peterson, R. L., M. J. Howarth, and D. P. Whittier
1981. Interactions between a fungal endophyte and gametophyte cells in *Psilotum nudum. Can. J. Bot.* 59:711–720.

Pitot, A.
1950. Sur l'anatomie de *Psilotum triquetrum* Sw. *Inst. Franc. d'Afrique Noire* (Paris) 12:315–334.

Reed, C. F.
1966. Index Psilotales. *Sociedade Broteriana, Boletin* 40:71–96.

Roth, I.
1963. Histogenese der Luftsprosse und Bildung der "dichotomen" Verzweigungen von *Psilotum nudum. Advan. Frontiers Plant Sci.* 7:157–180.

Rouffa, A. S.
1967. Induced *Psilotum* fertile-appendage aberrations. Morphogenetic and evolutionary implications. *Can. Jour. Bot.* 45:855–861.
1971. An appendageless *Psilotum.* Introduction to aerial shoot morphology. *Amer. Fern Jour.* 61:75–86.
1978. On phenotypic expression, morphogenetic pattern and synangium evolution in *Psilotum. Amer. Jour. Bot.* 65:692–713.

Savory, H. J.
1949. A botanical discovery, *Psilotum. Nigeria* 30:317.

Siegert, A.
1964. Morphologische, entwicklungsgeschichtliche und systematische Studien an *Psilotum triquetrum* Sw. I. Allgemeiner Teil. Erstarkung und primäres Dickenwachstum der Sprosse. *Beitr. Biol. Pflanzen* 40:121–157.

Smith, G. M.
1955. *Cryptogamic Botany. Vol.II. Bryophytes and Pteridophytes,* 2d edition. McGraw-Hill, New York.

Solms-Laubach, H. Grafen zu.
1884. Der Aufbau des Stockes von *Psilotum triquetrum* und dessen Entwicklung aus der Brutknospe. *Ann. Jard. Bot. Buitenzorg.* 4:139–194.

Sykes, M. G.
1908. The anatomy and morphology of *Tmesipteris. Ann. Bot.* 22:63–89.

Tryon, R. M., and A. F. Tryon
1982. *Ferns and Allied Plants.* Springer-Verlag, New York.

Tse, A., and G. H. N. Towers
1967. The occurrence of psilotin in *Tmesipteris, Phytochemistry* 6:149.

Verdoorn, F.
1938. *Manual of Pteridology.* Martinus Nijhoff, The Hague.

Wagner, W. H., Jr.
1977. Systematic implications of the Psilotaceae. *Brittonia* 29:54–63.

Whittier, D. P.
1973. Germination of *Psilotum* spores in axenic culture. *Can. J. Bot.* 51:2000–2001.
1975. The origin of the apical cell in *Psilotum* gametophytes. *Amer. Fern J.* 65:83–86.

Whittier, D. P., and R. L. Peterson
1980. Archegonial opening in *Psilotum. Can. J. Bot.* 58:1905–1907.

Zimmerly, B. C., and H. P. Banks
1950. On gametophytes of *Psilotum.* (Abstract) *Amer. Jour. Bot.* 37:668.

Zimmermann, W.
1959. *Die Phylogenie der Pflanzen,* 2d edition. G. Fischer, Stuttgart.

CHAPTER 9

Lycophyta

THE Lycophyta is a well-defined group of vascular plants consisting of fossil and living representatives. The known history of this group extends from the Paleozoic Era to the present. There are five living genera with more than 1,000 living species which occur in various parts of the world under varied climatic conditions. The living genera consist of the "ground pine" or club moss *Lycopodium* (Figs. 9-2, 9-3, 9-4); the spikemoss *Selaginella* (Fig. 9-19); the small, tuberous plant *Phylloglossum* (Fig. 9-1), which is greatly restricted in its distribution; the quillwort *Isoetes* (Fig. 9-49); and *Stylites* (Amstutz, 1957) found growing high in the mountains of Peru (Fig. 9-59, A, B).

All of these genera are small plants: some are erect, some live as epiphytes, others grow as creepers on the ground, and yet others produce underground rhizomes. In contrast with these plants of modest stature, many of the ancient lycopods *(Lepidodendron)* were good-sized trees, and their vegetative structures and spores constitute an important part of coal (see Figs. 9-40; 9-41). The importance of this assemblage of vascular plants cannot be measured in terms of the present economic value of living members, but rather by the morphological unity of the entire group and its value in the interpretation of phylogenetic trends in vascular plants.

The vegetative sporophyte is differentiated into a shoot system, consisting of stems and leaves, and a root system. Reminiscent of the Psilophyta, the shoot system of many forms is isotomously branched or modified by anisotomous branching (Chapter 3). Occasionally the axis may be unbranched (as in *Phylloglossum*). The arrangement of leaves is fundamentally helical with modifications (opposite, whorled) characteristic of certain species. Leaves of most living genera are relatively small, whereas those of certain extinct forms were considerably larger. Whatever the arrangement or form of the leaves, each one is generally traversed by a single unbranched vascular bundle. Such a leaf is designated a *microphyll* (Chapter 3). Each leaf of certain genera, such as *Selaginella, Lepidodendron, Isoetes,* and *Stylites* has a curious tonguelike appendage on its adaxial side termed the *ligule* (Figs. 9-25, A; 27; 28 C, D; 50; 54).

The vascular cylinder of the stem in most living species is protostelic. The primary xylem is generally exarch in development and consists primarily of tracheids with scalariform pitting. Whether the vascular cylinder is a protostele or a siphonostele, there are no breaks in the vascular tissue at the point of departure of leaf traces. No leaf gaps exist (Chapter 3). In a stem with a siphonostele a branch gap is present in the vascular cylinder of the main axis only at the level of divergence of the branch. In most living genera the roots, arising from rhizomes (e.g., in *Lycopodium*), branch dichotomously. In *Isoetes* there is a definite, perennial, root-producing meristem. Although the formation of secondary tissues was very common in ancient arborescent lycopods, this feature is characteristic of only two living genera—*Isoetes* and *Stylites*. A feature that unifies the entire group is the position of the eusporangium. Each sporophyll has a single sporangium which is either attached to the adaxial basal region of the sporophyll or is located in its axil. The conditions of homospory and heterospory are coexistent in the group; heterosporous forms always produce endosporic gametophytes, whereas homosporous forms produce only exosporic gametophytes.

Classification

LYCOPHYTA: Sporophyte differentiated into leaf, stem, root and eusporangium; microphylls ligulate or eligulate; typically one sporangium attached to or associated with each sporophyll; no leaf gaps; exarch xylem predominates; protostelic or siphonostelic; some have secondary growth.

LYCOPODIALES: Living and extinct plants; sporophytes with primary growth only, no vascular cambium; leaves eligulate; majority have definite strobili; homosporous.

LYCOPODIACEAE: Living and extinct plants; herbaceous; many with definite strobili; exosporic gametophytes; biflagellate sperms in the living genus *Lycopodium*.

Lycopodium, Phylloglossum, Lycopodites (extinct).

EXTINCT HERBACEOUS DEVONIAN LYCOPODS: Of Devonian age; low, herbaceous plants; dichotomously branched upright shoots from rhizomes; vascular cylinder cylindrical to lobed; some leaves with one to four sporangia, located near axil or on adaxial side of leaf; no definite, compact strobili; homosporous.

Baragwanathia, Leclercqia, Drepanophycus, Protolepidodendron, Asteroxylon.

SELAGINELLALES: Living and extinct plants; with primary growth only; no vascular cambium; microphyllous with ligule; definite strobili formed; heterosporous; gametophytes endosporic; sperms biflagellate in living members.

SELAGINELLACEAE: Characteristics as in Selaginellales.

Selaginella, Selaginellites (extinct).

LEPIDODENDRALES: Extinct plants; treelike and most, if not all, with secondary growth; microphyllous with ligule; large root stocks (rhizophores) formed; heterosporous; sporophylls grouped into strobili; some forming seedlike structures.

Selected genera: *Lepidodendron, Stigmaria* (form genus for rhizophores), *Sigillaria, Lepidostrobus* (a form genus for strobili), *Lepidocarpon* (form genus for seedlike structures).

ISOETALES: Living and extinct plants; sporophytes with cormlike stems; secondary growth; perennial root-producing meristem; ligulate microphylls; heterosporous; endosporic gametophytes; sperms multiflagellate in living members.

ISOETACEAE: Characteristics as in Isoetales.

Isoetes, Isoetites (extinct), *Stylites.*

PLEUROMEIALES: Extinct plants; upright unbranched stem with ligulate microphylls grouped at its upper end; upper end of axis terminates in a strobilus; rhizophore; heterosporous.

PLEUROMEIACEAE: Characteristics as in Pleuromeiales.

Pleuromeia, Nathorstiana.

Homosporous Forms in the Lycophyta

Lycopodiales—Lycopodiaceae

The family Lycopodiaceae includes two living genera, *Lycopodium* and *Phylloglossum*. The former, commonly termed club moss, is worldwide in distribution. Most of the species (about 400) of

Lycopodium are tropical, but others occur in temperate and arctic regions of the world. *Phylloglossum,* a highly reduced and specialized monotypic plant, is restricted to Australasia (Fig. 9-1). *Lycopodites,* a fossil lycopod from the Carboniferous Period to the Recent Epoch, resembled the modern club moss in many respects.

LYCOPODIUM. Although species of *Lycopodium* do not usually form a conspicuous part of the flora of temperate regions, this genus is very diversified in growth habit and abundantly represented in the American tropics (Haught, 1960).

Some species are erect shrubby plants (Figs. 9-2; 9-3, B), others have a trailing or creeping habit (Fig. 9-4, A, C), and still others grow as epiphytes. Some of the terrestrial prostrate types form "fairy rings" in open, undisturbed areas. Active rhizomatous growth takes place at the margins of such a circle, while that part of the colony produced in previous years decays. The ring may be in the shape of a circle, and increase in diameter as a function of time, resulting in an exponential curve of surprising exactness. One ring, measuring 11.25 meters in diameter in 1964, was estimated to have originated in 1839 (Van Soest, 1964).

FIGURE 9-2 *Lycopodium* sp. Note the sporangia (white structures) in the axils of certain leaves along the upper half of each branch and the clusters of gemmae on the uppermost portion of each shoot.

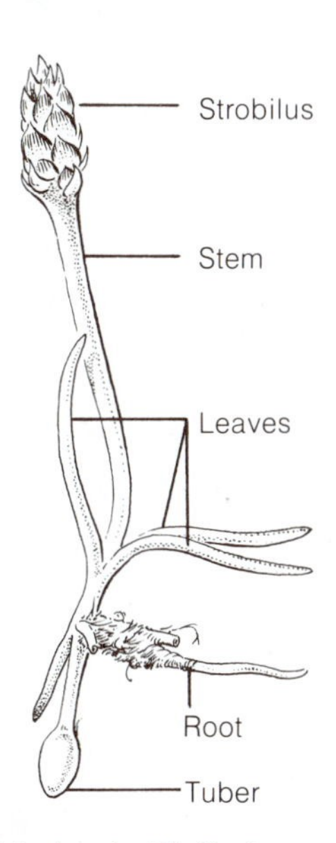

FIGURE 9-1 Habit sketch, *Phylloglossum drummondii.* The "tuber" is a vegetative reproductive body and is capable of developing into a typical plant under favorable environmental conditions.

It has been a convention to consider all club mosses as a single genus, *Lycopodium.* Some pteridologists (those who study lower vascular plants) believe that the variation in sporophytic and gametophytic features in the genus supports the establishment of at least subgenera within *Lycopodium,* or even for the elevation of subgenera to generic status. These concepts will be discussed later in the chapter after the reader becomes aware of the morphological variation in club mosses.

ORGANOGRAPHY. Whether a given species is erect or prostrate, branching is fundamentally dichotomous. The branches of a dichotomy may be equal (Fig. 9-3, A), or the branches of a dichotomy may be unequal, with one branch overtopping the other. The weaker branch system generally becomes determinate, often ending in one or more strobili (Fig. 9-4, C). This mode of branching is termed anisotomous, and it reaches its greatest development in

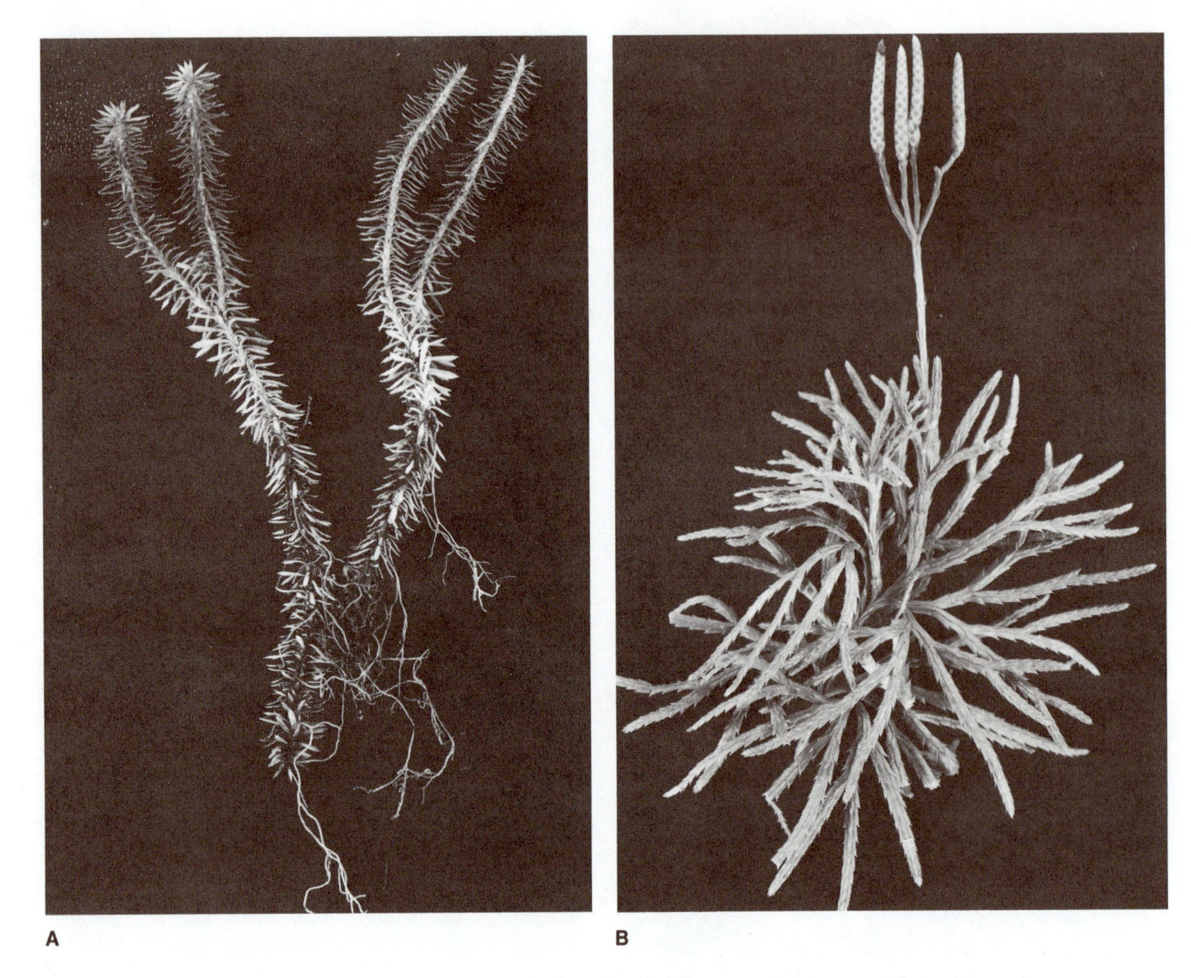

FIGURE 9-3 *Lycopodium.* **A,** *Lycopodium lucidulum;* note dichotomous branching of shoot, and roots. **B,** *Lycopodium digitatum;* note small, scalelike leaves and branch terminating in strobili.

forms with a prostrate rhizomelike main axis (Fig. 9-4, C). The leaves are microphylls which range in length from 2 to 20 millimeters or even up to 25 to 35 millimeters in a few species. Phyllotaxy is basically helical, but the arrangement may appear to be opposite or whorled, or even variable in different regions of the same plant (Bhambie, 1965). In some forms the leaf bases are decurrent (the leaf base is fused with and extends down the stem to varying degrees). In some species the leaves may be of two sizes (anisophylly), especially on lateral determinate branches.

Some species form vegetative reproductive structures—termed gemmae or bulbils—which become detached from the plant and grow into new sporophytes (Fig. 9-2). These structures arise in the positions of leaves and consist of a bud and preformed roots (Takeuchi, 1962). Gemmae have been interpreted as short, specialized branches resulting from anisotomous branching of the main stem axis (Stevenson, 1976). The factors that favor their formation are not well understood (Cutter, 1966).

Roots arise endogenously along the lower side of the stem in prostrate forms (Chapter 3; Fig. 9-4, C). In the upright forms roots may be initiated near the shoot tip and subsequently grow downward through the cortex, emerging at the base of the plant (Figs. 9-3, A; 9-5, A). After the root emerges from the stem it may branch freely in a dichotomous fashion.

FIGURE 9-4 *Lycopodium.* **A,** *Lycopodium inundatum,* showing prostrate rhizomes and upright fertile shoots terminated by strobili. **B,** *Lycopodium obscurum,* portion of upright branched shoot and terminal strobilus. **C,** *Lycopodium clavatum,* strongly rhizomatous species with determinate, fertile side branches; a root is evident along lower edge of the main axis.

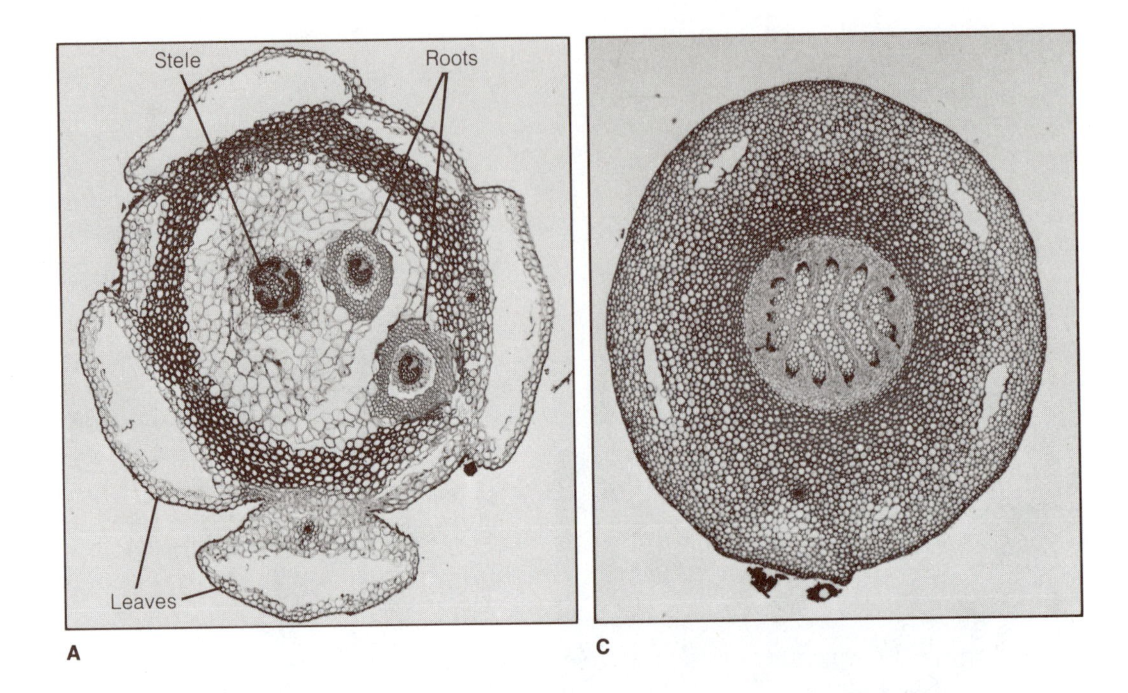

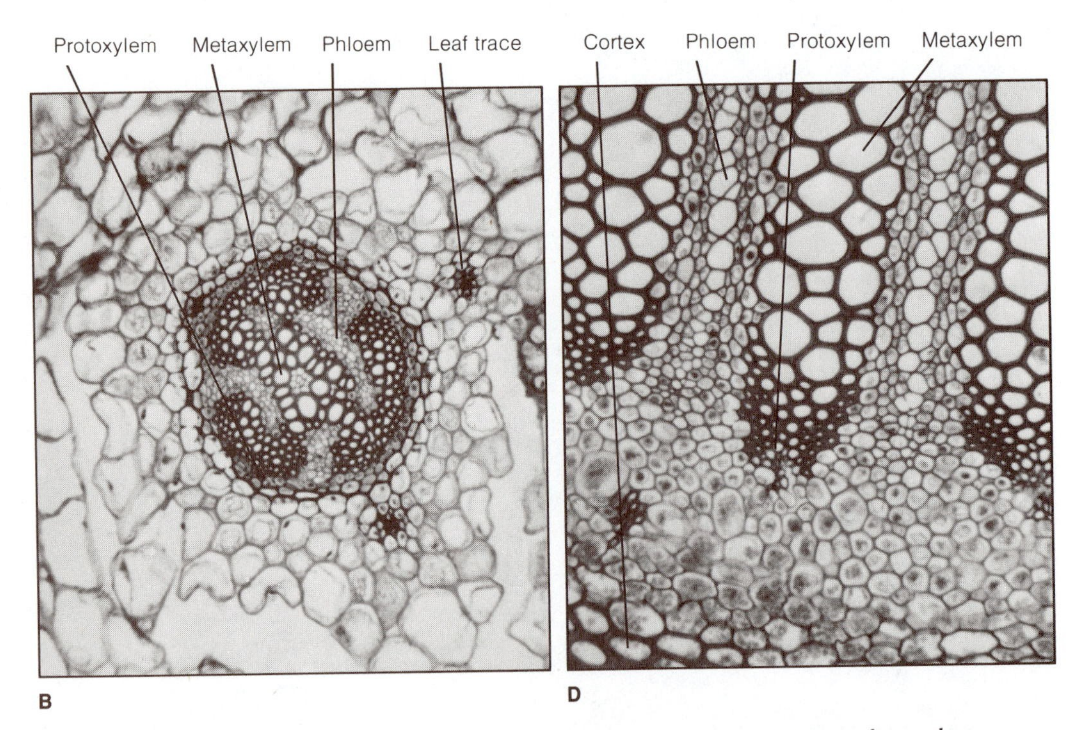

FIGURE 9-5 Stem anatomy in *Lycopodium*. **A,** transection, stem of *Lycopodium selago;* **B,** details of actinostele in *Lycopodium selago;* **C,** transection, stem of *Lycopodium* sp. showing plectostele; **D,** details, portion of stele in **C.**

Sporangia always occur singly on the adaxial surface of the sporophylls or in their axils. The sporophylls may be aggregated into definite strobili and may be quite different from vegetative leaves (Figs. 9-3, B; 9-4, B, C). In other species, however, "fertile" areas alternate with "sterile" regions along the stems, the sporophylls resembling ordinary foliage leaves (Figs. 9-2; 9-6, A).

STEM ANATOMY. The outermost layer of the stem is a uniseriate epidermis. The cortex is highly variable in thickness and structure (Fig. 9-5, A, C). In some species it remains parenchymatous, and in others the cells of specific regions undergo sclerification. Large air space systems may be present that extend into the leaves. Surrounding the vascular cylinder is the pericycle that may be two or three cells wide, or constitute a broad zone of parenchyma. An endodermis with casparian strips is not evident. The transition from pericycle to cortex in some species is abrupt because the walls of the cortical cells are thick (Fig. 9-5, D).

With the exception of the ferns, in no lower vascular plants is there such variation in the pattern of primary xylem and phloem in stems as is found in *Lycopodium*. The same species and even the same individual may show great variation during ontogeny (Wardlaw, 1924). In the mature plant body the vascular cylinder may be *actinostelic* with the primary phloem occupying the regions between the flanges of primary xylem (Fig. 9-5, B). In other species the primary xylem and phloem form strands of tissue which in transverse section appear as alternating bands of xylem and phloem; this type of vascular cylinder is designated a *plectostele* (Fig. 9-5, C, D). In still other species the central mass of xylem may be so modified as to form numerous strands of xylem and phloem (Ogura, 1938). It should be remembered that the seemingly isolated strands of xylem or phloem actually are interconnected. This can be demonstrated if their course if followed throughout the stem.

Ontogenetic studies have shown that the young sporophyte in most species is actinostelic. As growth of the sporophyte continues and the stem increases in size there is generally a change in the pattern of xylem and phloem. The actinostelic condition may persist with the formation of more protoxylem poles, or any of the configurations described above may result. In the smaller branches there may be a return to an actinostelic arrangement with only a few protoxylem poles.

Xylem maturation generally has been accepted to be strictly exarch in *Lycopodium,* but the results of a reinvestigation of three species would suggest that xylem development is at least sometimes and possibly always mesarch in indeterminate branches (Wilder, 1970). Mesarchy is, however, inconspicuous in that there are only a few tracheids formed in the centrifugal direction from a protoxylem pole. The bulk of the xylem cylinder, of course, consists of tracheids of the metaxylem, the larger of which have scalariform or circular bordered pits.

The phloem consists of sieve cells and parenchyma. The sieve cells are elongate with sieve areas distributed over the lateral walls as well as on the long, oblique end walls. The pores of the sieve areas of one investigated species are not lined with the carbohydrate callose, the presence of which is so characteristic of vascular plants in general (Warmbrodt and Evert, 1974). The sieve cells of *Psilotum* also lack callose (Chapter 8, p. 92).

The apical meristem of the shoot tip is reported to consist of a group of apical cells (Turner, 1924; Härtel, 1938; Freeberg and Wetmore, 1967; Nougarède and Loiseau, 1963) which by periclinal and anticlinal divisions contribute to the three primary meristematic tissues: protoderm, ground meristem, and procambium. The derivatives of these three primary tissues differentiate into epidermis, cortex, and vascular tissue, respectively. The centrally located procambium extends very close to the shoot apex, a feature characteristic of many lower vascular plants (Wetmore, 1943; Freeberg and Wetmore, 1967). The cells of the procambium are elongate (Fig. 9-7, A), and divide longitudinally, and frequently in the transverse plane.

To understand vascular differentiation, an examination of transverse stem sections taken at successive levels from the apex is essential. Near the tip the future vascular cylinder is represented by a compact core of procambial cells. Very early, however, within the procambial cylinder of a plectostele, for example, there is the centrifugal blocking-out of the future stele. The first procambial cells to differentiate cytologically are the future

A

B

FIGURE 9-6 *Lycopodium lucidulum.* **A,** fertile region of branch showing sporangia in axils of leaves (sporophylls). **B,** enlargement, as seen with the scanning electron microscope; note line of dehiscence across top of each sporangium. [**B** courtesy of Dr. R. H. Falk.]

cells of the metaxylem and metaphloem, followed by protoxylem and protophloem (Fig. 9-7, B, C). The future pattern may be established as close as 0.5 millimeters from the shoot apex. This initial centrifugal blocking-out is followed, at a lower level, by a stage of centripetal cellular maturation, the first elements to mature being the tracheids of the protoxylem (Fig. 9-7, D), followed by sieve cells of the protophloem. Maturation then proceeds centripetally until all metaxylem and metaphloem elements are mature. Complete maturation of the vascular cylinder may be complete only at a distance of 4 to 6 centimeters below the shoot apex (Freeberg and Wetmore, 1967).

Although the foregoing description emphasizes radial differentiation and maturation, procambial cells also are elongating and maturation occurs both longitudinally and radially in the stem.

LEAF ANATOMY. Initiation of the leaf may occur in a single superficial cell on the flank of the apical meristem, or in several superficial cells for certain species (Bhambie (1965). Growth in length, lateral extension of the lamina, and maturation of tissues produce the mature leaf. The growth of a leaf is associated with the development of a procambial strand into its base from the differentiating vascular tissue of the stem (Härtel, 1938). Differentiation of cells within this original procambial tract produces a vascular bundle, termed a leaf trace in its course through the cortex of the stem, and an unbranched vein within the leaf itself. Leaf traces are attached to lateral flanges or edges of the protostele of the stem axis (Fig. 9-7, C). Mature leaves range from minute scales to larger types that are lanceolate to ovate in outline, and generally lack definable petioles. Stomata may occur on both leaf surfaces or be re-

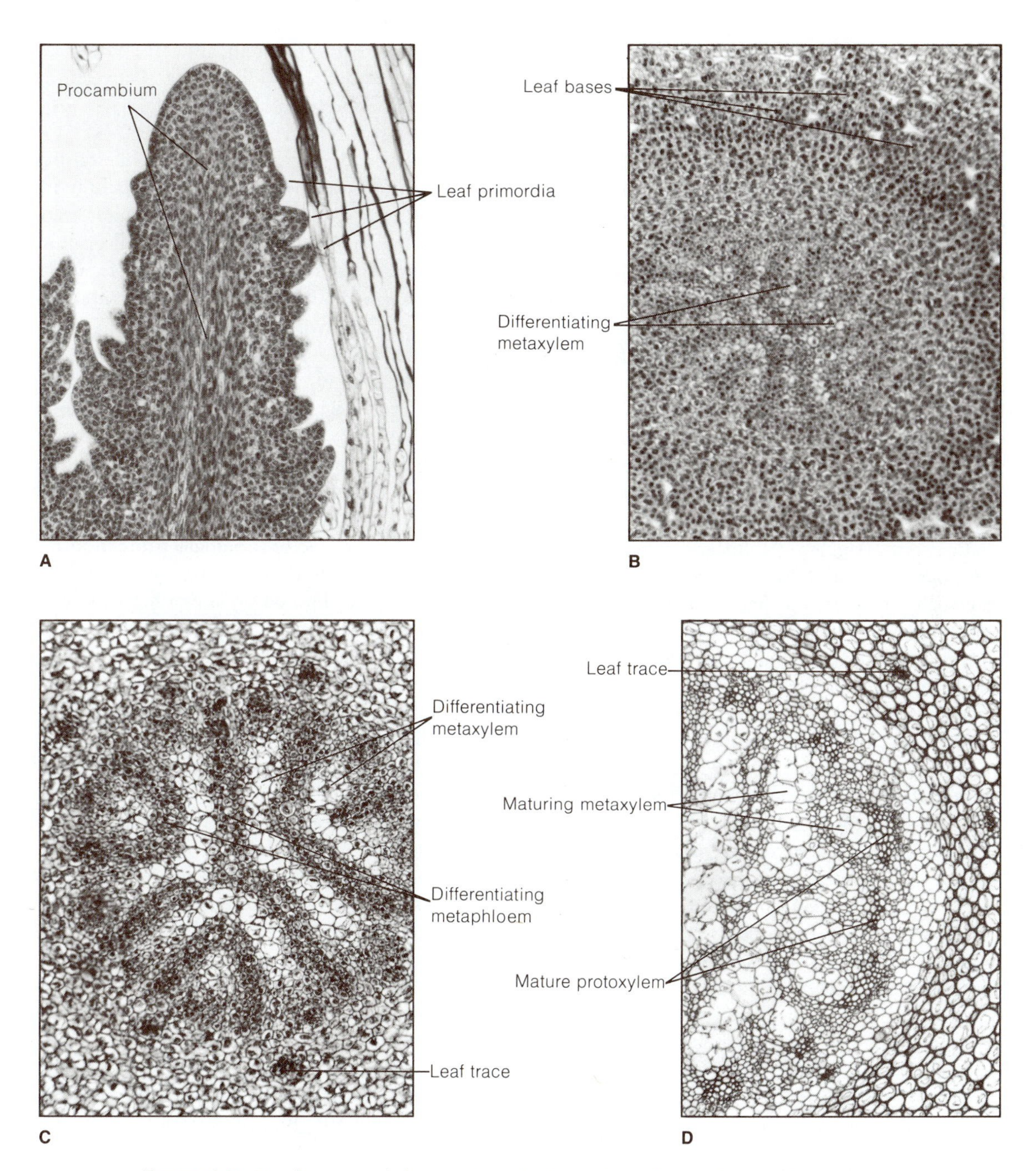

FIGURE 9-7 Development of the vascular cylinder (stele) in *Lycopodium*. **A,** longisection of shoot tip; **B-D,** transections of stem near shoot tip (**B**) and at increasing distances from the shoot apex (**C, D**). See text for details.

stricted to one (Chu, 1974). The mesophyll in many species is composed of more or less isodiametric cells with a conspicuous intercellular air space system.

The guard cells are formed directly from the original protodermal initial (Bhambie, 1965; Pant and Mehra, 1964).

ROOT. Except for the ephemeral primary root of the young sporophyte, the roots of actively growing plants arise from the stem very near the growing tip. These roots, which arise endogenously from the stem pericycle (Roberts and Herty, 1934), do not break through the cortex and epidermis immediately but often traverse the cortex for some distance before emerging. Since roots arise acropetally along the stem, many roots may be found in the stem cortex of the aerial portion of erect and epiphytic species (Fig. 9-5, A). In certain species (e.g., *Lycopodium pithyoides*) as many as 52 roots may be counted in the cortex at one level (Stokey, 1907). Only near the base of the stem do these roots emerge. In prostrate forms the roots take a more direct course from the stem axis to the exterior. After emerging from the stem, a root branches dichotomously, often with great regularity.

In propagating species of *Lycopodium* it is important to (1) obtain a portion of the plant with intact roots, or (2) use the upper portion of a shoot (since roots are initiated near the tip), or (3) secure a portion of the stem with arrested roots which emerge from the stem cortex on contact with a moist surface. Arrested roots may be identified as mounds on the under side of the stem of a prostrate form (Roberts and Herty, 1934).

Stokey (1907) has reported that four distinct groups of initials are present in the root apical meristem: a calyptrogen giving rise to the root cap; a tier of initials contributing to the developing protoderm; a group of initials giving rise to the cortex; and a set of initials for the vascular cylinder. These observations, however, need verification. Procambial differentiation and maturation result in a xylem strand that is crescent shaped (as seen in transverse section) and that partially surrounds a strand of phloem. Near the point of attachment to the rhizome a root may be polyarch (having several protoxylem poles) and maturation is exarch (Pixley, 1968). At this level the vascular cylinder (Fig. 9-8) may resemble that of the stem, and, except for size, it is sometimes difficult to distinguish between the two organs on the basis of stelar anatomy.

SPORANGIUM. One of the definitive characteristics of the Lycophyta is the association of one sporangium with each sporophyll; each sporangium is located on the adaxial side of a sporophyll or in its axil. In certain species of *Lycopodium* (for example, *L. lucidulum and L. selago*) the sporophylls are similar to vegetative leaves (Figs. 9-2; 9-6, A). No definite strobili are formed, but rather there are "fertile" areas on the stem alternating with vegetative or "sterile" regions. In species considered to be more specialized, the sporophylls are aggregated into definite conelike structures or strobili; the sporophylls of such cones may be unlike vegetative leaves in size, shape, and color and exhibit other specializations related to sporangial protection and spore dispersal. These strobili may occur on leafy stems or may be elevated on lateral branches with very small, scalelike leaves unlike those of the vegetative shoot (Figs. 9-3, B; 9-4, B, C).

Developmentally the sporangium is of the eusporangiate type originating from a group of superficial cells which divide periclinally (Fig. 9-9). The outer cells of such divisions form the multilayered wall, and the inner derivatives the sporogenous cells. The innermost layer of the sporangial wall functions as the tapetal layer.

Mature sporangia of most species are reniform (kidney shaped), their color ranges from yellow to orange, and they have a short stalk. There are some interesting relationships among position of mature sporangia, line of dehiscence, and specialization of the sporophyll. In certain species (e.g., *L. lucidulum*) the mature sporangium is axillary to a relatively unmodified sporophyll. Dehiscence is longitudinal or transverse to the long axis of the sporophyll (Fig. 9-6, B). In other species with definite strobili, the mature sporangia are foliar in position, and the sporophylls are imbricated and have abaxial extensions (Fig. 9-10, B). Dehiscence is modified, the opening being between the sporophyll and the abaxial extension of the sporophyll directly above. In still other species the sporangia are axillary, protected by sporophyll modifications, and open in a similar manner. Whether the sporangium is protected or not, the line of dehiscence

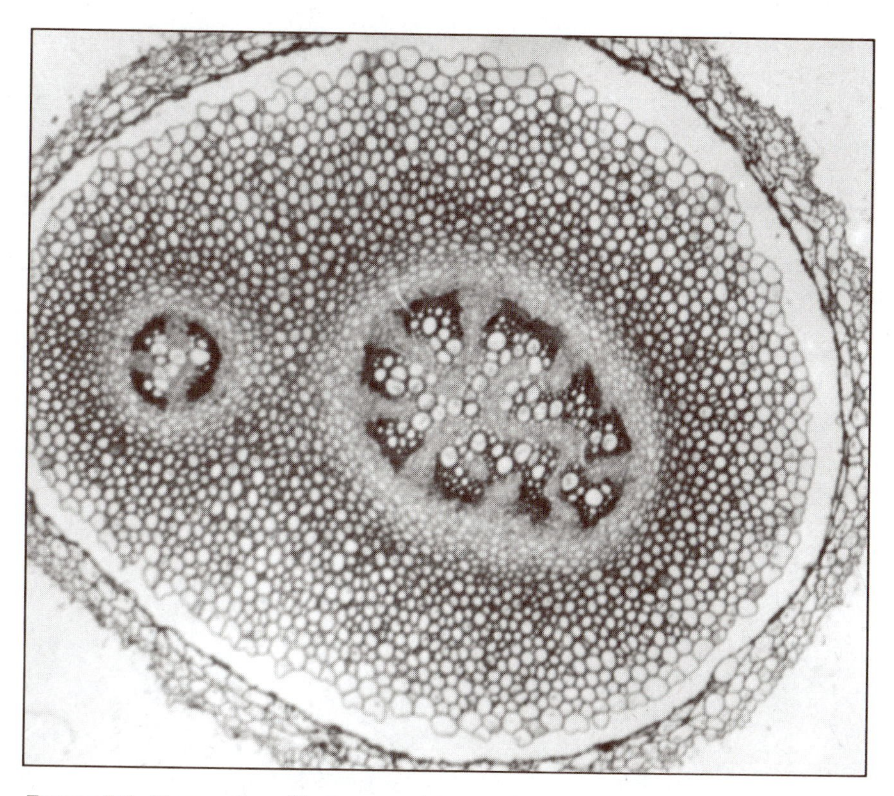

FIGURE 9-8 Transection of large root of *Lycopodium* sp. showing general similarity in organization of stele to that of a *Lycopodium* stem.

occurs in such a position as to insure efficient dispersal of spores (Sykes, 1908).

Meiosis occurs in the sporocytes, resulting in spore tetrads. The mature spores are yellow. The spore wall consists of an inner layer, called the intine, and an outer layer, the exine that displays an ornamentation that varies with the species. A triradiate ridge, present on the inner (proximal) face of each spore, is indicative of the mutual contact between members of a spore tetrad (Fig. 9-11, A, B). Spore morphology is useful in delimiting subgroups within the genus. (See Wilce, 1972, for illustrations of differences in spore wall ornamentation.)

The spores of certain species of *Lycopodium* are collected and sold as "lycopodium powder." This powder has been used in the manufacture of fireworks, but its use as a dusting powder on surgical gloves and pills has been discouraged; apparently the spores of *L. clavatum* cause inflammations in operative and other wounds (Whitebread, 1941).

THE GAMETOPHYTE. Depending on the species of *Lycopodium*, the spores may germinate immediately or after a delay of several years. A gametophyte plant of the first type *(L. cernuum, L. inundatum)*, generally found on the surface of the substrate, is ovoid to axial-dorsiventral, with short green aerial branches; the entire plant may not be over 3 millimeters long (Fig. 9-12, B). Rhizoids occur on the colorless basal portion. An endophytic fungus, entering the gametophyte plant early in development, is present in most species, occupying a definite region within the gametophyte. The sex organs generally occur near the bases of the aerial lobes. The time interval between spore germination and appearance of sex organs may vary from eight months to one year (Treub, 1884; Chamberlain, 1917; Eames, 1942).

After spore germination and when 6 to 8 cells have been formed, gametophytes of the second type may enter into a rest period of a year or more.

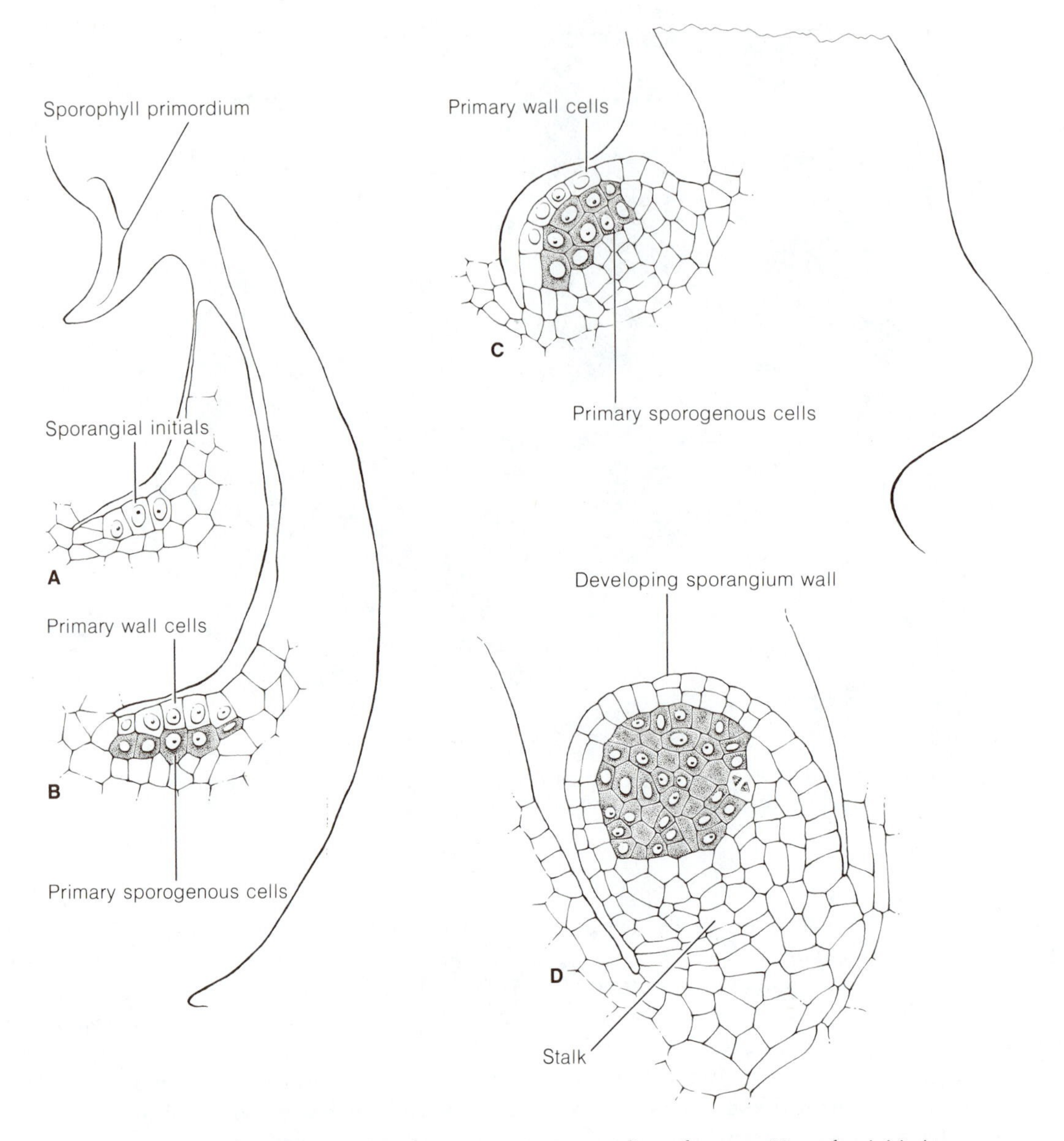

FIGURE 9-9 Ontogeny of the sporangium in *Lycopodium clavatum.* Note that initiation of the sporangium takes place in superficial cells by periclinal divisions, setting aside primary wall and primary sporogenous cells **(A, B)**. The tapetum ultimately arises from inner cells of the sporangium wall.

Apparently, further development is dependent on the entrance of a fungus. If this infection does not occur, all further growth ceases (Bruchmann, 1910). Physiologically, the fungus must supply certain substances vital for proper growth of the gametophyte plant. Subsequent development to a stage in which mature sex organs are present may require ten years or more (Eames, 1942). Development takes place beneath the surface of the ground or within a layer of humus. The gametophyte (e.g., *L. clavatum*) becomes disc shaped, with a convolute margin resembling a "walnut meat" (Fig. 9-12, A).

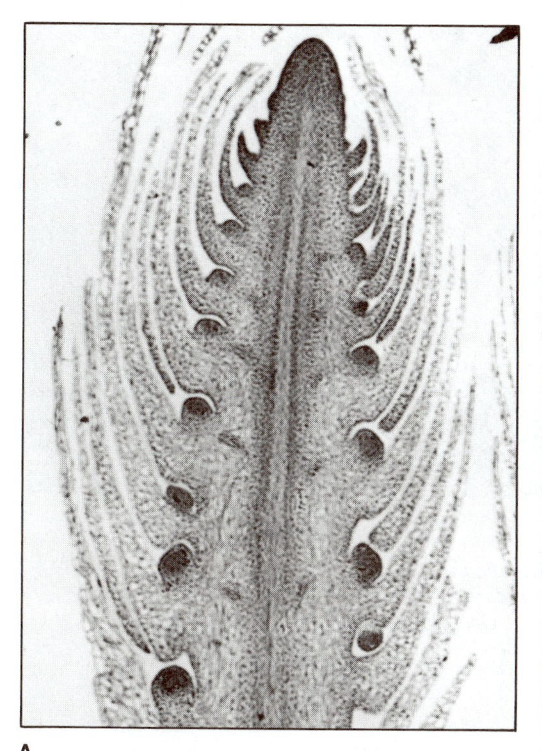

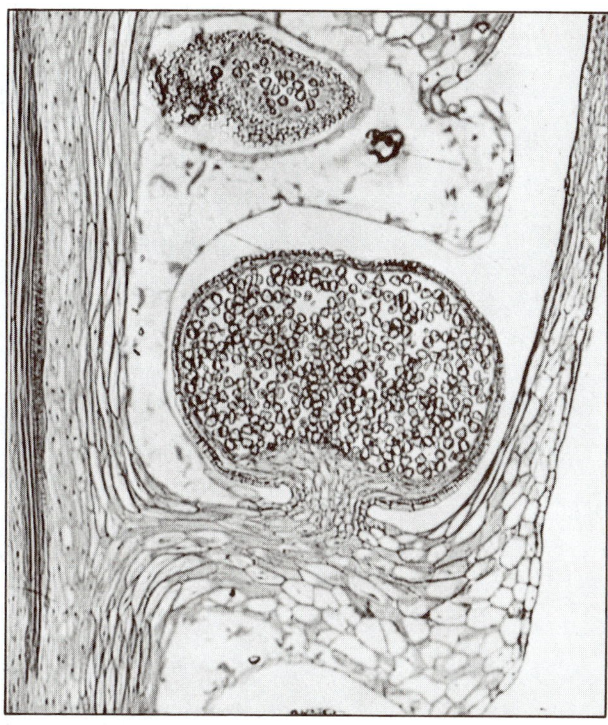

FIGURE 9-10 **A,** longisection of entire young strobilus of *Lycopodium clavatum;* developing sporangia can be seen near the bases of sporophylls. **B,** a mature sporangium of *Lycopodium* sp. attached to sporophyll; note numerous spores and the overarching abaxial extension of the sporophyll above; the cone axis is to the left.

But in other species the gametophyte may be cylindrical and branched, or assume the shape of a tiny carrot (Fig. 9-12, C). All of the subterranean gametophytes are colorless or yellowish to brown, developing chlorophyll only in those portions that become exposed near the surface (Spessard, 1922).

The subterranean forms are long lived, increasing in size by a marginal ring of meristematic tissue. Old gametophytes may be up to 2 centimeters in length or width.

In species whose gametophytes are of the green, annual type, antheridia and archegonia are generally intermingled near the bases of the upright lobes, whereas in the subterranean forms the sex organs are segregated into definite groups (Fig. 9-13, A) except in certain species (Spessard, 1922). In the course of development, antheridia generally appear first near the middle of the crown of the gametophyte. Initiation of archegonia and more antheridia then occurs in the immediate derivatives of the meristematic ring (Fig. 9-13, A).

The dependence of *Lycopodium* species on the infection of the gametophyte by a fungus presents an interesting physiological problem. It has been possible to culture gametophytes, particularly the annual type, to maturity by sowing the spores on soil taken from the original habitat (see Koster, 1941). Wetmore and Morel (1951a) were able to culture to maturity, in the laboratory under sterile conditions, the gametophyte of *L. cernuum* (a green annual type that in nature is associated with a fungus). After the spore coat had been sterilized with calcium hypochlorite, the spores were sown on a culture solution containing minerals and glucose. In some cultures the upright green branches became club shaped, while in others a filamentous "pin-cushion" type resulted (Figs. 9-14; 9-15, C). After six months of continued growth, under regu-

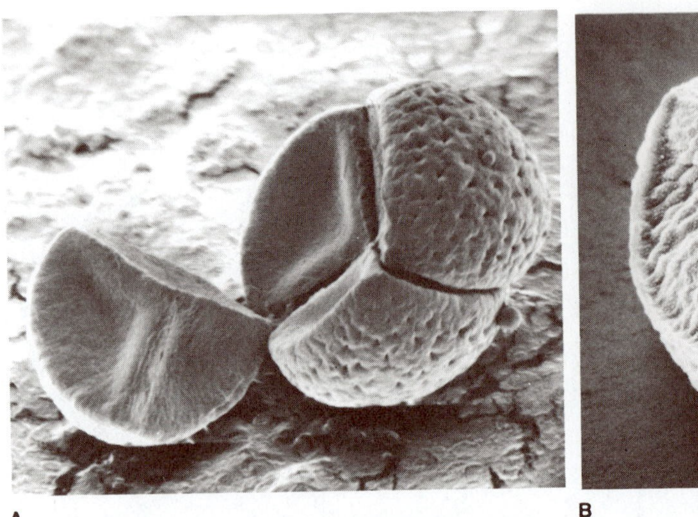

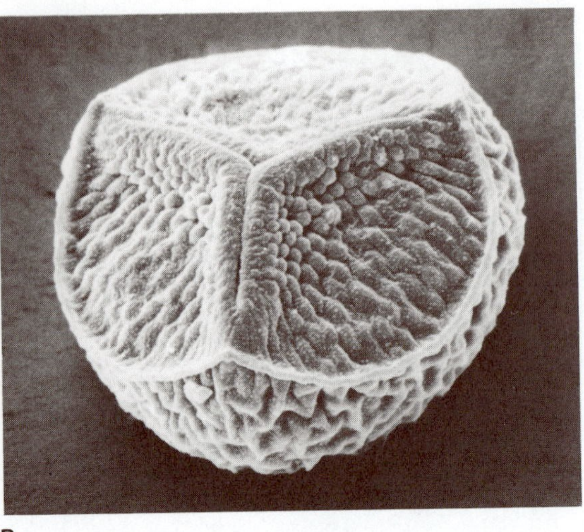

FIGURE 9-11 Scanning electron micrographs of *Lycopodium* spores. **A,** spore tetrad of *Lycopodium reflexum* (×1650). **B,** *Lycopodium inundatum;* note prominent triradiate ridge and contact faces with the other three spores (×1250). [Courtesy of Dr. G. Breckon.]

lated conditions, antheridia and archegonia were formed and many sporophytes developed (Fig. 9-14).

In 1957 Freeberg and Wetmore reported that they were able to germinate the spores of *Lycopodium selago* and *L. complanatum var. flabelliforme (L. digitatum).* Under natural conditions the gametophytes of both species are subterranean and long lived. However, under artificial cultural conditions the gametophytes were green and similar to those of *L. cernuum* discussed previously. On the basis of these results, some pteridologists expressed the belief that the gametophyte would be of little systematic value if the form of the gametophyte could be changed by simply altering the conditions for growth. Approximately 20 years later Bruce (1976a) made a detailed study of the sporophytes that developed on the gametophytes of all three species and concluded that all of the cultures were of *L. cernuum.* Apparently some spores of *L. cernuum* had been inadvertently introduced into the culture tubes, and they were the only spores that germinated. Pteridologists have now returned to the conviction that the form and physiology of the gametophyte can be of importance in the taxonomy and systematics of *Lycopodium.* In support of this belief Whittier (1977, 1981) was able to germinate spores *in vitro* and obtain mature gametophytes of *L. obscurum* and *L. digitatum.* Spores will germinate only if the culture tubes are placed in the dark for six or more months. The resulting gametophytes of *L. digitatum,* devoid of the endophytic fungus, are similar in form to those from nature. They are carrot shaped with a tapering base, and have a constricted neck below a cap like portion (Fig. 9-15, D). Although the endophytic fungus is not present, a layer of radially elongate cells is present that matches the region occupied by the endophytic fungus in nature. Gametophytes do turn green upon exposure to light after the required dark period, but the form remains the same.

The ontogeny of gametangia in *Lycopodium* has been described in detail in Chapter 5. A remarkable similarity in development exists in the early stages of ontogeny of the sex organs—namely initiation in a single superficial cell by a periclinal division, which sets aside the sterile jacket cell and the primary spermatogenous cell of the antheridium, and a division which forms the primary cover cell and the central cell of the young archegonium. The latter cell is the progenitor of the axial row (Fig. 9-13, B, C–E). At maturity an antheridium consists of a sterile jacket, one cell thick, enclosing many spermatids. Each one matures into a biflagellate sperm

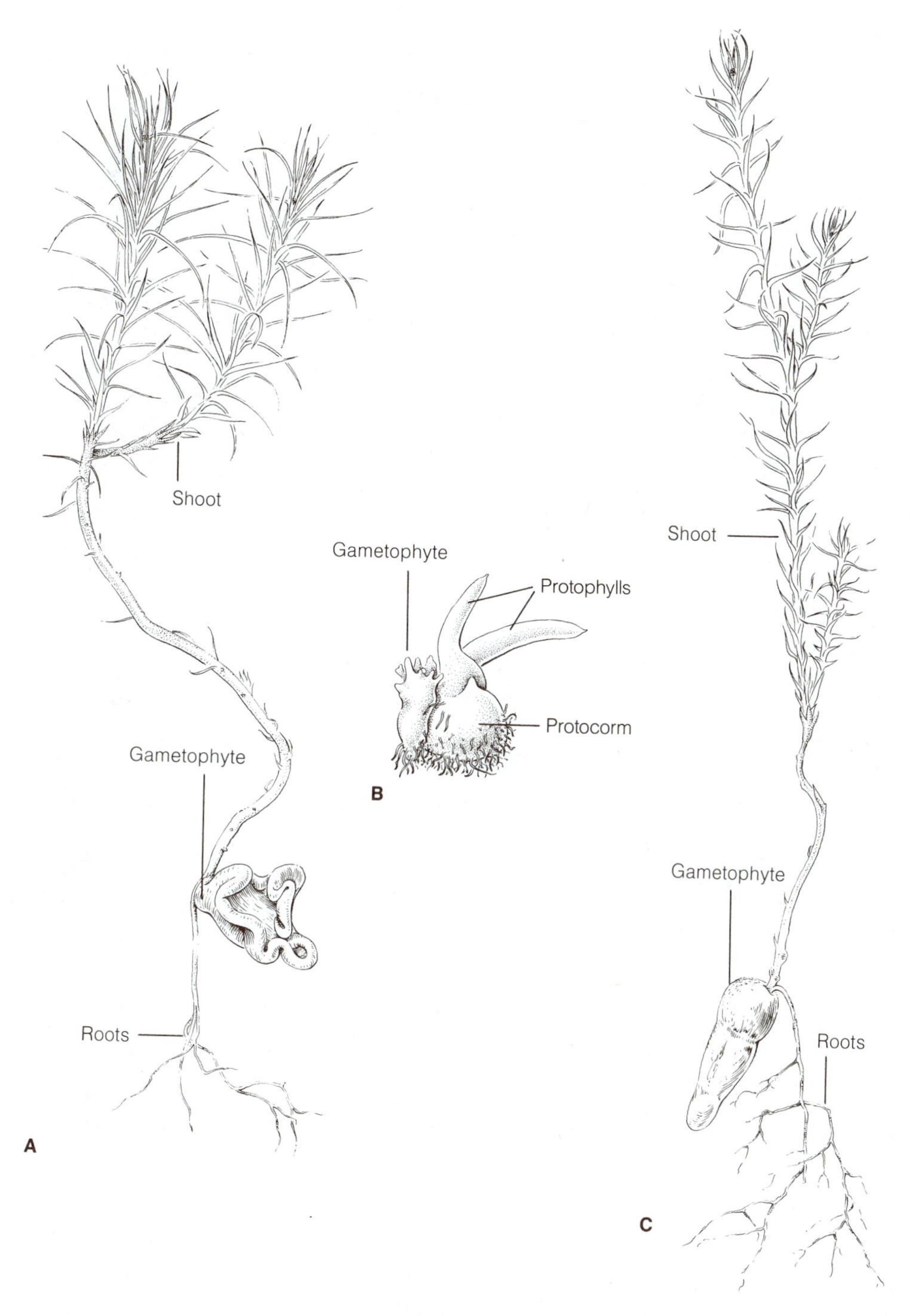

FIGURE 9-12 Gametophytes of *Lycopodium*. **A,** subterranean gametophyte of *Lycopodium clavatum,* with attached sporophyte; **B,** the subaerial or terrestrial type, *Lycopodium laterale;* **C,** subterranean type, *Lycopodium complanatum.* [**A** and **C** drawn from specimens supplied by Dr. A. J. Eames; **B** redrawn from Chamberlain, *Bot. Gaz.* 63:51, 1917.]

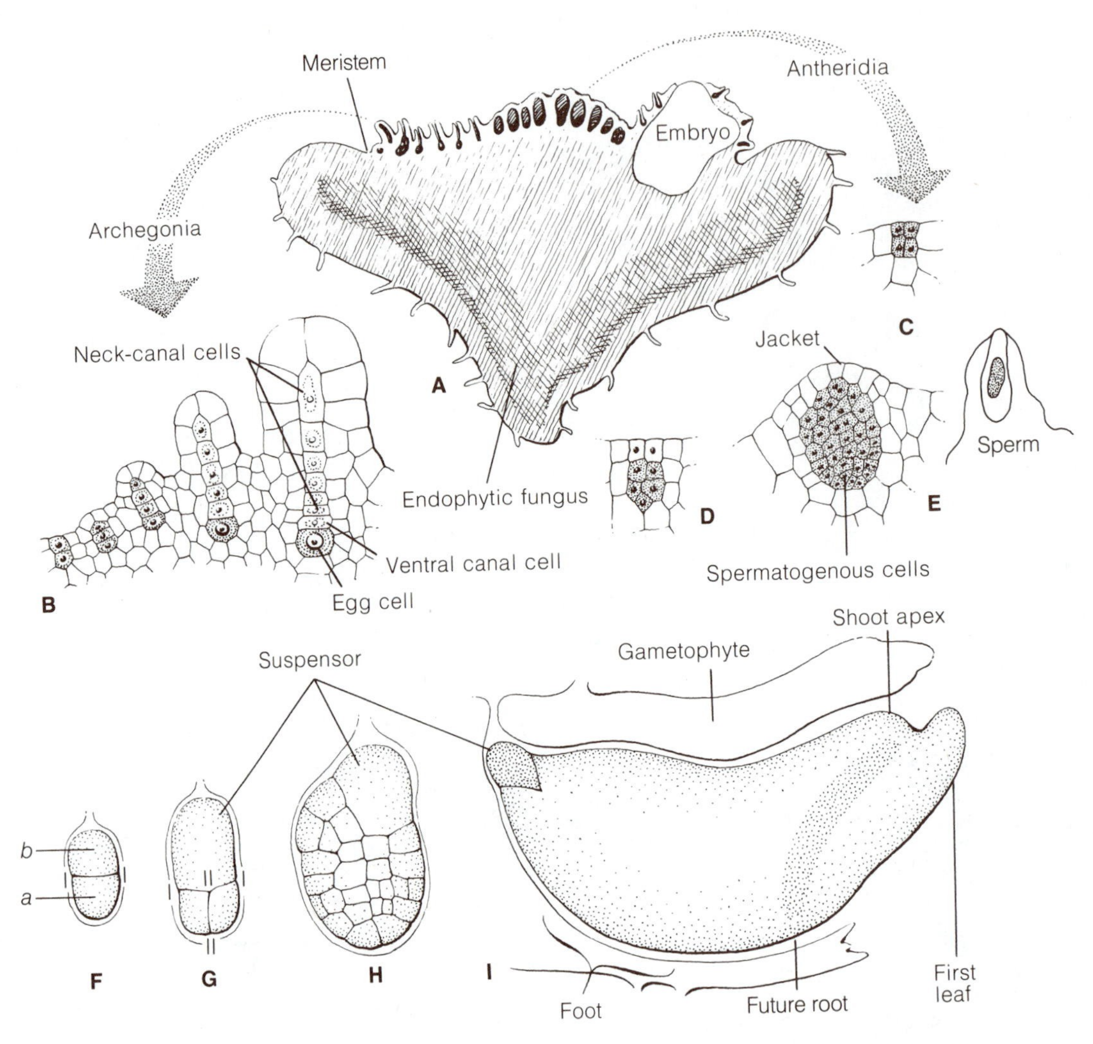

FIGURE 9-13 **A,** longisection of the gametophyte of *Lycopodium clavatum* showing the position of antheridia and archegonia, and one embryo; **B,** stages in development of an archegonium, *Lycopodium selago;* **C-E,** stages in ontogeny of an antheridium, *Lycopodium clavatum;* **F-I,** development of the embryo, *Lycopodium selago* (mouth of the archegonium is directed toward the top of the page). (Consult text for details.) [**A** redrawn from *Syllabus der Pflanzenfamilien* by Engler and Gilg, Berlin: Gebrüder Borntraeger, 1924; **B, G, H, I** redrawn from Bruchmann, *Flora* 101:220, 1910; **C-E** adapted from *Morphology of Vascular Plants. Lower Groups* by A. J. Eames. McGraw-Hill, New York, 1936.]

which closely resembles the sperms of certain algae. A sperm is a blunt-ended, fusiform cell, 8 to 10 micrometers long by 4 to 5 micrometers wide. The two flagella, each about 38 micrometers long, trail behind the cell as it swims. In its ultrastructure the sperm has a large coiled mitochondrion closely associated with a multilayered structure with microtubules forming a supportive spline. A large amyloplast, containing starch grains, occurs at the posterior end. The basal body of each flagellum lies close to the multilayered structure (Robbins and Carothers, 1978; Fig, 9-16).

FIGURE 9-14 *Lycopodium cernuum.* In vitro cultures of gametophytes with young sporophytes attached, in two culture tubes at the left. [Courtesy of Dr. Ralph H. Wetmore.]

The archegonia of surface-living, green, short-lived gametophytes have only three or four tiers of neck cells (Treub, 1884) and usually one neck-canal cell, whereas, according to Spessard (1922), archegonia of the subterranean forms have long necks with six or more neck-canal cells (Fig. 9-13, B). In either type the venter is embedded in the gametophyte tissue. In certain forms a doubling of the axial row may occur (Spessard, 1922), and archegonia may be formed with exceedingly long necks.

With the degeneration of the neck-canal cells and the ventral canal cell a passageway is created for the entrance of the motile biflagellate sperm, which reach the archegonium by swimming through a film of water on the surface of the gametophyte. Free citric acid or salts of citric acid may play a role in the attraction of sperm to the archegonia (Bruchmann, 1909; Doyle, 1970).

Despite the fact that the gametophytes of *Lycopodium* are bisexual, the results of an electrophoretic study of enzymes from the sporophytes of several to many colonies of three species indicate that the rates of intragametophyte selfing (self-fertilization) are very low. The gametophytes of these three species predominantly cross fertilize. The mechanism(s) promoting cross fertilization are unknown (Soltis and Soltis, 1988).

THE EMBRYO. Embryogeny is correlated, to some degree, with the type of gametophyte, but closer examination reveals a common basic plan. To gain an understanding of embryogeny, we will begin with a species possessing an underground gametophyte (Fig. 9-13, F-I). The embryo in *Lycopodium* is endoscopic, that is, the future shoot apex is directed away from the mouth of the archegonium. The first division of the zygote is transverse to the long axis of the archegonium, setting aside an apex, cell *a*, and a base, cell *b* (Fig. 9-13, F). Cell *b* undergoes no further divisions and becomes a suspensor. In our example, cell *a* then divides at a right angle to the original wall (Fig. 9-13, G, wall II–II). Additional divisions produce a multicellular embryo (Fig. 9-13, H). At about this developmental stage the future shoot apex grows laterally and upward, and a foot develops along the lower side of the embryo. The root is variable in position, but commonly arises between the first leaf and foot. With continued growth, the shoot tip emerges from the gametophytic tissue Fig. 9-13, I). The foot enlarges and maintains close connection with the gametophyte, acting as an haustorial structure until the sporophyte becomes physiologically independent. Sexually mature gametophytes may continue to live for some time, supporting one or more young sporophytes in various stages of development.

In certain species that have green, surface-living gametophytes, a foot is formed as well as a spherical parenchymatous body, termed a protocorm. No roots are produced on the protocorm, but leaflike structures — protophylls — arise on the upper surface, and rhizoids occur on the lower surface (Fig. 9-12, B). Only later does a shoot apical meristem become organized in cells of the protocorm, and a "normal" type of shoot is produced. In *L. carolinianum* a foot is formed but the protocorm stage is

FIGURE 9-15 **A, B,** *Lycopodium carolinianum.* **A,** gametophyte showing position of meristem (M) and spore (white arrow; ×40). **B,** gametophyte (G) with attached sporophyte showing protophyll (P) and four leaves (×20). **C, D,** gametophytes grown in sterile culture. **C,** *Lycopodium cernuum;* gametophyte is green under natural conditions and in artificial culture (×1.2). **D,** *Lycopodium digitatum;* gametophyte is subterranean in nature, but becomes green in culture when exposed to light (×7.8). [**A, B** from Bruce, *Amer. Jour. Bot.* 66:1156–1163, 1979; **C** courtesy of Dr. J. A. Freeberg; **D** courtesy of Dr. Dean P. Whittier.]

absent. The emerging sporophyte forms a single protophyll (Fig. 9-15, A, B) and a rhizome apex (Bruce, 1979).

Generic and Subgeneric Concepts

As mentioned previously there have been recommendations over the years to establish subgenera or genera of *Lycopodium.* One early suggestion was to recognize two subgenera—one, *Urostachya,* in which well defined strobili are not formed (e.g., *L. lucidulum, L. selago*) and branching is essentially isotomous. The second subgenus, *Rhopalostachya,* includes species that have definite cones and branching is anisotomous (e.g., *L. digitatum, L. clavatum*). In another system two families are recognized—the Urostachyaceae with one genus and the Lycopodiaceae with three genera. All four genera are said to have different basic chromosome numbers (Löve and Löve, 1958), and the chemistry of their flavones (Voirin and Jay, 1978), phenolics, and lignins differ (Towers and Maass, 1965).

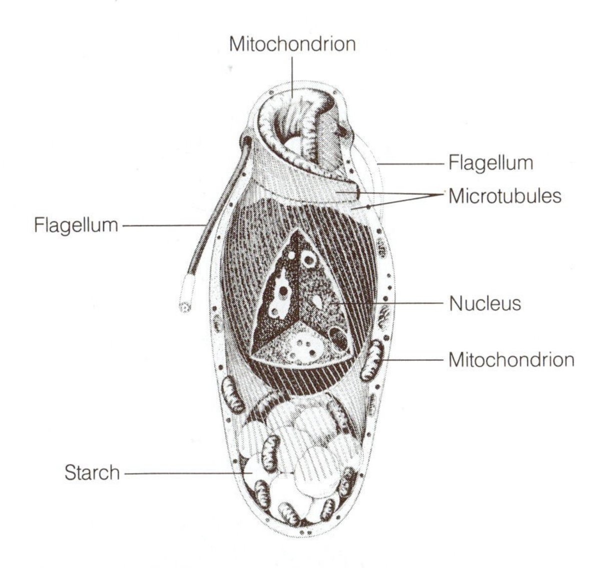

FIGURE 9-16 Perspective diagram of a mature sperm of *Lycopodium cernuum*, with wedge of nucleus removed to show internal organization (approx. ×7,000). [Courtesy of Dr. Robert R. Robbins.]

Other pteridologists believe that the one genus *Lycopodium* (the species of which are easily recognized) should be retained until more information is available on aspects such as gametophytes, life histories, and chromosome numbers (Hauke, 1969). In recent years more information has become available on the gametophyte generation (see p. 115). Chromosome counts of a limited number of species on a worldwide basis range from $n = 23$ to $n = 264$ (Löve et al., 1977). These counts indicate a history of hybridity and possibly polyploidy, although few multiple series (replications of entire chromosome sets) are evident.

Space does not permit a review of all proposed subgenera, but one system of classification, applied more specifically to North American species, will be presented. The classification is based upon the totality of morphological characteristics and chromosome numbers (Beitel, 1979; Wagner and Wagner, 1980; Beitel and Wagner, 1982).

Subgenera

Huperzia (e.g., *H. selago, H. lucidulum;* (Fig. 9-3, A). Little distinction between rhizome and upright branch system as a result of isotomous branching; leaves spirally arranged; sporophylls unmodified or slightly modified, arranged in zones on the stem or in pendant tassels; gametophytes rod shaped, branched or unbranched, subterranean or buried in moss or humus on trees. A known chromosome number of $n = 67$ may be indicative of an aneuploid series based upon multiples of the theoretical base number $x = 11$ plus one chromosome $[(6 \times 11) + 1]$.

Lycopodium [(in limited sense) e.g., *L. clavatum, L. obscurum;* Fig. 9-4, C]. Anisotomous branching; prostrate rhizome and upright terminal branches; leaves spirally arranged; sporophylls aggregated into definite strobili; gametophytes disc shaped and subterranean; chromosome base number of $n = (3 \times 11) + 1 = 34$.

Diphasiastrum (e.g., *L. complanatum, L. digitatum;* Fig. 9-3, B). Anisotomous branching; creeping rhizome and upright evergreen branches; majority have stems with four rows of leaves, fused with stem most of their length, forming flattened branches; definite strobili; gametophytes subterranean, cone shaped like tiny carrots; chromosome base number $n = (2 \times 11) + 1 = 23$.

Lycopodiella (e.g., *L. inundatum, L. carolinianum, L. cernuum;* Fig. 9-4, A). Considerable variation in growth habit from an erect, much-branched plant to creeping rhizomes with upright fertile branches that bear spirally arranged leaves; only the swollen rhizome tips in some species remain alive at the end of the growing season; gametophytes, where known, resemble tiny, green pincushions or are axial-dorsiventral with upright green lobes; chromosome base number $n = (7 \times 11) + 1 = 78$ appears to be representative, although there is considerable variation in the group, especially for *L. cernuum* and *L. carolinianum* (Löve et al., 1977; Bruce, 1976b).

One note of caution should be cited with respect to high chromosome numbers in *Lycopodium* (sensu lato) and the role that possible repeated episodes of polyploidy played in evolution. Soltis and Soltis (1988) pointed out that if lycopod species with high chromosome numbers are truly highly polyploid, they should possess many isozymes (different forms of an enzyme encoded by different gene loci). They found in their study of eight species (in three subgenera) that the number of isozymes present was typical of diploid seed plants for all but one enzyme. They conclude that there is no genetic evidence for widespread polyploidy in lycopods and no evidence to support the low, theoretical base numbers suggested for these plants, e.g., $x = 11, 12$. They favor the concept that the ancient ancestors of lycopods and other homosporous pteridophytes initially had high chromosome numbers.

Some Herbaceous Devonian Lycopods

A plant with definite lycopod characteristics is *Baragwanathia* from the Lower Devonian of Australia (Jaeger, 1962). It has been reported also from the Silurian, however this record needs verification. The plant branched dichotomously, and the stem was covered with helically arranged leaves, 0.5 to 1 millimeter wide and as long as 4 centimeters. Sporangia are associated with some leaves, but it is unknown (because of poor preservation) whether they were on the adaxial side of leaves or near the axils of leaves.

The genus *Protolepidodendron* from the Lower to Middle Devonian had leaves that forked near the tip. The plant was rhizomatous with upright dichotomous branches bearing helically arranged leaves, some of which had solitary sporangia located on the adaxial side of leaves (Fig. 9-17, A). Similar types of sporophylls have been reported for the new genus *Estinnophyton;* this genus, however, had one pair or two pairs of short, stalked sporangia on the adaxial side of the forked sporophylls (see Gensel and Andrews, 1984).

Another lycopod from throughout the Devonian is *Drepanophycus* (Fig. 9-17, C). As reconstructed, it had a rhizome with upright dichotomous branches with stiff, curved leaves. Stalked sporangia in some forms occurred singly on some leaves midway between the enlarged leaf base and tip (Grierson and Hueber, 1968).

The Lower Devonian genus *Asteroxylon* was formerly classified with the "psilophytes" because it was assumed to have had terminal sporangia. It was transferred to the Lycophyta because vascularized, stalked sporangia were borne laterally on stems, interspersed among leaves (Lyon, 1964). *Asteroxylon mackiei* grew to a height of 0.5 meters. The plant had a naked, dichotomously branched subterranean rhizome. Small branch systems probably functioned in anchorage and absorption. The tips of some dichotomies grew upright, bearing numerous, closely appressed, small leaves. The main upright axes exhibited irregular or anisotomous branching, whereas smaller side-branch systems displayed more regular dichotomous growth (Fig. 9-17, B).

The stem of *Asteroxylon* had an epidermis with thick outer walls, which was interrupted in places by stomata. The cortex was differentiated into outer, middle, and inner portions (Fig. 9-18). Occupying the central region of the stem was the vascular cylinder, which may be designated as an actinostele (Chapter 3). Primary xylem in the form of a fluted cylinder occupied the center of the stem. Protoxylem occurred near the extremities of the lobes but was surrounded on all sides by metaxylem, making the xylem mesarch in development. "Leaf" traces, departing from the vicinity of the lobes, passed obliquely through the cortex and ended abruptly near the base of each leaflike appendage. These traces were concentric; that is, each trace consisted of primary xylem surrounded by primary phloem. In one species, *Asteroxylon elberfeldense,* the vascular cylinder was siphonostelic in lower portions of the aerial stem.

Leclercqia, a recently described lycopod from Middle Devonian rocks, is the best preserved and most completely known herbaceous lycopod (Banks et al., 1972). Its leaves were 4.0 to 6.5 millimeters long and were five partite: an elongate tip and two lateral portions each of which was forked. Tracheids have been identified in the midvein. A branch vein extended to the base of each lateral division but did not enter the two segments. The stem had an exarch protostele with 14 to 18 protoxylem poles. There was one adaxial sporangium on each sporophyll and the plant was probably homosporous. In 1979, Grierson and Bonamo discovered the presence of a ligule on vegetative leaves and sporophylls. This discovery constitutes the first record of the ligule in a pre-Carboniferous, herbaceous, homosporous lycopod. It has long been thought there is a strict correlation between the ligulate condition and heterospory as characterized by the Carboniferous Lepidodendrales and the present day genera *Selaginella* and *Isoetes.* No doubt *Leclercqia* will be brought into future discussions on the phylogenetic interrelationships of Devonian and Carboniferous lycopods (see Stewart, 1983).

FIGURE 9-17 Diagrammatic reconstructions of extinct lycopods from the Devonian Period. **A,** *Protolepidodendron scharyanum;* sporophylls and sporangia enlarged above; **B,** *Asteroxylon mackiei;* **C,** *Drepanophycus spinaeformis.* [**A** redrawn from Kräusel and Weyland, *Senckenbergiana* 14:391–403, 1935; **B** redrawn from Kidston and Lang, *Trans. Roy Soc. Edinb.* 52, Part IV, 1921; **C** redrawn from Kräusel and Weyland, *Palaeontographica* 80(B):171–190, 1935.]

Some of the Devonian lycopods just described were probably ancestral forms in the line leading to *Lycopodites* (Middle Devonian-Carboniferous species) and then to the extant genus *Lycopodium*. The Devonian types were, in turn, undoubtedly derived from members of the Zosterophyllophyta — one of the divisions of early vascular plants (Chapter 7).

Heterosporous Groups in the Lycophyta

Selaginellales — Selaginellaceae: *Selaginella*

GENERAL CHARACTERISTICS. The genus *Selaginella*, the small club moss or spike moss, is widely

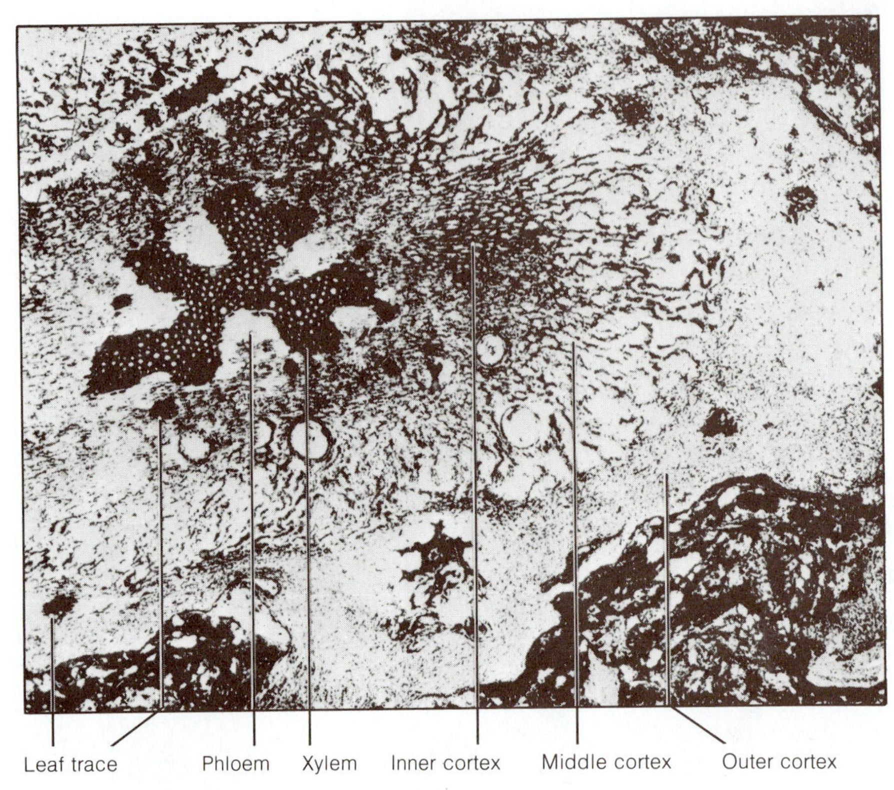

FIGURE 9-18 Transection, stem of *Asteroxylon mackiei.* [From Kidston and Lang, *Trans. Roy. Soc. Edinb.*, Vol. 52, 1920–21.]

distributed over the earth. But even though the genus includes about 700 species, it does not form a conspicuous part of the world's vegetation. Whereas many species of *Lycopodium* may be relatively large and coarse, most species of *Selaginella* are small and delicate. It is in the tropics that *Selaginella* is most abundantly represented, often being the dominant element of the forest floor in mesophytic tropical woodlands (Haught, 1960). Some species grow where climates are cold, and many others inhabit temperate regions, growing in damp areas or even occupying exposed rocky ledges. The genus is especially well represented on the eastern slopes of the Andes from Colombia to Bolivia (Tryon and Tryon, 1982). One species, *Selaginella lepidophylla,* caespitose in habit, has adapted to existence on a Mexican desert and in the arid regions of southwestern United States. The entire plant forms a tight ball during periods of drought; in the presence of moisture the branches expand and lie flat on the ground. This species is commonly known as the "resurrection plant." *Selaginella* is a greenhouse favorite, and is often used as a border plant. Species of this genus growing together in a greenhouse present an array of color shades—dark to light green, bluish—and some are iridescent.

GROWTH FORM. In growth habit there is considerable variation, although most species can be referred to two or perhaps three growth types. Some species are erect or often form tufts or mounds (Fig. 9-19, A). The leaves are helically arranged and are of the same size and shape (isophylly). In other species the plant may be flat, creeping along the surface of the ground or scrambling over shrubs (Fig. 9-19, B). Still others have a strongly developed rhizomatous stem with large, frondlike side branches which stand erect (Fig. 9-20, C). In the last two types anisophylly (the production of small and large leaves) is a prominent feature and the stems have dorsiventral symmetry (Fig. 9-21). Branching generally is considered to occur by a more or less equal bifurcation of the shoot apex in the establishment of two shoot axes, one of which overtops the other, resulting in

FIGURE 9-19 Two species of *Selaginella* showing contrasting growth forms. **A,** *Selaginella watsoni*, ascending branches with leaves of uniform size (isophylly). **B,** *Selaginella kraussiana*, a creeping or scrambling type; note that roots arise at points of branching. [B courtesy of Susan Larson.]

anisotomous branching. However, Hagemann (1980) has shown that a new branch in *Selaginella speciosa* arises in a lateral position on the dome-shaped apical meristem. Both the original axis as well as the new branch continue to grow. The branching pattern resembles the anisotomous type.

Peculiar leafless, proplike structures, originating from the stem at points of branching, have been termed "rhizophores," (see Figs. 9-19, B; 9-20, C; also Fig. 9-22, B); more will be said of their morphological interpretation in a later section of this chapter.

STEM ANATOMY. The outer cell walls of the epidermis are cutinized. Stomata are said to be lacking. In many species there are several layers of thick-

A
B
C

FIGURE 9-21 *Selaginella kraussiana,* showing two rows of small dorsal leaves and two rows of larger ventral leaves (anisophylly). [Courtesy of Susan Larson.]

walled cells beneath the epidermis, which merge gradually with thin-walled chlorophyllous cells of the inner cortex. In most species the trailing stem or prostrate rhizome is protostelic. Plants with a radial symmetry may have a simple, cylindrical protostele in the stem, whereas dorsiventral species may have two or more vascular strands that are either circular or ribbon-shaped as seen in transverse section (Fig. 9-22, A). The ribbon-shaped protostele in the rhizome (Fig. 9-22, D) may be replaced in the upright branches by a number of vascular bundles (meristeles). It should be emphasized that the vascular system of a shoot is interconnected and the term meristele is used only for convenience in describing a portion of the vascular system as seen in transverse section. An even more complex stelar pattern has been described for the tropical species *Selaginella exalta.* In this species the stele of the very large erect stem is a three-lobed plectostele, termed an "actinoplectostele" (Mickel and Hellwig, 1969).

In still other species the rhizome may be solenostelic, and the upright branches may have as many as 10 to 15 separate meristeles. Experimentally it has been shown that if such an upright branch is placed in a horizontal position, the newly developed portion of the shoot will become solenostelic (Wardlaw, 1924). Whatever the stelar configuration, the one central vascular cylinder or each meristele is supported in a large air-space system by radially elongated endodermal cells designated *trabeculae* (Figs. 9-22, A; 9-23; 9-24). These cells have the characteristic casparian strips. If the air-space system is large, each support or trabecula may consist of several cortical cells as well as the endodermal cell.

Regardless of stelar organization, the primary xylem is exarch in development, and the metaxylem consists primarily of tracheids with scalariform pitting. Ribbon-shaped steles may have more than one protoxylem pole (Zamora, 1958). Many years ago (Duerden, 1934) certain species were shown to possess vessels. The phloem of *Selaginella* consists of sieve cells and parenchyma. The end walls of the sieve cells are usually slightly oblique and scattered single "sieve pores" occur on the lateral and end walls rather than having typical sieve areas, i.e., groups of pores clustered in thin-walled areas (Lamoureux, 1961). Each mature sieve cell contains a degenerated nucleus and "refractive spherules" which are probably plastids (Burr and Evert, 1973).

LEAF ANATOMY. Leaves of all species are small, attaining a length of a few millimeters. In form the leaves may be ovate, lanceolate, or orbicular with one vein running nearly the length of the leaf. Exceptions to this condition have been reported for two species that have a branched venation system similar to megaphylls (Wagner, et al., 1982). Although helical arrangement is a common feature in *Lycopodium,* most species of *Selaginella* have leaves that are arranged in four rows along the stem (Fig. 9-21). There are two rows of small leaves on the

FIGURE 9-20 **A, B,** branches of two species of *Selaginella* showing strobili at tips of determinate branches. **C,** *Selaginella martensii,* showing large ascending or erect branches and numerous proplike roots.

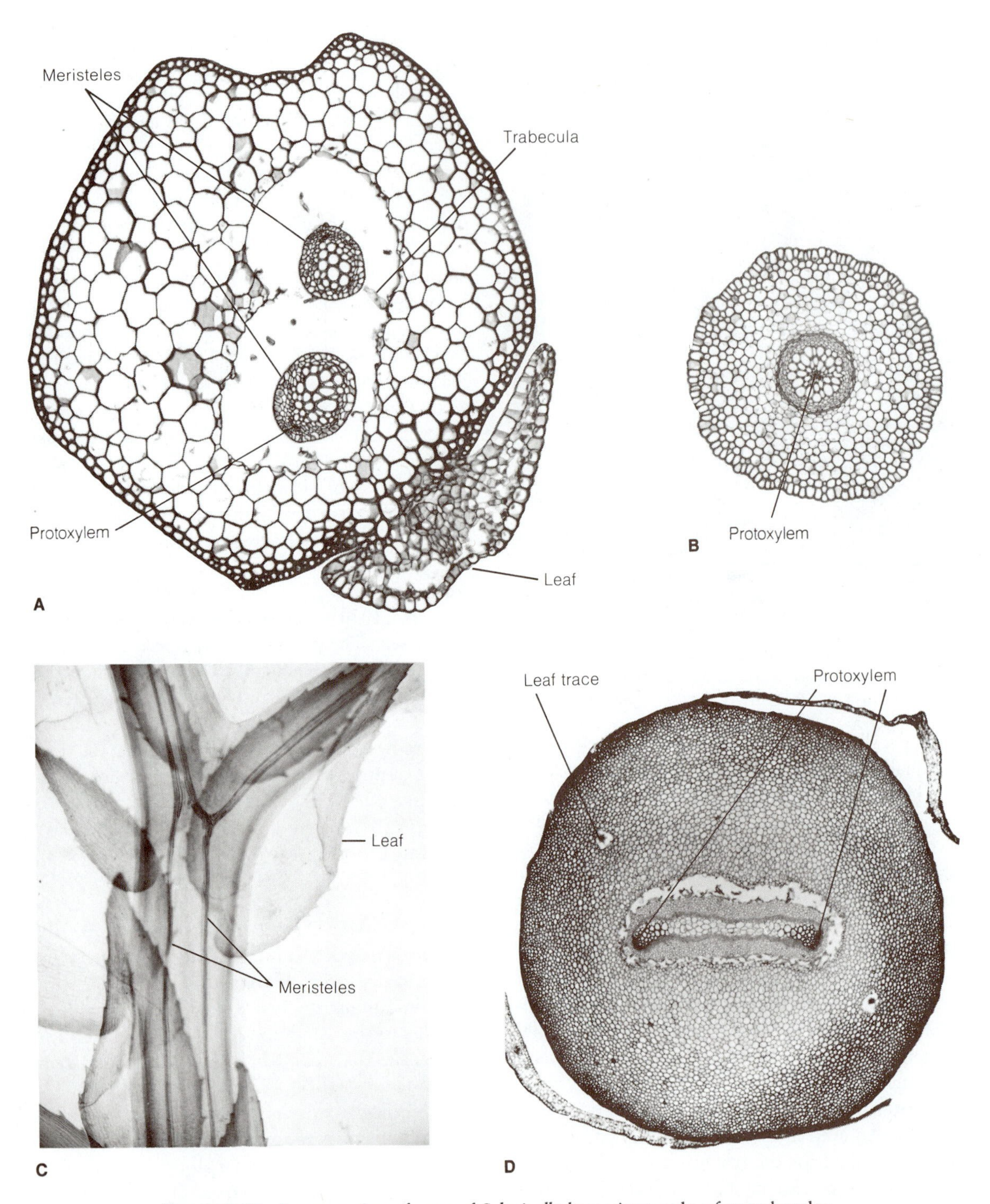

FIGURE 9-22 **A,** transection of stem of *Selaginella kraussiana;* only a few trabeculae appear in a transection of the stem; **B,** transection, rhizophore of *Selaginella* sp.; **C,** "cleared" and stained shoot of *Selaginella kraussiana* showing course of the vascular strands (meristeles) at a dichotomy of the stem; **D,** transection, stem of *Selaginella pallescens* near the base of a large branch.

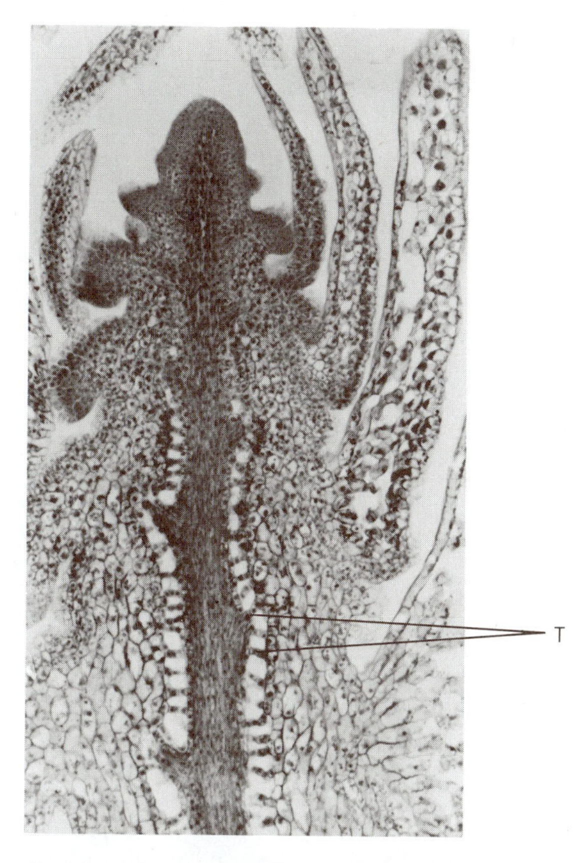

FIGURE 9-23 Longitudinal section of *Selaginella* sp. stem showing development of numerous trabeculae supporting the vascular cylinder in a large air-space system. T, trabeculae.

dorsal side of the stem, and two rows of larger leaves on the ventral side or in a lateral position. A small tonguelike structure, the ligule, is on the adaxial side of each leaf near the base. Anatomically the mature leaf may vary considerably. The cells of the two epidermal layers may be similar, or in some species they may be somewhat different (Hsü, 1937). Some species have bristles or short hairs extending out from the epidermis. The mesophyll may consist of a distinct palisade layer and spongy parenchyma, or the entire mesophyll may be a reticulum of lacunate parenchyma. Generally stomata are on the abaxial surface, although in certain species they are present on both surfaces.

In certain investigated species the shoot is terminated by an apical cell (Barclay, 1931; Hsü, 1937). The apical cell is tetrahedral and appears as an inverted pyramid in longitudinal sections of the shoot apex (Fig. 9-24). Through divisions of this apical cell, derivative cells are produced on the three "cutting" surfaces. Each of these cells (segments) undergoes a periclinal division, forming an outer and an inner cell. The outer cell, by further divisions, will produce the epidermis and cortex. Endodermis, pericycle, and vascular tissues are derived ultimately from derivatives of the inner cell (Fig. 9-24). Certain other species are reported to have one three-sided cell (Hagemann, 1980), or two adjoining apical initial cells at the shoot tip (Williams, 1931), or a group of apical initials (Bhambi and Puri, 1963).

Leaves have their origin in superficial cells located along the flanks of the apical meristem. A developmental study was made of anisophylly in *S. martensii* to determine if the smaller dorsal leaf is only an arrested form of the larger ventral leaf. Each pair of leaves—dorsal and ventral—originate at about the same time from the apical meristem, but the ventral leaf primordium is larger at the time of inception. The two can be distinguished when the leaves are only 0.1 millimeter in length. The subsequent pattern of histogenesis of both types of leaf is similar, but the smaller dorsal leaf is distinguished primarily by precocious maturation of tissues (Dengler, 1983a, b).

Early in its development a leaf is traversed by a procambial strand which is continuous with the vascular cylinder of the stem. Procambial cells eventually differentiate into primary xylem and primary phloem of the leaf vascular bundle.

A ligule (from Latin, *ligula,* "a small tongue"), located on the adaxial side of each vegetative leaf and sporophyll, makes its appearance through periclinal divisions in two or more short rows of surficial cells (Fig. 9-28). At maturity, a ligule is a surprisingly complex structure, considering its small size and short life. The ligule may be to some extent sunken in the leaf and has a basal sheath of cells with casparian strips and an adjacent group of large, vacuolated cells termed a *glossopodium* (from the Greek words *glossa,* meaning "tongue", and *podion,* meaning "foot"). The sheath cells resemble endodermal cells and may well perform a regulatory role in the movement of water and dissolved substances. The frequent development of tracheidlike cells (transfusion tissue) between the ligular sheath

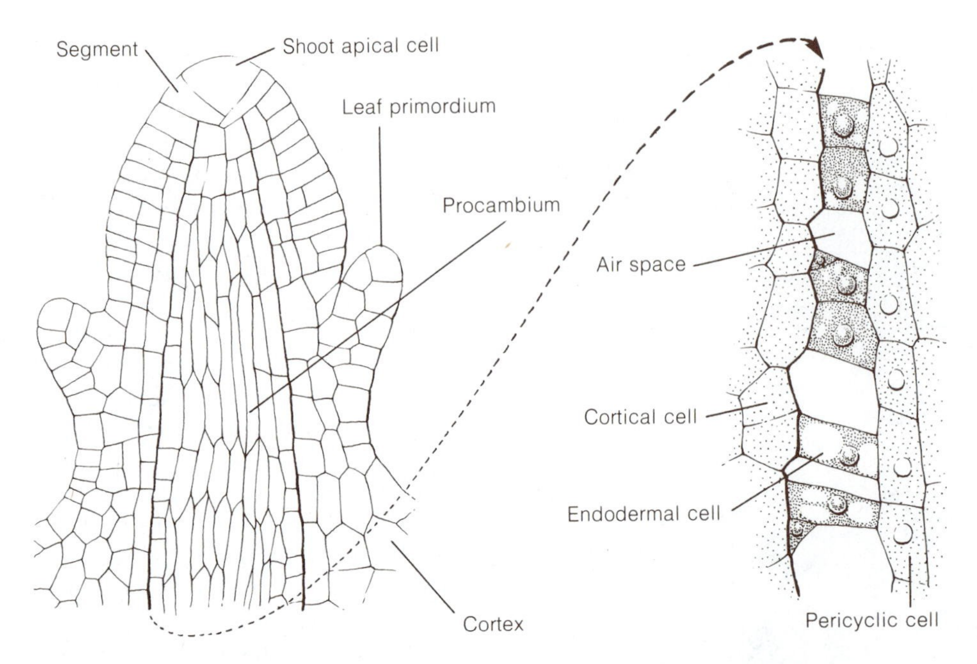

FIGURE 9-24 Stem development in *Selaginella sinensis.* Longisection of the shoot tip (left), and early development of trabeculae (endodermal cells) and of the air-space system surrounding the vascular cylinder (right); endodermal cells become separated from one another and undergo radial extension. [Redrawn from Hsü, *Bull. Chinese Bot. Soc.* 3:75, 1937.]

and adjacent vein of the microphyll strengthens the concept of direct conduction of substances to the base of the ligule.

A ligule develops precociously soon after initiation of the microphyll, and it has been assumed that it might function as an excretory structure in keeping the young leaf primordium (and sporangium of a sporophyll) moist during early development. There is now convincing evidence that the ligule in certain species, during early development, secretes a mucilage consisting of carbohydrates and proteins which also coats the apical meristem (Bilderback, 1987; Fig. 9-25, A, B).

Ligules also occur in *Isoetes* and *Stylites.* Phylogenetically, the origin and homology of the ligule are obscure problems. But the antiquity of ligules is shown by their presence in extinct arborescent lycopods of the Carboniferous. More recently, the ligule was shown to have been present in the herbaceous, homosporous Devonian lycopod, *Leclercqia* (see p. 124).

THE ROOT. Except for the primary root of the young sporophyte, the roots of most species arise at points of branching of the stem. Each root takes its origin from a meristem, termed an angle meristem, but the root generally remains visibly unbranched until it contacts the substratum. Traditionally, the leafless axis has been interpreted as stemlike (rhizophore), giving rise to roots at its distal end upon reaching the soil or humus. The young, initially unbranched root has no definite root cap, and occasionally develops into a leafy shoot under natural conditions or can be induced to do so under experimental conditions (Williams, 1937; Cusick, 1954).

Detailed histogenetic studies of root development have been made on three species of *Selaginella.* In *S. wallacei* (an isophyllous species) the root initially lacks a root cap, but very early the apical cell gives rise to root cap cells. Soon after root cap formation, the root apical meristem divides in preparation for branching. The apical meristem may continue to divide in an isotomous manner, but actual branching is not evident externally until the axis comes into contact with the soil (Webster and Steeves, 1964). Root development is essentially the same for the anisophyllous species, *S. kraussiana.* A root may become 4 to 5 centimeters in length be-

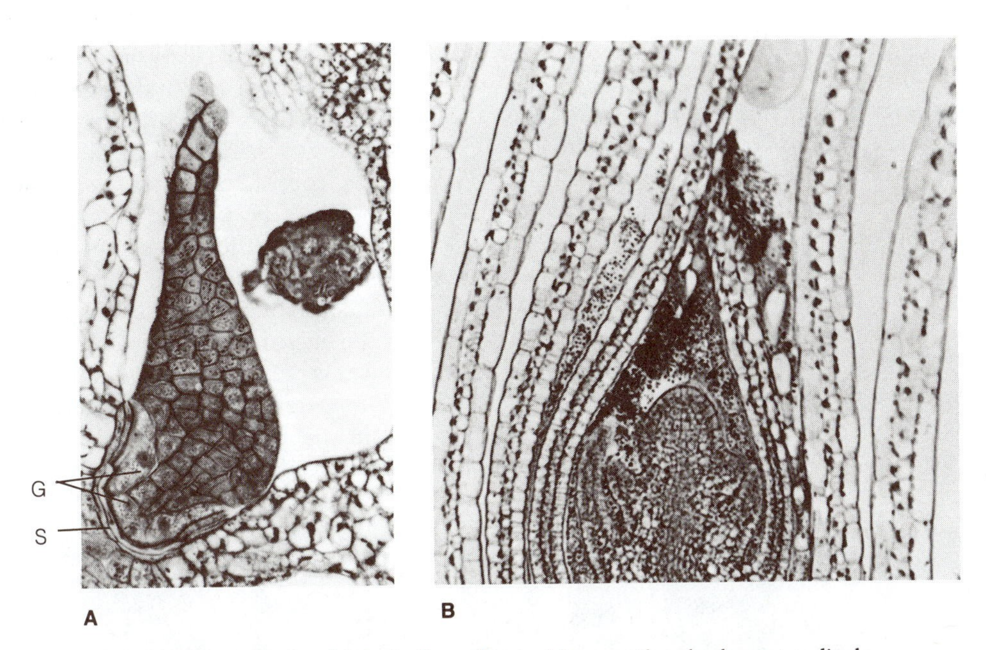

FIGURE 9-25 **A,** ligule of *Selaginella wallacei* with a mucilage body next to ligule. **B,** stem tip of *Selaginella kraussiana* grown in sterile culture with sucrose; mucilage surrounds the shoot apex and young leaves. G, glossopodium; *S*, sheath. [Courtesy of Dr. D. E. Bilderback.]

fore branching is evident externally. A root cap is lacking in *S. martensii* (an anisophyllous species) even when the root is several to many centimeters in length. Although not always evident to the eye, branching of the apical meristem has already occurred upon reaching the soil. Root cap formation then takes place and the dichotomously branching roots become evident (Webster and Steeves, 1967).

Physiological support for the conclusion that the leafless axis ("rhizophore") is in fact a root, rather than a stem, has come from experiments on auxin transport in *S. willdenovii*. In this species there are two angle meristems: one on the ventral side (lower side) of the flabellate shoots and one on the dorsal side. A root is formed from the ventral meristem and a leafy shoot may develop from the dorsal meristem. Experimental results have shown that auxin transport in the root is acropetal (toward the root tip) rather than basipetal (Wochok and Sussex, 1974). This acropetal transport is similar to that in the roots of angiosperms. Basipetal transport is the prevailing condition in shoots. When grown on a basic medium supplemented with 0.5 *M* naphthaleneacetic acid (NAA), 1-millimeter tips of roots less than 20 millimeters long continued to grow as roots (Wochok and Sussex, 1976). Twenty percent of the roots grown only on the basic medium or in the presence of triiodobenzoic acid (TIBA), an antagonist of auxin, developed into leafy shoots.

In summary, recent evidence supports the conclusion that the rhizophore is, in fact, a root. The occasional formation of a leafy shoot from an angle meristem that otherwise normally develops into a root perhaps might occur because in a primitive plant such as *Selaginella* the fate of a meristem may not be so precisely determined as in seed plants (Webster and Steeves, 1967; Mickel and Hellwig, 1969).

However, the seemingly unending debate on the homology of the rhizophore may not be over. An electrophoretic analysis of polypeptides from stem, leaf, root, and rhizophore of *S. kraussiana* revealed that the polypeptides of the rhizophore more closely resemble those of the stem rather than subterranean roots (Jernstedt and Mansfield, 1985).

THE STROBILUS. Unlike *Lycopodium*, all species of *Selaginella* form strobili or cones. Strobili occur terminally on side branches, although in some forms the apical meristem of the cone may continue meristematic activity, producing vegetative leaves. All sporophylls of a strobilus are generally alike (although not differing from vegetative leaves as much as in certain species of *Lycopodium*) and are arranged in four distinct rows. The sporophylls may fit tightly together, or the entire strobilus may be lax or an open type of cone (Figs. 9-20, A, B; 9-26, A).

Because *Selaginella* is heterosporous, sporangia are of two types: microsporangia and megasporangia (Figs. 9-26, B; 9-27). The sporophylls associated with these two types of sporangia are termed, respectively, microsporophylls and megasporophylls. The one mature sporangium associated with each sporophyll is generally axillary in position, although its origin may be from cells of the axis or the base of the sporophyll. There is variation in distribution of sporangia within the strobili of different species. Strobili may consist entirely of microsporangia or of megasporangia. However, the mixed condition is more common. The lower portion of a strobilus may consist of megasporangia and the upper portion of microsporangia, or the two types of sporangia may be mixed indiscriminately. A common arrangement is two vertical rows of each type (Fig. 9-26, B). In certain species (e.g., *Selaginella kraussiana*) only one megasporangium is present at the base of each strobilus.

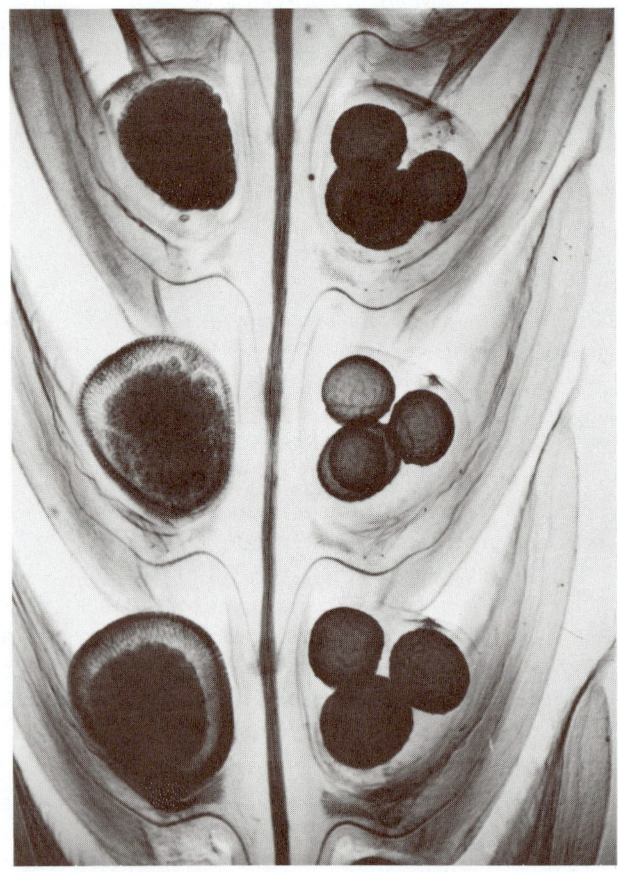

FIGURE 9-26 Strobili of *Selaginella*. **A,** two enlarged, compact strobili comprising four rows of sporophylls; **B,** one-half of a strobilus that has been "cleared" and stained; microsporangia to the left and megasporangia to the right; note the four megaspores in each megasporangium, also the vascular bundle in each sporophyll that passes beneath the sporangium.

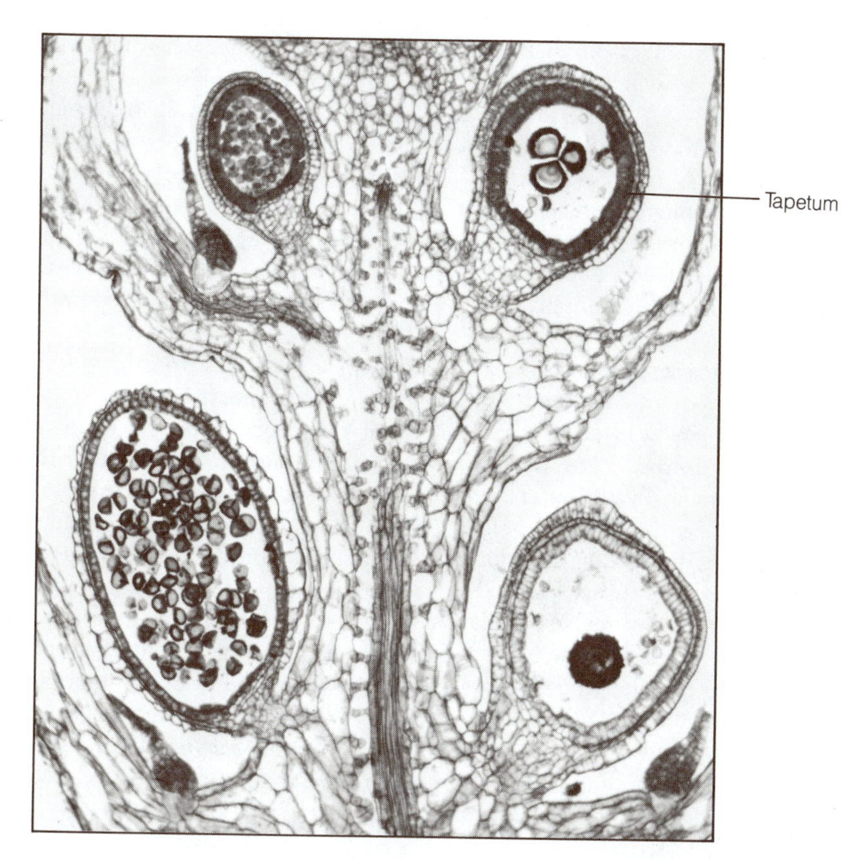

FIGURE 9-27 Portion of longisection of strobilus of *Selaginella* sp. showing late stages in development of sporangia. At upper left, a microsporangium with microsporocytes; note median sectional view of ligule. At lower left, a mature microsporangium with numerous microspores. Three megaspores, surrounded by degenerating sporocytes, are evident in megasporangium at upper right. The megasporangium at lower right is nearly mature, and at this level of section only a single megaspore is seen. (Consult Fig. 9-28 for details of early ontogeny of sporangium in *Selaginella*.)

The type and distribution of sporangia within the strobili of *Selaginella* have never been submitted to a detailed analysis by using mass collections. However, Horner and Arnott (1963) examined the pattern of distribution of megasporangia and microsporangia within strobili of species related taxonomically and geographically. They recognized three major patterns of sporangia distribution in 30 North American species of *Selaginella:* Pattern I—strobili having a basal megasporangiate zone with an upper zone of microsporangia; Pattern II—strobili having two rows of microsporangia and two rows of megasporangia; Pattern III—strobili that are wholly megasporangiate. It was concluded that sporangial arrangement is a useful taxonomic tool in *Selaginella*. In general, a natural group or series of species is all of one type. Furthermore, Horner and Arnott (1963) concluded that Pattern I is more primitive since this arrangement was exhibited by Carboniferous tree lycopods and fossil species of *Selaginellites*. From this type other patterns may have evolved.

Mature microsporangia are generally obovoid or reniform and reddish to bright orange. Megasporangia are larger than microsporangia and frequently are lobed, conforming in outline to the large spores within them. The megasporangia are characterized by lighter colors: whitish-yellow or light orange.

The site of sporangial initiation, of microsporangia or megasporangia, is in superficial cells of the axis, directly above the sporophyll, or in cells near the base of the sporophyll on the adaxial side. Whether two, three, or more superficial initials are involved, periclinal divisions in these initials separate an outer tier of cells — the primary wall cells — and an inner tier — the primary sporogenous cells (Fig. 9-28, A, B). By repeated anticlinal and periclinal divisions of the primary wall cells, a two-layered sporangial wall is formed. The primary sporogenous cells divide periclinally, the outer layer of cells eventually becoming the tapetum (Figs. 9-27; 28,C, D); the inner cells, by dividing in various planes, produce the sporogenous tissue (Fig. 9-28, C). Undoubtedly cells located near the base of the sporogenous tissue, not identified with the original periclinal divisions, serve to complete the continuity of the tapetum (Figs. 9-27; 9-28, D).

A sporangium at this stage consists of an immature sporangial wall of two layers, a short stalk, and a conspicuous tapetal layer enclosing sporocytes which normally round off and separate from each other prior to the meiotic divisions. Up to this stage, microsporangia and megasporangia are indistinguishable, although one study showed that a pair of sporangia (microsporangium and megasporangium) at the same node in the cone exhibit different growth rates up to the premeiotic stage (French, 1972). As development continues, the two types become clearly defined. If a sporangium is to become a microsporangium, a large percentage of the sporocytes undergo meiosis to form tetrads of microspores (Fig. 9-27).

In a potential megasporangium the functional megasporocyte becomes distinct from the nonfunctional ones prior to meiosis. Nonfunctional megasporocytes develop large vacuoles and accumulate starch while the functional sporocyte retains a dense cytoplasm (is rich in RNA) and lacks starch. The functional megasporocyte, encased in a thick coat of callose (Horner and Beltz, 1970), is generally in the central region of the sporocyte mass and undergoes meiosis, forming four megaspores; the nonfunctional sporocytes ultimately degenerate.

What determines why one sporangium will become microsporangiate and another megasporangiate? The earliest histological feature that distinguishes a megasporangium from a microsporangium is the number of potential sporocytes produced. In a megasporangium there are 100 to 150 fewer sporocytes (French, 1972) which means that the sporogenous cells failed to undergo a final *mitotic* division. In some interesting experiments, Brooks (1973) has shown that repeated spraying of *Selaginella wallacei* with Ethephon markedly modifies the determination of sporangia. After Ethephon is absorbed it decomposes, releasing ethylene. After 18 months, 98 percent of the strobili were entirely megasporangiate. Ethylene may inhibit the last mitotic division of sporogenous cells, thus resulting in the production of megasporangia.

Frequently one or more megaspores do not mature, or in certain species more than one megasporocyte is functional, resulting in the production of eight, twelve, and even more megaspores.

The difference in size between microspores and megaspores in *Selaginella* is dramatic (Fig. 9-29, A). Spore size and morphology are useful in the taxonomy of the genus.

CHROMOSOME NUMBERS. Chromosome numbers are known for only about 10 percent of the species of *Selaginella*. The small size of chromosomes (1 micrometer or less long in some species) and difficulties in obtaining suitable cytological preparations may have contributed to the lack of information on the genus. On the basis of a limited amount of information, some investigators recognize at least four basic or primary chromosome numbers ($x = 7, 8, 9, 10$; Löve et al., 1977), with actual chromosome counts of $2n = 14, 16, 18, 20$, with a few being $2n = 50$ to 60 (Jermy et al., 1967). Another investigator concluded that perhaps there are only two basic numbers, $x = 9$ and 10, although 10 may be basic and 9 an aneuploid derivative (Kuriachan, 1963). A majority of the actual counts are indeed $n = 9$ or 10. *Selaginella* remains a genus that has not undergone increases in chromosome numbers and polyploidy apparently has played no major role in evolution.

GAMETOPHYTES. In *Selaginella* development of microspores and megaspores generally begins while

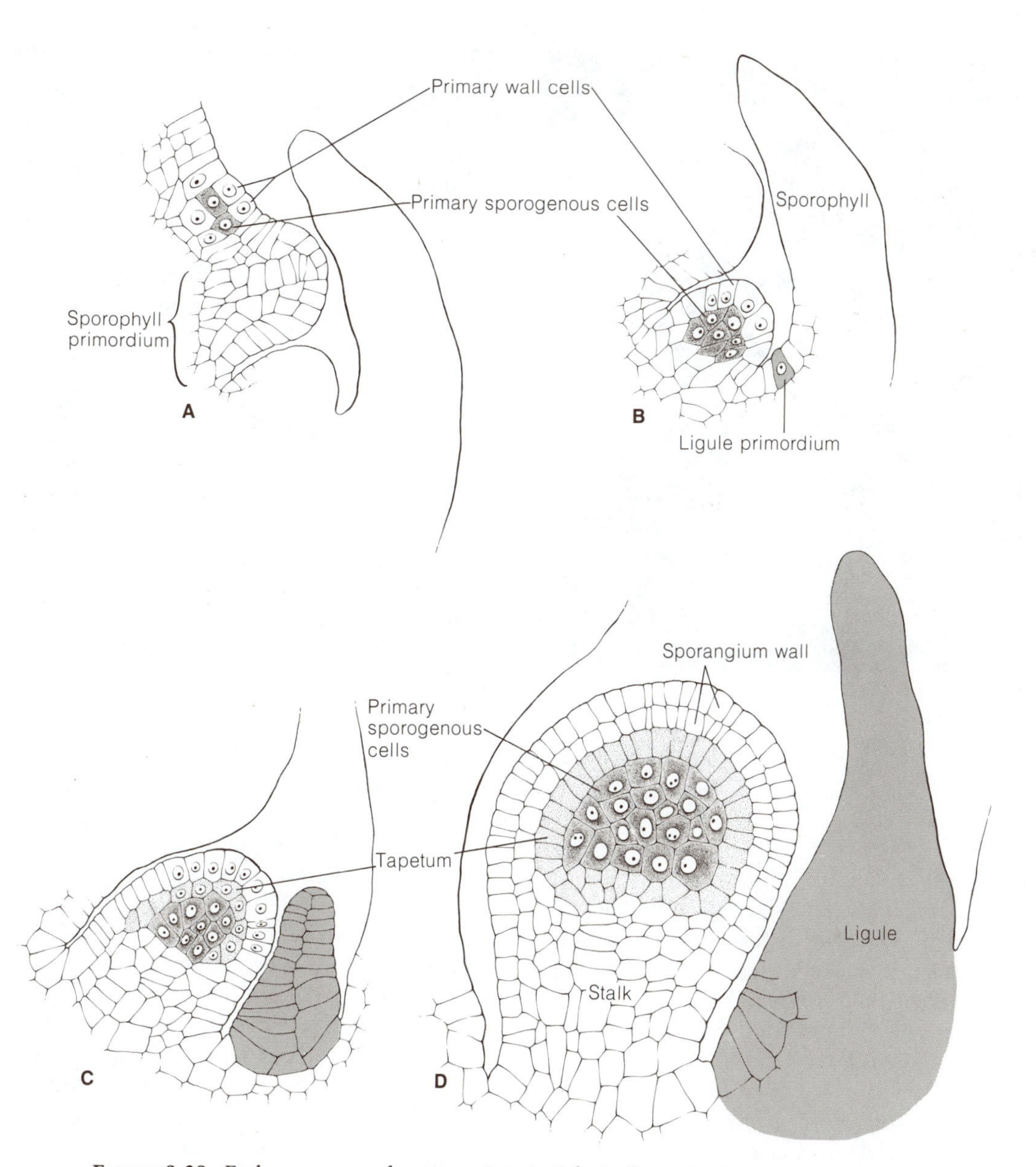

FIGURE 9-28 Early ontogeny of a sporangium in *Selaginella* sp. **A, B,** periclinal divisions in superficial cells separate primary wall and primary sporogenous cells. **C, D,** the tapetum is formed from outer sporogenous cells. Note precocious development of ligule. Whether the sporangium in **D** would have become a microsporangium or a megasporangium is not evident morphologically at this stage of development. However, physiological specialization may have occurred.

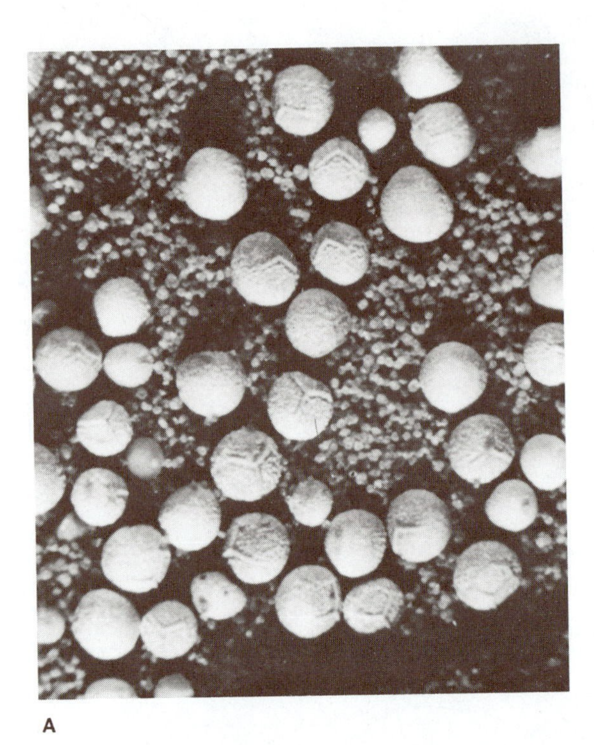

A

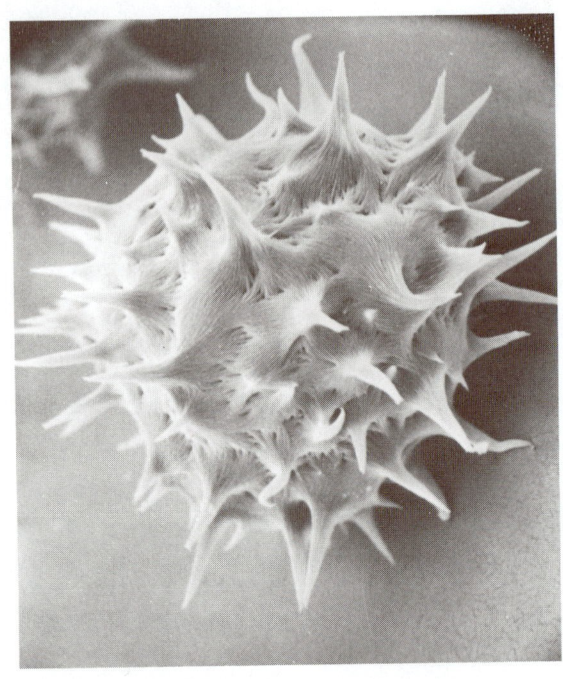

B

FIGURE 9-29 *Selaginella* spores. **A,** spores of *Selaginella* sp. illustrating dramatic difference in size of large megaspores and numerous small microspores (×35). **B,** scanning electron micrograph of microspore of *Selaginella kraussiana* (×875). [**A,** courtesy of Dr. E. G. Cutter; **B,** courtesy of Dr. R. H. Falk.]

they are still within their respective sporangia. In a microsporangium at a late stage of development the radial and inner tangential walls of the outer layer of the sporangium thicken; the inner wall layer becomes stretched and crushed; the tapetum may still be recognizable. Within the microsporangium the microspores (which may still be held together in tetrads) are thick-walled tetrahedral cells with various types of wall ornamentations such as spines or knobs. The first division in a microspore results in the formation of a small vegetative cell ("prothallial cell") and a large potentially meristematic cell (termed the "antheridial initial"; Fig. 9-30). The antheridial initial divides, and by several additional cell divisions a sterile jacket is established that encloses four primary spermatogenous cells, all within the original microspore wall. Dispersal of the spores (with the enclosed partially developed microgametophytes) may occur at this time or earlier by mechanisms characteristic of species.

In some species a microsporangium dehisces but there is no active mechanism for active spore ejection. In a study of 53 species of *Selaginella* Koller and Scheckler (1986) demonstrated that 21 of them, mainly from xeric environments, have the "passive type" of spore dispersal. The microsporangium dehisces longitudinally and the spores simply fall over the sides of the sporangium and sift down through the strobilus or are carried away on wind currents. The spores in certain species are forcibly discharged. In one type, a sporangium dehisces longitudinally into two valves (halves) and the valves continue to snap back and forth several times (Fig. 9-31, A). This is the "spore-ejector" type. In the third type, at the time of dehiscence, the sporangial wall is ruptured and becomes reflexed (bends back on itself), and the entire sporangium with most of the spores is ejected. Sporangia are ejected up to 16 to 20 centimeters away from the strobilus in certain species. Upon landing, a sporangium continues to undergo snapping motions causing the spores to be dispersed (Fig. 9-31, B, C). When movement stops, a sporangium usually remains in an open position, and may consist of two

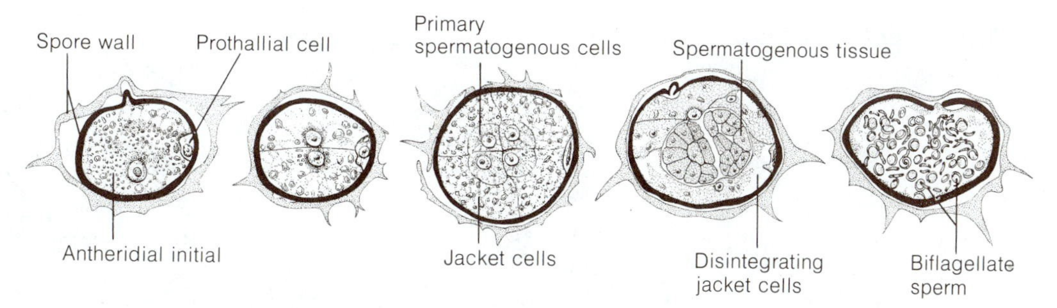

FIGURE 9-30 Development (from left to right) of the microgametophyte in *Selaginella kraussiana*. Early development may occur within the microsporangium prior to sporangial dehiscence. [Redrawn from Slagg, *Amer. Jour. Bot.* 19:106, 1932.]

lobes and a tonguelike portion or variations of this configuration (Fig. 9-31, D). This type of dehiscence is termed the "sporangium-ejector" type. In the two active types of spore dispersal, an annulus of thick-walled cells is involved in dehiscence, functioning much like the annulus of fern sporangia. (See Chapter 13, p. 262 for a discussion of the mechanism.)

The type of spore dispersal may well become useful in the taxonomy and systematics of *Selaginella*. Although only a few species have been studied in detail, it appears that the sporangium-ejector type may prove to be characteristic of the series *Articulatae* in the genus *Selaginella* (Somers, 1982; Koller and Scheckler, 1986).

According to Slagg (1932), the primary spermatogenous cells undergo several divisions, forming 128 or 256 spermatocytes which, on disintegration of the jacket cells and rupture of the spore wall along the triradiate ridge, are liberated as free-swimming biflagellate sperm (Fig. 9-30). Each mature swimming sperm (in *S. kraussiana*) is about 25 micrometers long and very narrow (0.25 micrometers at the anterior end to 0.50 micrometers at the posterior end). One flagellum is attached at the anterior end and the other (posterior flagellum) at about the middle of the sperm. Each flagellum is about 30 micrometers long (Fig. 9-32). In the sperm's ultrastructure a very long mitochondrion (12 micrometers) occupies the anterior half, and the nucleus most of the posterior half. A partial sheath of microtubules extends the entire length of the sperm, functioning perhaps as a support mechanism (Robert, 1974).

Just as early stages in formation of the microgametophyte begin while the microspores are still within the microsporangium, megagametophyte development begins while the megaspores are in the megasporangium. After meiosis in a megasporocyte the resulting megaspores soon develop a thick, layered cell wall. The outer layer, referred to as exospore (or exine) becomes thick, developing spines or ridges in a pattern useful in the identification of species (Figs. 9-33, A; 9-34). Beneath this layer is a thinner layer, the endospore (intine). In some species a third layer (mesospore) develops between the exospore and endospore. The exospore displays a rather unique organization in that there may be packets of aligned granules, each packet forming a polyhedron. In some investigated species there are heavy deposits of silica in the outer parts of the wall (Martens, 1960a, b; Stainier, 1965; Tryon and Lugardon, 1978). Very early a conspicuous vacuole develops within the cytoplasm. In material processed for microscopy the intine may appear to be separate from the exine, however, Pieniązek (1938) reported that the apparent differential growth of exine and intine is an artifact. He states that the spore-wall layers remain in contact throughout development if the spores are not subjected to unusual physical or chemical changes (e.g., use of chemical preservatives). Concomitant with the enlargement of the vacuole the megaspore nucleus divides, followed by additional nuclear divisions without cell-wall formation (referred to as a "free" nuclear phase). This results in a thin layer of multinucleate cytoplasm surrounding a large vacuole. The cytoplasm is rich in lipid globules and protein bodies. The formation of cell walls around nuclei begins initially at the apical end beneath the triradiate ridge (Fig. 9-34). In some species cell wall formation is a continuous process proceeding basi-

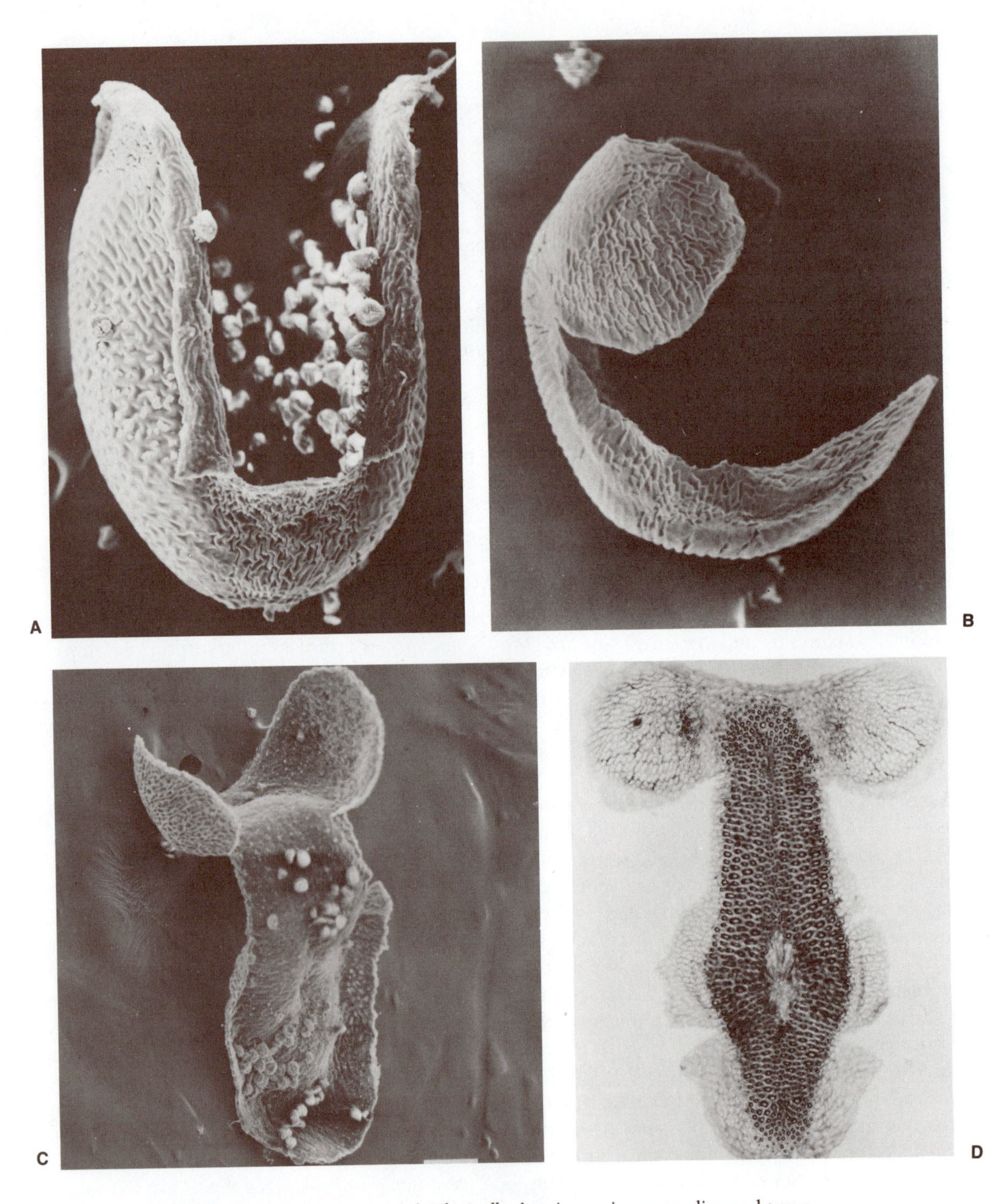

FIGURE 9-31 Microsporangia of *Selaginella* showing active spore dispersal types. **A,** *S. hoffmanii:* spore-ejector type (×160). **B, C,** sporangium-ejector type; **B,** *S. diffusa:* partially reflexed sporangium (×120); **C,** *S. galeottii:* reflexed sporangium; some spores still present (×86). **D,** *S. diffusa:* after dehiscence and completion of snapping motions; whole stained mount; annulus consists of thick-walled cells; region of former attachment to the sporophyll indicated by the lighter staining cells toward the lower end of the annulus. ×86 [Courtesy of Dr. A. L. Koller.]

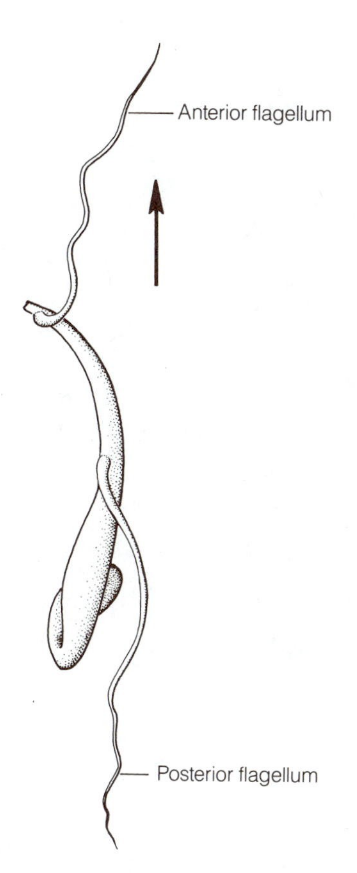

FIGURE 9-32 Biflagellate sperm of *Selaginella kraussiana* (approx. ×1,500). [Based on Robert, *Ann. Sci. Nat. (Bot.),* Sér. 12, 15:65–118.]

petally from the apical region, until the megagametophyte is entirely cellular or a multinucleate storage vesicle may remain at the basal end. In others, e.g., *S. kraussiana,* soon after cellularization begins, a conspicuous arching diaphragm (wall) connected to the intine is formed beneath the apical patch of tissue (Fig. 9-35). Cell walls do form around some nuclei in the storage vesicle below the diaphragm. The latter is perforated so that there is continuity between the two regions (Robert, 1971a, b).

Apparently there is considerable variation in the genus as to when sporangial dehiscence occurs and the stage of megagametophyte development. Dehiscence may occur at any time before or during the cellular stage. The final stages of megagametophyte development and fertilization take place while the megaspore with its enclosed megagametophyte rests on moist soil or humus. The megagametophyte increases in size resulting in the splitting open of the megaspore wall along the triradiate ridge. The exposed gametophyte may develop tufts of rhizoids which probably play a role in absorption of water, and possibly in anchorage (Fig. 9-38, A).

Archegonia make their appearance in the apical region (Fig. 9-35). Each archegonium develops from a single superficial cell and at maturity consists of eight neck cells, arranged in four rows of two cells each. There is one neck-canal cell, a ventral canal cell, and the egg. Only the terminal neck cells extend beyond the surface of the gametophytic tissue. The microgametophytes complete their development while situated on the exposed megagametophyte or in close proximity to it. After the biflagellate sperm are liberated they swim to the archegonia in a thin film of dew or rain water.

EXPERIMENTAL STUDY OF THE MEGAGAMETOPHYTE. Wetmore and Morel (1951b) cultured the female gametophytes of two species of *Selaginella.* On the culture medium the gametophytes remain alive for six months, and if vitamins are added large masses of undifferentiated tissue are produced, which are covered with rhizoids and archegonia.

THE EMBRYO. After fertilization, the sporophyte generation is established. The first division of the zygote is transverse, separating a suspensor cell (that cell toward the archegonial neck) and the embryonic cell (labeled "apex" in Fig. 9-36, A) that will form the remainder of the embryo (see Chapter 6).

The embryo is endoscopic. The suspensor may remain undivided or form several cells. A shoot apical cell is established as a result of longitudinal and oblique divisions within the original embryonic cell. At approximately this developmental stage the embryo proper undergoes a 90-degree turn. The first pair of leaves are formed laterally. A foot is produced on the lower side, and the primary root is formed between the suspensor and foot (Fig. 9-36). Variation may exist with respect to the origin and position of foot and root (Chapter 6, Fig. 6-2).

By continued growth of shoot and root, the young sporophyte emerges from the gametophytic tissue (Fig. 9-37). That portion of stem below the first leaves elongates rapidly (Fig. 9-38, B), and in many species the first branching of the shoot takes place immediately above the first pair of leaves (Fig.

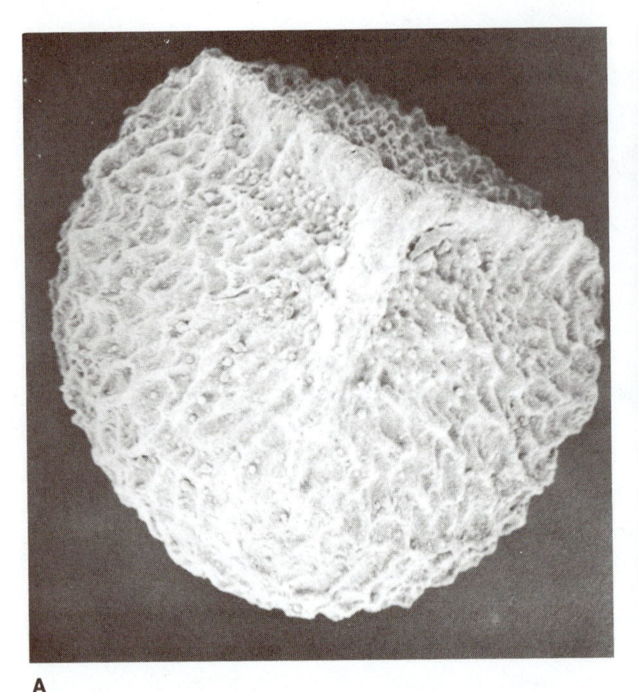

A

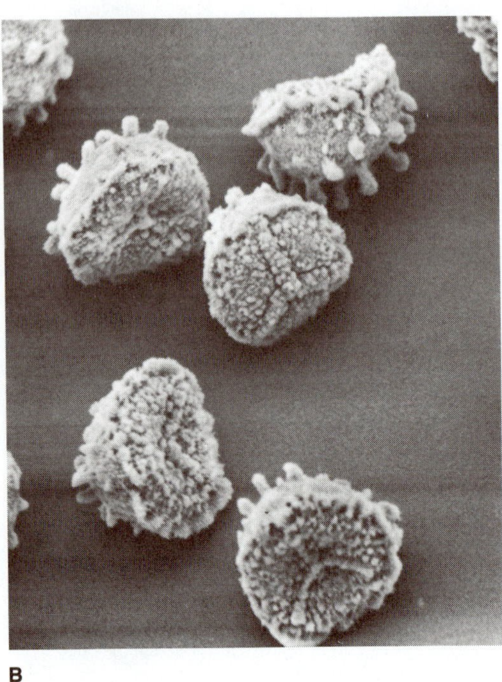

B

FIGURE 9-33 Scanning electron micrographs of spores of *Selaginella flabellata*. **A,** megaspore (×200). **B,** microspores (×650). [Courtesy of Dr. R. H. Falk.]

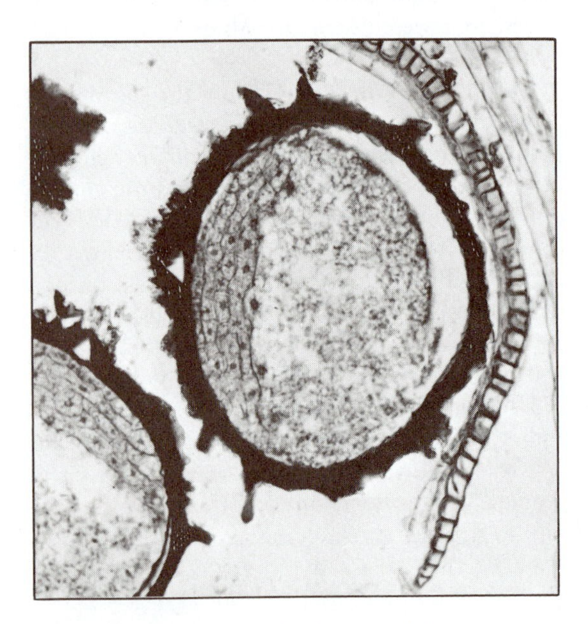

FIGURE 9-34 Section of developing cellular megagametophyte within thick megaspore wall of *Selaginella* sp. A large storage vesicle is present beneath the cellular tissue. The triradiate ridge is indicated by a triangular space (white) at the upper end of the spore (to the left). The developing endosporic gametophyte is still enclosed within the megasporangium.

9-38, C). The primary root grows downward and enters the soil. The foot remains in close contact with the nutritive tissue of the gametophyte contained within the megaspore wall. The complete separation of the sporophyte from the gametophyte may not occur until the sporophyte has undergone considerable growth (Fig. 9-38, C).

Most of our knowledge of embryo ontogeny is based on very old studies (e.g., Bruchmann, 1912). The descriptions may well be accurate, but we need confirmation with modern techniques and equipment.

Webster (1979) has developed a controlled artificial crossing technique for *Selaginella* for inheritance studies. *Selaginella kraussiana,* a common greenhouse plant, is green. *Selaginella kraussiana* var. *aurea* is a pigment-deficient cultivar with yellow-green foliage. When selfed, it produces green, yellow-green, and white (lethal) young sporophytes in a 1 : 2 : 1 ratio. These results together with results of additional crosses have shown that the *aurea* character is controlled by a single nuclear gene with two partially dominant alleles (Webster and Tanno, 1980).

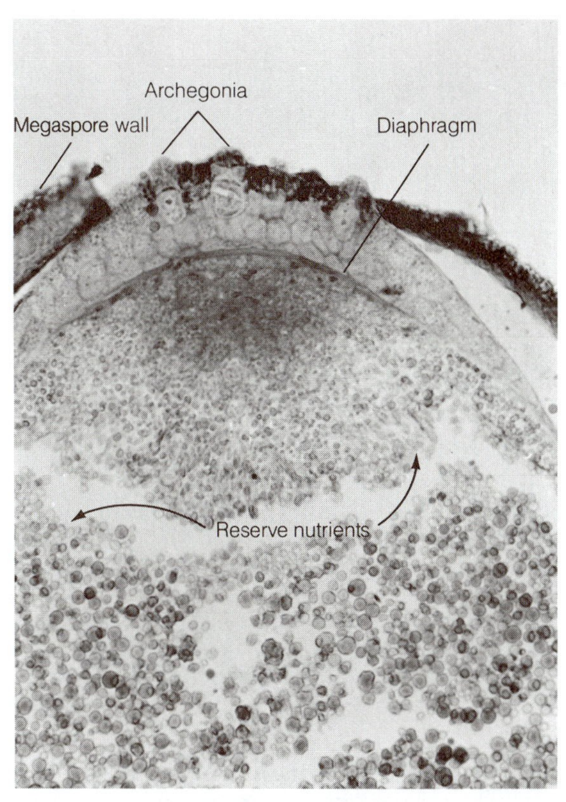

FIGURE 9-35 Section of megagametophyte of *Selaginella kraussiana* surrounded by megaspore wall. Note diaphragm that separates upper (apical) portion of gametophyte from the lower portion.

Selaginella martensii f. *albovariegata* is a variegated sport which produces white tissue in an irregular fashion. Reciprocal crosses between the green *S. martensii* and f. *albovariegata* revealed maternal inheritance of variegation. Additional crosses, involving progeny of selected reciprocal crosses, also indicated a lack of direct nuclear influence on variegation. Character expression and inheritance can be accounted for on the basis of random sorting of normal and defective cytoplasmic factors at cell division (Tanno and Webster, 1982a, b).

Carboniferous Relatives of *Selaginella*

Some herbaceous plants from the Carboniferous Period resembling *Selaginella* have been known for many years. The plants were small and were clearly heterosporous. Some of them were described as *Selaginellites*, others as *Selaginella*. One paleobotanical case history is of interest. Most species thought to be herbaceous and *Selaginella*-like, were described from compressions in a poor state of preservation. But in 1954 a ligulate lycopod was described from the Carboniferous, which had small stems and spirally arranged leaves. This plant was given the generic name *Paurodendron*. The stems were in an excellent state of preservation (Fry, 1954). In 1966 another plant of the same genus was found that had a basal root-producing portion, termed a rhizomorph (Fig. 9-39), that had secondary growth (Phillips and Leisman, 1966). Complete knowledge of the plant was obtained in 1969 when bisporangiate strobili (microsporangia and megasporangia) were found attached to stems. The entire plant was then described as *Selaginella fraiponti* because it resembled to a remarkable degree the modern-day species *Selaginella selaginoides*, which has a centralized basal root system (Schlanker and Leisman, 1969). This is of great morphological interest because it indicates that *Selaginella* existed in the Carboniferous, and provides information on possible ancestral forms of *Selaginella*. However, based upon a study of the meristem of the rhizomorph of *Paurodendron*, Rothwell and Erwin (1985) have aligned the genus with the Isoetales rather than with the Selaginellales.

Lepidodendrales

In addition to the low-growing lycopods of the Carboniferous Period, there were also large, tall trees with lycopsid features. The first evidence of these forms are found in rocks of the Upper Devonian, although they were not as tall as those in the Upper Carboniferous. These trees must have formed a conspicuous element of the Carboniferous "coal" swamps (Fig. 9-40) and are some of the best known fossils. They were arborescent, heterosporous, and ligulate, and had distinctive persistent leaf bases on the stems.

LEPIDODENDRON. Well-preserved material, especially of the trunks, is common in coal beds in Great Britain and central United States. Many of these dendroid plants attained a height of 30 to 35 meters or more and were at least 1 to 2 meters in diameter at the base. The trunks were unbranched for 10 to

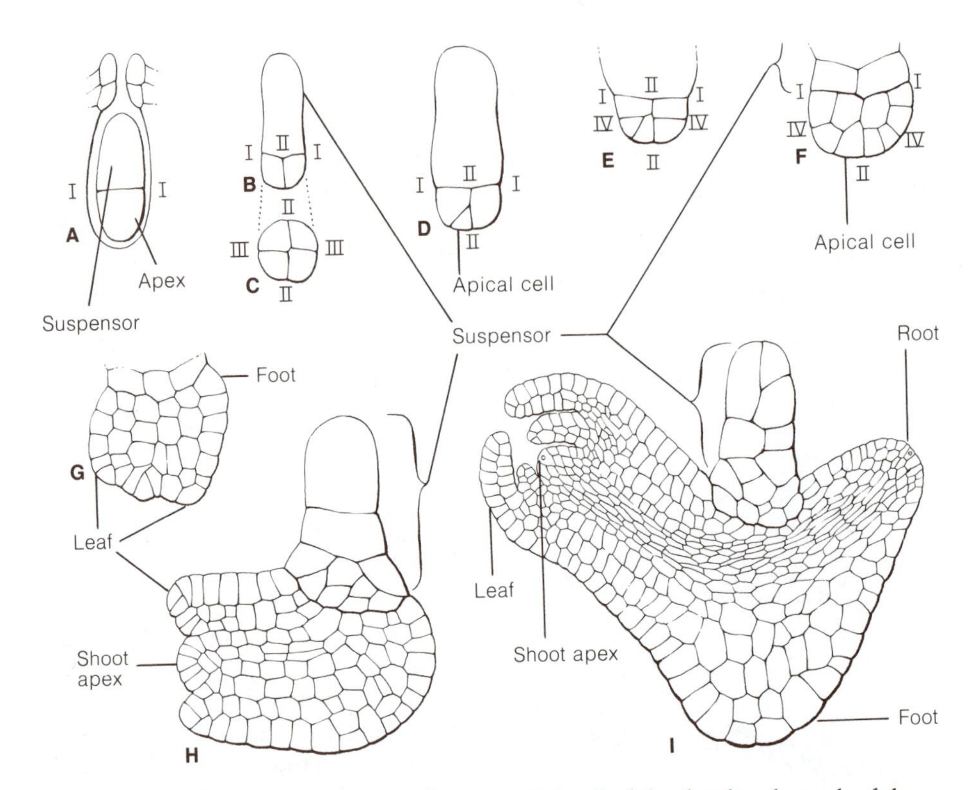

FIGURE 9-36 Embryogeny in *Selaginella martensii.* In all of the sketches the neck of the archegonium is directed toward the top of the page. The walls resulting from early cleavages are indicated as I-I, II-II, etc. An apical polar view of **B** is shown in **C.** (For details consult text.) [**B-I** redrawn from Bruchmann, *Flora* 99:12, 1909.]

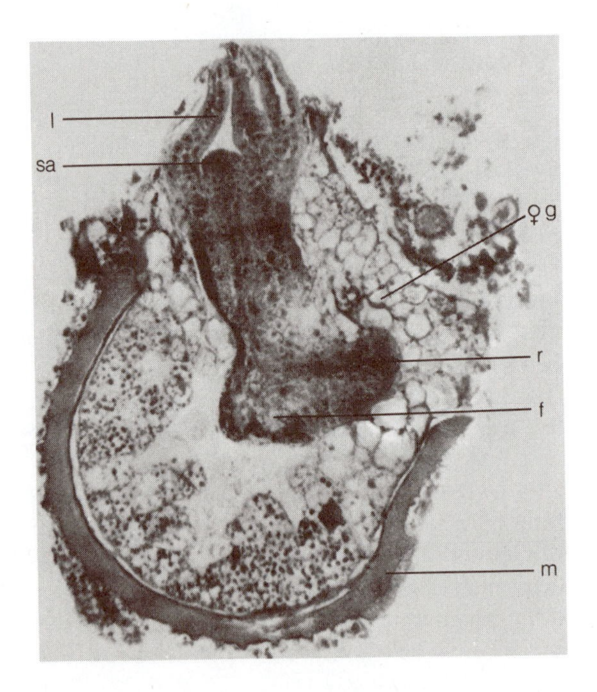

FIGURE 9-37 *Selaginella* sp. Section of megaspore, enclosed female gametophyte, and young sporophyte. f, foot; 1, leaf; ♀g, female gametophyte; m, megaspore wall; r, first root; sa, shoot apex. [From Bold et al., *Morphology of Plants and Fungi,* Harper and Row, New York 1980.]

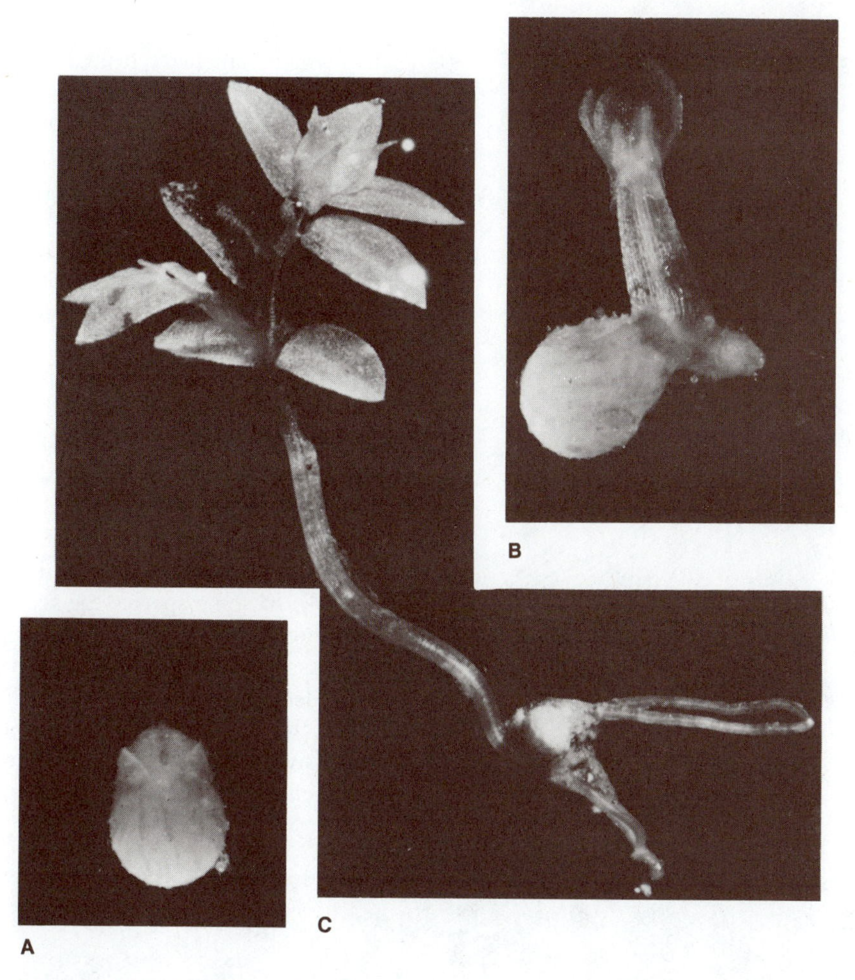

FIGURE 9-38 Megagametophyte and young sporophyte of *Selaginella kraussiana*. **A**, gametophyte tissue protruding through cracked spore wall; **B**, young sporophyte attached to gametophyte, showing root (to the right), stem, and first pair of leaves (oriented in the plane of the page); **C**, older sporophyte (megaspore wall with enclosed gametophyte is still visible at the juncture of stem and root).

20 meters or more and then terminated in an isotomously or anisotomously branched, umbrellalike shoot system in which the branches were clothed with spirally arranged, linear, or awl-shaped leaves (Fig. 9-41, A). In some species the branched portion consisted of two systems of branches, each of which branched dichotomously. It is believed that the helically arranged leaves on the trunks were grasslike, reaching lengths of several to 78 centimeters. Upon abscission of a leaf blade, a distinctive base (leaf cushion) remained attached to the stem (Fig. 9-42, A). The shorter leaves on the upper crown were persistent. The form of the leaf cushions and the details on the face of the scar left by the abscised leaf blades are useful in separating species in the genus. An examination of a leaf scar reveals a vascular bundle scar flanked by two small scars, each of which is termed a *parichnos* (from the Greek word, meaning "footprint") (Fig. 9-42, A). In the living condition each parichnos consisted of a strand of loosely organized parenchyma that extended from the stem cortex into the leaf blade, functioning presumably as aerating tissue. On some lepidodendrid leaf cushions there are two additional parichnos channels below the leaf scar (Fig. 9-42, B). The base of the trunk was divided into

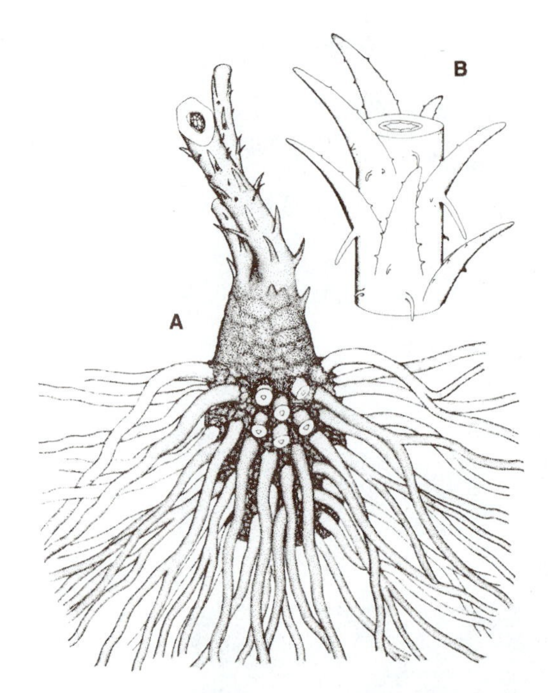

FIGURE 9-39 *Paurodendron (Selaginella) fraiponti.* **A,** reconstruction of lower portion of plant. **B,** enlarged portion of aerial stem. [From Phillips and Leisman, *Amer. Jour. Bot.* 53:1086–1100, 1966.]

four large rootlike structures, termed rhizophores, each of which branched repeatedly in dichotomous fashion (Figs. 9-41, A, B; 9-43).

Even though the trunk of one of these trees was very large, it had very little vascular tissue. Whether the primary vascular cylinder was protostelic or siphonostelic, the primary xylem was exarch in development. A vascular cambium produced a narrow cylinder of uniform secondary xylem, consisting of scalariform-pitted tracheids and uniseriate rays. Apparently the vascular cambium did not produce secondary phloem; only primary phloem has been identified (Eggert and Kanemoto, 1977). The entire vascular cylinder, even in large stems, was never more than several centimeters in diameter.

The bulk of the stem consisted of primary and secondary cortex (periderm). The primary cortex consisted of inner, middle, and outer tissue zones (Fig. 9-44). In some species the secondary cortex was produced centripetally from a ring of meristematic tissue (phellogen) which had its origin through dedifferentiation of outer primary cortical cells (Fig. 9-44). In others there was the production of successive layers of secondary cortex from meristematic cells derived from more deeply lying pri-

FIGURE 9-40 Reconstructions of plants in a Carboniferous swamp; trunks of lepidodendrids at left foreground with seed ferns behind them; *Equisetum*-like plant (a calamite) at right foreground. [Courtesy of Field Museum of Natural History, Chicago.]

FIGURE 9-41 Suggested reconstructions. **A,** *Lepidodendron* sp.; **B,** *Sigillaria elegans.* Note strobili and the large rhizophores with attached rootlets at base of trunks. *Form* or *organ* genera exist for all basic parts of the plants. (Consult text for pertinent information.) [Modified from *Handbuch der Paläobotanik* by M. Hirmer. R. Oldenbourg, Munich. 1927.]

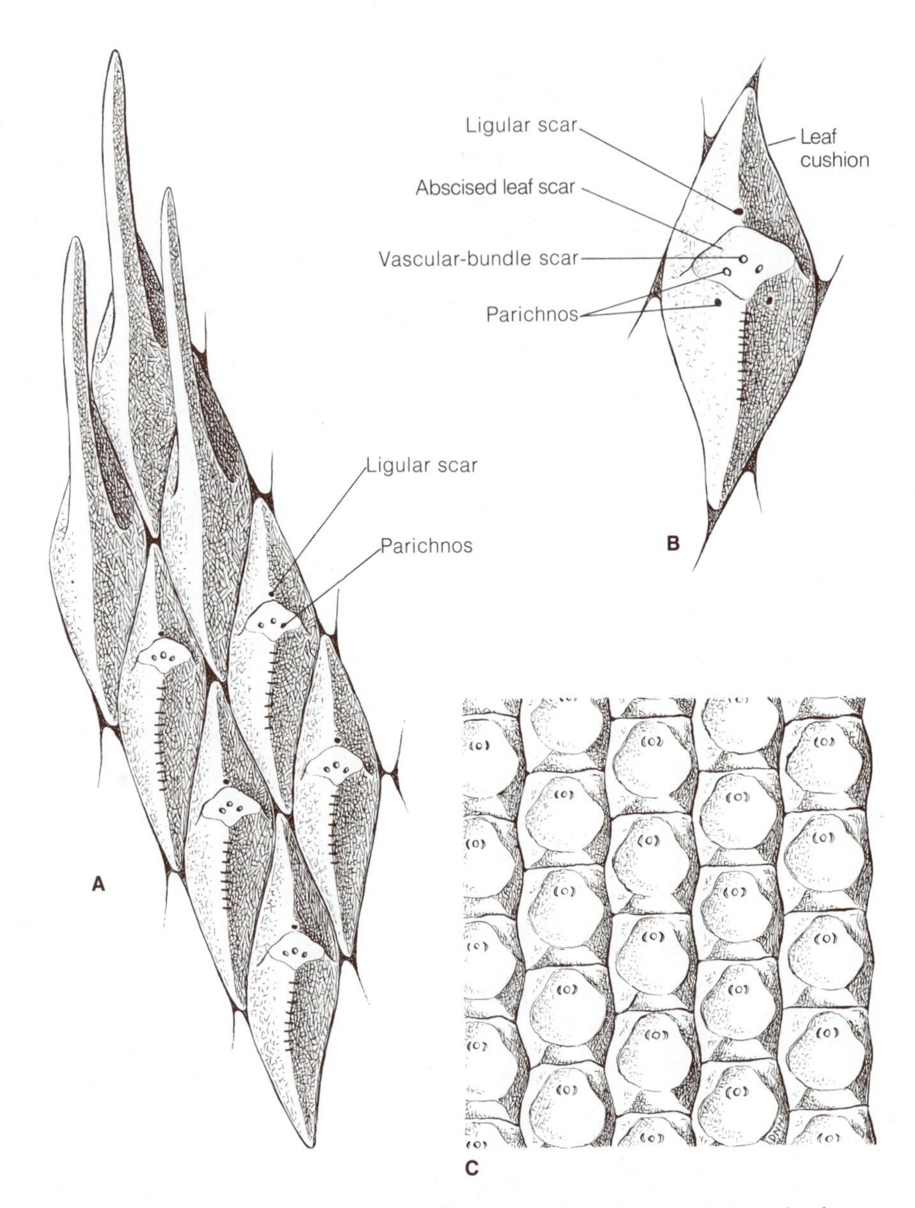

FIGURE 9-42 **A,** diagrammatic representation, portion of the surface of a branch of an arborescent lycopod (*Lepidodendron* sp.) showing three attached leaves and the scars left by the abscission of five others. **B,** one leaf scar of a lepidodendrid showing two sets of parichnos scars. **C,** surface of the stem of *Sigillaria* sp.

mary cortical cells. The development of secondary cortex brought about the separation of the outer primary cortical cells from that portion nearer the vascular cylinder. Also, the massive production of secondary cortical tissue resulted in the separation of leaf cushions and decortication of the lower half or more of a trunk (Fig. 9-41, A, B).

The small amount of xylem produced in relation to the large size of the trunk is remarkable. The secondary cortex did consist partly of fiber cells in some species, but this would not have resulted in the strength provided by secondary xylem of modern trees. The disparity between tree size and strengthening tissue may have literally led to the

FIGURE 9-43 Tree stump of a lepidodendrid in "Fossil Grove," Victoria Park, Glasgow, Scotland. The basal dichotomously branched lobes, to which rootlets were attached, are designated *Stigmaria* (an organ genus). The fossil is a cast of the original tree. Stumps, which measure 15 to 40 inches at their widest diameter, were exposed by carefully removing the hard rock that encased them. [Photograph courtesy of Dr. E. G. Cutter.]

downfall of the tree-lycopods and their disappearance from the world's flora toward the end of the Carboniferous.

Other genera coexisted with *Lepidodendron* in the Upper Carboniferous. They likewise were tall trees and were somewhat similar to *Lepidodendron,* but enough is known about them to assign them to different genera.

Lepidophloios was similar in general growth habit but differed from *Lepidodendron* primarily in the shape and organization of leaf cushions.

Sigillaria was unbranched with one or two dichotomies at the distal end (Fig. 9-41, B). The grass-like leaves of the distal branches were generally longer (up to 1 meter) than in the other two genera. The leaves were helically arranged, but the leaf cushions appear to lie in vertical rows. The leaf-blade scars were hexagonal to oval (Fig. 9-42, C). Strobili were borne terminally on short lateral branches at the base of the leafy crowns or intermingled with the leaves (Fig. 9-41, B). Where known, the vascular cylinder was siphonostelic with exarch primary xylem, surrounded by secondary xylem (Fig. 9-45, A, B).

ONTOGENY. Preservation of tree-lycopods is so excellent that the ontogeny of a plant can be deduced from fossil remains. As can be seen in the reconstruction illustrated in Fig. 9-46, the main trunk tapered slightly from the base to the initial dichotomy. Branches resulting from the first dichotomy were reduced in size. With each successive dichotomy there was a progressive decrease in size of branches which led finally to determinate growth, often in the production of strobili. *Lepidodendron* had a definitive life span. Eggert (1961) was able to correlate the successive decrease in size of branches with a decrease in complexity of the vascular cylinder. The xylem was protostelic at the base, siphonostelic with a relatively wide pith near the level of initial branching, and ultimately protostelic in the tiny terminal branches (Fig. 9-46). Associated with this progression, there was a decrease

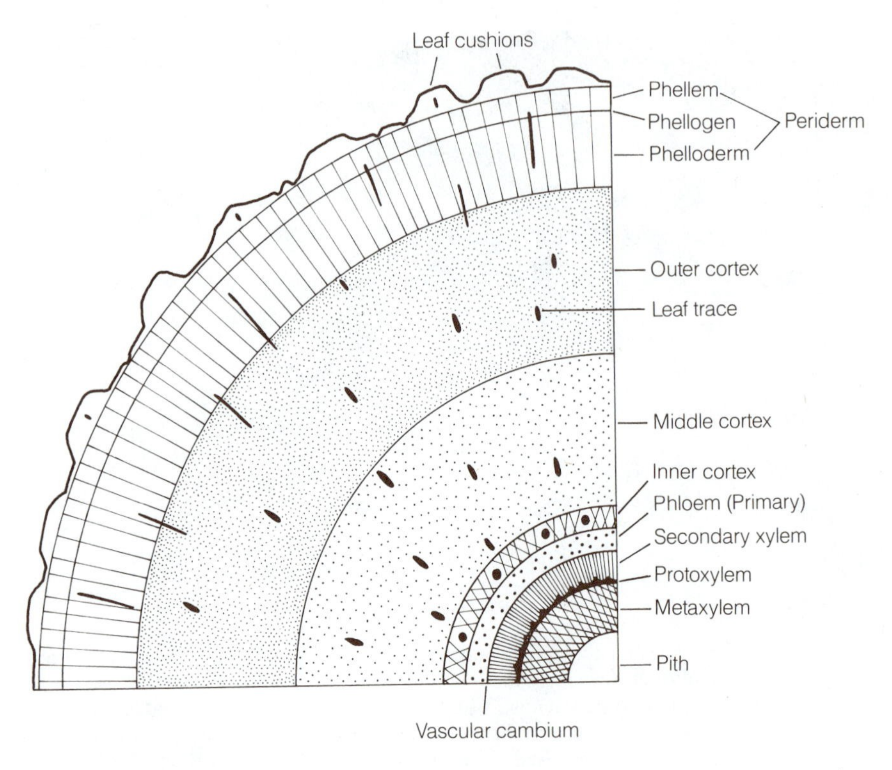

FIGURE 9-44 Diagrammatic representation of a transverse section of one type of *Lepidodendron* stem.

in size of the secondary xylem cylinder, as well as in the amount of cortical tissue produced. (See also Andrews and Murdy, 1958, and Delevoryas, 1964, for additional discussions.)

ORGAN GENERA. From the account in the previous section the reader might conclude that the discovery of intact remains of these arborescent genera is indeed remarkable. However, the story is not that simple. To uncover a fossil plant with all parts intact is the dream of paleobotanists; unfortunately, at best only unconnected portions or separate organs of large plants are generally discovered. This is understandable if we consider the amount of breakage possible during transport by water, for example, to sites of final preservation. The ontogeny of a plant, as discussed previously, has a bearing on the problem because stem structure, for example, may vary at different levels. Also, some structures (strobili, spores, etc.) may have been formed either at regular intervals or only near the end of the life of a plant. This dilemma, appreciated by early paleobotanists, led to the establishment of *organ genera* or *form genera* which has resulted in the accumulation of numerous, but necessary, genera for organs such as leaves, stems, and strobili. For example, the genus *Lepidodendron* originally referred to portions of the stem of lycopods with the distinctive leaf-cushion, leaf-scar pattern. Fossil leaves with one unbranched vein are placed in the organ genus *Lepidophylloides*. *Sigillariophyllum* is the name applied to leaves that are vascularized occasionally by two veins.

The massive underground rootlike organ is known as *Stigmaria*, another organ genus (Fig. 9-43). These dichotomizing axes (rhizophores) bore "stigmarian" rootlets in a helical pattern. Paolillo (1982) has suggested that the rootlets were derived from cell derivatives of an apical meristem rather than from a secondary meristem as in the corm of *Isoetes* (see p. 159). Usually the rootlets are found detached; only the scars are seen on the rhizophores. The internal structure of these rootlets is interesting because it resembles to a remarkable

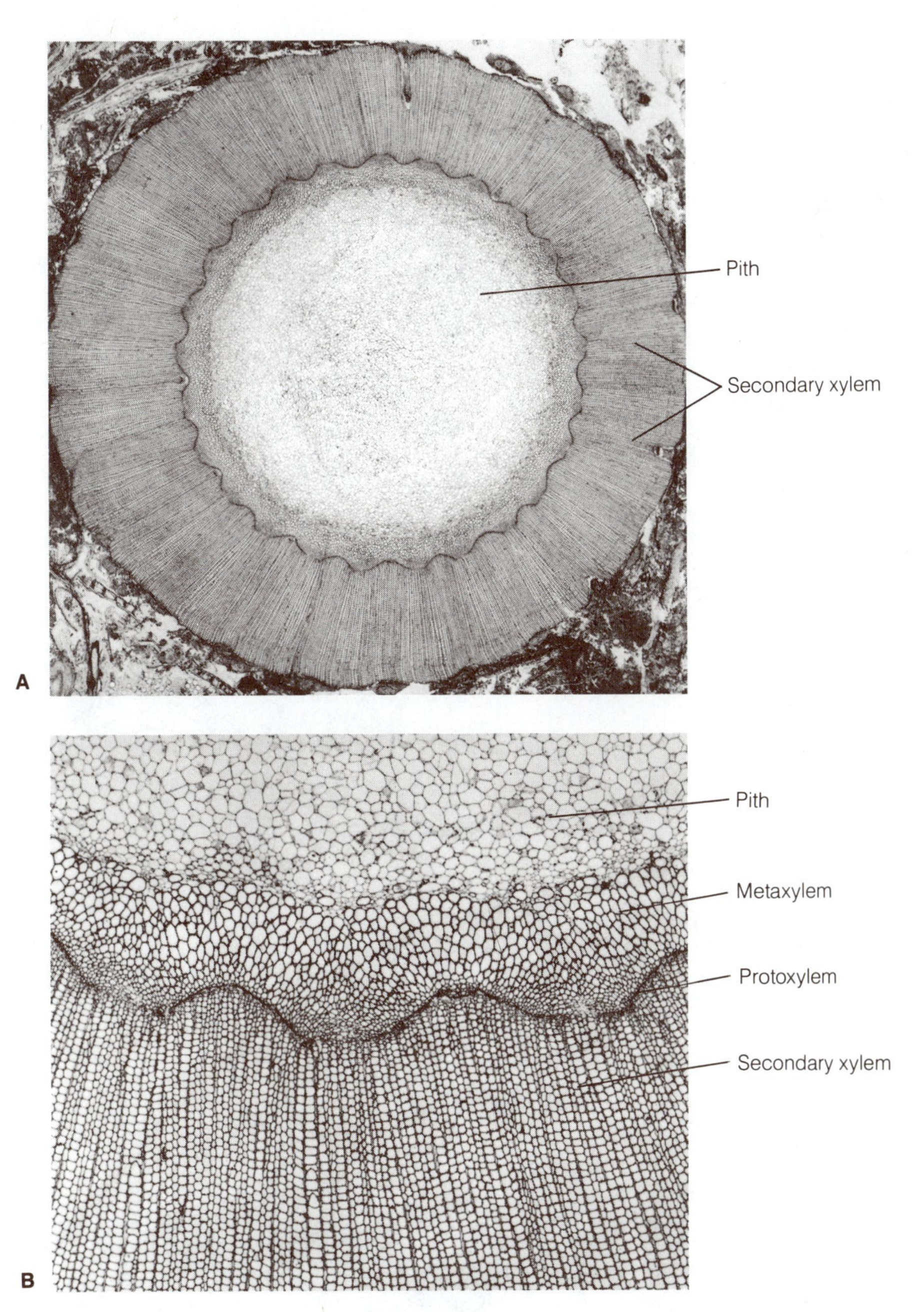

FIGURE 9-45 **A,** transverse section of a siphonostele of *Sigillaria approximata*. **B,** details of a section from **A.** [Courtesy of Dr. T. Delevoryas.]

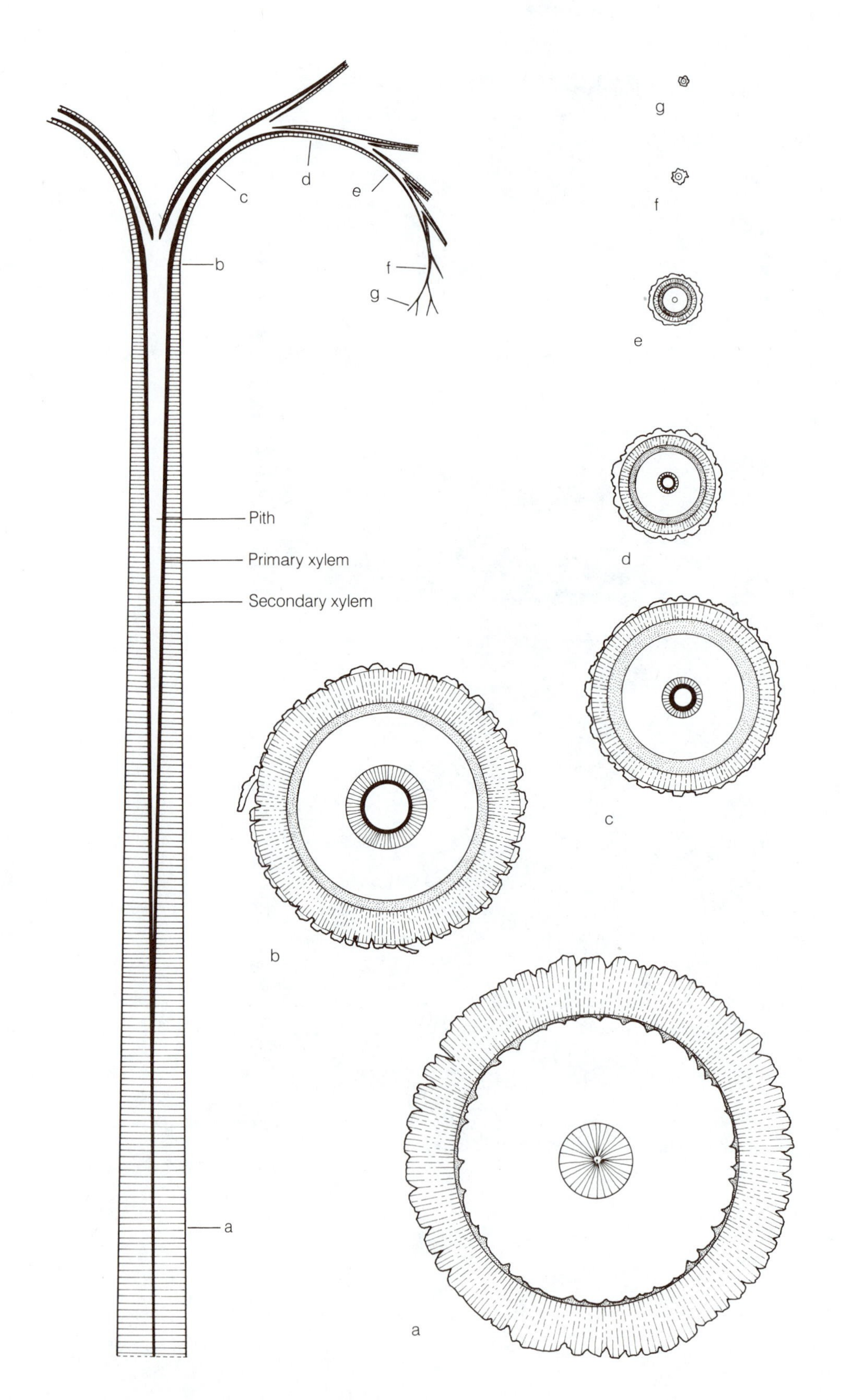

g
f
e
d
c
b
a
Pith
Primary xylem
Secondary xylem

degree that of the roots of *Isoetes;* in both, the root has a monarch vascular strand supported by a flange of the cortex in a large central air cavity (Fig. 9-53).

By continuing comparative studies, and with due consideration given to ontogeny, a better picture is emerging as to what constitute natural or biological species in *Lepidodendron.* For example, DiMichele (1981) has been able to document the occurrence of five distinct species that grew in the coal swamps of the Upper Carboniferous in Europe and America.

The strobili of *Lepidodendron* were either bisporangiate (having microsporangia and megasporangia in the same cone) or monosporangiate. The organ genus *Lepidostrobus* has been applied to both types. The cones were large (8 to 20 centimeters or even 35 centimeters long) with helically arranged overlapping ligulate sporophylls (Fig. 9-47). In bisporangiate cones the microsporangia, containing large numbers of microspores, occurred in the upper portion of the strobilus, and the megasporangia, containing fewer megaspores, were present in the lower portion. One sporangium was present on the adaxial side of each sporophyll, similar to *Lycopodium* and *Selaginella.* The spores were shed from the sporangia and the endosporic gametophytes developed much like those of extant *Selaginella* and *Isoetes.* Well preserved megagametophytes with archegonia, surrounded by the megaspore wall, have been described (Brack-Hanes, 1978) which resemble those of *Isoetes.* Microgametophytes have also been found which resemble those of *Selaginella* in a three-celled developmental stage. These preparations also reveal structures that have been interpreted as chromosomes (Brack-Hanes and Vaughn, 1978).

Some members of the Lepidodendrales progressed to an evolutionary level of producing seed-like structures. *Lepidocarpon* is an organ genus represented by monosporangiate strobili in which typically only one functional megaspore, with its enclosed megagametophyte, was retained within the sporangium (Fig. 9-48, A, B). Conclusive evidence of embryos in *Lepidocarpon* was produced in 1975 by Phillips et al., and further documented by Phillips in 1979. The embryos were vascularized, and post-embryonic stages were obtained showing the establishment of aerial stem and basal rhizophore axes. It was also shown that some of the megagametophytes described by earlier workers were in reality nonvascularized embryos that did not germinate. Another feature of considerable evolutionary interest is the enclosure of the sporangium and enclosed megagametophyte by two lateral extensions (lateral laminae) of the sporophyll, except for a slitlike opening along the top. The entire structure was very much like a seed, the lateral laminae functioning as an integument. The sporophylls were shed from the strobilus, and it is thought that each unit could float on water; hence, there was a mechanism for aquatic dispersal. Fertilization could occur during the flotation period or when the sporophyll unit came to rest on the floor of the muddy swamp. Fertilization probably was effected much like that in *Selaginella.*

Isoetales – Isoetaceae: *Isoetes*

The genus *Isoetes* is a most interesting and enigmatic vascular plant. The plant body of all species is relatively small with a greatly shortened axis, and has tufts of leaves and roots (Fig. 9-49).

The first popular names recorded for *Isoetes* were "quillwort" and "Merllyn's Grass." The former name is frequently applied to the genus in America; in Europe it is still known as Merlin's Grass. Economically the genus is relatively unimportant today, but there are past records of the plants being eaten occasionally in Europe (Pfeiffer, 1922). Great quantities of starch and oil are present in the plant body. Birds, pigs, muskrats, and ducks may eat the fleshy plant body, and cattle often graze on the leaves.

Isoetes includes about 100 described species, although some taxonomists recognize only 50 to 70 as being valid species. The genus is worldwide in distribution, occurring, for example, in America, Europe, Asia, India, and Australia. In America it occurs from coastal Alaska, southward through the United States, Mexico, and most of South America.

FIGURE 9-46 **A,** diagram of a longitudinal section of the vascular system of an arborescent lycopod showing the distribution of primary and secondary xylem from base to ultimate branches. Levels *a–g* indicate level of transverse sections *a–g.* Note that the axis was protostelic at the base, becoming siphonostelic and then once again protostelic in the small terminal branches. [Redrawn from Eggert, *Palaeontographica* 108B:43, 1961.]

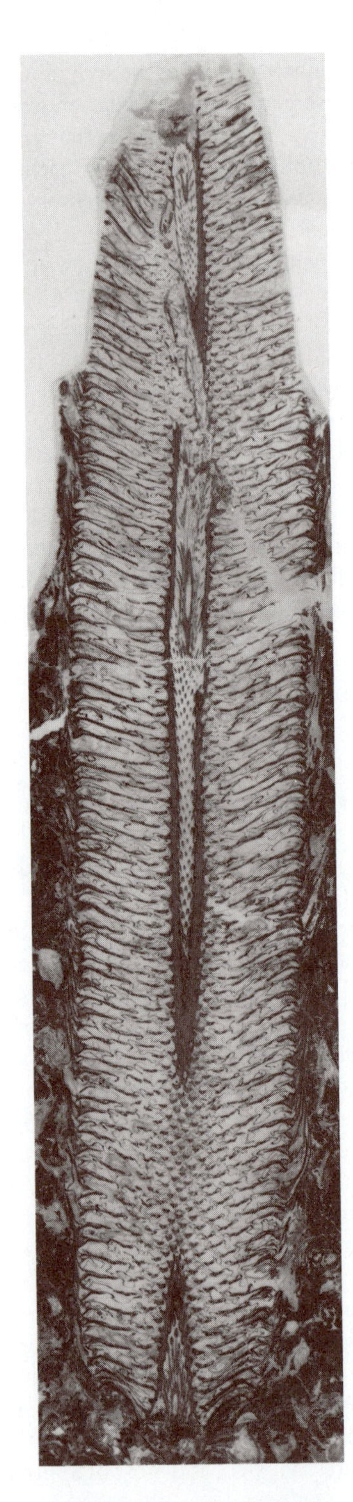

FIGURE 9-47 *Lepidostrobus,* longitudinal section of a large cone of a lepidodendrid. The sporophylls were helically arranged as can be seen in the lower part of the cone where the sporophylls have been cut transversely. The actual specimen is 35 centimeters long.

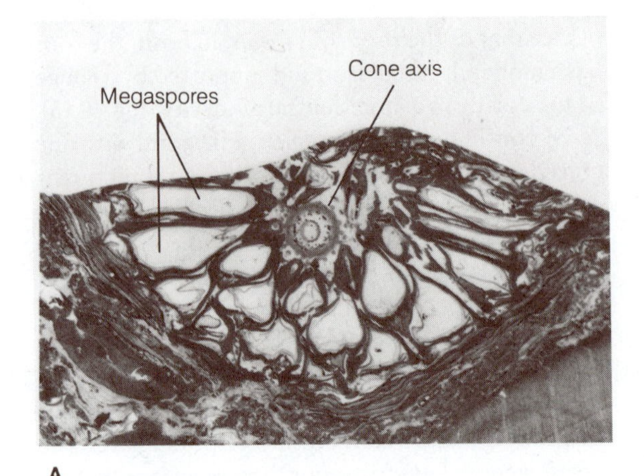

A

Lateral laminae of sporophyll
Megasporangium wall
♀
Vascular bundle of sporophyll

B

FIGURE 9-48 *Lepidocarpon.* **A,** transverse section of a portion of a fossilized cone showing the cone axis, sporophylls, and large megaspores (one per megasporangium) (×1). **B,** transverse section of one megasporophyll showing structural features and the developing megagametophyte (♀) (×4).

It grows in lakes, ponds, rivers, swampy areas, and in ephemeral pools (e.g., vernal pools) that become dry during part of the year (Tryon and Tryon, 1982). It also grows in lakes that are covered with ice during winter months.

ORGANOGRAPHY. The axis is a short, erect structure commonly referred to as a corm. The basal portion of a young plant is usually two-lobed, and may remain so during the life of the plant, but the corm may become three, or rarely, four-lobed in older plants of certain species by the development of additional furrows (Karrfalt and Eggert, 1977a). Along the sides of the grooves or furrows are numerous roots. The upper part of the axis is covered with a dense cluster of leaves which have broad overlapping bases (Figs. 9-49; 9-50, A-C). The shoot apex is completely hidden in a depression by the tightly overlapping leaves (Fig. 9-51, A). The sides of mature plants become very rough with layers of sloughing tissues. Each increment includes leaf bases and severed roots of previous growing seasons.

ANATOMY OF THE CORM. Any morphological interpretation of the corm of this peculiar plant must necessarily be based partially on internal structure. The corm has been described variously as an erect rhizome; as a stock; as a stem; as a stem combined with a stigmarian type of rhizophore; and as an upper leaf-bearing part, the stem, and a lower root-bearing part, the rhizomorph. (See Paolillo, 1963, for a review of the subject.) If a mature plant is cut longitudinally in the plane of the basal groove or furrow (considering that the plant is two-lobed), the shoot apex is seen at the bottom of a depression with surrounding leaf bases and leaf traces (Fig. 9-52, A). In a well-established plant the shoot apex is a low dome or a small cone, about twice as broad as high; a group of cells occupy the distal region of the cone (Paolillo, 1963; Michaux, 1966). A single cell may dominate the apical group, but generally there is no conspicuous apical cell. However, Karrfalt (1977) has described apices of older plants in which the arrangement of cells appears to be related developmentally to a tetrahedral apical cell with three initiating faces. The xylem core is in the outline of an anchor—cylindrical in the upper part with the lower part extended horizontally with upturned arms. Roots and root primordia are evident below the xylem core.

Secondary growth is a characteristic feature of *Isoetes* and the specialized cambium comprises two parts: (1) a *lateral meristem* that gives rise to vascular tissue (prismatic layer), centripetally, and to secondary cortical tissues, centrifugally (Figs. 9-51, B; 9-52, B); (2) a *basal meristem,* continuous with the lateral meristem, which adds to the xylem core and produces basally the surrounding ground tissue, in which root primordia become organized (Figs. 9-51, A, B; 9-52, A, B).

FIGURE 9-49 Entire plant of *Isoetes* sp. showing crown of tightly packed leaves (sporophylls), short "corm" (dark) and roots.

Both lobes of the corm are evident in a median longitudinal section cut at right angles to the furrow (Figs. 9-51, B; 9-52, B).

A transverse section through the leaf-bearing portion of a corm reveals the xylem core surrounded by the cylindrical prismatic layer and lateral meristem (Fig. 9-52, C). At a lower level the lateral meristem is evident and the basal meristem appears at two locations, reflecting the curved contour of the xylem core (see Fig. 9-52, C and D).

Of the many debatable features of *Isoetes,* interpretation of the secondary vascular tissue has probably caused the most discussion. The cells of this tissue, which are derived from a definite storied cambium (lateral meristem), are extremely short, being not much taller than they are wide. The form of these cells, combined with the difficulty of interpretation, led to the general acceptance of the noncommittal term "prismatic layer" for this part of the axis. Nevertheless, this tissue has been inter-

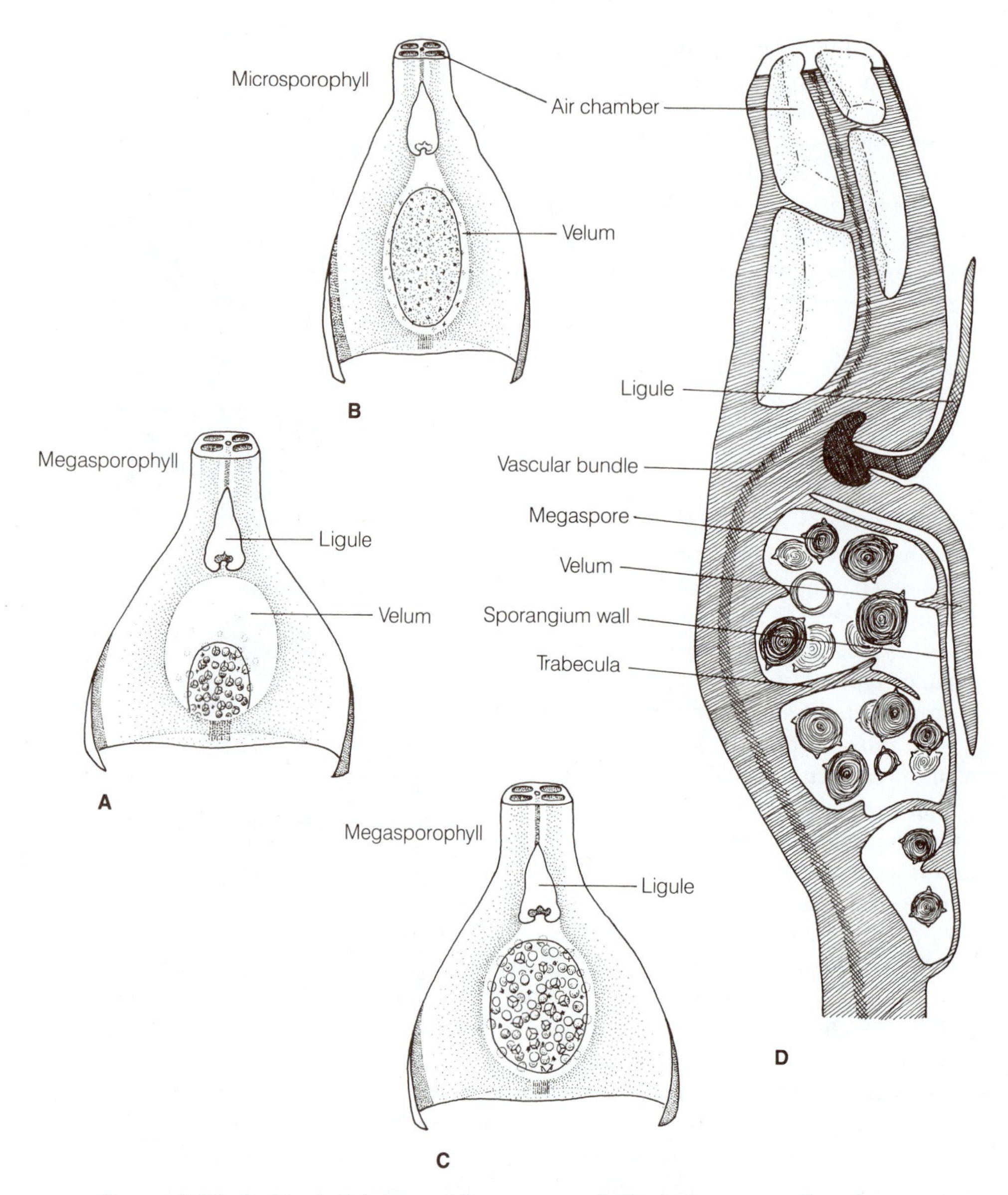

FIGURE 9-50 **A–C**, adaxial views of *Isoetes* sporophylls. **A**, *Isoetes* sp.; the velum covers a greater portion of the megasporangium. **B, C**, *Isoetes howellii*. **B**, microsporophyll showing microsporangium (spores are shown as black dots within the sporangium, the ends of trabeculae as larger black asteroids); the velum covers only a small portion of the sporangium. **C**, megasporophyll, showing megasporangium and megaspores. **D**, longisection of a megasporophyll comparable to that in **A** (markings on spore walls are entirely schematic).

preted as secondary xylem, as secondary phloem, as a secondary tissue containing both tracheids and sieve elements, and as a tissue composed of (1) occasional tracheids, (2) considerable unmodified parenchyma, and (3) specialized parenchyma cells which are concerned with conduction. (See Paolillo, 1963, for a review of older literature.)

Paolillo (1963) provided a detailed study of the secondary vascular tissue (prismatic layer) of *Isoetes howellii*. In young plants of this species the deriva-

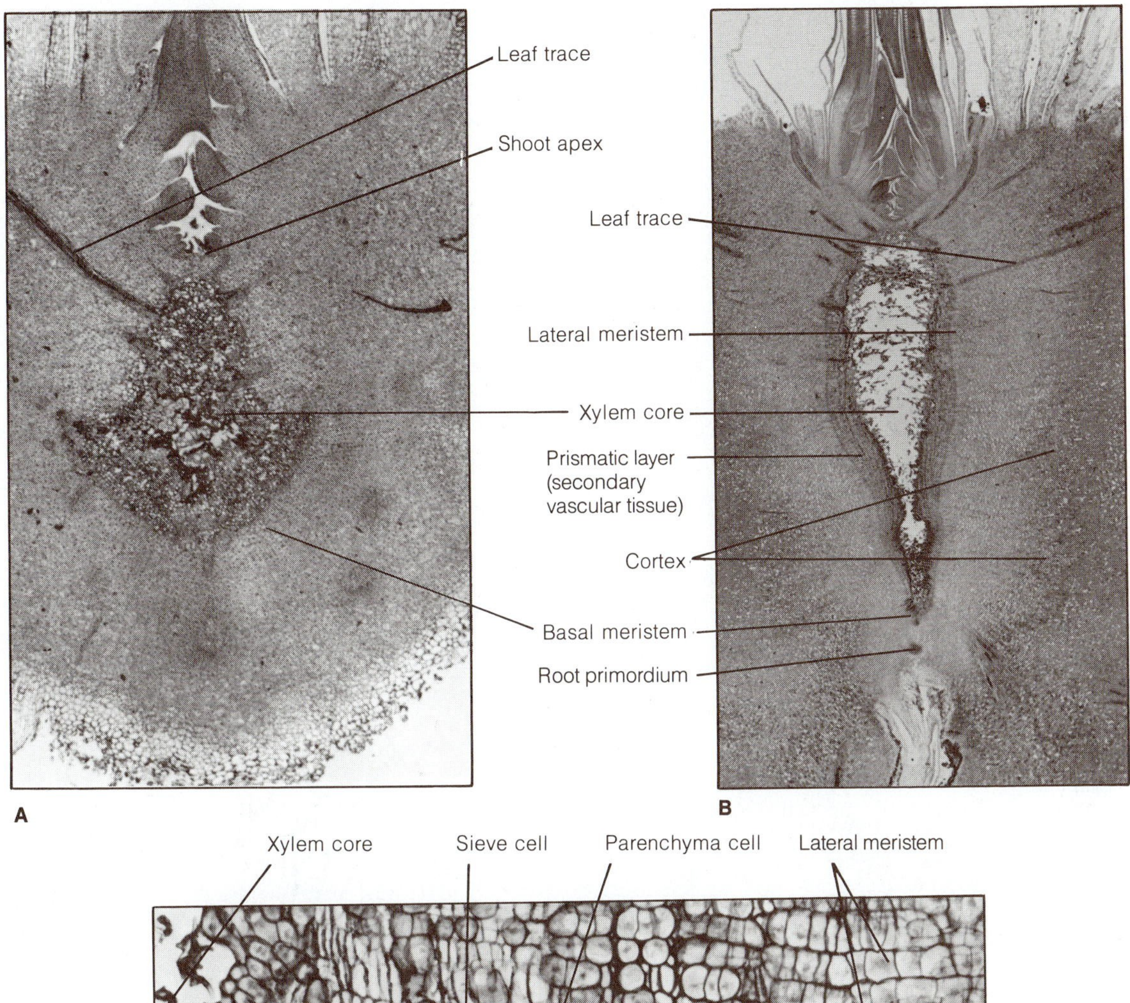

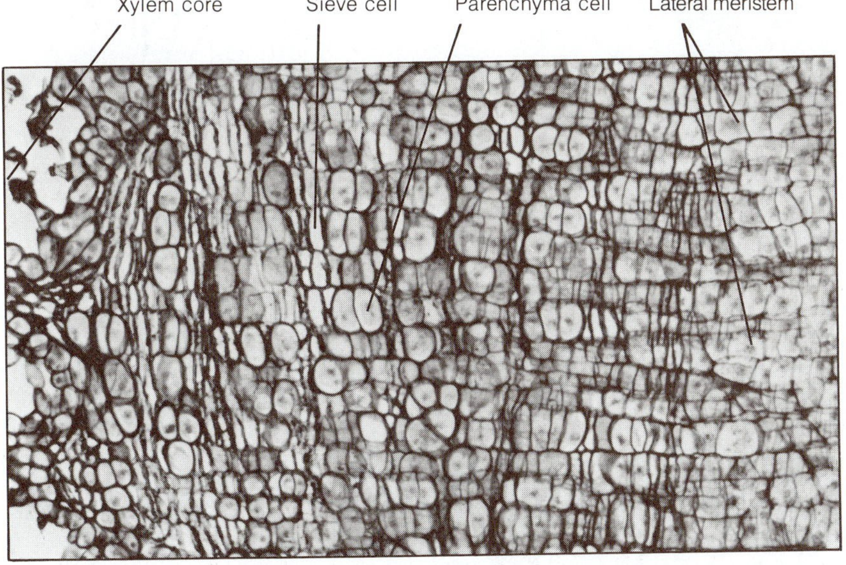

FIGURE 9-51 Anatomy of the corm in *Isoetes howellii*. **A,** longisection, in the plane of the basal groove; **B,** median longisection, at right angles to the basal groove (several root traces can be seen at lower right); **C,** a portion of the secondary vascular tissue and adjacent tissues.

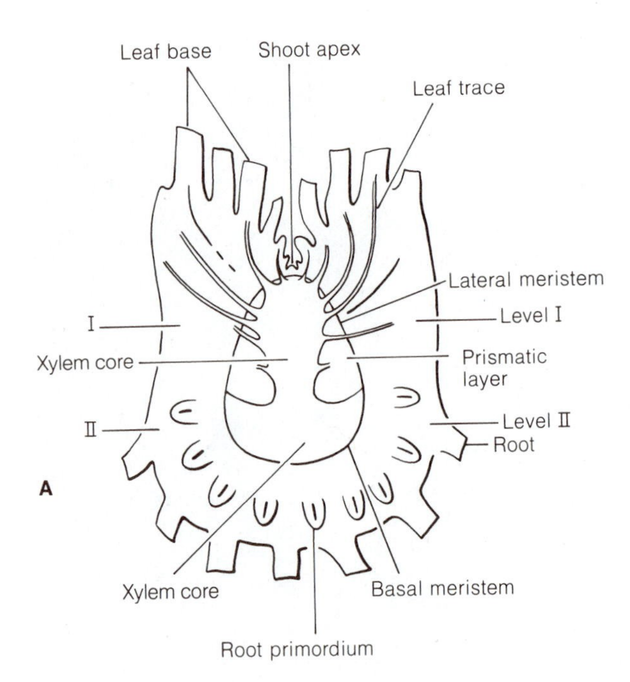

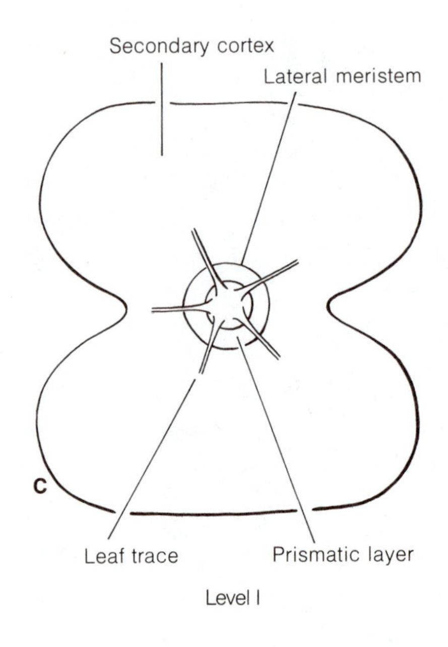

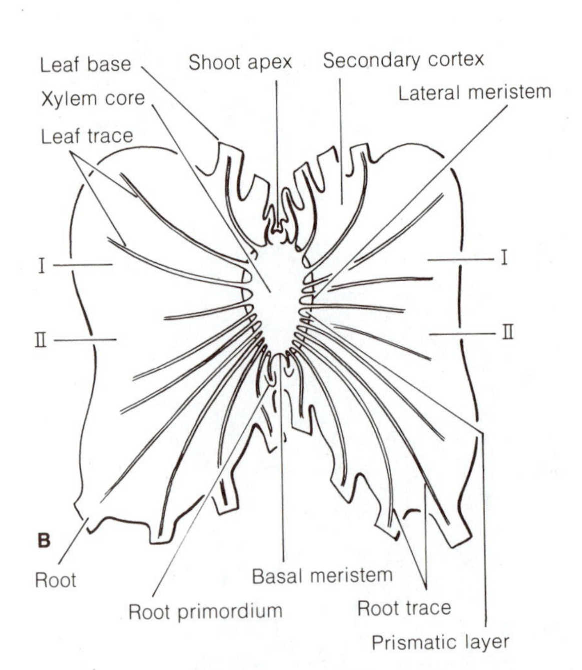

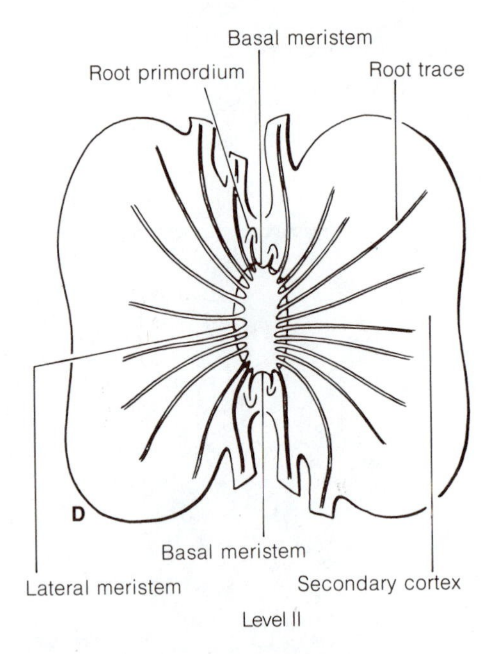

FIGURE 9-52 *Isoetes howellii.* **A, B,** longitudinal sections of two-lobed plants in the plane of the furrow or groove (**A**) and perpendicular to the furrow (**B**); **C,** transverse section at high level (I-I in **A** and **B**); **D,** transverse section at low level (II-II in **A** and **B**). [Redrawn from Paolillo, *Illinois Biol. Monogr.* 31, 1963.]

tives of the lateral meristem differentiate as sieve elements. In older (larger) plants the inner derivatives of the lateral meristem differentiate as layers of sieve elements, alternating with layers of parenchyma (Fig. 9-51, C). Some or many of the cells in the parenchyma layers may differentiate as tracheids. A comparable type of anatomy occurs in *I. taiwanensis* (Chiang and Chen, 1986).

Recognition of sieve elements is made possible by a combination of features: sieve pores or areas, deposition of callose, and generally thicker walls than those of parenchyma cells (Esau et al., 1953). A degenerated nucleus persists in a functioning sieve element, comparable to other lower vascular plants. In older plants some or many of the cells in the layers of parenchyma may differentiate as tracheids. An ultrastructural study of *I. muricata* detected very little callose lining the pores, although with the approach of dormancy the pores become occluded with copious amounts of callose. The dormancy callose begins to disappear with the resumption of growth, and some sieve elements near the lateral meristem are reactivated before the lateral meristem becomes functional (Kruatrachue and Evert, 1977). This type of reactivation is similar to that of certain woody dicotyledons: *Vitis* (grapevine) and *Tilia* (linden).

The lateral meristem also produces cells centrifugally which retain their meristematic activity, but ultimately mature into parenchyma cells—constituting a broad secondary cortex (Figs. 9-51, B; 9-52, B–D). The secondary cortex is added to regularly and the outer part sloughs off from the "shoulders" of the corm.

The basal meristem, as noted earlier, contributes to the xylem core and also forms ground tissue distally in which root primordia become organized. The roots eventually penetrate the ground tissue. Roots are produced on either side of the furrow—forming series (rows roughly parallel with the furrow) and orthostichies (rows running at right angles to the furrow). Series are more easily seen near the furrow. (See Paolillo, 1963, Karrfalt and Eggert, 1978, for discussions.)

Thus far, organography, basic organization, and functioning of lateral and basal meristems (collectively constituting a cambium) have been described. What constitutes primary growth (and tissues) of the corm? Leaves originate in the surface layer near the base of the small apical cone; the subapical cells produced internally enlarge and divide, producing radiating files. It is within this tissue that axial procambium as well as leaf traces differentiate.

The primary plant body of the shoot, then, consists of a core of primary xylem, primary phloem, primary cortex, leaf traces, and leaves. The lateral meristem arises outside of the primary phloem and gives rise to secondary vascular tissue (prismatic layer) from its inner face and secondary cortex from its outer face. The origin of the lateral meristem outside the primary phloem is quite different from that of the vascular cambium in most woody seed plants, that is, between the primary xylem and phloem.

Paolillo (1963) regards the basal meristem as part of the cambium because the two meristems (lateral and basal) originate together and, during ontogeny, portions of the lateral meristem may be added to the basal meristem by conversion of initials from one function to another. Earlier, it will be remembered, the term xylem core was used to describe the central core of tissue. This term purposely was used because the origin of the central xylary tissue is from two quite different sources—primary xylem derived from the procambium (a primary meristem) of the upper part of the plant axis and the internally produced cells of the basal meristem (part of the cambium). However, Karrfalt and Eggert (1977b, 1978) have presented evidence for their belief that the basal meristem also should be considered a primary meristem.

In summary, one might ask: How is the mature plant form in *Isoetes* related to meristematic activity? Longitudinal growth, accomplished through the functioning of the apical meristem of the shoot apex and of the basal meristem, is very slow, as evidenced by the extremely short axis and the crowding of appendages. The lateral meristem, which surrounds the upper portion of the vascular cylinder and encloses the sides of the basal portion of the xylem core, gives rise yearly to increments of secondary vascular tissue and a large amount of secondary cortical tissue. This accounts for the fact that older plants appear to be more broad than tall.

ROOTS. Roots branch dichotomously (except in one species) after they emerge from the corm ground tissue. For *Isoetes howellii* the root apical meristem is reported to consist of a layer of initials that gives rise to the cells of the outer cortex, epidermis, and root cap, and a group of initials common to the inner cortex and procambium (Paolillo, 1963).

A transverse section of a mature functioning root reveals a simple type of root of unusual interest. It consists of a cylindrical cortex which surrounds a large air cavity, and a vascular cylinder which is supported in a flange of the cortex in the cavity (Fig. 9-53, A). The primary xylem and phloem are collateral in arrangement, the phloem being oriented toward the cavity. An endodermis, with the usual casparian strips, is present around the primary vascular tissues. The air cavity is formed by a breakdown of cortical cells throughout the length of that portion of the root which has emerged from the corm. Histologically the root of *Isoetes* resembles very closely an appendage on the rhizophore of *Stigmaria* (Fig. 9-53, B). The similarity is conclusive enough to support the belief of some botanists that a phylogenetic relationship between *Isoetes, Stylites,* and some members of the Lepidodendrales is certain.

THE LEAF. Actually each foliar appendage is a potential sporophyll, either a microsporophyll or a megasporophyll. (The terms leaf and sporophyll will be used interchangeably.) Each sporophyll has a thickened and expanded base, with a tapering upper portion which is awl-shaped and pointed. In young plants leaf arrangement is distichous (leaves arranged in two rows), but soon becomes helical. With *Isoetes tegetiformans* (Rury, 1978) the distichous arrangement persists (Fig. 9-58). Leaves may be a few centimeters long or 50 centimeters or more *(Isoetes japonica)*. The lower parts of the leaves may be buried in the soil and lack chlorophyll, and are commonly a glistening white. Some species have black leaf bases. In many species most of the leaves die and decay at the termination of the growing season. In permanently aquatic species leaves may remain on the plant for some time. The upper portion of a leaf is traversed longitudinally by four large air chambers (Fig. 9-50, B, D) that may be partitioned into compartments by transverse tissue diaphragms — the possession of large air cavities is a feature common to many water plants. Running throughout the length of the leaf (a microphyll) is an unbranched vascular bundle.

Located on the adaxial side near the base of each leaf is a sporangium (Fig. 9-50, A-C). A ligule is present just above the sporangium. Covering or partially covering the sporangium is a protective flap of leaf tissue, the velum (Fig. 9-50). All sporophylls generally have normal sporangia except several of the late-formed sporophylls of a growing season.

EXPERIMENTAL MORPHOLOGY AND PHYSIOLOGY. Expression of form in plants is determined largely by the occurrence and distribution of endogenous growth regulators. Perhaps the small, short corm-like form of *Isoetes* is due to a lack of certain endogenous regulating substances. In one experiment the investigator tested the effects of indoleacetic acid and gibberellic acid on *Isoetes*. Applied indoleacetic acid is known to stimulate or inhibit certain growth functions, and treatment with gibberellic

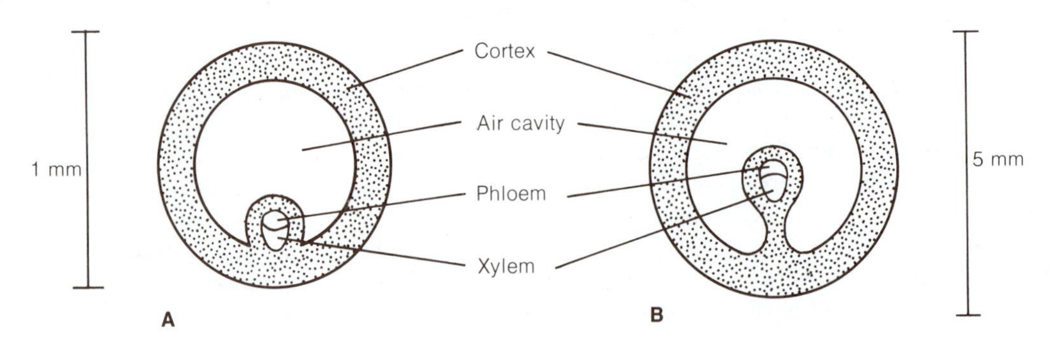

FIGURE 9-53 Schematic representations. **A,** transection of root of *Isoetes macrospora;* **B,** a stigmarian appendage of *Stigmaria.* [Redrawn from Stewart, *Amer. Jour. Bot.* 34:315–324. 1947.]

acid is known to result in stem elongation in many plants. Except for some proliferation of cortical tissue in the upper part of the corm, stimulated by the indoleacetic acid, there were no other obvious effects on growth (Zinda, 1966). Additional experiments, using combinations of certain growth regulators, would be of great interest to the better understanding of control of form in *Isoetes.*

Isoetes howellii has been shown to have crassulacean acid metabolism (CAM). It has been shown that when plants are completely submerged in vernal pools, dissolved free carbon dioxide is often totally depleted by noon on sunny days. The submerged leaves fix carbon dioxide into malic acid at night. During daylight hours, the malic acid is broken down and the carbon dioxide is utilized in photosynthesis. As the water level in vernal pools drops and the plants become emergent, the CAM mechanism is largely lost (Keeley et al., 1983).

SPOROPHYLL AND SPORANGIUM DEVELOPMENT. Sporophylls originate in a position lateral to the apical cone. Very soon after a sporophyll primordium is initiated, a ligule is produced on the adaxial face of the sporophyll. By repeated divisions the ligule soon overtops the leaf apex. The ligule eventually becomes tongue shaped (Fig. 9-50, A–C) and shows a high degree of histologic specialization comparable to ligules in *Selaginella* (p. 131–132).

In *Isoetes,* the glossopodium is deeply embedded in the leaf tissue and consists of a transverse bar and a pad that connects two side arms (Fig. 9-54); each of the arms may become anchor shaped (Sharma and Singh, 1984). In a median longitudinal section of a leaf only the pad is evident (Fig. 9-50, D). The sporangium and velum have their origin through periclinal divisions in surface cells below the ligule. The velum is first to take form, and is followed by growth of the sporangium. As a result of the first periclinal divisions, which localize the sporangial position, a central mass of potentially sporogenous cells is separated from outer layers of peripheral cells. During development, the cells of the outer layers may continue to add derivatives to the sporogenous mass. Ultimately the outer three or four peripheral layers constitute the sporangium wall. The sporogenous cells divide in all planes.

Microsporangia and megasporangia are indistinguishable during early stages of development, and it is only after the potential microsporocytes or megasporocytes become apparent that the two types of sporangia can be distinguished. In a microsporangium irregular groups of deeply staining cells ultimately become the microsporocytes, and cer-

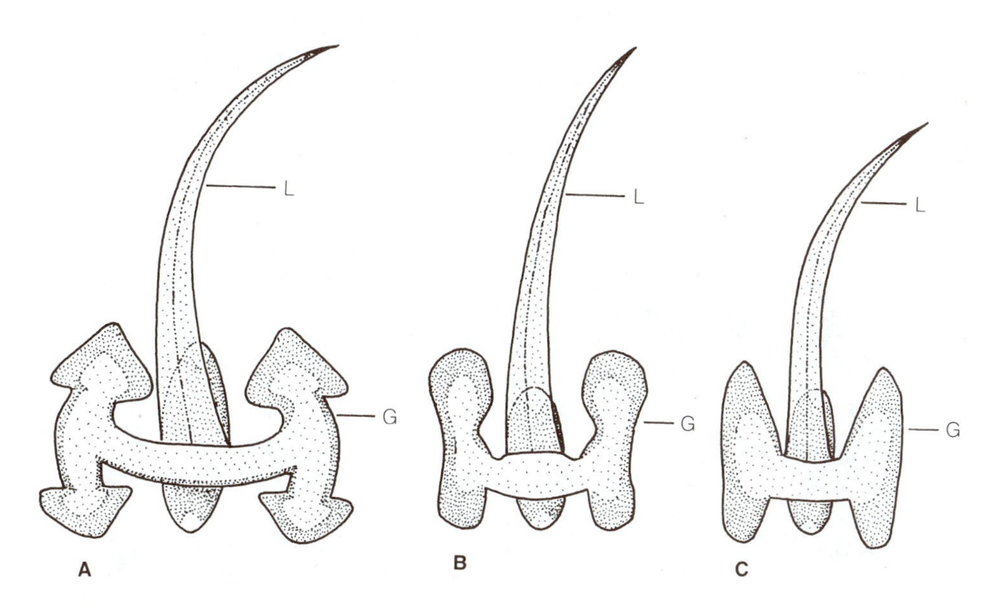

FIGURE 9-54 Reconstructions of *Isoetes* ligules. **A,** *Isoetes coromandelina.* **B,** *I. rajasthanensis.* **C,** *I. reticulata.* G, glossopodium; L, free portion of ligule. [Redrawn from Sharma and Singh, *Amer. Fern Jour.* 74:22–28, 1984.]

tain bands of lightly staining cells (originally potentially sporogenous) become the trabeculae, which traverse the sporogenous mass but do not divide it into compartments or locules (Fig. 9-55). Covering the trabeculae is a tapetum which may be biseriate or multilayered, the cells of which are derived also from potentially sporogenous cells. The tapetum of the trabeculae is continuous with a tapetal layer lining the sporangium wall. The microsporocytes separate from each other prior to meiosis. Estimates of the number of bifacial (monolete) microspores produced by a single sporangium range from 300,000 to 1,000,000. The spore number of each sporangium in *Isoetes* is probably greater than it is in any other vascular plant.

In a megasporangium, even before the trabeculae are distinguishable, certain cells are greatly enlarged over their neighbors and will become the megasporocytes. Not all of the enlarged cells will become megaspore mother cells, but some will degenerate and be resorbed. It is only after megaspore mother cells are in evidence that trabeculae become apparent. As a result of meiosis, each sporangium contains approximately 100 to 300 tetrahedral (trilete) megaspores which may range from 200 to 900 micrometers in diameter (Fig. 9-50, D). Megaspores may be white, gray, or black. The wall may be smooth or have a distinctive ornamentation. Monolete microspores are small, from 20 to 40 micrometers long, and have various characteristic wall patterns.

Sporangia are indehiscent, and liberation of spores is brought about by decay of sporophylls in the fall or winter seasons in cooler latitudes. Certain species of *Isoetes* growing in vernal pools, in which the corm is entirely buried, have a special means of exposing the sporangial contents. In these cases decaying sporophylls of the previous season are forced up by the expansion of mucilage cells at the base of the sporophylls, whereupon the spores are brought to the surface (Osborn, 1922). Distribution of spores is by wind, disturbance of mud in which certain species grow, or wave action in lakes; also, earthworms have been reported as carriers (Duthie, 1929).

CYTOLOGY AND SYSTEMATICS. The chromosomes of *Isoetes* range in length from 1 to 2 micrometers,

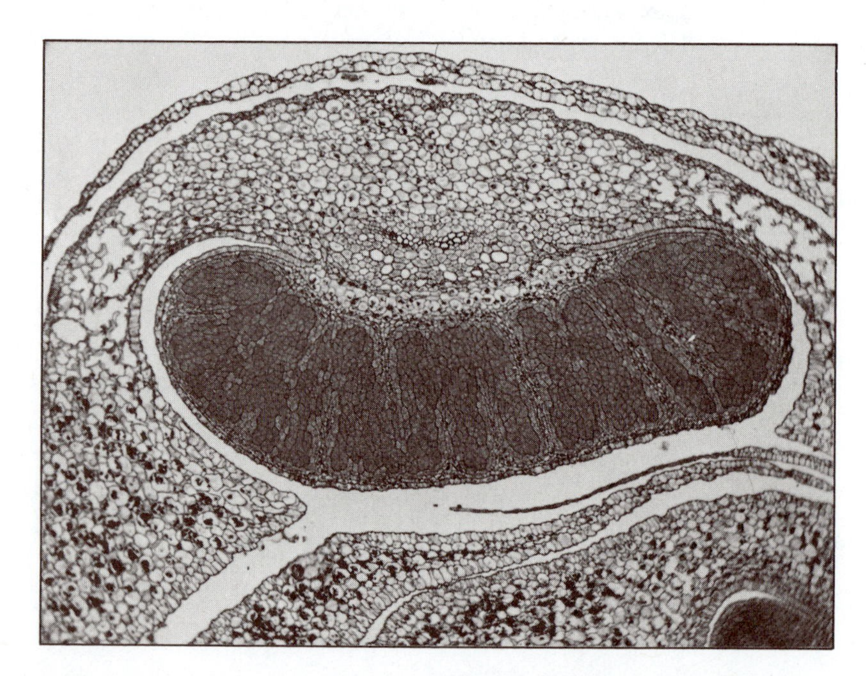

FIGURE 9-55 Transection of microsporophyll and developing microsporangium of *Isoetes howellii*. Trabeculae can be seen traversing the mass of potential microsporocytes. The bases of other sporophylls are evident.

or less, up to 7 to 8 micrometers. The basic chromosome number for the genus is $x = 11$, and polyploidy is common in the genus. A study of six species in northeastern North America revealed diploids $2n = 22$, tetraploids $2n = 44$, and one decaploid $2n = 110$ (Kott and Britton, 1980). A triploid $2n = 33$ has been reported for an apogamous (without fertilization) species from India (Abraham and Ninan, 1958), and a hexaploid $2n = 66$ for *I. japonica* from Japan (Löve et al., 1977). Chromosome counts of $2n = 22$, 44, and 132 have been reported for several neotropical species (Hickey, 1984).

Characters such as spore color and morphology, extent of velum cover, and leaf length have, for example, been used to separate species. The results of certain population studies indicate that these features are quite variable and become greatly modified under certain environmental conditions. For example, *Isoetes* populations associated with granite outcrops of the southeastern Piedmont in the United States range in morphology from those that can be classified as *I. melanospora,* through a series of intermediate populations, to those classified as *I. piedmontana.* Also, the populations have similar polyphenolic patterns (Matthews and Murdy, 1969).

Interspecific hybridization is now known to be common in *Isoetes* and probably accounts for much of the confusion in the past over the delimitations of species. Under natural conditions, species are often isolated geographically or ecologically and they can be distinguished morphologically from one another. However, the ease by which interspecific hybridization can occur between species of different sections of the genus has been demonstrated recently for four species. Crosses were set up in such a manner that the megaspores of each species were brought into contact with the microspores of every other species, under rigorous experimental conditions. Progeny were produced from all crosses (Boom, 1980). Experiments such as this may help to explain the several to many described varieties for certain species when populations impinge on each other or overlap geographically.

GAMETOPHYTES. Spores may germinate immediately after being shed from the sporangium, but generally germination does not take place until winter in warm-climate species or spring in cold-climate species, after decay of the sporophylls. The gametophytes, as in *Selaginella,* are endosporic. The microgametophyte is retained entirely within the spore wall, though a portion of the megagametophyte may be exposed.

MICROGAMETOPHYTE DEVELOPMENT. The first division of the microspore forms a small prothallial cell, and a large cell that is interpreted as an antheridial initial. By several divisions the antheridial initial is subdivided into a layer of jacket cells and a total of four spermatogenous cells (the actual spermatids, in this case). After a developmental period of about two weeks, the spore wall cracks along the flat surface. The jacket cells and prothallial cell degenerate, and four multiflagellate sperm are liberated (Liebig, 1931). The general scheme of development is comparable to that in *Selaginella* (Fig. 9-30). The possession of sperm with many flagella contrasts strikingly with the biflagellate condition of the sperm of other genera (*Lycopodium* and *Selaginella*) of the Lycophyta.

MEGAGAMETOPHYTE DEVELOPMENT. A mature megaspore contains a considerable amount of stored food surrounded by the spore wall. The primary nucleus is quite large and may be at the base or apex (toward the triradiate ridge) (Campbell, 1891; LaMotte, 1933). A period of free nuclear division ensues. Wall formation then takes place rapidly around nuclei at the apical end, proceeding basipetally and centripetally at a slower rate. Stored material is prominent in the basal end of the developing megagametophyte. No large central vacuole, as in *Selaginella,* is present. The megagametophyte may not become entirely cellular until the embryo is quite advanced in development (LaMotte, 1933). With an increase in volume of the megagametophyte, the megaspore wall breaks along the triradiate ridge (Fig. 9-56). The first archegonium appears at the apex, and at maturity consists of four tiers of neck cells, a neck-canal cell, a ventral canal cell, and an egg cell. If fertilization does not occur, many more archegonia may be formed among rhizoids that extend above the surface of the gametophyte (LaMotte, 1933).

THE EMBRYO. After fertilization, the first division of the zygote is transverse or oblique to the long

FIGURE 9-56 Megagametophytes of *Isoetes howellii* protruding through cracked spore walls. In each case, the megaspore wall was ruptured along the triradiate ridge, and three portions of the wall are visible. The dark areas are archegonia, and one tier of four neck cells is apparent in an archegonium of the gametophyte to the right.

axis of the archegonium (Fig. 9-57, A). The embryo becomes globose, and in some instances quadrants (Fig. 9-57, B) are recognizable (Campbell, 1891; LaMotte, 1937). The upper half (hypobasal) of the embryo (toward the neck of the archegonium) gives rise to the foot and root, and the lower half (epibasal) produces the first leaf and shoot apex. The embryo is thus endoscopic, but no suspensor is formed as in *Lycopodium* and *Selaginella*. Interpretation of subsequent embryonic development is indeed difficult, though the following description may represent a reasonably accurate account (LaMotte, 1933, 1937).

That portion of the embryo which will become the foot grows downward obliquely and into the storage tissue of the megagametophyte. At the same time, that portion of the embryo which will produce the first leaf and shoot apex grows laterally or perpendicular to the long axis of the archegonium. The primary root grows in the same plane but in the direction opposite from the leaf. Reorientation of the embryo is thereby achieved (Fig. 9-57, C, D). With further development the first leaf of the sporophyte breaks through the gametophytic sheath; the root emerges and turns downward. The young sporophyte may become firmly established on the substratum, but it remains attached for some time to the gametophyte and surrounding megaspore wall.

NEWLY DISCOVERED FORMS. Our study of *Isoetes* would be incomplete without a discussion of the recently described *Isoetes tegetiformans* (Rury, 1978), a mat-forming species that occurs in Georgia in the United States, and *Stylites* that occurs in Peru.

The corm of *I. tegetiformans* is bilobed, and the lobes exhibit intercalary extension from the median furrow (Fig. 9-58). Phyllotaxy is distichous (two rows of leaves) throughout the life of a plant, contrary to the usual helical arrangement in adult forms of other *Isoetes* species. The stem vasculature is a sympodium of leaf and root traces. Another distinctive feature is the presence of three rows of nondichotomizing roots. Also, adventitious buds arise on the elongate lobes of the corm.

The family Isoetaceae was considered to be monotypic until 1957 when another genus, *Stylites*, was established for plants which were found growing in dense cushions around the boggy margins of a small glacial lake high in the Andes of Peru (Amstutz, 1957). The plants resemble certain species of *Isoetes*, but the corms are much more elongate, up to several centimeters long, and branch dichotomously (Fig. 9-59, A, B).

The morphology, anatomy, and life history of the originally described species *(Stylites andicola)* and of another species *(S. gemmifera)* are now well known (Rauh and Falk, 1959a, b). Secondary growth occurs, as in *Isoetes*, but the root-producing meristem is upturned and the nondichotomizing roots, generally present only on one side of each axis, appear to be more lateral in position than basal. In *S. gemmifera* adventitious buds can arise at the base of sporophylls as has been reported for

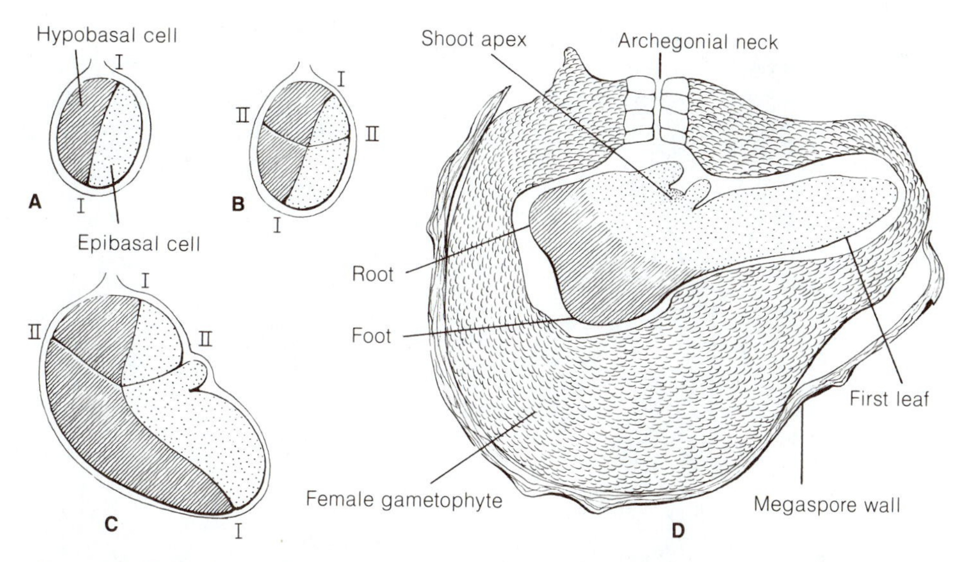

FIGURE 9-57 Early cleavages and subsequent growth of the embryo in *Isoetes lithophila*. **A,** the first cleavage (I-I) in this species is usually as shown, only rarely being at right angles to the neck of the archegonium; **B,** quadrant stage; **C, D,** later stages, each segment being multicellular, but not indicated by cells. More rapid growth occurs in the lower two quadrants, resulting in an apparent rotation of the embryo. [Redrawn from La Motte, *Ann. Bot.* n.s. 1:695, 1937; *Amer. Jour. Bot.* 20:217, 1933.]

FIGURE 9-58 *Isoetes tegetiformans.* Arrows indicate the adventitious buds that arise on the elongate lobes of the corm. [Courtesy of Dr. P. M. Rury.]

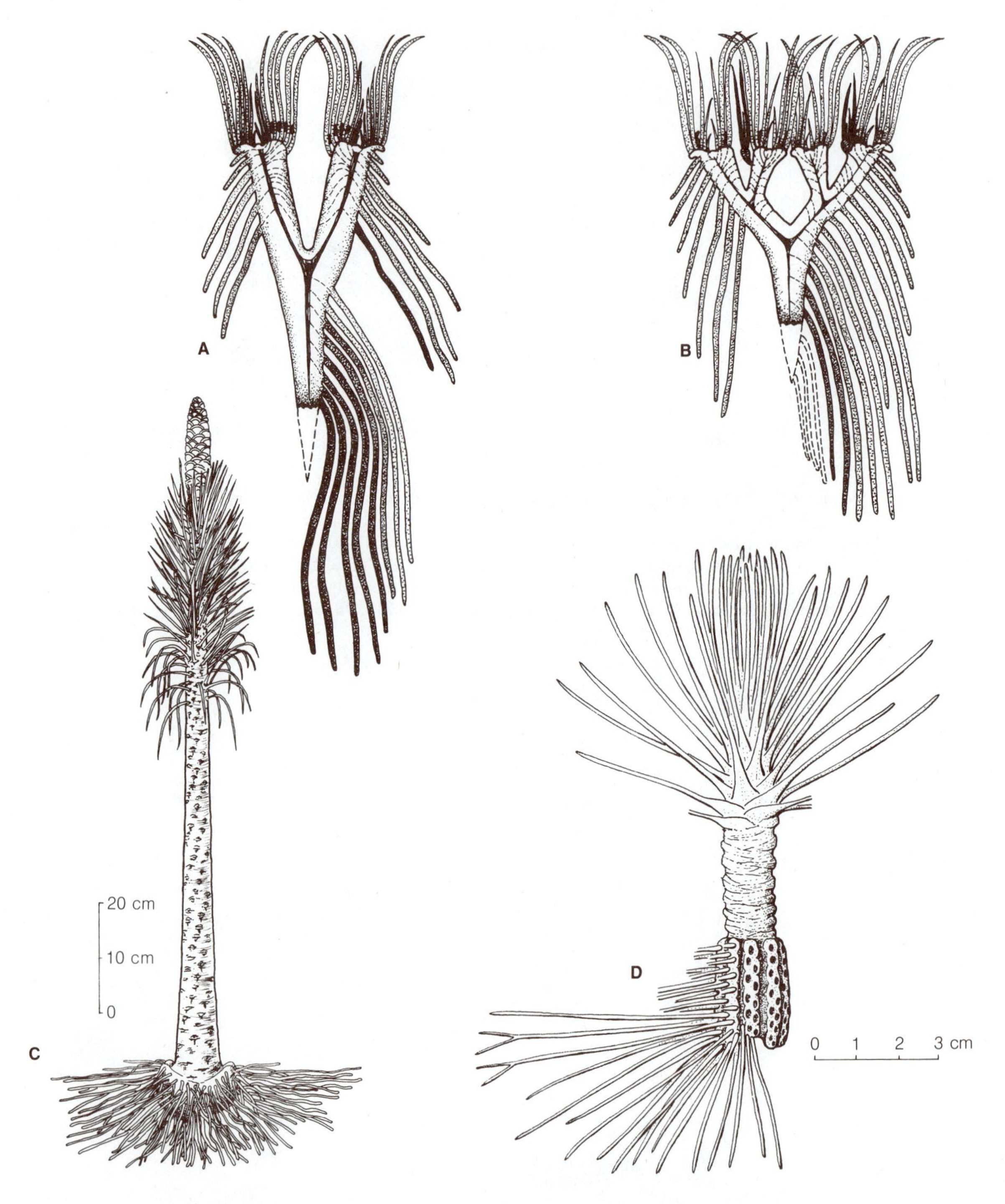

FIGURE 9-59 Schematic representation of growth form and type of branching in *Stylites andicola* (**A**) and *Stylites gemmifera* (**B**). **C**, reconstruction of *Pleuromeia sternbergii*. **D**, reconstruction of *Nathorstiana arborea*. [**A**, **B** redrawn from Rauh and Falk, *Sitzb. Heidelb. Akad. Wiss. Jahrgang* 1–83, 1959; **C** redrawn from Hirmer, *Palaeontographica* 78(B):47–56, 1933; **D** redrawn from *Paläobiologie der Pflanzen* by K. Magdefrau. Gustav Fischer Verlag, Stuttgart. 1968.]

certain species of *Isoetes*. An interesting physiological feature of *S. andicola* has been reported by Keeley et al. (1984). *Stylites* lacks stomata and derives nearly all of its photosynthetic carbon through its roots. This species possesses characteristics of crassulacean acid metabolism (CAM) comparable to *Isoetes howellii* (p. 161).

A reinvestigation of *Isoetes triquetra* from the Andes of Colombia revealed many similarities to *Stylites* (Kubitzki and Borchert, 1964). Enough information has been accumulated to convince some botanists that the validity of the genus *Stylites* should be questioned. Perhaps the two species of *Stylites* should be considered only as extremes in the morphological variation of *Isoetes*, or, at best, *Stylites* should be recognized only as a subgenus of *Isoetes*. (For discussions, see Rury, 1978, and Gomez, 1980.)

Pleuromeiales

A representative member of the order Pleuromeiales is the extinct Triassic genus *Pleuromeia* (Fig. 9-59, C). The plant body consisted of an erect, unbranched stem about 1 meter or more in height and 10 centimeters in diameter. The stem axis terminated in a strobilus composed of closely overlapping sporophylls with apparently adaxial sporangia (Emberger, 1968; Delevoryas, 1962). Spirally arranged vegetative leaves were present beneath the strobilus. A very remarkable structural feature of the genus was the enlargement of the stem base. This basal part was divided commonly into four lobes (a lobed rhizophore) upon which roots were produced in an orderly fashion.

On the basis of a careful morphological study of *Pleuromeia*, Paolillo (1982) has presented evidence that root initiation was similar to that of *Isoetes*. Anatomically, the roots were of the "stigmarian" type (Fig. 9-53, B), and similar to *Isoetes*.

Another genus, *Nathorstiana*, of the Cretaceous, is often included in this order. The plants were probably about 10 to 30 centimeters tall, had a crown of leaves, and had an enlarged lobed basal rhizophore (Fig. 9-59, D). Paolillo (1982) has suggested that root initiation may have been comparable to that of *Isoetes* although, on the basis of a study of young unlobed rhizophores of *Nathorstiana*, Karrfalt (1984) and Rothwell (1984) believe that the basal meristems of the lobed types of *Nathorstiana* and of *Isoetes* are primary meristems rather than secondary meristems. According to their view a homology exists among the root-producing meristems of the rhizophores of lepidodendrids, *Nathorstiana*, and *Isoetes*.

Considerable evidence points toward a phylogenetic relationship among the three genera. The *Isoetes*-type plant body is probably a miniature form of the *Pleuromeia-Nathorstiana* type, brought about by the telescoping of the entire axis. Plants with certain features of a modern *Isoetes* occurred in the Triassic; a more convincing fossil, *Isoetites serratus*, has been described from the Upper Cretaceous. (See Taylor, 1981, for additional details.)

REFERENCES

Abraham, A., and C. A. Ninan
1958. Cytology of *Isoetes*. Curr. Sci. 27:60–61.

Amstutz, E.
1957. *Stylites*, a new genus of Isoetaceae. *Ann. Missouri Bot. Gard.* 44:121–123.

Andrews, H. N., Jr., and W. H. Murdy
1958. *Lepidophloios*—and ontogeny in arborescent lycopods. *Amer. Jour. Bot.* 45:552–560.

Banks, H. P., P. M. Bonamo, and J. D. Grierson
1972. *Leclercqia complexa* gen. et sp. nov., a new lycopod from the late Middle Devonian of eastern New York. *Rev. Palaeobot. Palynol.* 14:19–40.

Barclay, B. D.
1931. Origin and development of tissues in stem of *Selaginella Wildenovii*. *Bot. Gaz.* 91:452–461.

Beitel, J.
1979. Clubmosses *(Lycopodium)* in North America. *Fiddlehead Forum, Bull. Amer. Fern Soc.* 6(5):1–8.

Beitel, J. M., and F. S. Wagner
1982. The chromosomes of *Lycopodium lucidulum*. *Amer. Fern Jour.* 72:33–35.

Bhambie, S.
1965. Studies in pteridophytes. V. The development, structure and arrangement of leaves in some species of *Lycopodium*. *Proc. Indian Acad. Sci.* (Sect. B) 61:242–252.

Bhambie, S., and V. Puri
1963. Shoot apex organization in Lycopodiales. *Mem. Indian Bot. Soc.* 4:55–63.

Bilderback, D. E.
1987. Association of mucilage with the ligule of several species of *Selaginella*. *Amer. Jour. Bot.* 74:1116–1121.

Boom, B. M.
1980. Intersectional hybrids in *Isoetes*. *Amer. Fern Jour.* 70:1–4.

Brack-Hanes, S. D.
1978. On the megagametophytes of two lepidodendracean cones. *Bot. Gaz.* 139:140–146.

Brack-Hanes, S. D., and J. C. Vaughn
1978. Evidence of Paleozoic chromosomes from lycopod microgametophytes. *Science* 200:1383–1385.

Brooks, K. E.
1973. Reproductive biology of *Selaginella*. I. Determination of megasporangia by 2-chloroethylphosphonic acid, an ethylene-releasing compound. *Plant Physiol.* 51:718–722.

Bruce, J. G.
1976a. Gametophytes and subgeneric concepts in *Lycopodium*. *Amer. Jour. Bot.* 63:919–924.
1976b. Comparative studies in the biology of *Lycopodium carolinianum*. *Amer. Fern Jour.* 66:125–137.
1979. Gametophyte and young sporophyte of *Lycopodium carolinianum*. *Amer. Jour. Bot.* 66:1156–1163.

Bruchmann, H.
1909. Von der Chemotaxis der *Lycopodium*—Spermatozoiden. *Flora* 99:193–202.
1910. Die Keimung der Sporen und die Entwicklung der Prothallien von *Lycopodium clavatum* L., *L. annotinum* L., und *L. Selago* L. *Flora* 101:220–267.
1912. Zur Embryologie der Selaginellaceen. *Flora* 104:180–224.

Burr, F. A., and R. F. Evert
1973. Some aspects of sieve-element structure and development in *Selaginella kraussiana*. *Protoplasma* 78:81–97.

Campbell, D. H.
1891. Contributions to the life-history of *Isoetes*. *Ann. Bot.* 5:231–258.

Chamberlain, C. J.
1917. Prothallia and sporelings of three New Zealand species of *Lycopodium*. *Bot. Gaz.* 63:51–65.

Chiang, S.-H. T., and S.-Y. Chen
1986. Components of vascular tissue in the corm of *Isoetes taiwanensis*. *Taiwania* 31:1–14.

Chu, M. C.-Y.
1974. A comparative study of the foliar anatomy of *Lycopodium* species. *Amer. Jour. Bot.* 61:681–692.

Cusick, F.
1954. Experimental and analytical studies of Pteridophytes XXV. Morphogenesis in *Selaginella Willdenovii*. II. Angle-meristems and angle-shoots. *Ann. Bot.* n. s. 18:171–181.

Cutter, E. G.
1966. Patterns of organogenesis in the shoot. Pp. 220–234 *in* Cutter, E. G. (ed.) *Trends in Plant Morphogenesis*. Longmans, Green, London.

Delevoryas, T.
1962. *Morphology and Evolution of Fossil Plants*. Holt, Rinehart, and Winston. New York.
1964. Ontogenetic studies of fossil plants. Phytomorphology 14:299–314.

Dengler, N. G.
1983a. The developmental basis of anisophylly in *Selaginella martensii*. I. Initiation and morphology of growth. *Amer. Jour. Bot.* 70:181–192.
1983b. The developmental basis of anisophylly in *Selaginella martensii*. II. Histogenesis. *Amer. Jour. Bot.* 70:193–206.

DiMichele, W. A.
1981. Arborescent lycopods of Pennsylvanian age coals: a *Lepidodendron*, with description of a new species. *Palaeontographica* 175B:85–125.

Doyle, W. T.
1970. *The Biology of Higher Cryptogams*. Macmillan, London.

Duerden, H.
1934. On the occurrence of vessels in *Selaginella*. *Ann. Bot.* 48:459–465.

Duthie, A. V.
1929. The method of spore dispersal of three South African species of *Isoetes*. *Ann. Bot.* 43:411–412.

Eames, A. J.
1942. Illustrations of some *Lycopodium* gametophytes. *Amer. Fern Jour.* 32:1–12.

Eggert, D. A.
1961. The ontogeny of Carboniferous arborescent Lycopsida. *Palaeontographica*. 108(B):43–92.

Eggert, D. A., and N. Y. Kanemoto
1977. Stem phloem of a Middle Pennsylvanian *Lepidodendron*. *Bot. Gaz.* 138:102–111.

Emberger, L.
1968. *Les Plantes Fossiles dans leurs Rapports avec*

les Végétaux Vivants, 2d edition. Masson et Cie., Paris.

Esau, K., V. I. Cheadle, and E. M. Gifford, Jr.
1953. Comparative structure and possible trends of specialization of the phloem. *Amer. Jour. Bot.* 40:9–19.

Freeberg, J. A., and R. H. Wetmore
1957. Gametophytes of *Lycopodium* as grown *in vitro. Phytomorphology* 7:204–217.
1967. The Lycopsida—a study in development. *Phytomorphology* 17:78–91.

French, J. C.
1972. Dimensional correlations in developing *Selaginella* sporangia. *Amer. Jour. Bot.* 59:224–227.

Fry, W. L.
1954. A study of the carboniferous lycopod, *Paurodendron,* Gen. Nov. *Amer. Jour. Bot.* 41:415–428.

Gensel, P. G., and H. N. Andrews
1984. *Plant Life in the Devonian.* Praeger, New York.

Gomez, P. L. D.
1980. Vegetative reproduction in a Central American *Isoetes* (Isoetaceae). Its morphological, systematic, and taxonomical significance. *Brenesia* 18:1–14.

Grierson, J. D., and P. M. Bonamo
1979. *Leclercqia complexa:* earliest ligulate lycopod (Middle Devonian). *Amer. Jour. Bot.* 66:474–476.

Grierson, J. D., and F. M. Hueber
1968. Devonian lycopods from northern New Brunswick. In D. Oswald (ed.), *International Symposium on the Devonian System,* pp. 823–836.

Hagemann, W.
1980. Über den Verzweigungsvorgang bei *Psilotum* und *Selaginella* mit Anmerkungen zum Begriff der Dichotomie. *Plant Syst. Evol.* 133:181–197.

Härtel, K.
1938. Studien an Vegetationspunkten einheimischer Lycopodien. *Beitr. Biol. Pflanzen* 25:125–168.

Haught, O.
1960. Lycopods in the American Tropics. *Castanea* 25:127–129.

Hauke, R. L.
1969. Problematic groups in the fern allies and the treatment of subspecific categories. *BioScience* 19:705–707.

Hickey, R. J.
1984. Chromosome numbers of Neotropical *Isoetes. Amer. Fern Jour.* 74:9–13.

Horner, H. T., Jr., and H. J. Arnott
1963. Sporangial arrangement in North American species of *Selaginella. Bot. Gaz.* 124:371–383.

Horner, H. T., Jr., and C. K. Beltz
1970. Cellular differentiation of heterospory in *Selaginella. Protoplasma* 71:335–361.

Hsü, J.
1937. Anatomy, development and life history of *Selaginella sinensis.* I. Anatomy and development of the shoot. *Bull. Chinese Bot. Soc.* 3:75–95.

Jaeger, H.
1962. Das Alter der ältesten bekannten Landpflanzen (*Baragwanathia*-flora) in Australien auf Grund der begleitenden Graptolithen. *Paläontol. Zeit.* 36:7.

Jermy, A. C., K. Jones, and C. Colden
1967. Cytomorphological variation in *Selaginella. Jour. Linn. Soc. Bot.* 60:147–158.

Jernstedt, J. A., and M. A. Mansfield
1985. Two-dimensional gel electrophoresis of polypeptides from stems, roots, leaves, and rhizophores of *Selaginella kraussiana. Bot. Gaz.* 146:460–465.

Karrfalt, E. E.
1977. An apical cell in the shoot apex of *Isoetes tuckermanii. Amer. Fern Jour.* 67:68–72.
1984. Further observations on *Nathorstiana* (Isoetaceae). *Amer. J. Bot.* 71:1023–1030.

Karrfalt, E. E., and D. A. Eggert
1977a. The comparative morphology and development of *Isoetes* L. I. Lobe and furrow development in *I. tuckermanii* A. Br. *Bot. Gaz.* 138:236–247.
1977b. The comparative morphology and development of *Isoetes* L. II. Branching of the base of the corm in *I. tuckermanii* A. Br. and *I. nuttallii* A. Br. *Bot. Gaz.* 138:357–368.
1978. The comparative morphology and development of *Isoetes* L. III. The sequence of root initiation in three- and four-lobed plants of *I. tuckermanii* A. Br. *Bot. Gaz.* 139:271–283.

Keeley, J. E., R. P. Mathews, and C. M. Walker
1983. Diurnal acid metabolism in *Isoetes howellii* from a temporary pool and a permanent lake. *Amer. Jour. Bot.* 70:854–857.

Keeley, J. E., C. B. Osmond, and J. A. Raven
1984. *Stylites,* a vascular land plant without stomata absorbs CO_2 via its roots. *Nature* 310:694–695.

Koller, A. L., and S. E. Scheckler
1986. Variations in microsporangia and microspore dispersal in *Selaginella. Amer. J. Bot.* 73:1274–1288.

Koster, H.
1941. New *Lycopodium* gametophytes from New Jersey. *Amer. Fern Jour.* 31:53–58.

Kott, L. S., and D. M. Britton
1980. Chromosome numbers for *Isoetes* in northeastern North America. *Can. Jour. Bot.* 58:980–984.

Kruatrachue, M., and R. F. Evert
1977. The lateral meristem and its derivatives in the corm of *Isoetes muricata. Amer. Jour. Bot.* 64:310–325.

Kubitzki, K., and R. Borchert
1964. Morphologische Studien an *Isoëtes triquetra* A. Braun und Bermerkungen über das Verhältnis der Gattung *Stylites* E. Amstutz zur Gattung *Isoëtes* L. *Ber. Deutsch. Bot. Ges.* 77:227–233.

Kuriachan, P. I.
1963. Cytology of the genus *Selaginella. Cytologia* 28:376–380.

LaMotte, C.
1933. Morphology of the megagametophyte and the embryo sporophyte of *Isoetes lithophila. Amer. Jour. Bot.* 20:217–233.
1937. Morphology and orientation of the embryo of *Isoetes Ann. Bot.* n.s. 1:695–716.

Lamoureux, C. H.
1961. Comparative studies on phloem of vascular cryptogams. Ph.D. dissertation. University of California, Davis.

Liebig, J.
1931. Ergänzungen zur Entwicklungsgeschichte von *Isoëtes lacustre* L. *Flora* 125:321–358.

Löve, A., and D. Löve
1958. Cytotaxonomy and classification of lycopods. *Nucleus* (Calcutta) 1:1–10.

Löve, A., D. Löve, and R. E. G. Pichi-Sermolli
1977. *Cytotaxonomical Atlas of the Pteridophyta.* J. Cramer, Vaduz.

Lyon, A. G.
1964. Probable fertile region of *Asteroxylon mackiei* K. and L. *Nature* 203:1082–1083.

Martens, P.
1960a. Sur une structure microscopique orientée, dans la paroi mégasporale d'une Sélaginelle. *Compt. Rend. Acad. Sci.* (Paris) 250:1599–1602.
1960b. Nouvelles observations sur la structure des parois mégasporales de *Selaginella myosurus* (Sow.) Alston. *Compt. Rend. Acad. Sci.* (Paris) 250:1774–1775.

Matthews, J. F., and W. H. Murdy
1969. A study of *Isoetes* common to the granite outcrops of the southeastern Piedmont, United States. *Bot. Gaz.* 130:53–61.

Michaux, N.
1966. Structure et fonctionnement du méristème apical de l'*Isoetes setacea* Lam. *C. R. Acad. Sci.* (Paris) 263:501–504.

Mickel, J. T., and R. L. Hellwig
1969. Actino-plectostely, a complex new stelar pattern in *Selaginella. Amer. Fern Jour.* 59:123–134.

Nougarède, A., and J.-E. Loiseau
1963. Étude morphologique des rameaux du *Lycopodium selago* L.; structure et fonctionnement de l'apex. *C. R. Acad. Sci.* (Paris) 257:2698–2701.

Ogura, Y.
1938. Anatomie der Vegetationsorgane der Pteridophyten. *Handbuch der Pflanzenanatomie.* Gebrüder Borntraeger, Berlin.

Osborn, T. G. B.
1922. Some observations on *Isoetes Drummondii,* A. Br. *Ann. Bot.* 36:41–54.

Pant, D. D., and B. Mehra
1964. Development of stomata in some fern allies. *Proc. Nat. Inst. Sci. India* 30(B):92–98.

Paolillo, D. J., Jr.
1963. *The Developmental Anatomy of Isoetes.* (Illinois Biological Monographs. No. 31) University of Illinois Press, Urbana.
1982. Meristems and evolution: developmental correspondence among the rhizomorphs of the lycopsids. *Amer. Jour. Bot.* 69:1032–1042.

Pfeiffer, N. E.
1922. Monograph of the Isoetaceae. *Ann. Missouri Bot. Gard.* 9:79–232.

Phillips, T. L.
1979. Reproduction of heterosporous arborescent lycopods in the Mississippian-Pennsylvanian of Euramerica. *Rev. Palaeobot. Palynol.* 27:239–289.

Phillips, T. L., M. J. Avcin, and J. M. Schopf
1975. Gametophytes and young sporophyte development in *Lepidocarpon. Bot. Soc. Amer. Abstract,* Corvallis, Oregon, p. 23.

Phillips, T. L., and G. A. Leisman
1966. *Paurodendron,* a rhizomorphic lycopod. *Amer. Jour. Bot.* 53:1086-1100.

Pieniązek, S. Al.
1938. Über die Entwicklung und das Wachstum der Makrosporenmembranen bei *Selaginella. Sprawozdania Towarzystwa Nauk. Warszawskiego Wydziat 4 (Compt. Rend. Soc. Sci. Varsovie Cl. 4)* 31:211-230.

Pixley, E. Y.
1968. A study of the ontogeny of the primary xylem in the roots of *Lycopodium. Bot. Gaz.* 129:156-160.

Rauh, W., and H. Falk
1959a. *Stylites* E. Amstutz, eine neue Isoëtacee aus den Hochanden Perus. 1. Teil: Morphologie, Anatomie und Entwicklungsgeschichte der Vegetationsorgane. *Sitzb. Heidelb. Akad. Wiss. Jahrgang* 1959, 1-83.
1959b. *Stylites* E. Amstutz, eine neue Isoëtacee aus den Hochanden Perus. 2. Teil. Zur Anatomie des Stammes mit besonderer Berücksichtigung der Verdikungsprozesse. *Sitzb. Heidelb. Akad. Wiss. Jahrgang* 1959, 87-160.

Robbins, R. R., and Z. B. Carothers
1978. Spermatogenesis in *Lycopodium:* the mature spermatozoid. *Amer. Jour. Bot.* 65:433-440.

Robert, D.
1971a. Le gamétophyte femelle de *Selaginella kraussiana* (Kunze) A. Br. I. Organisation générale de la mégaspore. Le diaphragme et l'endospore. Les reserves. *Rev. Cytol. Biol. Vég.* 34:93-164.
1971b. Le gamétophyte femelle de *Selaginella kraussiana* (Kunze) A. Br. II. Organisation histologique du tissu reproducteur et principaux aspects de la dédifférenciation cellulaire préparatoire à l'oogénèse. *Rev. Cytol. et Biol. Vég.* 34:189-232.
1974. Étude ultrastructurale de la spermiogenèse, notamment de la différenciation de l'appareil nucléaire, chez le *Selaginella kraussiana* (Kunze) A. Br. *Ann. Sci. Nat. Bot.* (Paris), Sér. 12, 15:65-118.

Roberts, E. A., and S. D. Herty
1934. *Lycopodium complanatum* var. *flabelliforme* Fernald: its anatomy and a method of vegetative propagation. *Amer. Jour. Bot.* 21:688-697.

Rothwell, G. W.
1984. The apex of *Stigmaria* (Lycopsida), rooting organ of Lepidodendrales. *Amer. J. Bot.* 71:1031-1034.

Rothwell, G. W., and D. M. Erwin
1985. The rhizomorph apex of *Paurodendron;* implications for homologies among the rooting organs of Lycopsida. *Amer. Jour. Bot.* 72:86-98.

Rury, P. M.
1978. A new and unique, mat-forming Merlin's-grass *(Isoetes)* from Georgia. *Amer. Fern Jour.* 68:99-108.

Schlanker, C. M., and G. A. Leisman
1969. The herbaceous Carboniferous lycopod *Selaginella fraiponti* comb. nov. *Bot. Gaz.* 130:35-41.

Sharma, B. D., and R. Singh
1984. The ligule in *Isoetes. Amer. Fern Jour.* 74:22-28.

Slagg, R. A.
1932. The gametophytes of *Selaginella Kraussiana.* I. The microgametophyte. *Amer. Jour. Bot.* 19:106-127.

Soltis, D. E., and P. S. Soltis
1988. Are lycopods with high chromosome numbers ancient polyploids? *Amer. J. Bot.* 75:238-247.

Soltis, P. S., and D. E. Soltis
1988. Estimated rates of intragametophytic selfing in lycopods. *Amer. J. Bot.* 75:248-256.

Somers, P.
1982. A unique type of microsporangium in *Selaginella* series Articulatae. *Amer. Fern Jour.* 72:88-92.

Spessard, E. A.
1922. Prothallia of *Lycopodium* in America. II. L. *lucidulum* and L. *obscurum* var. *dendroideum. Bot. Gaz.* 74:392-413.

Stainier, F.
1965. Structure et infrastructure des parois sporales chez deux Sélaginelles. *(Selaginella myosurus* et *S. kraussiana). Cellule* 65:222-244.

Stevenson, D. W.
1976. Observations on phyllotaxis, stelar morphology, the shoot apex and gemmae of *Lycopodium lucidulum* Michaux (Lycopodiaceae). *Bot. J. Linnean Soc.* 72:81-100.

Stewart, W. N.
1983. *Paleobotany and the Evolution of Plants.* Cambridge University Press, Cambridge.

Stokey, A. G.
1907. The roots of *Lycopodium pithyoides. Bot. Gaz.* 44:57-63.

Sykes, M. G.
1908. Notes on the morphology of the sporangium-

bearing organs of the Lycopodiaceae. *New Phytol.* 7:41–60.

Takeuchi, K.
1962. Study on the development of gemmae in *Lycopodium chinense* Christ and *Lycopodium serratum* Thunb. *Jap. Jour. Bot.* 18:73–85.

Tanno, J. A., and T. R. Webster
1982a. Variegation in *Selaginella martensii* f. *albovariegata.* I. Expression and inheritance. *Can. J. Bot.* 60:2375–2383.
1982b. Variegation in *Selaginella martensii* f. *albovariegata.* II. Plastid structure in mature leaves. *Can. J. Bot.* 60:2384–2393.

Taylor, T. N.
1981. *Paleobotany: An Introduction to Fossil Plant Biology.* McGraw-Hill, New York.

Towers, G. H. N., and W. S. G. Maass
1965. Phenolic acids and lignins in the Lycopodiales. *Phytochemistry* 4:57–66.

Treub, M.
1884. Études sur les Lycopodiacées. *Ann. Jard. Bot. Buitenzorg.* 4:107–138.

Tryon, A. F., and B. Lugardon
1978. Wall structure and mineral content in *Selaginella* spores. *Pollen et Spores* 20:315–339.

Tryon, R. M., and A. F. Tryon
1982. *Ferns and Allied Plants.* Springer-Verlag, New York.

Turner, J. J.
1924. Origin and development of vascular system of *Lycopodium lucidulum. Bot. Gaz.* 78:215–225.

Van Soest, J. L.
1964. Estimation of the age of a fairy circle *(Lycopodium complanatum* L. var. *chamaecyparissus)* (A. Br.) Döll. *Acta Bot. Neer.* 13:623.

Voirin, B., and M. Jay
1978. Apport de la biochimie flavonique à la systematique du genre *Lycopodium. Biochem. Syst. Ecol.* 6:95–97.

Wagner, W. H., Jr., J. M. Beitel, and F. S. Wagner
1982. Complex venation patterns in the leaves of *Selaginella:* megaphyll-like leaves in lycophytes. *Science* 218:793–794.

Wagner, W. H., Jr., and F. S. Wagner
1980. Polyploidy in Pteridophytes. In W. H. Lewis (ed.), *Polyploidy: Biological Relevance.* Plenum Press, New York, pp. 199–214.

Wardlaw, C. W.
1924. Size in relation to internal morphology. No. 1. Distribution of the xylem in the vascular system of *Psilotum, Tmesipteris,* and *Lycopodium. Trans. Roy. Soc. Edinb.* 53:503–532.

Warmbrodt, R. D., and R. F. Evert
1974. Structure and development of the sieve element in the stem of *Lycopodium lucidulum. Amer. Jour. Bot.* 61:267–277.

Webster, T. R.
1979. An artificial crossing technique for *Selaginella. Amer. Fern J.* 69:9–13.

Webster, T. R., and T. A. Steeves
1964. Developmental morphology of the root of *Selaginella kraussiana* A. Br. and *Selaginella wallacei* Hieron. *Can. J. Bot.* 42:1665–1676.
1967. Developmental morphology of the root of *Selaginella martensii* Spring. *Can. Jour. Bot.* 45:395–404.

Webster, T. R., and J. A. Tanno
1980. Inheritance of pigment deficiencies and ultrastructural defects in plastids of *Selaginella kraussiana* var. *aurea. Can. J. Bot.* 58:1929–1937.

Wetmore, R. H.
1943. Leaf stem relationships in the vascular plants. *Torreya* 43:16–28.

Wetmore, R. H., and G. Morel
1951a. Sur la culture *in vitro* de prothalles de *Lycopodium cernuum. Compt. Rend. Acad. Sci.* (Paris) 233:323–324.
1951b. Sur la culture du gametophyte de Sélaginelle. *Compt. Rend. Acad. Sci.* (Paris) 233:430–431.

Whitebread, C.
1941. Beware of "*Lycopodium*"! *Amer. Fern Jour.* 31:100–102.

Whittier, D. P.
1977. Gametophytes of *Lycopodium obscurum* as grown in axenic culture. *Can. Jour. Bot.* 55:563–567.
1981. Gametophytes of *Lycopodium digitatum* (formerly *L. complanatum* var. *flabelliforme*) as grown in axenic culture. *Bot. Gaz.* 142:519–524.

Wilce, J. H.
1972. Lycopod spores, I. General spore patterns and the generic segregates of *Lycopodium. Amer. Fern Jour.* 62:65–79.

Wilder, G. J.
1970. Structure of tracheids in three species of *Lycopodium. Amer. Jour. Bot.* 57:1093–1107.

Williams, S.
1931. An analysis of the vegetative organs of *Selaginella grandis* Moore, together with some

observations on abnormalities and experimental results. *Trans. Roy. Soc. Edinb.* 57:1–24.
1937. Correlation phenomena and hormones in *Selaginella. Nature* 139:966.

Wochok, Z. S., and I. M. Sussex
1974. Morphogenesis in *Selaginella.* II. Auxin transport in the root (rhizophore). *Plant Physiol.* 53:738–741.
1976. Redetermination of cultured root tips to leafy shoots in *Selaginella willdenovii. Plant Sci. Letters* 6:185–192.

Zamora, P. M.
1958. Anatomy of the protoxylem elements of several *Selaginella* species. *Philippine Jour. Sci.* 87:93–114.

Zinda, D. R.
1966. A preliminary report on the comparative morphology of the shoot apex of *Isoetes macrospora* Dur. and some effects of experimentally applied indole-3-acetic acid and gibberellic acid. *Jour. Minnesota Acad. Sci.* 33:107–116.

CHAPTER 10

Sphenophyta

THE Sphenophyta is a well-defined group of living and extinct plants. The oldest fossil remains are from the Upper Devonian, and during the Carboniferous Period the group attained almost worldwide distribution and significant diversity of growth form. Arborescent and smaller sphenopsids coexisted during this period; however, by the end of the Jurassic Period (Mesozoic Era) only a few representatives remained, and they were relatively small plants. Today only one genus, *Equisetum,* with fifteen species, remains of this once conspicuous and diversified group. There is some evidence that *Equisetum* itself may have been present in the Carboniferous and may not have undergone any significant change since then. If this is so, *Equisetum* may be one of the oldest living genera of vascular plants in the world today.

The most conspicuous external morphological feature of the group as exemplified by *Equisetum* is the subdivision of the shoot axis into definite nodes and internodes—that is, the stem is jointed (Figs. 10-1, A, B; 10-2; 10-3). At the nodes are whorls of relatively small leaves that alternate with branches. In addition to the stem joints there are definite, easily observed longitudinal ridges on the stem. The reproductive structures of the sporophyte are terminal cones or strobili. The strobilus consists of an axis with whorls of stalklike structures (sporangiophores) with attached groups of eusporangia (Figs. 10-11, A, B; 10-12, A).

Classification

Hyenia, a fossil from the Middle Devonian, has been described as a possible early sphenopsid. The genus is now considered to be better aligned with the early ferns (order Cladoxylales) based upon the lack of the characteristic whorled arrangement of appendages, type of branching, and lack of any knowledge of its stem anatomy (Schweitzer, 1972). *Calamophyton,* a companion of *Hyenia,* also was removed from the sphenopsids (Leclercq and Schweitzer, 1965). However, Stewart (1983) believes that the two genera represent a "transitional group not far removed from their trimerophyte ancestors and sharing characteristics with the sphenopsids and Cladoxylales." Also, the extinct arborescent sphenopsids of the Carboniferous were previously placed in the order Calamitales. Paleobotanists are now inclined to combine the Calamitales with the Equisetales, based upon many morphological similarities.

SPHENOPHYTA: Living and extinct plants; sporophyte differentiated into stem, leaf, root, and eusporangium; jointed stems; monopodial

A

B

FIGURE 10-1 *Equisetum telmateia* subsp. *braunii*. **A,** colony growing along roadside embankment, Berkeley, California; **B,** series of plants arranged to show stages in the growth and expansion of vegetative shoots. [Courtesy Mr. Louis Arnold.]

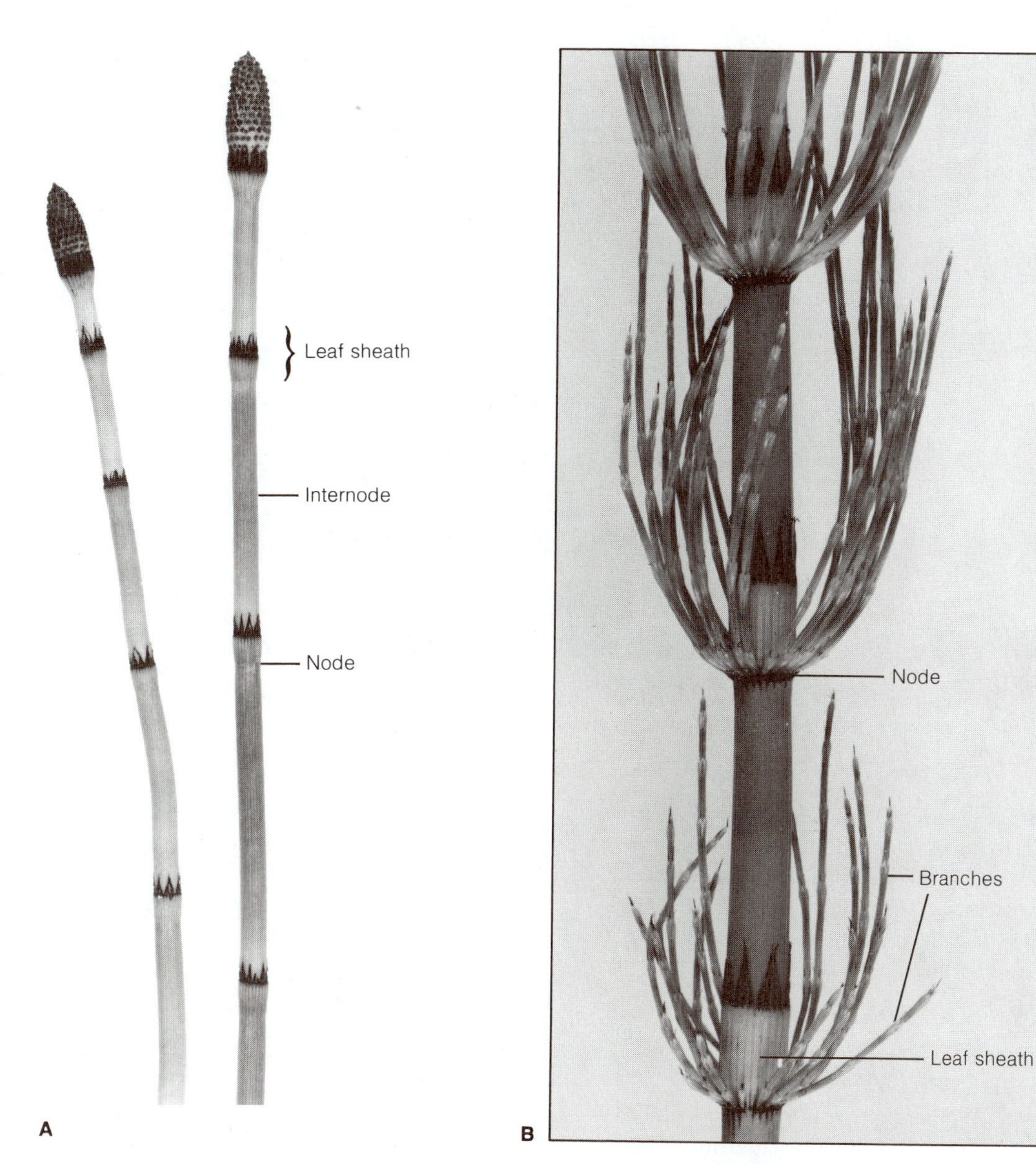

FIGURE 10-2 *Equisetum* shoots. **A,** *Equisetum hyemale,* portions of two unbranched shoots with terminal strobili. **B,** *Equisetum telmateia* subsp. *braunii,* portion of sterile (vegetative) shoot showing leaf sheaths and lateral branches.

branching; stem protostelic or siphonostelic; xylem exarch or endarch; secondary growth in some forms; sporangia borne on specialized stalks (sporangiophores) that are organized into strobili; mostly homosporous, some heterosporous (extinct).

EQUISETALES: Upper Devonian to present; living and extinct plants; ribbed stems; whorls of microphylls that are fused to form a sheath around the stem, or fused only at their bases; stems siphonostelic with endarch xylem; conspicuous protoxylem lacunae (carinal canals); internodal pith cavities; spores with elaters.

EQUISETACEAE: Permian to present; no secondary growth; microphylls fused to

FIGURE 10-3 *Equisetum scirpoides.* A small species having unbranched aerial shoots with terminal strobili; grown under greenhouse conditions.

form a sheath around stem; strobili consisting only of whorled sporangiophores; strobili terminal on main vegetative stem, or occasionally on branches, or on a specialized strobiliferous stem; homosporous.

Equisetum, Equisetites (extinct).

CALAMITACEAE: Extinct plants from Carboniferous to Lower Permian; arborescent; secondary growth; microphylls fused at their bases to form a collar around the stem at the node; strobili consisting of (1) whorls of closely associated sporangiophores and bracts, or (2) alternate whorls of sterile bracts and sporangiophores; some heterosporous.

Calamites (pith casts, stem impressions); *Arthropitys* and *Calamodendron* (stem petrifactions); *Annularia* and *Asterophyllites* (leaves); *Palaeostachya* and *Calamostachys* (strobili).

SPHENOPHYLLALES: Extinct plants from Carboniferous to Lower Permian; small plants, either upright or trailing; stem protostelic, exarch xylem; secondary growth; leaves commonly with dichotomous venation; strobili composed of whorls of bracts and sporangiophores, the latter generally partially fused to bracts.

Sphenophyllum(stems, leaves, roots); *Bowmanites* and *Sphenostrobus* (strobili).

PSEUDOBORNIALES: Extinct plants from Upper Devonian; arborescent; three orders of branching; leaves in whorls on ultimate branches; strobili terminal on first order branches in upper portion of plant.

Pseudobornia ursina.

Equisetales – Equisetaceae: *Equisetum*

The sporophytes of *Equisetum,* with their characteristic jointed stems and rough texture, have several names — "horsetails" and "scouring rushes" are the most popular. During the American colonial period, horsetails were used as scouring agents, as they probably are today in some regions of the world. The American Indians used the stems of

Equisetum as an abrasive for polishing bows and arrows (Lloyd, 1964).

The horsetails are worldwide in distribution today, except for Australia and New Zealand. They generally grow in wet or damp habitats, being particularly common along the banks of streams or irrigation ditches (Fig. 10-1, A); some have become adapted to dry or mesophytic conditions, for at least a part of the year. In some localities they are a serious weed problem for farmers and a matter of concern for livestock owners because of the poisonous substances in the stems of some species. Horses are especially susceptible to their toxins. However, a certain species in Costa Rica is used medicinally, as a treatment for human kidney trouble (Hauke, 1969c). American Indians prepared infusions of certain species for treatment of various medical problems—to counteract diarrhea, to rid the hair of vermin, and as an eye wash (Lloyd, 1964).

Certain species are indicators of the mineral content of the soil in which they grow (Vogt, 1942). These plants accumulate minerals, including gold, up to 4½ ounces per ton (Benedict, 1941). This source of mineral information is therefore of some value in prospecting for new ore deposits.

The horsetails, despite the range of variation in branching and shoot dimorphism, are usually not mistaken for some other group, and most botanists recognize the one genus *Equisetum*. However, some specialists have proposed that at least two genera be recognized. Consequently, the genera *Equisetum* and *Hippochaete*, or even three genera, have been described. Hauke (1969a) has pointed out that there are some real and consistent differences between the two major groups of species, but that they are minor, and the number of similarities outweigh the differences. Therefore, he could see no gain in establishing two genera instead of one. Two subgenera, however, recognize the differences (Hauke, 1974). Table 10-1 summarizes the main differences between the subgenera. (For earlier literature and taxonomic analyses, see Rothmaler, 1944, and Hauke, 1961, 1962a, b, c, 1963, 1978).

Table 10-1 Subgenera of *Equisetum*

	Equisetum subgenus *Equisetum*
1	Aerial stems generally branched, annual or short lived.
2	Strobili borne on chlorophyllous vegetative stems or on nonchlorophyllous stems, lacking apiculum (not pointed).
3	Stomata on stems scattered in the furrows between the ridges, or in bands one to three stomata wide.
4	Stomata flush with the epidermis.
5	Upright, platelike lamellae on gametophytes.
6	Projecting antheridia, generally with more than two opercular cells.
	Examples: *E. arvense, telmateia, sylvaticum*
	Equisetum subgenus *Hippochaete*
1	Aerial stems generally unbranched, and may function for several seasons.
2	Strobili borne on chlorophyllous vegetative shoots, apiculate.
3	Stomata on stems in one or two longitudinal rows in the furrows between the ridges.
4	Stomata sunken below surface of the epidermis.
5	Upright, columnlike lamellae on gametophytes.
6	Antheridia sunken, with two opercular cells.
	Examples: *E. hyemale, scirpoides, laevigatum*

Organography

The shoot system of *Equisetum* consists of an aerial portion and an underground rhizome system that exhibit monopodial branching. The sporophyte is perennial, at least the rhizome is, even though the aerial shoots may die back during part of the year. In some species the rhizomes may reach soil depths of 2 to 3 meters. Some species are small, particularly those of arctic and alpine regions. Some species grow near or within the Arctic Circle in Europe, Siberia, and North America. A South American species, *E. giganteum*, may reach a height of 5 meters; the stems are relatively small in diameter, and supported partly by surrounding tall grass.

The stem of *Equisetum* has a very rough texture, and an irregular surface pattern that is especially evident in scanning electron micrographs of the stem surface (Fig. 10-4). Stomata occur in one or two longitudinal rows either in the furrows between the ridges, scattered in the furrows, or in bands one to three stomata wide. The extremely rough texture results from the deposition of silica as discrete knobs (Fig. 10-4, D) and rosettes on the epidermal surface (e.g., in *Equisetum arvense)* or in a uniform pattern (Fig. 10-4, A, B) on and in the

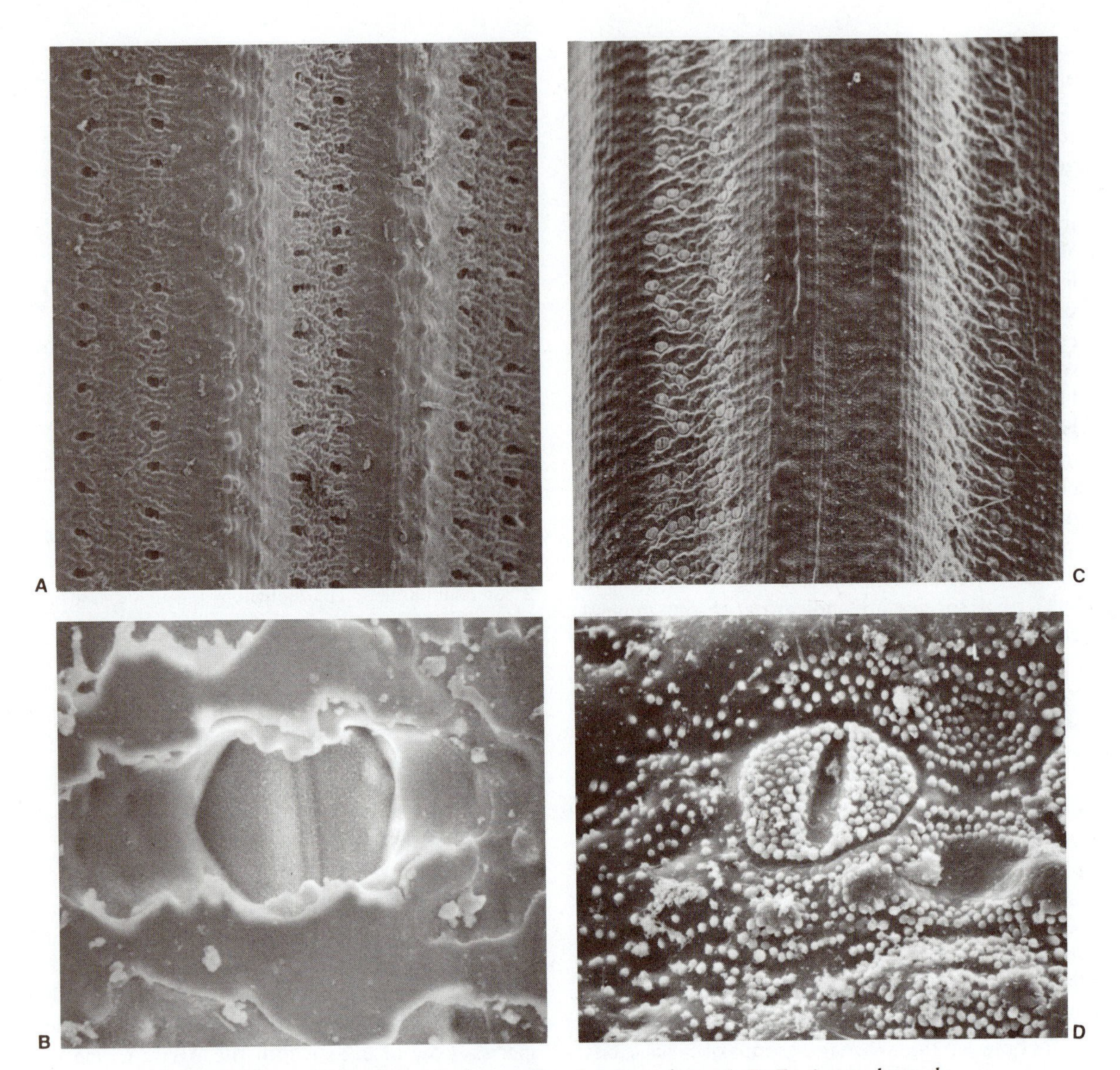

FIGURE 10-4 Scanning electron micrographs of stem surfaces. **A, B,** *Equisetum hyemale,* showing two rows of stomata in the valleys between the ribs (**A**), and one sunken stoma (**B**). **C, D,** *Equisetum telmateia* subsp. *braunii,* showing bands of stomata (**C**), and one stoma (**D**). (**A, C** × 50; **B, D** × 700)

entire outer epidermal cell walls (e.g., in *Equisetum hyemale,* Kaufman, et al., 1971). Silica is an essential element for normal growth of Equisetum (Chen and Lewin, 1969) and may play an important role in maintaining erectness of the plant, compensating for the very low lignin content of the cell walls. Other functions have been proposed: protection against pathogens and predators and prevention of excessive water loss (Kaufman et al., 1971). Silicon is necessary for the completion of the life cycle. Sporophytes, produced on gametophytes, formed viable spores only if the artificial culture medium contained at least 20 milligrams per liter of silicon (Hoffman and Hillson, 1979).

The shoot system of all species of *Equisetum,* comprising an underground rhizome and an aerial shoot, possesses the same fundamental organization. The stem is divided into definite nodes and internodes, and leaves are attached at nodes, united at least for a part of their length, forming a sheath around the stem (Fig. 10-2). The number of leaves per node varies according to the species and the position on the stem in most species. Where adjacent nodes have the same number of leaves in the leaf sheath, there is a very precise type of symmetry. The number of leaves at a node corresponds to the number of ribs on the internode below. Opposite each rib is a vascular bundle which enters the leaf that is on a line with the rib (Figs. 10-7, A; 10-10). However, careful examination of an aerial shoot reveals that the number of leaves per node generally increases from the base for some distance and then decreases. The type of vascularization just described is not observed for this growth pattern; adjustments in vascular connections occur such that some leaves of a whorl may have two vascular bundles instead of one (Bierhorst, 1959).

An internode may continue to grow for some time through the activity of an intercalary meristem located at its base. Lateral branches, when evident, are attached at the nodes and alternate with the leaves. During development, each branch breaks through the lower part of the leaf sheath in reaching the exterior. Whether a species is highly branched or not, branch and root primordia are formed at each node (Fig. 10-5). In the rhizome a root primordium develops into a root; a branch primordium may develop into an erect aerial shoot, remain as an arrested bud, develop into a new rhizome, or become abortive. In aerial shoots (if branching is a characteristic of the species) the branch primordia develop while the roots remain in an arrested condition unless the stems become procumbent and come to lie on a moist surface. This information can be used in vegetative propagation. A plant may be propagated vegetatively also in some species from arrested tuberous branches which occur at the nodes of rhizomes. The propensity for vegetative propagation is the major factor in *Equisetum* becoming a weed that is difficult to control in some localities. A small detached branch or tuber could be the start of the colonization of a very large area under favorable conditions for growth. For a farmer or cattle rancher to eradicate the last bit of rhizome is almost an impossible task!

The rhizome, and the aerial stem, can be pulled apart into internodal lengths or "pipes." In an interesting study, Treitel (1943) found that before a rhizome actually breaks under tension its elasticity closely compares with that of muscle or rubber.

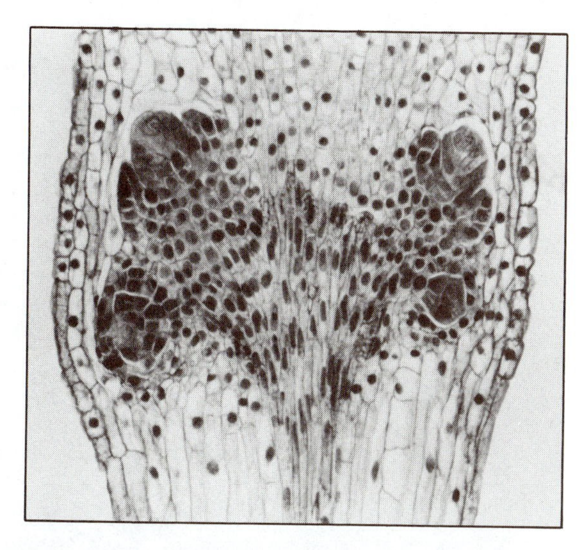

FIGURE 10-5 *Equisetum scirpoides.* Stem longitudinal section showing bud and root primordia at a node. Bud primordia toward top of figure.

Abnormal growths or monstrosities (teratological forms) are always of interest. Certain shoots may have unusually short internodes. Other shoots may have flexuous (snakelike) stems or have continuous spiral leaf sheaths, exhibit dichotomous branching, or produce a vegetative shoot beyond the usual terminal cone (Schaffner, 1933). All of these abnormalities undoubtedly are the result of an unbalanced growth-regulating system.

Strobili may occur terminally on many of the main vegetative axes of a highly branched species (*E. myriochaetum*), but more often a single strobilus terminates the main axis, whether the plant is branched or not (Fig. 10-2, A). Shoot dimorphism or segregation of function occurs in certain species: some shoots are green and purely vegetative (Figs. 10-1, A; 10-2, B), and others are fertile, unbranched (Fig. 10-6), brownish in color, and have terminal strobili (e.g., *E. arvense*). Still other species (e.g., *E. sylvaticum*), in addition to having purely green vegetative shoots, produce brownish, fertile shoots which may develop chlorophyll after the spores are shed; green branches then grow out from the nodes of the stem.

FIGURE 10-6 *Equisetum arvense.* Photograph taken in early spring showing unbranched fertile shoots terminated by strobili; some branched sterile (vegetative) shoots are evident to the right. [Courtesy of Dr. Richard L. Hauke.]

Anatomy of the Mature Stem

The stem of *Equisetum,* as seen in transverse section, has prominent ridges. The epidermal cell walls are thick and have a generous deposition of siliceous material. Distribution of stomata in the furrows between the ridges varies according to the subgenus (Table 10-1). In the subgenus *Hippochaete,* the stomata are sunken, consisting of two guard cells and overarching subsidiary cells; the latter have ridges on their inner walls (Hauke, 1957). Beneath the epidermis is the cortex, which exhibits a varied type of organization. The outer cortex is composed of collenchyma (Brown, 1976) and thinner-walled chlorenchyma. The distributional pattern of the two types of tissue varies according to species. Collenchyma is often excessively developed opposite the ribs (e.g., *E. hyemale,* Fig. 10-7, A), and a large longitudinal air space (vallecular canal) is present between the ridges.

Opposite each ridge is a vascular bundle of unusual interest. A transverse section of the internode of a mature stem reveals a protoxylem lacuna (carinal canal) associated with a vascular bundle (Fig. 10-7, A). The carinal canal is formed during elongation of an internode by the separation and disruption of protoxylem elements. In a mature stem one or more late-maturing tracheids can be seen along the edge of the carinal canal (Fig. 10-7, B). The parenchyma cells lining the canal develop thick walls, and the canal itself functions in water conduction (Bierhorst, 1958a), as shown by experiments with dyes. Opposite the carinal canal is the primary phloem flanked on each side by strands of xylem (Fig. 10-7, B), the ontogeny of which has been variously interpreted. In the past these strands have been considered to be the metaxylem of the bundle (Golub and Wetmore, 1948a, b), but Bierhorst (1958a) considers this portion of the xylem to be separate and distinct (developmentally) from the carinal group; he refers to the flanking strands as "lateral xylem." Authors do agree that, despite some irregularities, maturation of the tracheary elements in the lateral xylem is centrifugal. Also, vessels have been reported to occur in the lateral xylem of the rhizomes of five species (Bierhorst, 1958b).

Protophloem elements of the centripetally differentiating phloem are small in diameter, whereas those of the metaphloem are large (Fig. 10-7, B). End walls of the metaphloem sieve elements are

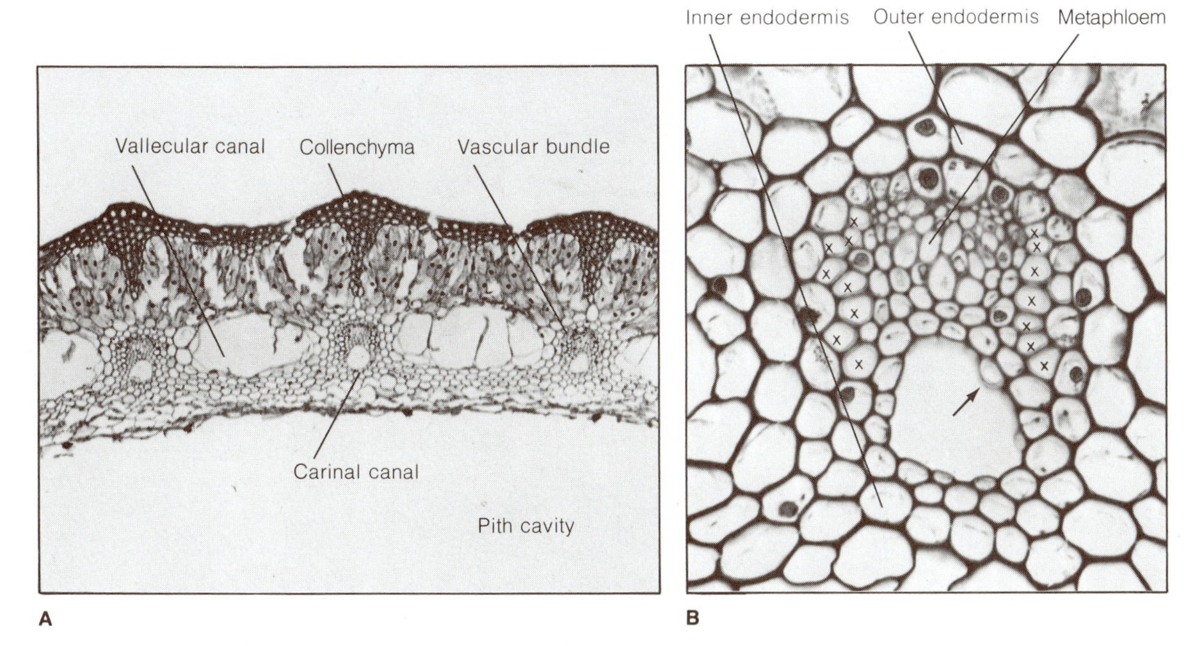

FIGURE 10-7 Stem anatomy in *Equisetum hyemale.* **A,** transection, portion of stem (note that carinal canals are opposite ridges on stem); **B,** structure of vascular bundle; one protoxylem tracheid is visible along edge of carinal canal at right (arrow); cross marks designate tracheids of the metaxylem(lateral xylem).

transverse to slightly oblique with callose-lined pores that are larger than those occurring on lateral walls (Lamoureux, 1961). Structure of a sieve element is much like a sieve-tube member of an angiosperm.

Depending on the species, one endodermal layer may be present outside the cylinder of vascular bundles (Fig. 10-8, A), or there may be, in addition, an inner endodermis (Fig. 10-8, B). In other species, an endodermis may completely surround each vascular bundle (Fig. 10-8, C). A pericycle may or may not be evident. Sometimes the pericycle cells are filled with starch. The vascular cylinder of *Equisetum* has been described as an ectophloic siphonostele of the "*Equisetum* type," or as a "perforated" ectophloic siphonostele (Schmid, 1982).

The stem nodes are unusual in that the xylem is extensively developed as a conspicuous circular ring. There are no carinal or vallecular canals. In addition, a nodal plate of pith tissue separates one internode from another.

The tips of underground rhizomes are covered by scalelike leaf sheaths, which may secrete mucilage from their abaxial epidermal cells. Trichomes may also be present on the adaxial side of the sheaths. There is collenchyma in the outer cortex. Vallecular canals and protoxylem lacunae are present. An endodermis around each vascular bundle is common (Fig. 10-8, C). There is often no consistency in the arrangement and number of endodermal layers in the aerial stem and rhizome of the same plant (Fig. 10-8, A, B).

Development of the Shoot

The shoot apical meristem is dominated by a conspicuous tetrahedral cell, the apical cell, that has three lateral ("cutting") faces that are directed downward. The fourth side is the rounded cap (Fig. 10-9, A, B). The apical cell cuts off daughter cells (segments) in a continuous sequence, so that three tissue regions are established (Gifford and Kurth, 1983). Each segment divides anticlinally. By subsequent divisions of these cells, a pith meristem and a lateral circular bulge are formed. The lateral bulge is

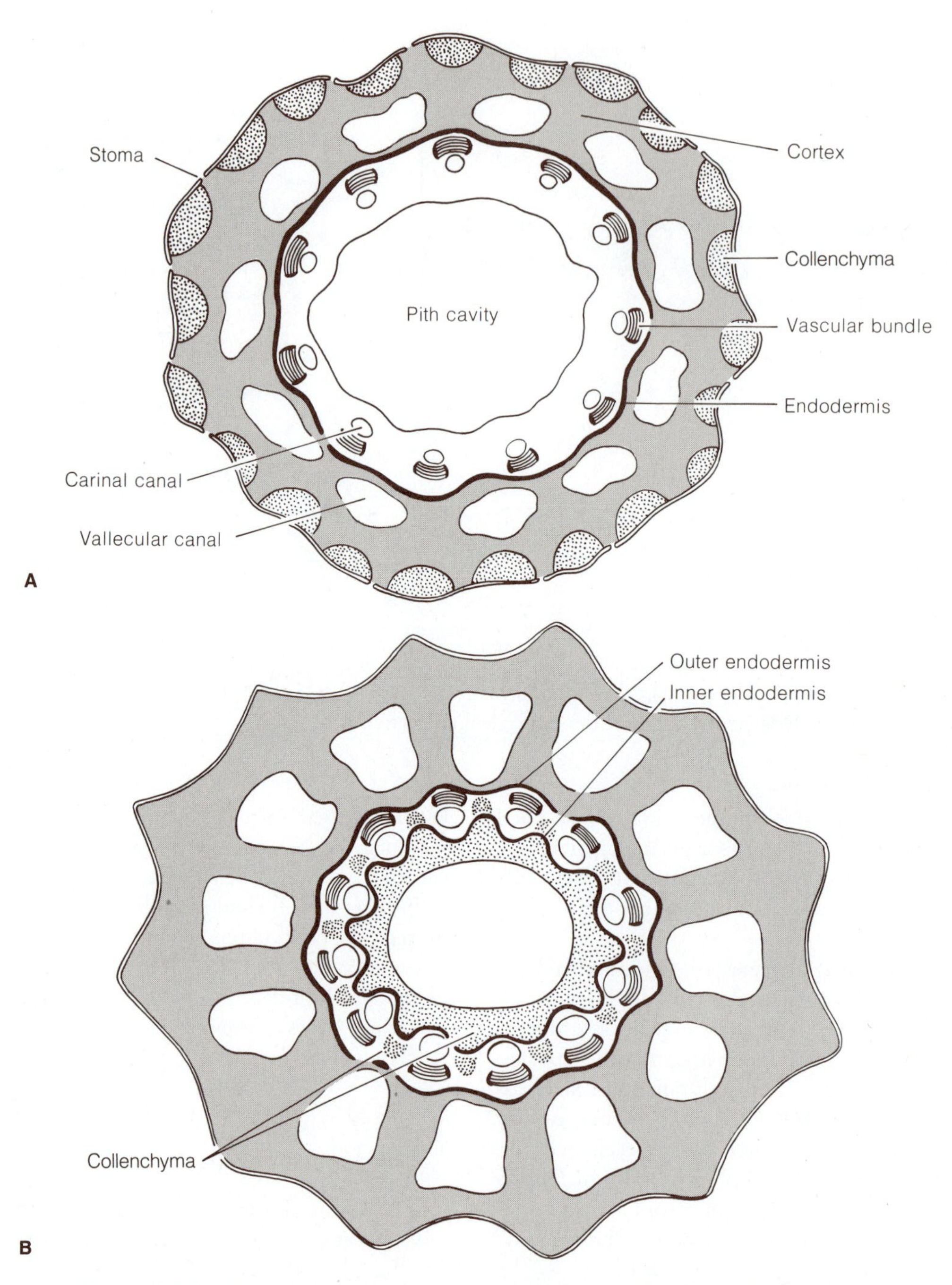

FIGURE 10-8 Schematic representations of variations in configuration and position of the endodermis in two species of *Equisetum*. **A,** aerial stem, *Equisetum sylvaticum;* **B,** rhizome, *Equisetum sylvaticum;* **C,** rhizome, *Equisetum hyemale.*

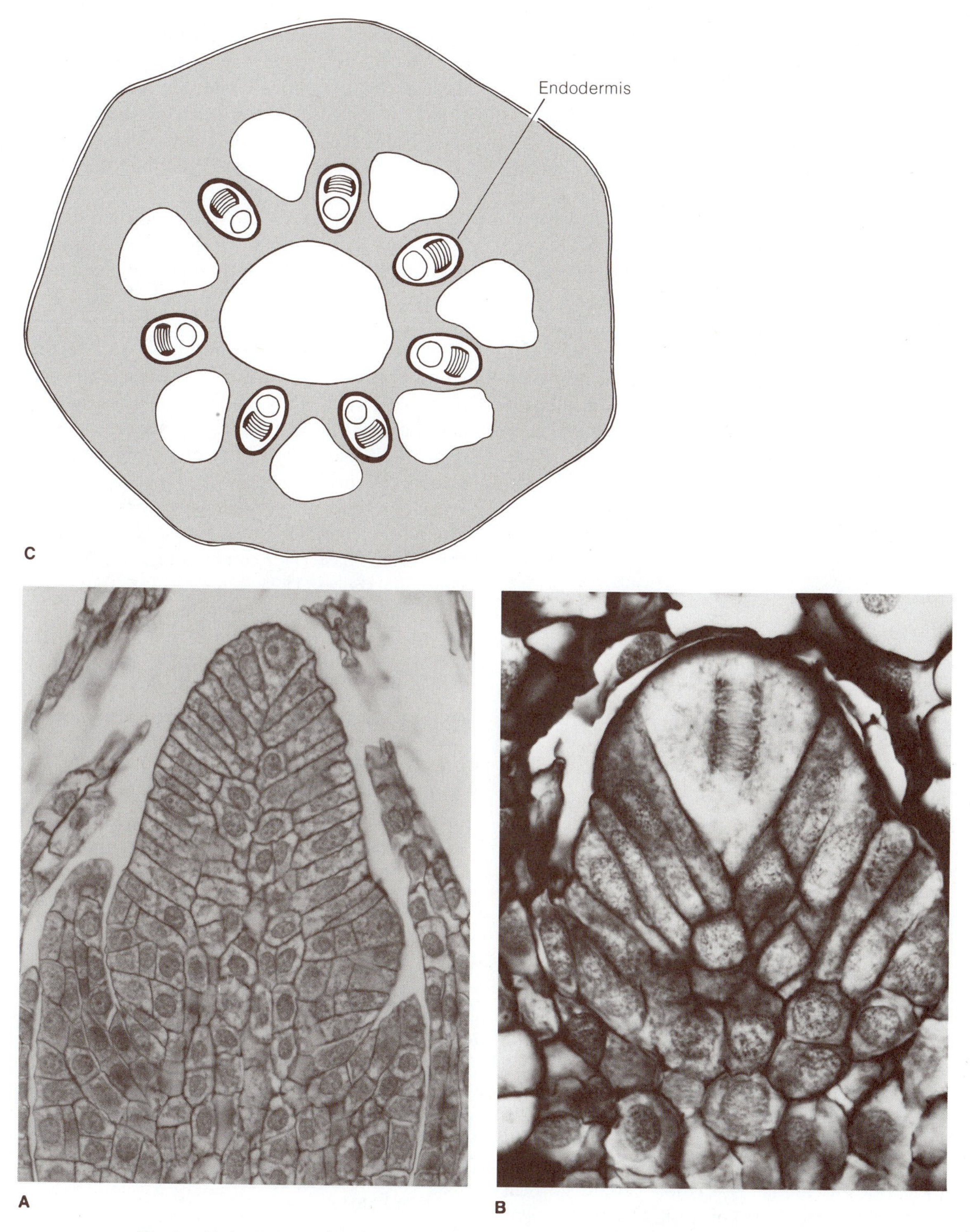

FIGURE 10-9 **A, B,** median longitudinal sections, shoot tips of *Equisetum;* note prominent apical cell in each, and the initiation of leaf primordia at the base of the apical cone in (**A**); apical cell in (**B**) is in anaphase. [**B** courtesy of Dr. D. W. Bierhorst.]

reported to consist of five to six layers in *E. arvense* (Golub and Wetmore, 1948a). Cells of the upper two layers will produce the future node and leaves; the lower layers will form the future intercalary meristem of the internode.

Very early in development, leaf-sheath initials can be distinguished as certain regularly placed surface cells of the upper two layers. A leaf then grows in length and widens at its base, eventually becoming fused to varying degrees with adjacent leaves, thus forming the characteristic leaf sheath (Fig. 10-2, A, B). At the time leaf initials are apparent, procambial cells become differentiated in the nodal and internodal regions near the pith. The results of one study (Golub and Wetmore, 1948b) indicate that the procambium in *E. arvense* is acropetally continuous throughout the shoot. (For additional information on mitotic activity, early shoot histogenesis, and determination of DNA content of various cells of the apical meristem, see Gifford and Kurth, 1983).

Vascularization

There has been considerable controversy regarding the longitudinal differentiation and maturation of xylem and phloem (see Golub and Wetmore, 1948b). In *E. arvense,* the differentiation of protoxylem and protophloem within a procambial strand begins at about the fourth node. Differentiation then proceeds acropetally into a developing leaf and basipetally into the internode below. In the elongation of older internodes the protoxylem and protophloem elements become stretched and torn because the intercalary meristem, located at the base of each internode, does not contribute new cells to the procambial strand. It is during this period of rapid elongation that protoxylem lacunae (carinal canals) are formed. After considerable elongation of an internode, the differentiation of the lateral xylem and metaphloem proceeds basipetally from a node through the internode below, and continuity is finally achieved with the same tissues in the node below.

During internodal elongation certain cortical cells separate from each other and form vallecular canals. The pith cavity is reported to be formed by mechanical tearing of cells (Golub and Wetmore, 1948b).

Following a vascular bundle up through an internode reveals a trichotomy of the protoxylem at the level of the leaf attachment. The median bundle enters the scale leaf, the two laterals diverge right and left, and each is joined laterally with an adjacent strand to form one of the vascular bundles of the next higher internode (Fig. 10-10).

Branches occur at nodes on radii alternating with those of the leaves. At about the sixth node from the shoot tip, a lateral branch is initiated from a single surface mother cell. An apical cell is soon established, and a branch bud with whorled leaves is formed. At about the time the first or second whorl of leaves is formed, a large cell appears near the basal end of each branch bud. This is the root apical cell which may give rise to several segments (Fig. 10-5). The branch buds may continue to develop if the species is normally branched; the roots remain arrested in development except on underground rhizomes. Of course, some of the buds on the rhizome have the potential to grow upward as new aerial shoots. The vascular tissue at the base of the branch shoot is in the form of a continuous cylinder (siphonostele) and is in continuity with the xylem and phloem of the nodal region. There is tissue continuity between the pith of the parent axis and the pith of the branch axis. These interruptions in the vascular tissue at the nodes constitute branch gaps (Jeffrey, 1899).

The leaves at each node are united to varying degrees. The thick-walled cells of the abaxial epidermis have various types of ornamentation on their outer tangential walls. Internally the mesophyll is lacunate and is traversed by one small median vascular bundle. Mesarch xylem has been reported by several workers.

One interesting feature of the *Equisetum* leaf is the presence of specialized "water stomata" (hydathodes) on the adaxial surface of the leaf along the midvein region (Johnson, 1937). *Equisetum* secretes water as it is associated with the conditions of high moisture around the roots and a saturated atmosphere.

The Root

The primary root is ephemeral; all other roots in *Equisetum* arise at the nodes of stems. Cells of the outer root cortex often have thick walls, those of

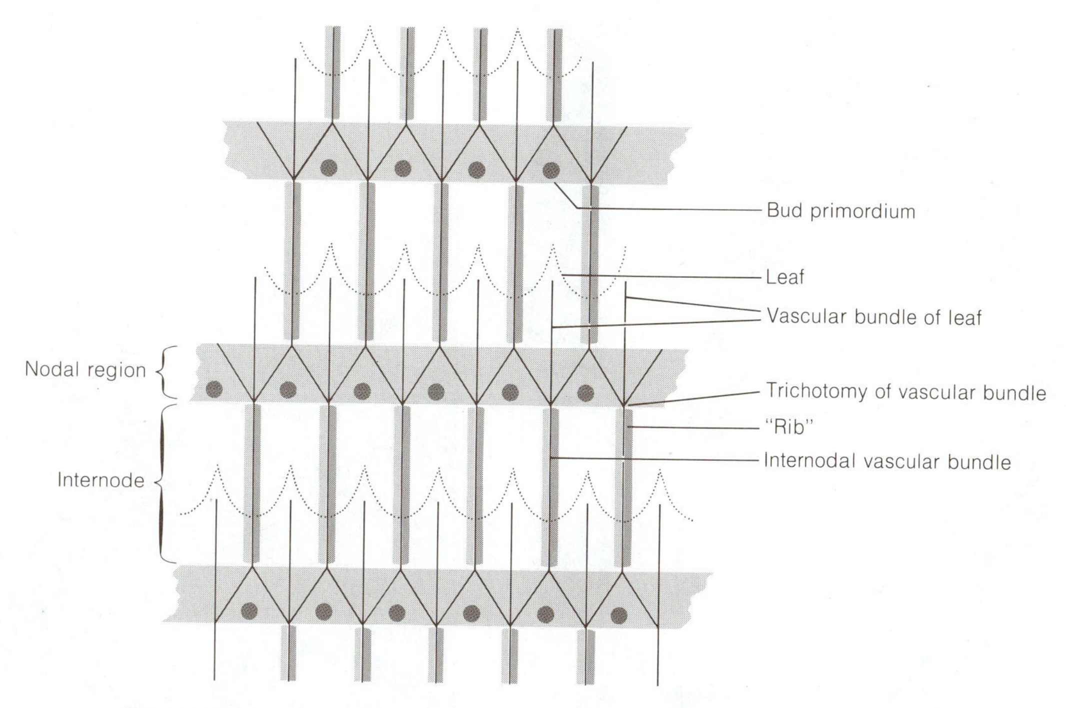

FIGURE 10-10 Schematic representation of a portion of the vascular system of *Equisetum,* shown in relationship to the "ribs," leaves, and buds.

the inner cortex are thinner. The xylem is triarch or tetrarch, or, in smaller roots, may be diarch. Small roots, as seen in transverse section, may have one large metaxylem element in the center of the root (Walton, 1944). The cells of the endodermis and pericycle occur on the same radii, indicating a common origin. Lateral roots have their origin in the endodermis.

A tetrahedral cell is the most prominent feature of the root apical meristem. A root cap is formed distally; the apical cell also produces segments laterally (proximally), each of which by two divisions gives rise to three cells, which in turn are the origin of the vascular tissue, cortex (including endodermis and pericycle), and epidermis (Johnson, 1933). Historically, the root apical cell in lower vascular plants was considered to be the ultimate source of cells of the root. In recent years evidence has been presented for the belief that the apical cell is active in cell division only during early stages of development. It then ceases to divide and the nucleus may become endopolyploid. Some surrounding cells assume the role of initials (Avanzi and D'Amato, 1967; D'Amato, 1975). Evidence from a more recent investigation on *E. scirpoides* supports the original tenet that the apical cell is active mitotically throughout root growth; it is not quiescent and does not become endopolyploid (Gifford and Kurth, 1982).

The Strobilus

The strobilus terminates an axis, whether it be on a vegetative stem or a strictly fertile nonchlorophyllous axis. The strobilus is composed of an axis with whorls of stalked, peltate structures termed sporangiophores (Fig. 10-11, A). Each sporangiophore is umbrellalike in shape, with pendant sporangia (five to ten in number) attached to the underside of the polygonal, disk-shaped shield (Fig. 10-12, A). The flattened tips of the sporangiophores fit closely together, providing protection for the sporangia during development. At maturity the cone axis elongates, separating the sporangiophores, and the sporangia open by a longitudinal cleft that is formed down the inner side of each sporangium. Additional protection during early development is provided by a rudimentary leaf sheath,

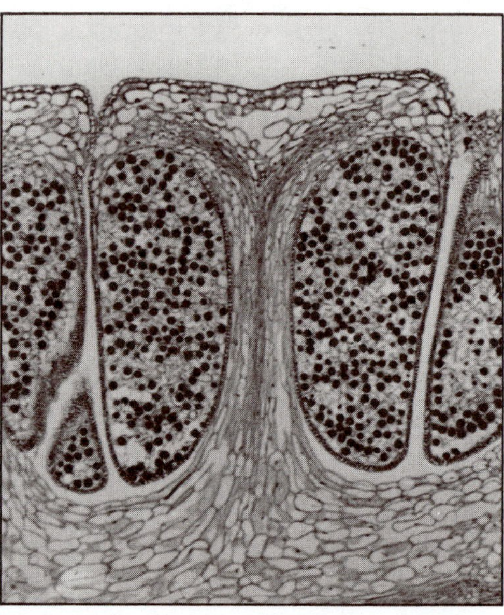

FIGURE 10-11 *Equisetum telmateia* subsp. *braunii*, **A,** fertile shoots (the strobili are at different stages of maturity; the youngest is to the left); **B,** median longisection of one sporangiophore showing the vascular bundle in the stalk and its mode of branching (crowded spores with elaters are evident in the sporangia).

the so-called annulus, at the base of the cone in some species.

The vascular cylinder of the cone axis is a network of interconnected vascular bundles. No large canals are formed as described for the stem. Vascular bundles (sporangiophore traces) diverge from the vascular cylinder at regular intervals and enter the successive whorls of sporangiophores. At the distal end of the sporangiophore the bundle is branched; each strand is recurved and ends near the base of a sporangium (Fig. 10-11, B).

Early in development, whorls of sporangiophore primordia arise in an acropetal manner on the flanks of the meristematic cone axis (Fig. 10-13). After enlargement of the sporangiophore primordium, sporangia are initiated in single superficial cells around the rim of the sporangiophore. The sporangium initial divides periclinally, setting aside an inner and an outer cell. The inner cell, by further divisions in various planes, produces sporogenous tissue (see Chapter 4, also Fig. 10-14). The outer cell, by anticlinal and periclinal divisions, gives rise to irregular tiers of cells, the inner of which may also become sporogenous; the outer tiers become the future sporangial wall cells. Superficial cells adjacent to the original initials may also contribute to the development of the sporangium. Before the sporocytes separate and round off prior to the meiotic divisions, two to three layers of cells adjacent to the sporogenous mass differentiate as the tapetum. In addition, not all of the sporogenous cells function as sporocytes; many degenerate and their cytoplasm, together with that of the tapetum, forms a multinucleate nourishing substance which occupies the spaces between the sporocytes.

The results of one study indicate that the first

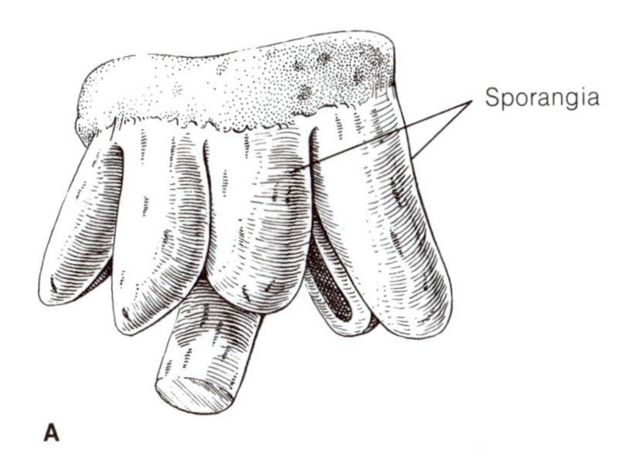

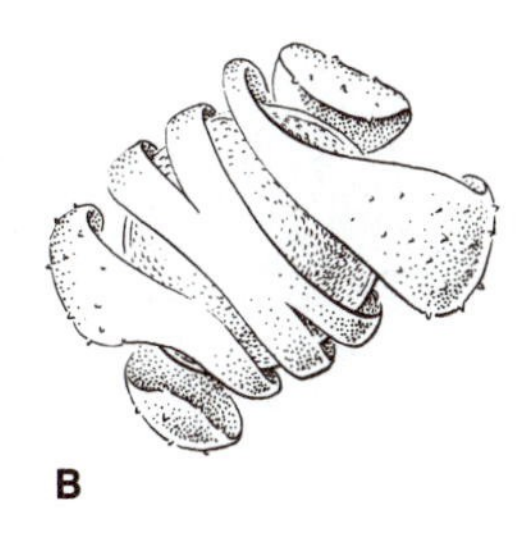

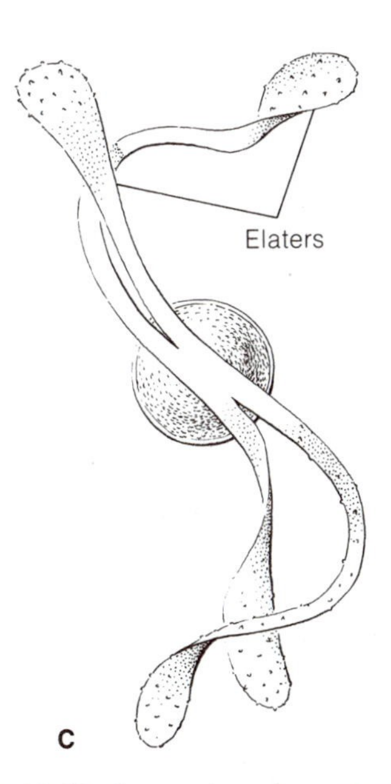

FIGURE 10-12 Sporangia and spores of *Equisetum.* **A,** a single sporangiophore with pendant sporangia; **B,** a mature spore whose coiled elaters indicate that it is moist; **C,** a spore whose uncoiled elaters indicate that it is dry.

FIGURE 10-13 Scanning electron micrograph of a young strobilus of *Equisetum hyemale,* showing whorls of developing sporangiophores (×50).

sporangia to mature in a strobilus are sporangia situated in the widest part of the cone. Furthermore, within a single sporangium the sporocytes may be in various stages of meiosis. This may be related to the fact that the sporocytes are separated into pockets surrounded by the multinucleate plasma (Manton, 1950).

After the meiotic divisions have taken place, the spore tetrads separate from one another, and each spore becomes spherical. The spore wall is said to be laminated. The outer layer is deposited on the spore in the form of four bands, derived presumably from the breakdown products of the nonfunctional sporocytes and tapetal cells. The four bands are attached to the spore wall at a common point and remain tightly coiled around the spore until the sporangium is completely mature. The spores are filled with densely packed chloroplasts, a feature relatively uncommon to other lower vascular plants.

At maturity a sporangium consists of an outer wall composed of two layers of cells, the inner of which is generally compressed; the cells of the outer layer develop helical thickenings similar to tracheids which presumably are involved in sporangial dehiscence. Internal to the wall is the mass of

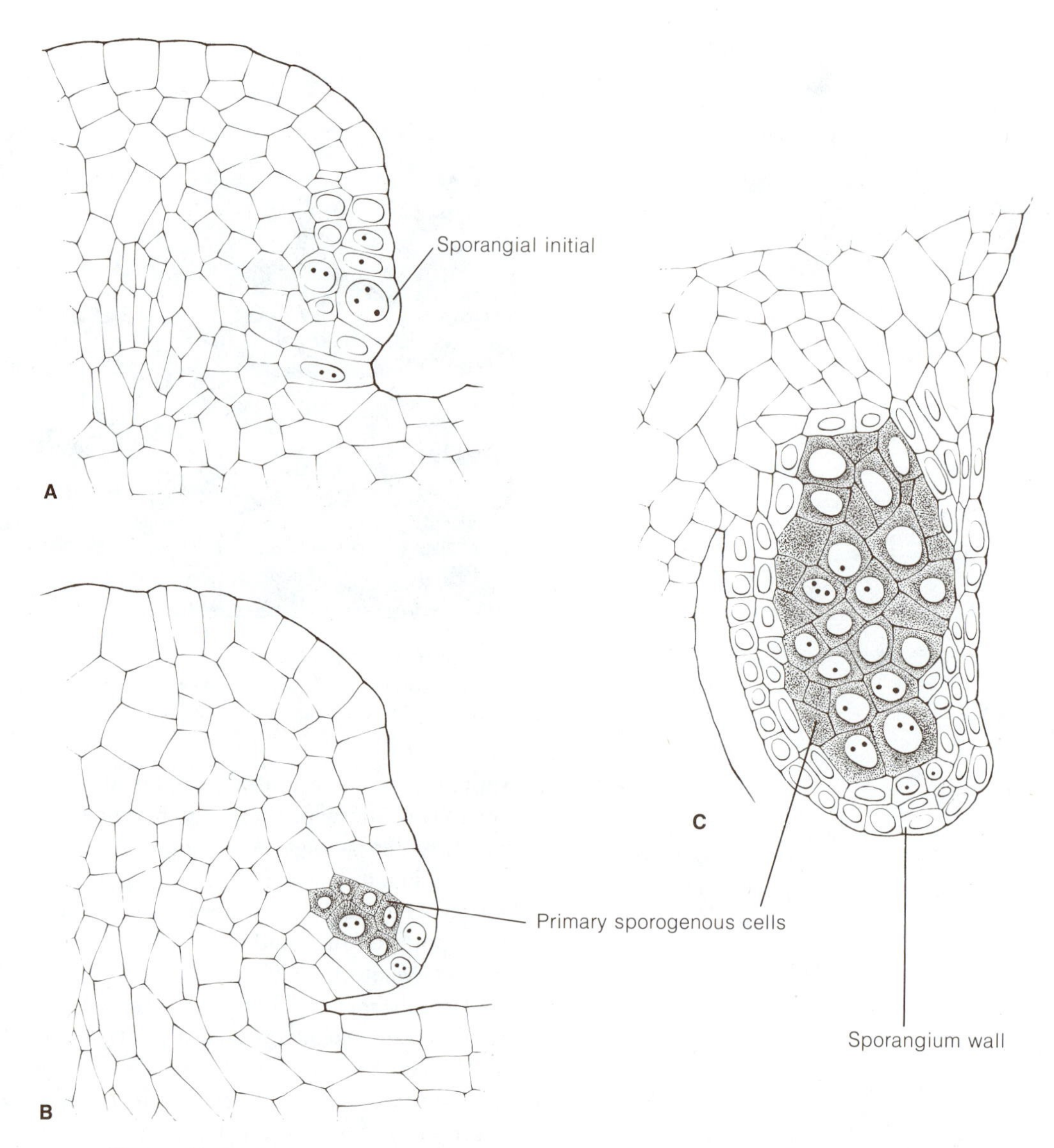

FIGURE 10-14 Early ontogeny of the sporangium in *Equisetum hyemale.* Initiation occurs in a single superficial cell (**A**), although lateral derivatives of the sporangial initial as well as adjacent superficial cells may contribute to the formation of primary sporogenous cells and primary wall cells. (See text for details.)

spores. At the time of dehiscence the free ends of the four bands, called elaters, separate from the spore wall. The elaters are hygroscopic, uncoiling as their water content decreases and recoiling with the addition of moisture (Fig. 10-12, B, C). It has been assumed that through this action the elaters assist in the dehiscence process and also bring about the dispersal of spores in large clumps from the sporangium.

Chromosome Numbers and Hybridity

In the species that have been studied critically, the chromosome number is $n = 108$ (Manton, 1950; Bir, 1960). Chromosomes vary in size and shape between the two subgenera (*Equisetum* and *Hippochaete;* Table 10-1), but the uniformity of the chromosome number may be indicative of a highly derived level and the extinction of lower numbers

(Tryon and Tryon, 1982). The high chromosome number could be indicative of a high degree of polyploidy. However, the data from enzyme electrophoresis of three species indicate that *Equisetum* possesses, in general, the number of isozymes typical of diploid angiosperms and gymnosperms (Soltis, 1986). Thus, for most enzymes examined, these plants are genetically diploid despite the high chromosome number. Hybridization is common in both subgenera, and this is reflected in the observed peculiarities at meiosis and the production of abortive spores. Hybridization has been responsible for much confusion in the taxonomy of the genus. In an intensive study of the subgenus *Hippochaete*, Hauke (1962c) described six hybrids between various pairs of seven species. Five hybrids in the subgenus *Equisetum* occur in North America (Hauke, 1983); also, artificial hybrids have been produced (Duckett, 1979b; Duckett and Page, 1975). Even if a hybrid produced nonviable spores, the sporophyte resulting from the initial cross could propagate itself very efficiently through fragmentation and growth of rhizomes (Hauke, 1969a).

Gametophyte Generation

Under natural conditions the gametophytic plants may be found growing in damp areas, on

A

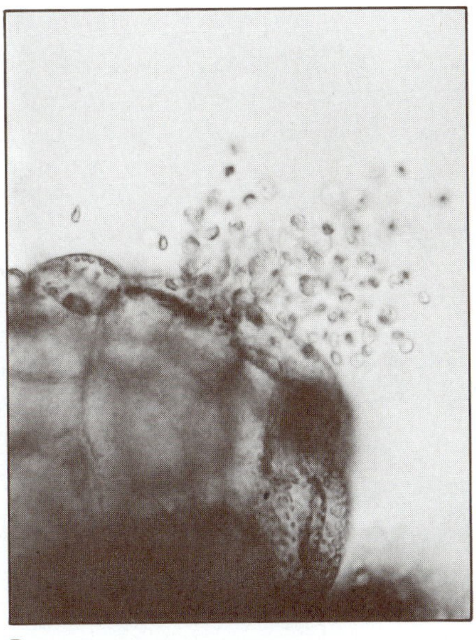

B

FIGURE 10-15 **A,** gametophytes of *Equisetum* sp. grown on simple inorganic nutrient medium; young sporophytes attached to gametophytes can be seen toward upper edge of figure. **B,** liberation of sperms from an antheridium.

mud, along creek banks, and even on the damp floors of abandoned mines and quarries. Mature plants may range in diameter from a few millimeters to 1 centimeter or more. A tropical species (Kashyap, 1914) and one from California (Mesler and Lu, 1977) may be even as large as 3 to 3.5 centimeters in diameter. An older plant may be uniform in outline; viewed from above, it resembles a miniature pin cushion (Fig. 10-15, A).

The time interval between shedding of spores and germination is quite critical. Spores germinate very readily if they land in a suitable environment; however, the limiting factor seems to be the amount of available water. If the spores do not germinate at once their viability decreases rapidly.

The first division of the spore results in two cells unequal in size. The smaller cell, which has fewer chloroplasts, elongates and forms a rhizoid (Fig. 10-16, A, B). The larger cell may divide transversely, or the division may be perpendicular to the original wall.

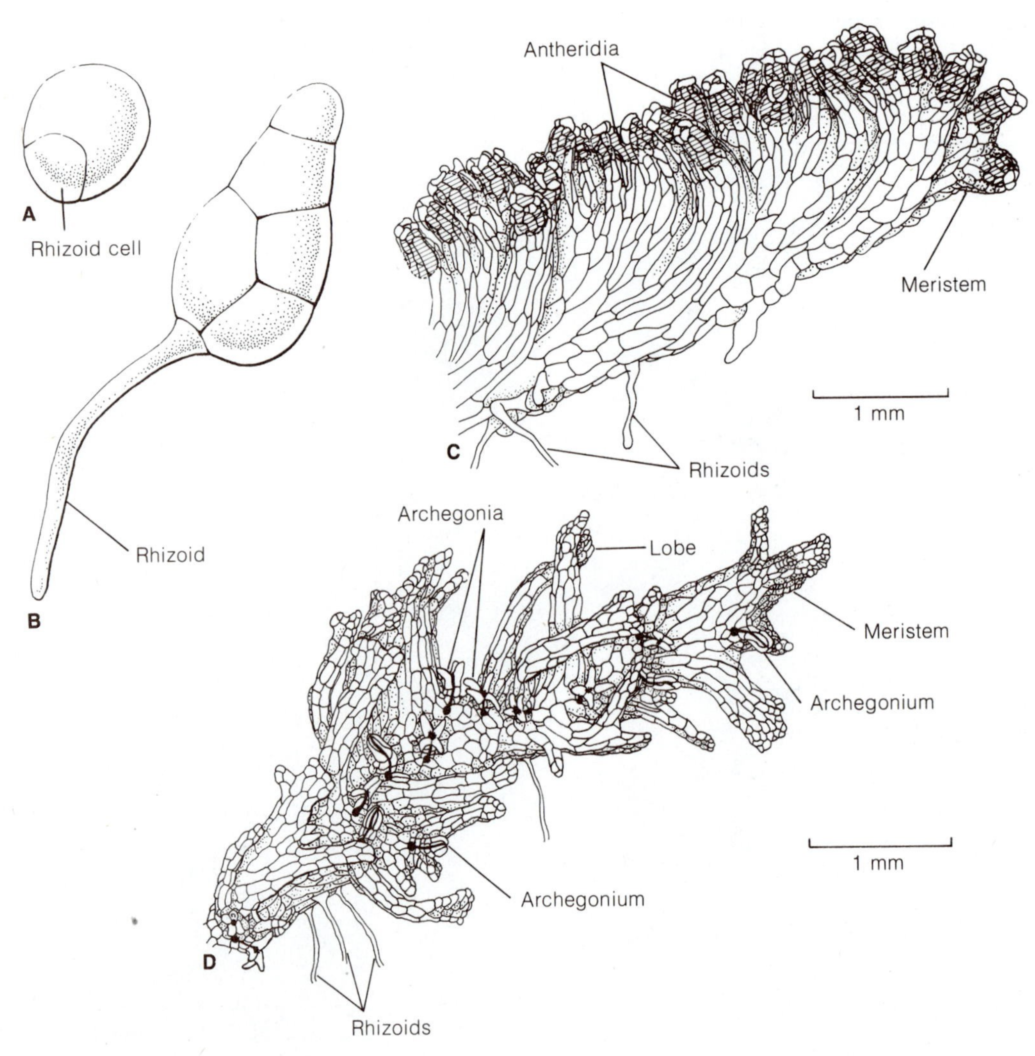

FIGURE 10-16 **A, B,** spore germination and early development of the gametophyte in *Equisetum hyemale;* **C,** apical portion, branch of male gametophyte of *Equisetum arvense;* **D,** branch of initially female plant. (See text for details of development.) [C, D redrawn from Duckett, *Bot. Jour. Linn. Soc.* 63(4):327–352, 1970.]

Single spore cultures of five species in subgenus *Equisetum (E. arvense, E. telmateia, E. fluviatile, E. palustre,* and *E. sylvaticum)* reveal that within 15 to 20 days after germination a developing gametophyte becomes organized into two quite different regions: (1) a basal region with rhizoids on the lower surface and (2) an upright, bright green, branched region. After 15 to 25 days in culture a localized meristem develops that gives rise to new basal cells of the "cushion" and produces derivatives which contribute to the development of upright, platelike lobes (Duckett, 1970a). Subsequent development and morphology is dependent on the type of sex organ produced.

Sexual behavior in *Equisetum* has long been a standing controversy. Some investigators reported that for the species they studied the gametophytes were either unisexual or bisexual, or were initially unisexual and then became bisexual. Heterospory was claimed to support sexual differentiation, but subsequent research failed to support this contention (Duckett, 1970b).

According to Duckett (1970a), *Equisetum* gametophytes generally are either male or female initially and the first sex organs to appear are antheridia on male plants. In the five species mentioned earlier, the male gametophytes are smaller and grow more slowly than the females. The initially female gametophytes eventually produce antheridia, generally in great numbers. The proportions of males to bisexual gametophytes are much the same within each species, but vary greatly between species.

MALE GAMETOPHYTES. The basal or cushion meristem, after producing a few upright lobes, suddenly begins forming antheridial tissue, and the gametophyte often has a distinctive coloration (Fig. 10-16, C). Old males may have several hundred antheridia.

INITIALLY FEMALE GAMETOPHYTES. Archegonia are found on the cushion and at maturity lie at the bases of the upright lobes (Fig. 10-16, D). Female gametophytes sooner or later produce antheridia. When they do, archegonia and antheridia are both present, but in older gametophytes the number of antheridia produced increases. If a culture of a bisexual gametophyte (derived from one spore) is flooded with water, homozygous sporophytes are produced (also see Sporne, 1964).

Early sex determination, at least in some species, appears to be related to environmental conditions —temperature, light, and humidity as well as the supply of nutrients. (See discussions in Schratz, 1928; Wollersheim, 1957a, b; Hauke, 1967; Duckett, 1970a, 1972.) In other experiments designed to explore sexuality in *Equisetum,* Duckett (1977) subcultured fragments of male gametophytes. Some of the fragments remained male throughout successive subcultures. Others produced archegonia, but subsequently produced antheridia in increasing numbers, supporting the contention that *Equisetum* gametophytes are potentially bisexual. In contrast to these results Hauke (1969b) reported that gametophytes of *E. bogotense* were unisexual and never changed. Duckett (1972) also found that in *E. ramosissimum* and *E. variegatum* the initially male gametophytes never became bisexual. However, for *E. giganteum* (subgenus *Hippochaete*) Hauke (1969b) reported that the gametophytes were normally bisexual, with archegonia and antheridia appearing at the same time. The pattern in *E. giganteum* was considered by Hauke to be primitive and could be correlated with primitive features of the sporophyte of the species; Duckett (1979a), however, believes that our knowledge of gametophytes does not provide definitive information to support the view that the subgenus *Hippochaete* is a more primitive group.

In the absence of clear-cut heterospory, it seemed unlikely to Duckett (1970b) that any exact sex-determining mechanism is operative in *Equisetum;* rather, sexuality is controlled by a complex set of interactions, although Hauke (1977) believes that all the available information points to a genetic basis for sex in *Equisetum* spores.

Although the gametophytes of most species of *Equisetum* are potentially bisexual, a study of sexuality in *E. arvense* revealed that intragametophytic self-fertilization is infrequent. This conclusion is based upon genetic data from enzyme electrophoresis (Soltis et al., 1988).

GAMETANGIA. Antheridia first appear on the lobes of the male gametophyte and they may occur by the hundreds on the rounded, compact surface of old bisexual gametophytes at the "pincushion" stage. An antheridium is initiated by a periclinal division in a superficial cell which gives rise to a cover or jacket cell and a primary spermatogenous cell. The jacket cell divides anticlinally, and at maturity the jacket layer may consist of two or more cells. The primary spermatogenous cell divides, and the derivative cells then undergo a series of synchronized divisions, forming spermatids. At maturity the spermatids escape through a pore (Fig. 10-15, B) created by the separation of the jacket or opercular cells (Hauke, 1968; Duckett, 1972, 1973; Bilderback et al., 1973). A sperm escapes after the spermatid wall breaks; each sperm has about 120 flagella attached to a lamellar band (Duckett, 1975; Duckett and Bell, 1977). The flagella beat with a continuous, traveling, helical wave, propelling a sperm in a unidirectional, helical path (Fig. 10-17; Bilderback et al., 1973).

An archegonium follows the common type of ontogeny displayed in other lower vascular plants (Chapter 5, Fig. 1). At maturity an archegonium may have a projecting neck frequently comprising three tiers of neck cells arranged in four rows, two adjacent neck-canal cells, often of unequal size, a ventral canal cell, and an egg at the base of the embedded venter (Chatterjee and Ram, 1968;

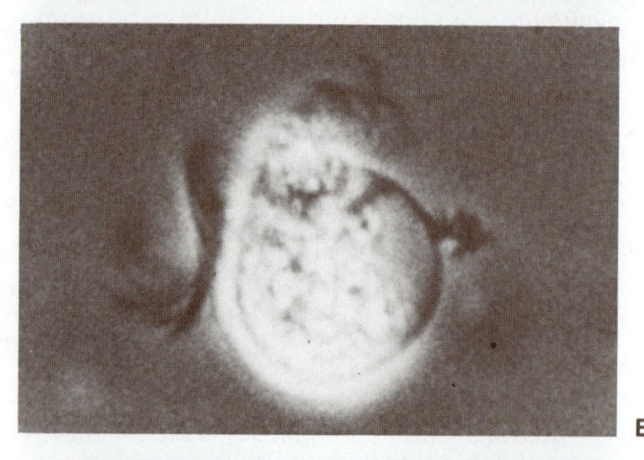

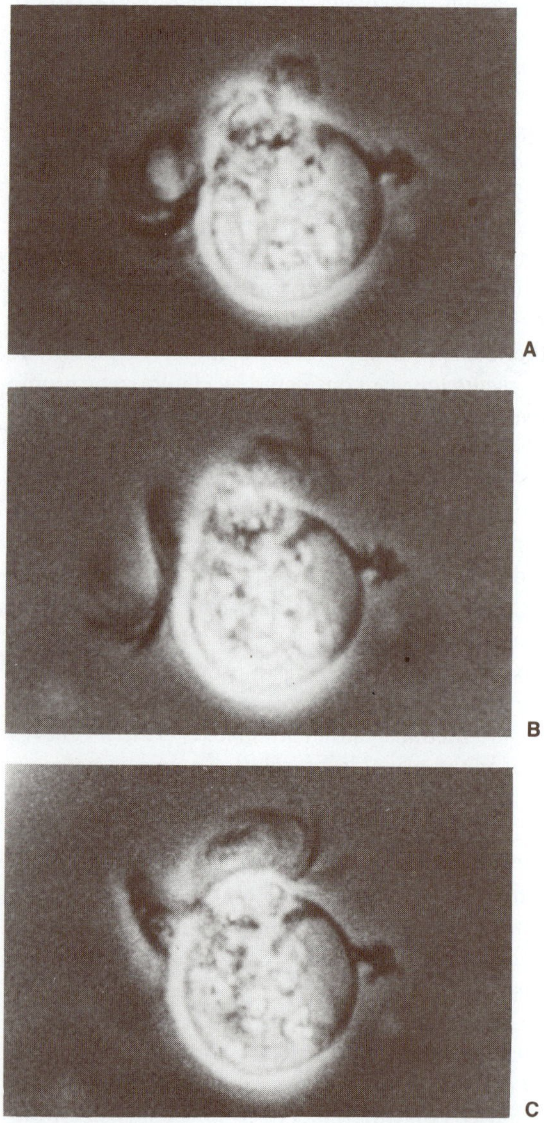

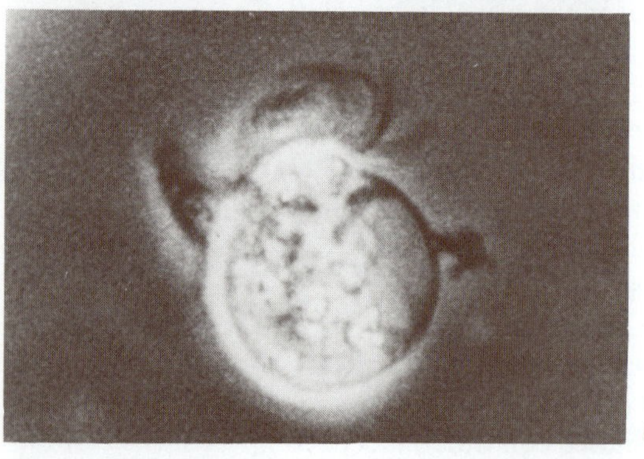

FIGURE 10-17 **A–C,** a sequential series of a swimming sperm of *Equisetum hyemale,* illustrating the helical nature of the flagellar beat (× 1500). [Courtesy of Dr. David E. Bilderback.]

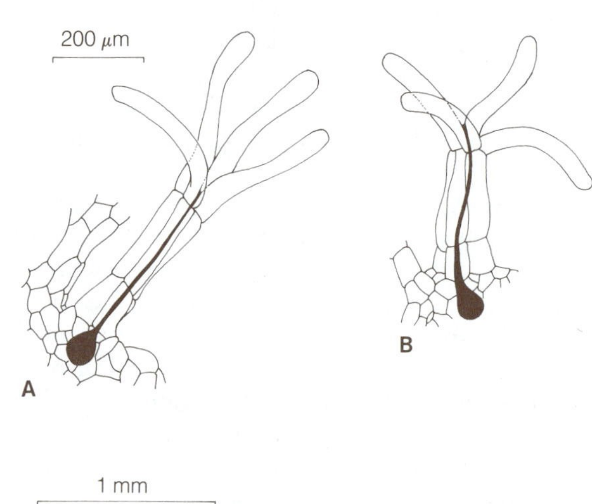

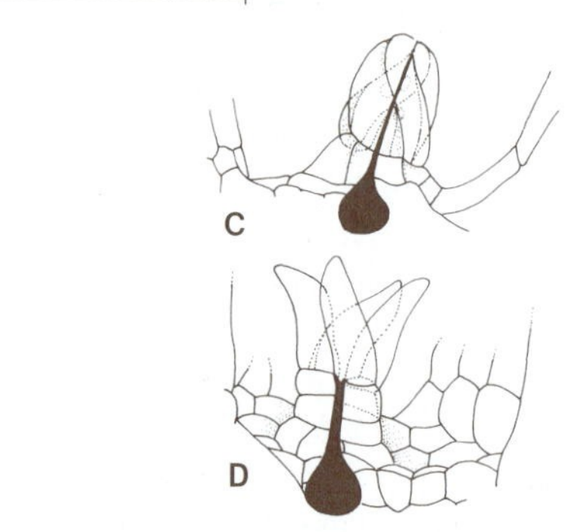

FIGURE 10-18 **A, B,** archegonia of *Equisetum fluviatile* showing greatly elongated terminal tier of neck cells. **C, D,** archegonia, *E. scirpoides,* illustrating the twisted form of the terminal tier of neck cells (**C**), and their spreading apart at archegonial maturity (**D**). [**A, B** redrawn from Duckett, *Bot. Jour. Linn. Soc.* 66:1–22, 1973; **C, D** redrawn from Duckett, *Bot. Jour. Linn. Soc.* 79:179–203, 1979.]

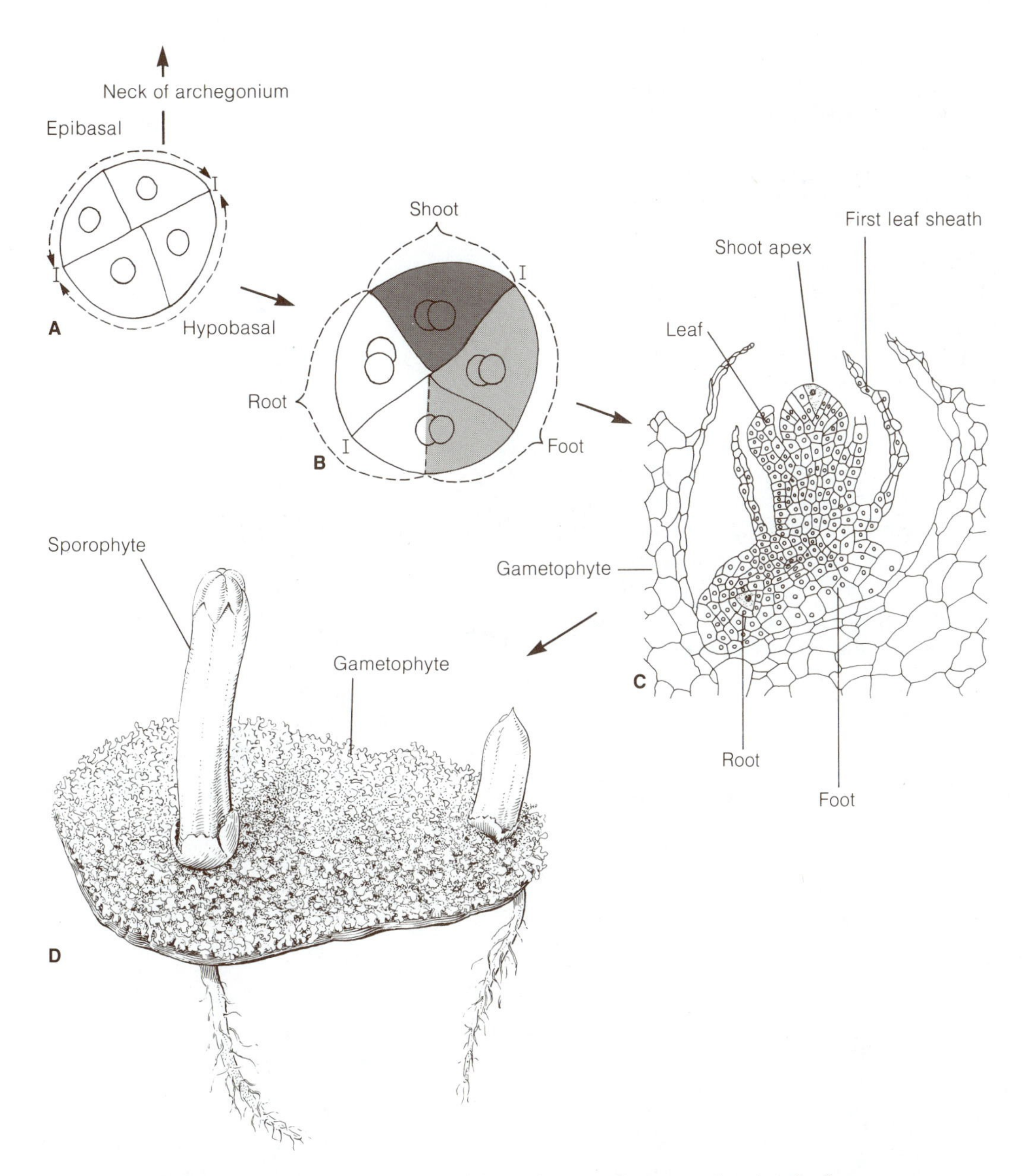

FIGURE 10-19 **A–C,** development of the embryo in *Equisetum* showing the first cleavage of the zygote (I–I in **A**) and derivation of shoot, root, and foot from the eight-celled embryo (**B**); **D,** dorsiventral gametophyte of *Equisetum laevigatum* with two attached sporophytes. [**A, B** redrawn from Laroche, *Rev. Cytol. Biol. Veg.* 31:155–216; **C** redrawn from *Cryptogamic Botany,* Vol. II, 2d edition, by G. M. Smith. McGraw-Hill, New York. 1955: **D** redrawn from Walker, *Bot. Gaz.* 71:378–391, 1921.]

Duckett, 1973). During egg maturation and degeneration of the ventral and neck-canal cells, the terminal tier of neck cells elongate considerably. In some species the subterminal neck cells may also elongate (Fig. 10-18, A, B). The terminal tier of neck cells are often twisted initially but then separate from each other and spread apart in an arching manner (Duckett, 1973; Hauke, 1980; Fig. 10-18, C, D).

To achieve fertilization the gametophytes must

be covered by at least a thin layer of water in which the motile sperm swim to the archegonia. Mature spermatids are probably not released until the proper osmotic conditions are achieved. When gametophytes are flooded, water probably enters the cells of the mature antheridia because of an osmotic gradient. This would bring about swelling of the spermatids and opercular cells and is possibly the causal factor in the opening of the antheridium. The spermatid cells have elaterlike structures which may aid discharge and dispersal (Bilderback et al., 1973).

The Embryo

Observations on fertilization have indicated that numerous sperm may enter the archegonium, and even penetrate the egg cytoplasm, but only one sperm actually fuses with the egg nucleus (Chatterjee and Ram, 1968). The first division of the zygote has been described as transverse (Sadebeck, 1878) for certain species; the first division in *E. arvense* has been described as being oblique (Laroche, 1968), setting aside an upper cell, the epibasal cell, and a lower cell, the hypobasal cell. The embryo is therefore exoscopic in polarity (Chapter 6; Fig. 10-19).

FIGURE 10-20 *Equisetum* gametophyte with attached sporophytes (× 4).

There are conflicting accounts of subsequent embryogeny in *Equisetum,* which may indicate variation in the genus. Earlier descriptions were presented at a time when embryogeny was considered to be a precise series of unalterable events. Assign-

FIGURE 10-21 Reconstruction of *Pseudobornia ursina* from the Devonian Period. [From Schweitzer, *Palaeontographica* 120B:116–137, 1967.]

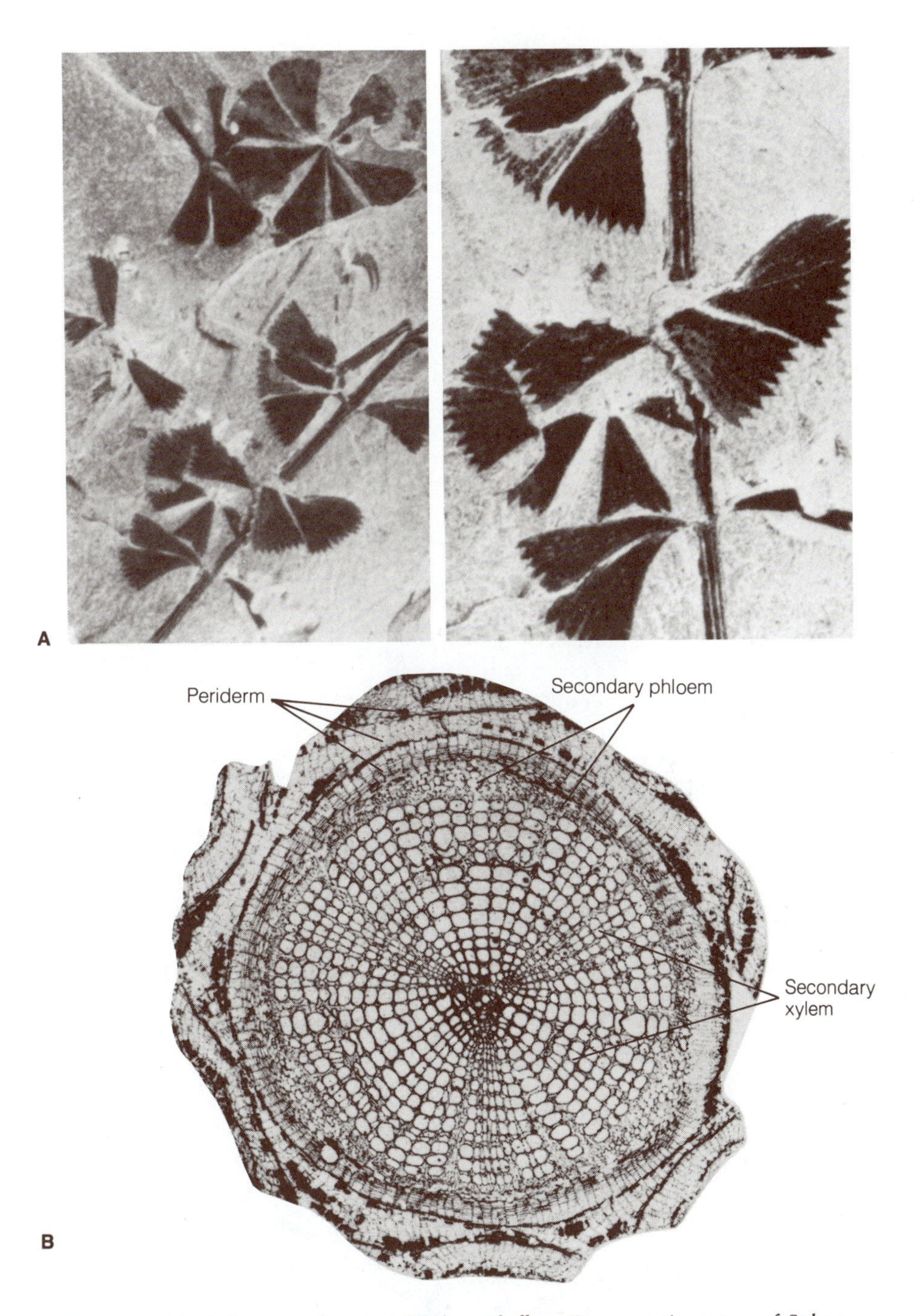

FIGURE 10-22 **A,** leaves and stems of *Sphenophyllum*. **B,** transection, stem of *Sphenophyllum plurifoliatum*. [**A** from Remy and Remy, *Pflanzenfossilien*, 1959; **B** from Eggert and Gaunt, *Amer. Jour. Bot.* 60:755–770, 1973.]

ment of the first leaf, stem (future shoot apex), root, and foot to definite segments of the embryo in the quadrant stage was described by Sadebeck (1878). The first leaf and the shoot apex were said to develop from the epibasal portion, the first root and the foot from different quadrants of the hypobasal portion. Later workers reported variations in segmentation and origin of fundamental organs. Until more complete studies are made of more species, the following general outline of development may be taken as representative for *E. arvense* (Laroche, 1968) and some other species.

Following the first oblique division, the epibasal and hypobasal cells divide at right angles to the original wall. This establishes the quadrant stage (Fig. 10-19, A). After subsequent divisions in the four cells, the future shoot apex is organized from derivatives of one quadrant of the epibasal hemisphere. The foot takes its origin from one quadrant of the hypobasal hemisphere and a portion of the other adjacent hypobasal quadrant. The first root is organized from one of the epibasal quadrants and a portion of the subjacent hypobasal quadrant (Fig. 10-19, B). The shoot then grows rapidly, forming whorls of three or four scalelike leaves (Fig. 10-19, C). Later the root penetrates the gametophytic tissue in reaching the soil or substratum (Fig. 10-19, D). Additional erect shoots are formed from buds on the primary axis of the sporophyte. Large, mature gametophytes may support several sporophytes (from multiple fertilizations) in varying stages of development (Fig. 10-20).

Fossil Sphenopsids

Pseudoborniales — *Pseudobornia*

The order Pseudoborniales is represented by one species from the Upper Devonian, *Pseudobornia ursina,* collected from Bear Island, south of Spitzenberg, and from northeastern Alaska. *Pseudobornia* was a monopodial branched tree about 15 to 20 meters in height (Fig. 10-21). The basal portion of the trunk was up to 60 centimeters in diameter. There were three orders of branching, with one or two branches at the nodes of the main axis; the ultimate branches bore leaves in whorls of four, each of which dichotomized two or three times and were finely dissected along their margins. The strobili at the ends of first order branches were about 30 centimeters long. Strobili consisted of whorled bracts and sporangiophores; the sporangiophores were upturned, divided into two segments, and bore about thirty terminal sporangia (Schweitzer, 1967). The affinities of *Pseudobornia* are obscure. Organization of the reproductive structures would appear to support a relationship with the Sphenophyllales (Stewart, 1983).

Sphenophyllales

The group Sphenophyllales was common in the Carboniferous Period; it may have existed as early as the Upper Devonian, but disappeared by the Per-

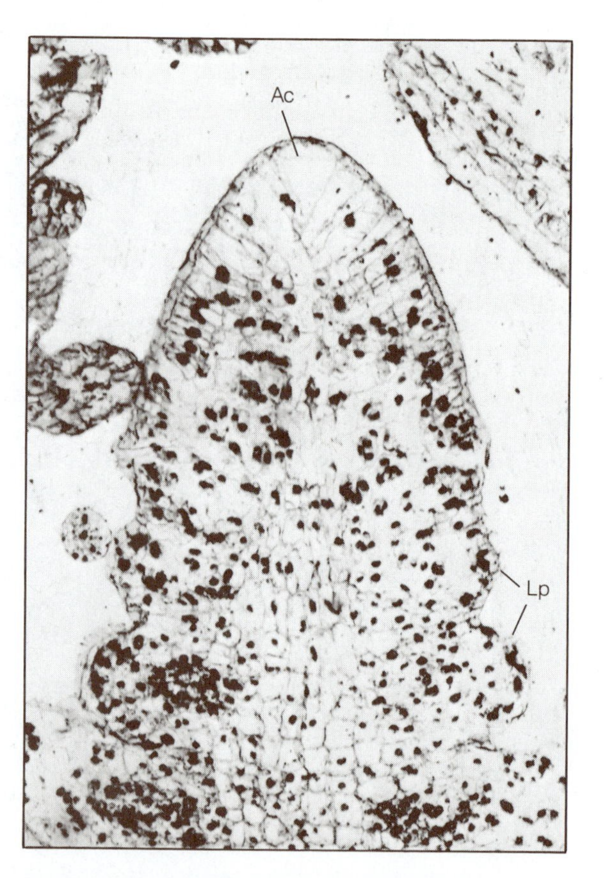

FIGURE 10-23 Median longitudinal section, shoot apex of *Sphenophyllum* showing the large inverted pyramidal apical cell (Ac) and derivative cells. Lp, leaf primordia. [Courtesy of Dr. T. N. Taylor.]

mian. These plants were probably less than 1 meter tall and are variously depicted as erect and self-supporting, or supported by surrounding vegetation. The generic name *Sphenophyllum* was originally applied to leaf compressions, but now includes extremely well-preserved stems, roots, and leaves. Similar to other sphenopsids, the stems were jointed with whorls of leaves at each node. Unlike other groups, however, the leaves were commonly wedge shaped, the distal margins toothed or deeply notched, with a dichotomously branched venation system (Fig. 10-22, A). The stem had an exarch protostele which was usually triarch (Fig. 10-22, B). In larger stems an abundant amount of secondary xylem was produced consisting of elongate tracheids (in excess of 8 millimeters) with circular to elliptical pits. Between the tracheids were longitudinal strands of narrow parenchymatous cells which become confluent in the outer part of the xylem to form a type of multiseriate ray. Eggert and Gaunt (1973) have shown that secondary phloem was formed in *Sphenophyllum plurifoliatum*. *Sphenophyllum* appears to be the only nonseed plant in which a well-documented bifacial vascular cambium has been recorded. A periderm was produced by a phellogen that probably arose early in development in tissue believed to be the pericycle.

A remarkable discovery was made of beautifully preserved shoot tips of *Sphenophyllum* by Good and Taylor (1972). The shoot apices resemble in every way those of *Equisetum*, including the prominent apical cell, derivatives, young leaves, and the initiation of intercalary meristems (Fig. 10-23). Because *Sphenophyllum* is protostelic, it formed a cyl-

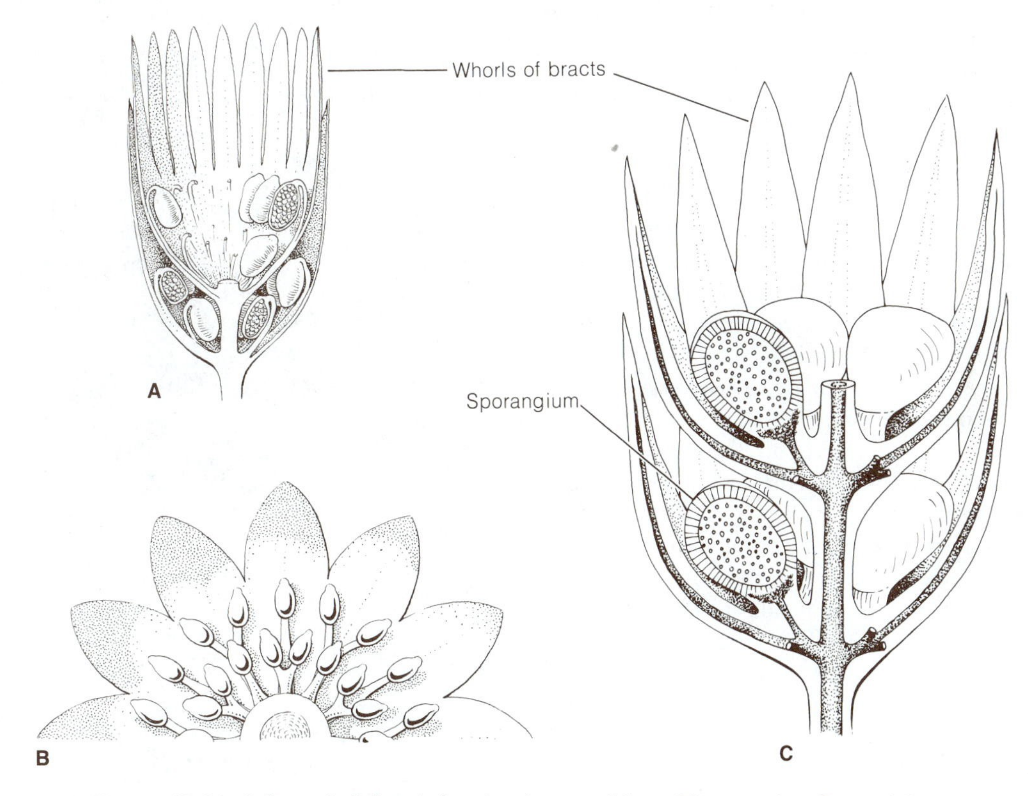

FIGURE 10-24 Sphenophyllales. **A,** longisection, strobilus of *Bowmanites dawsoni;* **B,** same as **A,** viewed from above (stalked sporangia were fused with the laterally joined bracts to variable degrees); **C,** strobilus of *Litostrobus iowensis* as seen in longisection. [**A** redrawn from *An Introduction to the Study of Fossil Plants* by J. Walton. Adam and Charles Black, London, 1953; **B** redrawn from *Handbuch der Paläobotanik* by M. Hirmer. R. Oldenbourg, Munich, 1927; **C** redrawn from Reed, *Phytomorphology* 6:261–272, 1956.]

inder of procambium in the center of the axis, rather than a pith meristem as in *Equisetum.*

Strobili consisted of alternating whorls of bracts and sporangiophores. Commonly there were one to two sporangiophores per bract which were fused to varying degrees with the adaxial sides of the bracts (Fig. 10-24, A–C).

Equisetales — Calamitaceae

Members of this family were trees reaching heights up to 20 meters and constituted a significant part of the Upper Carboniferous flora. Arborescent sphenopsids also occurred in the Upper Devonian to Lower Carboniferous and are placed in the family Archaeocalamitaceae, but we will discuss only the Calamitaceae.

If a complete calamite plant could be assembled, the plant body would consist of an aerial branch system and an underground rhizome system similar to those of *Equisetum* (Fig. 10-25). Roots, as well as aerial branches, originated at nodes along the rhizome. The aerial, articulated shoot exhibited limited or extensive branching depending upon recognized subgenera. The stems were jointed, as in *Equisetum,* but had a smooth outer surface. Whorls of linear, lanceolate, or spatulate microphylls occurred at the nodes (Fig. 10-26, A). These leaves were larger than those of *Equisetum* and were often fused at the base to form a collar around the base of an internode. In contrast to *Sphenophyllum,* the leaves had only a single unbranched vein, similar to *Equisetum.* Strobili were at the tips of side branches, not terminally on the main axis as is common in *Equisetum.*

Form or organ genera are common in the Calamitaceae. There are genera for casts of pith cavities (e.g., *Calamites*), permineralizations of stems *(Arthropitys, Calamodendron),* leaf whorls *(Annularia, Asterophyllites),* and strobili *(Calamostachys, Palaeostachya).* Not all organ genera are listed here. It is conventional to use the generic name *Calamites* in speaking of entire plants (Fig. 10-25).

The remains of a calamite stem are most commonly of two types: pith casts and permineralizations. Pith casts were formed by sediments infiltrating pith cavities and hardening there. When the surrounding organic material decayed, an image of the inner surface of the vascular cylinder remained. On the surface of a pith cast there are longitudinal furrows and ridges (Fig. 10-26, B); the furrows mark the positions of the primary xylem and can be correlated with a stem seen in transverse section (compare Figs. 10-26, B and 10-27). The permineralized stems reveal a remarkable similarity to those of *Equisetum,* except for the cylinder of secondary xylem.

Annularia is a genus for whorls of leaves attached to small stems (Fig. 10-26, A). Strobili consisted of alternate whorls of bracts and sporangiophores (*Calamostachys;* Fig. 10-28, D, E) or whorls of sporangiophores that were situated in the

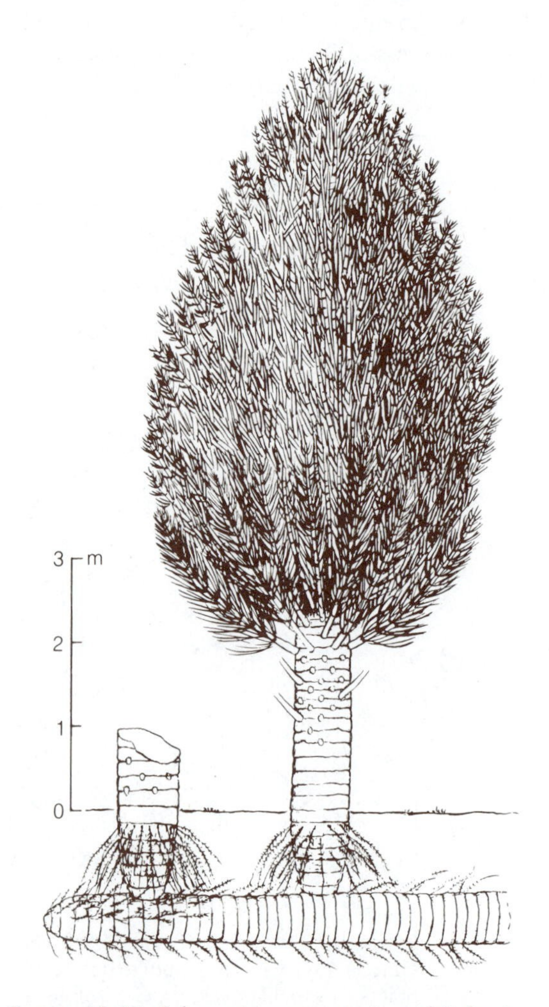

FIGURE 10-25 Schematic representation of *Calamites* showing rhizome and a tall, upright, branched portion. [Redrawn from *Les Plantes Fossiles* by L. Emberger. Masson et Cie, Paris, 1968.]

FIGURE 10-26 **A,** *Annularia radiata,* stem with whorls of leaves. **B,** pith cast of a calamite; note prominent nodes and internodes, and the presence of branch scars at the nodes. [**A** from R. E. Janssen, *Leaves and Stems from Fossil Forests,* Popular Series, Vol. I. Illinois State Museum, Springfield, Illinois, 1965.]

axils of sterile bracts (*Palaeostachya;* Fig. 10-28, A–C). Some calamites were heterosporous, and in one species each megasporangium retained only one megaspore (Baxter, 1963).

Equisetales — Equisetaceae: *Equisetites*

Certain sphenopsids of the Carboniferous resembled the living genus *Equisetum* and are placed in the genus *Equisetites.* However, most of the species are known from the early Mesozoic and later. The generic name *Equisetum* is used for Cenozoic fossils that are in every way similar to living species of the genus.

The reader is referred to books on paleobotany for descriptions of additional extinct genera (Taylor, 1981; Stewart, 1983).

An Appraisal

In this book we have recognized three orders in the Sphenophyta: Pseudoborniales, Sphenophyllales, and Equisetales. The latter order is comprised of two families: Equisetaceae and Calamitaceae. Some authors recognize five orders: Hyeniales, Pseudoborniales, Sphenophyllales, Calamitales, and Equisetales, although the Hyeniales are considered to be

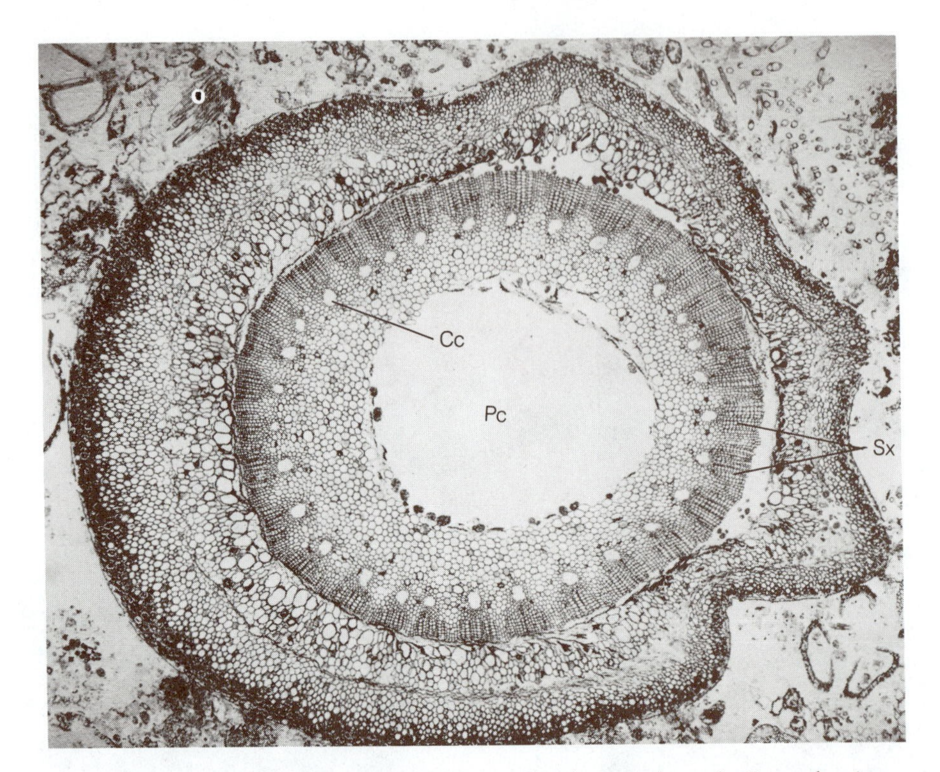

FIGURE 10-27 Transverse section of the stem of *Arthropitys* from the Pennsylvanian Epoch of the Carboniferous Period. Cc, carinal canal; Pc, pith cavity; Sx, secondary xylem. [Courtesy of Dr. T. L. Phillips.]

primitive ferns by some paleobotanists. Additionally, the Sphenophyta is designated, for example, as the Articulatae or Arthrophyta by other authors in reference to the jointed nature of the stems.

The Sphenophyllales is quite distinct from other members of the Sphenophyta in leaf morphology, possession of a protostele, and the close association of sporangiophores with bracts. In some instances the sporangiophore-bract complex is exceedingly intricate.

The general morphology of *Equisetum* and that of an extinct calamite is similar except for the great size of a calamite and the formation of secondary xylem. Morphologists are confronted with the question: Is the strobilar organization in *Equisetum* and the fossil *Equisetites* essentially primitive or is it specialized? In both genera a strobilus consists of sporangiophores without bracts. If *Equisetum* does truly represent a highly reduced calamite, then strobilar organization has involved the loss of bracts in the strobilus. In this case, the *Equisetum* strobilus would be specialized and reduced—by the loss of bracts in the course of evolution. Another way of looking at the problem is to consider sporangiophores and bracts as equivalent or serially homologous structures—perhaps both fertile in the past—the bracts having lost their reproductive structures (sporangia) in the course of time. Page (1972) has presented evidence for such a concept based upon the presence of occasional leafy sporangiophores and the occurrence of gradual transitions between typical sporangiophores to leafy ones in strobili of living species. The leaf has not evolved through a process of planation and webbing as proposed by the telome theory (Zimmermann, 1959), but has evolved by reduction and loss of dichotomies (Fig. 10-29).

It seems reasonable to assume, on the basis of morphology and anatomy, that there is an evolutionary relationship between *Equisetum* and the

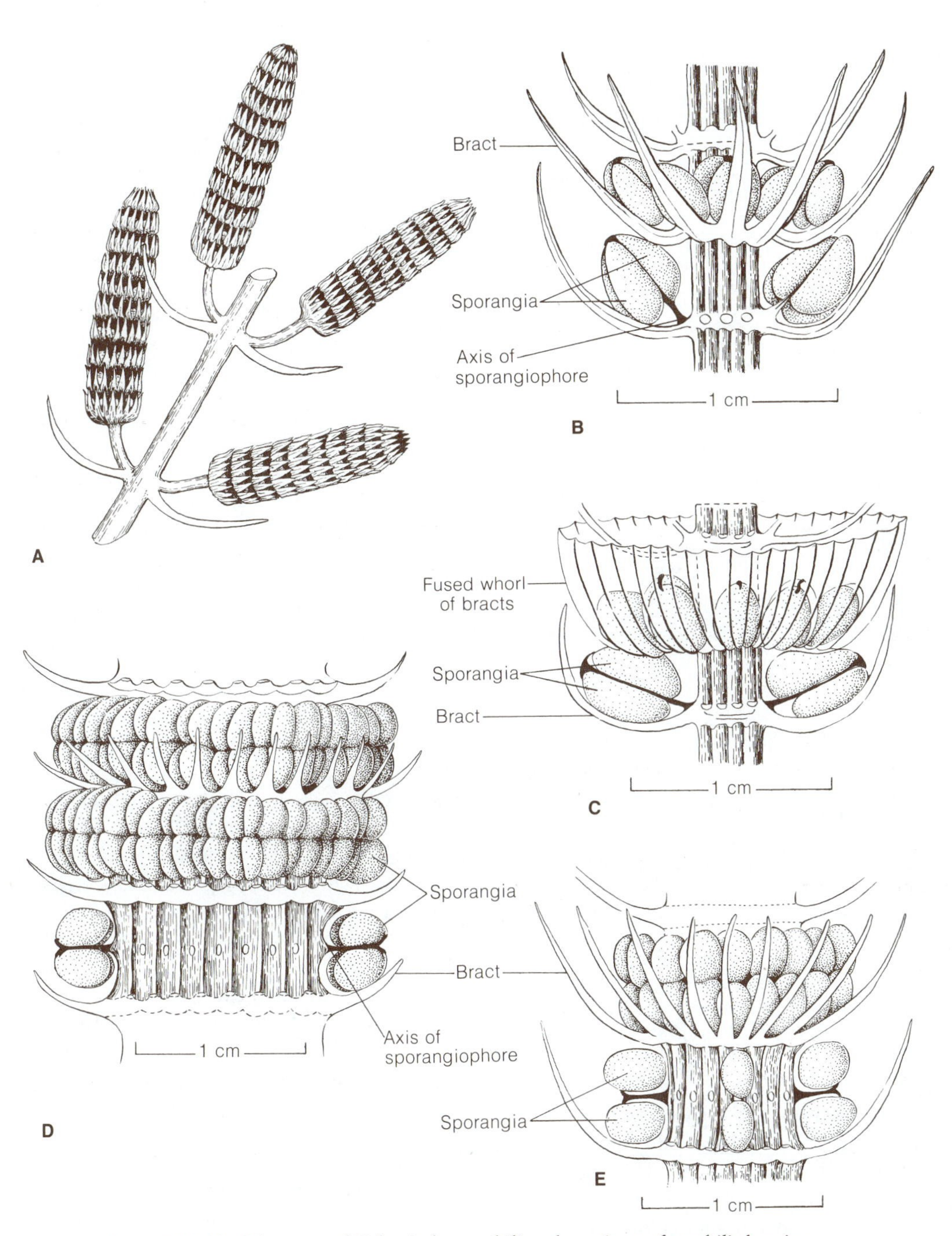

FIGURE 10-28 Diagrams of Calamitales strobili and portions of strobili showing position of sporangiophores and bracts. **A**, *Palaeostachya decacnema;* **B**, *Palaeostachya ovalis;* **C**, *Palaeostachya aperta;* **D**, *Calamostachys interculata:* **E**, *Calamostachys longibracteata.* [**A** redrawn from Delevoryas, *Amer. Jour. Bot.* 42:481–488, 1955; **B–E** redrawn from Abbott, *Palaeontogr. Amer.* 6:1–49, 1968.]

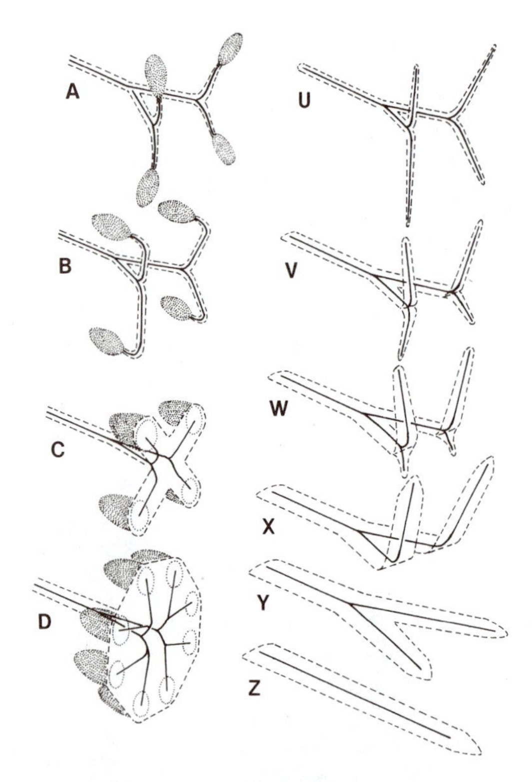

FIGURE 10-29 Diagram illustrating the suggested evolutionary stages of the *Equisetum* appendages: **A–D,** sporangiophores; **U–Z,** leaf. Broken lines indicate the outline of the appendage, stippled areas the sporangia and solid lines the vascular supply. **A** and **U** are the primitive appendages orientated in their probable planes of growth on a vertical shoot and differ from one another only in the presence or absence of sporangia. **D** and **Z** represent the morphology of sporangiophore and leaf of modern *Equisetum,* respectively. **B–C** and **V–Y** are the suggested intermediate evolutionary steps. [Redrawn from Page, *Bot. Jour. Linn. Soc.* 65:359–397, 1972.]

Calamitaceae. They probably shared a common ancestry with one line leading to the herbaceous, bractless *Equisetum*-type; the second line led to the arborescent, woody Calamitaceae with strobili having both bracts and sporangiophores.

REFERENCES

Avanzi, S., and F. D'Amato
1967. New evidence on the organization of the root apex in leptosporangiate ferns. *Caryologia* 20:257–264.

Baxter, R. W.
1963. *Calamocarpon insignis,* a new genus of heterosporous, petrified calamitean cones from the American Carboniferous. *Amer. Jour. Bot.* 50:469–476.

Benedict, R. C.
1941. The gold rush: a fern ally. *Amer. Fern Jour.* 31:127–130.

Bierhorst, D. W.
1958a. The tracheary elements of *Equisetum* with observations on the ontogeny of the internodal xylem. *Bull. Torrey Bot. Club* 85:416–433.
1958b. Vessels in *Equisetum. Amer. Jour. Bot.* 45:534–537.
1959. Symmetry in *Equisetum. Amer. Jour. Bot.* 46:170–179.

Bilderback, Diane E., D. E. Bilderback, T. L. Jahn, and J. R. Fonseca
1973. The release mechanism and locomotor behavior of *Equisetum* sperm. *Amer. Jour. Bot.* 60:796–801.

Bir, S. S.
1960. Chromosome numbers of some *Equisetum* species from the Netherlands. *Acta Bot. Neer.* 9:224–234.

Brown, J. T.
1976. Observations on the hypodermis of *Equisetum. South African Jour. Sci.* 72:303–305.

Chatterjee, J., and H. Y. M. Ram
1968. Gametophytes of *Equisetum ramosissimum* Desf. subsp. *ramosissimum.* I. Structure and development. *Bot. Notis.* 121:471–490.

Chen, C., and J. Lewin
1969. Silicon as a nutrient element for *Equisetum arvense. Can. Jour. Bot.* 47:125–131.

D'Amato, F.
1975. Recent findings on the organization of apical meristems with single apical cell. *G. Bot. Ital.* 109:321–334.

Duckett, J. G.
1970a. Sexual behavior of the genus *Equisetum,* subgenus *Equisetum. Bot. Jour. Linn. Soc.* 63:327–352.
1970b. Spore size in the genus *Equisetum. New Phytol.* 69:333–346.
1972. Sexual behaviour of the genus *Equisetum,* subgenus *Hippochaete. Bot. Jour. Linn. Soc.* 65:87–108.
1973. Comparative morphology of the gametophytes of the genus *Equisetum* subgenus *Equisetum. Bot. Jour. Linn. Soc.* 66:1–22.

1975. Spermatogenesis in pteridophytes. In J. G. Duckett and P. A. Racey (eds.), *The Biology of the Male Gamete*, pp. 97–127. *Biolog. Jour. Linnean Soc.* 7, Suppl. 1. Academic, London.
1977. Towards an understanding of sex determination in *Equisetum:* an analysis of regeneration in gametophytes of the subgenus *Equisetum*. *Bot. Jour. Linn. Soc.* 74:215–242.
1979a. Comparative morphology of the gametophytes of *Equisetum* subgenus *Hippochaete* and the sexual behavior of *E. ramosissimum* subsp. *debile,* (Roxb.) Hauke, *E. hyemale* var. *affine* (Engelm.) A. A., and *E. laevigatum* A. Br. *Bot. Jour. Linn. Soc.* 79:179–203.
1979b. An experimental study of the reproductive biology and hybridization in the European and North American species of *Equisetum*. *Bot. Jour. Linn. Soc.* 79:205–229.

Duckett, J. G., and P. R. Bell
1977. An ultrastructural study of the mature spermatozoid of *Equisetum*. *Philos. Trans. Roy. Soc. London, B, Biol. Sci.* 277:131–158.

Duckett, J. G., and C. N. Page
1975. In C. A. Stace (ed.), *Hybridization and the Flora of the British Isles,* pp. 99–103. Academic, London.

Eggert, D. A., and D. D. Gaunt
1973. Phloem of *Sphenophyllum*. *Amer. Jour. Bot.* 60:755–770.

Gifford, E. M., Jr., and E. Kurth
1982. Quantitative studies of the root apical meristem of *Equisetum scirpoides*. *Amer. Jour. Bot.* 69:464–473.
1983. Quantitative studies of the vegetative shoot apex of *Equisetum scirpoides*. *Amer. Jour. Bot.* 70:74–79.

Golub, S. J., and R. H. Wetmore
1948a. Studies of development in the vegetative shoot of *Equisetum arvense* L. I. The shoot apex. *Amer. Jour. Bot.* 35:755–767.
1948b. Studies of development in the vegetative shoot of *Equisetum arvense* L. II. The mature shoot. *Amer. Jour. Bot.* 35:767–781.

Good, C. W., and T. N. Taylor
1972. The ontogeny of Carboniferous articulates: the apex of *Sphenophyllum*. *Amer. Jour. Bot.* 59:617–626.

Hauke, R. L.
1957. The stomatal apparatus of *Equisetum*. *Bull. Torrey Bot. Club* 84:178–181.
1961. A resume of the taxonomic reorganization of *Equisetum,* subgenus *Hippochaete*. I. *Amer. Fern Jour.* 51:131–137.
1962a. A resume of the taxonomic reorganization of *Equisetum,* subgenus *Hippochaete*. II. *Amer. Fern Jour.* 52:29–35.
1962b. A resume of the taxonomic reorganization of *Equisetum,* subgenus *Hippochaete*. III. *Amer. Fern Jour.* 52:57–63.
1962c. A resume of the taxonomic reorganization of *Equisetum,* subgenus *Hippochaete*. IV. *Amer. Fern Jour.* 52:123–130.
1963. A taxonomic monograph of the genus *Equisetum,* subgenus *Hippochaete*. *Beihefte Nova Hedwigia* 8:1–123.
1967. Sexuality in a wild population of *Equisetum arvense* gametophytes. *Amer. Fern Jour.* 57:59–66.
1968. Gametangia of *Equisetum bogotense*. *Bull. Torrey Bot. Club* 95:341–345.
1969a. Problematic groups in the fern allies and the treatment of subspecific categories. *BioScience* 19:705–707.
1969b. Gametophyte development in Latin American horsetails. *Bull. Torrey Bot. Club* 96:568–577.
1969c. The natural history of *Equisetum* in Costa Rica. *Rev. Biol. Trop.* 15:269–281.
1974. The taxonomy of *Equisetum*—an overview. *New Botanist* 1:89–95.
1977. Experimental studies on growth and sexual determination in *Equisetum* gametophytes. *Amer. Fern Jour.* 67:18–31.
1978. A taxonomic monograph of *Equisetum* subgenus *Equisetum*. *Nova Hedwigia* 30:385–455.
1980. Gametophytes of *Equisetum diffusum*. *Amer. Fern Jour.* 70:39–44.
1983. Horsetails *(Equisetum)* in North America. *Fiddlehead Forum, Bull. Amer. Fern Soc.* 10: 39–42.

Hoffman, F. M., and C. J. Hillson
1979. Effects of silicon on the life cycle of *Equisetum hyemale* L. *Bot. Gaz.* 140:127–132.

Jeffrey, E. C.
1899. The development, structure, and affinities of the genus *Equisetum*. *Mem. Boston Soc. Natur. Hist.* 5:155–190.

Johnson, M. A.
1933. Origin and development of tissues in *Equisetum scirpoides*. *Bot. Gaz.* 94:469–494.
1937. Hydathodes in the genus *Equisetum*. *Bot. Gaz.* 98:598–608.

Kashyap, S. R.
1914. The structure and development of the prothallus of *Equisetum debile,* Roxb. *Ann. Bot.* 28:163-181.

Kaufman, P. B., W. C. Bigelow, R. Schmid, and N. S. Ghosheh
1971. Electron microprobe analysis of silica in epidermal cells of *Equisetum. Amer. Jour. Bot.* 58:309-316.

Lamoureux, C. H.
1961. Comparative studies on phloem of vascular cryptogams. Ph.D. dissertation, University of California, Davis.

Laroche, J.
1968. Contributions à l'étude de l'*Equisetum arvense* L. II. Etude embryologique. Caractères morphologiques, histologiques et anatomiques de la première pousse transitoire. *Rev. Cytol. Biol. Veg.* 31:155-216.

Leclercq, S., and H.-J. Schweitzer.
1965. *Calamophyton* is not a sphenopsid. *Bull. l'Académie Roy. Belgique, Sciences.* 51:1395-1403.

Lloyd, R. M.
1964. Ethnobotanical uses of California pteridophytes by western American Indians. *Amer. Fern Jour.* 54:76-82.

Manton, I.
1950. *Problems of Cytology and Evolution in the Pteridophyta.* Cambridge University Press, London.

Mesler, M. R., and K. L. Lu
1977. Large gametophytes of *Equisetum hyemale* in northern California. *Amer. Fern Jour.* 67: 97-98.

Page, C. N.
1972. An interpretation of the morphology and evolution of the cone and shoot of *Equisetum. Bot. Jour. Linn. Soc.* 65:359-397.

Rothmaler, W.
1944. Pteridophyten-Studien. I. *Repert. Spec. Nov. Regni Veg.* 54:55-82.

Sadebeck, R.
1878. Die Entwicklung der Keimes der Schachtelhalme. *Jahrb. Wiss. Bot.* 11:575-602.

Schaffner, J. H.
1933. Six interesting characters of sporadic occurrence in *Equisetum. Amer. Fern Jour.* 23:83-90.

Schmid, R.
1982. The terminology and classification of steles: historical perspective and the outlines of a system. *Bot. Rev.* 48:817-931.

Schratz, E.
1928. Untersuchungen über die Geschlechterverteilung bei *Equisetum arvense. Biol. Zentralbl.* 48:617-639.

Schweitzer, H.-J.
1967. Die Oberdevon-flora Bäreninsel. l. *Pseudobornia ursina* Nathorst. *Palaeontographica* 120 (B):116-137.
1972. Die Mitteldevon — Flora von Lindlar (Rheinland). 3. Filicinae — *Hyenia elegans* K. & W. *Palaeontographica* 137(B):154-175.

Soltis, D. E.
1986. Genetic evidence for diploidy in *Equisetum. Amer. J. Bot. 73:908-913.*

Soltis, D. E., P. S. Soltis, and R. D. Noyes
1988. An electrophoretic investigation of intragametophytic selfing in *Equisetum arvense. Amer. J. Bot.* 75:231-237.

Sporne, K. R.
1964. Self-fertility in a prothallus of *Equisetum telmateia,* Ehr. *Nature* 201:1345-1346.

Stewart, W. N.
1983. *Paleobotany and the Evolution of Plants.* Cambridge University Press, Cambridge.

Taylor, T. N.
1981. *Paleobotany: An Introduction to Fossil Plant Biology.* McGraw-Hill, New York.

Treitel, O.
1943. The elasticity, breaking stress, and breaking strain of the horizontal rhizomes of species of *Equisetum. Trans. Kansas Acad. Sci.* 46: 122-132.

Tryon, R. M., and A. F. Tryon
1982. *Ferns and Allied Plants.* Springer-Verlag, New York.

Vogt, T.
1942. Geokjemisk og geobotanisk malmleting. III. Litt om planteveksten ved Rorosmalmene. [Geochemical and geobotanical ore prospecting.] III. Some notes on the vegetation at the ore deposits at Roros. *K. Norske Vid. Selsk. Forh.* 15:21-24. (In Norwegian, Engl. summary).

Walton, J.
1944. The roots of *Equisetum limosum* L. *New Phytol.* 43:81-86.

Wollersheim, M.
1957a. Untersuchungen über die Keimungsphysiologie der Sporen von *Equisetum arvense* und *Equisetum limosum Zeit. Bot.* 45:145–159.
1957b. Entwicklungsphysiologische Untersuchungen der Prothallien von *Equisetum arvense* und *Equisetum limosum* mit besonderer Berücksichtigung der Frage nach der Geschlechtsbestimmung. *Zeit. Bot.* 45:245–261.

Zimmermann, W.
1959. *Die Phylogenie der Pflanzen,* 2d edition. Gustav Fischer Verlag, Stuttgart.

CHAPTER 11

Filicophyta: Ferns

GENERAL MORPHOLOGY; EARLY FERNS

FOR the groups of plants discussed thus far (Chapters 7 through 10) the *microphyll,* with some exceptions (e.g., in *Sphenophyllum*), is the prevailing type of foliar appendage. Most of the primary vascular tissue of microphyllous plants is cauline; the leaf traces are small strands of vascular tissue that separate from the periphery of the stele at the nodes. Ferns, on the contrary, have *megaphylls.* A megaphyll, whether large or small, usually has a branched venation system, and its leaf-trace system in ferns is associated with a *leaf gap* in the vascular cylinder of the stem unless, of course, the stem is protostelic. (Refer to Chapter 3 for a more complete discussion of microphylls and megaphylls). In megaphyllous plants, the form of the primary vascular system of the stem is markedly affected by the development of the larger leaves and their associated leaf traces. It should be pointed out, however, that regardless of the stelar pattern of the shoot axis, the primary vascular system of the fern *root* is exarch and radial in organization as it is in all other vascular plants. This remarkable conservatism of the root seems to be correlated with the absence of foliar appendages, and thus lends further support to the idea that the vascularization of stems in most ferns and seed plants is significantly correlated with the development of megaphylls and their leaf traces.

Today approximately 11,000 species of ferns are widely distributed over the earth's surface. Some species are restricted to narrow environmental niches and are endemic to certain localities. The common bracken fern *(Pteridium aquilinum),* on the other hand, is worldwide in distribution in the tropics and temperate zones and is a troublesome weed in some regions. Ferns are quite numerous and are most diverse in the tropical rain forests, many of them becoming trees 6 to 12 meters high, or growing as epiphytes. However, even desert areas and mountains of the temperate regions may have a fern population.

What general characteristics do we associate with a common field, garden, or house fern? Naturally we think of a large fern leaf, commonly called a *frond* in everyday usage—a term also used by many fern specialists. The fern frond may be a simple expanded blade or lamina with a petiole or *stipe*—the latter term used by some students of fern morphology —or, which is more common, the frond may have incisions in the blade, resulting in a pinnatifid leaf from the Latin (*pinnatus,* meaning "featherlike," or "with parts arranged along the two sides of an axis"). The pinnate plan of organization reaches its highest degree of development in pinnately compound leaves. In the latter type of organization the petiole (stipe) is devoid of any expanded blade, and its continuation as the main axis of the frond is called the *rachis*. Attached to the rachis by petiolules, and approximately opposite each other, are pairs of leaflets, each called a *pinna* (plura, *pinnae*). Each pinna may likewise by subdivided into pairs of *pinnules,* and there may be further subdivisions. Thus, a frond may be once pinnate, bipinnate, tripinnate, and so on (Fig. 13-1). These plans of organization do not describe all of the variations of pattern in fern fronds, but do describe those of a large number of ferns.

In most ferns the stem is an underground rhizome and is not apparent except in stocky, erect species. The large trunks of tropical ferns, however, compare in size with the trunks of moderately large palms. Roots usually are apparent at the lower part of an aerial stem or arise from the lower surface of a rhizome, often characteristically related to each leaf (Fig. 3-6).

Young fern fronds expand by unrolling and are often referred to as "fiddleheads," "monkey tails," or "croziers" (Figs. 11-1, 11-2).

Sporangia

Ferns have brownish to black splotches on the lower surface (abaxial side) of the frond. Each "spot" is technically a *sorus,* i.e., a collection of sporangia that is, in some species, protected by an outgrowth from the leaf surface called an *indusium* (Fig. 13-9).

In contrast with the adaxial, solitary sporangia of the Lycophyta, the sporangia of higher ferns are either marginal or, more commonly, on the abaxial surface of the fertile pinnae (Fig. 13-7). Abaxial sporgania, frequently fused into synangia, are also typical of certain members of the Marattiaceae (Fig. 12-15, B). In the Ophioglossaceae, often regarded as the most primitive family in the ferns, the sporangia occur singly on the upper part of the fertile leaf segment *(Botrychium)* or are embedded in tissue of the fertile segment or spike of *Ophioglossum* (Figs. 12-2, B; 12-11). In some members of the coenopterid ferns (extinct) individual sporangia were terminal on frond segments (Fig. 11-4, A). In general, three sporangial positions occur in the Filicophyta: terminal, marginal, and abaxial. The abaxial position is common in the more specialized ferns (Filicales) and serves to help demarcate this group of ferns.

From the standpoint of the structure and method of development of their sporangia, the Filicophyta are either *eusporangiate* or *leptosporangiate.* Eusporangiate development is characteristic of the more primitive orders of ferns (i.e., Ophioglossales and Marattiales). The more specialized ferns (Filicales, Marsileales, Salviniales) are remarkable among all vascular plants by virtue of the presence of leptosporangia, one of the most distinctive morphological features of these plants. As pointed out in Chapter 4, the leptosporangium is evidently an extreme modification of the more archaic eusporangium.

With reference to the kinds of spores produced, the ferns are characterized by both homosporous and heterosporous types. *Homospory* is typical of most ferns; *heterospory* occurs in certain "water ferns" (Marsileales and Salviniales). There is no strict correlation between sporangial type and type of spores produced.

Gametophytes and Embryos

The type and relative prominence of the gametophytes are closely correlated with the conditions of homospory and heterospory in the ferns. Thus, in the homosporous ferns the gametophyte is a freely developed, independent plant *(exosporic)* that is photosynthetic or (as in many of the eusporangiate groups) subterranean and associated with a fungus. In contrast, the male and female gametophytes of heterosporous ferns are *endosporic* and much smaller than those of homosporous ferns. All known living ferns produce multiflagellate sperm.

No single type of polarity characterizes the embryogeny of ferns. The pattern in which the first division of the zygote is longitudinal, and hence results in a *lateral orientation* of the apical and basal poles, characterizes the largest number of ferns (see Chapter 6). *Exoscopic* and *endoscopic* embryos also are found in ferns. The intermediate, called "prone," type of embryo, devoid of a suspensor, is

FIGURE 11-1 Tree ferns growing in Golden Gate Park, San Francisco, California. Young leaves, which exhibit circinate vernation, are seen in various stages of growth. The mature fronds are large and compound pinnate. [Courtesy Dr. T. E. Weier, from *Botany. An Introduction to Plant Science.* 2d edition, by W. W. Robbins, T. E. Weier, and C. R. Stocking. Wiley, New York, 1957.]

FIGURE 11-2 Greatly enlarged crozier, or fiddlehead, of a fern. Note that the pinnae and further subdivisions exhibit circinate vernation. [Courtesy Dr. T. E. Weier.]

another feature that separates the highly specialized leptosporangiate ferns (except for a few species) from other vascular plants.

Early Ferns

The ferns are well known from a fossil record extending back to the Lower Devonian. Some of these putative ferns were not too different from those of certain trimerophytes from which they probably took their origin. Ferns increased in number and diversity in the Upper Carboniferous and have persisted to the present day. However, some groups did not leave any direct descendents in today's flora. Even so, the Filicophyta, especially the Filicales, is a highly diversified and successful group at present, having overcome the rigors of existence in a changing world much better than their frequent associates, the lycopods and horsetails.

The search for earliest fern records immediately becomes a complicated study, but one not without some degree of hope. The bulk of Paleozoic fern foliage, originally thought to be exclusively that of spore-producing ferns and to which form genera (see Chapter 9) were assigned, was shown to represent actually the leaves of a great many seed-producing ferns — the Pteridospermophyta (Fig. 15-1). Identification keys that use shape, method of attachment of pinnules, and type of venation have been established for Paleozoic fern leaves. In certain instances a "natural plant" can be synthesized from the form genus for foliage and from the numerous form genera for other parts of the plant which were originally found as isolated fragments.

The problem, then, in tracing the history of the Filicophyta is to separate those fossil forms that may represent morphological steps in the evolution of the fern type of organization, but at the same time to recognize that seed ferns may have shared a common ancestry with the Filicophyta.

As a result of extensive studies on ferns, Bower (1935) proposed certain features that would, in his opinion, characterize a primitive fern. Bower's fern archetype was an upright, dichotomizing plant, if branched at all, in which the distinction between leaf and axis was either absent or ill defined. The leaf, where recognizable, was long-stalked and dichotomously branched with the shanks of the dichotomies free from one another. Sporangia were relatively large, solitary, and located at the distal ends of the subdivisions of leaves. The sporangial wall was thick, opening by a simple dehiscence mechanism, and the sporangia contained only one type of spore.

Cladoxylales

One assemblage of ancient fernlike plants is the order Cladoxylales (sometimes elevated to the rank of class). These Paleozoic plants [Lower Devonian (Emsion) into the Lower Carboniferous] are of interest because of their complex stelar configurations and leaf forms. Most species were relatively small plants with irregularly branched main axes and dichotomously branched smaller ones. The

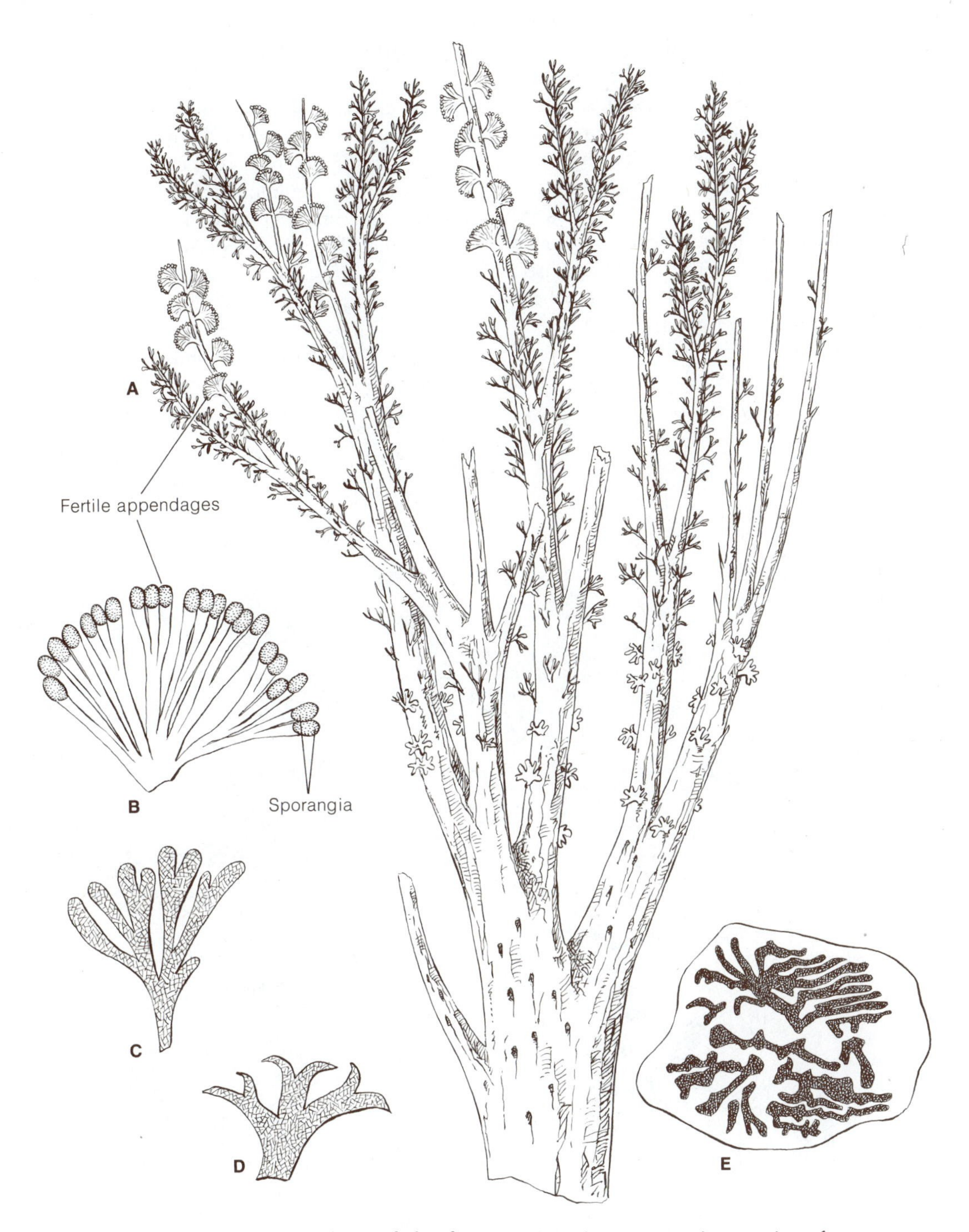

FIGURE 11-3 Cladoxylales—*Cladoxylon scoparium*. **A**, reconstruction, portion of a plant; **B**, fertile appendages: **C**, **D**, sterile appendages from upper (**C**) and lower (**D**) portions of the plant; **E**, xylem of stem as seen in transverse section. [Redrawn from Kräusel and Weyland, *Abh. Senckenberg. Naturforsch. Ges.* 40:115–155, 1926.]

smaller appendages were forked and the fertile leaves were flat, fan-shaped, and bore terminal sporangia on partially fused ultimate segments (Fig. 11-3). The vascular tissue in the stem was organized as a system of interconnected strands (each strand, on many specimens, being elongated radially as seen in transverse section of a stem). Some species are reported to have had secondary xylem around each primary vascular bundle, a feature uncharacteristic of ferns. However, some paleobotanists believe that the tissue consisted of regularly aligned metaxylem cells that were not derived from a vascular cambium.

Coenopterid Ferns

The extinct coenopterids is a fascinating and enigmatic group of ferns. Some specimens have been identified as being of Upper Devonian, but the group was more common in the Carboniferous, and finally disappeared in the Permian. It has long been known that these ferns represent a rather artificial assemblage of plants. Originally, the ordinal name Coenopteridales was established for the entire group, but now three orders are commonly recognized. On the basis of frond morphology and sporangial characteristics, some paleobotanists

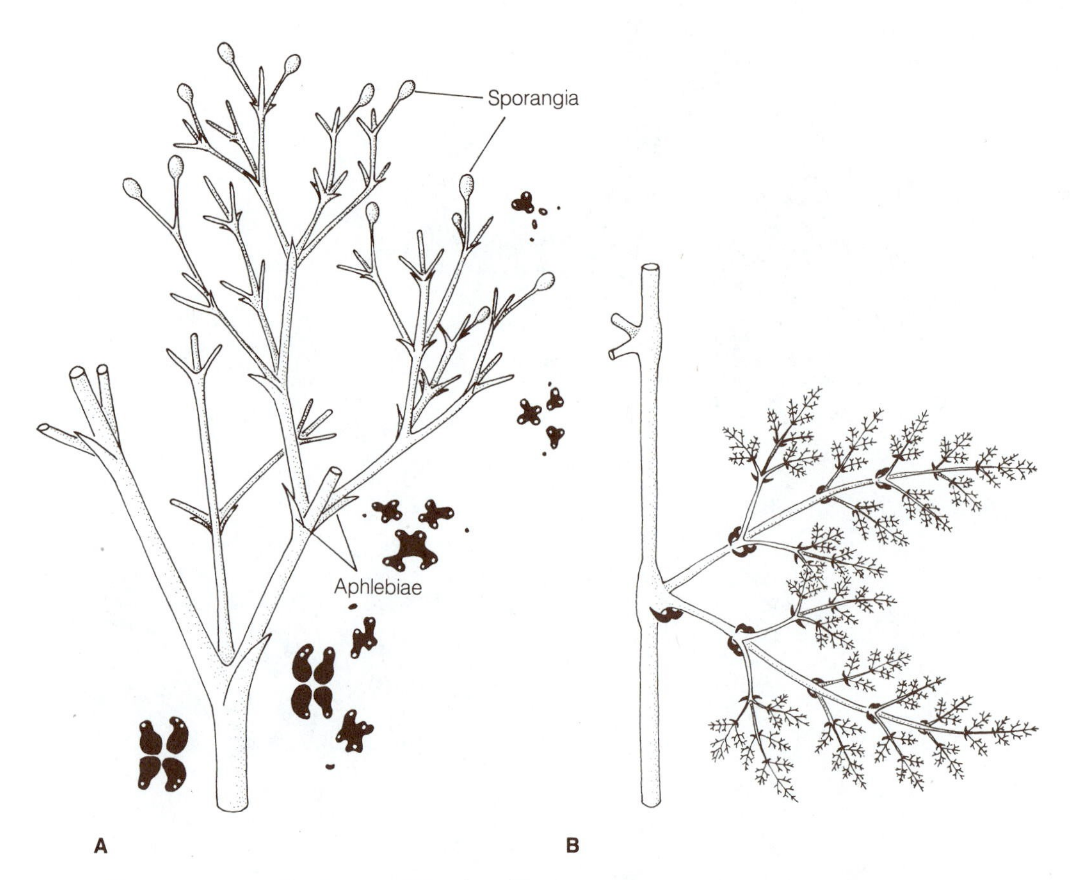

FIGURE 11-4 **A**, *Stauropteris oldhamia.* Idealized reconstruction, portion of the aerial system. Transverse sectional configurations of the xylem at successively higher levels in the branching system shown at left and right. Protoxylem indicated in white; the small vascular strands belong to aphlebiae. **B**, *Stauropteris burntislandica.* Diagrammatic reconstruction of part of a frond; aphlebiae shown in black. [**A** from Eggert, *Mem. Torrey Bot. Club* 21:38, 1964; **B** from Surange, *Philos. Trans. Roy. Soc.* (London), 237(B):73, 1952.]

(Taylor, 1981) have transferred certain families to the Filicales (leptosporangiate ferns).

In some coenopterids, there was a general lack of distinction between stem axis and the frond petiole (stipe) at the level of frond attachment. Stem and petiole, however, generally had quite different vascular anatomy. Certain coenopterid ferns exhibited primitiveness by the fact that the stem was commonly protostelic, by the three-dimensional branching of the frond, and by the terminal position of sporangia on ultimate frond segments. Specialization led to (1) the development of siphonostelic stems in certain genera, (2) the formation of elaborate vascular cylinders in the frond, and (3) planate megaphylls with sporangia located near the margins on the abaxial side of ultimate

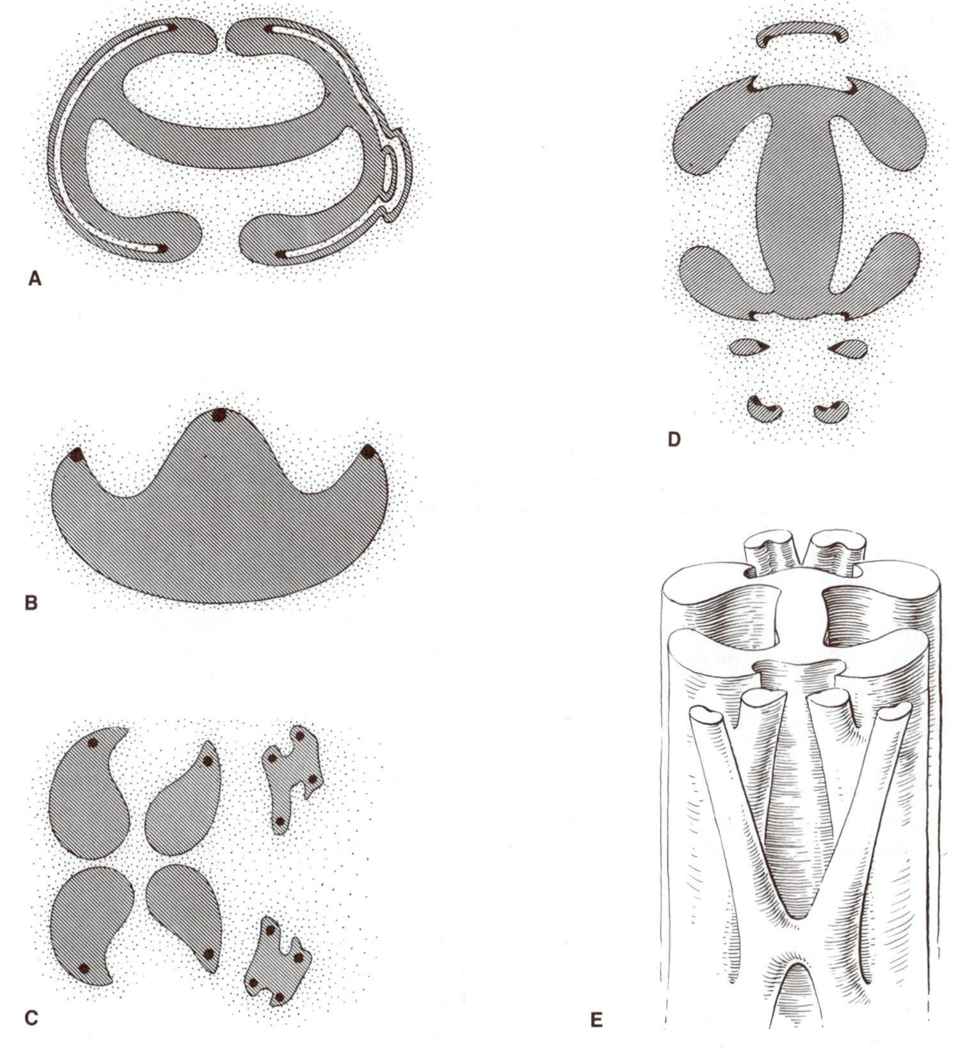

FIGURE 11-5 Petioles (stipes, phyllophores) of the coenopterid ferns. **A–D,** xylem cylinders only, as seen in transection (solid black areas represent positions of protoxylem). **A,** *Ankyropteris;* **B,** *Botryopteris;* **C,** *Stauropteris;* **D,** *Etapteris;* **E,** stereogram of xylem cylinder in *Etapteris.* Smaller, peripheral strands in all diagrams are traces departing to subdivisions of the frond. [**A, D, E** redrawn from *Handbuch der Paläobotanik* by M. Hirmer. R. Oldenbourg, Munich, 1927; **B, C** redrawn from Anatomie der Vegetationsorgane der Pteridophyten by Y. Ogura. In *Handbuch der Pflanzenanatomie.* Gebrüder Borntraeger, Berlin, 1938.]

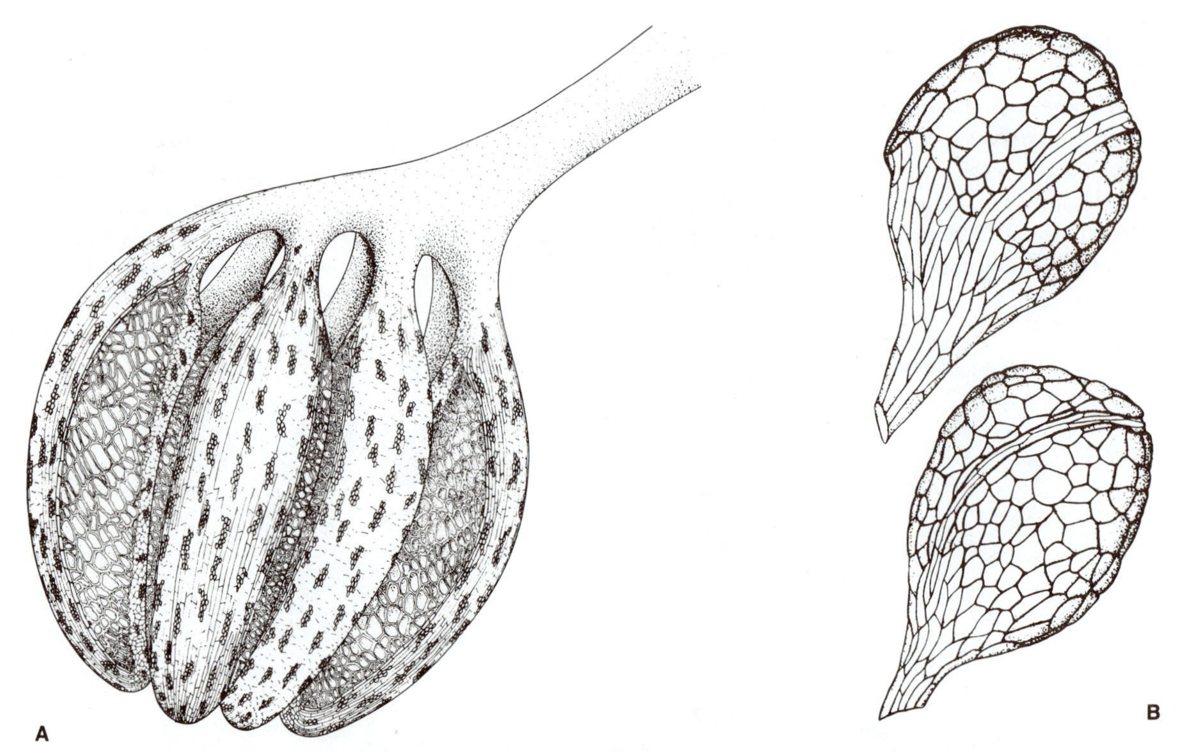

FIGURE 11-6 **A**, cluster of banana-shaped sporangia of *Biscalitheca musata;* annuli are lateral in position with thick cell walls; **B**, sporangia of *Botryopteris globosa;* note that the line of dehiscence runs over the top of the sporangia. [**A** from Phillips and Andrews, *Paleontology* 11:104–115, 1968; **B** from Murdy and Andrews, *Bull. Torrey Bot. Club* 84:252–267, 1957.]

laminate segments. Tracheids were scalariformly pitted in some (considered to be the primitive condition); others had elliptical to circular bordered pits. (See Eggert, 1964, and Phillips, 1974, for more detailed discussions).

Space does not permit a complete discussion of these ferns; only representative genera will be described with notations as to which order and family each belongs.

STAUROPTERIS **(Stauropteridales—Stauropteridaceae).** The two species assigned to this genus were apparently small, bushy plants that exhibited three-dimensional branching (Figs. 11-4, A, B). No laminar tissue was formed and sporangia were terminal on ultimate branches. At each level of branching the pairs of branches were associated with a pair of scalelike appendages termed aphlebiae (Figs. 11-4, A, B). The vascular cylinder was four-lobed, or the lobes were separated by parenchyma (Fig. 11-4, A). There were no distinctive fronds in *S. oldhamia* which suggests that the entire plant may have consisted of a three-dimensional branch system. The frond of *S. burntislandica* consisted of alternate pairs of branches on opposite sides of all axes (Fig. 11-4, B). *Stauropteris oldhamia* was homosporous, but heterospory has been documented for *S. burntislandica* (Surange, 1952) which is rather unique for a fern from the Carboniferous.

ZYGOPTERIS **(Zygopteridales-Zygopteridaceae).** Representatives of this order existed in the Upper Devonian, but were more common in the Carboniferous. Some species of *Zygopteris* were rhizomatous, branched dichotomously, and bore fronds in two lateral ranks. The vascular tissue of the frond axis or *phyllophore* was in the form of the letter H as seen in transverse section. The form genus *Etapteris* is applied to fronds (Fig. 11-5, D, E). The term phyllophore (leaf-bearing structure) traditionally has been used to designate the petiole and rachis of a frond, coined at a time when paleobotanists were

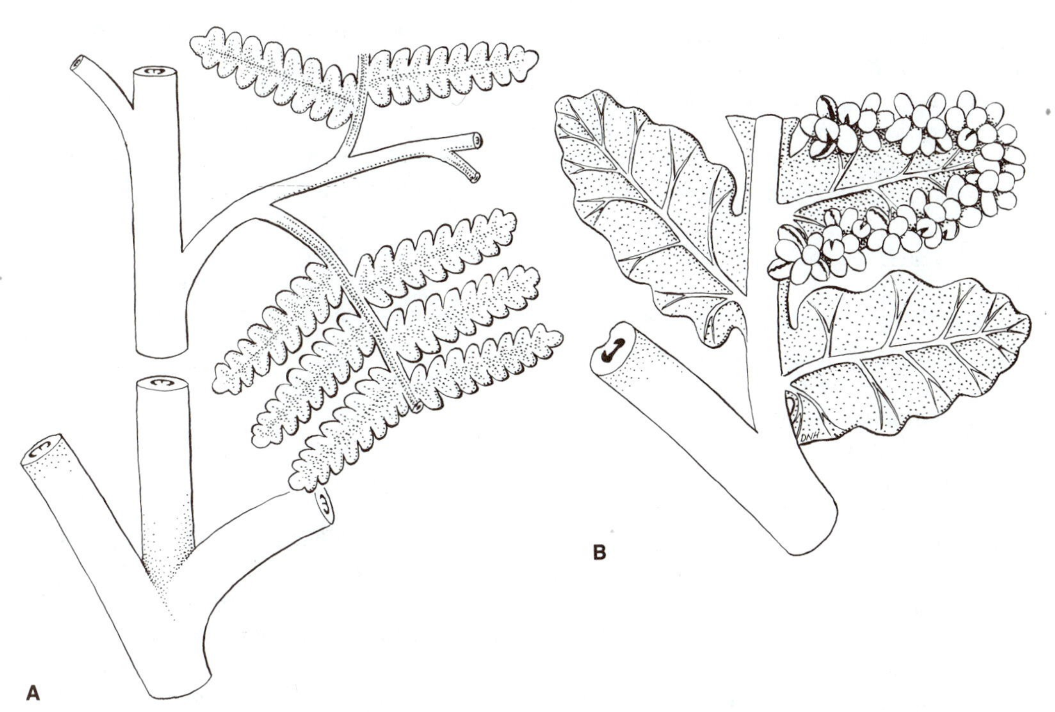

FIGURE 11-7 Reconstructions of portions of fronds of two coenopterid genera. **A,** *Botryopteris forensis,* distal part of a frond; **B,** *Ankyropteris,* portion of fertile frond (form genus *Tedelea*) showing sporangia on abaxial side of one leaf segment. [**A** redrawn from Delevoryas and Morgan, *Amer. Midland Natur.* 52:374–387, 1954; **B** redrawn from Eggert and Taylor, *Palaeontographica* 118(B):52–73, 1966.]

uncertain as to what constituted a frond in coenopterid ferns. The ultimate segments (pinnules) of the frond were planated in some; clusters of elongate, banana-shaped sporangia on segments of fertile fronds have been described and given the name *Biscalitheca musata* (Fig. 11-6, A).

ANKYROPTERIS (Zygopteridales — Tedeleaceae). In *Ankyropteris* the fronds were helically arranged on the stem. The fronds were pinnately compound, with planated pinnae and small laminate pinnules, remarkably similar to modern-day ferns (Fig. 11-7, B). Unlike other coenopterids and the majority of ferns, branching was axillary — new shoots developing in the axils of fronds (Phillips, 1974). Xylem in the phyllophore was in the form of the letter H or resembled two anchors joined by a bar as seen in transverse section (Fig. 11-5, A). Fertile fronds were discovered and given the name *Tedelea* (Eggert and Taylor, 1966). Sporangia occurred in clusters near the margins on the lower surface of pinnules at the termination of lateral veins (Fig. 11-7, B). Sporangial structure was similar to certain extant ferns. On the basis of leaf morphology and sporangial characteristics, *Ankyropteris* has been transferred to the Filicales by some paleobotanists (Mickle, 1980; Taylor, 1981). *Ankyropteris brongniartii* is now recognized as the only species, with all other species of *Ankyropteris* and *Tedelea* being placed in synonomy (Mickle, 1980).

BOTRYOPTERIS (Coenopteridales — Botryopteridaceae). Ten or more species have been described that had helically arranged fronds on the stems; there was considerable variation in frond morphology. In certain species the lower (proximal) portion of the frond was branched three-dimensionally while the upper (distal) portion was planate and the ultimate segments were laminate (Fig. 11-7, A). The xylem cylinder in the petiole and rachis typically was in the form of the Greek letter omega (ω), as seen in transverse section (Fig. 11-5, B). The sporangia were borne terminally on ultimate segments of a much branched fertile portion of a frond, resulting in the formation of large masses of sporangia. The sporangia (Fig. 11-6, B) resemble those of the extant fern family Osmundaceae, and thus *Botryopteris* is considered to be a taxon more closely aligned with the Filicales (Mickle, 1980; Taylor, 1981). However, as Stewart (1983) has pointed out, all coenopterid ferns differ to varying degrees from the Filicales in their vegetative morphology, and all are extinct.

In summary, while some of the coenopterid ferns may have been ancestral to some modern-day ferns, certain ones most probably represent end points in evolutionary experimentation. The Trimerophytophyta (Chapter 7) remains as the group most likely to have been ancestral to the coenopterid ferns through the elaboration of lateral branched axes with terminal sporangia, and finally by the shifting of sporangia to the abaxial side of laminate pinnules in some taxa.

Classification

The following outline of classification will serve as a guide for discussions of fern orders found in Chapters 12 and 13. The extinct Cladoxylales and coenopterid ferns are omitted.

FILICOPHYTA: Living and extinct ferns; plants showing a conspicuous alternation of generations (in modern representatives at least); sporophyte most conspicuous and often elaborately developed; megaphylls present; stems protostelic or siphonostelic, often with complex vascular cylinders; sporangia terminal on ultimate axes, terminal on veins, marginal, or on abaxial surface of fronds; eusporangiate, or more commonly leptosporangiate in living species; homosporous, a few heterosporous; gametophytes: (1) majority exosporic and green, (2) others exosporic, nonchlorophyllous, and subterranean, and (3) endosporic (restricted to heterosporous types); multiflagellate free-swimming sperm; embryo exoscopic, endoscopic, or intermediate.

OPHIOGLOSSALES (Chapter 12): Living ferns, meager fossil record; sporophyte axis usually short and fleshy; stem siphonostelic; fronds simple or pinnately compound; vernation noncircinate to weakly circinate *(Botrychium);* each fertile frond consists of a fertile segment or spike, bearing sporangia, and a sterile or vegetative segment; eusporangiate, producing great quantities of spores per sporangium; homosporous; gametophytes subterranean, bisexual, nonchlorophyllous, tuberous or wormlike with endophytic fungus; embryo exoscopic or endoscopic.

OPHIOGLOSSACEAE: Characteristics as in Ophioglossales.

Ophioglossum, Botrychium, Helminthostachys.

MARATTIALES (Chapter 12): Living and extinct ferns; sporophyte stem in most erect and short, or may be dorsiventral; stem with complex dictyostelic vascular cylinder; fronds commonly large, simple pinnate to tripinnate, and circinate in vernation; paired, clasping stipules at base of each leaf; eusporangia free and grouped into elongate sori or united into synangia on abaxial surface of fronds; many spores formed per sporangium; homosporous; gametophyte terrestrial, green, cordate to ribbon shaped, bisexual, and with endophytic fungus; endoscopic embryo.

MARATTIACEAE: Characteristics as in Marattiales.

Representative genera: *Angiopteris, Marattia. Psaronius* (extinct), *Eoangiopteris, Scolecopteris* (form genera for sori).

FILICALES (Chapter 13): Living and extinct plants, of diverse growth habits and habitats; stems vary from protostelic to intricately dictyostelic; fronds simple to compound pinnate; sporangia scattered or grouped into sori; sori marginal or on abaxial side of fronds; sori with or without a protective structure, the indusium; leptosporangiate, most with a definite dehiscence mechanism, the annulus; spores numerous to few per sporangium, tetrahedral or bilateral; homosporous; gametophytes primarily green, exosporic, commonly thalloid, some filamentous; embryo "prone."

MARSILEALES (Chapter 13): Living and extinct ferns; grow on damp soil or submerged in water; solenostelic; lamina consisting of four pinnae, bipinnate, or without pinnae; circinate vernation; sori enclosed in sporocarps; leptosporangiate; heterosporous; microsporangia and megasporangia in same sporocarp; endosporic gametophytes.

MARSILEACEAE: Characteristics as in Marsileales.

Marsilea, Regnellidium, Pilularia

SALVINIALES (Chapter 13): Living and extinct ferns; small plants that float on surface of water; sori enclosed in specialized indusia ("sporocarps"); leptosporangiate; heterosporous; microsporangia and megasporangia in separate sporocarps; endosporic gametophytes.

SALVINIACEAE: Characteristics as in Salviniales.

Azolla, Salvinia.

Before proceeding with a detailed account of the living orders of ferns, it is important for the reader to have a clear idea of the morphological features that are used for comparative purposes. It was the celebrated British morphologist F. O. Bower who realized the importance of exploring and exploiting the totality of morphological features before a reasonable phylogeny of ferns could be achieved. Bower (1923, 1935) concluded that there are at least twelve major morphological and anatomical criteria that should be utilized. These are listed here because discussions of these points are unavoidable in ferns, and because most of them are of great importance in later discussions.

1. External morphology and habit of plant
2. Apical meristem organization
3. Architecture and venation of the leaf
4. Vascular system of the shoot
5. Morphology of hairs and scales
6. Position and structure of the sorus
7. Protection of the sorus by an indusium
8. Development and mature structure of the sporangium including form of and markings on spores
9. Number of spores produced
10. Morphology of the gametophyte
11. Position and structure of sex organs
12. Embryology of the sporophyte

Additional information that has a bearing on the systematics and phylogeny of ferns can be found in the following list of selected publications: anatomy (White, 1963, 1979; Ogura, 1972), cytology and cytogenetics (Manton, 1950, 1959; Abraham et al., 1962; Fabbri, 1963; Löve et al., 1977; Walker, 1979), experimental physiology and morphogenesis (Wardlaw, 1952, 1968a, b; Wetmore and Wardlaw, 1951), and gametophytes (Nayar and Kaur, 1971).

REFERENCES

Abraham, A., C. A. Ninan, and P. M. Mathew
1962. Studies on the cytology and phylogeny of the pteridophytes. VII. Observations on one hundred species of South Indian ferns. *Jour. Indian Bot. Soc.* 41:339–421.

Bower, F. O.
1923. *The Ferns.* Vol. I. Cambridge University Press, London.
1935. *Primitive Land Plants.* Macmillan, London.

Eggert, D. A.
1964. The question of the phylogenetic position of the Coenopteridales. *Mem. Torrey Bot. Club* 21:38–57.

Eggert, D. A., and T. N. Taylor
1966. Studies of Paleozoic ferns: on the genus *Tedelea* gen. nov. *Palaeontographica* 118(B):52–73.

Fabbri, F.
1963. Primo supplemento alle tavole cromosomiche delle Pteridophyta di Alberto Chiarugi. *Caryologia* 16:237–335.

Löve, A., D. Löve, and R. E. G. Pichi-Sermolli
1977. *Cytotaxonomical Atlas of the Pteridophyta.* Cramer, Vaduz.

Manton, I.
1950. *Problems of Cytology and Evolution in the Pteridophyta.* Cambridge University Press, London.
1959. Chromosomes and fern phylogeny with special reference to "Pteridaceae". *Jour. Linn. Soc. London Bot.* 56:73–91.

Mickle, J. E.
1980. *Ankyropteris* from the Pennsylvanian of eastern Kentucky. *Bot. Gaz.* 141:230–243.

Nayar, B. K., and S. Kaur
1971. Gametophytes of homosporous ferns. *Bot. Rev.* 37:295–396.

Ogura, Y.
1972. Comparative Anatomy of Vegetative Organs of the Pteriodphytes, 2d edition, *Encyclopedia of Plant Anatomy,* Vol. 7, part 3. Gebrüder Borntraeger, Berlin.

Phillips, T. L.
1974. Evolution of vegetative morphology in coenopterid ferns. *Ann. Missouri Bot. Garden* 61:427–461.

Stewart, W. N.
1983. *Paleobotany and the Evolution of Plants.* Cambridge University Press, Cambridge.

Surange, K. R.
1952. The morphology of *Stauropteris burntislandica* P. Bertrand and its megasporangium *Bensonites fusiformis* R. Scott. *Phil. Trans. Roy. Soc. London* 237(B):73–91.

Taylor, T. N.
1981. *Paleobotany: An Introduction to Fossil Plant Biology:* McGraw-Hill, New York.

Walker, T. G.
1979. The cytogenetics of ferns. In A. F. Dyer (ed.), *The Experimental Biology of Ferns,* pp. 87–132. Academic, London.

Wardlaw, C. W.
1952. *Phylogeny and Morphogenesis.* Macmillan, London.
1968a. *Essays on Form in Plants.* University Press, Manchester, England.
1968b. *Morphogenesis in Plants: A Contemporary Study.* Methuen, London.

Wetmore, R. H., and C. W. Wardlaw
1951. Experimental morphogenesis in vascular plants. *Ann. Rev. Plant Physiol.* 2:269–292.

White, R. A.
1963. Tracheary elements of the ferns. II. Morphology of tracheary elements; Conclusions. *Amer. Jour. Bot.* 50:514–522.
1979. Experimental investigations of fern sporophyte development. In A. F. Dyer (ed.), *The Experimental Biology of Ferns,* pp. 505–549. Academic, London.

CHAPTER 12

Filicophyta

EUSPORANGIATE FERNS: OPHIOGLOSSALES, MARATTIALES

We began our discussion of Filicophyta in the previous chapter with a general description of the ferns. In this chapter, we turn to the eusporangiate ferns of the families Ophioglossales and Marattiales.

Ophioglossales — Ophioglossaceae

Over the years, the fossil record of the Ophioglossales has been sketchy to nonexistent. Spores have been described that could be from plants of the order, but they are similar to those of other ferns, and hence did not constitute definitive evidence. In 1974 Chandrasekharam discovered remains of *Botrychium* (from the Paleocene) near Genesee, Alberta, Canada. The specimens consist of fertile portions and fragments of laminae. This report was subsequently confirmed by G. W. Rothwell (pers. comm.). Some morphologists consider that certain member(s) of the coenopterids may have been ancestral to the Ophioglossales. Others believe that the order originated more directly from the Trimerophytophyta.

The family comprises three recognized genera, of which two are more commonly seen and known by botanists. One genus, *Botrychium* (grape fern, moonwort), having about 25 or more species, is nearly worldwide in distribution. There has always been considerable disagreement among pteridologists with respect to the taxonomy of *Botrychium*. In addition to species, numerous varieties and subspecies have been described. In an effort to better define species of the New World, Wagner and Wagner (1983) have made mass collections of fronds and have adopted the "Genus Community Method" of analysis. *Ophioglossum* (adder's tongue), with 30 or more species, is widely spread throughout the world but is more abundant in the tropics. Although members of the Ophioglossaceae are typically terrestrial, there are some tropical, epiphytic species of *Ophioglossum*. Commonly the stem is short and erect and has a frond that is divided into a more or less flattened, vegetative portion (sterile segment) and a sporangium-bearing portion (fertile segment). However, the genera *Ophioglossum* and *Botrychium* can be separated easily (with a few exceptions) by examining the fronds. The lamina in *Ophioglossum* is typically simple in outline with reticulate venation; it is pinnate in plan in *Botrychium*, with open dichotomous venation. One other genus, *Helminthostachys*, which is native to the Indo-Malayan regions, can be recognized by its creeping rhizome and ternately compound leaves, which have an open dichotomous venation system.

Botrychium (Grape Fern, Moonwort)

The plant axis is typically a short, stocky, subterranean rhizome from which roots arise at the bases of leaves (Fig. 12-1). Generally one frond matures each year in temperate-climate species; the decaying leaves of previous seasons surround the base of the plant. Each frond has a sheath at its base that encloses the next younger leaf. Although only one leaf may mature each year, there are several immature leaves of future seasons in varying stages of development within the terminal bud. Traditionally, the frond has been described as noncircinate in vernation. Stevenson (1975) however, has demonstrated that large leaves of *Botrychium multifidum* are circinate during early development but not as pronounced as in the Filicales (Figs. 11-1; 11-2). The frond is fleshy and consists of two parts: the vegetative lamina (blade), which usually exhibits a pinnate pattern of branching, and a fertile segment. The fertile segment likewise is constructed on the pinnate plan, bearing eusporangia in two rows on ultimate axes (Fig. 12-2, A, B). There is some degree of correlation between the amount of pinnate branching of the fertile and sterile portions of the frond. In certain species the vegetative lamina may be multipinnate and the fertile segment may be of the same degree of branching; in others, the blade may be simple and entire with the fertile segment showing the same degree of reduction. This reduction series is considered to be an evolutionary sequence. Morphological interpretation of the frond in *Botrychium* is a controversial subject to be discussed later in this chapter.

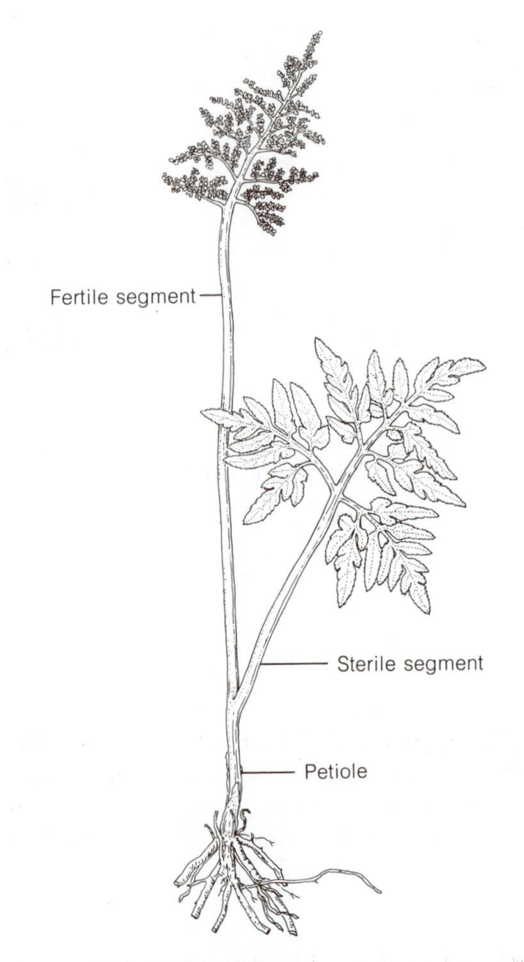

FIGURE 12-1 Habit sketch of sporophyte of *Botrychium dissectum* var. *obliquum,* showing pinnate fertile and sterile segments of one leaf.

ANATOMY OF THE STEM. Growth of the shoot is reported to be initiated by an apical cell that gives rise to derivatives (segments). The segment cells divide, and then by continued cell divisions tissues are provided for leaf initiation at the surface of the shoot apex as well as for the production of cells basally that will ultimately differentiate into tissues of the stem. By using more modern histological methods, Stevenson (1976) has described a zonate type of organization in *Botrychium multifidum.* The apex is reported to consist of (1) a surface zone of initials (in which a prominent apical cell may or may not be evident), (2) a zone of subsurface initials, and (3) a cup-shaped zone that is subdivided into a peripheral zone and a pith meristem.

The primary vascular system of *Botrychium* is considered to be a sympodium of anastomosing, collateral leaf traces in which the primary xylem is endarch. A unifacial vascular cambium is formed between the primary xylem and phloem; only secondary xylem is produced (Stevenson, 1980). Secondary vascular tissues are not found in any other living fern. In stelar terminology, the vascular cylinder of *Botrychium* is an ectophloic siphonostele. The leaf gaps often do not overlap; consequently, a continuous vascular cylinder or a cylinder with one

FIGURE 12-2 *Botrychium multifidum.* **A,** growth habit; **B,** enlarged view, portion of fertile segment showing sporangia; line of dehiscence can be seen along top of some sporangia.

gap may be present in a stem as seen in transverse section (Fig. 12-3, A). The walls of sieve cells of the phloem are conspicuously thickened except for the sieve areas. However, the presence of callose has not been detected in stem phloem (Esau et al., 1953) or in the phloem of the petiole or rachis (Evert, 1976). A periderm is generally present at the periphery of the broad cortex, another characteristic not found in extant ferns.

THE ROOT. Attached near the bases of leaves are fleshy roots that branch sparsely and are devoid of root hairs. The roots are mycorrhizal, and the fungus probably functions in absorption of water

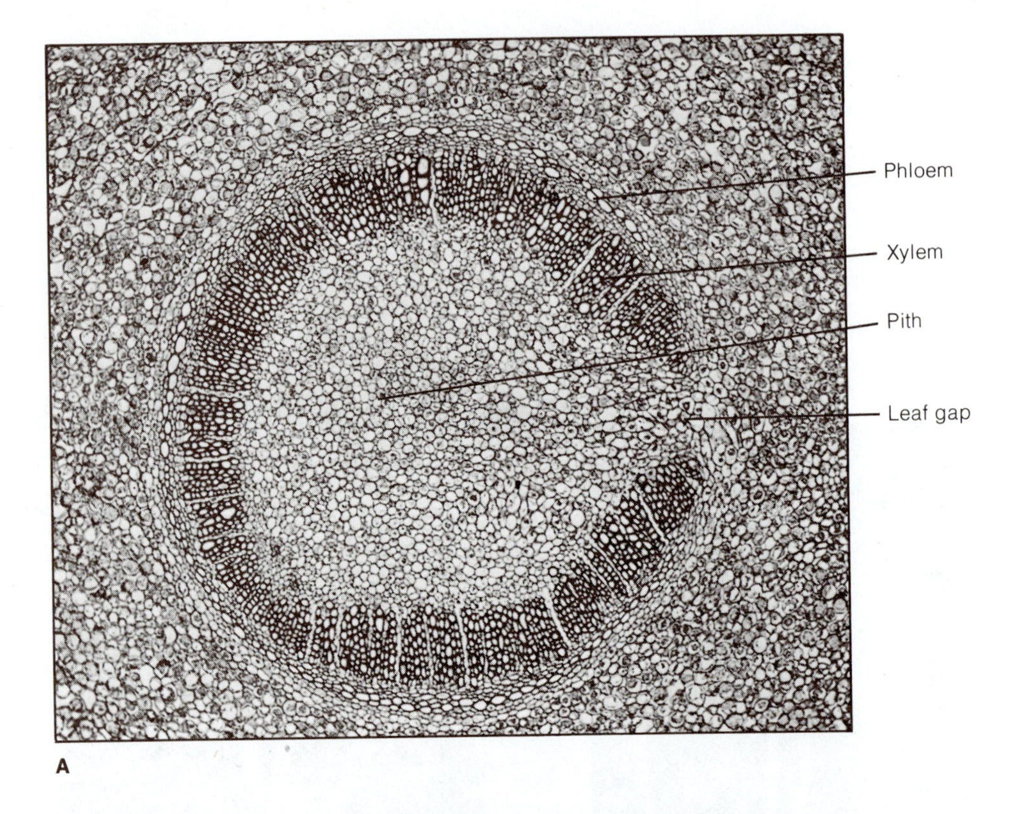

A

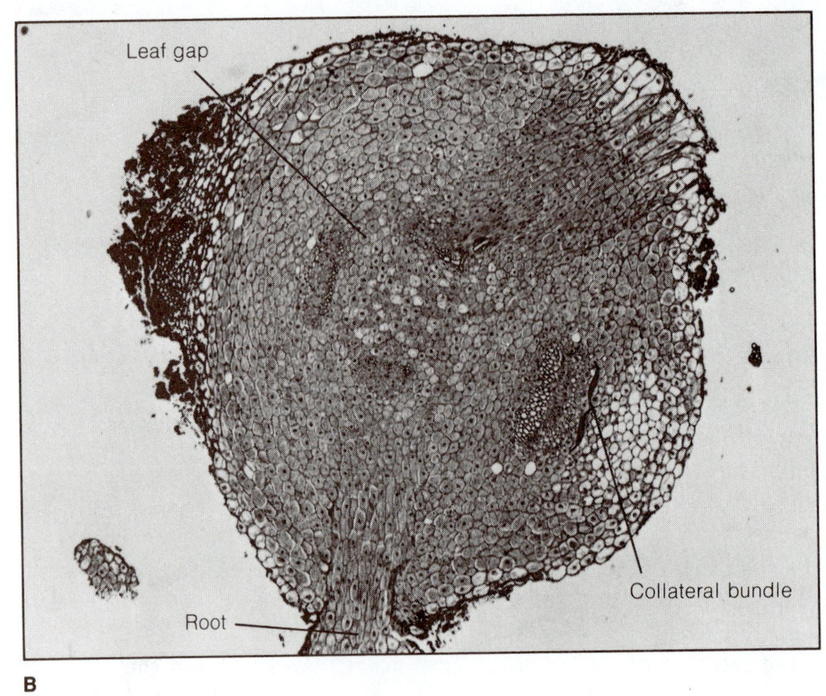

B

FIGURE 12-3 Stem anatomy in the Ophioglossales. **A,** transection, stem of *Botrychium* sp. showing ectophloic siphonostele and secondary xylem. **B,** transection, stem of *Ophioglossum* sp. showing ectophloic siphonostele.

and minerals. The vascular cylinder usually has two to five protoxylem poles. It is reported that a tetrahedral apical cell is present in root tips, but divisions that give rise to the root cap may not be as precise as they are in most leptosporangiate ferns (Bower, 1926).

THE FROND (LEAF). The vegetative portion of the frond in *Botrychium* has an open dichotomous venation system and a rather homogeneous mesophyll organization. Transcending the mere knowledge of leaf structure is the question of phylogeny of the frond. According to one view (Chrysler, 1925) the fertile portion represents two pinnae which have become fused during the course of evolution and now stand erect. This hypothesis is based upon (1) a study of the branched vascular system of the frond and (2) the occasional presence in some species of fertile pinnae occupying the position of otherwise normal vegetative pinnae.

In 1942 Zimmermann — the originator of the telome theory — published a reinterpretation of the frond. This concept is also based upon an analysis of the branched venation system. The repeated bifurcations of the vascular system are interpreted as being indicative of a once freely dichotomous branch system. The first dichotomy is at the level below the union of the fertile portion and vegetative lamina (or may occur at the level of the departing leaf trace from the stem vascular cylinder). Thus, this interpretation adds weight to the idea that the frond complex phylogenetically is a modified, reduced dichotomous branch system, reminiscent of a coenopterid type of fern. Chrysler (1945) subsequently accepted this interpretation.

As a natural extension of this concept some authors consider the petiole or stipe of the "leaf complex" of present-day species of *Botrychium* to be cauline; they believe that the sterile and fertile segments are phylogenetically equivalent to fronds and are therefore homologous (Nozu, 1950, 1955; Nishida, 1952, 1957).

THE SPORANGIUM. The fronds of several future seasons are in varying stages of development within the bud. Therefore the stages of development of the fertile spikes are also variable. Eusporangia are initiated from one or from several superficial cells (Bower, 1935) in an acropetal manner on the pinnae. If only one initial superficial cell is evident early in development, sooner or later adjacent superficial cells may divide periclinally, resulting in the separation of sporogenous cells and wall cells (Chapter 4). The tapetum becomes several layers thick, and its cells break down very early, their contents permeating the spaces between the sporocytes.

At maturity each sporangium is large and has a vascular strand that extends to the base of the capsule. Each sporangium, which dehisces by means of a terminal slit, produces 2,000 or more tetrahedral spores.

CHROMOSOME NUMBERS. The results of several studies have shown that the haploid number of chromosomes in a large number of species is 45. Some species are presumably tetraploid with $n=90$ (Löve et al., 1977).

GAMETOPHYTE. Mature gametophytes of *Botrychium* are subterranean, tuberous or somewhat elongate to button shaped, and have numerous rhizoids (Figs. 12-4, A; 12-5, A). The gametophytes vary from 1 to 3 millimeters to 5 to 6 millimeters in length and even reach 1.5 to 2 centimeters in some species. They have an associated endophytic fungus, the presence of which presumably is essential for continued growth under natural conditions. Typically the gametophytes have a "dorsal ridge" in which antheridia are embedded (Fig. 12-4, A, B). Archegonia generally are along each side of the dorsal ridge (Nozu, 1954; Bierhorst, 1958; Daigobo, 1979). An apical meristem located at the anterior end, toward the dorsal side, gives rise to new tissue in which more antheridia and archegonia are initiated. Antheridia are sunken and produce multiflagellate sperms. Archegonia have protruding necks at maturity, and they may have a binucleate neck-canal cell, a ventral canal cell, and an egg cell.

Although *Botrychium* spores were reported to germinate on soil or humus, presumably in the light, recent investigations on culturing gametophytes on agar under sterile conditions have revealed a dark requirement for germination. Whittier (1972, 1973) found that the spores of *B. dissectum* would germinate only after the spores were subjected to a minimum of 3 to 4 weeks of darkness. The morphology of the gametophytes fits the description of the species from nature, except for the absence of the en-

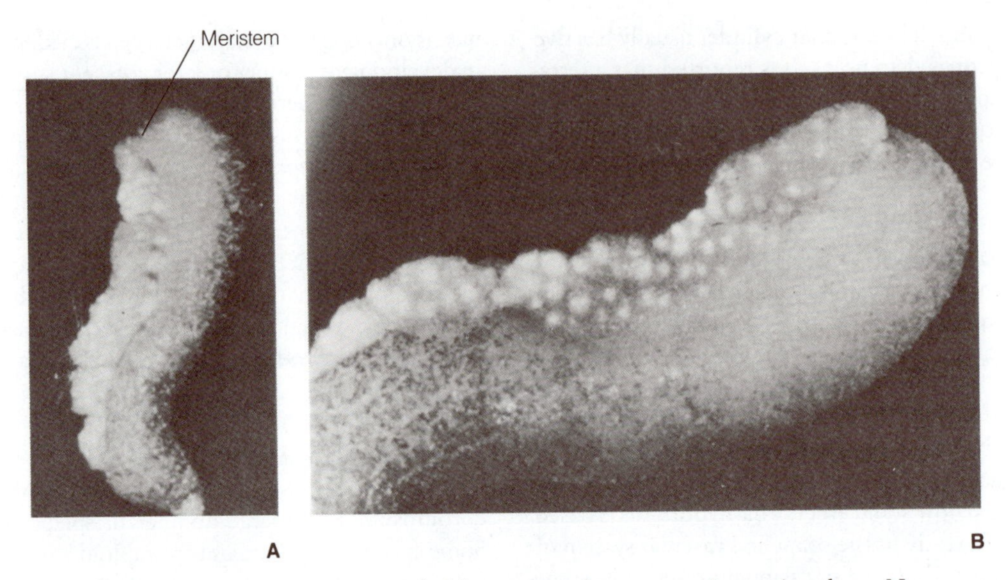

Figure 12-4 **A, B,** *Botrychium multifidum* gametophytes grown in axenic culture. Note antheridia on dorsal ridges (× 15). [From Gifford and Brandon, *Amer. Fern Jour.* 68:71, 1978.]

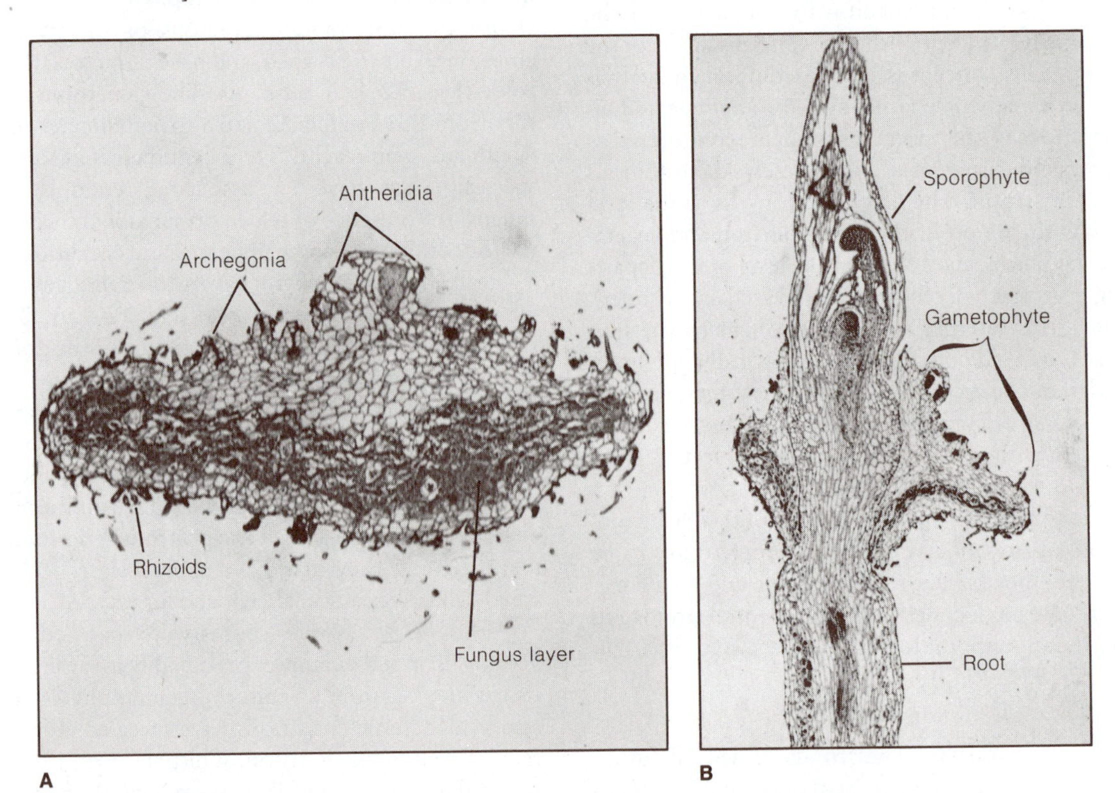

Figure 12-5 **A,** section, gametophyte of *Botrychium* sp. showing antheridial ridge and subjacent archegonia; **B,** section, young sporophyte attached to gametophyte of *Botrychium* sp. The sporophyte has several immature leaves in the bud; the first root is large and is toward the bottom of the page.

dophytic fungus. A variable dark requirement also was demonstrated for other species (Whittier, 1981). Comparable results were obtained for *B. multifidum* by Gifford and Brandon (1978). The gametophytes of *B. multifidum,* previously unknown from nature, were found in Japan (Daigobo, 1979) and were similar in morphology to those grown in culture.

The gametophytes at sexual maturity are bisexual and intragametophytic fertilization (self fertilization) should be possible. A high level of selfing (inbreeding) has been demonstrated for two species of *Botrychium,* based upon data from enzyme electrophoresis (See Soltis and Soltis, 1987).

THE EMBRYO. There is some variation in the early development of the embryo in *Botrychium,* but the first division of the zygote is transverse or slightly oblique. Depending on the species the embryo is either exoscopic or endoscopic. If endoscopic, a suspensor may or may not be formed. Quadrants are usually formed, but the derivation of future parts of the embryo is not easily determined. Critical studies of embryogeny, using modern techniques, have not been undertaken, due in part to the difficulty of obtaining suitable material. The relationship of a young sporophyte to the gametophyte is shown in Figs. 12-5, B and 12-6. Additional information on embryogeny in *Botrychium* can be found in the following references: Jeffrey, 1897; Lyon, 1905; Nishida, 1955; Wardlaw, 1955; Rao, 1962; Bierhorst, 1971.

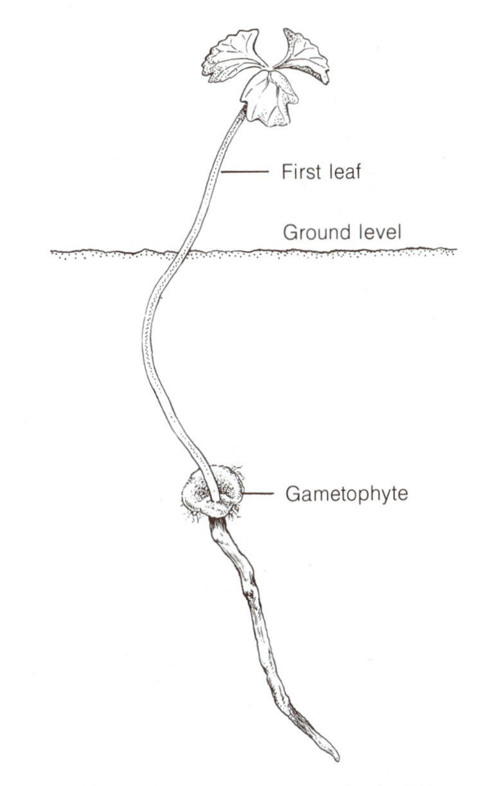

FIGURE 12-6 Subterranean gametophyte of *Botrychium dissectum* with attached sporophyte. [Drawn from specimen, courtesy Dr. A. J. Eames.]

Ophioglossum (Adder's Tongue)

Most species are terrestrial; the exceptions are certain tropical epiphytic forms. The general organization of the plant body is similar to that of *Botrychium,* namely, there is a short, sometimes globose rhizome, roots, and generally the formation of a single mature frond each year in temperate species (Fig. 12-7). Here also, a leaf sheath encloses the next younger leaf in the bud. The vegetative lamina in the majority of species is simple and entire in outline, or it may be lobed in an epiphytic species (Figs. 12-7; 12-8; 12-9). In contrast to *Botrychium,* leaf venation is of the reticulate type. In general a fertile spike or segment consists of an axis with two lateral rows of embedded sporangia (Fig. 12-11). In the tropical, epiphytic species *Ophioglossum palmatum,* there may be several fertile spikes inserted along the margins of the basal part of the lamina and on the upper part of the petiole (Fig. 12-9). A phylogenetic reductional series has been proposed for *Ophioglossum,* as in *Botrychium,* beginning with the large epiphytic species and culminating with the minute and inconspicuous types a few centimeters in height (Eames, 1936). However, Wagner (1952) believes that the large epiphytic types may be more advanced.

ANATOMY OF THE SHOOT. There are reports of the presence of a definite apical cell at the shoot tip (Bower, 1926; Bierhorst, 1971). For *Ophioglossum lusitanicum* subsp. *lusitanicum,* Gewirtz and Fahn (1960) reported the occurrence of a group of initials. The vascular cylinder in the upper part of a well-developed plant, enclosed by fleshy storage cortical tissue, is an ectophloic siphonostele with overlapping leaf gaps (Fig. 12-3, B). The vascular

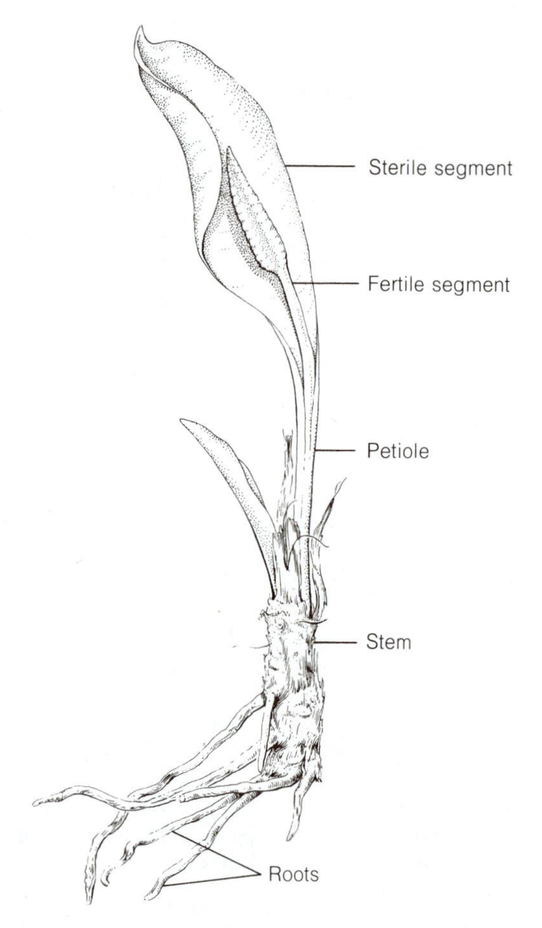

FIGURE 12-7 Sporophyte of *Ophioglossum lusitanicum* var. *californicum*. Each leaf consists of a simple sterile segment and a compact spikelike fertile segment. One leaf or frond matures every year. The stem is a short upright rhizome bearing roots (× 3). [Specimen supplied by Dr. W. H. Wagner, Jr.]

bundles are collateral (consisting of radially aligned primary phloem and xylem) comparable to the primary state of development in *Botrychium*. No secondary xylem is produced. In *O. petiolatum* the leaf traces are attached to the vascular system in the stem at the margins of the leaf gaps rather than at the base. In some atypical stems there may be traces to two or three leaves that are associated with a single common gap (Webb, 1975, 1981). Other species of *Ophioglossum* have a much more complex stelar organization. The occurrence and position of an endodermis with casparian strips are seemingly variable in *Ophioglossum* (see Gewirtz and Fahn, 1960; Webb, 1975).

The base of the frond is traversed by several vascular bundles which are subdivisions of a single leaf trace (in some species). Vascular bundles on the adaxial side of the petiole are continuous into the fertile segment; others traverse the vegetative lamina and are branched, forming a reticulate venation system. The morphological interpretation of the fertile spike with its inherent problems is comparable to the situation in *Botrychium*.

THE ROOT. The fleshy, generally unbranched mycorrhizal roots often have a relationship to the leaves. A root arises endogenously and its vascular strand may be attached to a vascular bundle of the stem (Gewirtz and Fahn, 1960) or near the margin of a leaf gap. In the latter situation, a small "root gap" or perforation occurs in the vascular system of the stem (Webb, 1975). Growth of the root is initiated by a tetrahedral cell; the root becomes monarch or diarch, or more protoxylem poles are formed in some species. For more information on general anatomy of stem and roots see Petry, 1914; Joshi, 1940; Chrysler, 1941; Sen, 1968; Gewirtz and Fahn, 1960; and Webb, 1975, 1981.

Vegetative propagation in *Ophioglossum* is especially interesting in that new shoots in many species take their origin from roots (Fig. 12-10). The bud meristem originates from recent derivatives of the root apical cell (Peterson, 1970). The parent root continues to grow as well as newly initiated roots and more buds are formed. In this fashion a large colony can develop from an original plant. Buds also can be induced experimentally to arise from the pith of decapitated shoots (Wardlaw, 1953, 1954).

THE SPORANGIUM. Eusporangia, which originate from superficial cells along two sides of the young fertile spike, are in varying stages of development before the frond emerges for the season. The sporangia always remain embedded in tissues of the fertile spike; small vascular strands are present between the sporangia and often are turned toward them (Fig. 12-11). At maturity numerous tetrahedral spores are formed which are liberated through a slit in the sporangial wall, perpendicular to the wide surface of the fertile spike.

CHROMOSOME NUMBERS. The largest number of chromosomes of any living organism is in *Ophio-*

FIGURE 12-8 *Ophioglossum petiolatum.* Group of plants produced by means of vegetative reproduction (see Fig. 12-10). Note sterile and spikelike fertile segments.

glossum reticulatum: $2n$ = about 1260 (Abraham and Ninan, 1954; Ninan, 1958). As pointed out by Stebbins (1971), it is nothing short of miraculous that so many chromosomes could sort themselves out at the time of meiosis, each one "seeking" a specific mate in the formation of hundreds of bivalents in each sporocyte. The lower end of the range of chromosome numbers in the genus is $2n = 240$ with some intermediates (Löve et al., 1977). One must conclude that there has been an extremely high degree of polyploidy in these presumably ancient ferns.

EXPERIMENTAL STUDIES ON THE DEVELOPMENT OF THE FERTILE SEGMENT. A study of the mechanism of elongation of the *fertile segment* in *Ophioglossum petiolatum* was undertaken by Peterson and Cutter (1969a). In this species the fertile segment undergoes considerable elongation during development. The fertile segment originates as a primordium on the adaxial side of the sterile segment and exhibits apical growth initially. Apical growth ceases rather early in ontogeny at a time when the sporangial region becomes organized. Meristematic activity then becomes more pronounced in an intercalary meristem located at the base of the sporangial region. Marking experiments (using India ink) indicated that subsequent growth of the upper portion of the stalk of the fertile segment accounted for most of the final length of the stalk. These results were confirmed by the localization of mitoses, cell length, and the use of ^{3}H-thymidine to label DNA.

Additional experiments were performed to determine the causal basis for elongation of the fertile segment. Surgical removal of the sterile segment of the leaf had no effect on the elongation of the fertile segment. Excision of the sporangial portion early in development resulted in cessation of elongation of the stalk of the fertile segment. Replacing the excised sporangial portion with auxins resulted in elongation of the stalk, but gibberellic acid alone or kinetin alone had very little effect. These studies have shown the controlling influence that sporo-

FIGURE 12-9 *Ophioglossum palmatum.* Epiphytic, lobed species from a montane forest, Ecuador. Note upright fertile segments along petiole and base of the lamina (at top of the figure). [From *Ferns and Allied Plants,* by R. M. Tryon and A. F. Tryon. Springer-Verlag, New York. 1982.]

genous tissue in general may have on adjacent tissues (Peterson and Cutter, 1969b).

THE GAMETOPHYTE. There are reports in the older literature of successful attempts to germinate *Ophioglossum* spores, but the conditions for germination were not always well defined. Just as for *Botrychium,* Whittier (1981, 1983) has demonstrated a dark requirement for germination in *O. engelmannii.* One year in the dark was necessary for germination in axenic culture. Gametophytes from nature are reported to be slow growing, and some may live for several years. Gametophytes are subterranean, without chlorophyll, and contain an endophytic fungus that enters gametophytic cells soon after germination and is necessary for sustained growth. Mature gametophytes may be irregularly cylindrical and, according to the species, unbranched or branched (Figs. 12-12; 12-13, A). They may be 1 millimeter or less in diameter and 2 to 6 millimeters in length (Gewirtz and Fahn, 1960; Mesler et al., 1975); in some species they may be 5 to 6 centimeters long and resemble roots of the same species. For the epiphytic species *O. palmatum* (Mesler, 1975) the gametophytes are fundamentally cylindrical and much branched; the branches are often oriented in a single plane (Fig. 12-13, B). Growth is initiated by a group of cells at the surface of the apical meristem rather than by an apical cell as reported for some other species (see Whittier, 1983). The gametophyte of *O. crotalophoroides* is interesting in that it is globose or approximately hemispherical, ranging from 1 to 3 millimeters in diameter (Mesler, 1973, 1976), and resembling to some extent those of *Botrychium.*

Antheridia and archegonia are initiated in derivatives of apical meristems or from meristematic cells on the periphery in *O. crotalophoroides.* Gametangia in most species are scattered and intermingled over the surface of the gametophyte and resemble those of *Botrychium.* Antheridia give rise to numerous multiflagellate sperm.

THE EMBRYO. As is true of most groups of lower vascular plants, the first division of the zygote in *Ophioglossum* is transverse, that is, perpendicular to the long axis of the archegonium. The embryo is exoscopic in polarity (Fig. 6-1). A cell division at the apical pole (epibasal cell) and at the basal pole (hypobasal cell) results in the formation of four cells—the quadrant stage of embryogeny. Subsequent cell divisions are irregular, but it is accepted with little doubt that the first leaf and future shoot apex are derived from the epibasal portion of the embryo and the foot is derived from the hypobasal portion. Origin of the root is uncertain, but seemingly it arises near the middle of the embryo and enlarges rapidly in a lateral direction before other parts of the embryo become conspicuous (Fig. 12-13, A).

Appraisal of the Ophioglossales

In this book we have included the Ophioglossales as an order in the Filicophyta along with other ferns. Some botanists, however, consider the members of the Ophioglossales to be sufficiently different from other ferns to warrant making them a separate division. Some of the important attributes

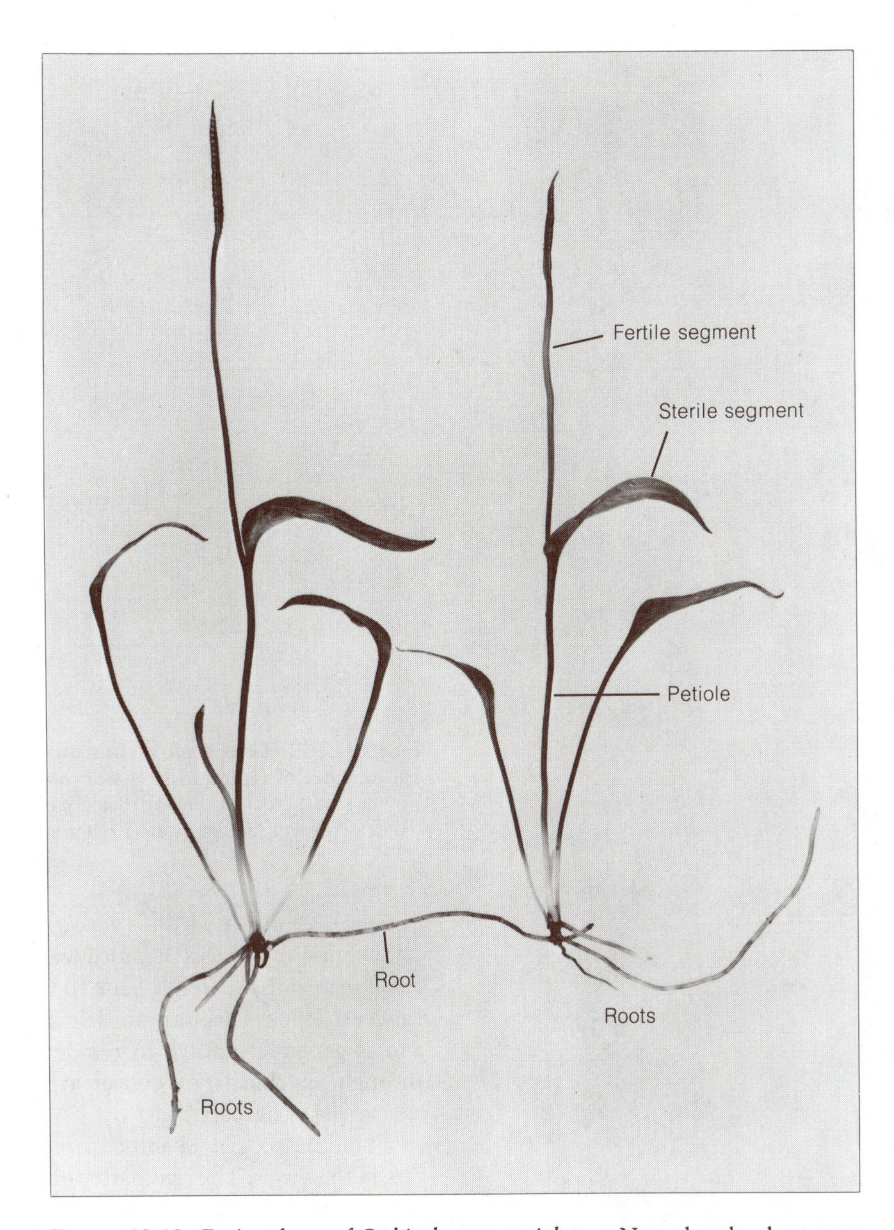

FIGURE 12-10 Entire plants of *Ophioglossum petiolatum*. Note that the short erect stems arise from roots; only one leaf is fertile on each shoot.

for consideration are: (1) the peculiar fertile segment, (2) collateral vascular bundles, (3) roots with endophytic fungus and without root hairs, (4) noncircinate vernation of leaves (although see section on *Botrychium*), (5) lack of sclerenchyma in the plant body, and (6) subterranean gametophytes with an associated fungus. Although some of these features are not confined to the Ophioglossales, no other group of vascular plants has all of these attributes. The Ophioglossales is undoubtedly an ancient group, but the meager number of fossils continues to shroud its phylogeny.

Marattiales — Marattiaceae

This group of plants more closely resembles the Filicales than the Ophioglossales. Many of them

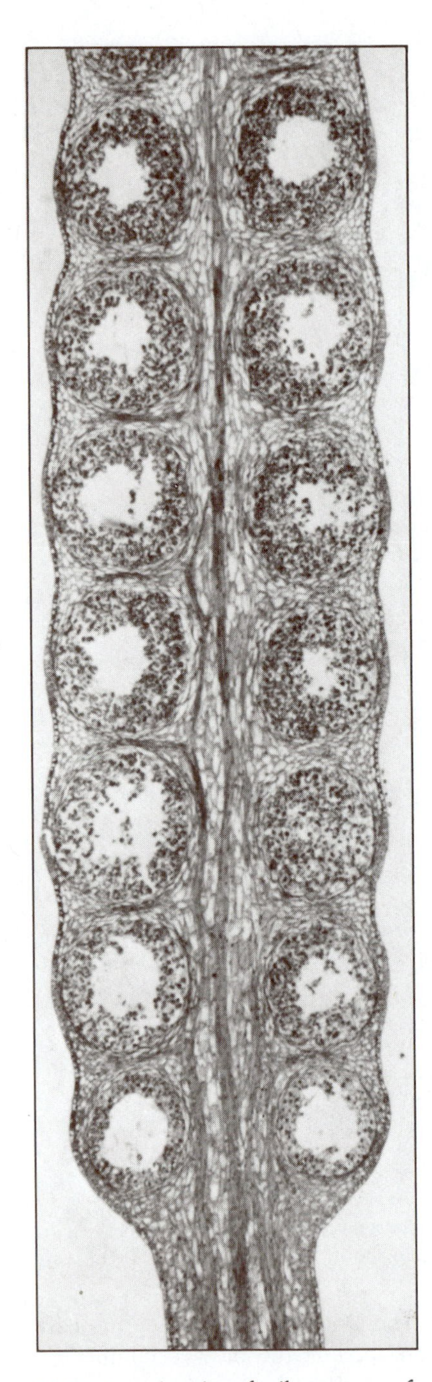

FIGURE 12-11 Longisection, fertile segment of *Ophioglossum lusitanicum* var. *californicum*, showing two rows of embedded sporangia. Branches of the main vascular system are evident between sporangia at left.

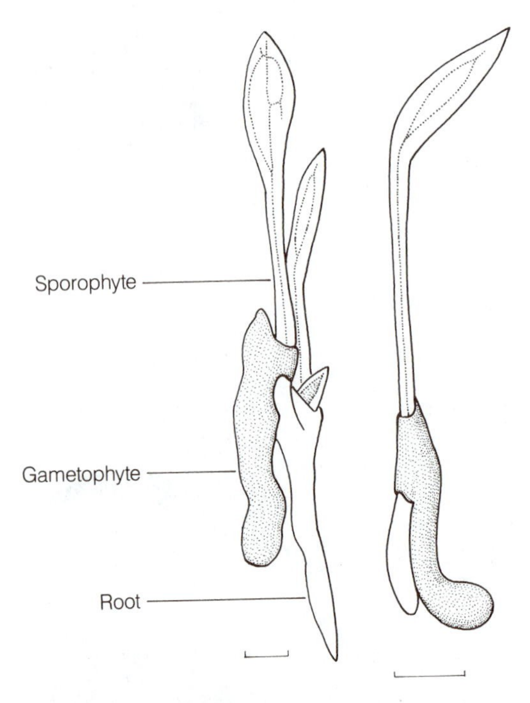

FIGURE 12-12 Gametophytes and attached young sporophytes of *Ophioglossum nudicaule;* the gametophytes stand erect in the soil. Scale bars = 1 millimeter. [Redrawn from Mesler et al., *Phytomorphology* 25:156, 1975.]

possess large pinnate fronds with sporangia located on their lower surfaces. It is an ancient group with a fossil record that extends back to the transition of Lower to Upper Carboniferous. It is a tropical order and is generally known in temperate zones only through specimens in conservatories and dried specimens in herbaria.

There are six genera and perhaps 200 living species in the order. The two better-known genera are *Angiopteris* (about 100 species) and *Marattia* (about sixty species); the former is distributed throughout the tropics of southeastern Asia, the latter is pantropical. In growth habit the two genera typically have upright, unbranched, fleshy tuberous stems or short trunks bearing large, pinnately compound leaves (circinate in vernation) and thick fleshy roots (Fig. 12-14). A pair of clasping fleshy stipules is present at the base of each leaf, covering part of the stem; they persist, along with leaf bases, even after the fronds abscise. New plantlets can arise from the stipules. On the lower surface (abaxial side) of the fronds, which may be up to 5 or 6 meters long, sporangia occur along veins; venation in the ultimate segments is of the open dichotomous type. In *Angiopteris* a sorus consists of sporangia crowded together in two rows along a vein; each sporangium dehisces by a longitudinal slit on the side facing the other row of sporangia (Figs. 12-15, A; 12-16). In *Marattia* the two rows of sporangia are united into a compact soral group surrounded by a common wall (Fig. 12-15, B). This structure is termed a *synangium*. At maturity the synangium opens, much like a clam shell, exposing the sporangia which dehisce by longitudinal slits.

In most other genera the stems are trailing, dorsiventral rhizomes. The fronds may be simple to once pinnate *(Danaea)*, or palmately compound with reticulate venation and scattered circular or ringlike synangia *(Christensenia)*.

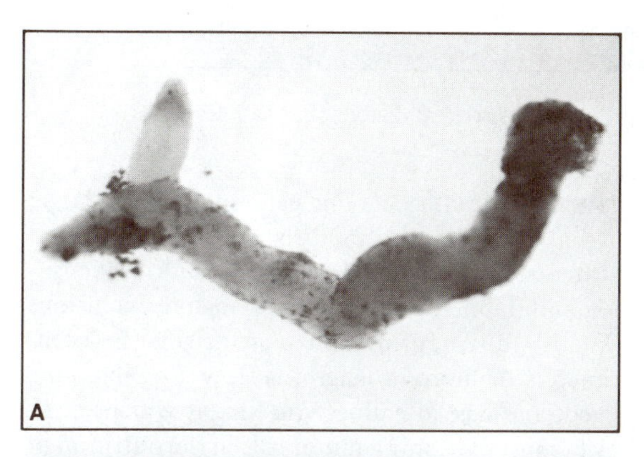

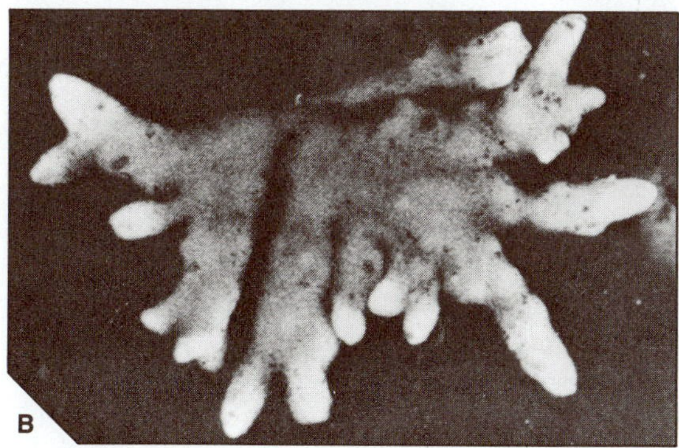

FIGURE 12-13 Gametophytes of *Ophioglossum*. **A,** *O. vulgatum;* precocious root of young attached sporophyte is seen at left (× 7). **B,** *O. palmatum;* branching confined essentially to a single plane (× 10). [Specimen for **A** supplied by Dr. D. Stevenson; **B** from Mesler, *Amer. Jour. Bot.* 62:982, 1975.]

ANATOMY OF THE SHOOT. The apical meristem of a mature adult plant is reported to have a group of apical initials or an apical cell that is not regular or precise in its divisions (Charles, 1911; Bower, 1923). As mentioned previously, the stems are fleshy and have numerous mucilage canals and tannin-filled cells throughout the plant body. The stem is protostelic at the base of the plant. Higher up, the stele is an amphiphloic siphonostele with overlapping leaf gaps. At even higher levels in older plants the stele is a complex polycyclic dictyostele consisting of two or more concentric vascular cylinders (Fig. 12-17). It should be emphasized that the inner cylinders are continuous with the outer cylinder at lower levels in the stem. In the petiole the leaf trace becomes subdivided in the formation of concentric cylinders of vascular bundles; the cylinders decrease in number higher up in the frond rachis. Root primordia have their origin in the pericycle of stem vascular bundles of the outer and inner vascular cylinders. After their initiation, the large roots "bore" their way through the stem cortex and then grow between the leaf bases, generally being evident on the surface of the stem before entering the soil.

The leaves of marattiaceous ferns are circinate in vernation during development. The ultimate frond segments in most species have a single midvein with lateral dichotomous veins. The mesophyll in most forms is differentiated into an adaxial palisade tissue and an abaxial spongy mesophyll. Stomata occur on the abaxial surface. Mucilage cavities, hypodermal sclerenchyma, or collenchyma are often present in the petiole.

THE ROOT. In the primary root and the first-formed roots on the stem there is a definite apical cell; later-formed roots are reported to have a group of about four equivalent initials (West, 1917). Roots become large and fleshy and contain mucilage cavities. Typically the vascular cylinder is polyarch—a feature not generally found in other ferns.

THE SPORANGIUM. Sporangia are of the eusporangiate type and commonly originate from mounds of tissue paralleling the veins of developing fronds. At maturity each sporangium has a broad base and a sporangial wall that consists of several layers of

FIGURE 12-14 *Angiopteris evecta,* showing short, tuberous stem and large bipinnate leaves.

cells. When mature the sporangia may be separate from each other *(Angiopteris),* or the sporangial walls may become confluent during development so that each sporangium is actually a pocket or loculus in a compact structure, the synangium (Fig. 12-15, B). Dehiscence of individual sporangia in a synangium is brought about by the drying out of wall cells, which results in longitudinal splitting of each sporangium (after the halves of the synangium separate in *Marattia*), or by the formation of a pore at the tip of each sporangium as in *Danaea.* Spore output is large; spore numbers range from a minimum of 1,000 up to a maximum of 7,000 spores formed by each sporangium (Bower, 1935). In *Marattia* the spores are small, bilateral (monolete); those of *Angiopteris* are tetrahedral (trilete).

CHROMOSOME NUMBERS. Chromosome numbers are known for three genera. The theoretical base number for *Danaea* and *Angiopteris* is $x = 10$, with actual counts of $n = 40, 80$. For *Marattia,* $x = 13$ is the probable base number with actual counts of $n = 78, 156$ (Löve et al., 1977). As with some other ferns, polyploidy apparently has been operative in the order.

THE GAMETOPHYTE. The gametophyte is a large, green, dorsiventral ribbon-shaped or heart-shaped structure with a prominent ventral midrib or cushion and thinner, ruffled, lateral winglike extensions (Fig. 12-18). The gametophyte, which may be 2 centimeters or more in length, is slow growing, long lived, and has an endophytic fungus which, however, must play only a minor role in the nutrition of the gametophyte because of the presence of chlorophyll. There are absorbing rhizoids along the ventral midrib. Gametangia of *Angiopteris* show the following pattern of distribution: antheridia are on the ventral surface but may occur on the dorsal side (Nozu, 1956); archegonia are restricted to the projecting ventral midrib (Haupt, 1940). Both antheridia and archegonia are sunken, and the main stages in their development are shown in Fig. 12-19. Mature sperms are multiflagellate.

THE EMBRYO. Just as in the Ophioglossales, the first division of the zygote is transverse, resulting in a two-celled embryo. Details of embryogeny in the Marattiales are not too well known, but in contrast to the Ophioglossales, the Marattiales exhibit only endoscopic polarity (Chapter 6, Fig. 6-1). The fu-

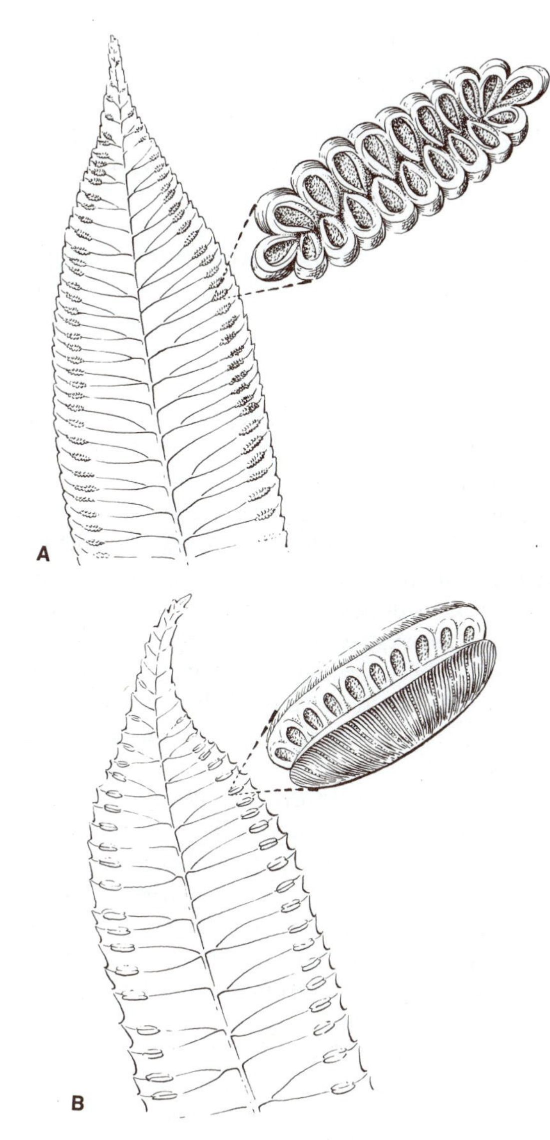

FIGURE 12-15 Abaxial views of fertile pinnae. **A,** *Angiopteris;* **B,** *Marattia.* One sorus of each is enlarged. The sporangia in *Marattia* form a definite synangium. [Pinnae redrawn from *The Ferns,* Vol. II, by F. O. Bower. Cambridge University Press, London. 1926.]

FIGURE 12-16 Section, lamina and sorus of *Angiopteris* sp. Note the presence of numerous spores within each thick-walled sporangium.

FIGURE 12-17 *Angiopteris evecta.* Transverse section of stem, showing polycyclic dictyostele. Root traces enclosed in clear areas. [Redrawn from *The Ferns,* Vol. II, by F. O. Bower. Cambridge University Press, London. 1926.]

ture shoot apex and first leaf have their origin from the cell (epibasal cell) directed away from the neck of the archegonium. The cell (hypobasal cell) toward the archegonial neck gives rise to a multicellular foot. The root meristem, appearing late in embryogeny, is endogenous in origin, and in one study it was reported to be derived from the epibasal portion of the embryo. With subsequent growth

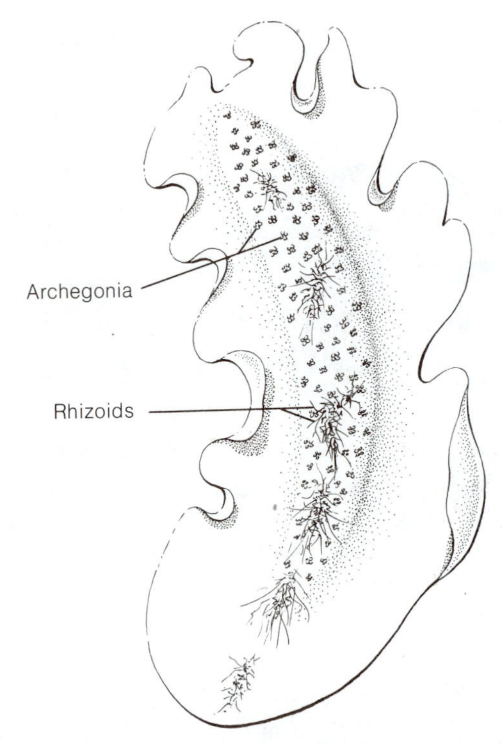

FIGURE 12-18 Ventral surface of ribbon-shaped gametophyte of *Marattia douglasii.* Most of the antheridia (not shown in drawing) are on the ventral surface (× 5). [Drawn from specimen supplied by Dr. W. H. Wagner, Jr.]

of the embryo the young shoot grows up through the gametophyte, emerging from the upper surface. The presence of a suspensor has been reported in some species of *Angiopteris* (Land, 1923) and other genera (Campbell, 1940). The first vascular bundle of the embryo is continuous between the root and first leaf. The vascular bundle of the next leaf is joined to the first vascular strand.

FOSSIL MEMBERS OF THE MARATTIALES. Considerable information is now available on the extinct Carboniferous swamp relatives of the living members of the Marattiales. They were trees, large at the base, tapering toward the top and ending in a collection of large pinnately compound fronds which were probably very graceful in appearance (Fig. 12-20). There are form genera for these plants, but the generic name *Psaronius* is generally used in descriptions of these trees that may have reached 10 meters or more in height. The stems were relatively small in diameter and probably could not have supported the plant except for the presence of large masses of roots that formed a mantle on the stem. These roots, originating high on the stem, grew downward and produced lateral roots, forming a compact inner zone and a peripheral zone of more loosely arranged roots (Fig. 12-21). Roots of the inner zone were bound together by proliferative tissue derived in part from the stem and from the cortex of the closely appressed roots (Ehret and Phillips, 1977). Some roots are known to have been 2 centimeters in diameter—the largest for any known fern. The mantle of roots at the base of a large tree trunk could be up to 0.5 to 1 meter in diameter in some specimens. Present-day tree ferns resemble *Psaronius* in general growth habit (Fig. 11-1). The anatomy of a mature root was similar to that of *Marattia* and *Angiopteris.* Well-preserved root tips were discovered that show an apical organization remarkably similar to present-day marattialean species (Ehret and Phillips, 1977). Phyllotaxy was variable: it was helical at all levels of the stem (Rothwell and Blickle, 1982) or it was helical in the basal part of the stem, becoming whorled higher up (Morgan, 1959). Some species were distichous—two longitudinal rows of leaves, one on each side of the stem (DiMichele and Phillips, 1977). Phyllotaxy can be deduced by examining the arrangement of leaf scars from compression-impression fossils of stems (Fig. 12-22); different types are placed in form genera when internal anatomy of the stem is not known.

The stems of *Psaronius* were obconical in growth habit—that is, the stem was small in diameter at the base, increasing in width at higher levels, although the mantle of roots gives the impression that the stem was larger at the base of the plant (Fig. 12-21). Correlated with this change was a modification of the stele. It is assumed that a young plant was protostelic (this would be the stelar type at the base of an older plant). As the plant continued to grow, there was a transition to the siphonostelic condition, more specifically to a monocyclic and then to a dicyclic dictyostele (Stidd, 1974). At higher levels more cylinders of vascular tissue were formed in the production of a complex, concentric polycyclic dictyostele (Fig. 12-23, A). Each new vascular cylinder was attached to the inner face of the previously formed cylinder. What has just been de-

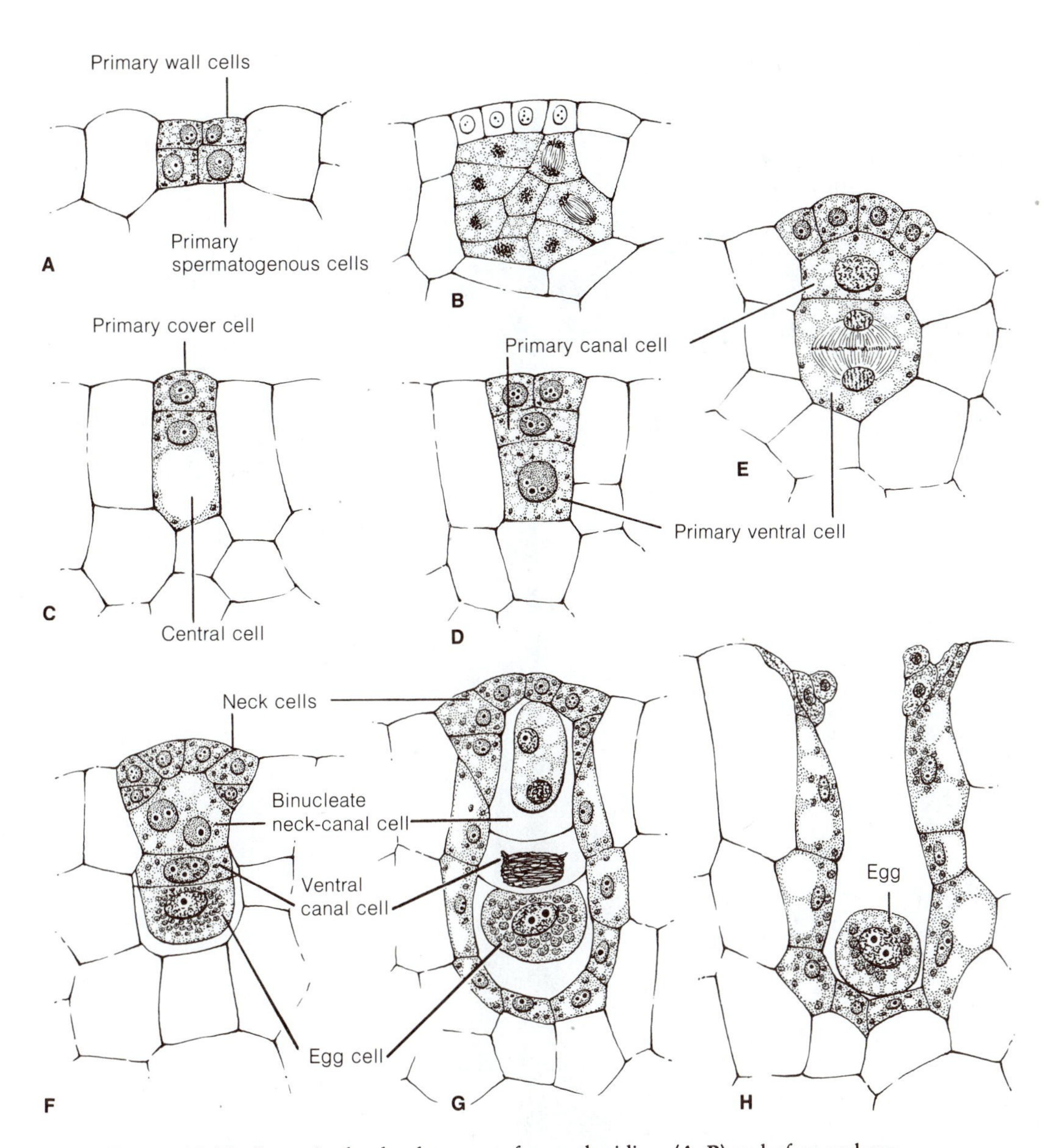

FIGURE 12-19 Stages in the development of an antheridium (**A, B**) and of an archegonium (**C–H**) of *Angiopteris evecta*. [Redrawn from Haupt, *Bull. Torrey Bot. Club* 67:125, 1940.]

FIGURE 12-20 Reconstruction of the tree fern *Psaronius,* considered to have been approximately 6 to 8 meters tall. [Redrawn from Morgan, *Illinois Biol. Monogr.* 27:1–108, 1959.]

FIGURE 12-21 *Psaronius magnificus.* Transverse section of trunk showing the small central stem, prominent inner zone of compact roots, and peripheral free roots (× 0.3). [From Rothwell and Blickle, *Jour. of Paleontology* 56:459, 1982.]

scribed is an overview of a possible sequence. The stem of *P. simplicicaulis* was probably smaller than most Upper Carboniferous forms; it had distichous leaf arrangement, and the stele remained as a monocyclic dictyostele (DiMichele and Phillips, 1977). *Psaronius magnificus* had helically arranged leaves at all levels and the stem had three to six stelar cycles (Rothwell and Blickle, 1982). The dictyosteles of other species were amazingly complex, consisting of several concentric cycles (Morgan, 1959) as shown in Fig. 12-23, B.

The xylem in each vascular bundle (meristele) was endarch in development; phloem surrounded the xylem, but an endodermis with casparian strips has not been identified (Smoot and Taylor, 1981). No secondary vascular tissues were formed. This type of anatomy, except for the occasional appearance of a typical endodermis, also characterizes the living members of the Marattiales as well as the occurrence of polycyclic dictyosteles in most of them. Additional information on anatomy can be found in the treatise by Ogura (1972).

There is a wealth of information concerning sporangial structures of the Upper Carboniferous marattialean ferns. Pinnae have been found with

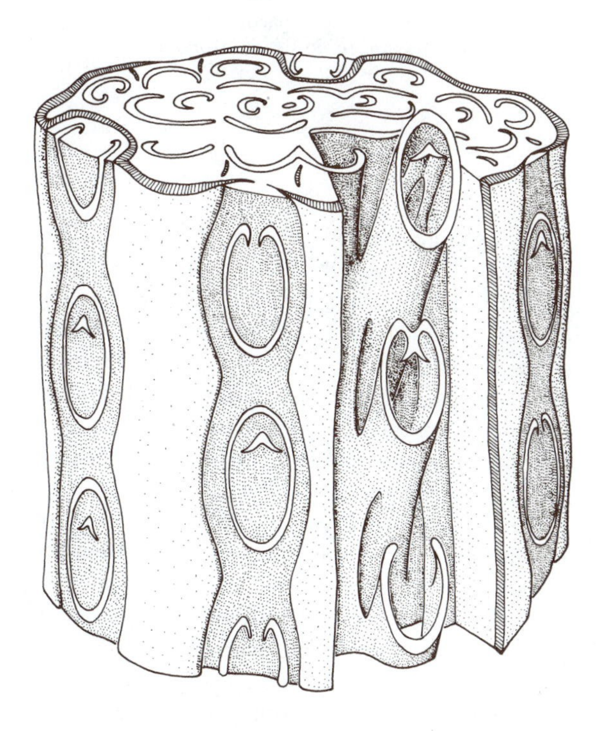

FIGURE 12-22 Reconstruction of a *Psaronius* stem showing origin of leaf traces (in cut-away portion) in a helical phyllotaxy. The vascular system of the stem is a polycyclic dictyostele. [Redrawn from Stidd, *Ann. Missouri Bot. Gard.* 61:388, 1974.]

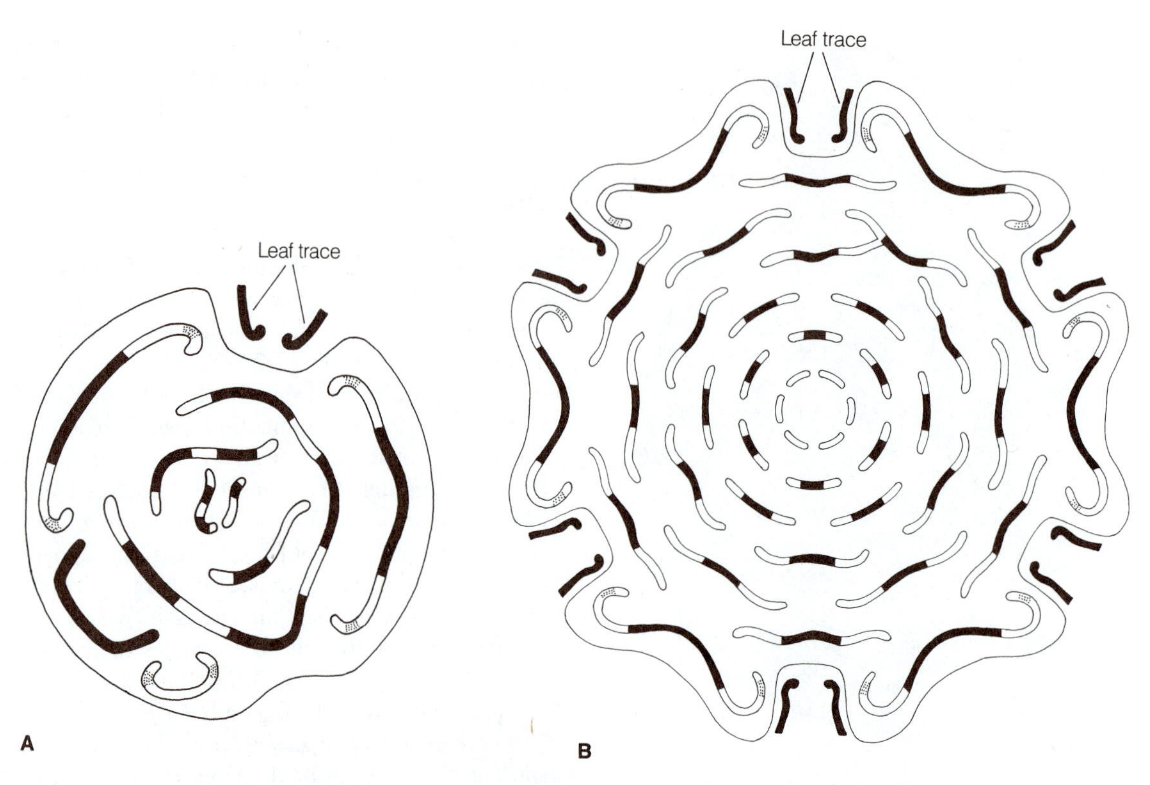

FIGURE 12-23 *Psaronius blicklei.* Diagrammatic reconstructions of stem transverse sections showing polycyclic dictyosteles. **A,** stem with helical phyllotaxy. **B,** large stem with whorled phyllotaxy. Leaf traces and potential leaf-trace tissue shown in solid black; cauline vascular tissue unshaded; stippling represents cauline vascular tissue that will contribute to departing leaf traces at higher levels. Compare with Fig. 12-22. [From Morgan, *Illinois Biol. Monogr.* 27:1–108, 1959.]

sori on the abaxial side of pinnules (Fig. 12-24) which can be compared favorably with extant genera. In some, the sporangia are fused to varying degrees, forming synangia. The sorus of the form genus *Acaulangium* consisted of a group of four to six sporangia. The sporangial bases extended into a common basal pad of tissue and the sporangia were laterally attached prior to dehiscence (Millay, 1977). *Scolecopteris* was essentially the same (Fig. 12-25, A) except for a well-defined pedicel supporting the synangium (Mamay, 1950). In other forms the sporangia occurred in two rows along a vein much like the present-day *Angiopteris.* In *Eoangiopteris goodii* there were up to nineteen sporangia in a sorus. The sporangia were embedded in an elongate parenchymatous pad and the sporangia were attached for a short distance distal to the base (Millay, 1978). *Eoangiopteris andrewsii* was similar except that there were fewer sporangia in a sorus (Fig. 12-25, B).

The marattialean flora declined after the Upper Carboniferous, but fossils have been described from the Mesozoic (Triassic and Jurassic) and Cenozoic (Eocene) that were essentially similar to certain modern-day genera. There is very little doubt about the naturalness of the order. Marattialean types appeared in the transition from Lower to Upper Carboniferous, reached a high point in the world flora during the Upper Carboniferous, but derivative forms persisted to the present time. For additional information consult recent textbooks on paleobotany (Taylor, 1981; Stewart, 1983).

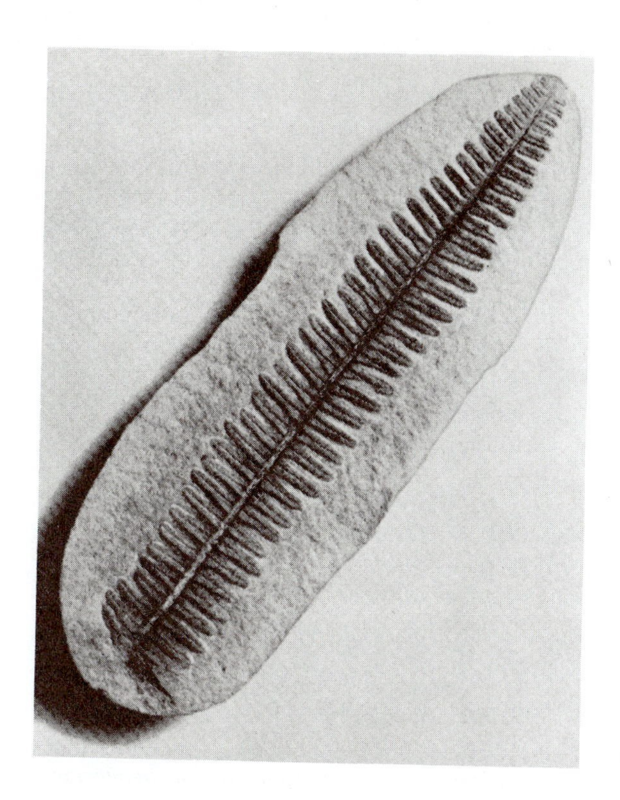

FIGURE 12-24 *Pecopteris*. Form genus for pinnules of the Carboniferous marattialean fern *Psaronius*. [From *Leaves and Stems from Fossil Plants* by R. E. Janssen. Illinois State Museum, Springfield, Ill.]

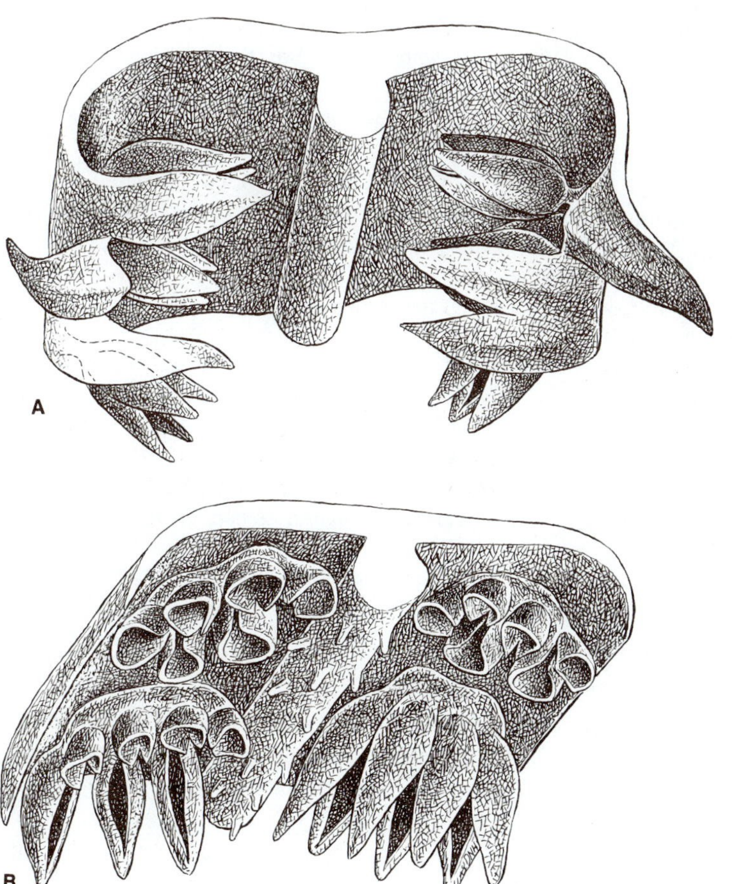

FIGURE 12-25 **A,** *Scolecopteris incisifolia:* reconstruction of lower surface of a fertile pinnule showing groups of sporangia, each group attached by a common pedicel. **B,** *Eoangiopteris andrewsii:* lower surface of a portion of a pinnule with four groups of sporangia; upper two groups sectioned transversely. [Redrawn from Mamay, *Ann. Missouri Bot. Gard.* 37:409–476, 1950.]

REFERENCES

Abraham, A., and C. A. Ninan
1954. Chromosomes of *Ophioglossum reticulatum* L. *Curr. Sci.* 23:213–214.

Bierhorst, D. W.
1958. Observations on the gametophytes of *Botrychium virginianum* and *B. dissectum*. *Amer. Jour. Bot.* 45:1–9.
1971. *Morphology of Vascular Plants.* Macmillan, New York.

Bower, F. O.
1923. *The Ferns,* Vol. I. Cambridge University Press, London.
1926. *The Ferns,* Vol. II. Cambridge University Press, London.
1935. *Primitive Land Plants.* Macmillan, London.

Campbell, D. H.
1940. *The Evolution of the Land Plants (Embryophyta).* Stanford University Press, Stanford, California.

Chandrasekharam, A.
1974. Megafossil flora from the Genesee locality, Alberta, Canada. *Palaeontographica.* 147B: 1–41.

Charles, G. M.
1911. The anatomy of the sporeling of *Marattia alata. Bot. Gaz.* 51:81–101.

Chrysler, M. A.
1925. *Botrychium lanuginosum* and its relation to the problem of the fertile spike. *Bull. Torrey Bot. Club* 52:127–132.
1941. The structure and development of *Ophioglossum palmatum. Bull. Torrey Bot. Club* 68:1–19.
1945. The shoot of *Botrychium* interpreted as a series of dichotomies. *Bull. Torrey Bot. Club* 72:491–505.

Daigobo, S.
1979. Observations on the gametophytes of *Botrychium multifidum* from nature. *Jour. Jap. Bot.* 54:169–177.

DiMichele, W. A., and T. L. Phillips
1977. Monocyclic *Psaronius* from the Lower Pennsylvanian of the Illinois Basin. *Can. Jour. Bot.* 55:2514–2524.

Eames, A. J.
1936. *Morphology of Vascular Plants. Lower Groups.* McGraw-Hill, New York.

Ehret, D. L., and T. L. Phillips
1977. *Psaronius* root systems—morphology and development. *Palaeontographica* 161(B):147–164.

Esau, K., V. I. Cheadle, and E. M. Gifford, Jr.
1953. Comparative structure and possible trends of specialization of the phloem. *Amer. Jour. Bot.* 40:9–19.

Evert, R. F.
1976. Some aspects of sieve-element structure and development in *Botrychium virginianum. Israel Jour. Bot.* 25:101–126.

Gewirtz, M., and A. Fahn
1960. The anatomy of the sporophyte and gametophyte of *Ophioglossum lusitanicum* L. ssp. *lusitanicum. Phytomorphology* 10:342–351.

Gifford, E. M., Jr., and D. D. Brandon
1978. Gametophytes of *Botrychium multifidum* as grown in axenic culture. *Amer. Fern Jour.* 68:71–75.

Haupt, A. W.
1940. Sex organs of *Angiopteris evecta. Bull. Torrey Bot. Club* 67:125–129.

Jeffrey, E. C.
1897. The gametophyte of *Botrychium virginianum. Trans. Canadian Inst.* 5:265–294.

Joshi, A. C.
1940. A note on the anatomy of the roots of *Ophioglossum. Ann. Bot.* n.s. 4:663–664.

Land, W. J. G.
1923. A suspensor in *Angiopteris. Bot. Gaz.* 75:421–425.

Löve, A., D. Löve, and R. E. G. Pichi-Sermolli
1977. *Cytotaxonomical Atlas of the Pteridophyta.* J. Cramer, Vaduz.

Lyon, H. L.
1905. A new genus of Ophioglossaceae. *Bot. Gaz.* 40:455–458.

Mamay, S. H.
1950. Some American Carboniferous fern fructifications. *Ann. Missouri Bot. Gard.* 37:409–476.

Mesler, M. R.
1973. Sexual reproduction in *Ophioglossum crotalophoroides. Amer. Fern Jour.* 63:28–33.
1975. The gametophytes of *Ophioglossum palmatum* L. *Amer. Jour. Bot.* 62:982–992.
1976. Gametophytes and young sporophytes of *Ophioglossum crotalophoroides* Walt. *Amer. Jour. Bot.* 63:443–448.

Mesler, M. R., R. D. Thomas, and J. G. Bruce
1975. Mature gametophytes and young sporophytes of *Ophioglossum nudicaule. Phytomorphology* 25:156–166.

Millay, M. A.
1977. *Acaulangium* gen. n., a fertile marattialean

from the Upper Pennsylvanian of Illinois. *Amer. Jour. Bot.* 64:223–229.
1978. Studies of Paleozoic marattialeans: the morphology and phylogenetic position of *Eoangiopteris goodii* sp. n. *Amer. Jour. Bot.* 65:577–583.

Morgan, J.
1959. The morphology and anatomy of American species of the genus *Psaronius*. *Illinois Biol. Monogr.* 27:1–108.

Ninan, C. A.
1958. Studies on the cytology and phylogeny of the Pteridophytes. VI. Observations on the Ophioglossaceae. *Cytologia* 23:291–316.

Nishida, M.
1952. Dichotomy of vascular system in the stalk of Ophioglossaceae. *Jour. Jap. Bot.* 27:165–171.
1955. The morphology, gametophyte, young sporophyte and systematic position of *Botrychium japonicum* Und. *Phytomorphology* 5:449–456.
1957. Studies on the systematic position and constitution of Pteridophyta. 10. A further investigation on the vascular dichotomy in the phyllomophore of Ophioglossales, with special references to phylogeny. *Jour. Coll. Arts Sci. Chiba Univ.* 2:179–211.

Nozu. Y.
1950. On the so-called petiole of *Botrychium*. *Bot. Mag.* (Tokyo) 63:4–11.
1954. The gametophyte and young sporophyte of *Botrychium japonicum* Und. *Phytomorphology* 4:430–433.
1955. Anatomical and morphological studies of Japanese species of the Ophioglossaceae. I. Phyllomophore. *Jap. Jour. Bot.* 15:83–102.
1956. Notes on gametophyte and young sporophyte of *Angiopteris subopposititfolia* de Vries. *Bot. Mag.* (Tokyo) 69:474–480.

Ogura, Y.
1972. Comparative Anatomy of Vegetative Organs of the Pteridophytes. *Encyclopedia of Plant Anatomy*, 2d edition, Vol. 7, part 3. Gebrüder Borntraeger, Berlin.

Peterson, R. L.
1970. Bud development at the root apex of *Ophioglossum petiolatum*. *Phytomorphology* 20:183–190.

Peterson, R. L., and E. G. Cutter
1969a. The fertile spike of *Ophioglossum petiolatum*. I. Mechanism of elongation. *Amer. Jour. Bot.* 56:473–483.
1969b. The fertile spike of *Ophioglossum petiolatum*. II. Control of spike elongation and a study of aborted spikes. *Amer. Jour. Bot.* 56:484–491.

Petry, L. C.
1914. The anatomy of *Ophioglossum pendulum*. *Bot. Gaz.* 57:169–192.

Rao, L. N.
1962. Life-history of *Botrychium lanuginosum* Wall. ex. Hook et Grev. *Proc. Indian Acad. Sci.* (Sect. B) 55:48–64.

Rothwell, G. W., and A. H. Blickle
1982. *Psaronius magnificus* n. comb., a marattialean fern from the Upper Pennsylvanian of North America. *Jour. Paleontology* 56:459–468.

Sen, U.
1968. Morphology and anatomy of *Ophioglossum reticulatum*. *Can. Jour. Bot.* 46:957–968.

Smoot, E. L., and T. N. Taylor
1981. Phloem histology of a Lower Pennsylvanian *Psaronius*. *Palaeobotanist* 28–29:81–85.

Soltis, D. E., and P. S. Soltis
1987. Polyploidy and breeding systems in homosporous Pteridophyta: a reevaluation. *Amer. Naturalist* 130:219–232.

Stebbins, G. L.
1971. *Chromosomal Evolution in Higher Plants.* Addison-Wesley, Reading, Mass.

Stevenson, D. W.
1975. Taxonomic and morphological observations on *Botrychium multifidum* (Ophioglossaceae). *Madroño* 23:198–204.
1976. The cytohistological and cytohistochemical zonation of the shoot apex of *Botrychium multifidum*. *Amer. Jour. Bot.* 63:852–856.
1980. Ontogeny of the vascular system of *Botrychium multifidum* (S. G. Gmelin) Rupr. (Ophioglossaceae) and its bearing on stelar theories. *Bot. Jour. Linn. Soc.* 80:41–52.

Stewart, W. N.
1983. *Paleobotany and the Evolution of Plants.* Cambridge University Press, Cambridge.

Stidd, B. M.
1974. Evolutionary trends in the Marattiales. *Ann. Missouri Bot. Gard.* 61:388–407.

Taylor, T. N.
1981. *Paleobotany: An Introduction to Fossil Plant Biology.* McGraw-Hill, New York.

Wagner, W. H., Jr.
1952. Types of foliar dichotomy in living ferns. *Amer. Jour. Bot.* 39:578–592.

Wagner, W. H., Jr., and F. S. Wagner
1983. Genus communities as a systematic tool in the study of New World *Botrychium* (Ophioglossaceae). *Taxon* 32:51–63.

Wardlaw, C. W.
1953. Endogenous buds in *Ophioglossum vulgatum* L. *Nature* 171:88–89.
1954. Experimental and analytical studies of pteridophytes. XXVI. *Ophioglossum vulgatum:* Comparative morphogenesis in embryos and induced buds. *Ann. Bot.* n.s. 18:397–406.
1955. *Embryogenesis in Plants.* Methuen, London.

Webb, E.
1975. Stem anatomy and phyllotaxis in *Ophioglossum petiolatum. Amer. Fern Jour.* 65:87–94.
1981. Stem anatomy, phyllotaxy, and stem protoxylem tracheids in several species of *Ophioglossum.* I. *O. petiolatum* and *O. crotalophoroides. Bot. Gaz.* 142:597–608.

West, C.
1917. A contribution to the study of the Marattiaceae. *Ann. Bot.* 31:361–414.

Whittier, D. P.
1972. Gametophytes of *Botrychium dissectum* as grown in sterile culture. *Bot. Gaz.* 133:336–339.
1973. The effect of light and other factors on spore germination in *Botrychium dissectum. Can. Jour. Bot.* 51:1791–1794.
1981. Spore germination and young gametophyte development of *Botrychium* and *Ophioglossum* in axenic culture. *Amer. Fern Jour.* 71:13–19.
1983. Gametophytes of *Ophioglossum engelmannii. Can. Jour. Bot.* 61:2369–2373.

Zimmermann, W.
1942. Die Phylogenie des Ophioglossaceen-Blattes. *Ber. Deutsch. Bot. Ges.* 60:416–433.

CHAPTER 13

Filicophyta

LEPTOSPORANGIATE FERNS
FILICALES, MARSILEALES, SALVINIALES

IN this chapter, we continue our survey of Filicophyta with a discussion of the leptosporangiate ferns Filicales, Marsileales, and Salviniales.

Filicales

Members of the group Filicales are often called the "true" ferns and are the types most commonly encountered in home and public gardens.

Economic Importance

Aside from the purely aesthetic value of ferns, they do have moderate economic importance by providing food for certain groups of peoples. Young leaf tips of some ferns are eaten throughout Malaysia and adjacent regions, and in Japan. In fact, in these areas certain ferns are grown commercially with a value sometimes exceeding that of rice (Copeland, 1942). Not only are ferns enjoyed as food in such tropic regions, but "fiddlehead greens" are shipped from Maine to restaurants in New York. One gourmet reputedly remarked, "They taste, simply and beautifully, like the soul of spring." This edible fern, *Matteuccia struthiopteris* — known as the ostrich fern — is distributed throughout most of the Northern Hemisphere, but its consumption as a spring vegetable has been restricted largely to the Maritime Provinces of Canada and to neighboring Maine in the United States. American Indians were eating the fern at the time of the arrival of European colonists and later sold the fiddleheads to colonists at local markets, a practice that continued into the early twentieth century. Subsequently, fiddleheads were sold as canned food, but the advent of fresh, frozen, packaged fiddlehead greens has largely replaced the canning process. One company now packages about 50 to 100 tons per season. Freezing apparently enhances the taste. For consumption, the croziers are steamed for five minutes and served with butter and lemon juice, or they may be marinated, used in salads, or fried. Attempts are now being made to establish ostrich fern plantations. Wild populations of the fern are abundant, but the logistics of collecting and processing the fiddleheads have revived an interest in establishing permanent fern plantations. (See von Aderkas, 1984, for an interesting historical account of the use of the ostrich fern.)

During devastating famines in Europe in past times, precious stores of barley and rye were conserved by mixing small amounts of the grains with dry, ground-up male fern *(Dryopteris filix-mas)* or bracken fern — the mixture being termed "*pain de fougère*" in France (Coquillat, 1950). The bracken fern *(Pteridium aquilinum)* probably has had more economic impact on human affairs than any other fern. It is worldwide in distribution; twelve varieties have been described. The young leaves (croziers) are considered a delicacy in Japan, eaten in season or canned. In Britain in times past bracken has been used as fuel, thatch, litter for animals, as compost, as a source of potash for glass and soap making, and for the production of bread-type food in times of severe famine. The tissues contain toxins that produce a variety of symptoms in livestock, sometimes resulting in sudden death. Also, potent carcinogens have been isolated from bracken. Bracken has had a tremendous impact on agriculture in Scotland during the past two to three centuries. With the clearing of land — by felling trees and burning other vegetation — bracken can rapidly colonize the open areas almost to the exclusion of forage plants. A survey of Scotland in 1970 showed that about 400,000 acres of potential rangeland were infested by bracken. The fern can be controlled to some extent by cutting and burning, but it cannot be eradicated by either method because the pernicious underground rhizomes are not killed. Certain herbicides can now be used quite successfully, but often the infested areas then become overrun by other weeds.

Many ferns, including bracken, produce large amounts of naturally occurring insecticides in the form of molting hormones that interfere with the normal molting process of insects. These hormones as well as toxic compounds probably have resulted in the relative immunity of ferns to devastating insect attacks, which may have been important in the co-evolution of ferns and seed plants. For additional information the reader is referred to the interesting collection of articles edited by Perring and Gardiner (1976).

Commercially, ferns are used extensively by florists in bouquets and floral arrangements. Ferns with good keeping qualities are the American shield fern (fancy fern) and *Rumohra* (leather fern). In greenhouses and conservatories orchids often are grown on pieces of the trunk and roots of tree ferns or on *Osmunda* "fiber." In some regions of the world the fibrous material obtained from ferns has been and perhaps still is used as a stuffing for mattresses.

The male fern *(Dryopteris filix-mas)* has been used medicinally in the treatment of tapeworm since its beneficial effect was discovered by Dioscorides in the days of Nero. Two more uses of the male fern might be cited: tissues of the fern have been substituted for hops in brewing ale, and juices of the fern are said to have been used by witches in concocting love philters!

Geographical Distribution and Growth Habit

There are approximately 10,000 species in the Filicales and about 300 genera. The number of recognized families varies widely depending upon the morphological characters used to define families. The Filicales reach their greatest numbers and are most diversified in the tropics, where both epiphytes and tree ferns are common. The two main regions of distribution are southern Asia to Malesia and the American tropics, i.e., southern Florida, Caribbean region, Mexico, Central America, and northern South America including Brazil. Some mountains in the tropics may have hundreds of species. As an example, a certain mountain in Borneo has at least 437 species (Copeland, 1939). The ferns are not restricted to the tropics, but their numbers decrease with increasing latitude and decreasing moisture. In temperate regions ferns are largely terrestrial with a short, erect stem or commonly a prostrate rhizome with no aerial stem, the leaves being the only structures visible above ground. Ferns constitute only about one-fiftieth of the total species of vascular plants in California, whereas in Guam one-eighth of the total species are ferns.

Morphology and Anatomy of the Shoot

It is impossible to describe all of the morphological variations in the anatomy of the fern sporophyte. We can only look at certain common features, leaving complete surveys to monographic treatments. However, near the end of the chapter the reader will find summaries of the important characters of selected families and subfamilies and their interrelationships.

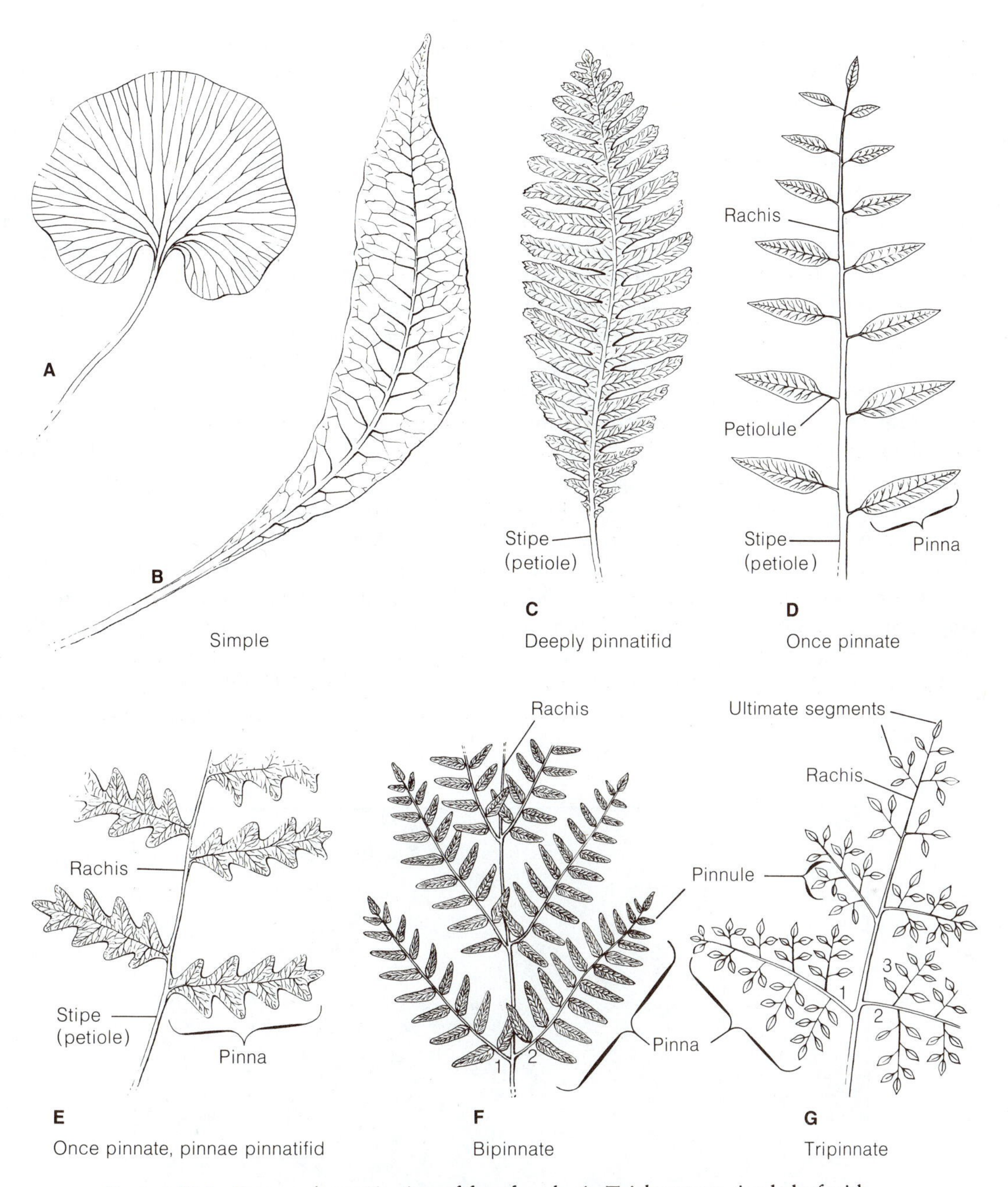

FIGURE 13-1 Form and organization of fern fronds. **A,** *Trichomanes,* simple leaf with open dichotomous venation; **B,** *Paraleptochilus,* simple leaf with pinnate reticulate venation; **C,** *Trichomanes;* **D–G,** schematic only, numbers 1, 2, 3 indicate degree of compoundness. [**A, C** redrawn from *The Ferns,* Vols. I, II, respectively, by F. O. Bower. Cambridge University Press, London, 1923, 1926; **B** redrawn from E. B. Copeland, *Genera Filicum.* A Chronica Botanica Publication. Copyright 1947, The Ronald Press Company.]

If the aerial stem or rhizome of a fern is short and erect, the plant appears to be a collection of leaves. If, on the other hand, the rhizome is prostrate, the leaves tend to be somewhat farther apart, and nodal and internodal regions can be more easily seen. Branching of a prostrate rhizome in some ferns is dichotomous, but it is usually irregular by the activation of preformed buds at varying distances from the rhizome apex, often as a result of injury to the main axis. In *Hypolepis repens,* buds formed on petiolar bases may develop into rhizomes (Gruber, 1981). This type of branching results in the rapid occupation of favorable environments. The majority of filicalean ferns have pinnate leaves, but many have simple ones (Fig. 13-1). Members of all families of ferns possess epidermal appendages on the leaf and stem and frequently on the root. These appendages may be only simple hairs, or develop into large chaffy scales called paleae (see Fig. 13-6, E). Roots arise endogenously from the stem. Depending upon the taxon, roots may occur along a rhizome, be restricted to the nodes, or arise at the bases of leaves or at the base of lateral buds (Ogura, 1972).

INITIATION AND DEVELOPMENT OF THE LEAF. At the tip of the shoot apical meristem is an apical cell which varies in shape according to the species. An apical cell with three sides (or faces) from which cells (segments) are cut off on two sides has been reported by some investigators. In the majority of described species, however, the apical cell is a tetrahedral, pyramidal cell; new cells are initiated from the three oblique, basal faces. The fourth side is in the form of a rounded cap at the surface of the shoot apex (Fig. 13-2). The apical cell traditionally has been considered to be the *ultimate* source of cells of the shoot, but the apical cell, if present, is not always easily identified in sectioned material. This difficulty led to a zonate concept of fern apices in which the apical cell may or may not be apparent in a group of surface initials (McAlpin and White, 1974; Stevenson, 1976).

Leaves are initiated from surface cells of the apical meristem at rather specific distances from the apical cell. One cell among a group of meristematic cells functions as the initial of a new leaf. This cell functions as the apical cell of the leaf, giving rise to derivatives which in turn are meristematic. The api-

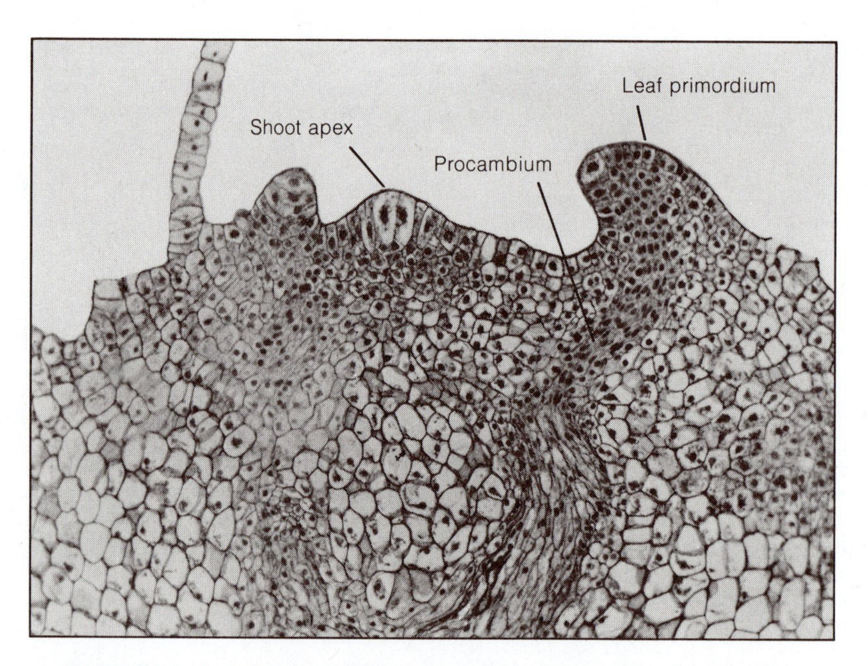

FIGURE 13-2 Longisection of shoot apex and leaf primordia of *Asplenium bulbiferum*. Note procambial tissue extending into leaf primordium at right. [Photomicrograph courtesy of K. Esau.]

cal cell may be tetrahedral during early leaf development in primitive ferns, but more commonly it is three-sided, with new segments cut off on the two lateral faces. This type of cell is often referred to as being "bilateral" or lenticular (Bierhorst, 1977). The leaf then undergoes a period of apical growth which may continue for varying lengths of time depending on the ultimate size of the frond. A developing vascular bundle may be evident in a leaf that is only a fraction of a millimeter in height. The lamina or pinnae (if the leaf is compound) are formed as lateral outgrowths on the original peglike structure; the latter becomes the stipe (petiole) and rachis. The blade portions of the leaf are formed by the activity of marginal meristems (Fig. 13-3). The derivatives of the marginal initials continue to divide in a definite pattern, resulting in the formation of a characteristic lamina for each species (Zurakowski and Gifford, 1988). During the later phases of growth the typical fern leaf displays circinate vernation, not only of the main axis but also of its subdivisions. Circinate vernation is undoubtedly related in part to the presence of naturally occurring growth hormones (see Steeves and Wetmore, 1953, and Briggs and Steeves, 1959, for detailed discussions).

SIZE, FORM, AND TEXTURE OF THE LEAF. The majority of ferns have pinnatifid, pinnate, or simple leaves (Fig. 13-1). In size, fronds may vary from the enormous, almost branchlike leaves of tree ferns to the small leaves of certain water ferns. In living forms the large, compound, pinnate leaves are considered more primitive, whereas the small, simple leaves have been reduced in size during evolution. The latter type is therefore interpreted as a derived form. There is considerable variation in the texture of fern leaves: they may be thick and leathery, crisp, or very delicate (as in the so-called filmy ferns).

A survey of living ferns suggests two principal trends in the evolution of leaf form, namely, (1) the evolution of a simpler leaf form from a more complex one and (2) the evolution of all other principal leaf forms from the pinnate-determinate type (R. M. Tryon, 1964) (Fig. 13-4, A). Approximately 85 percent of the species and genera of ferns have the pinnate-determinate type. Specialization has led, for example, to the helicoid, radiate, and furcate types (Fig. 13-4, C–E). Tryon has concluded that there have been two principal adaptive evolutionary developments, one leading to the simplification of the lamina and reduction of leaf surface, such as in ferns of tropical xeric environments (adequate soil moisture available only seasonally) and the other to the climbing types with pinnate leaves of indeterminate growth (Fig. 13-4, F).

VENATION OF THE LEAF. The stipe, rachis, and axes of frond subdivisions have prominent vascular bundles. The smaller veins of the lamina often have an open dichotomous type of venation. In some ferns the venation pattern is a reticulate type (Fig. 13-5).

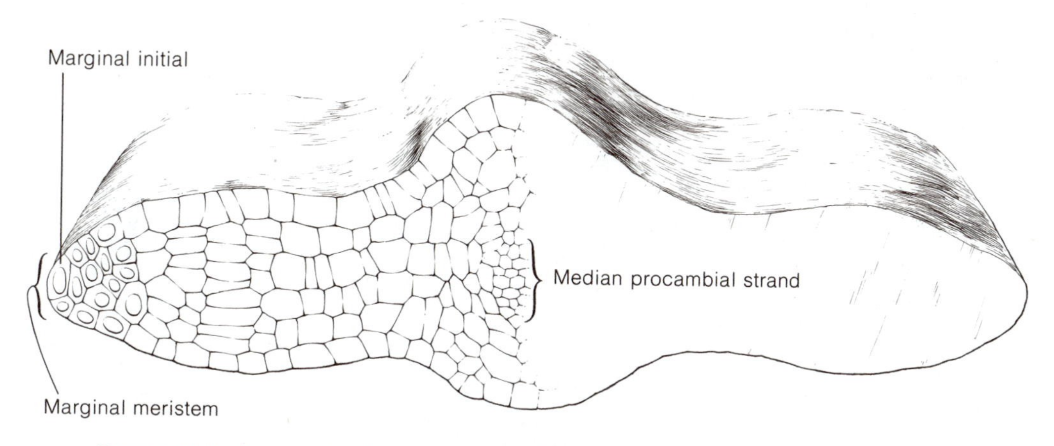

FIGURE 13-3 Transectional view of a young developing lamina of *Dicksonia*. The epidermal and mesophyll cells have their origins from derivatives of the marginal initials. Only one cell of a row of marginal initials is shown.

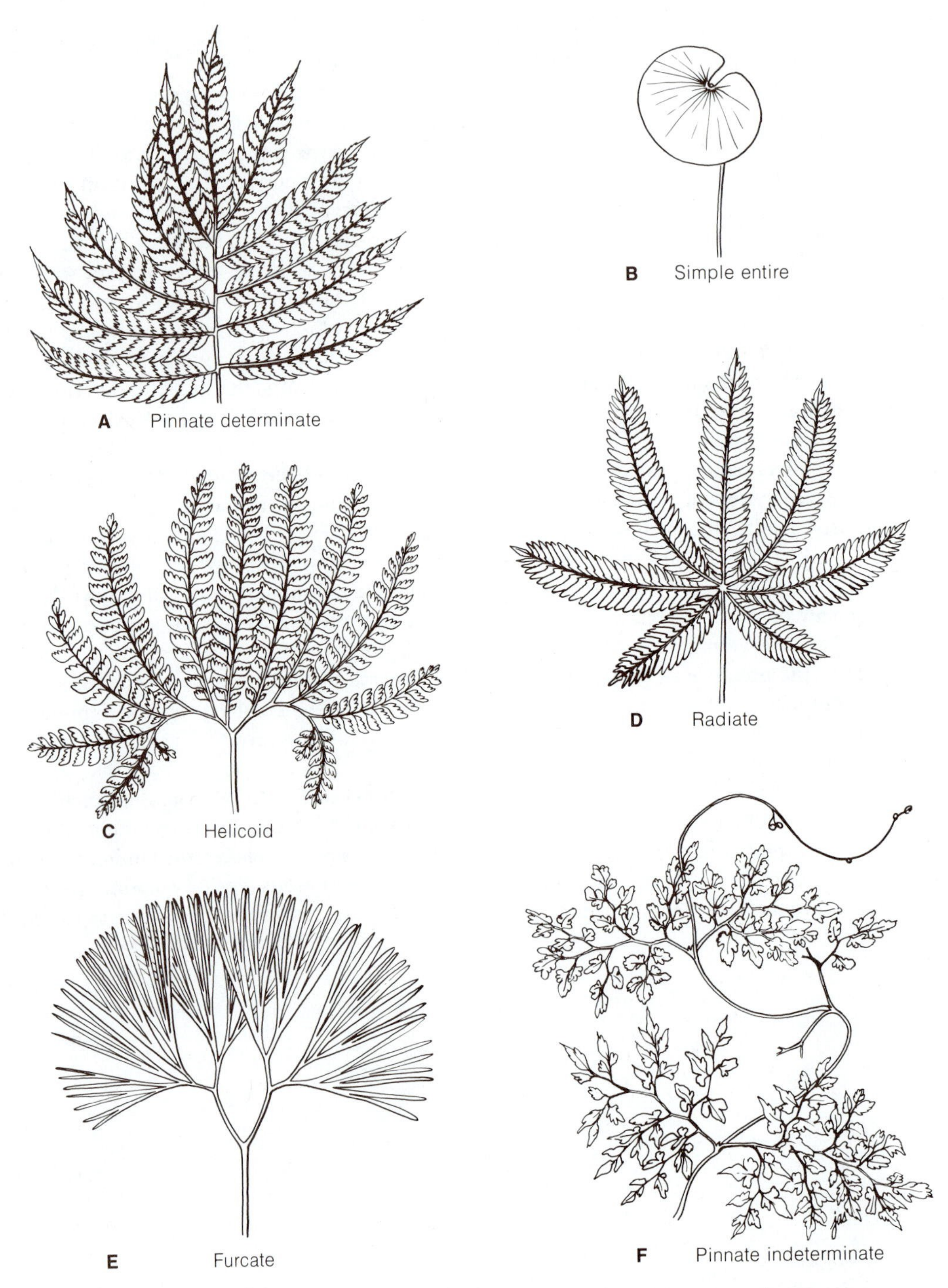

FIGURE 13-4 Diversity and evolution of leaf form in ferns. **A,** *Adiantum pulverulentum:* lamina has pinnately arranged parts and determinate growth; **B,** *Adiantum reniforme:* lamina is simple and not lobed or laciniate; **C,** *Adiantum pedatum:* parts of lamina are borne on only one side of rachis branches; **D,** *Cheilanthes radiata:* parts of lamina radiate from one point; **E,** *Schizaea dichotoma:* parts of lamina branch in a dichotomous manner; **F,** *Lygodium japonicum:* leaf has indefinite growth; the parts are arranged in a pinnate manner. [Based on Tryon, *Mem. Torrey Bot. Club* 21:73–85, 1964.]

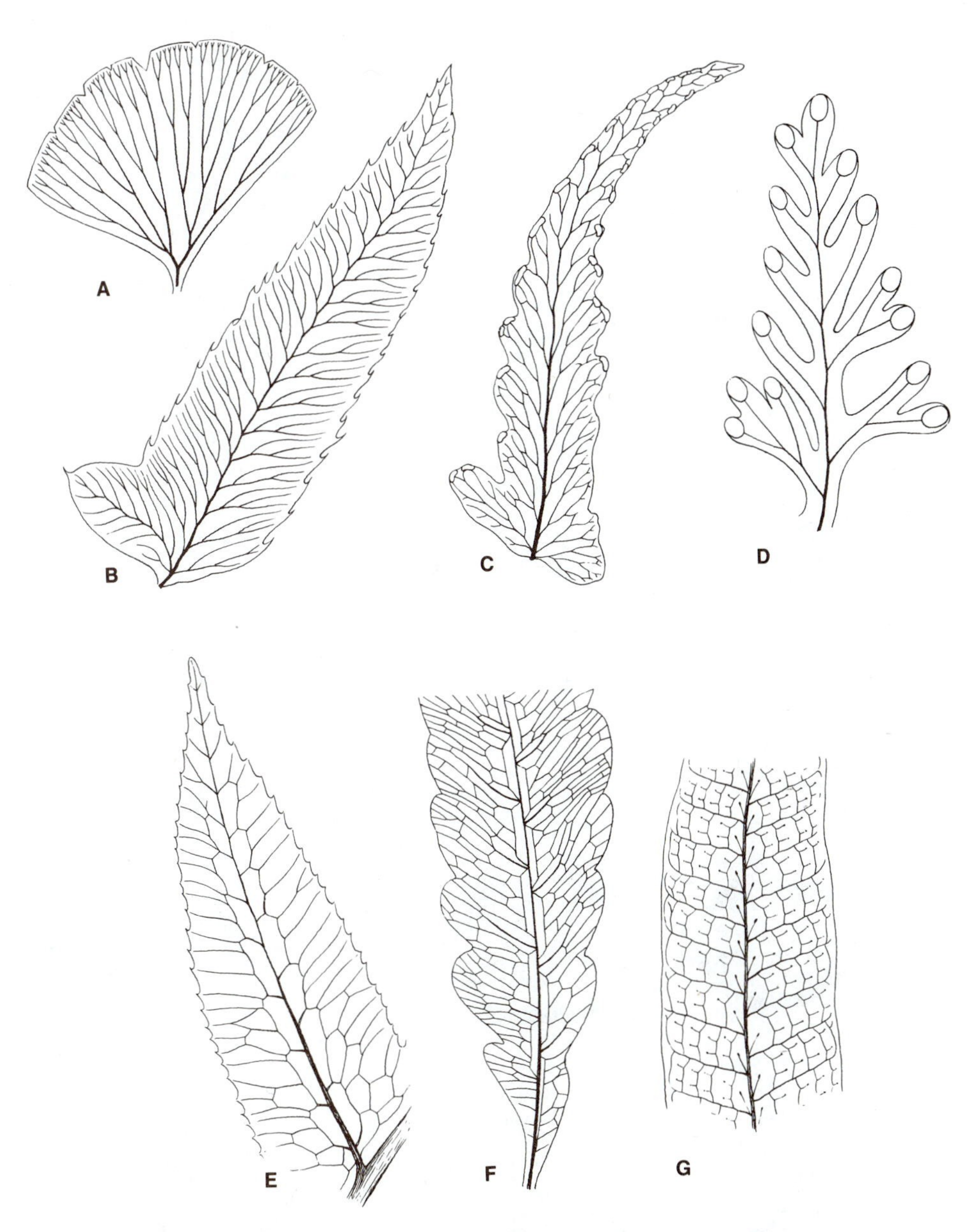

FIGURE 13-5 Leaf venation in ferns, illustrating the basic type of open dichotomous venation system in some ferns as contrasted with the highly complex reticulate patterns in others. **A,** *Adiantum;* **B,** *Polystichum;* **C,** *Diellia;* **D,** *Davallia;* **E,** *Woodwardia;* **F,** *Onoclea;* **G,** *Polypodium.* [**A** redrawn from *Organographie der Pflanzen. Dritte Auflage. Zweiter Teil* by K. Goebel. Gustav Fischer, Jena, 1930; **B** redrawn from *Morphology of Vascular Plants. Lower Groups* by A. J. Eames. McGraw-Hill, New York, 1936; **C, D** redrawn from Wagner, *Univ. Calif. Publ. Bot.* 26:1, 1952; **E–G** redrawn from *The Ferns,* Vols. I, II, III, by F. O. Bower. Cambridge University Press, London, 1923, 1926, 1928.]

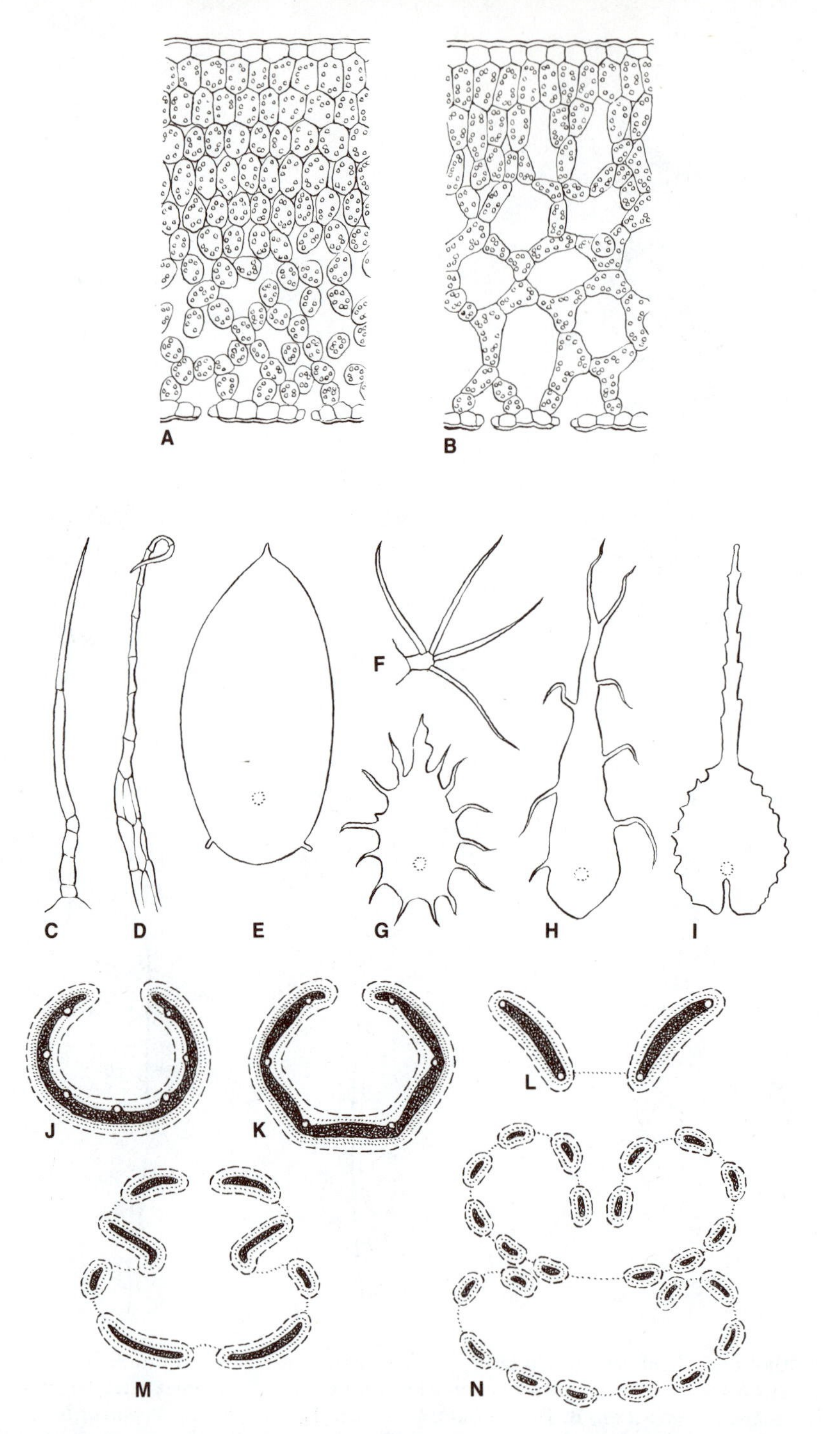

Figure 13-6 **A, B,** representation of internal leaf structure of ferns; **C–I,** representative types of dermal hairs and scales in ferns; the large peltate scales—**E, G–I**—are the more specialized forms; **J–N,** vascular cylinders of frond stipes as seen in transection; the primitive condition is seen in **J,** and a specialized condition in **N;** xylem is black, phloem stippled, protoxylem represented by white dots in **J–L. A, B, E, G–I,** *Polypodium;* **C,** *Matonia;* **D,** *Dipteris;* **F,** *Trichomanes;* **J,** *Osmunda;* **K,** *Gleichenia;* **L,** *Asplenium;* **M,** *Histiopteris;* **N,** *Cyathea.* [**A, B, E, G–N** redrawn from Anatomie der Vegetationsorgane der pteridophyten by Y. Ogura. In *Handbuch der Pflanzenanatomie.* Gebrüder Borntraeger, Berlin, 1938; **C, D, F** redrawn from *The Ferns,* Vol. I, by F. O. Bower. Cambridge University Press, London, 1923.]

Vein fusions are considered an evolutionary advancement. The first steps toward vein fusions are expressed by the appearance of marginal loops which connect the tips of vein dichotomies, or by the formation of a series of meshes adjacent to the midrib (Fig. 13-5, E) or throughout a leaf segment (Fig. 13-5, F).

ANATOMY OF THE LEAF. The typical fern leaf is dorsiventral. Epidermal cells have thickened, outer tangential walls and usually contain chloroplasts, a feature not shared by most other vascular plants. Stomata generally occur on the abaxial surface of leaves. The two guard cells of a stoma (guard cells and pore) may originate directly by cell division of a protodermal cell. In other ferns, a protodermal cell divides twice or three times; one of the derivatives functions as the initial of the guard cells, the others become surrounding subsidiary cells which apparently participate in the osmotic changes involved in the movement of the guard cells in the opening and closing of the pore. Stomatal types are of taxonomic value in determining relationships among ferns. The mesophyll may be uniform in organization, consisting of homogeneous parenchyma with chloroplasts, or the cells may be organized into definite adaxial palisade and abaxial spongy parenchyma layers (Fig. 13-6, A, B). Considerable evolutionary importance is attached to the form and arrangement of the vascular tissue in the stipe and rachis. There may be a simple arc, as seen in transverse section, or the vascular bundle may be convoluted, or the vascular tissue may consist of a cylinder of separate bundles (Fig. 13-6, J–N). The last arrangement is considered to be derived. As an added complication, the pattern of the vascular tissue may change in passing from the stipe through the rachis to the axes of the pinnae and pinnules.

THE SORUS. In the vast majority of ferns the foliage leaves serve in both photosynthesis and reproduction. In some species there may be a distinct separation of these functions: certain leaves function in photosynthesis, whereas others are strictly "sporophylls" and nonphotosynthetic. In other species there is an intermediate arrangement. For example, sporangia may be restricted to certain specific portions of a photosynthetic leaf. For the most part, sporangia are crowded into compact groups on leaves, each group being termed a *sorus* (Fig. 13-7). Sori may be circular or linear in outline. If sporangia are not grouped into definite sori they may form marginal tassels along narrow, reduced leaf segments or be scattered over the lower surface of expanded leaves, along and sometimes between veins. In those species having definite sori the sorus is along or near the frond margin or on the abaxial side of the frond.

ORGANIZATION OF THE SORUS. Both marginal sori and superficial sori (that is, those on the abaxial side of the frond) are most commonly found over a vein or at the terminus of a vein (Fig. 13-7, C, F). That portion of the leaf surface to which sporangia are attached is termed the *receptacle* (Fig. 13-12). It may be a slight protuberance, a definite bulge, or an elongated cone. It is from the superficial cells of the receptacle that sporangia originate while the leaf is still in a very young developmental stage. Undoubtedly circinate vernation provides protection for the delicate sporangia during their ontogeny.

Sporangia may or may not be protected by a covering termed an *indusium*. If a sorus lacks an indusium it is a naked sorus (exindusiate) (Figs. 13-7, E; 13-8, D). If an indusium is present it may be formed by adaxial and abaxial extensions of the lamina; this results in the formation of a cup or pouchlike structure (Fig. 13-7, B). In some species a reflexed marginal portion of the lamina itself is associated with an indusium, forming a pouchlike structure (Figs. 13-9, B; 13-10, A). In forms that have sori on the abaxial surface some distance from the leaf margin, the indusium is an outgrowth from the epidermis of the lamina or of the receptacle. The form of the indusium is variable: it may be a delicate, linear flap attached along one side only (unilateral indusium); it may be horseshoe shaped or circular and elevated (peltate); it may be cup-shaped; it may be a collection of scalelike structures overarching the sporangia; or the leaf margin may be turned back upon itself (the so-called false indusium), with the sorus borne on it (see Figs. 13-7, 13-8, 13-9, 13-10, 13-11, 13-12 for types).

DEVELOPMENT AND STRUCTURE OF THE SPORANGIUM. In the Filicales (leptosporangiate ferns) a single surface cell functions as the sporangial initial (Fig. 13-13, A, B). The sporangium grows for a time

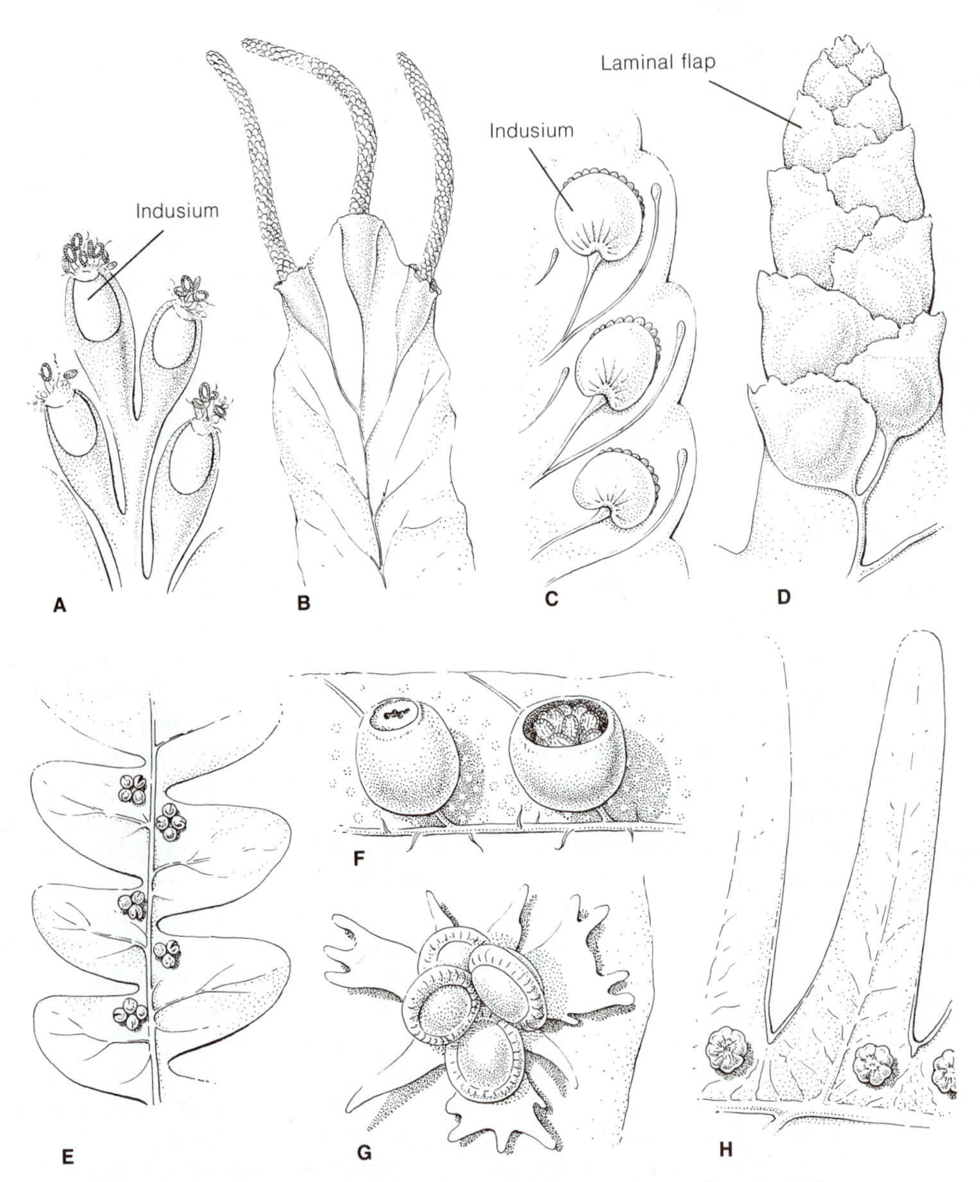

FIGURE 13-7 Variation in position and form of fern sori. **A,** *Davallia,* pouchlike indusium, joined with lamina, open at laminal margin; **B,** *Trichomanes,* marginal, receptacle elongate; **C,** *Nephrolepis,* indusium attached at one side; **D,** *Lygodium,* each sporangium covered by a laminal flap; **E,** *Gleichenia,* superficial position, no indusium; **F,** *Cyathea,* cup-shaped indusium; **G,** *Woodsia,* basal membranous indusial segments; **H,** *Matonia,* peltate indusium. [**C** redrawn from *The Ferns,* Vol. III, by F. O. Bower. Cambridge University Press, London, 1928; **F** adapted from *Morphology of Vascular Plants. Lower Groups* by A. J. Eames. McGraw-Hill, New York, 1936.]

A

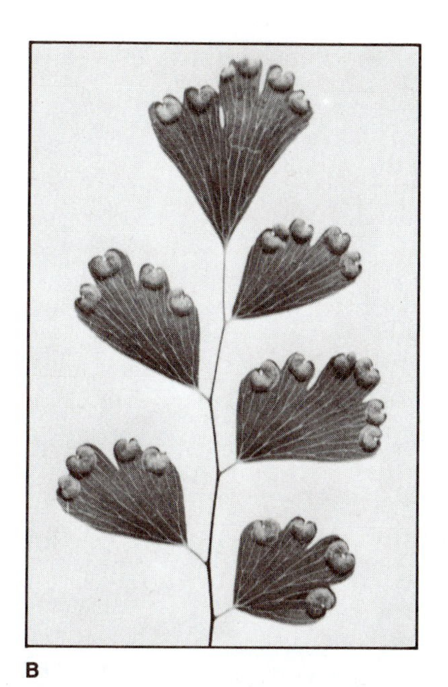

B

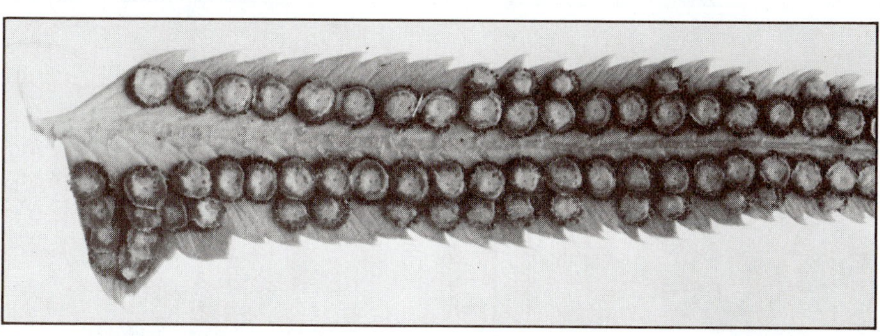

C

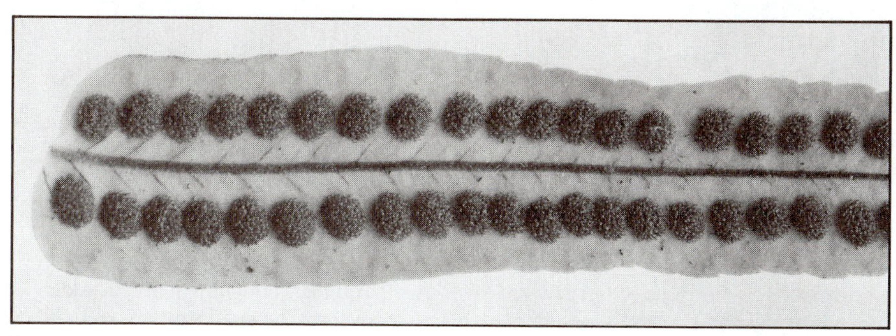

D

FIGURE 13-8 Morphology of fern sori. **A,** *Dryopteris,* reniform indusia; **B,** *Adiantum,* each sorus occurs on a reflexed portion of the lamina ("false indusium"); **C,** *Polystichum,* peltate indusia (sporangia can be seen extending beyond the margin of each indusium); **D,** *Polypodium,* naked sori ("exindusiate").

A

B

C

D

FIGURE 13-9 Portions of fertile fern leaves. **A,** *Pteridium,* young pinna at left, older pinna at right (the reflexed margin of the lamina covering the coenosorus is evident at left); **B,** *Dicksonia,* each "indusium" consists of a reflexed laminal flap tightly joined with true abaxial indusium (also see Fig. 13-10, A); **C,** *Lygodium,* individual sporangia covered by a flap of leaf tissue; **D,** *Asplenium,* unilateral indusia.

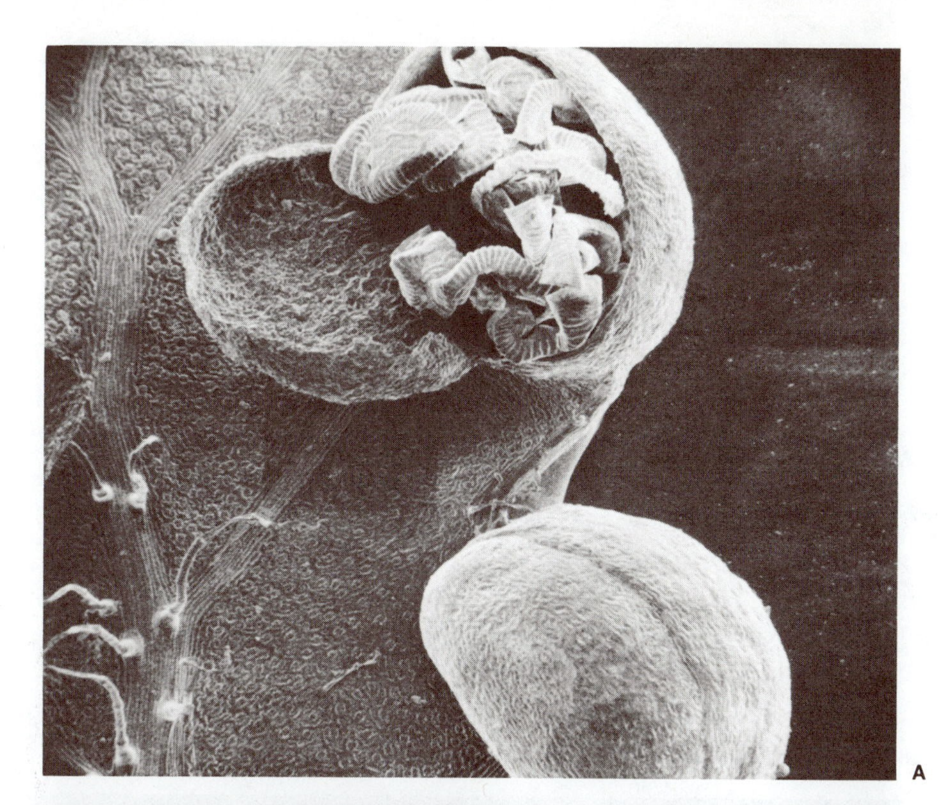

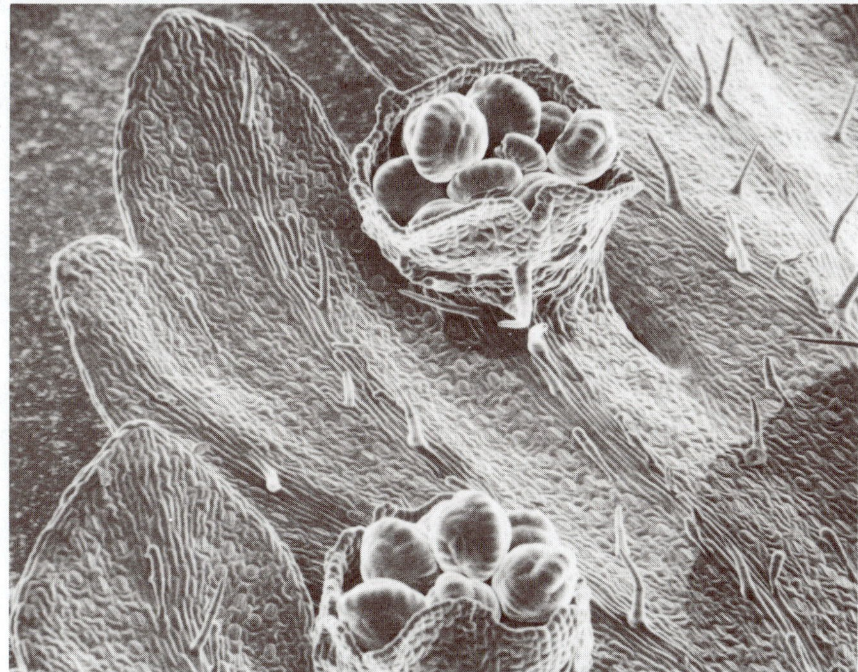

FIGURE 13-10 Scanning electron micrographs of fern sori. **A,** *Dicksonia antarctica:* the sporangia are protected by the reflexed margin of the lamina, which is tightly appressed to the indusium (sorus toward the lower edge of the figure); the indusium is depressed in the sorus toward the upper edge of the figure, revealing the sporangia; note veins and numerous stomata on abaxial side of lamina (×50); **B,** *Dennstaedtia cicutaris:* indusia are cup shaped and attached along the margin of the lamina (×110). [Courtesy of Dr. R. H. Falk.]

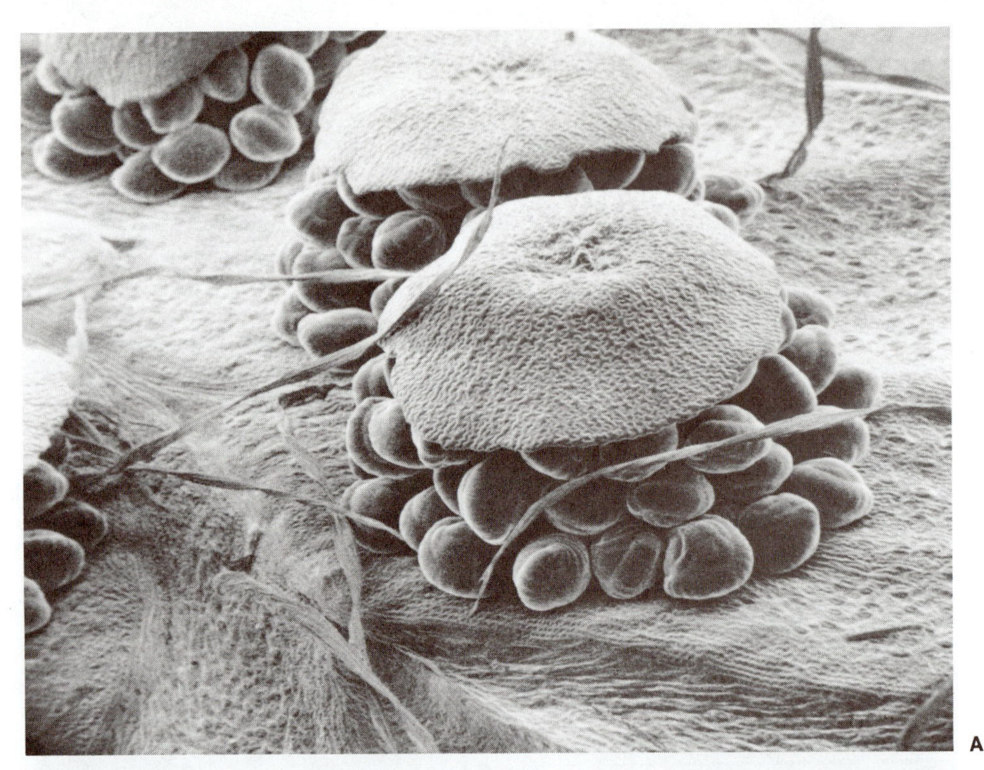

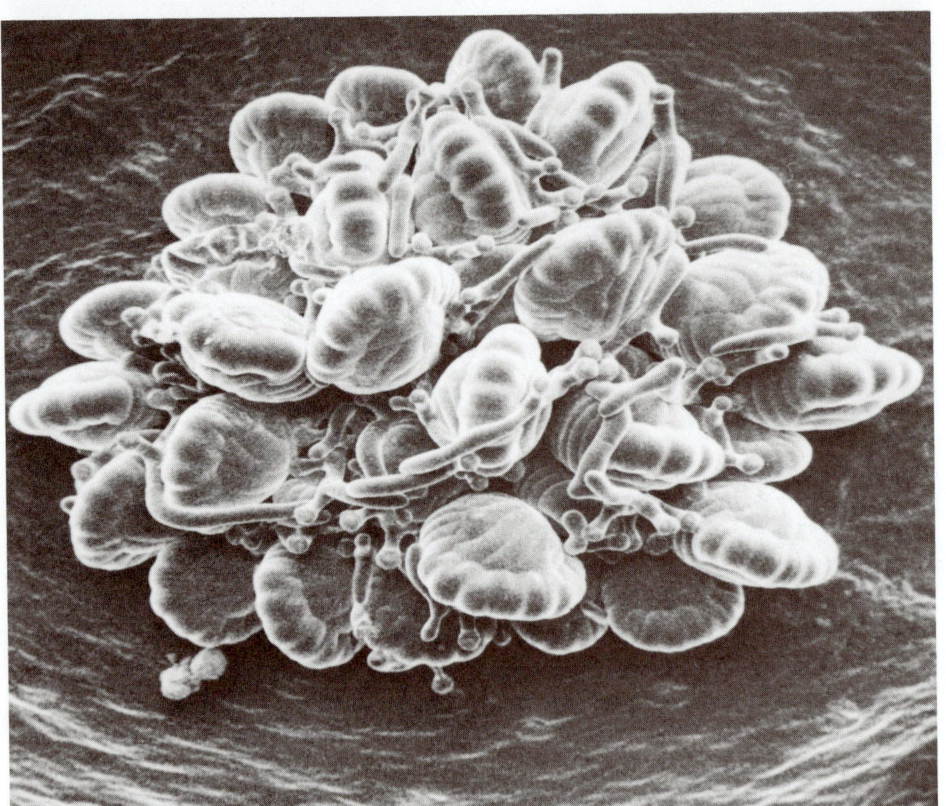

FIGURE 13-11 Scanning electron micrographs of fern sori. **A,** *Polystichum setosum:* note peltate indusia (×35); **B,** *Polypodium heterophyllum:* note prominent annuli and branched paraphyses (×80). [Courtesy of Dr. R. H. Falk.]

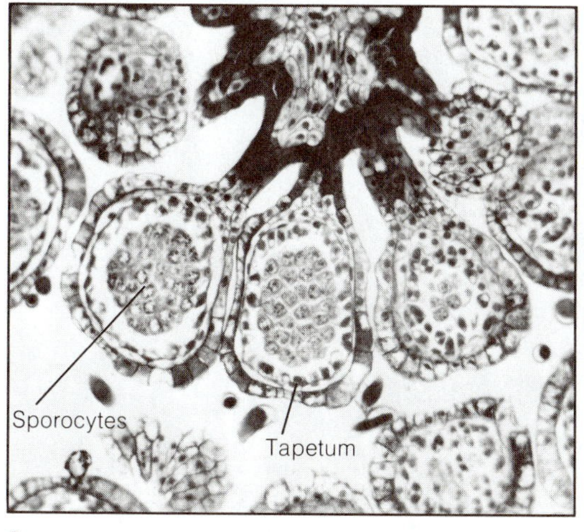

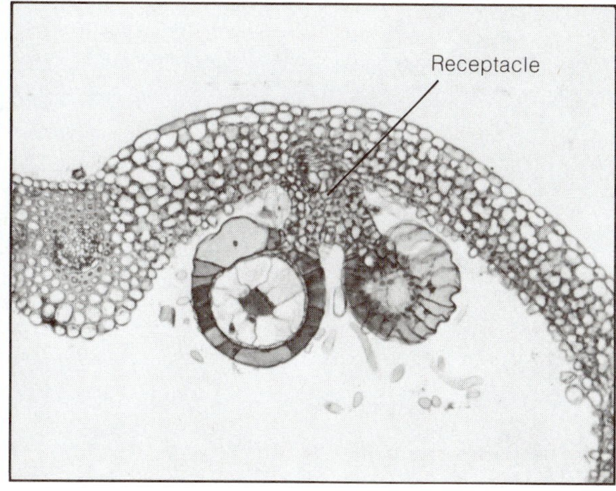

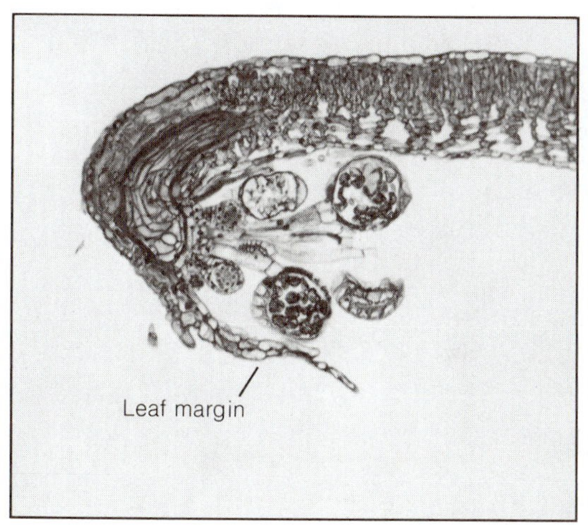

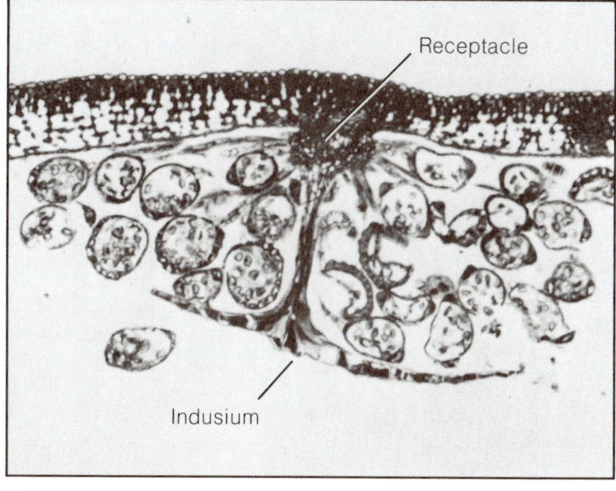

FIGURE 13-12 Sori and sporangia of ferns as seen in sectional view. **A,** *Osmunda,* section of leaf segment showing large sporangia, each one with many sporocytes and a massive tapetum; **B,** *Gleichenia,* showing large sporangia and lack of indusium; **C,** *Pteris,* intramarginal sorus with protective reflexed leaf margin (maturation of sporangia is of the mixed type); **D,** *Polystichum,* peltate indusium.

by the activity of an apical cell and by divisions in subjacent cells. The apical cell eventually forms a jacket cell and an internal cell. The latter cell gives rise to the tapetal initials and then to the sporocytes (Fig. 13-13). The wall remains one cell in thickness throughout development.

The two features that are regularly considered in comparing sporangia are final length of the stalk, and the number of cells (or rows of cells) making up the stalk. Short, thick stalks are considered primitive, and long, delicate stalks (frequently consisting of three rows of cells or even one) are derived.

The sporangial wall is typically one cell in thickness at maturity. The main point of interest, however, is in the means of dehiscence. In the Filicales there are various methods of dehiscence, depending

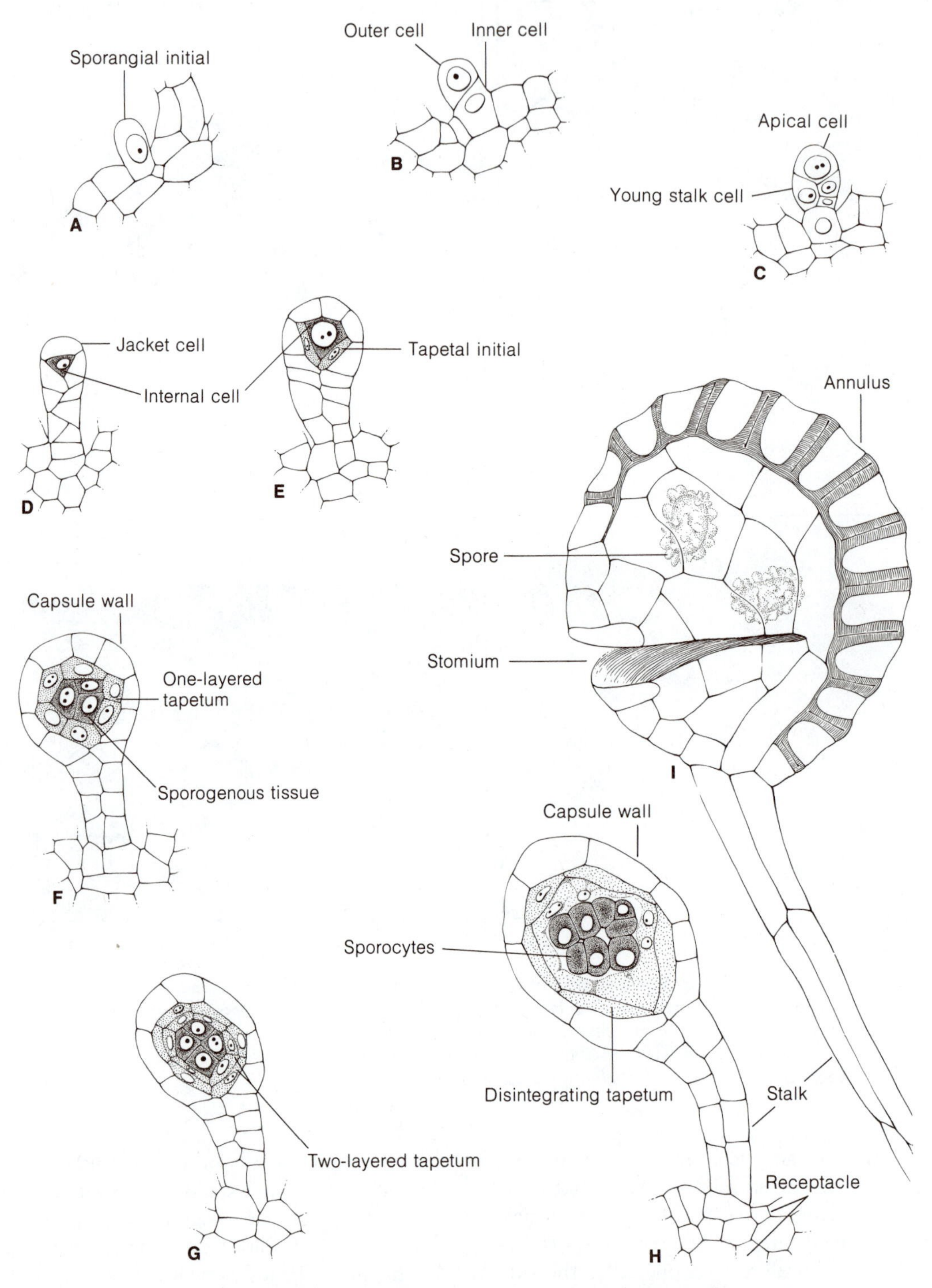

FIGURE 13-13 Stages in the development of a leptosporangium. Dehiscence has occurred in **I. A–H,** *Polypodium;* **I,** *Polystichum.*

on the position of the annulus (thick-walled cells). In the Osmundaceae the annulus, located to one side, is responsible for the formation of a cleft that runs over the top of the sporangium and down the opposite side. The sporangium opens like a clam. In other ferns the annulus may form a cap at the distal end of the capsule (Fig. 13-14, B), be obliquely placed (Fig. 13-14, E), or run over the top of the capsule in line with the stalk (in a vertical or longitudinal position). These three positions result in longitudinal, oblique, and transverse dehiscence, respectively (Fig. 13-14).

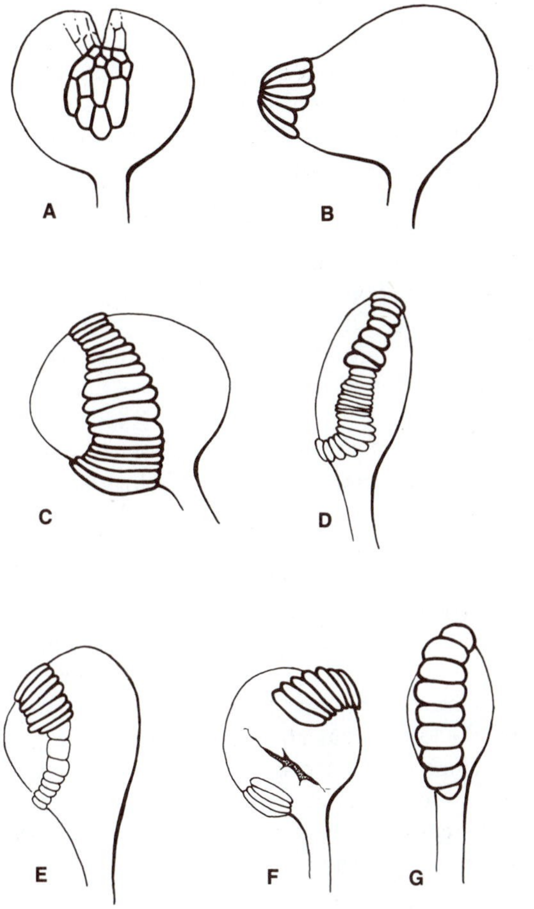

FIGURE 13-14 Variation in position of the annulus in leptosporangia. **A,** *Todea,* annulus subapical or lateral, which results in longitudinal dehiscence; **B,** *Lygodium,* annulus apical; **C,** *Gleichenia,* annulus oblique; **D,** *Plagiogyria,* annulus oblique; **E,** *Loxsoma,* annulus oblique, not all cells thickened; **F,** *Hymenophyllum,* annulus oblique, oblique dehiscence; **G,** *Leptochilus,* annulus vertical, which results in transverse dehiscence. [Redrawn from *Primitive Land Plants* by F. O. Bower. Macmillan, London, 1935.]

MATURATION OF SPORANGIA WITHIN A SORUS. The simplest way in which fern sporangia are borne is singly along the margins of leaf segments—each with a vascular bundle leading to its base (as in some extant ferns such as *Osmunda*). This arrangement is a primitive condition. In the majority of Filicales, sporangia are aggregated to form sori. A sorus in which all of the sporangia originate, grow, and mature at the same time is termed a "simple" sorus (Fig. 13-15).

If sporangia are initiated over a period of time in a definite sequence the "gradate" sorus is produced. The order of sporangial initiation and development is basipetal; the oldest sporangium is near the summit of a receptacle with successively younger sporangia toward the base (Fig. 13-15). When the fossil record is considered and compared

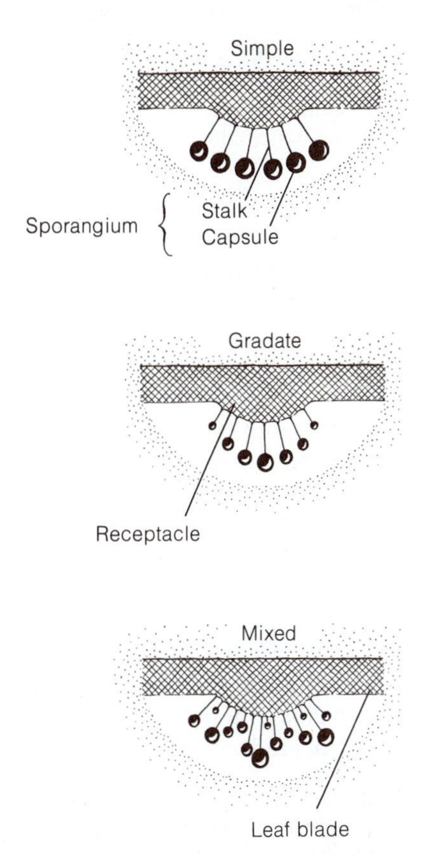

FIGURE 13-15 Three types of sporangial maturation in sori of leptosporangiate ferns.

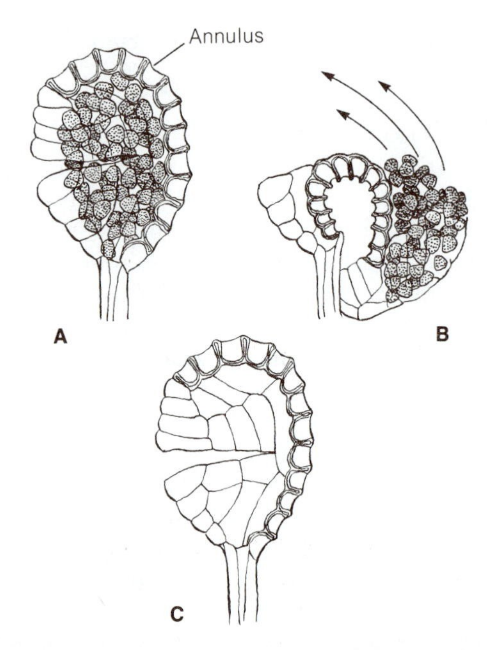

FIGURE 13-16 Behavior of a fern sporangium during drying and dispersal of spores. See text for discussion. [Redrawn from *Plant Physiology,* 1st edition, by B. S. Meyer and D. B. Anderson, © 1939 by Litton Educational Publishing, Inc. Reprinted by permission of Van Nostrand Reinhold Co.]

with that of living ferns having the simple type of sorus, the gradate maturation is clearly a derived type. It must be emphasized that not all ferns with gradate sori are necessarily closely related. Gradate maturation represents an evolutionary level of specialization that has been achieved by different species.

The most advanced evolutionary level of development is achieved in the sorus that has intermingled sporangia, all in different stages of growth (Fig. 13-15). This is the "mixed" sorus. The more highly specialized and evolved families or subfamilies have this mode of soral development. Bower (1923, 1935) emphasized the physiological importance of extending sporangial development over a longer period of time during the reproductive phase of the sporophyte. Neither should the adaptive value of prolonged spore production be underestimated.

SPORANGIAL DEHISCENCE. The dehiscence of the fern sporangium is ingenious and has attracted much attention from interested students. The annulus is the structural feature associated with dehiscence and the forceful ejection of spores. As a sporangium matures, water is lost from the cells of the annulus by evaporation. There is a powerful adhesion between the cell walls and water. The continued loss of water from each cell of the annulus results in the thin outer tangential wall of each cell being drawn inward, while the ends of the radial walls are pulled toward each other. This results in the tearing open of the sporangium on the weak side and eventually in the complete inversion in the position of the annulus. The annulus is now under tremendous tension. Water continues to evaporate. Eventually the cohesive force of the water in cells of the annulus is exceeded and the annulus returns suddenly to approximately its original position. In the process the spores are thrown out forcefully for a distance of a centimeter or so (Fig. 13-16). The tensions built up prior to dehiscence are equivalent to about 300 or more atmospheres of pressure.

A SURVEY OF SORAL TYPES AND THEIR DEVELOPMENT. Variations in soral morphology are endless. At best only a few of the more common soral types that illustrate a generalized type or represent steps in a possible evolutionary series can be described.

Sori on fern leaves may occupy (1) a marginal position, (2) an intramarginal (near the margin) position, or (3) the abaxial or "superficial" position in which sori are at some distance from the leaf margin on the lower (abaxial) side of a frond. During leaf development, the origin and final position of a sorus are correlated with the activity of marginal meristems and their derivatives (Wardlaw, 1958, 1962; Wardlaw and Sharma, 1961).

SORI OF MARGINAL POSITION. For a sorus to qualify as truly marginal the receptacle must originate strictly from the margins of the developing pinnae or pinnules (Fig. 13-17, A, B). An indusium, if

FIGURE 13-17 A, B, *Hymenophyllum,* transverse sections through margins of fertile portions of leaves showing stages in development of marginal sorus; C–E, *Cryptogramma,* sections through leaf margins showing development of intramarginal sorus. Margin of lamina becomes extended as a protective flap (E). [B redrawn from *Cryptogamic Botany,* Vol. II, by G. M. Smith. McGraw-Hill, New York. 1938; C–E redrawn from Wardlaw, *Ann. Bot.* 25:481, 1961.]

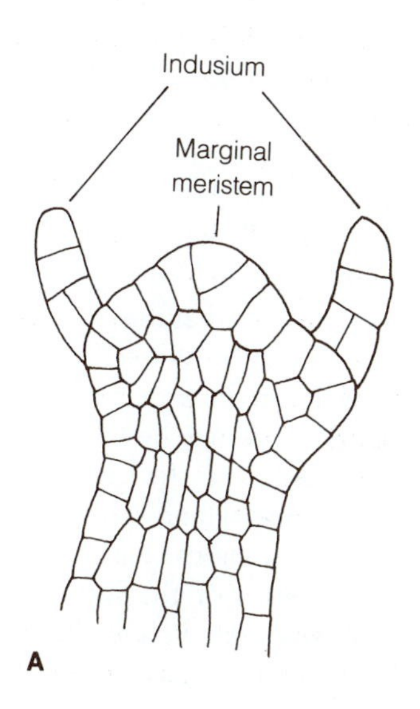
Indusium
Marginal meristem
A

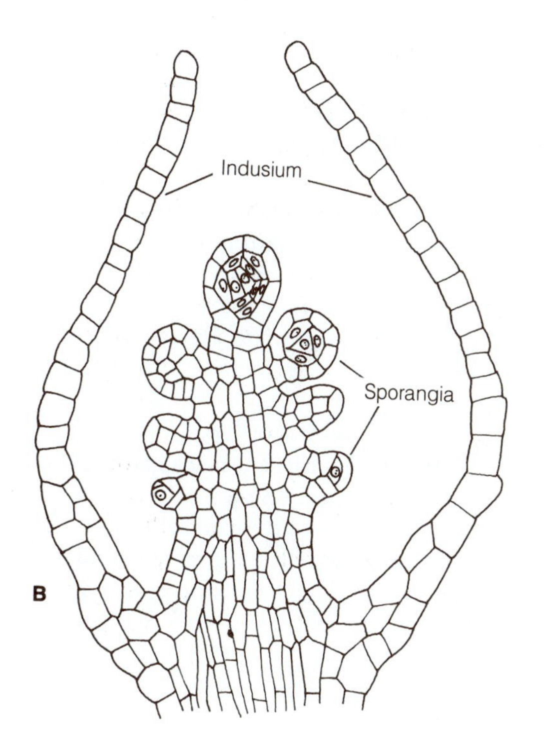
Indusium
Sporangia
B

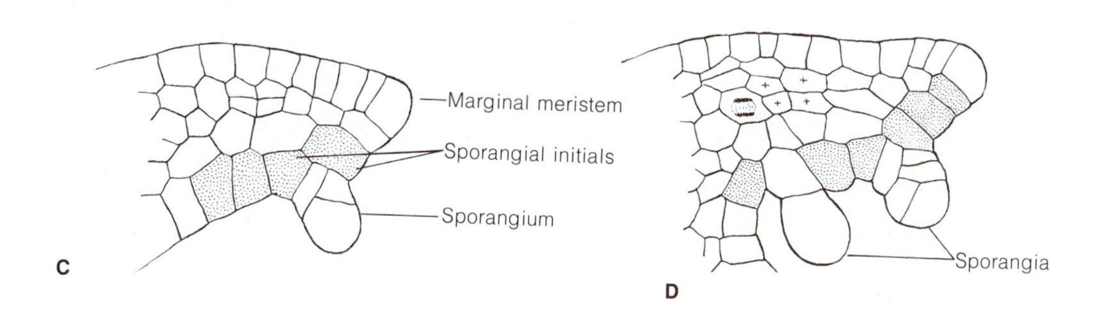
Marginal meristem
Sporangial initials
Sporangium
C
Sporangia
D

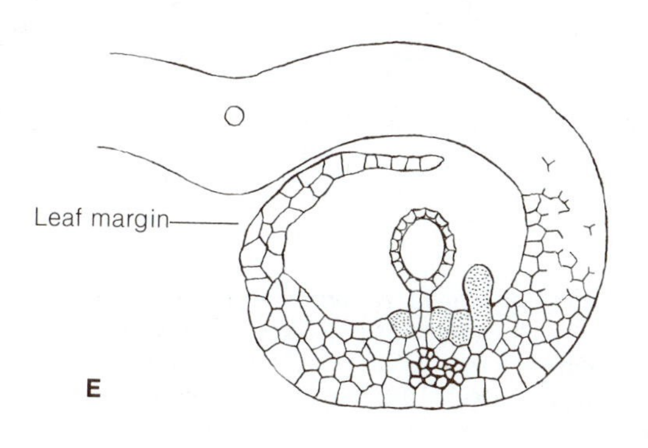
Leaf margin
E

present, is formed by submarginal outgrowths around the receptacle. The indusium in its final form may be funnel shaped or two-lipped. The receptacle is actually a continuation of the marginal meristem and retains its activity by producing sporangial initials. Wardlaw (1958, 1962) has recommended that the term "sporogenous meristem" should replace the term receptacle, since it is more descriptive of its function. Examples of the marginal type are *Hymenophyllum* and *Trichomanes* (Fig. 13-7, B).

SORI OF INTRAMARGINAL POSITION. The sori of many ferns may be described as being marginal because of how they appear on the adult lamina, but actually they originate near the margin during ontogeny. In these ferns, the cells at the margin suddenly or gradually lose their meristematic potentialities and become parenchymatous, often forming a thin extension (Fig. 13-17, C–E). Submarginal cells on the abaxial side of the developing lamina near the margin continue their meristematic activity—forming the receptacle or sporogenous meristem. Associated with this change in activity, an inner or abaxial indusial flap also may be formed as a thin outgrowth from surface cells near the receptacle. The final width and length of the receptacle depends upon the length of time that the sporogenous meristem is active and upon how much of the laminal margin participates in the processes. In some instances individualized sori may be formed along veins or at the tips of veins near the leaf margin; in other instances the receptacle (and sporangia) may extend parallel with and along the entire length of the leaf segment. Examples of the intramarginal type are: *Pteris, Cryptogramma, Pellaea, Pteridium,* and *Dicksonia* (the last two also have a true abaxial indusium).

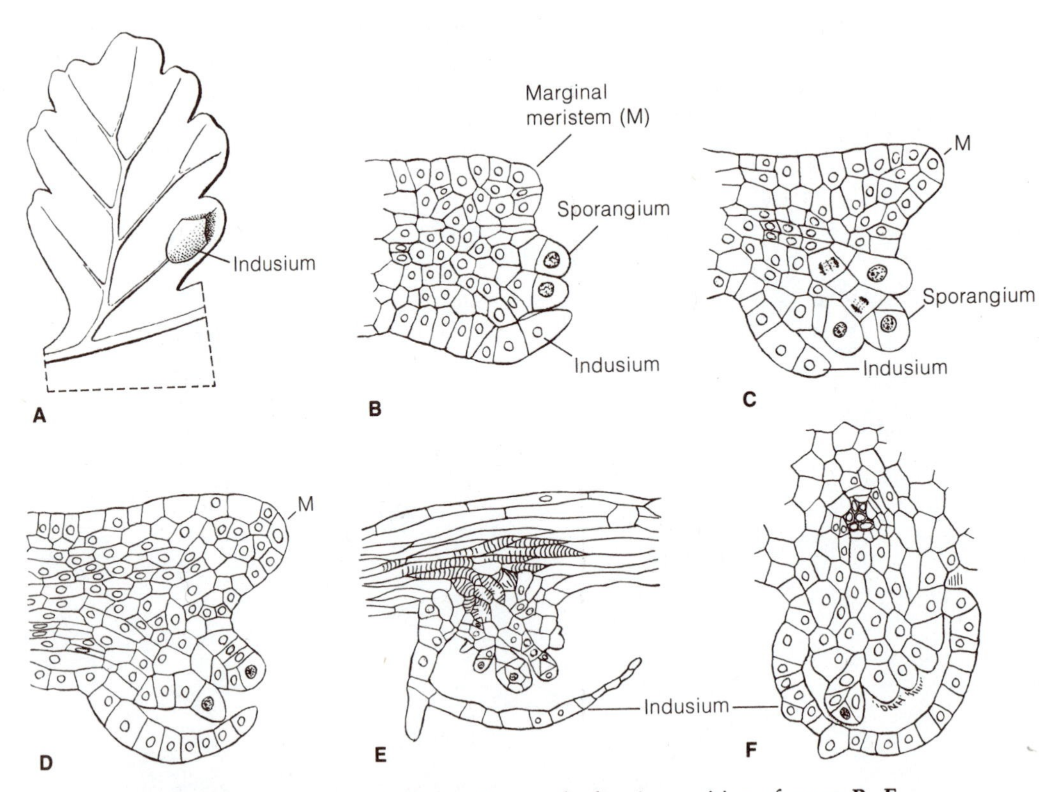

FIGURE 13-18 *Cystopteris bulbifera.* **A,** pinnule showing position of sorus; **B–E,** transverse sections through margins of pinnules, showing development of sorus; **F,** developing sorus as seen in a section that cuts across a vein. [Redrawn from Palser and Barrick, *Bot. Gaz.* 103:172, 1941, University of Chicago Press.]

SORI OF ABAXIAL OR "SUPERFICIAL" POSITION. In ferns whose sori are in the abaxial position, a young receptacle has its inception from submarginal cells on the abaxial side rather early during marginal growth of the lamina. However, in contrast to the intramarginal type of development, the margin of the lamina remains actively meristematic and adds new tissue to the lamina. Therefore, as growth of a young leaf segment proceeds, the young sorus occupies positions progressively farther from the margin (Figs. 13-18, 13-19). An indusium, if characteristic of the species, is formed from superficial cells near the developing receptacle and eventually overarches it (Fig. 13-18). The indusium, in its final form, varies in shape from elongate and attached along one of the longer sides (unilateral indusium), to half-moon shaped or reniform and attached at the sinus (*Dryopteris,* Fig. 13-8, A). In some species the indusium is an outgrowth from the top of the receptacle, resulting in the formation of a stalk and a radially symmetrical cap (peltate type; Figs. 13-8, C; 13-11, A; 13-12, D).

ANATOMY OF THE STEM. A variety of structures and tissues take their origin from the meristematic derivative cells of the apical cell: leaves, the protoderm of leaves and of the stem, epidermal hairs and dermal appendages, ground meristem, and procambium. From available studies it appears that procambium in ferns develops in relation to the appearance of leaves. (For a review see Esau, 1954.) Very soon after a leaf is initiated a procambial strand becomes differentiated at its base and soon becomes apparent in the leaf itself (Fig. 13-2). The development of procambium is acropetal and apparently always in continuity with existing procambium at a lower level in the stem. Within a procambial strand, phloem differentiation also

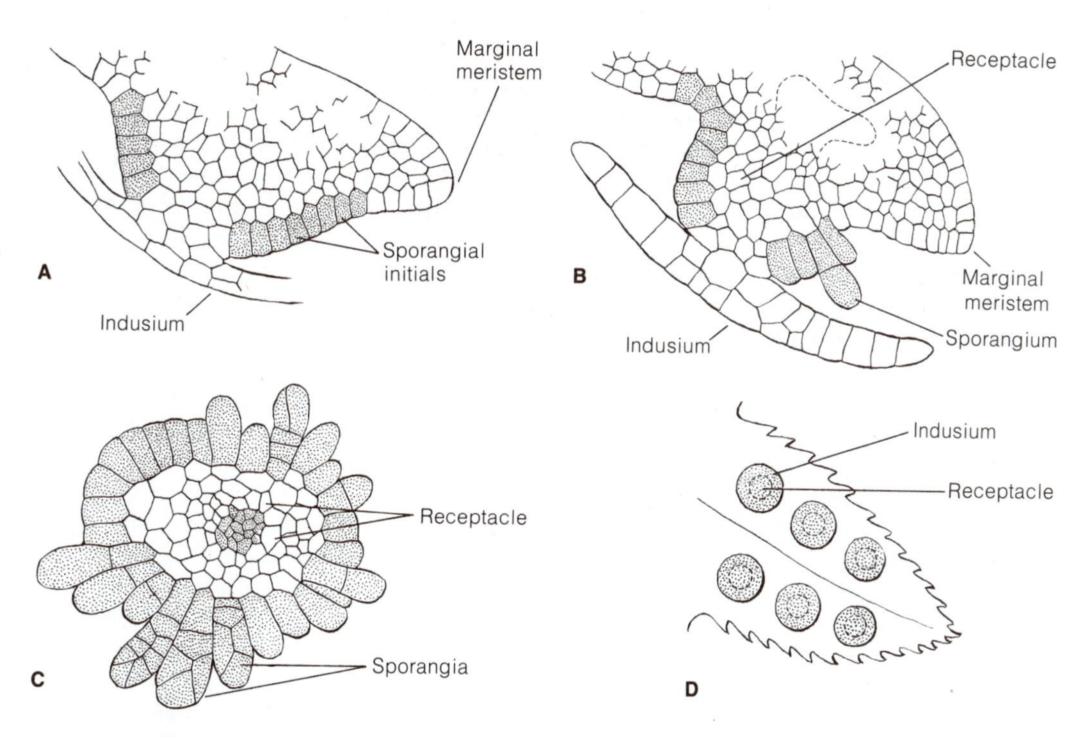

FIGURE 13-19 *Polystichum aculeatum.* **A, B,** transverse sections through margins of fertile leaf segments showing origin and development of the sorus; note displacement of sorus from marginal region by continued activity of the marginal meristem. **C,** transverse section of a young sorus at a level below the indusium; note origin of sporangia from surface cells (sporogenous meristem) of the receptacle. **D,** schematic representation showing position of mature sori on abaxial side of a fertile leaf segment. [Redrawn from Wardlaw, *Ann. Bot.* 25:482, 1961.]

occurs in an acropetal direction (upward into the leaves). In some ferns there is evidence that the differentiation of tracheary elements in the xylem may be discontinuous before the leaf has its normal amount of mature xylem. In general, though, the wave of vascular differentiation is in an acropetal direction.

Throughout the Filicales, including the fossil forms, there is no indication of cambial activity resulting in the formation of secondary vascular tissues. There is, however, the development of considerable sclerenchyma in the axes of some ferns. The radial maturation of xylem in a stem vascular bundle is typically exarch or mesarch. The large tracheids of the metaxylem generally have tapered ends and scalariform pitting. It has been demonstrated that certain filicalean ferns (e.g., *Pteridium aquilinum*) possess vessels; scalariform perforations instead of pits are present in the oblique end walls between two vessel members (Bliss, 1939). Vessels with simple perforations also have been reported to occur in the nodal regions of the rhizome of *Actiniopteris radiata* (Singh et al., 1978). Vessels also occur in the roots and rhizomes of certain water ferns (see p. 306).

The length of tracheary elements varies with location in the plant, age of the plant, length of an internode, polyploidy, and habitat (White, 1963a, b). In spite of these variations the following changes appear to be correlated with specialization from the primitive condition for each feature: (1) shortening of tracheary elements, (2) increase in the occurrence of a modified type of scalariform pitting (opposite and alternate), (3) increase in the occurrence of slightly oblique to transverse end walls, and (4) sporadic occurrence of vessels, e.g., in *Pteridium* and *Marsilea* (see p. 306).

TYPES OF VASCULAR CYLINDERS. Just as the protostele is considered primitive and of common occurrence in other lower vascular plants, so have certain members of the Filicales retained this type of stele. For example, it occurs in the living genera *Gleichenia* and *Lygodium,* and in the family Hymenophyllaceae. Most species of *Gleichenia* are protostelic but of a rather special type (Fig. 13-20, A). The bulk of the central column is primary xylem in which tracheids of the metaxylem are interspersed with parenchyma. This type of stele has been designated as a protostele with a "mixed pith" (Schmid, 1982). Protoxylem occupies definite loci near the periphery of the xylem. External to the xylem is a cylinder of phloem consisting of relatively large, thin-walled sieve cells and smaller parenchyma cells. The vascular cylinder is enclosed by a rather wide pericycle several cell layers thick which, in turn, is surrounded by the endodermis. Frequently the endodermis and particularly the casparian strips are not well defined. Topographically the endodermis is located at the boundary between thick-walled cortical cells and thin-walled pericyclic cells. The departure of a leaf trace does not materially affect the outline of the vascular cylinder of the stem, and there is no leaf gap. The significance of this type of protostele in the evolution of the siphonostele has already been considered in Chapter 3.

The most commonly accepted viewpoint is that there are two types of siphonosteles: the *ectophloic siphonostele* (outer cylinder of phloem only), and the *amphiphloic siphonostele* (inner and outer cylinders of phloem) (see Chapter 3). *Osmunda* is one of the ferns commonly used to demonstrate an ectophloic siphonostele (Fig. 13-20, B). *Osmunda* has a short erect stem with a crown of leaves. The stem is completely invested by leaf bases. A transverse section of an older shoot reveals the crowding of the leaves and the presence of a vascular strand in each leaf base. Centripetal to the leaf bases is the outer cortex, consisting of thick-walled sclerenchyma with embedded leaf traces. The inner cortex is parenchymatous, the cells often packed with starch grains. The pith occupies a generous amount of the axis. Along the outer edge of the pith is a cylinder of more-or-less separate xylem strands, often U shaped in outline with the open side toward the pith. Protoxylem is in the sinus formed by the two arms of metaxylem. For this reason the xylem is considered to be mesarch. Surrounding the xylem is a zone of parenchyma, several cells in thickness. External to this layer is the phloem with its relatively large sieve cells, often elongated in an oblique longitudinal direction. This is particularly true of the protophloem. Parenchyma cells also are interspersed among the sieve cells. External to the phloem is the *pericycle,* which, in turn, is surrounded by tannin-filled *endodermal* cells. The outline of the main vascular cylinder may be some-

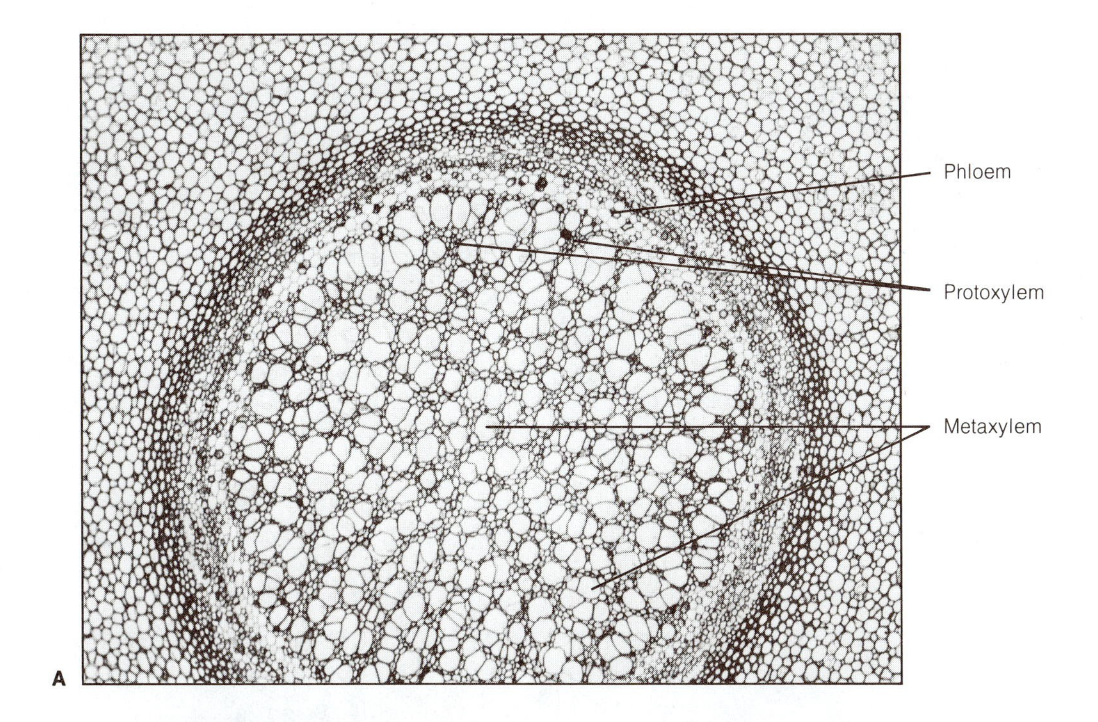

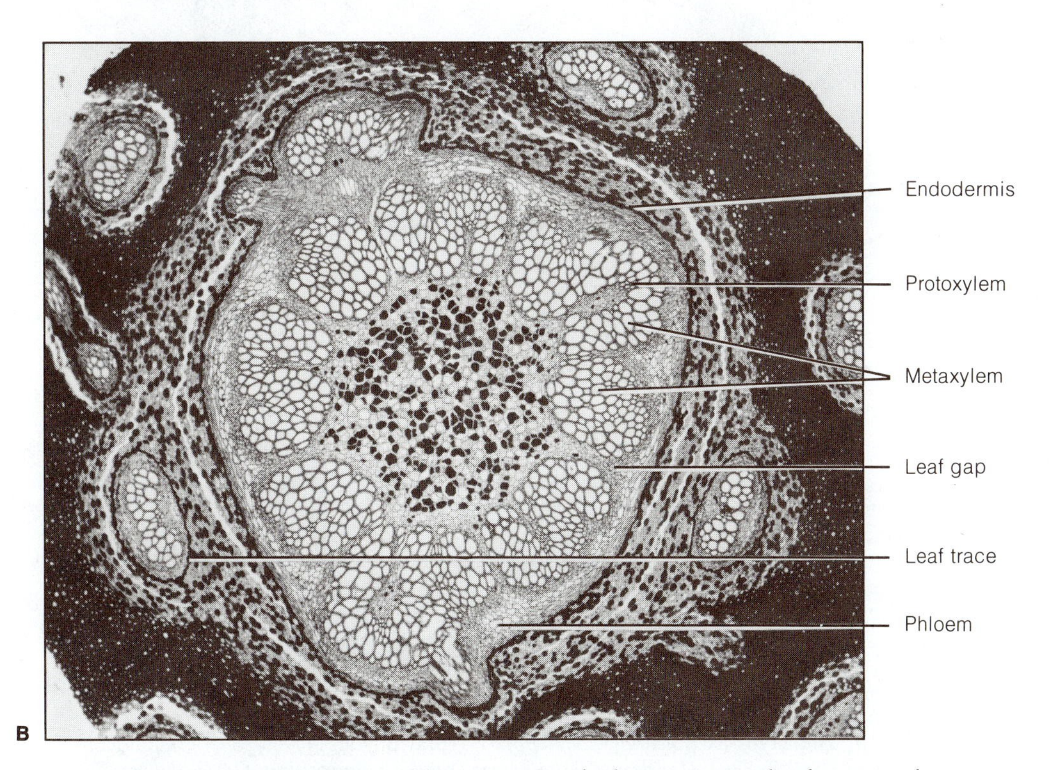

FIGURE 13-20 Transections of fern stems. **A,** *Gleichenia,* an example of a protostele with a "mixed pith" (consisting of metaxylem tracheids intermingled with parenchyma); **B,** *Osmunda,* example of an ectophloic siphonostele.

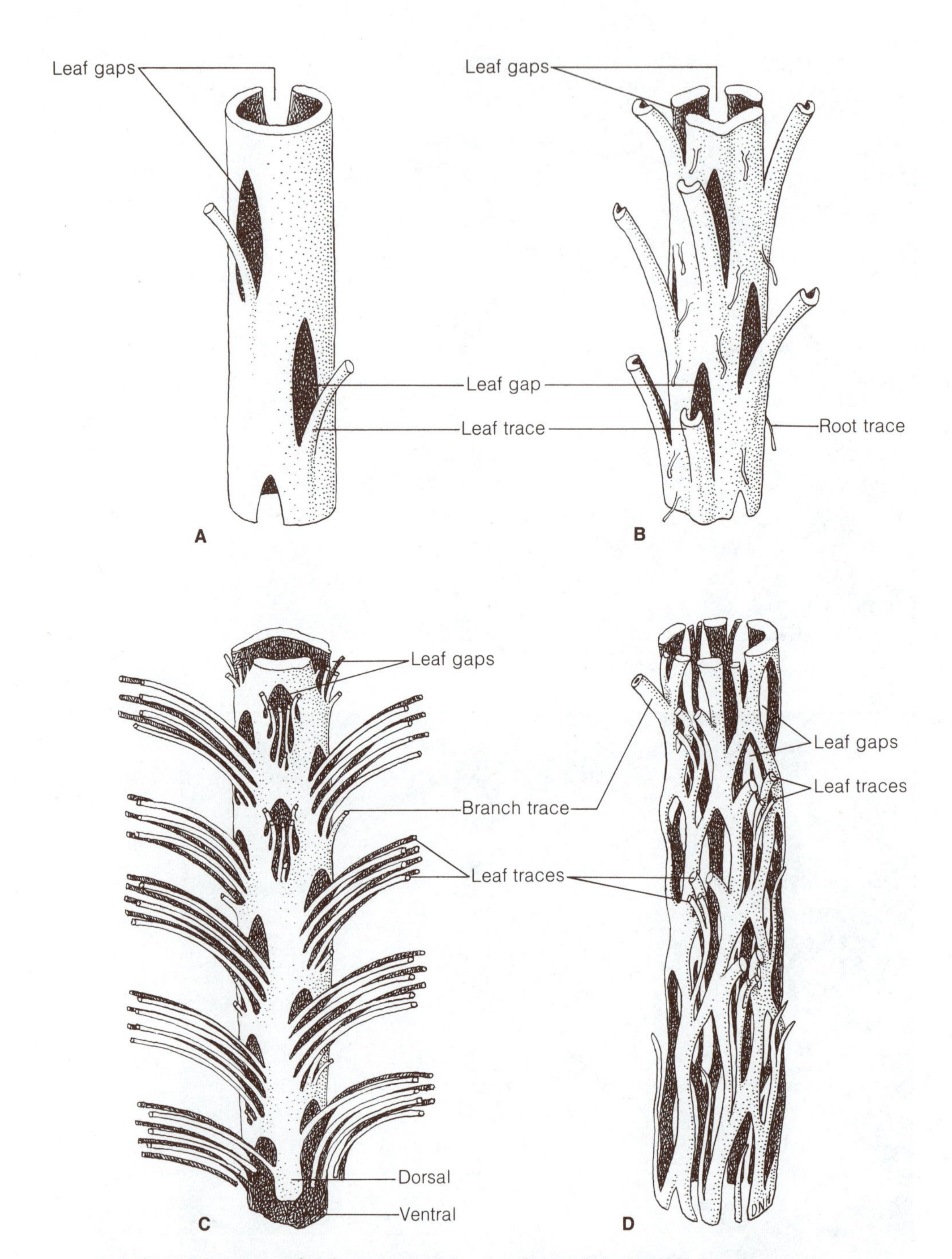

FIGURE 13-21 Types of stelar organization. **A,** solenostele, schematic; **B,** dictyostele *(Cheilanthes tenuifolia);* **C,** dictyostele, rhizome has dorsiventral organization, leaves arising on dorsal side; some leaf traces are attached along edge of a leaf gap *(Bolbitis diversifolia);* **D,** dictyostele *(Oleandra wallichii).* [**B** redrawn from Nayar, *Jour. Linn. Soc. London Bot.* 58:449, 1963; **C** redrawn from Nayar and Kaur, *Jour. Linn. Soc. London Bot.* 59:127, 1965; **D** redrawn from Nayar, Bajpai, and Chandra, *Jour. Linn. Soc. London Bot.* 60:265, 1968.]

what irregular, due to the divergence of leaf traces. Each leaf trace is confronted by a leaf gap which may extend longitudinally and obliquely for some distance in the xylem. The phloem, however, forms a continuous cylinder. In some species tracheids may be intermingled with parenchymatous pith cells, and in one species *(Osmunda cinnamomea)* an internal endodermis has been reported.

Ferns with vascular cylinders of the types just described have also been encountered as fossils. The stem anatomy of a late Paleozoic fern, *Thamnopteris,* closely resembled that of *Osmunda.* In certain species of *Thamnopteris,* however, the pith consisted of intermingled parenchyma and tracheids surrounded by a compact cylinder of xylem, without leaf gaps. The vascular cylinder was a mixed-pith protostele. Some species of the Mesozoic genus *Osmundacaulis* had stems that were even more similar in anatomy to those of *Osmunda* (see p. 266).

Without doubt many more ferns have the amphiphloic siphonostele than have the ectophloic type. Fundamentally there are two types of amphiphloic siphonosteles: the *solenostele,* and the *dictyostele.* The solenostele, without overlapping leaf gaps (Fig. 13-21, A), frequently occurs in many creeping ferns that have conspicuous internodes. Common examples are certain species of *Adiantum* and *Dennstaedtia* (hay-scented fern). A transverse section within an internodal region reveals a complete cylinder of vascular tissue surrounded by the cortex and enclosing a pith that is often sclerified (Fig. 13-22, A). Starting from the outer edge of the vascular cylinder (stele), the outer endodermis can usually be distinguished by phenolic substances in its cells. Internal to the endodermis is the pericycle (one or more layers of cells in width) followed by the phloem (outer phloem). As usual the ring of xylem is most conspicuous, being composed of scalariformly pitted tracheids. Inner phloem, pericycle, and inner endodermis follow, in that order, in progressing toward the pith. At a node the seemingly closed cylinder of vascular tissue is broken by the presence of a leaf gap. The cylinder again appears closed within the next higher internode. This type of stele is not an entirely new innovation because species of *Adiantum* have been found from the Tertiary. The creeping water fern, *Marsilea,* is also solenostelic with a highly lacunate cortex typical of many hydrophytes (Fig. 13-51). Solenosteles may become highly elaborate, as in *Matonia* where there is a variable number of concentric, amphiphloic cylinders *(polycyclic solenostele);* the leaf-trace complex also may be a collection of concentric cylinders.

The most specialized type of siphonostele in ferns is the dictyostele. The dictyostele is an amphiphloic tube of vascular tissue in which parenchymatous leaf gaps overlap. A transverse section made at any level of a well-developed stem shows a ring of separate vascular bundles. These bundles vary considerably in size and shape, but each one consists of primary xylem enclosed by primary phloem. These strands are termed concentric, amphicribral vascular bundles, or *meristeles* (Fig. 13-22, B; also see Chapter 3, p. 45). The parenchymatous zones between meristeles are leaf gaps (Fig. 13-22, B). Most of the Filicales are fundamentally dictyostelic and of the type just described (Figs. 13-21, B–D; 13-25, B). This is true of upright as well as creeping species. If the leaf gaps are long and extensive and more than one leaf trace is associated with each gap, or if there are breaks in the cylinder not associated with leaves, the vascular cylinder becomes a network of vascular strands. In one transverse section it is difficult if not impossible in these cases to distinguish the actual limits of the gap. We are describing the so-called *dissected* or *perforated* dictyostele. *Polypodium* is an example of a common fern with this type of stele (Fig. 13-23, A). Inner vascular cylinders may be present in some dictyostelic species.

The rhizome of *Pteridium aquilinum* (bracken fern) has "medullary strands" within an outer vascular cylinder (Fig. 13-23, B). Depending upon the variety of *P. aquilinum,* the leaves may be borne on both long and short lateral shoots (O'Brien, 1963) or only on short lateral shoots (Webster and Steeves, 1958). Therefore, the parenchymatous regions between vascular bundles (meristeles) of a rhizome of the latter type, which is devoid of leaves, cannot be regarded as leaf gaps; their morphological significance is unknown. Schmid (1982) has designated this type of stele as a perforated solenostele. Sclerenchyma may entirely or partially enclose the inner strands, or it may occur as two separate bands as seen in transectional view. When a lateral bud grows out, the vascular tissue of the new rhizome

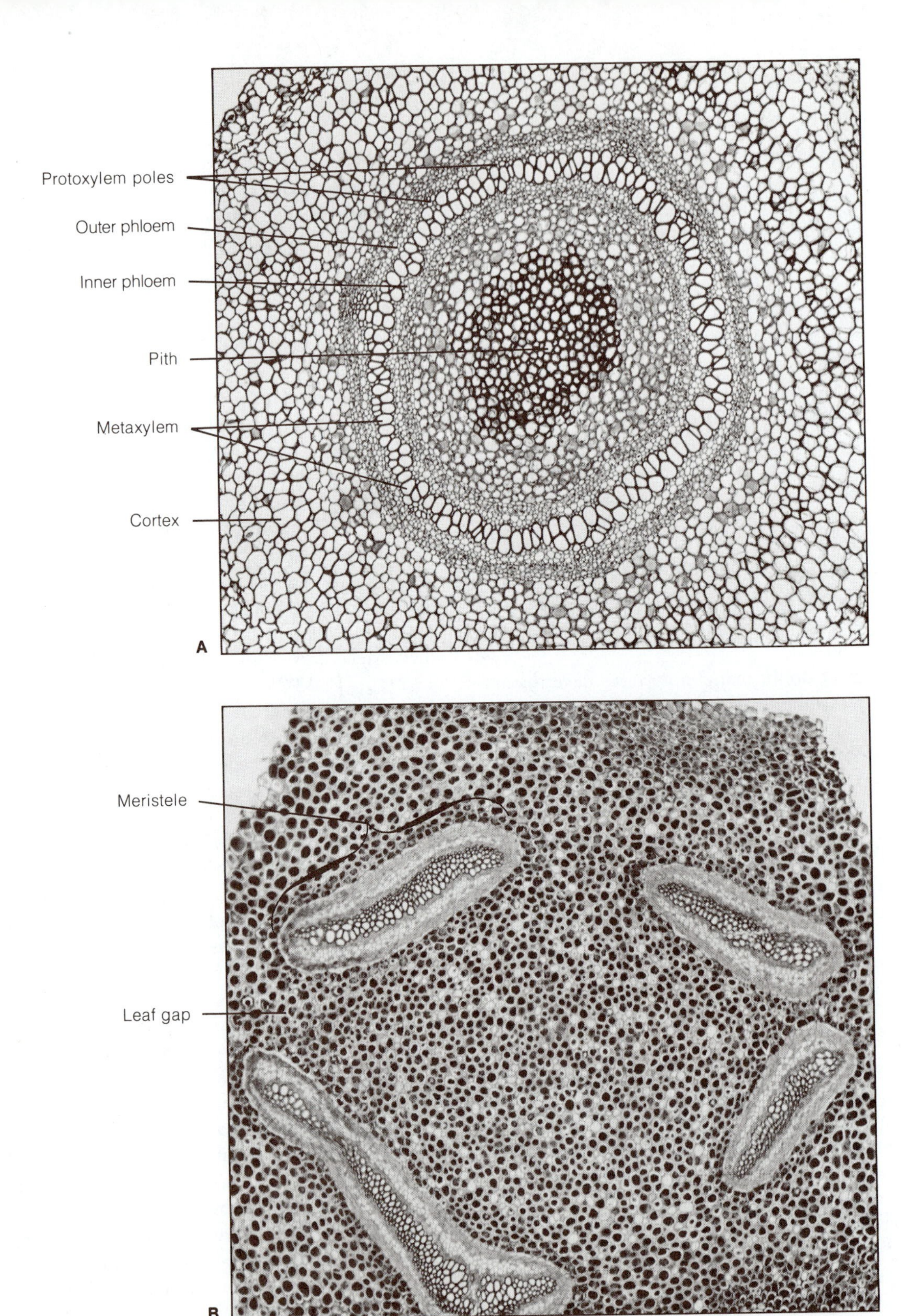

FIGURE 13-22 Transections of fern steles. **A,** *Dennstaedtia,* solenostele, phloem, and endodermis occur on both sides of the xylem; section was made at a level between leaf gaps; **B,** *Phyllitis,* dictyostele.

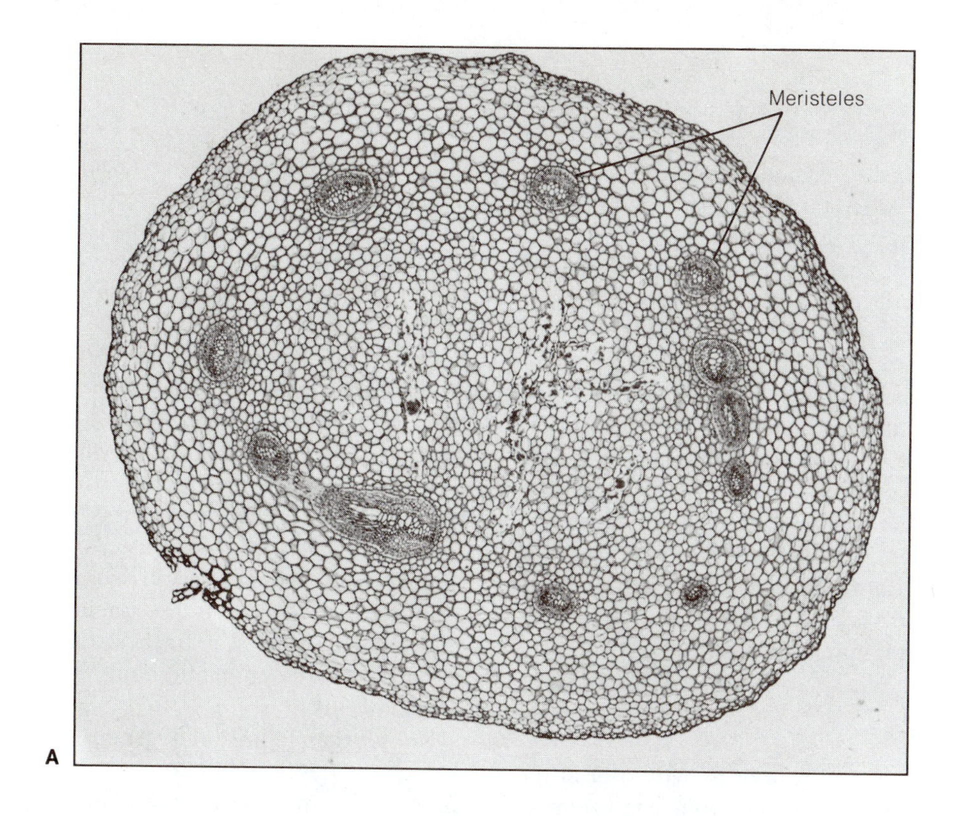

FIGURE 13-23 Transections of fern rhizomes. **A,** *Polypodium,* example of a dissected dictyostele; **B,** *Pteridium,* a stele with outer and inner vascular cylinders and intervening sclerenchyma.

assumes the same configuration as that of the parent rhizome. The structural details of one vascular bundle are shown in Fig. 13-24.

More highly complicated vascular structures have been described for ferns, namely, species with exceedingly long and large leaf gaps with numerous leaf traces attached to the sides of the gaps, and species with small innumerable meristeles, arranged in intricate patterns and often associated with leaf gaps in various ways. In concluding this section it should be pointed out that critical ontogenetic work, utilizing modern techniques, still remains to be done on fern vasculation before the results of earlier works can be fully accepted. Evolution of the stele in ferns is discussed in a later section on special problems in fern morphology.

The Root

Except for the first root or primary root of the embryo, all roots of a fern arise from the rhizome or upright stem, near leaf bases or below them. They are usually small in diameter, monopodially branched, and dark in color. The outer investment of fibrous material on the lower portion of the trunks of tree ferns is a mat of tangled, living, dead, and dying roots (Fig. 13-25, A). The actual stem axis may be only a relatively small cylinder within the covering of roots.

At the level of mature primary tissues in the root the epidermal cells are usually thin walled (some cells may form root hairs). A generous portion of the inner cortex may be sclerenchymatous, or only the cells of a single layer around the endodermis may be thickened. The endodermis is always well defined, and each cell has the usual casparian strip; at maturity the cell walls vary in thickness according to the species. Fern roots are commonly diarch to tetrarch with very few tracheids in the xylem. Xylem is exarch in maturation. No secondary growth occurs.

Roots generally arise endogenously from cells of the stem pericycle. They force their way through the cortex to the outside. Early in their development a conspicuous apical cell makes its appearance and functions throughout active apical growth of

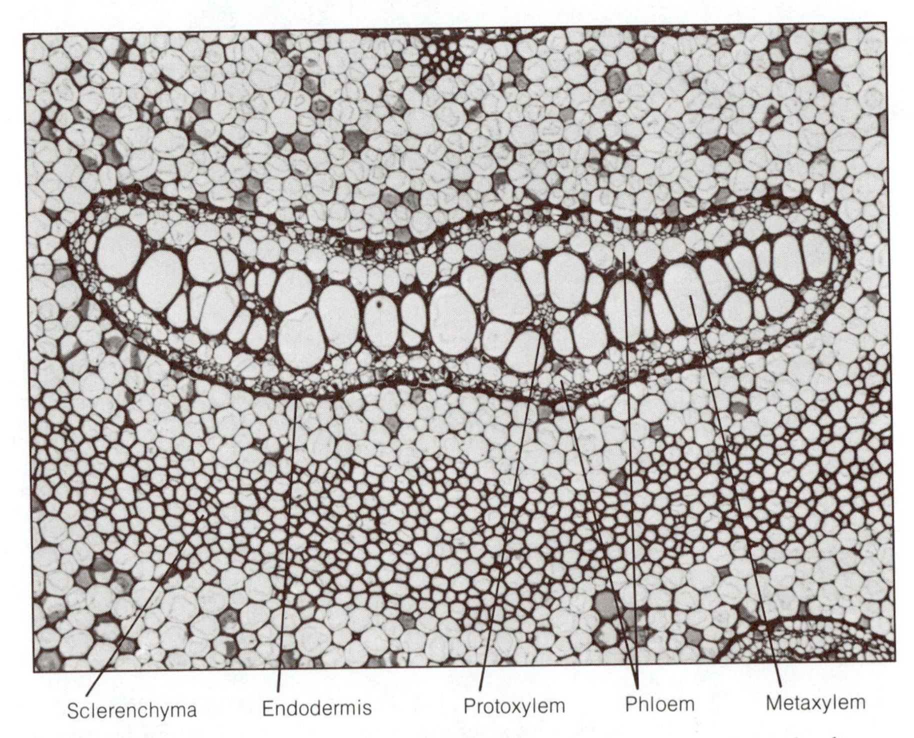

FIGURE 13-24 Transection of rhizome of *Pteridium* showing structural details of one meristele (vascular bundle).

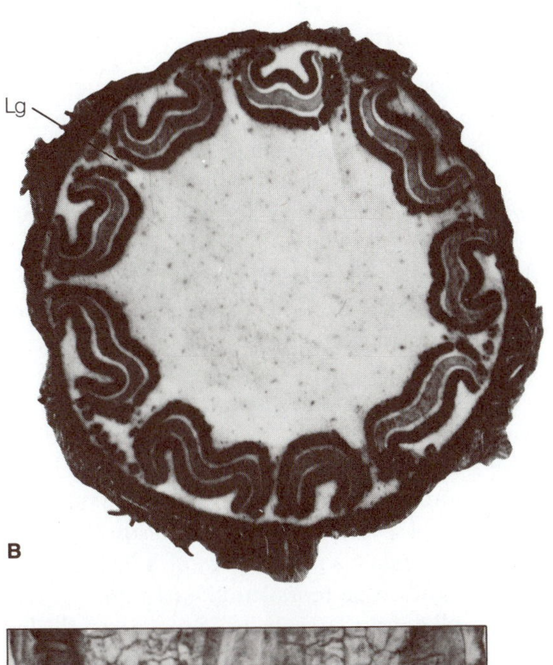

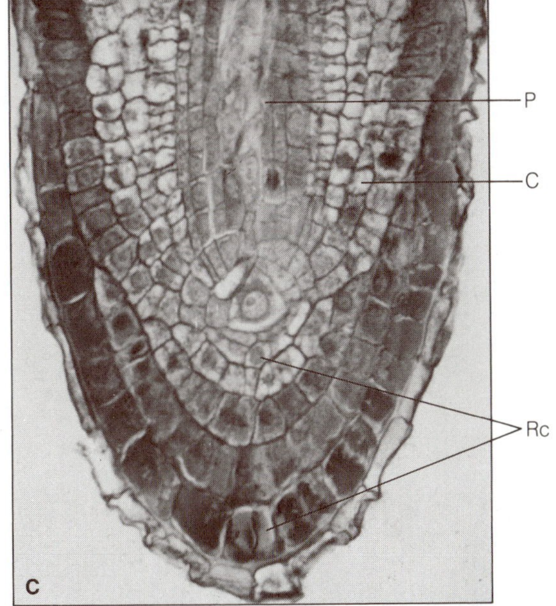

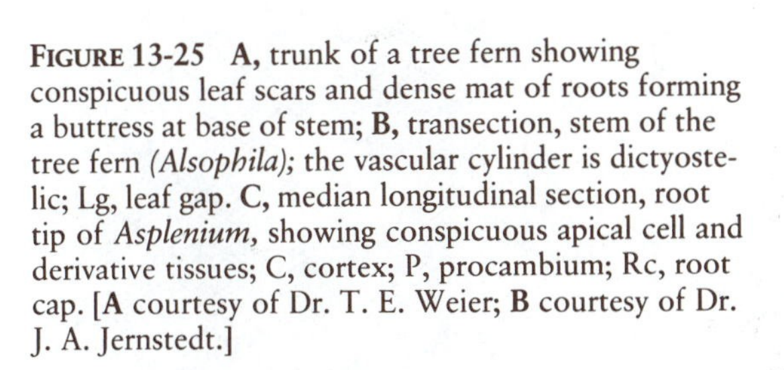

FIGURE 13-25 **A,** trunk of a tree fern showing conspicuous leaf scars and dense mat of roots forming a buttress at base of stem; **B,** transection, stem of the tree fern *(Alsophila);* the vascular cylinder is dictyostelic; Lg, leaf gap. **C,** median longitudinal section, root tip of *Asplenium,* showing conspicuous apical cell and derivative tissues; C, cortex; P, procambium; Rc, root cap. [**A** courtesy of Dr. T. E. Weier; **B** courtesy of Dr. J. A. Jernstedt.]

the root (Fig. 13-25, C). In most investigated ferns the apical cell is pyramidal in shape, cutting off cells to the main body of the root on three sides and contributing segments to the root cap on the fourth side. An account of recent research on the mitotic activity of the apical cell and root histogenesis is presented later in this chapter (p. 316). The three primary meristems—protoderm, ground meristem, and procambium—have their origin from derivative cells of the apical cell, and they, in turn, give rise to the epidermis, cortex, and vascular cylinder, respectively. Lateral roots originate from cells of the endodermis, in contrast to higher plants in which the pericycle is usually the initiating layer (Esau, 1965; Guttenberg, 1966; Ogura, 1972).

The Gametophyte

The origin of the gametophyte is normally from a haploid cell, the spore. Members of the Filicales are homosporous and have exosporic gametophytes, with the exception of *Platyzoma microphyllum* which exhibits "incipient heterospory" (A. F. Tryon, 1964; Tryon and Vida, 1967). Because there is a renewed interest in ferns, our knowledge of the range in variability of the fern gametophyte has been greatly extended (Atkinson and Stokey, 1964; Nayar and Kaur, 1971).

Fern spores are either *tetrahedral* or *bilateral* in symmetry. If tetrahedral, the spore has a triradiate ridge on the proximal face (side in contact with other spores of a tetrad) and the spore is said to be trilete (Fig. 13-26, A). If bilateral, a distinctive portion of the exine is in the form of a line and the spore is said to be monolete (Fig. 13-26, B). The spore wall consists essentially of two parts; the inelastic outer layer, the *exine* (exospore), and the inner layer, the *intine* (endospore). In the majority of species there may be an additional outer envelope termed the *perispore* or perine. Sporopollenin occurs in the exine and perispore.

SPORE GERMINATION. Spores of most ferns will germinate if provided with moisture and maintained at a pH of 4.5 to 7 and in the temperature range 15 to 30° C. Light in the visible spectrum is probably the most critical factor required for germination. In general, fern spores do not germinate in the dark, but there are exceptions. It has long been known that long and short wavelengths of light have different effects on fern-spore germination. Red light promotes germination whereas blue light is relatively ineffective or is inhibitory. Far-red light alone also exerts an inhibitory effect (Mohr, 1963). There are exceptions to the generalizations just made. For example, *Pteridium aquilinum* is indifferent to wavelength and germinates under light of all wavelengths as well as in total darkness. The spores of *Anemia phyllitidis* will not normally ger-

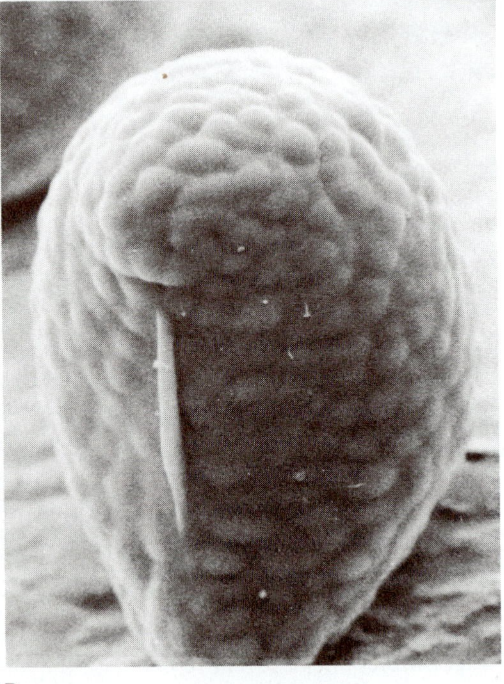

FIGURE 13-26 Two representative types of fern spores. **A,** *Osmunda regalis* (trilete type, ×1450); **B,** *Scyphularia pentaphylla* (monolete type, ×1500). [Courtesy of Dr. R. H. Falk.]

minate in total darkness, but they will if gibberellic acid is added to the culture medium (Schraudolf, 1962).

Fern spores contain varying amounts of storage proteins, lipids, and carbohydrates (generally as sucrose). The spores of certain species contain chlorophyll, others are nonchlorophyllous. The difficulty of sectioning fern spores because of the hard cell wall originally restricted histochemical studies, but the use of plastic embedding techniques has overcome this problem. Germination involves the enzymatic breakdown of stored materials that will be used in synthetic processes. Also, the occurrence of nonnuclear DNA has been reported in germinating nonchlorophyllous spores, probably associated with the biogenesis of chloroplasts (see Gantt and Arnott, 1965; Raghavan, 1980; and DeMaggio and Stetler, 1985, for discussions of the subject).

There are variations in the sequence of wall formation as well as in polarity during spore germination (Fig. 13-27). In one type (*Osmunda* type) the first division of the spore results in the formation of a small cell toward the proximal face and a larger

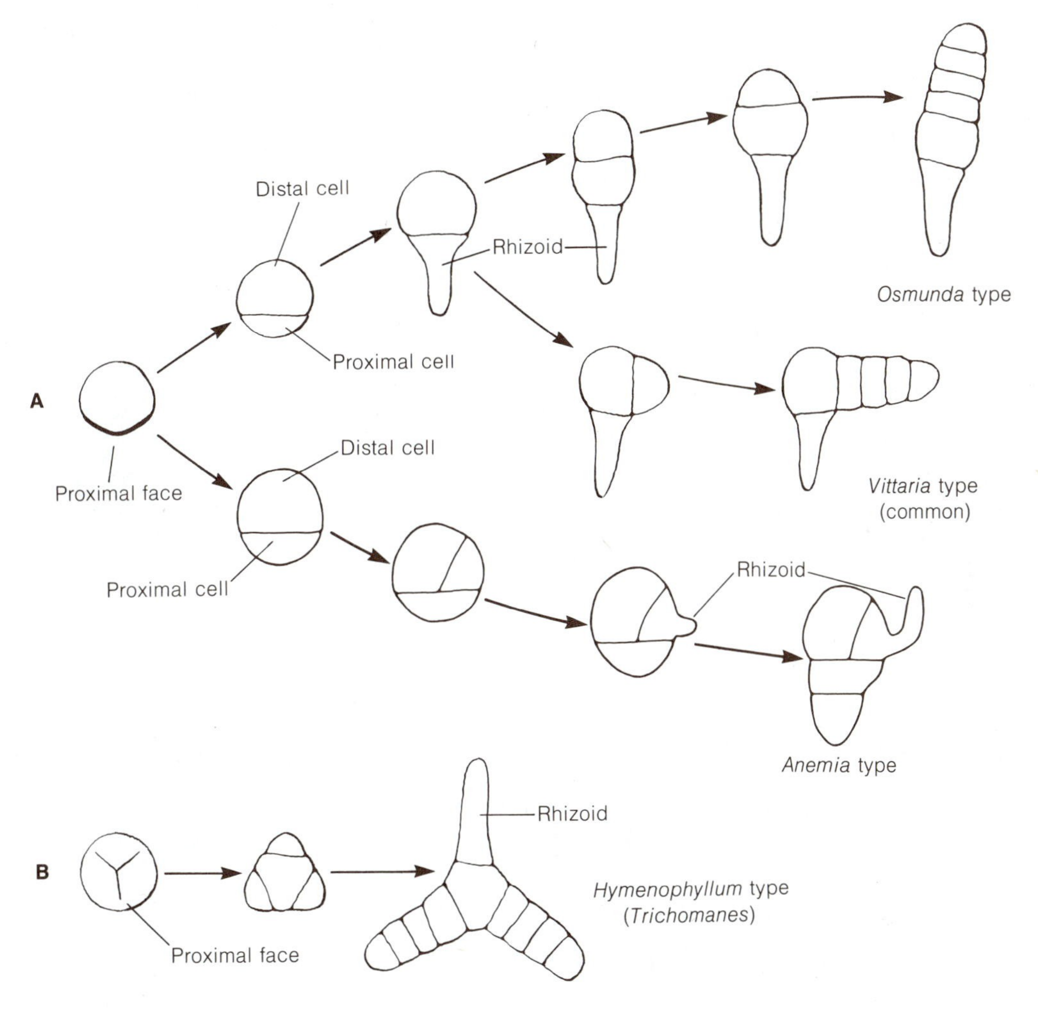

FIGURE 13-27 Schematic representation of certain types of spore germination in homosporous leptosporangiate ferns. The proximal face (where spore was in contact with others in a tetrad) is shown schematically. Consult text for details. [Redrawn from Nayar and Kaur, *Bot. Rev.* 37:301, 1971; *Anemia* type redrawn from Raghavan and Huckaby, *Amer. J. Bot.* 67:653, 1980.]

one toward the distal face. The smaller cell will become a rhizoid and the larger one is the initial of the green prothallial filament. Additional transverse divisions in the initial and its derivatives produces a filament of several cells. A variation of this type and one that is common in the Filicales is shown in Fig. 13-27 (*Vittaria* type). In this type the prothallial initial divides at right angles to the first wall, and one of the two cells proceeds to form a filament. In a third type (*Anemia* type), the spore divides into a small proximal cell, which is the initial of the prothallial filament. The larger distal cell, by an unequal division, gives rise to a rhizoid initial (Fig. 13-27). Some ferns are tripolar in germination (Fig. 13-27). Additional examples can be found in the review by Nayar and Kaur (1971) and in the article by Raghavan and Huckaby (1980).

DEVELOPMENT OF THE GAMETOPHYTE. Subsequent to spore germination each species tends to have a particular, established pattern of development; also, many larger groups of ferns share a

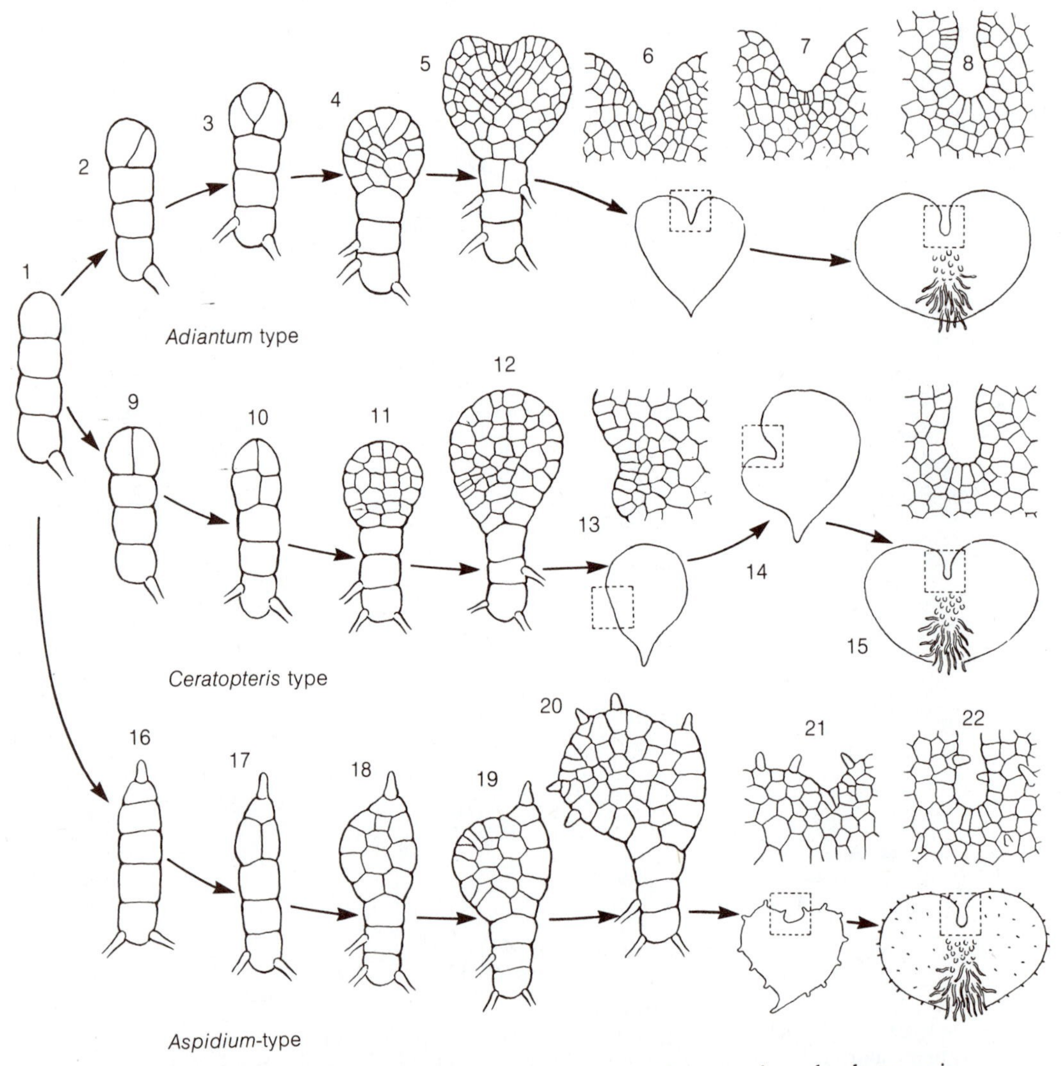

FIGURE 13-28 Schematic representation of three types of gametophyte development in homosporous, leptosporangiate ferns. The arrows indicate successive stages in each type of development as follows: numbers 1–8, *Adiantum* type; 1, 9–15, *Ceratopteris* type; 1, 16–22, *Aspidium* type. [Redrawn from Nayar and Kaur, *Bot. Rev.* 37:304, 1971.]

common pattern (Nayar and Kaur, 1971). In general, seven different patterns are recognized. We will describe three patterns of development that lead to the common cordate type of gametophyte.

In the *Adiantum* type of gametophytic development, divisions in the terminal cell of the prothallial filament result in the early establishment of a wedge-shaped apical cell. Cells are produced laterally from the apical cell, and these in turn remain meristematic. The apical portion becomes notched, and the apical cell then divides transversely and derivative cells continue to divide. Soon a thick midrib is formed by the cell divisions in the plane of the flat thallus. Ultimately the gametophyte becomes cordate with a midrib and thin lateral wings (Fig. 13-28, 1-8).

In the *Ceratopteris* type of gametophytic development a broad plate of cells is formed, but without a definite apical cell. Meristematic activity is restricted to a group of cells along the margin of the gametophyte. A notch is produced laterally, and one wing may become larger than the other, although the gametophyte may eventually become symmetrically cordate (Fig. 13-28, 9-15).

The early formation of a trichome (hair) from the terminal cell of the prothallial filament restricts apical growth in *Aspidium*-type development (Fig. 13-28, 16-22). A broad plate of cells is formed behind the anterior end of the developing gametophyte. An apical cell becomes established laterally but eventually the thallus becomes cordate (Fig. 13-29, D). Additional trichomes are formed from marginal and superficial cells.

The types of gametophyte development just described are considered representative of the Filicales, although development can vary, even within species of a section of a genus (e.g., *Pellaea*, Pray, 1968).

FORM OF MATURE GAMETOPHYTES. As mentioned earlier the cordate type of gametophyte is the most common in the Filicales. The plant generally lies flat on the substrate, but in some species the wings are raised or have ruffled margins (Figs. 13-29; 13-30).

Examples of other types of gametophytes, which are rare, are (1) strap or branched, ribbon shaped (Fig. 13-30, B, E), (2) filamentous (Fig. 13-30, D), (3) tuberous and subterranean, and (4) more-or-less filamentous, with the diameters of some axes comprising several cells (Fig. 13-30, F).

GAMETANGIA. A mature gametophyte of the Filicales typically has both antheridia and archegonia; it is bisexual. In the dorsiventral, cordate types, the gametangia generally are restricted to the ventral side. Archegonia are present near the notch on the so-called archegonial pad that is more than one cell in thickness. Antheridia are generally formed toward the posterior end of the gametophyte, situated among rhizoids (Fig. 13-31), but they may occur on the wings or toward the notch. From all available evidence, it is clear that antheridial initiation is controlled by a mechanism involving the production of naturally occurring hormones. The physiology of sexuality in ferns is discussed on p. 288.

The mature antheridium of a specialized (advanced) homosporous, leptosporangiate fern is composed of three jacket cells — a basal cell, a ring cell, and a cap cell — enclosing the spermatids (Fig. 13-31, I). (The ontogeny of this type of antheridium is described in Chapter 5, p. 63.) Not all filicalean ferns have this type of antheridial organization. There are, for example, variations in the number of jacket cells and spermatids; these variations are described in later sections.

The venter of the archegonium is normally embedded in gametophytic tissue. It has a protruding neck comprising four rows of from five to seven tiers of cells that enclose a binucleate neck-canal cell, a ventral canal cell, and the egg cell (Fig. 13-31, B – F). (For a detailed account of archegonial development, see Chapter 5, p. 62.)

FIGURE 13-29 Gametophytes of *Pteridium aquilinum* grown on nutrient agar (×6).

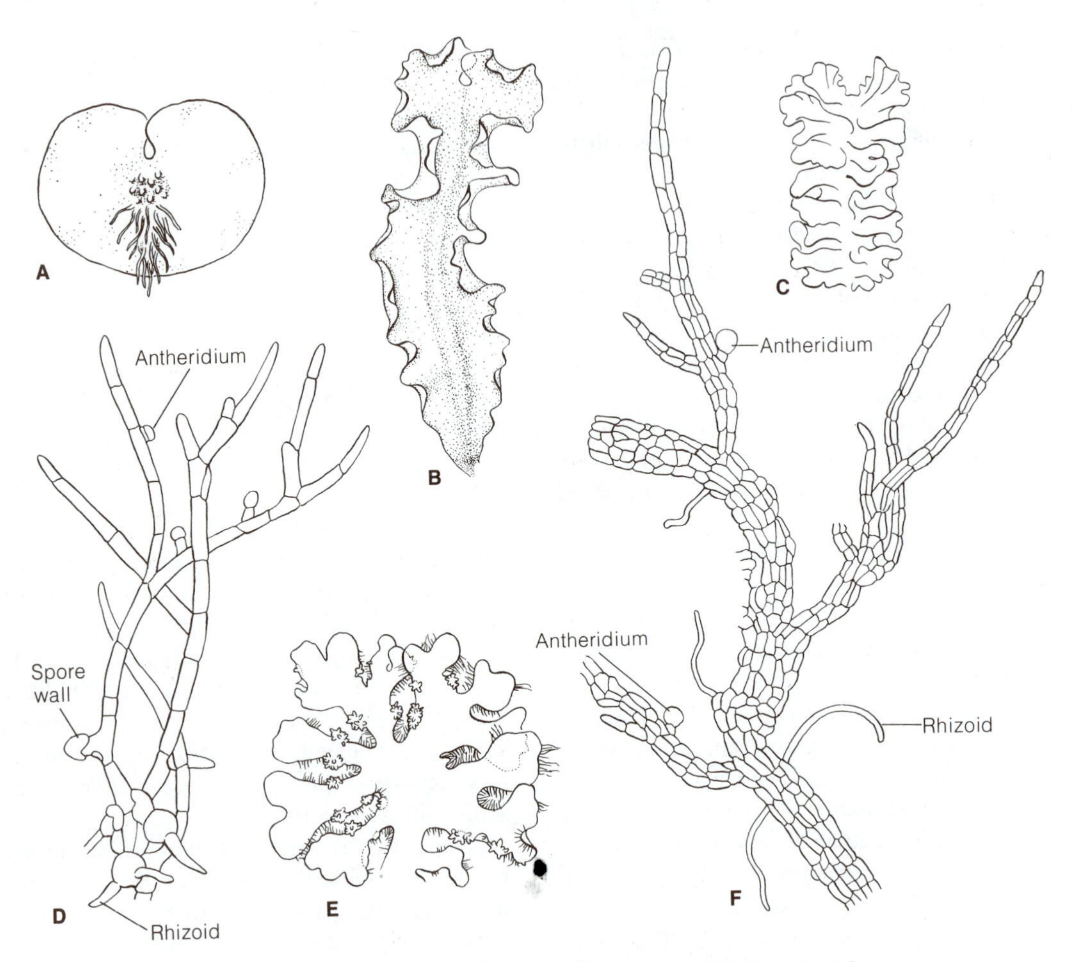

FIGURE 13-30 Representative types of gametophytes in the Filicales. **A,** *Adiantum caudatum;* **B,** *Blechnum brasiliense;* **C,** *Dipteris conjugata;* **D,** *Schizaea pusilla;* **E,** *Mecodium flabellatum;* **F,** *Lophidium (Schizaea) dichotoma.* See text for explanation of types. [Redrawn from Nayar and Kaur, *Bot. Rev.* 37:295–396, 1971.]

Dynamics of Fertilization and Embryogeny

When a gametophyte is flooded with water, an antheridium shows signs of an increased internal pressure followed by the loosening of the cap cell and the escape of the spermatids. In a few seconds, or at the most a few minutes, the spermatids are metamorphosed into multiflagellate, swimming sperm. A sperm resembles a short corkscrew, with the flagella attached to a lamellar band in the anterior gyres of the helix. The flagella beat in a coordinated wavelike fashion, thrusting the sperm forward and also causing it to rotate (Fig. 13-32).

When a gametophyte is immersed in water, the impending opening of an archegonium is heralded by the enlargement of the distal end and the separation of apical neck cells. This allows a tiny stream of mucilaginous material or slime to be released, which is followed by the forceful release of definable cytoplasmic bodies—presumably they are entities of the axial row except for the basal egg cell (Ward, 1954). Following this phase, the upper tiers of neck cells split apart. Swimming sperm are at first not attracted to archegonia, but later, when the slime is diluted, they change direction chemotactically and swim toward an archegonium. In *Phlebodium aureum,* from three to five sperm may swim into an open archegonium and occupy the ventral cavity at the same time. However, only one sperm fertilizes the egg (Fig. 13-31, H).

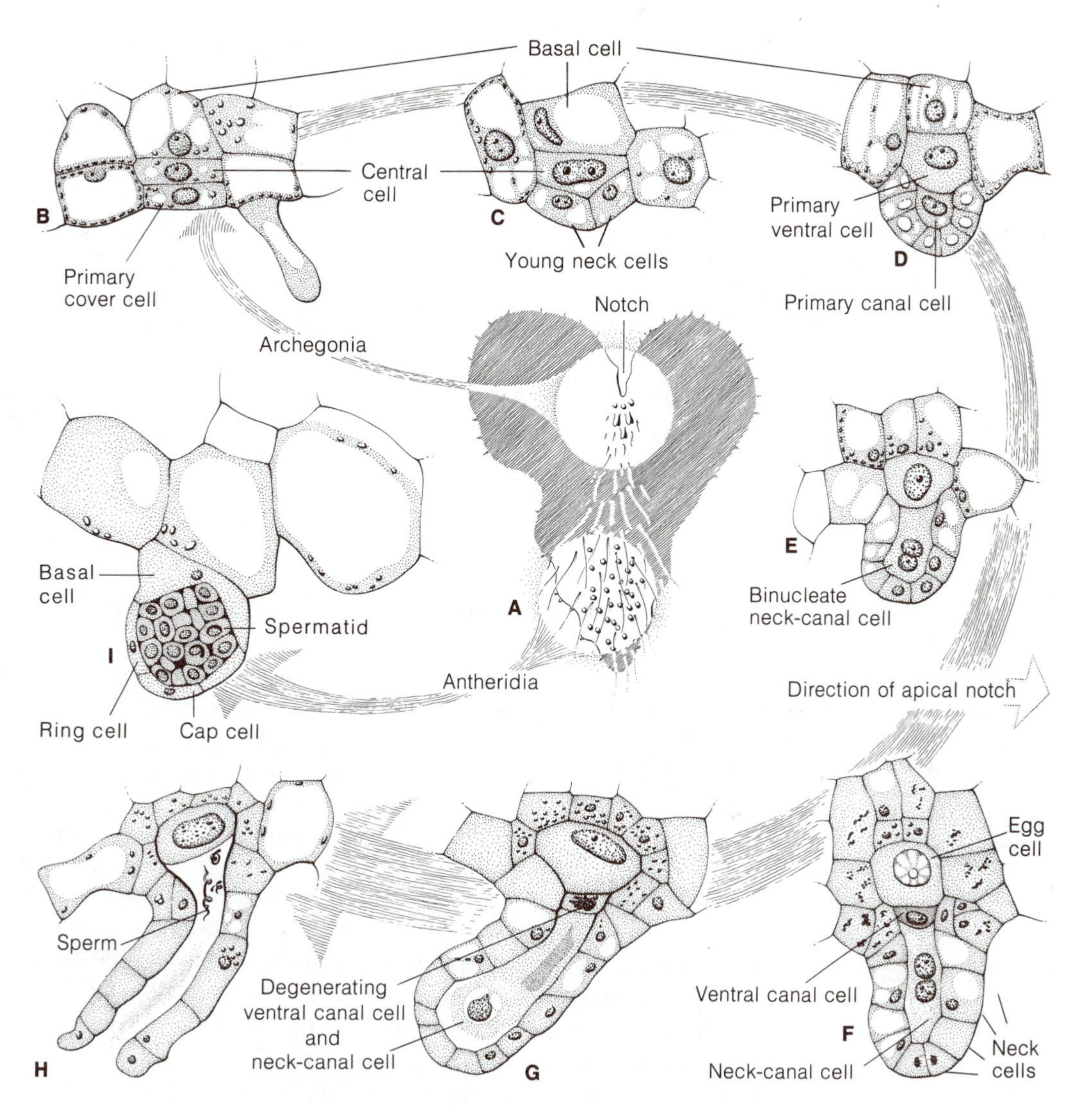

FIGURE 13-31 Gametophyte and gametangia of a fern in the Filicales. **A,** form of a sexually mature, heart-shaped gametophyte; **B–H,** stages in the development of an archegonium; **I,** a nearly mature antheridium. [From slides prepared by Dr. F. V. Ranzoni.]

The first division of the zygote is reported to take place anywhere from one hour to ten days after fertilization. In *Phlebodium aureum* it is five days (Ward, 1954). Even before the zygote divides, cells of the surrounding gametophyte divide and form a partially ensheathing calyptra. The first division wall of the zygote is generally parallel to the long axis of the archegonium (Fig. 13-33, A). This initial division separates an anterior cell or hemisphere that is directed toward the notch of the gametophyte and a posterior cell or hemisphere that is directed away from the notch. The former cell is the apical pole or epibasal cell, and the latter is the basal pole or hypobasal cell. The zygote in the Filicales is said to be prone in orientation. Each of the two cells divides, and the new cell wall in each is perpendicular to the original wall, resulting in the development of a four-celled embryo or "quadrant stage" of embryogeny (Fig. 13-33, B). Each of the four cells then undergoes (usually by synchronized divisions) more divisions (Fig. 13-33, C). According to the classical descriptions of fern embryogeny, the pri-

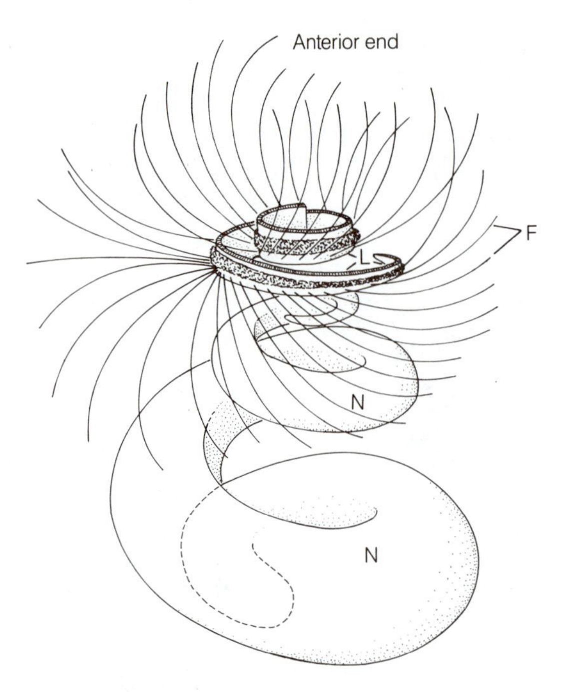

FIGURE 13-32 Diagrammatic reconstruction of the relationship between the nucleus and lamellar band in a mature sperm of *Pteridium*. F, flagella; L, lamellar band; N, nucleus. Other cellular details omitted. [Redrawn from J. G. Duckett in *The Biology of the Male Gamete* edited by J. G. Duckett and P. A. Racey. Academic, London. 1974.]

mary organs may be traced back to specific segments of the quadrant stage. The outer anterior quadrant or cell reportedly gives rise to the first leaf, the inner anterior quadrant to the future shoot apex. The primary root originates from the outer posterior quadrant, the foot from the inner posterior quadrant.

Some workers have questioned the preciseness of fern embryogeny (particularly of later stages) as presented in classical studies. In *Gymnogramme sulphurea* (Vladesco, 1935) the foot is reported to be derived from certain derivatives of the original inner anterior quadrant as well as from the inner posterior quadrant (Fig. 13-33, D). The setting aside of a shoot apical cell is delayed, and frequently it is a cell located very near the equatorial plane of the embryo and not a centrally located derivative of the inner anterior quadrant. According to another report (Ward, 1954) the inner two quadrants in *Phlebodium aureum* give rise to the foot, and the first leaf and root, appearing in that order, have their origin as described earlier. The origin of the future shoot apical meristem cannot be assigned to any definite quadrant, but the shoot apex arises from derivative cells of the inner and outer anterior quadrants midway between the organized foot and first leaf.

After whatever manner the organs are delimited, the first root and leaf begin to grow rapidly and pierce the calyptra (Fig. 13-34, B). The first leaf eventually grows forward and upward through the notch. After the first leaf has unfolded, more leaves are formed by the shoot apical meristem, and roots are produced on the developing stem axis (Fig. 13-35). Sooner or later, the gametophyte degenerates and dies. Leaves of young sporophytes often do not resemble those of adult plants.

Special Problems in Fern Morphology

PHYLETIC SLIDE OF SPORANGIA AND SORI. The most primitive position of sporangia in ancient ferns is thought by many morphologists to have been terminal if the megaphyll represents essentially a modified branch system. A terminal position of sporangia on dichotomizing axes is well illustrated by the extinct Rhyniophyta (Chapter 7). By means of anisotomous branching (overtopping) a pseudomonopodial branch system resulted in which sporangia were terminal on lateral determinant branch systems, as in the Trimerophytophyta (Chapter 7). By the processes of planation and development of laminar tissue (webbing), sporangia came to occupy a marginal position on a leaf segment. All of these processes are embodied in the telome theory (Chapter 3, p. 31). In some coenopterid ferns from the Upper Devonian to Carboniferous, sporangia were terminal on ultimate segments of fronds or, in others, were abaxial on leaf segments near the margin (Chapter 11). In the course of evolution it is not too difficult to visualize how the abaxial position could be achieved. During ontogeny sporangia could originate near the margin, and by continued marginal growth of the leaf segment the individual sporangia and/or sori would be left behind, now occupying an abaxial position—hence a "phyletic slide" to the lower leaf surface.

A "true" indusium is a prominent and common feature in the majority of Filicales. Cup-shaped or tubular indusia may occur in ferns in which the

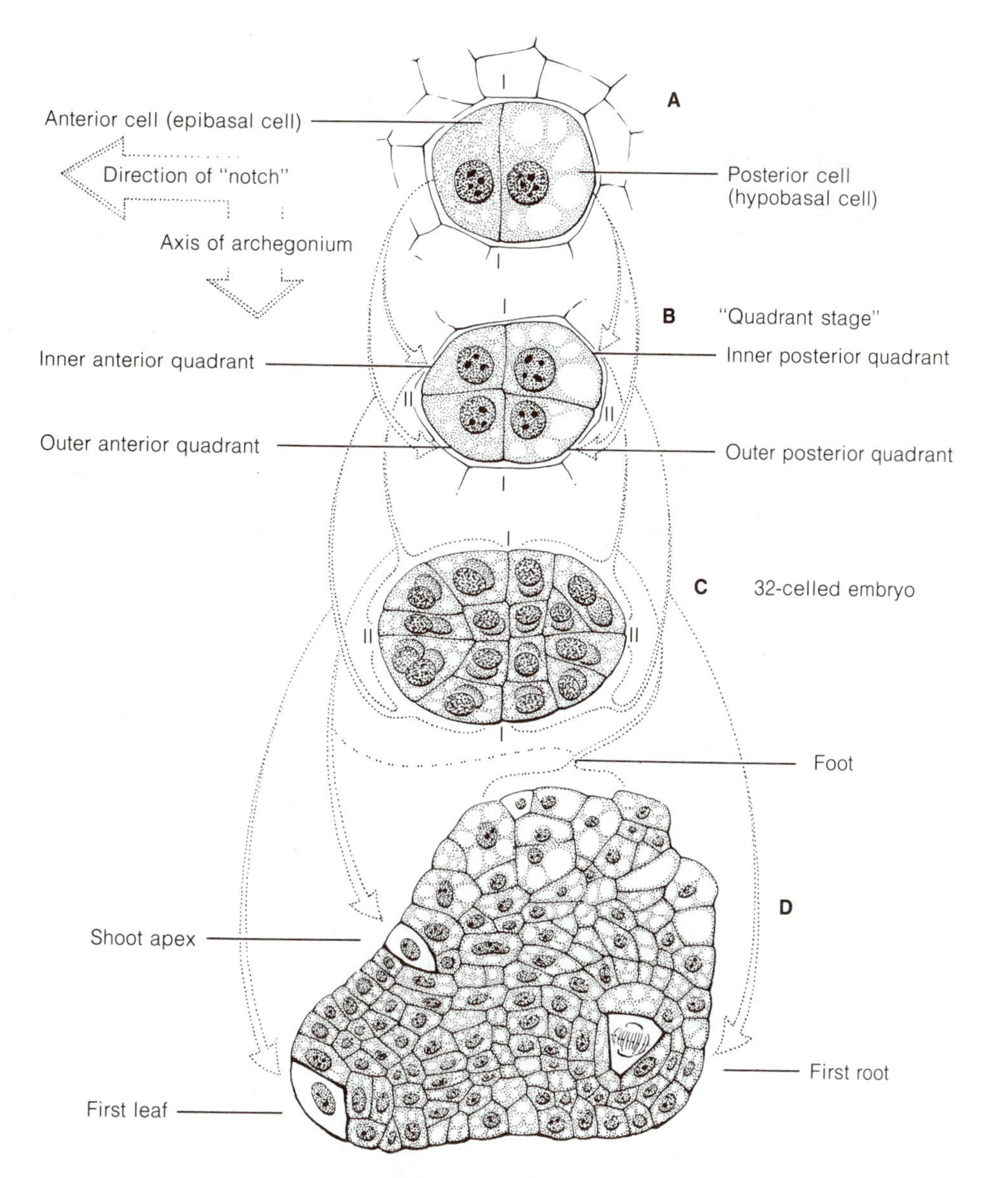

FIGURE 13-33 Developmental stages in the growth of a leptosporangiate fern embryo *(Gymnogramme sulphurea)*. [Redrawn from Vladesco, *Rev. Gen. Bot.* 47:513, 1935.]

soral receptacle is truly marginal (e.g., *Hymenophyllum, Trichomanes*), submarginal *(Dicksonia, Davallia)*. Did the indusium arise *de novo* in various ways in several phyletic lines in the Filicales, or was there some type of basic underlying developmental plan that could account for the various indusial types? According to one theory the primitive soral position was marginal, an assumption that has some support from extinct ferns. In this type the receptacle was a continuation of the leaf margin, and the indusium arose as outgrowths or flaps from the adaxial and abaxial leaf surfaces. In the course of evolution the adaxial flap became modified as the new leaf margin and became vascularized. The abaxial flap remained as the indusium. Subsequently the sorus became positioned farther and farther away from the leaf margin and the indusium could have assumed the various shapes encoun-

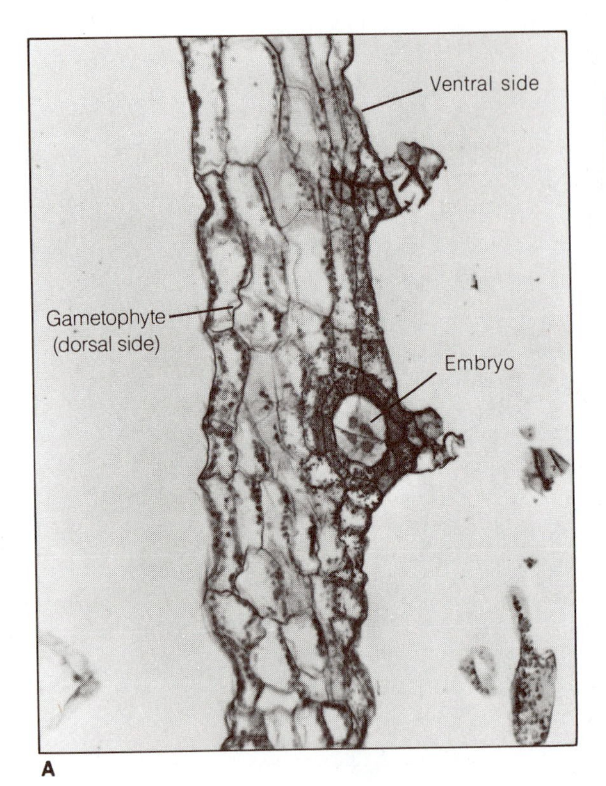

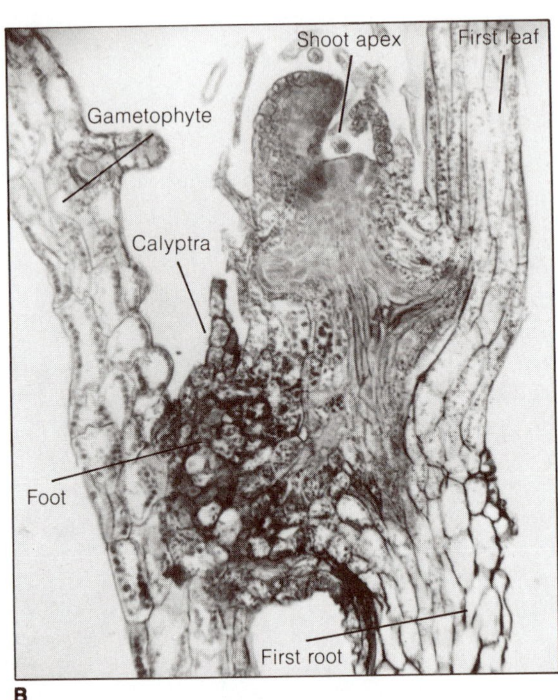

FIGURE 13-34 Vertical sections of gametophytes of *Adiantum* sp. showing attached eight-celled embryo (four cells visible) in **A,** and young developing sporophyte in **B.**

tered in various lines of extant ferns (Fig. 13-36). The "phyletic slide" of the sorus to the abaxial side is based upon countless modified ontogenies during evolution leading to soral positions and types that presumably had adaptive value. Whether such transitions took place is open to question, probably never to be proved.

EVOLUTIONARY AND ONTOGENETIC IMPLICATIONS OF STEM ANATOMY. On the basis of comparative morphology of both living and extinct plants, the protostele, with little doubt, is the primitive type of vascular cylinder. This is true not only in ferns but also in other groups of vascular plants. Any departure from a simple protostele (haplostele) is commonly regarded as a derived condition.

The value of studies based on the comparative anatomy of mature stems should not be underestimated, but most ferns, during their early ontogeny, have a vascular pattern quite different from that of the adult plant. For example, the base of a fern stem (older portion) may be protostelic, whereas at a higher level of the same stem (younger portion) the vascular cylinder may be solenostelic or dictyostelic. Of course this change is ontogenetic, but it has been interpreted by some morphologists as representing an evolutionary recapitulation of stelar types. Rightfully, these same morphologists realize the importance of comparing the steles of plants of comparable age. Therefore the vascular cylinders of well-established adult plants can be compared with some degree of assurance. Similarity of vascular cylinders, however, cannot be used in itself as an indication of close relationship between ferns. The attainment of similar levels of structural specialization may be reached in several evolutionary lines. We are dealing here with convergent evolution or homoplastic development (Bower, 1923).

The shoots of many plants, particularly herbaceous species, become obconical during growth, and, as a result, the stem at an upper level may be several times larger in diameter than the base. This

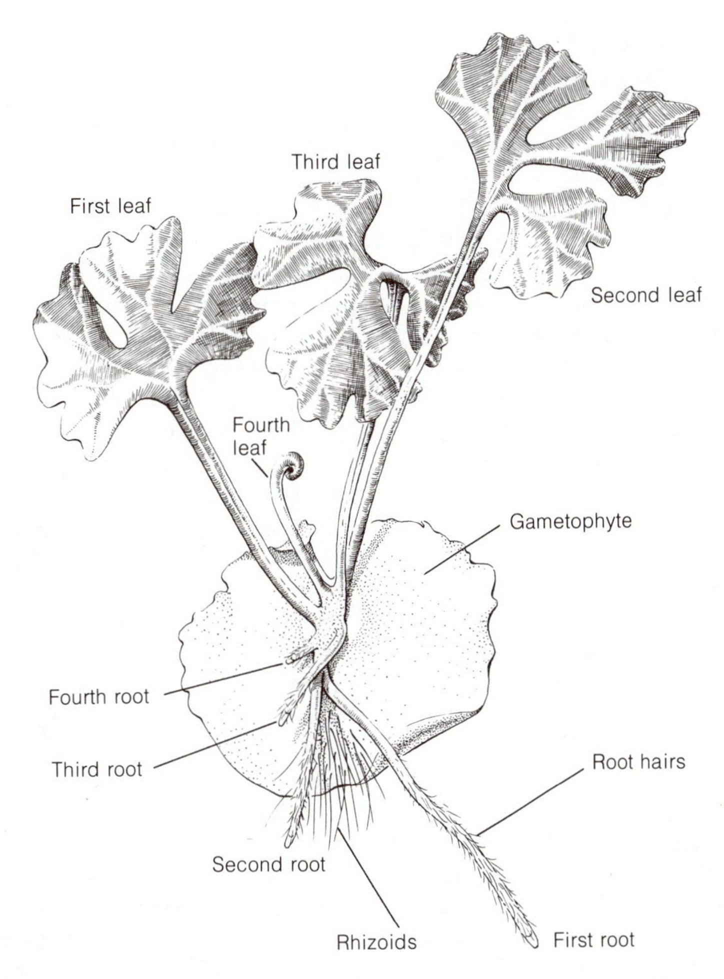

FIGURE 13-35 Ventral (lower) side of gametophyte of fern with attached young sporophyte. Note circinate vernation of youngest (fourth) leaf.

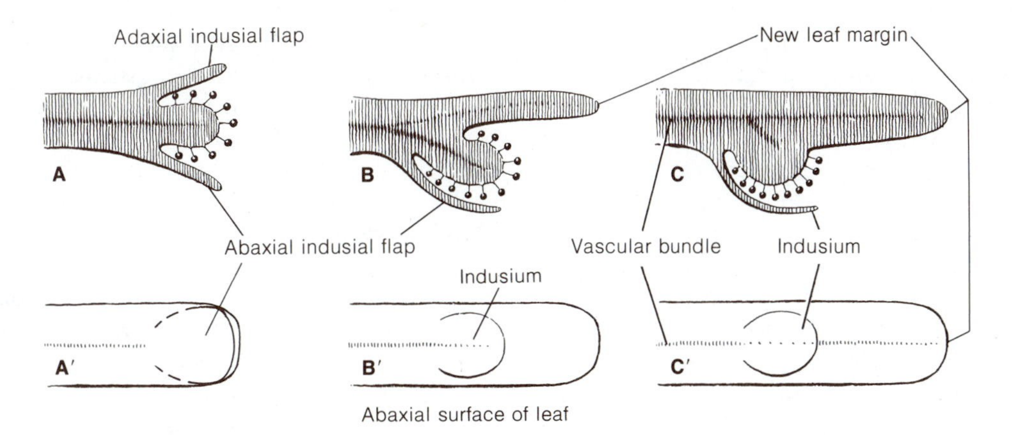

FIGURE 13-36 Possible evolutionary stages in the movement of the marginal sorus to the abaxial side of the fern leaf (the "phyletic slide"). **A–C**, transverse sections, hypothetical leaf lamina; **A′–C′**, abaxial surface views. See text for details.

phenomenon is particularly well displayed in ferns. Associated with this increase in size is generally a change in the form of the vascular cylinder. Using Galileo's principle of similitude, Bower calculated that a mere increase in size or volume of a smooth cylinder of dead tracheids is not compensated for by a comparable increase in surface of the cylinder. The volume increases as the cube of the linear dimensions, and the surface increases as the square of the linear dimension, if the same shape is maintained. Bower believed that the proper relationship between dead conducting elements and living parenchyma must be maintained. This can be accomplished, then, in any one of several ways or by combinations of these methods: (1) formation of a central pith, (2) formation of flanges or irregular lobes on an otherwise smooth protostele, (3) formation of a highly intricate reticular vascular cylinder (e.g., dictyosteles), and (4) presence of living cells among dead elements. Whatever physiological advantages are achieved by these methods they appear to have been favorably acted upon through natural selection.

There has been a renewed interest by morphologists in the United States and Great Britain in the problem of size and form. In England, Wardlaw was the leader of a group that approached morphological problems experimentally. Bower was impressed with the phylogenetic implications of stelar morphology, but he was aware also of the plasticity and variability of the vascular cylinder. Wardlaw (1949) has shown in *Dryopteris* that if the shoot apex is isolated from existing vascular tissue by vertical incisions, the vascular cylinder of the new "plug" of tissue will be solenostelic instead of dictyostelic for a short distance. Eventually the dictyostelic condition is achieved again. If the incisions are made very close to the apical cell, a protostele is produced. Wardlaw interprets these results as an indication of the importance of nonhereditary factors in determining shoot organization. According to Wardlaw, the availability of nutrients (an extrinsic factor) perhaps has more of an influence on growth than do hereditary factors. Starvation experiments also support this general idea (Bower 1923).

EXPERIMENTAL MORPHOLOGY: SPOROPHYTE. What are the methods or techniques open to the experimental morphologist? Changes in plants can be brought about by culturing isolated tissues and organs, by applying chemicals directly to the intact surfaces, and through surgical manipulations. Experimental morphologists, then, attempt to evoke changes in plant form and development under carefully controlled conditions. When changes occur, they seek to relate these changes to the conditions of the experiment, thereby hoping to arrive at a more complete understanding of the basic organization and development of the organism.

THE SHOOT APICAL MERISTEM. In *Dryopteris* there are three different types of structures derived from the apical meristem: leaves, scales, and bud primordia. Leaves arise within the basiscopic margin of the apical cone; scales originate on the margin; buds have their origin on the broad subapical region (Fig. 13-37). The shoot apex usually exerts physiological dominance over buds or bud primordia. That this is controlled by a substance (termed a growth hormone) diffusing basipetally has been demonstrated experimentally. For example, if the tip of a main shoot is removed, dormant buds or bud primordia will begin to grow. If, as in the ferns *Matteuccia* and *Onoclea,* the cut surface of the stem is smeared with indoleacetic acid in lanolin, the development of lateral buds does not take place (Wardlaw, 1946). The applied growth-regulating substance performs the same function as the natural growth hormone that is produced at the tip of the intact rhizome.

The shoot apical meristem of vascular plants has long been of interest to morphologists. We might ask if the apical meristem is a self-determining region or if its characteristics and activities are controlled by older, more mature tissues below it. Wardlaw (1949) demonstrated that if the conical apical meristem of *Dryopteris* is isolated from adjacent tissues by four deep longitudinal incisions, it is still capable of growing and giving rise to a leaf-bearing axis. The initial incisions isolated the apical meristem on a plug of parenchymatous tissue. The newly formed vascular tissue of the "isolated" shoot never became continuous with that below. It has been concluded that the shoot apex is a self-determining region, and, except for nutrients, is not dependent upon older tissues. Although a plug of tissue consisting of the apical meristem and subjacent pith can be isolated from mature tissues, it may be difficult to isolate surgically an apex on a plug

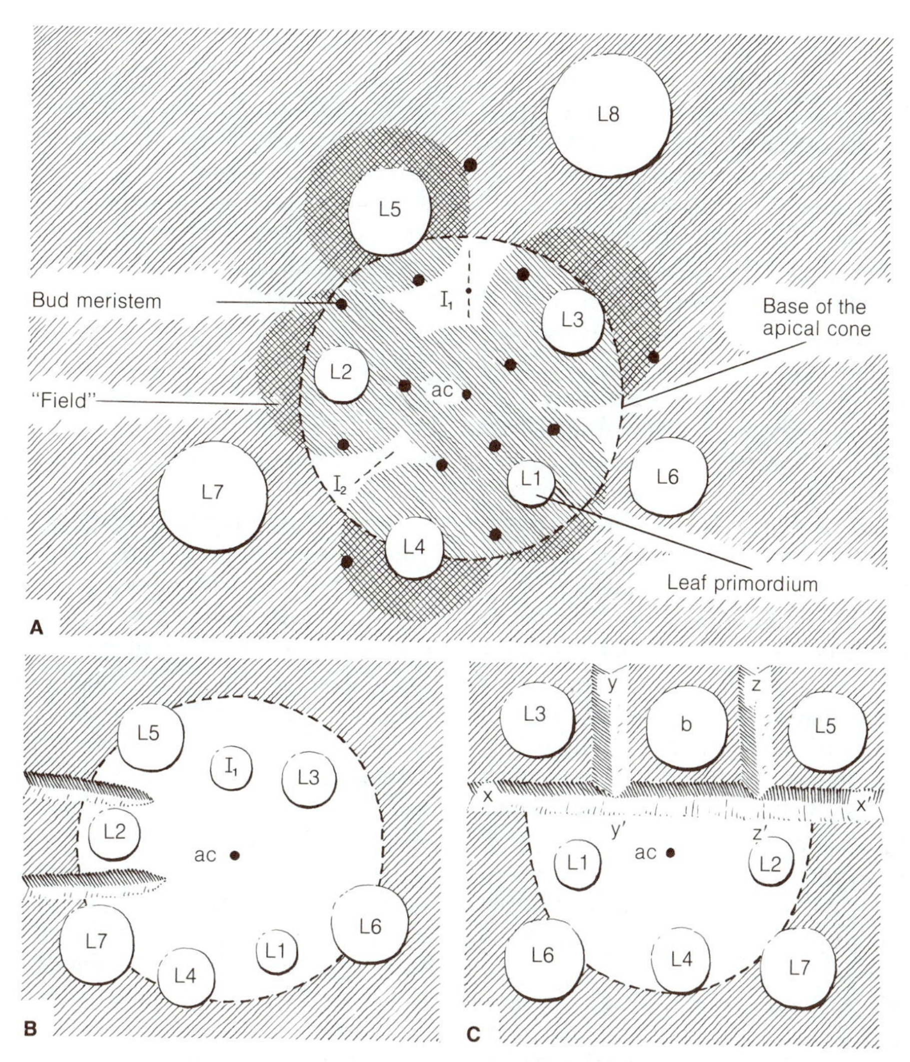

FIGURE 13-37 **A,** surface view of apical cone, shoot of *Dryopteris* (consult text for explanation); **B,** isolation of young leaf primordium (L2) by longitudinal radial incisions; C, isolation of I_1, by tangential incision (X–X′) and radial incisions (Y–Y′, Z–Z′), which resulted in formation of a bud (b). ac, apical cell; b, bud; I_1, position of first presumptive leaf; I_2, position of a second presumptive leaf. [Redrawn from *Phylogeny and Morphogenesis* by C. W. Wardlaw. Macmillan, London. 1952.]

that did not contain incipient vascular tissue (Larson, 1975).

Additional evidence for the autonomous nature of the shoot apical meristem was provided by in vitro culture techniques. Wetmore (1950, 1954) was able to excise the shoot apical meristems of *Adiantum* and grow normal plants from them on a mineral agar medium. The addition of auxin, yeast extract, or autoclaved coconut milk speeded up the rate of growth without changing the developmental pattern.

The question was raised: Does the shoot apex (more specifically the apical cell) determine the position of leaves? In a test experiment great care was taken to puncture only the apical cell of *Dryopteris,* although there was some necrosis of adjacent cells; existing primordia and the loci of yet uninitiated leaves were carefully avoided in order not to injure the tissues in any way. After the operation, the existing leaf primordia continued to grow, and new ones were formed in a normal pattern until all the space on the apical cone was occupied before the termination of shoot growth. Apparently the apical cell does not determine leaf position, nor is it responsible for leaf initiation (Wardlaw, 1949).

In a more refined experiment on *Osmunda,* Kuehnert and Miksche (1964) were able to destroy the apical cell in a very precise manner by use of a microbeam; a new growth center developed near the apical cell and assumed control of development.

As a working hypothesis, Wardlaw assumed that the apical cell and all young leaf primordia constitute growth centers, each with its own physiological "field" (Fig. 13-37, A). This field restricts the initiation of new primordia to within a certain distance from its center. A new or presumptive leaf (I_1) arises in the first available space on the apical meristem—outside the fields of older adjacent leaves.

If a young primordium (e.g., L2) was isolated from adjacent primordia by deep longitudinal *radial* incisions, it soon became larger than older leaves (Fig. 13-37, B). If an incision was made in the position of a presumptive leaf (I_1), the next leaf in the phyllotactic pattern in that vicinity was shifted toward the incision. These experiments are reputed to show that leaf primordia have inhibitory effects on each other.

In other experiments, when I_1 (or a L1 primordium and sometimes L2 or even L3) was isolated by a deep incision on the *adaxial* side of the primordium, a bud and not a leaf usually formed in that position. The same effect was achieved by making both radial and tangential incisions (Fig. 13-37, C). When, however, shallow cuts were made, normal leaves were formed. In the instances in which buds developed from potential leaf primordia it would appear that such primordia are undetermined in their early development. According to Cutter (1954, 1956) the course of development of a leaf primordium of *Dryopteris* can be changed unless its fate has become determined by the appearance of a large lenticular apical cell at the tip of the leaf primordium (or by some other coordinated event). Also, all of these experiments emphasize the importance of "prevascular tissue" (a term used by Wardlaw) in early development. *Deep* incisions would sever the "prevascular" tissue but *shallow* incisions would not disrupt it. The controlling effects of the apical meristem are probably transmitted through prevascular tissue and a disruption of that tissue would affect the future development of primordia. (Additional information on these experiments and others can be found in the following selected publications: Wardlaw, 1952, 1965, 1968; Wardlaw and Cutter, 1954, 1956; Cutter, 1965; White, 1979; Halperin, 1978.)

Experimental studies were made of leaf primordia of *Osmunda cinnamomea.* In this fern a bud may form on the adaxial face of a young primordium. These buds are not cauline or axillary in nature. A bud is produced only when a primordium has been isolated from the rest of the shoot system. The buds form on primordia as young as P2 when they are cultured singly or isolated from the parent shoot apical meristem by a barrier (a thin mica chip). If P4 is cultured on the same plug of tissue as the shoot apical meristem, the primordium never develops a bud. When P4 is isolated from the shoot apex by a sterile mica chip, 75% of the primordia develop buds; 62.5% of singly cultured P4 primordia also had the capacity to produce buds. If the shoot apical meristem and P4 were excised and cultured together as a paired unit, buds were formed only 30% of the time. Conclusion: young leaf primordia, when removed from the influence of the shoot apical meristem, are capable of forming buds that eventually produce leaves and roots. What are the potentialities of P1? When P1 is in physical contact with the shoot apex it always develops as a leaf. When physiological continuity is broken between the shoot apex and P1, 37% develop as buds. Thus, P1 is described as an undetermined mass of meristematic tissue capable of developing either as a leaf or as a shoot (Haight and Kuehnert 1969). (Additional information on the subject can be found in the following publications: Steeves, 1961, 1962; Hicks and Steeves, 1969; White, 1971, 1979; Kuehnert, 1972).

EXPERIMENTAL MORPHOLOGY: CONTROL OF GAMETOPHYTIC GROWTH FORM. During the last fifteen years there has been a renewed interest in the fern gametophyte as an experimental object. The common type of fern gametophyte is autotrophic, relatively small, and can be grown in large numbers and manipulated with considerable ease.

Most ferns germinate to form a filament of a few cells. This is referred to as one-dimensional (1-D) growth. Sooner or later cells divide in another plane (longitudinal division) and a plate of cells (Fig. 13-38, A) is formed (two-dimensional, or 2-D, growth). What are the causal factors responsible for this change in growth pattern?

The transition from 1-D growth to 2-D growth in normal gametophyte development is in response to light—a photomorphogenic response. Light quality (wavelength) is the important factor, but light quantity also plays a role. Blue light is neces-

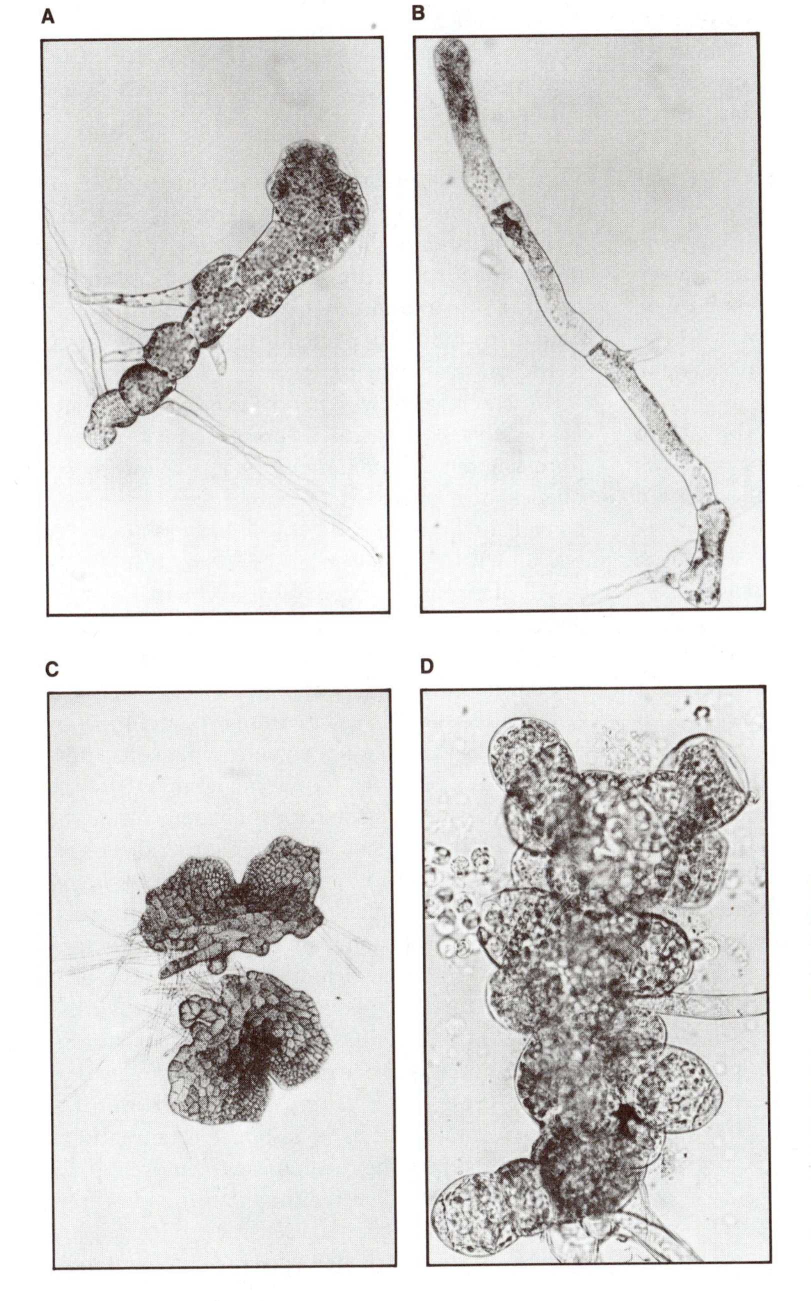

FIGURE 13-38 **A, B,** young gametophytes of *Pteridium aquilinum* (of same age) grown on nutrient agar and illuminated with white light (**A**) and red light (**B**). **C, D,** fourteen-day-old cultured gametophytes of *Onoclea sensibilis;* **C,** control plants grown on nutrient agar without antheridial factor; **D,** gametophyte grown from time of spore germination on medium containing antheridial factor *(Pteridium);* note large number of antheridia, and the release of sperm in one. (**C** and **D** are not at same magnification.) [Courtesy of Dr. E. G. Cutter.]

sary for 2-D growth. Under far-red and red light, a gametophyte generally remains filamentous (Fig. 13-38, B) although there are exceptions (see below). This all suggests that a photoreceptor-pigment is involved in the process. Ohlenroth and Mohr (1964) suggested a working model in which blue light is absorbed by a photoreceptor-pigment, resulting in the production of a reaction product that enters the nucleus and evokes a gene response. A messenger RNA with new information is produced, and then specific proteins (enzymes) concerned with 2-D growth are formed in the cytoplasm. However, the isolation and characterization of a blue light-absorbing pigment have yet to be achieved, nor has a specific messenger RNA been identified.

With the transition to 2-D growth, changes in protein and RNA content occur. These changes are, however, thought to be only correlated with the photomorphogenetic transition from 1-D to 2-D growth and are not the cause of the transition (Howland and Edwards, 1979).

Another explanation for the change from 1-D to 2-D growth, which entails changing rates of cell division and cell elongation, has been suggested (Sobota and Partanen, 1966, 1967; Sobota, 1970). This hypothesis assumes that no single mechanism alone is responsible for the change. Sobota and Partanen cite Errera's law which states that a cell so divides in a plane that the new cell wall has minimal surface area. Exceptions do exist to the rule, but in the fern gametophyte the plane of division would depend upon the length of a cell. If a cell of the germination filament (grown under red light) was quite long when it divided, the new cell wall would be formed transverse to the long axis of the filament. If there was a division before the cell had greatly elongated (as is the case under blue light), the new cell wall could be either transverse or longitudinal. This concept does have merit because when red light-grown filamentous gametophytes are transferred to blue light, the terminal cell swells within several hours and becomes bulbous. When the cell divides, the orientation of the cell plate may vary in the now isodiametric apical cell. If the cell divides longitudinally or obliquely, 2-D growth is initiated.

In general, external factors that favor and promote a high sustained rate of cell division, rather than cell elongation, influence the attainment and continuance of 2-D growth. Blue light favors 2-D growth, but, as Sobota and Partanen (1966) have shown, 2-D growth can be maintained in *Pteridium* under red light if the intensity is high enough. Also, the gametophytes of the Japanese climbing fern *(Lygodium japonicum)* are an exception to the general rule; gametophytes become 2-D in both red and blue light but retain a filamentous morphology in far-red light (Raghavan, 1973). The reader is referred to the following reviews for additional reading: Miller, 1968; Raghavan, 1974; Howland and Edwards, 1979.

SEXUALITY AND GENETICS. The gametophytes of filicalean ferns at maturity are commonly bisexual —having both archegonia and antheridia on the same individual. Spores germinated on agar or on another substratum often yield antheridial, archegonial, or bisexual plants. Antheridia are often found on small gametophytes.

Research has shown that the expression of maleness is correlated with the presence of a secreted hormone termed *antheridiogen.* The following sequence of events that probably occur in a mass culture of bracken fern *(Pteridium aquilinum)* is based on reviews by Voeller (1971) and Näf (1979).

All meristic gametophytes (those with an organized meristem) develop into heart-shaped gametophytes. Ameristic gametophytes (without organized meristem) are of irregular shape and bear only antheridia. All meristic gametophytes form archegonia as they mature, some forming antheridia initially. In a population of germinating spores the most rapidly growing meristic individuals attain the archegonial phase without a prior antheridial phase (about 20 percent under artificial culture). The remaining meristic individuals have a prior antheridial phase before archegonia are formed. Observations also have shown that all individuals form antheridia if exposed to antheridiogen at an early stage, and all meristic individuals have the capacity for antheridiogen synthesis. An hypothesis was formulated to explain these observations as follows: gametophytes form antheridia in response to antheridiogen secreted into the medium by the more rapidly growing individuals of a population, but they, themselves, have become insensitive to it. This hypothesis was strengthened with the demonstration that all isolated individuals attain the archegonial phase without a prior antheridial phase. In *Pteridium,* the formation of antheridia prior to the attainment of the archegonial phase appears to occur as an interaction among gametophytes. Some variations and exceptions to this general scheme have been reported (Schedlbauer, 1976; Näf, 1979).

Onoclea sensibilis (sensitive fern) has been used as an assay organism for antheridiogens because the gametophytes often do not become male on an agar medium except after prolonged culture. This prompted Rubin and Paolillo (1983) to wonder how *Onoclea* reproduces sexually in nature. From their experiments they found that gametophytes grown on soil quickly (in from nine to thirteen days) became male, but this stage was delayed until nearly three weeks after germination if the gametophytes were grown on agar. They believe that soil rather than agar should be used in experiments of sexuality in ferns. They also observed that gametophytes developing from spores that germinated on agar or soil more than four weeks after sowing did not become antheridial. They would be expected to become antheridial if antheridiogen were present. They concluded that "antheridiogens are not the basis for the fundamental control of maleness. Instead, they are modifiers of the basic control, which is still to be elucidated."

Antheridiogen produced by one species of ferns is effective on others (Fig. 13-38, C, D). For example, the antheridiogen produced by *Pteridium aquilinum* (bracken fern) is effective on certain species of twenty genera in four families. Members of certain groups, however, apparently have their own specific antheridiogen—sometimes called antheridiogen-B. One such group is the family Schizaeaceae. Gibberellic acid also can substitute for antheridiogen-B (Fig. 13-39, A, B).

Despite the present state of our knowledge of sexuality in homosporous ferns, it is apparent that there are genetic and physiological mechanisms that favor intergametophytic fertilization and hence heterozygosity. Self fertilization results in homozygosity with all of the bad features inherent in such a system. Klekowski (1971) has shown that single spore cultures (from same parent sporophyte) may reveal lethal genes in that sporophytes fail to form. However, placing portions of two "sporophyteless" hermaphroditic gametophytes of different origin together in culture results in the formation of one or more sporophytes.

Homosporous ferns have high chromosome numbers (96 percent have a basic chromosome number higher than 27). Klekowski and Baker (1966) think that in homosporous ferns this is important in creating and maintaining genetic variation in the face of the homozygotizing effects of possible habitual self fertilization in hermaphroditic gametophytes.

In a refinement of this concept Klekowski (1973) proposed the following: meiosis in *polyploid* homosporous ferns permits pairing between duplicated chromosomes within polyploid sets (*homoeologous* pairing) rather than restricting pairing to homologs; this allows the sporocytes of even a homozygous sporophyte (as a result of intragametophytic fertilization) to give rise to genetically heterogeneous spores.

Recent research has shown that intergametophytic fertilization and heterozygosity are probably more common in homosporous ferns than originally suspected. *Bommeria hispida* is obligately outcrossing (fertilization is intergametophytic) when gametophytes are grown in culture. An electrophoretic analysis of enzymes of gametophytes and naturally occurring sporophyte populations revealed a high degree of heterozygosity (Haufler and Soltis, 1984). Gastony and Gottlieb (1982, 1985) have shown that gametophytes and sporophytes of *Pellaea andromedaefolia* exhibit heterozygous electrophoretic patterns for several enzyme systems. They concluded, however, that the genetic variation in sexual homosporous ferns is produced by cross fertilization of genetically different gametes and may not be the result from pairing between homoeologous chromosomes as proposed by Klekowski (1973). There is a considerable amount of information on fern breeding systems; the reader is referred to the following review articles: Lloyd, 1974; Walker, 1979; Klekowski, 1979; Soltis and Soltis, 1987.

EVOLUTIONARY TRENDS IN GAMETOPHYTE MORPHOLOGY. It is unfortunate that there is no fossil record of fern gametophytes. If one considers the eusporangiate ferns that are thought to have retained many primitive or less advanced characteristics, we find that the gametophytes of the

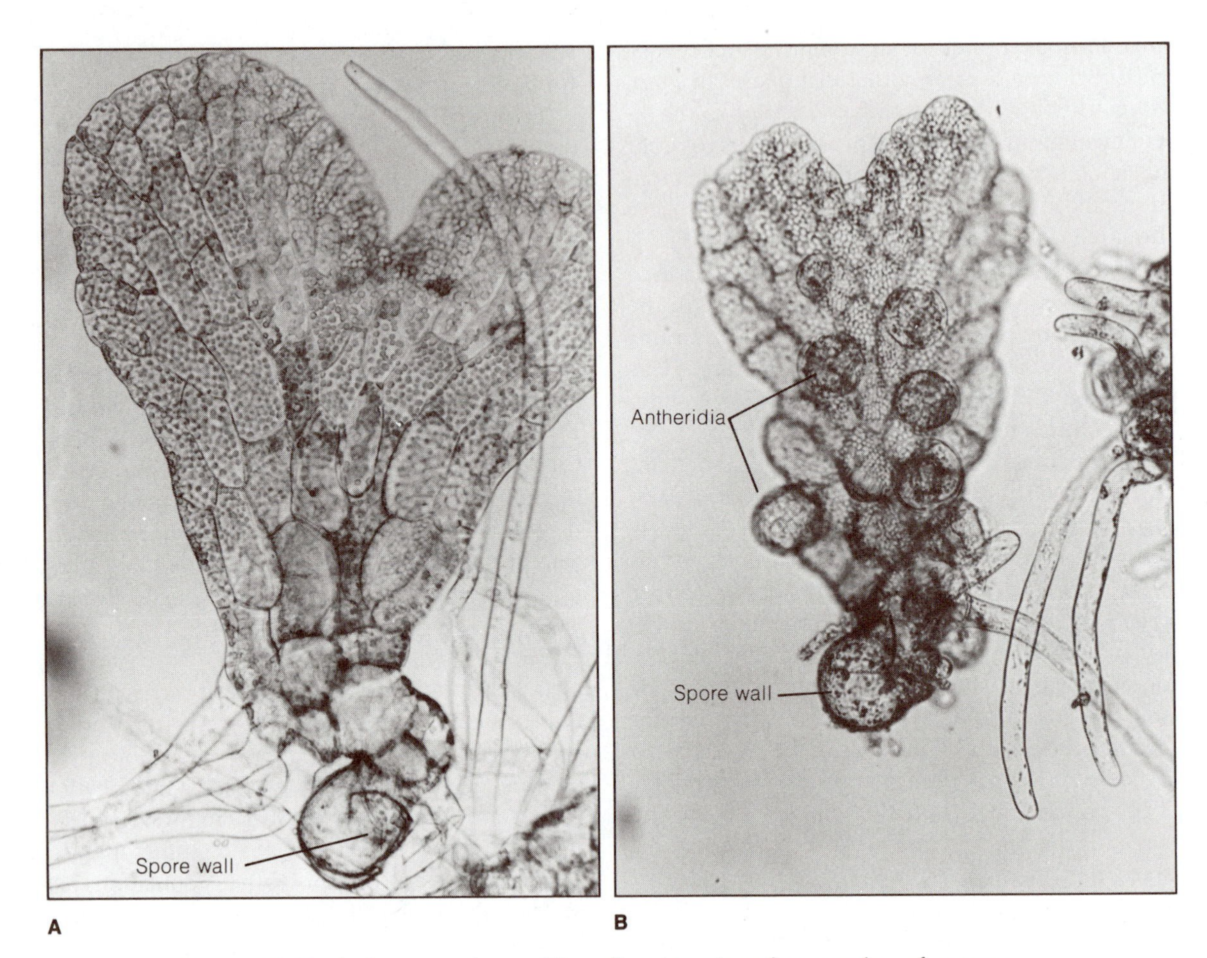

FIGURE 13-39 **A, B,** gametophytes of *Lygodium japonicum* fourteen days after spore germination; **A,** gametophyte grown on simple nutrient agar under white light without gibberellic acid added; **B,** grown as in **A,** but with gibberellic acid added to the culture medium. [Courtesy of Dr. E. G. Cutter.]

Ophioglossales are subterranean, tuberous to elongate; those of the Marattiales are surficial, green, and cordate to elongate with a prominent midrib, and the wings are more than one cell thick except at the extreme margins. In both types a gametophyte develops slowly and lives for a long time. In the Filicales the gametophyte has undergone evolution along different lines of specialization. The surficial, elongate, or cordate type with a prominent midrib characterizes the Osmundaceae, Gleicheniaceae, and Matoniaceae. Most of the families and subfamilies in the Filicales that are considered to be more highly advanced or specialized have the typical cordate type of green gametophyte, but it matures rapidly, is short lived, and lacks a prominent midrib. The wings are delicate, generally consisting of only one layer of cells (e.g., *Pteridium, Adiantum*). In others, specialization has led to branched, ribbonlike forms without a midrib *(Hymenophyllum)* to branched, filamentous types (*Trichomanes* and certain species in *Schizaea*). The subterranean type does occur in some species of *Schizaea* and in *Actinostachys,* both in the family Schizeaceae (Fig. 13-30, F).

Phylogenetically, the antheridium has undergone certain changes. The less advanced type of antheridium consists of a jacket composed of several to many cells that enclose numerous spermatids (e.g., in the Osmundaceae, Gleicheniaceae). Specialization has led to the development of a small antheridium with only three jacket cells enclosing sixteen or thirty-two spermatids.

In more primitive taxa of the Filicales, the archegonial necks are rather long (seven to twelve tiers of cells), whereas in advanced ferns the neck consists of only four to five tiers and the neck curves away from the apex (notch) of the gametophyte.

In summary, gametophyte morphology is playing an ever increasing role in determining phylogenetic trends in the Filicales.

EXPERIMENTAL EMBRYOLOGY. Each phase of the fern life cycle—gametophytic and sporophytic—begins with a single-cell spore or zygote. Commonly a spore has a reduced number of chromosomes (haploid, n) and a zygote has a double set of chromosomes (diploid, $2n$) as a result of fertilization. It might be thought that the profound differences between gametophyte and sporophyte are related to chromosome numbers. However, the many well-documented cases of apogamy and apospory (see p. 293) have shown that the difference in chromosome numbers cannot fully account for the differences in morphology of the two generations. For example, in apogamously reproducing ferns both generations have the same chromosome number.

Many years ago the eminent British botanist W. H. Lang postulated that physical and nutritional conditions influence the early development of the spore and fertilized egg. He suggested that the restraints exerted upon the "encapsulated" zygote by the archegonium and surrounding gametophytic cells influence its characteristic growth; the spore, on the other hand, germinates in a "free" environment.

Since Lang's time some experiments have been undertaken to test his ideas. Incisions in the gametophyte around a fertilized archegonium of *Phlebodium (Polypodium) aureum* did effect a partial "release" of the young embryo. The course of embryogenesis was modified, and an unusual sequence of stages occurred, but ultimately a normal sporophyte was produced (Ward and Wetmore, 1954).

In other experiments in which pure cultures of the fern *Thelypteris palustris* were used, departures were noted from the normal embryogenesis as a result of various surgical and isolation techniques (Jayasekera and Bell, 1959). By delicate and precise surgical techniques an archegonium and a small pad of gametophytic tissue were removed three or five days after fertilization and transferred to a mineral-agar culture medium. The course of embryo development was not altered materially except that there was a delay in the emergence of the embryo and in the appearance of the first root. In other experiments unfertilized archegonia were removed, inseminated in a suspension of sperm, and then transferred to the agar medium. If the small fragment of tissue contained the apical notch region of an actively growing gametophyte, embryogeny proceeded normally but at a slower pace (Fig. 13-40, A, B). If the fragment did not include an active apical notch region, development was delayed, especially the development of the root (Fig. 13-40, C, D). If indoleacetic acid was applied to a fragment that did not contain the apical notch region, the root was the first organ to emerge from the calyptra. Normally an embryonic leaf emerges first. Removal of the archegonial neck of a fertilized archegonium induced aberrations in that the embryo emerged from the upper (dorsal) side of the gametophyte and initially consisted of only leaves and shoot apices (Fig. 13-40, E). Roots appeared after fifty days. The investigators concluded that a supply of auxin, known to be produced by cells in the notch region (Albaum, 1938), may determine the plane of the first division and promote embryonic differentiation, especially of the root. The archegonial neck may place a restraint on the embryo, and, if it is removed, the orientation and future development of the embryo may be disturbed.

That the surrounding somatic tissue produces growth regulators which influence embryo development, especially of the root, was shown rather clearly by Rivières (1959). An isolated embryo of *Pteris longifolia* contained in an archegonium developed into a sporophyte, but the root did not develop.

DeMaggio (1963) pursued the idea that physical restraint, environment, and nutrition are important in the embryogeny of ferns. He isolated twenty-day-old embryos of *Todea barbara* and grew them on a simple nutrient medium. Growth was delayed but the embryos developed normally. Zygotes isolated prior to the first division lost their globular form, assuming a two-dimensional form much like a gametophyte. Generally, growth ceased after thirty days. He concluded that restraint is important and that there is an essential nutritional interaction between the gametophyte and the embryo during the early stages of development.

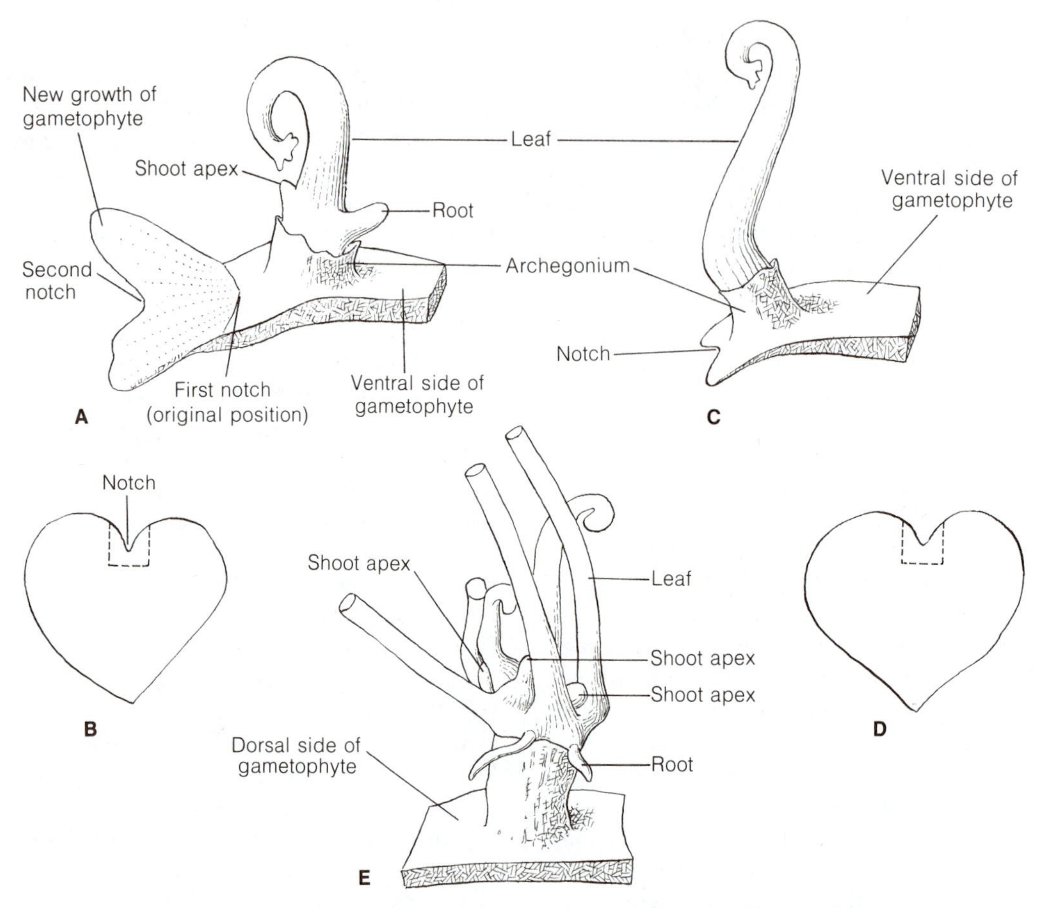

FIGURE 13-40 Experimental embryology in *Thelypteris palustris.* **A,** embryo twenty-two days old, the archegonium having been excised from the gametophyte prior to fertilization, but with the apical notch of the gametophyte attached and active; **B,** diagram showing portion of gametophyte excised. **C,** embryo twenty-six days old, the archegonium having been excised prior to fertilization, the apical notch of gametophyte attached but inactive; **D,** diagram showing portion of gametophyte excised. **E,** cylindrical body developing from zygote, seventy days after removal of archegonial neck on fifth day after fertilization. See text for details. [Redrawn from Jayasekera and Bell, *Planta* 54:1–14, 1959.]

It might be concluded that Lang's ideas about the importance of physical restraint and nutrition have been borne out by the experiments just described. Bell (1970) subscribes to the effects of physical restraint on the embryos of ferns, but, however, considers the egg (before fertilization) to be a rather specialized cell and quite different from a fern spore. On the basis of electron microscopy and autoradiography (i.e., use of radioactive isotopes), Bell (1979) has described some interesting cytological features of oogenesis (egg development) in *Pteridium.* During early stages of archegonial ontogeny there is an active phase of cytoplasmic vesiculation, presumably through the action of hydrolytic enzymes. At about the time the axial row of cells is present, there is a curious nucleo-cytoplasmic interaction. Evaginations from the nucleus are present that have their own envelope and contain some nucleolar material. These evaginations may still be visible six hours after fertilization, but eventually they are not recognizable. The ultimate fate of the evaginations is not known. There is no indication that DNA is present in the evaginations. Considerable RNA and protein synthesis occurs during early development, but synthesis declines dramatically with maturation of the egg. Undoubtedly all these changes are occurring in preparation for fertilization.

APOSPORY AND APOGAMY. The sporophyte of an homosporous fern forms sporangia that produce spores as a result of meiosis. These spores germinate, forming usually bisexual gametophytes. Antheridia and archegonia are formed on the gametophytes and self (intragametophytic) or intergametophytic fertilization occurs (Fig. 13-41). The embryo or new sporophyte is thus established. As sporophyte and gametophyte often have a natural means of vegetative propagation or can be propagated experimentally, theoretically each generation could be perpetuated indefinitely.

There are some interesting and striking deviations from the "normal" life cycle in ferns, some occurring naturally, others being induced experimentally in the laboratory. One of the processes in the alteration of the life cycle is termed *apospory*. In the process of apospory a gametophyte is formed from a vegetative cell or cells of a sporophyte. Apospory can result from the proliferation of leaf tissue in the formation of a gametophyte (Fig. 13-42). The gametophytes formed in this manner have the same chromosome number as the sporophyte. For *Adiantum pedatum*, Morel (1963) has shown that juvenile leaves of young sporophytes will proliferate into gametophytes in culture, but that this capacity is lost in older leaves.

Sheffield (1984, 1985) has described the ontogeny of aposporous gametophytes using scanning electron microscopy. When detached juvenile leaves of *Pteridium aquilinum* were placed on sim-

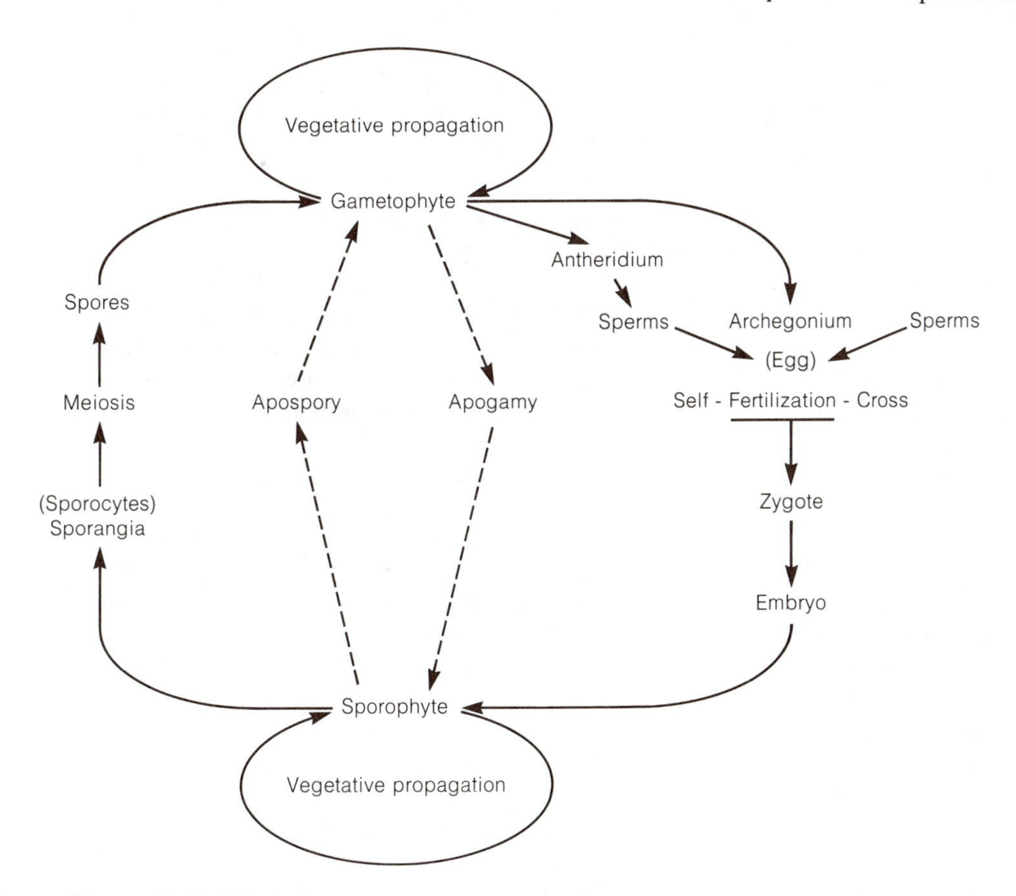

FIGURE 13-41 "Typical" generalized life cycle of a homosporous fern and possible deviations (apospory and apogamy) from the complete "sexual cycle."

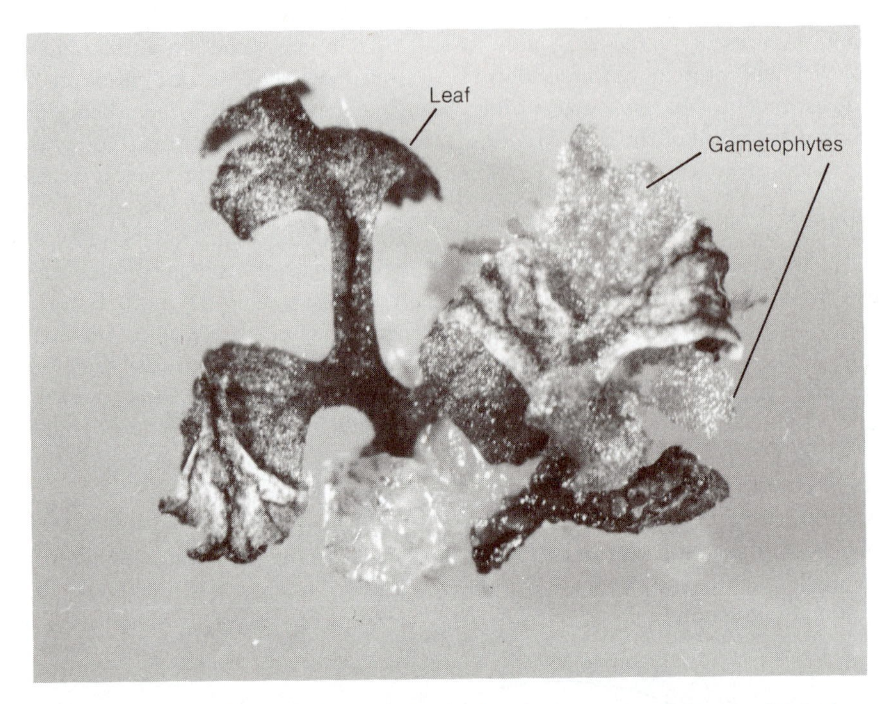

FIGURE 13-42 Production of gametophytes directly from juvenile leaves of *Pteridium aquilinum*. Leaves were placed on an agar medium that contained simple inorganic constituents. This is an example of apospory.

ple nutrient agar, gametophytes were initiated within three days from individual epidermal cells (Fig. 13-43) or from small groups of cells near the margins of the lamina. By five days multicellular outgrowths were formed. All epidermal cells except stomatal guard cells were capable of initiating aposporous outgrowths. Antheridia were commonly produced on older outgrowths. In one instance an antheridium developed from a leaf epidermal cell. The rapidity of the change from sporophytic phase to gametophytic is indeed remarkable. Experimentally, gametophytes also have been induced to form from the cut surface of fern petioles and rhizome of *Phlebodium aureum* (Ward, 1963).

Apogamy, the development of a sporophyte from somatic tissue of a gametophyte, is a feature of even more significance than apospory in the life cycles of some ferns. In species in which apogamy occurs regularly, antheridia and archegonia may be present, but the archegonia are nonfunctional. A cell or a group of cells of the gametophyte just behind the apical notch undergoes cell divisions, and from this mass of cells a new sporophyte is formed. Since only mitotic divisions are involved, both gametophyte and sporophyte have the same chromosome number.

If spores are produced regularly by the sporophyte of a fern exhibiting apogamy on a regular basis *(obligate apogamy)* some type of *compensating mechanism* must be involved because both generations have the same chromosome number. Cytological studies of sporogenesis have revealed the answers.

The most common type of compensating mechanism is as follows: The first division in each of eight sporocytes is mitotic, but cell-wall formation does not occur, and a "restitution" nucleus is formed that has a double set of chromosomes. The restitution nucleus then undergoes meiosis, resulting in the production of four spores with the same chromosome number as the original sporocyte (Manton, 1950). The total number of spores produced in each sporangium is thirty-two (8 sporocytes × 4 = 32 spores; Fig. 13-44, A). In a second type, the first meiotic division in each of sixteen sporocytes fails and a restitution nucleus is formed. The restitution nucleus then undergoes mitosis followed by cytokinesis, and two spores are

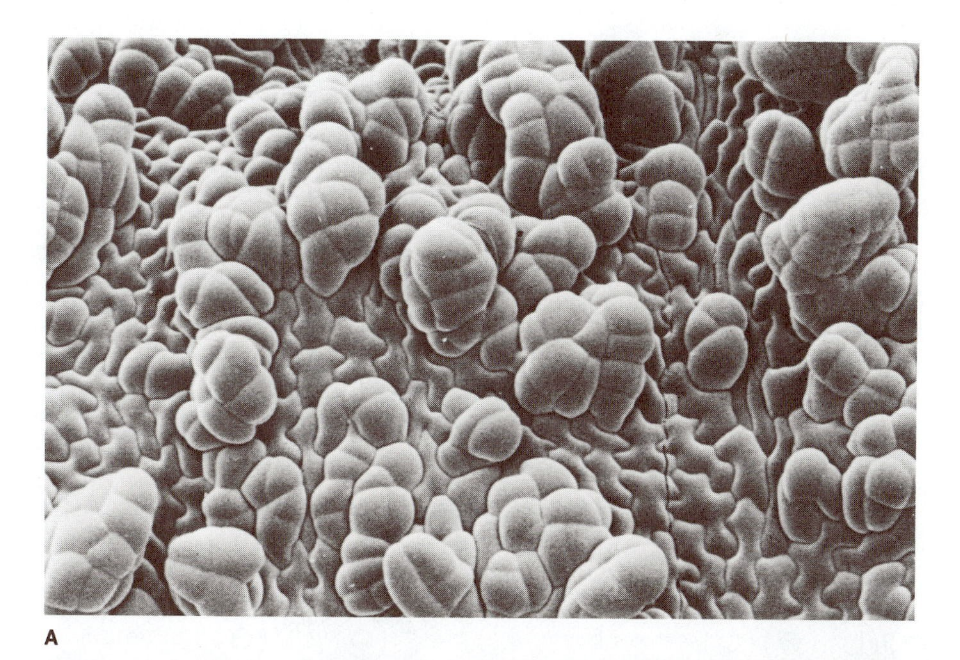

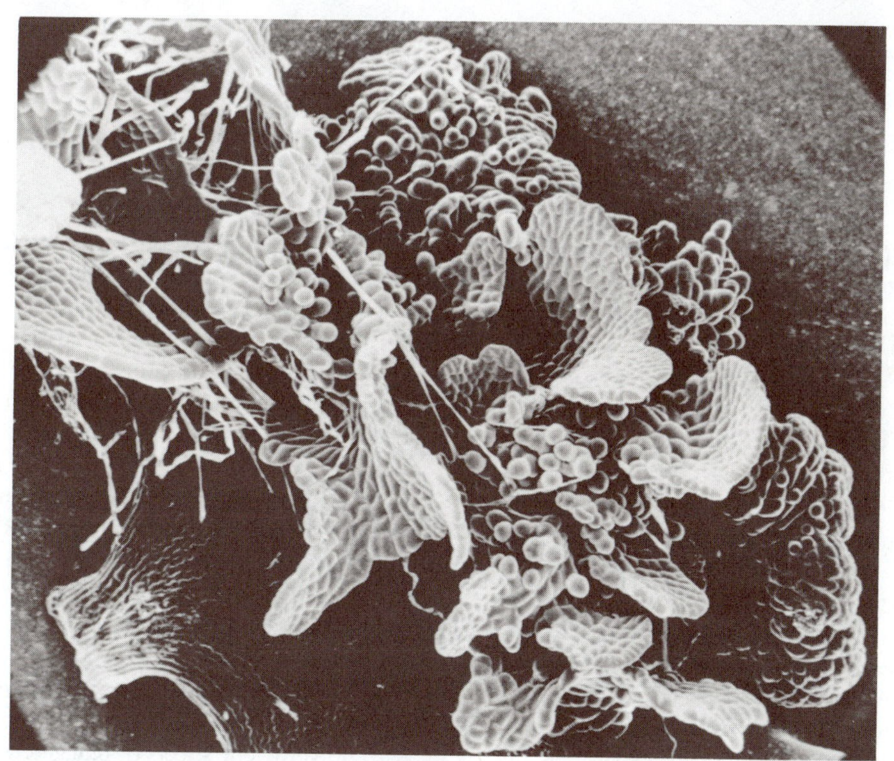

FIGURE 13-43 Scanning electron micrographs of aposporous gametophytes formed on juvenile leaves of *Pteridium aquilinum* that were placed on nutrient agar. **A,** numerous multicellular outgrowths from leaf surface (×130); **B,** later stages in development of dorsiventral gametophytes bearing antheridia; base of leaf-lamina at lower left (×42). [Courtesy of Dr. E. Sheffield.]

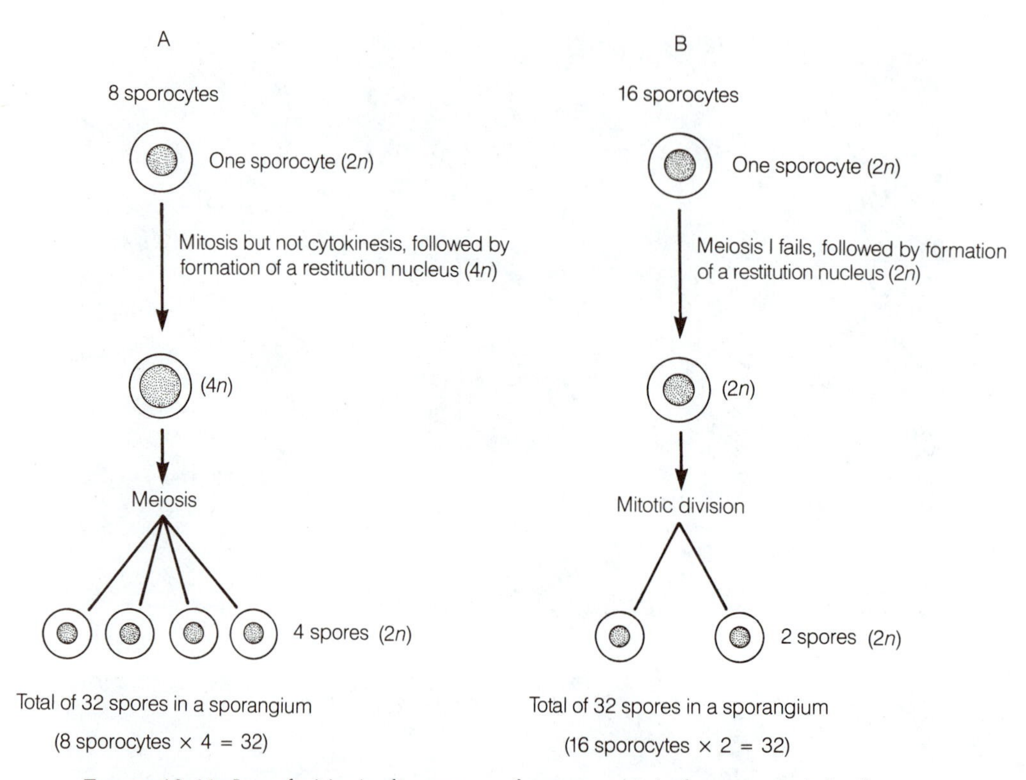

FIGURE 13-44 Irregularities in the process of sporogenesis in fern sporangia leading to the production of $2n$ spores. Consult text for explanation.

formed with the same number of chromosomes as in the original sporocyte (Braithwaite, 1964). The total number of spores formed in each sporangium is thirty-two (16 sporocytes × 2 = 32 spores; Fig. 13-44, B). A third type has been reported for one species in which sixteen sporocytes undergo only mitosis resulting in the formation of thirty-two spores with an unreduced chromosome number (Evans, 1964). However, the validity of the latter type has been questioned by Walker (1979, 1985).

A spore with an unreduced number of chromosomes, formed by one of the means outlined above, germinates and gives rise to a gametophyte. Then a sporophyte is formed apogamously and the cycle is complete without any change taking place in chromosome number of the two generations.

Apogamy can be induced experimentally in some ferns that otherwise have the ability to produce sporophytes by fertilization. For these experiments the spores were sown on nutrient agar and maintained under sterile conditions. For example, induced apogamy was achieved in fourteen strains of seven species in response to suitable concentrations of sugar (Whittier and Steeves, 1962). Through extensive studies on *Pteridium aquilinum* it was shown that naphthalene acetic acid and gibberellic acid increased apogamy when applied during the initiative phase of apogamy (Whittier, 1966). The apogamous response was also enhanced by high concentrations of phosphorus. The omission of any mineral element from the medium inhibited apogamy (Treanor and Whittier, 1968) in contrast to reports that low mineral levels induce apogamy. In two other species that are obligately apogamous, sucrose hastened the onset of apogamy (Whittier, 1964, 1965).

With the knowledge that apogamy is increased when gametophytes are grown in containers sealed with polyethylene sheets or plugged to reduce drying out, it seemed possible that gametophytes release a volatile that might influence apogamy. This volatile turned out to be the gas *ethylene*, known to cause a large number of different responses in plants (Elmore and Whittier, 1973). For *Pteridium* there was a massive production of apogamous sporophytes in response to a continuous flow of ethylene through the culture vessels. Also, when mercuric perchlorate, an absorber of ethylene, was placed in the closed containers, apogamy was reduced to a very low level. On the basis of these results and subsequent experiments, Elmore and Whittier (1975) demonstrated that apogamy is *induced* in response to ethylene during an eight-day induction period, but only in the presence of sucrose (4 percent was optimal). Once apogamy was induced, further growth of the apogamous buds was independent of ethylene, but still was dependent on a supply of sucrose. Just how these results from artificial culture apply to natural populations of gametophytes are obscure. Perhaps if gametophytes are growing under crowded conditions in a more or less confined space, ethylene might accumulate to the extent that it would induce apogamy. The reader should be reminded, however, that some ferns under natural conditions are obligately apogamous and the role played by ethylene, if any, is unknown.

ADDITIONAL DEVIATIONS FROM THE NORMAL LIFE CYCLE. Some ferns have deviated from the "normal" life cycle by growing and reproducing only as independent gametophytes. The gametophytes reproduce vegetatively by the production of filamentous multicellular propagules termed gemmae (Wagner and Sharp, 1963; Farrar, 1967, 1974; Farrar et al., 1983). Independent gametophytes assigned to three families have been described from the eastern United States. The gametophytes of *Trichomanes* (family Hymenophyllaceae) at their most northern habitat are more than 900 kilometers north from the nearest sporophyte of the genus. Also, *Vittaria* (family Vittariaceae) is represented only by gametophytes in temperate regions of the United States. At their northern limit the gametophytes are over 1,000 kilometers north of *V. lineata* in Florida, the only sporophyte-producing species of *Vittaria* in continental United States. Starch-gel electrophoresis of enzymes indicates that the independent gametophytes are distinct from *V. lineata* (Farrar, 1985). The gametophytes of *Trichomanes* and *Vittaria* have gametangia, but do not produce sporophytes. As the sporophyte-producing species are primarily tropical, it is proposed that the inability of independent gametophytes to produce sporophytes is the result of their long existence in habitats unsuitable for sporophyte growth. The gametophytes generally grow in deep crevices and grottoes of moist rock outcrops that have very low light levels (Farrar et al., 1983). The gemmae (Fig. 13-45) also are considerably tolerant to desiccation and freezing (Farrar, 1978). These independent gametophytes may be ancient tropical relicts of past ages that have lost the sporophyte phase of their life cycle through adaptation to a temperate climate.

Systematics and Phylogeny of the Filicales

It is beyond the scope of this book to give extensive descriptions of fern families and subfamilies. However, we will present some criteria that have been used to establish taxonomic groups and arrive at phylogenies. In the past, the scheme of the British pteridologist, F. O. Bower, was adopted by many botanists. Certain families recognized by Bower are still accepted today, e.g., Osmundaceae, Schizaeaceae, Hymenophyllaceae, Gleicheniaceae, Matoniaceae. The Polypodiaceae, as constituted by

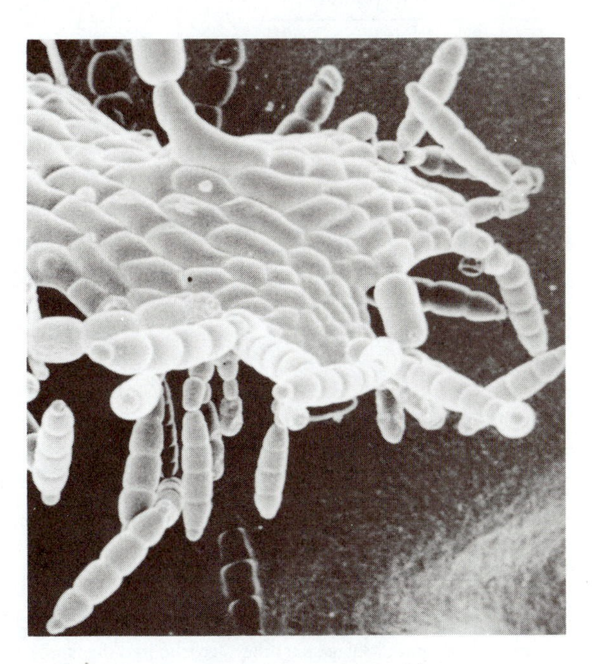

FIGURE 13-45 *Vittaria* gametophyte with numerous gemmae (×112). [Courtesy of Dr. E. Sheffield.]

Bower, was a large family comprising subgroups (e.g., davallioids, pteroids, blechnoids) that represent phylogenetic end points in the family (Table 13-1). Bower realized that the Polypodiaceae was polyphyletic. To qualify as a member of the Polypodiaceae a fern had to have certain morphological features: (1) a highly specialized sporangium consisting of a long delicate stalk and transverse dehiscence of the capsule, (2) a mixed type of soral maturation, and (3) a small specialized antheridium consisting of three jacket cells, generally producing relatively few sperms. About seven-eighths of the fern genera were included in this highly artificial family. In summary, the genera in the Polypodiaceae (in the sense of Bower) attained the same *evolutionary level,* although they were derived from different natural phylogenetic lines.

In more recent monographic treatments there is a tendency to establish more natural families out of the genera of the Polypodiaceae (*sensu* Bower). If this procedure is followed, the Polypodiaceae shrinks in number of genera. For example, asplenioids become members of the Aspleniaceae, and the blechnoids become the Blechnaceae.

Table 13-1 Fern classifications. A comparison of F. O. Bower's and E. B. Copeland's systems. See text for explanation

Bower	Copeland
FILICALES	FILICALES
OSMUNDACEAE	OSMUNDACEAE
SCHIZAEACEAE	SCHIZAEACEAE
HYMENOPHYLLACEAE	HYMENOPHYLLACEAE*
GLEICHENIACEAE	GLEICHENIACEAE
MATONIACEAE	MATONIACEAE
DICKSONIACEAE POLYPODIACEAE pteroids gymnogrammoids	PTERIDACEAE
davallioids	DAVALLIACEAE
asplenioids	ASPLENIACEAE
dryopteroids woodsioids onocleoids	ASPIDIACEAE
blechnoids	BLECHNACEAE
dipteroids	POLYPODIACEAE
CYATHEACEAE	CYATHEACEAE
MARSILEACEAE	MARSILEACEAE
SALVINIACEAE	SALVINIACEAE

*Copeland recognized 33 genera; others have limited the family to two genera.

An example of a system that reflects the recognition of several natural families within the originally large Polypodiaceae is that of Copeland (1947). Copeland attempted to make his taxonomic system phyletic with all members of common ancestry being included within a single family. He admitted that this approach often tends to make families large and difficult or impossible to define, although on the generic level it is possible to define limits. In his system, for example, the pteroids and gymnogrammoids of Bower's Polypodiaceae, as well as the family Dicksoniaceae, become members of one large family, the Pteridaceae (Table 13-1). A notable feature of a system proposed by Holttum (1947, 1949) is the establishment of a very large central family, the Dennstaedtiaceae (with eleven subfamilies). The family includes, among others, Bower's onocleoids, dryopteroids, asplenioids, and davallioids. The family is large, taxonomic limits are difficult to define, and there are several evolutionary lines within the family.

Other systems have been proposed. In one instance eighteen orders and thirty-eight families are recognized (Pichi-Sermolli, 1958). In another, chromosome numbers (especially basic numbers) become the overriding diagnostic feature (Abraham et al., 1962). In a recent, well-illustrated monograph, Tryon and Tryon (1982) recognize twenty-four families on a worldwide basis, twenty of which have genera in the New World. Their classification is based on character similarities with no attempt made to determine phyletic relationships.

W. H. Wagner, Jr., an eminent pteridologist, believes that proliferation in the numbers of families in the Filicales is unnecessary. Wagner recognizes the existence of seven families in the Filicales (Fig. 13-46). Wagner's classification system is based upon the ever-increasing use of the cladistic approach in determining evolutionary history (phylogeny). The ground plan–divergence method is one system of developing a cladogram or "phylogenetic tree." In this method one needs to select what appear to be reliable morphological characters and know the evolutionary direction of a character. This is generally done by making comparisons with nearest relatives outside the group that is being ana-

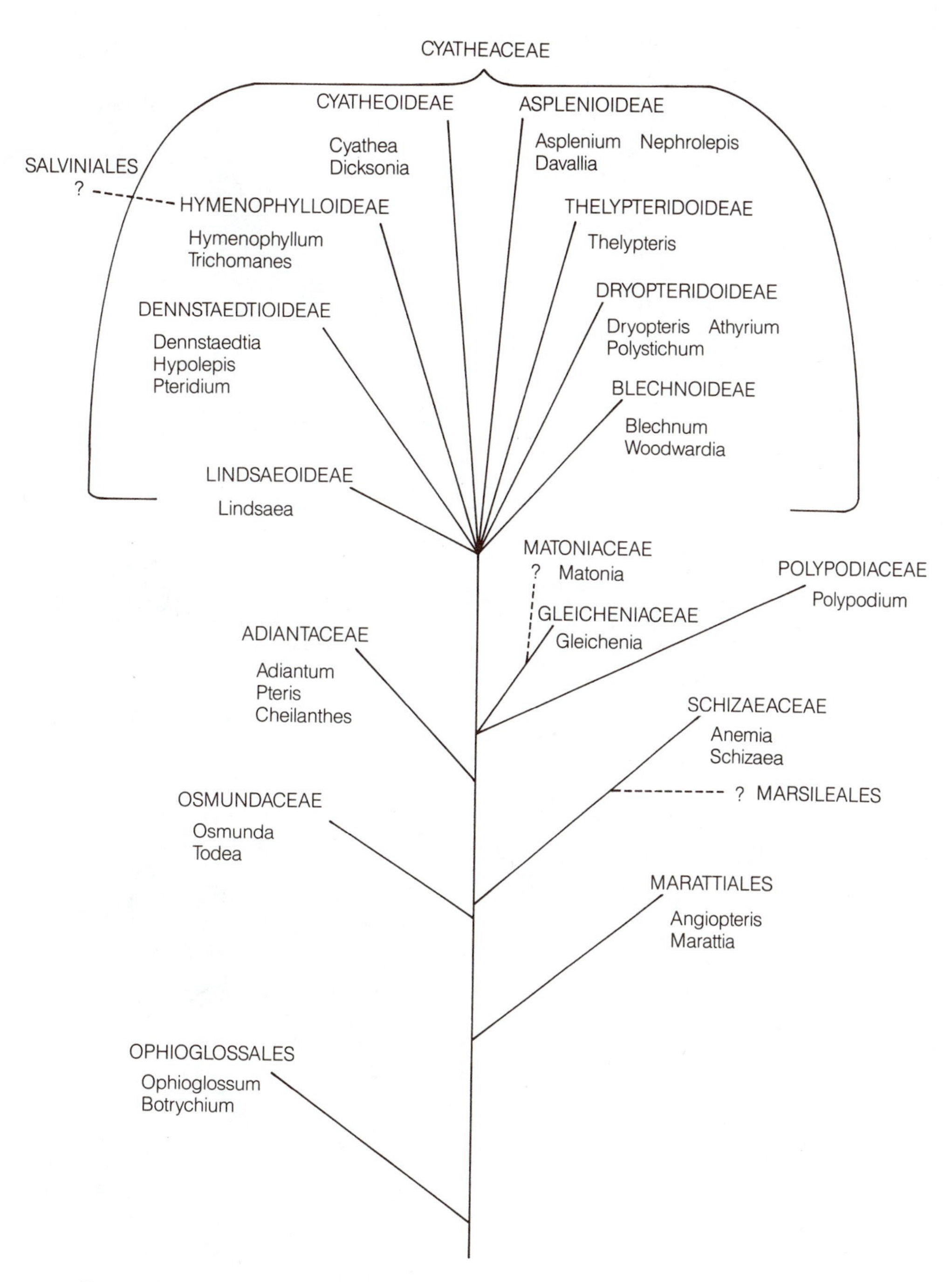

FIGURE 13-46 A phylogenetic scheme of ferns based on the ground plan-divergence method. See text for details.

FIGURE 13-47 Osmundaceae. **A,** *Osmunda cinnamomea,* sterile and fertile fronds; **B,** *Osmunda claytoniana;* **C,** *Todea barbara;* **D,** *Leptopteris hymenophylloides;* **E,** *Osmunda regalis;* **F,** *Osmunda cinnamomea,* transverse section of rhizome. See text for details. [Redrawn from Hewitson, *Ann. Missouri Bot. Gard.* 49:57–93, 1962.]

lyzed (outgroup comparison). If one can obtain a series of uniquely derived character states, one can reconstruct phylogeny. The ground plan is all the primitive states of the characters, and the ends of the lines (or branches) represent unique combinations of advanced character states. (See Wagner, 1980, for a review of the method.)

The phylogenetic tree in Fig. 13-46 is based on the ground plan–divergence method for analyzing living ferns. It will be noted that the eusporangiate orders are on the lower branches, followed by families of the Filicales considered to be unique and probably primitive in other systems: Osmundaceae, Schizaeaceae, Gleicheniaceae, and Matoniaceae. Except for the Adiantaceae and Polypodiaceae, all other derivative filicalean ferns are placed in the family Cyatheaceae for which eight subfamilies are recognized. The family name Cyatheaceae was chosen on the basis of priority. It was the first family name to be used (1827) in descriptions of ferns. Members of the Cyatheaceae commonly have true indusia. There are two trends among the subfamilies—those tending toward marginal sori and those tending toward abaxial sori (to the right in Fig. 13-46). It is interesting to note that some of the subfamilies are the derivative groups that Bower included in his classification. Representative genera are indicated for families and subfamilies, primarily those that have been used in the earlier part of the chapter to illustrate various fern characteristics.

OSMUNDACEAE. The Osmundaceae, consisting of three living genera, *Osmunda, Todea,* and *Leptopteris,* is represented in rocks of the Permian (e.g., *Thamnopteris*). Other relatives (e.g., *Osmundacaulis*) were common in the Mesozoic. For more than 100 million years the family has displayed a remarkable constancy in characters. *Osmunda* as a genus has probably existed for 70 million years, and the family was probably more abundant in the past and more widely distributed than at present. Extant species are terrestrial, generally stand erect, and have simple trichomes. Some individual plants of *Osmunda* may be over 100 years old. In *Osmunda cinnamomea* (cinnamon fern) there are two types of fronds—fertile and sterile (Fig. 13-47, A). In other species the frond consists of sterile and fertile regions (Fig. 13-47, B, E). In *Osmunda* large sporangia are attached along the margins of narrow leaf segments, often in clusters, but without a definite soral-type organization. In *Todea* the sporangia occur along veins on the abaxial side of the lamina (Fig. 13-47, C). The origin of a sporangium is not always easily referable to a single surficial initial, and the mature sporangium has many eusporangiate features (Chapter 4). A sporangium is large with a lateral annulus; upon dehiscence the sporangium opens much like a clam shell (Fig. 13-48), liberating a large number of spores (up to about 512). A chromosome number of $n = 22$ is uniform in the family.

A transverse section of an *Osmunda* stem reveals a large number of closely packed leaf bases and the vascular cylinder. The stele is an ectophloic siphonostele with overlapping leaf gaps (Figs. 13-47, F; 13-20, B). Some fossil genera from the Permian were strictly protostelic or protostelic with mixed pith (tracheids and parenchyma). Enough fossil forms are known that a stelar evolutionary sequence can be traced from the protostele to the

FIGURE 13-48 Scanning electron micrograph of a cluster of mature sporangia of *Osmunda regalis;* dehiscence is longitudinal (×20). [Courtesy of Dr. R. H. Falk.]

Osmunda type of today. From all evidence it appears that the siphonostelic organization of extant species has been derived (in an evolutionary sense) through developmental changes that occurred in the central part of the stele. Cells differentiated into parenchyma cells rather than tracheids, resulting in the formation of a pith. This change was correlated with the formation of leaf gaps.

The gametophyte is a large green thallus with a conspicuous midrib. Long-lived gametophytes may reach 4 to 5 centimeters in length. Members of the Osmundaceae combine more eusporangiate and leptosporangiate characteristics in their morphology than do members of any other family in the Filicales. This is reflected in the establishment by some pteridologists (e.g., Pichi-Sermolli, 1958) of a separate order, Osmundales, for the group. Many morphologists consider that no other group of ferns has been derived from the Osmundaceae. (Additional information on the family, especially fossils, can be found in the following references: Hewitson, 1962; Miller, 1967; Tidwell and Rushforth, 1970.)

SCHIZAEACEAE. The family Schizaeaceae is considered to be ancient, dating perhaps from the Carboniferous Period based upon similarities in structure of sporangia. By the Jurassic to the Cretaceous Periods the family was well established. The family is primarily tropical or subtropical and of the New World, except for one genus. The stem is erect to decumbent or creeping in growth habit. The leaf varies from simple to dichotomously branched to pinnate. The pinnate leaf of *Lygodium* is distinctive in that it has almost unlimited apical growth and may attain the remarkable length of 5 to 10 meters, and is adapted to climbing on other vegetation (Fig. 13-4, F). The stem is protostelic (with mixed pith). Sporangia are not grouped into sori, but occur singly, each protected by a laminar outgrowth or flap (Figs. 13-7, D; 13-9, C). The annulus is apical, the sporangium dehiscing longitudinally (Fig. 13-14, B). *Anemia phyllitidis* is another representative species in which only the elongate basal pinnae of a leaf are fertile. Sporangia are large, with short stalks and produce a large number of spores. Each sporangium has an apical, girdling annulus which brings about longitudinal dehiscence (Fig. 13-49, A, B). Some members are protostelic, others are siphonostelic. Research in recent years has resulted in the recognition of at least three different types of gametophytes in the family. The common green, cordate type is characteristic of *Lygodium* and *Anemia*. The species of one section of *Schizaea* have green, filamentous gametophytes and others *(Schizaea dichotoma)* are subterranean and filamentous, but the larger axes are multicellular. For *Actinostachys* the gametophytes are of the subterranean, axial, tuberous type and are associated with an endophytic fungus.

Departures from the common type of embryogeny in filicalean ferns (Chapter 6; p. 70) have been reported by Bierhorst (1968, 1983). For *Actinostachys* and *Schizaea dichotoma,* the first division of the zygote is transverse instead of longitudinal. Also, no primary leaf, or even a root in *Actinostachys,* is formed during early embryogeny. These organs arise later in development of the young sporophyte. This type of development may be correlated with the subterranean nature of the gametophytes and the longer dependence of the sporophyte on the gametophyte for nutrition. Genera assigned to the Schizaeaceae share many common features, but some pteridologists, in an effort to point up the differences, separate the genera into two or three families or establish subfamilies for the Schizaeaceae.

Fertile leaves with sporangia *(Seftenbergia)* have been described, suggesting that relatives of the Schizaeaceae were present in the Carboniferous. Sporangia from the Upper Carboniferous are similar to those of the coenopterid *Ankyropteris* (Chapter 11). Sporangia resembling members of the Schizaeaceae were more numerous in the Jurassic, and by the Upper Cretaceous sporangia were about identical with *Anemia. Lygodium* was worldwide in distribution during the Eocene (Stewart, 1983).

POLYPODIACEAE. Members of the family Polypodiaceae are primarily epiphytes of the tropics, although the common *Polypodium vulgare* and related species occur in Europe and in the north temperate zone. A dictyostelic rhizome is the prevailing type in the family and the rhizome generally bears scales. Leaves may be simple, pinnatifid or once pinnate, and many have a complex system of anastomosing veins in the lamina. Sori are round to elongate but are *without indusia* (Fig. 13-8, D). Sori

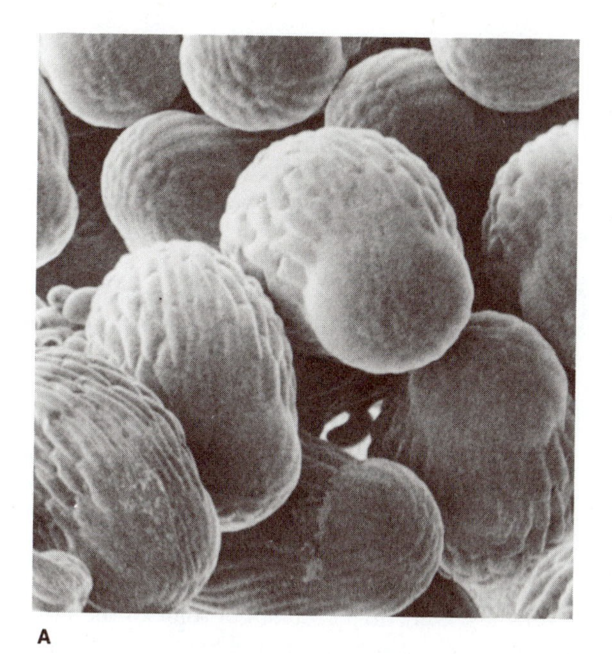

A

B

FIGURE 13-49 Schizaeaceae. *Anemia phyllitidis:* scanning electron micrographs of sporangia (**A**) and one sporangium which has dehisced, exposing the spores (**B**). An annulus is apical in position, as shown by rounded tips in **A.**

may be crowded on a portion of the frond, or, as in some genera such as *Platycerium* (staghorn fern), sporangia are scattered and not organized into definite sori. The annulus is vertical (longitudinal), and the spores are either bilateral (monolete) and non green, or tetrahedral-globose and green. The differences in spore morphology, along with other characters, are used in establishing subfamilies within the Polypodiaceae. Many pteridologists consider the family to have been derived from the Gleicheniaceae or from some common ancestral group.

ADIANTACEAE. Genera in the family Adiantaceae are cosmopolitan in distribution, ranging from the tropics to xerophytic conditions, often with the conspicuous development of glandular trichomes or scales on the leaves. In growth habit the stem may be erect, decumbent, to creeping. Leaf morphology ranges from palmately lobed to pinnate. Venation may be of the open dichotomous type or reticulate. The most distinctive feature of the family is the presence of sporangia in lines along the veins or at the ends of veins near the leaf margin. Sporangia are either unprotected or are covered by a more or less rolled or reflexed leaf margin ("false" indusium, Fig. 13-12, C). In some, the sporangia form a continuous line (coenosorus) or are organized in crowded groups protected by the leaf margin *(Pteris, Pellaea).* The annulus of the spore capsule is generally vertical (longitudinal), resulting in transverse dehiscence. Spores are mainly tetrahedral without perispore.

GLEICHENIACEAE. The ancient family Gleicheniaceae may have been derived from certain coenopterid ferns. There is some evidence for the existence of the family in the Carboniferous. The fossil genus *Gleichenites* has been reported from the Jurassic; more typical forms of the family have been described from the Upper Cretaceous.

Modern members are pantropical to subtropical in distribution. They are terrestrial with long creeping rhizomes and leaves that may clamber over other vegetation. They often form dense thickets at the edges of forests. Leaves have a rather unique architecture. They fork repeatedly, and a foliar bud is present in the sinus between the two axes at the level of branching. The bud of the main axis may continue to grow periodically, and more pinnae are formed. Leaves of some species may become 3 to 10 meters long. Side axes (pinnae) grow in a similar fashion but the leaf buds remain permanently arrested. The ultimate portions of the branch system develop laminar segments. Chrysler (1943) described this type of branching as a series of dichotomies in which the leaf bud is a short axis of one dichotomy; the other axis of a dichotomy grows in length, becoming part of the rachial system.

In some systems of classification two genera, *Gleichenia* and *Dicranopteris,* are recognized (Tryon and Tryon, 1982). In *Gleichenia* the ultimate lateral veins of leafy segments are unbranched or dichotomize once. Sori, consisting commonly of two to four large sporangia (Fig. 13-12, B), are in two rows, one on each side of the midrib (Fig. 13-7, E); there is no indusium. The ultimate veins in *Dicranopteris* dichotomize more than once, and sori consist of eight to fifteen sporangia. Dehiscence of a sporangium is brought about by the functioning of a transverse, to a transverse-oblique, girdling annulus. Each sporangium produces a large number of spores that are trilete or monolete depending on the species. Stems of most species are protostelic of the mixed-pith type (Fig. 13-20, A). Gametophytes are large, green, and dorsiventral with a conspicuous midrib. An antheridium is large and generally produces several hundred sperm. In considering all morphological characteristics, one can say that the family has retained many primitive features.

MATONIACEAE. The family Matoniaceae probably has closest affinity with the Gleicheniaceae, based on soral organization and structure of sporangia. Fossil members have been described from the Triassic and Cretaceous. The two extant genera are from Indonesia, Borneo, and New Guinea. In *Matonia* the frond axis is divided into two parts, each of which undergoes a series of unequal dichotomies. The leaf segments are pinnatifid, and sori are borne in a single row on each side of the midrib. Each sorus consists of a few large sporangia arranged in a ring around the receptacle which continues as the stalk of an umbrellalike indusium (Fig. 13-7, H). The sporangia of a sorus, which mature simultaneously, have short stalks and oblique annuli. The stele of a large rhizome is solenostelic with three concentric vascular cylinders (polycyclic). The spores are trilete that germinate to form thalloid green gametophytes that have conspicuous midribs and ruffled margins. Gametophyte morphology and the fact that the antheridial wall consists of several cells are suggestive of an affinity with the Gleicheniaceae.

CYATHEACEAE. According to the classification system adopted in this book, the vast majority of the filicalean ferns are placed in the family Cyatheaceae consisting of eight subfamilies. Members of this large family are characterized generally by the presence of "true" indusia, i.e., outgrowths from the leaf epidermis. In some species the indusium may be absent, probably representing a secondary loss of the structure in the course of evolution. Space does not permit a detailed analysis of each subfamily. Only brief sketches of some of the subfamilies will be presented, and reference will be made to genera discussed earlier in the chapter that typify certain morphological characters.

Cyatheoideae. Ferns in the subfamily Cyatheoideae are terrestrial, medium to very large tree ferns, or rhizomatous; the rhizomes are often ropelike. The vascular cylinders vary from solenosteles to polycyclic dictyosteles (Fig. 13-25, B). A sorus may be away from the margin, and the indusium cup shaped (*Cyathea;* Fig. 13-7, F). Or, the sorus may be marginal to submarginal (*Dicksonia;* Fig. 13-10, A). The annulus of the spore capsule is generally oblique in orientation. The spores are mostly tetrahedral, without perispore. Some pteridologists recognize two families or subfamilies based upon differences, for example, between *Cyathea* and *Dicksonia.*

Hymenophylloideae (filmy ferns). The filmy ferns are represented by two genera: *Hymenophyllum* and *Trichomanes.* These ferns are primarily of tropical rain forests or wet temperate regions where they grow as epiphytes. Leaves are simple, pinnatifid to pinnate, and the leaf blade is only one to two cells thick, except over veins—hence the name filmy ferns. Leaves are without stomata. The stem is protostelic. A sorus is marginal and the indusium is two lipped *(Hymenophyllum),* or the receptacle is elongate surrounded by a tubular indusium (*Trichomanes;* Fig. 13-7, B). The developmental order of short-stalked sporangia is basipetal (gradate type of maturation). The annulus of the sporangium is oblique and the spores are tetrahedral-globose and contain chlorophyll. Some pteridologists recognize thirty to forty genera in the group rather than two. A chromosome number of $n = 36$ is common in the subfamily.

Dennstaedtioideae. The stems of the ferns of subfamily Dennstaedtioideae vary from erect to long creeping rhizomes that are solenostelic or dictyostelic, and bear trichomes or scales or both. Fronds are primarily pinnate, rarely simple. Sori are marginal or submarginal. In marginal forms the indusium may be cup shaped (*Dennstaedtia;* Fig. 13-10, B). In others the sori are submarginal and the indusium is half-cup shaped, attached at its base and sides (e.g., *Microlepia*), or exindusiate protected by recurved leaf margins. The annulus is vertical or slightly oblique. Spores are tetrahedral-globose, or strongly lobed to monolete. The common bracken fern *(Pteridium aquilinum)* is in this subfamily. Sporangia extend in a line in a submarginal position, protected by a reflexed modified leaf margin (Fig. 13-9, A), and by a membranous, abaxial, true indusium.

Asplenioideae (spleenworts). In the subfamily Asplenioideae the stems may be erect, decumbent, or creeping, and are clothed with scales. Sori are abaxial, crescent shaped to elongate to very long, and borne on a vein. The sporangia are protected by a unilateral indusium, the open side of which usually faces toward the apex of the leaf segment (Fig. 13-9, D). Sporangia have the typical vertical annulus and the stalk consists of one row of cells. Spores are monolete with perispore. The genus *Asplenium,* comprising about 600 species, typifies the subfamily.

Dryopteridoideae (male fern, lady fern). Stems of ferns in the subfamily Dryopteridoideae are erect to creeping, clothed with non-overlapping scales, and are commonly dictyostelic. Fronds are simple to pinnate, sometimes becoming very large. Sori are abaxial and may be slightly elongate to lunate with a lateral indusium curved across the vein in a hook-shaped fashion *(Athyrium).* In others the sori are round, the indusium more or less reniform, attached to the receptacle near the sinus (*Dryopteris;* Fig. 13-8, A). In *Polystichum* and *Cyrtomium* the indusium is peltate and radially symmetrical (Fig. 13-8, C). Spores are commonly ellipsoidal, monolete, with perispore. The chromosome number is commonly $n = 40$ or 41.

Blechnoideae (chain ferns). The stems of the chain ferns are erect or decumbent to creeping, bearing scales, and solenostelic (polycyclic) or dictyostelic. The leaves may be deeply pinnatifid to pinnate. Sori occur in separated "loops" along major veins *(Woodwardia),* or sori may extend nearly the length of a leaf segment *(Blechnum),* the indusia opening inward toward the main vein of the leaf segment. Young foliage is commonly red, turning green. Spores are monolete with perispore.

Marsileales: *Marsilea, Regnellidium,* and *Pilularia*

This order is one of the two groups of "water ferns," both of which are heterosporous. In some earlier systems of classification this group constituted a family in the Filicales (the other group of water ferns constituted, in those systems, the family Salviniaceae). In other systems, the families have been placed in the order Hydropteridales—to emphasize the fact that they are aquatic as well as heterosporous. More recently the two orders Marsileales and Salviniales have been established to emphasize their structural differences.

There are three genera in the Marsileales—(family Marsileaceae): *Marsilea, Regnellidium,* and *Pilularia* (Figs. 13-50; 13-52, A, B). All are aquatic plants being generally rooted in mud or damp soil and often submerged in shallow water with just the leaf blades (*Marsilea* and *Regnellidium*) floating on the surface of the water. They are rhizomatous with roots arising from the lower side of the rhizome, usually at the nodes. *Marsilea,* with about fifty species, is primarily a genus of seasonally wet habitats with the production of reproductive structures during the drier periods (e.g., *Marsilea vestita*). *Regnellidium* (one species) is a South American plant, and *Pilularia* (about five species) is more widespread in distribution. Very little is known about the history of this line of ferns, but there is some evidence that a *Marsilea*-type plant was present in the Tertiary Period.

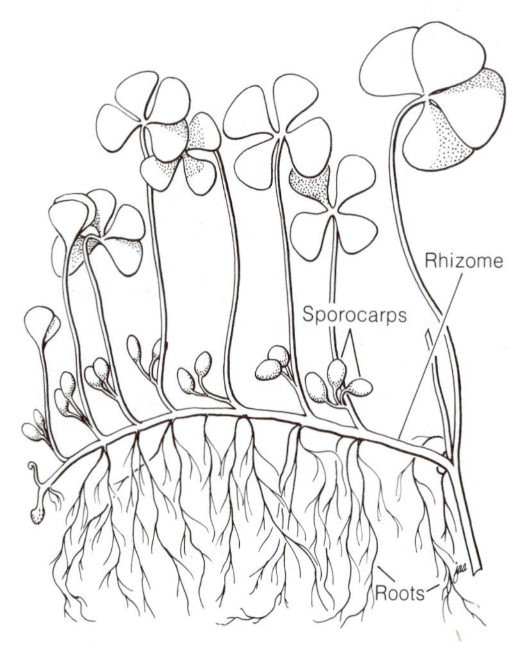

FIGURE 13-50 Sporophyte of *Marsilea quadrifolia*. [Redrawn from *Morphology of Vascular Plants: Lower Groups* by A. J. Eames. McGraw-Hill, New York. 1936.]

Marsilea

The lamina of *Marsilea* is quadrifid (Fig. 13-50) resembling a four-leaf clover and has, incidentally, been used on occasion as a substitute for the true four-leaf clover. Each pinna has basically dichotomous venation, but there are cross connections that unite the major vein system into a closed reticulum. A young leaf is circinate in vernation and the pinnae are folded together until late in development. At maturity the pinnae are extended perpendicular to the petiole. Leaves that remain totally submerged ("water form") differ morphologically from the "land form," and the two types have been studied experimentally (see p. 311). Reproductive structures, termed *sporocarps*, are found attached to or near petioles (Fig. 13-50).

ANATOMY OF THE RHIZOME. The tip of the rhizome is dominated by a conspicuous tetrahedral apical cell that is reported to give rise regularly to segments or merophytes that are the building units of the shoot. There is some evidence that the apical cell is active mitotically during early stages of growth, but the rate of division may decrease in older plants (Kuligowski-Andrès, 1978). The prostrate rhizome of *Marsilea* is solenostelic (Fig. 13-51). The inner portion of the cortex usually consists of a compact tissue; the outer portion is lacunate with large air spaces around radiating rows of parenchyma. The rhizomes of submerged plants generally have a thin-walled parenchymatous pith, whereas those growing on mud or damp soil have a sclerotic pith.

ROOTS. Roots arise from the lower side of the rhizome at the nodes. The apical meristem consists of a tetrahedral apical cell from which segments or merophytes are regularly produced from the three proximal "cutting" faces. Cell divisions occur in the merophytes giving rise to the primary meristems—protoderm, ground meristem, and procambium. Distally, the apical cell gives rise to segments from which the root cap develops. This concept of root histogenesis has been challenged in recent years. The apical cell is reported to be active mitotically only in the initiation and early organization of the root apex. Very soon the apical cell is reported to undergo endopolyploidy and/or endoreduplication rather than cell division (Avanzi and D'Amato, 1970; D'Amato and Avanzi, 1965). Kurth (1981), however, has provided compelling evidence that the apical cell in *Marsilea vestita* remains mitotically active even in roots 12 centimeters long, and there is no evidence that the apical cell becomes polyploid. Similar conclusions, based upon different data, were obtained by Vallade and Bugnon (1979) for *Marsilea diffusa*.

Internally a mature root is much like that of other ferns, but vessels do occur in the xylem. As mentioned previously (p. 266), vessels in the rhizome, petiole, and root were discovered many years ago in the bracken fern *(Pteridium aquilinum)*. Vessels have been found only in the roots of several *Marsilea* species, but in two species, *Marsilea drummondii* and *M. elata*, vessels occur in the rhizome as well (White, 1961; Bhardwaja and Baijal, 1977). *Marsilea* and *Pteridium* are rather far apart taxonomically, and the presence of vessels in these two genera indicates that there were at least two independent origins of vessels in ferns (see also the following section).

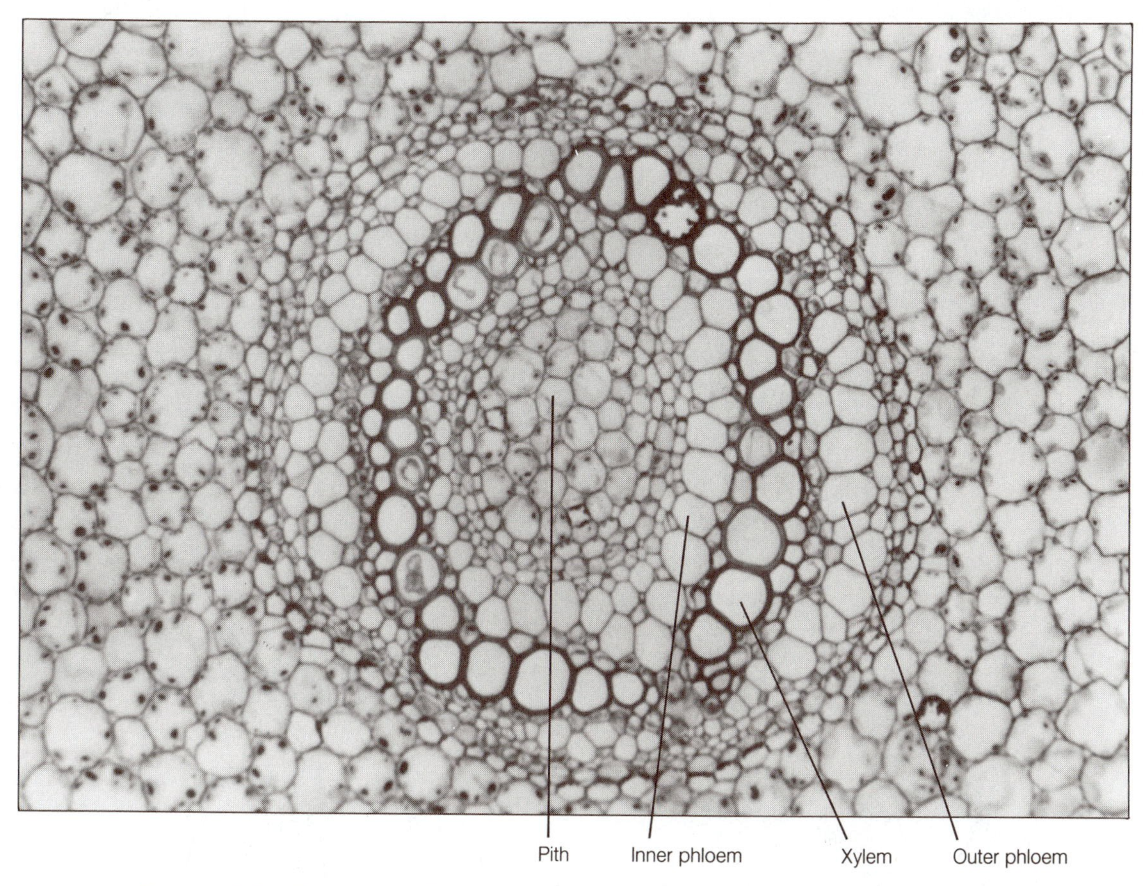

FIGURE 13-51 Transection of the rhizome of *Marsilea* sp. showing details of the solenostele.

Regnellidium and *Pilularia*

Regnellidium (Fig. 13-52, A), a monotypic genus, differs from *Marsilea* in having leaves that are bifid; venation is of the dichotomous type (Pray, 1962) although the entire system of veins is united by a marginal vein. Also, *Regnellidium* is the only known nonflowering plant that produces latex. It is formed in unbranched laticiferous ducts in the cortex of the rhizome, petiole, and in the lamina (Labouriau, 1952). *Pilularia,* a relatively small plant, has a filiform leaf that consists of only a petiole; no lamina is formed (Fig. 13-52, B). The petiole, just as is true of the leaves of *Marsilea* and *Regnellidium,* is circinate during development. The internal anatomy of the rhizomes of *Pilularia* is essentially the same as that of *Marsilea.* The leaf of *Pilularia* is a simple structure, but it probably became simple through evolutionary processes of reduction. In the course of time there has been a loss of the lamina. It is interesting to note that vessels have also been reported to occur in the root of *Regnellidium diphyllum* (Tewari, 1975).

Reproductive Cycle of *Marsilea*

Reproductive structures, termed *sporocarps* and containing sporangia, are hard, bean-shaped bodies. In most species they occur singly, but there are exceptions (Fig. 13-50). Sporocarps withstand desiccation and are reported to be viable even after twenty to thirty-five years. These specialized structures are thought to be laminar in evolutionary origin. The question does arise, however, of whether a sporocarp represents a simple, folded basal pinna,

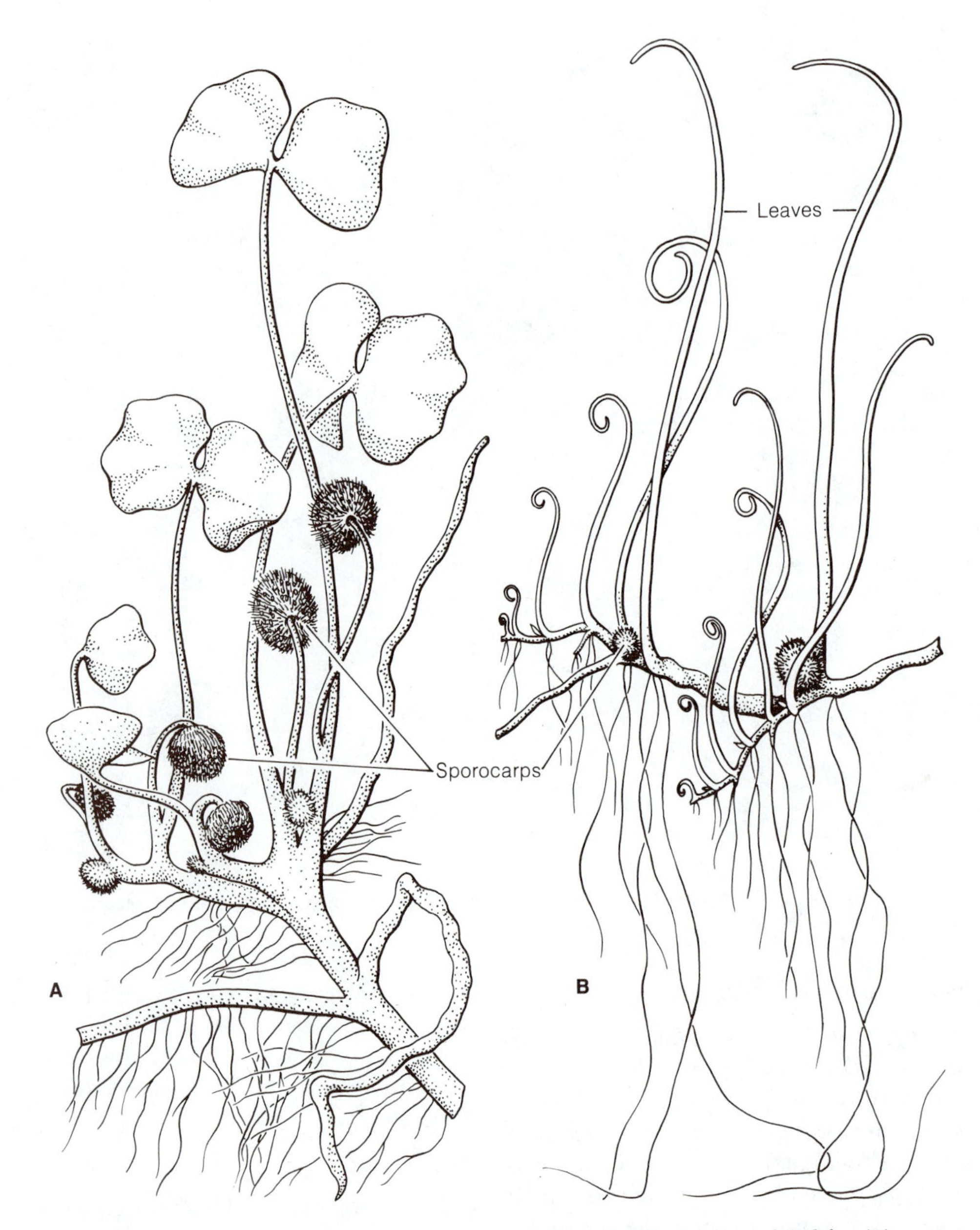

FIGURE 13-52 Sporophytes of *Regnellidium diphyllum* (A) and *Pilularia globulifera* (B); natural size. [Redrawn from *Morphology of Vascular Plants: Lower Groups* by A. J. Eames. McGraw-Hill, New York. 1936.]

or whether in evolution it has arisen from an entire pinnate leaf (for discussions see Eames, 1936; Puri and Garg, 1953; Smith, 1955). There is one main vein in the sporocarp, with lateral side branches that dichotomize (Fig. 13-53, A). Sori are on the inner side of each half of the sporocarp, oriented parallel to the lateral veins. Each sorus consists of a receptacle to which are attached megasporangia along its crest and microsporangia along the flanks (Fig. 13-53, B, C). An indusium covers each sorus and is a hoodlike structure which extends to the margin of the sporocarp (Fig. 13-53, B, C). Develop-

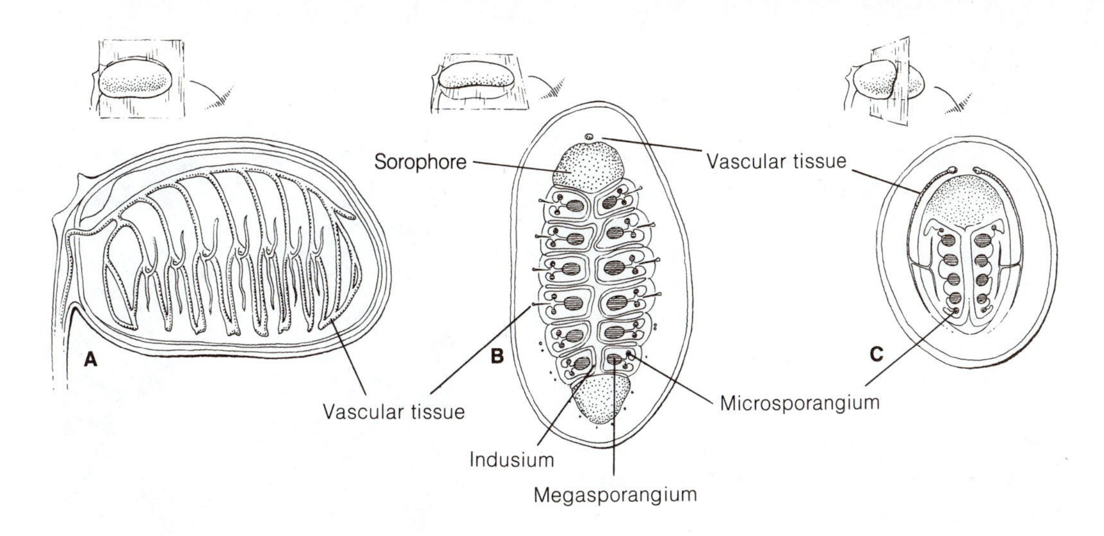

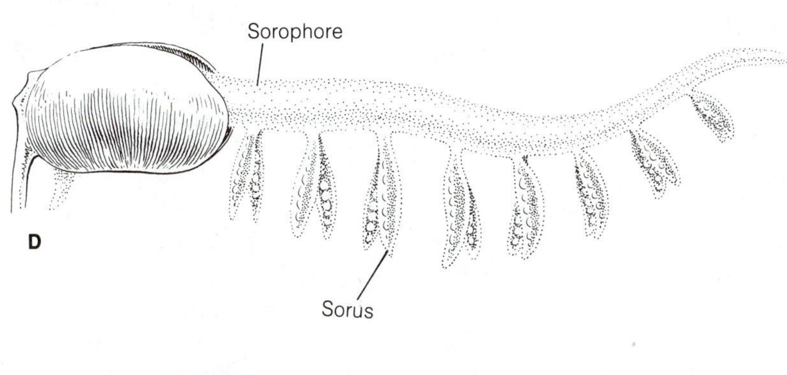

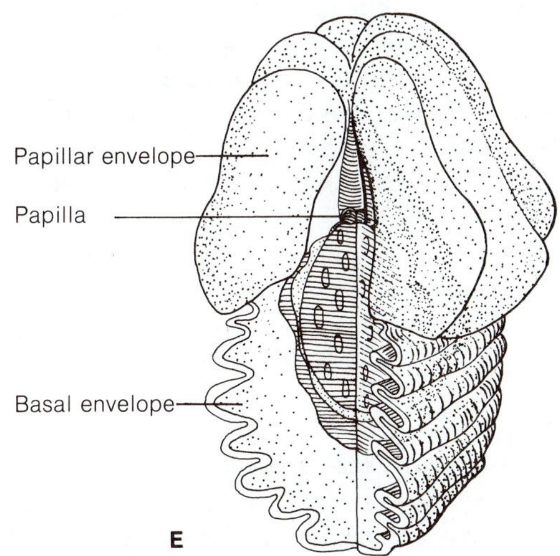

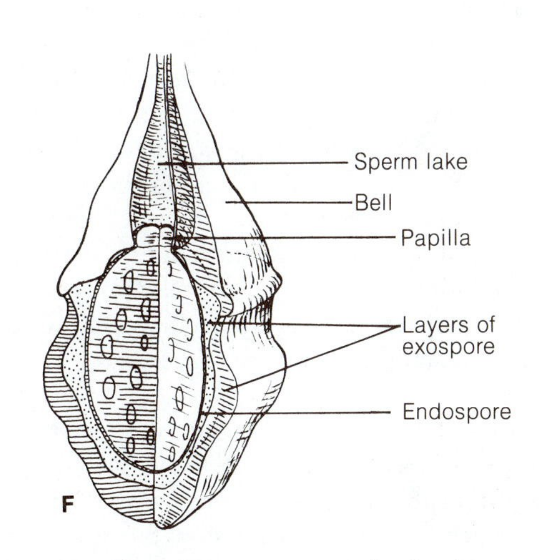

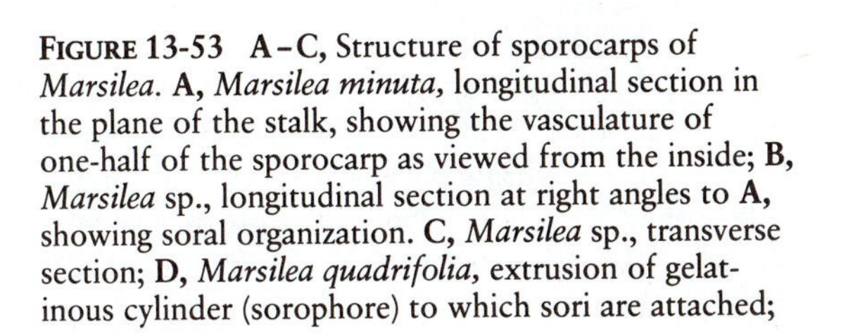

FIGURE 13-53 **A–C,** Structure of sporocarps of *Marsilea.* **A,** *Marsilea minuta,* longitudinal section in the plane of the stalk, showing the vasculature of one-half of the sporocarp as viewed from the inside; **B,** *Marsilea* sp., longitudinal section at right angles to **A,** showing soral organization. **C,** *Marsilea* sp., transverse section; **D,** *Marsilea quadrifolia,* extrusion of gelatinous cylinder (sorophore) to which sori are attached; **E, F,** *Marsilea vestita,* megaspores shortly after hydration, the papillar and basal envelopes removed in **F.** (See text for details.) [**A** redrawn from Puri and Garg, *Phytomorphology* 3:190, 1953; **B–D** adapted from *Morphology of Vascular Plants: Lower Groups* by A. J. Eames. McGraw-Hill, New York. 1936; **E, F** redrawn from Machlis and Rawitscher-Kunkel, *Amer. Jour. Bot.* 54:692, 1967.]

mentally each sorus is generally of the gradate type, with megasporangia being initiated first, followed by microsporangia. Sporangial development is of the leptosporangiate type. Only one spore in each megasporangium is functional after meiosis. The sporangium wall shows no sign of cellular specialization such as the formation of an annulus — hence there is no mechanism for dehiscence.

OPENING OF THE SPOROCARP AND DEVELOPMENT OF GAMETOPHYTES. A sporocarp is a hard structure and withstands desiccation. The spores are capable of germinating as soon as the sporocarp is mature, but do not do so until the sporocarp opens. In nature the sporocarps do not open until two or three years after their formation. This delay is probably due to the imperviousness of the hard sporocarp wall. Opening of the sporocarp can be hastened by cracking or scoring the stony covering. When the sporocarp is placed in water the tissues within begin to imbibe water and swell. As the sporocarp swells it splits open, and within a few minutes a long, wormlike gelatinous structure *(sorophore)* emerges, to which are attached the sori (Fig. 13-53, D). Cells of the sorophore, an inner region of the sporocarp wall, contain large amounts of mucilage. The mucilage carbohydrate is hydroscopic. Upon hydration, the sorophore cells swell in response to the expanding mucilage, and the sori are carried to the outside on the sorophore (Bilderback, 1978). The sorophore may become ten to fifteen times longer than the sporocarp. The indusial and sporangial walls are intact at first, but with the imbibition and swelling of gelatinous layers around the spores there is a general breakdown of the tissues, and the spores are released.

MALE GAMETOPHYTE. The numerous *microspores* begin to develop almost immediately (by cell divisions within the spore wall — endosporic type of development). After several divisions two groups of sixteen spermatids are formed, surrounded by jacket cells. Spermatid development and liberation of sperms are temperature dependent. For example, sperm discharge can be expected to occur in about six hours at 25° C. Prior to actual liberation of the sperms, the outer layer of the microspore, termed the *exospore,* ruptures along the triradiate ridge.

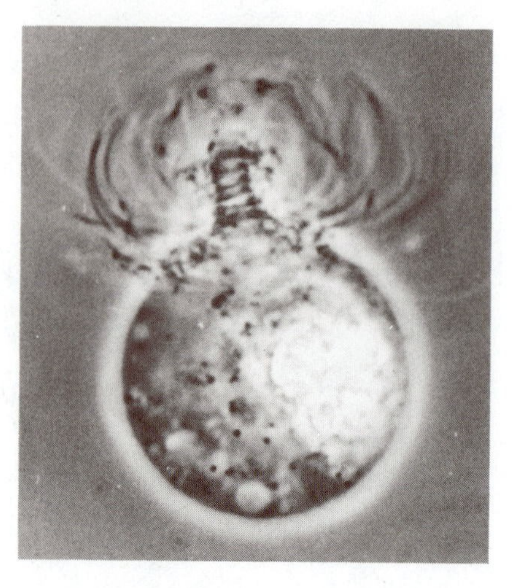

FIGURE 13-54 Swimming sperm of *Marsilea vestita* as seen with phase-contrast microscopy. Note the posterior cytoplasmic vesicle and the anterior flagella-bearing coil (×1650). [Courtesy of Dr. W. M. Laetsch.]

The spermatids, enclosed by the inner layer of the spore wall, the *endospore,* emerge. The endospore then breaks, liberating the two groups of spermatids, each of which becomes a motile sperm by dissolution of the spermatid membrane.

Each mature swimming sperm consists of a large posterior cytoplasmic vesicle, a nucleus and a coiled, flagella-bearing band at the anterior end (Fig. 13-54). A sperm may be active for about one hour after liberation at 22 to 25° C (Rice and Laetsch, 1967). When sperm are released in the vicinity of megaspores they become trapped in the gel around a megaspore. As a sperm loses its motility, it becomes less tightly coiled and the posterior cytoplasmic vesicle is lost (Myles, 1975).

FEMALE GAMETOPHYTE. Approximately five minutes after a dry megaspore is moistened, the layers of the spore wall imbibe water and expand to form a gelatinous mass around the megaspore. At the apical end (where the endospore wall layer expands to form a small papilla) the gelatinous mass is characterized by longitudinal folds; the comparable structure around the basal end has horizontal folds (Fig. 13-53, E). During this process the sporangial wall is shed. If a large concentration of sperm is present, these envelopes are rapidly destroyed (Machlis and Rawitscher-Kunkel, 1967). Within the envelopes just described is a thick gelatinous structure shaped like a bell whose interior is liquid. These are referred to as the "bell" and "sperm lake" (Fig. 13-53, F). The former constitutes the "funnel"—a term often used in the literature on *Marsilea.* If sperm are in the vicinity they become trapped in the gel or can be seen swimming in the sperm lake. The bell and subjacent layers of the exospore merge into a single gelatinous matrix a few hours after hydration. This is the stage generally illustrated in textbooks (Fig. 13-55, E).

The first cell division of the large megaspore occurs very soon after hydration—reportedly within two and a half hours in one instance (Demalsy-Feller, 1957). There is apparently some variation in the events that follow the first division, but very soon a group of cells is formed at the apical end (within the papilla) and a cell wall separates the group of cells from a very large nutritive basal cell. An archegonial initial is formed, and from it an archegonium is initiated consisting of an egg cell, a ventral canal cell, a neck-canal cell, and neck cells (Fig. 13-55, A, B). Growth of the archegonial complex ruptures the thin edospore-wall layer at the apical end and a conspicuous papilla or protuberance is formed. With a slight separation of the neck cells and disintegration of the ventral canal and the neck-canal cell, a passage is created for a sperm to enter from the sperm lake.

EMBRYO DEVELOPMENT. Following fertilization, which may occur as early as ten hours after hydration of the sporocarp, the first division of the zygote is longitudinal (in reference to the neck of the archegonium). Each cell of the two-celled embryo (Fig. 13-55, C) then divides transversely, resulting in the quadrant stage of the embryo. There are, however, reports that the first division may be transverse (Smith, 1955). Subsequent development is much like that of members of the Filicales in that the first root, leaf, foot, and shoot apex can be related to specific quadrants of the four-celled embryo (Fig. 13-55, D). Within four or five days the young sporophyte has a well-developed primary root and leaf (Fig. 13-55, F). If a mineral nutrient solution is provided, the sporophyte will continue to grow in a water environment, producing several "juvenile" type leaves before the quadrifid "adult" leaves develop.

Sterile cultures of sporophytes can be established rather easily using special methods that have been developed (Laetsch, 1967).

Land and Water Leaf Forms

Many species of plants exhibit differences in leaf form depending upon whether they are growing submerged in water or on land.

Marsilea is useful in a study of this phenomenon. Adult land and water forms of this genus are easily distinguished from each other. A "water form" has elongated internodes, short petioles, and leaves with a divided lamina, expanded in the plane of the petiole. A "land form" has short, thick internodes and long petioles, with the lamina quadrifid and expanded at right angles to the petiole (Fig. 13-56, A, B).

Allsopp (1963) has related these differences primarily to nutrition—specifically to the internal sugar concentration. Young sporophytes of *Marsilea drummondii* grown in vitro on 5 percent glucose resembled the land form; others grown on 1 to 2 percent glucose resembled the water form. The water form could be induced to develop in plants grown on 4 percent glucose—which normally would result in the land form—if gibberellic acid was added to the medium. This supplemented medium also accelerated growth. Inhibition of protein synthesis was reported to be associated with the development of land forms (White, 1966).

Gaudet (1963, 1965) described leaf development under different light regimes. Young sporophytes of *Marsilea vestita,* grown initially in darkness or far-red light and then transferred to continuous light, developed as land forms. Even sporophytes cultured in a medium that normally permits only the water form to develop responded to a transfer to continuous light in this way. Light of far-red wavelength was more effective than darkness in causing this conversion. Long wavelengths of far-red light do not penetrate as far as blue into water. Gaudet suggested that, therefore, the water form would develop under natural conditions.

This light effect would appear to be a plausible explanation of the phenomenon, but the results of some recent experiments tend to confound the

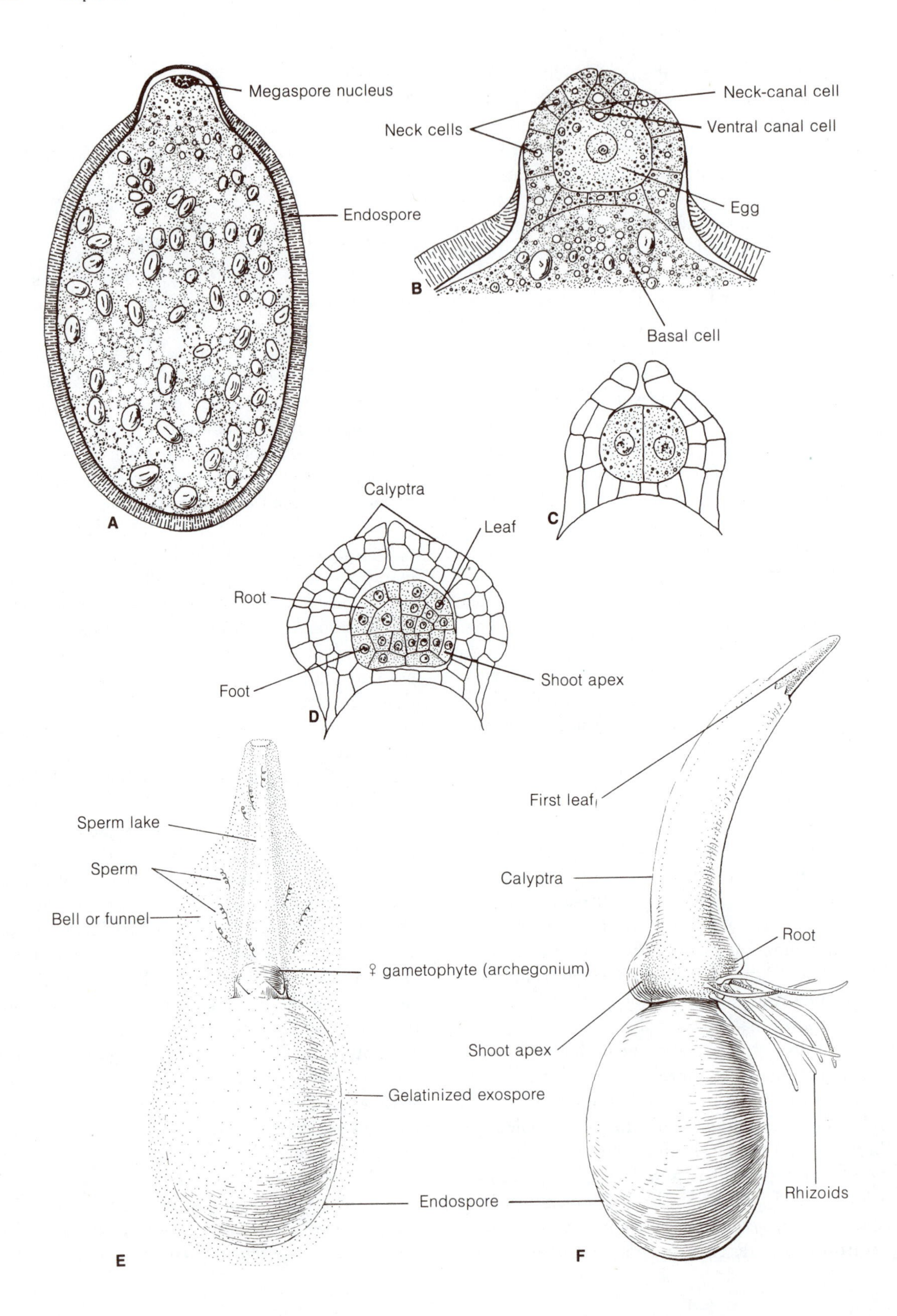
Megaspore nucleus
Endospore
A
Neck-canal cell
Neck cells
Ventral canal cell
Egg
B
Basal cell
C
Calyptra
Leaf
Root
Foot
Shoot apex
D
Sperm lake
Sperm
Bell or funnel
♀ gametophyte (archegonium)
Gelatinized exospore
Endospore
E
First leaf
Calyptra
Root
Shoot apex
Rhizoids
F

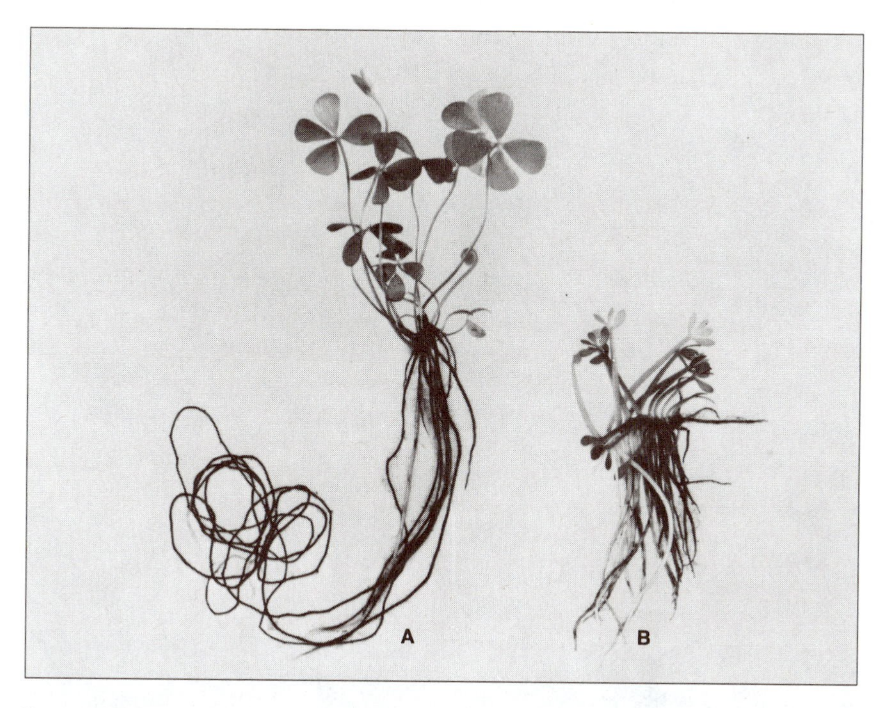

FIGURE 13-56 Sporophytes of *Marsilea* sp. **A,** grown in an inorganic nutrient solution with 4 percent sucrose added; **B,** grown as in **A,** but with the addition of 1 percent sucrose. The "land form" is exhibited in **A,** the "water form" in **B.**

final solution of the problem. By growing young plants of *Marsilea* on a solid substrate in an atmosphere with increased partial pressure of carbon dioxide, water forms were produced instead of land forms. The increased concentrations of carbon dioxide were effective whether the sporophytes were grown in darkness or light (Bristow and Looi, 1968). These studies throw some doubt on the role of internal physiological differences of the types described earlier. It is clear, as White (1971) has stated, that the possible causes are not as simple as first thought—i.e., nutrition and light quality.

Phylogeny

Some fern specialists suggest that the Marsileales should be treated as a family in the Filicales (Table 13-1), possibly related to the Schizaeaceae. The group is undoubtedly related in some fashion to the Filicales, but they have become very specialized through heterospory and the production of endosporic gametophytes. It seems reasonable, until contradictory paleobotanical evidence is forthcoming, to recognize the group as constituting a separate order.

FIGURE 13-55 **A–D,** *Marsilea quadrifolia.* **A,** longitudinal section of uninucleate megaspore, outer wall layers (exospore) removed; **B,** mature archegonium and portion of basal or nutritive cell; **C,** two-celled embryo; **D,** embryo enclosed by developing calyptra; **E,** *Marsilea vestita,* megaspore with enclosed megagametophyte surrounded by gelatinous sheath in which sperm are embedded; **F,** *Marsilea vestita,* external appearance of young sporophyte attached to megaspore and enclosed megagametophyte. [**A–D** redrawn from *Plant Morphology* by A. W. Haupt. McGraw-Hill, New York. 1953.]

A

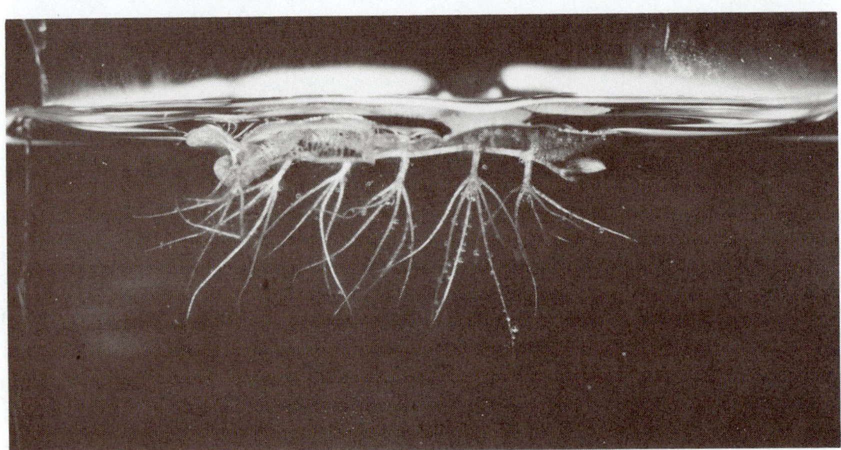

B

FIGURE 13-57 *Salvinia* sp. **A,** group of floating plants as viewed from above (about natural size); note that two of the three leaves at each node can be seen; **B,** a floating plant viewed from the side showing the branched, filiform third leaf at each node, which extends into the water.

FIGURE 13-58 *Azolla filiculoides.* A group of floating plants showing overlapping leaves and slender roots.

Salviniales: *Azolla* and *Salvinia*

The ferns in this order are referred to as "water ferns," and the descriptive term is probably more justified for them than for the Marsileales. Members of the Salviniales are small and float free in ponds and lakes. There are two genera—*Azolla* and *Salvinia* (Figs. 13-57; 13-58). Both genera are heterosporous, and bear sporangia in sporocarps which, in contrast to those in the Marsileales, are modified sori. The gametophytes are endosporic in development as in the Marsileales.

Salvinia, known to have been present in the Cretaceous Period, is primarily of tropical regions and consists of ten species. *Azolla,* with about six extant species, is widely distributed. Several biotypes are recognized for certain species. Species of *Azolla* are known from the Cretaceous and were widespread during the Tertiary (Hall and Swanson, 1968; Hall, 1969).

The leaves of *Salvinia* are in whorls of three on the floating rhizome. There are no roots, but one leaf of each whorl is submerged and is composed of dissected, hairlike filaments (Fig. 13-57, B). These filiform structures may serve the function of roots. The two floating leaves are concave and bear numerous waxy trichomes which can trap air if the plants become immersed.

The thin, floating stem of *Azolla* bears leaves and roots (Fig. 13-58). The crowded leaves are small and bilobed. The upper (dorsal) lobe is photosynthetic, while the lower (ventral) lobe is more delicate and often nearly colorless. Early in development, a pouch is formed in the upper lobe that becomes filled with an atmospheric nitrogen-fixing blue-green alga (cyanobacterium), *Anabaena azollae. Azolla* can be grown on culture media without an added source of nitrogen. *Azolla* has been grown for many years as a "green manure" in rice fields of Vietnam and China, and its use as a fertilizer is being explored for other rice-growing regions of the world (Lumpkin and Plucknett, 1980). When growing in mass, the plants may conceal the water below, and they may develop a distinctive red to reddish-orange anthocyanin in intense light (Holst, 1977). There are methods for obtaining aseptic

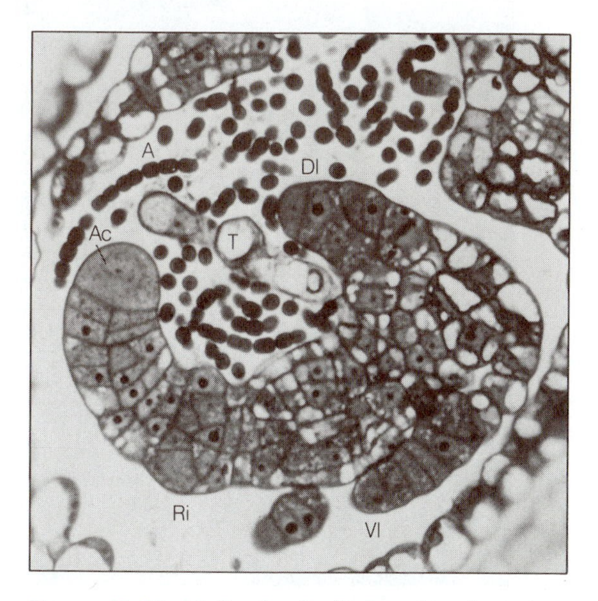

FIGURE 13-59 Median longitudinal section, shoot tip of *Azolla filiculoides.* A, *Anabaena azollae;* Ac, apical cell; Dl, dorsal lobe of leaf; Ri, root initial; T, trichome; Vl, ventral lobe of leaf.

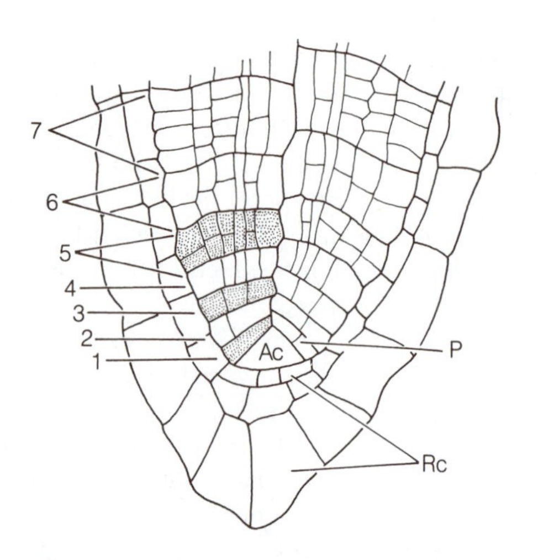

FIGURE 13-60 Median longitudinal section of the tip of a young root of *Azolla filiculoides* showing the apical cell (Ac) and merophytes 1 through 7. Each merophyte originates from one mitotic division of the apical cell. The first division within each merophyte is periclinal (P); transverse divisions first occur in merophyte 5. The root cap (Rc) originates from its own meristem (calyptrogen) very early in root ontogeny.

clones of *Azolla* for experimental purposes (Posner, 1967).

A three-sided apical cell is present at each shoot tip (Fig. 13-59) giving rise to derivatives (merophytes) from two "cutting" faces (Gifford and Polito, 1981a, b). Roots may be only two to three centimeters in length. The root apical cell is tetrahedral, but the root cap is produced by its own meristem (calyptrogen). Studies have shown that the apical cell may be active mitotically only during early root development, giving rise to merophytes or structural units (Fig. 13-60). Continued cell divisions and growth occur in these units as a root grows in length and then very soon becomes determinant (Gunning et al., 1978; Nitayangkura et al., 1980).

Reproductive Cycle

Space does not permit a detailed account of reproduction, but we will describe in a general way that of *Azolla*.

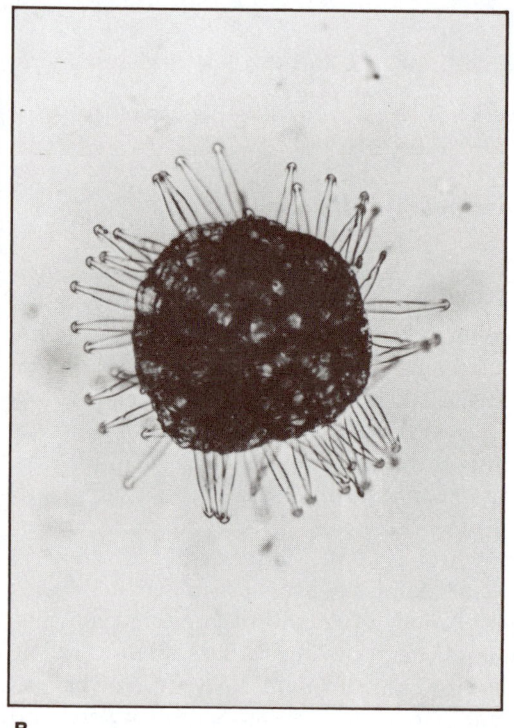

FIGURE 13-61 *Azolla filiculoides*. **A,** ventral side of a sporophyte showing four sporocarps. **B,** one microsporic massula showing barbed glochidia (greatly enlarged).

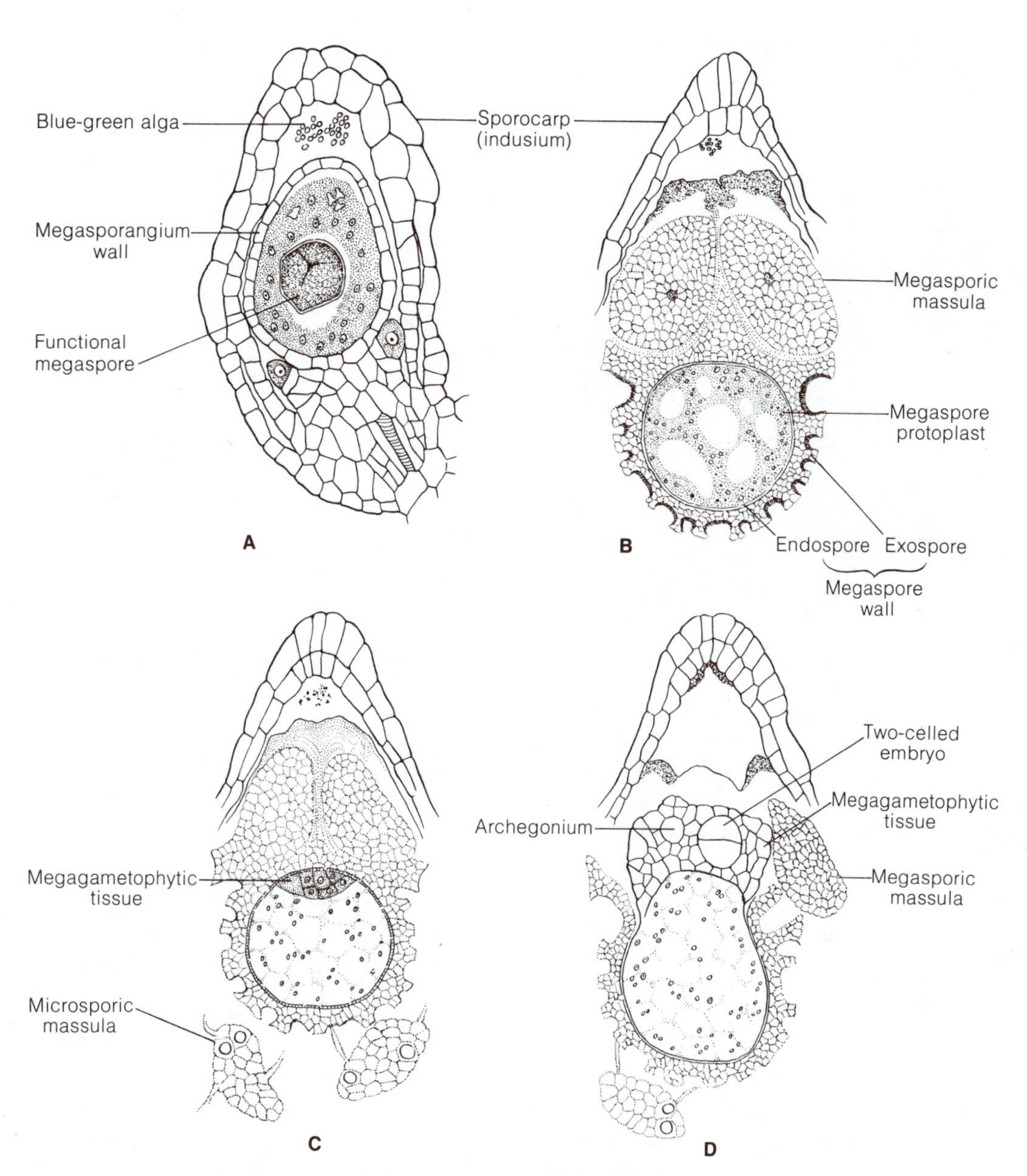

FIGURE 13-62 *Azolla filiculoides.* **A,** megasporangium and surrounding sporocarp wall (indusium); **B–D,** stages in development of megagametophyte and massulae. See text for details. [Redrawn from *Cryptogamic Botany,* Vol. II, 2d edition, by G. M. Smith. McGraw-Hill, New York. 1955.]

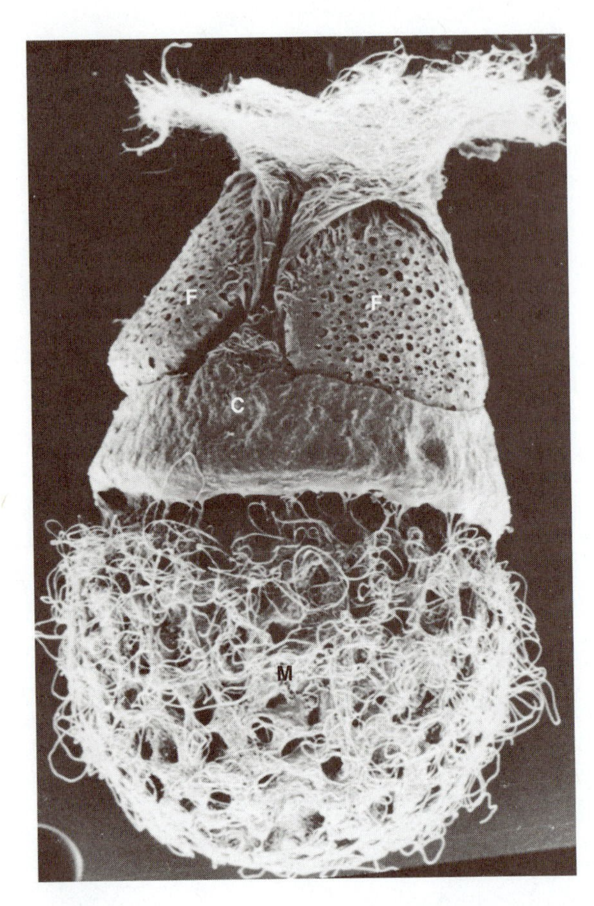

FIGURE 13-63 Scanning electron micrograph, megaspore apparatus of *Azolla filiculoides*. C, collar; F, float; M, megaspore with excrescence of exoperine wall-layer (×180). [Courtesy of Professor E. Cutter.]

Mature sporocarps of *Azolla* are of two sizes. The larger one contains the microsporangia surrounded by an indusium (Fig. 13-61, A). At maturity the megasporocarp contains one functional megasporangium. The differentiation of sporocarps occurs rather late in development, by the survival of either one megasporangium or many microsporangia. A sorus is gradate in development, the potential megasporangium being formed first. A mature megasporangium contains one functional megaspore, surrounded by a multinucleate plasmodium, which is derived from the breakdown of the tapetal cells (Fig. 13-62, A). The functional megaspore increases in size, and the plasmodium above the spore becomes organized into large masses, called *massulae* or "floats" (Fig. 13-62, B). The embedded megaspore soon forms a group of cells (megagametophyte) at its apical end (Fig. 13-62, C). With continued growth the spore wall breaks and the megagametophytic tissue is exposed. Archegonia are then formed at the surface. During the growth processes, the original sporangial wall and sporocarp wall (indusium) are ruptured (Fig. 13-62, D), liberating the megaspore and its surrounding floats (Fig. 13-62, D). The Old World species, *Azolla nilotica* and *A. pinnata,* have nine floats; the American species have three. The entire megaspore apparatus is complex, and the spore wall (sporoderm) consists of an exine and outer perine layers (Fig. 13-63). Some fossil *Azolla* species had numerous floats; the occurrence of three or nine floats in modern-day species appears to be the derived condition (Stewart, 1983).

All of the microspores within a microsporangium are functional (Fig. 13-64, A). During meiosis there is a breakdown of the tapetum into a multinucleate plasmodium, just as in a megasporangium. The spores move to the periphery of the sporangium and become encased in three or more massulae (Fig. 13-64, B, C). In some species of *Azolla,* elongate, hooked processes *(glochidia)* are formed on the massulae (Figs. 13-61, B; 13-64, D). These serve the purpose of attaching the "microsporic massulae" to the megaspore apparatus, thereby bringing the male gametophytes into the vicinity of a megametophyte and better assuring that fertilization will take place (Fig. 13-62, C). The microspores in the massulae germinate by undergoing several cell divisions in the production of sperm. The sperm then swim to the archegonia. The first division of the zygote is reported to be transverse (Fig. 13-62, D) to oblique followed by a longitudinal division in each of the two cells. The four portions of the embryo—first leaf, root, shoot apex, and foot—can be related to the four quadrants of the embryo much like those of other ferns.

As many of the details have been omitted from the account of this extraordinary sexual cycle in ferns, the interested reader should consult the pertinent research articles as well as other textbooks: Smith, 1955; Bonnet, 1957; Bierhorst, 1971; Konar and Kapoor, 1974; Fowler and Stennett-Willson, 1978; Bold et al., 1980.

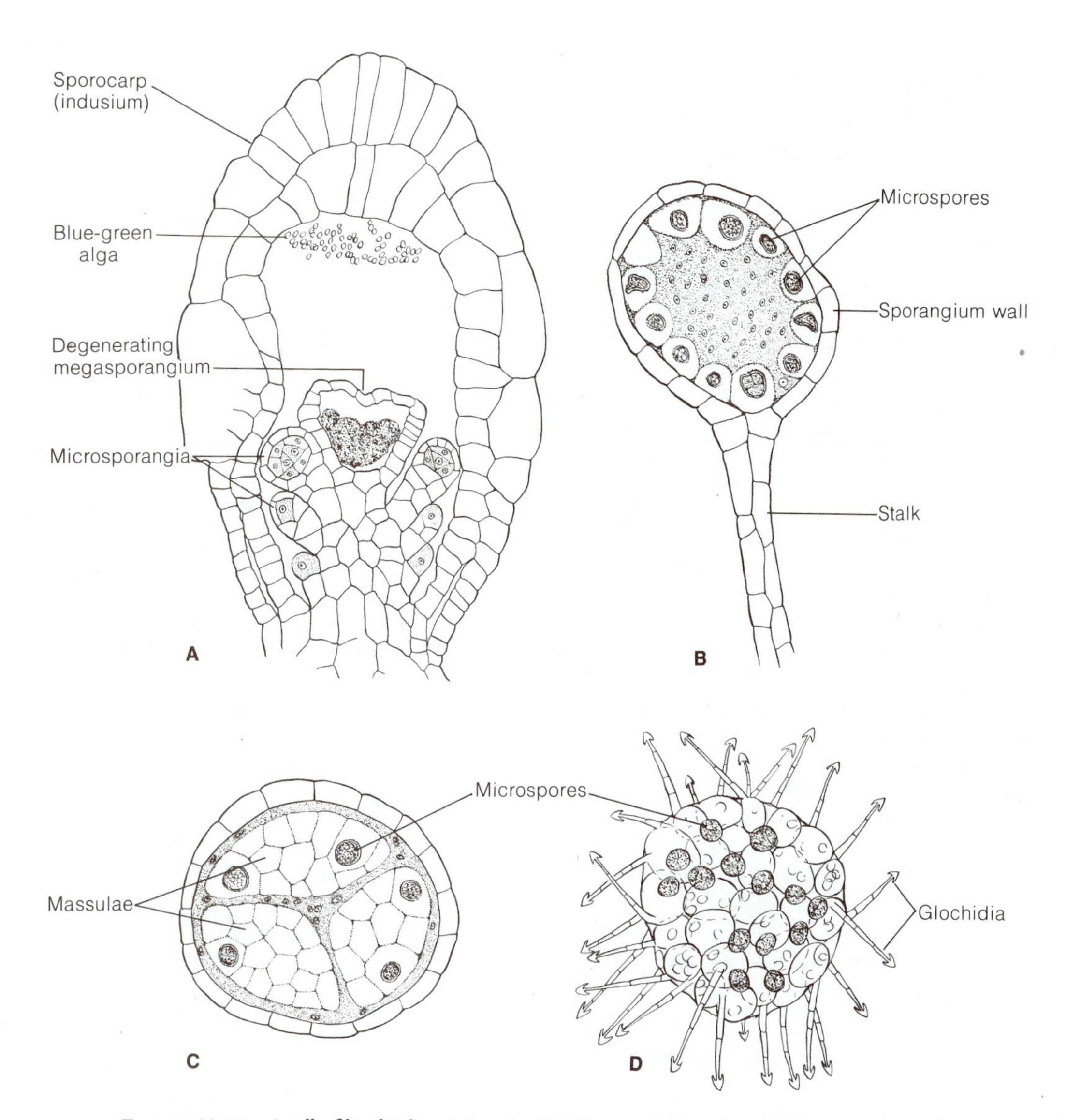

FIGURE 13-64 *Azolla filiculoides.* **A,** longitudinal section of young microsporangial sporocarp; **B, C,** stages in development of massulae within microsporangia; **D,** microsporic massula after liberation from microsporangium. [**A–C** redrawn from *Cryptogamic Botany,* Vol. II, 2d edition, by G. M. Smith. McGraw-Hill, New York. 1955; **D** redrawn from *Morphology of Vascular Plants: Lower Groups* by A. J. Eames. McGraw-Hill, New York. 1936.]

REFERENCES

Abraham, A., C. A. Ninan, and P. M. Mathew
1962. Studies on the cytology and phylogeny of the pteridophytes. VII. Observations on 100 spp. of South Indian ferns. *Jour. Indian Bot. Soc.* 41:339–421.

Albaum, H. G.
1938. Inhibitions due to growth hormones in fern prothallia and sporophytes. *Amer. Jour. Bot.* 25:124–133.

Allsopp, A.
1963. Morphogenesis in *Marsilea. Jour. Linn. Soc. London Bot.* 58:417–427.

Atkinson, L. R., and A. G. Stokey
1964. Comparative morphology of the gametophyte of the homosporous ferns. *Phytomorphology* 14:51–70.

Avanzi, S., and F. D'Amato
1970. Cytochemical and autoradiographic analyses on root primordia and root apices of *Marsilea strigosa.* (A new interpretation of the apical structure in cryptogams.) *Caryologia* 23:335–345.

Bell, P. R.
1970. The archegoniate revolution. *Sci. Progr.* (England) 58:27–45.
1979. Gametogenesis and fertilization in ferns. In A. F. Dyer (ed.), *The Experimental Biology of Ferns,* pp. 471–503. Academic, London.

Bhardwaja, T. N., and J. Baijal
1977. Vessels in rhizome of *Marsilea. Phytomorphology* 27:206–208.

Bierhorst, D. W.
1968. Observations on *Schizaea* and *Actinostachys* spp., including *A. oligostachys,* sp. nov. *Amer. J. Bot.* 55:87–108.
1971. *Morphology of Vascular Plants.* Macmillan, New York.
1977. On the stem apex, leaf initiation and early leaf ontogeny in filicalean ferns. *Amer. J. Bot.* 64:125–152.
1983. On the embryogeny of *Schizaea dichotoma. Amer. J. Bot.* 70:1057–1062.

Bilderback, D. E.
1978. The development of the sporocarp of *Marsilea vestita. Amer. J. Bot.* 65:629–637.

Bliss, M. C.
1939. The tracheal elements in the ferns. *Amer. Jour. Bot.* 26:620–624.

Bold, H., C. J. Alexopoulos, and T. Delevoryas
1980. *Morphology of Plants and Fungi,* 4th edition. Harper and Row, New York.

Bonnet, A. L.-M.
1957. Contribution à l'étude des Hydroptéridées. III. Recherches sur *Azolla filiculoides* Lamk. *Rev. Cytol. Biol. Veg.* 18:1–88.

Bower, F. O.
1923. *The Ferns,* Vol. I. Cambridge University Press, London.
1935. *Primitive Land Plants.* Macmillan, London.

Braithwaite, A. F.
1964. A new type of apogamy in ferns. *New Phytol.* 63:293–305.

Briggs, W. R., and T. A. Steeves
1959. Morphogenetic studies on *Osmunda cinnamomea* L. The mechanism of crozier uncoiling. *Phytomorphology* 9:134–147.

Bristow, J. M., and A. Looi
1968. Effects of carbon dioxide on the growth and morphogenesis of *Marsilea. Amer. Jour. Bot.* 55:884–889.

Chrysler, M. A.
1943. The vascular structure of the leaf of *Gleichenia.* I. The anatomy of the branching regions. *Amer. J. Bot.* 30:735–743.

Copeland, E. B.
1939. Antarctica as the source of existing ferns. *Pac. Sci. Congr. Proc.* 4:625–627.
1942. Edible ferns. *Amer. Fern Jour.* 32:121–126.
1947. *Genera Filicum* (The Genera of Ferns). Chronica Botanica Co., Waltham, Mass.

Coquillat, M.
1950. Au sujet du "pain de fougère" en Mâconnais. *Bull. Mens. Soc. Linn. Lyon* 19:173–175.

Cutter, E. G.
1954. Experimental induction of buds from fern leaf primordia. *Nature* 173:440–441.
1956. Experimental and analytical studies of Pteridophytes. XXXIII. The experimental induction of buds from leaf primordia in *Dryopteris aristata* Druce. *Ann. Bot.* n.s. 20:143–165.
1965. Recent experimental studies of the shoot apex and shoot morphogenesis. *Bot. Rev.* 31:7–113.

D'Amato, F., and S. Avanzi
1965. DNA content, DNA synthesis and mitosis in the root apical cell of *Marsilea strigosa. Caryologia* 18:383–394.

De Maggio, A. E.
1963. Morphogenetic factors influencing the development of fern embryos. *Jour. Linn. Soc. London Bot.* 58:361–376.

DeMaggio, A. E., and D. A. Stetler
1985. Mobilisation of storage reserves during fern spore germination. *Proc. Roy. Soc. Edinburgh* 86B:195–202.

Demalsy-Feller, M.-J.
1957. Etudes sur les Hydropteridales. V. Gamétophytes et gamétogenèse dans le genre *Marsilea. Cellule* 58:169–207.

Eames, A. J.
1936. *Morphology of Vascular Plants. Lower Groups.* McGraw-Hill, New York.

Elmore, H. W., and D. P. Whittier
1973. The role of ethylene in the induction of apogamous buds in *Pteridium* gametophytes. *Planta* 111:85–90.
1975. The involvement of ethylene and sucrose in the inductive and developmental phases of apogamous bud formation in *Pteridium* gametophytes. *Can. J. Bot.* 53:375–381.

Esau, K.
1954. Primary vascular differentiation in plants. *Biol. Rev.* 29:46–86.
1965. *Plant Anatomy,* 2d edition. Wiley, New York.

Evans, A. M.
1964. Ameiotic alternation of generations: a new life cycle in the ferns. *Science* 143:261–263.

Farrar, D. R.
1967. Gametophytes of four tropical fern genera reproducing independently of their sporophytes in the southern Appalachians. *Science* 155:1266–1267.
1974. Gemmiferous fern gametophytes—Vittariaceae. *Amer. J. Bot.* 61:146–155.
1978. Problems in the identity and origin of the Appalachian *Vittaria* gametophyte, a sporophyteless fern of the eastern United States. *Amer. J. Bot.* 65:1–12.
1985. Independent fern gametophytes in the wild. *Proc. Roy. Soc. Edinburgh* 86B:361–369.

Farrar, D. R., J. C. Parks, and B. W. McAlpin
1983. The fern genera *Vittaria* and *Trichomanes* in the northeastern United States. *Rhodora* 85:83–92.

Fowler, K., and J. Stennett-Willson
1978. Sporoderm architecture in modern *Azolla. Fern Gaz.* 11:405–412.

Gantt, E., and H. J. Arnott
1965. Spore germination and development of the young gametophyte of the ostrich fern *(Matteuccia struthiopteris). Amer. Jour. Bot.* 52:82–94.

Gastony, G. J., and L. D. Gottlieb
1982. Evidence for genetic heterozygosity in a homosporous fern. *Amer. J. Bot.* 69:634–637.
1985. Genetic variation in the homosporous fern *Pellaea andromedifolia. Amer. J. Bot.* 72:257–267.

Gaudet, J. J.
1963. *Marsilea vestita:* conversion of the water form to the land form by darkness and by far-red light. *Science* 140:975–976.
1965. The effect of various environmental factors on the leaf form of the aquatic fern *Marsilea vestita. Physiol. Plant.* 18:674–686.

Gifford, E. M., Jr., and V. S. Polito
1981a. Growth of *Azolla filiculoides. BioScience* 31:526–527.
1981b. Mitotic activity at the shoot apex of *Azolla filiculoides. Amer. J. Bot.* 68:1050–1055.

Gruber, T. M.
1981. The branching pattern of *Hypolepis repens. Amer. Fern Jour.* 71:41–47.

Gunning, B. E. S., J. E. Hughes, and A. R. Hardham
1978. Formative and proliferative cell divisions, cell differentiation, and developmental changes in the meristem of *Azolla* roots. *Planta* 143:121–144.

Guttenberg, H. von
1966. Histogenese der Pteridophyten. In W. Zimmermann (ed.), *Handbuch der Pflanzenanatomie,* Band VII, Teil 2. Gebrüder Borntraeger, Berlin.

Haight, T. H., and C. C. Kuehnert
1969. Developmental potentialities of leaf primordia of *Osmunda cinnamomea.* V. Toward greater understanding of the final morphogenetic expression of isolated set I cinnamon fern leaf primordia. *Can. J. Bot.* 47:481–488.

Hall, J. W.
1969. Studies on fossil *Azolla:* primitive types of megaspores and massulae from the Cretaceous. *Amer. Jour. Bot.* 56:1173–1180.

Hall, J. W., and N. P. Swanson
1968. Studies on fossil *Azolla: Azolla montana,* a Cretaceous megaspore with many small floats. *Amer. Jour. Bot.* 55:1055–1061.

Halperin, W.
1978. Organogenesis at the shoot apex. In W. R. Briggs, P. B. Green, and R. L. Jones (eds.), *Annual Review of Plant Physiology* 29:239–262.

Haufler, C. H., and D. E. Soltis
1984. Obligate outcrossing in a homosporous fern: field confirmation of a laboratory prediction. *Amer. J. Bot.* 71:878–881.

Hewitson, W.
1962. Comparative morphology of the Osmundaceae. *Ann. Missouri Bot. Gard.* 49:57–93.

Hicks, G. S., and T. A. Steeves
1969. In vitro morphogenesis in *Osmunda cinnamomea.* The role of the shoot apex in early leaf development. *Can. Jour. Bot.* 47:575–580.

Holst, R. W.
1977. Anthocyanins of *Azolla*. *Amer. Fern J.* 67:99–100.

Holttum, R.
1947. A revised classification of the leptosporangiate ferns. *Jour. Linn. Soc. London Bot.* 53:123–158.
1949. The classification of ferns. *Biol. Rev.* 24:267–296.

Howland, G. P., and M. E. Edwards
1979. Photomorphogenesis of fern gametophytes. In A. F. Dyer (ed.), *The Experimental Biology of Ferns*, pp. 393–434. Academic, London.

Jayasekera, R. D. E., and P. R. Bell
1959. The effect of various experimental treatments on the development of the embryo of the fern *Thelypteris palustris*. *Planta* 54:1–14.

Klekowski, E. J., Jr.
1971. Ferns and genetics. *BioScience* 21:317–322.
1973. Sexual and subsexual systems in homosporous pteridophytes: a new hypothesis. *Amer. J. Bot.* 60:535–544.
1979. The genetics and reproductive biology of ferns. In A. F. Dyer (ed.), *The Experimental Biology of Ferns*, pp. 133–170. Academic, London.

Klekowski, E. J., Jr., and H. G. Baker
1966. Evolutionary significance of polyploidy in the Pteridophyta. *Science* 153:305–307.

Konar, R. N., and R. K. Kapoor
1974. Embryology of *Azolla pinnata*. *Phytomorphology* 24:228–261.

Kuehnert, C. C.
1972. On determination of leaf primordia in *Osmunda cinnamomea* L. In M. W. Miller and C. C. Kuehnert (eds.), *The Dynamics of Meristem Cell Populations*, pp. 101–118. Plenum Press, N.Y.

Kuehnert, C. C., and J. P. Miksche
1964. Application of the 22.5 Mev deuteron microbeam to the study of morphogenetic problems within the shoot apex of *Osmunda claytoniana*. *Amer. J. Bot.* 51:743–747.

Kuligowski-Andrès, J.
1978. Contribution à l'étude d'une fougère, *Marsilea vestita* (Marsileacées), du stade embryon au stade sporophyte adulte. III. Le méristème apical du sporophyte juvénile: étude histochimique et histoautoradiographique. *Ann. Sci. Nat. Bot. Sér.* 12, 19:219–248.

Kurth, E.
1981. Mitotic activity in the root apex of the water fern *Marsilea vestita* Hook. and Grev. *Amer. J. Bot.* 68:881–896.

Labouriau, L. G.
1952. On the latex of *Regnellidium diphyllum* Lindm. *Phyton* 2:57–74.

Laetsch, W. M.
1967. Ferns. In F. H. Wilt and N. K. Wessels (eds.), *Methods in Developmental Biology*, pp. 319–328. Crowell, New York.

Larson, P. R.
1975. Development and organization of the primary vascular system in *Populus deltoides* according to phyllotaxy. *Amer. J. Bot.* 62:1084–1099.

Lloyd, R. M.
1974. Reproductive biology and evolution in the Pteridophyta. *Ann. Missouri Bot. Gard.* 61:318–331.

Lumpkin, T. A., and D. L. Plucknett
1980. *Azolla:* botany, physiology, and use as a green manure. *Econ. Bot.* 34:111–153.

Machlis, L., and E. Rawitscher-Kunkel
1967. The hydrated megaspore of *Marsilea vestita*. *Amer. Jour. Bot.* 54:689–694.

Manton, I.
1950. *Problems of Cytology and Evolution in the Pteridophyta*. Cambridge University Press, Cambridge.

McAlpin, B. W., and R. A. White
1974. Shoot organization in the Filicales: the promeristem. *Amer. J. Bot.* 61:562–579.

Miller, C. N., Jr.
1967. Evolution of the fern genus *Osmunda*. *Contr. Mus. Paleontol. Univ. Mich.* 21:139–203.

Miller, J. H.
1968. Fern gametophytes as experimental material. *Bot. Rev.* 34:361–440.

Mohr, H.
1963. The influence of visible radiation on the germination of archegoniate spores and the growth of the fern protonema. *Jour. Linn. Soc. London Bot.* 58:287–296.

Morel, G.
1963. Leaf regeneration in *Adiantum pedatum*. *Jour. Linn. Soc. London Bot.* 58:381–383.

Myles, D. G.
1975. Structural changes in the sperm of *Marsilea vestita* before and after fertilization. In J. G. Duckett and P. A. Racey (eds.), *The Biology of*

the Male Gamete, pp. 129–134. Academic, London.

Näf, U.
1979. Antheridiogens and antheridial development. In A. F. Dyer (ed.), *The Experimental Biology of Ferns,* pp. 435–470. Academic, London.

Nayar, B. K., and S. Kaur
1971. Gametophytes of homosporous ferns. *Bot. Rev.* 37:295–396.

Nitayangkura, S., E. M. Gifford, Jr., and T. Rost
1980. Mitotic activity in the root apical meristem of *Azolla filiculoides* Lam., with special reference to the apical cell. *Amer. J. Bot.* 67:1484–1492.

O'Brien, T. P.
1963. The morphology and growth of *Pteridium aquilnum* var. *esculentum* (Forst.) Kuhn. *Ann. Bot.* n.s. 27:253–267.

Ogura, Y.
1972. Comparative anatomy of vegetative organs of the Pteridophytes. In W. Zimmerman (ed.), *Handbuch der Pflanzenanatomie,* 2d edition, Band VII, Teil 3. Gebrüder Borntraeger, Berlin.

Ohlenroth, K., and H. Mohr
1964. Die Steuerung der Protein-synthese durch Blaulicht und Hellrot in den Vorkeimen von *Dryopteris filix-mas* (L.) Schott. *Planta* 62:160–170.

Perring, F. H., and B. G. Gardiner (eds.)
1976. The Biology of Bracken. *Bot. Jour. Linn. Soc.* 73: nos. 1–3. Academic, London.

Pichi-Sermolli, R. E. G.
1958. The higher taxa of the Pteridophyta and their classification. In O. Hedberg (ed.), *Systematics of Today,* pp. 70–90. Almquist and Wiksells, Uppsala.

Posner, H. B.
1967. Aquatic vascular plants. In F. H. Wilt and N. K. Wessells (eds.), *Methods in Developmental Biology,* pp. 301–317. Crowell, New York.

Pray, T. R.
1962. Ontogeny of the closed dichotomous venation of *Regnellidium. Amer. Jour. Bot.* 49:464–472.
1968. The gametophytes of *Pellaea* section *Pellaea:* dark-stiped series. *Phytomorphology* 18:113–143.

Puri, V., and M. L. Garg
1953. A contribution to the anatomy of the sporocarp of *Marsilea minuta* L. with a discussion of the nature of sporocarp in the Marsileaceae. *Phytomorphology* 3:190–209.

Raghavan, V.
1973. Photomorphogenesis of the gametophytes of *Lygodium japonicum. Amer. J. Bot.* 60:313–321.
1974. Control of differentiation in the fern gametophyte. *Amer. Scientist* 62:465–475.
1980. Cytology, physiology, and biochemistry of germination of fern spores. *Intl. Rev. Cytol.* 62:69–118.

Raghavan, V., and C. S. Huckaby
1980. A comparative study of cell division patterns during germination of spores of *Anemia, Lygodium* and *Mohria* (Schizeaceae). *Amer. J. Bot.* 67:653–663.

Rice, H. V., and W. M. Laetsch
1967. Observations on the morphology and physiology of *Marsilea* sperm. *Amer. Jour. Bot.* 54:856–866.

Rivières, R.
1959. Sur la culture in vitro d'embryons isolés de Polypodiacées. *Compt. Rend. Acad. Sci.* (Paris) 248:1004–1007.

Rubin, G., and D. J. Paolillo, Jr.
1983. Sexual development of *Onoclea sensibilis* on agar and soil media without the addition of antheridiogen. *Amer. J. Bot.* 70:811–815.

Schedlbauer, M. D.
1976. Fern gametophyte development: controls of dimorphism in *Ceratopteris thalictroides. Amer. J. Bot.* 63:1080–1087.

Schmid, R.
1982. The terminology and classification of steles: historical perspective and the outlines of a system. *Bot. Rev.* 48:817–931.

Schraudolf, H.
1962. Die Wirkung von Phytohormonen auf Keimung und Entwicklung von Farnprothallien. I. Auslösung der Antheridienbildung und Dunkelkeimung bei Schizaeaceen Gibberellinsäure. *Biol. Zentralbl.* 81:731–740.

Sheffield, E.
1984. Apospory in the fern *Pteridium aquilinum* (L) Kuhn. 1. Low temperature scanning electron microscopy. *Cytobios* 39:171–176.
1985. Cellular aspects of the initiation of aposporous outgrowths in ferns. *Proc. Roy. Soc. Edinburgh* 86B:45–50.

Singh, R., D. R. Bohra, and B. D. Sharma
1978. Vessels in the rhizome of *Actiniopteris radiata. Phytomorphology* 28:455–457.

Smith, G. M.
1955. *Cryptogamic Botany,* Vol. II, 2d edition. McGraw-Hill, New York.

Sobota, A. E.
1970. Interaction of red light and an inhibitor produced in the meristem of the fern gametophyte. *Amer. Jour. Bot.* 57:530–534.

Sobota, A. E., and C. R. Partanen
1966. The growth and division of cells in relation to morphogenesis in fern gametophytes. I. Photomorphogenetic studies in *Pteridium aquilinum. Can. Jour. Bot.* 44:497–506.
1967. The growth and division of cells in relation to morphogenesis in fern gametophytes. II. The effect of biochemical agents on the growth and development of *Pteridium aquilinum. Can. Jour. Bot.* 45:595–603.

Soltis, D. E., and P. S. Soltis
1987. Polyploidy and breeding systems in homosporous Pteridophyta: a reevaluation. *Amer. Naturalist* 130:219–232.

Steeves, T. A.
1961. A study of the developmental potentialities of excised leaf primordia in sterile culture. *Phytomorphology* 11:346–359.
1962. Morphogenesis in isolated fern leaves. *Symp. Soc. Study Develop. Growth* 20:117–151.

Steeves, T. A., and R. H. Wetmore
1953. Morphogenetic studies on *Osmunda cinnamomea* L.: some aspects of the general morphology. *Phytomorphology* 3:339–354.

Stevenson, D. W.
1976. Shoot apex organization and origin of the rhizome-borne roots and their associated gaps in *Dennstaedtia cicutaria. Amer. J. Bot.* 63:673–678.

Stewart, W. N.
1983. *Paleobotany and the Evolution of Plants.* Cambridge University Press, Cambridge.

Tewari, R. B.
1975. Structure of vessels and tracheids of *Regnellidium diphyllum* Lindman (Marsileaceae). *Ann. Bot.* 39:229–231.

Tidwell, W. D., and S. R. Rushforth
1970. *Osmundacaulis wadei,* a new osmundaceous species from the Morrison Formation (Jurassic) of Utah. *Bull. Torrey Bot. Club* 97:137–144.

Treanor, L. L., and D. P. Whittier
1968. The effect of mineral nutrition on apogamy in *Pteridium. Can. Jour. Bot.* 47:773–777.

Tryon, A. F.
1964. *Platyzoma*—a Queensland fern with incipient heterospory. *Amer. Jour. Bot.* 51:939–942.

Tryon, A. F., and G. Vida
1967. *Platyzoma:* a new look at an old link in ferns. *Science* 156:1109–1110.

Tryon, R. M.
1964. Evolution in the leaf of living ferns. *Mem. Torrey Bot. Club* 21:73–85.

Tryon, R. M., and A. F. Tryon
1982. *Ferns and Allied Plants.* Springer-Verlag, New York.

Vallade, J., and F. Bugnon
1979. Le rôle de l'apicale dans la croissance de la racine du *Marsilea diffusa. Rev. Cytol. Biol. Vég. Bot.* 2:293–308.

Vladesco, M. A.
1935. Recherches morphologiques et expérimentales sur l'embryogénie et l'organogénie des fougères leptosporangiées. *Rev. Gen. Bot.* 47:513–528; 564–588.

Voeller, B.
1971. Developmental physiology of fern gametophytes: relevance for biology. *BioScience* 21:266–270.

von Aderkas, P.
1984. Economic history of ostrich fern, *Matteuccia struthiopteris,* the edible fiddlehead. *Econ. Bot.* 38:14–23.

Wagner, W. H., Jr.
1980. Origin and philosophy of the Groundplan-divergence Method of cladistics. *Syst. Bot.* 5:173–193.

Wagner, W. H., Jr., and A. J. Sharp
1963. A remarkably reduced vascular plant in the United States. *Science* 142:1483–1484.

Walker, T. G.
1979. The cytogenetics of ferns. In A. F. Dyer (ed.), *The Experimental Biology of Ferns,* pp. 87–132. Academic, London.
1985. Some aspects of agamospory in ferns—the Braithwaite system. *Proc. Roy. Soc. Edinburgh* 86B:59–66.

Ward, M.
1954. Fertilization in *Phlebodium aureum* J. Sm. *Phytomorphology* 4:1–17.
1963. Developmental patterns of adventitious sporophytes in *Phlebodium aureum* J. Sm. *Jour. Linn. Soc. London Bot.* 58:377–380.

Ward, M., and R. H. Wetmore
1954. Experimental control of development in the

embryo of the fern, *Phlebodium aureum*. *Amer. Jour. Bot.* 41:428–434.

Wardlaw, C. W.
1946. Experimental and analytical studies of pteridophytes. VIII. Further observations on bud development in *Matteuccia struthiopteris*, *Onoclea sensibilis*, and species of *Dryopteris*. *Ann. Bot.* n.s. 10:117–132.
1949. Further experimental observations on the shoot apex of *Dryopteris aristata* Druce. *Phil. Trans. Roy. Soc. London* 223(B):415–451.
1952. *Phylogeny and Morphogenesis*. Macmillan, London.
1958. Reflections on the unity of the embryonic tissues in ferns. *Phytomorphology* 8:323–327.
1962. The sporogenous meristems of ferns: a morphogenetic commentary. *Phytomorphology* 12:394–408.
1965. The organization of the shoot apex. In W. Ruhland (ed.), *Encylopedia of Plant Physiology*, Vol. XV, pp. 966–1076. Springer-Verlag, New York.
1968. *Essays on form in plants*. University Press, University of Manchester.

Wardlaw, C. W., and E. G. Cutter
1954. Effect of deep and shallow incisions on organogenesis at the fern apex. *Nature* 174:734–735.
1956. Experimental and analytical studies of pteridophytes. XXXI. The effect of shallow incisions on organogenesis in *Dryopteris aristata* Druce. *Ann. Bot.* n.s. 20:39–56.

Wardlaw, C. W., and D. N. Sharma
1961. Experimental and analytical studies of pteridophytes. XXXIX. Morphogenetic investigations of sori in leptosporangiate ferns. *Ann. Bot.* n.s. 25:477–490.

Webster, B. D., and T. A. Steeves
1958. Morphogenesis in *Pteridium aquilinum* (L.) Kuhn. General morphology and growth habit. *Phytomorphology* 8:30–41.

Wetmore, R. H.
1950. Tissue and organ culture as a tool for studies in development. *Rep. Proc. 7th Int. Bot. Congr.* (Stockholm) p. 369.
1954. The use of "in vitro" cultures in the investigation of growth and differentiation in vascular plants. *Brookhaven Symp. Biol.* 6:22–40.

White, R. A.
1961. Vessels in roots of *Marsilea*. *Science* 133:1073–1074.
1963a. Tracheary elements of the ferns. I. Factors which influence tracheid length; correlation of length with evolutionary divergence. *Amer. Jour. Bot.* 50:447–455.
1963b. Tracheary elements of the ferns. II. Morphology of tracheary elements; conclusions. *Amer. Jour. Bot.* 50:514–522.
1966. The morphological effects of protein synthesis inhibition in *Marsilea*. *Amer. Jour. Bot.* 53:158–165.
1971. Experimental studies of the sporophytes of ferns. *BioScience* 21:271–275.
1979. Experimental investigations of fern sporophyte development. In A. F. Dyer (ed.), *The Experimental Biology of Ferns*, pp. 505–549. Academic, London.

Whittier, D. P.
1964. The effect of sucrose on apogamy in *Cyrtomium falcatum* Presl. *Amer. Fern Jour.* 54:20–25.
1965. Obligate apogamy in *Cheilanthes tomentosa* and *C. alabamensis*. *Bot. Gaz.* 126:275–281.
1966. The influence of growth substances on the induction of apogamy in *Pteridium* gametophytes. *Amer. Jour. Bot.* 53:882–886.

Whittier, D. P., and T. A. Steeves
1962. Further studies on induced apogamy in ferns. *Can. J. Bot.* 40:1525–1531.

Zurakowski, K. A., and E. M. Gifford
1988. Quantitative studies of pinnule development in the ferns *Adiantum raddianum* cv. Decorum and *Chleilanthes viridis*. *Amer. Jour. Bot.* 75:1559–1570.

CHAPTER 14

General Morphology of Gymnosperms and the Progymnospermophyta

THE term *gymnosperm* literally means "naked seed"; it designates an important characteristic of those groups of seed-bearing plants in which the seeds are not enclosed within a carpel, as in angiosperms, but are borne on sporophylls, scales, or comparable structures. The gymnosperms include ancient lines of seed-bearing plants. Their long evolutionary history — extending back at least 300 million years — contains many examples of organisms that flourished for a time and then became extinct as a result of changes in climate, topography, and biological competition. One of the most interesting and phylogenetically important of the extinct gymnosperms was an assemblage of plants decidedly fernlike in its foliage and general appearance and which possessed a primitive type of seed. Indeed, the pinnatifid leaves of some of these plants for a long time were classified as those of ferns, and it was not until seeds were found attached to the fronds that the unique nature of these organisms was fully appreciated. This group was well named the "seed ferns" or "pteridosperms" (Chapter 15).

In addition to the seed ferns, two other extinct orders of gymnosperms are well known from the fossil record: the Cordaitales (cordaites) and the Cycadeoidales. The cordaites formed large forests, reached considerable height, and differed from the seed ferns and the majority of cycadophytes by their simple leaves, which in some were as long as 1 meter (Chapter 17). The widely distributed extinct form genus *Callixylon* — notable for the beautiful preservation of the wood of the trunk and branches — was considered for many years a member of the Cordaitales until Beck (1960b) discovered well-known fossil fernlike fronds (form genus *Archeopteris)* attached to the stems. Beck then established a new major taxon, the Progymnospermopsida, for this group of plants. (See p. 334 for a discussion of the reconstruction of the progymnosperm *Archaeopteris*.)

The Cycadeoidales, which resembled certain modern-day cycads in their short or columnar, weakly branched trunks and often large, pinnately compound leaves, flourished during the Jurassic (Chapter 15). The cycadeoids were contemporaneous with extinct cycads (Cycadales) and the large dinosaurs; it is because of the abundance of cycadophytes in the Jurassic that this period is called the "age of cycads and cycadeoids."

Classification

Many years ago Chamberlain (1935) classified the gymnosperms into two major evolutionary groups —the "cycadophytes" and the "coniferophytes." The cycadophytes include gymnosperms with fern-like, pinnatifid leaves, globose or columnar trunks that are weakly branched and have large conspicuously developed pith and cortical zones in the stems. The cylinder of secondary xylem is relatively small and is composed of tracheids with abundant parenchyma; wood of this type has been termed *manoxylic* (Greek: "loose texture"). The cycadophytes are essentially a fossil group and comprise the extinct seed ferns (e.g., Fig. 15-4) and cycadeoids (e.g., Fig. 15-12); the only surviving representatives are the modern cycads (e.g., Figs. 15-16, 15-17, 15-18). Exceptions to these characteristics are known, especially in the seed ferns, but the term cycadophytes is useful when applied in a general way. In contrast, the coniferophytes are distinguished by their more profusely branched trunks, their simple leaves (needlelike, scalelike, or laminate) and the relatively small pith and cortex in the stem. The amount of secondary xylem of typical coniferophytes is massive and less parenchymatous than that of the cycadophytes; wood of this type is termed *pycnoxylic* (Greek: "thick, dense"). In this system the coniferophytes include the extinct Cordaitales and Voltziales, the extinct and living conifers, and the Ginkgoales. Because of their uncertain phylogeny, the Gnetophyta are usually not included in either the cycadophytes or coniferophytes, but have been regarded for many years as an enigmatic group. See Chapter 18 for possible relationships.

In recent years there has been a general tendency to establish several major taxa of extinct and living gymnosperms. In this book we recognize seven major taxa: (1) Progymnospermophyta (an extinct group that may have been ancestral to other gymnosperms), (2) Pteridospermophyta (extinct seed ferns), (3) Cycadophyta (cycads), (4) Cycadeoidophyta (extinct cycadlike plants), (5) Ginkgophyta (*Ginkgo*), (6) Coniferophyta (conifers), and (7) Gnetophyta (three genera of unique plants that have some similarities to angiosperms—the flowering plants). An expanded version of this classification, together with the occurrence of the taxa in geologic time, is shown in Table 14-1.

Although the fossil record has convinced many morphologists and paleobotanists that the "cycadophytes" and "coniferophytes" probably represent different lines of gymnosperm evolution, it is still uncertain whether the two lines arose from a common ancestor or originated independently. As we will show later, the answer to the problem perhaps lies in the continued study of the progymnosperms—a plexus of Middle Devonian

Table 14-1 The classification of gymnosperms used in this textbook. Only the major divisions and selected orders are listed, and their occurrence in geologic time is indicated.

Taxa	Geologic Time
Progymnospermophyta (Division)	Middle Devonian to Lower Carboniferous
Aneurophytales (Order)	
Archaeopteridales (Order)	
Pteridospermophyta	
Pteridospermales	Carboniferous to Permian
Glossopteridales	Permian to Triassic
Caytoniales	Triassic to Cretaceous
Cycadophyta	Permian to Recent
Cycadales	
Cycadeoidophyta	Triassic to Cretaceous
Cycadeoidales	
Ginkgophyta	Triassic to Recent
Ginkgoales	
Coniferophyta	
Cordaitales	Upper Carboniferous to Permian
Voltziales	Upper Carboniferous to Jurassic
Coniferales	Triassic to Recent
Gnetophyta	Permian (?) to Recent
Ephedrales	
Gnetales	
Welwitschiales	

to Lower Carboniferous plants which combined, in a remarkable way, free-sporing reproduction with a gymnospermous type of wood anatomy.

A detailed account of selected representatives of gymnosperms will be presented in Chapters 15 through 18. However, it is desirable to describe the general nature and geographical distribution of the surviving gymnosperms. The largest group, containing the most familiar gymnosperms, is the Coniferophyta. Examples of this order are pines, spruces, firs, cedars, and junipers. Some conifers are not only among the largest plants on earth, but are also organisms with a lifespan exceeding 3,000 years (e.g., *Sequoiadendron giganteum* and *Pinus longavea*). The Coniferophyta are worldwide in distribution, and many of them form extensive forests in the northern and southern hemispheres. In contrast, the living representatives of the Ginkgophyta and Cycadophyta are veritable "living fossils." *Ginkgo biloba,* for example, is the sole living member of the Ginkgophyta and exists in the wild state only in certain mountains in southeastern China. The living cycads are also relicts from a past age and are confined to limited areas in the subtropics and tropics. The cycads only rarely form continuous stands or represent abundant components of a given flora. The final living group of the gymnosperms is the Gnetophyta. *Welwitschia* is a monotypic genus restricted to certain desert regions in southwest Africa. The other two genera, *Ephedra* and *Gnetum,* are more diversified taxonomically and have a wider range of distribution. *Ephedra*—usually a shrub or small tree—occurs in tropical and temperate Asia and certain temperate regions of North and South America. *Gnetum,* which is most commonly a vine, is found in tropical areas in Asia, Africa, and South America. The Gnetophyta has frequently been regarded as a group bearing certain relations to the angiosperms (see Chapter 18).

Ontogeny and Structure of the Seed

The precursor of a gymnosperm seed is the *ovule,* a structure often characterized as "an integumented megasporangium." Phylogenetically, however, it is not yet certain whether primitive types of ovules originated from sporangia of homosporous or heterosporous plants (see Doyle, 1953). Furthermore, delineating the morphology and phyletic origin of the integument are still elusive problems, although considerable progress has been made toward understanding the evolutionary history of the integument of seeds of the pteridosperms (see Chapter 15).

Despite the present state of knowledge of the phylogeny of seeds, considerable information exists on the ontogeny and structure of the seeds of living gymnosperms. In simplest ontogenetic terms, a gymnospermous seed results from the fertilization of the egg cell of the female gametophyte contained within the ovule. The diploid zygote produces the embryo, which remains embedded in the nutritive tissue of the female gametophyte, and the integument forms the seed coat. The entire structure becomes detached from the parent sporophyte and ultimately germinates to produce a new plant. From the standpoint of alternation of generations (or phases in the life cycle), the seed is thus a remarkable combination of two sporophytic generations and one gametophytic phase (Fig. 14-2, D). The seed coat is diploid and represents a part of the previous sporophyte; the nutritive tissue is the haploid female gametophyte, and the embryo is the new diploid sporophyte generation.

It is now appropriate to discuss the main stages in the development of an ovule into a seed. Reference to the generalized diagram of the life cycle of a hypothetical gymnosperm, given in Fig. 14-1, will greatly aid in understanding the following discussion.

The Ovule and Megasporogenesis

One of the essential prerequisites for the development of a seed is the production of two different spore types, i.e., microspores and megaspores. Unfortunately, these names suggest a direct comparison, or homology, with the large megaspore and the much smaller microspore of lower vascular plants. Several investigations, however, have revealed that the size relations between microspores and megaspores in the gymnosperms tend to be the *reverse* of those characteristic of lower heterosporous plants. Very commonly the microspore is the larger of the two spore types, or the two kinds of spores are approximately equal in size. On the basis of their

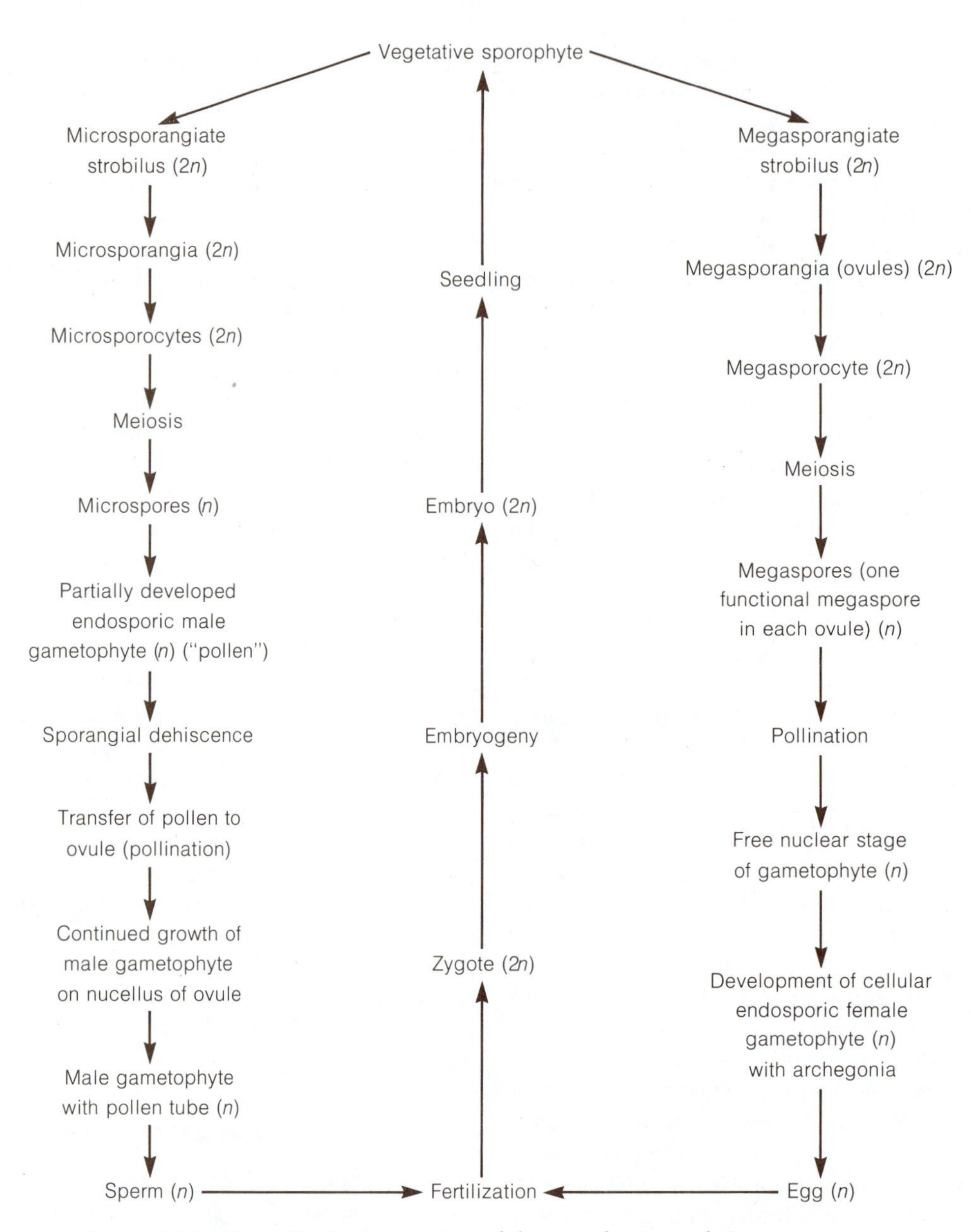

FIGURE 14-1 Generalized representation of the reproductive cycle in gymnosperms.

studies on spore dimensions, Thomson (1927, 1934) originally proposed "pollen spores" and "seed spores" as substitutes, respectively, for microspores and megaspores, whereas Doyle (1953) suggested as alternative terms "androspores" and "gynospores." Although it can be argued that there is considerable merit in these proposals, the terms *microspore* and *megaspore* are so widely used in morphological literature that it seems preferable to retain them in this book. Simply put, megaspores give rise to female gametophytes and microspores to male gametophytes, regardless of spore size. But it must be emphasized that this is done with the understanding that these terms do not necessarily imply a *direct phylogenetic connection* between *modern* gymnosperms and any group of living, heterosporous vascular plants with free spores.

Much more fundamental than a mere distinction in size is the marked difference in the *behavior* of the two spore types in the gymnosperms. The numerous microspores develop in distinct microsporangia of the eusporangiate type, and the endosporic male gametophytes may have already begun their development at the time of dehiscence of the sporangium (Fig. 17-31). In contrast, the usually solitary functional megaspore arises from deep within the tissue of the ovule and is never released by dehiscence from this type of sporangium. On the contrary, the megaspore enlarges and gives rise to an endosporic female gametophyte which is nourished and sheltered by the enveloping tissues of the ovule. This characteristic *retention of the megaspore* and its growth and development into a female gametophyte within its own sporangium represent fundamental prerequisites for seed development. The structure of the gymnospermous ovule differs in a number of respects from a typical megasporangium. Figure 14-2, A depicts somewhat diagrammatically an ovule in median longisectional view. The main body of the ovule is known as the *nucellus* and consists of parenchymatous tissue. Deeply embedded within the nucellus is a linear tetrad of four well-defined *haploid megaspores* originating by meiosis from a single *megasporocyte*. The lowermost of these megaspores (i.e., the one farthest away from the micropyle) will enlarge and give rise to the female gametophyte, whereas the remaining spores above it will degenerate (Fig. 14-2, B). Surrounding the upper free end of the nucellus is a collar or rimlike integument with a small central opening or *micropyle*. The micropyle in gymnospermous ovules provides the means of entrance to the interior of the ovule for the endosporic male gametophytes (pollen). Except for the Gnetophyta, which are frequently said to develop ovules with two integuments, all other living gymnosperms consistently form a single integument that, in the majority of gymnosperms, is free from the nucellus only near the micropylar end of the ovule. The lower portion of the ovule, where integument and nucellus are firmly joined, is termed the *chalaza*.

Although the nucellus of the ovule clearly appears to be *functionally equivalent* to a megasporangium, the integument is an accessory structure not found associated with the megasporangia of lower heterosporous vascular plants. It has been conjectured that the integument represents phylogenetically an indusiumlike structure and that the nucellus may represent the only surviving sporangium of a hypothetical fern sorus. This theory has been abandoned in favor of the concept that the integument represents, phylogenetically, the fusion of separate integumentary lobes that surrounded the nucellus (see Chapter 15 for details).

The anatomical features of the integument have been rather thoroughly studied, and certain facts deserve mention. Throughout the gymnosperms the integument is histologically differentiated into three zones or layers: an outer fleshy layer, a middle "stony" or sclerenchymatous layer, and an inner fleshy layer (Fig. 14-2, A). The degree of development of each of these layers during seed ontogeny varies within the different groups of the gymnosperms: in some (e.g., many conifers) the outer fleshy layer is rudimentary, whereas in others (e.g., the cycads and *Ginkgo*) this layer is thick, and may be conspicuously pigmented in the mature seed. In all, however, the inner fleshy layer tends to collapse and in the mature seed appears as a papery layer lining the inner surface of the stony layer. Further indication of the histological complexity of the integument is shown in many gymnosperms by the development of a vascular system (Fig. 15-25, D). In the cycads vascular strands traverse *both* the outer and inner fleshy layers, but in *Ginkgo* only the inner bundle system is developed. In members of the pine family both sets of vascular bundles are lacking. These varying conditions are of interest when compared with the structure of certain types of Paleozoic seeds in which the nucellus itself, as well as the outer fleshy layer of the integument, contained vascular tissue (Fig. 15-7). Therefore it seems reasonable to conclude that there has been a general tendency in seed evolution to eliminate vascular strands first from the nucellus and then from one or both fleshy layers of the integument.

The Female Gametophyte

The first phase of development of the gymnospermous female gametophyte is characterized by an extensive series of free nuclear divisions (Fig.

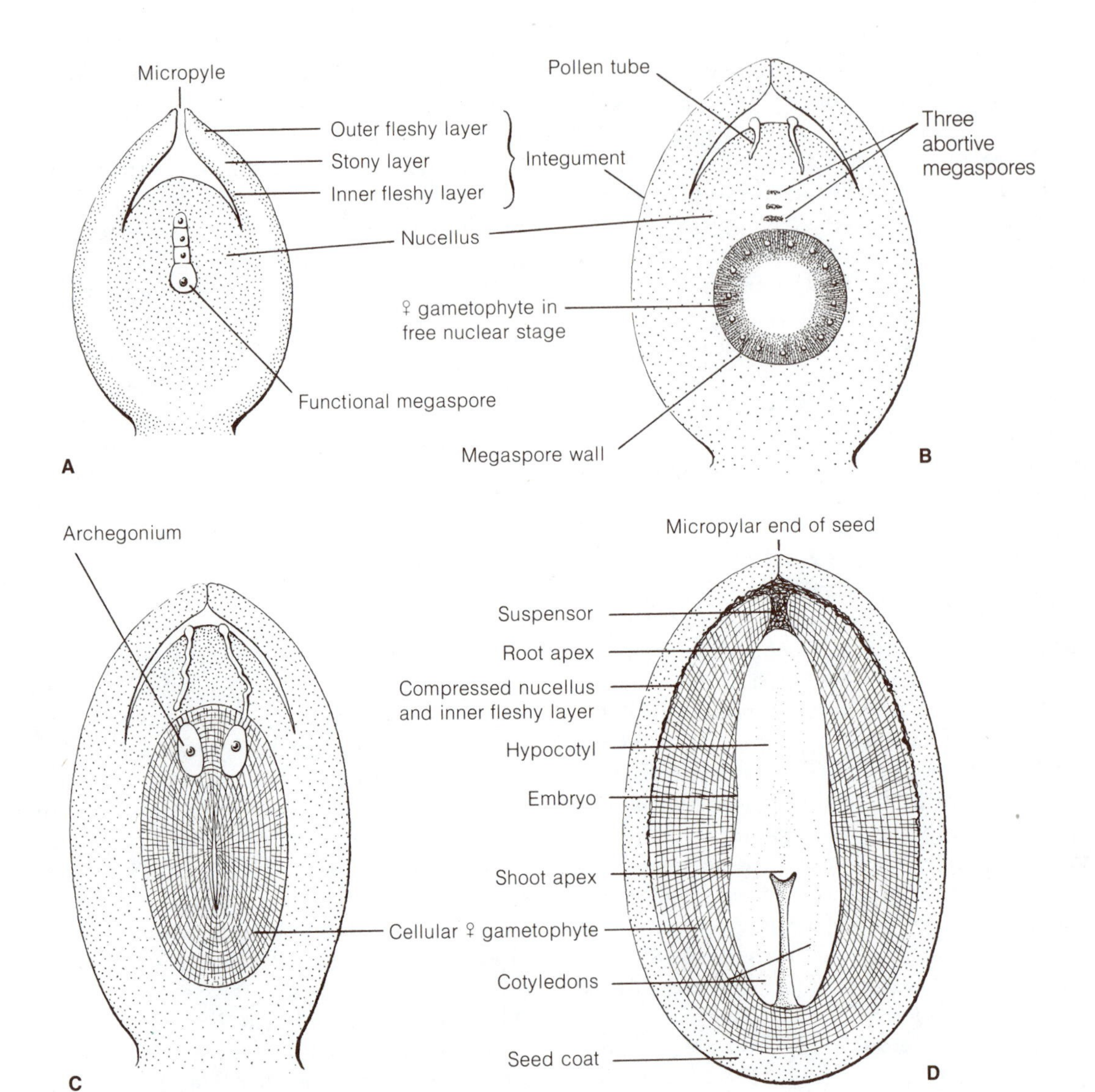

FIGURE 14-2 The processes and structures concerned in the development of a seed in gymnosperms. **A,** longisection of ovule, showing linear tetrad of megaspores, the lowermost of which will develop into the female gametophyte; **B,** female gametophyte in free nuclear stage of development (note early phases in growth of pollen tubes in nucellus); **C,** cellular female gametophyte with two archegonia (note that a pollen tube has reached the archegonium at the right); **D,** mature seed, consisting of seed coat, remains of the nucellus, female gametophytic tissue, and endoscopically oriented embryo.

14-2, B). Ultimately wall formation begins at the periphery and proceeds centripetally until the entire gametophyte consists of cells that later become filled with reserve food materials (Fig. 14-2, C). Throughout the ontogeny of the gametophyte this structure is encased by the well-defined *megaspore wall,* which appears in some species to increase in thickness during the development of the ovule. It may attain a thickness of 9 or 10 micrometers in the adult seeds of some of the cycads. The presence of a conspicuous megaspore wall surrounding the female gametophyte is one of the most important and definitive features common to gymnospermous seeds; it may serve to indicate the phylogenetic connection between the female gametophytes of free-sporing plants and those of primitive gymnosperms.

During or following the process of wall formation, certain individual superficial cells of the gametophyte—usually those near its micropylar end—give rise to archegonia (Fig. 14-2, C). The number of archegonia varies considerably, ranging from usually two in *Ginkgo* to many, as in some of the conifers. All living gymnosperms, with the exception of *Gnetum* and *Welwitschia,* are thus archegoniate plants, despite the fact that the female gametophyte is no longer a freely exposed sexual plant. (See Chapter 5 for a general discussion of the Archegoniatae.)

Pollination, the Formation of Pollen Tubes, and Fertilization

Since the ovules of gymnosperms are exposed, the process of pollination consists in the transferral (usually by wind) of the partly developed endosporic male gametophyte (pollen grain) to the micropylar end of the ovule. In many gymnosperms, the pollen is said to adhere to a drop of fluid (the so-called "pollination drop") which exudes from the micropyle. According to Chamberlain (1935), the retraction of the pollen drop causes the pollen grains to be drawn through the micropylar canal to the nucellus. This explanation, however, is not supported by the detailed studies of Doyle and O'Leary (1935) on a number of conifers. In *Pinus,* for example, they found that the pollen *at first* adheres to the projecting arms of the integument, which thus function somewhat like a stigma. Subsequently, a small amount of liquid is secreted by the ovule, filling the micropylar canal and spreading as a film on the projecting arms. Pollen grains float in this liquid, and then, since the ovules of *Pinus* are inverted, the grains begin to *rise upwardly in the liquid,* traversing the micropylar canal and finally reaching the nucellus. In Douglas fir *(Pseudotsuga),* however, no fluid is secreted by the ovule, and the pollen grains, "trapped" within the lobes of the integument, germinate and form pollen tubes which elongate and finally reach the surface of the nucellus (see Allen and Owens, 1972).

In some gymnosperms, a well-defined recess, termed the pollen chamber, is developed at the free end of the nucellus (Figs. 15-7, A; 15-25, D; and 18-13, B). Pollen chambers were well developed in Paleozoic gymnosperms; this is true today of the living cycads, *Ginkgo,* and *Ephedra.*

One of the distinctive morphological characters of *living* gymnosperms is the production of a more-or-less tubular outgrowth of the male gametophyte known as the pollen tube (Fig. 14-2, B, C). In *Ginkgo* and the cycads the pollen tube is largely haustorial in function, and grows, often for several months, like a fungal hypha *laterally* into the tissue of the nucellus; its function seems to be the absorption of food materials that are used by the gametophytic cells at the lower end of the tube. At the time of fertilization the basal end of the pollen tube bursts, liberating two large flagellated sperm together with some liquid into the cavity (archegonial chamber) directly above the female gametophyte (Fig. 15-25, E). One or both sperm may then enter an archegonium, and one of them fertilizes the large egg cell. In contrast, the pollen tubes of the conifers are sperm carriers, and after growing downward through the intervening nucellar tissue they convey the nonflagellated male gametes directly to the archegonium (Fig. 14-2, C).

The term "siphonogamous" has been used to collectively designate plants in which the sperm are directly conveyed to the egg by means of a pollen tube. Lower vascular plants, by contrast, have been designated as "zooidogamous" because the motile flagellated sperm are freely liberated from the antheridium into water through which they must swim (often for some considerable distance) to reach and fertilize the eggs. The evolutionary steps in the transition from zooidogamy to siphonogamy are by no means entirely clear, but the development of sperm-carrying pollen tubes, typical of higher gymnosperms and of all angiosperms, was surely a significant achievement. By means of the pollen tube the considerable chances and hazards of the aquatic zooidogamous method were eliminated, and much greater assurance of fertilization was made possible. Among living gymnosperms, the cycads and *Ginkgo* appear to have a primitive type of pollen tube which serves primarily as a haustorium. It seems significant in this connection that these are the only known living seed plants to have retained the flagellated type of sperm.

Embryogeny and the Maturation of the Seed

One of the most distinctive features of seed development in the majority of living gymnosperms is a period of free nuclear divisions at the beginning of embryogeny. The only known exceptions are found in *Welwitschia, Gnetum,* and *Sequoia sempervirens* in which, as in angiosperms (aside from *Paeonia*) and all lower vascular plants, the first division of the zygote is followed directly by the development of a wall separating the two cells. In most gymnosperms, beginning with the first mitotic division of the nucleus of the fertilized egg, there is a more-or-less protracted phase of nuclear multiplication unaccompanied by the formation of walls. This initial phase of embryogeny thus resembles the early stage of development of the female gametophyte. The number of free diploid nuclei formed in the young embryo of gymnosperms varies widely. In some of the cycads (e.g., *Dioon edule*), about 1,000 free nuclei are formed before walls begin to appear, whereas in *Pinus* the number is four.

The phylogenetic significance of the period of free nuclear divisions in gymnosperm embryogenesis presents a difficult problem. With reference to the cycads, Chamberlain (1935) maintained that "a free nuclear period arose as a consequence of the enlarging egg. The mass of protoplasm became so large that the early mitotic figures could not segment it." From both a phylogenetic and morphogenetic viewpoint, however, it is not clear (1) why there is so much variation in the *extent* of the free nuclear period among various gymnosperm taxa or (2) why in *Sequoiadendron giganteum* (giant sequoia) the proembryo is initiated by free nuclear divisions, whereas in the related *Sequoia sempervirens* (coast redwood) the first division of the zygote is accompanied by wall formation (Buchholz, 1939).

Following the free nuclear phase, the embryo in gymnosperms becomes cellular and gradually differentiates into a suspensor, shoot apex, cotyledons, hypocotyl, and radicle. From the standpoint of polarity, the embryo of gymnosperms is strictly endoscopic with the shoot end directed away from the micropyle (Fig. 14-2, D).

In most of the gymnosperms there is a marked tendency toward the condition of *polyembryony,* meaning the formation of several embryos in a single gametophyte. This is possible because more than one archegonium is commonly fertilized, and hence several zygotes may be produced. But more remarkable is the process of cleavage polyembryony. In this process, certain cells of the young embryo become separated from one another and give rise to a system of four or more distinct embryos. In some of the conifers both types of polyembryony may occur in the same developing seed. Physiological competition between the various embryos usually results in the elimination of all but one, which continues its differentiation and becomes the dominant embryo in the fully developed seed.

During the last phases of embryogeny the nucellar tissue of the ovule becomes disorganized and frequently persists only as a paperlike cap of dry tissue at the micropylar end of the seed (Fig. 14-2, D). Further histological maturation of the various layers of the integument continues, and the stony layer becomes an extremely hard, resistant shell which effectively encloses and mechanically protects the female gametophyte and the embryo. Except for the cycads and *Ginkgo,* the detached seed of gymnosperms remains dormant for some time. Under favorable conditions the embryo resumes growth and, rupturing the seed coat, develops into a new sporophyte plant (Chapter 17).

Progymnosperms and the Origin of Gymnosperms

Because the gymnosperms include such ancient lines of seed-bearing plants, understanding their origins depends on the analysis of the fragmentary fossil record. The formidable nature of the task that confronts paleobotanists is aggravated by the fact that many gymnosperm fossils—like the remains of other, more primitive vascular plants—are often isolated portions of leaves, strobili, sporangia, spores, stems, and roots. Many of these separate parts have been provisionally assigned generic names, but efforts to reconstruct a "complete plant" from the synthesis of such "organ genera" are often theoretical and subject to drastic revision as new paleobotanical discoveries are made. The most significant progress in understanding, however, is made when two or more "organ genera" can be demonstrated to be physically interconnected. One of the most remarkable discoveries of this sort was made by Beck (1960a, 1960b) who found two well-known organ genera, *Archaeopteris* (the "frond" of a presumed fern) and *Callixylon* (the stem of a gymnospermous tree) in organic connection (Fig. 14-3, A). The plant represented by the interconnection of *Archaeopteris* and *Callixylon* was placed by Beck (1960b) in a new class of vascular plants, the Progymnospermopsida. As relatively advanced representatives of this group, *Archaeopteris* combined the anatomy typical of gymnosperms (i.e., secondary xylem tracheids with circular bordered pits) with fernlike fronds and so-called pteridophytic reproduction (i.e., reproduction by dispersed spores, rather than by seeds). According to Beck's (1970) interpretation, "the progymnosperms are of great significance because they seem to be the immediate ancestors of seed plants."

Following the original reconstruction of *Archaeopteris*, Beck (1962, 1964, 1970, 1971, 1976, 1981) and others (Carluccio et al., 1966; Banks, 1968) have considerably advanced our knowledge of its morphology, anatomy, and phylogenetic relationships.

Historical Résumé

Prior to Beck's (1960b) investigations, *Callixylon* and *Archaeopteris* were two of the most widespread organ genera of the Upper Devonian. The generic name *Archaeopteris* was given to compressions of large fernlike "leaves" which Arnold (1947) considered "the most widely distributed fern in the Upper Devonian." According to Beck (1960a), *Archaeopteris* is a common fossil in eastern North America and has been collected from Gaspé Peninsula, Quebec, to southwestern Pennsylvania; Beck found the specimen that he used in his original reconstruction in beds of Upper Devonian age near Sidney, New York. *Callixylon,* on the other hand, is a name originally applied to petrifications of certain Devonian trees with gymnospermous secondary xylem. Stumps of *Callixylon* 1.5 meters in diameter have been reported, and logs 6 to 9 meters long or more have been uncovered in Kentucky and Texas.

The specimen, which revealed the organic connection between the frond of *Archaeopteris* and the stem of *Callixylon,* consisted of a small axis "to which are attached several fragments of bipinnately compound leaves, one of which bears both fertile and vegetative parts" (Beck 1960b). In accordance with the rule of priority, established by the International Code of Nomenclature, Beck used the earlier generic name *Archaeopteris* for the entire plant represented by the two organ genera (*Archaeopteris* and *Callixylon*); the closest affinity of his specimen was with *Archaeopteris macilenta.*

Obviously, the critical point in Beck's discovery was his demonstration that the axis of his specimen represents a portion of the stem of *Callixylon.* This was shown by the presence of well-defined *groups* of circular bordered pits on the radial walls of the tracheids of the secondary xylem (Fig. 14-4), a feature which had long been considered distinctive of the wood of *Callixylon* (see Beck 1960b, 1970) and Beck et al., 1982, for a detailed analysis of the secondary xylem of *Callixylon*).

Using data based on previous descriptions of the habit of *Callixylon* by Arnold (1931), *Archaeopteris* was considered by Beck (1962) to have been a large tree-attaining a height of 18 meters or more — with a crown of branches bearing large bipinnately compound leaves (Fig. 14-3, A). The leaves were regarded as megaphylls, comparable to the leaves of ferns and certain pteridosperms (extinct seed ferns).

A very different interpretation of the "frond" of *Archaeopteris* was subsequently proposed by Carluccio et al. (1966). They found that the so-called rachis of the frond has a radially symmetrical vascular system like the stem and that the axes of the lateral "pinnae" had a similar vasculature. Using these facts, Carluccio et al. concluded that the "frond" of *Archaeopteris* is not a pinnately compound leaf but a flattened leafy *branch system.* Ac-

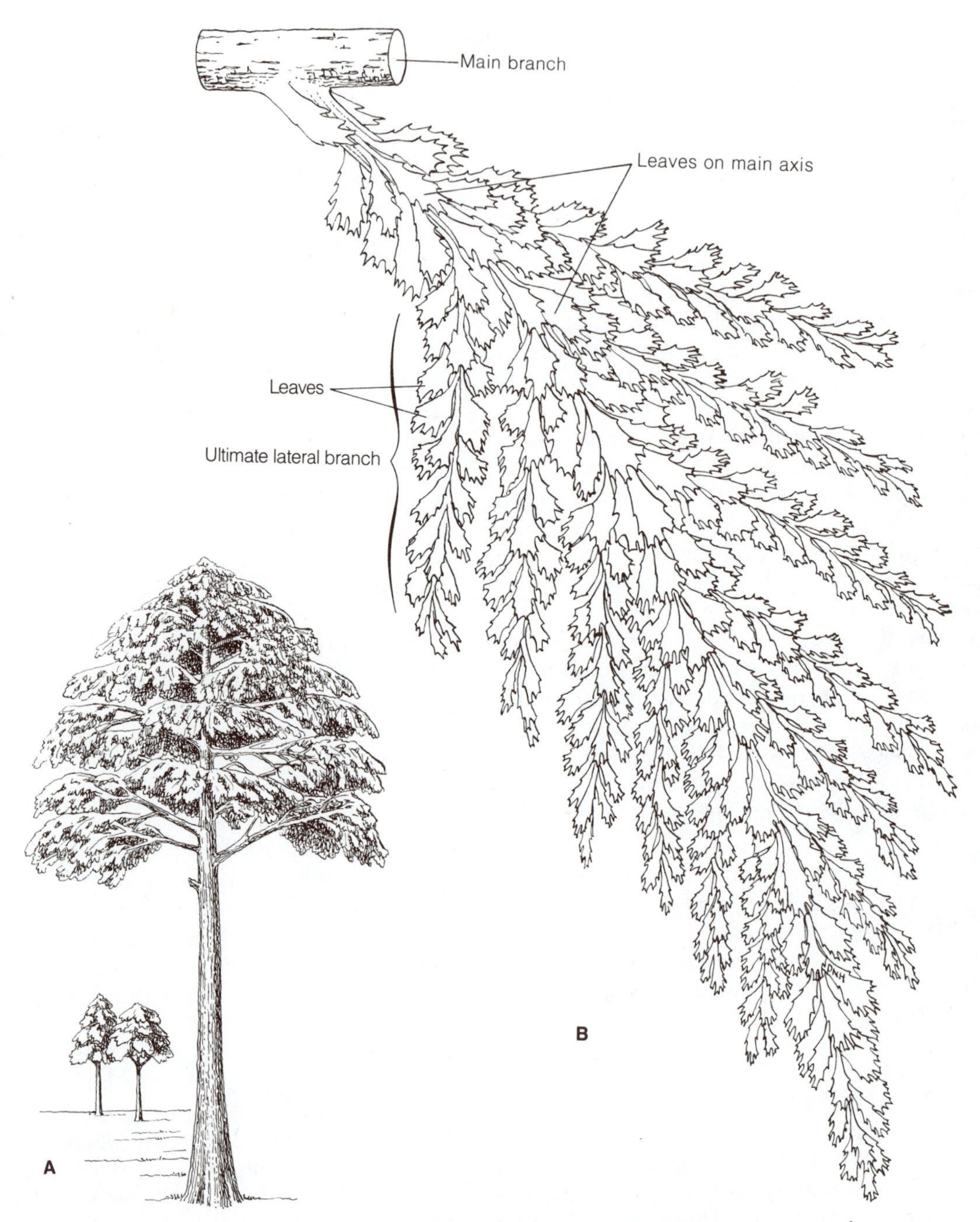

FIGURE 14-3 Reconstructions of *Archaeopteris*. **A,** growth habit; **B,** reconstruction of a complete vegetative, lateral branch system attached to a larger axis. [**A** redrawn from Beck, *Amer. J. Bot.* 49:373, 1962; **B** redrawn from Beck, *Amer. J. Bot.* 58:758, 1971.]

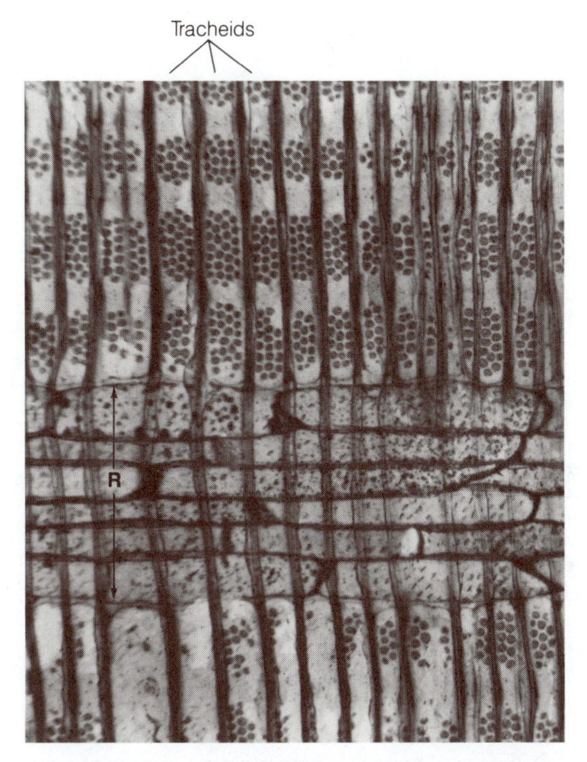

FIGURE 14-4 Radial longitudinal section, secondary wood of *Callixylon newberryi* showing groups of pits on tracheid walls. R, wood ray. [Courtesy of Dr. C. B. Beck.]

cording to this interpretation the actual leaves, which in *Archaeopteris macilenta* are broadly laminate, are attached to the ultimate lateral branches (Fig. 14-3, B). Beck (1970) subsequently adopted this revised interpretation. In some species the leaves are more finely divided. Fertile, ultimate appendages are considered homologous with vegetative leaves and are laminate, often dichotomously branched, distally. The ultimate fertile shoots exhibit the characteristics of strobili (Beck 1981). Numerous sporangia, which dehisced longitudinally, occurred on the adaxial side of sporophylls (Fig. 14-5).

Vascular Anatomy

The conclusions reached by Beck (1971) from his study of the external morphology of compression-type specimens were supported by his investigations of stelar anatomy. A reconstruction of the vasculature of a main axis is shown in Fig. 14-6. The primary xylem forms a cylinder of nine ribs, each of which is represented by an axial (or sympodial) vascular bundle; primary xylem is mesarch. Branch traces and leaf traces diverge from the axial bundles as follows: two of the nine stelar bundles give rise to traces that vascularize the members of the two distichous rows of lateral, determinate branches (Fig. 14-6, labeled brtr). The remaining seven "ribs," or axial bundles — three on one side of the stele and four on the opposite side — are the points of origin of the traces that extend into the helically arranged leaves of the main axis (Fig. 14-6, labeled ltr). Each leaf trace follows a steep, obliquely radial course through the cortex of the axis and then becomes divided into two strands as it enters the base of a leaf (Fig. 14-6, labeled lb, leaves 17 and 18). The further branching of these two strands produces the dichotomous venation characteristic of *Archaeopteris* leaves. The stele is siphonostelic but there are no leaf gaps (parenchymatous regions) directly above a departing leaf trace. Secondary xylem (sx) was produced, of the type described earlier; secondary phloem also was formed, but more completely described for an earlier (older) progymnosperm in the order Aneurophytales (Wight and Beck, 1984). The production of secondary phloem and xylem is another unique feature among Devonian plants and serves to strengthen the case for a phylogenetic relationship between the progymnosperms and gymnosperms.

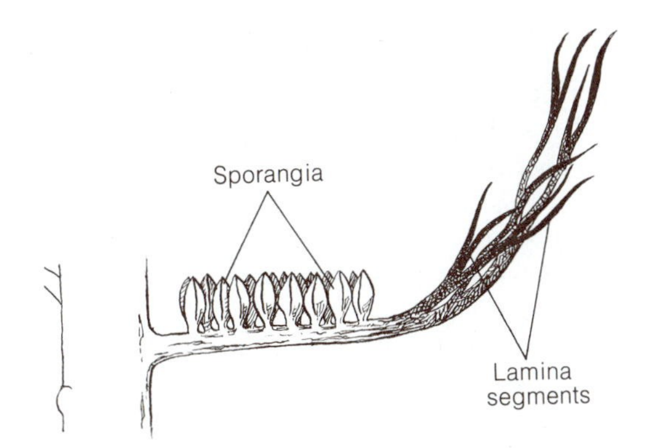

FIGURE 14-5 Reconstruction of a fertile appendage of *Archaeopteris*. Note the two rows of adaxial sporangia. [Redrawn from Carluccio et al., *Amer. Jour. Bot.* 53:719, 1966.]

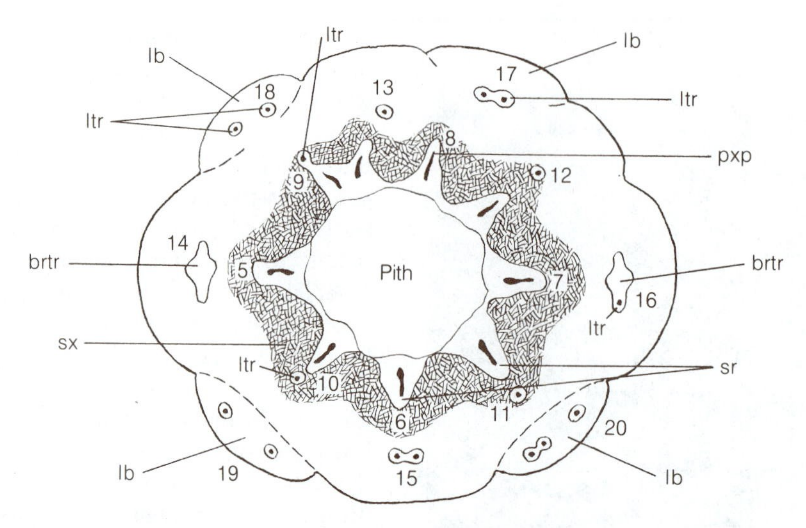

FIGURE 14-6 Reconstruction of the vascular anatomy of the main axis of *Archaeopteris macilenta.* The stele consists of nine "ribs" (sr) and is enclosed by a cylinder of secondary xylem (sx) indicated by hatching. The helical sequence in the divergence of branch traces (brtr) and leaf traces (l tr) is indicated by numbers 5 to 20; higher numbers indicate older traces, lower numbers younger traces. l b, leaf base; px p, protoxylem pole. See text for further explanation. [Redrawn from Beck, *Amer. Jour. Bot.* 58:758, 1971.]

Phylogenetic Significance of the Progymnosperms

The external morphology and vascular anatomy of *Archaeopteris* support the concept that the so-called frond of this progymnosperm represents a complex branch system to which leaves are attached. If this interpretation is correct, it seems doubtful that the *Archaeopteris* branch system typifies a stage in the evolution of the megaphyllous type of leaf characteristic of either the ferns or of such ancient gymnosperms as the seed ferns (pteridosperms). Beck (1971) noted the striking resemblance between the leafy branch systems of *Archaeopteris* and the frondlike branch systems of such living conifers as *Metasequoia* and *Taxodium.* Some of the Permian conifers, such as *Carpentieria,* also closely resembled *Archaeopteris* even in details of leaf morphology. Heterospory has been demonstrated in several species, including *A. macilenta* (see Beck, 1976, 1981). In another species, *Archaeopteris halliana,* the fertile leaves were interspersed among vegetative leaves on ultimate branches. Each fertile leaf was dichotomously branched with erect sporangia located on the adaxial surface. The sterile leaves were fan shaped with dichotomous venation (Fig. 14-7). This species also was heterosporous (Phillips et al., 1972).

The Progymnospermophyta, as originally conceived by Beck (1960b), includes not only *Archaeopteris* but also other Devonian genera with secondary xylem and free-sporing reproduction. Figure 14-8 shows representative genera included in the order Aneurophytales of the Progymnospermophyta. Branching in this order was mainly three dimensional, that of vegetative branches being either helical or decussate. Sporangia were borne terminally. The primary xylem was a ribbed protostele except in ultimate appendages where it was terete. It is particularly interesting to note that *Protopteridium,* now designated *Rellimia* (Bonamo, 1977), was formerly regarded as a primitive fern (Fig. 14-8, A). It had a three-lobed protostele with mesarch protoxylem in each lobe, and sporangia were borne terminally on ultimate axes. In *Tetraxylopteris* branching was decussate, and sporangia were borne terminally on axes arranged in an irregularly pinnate manner (Bonamo and Banks, 1967; Fig. 14-8, C). The protostele was a four-lobed actinostele, and secondary xylem has been described. The

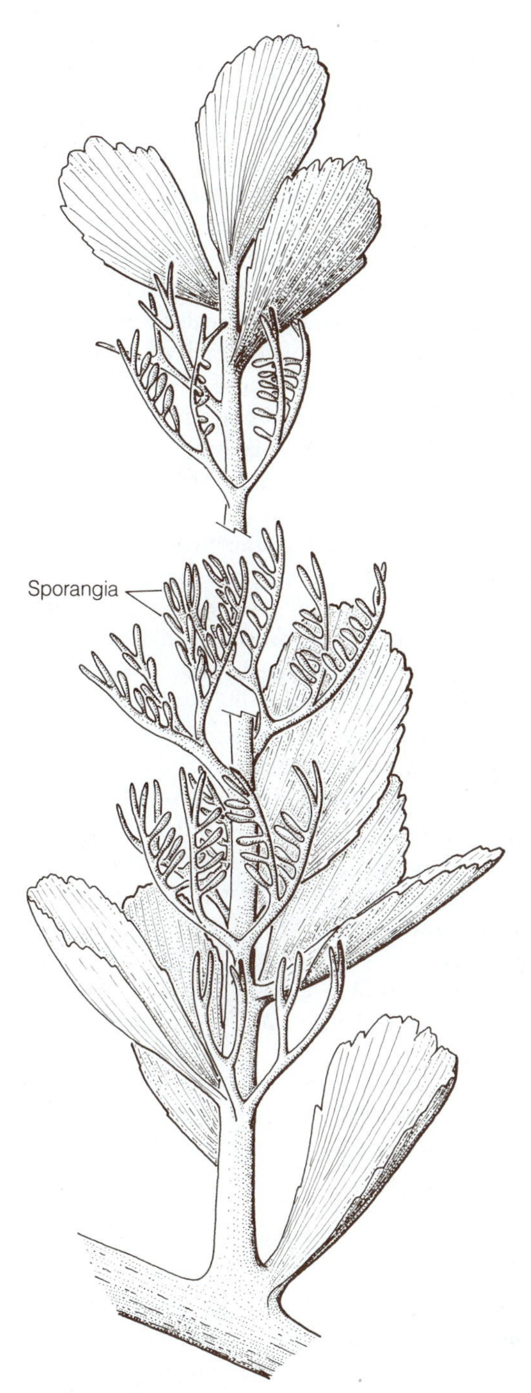

FIGURE 14-7 *Archaeopteris halliana* (Archaeopteridales). Ultimate branch, showing fertile and sterile leaves. [Redrawn from Phillips et al., *Palaeontographica* 139B;47–71, 1972.]

FIGURE 14-8 Examples of genera included in the Aneurophytales. **A,** *Protopteridium (Rellimia) hostinense;* **B,** *Aneurophyton;* **C,** *Tetraxylopteris schmidtii,* showing part of a sporangial complex. [**A, B** redrawn from Banks, pp. 73–107, in *Evolution and Environment* edited by E. T. Drake. Yale University Press, New Haven. 1968; **C** redrawn from Bonamo and Banks, *Amer. J. Bot.* 54:755–768, 1967.]

discovery of members of the Aneurophytales in the same fossil bed with the most ancient seeds led Rothwell and Erwin (1987) to propose that members of this order are more likely to be ancestral to the gymnosperms than the Archaeopteridales.

If *Archaeopteris* is neither a primitive fern nor the progenitor of the ferns, what evidence supports the possible evolution of coniferophytic gymnosperms from the progymnosperms? As stated earlier, *Archaeopteris* (including the form genus *Callixylon*) produced secondary xylem similar to extant conifers. Also, the investigations of Namboodiri and Beck (1968a, b, c) have suggested that an understanding of evolutionary relationships lies in the correct interpretation of the structure and evolution of the stele in progymnosperms and gymnosperms.

However, before describing the investigations of Namboodiri and Beck, it is necessary to review briefly the concept of Jeffrey (1917). He believed that the first step in stelar evolution from the protostele to a tubular siphonostele was the enclosure of cortical tissue within the stele and the formation of parenchymatous regions, *leaf gaps,* above the point of departure of leaf traces. The tubular nature of the stele would be obscured if the leaf gaps overlapped as in many ferns (see Chapter 13). By this reasoning, Jeffrey concluded that the vasculature of seed plants was derived phylogenetically from ferns. The presence of the leaf gap was one of the definitive characters used by Jeffrey in establishing the "Pteropsida," a major taxon under which he grouped ferns, gynmosperms, and angiosperms. This system of classification has generally been abandoned by botanists. (For details of his theory the reader is referred to Jeffrey's 1917 publication.)

According to Namboodiri and Beck (1968a, b, c) the evolution of the stele in progymnosperms and gymnosperms has never involved the formation of leaf gaps (in the sense of Jeffrey). Beck (1964) concluded that "stelar evolution in ferns and gymnosperms comprises two entirely separate lines." (Also, see Rothwell, 1976.) Let us now examine the nature of the evidence supporting the conclusions of Namboodiri and Beck.

Figure 14-9 represents diagrammatically the main steps in the evolution of the primary vascular system of progymnosperms and gymnosperms. The most primitive stage was a protostele, the ribs of which represent the points of origin of the radially diverging traces to the "appendages" (Fig. 14-9, A). This kind of stele is illustrated by certain primitive progymnosperms such as *Aneurophyton* and *Tetraxylopteris*. The next stage in stelar evolution was characterized by the gradual dissection of the protostele into "longitudinal columns," a process accompanied by "medullation," i.e., the origin of a central pith region (Fig. 14-9, B). Further evolutionary specialization led to a more definitive system of discrete *sympodial bundles,* each of which, at the node, divided *tangentially* into two *radially aligned* strands, an outer leaf trace and an inner "reparatory strand"; the latter continued its upward course, and a leaf gap was not formed (Fig. 14-9, C). This is the type of primary vasacular system that Beck (1971) found in the main axis of the "frond" of *Archaeopteris* (see Fig. 14-6). The next step is a change in the nature of the division of a sympodial bundle. A sympodial bundle divides radially to form two bundles, initially placed side by side along the same tangential plane. One bundle is the departing leaf trace and the other is the reparatory strand, continuing upward as the sympodial bundle (Fig. 14-9, D). This mode of separation of leaf traces also occurred without the formation of leaf gaps. The spaces between sympodial bundles can be described as interfasicular regions. This type of stele is illustrated by the Carboniferous seed fern, *Lyginopteris*. According to Beck (1970) the stelar system of *Lyginopteris oldhamia,* "in all of its major features, including the clear absence of a leaf gap, is identical with that of conifers."

Figure 14-9, E depicts the eustele typical of living conifers with helically arranged leaves. As in progymnosperms and *Lyginopteris,* leaf gaps are absent, and, in addition, the sympodial bundles at various levels become closely approximated but remain unfused throughout their course in the stem. The vascular cylinder of the shoot of *Abies* (fir) is shown in Fig. 14-10 as if it were split open and laid flat with the outer surface facing the observer (from Namboodiri and Beck, 1968a). There are thirteen sympodia numbered with Roman numerals in Fig. 14-10 that follow an undulating course through the stem. The divergence of successive leaf traces along each sympodium occurs at intervals of thirteen nodes; for example, leaf trace 14 stands above leaf trace 27 on the same sympodium.

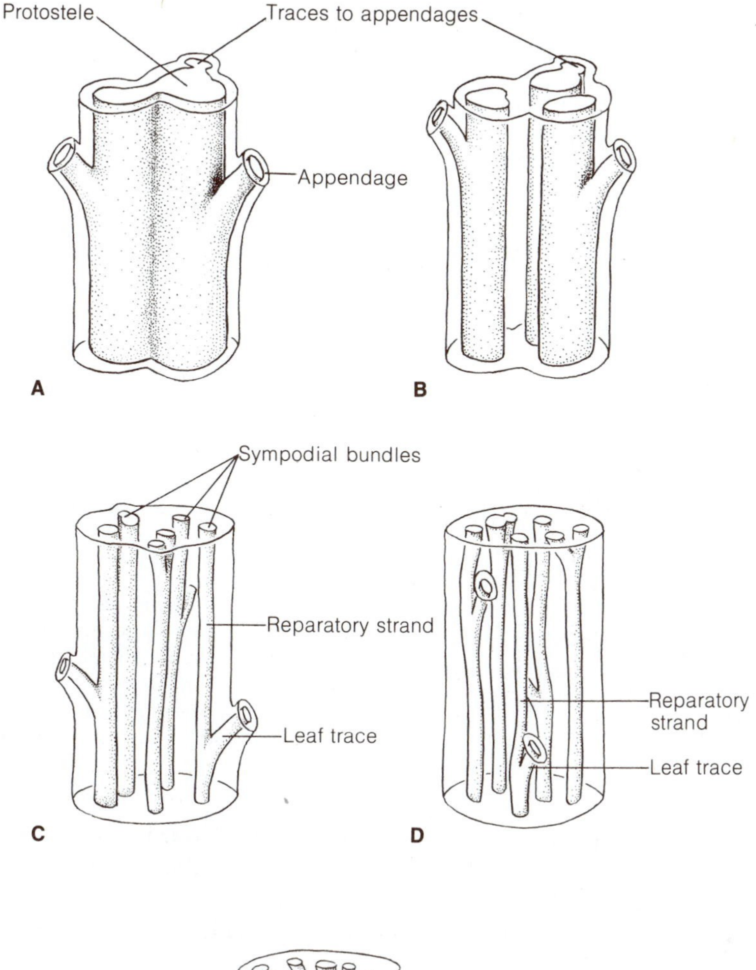

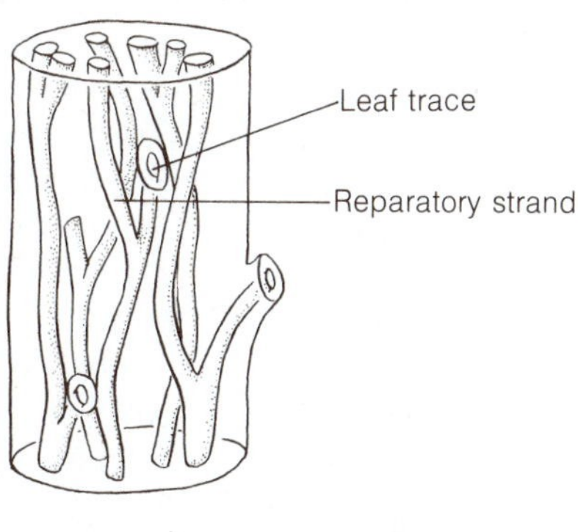

FIGURE 14-9 Diagrams showing the probable evolutionary development of the primary vascular systems of progymnosperms and gymnosperms: **A,** the primitive protostelic condition; **B,** division of protostele into longitudinal columns; **C,** vascular system composed of discrete sympodial bundles, each of which, at a nodal level, divides tangentially into a reparatory strand and a leaf trace, as illustrated in *Archaeopteris:* **D,** vascular system in which the reparatory strand and leaf trace are formed by the radial division of the sympodial bundle, as illustrated in *Lyginopteris,* a seed fern; **E,** primary vascular system characteristic of living conifers with helical phyllotaxis. See text for further explanation. [Redrawn from Namboodiri and Beck, *Amer. Jour. Bot.* 55:464, 1968.]

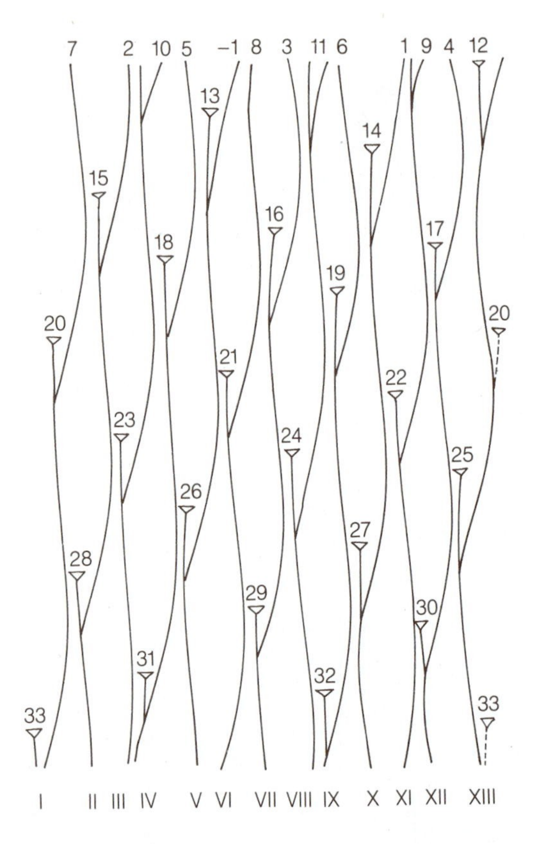

FIGURE 14-10 Primary vasculature of *Abies concolor* split open showing thirteen vertical sympodia numbered using Roman numerals at bottom; leaf traces (shown terminating in triangles) are numbered in their ontogenetic sequence, younger ones at top. [Redrawn from Namboodiri and Beck, *Amer. J. Bot.* 55:447–457, 1968.]

The salient feature of the concept of stelar evolution proposed by Namboodiri and Beck (1968c) concerns the similarity between the general mode of ontogenetic origin and divergence of leaf traces in progymnosperms and conifers. In both groups, leaf traces are described as arising directly from sympodial bundles of the eustele without the formation of leaf gaps (that is, the formation of leaf gaps or "perforations" in an otherwise tubular siphonostele as described by Jeffrey). In other words, the stele in gymnosperms evolved initially from a "dissected" protostele.

In stems of seed plants in which the primary vascular cylinder consists of sympodial bundles and departing leaf traces (eustele), the recognition of "leaf gaps" is rather uncertain because the parenchyma confronting the diverging leaf trace is confluent with adjacent interfascicular regions of the stele. Frequently it is only after some secondary growth has occurred that the "gap" becomes evident because secondary xylem formation is delayed in the interfascicular region confronting the departing leaf trace. In transectional view, the foliar gaps are then strictly correlated *in position* with the departing leaf traces. It is this aspect of anatomy that has made possible the recognition of various types of nodal anatomy characteristic of woody dicotyledons and conifers (see Chapters 17 and 19). From purely descriptive and practical standpoints, the term leaf gap is a useful one as long as its application is tempered with the realization that the steles in ferns and seed plants have probably evolved independently.

From the comprehensive studies on stelar anatomy, the origin of gymnosperms from ferns—as postulated by Jeffrey (1917)—no longer appears to be a useful hypothesis. Ancestors of the gymnosperms are to be found in the progymnosperms—one line leading to the seed ferns, the other to the coniferophyte assemblage. The ferns probably had their origin from some group of ancient plants of the *Psilophyton* type within the Trimerophytophyta (Fig. 14-11).

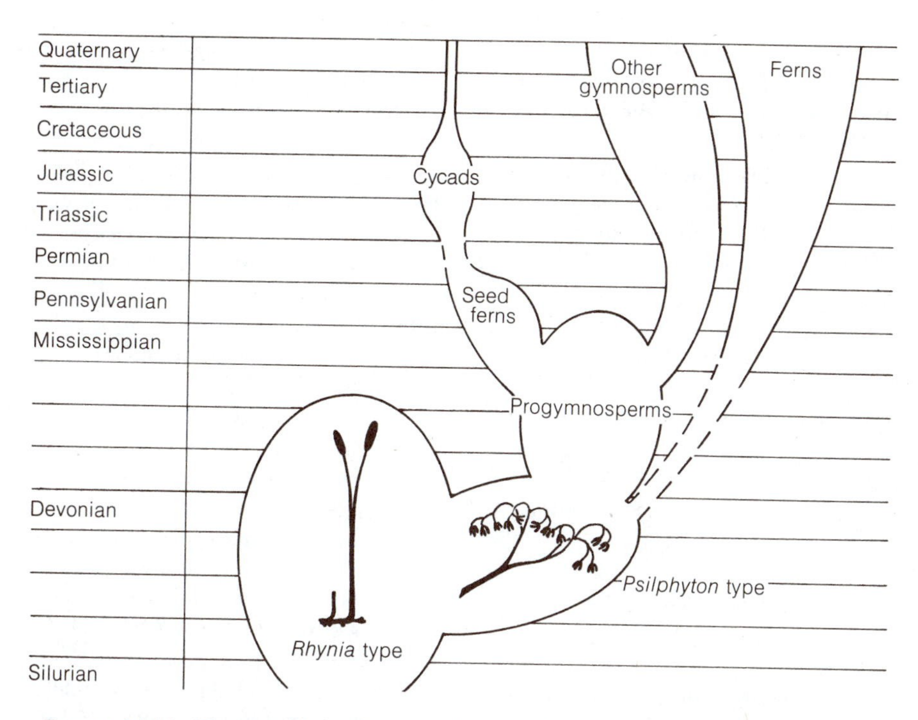

FIGURE 14-11 The possible evolutionary history of ferns, progymnosperms and gymnosperms. [Modified from *Evolution and Plants of the Past* by H. P. Banks. Wadsworth, Belmont, Calif. 1970.]

REFERENCES

Allen, G. S., and J. N. Owens
1972. *The Life History of Douglas Fir.* Information Canada, Ottawa.

Arnold, C. A.
1931. On *Callixylon newberryi* (Dawson) Elkins et Wieland. *Contr. Mus. Paleontol. Univ. Mich.* 3:207–232.
1947. *An Introduction to Paleobotany.* McGraw-Hill, New York.

Banks, H. P.
1968. The early history of land plants. *In* E. T. Drake (ed.), *Evolution and Environment.* Yale University Press, New Haven.

Beck, C. B.
1960a. Connection between *Archaeopteris* and *Callixylon. Science* 131:1524–1525.
1960b. The identity of *Archaeopteris* and *Callixylon. Brittonia* 12:351–368.
1962. Reconstructions of *Archaeopteris* and further consideration of its phylogenetic position. *Amer. Jour. Bot.* 49:373–382.
1964. The woody, fern-like trees of the Devonian. *Mem. Torrey Bot. Club* 21:26–37.
1970. The appearance of gymnospermous structure. *Biol. Rev.* 45:379–400.
1971. On the anatomy and morphology of lateral branch systems of *Archaeopteris. Amer. Jour. Bot.* 58:758–784.
1976. Current status of the Progymnospermopsida. *Rev. Palaeobot. and Palynol.* 21:5–23.
1981. *Archaeopteris* and its role in vascular plant evolution. In K. J. Niklas (ed.), *Paleobotany, Paleoecology, and Evolution,* Vol. 1, pp. 193–230. Praeger, New York.

Beck, C. B., K. Coy, and R. Schmid
1982. Observations on the fine structure of *Callixylon* wood. *Amer. J. Bot.* 69:54–76.

Bonamo, P. M.
1977. *Rellimia thomsonii* (Progymnospermopsida) from the Middle Devonian of New York State. *Amer. J. Bot.* 64:1272–1285.

Bonamo, P. M., and H. P. Banks
1967. *Tetraxylopteris schmidtii:* its fertile parts and its relationships within the Aneurophytales. *Amer. J. Bot.* 54:755–768.

Buchholz, J. T.
1939. The embryogeny of *Sequoia sempervirens* with a comparison of the Sequoias. *Amer. Jour. Bot.* 26:248–257.

Carluccio, L. M., F. M. Heuber, and H. P. Banks
1966. *Archaeopteris macilenta,* anatomy and morphology of its frond. *Amer. Jour. Bot.* 53:719–730.

Chamberlain, C. J.
1935. *Gymnosperms. Structure and Evolution.* University of Chicago Press, Chicago.

Doyle, J.
1953. Gynospore or megaspore—a restatement. *Ann. Bot.* n.s. 17:465–476.

Doyle, J., and M. O'Leary
1935. Pollination in *Pinus. Sci. Proc. Roy. Dublin Soc.* 21:181–190.

Jeffrey, E. C.
1917. *The Anatomy of Woody Plants.* University of Chicago Press, Chicago.

Namboodiri, K. K., and C. B. Beck
1968a. A comparative study of the primary vascular system of conifers. I. Genera with helical phyllotaxis. *Amer. Jour. Bot.* 55:447–457.
1968b. A comparative study of the primary vascular system of conifers. II. Genera with opposite and whorled phyllotaxis. *Amer. Jour. Bot.* 55:458–463.
1968c. A comparative study of the primary vascular system of conifers. III. Stelar evolution in gymnosperms. *Amer. Jour. Bot.* 55:464–472.

Phillips, T. L., H. N. Andrews, and P. G. Gensel
1972. Two heterosporous species of *Archaeopteris* from the Upper Devonian of West Virginia. *Palaeontographica* 139(B):47–71.

Rothwell, G. W.
1976. Primary vasculature and gymnosperm systematics. *Rev. Palaeobot. Palynol.* 22:193–206.

Rothwell, G. W., and D. M. Erwin
1987. Origin of seed plants: an aneurophyte/seed fern link elaborated. *Amer. J. Bot.* 74:970–973.

Thomson, R. B.
1927. Evolution of the seed habit in plants. *Proc. Trans. Roy. Soc. Canada* 21:229–272.
1934. Heterothally and the seed habit *versus* heterospory. *New Phytol.* 33:41–44.

Wight, D. C., and C. B. Beck
1984. Sieve cells in phloem of a Middle Devonian progymnosperm. *Science* 225:1469–1470.

CHAPTER 15

Pteridospermophyta (Seed Ferns), Cycadeoidophyta (Cycadeoids), and Cycadophyta (Cycads)

IN this chapter we will discuss the living and extinct representatives of the cycadophyte lines of gymnosperm evolution. The cycadophytes are distinguished in general by their large, pinnately compound leaves, columnar or sparingly branched trunks, and manoxylic secondary xylem. The only surviving cycadophytes are the modern cycads, Cycadophyta, which will receive the major attention in this chapter. First, however, we will consider the extinct Pteridospermophyta, or "seed ferns," and later in the chapter we will discuss the extinct Cycadeoidophyta. Both of these taxa are of exceptional evolutionary interest. The seed ferns provide important clues about the nature of the seed in ancient gymnosperms. The Cycadeoidophyta, although sharing some features in habit and anatomy with modern cycads, appear to represent a separate "blind end" in cycadophyte evolution, although they have been cited periodically as the progenitors of angiosperms.

Pteridospermophyta

Pteridospermales

The Paleozoic pteridosperms extended from the Carboniferous into the Permian. They combined in a most remarkable way, the general habit and foliage of ferns with the formation of gymnosperm-type seeds (Figs. 15-l, 15-3, 15-4). The demonstration by Oliver and Scott (1905) that many of the presumed "ferns" of the Carboniferous were in reality seed-bearing plants was an outstanding achievement in the history of paleobotany and much study has been devoted subsequently to the comparative morphology and systematics of these interesting plants.

This book discusses two families in the order Pteridospermales — the Lyginopteridaceae and the Medullosaceae. A more thorough treatment of these families is beyond the scope of our discussion.

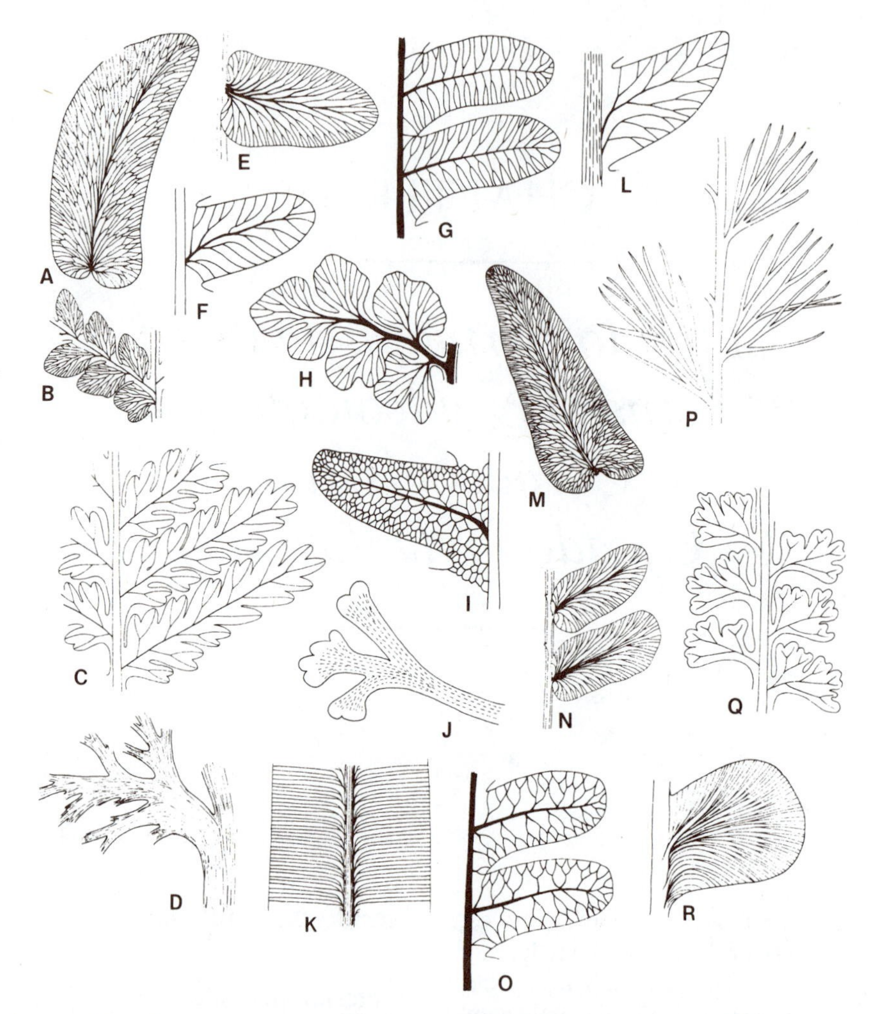

FIGURE 15-1 Representative examples of Paleozoic foliage, many of which have been shown to be the pinnules of seed-fern fronds. **A,** *Linopteris;* **B,** *Mariopteris;* **C,** *Sphenopteris;* **D,** *Aphlebia;* **E,** *Mixoneura:* **F,** *Pecopteris;* **G,** *Alethopteris;* **H,** *Eusphenopteris;* **I,** *Lochopteris;* **J,** *Kankakeea;* **K,** *Taeniopteris;* **L,** *Lescuropteris;* **M,** *Reticulopteris;* **N,** *Neuropteris;* **O,** *Lochopteridium;* **P,** *Rhodea;* **Q,** *Alloiopteris;* **R,** *Odontopteris.* [From *Paleobotany: An Introduction to Fossil Plant Biology* by T. N. Taylor. McGraw-Hill, New York. 1981.]

Therefore, we will adopt the type-method approach to describe vegetative and reproductive structures.

Lyginopteridaceae

Lyginopteris was established as a form genus for stems from the Carboniferous Coal Measures of Britain and Europe but now applies to an entire plant. *Lyginopteris* is thought to have been somewhat vinelike with large fronds, and probably supported by surrounding vegetation. The stem was from 3 to 4 centimeters in diameter and was eustelic; the primary xylem strands were mesarch in development, comparable to progymnosperms (see p. 337). Tracheids of the relatively small amount of manoxylic xylem was formed with multiseriate bordered pits on the radial walls of tracheids. The outer cortex consisted of a network of fibrous strands (Fig. 15-2). The microsporangiate structures were borne on laminar segments or were grouped terminally on axes of otherwise planated fronds; the sporangia were fused to form synangia (Fig. 15-3C). Microspores, often referred to as "prepollen," were trilete and resembled those of certain ferns. Ovules of *Lyginopteris,* assigned to the form genus *Lagenostoma,* were small — 5.5 millimeters long by 4.4 millimeters in diameter. An interesting feature was the development of a *cupule* around the ovule (Fig. 15-3, D). The cupule was divided distally into eight to ten vascularized lobes. Capitate glands were present on the outer surface of the cupule which may have served to attract insects. The single integument was vascularized and fused with the nucellus except at the distal end, much like living gymnosperms. There was a central column of nucellar tissue, surrounded by the so-called lagenostome or *salpinx* (Greek, meaning "trumpet"), also of nucellar origin; the space between the two was the pollen chamber in which spores of the trilete

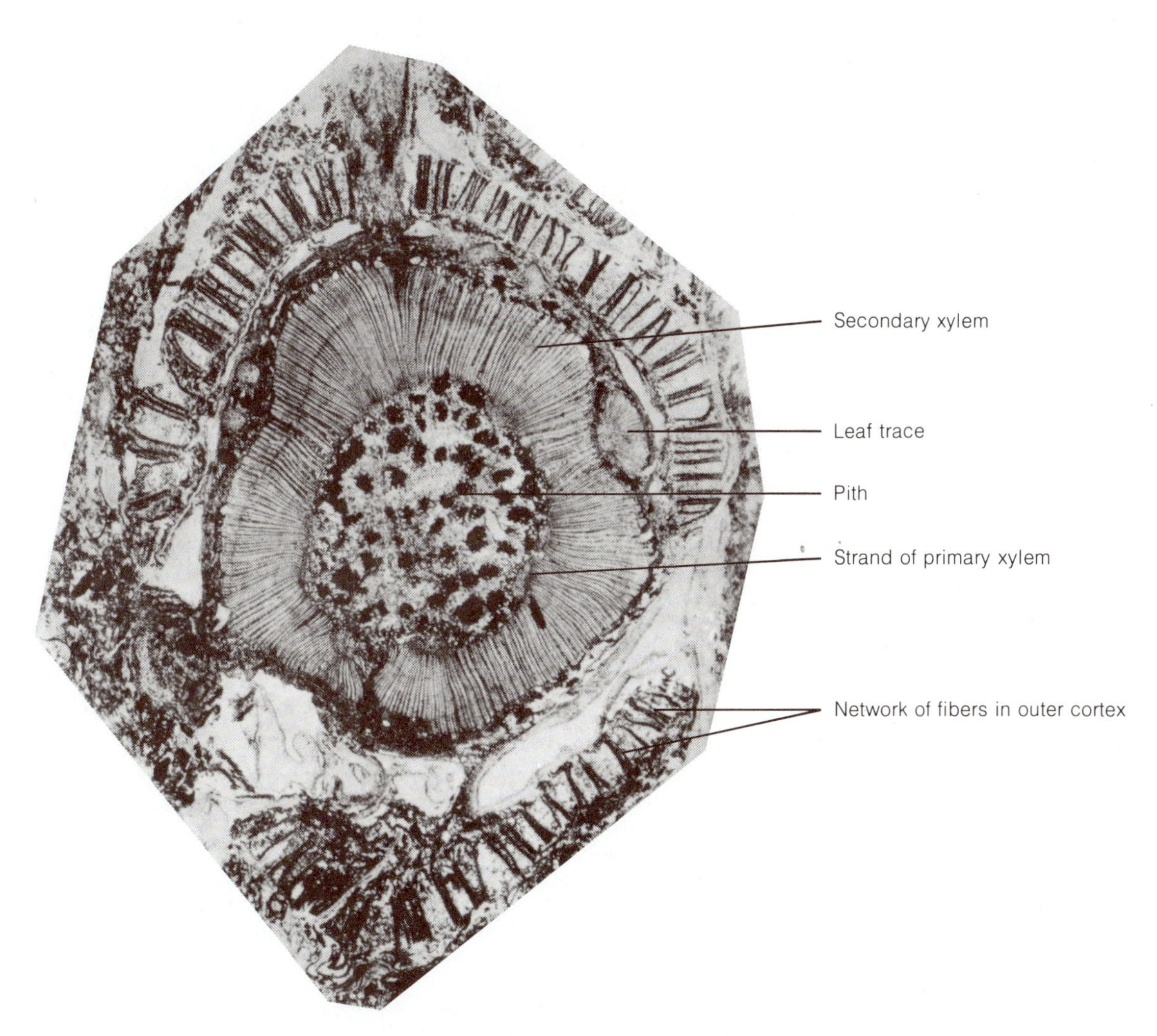

FIGURE 15-2 Transection of stem of *Lyginopteris oldhamia,* showing primary vascular strands, leaf traces, and the well-developed cylinder of secondary xylem. Note the characteristic network of fibrous strands in the outer cortex. [From *An Introduction to the Study of Fossil Plants* by J. Walton. Adam and Charles Black, London. 1953.]

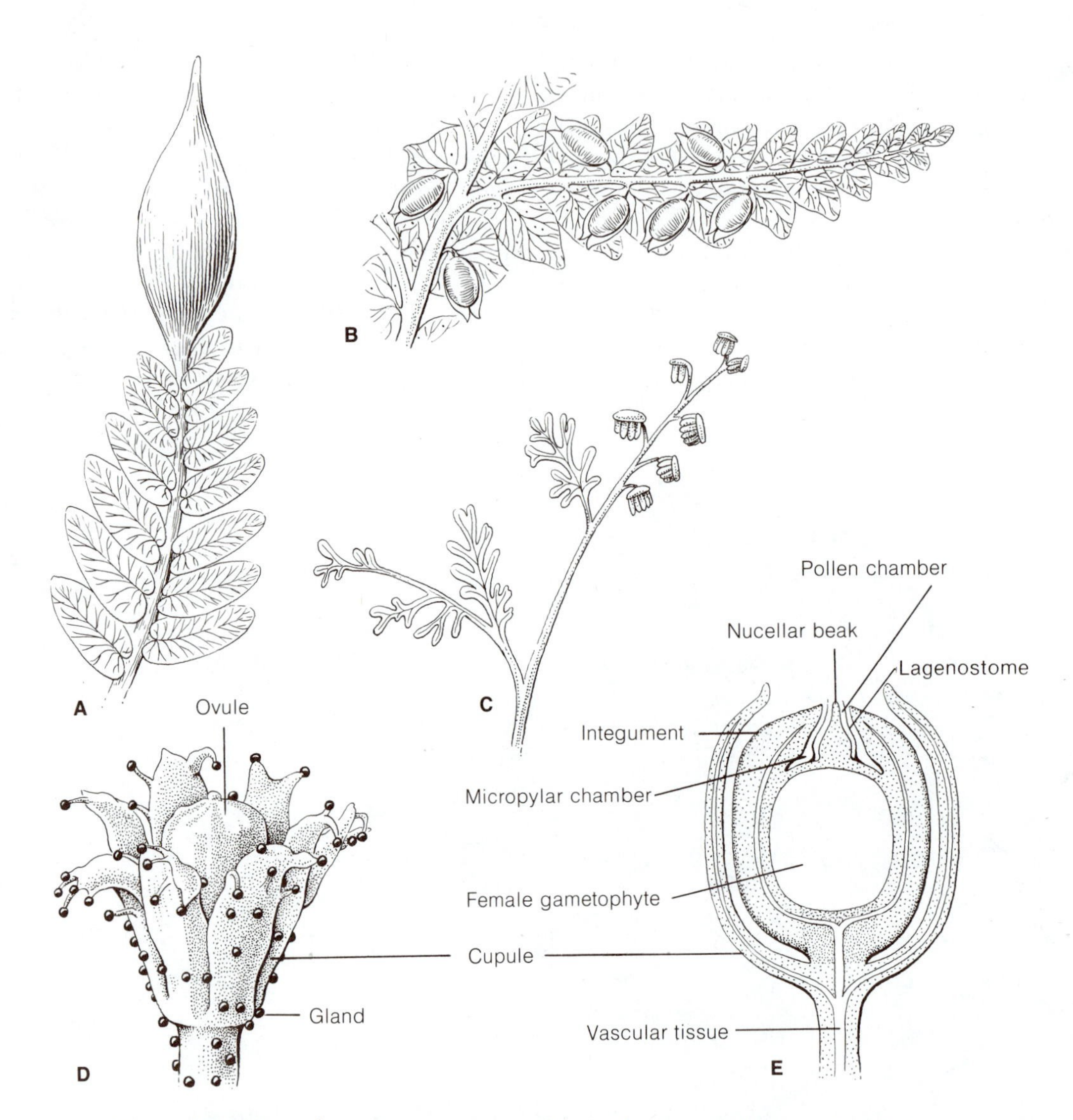

FIGURE 15-3 Reproductive structures in seed ferns. **A,** pinna with terminal seed of *Neuropteris heterophylla:* **B,** *Emplectopteris triangularis,* seeds attached to subdivisions of leaf; **C,** portion of fertile leaf of *Crossotheca* bearing clusters of pendant microsporangia; **D,** reconstruction of the ovule and cupule of *Lagenostoma lomaxi:* **E,** longisection of ovule of *Lagenostoma* showing details of internal structure. [**A** redrawn from *Textbook of Paleobotany* by W. C. Darrah. Copyright D. Appleton-Century Co., New York, 1939, by permission Appleton-Century-Crofts, Inc; **B** redrawn from *Ancient Plants and the World They Lived In* by H. N. Andrews. Comstock, New York. 1947; **C** and **D** from *Les Plantes Fossiles dans leurs Rapports avec les Végétaux Vivants* by L. Emberger. Masson et Cie, Paris. 1944; **E,** adapted from *An Introduction to Paleobotany* by C. A. Arnold. McGraw-Hill, New York. 1947.]

type have been found (Fig. 15-3, E). (Additional information can be found in Taylor and Millay, 1981.)

Medullosaceae

Members of this family, known from the Lower Carboniferous to the Permian, are generally depicted as rather tall trees (3 to 8 meters high) with alternately arranged compound leaves, resembling modern tree ferns (Fig. 15-4). Leaves are referable to two or more form genera (e.g., *Alethopteris, Neuropteris;* Fig. 15-1). *Medullosa* was used originally as a form genus for stem remains, but is now generally used for the entire plant. A transection of

FIGURE 15-4 Reconstruction of a specimen (3.5 to 4.5 meters high) of *Medullosa noei.* [Redrawn from Stewart and Delevoryas, *Bot. Rev.* 22:45, 1956.]

a medullosan stem may reveal the presence of two or more vascular cylinders, each consisting of primary xylem surrounded by a cylinder of manoxylic secondary xylem (Fig. 15-5, A, B). For many years each of these seemingly separate cylinders were termed "steles," and the stem was said to be "polystelic." According to a now more generally accepted concept, the stem is not polystelic, but rather has evolved in the course of evolution from the dissection of a single protostelic vascular cylinder. The individual vascular cylinders are defined as vascular *segments*, not "steles." All of the vascular segments, collectively, comprise a *eustele. Sympodia* are represented by protoxylem strands in the vascular segments rather than by the vascular segments themselves (Fig. 15-5, A, B). Steps in the possible evolution of the *Medullosa*-type eustele are shown in Fig. 15-5, C–E. Leaf gaps are absent. It should be noted, however, that the eustele in progymnosperms, gymnosperms, and angiosperms lack the formation of secondary xylem toward the pith.

All medullosan microsporangiate structures ("pollen organs"), whether simple or complex, were synangiate and consisted of tubular sporangia. In *Codonotheca* each unit consisted of several sporangia fused at their bases (Fig. 15-6, A). In the form genus *Whittleseya* (attached to *Neuropteris*-type foliage) many sporangia were joined laterally, forming a bell-shaped structure (Fig. 15-6, B). In the compound synangiate pollen organ of *Dolerotheca formosa (Bernaultia formosa)* there were up to 1200 sporangia in the entire structure, which may be 4.5 cm in its maximum dimension (Rothwell and Eggert, 1986). Microspores in the Medullosaceae

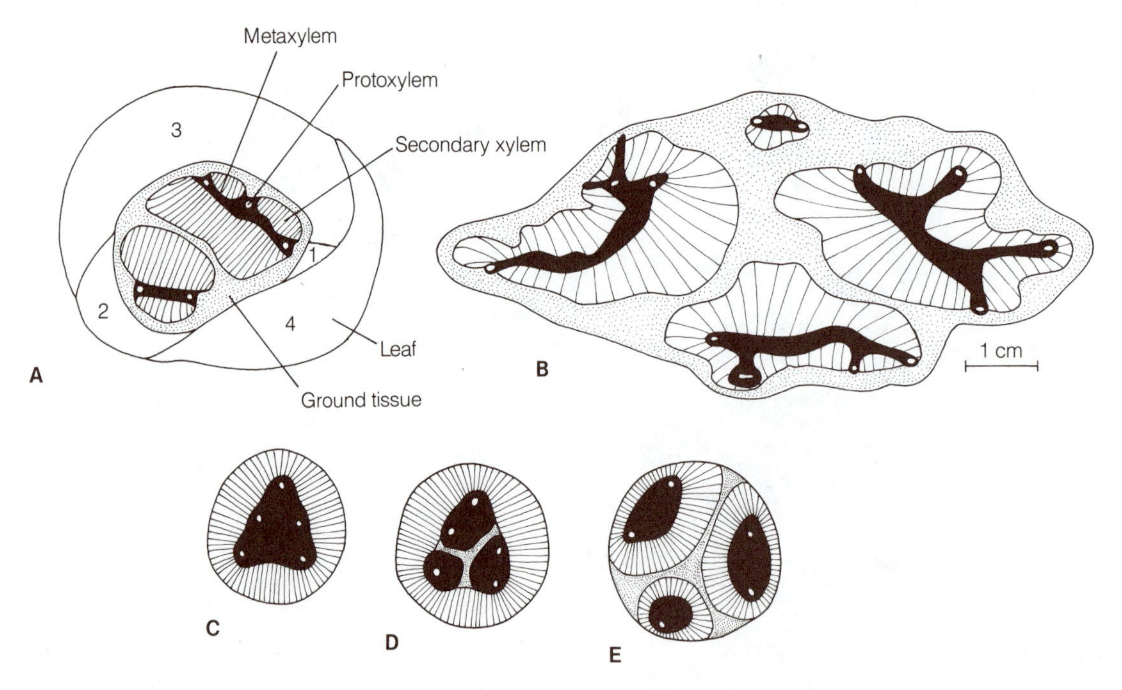

Figure 15-5 **A, B,** transverse sections of medullosan stems. **A,** *Medullosa noei.* At this level the vascular cylinder consists of two vascular "segments" and five sympodia (corresponding to five protoxylem poles) surrounded by ground tissue and four leaves (numbered 1 to 4); **B,** *Medullosa primavea.* Vascular cylinder comprising five vascular segments and thirteen sympodia; scale bar = approx. 1 centimeter. **C–E,** hypothetical sequence of how vascular segments could have been derived by dissection of the protostele. [**A, B** redrawn from Basinger, Rothwell, and Stewart, *Amer. J. Bot.* 61:1002–1015, 1974: **C–E,** redrawn from Stewart, *Birbal Sahni Institute of Paleobotany,* Lucknow, India, pp. 1–13, 1973.]

FIGURE 15-6 Representative medullosan pollen-producing structures. **A,** *Codonotheca* sp., cluster of synangia, each consisting of four partially fused, elongate microsporangia; **B,** synangium *(Whittleseya media)* attached to *Neuropteris*-type foliage; the synangium consists of numerous fused elongate microsporangia. [Redrawn from *Paleobotany and the Evolution of Plants* by Stewart. Cambridge University Press, Cambridge. 1983.]

were often monolete on the proximal surface, and germination probably occurred along this suture. The seeds were large, and, as illustrated by *Pachytesta,* the integument was completely free from the nucellus, unlike *Lagenostoma* (compare Figs. 15-3, E and 15-7). A vascular system was present in both the outer layer (sarcotesta) of the integument and the nucellus (Fig. 15-7, B).

Space does not permit a complete discussion of the Pteridospermales, but we should mention the interesting discovery of pollen grains in the family Callistophytaceae. Pollen was shed in an endosporic, multicellular condition, remarkably similar to pollen grains of living gymnosperms. Also, pollination-drops and pollen tubes have been identified (Rothwell, 1977; Millay and Eggert, 1974).

Mesozoic Seed Ferns

The two orders Caytoniales (Upper Triassic to Lower Cretaceous) and Glossopteridales (primarily of the Permian and Triassic) are rather enigmatic groups. Some paleobotanists align them with the Paleozoic seed ferns, others believe them to be of unknown affinity; still others consider them as possible preangiosperms or at least exhibiting angiospermlike reproductive characteristics.

Caytoniales

Plants assigned to the order Caytoniales had either pinnately or palmately divided leaves based upon compression-impression fossils. Very little is known about the internal structure of stems. Microsporophylls were generally branched, the sporangia aggregated on ultimate segments or fused into elongate, pendent synangia (Fig. 15-8, C, D). The discovery by Thomas (1925) of ovules enclosed in a cupulelike structure created quite a stir in botanical circles at the time. These cupules were borne along a bilaterally symmetrical megasporophyll (Fig. 15-8, A). Each cupule was recurved with a liplike opening near the stalk which became closed at maturity. Numerous small ovules were contained within the cupule (Fig. 15-8, B). Thomas originally described the lipped portion as a "stigmatic surface" comparable to the stigma of angiosperms, because pollen grains were observed on it. The cupule was interpreted as a "fruit." Subsequently, pollen grains were discovered inside the cupules, which detracted from the idea that a cupule was a fruit comparable to that of angiosperms. Nevertheless, the formation of cupules does represent a method whereby ovules could be enclosed and protected during development.

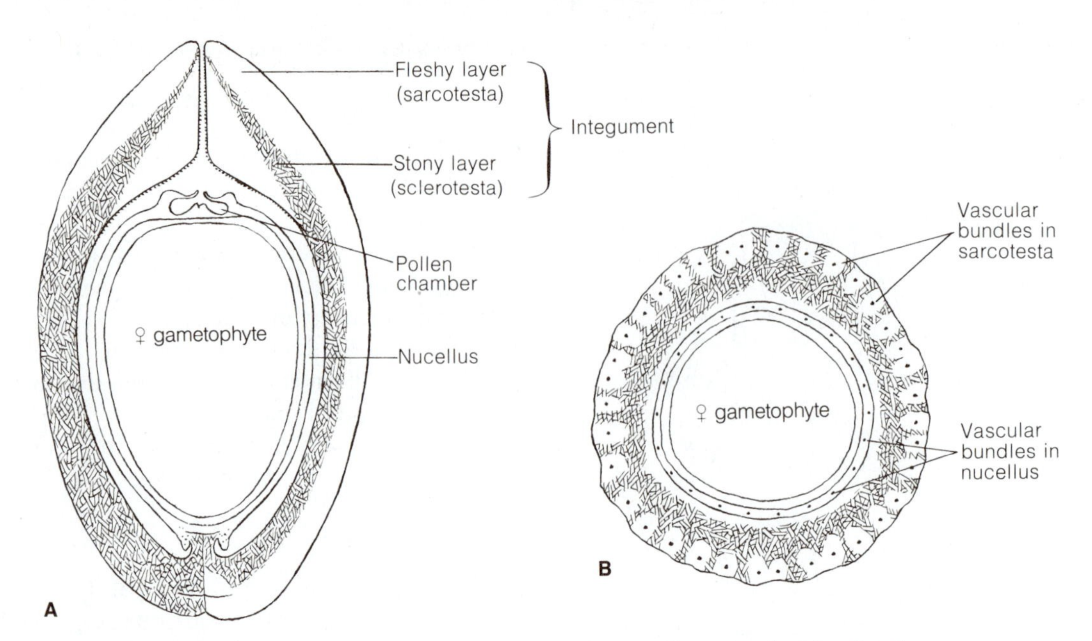

FIGURE 15-7 Reconstruction of the seed of *Pachytesta illinoensis* showing the absence of union between the integument and the nucellus. **A,** longitudinal view; **B,** transectional view. Note numerous vascular bundles in both the integument and the nucellus. [Redrawn from Taylor, *Palaeontographica* 117(B):1, 1965.]

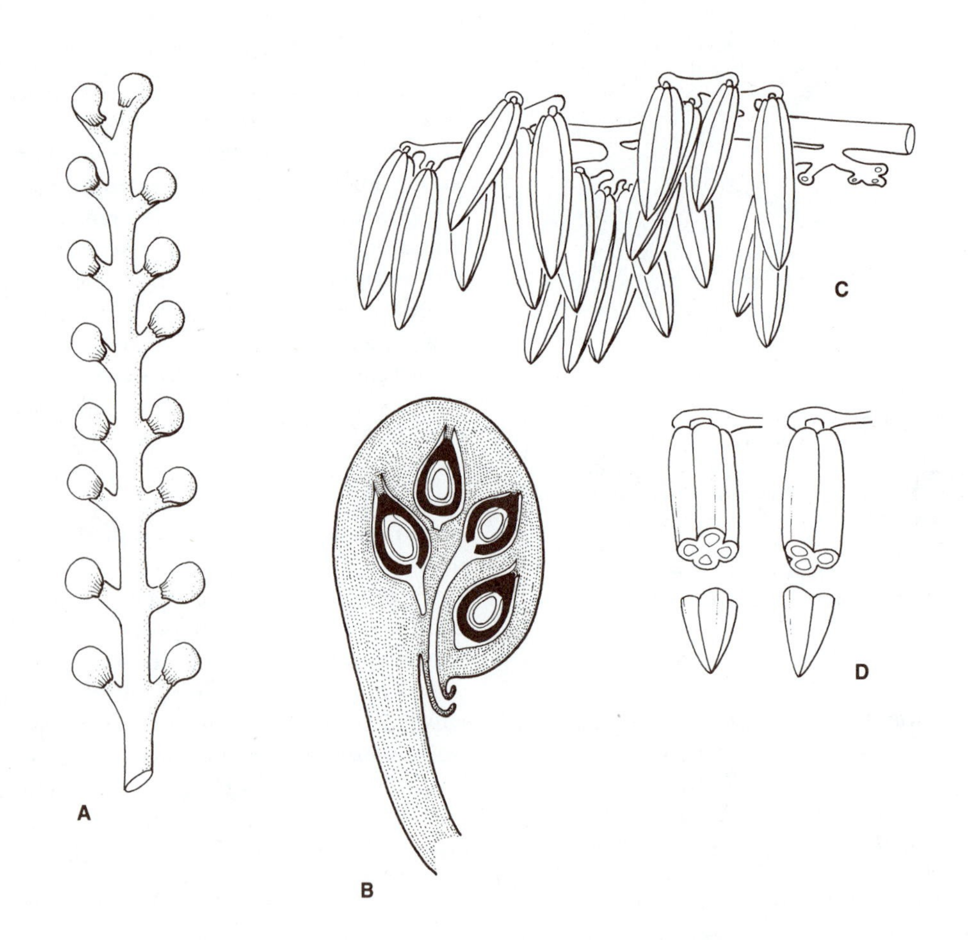

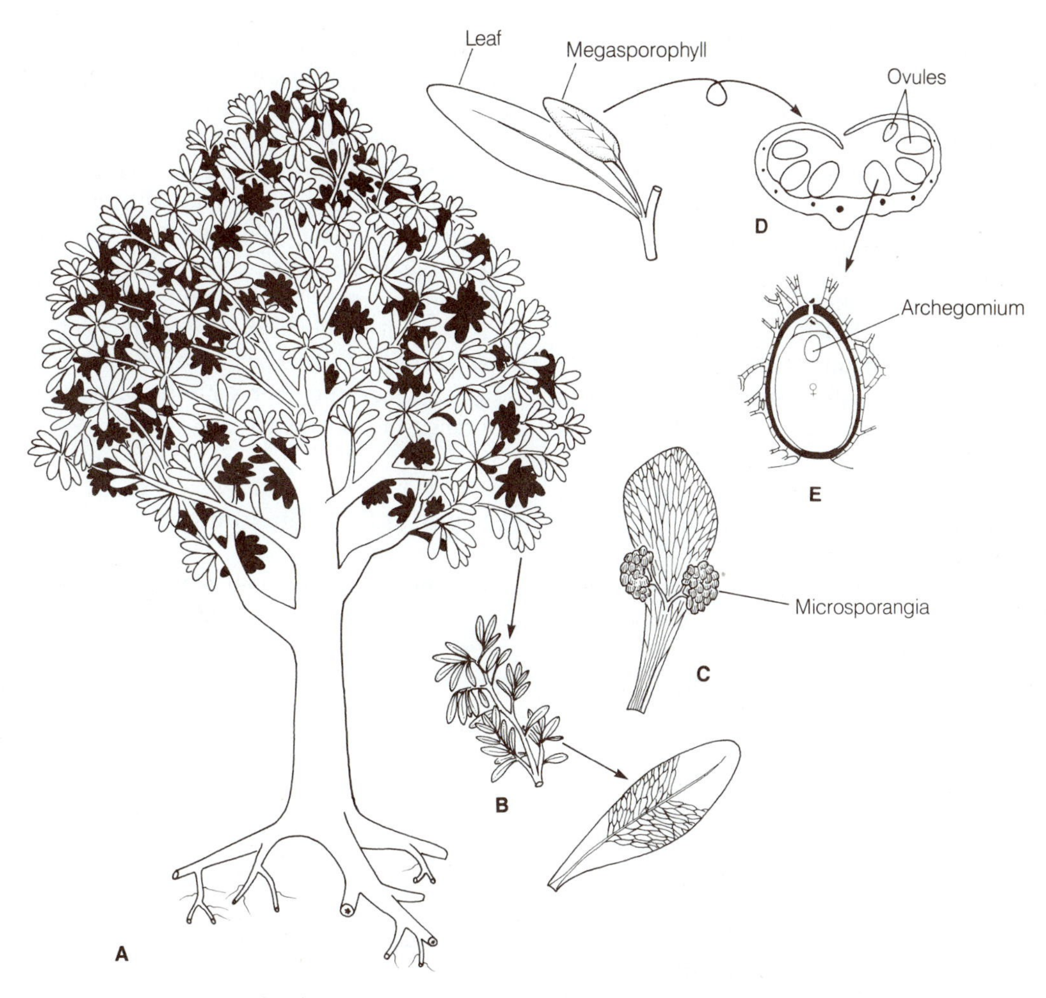

FIGURE 15-9 Stylized reconstruction of a *Glossopteris* tree. **A,** growth habit; **B,** leafy shoot and leaf; **C,** microsporophyll with two clusters of microsporangia (pollen sacs); **D,** inverted megasporophyll, or seed-bearing structure; **E,** longitudinal section of ovule showing position of female gametophyte (♀) and one archegonium. [Redrawn from Gould and Delevoryas, *Alcheringa* 1:387–399, 1977.]

FIGURE 15-8 Reproductive structures of Caytoniales. **A,** megasporophyll of *Caytonia nathorsti* showing two rows of cupules. **B,** *Caytonia thomasi,* longitudinal section of a cupule showing position of ovules. **C, D,** *Caytonanthus kochi,* portion of microsporophyll (**C**) and sectioned synangia showing microsporangia (**D**). [**A** redrawn from Thomas, *Phil. Trans. Roy. Soc.,* London, 213(B):299–363, 1925; **B** redrawn from Harris, *New Phytol.* 32:97–114, 1933; **C, D** redrawn from *Studies in Paleobotany* by H. N. Andrews, Jr. Wiley and Sons, New York. 1961.]

Glossopteridales

Fossils of the Glossopteridales have been found principally in India, South America, South Africa, Australia, and Antarctica and were the main component of the flora of the southern continent of Gondwanaland of the Permian. The glossopterids were seed-bearing plants and have been classified by various authors as seed ferns, cycads, cordaites, and angiosperms. Lacking definitive evidence that the glossopterids represent the precursor-group of the angiosperms, many paleobotanists retain the glossopterids as a specialized group of seed ferns; however, the method of enclosure of ovules suggests how a carpel may have evolved.

Numerous species of *Glossopteris* have been described, some of which may be only variants of a single species or represent ontogenetic stages in the growth of a single species. Gould and Delevoryas (1977) have described *Glossopteris* as a tree with pycnoxylic wood, unlike other seed ferns (Fig. 15-9). Leaves were arranged in tight spirals (helical phyllotaxy). In form, leaves were often lanceolate or spatulate with a prominent midrib and reticulate venation. Microsporangiate structures have been discovered; in one example it was leaflike with two stalked clusters of microsporangia (Fig. 15-9, C). Over the years there has been disagreement, often heated, on the correct interpretation of ovulate structures. In general, one or more of the ovulate structures were stalked and attached near the base of a leaf or in its axil. An entire structure has been termed a megasporophyll, capitulum, or cupule. Several to many ovules were attached to the capitulum and the margins enclosed the ovules to varying degrees (Figs. 15-9, D, E; 15-10, B). When well preserved the ovules are gymnospermous, showing the enclosed female gametophyte with one archegonium (Fig. 15-9, E; Gould and Delevoryas, 1977). In one form there were several stalked, cupulelike structures, each containing one ovule (Fig. 15-10, A). It should be mentioned that some of the "glossopterids" have been aligned with the Pteridospermales rather than with the Glossopteridales (Surange and Chandra, 1975).

Origin of Seed Integument

In Chapter 14 we pointed out that the nature and origin of the integument of the seed in gymno-

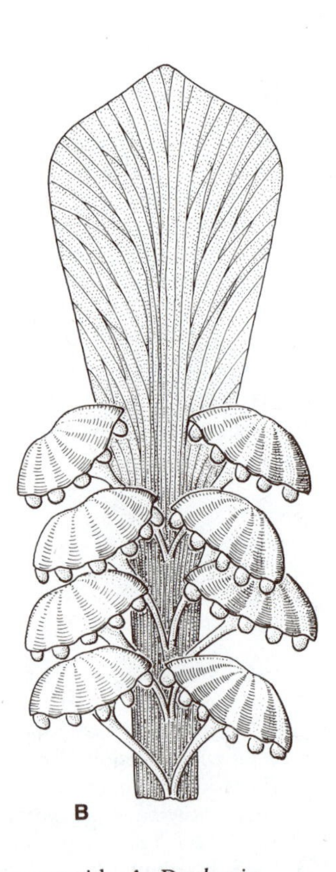

FIGURE 15-10 Reconstruction of ovulate structures of glossopterids. **A,** *Denkania indica;* six cupulelike structures attached to petiole of a leaf. **B,** *Lidettonia mucronata;* seeds attached on the lower surfaces of capitular disks. [**A, B** redrawn from Surange and Chandra, *Palaeontographica* 149(B):153–180, 1975.]

sperms represent difficult morphological problems. Perhaps a partial answer is provided by studies on the seeds of certain pteridosperms of the Carboniferous. The nucellus of the seed, *Genomosperma kidstonii,* in place of being enclosed by and partially joined with an integument, is surrounded basally by a whorl of eight free filamentous processes (Fig. 15-11, A). A more advanced condition toward the formation of an integument is represented by the seed of *Genomosperma latens,* in which the eight integumentary "lobes" are apically appressed—forming what Andrews (1963) regards as a "rudimentary micropyle"—and joined with each other for about a third of their length (Fig. 15-11, B). Fusion of the integumentary lobes is almost complete in the seed of *Eurystoma,* and complete union is shown by the integument of the seed of *Stamnostoma* (Fig. 15-11, C–D). Much remains to be learned about Paleozoic seeds, including the nature and origin of cupulate types. The oldest record of the seed-bearing gymnosperms in the Devonian was reported by Pettitt and Beck (1968) from the Upper Devonian; the seed was named *Archaeosperma arnoldii.* More recently a cupulate form was discovered from the Upper Devonian of West Virginia that is somewhat older than *Archaeosperma* (Gillespie et al., 1981).

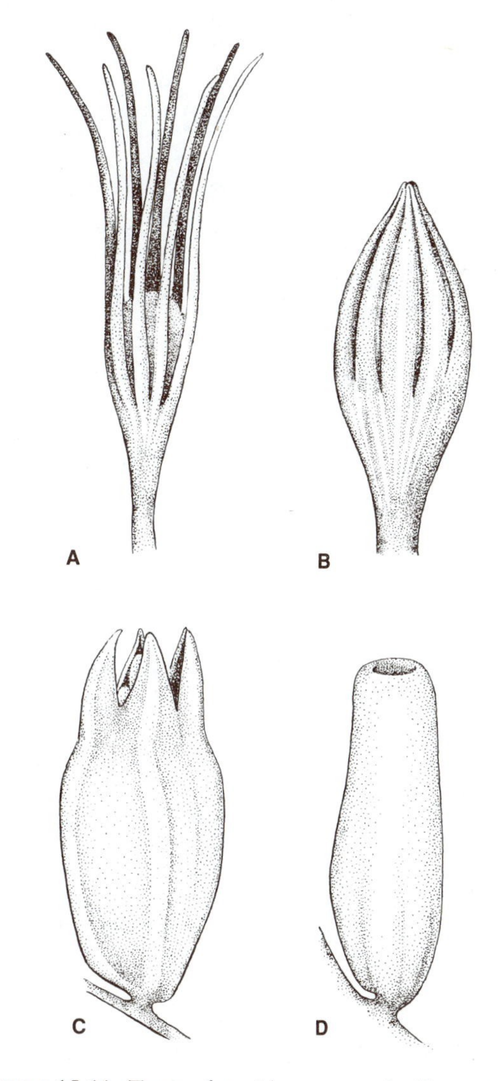

FIGURE 15-11 Types of pteridosperm seeds, illustrating the theoretical origin and phylogenetic development of the integument. **A,** *Genomosperma kidstoni:* **B,** *Genomosperma latens:* **C,** *Eurystoma angulare:* **D,** *Stamnostoma huttonense.* [Redrawn from Andrews, *Science* 142:925, 1963. Copyright © 1963 by the American Association for the Advancement of Science.]

Cycadeoidophyta

The extinct group Cycadeoidophyta was coexistent with extinct cycads in the Jurassic and Cretaceous. The variation in leaf morphology was quite similar in the two groups, but they can be separated from those of the Cycadophyta because the stomata were *syndetocheilic* (subsidiary cells and guard cells originate from the same initial cell). In growth habit some of the cycadeoids are depicted as having columnar trunks; in others the stem is squat and globose. This is the same variation present in the Cycadophyta. The pith was large in diameter. When preserved, the primary xylem was endarch and surrounded by a cylinder of secondary xylem, but the stems lacked the girdling leaf traces characteristic of cycads. Two families are recognized—the Williamsoniaceae and Cycadeoidaceae. Genera in the Williamsoniaceae had slender branched stems, although *Williamsonia sewardiana* has been depicted as assuming the form of a columnar cycad (Fig. 15-12, A). Leaves were simple or pinnate. Strobili were either monosporangiate or bisporangiate (both microsporophylls and ovulate structures in the same strobilus). The cone of *Williamsoniella* consisted of a central axis (receptacle) covered with long-stalked ovules with projecting micropyles. Interspersed among the ovules were sterile (interseminal) scales; surrounding the ovule-bearing portion were wedge-shaped microsporophylls, each

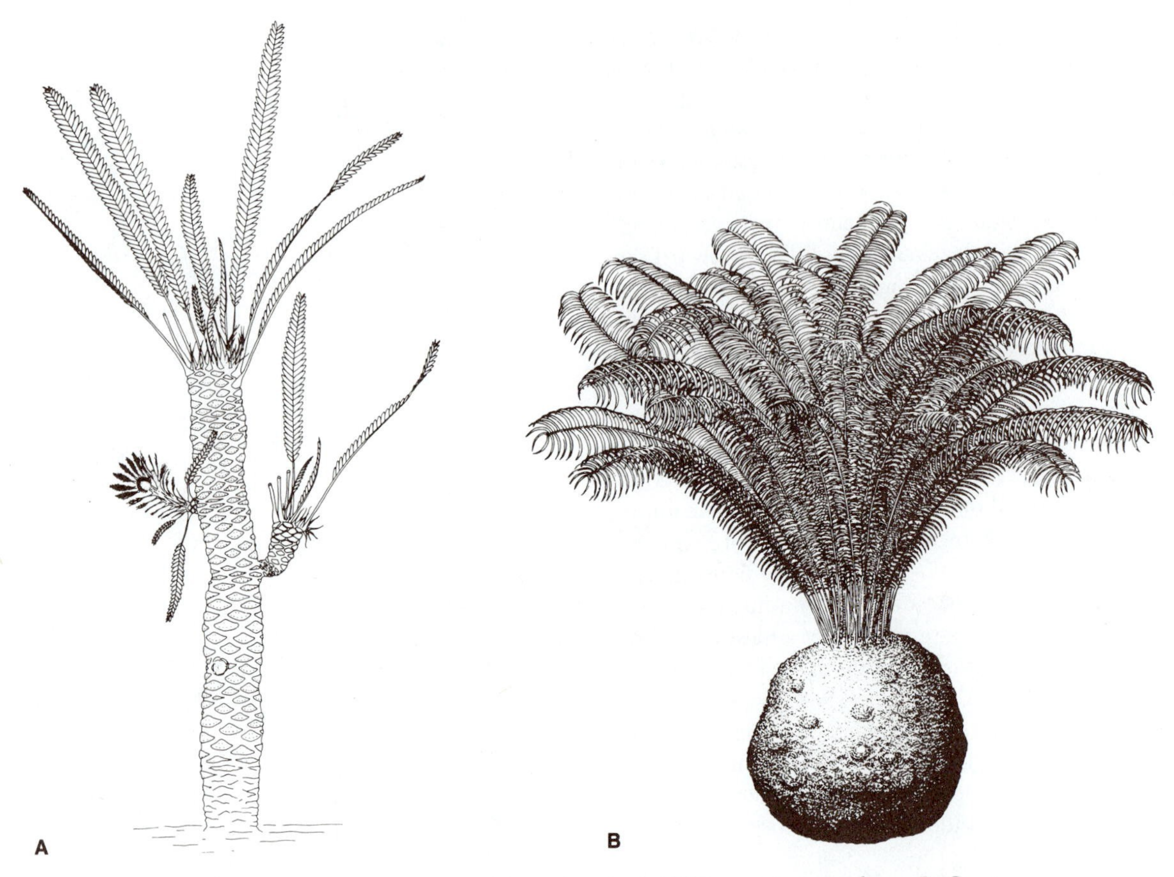

FIGURE 15-12 Reconstructions of cycadeoids. **A,** *Williamsonia sewardiana.* **B,** *Cycadeoidea.* [**A, B** from Delevoryas, *Proc. North Amer. Paleontological Convention,* Part L, 1660, 1971.]

with embedded synangia. The entire fertile axis was enclosed by bracts which probably provided protection during ontogeny (Harris, 1944; Fig. 15-13).

In contrast to the family Williamsoniaceae, the short globose trunks of *Cycadeoidea* (Fig. 15-12, B) have attracted much popular attention, and, before they had been studied by paleobotanists, they were frequently thought to represent fossil beehives, wasps' nests, or corals (Fig. 15-14). Remains of cycadeoids from the Jurassic to the Cretaceous have been found in such widely separated areas as England, the continent of Europe, and India. In North America the cycadeoids were widespread, occurring generally across the United States from the east coast to Arizona, New Mexico, and California. In North America, the most famous and productive localities are the Black Hills of South Dakota. Hundreds of fossil trunks from the Black Hills provided the critical material upon which G. R. Wieland (1906, 1916) based his classical morphological and taxonomic studies. Anatomy of the trunk was of the cycadean type: wide pith, cylinder of manoxylic secondary xylem, although the leaf traces were not of the girdling type as in cycads.

As shown in Fig. 15-12, B, the strobili of *Cycadeoidea* were borne laterally among the persistent leaf bases that formed the armor of the trunk. From all available evidence the strobili of cycadeoids were bisporangiate. The strobilus consisted of a basal collection of bracts surrounding the reproductive parts. Hundreds of stalked ovules, interspersed with sterile scales occurred on a terminal cone-shaped receptacle, which in turn was surrounded by a whorl of microsporophylls (Fig. 15-15). In Wie-

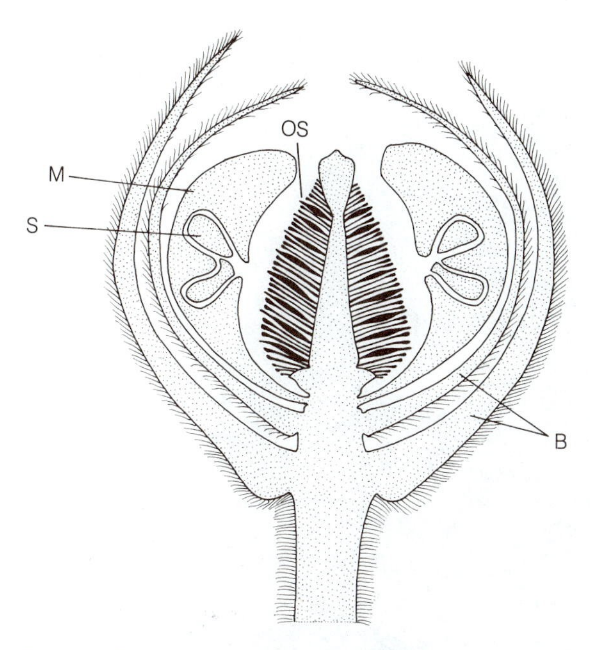

FIGURE 15-13 Diagrammatic representation of a longitudinal section of strobilus of *Williamsoniella.* B, bracts; M, microsporophyll; OS, stalked ovules and sterile scales; S, synangium. [Redrawn from Harris, *Philos. Trans. Roy. Soc.,* London 231B:313, 1944.]

land's (1906) original description the microsporophylls, when mature, were depicted as expanded, pinnate, small frondlike structures, each pinna bearing two rows of synangia. However, the morphology of the pollen-forming organs has been reinterpreted in recent years (Delevoryas, 1963, 1968a; Crepet, 1972, 1974). Each microsporophyll is considered to be basically pinnate; the sporophylls were recurved and joined laterally, enclosing the ovulate receptacle. Each pinna bore two rows of synangia. It is doubtful that the individual units were expanded as free sporophylllike structures because the presence of a dome of parenchymatous tissue would restrict unfolding (Fig. 15-15). If the microsporophyll units never expanded, one wonders how wind pollination would occur. Crepet (1974) has provided evidence that microsporophylls may have disintegrated in place and the pollen was released in close proximity to the ovules. Another possibility is that pollination may have been achieved by insects; Crepet (1972, 1974) has discovered bore holes in cones that might support this conclusion. Similar bore holes made by "snout weevils" occur in the living cycad, *Zamia furfuracea.* The insects bring about pollination in the absence of airborne pollen (Norstog et al., 1986; Norstog, 1987). These two mechanisms would result primarily in selfing without the benefit of outcrossing; which may have contributed to the demise of *Cycadeoidea.*

It seems to be generally agreed that the cycadeoids represent a terminal line in gymnosperm evolution. Cycadeoids differ so markedly in cone structure from any known cycad that it would be impossible to imagine any *direct* phylogenetic connection between the two groups. Furthermore, the stomata of the Cycadeoidales are *syndetocheilic* in contrast to the *haplocheilic* stomata of the Cycadales. (For a description of the ontogeny of the two types, see Chapter 17, p. 414.) This distinction between the two orders has proved extremely helpful in the correct identification of isolated fossil fronds of the two groups. The evolutionary origin of the Cycadeoidales is still an unsettled question although, most likely, ancestral forms would seem to be the pteridosperms (Delevoryas, 1968b).

Cycadophyta

According to paleobotanical studies, modern cycads appear to represent the surviving members of a former larger cycad flora that extended over much of the earth during the Mesozoic Era. Cycads as a whole have existed over a period of at least 250 million years. The best well preserved fossils come from the Upper Triassic and Jurassic. Cycad remains have been discovered in such widely separated regions of the world as Siberia, Manchuria, Oregon, Alaska, several islands in the Arctic Ocean, Greenland, Sweden, England, central Europe, India, Australia, and the Antarctic continent. The distribution of the cycads was probably at its maximum during the Jurassic, coinciding with the age of the giant dinosaurs.

The most comprehensive account of the cycads — based on studies in both the field and laboratory — was given by Chamberlain (1919, 1935), and his comparative studies have profoundly influenced our present understanding of their reproductive morphology and relationships. Chamberlain (1919) recognized nine genera, which he placed in the sin-

FIGURE 15-14 Trunk of *Cycadeoidea marylandica,* the earliest described American specimen of a fossil cycadeoid. The specimen, according to Wieland, was discovered in 1860, in iron-ore beds of the Potomac formation of Maryland, and is represented here about one-fifth its natural size. Note the numerous strobili distributed among the persistent leaf bases which constitute the armor of the trunk. [From *American Fossil Cycads,* Vol. I, by G. R. Wieland. Carnegie Institution, Washington, D.C. 1906.]

gle family Cycadaceae under the order Cycadales. Johnson (1959) added the genus *Lepidozamia* and recognized three separate families of cycads: the Cycadaceae and Stangeriaceae, each represented by a single genus, and the Zamiaceae, which includes the remaining eight genera. Stevenson (1981) now recognizes another family (Boweniaceae; genus *Bowenia*) and has recently discovered a new genus, *Chigua,* endemic to South America (Stevenson, 1988a).

The pattern of present distribution of the eleven genera and about 160 species is peculiar and difficult to interpret from a phytogeographical point of view. In the Western Hemisphere there are four genera: *Microcycas* (a monotypic genus restricted to western Cuba), *Ceratozamia* and *Dioon* (endemic to Mexico), *Chigua* (endemic to northern Colombia), and *Zamia,* which occurs in southern Florida (the only genus of cycads native to the United States), the West Indies, Mexico, Central

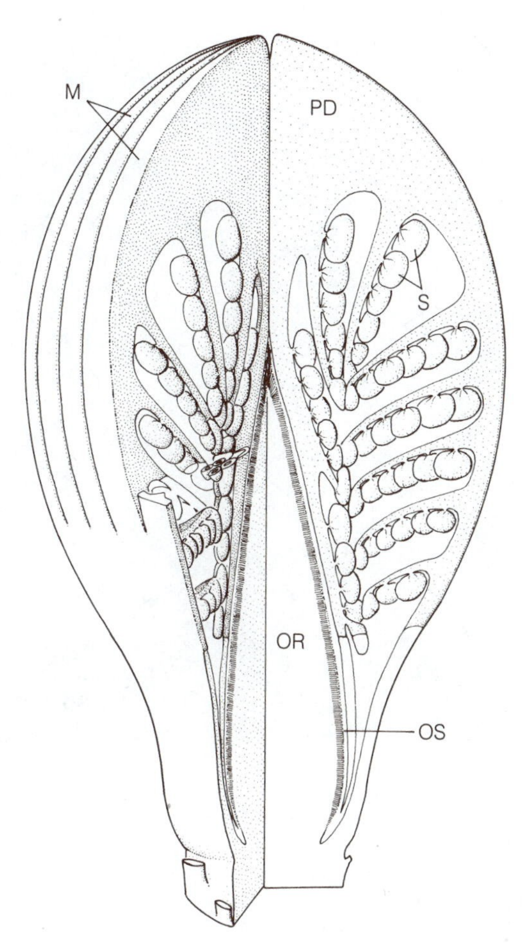

FIGURE 15-15 Diagrammatic reconstruction of the bisporangiate strobilus of *Cycadeoidea*. At the right, one of the microsporophylls is shown in median radial sectional view. M, microsporophylls; PD, parenchymatous dome; OR, ovulate receptacle; OS, numerous stalked ovules and sterile scales. [Redrawn from Crepet, *Palaeontographica* 148B:144, 1974.]

America, and northern and western South America. In the Eastern Hemisphere, the important cycad areas are: Australia, where the endemic genera *Bowenia, Macrozamia,* and *Lepidozamia,* as well as *Cycas,* are found, and Africa, which is the home of *Stangeria* and *Encephalartos.* The relatively large genus *Cycas* occurs not only in Australia but is also represented by species in India, China, the southern islands of Japan, Madagascar, East Africa, and certain islands of the south Pacific Ocean. If we attempt, speculatively, to reconstruct the paths of past migration which led to the present geographical distribution of the living cycads, it becomes very difficult to explain the present isolation of such monotypic genera as *Stangeria* (Africa) and *Microcycas* (Cuba). Such isolated but morphologically well-defined genera are perhaps "leftovers," as Arnold (1953) suggested, of a former, more extensive cycad population. Another explanation for the present distribution may be continental drift.

Because of their large, attractive palmlike leaves, many of the cycads are widely grown as ornamentals. The commonest species in cultivation throughout most of the Northern Hemisphere is the so-called Sago Palm *(Cycas revoluta).* This form requires greenhouse protection in regions of cold winters but is quite hardy outdoors in Southern California, parts of Louisiana, and Florida (Fig. 15-16, A). The most extensive outdoor collections of cycads in the United States are found in the Fairchild Tropical Garden, Miami, Florida, and the Huntington Botanical Gardens, San Marino, California. The growth and general aspects of the species found at the latter institution are interestingly described and photographically illustrated in a book by Hertrich (1951).

Vegetative Organography and Anatomy

The general habit of cycads ranges from types in which the stem is tuberous and partly or wholly subterranean (species of *Cycas, Zamia, Macrozamia,* and *Encephalartos;* and the genera *Stangeria* and *Bowenia*) to one species of *Zamia* which is an epiphyte, to relatively tall plants with the general aspect of tree ferns or palms. The latter category is illustrated by the Cuban cycad *Microcycas calocoma* (Fig. 15-17). The columnar trunk of this inappropriately named genus was found by Foster and San Pedro (1942) to attain a height of about 9 meters as measured from the ground to the base of the terminal crown of leaves. According to Chamberlain (1919), the tallest of the cycads is *Macrozamia hopei,* a native of Queensland, Australia, which may reach a height of 18 meters. The same author records specimens of the Mexican cycad *Dioon spinulosum,* which measured 15 meters in height. Since cycads in general are conspicuously sluggish in their rate of growth, large arborescent types probably are extremely old. Estimates of the age of

A

B

FIGURE 15-16 Habit and general organography of cycads. **A,** specimen of *Cycas revoluta* growing in cultivation in Pasadena, California (note the crowded zones of persistent leaf bases which comprise the armor of the trunk); **B,** specimen of *Dioon spinulosum* growing in the Conservatory at Golden Gate Park, San Francisco, California (note the enormous size of the pendulous megasporangiate cone and the conspicuous armor of leaf bases). [**B** courtesy of Dr. T. E. Weier.]

large specimens have been based on a study of the persistent leaf bases which constitute the characteristic armor of the trunk (Fig. 15-16). If the average number of leaves produced each year can be determined, this number divided into the total number of leaf bases on the entire trunk yields some approximation of the age of the plant. By this method Chamberlain concluded that a plant of *Dioon edule* with a trunk only 2 meters in height was at least 1,000 years old. Age determinations based on the study of the leaf armor, however, are probably conservative because the sluggish period of development of the seedling and the prolonged periods of dormancy of older plants are not taken into consideration. In some genera that have subterranean fleshy, tuberous stems, leaf bases do not persist and the stems are contractile (Stevenson, 1980c).

Cycads as a whole, particularly the arborescent types, are unbranched, but there are exceptions—in species having the tuberous habit of growth as well as those of the columnar habit (Fig. 15-18) (Swamy, 1948). In *Cycas revoluta,* for example, buds commonly develop from various areas of the trunk, and their irregular expansion into shoots results in the production of irregularly or grotesquely branched individuals. These buds, which apparently originate from the living tissue of the persistent leaf bases, are capable of rooting and producing new plants if detached from the trunk. In other instances branching is clearly the result of the injury or destruction of the main terminal bud. In these

FIGURE 15-17 A microsporangiate plant of *Microcycas calocoma* growing in the colony at Cayo Ramones, Province of Pinar del Rio, Cuba. The trunk of this large specimen measured 8.5 meters in height and was 113 centimeters in circumference near the base. [From Foster and San Pedro, *Mem. Soc. Cubana Hist. Natur.* 16:105, 1942.]

instances the buds arise endogenously from internal organized meristems such as the vascular cambium (Stevenson, 1980b). Striking examples of this were discovered by Foster and San Pedro (1942) in *Microcycas.* In this genus adventitious buds arise following injury (often as a result of wind) to the upper part of the trunk, and may give rise to large candelabra-shaped branches.

STEM. The primary vascular cylinder is eustelic. As viewed in transverse section, the stem is distinguished by several characteristics. First, the cortex and pith are exceptionally large, as compared with the relatively narrow diameter of the vascular cylinder (Fig. 15-19, A). For example, the diameter of the pith in large trunks of *Cycas revoluta* may measure as much as 15 centimeters. Secondly, the course of the vascular strands or traces which supply each of the leaves is complex. Since the various strands entering a given leaf extend horizontally through the cortex before entering the leaf base, a girdling arrangement is seen in thick or cleared sections of the stem. Although secondary growth by means of a vascular cambium occurs in the stems of many genera, the amount of secondary vascular tissues produced is relatively small, and the xylem is traversed by numerous broad parenchymatous rays. Seward (1917) designated the loose-textured wood of cycads as "manoxylic," in contrast with the more compact and dense xylem of conifers, which he termed "pycnoxylic." Moreover, although concentric and definable growth rings may be seen in certain genera, they do not appear to represent true annual increments or annual rings. According to Chamberlain's study of *Dioon spinulosum,* the rings correspond to those periods in development when new crowns of leaves were produced by the terminal bud.

LEAVES. Cycads are distinguished from all other living gymnosperms by their pinnate leaves, which, in certain genera, may reach a length of about 1.5 meters. The genus *Bowenia* is exceptional in having large bipinnate leaves; in all other living cycads the leaf develops a single series of leaflets (Figs. 15-16, 15-20). Except during the juvenile phases of development, the leaves in most of the genera are produced in crowded clusters, or crowns, which thus

FIGURE 15-18 **A,** large specimen of *Encephalartos transvenosus*, a cycad native to Africa. Note huge ovulate cone at tip of the single lateral branch. **B,** ovulate cones of *Encephalartos umbeluziensis*. [From Dyer, *Cactus Succulent Jour.* 44:209, 1972.]

B

A

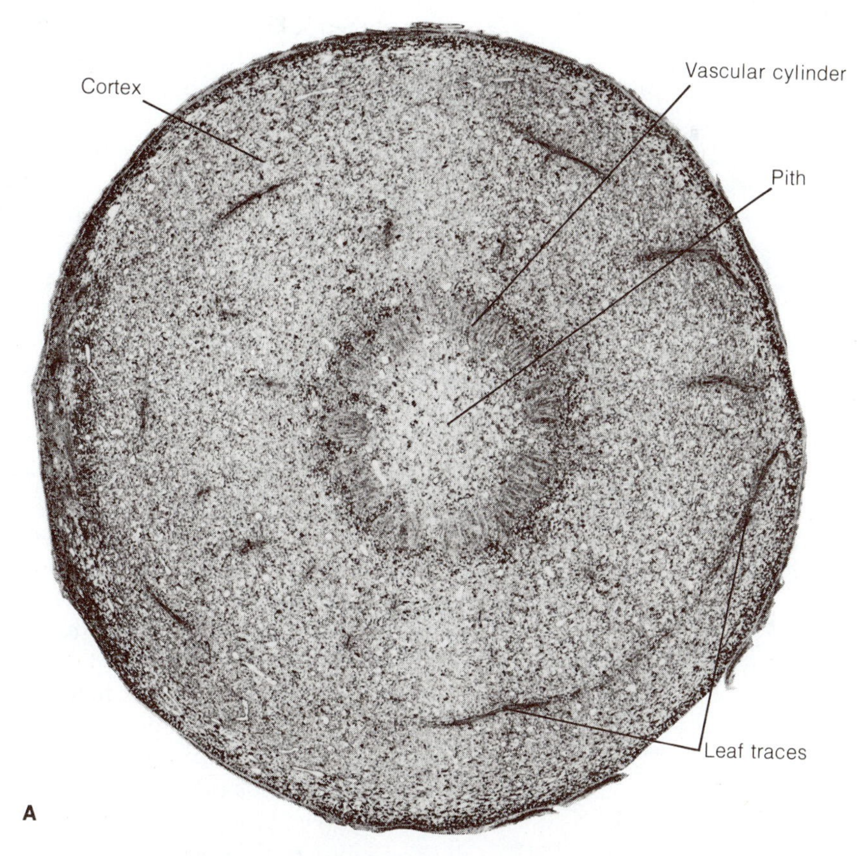

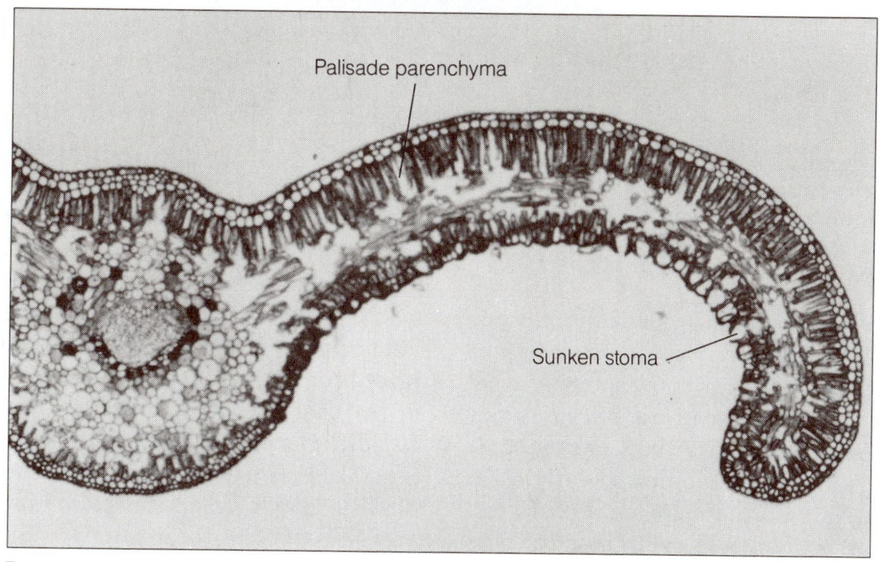

FIGURE 15-19 **A,** transection of stem of *Zamia sp.* demonstrating the large pith and cortex and the characteristic girdling leaf traces (× 4). **B,** transection, portion of leaf of *Cycas.*

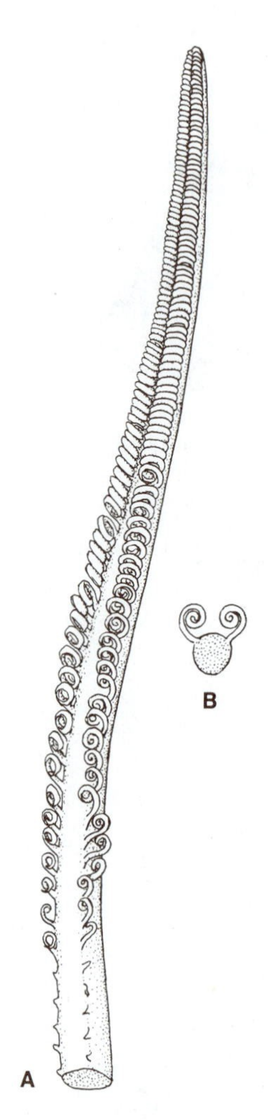

FIGURE 15-20 Circinate vernation in *Cycas revoluta*. **A,** adaxial view of foliage leaf showing the acropetal unrolling of the rows of circinate pinnae; **B,** transection of rachis of leaf showing two circinate pinnae.

impart a palmlike aspect to many of the arborescent types (Figs. 15-16, A; 15-17). Prior to the expansion of a crown of new foliage leaves the latter are overlaid and protected by an extensive imbricated series of tough bud scales. These structures, like the bases of the foliage leaves, persist for many years on the trunk and thus further contribute to the characteristic leaf armor. In the aboveground columnar-type cycads, the new leaves are erect in *vernation* (form of the leaf in the bud) and the leaflets are flat during expansion of the leaf. In the genus *Cycas* the leaf axis is erect or only slightly bent, but vernation of the young leaflets is distinctly circinate (coiled) as in many ferns. Figure 15-20 shows that the unrolling of successive coiled pinnae (leaflets) proceeds from the base upward. In the subterranean tuberous species, young leaves may be bent or folded *(Zamia)* or display true circinate vernation *(Bowenia)* which may be correlated with subterranean existence of the plant, providing protection until the leaf has emerged through the soil (Stevenson, 1980b).

There are various types of venation patterns in the leaflets or pinnae of cycad leaves, and some of these patterns provide consistent characters useful in the recognition of genera. The genus *Cycas* is unique because each pinna is traversed by only a single prominent midvein and lateral veins are absent (Fig. 15-19, B). In contrast, the leaflets of *Stangeria* develop a pinnate type of venation consisting of a midvein from which radiate lateral veins that are closely spaced and dichotomously branched. Many of the veinlets in the pinnae of *Stangeria* anastomose near the margin, forming conspicuous closed loops as is shown in Fig. 15-21, B. In the remaining eight genera, the pinnae are devoid of a distinct midvein, and the venation consists of a series of longitudinal veins which branch dichotomously at various levels in their course through the pinnae (Fig. 15-21, A). Although Troll (1938) called attention to the vein anastomoses in the pinnae of *Ceratozamia mexicana* and *Stangeria,* this aspect of leaf venation in the cycads has received much less attention than it deserves. The genus *Zamia,* for example, provides striking examples of the union between the branches derived from the dichotomy of two adjacent veins. This distinctive type of anastomosis is evidently a very primitive kind of vein union since it occurs not only in many ferns but also in the dichotomously veined lamina of *Ginkgo biloba* (Arnott, 1959). It is interesting, from a taxonomic point of view, that although Johnson (1959) placed considerable emphasis on venation patterns in his characterization of the three families of cycads, he failed to mention the occurrence in some genera of vein anastomoses. It seems entirely possible that more detailed studies will reveal that the presence or absence of anastomoses has value in the separation of cycad taxa at both the generic and specific levels (see Brashier, 1968).

Certain interesting aspects of leaf histology may be briefly noted at this point. In *Cycas,* as previously mentioned, each pinna is vascularized by a

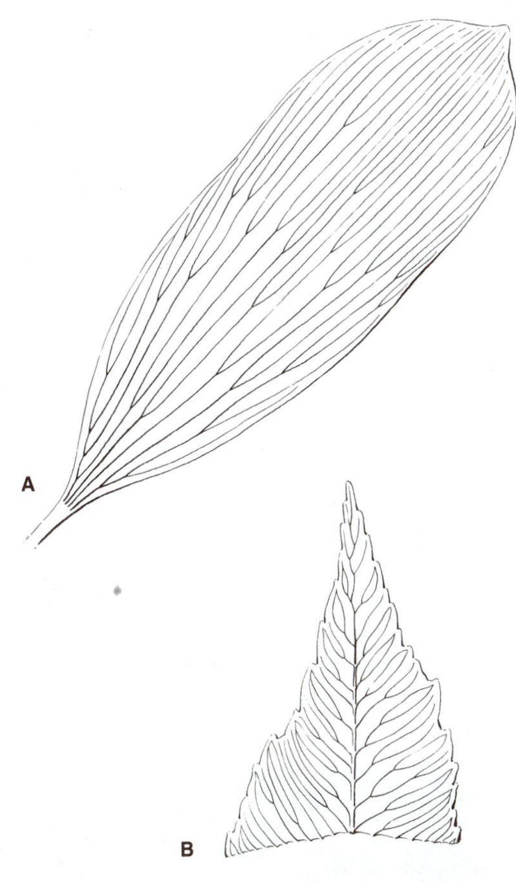

FIGURE 15-21 Dichotomous venation in leaflets of cycads. **A,** *Zamia wallisii:* **B,** *Stangeria schizodon,* tip of leaflet (note midvein, dichotomously branched lateral veins, and union of certain vein endings). [From *Vergleich. Morphologie der höheren Pflanzen* by W. Troll. Gebrüder Borntraeger, Berlin. 1938.]

single bundle which extends through the midrib region. Although lateral veins are absent, a sheet of so-called accessory transfusion tissue extends from each side of the midrib to the margins of the pinna. Some of the cells in this transfusion tissue resemble tracheids in their elongated form and wall structure and perhaps facilitate the translocation of material between the midvein and the mesophyll of the pinna. (See Brashier, 1968, for further details regarding transfusion tissue in the pinnae of cycad leaves.) Of additional interest, *Cycas* has *both* endarch and exarch xylem in the midrib bundle of the pinnae as well as in the vascular strands in the leaf rachis.

APOGEOTROPIC ROOTS. According to Chamberlain (1935), all cycads form *apogeotropic roots* which grow upwardly and, near the soil surface, branch dichotomously, forming peculiar coralloid masses of rootlets. Coralloid rootlets may form in seedlings but are particularly conspicuous near the base of the trunk of well-developed cultivated specimens (e.g., in *Cycas revoluta*), especially if the plants are grown in tubs.

One of the distinctive features of coralloid roots is the presence in the root cortex of a well-defined cylindrical zone inhabited by the blue-green alga (cyanobacterium) *Anabaena cycadeae.* Entry of the alga is through a break in the root epidermis (Nathanielsz and Staff, 1975a). Associated with initial entry of the algal cells, large amounts of mucus are apparently secreted by algal-zone cells of the host and deposited in intercellular spaces of the cortex. Storey (1968) found that "somatic reduction" of chromosomes occurs in many cells of the developing "algal zone" which results in the formation of small cells among the larger diploid cells. The small cells form the slime that apparently provides a pathway for the establishment of the colonies of blue-green algae. The algal cells also may actually penetrate cortical cells, filling much of the host cell cavity (Nathanielsz and Staff, 1975b).

From all available information the initial entrance of the blue-green alga is not responsible for the formation of coralloid roots. Some interesting in vitro experiments have been performed to determine the factors that induce apogeotropic roots, and nodulation leading to the formation of coralloid roots. In some cycads, nodulation occurs without any specific external stimulus (Lamont and Ryan, 1977; Webb, 1983b). In others, root nodulation occurs on seedling roots when exposed to light. In *Zamia pumila,* for example, apogeotropism was more frequent at low light intensities, but nodulation increased at higher light intensities (Webb, 1982, 1983a). Nodulation does not occur in dark-grown seedlings; the roots were geotropic and became much longer than roots of seedlings grown in light. Under natural conditions, as Webb has pointed out, it is possible that low levels of light stimulate apogeotropic growth near the soil surface. As the roots grow upward, they may respond to higher intensities of illumination by becoming nodulated and be more easily colonized by blue-green algae. The biological role of the algae has not been determined, but nitrogenous substances may be passed on to the host plant because *Anabaena* is known to fix atmospheric nitrogen.

The Reproductive Cycle

STROBILI. All carefully investigated species of cycads are strictly dioecious with microsporangiate and megasporangiate strobili being borne on separate plants. Chamberlain stated emphatically that "in thirty years of study in the field and in greenhouses I have never seen anything to indicate that the cycads are not absolutely dioecious." Dioecism in cycads is based apparently upon an XY type of sex determination. In a megasporangiate plant (female) of *Cycas pectinata* each chromosome of one pair of the eleven pairs of chromosomes ($2n = 22$) has a satellite (XX condition). In a microsporangiate plant (male) the pair is heteromorphic (XY); only one chromosome has a satellite (Abraham and Mathew, 1962). However, Marchant (1968) found no compelling cytological evidence for explaining dioecy in cycads. Chromosome numbers $2n = 16$, 18, 22, 24 are common in cycads, with most genera and species being $2n = 18$ (Norstog, 1980).

In all genera, with the exception of *Cycas,* both types of strobili are compact conelike structures with determinate growth. The most common explanation of the process of cone production in cycads is by sympodial branching. According to Chamberlain (1935), for example, the first cone produced by *Dioon* is terminal and a new vegetative meristem, developing laterally at the base of this cone, gives rise to the next crown of leaves. Stevenson (1988b) has produced evidence that vegetative and cone axes are formed by dichotomous (anisotomous) branching. One branch of a dichotomy exhibits continuous growth as a strobilus, while the other branch is at first dormant but later begins vegetative growth. After some years certain cycads produce multiple cones. In these cases the vegetative apex is centrally located and produces several

A

B

FIGURE 15-22 Strobili of *Cycas revoluta.* **A,** microsporangiate strobilus (cultivated specimen, Pasadena, California); **B,** megasporangiate strobilus (note pinnatifid form of megasporophylls). [**B** courtesy of Dr. T. E. Weier.]

to many cone primordia, equidistant from each other, around the periphery of the shoot apex (Stevenson, 1988b).

Cycas is unique in that the megasporophylls are produced, like the foliage leaves, in a relatively loose crown surrounding the shoot apex of the terminal bud (Fig. 15-22, B). After seed maturation, a new crown of foliage leaves expands above the cluster of megasporophylls; the latter persist as dead or dying structures below this new foliage. The microsporangiate strobilus of *Cycas,* however, is a compact, determinate cone as in the other genera (Fig. 15-22, A).

According to Chamberlain, "the largest cones that have ever existed are found in the living cycads." Apparently the largest dimensions and greatest weight are characteristic of the megasporangiate strobili of certain species of *Encephalartos, Dioon,* and *Macrozamia* (Figs. 15-16, B; 15-18). Weights up to 40 kilograms have been recorded for *Encephalartos,* and Chamberlain describes the seed cones of *Macrozamia denisoni* as measuring 60 centimeters in length, nearly 30 centimeters in diameter at the base, and with a weight of 25 to 30 kilograms. In comparison with such gigantic cones, the cones of the more familiar gymnosperms, such as pine, indeed seem insignificant in size.

SPOROPHYLLS AND SPORANGIA. The microsporophylls in the cycads, although varying in size and form, all are rather thick, scalelike structures that bear the microsporangia on their lower or abaxial surfaces (Fig. 15-23, B). This resemblance to the abaxial position of sporangia typical of many ferns is strengthened by the fact that the microsporangia are arranged in somewhat definite soral clusters.

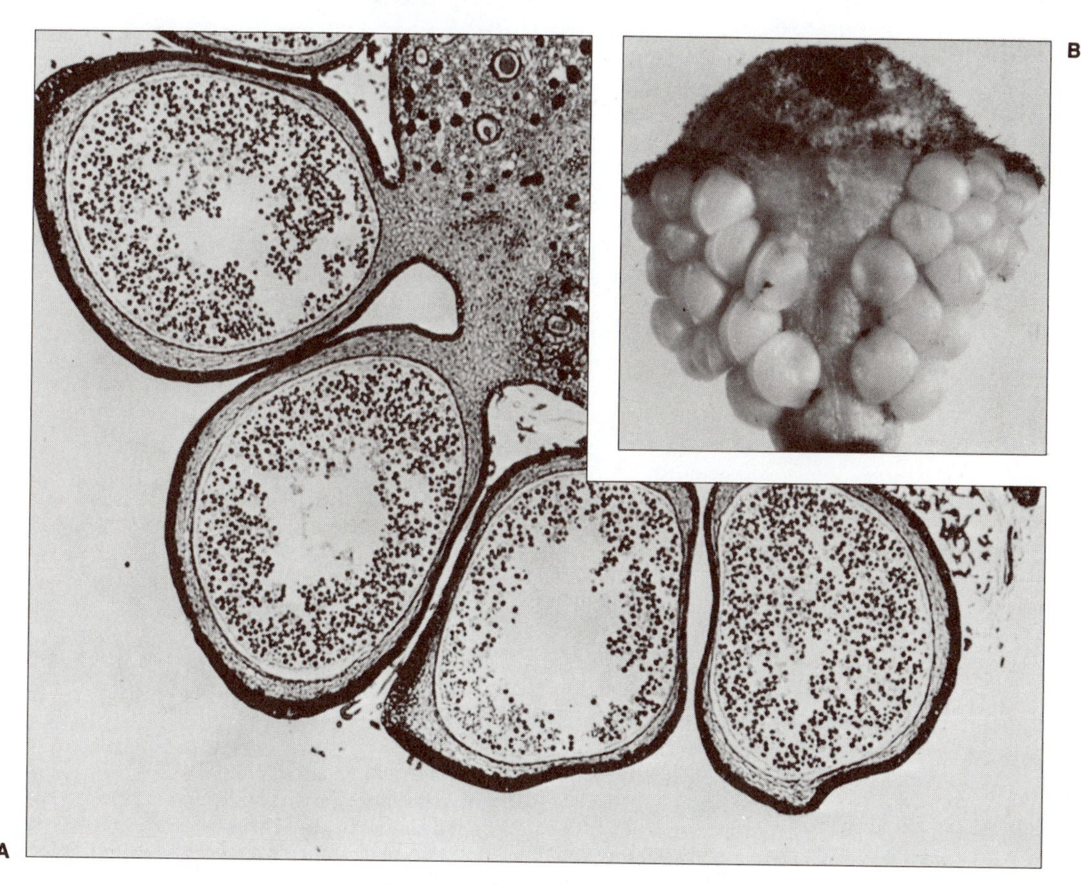

FIGURE 15-23 *Zamia* sp. **A,** transection of microsporophyll; **B,** abaxial view of microsporophyll, showing soral clusters of microsporangia. Note thick walls of the stalked microsporangia and the abundant microspores produced in each of them.

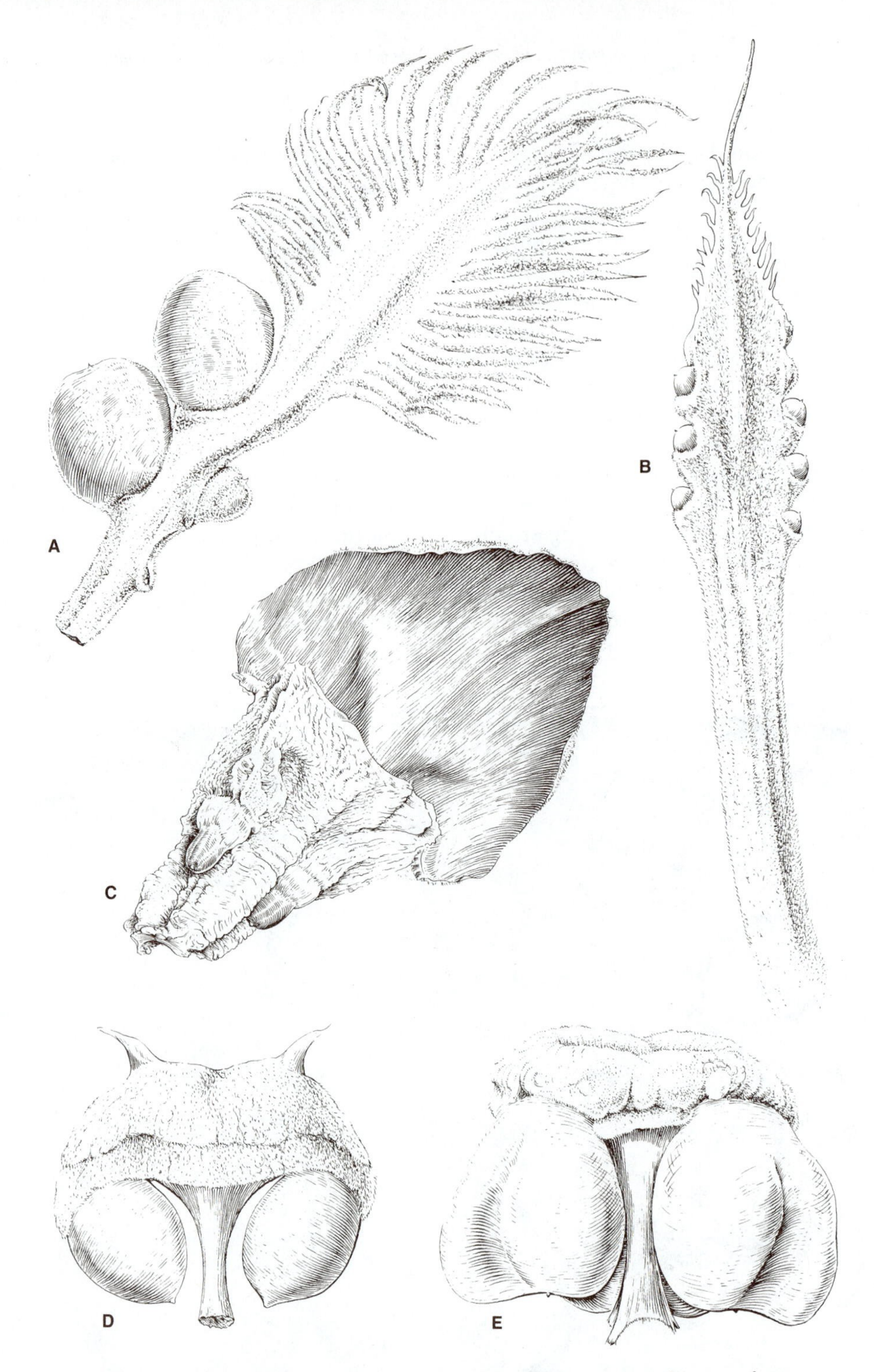

FIGURE 15-24 Megasporophylls in cycads. **A,** *Cycas revoluta,* pinnatifid type of megasporophyll with developing seeds; **B,** *Cycas circinalis,* note very rudimentary pinnae; C, *Dioon edule,* megasporophyll with expanded lamina tip and two ovules; **D,** *Ceratozamia,* peltate type of megasporophyll bearing two ovules (the two spines at the top of the sporophyll are characteristic for this genus); **E,** *Zamia,* peltate type of megasporophyll with two ovules. [**B** redrawn from *Die natür. Pflanzenfamilien* by A. Engler and K. Prantl. Wilhelm Englemann, Leipzig. 1926; **D** redrawn from *Syllabus der Pflanzenfamilien* by A. Engler and E. Gilg. Gebrüder Borntraeger, Berlin. 1924.]

The number of microsporangia borne by a single sporophyll varies from more than a thousand in *Cycas media* to several dozen in *Zamia floridana.* Although much remains to be done in the study of the origin and early ontogeny of the microsporangium, it is known that its structure at maturity is typically eusporangiate with a wall several cell layers in thickness, a tapetum, and numerous small microspores (Fig. 15-23, A). The surface cells of the microsporangium are large and very thick walled (except at the point where dehiscence will occur) and have collectively been regarded by Jeffrey (1917) as an annulus.

The megasporophylls vary considerably in size and form; in many cases their shape is of great systematic value in the characterization of genera or even species. Two extreme types occur. In *Zamia, Microcycas,* and *Ceratozamia,* for example, the megasporophylls are peltate, scalelike organs, each bearing two ovules (Fig. 15-24). In marked contrast, the megasporophylls of *Cycas revoluta* are conspicuously pinnatifid, leaflike structures that bear six to eight ovules laterally arranged on the sporophyll axis below the terminal group of rudimentary pinnae (Fig. 15-24). In other genera, such as *Dioon* and *Encephalartos,* the sporophylls, although essentially scalelike in form, may show extended tips or marginal serrations suggestive of reduced leaf blades. The common interpretation of this range in sporophyll form and ovule number is to regard the condition in *Cycas revoluta* as primitive and the other conditions as being the result of varying degrees of suppression and ultimate elimination of a definable lamina in the megasporophyll (Fig. 15-24). However, recent studies of fossil cycads have cast some doubts on this interpretation (see p. 381).

The ovules of cycads are oriented horizontally with respect to the cone axis. In the genus *Cycas* the micropyles of the ovules are turned obliquely outward and because of the loose arrangement of the pinnatifid megasporophylls are a striking example of veritable naked ovules (Fig. 15-24). In the other genera the micropyles of the paired ovules of each sporophyll are directed inward toward the axis of the strobilus and, except during the brief period when pollination occurs, are not directly exposed to the air (Fig. 15-24).

On the whole, cycad ovules are large as compared with the ovules of other gymnosperms. According to Chamberlain, the ovules of *Cycas circinalis* and *Macrozamia denisoni* may reach 6 centimeters in length; in other genera the ovules are smaller, the most diminutive being those of *Zamia pumila* which measure only 5 to 7 millimeters long. Structurally the cycad ovule consists of the integument laterally joined with the massive nucellus except near the micropylar end of the ovule (Fig. 15-25, D). During its ontogeny, the integument becomes histologically differentiated into three layers: an outer and an inner fleshy layer and a middle stony layer, which is sclerified and hard in texture. A very interesting feature of the ovule in cycads is the development of two separate vascular systems. The outer system extends as a series of unbranched veins through the outer fleshy layer, whereas the inner system, consisting of numerous dichotomously branched strands, vascularizes the inner fleshy layer (Fig. 15-25, D, E). It has been reported that in several genera, some of the branches of the inner vaescular system may terminate in the base of the free part of the nucellus (Kershaw, 1912; Reynolds, 1924; Shapiro, 1951). The inner layer of the integument usually breaks down during seed development, but the outer fleshy layer persists and, in the ripe seed, may be bright red *(Encephalartos, Zamia),* salmon pink *(Microcycas),* or nearly white *(Dioon, Ceratozamia).*

GAMETOPHYTES. Our knowledge of the gametophytes, fertilization, and embryogeny in cycads is still somewhat fragmentary, and the brief discussion of these topics here necessarily represents a composite picture derived from the available data. In many respects the investigations of Chamberlain, especially on *Dioon.* are classical and have provided the firm basis for most of the subsequent investigations on other species.

MALE GAMETOPHYTE. The development of the endosporic male gametophyte of cycads proceeds through the early stages before the pollen grains are shed from the microsporangium. The mitotic division of the microspore nucleus forms two cells: a small basal *prothallial cell* and a larger meristematic initial. The latter functions as an *"antheridial initial,"* and its division yields a *generative cell,* which lies next to the prothallial cell, and a *tube cell* comprising a nucleus and its associated cytoplasm Fig.

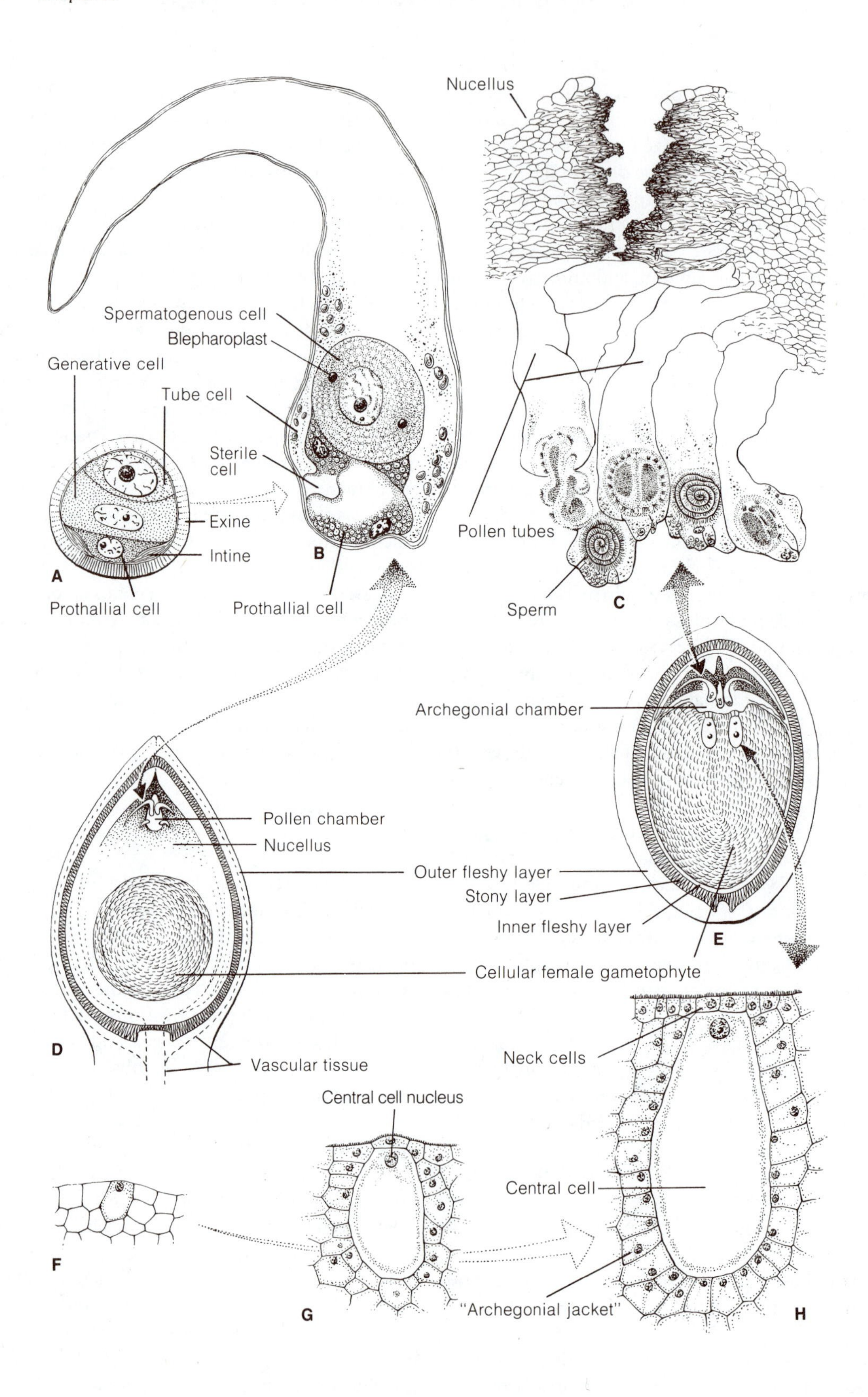
Nucellus
Spermatogenous cell
Blepharoplast
Generative cell
Tube cell
Sterile cell
Exine
Intine
A
B
Prothallial cell
Prothallial cell
Pollen tubes
Sperm
C
Archegonial chamber
Pollen chamber
Nucellus
Outer fleshy layer
Stony layer
Inner fleshy layer
E
Cellular female gametophyte
D
Vascular tissue
Neck cells
Central cell nucleus
Central cell
F
G
"Archegonial jacket"
H

15-25, A). At this three-celled stage, the partially developed male gametophyte, enclosed within the original spore wall, is released from the microsporangium as a pollen grain.

A survey of 29 species of ten genera in the cycads revealed that the pollen grains are elongated, boat-shaped, longitudinally monosulcate, and bilaterally symmetrical. The sulcus is an elongate furrow on the surface of the pollen grain through which the pollen tube emerges following pollination. The surface features of the grains do provide information on relationships within the order and may be related to the mode of pollination (Dehgan and Dehgan, 1988).

The manner in which the pollen of such dioecious plants as the cycads reaches the megasporangiate cones needs more careful study in the field. Although Chamberlain (1935) maintained that all cycads are probably wind pollinated, there have been reports of insect pollination in such genera as *Encephalartos* and *Macrozamia* (see Rattray, 1913, and Baird, 1939). Norstog and Stevenson (1980) found that "snout weevils" spend part of their lives within the pollen cones of *Zamia furfuracea,* first as larvae and later as pupae. The metamorphosed adults bore tunnels through the microsporophylls to the outside. In so doing they become dusted with pollen from the microsporangia. The weevils enter ovulate cones between the megasporophylls, bringing about pollination (Norstog et al., 1986). Norstog (1987) gives an interesting account of pollination in cycads and the possible coevolution of cycads and their insect pollinators.

Whatever may be the exact mechanism of pollen transfer, the pollen grains, on reaching the ovules, are said by Chamberlain to be drawn into the pollen chamber by the "retraction" of the pollination droplet that exudes from the micropylar end of the

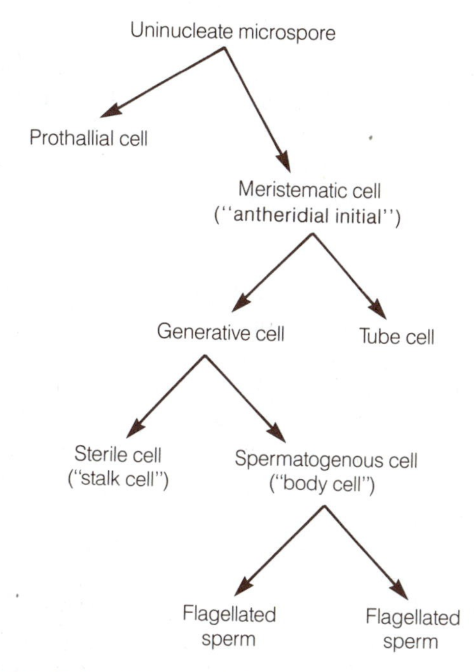

FIGURE 15-26 Diagram of the successive stages in development of the male gametophyte in cycads. See text for details.

ovule. Within the pollen chamber, pollen tubes begin to form, and the male gametophyte completes its development in the manner represented diagrammatically in Fig. 15-26.

According to Chamberlain's studies, the interval between pollination and fertilization varies from four to six months. During this interval, a pollen tube, produced by the tubular extension of the upper portion of the pollen-grain wall (i.e., the end *opposite* the basal prothallial cell), invades the tissue of the nucellus (Fig. 15-25, D, E). In some cycads, this haustorial portion of the pollen tube may become irregularly branched. Concomitant with the

FIGURE 15-25 The reproductive cycle in cycads. **A–C,** development of male gametophyte in *Cycas.* **A,** the three-celled stage at time of shedding; **B,** the generative cell has divided forming the spermatogenous cell and the sterile cell. This stage occurs after pollination and during the invasion of the nucellus by the pollen tube; **C,** reconstruction of adjacent serial sections showing five pollen tubes with sperm in various stages of development, hanging from the nucellus in the archegonial chamber (note the spirally arranged band of flagella on two of the mature sperm); **D,** diagram of longisection of ovule of *Dioon edule* early stage in development of pollen tubes and the cellular stage of the female gametophyte; **E,** diagram of longisection of ovule of *Dioon edule* showing pollen tubes hanging downward in the archegonial chamber and the position of the archegonia in the female gametophyte; **F–H,** ontogeny of archegonium in *Dioon edule.* [**A–C** redrawn from Swamy, *Amer. Jour. Bot.* 35:77, 1948; **D–H** redrawn from *Gymnosperms. Structure and Evolution* by C. J. Chamberlain. The University of Chicago Press, Chicago. 1935.]

haustorial growth of the upper part of the tube, the lower swollen prothallial end of the tube grows downward and the pollen chamber becomes enlarged by the continued breakdown of the adjacent nucellar tissue (Fig. 15-25, D). Finally the pollen tubes become freely suspended in the archegonial chamber above the exposed female gametophyte (Fig. 15-25, E).

During the growth of the pollen tube, the generative cell divides, producing a *sterile cell* located next to the prothallial cell and a terminal *spermatogenous cell* (Fig. 15-25, B). These terms proposed by Sterling (1963) seem preferable to the older terminology in which the two cells derived from the division of the generative cell were designated as the "stalk cell" and the "body cell" (see Fig. 15-26). The term stalk cell suggests an unproved homology with the stalk of the antheridium of bryophytes and certain leptosporangiate ferns. Furthermore, the so-called stalk cell in the majority of cycads is nonfunctional — and hence "sterile" — although in *Microcycas,* according to Downie (1928), the stalk cell divides repeatedly, forming a linear series of body cells each of which forms two sperm. It is also evident that the term "body cell" should be replaced by the more explicit term "spermatogenous cell," which clearly defines the basic role of this cell, i.e., the production of a pair of gametes. Prior to the division of a spermatogenous cell, two organelles (blepharoplasts) are evident in the cytoplasm on opposite sides of the nucleus (Fig. 15-25, B). They are the organizing centers for the eventual production of basal bodies of the multiflagellate sperm (see Chapter 16 on *Ginkgo* for more details). Figure 15-25, C, based on the reconstruction of several serial sections, depicts a group of suspended pollen tubes just prior to the release of the sperm. A more detailed description of the structure and behavior of the remarkable flagellated sperm will be given in the discussion of fertilization in cycads.

FEMALE GAMETOPHYTE. As stated in the previous chapter, the female gametophyte in gymnosperms arises by the enlargement and division of a single functional megaspore which is deeply situated within the nucellar tissue of the ovule (Fig. 14-2, A). However, because of the many technical difficulties of securing young ovules at the critical stages, our present knowledge of the details of megasporogenesis in the cycads is both meager and contradictory. One of the most thorough investigations was made by Smith (1910), who found that in *Zamia floridana* a linear tetrad of megaspores is produced from the megasporocyte; the three upper (toward the micropyle) megaspores degenerate, and the lowermost megaspore enlarges and gives rise to the female gametophyte. In certain other cycad genera, it has been reported that a row of three rather than four cells is produced during megasporogenesis (Maheshwari and Singh, 1967). In this case the uppermost cell in the series is an *undivided dyad cell,* and the two cells below it are megaspores derived from the division of the lower dyad cell. As in *Zamia,* however, only the lower of the two megaspores is functional and the cells above it degenerate (DeSloover, 1961).

Following the enlargement of the functional megaspore, the female gametophyte is initiated by a series of synchronized free nuclear divisions. The extremely numerous nuclei that are formed during this stage in development become arranged in the peripheral layer of cytoplasm which is found between the megaspore wall and the large central vacuole of the coenocytic gametophyte. Because of the very delicate structure of the gametophyte, it is extremely difficult to estimate accurately the number of free nuclei. Chamberlain (1935) stated that about 1,000 free nuclei are present at the end of the coenocytic period of gametophyte development in *Dioon edule.* Much larger numbers of nuclei — some greater than 3,000 — were reported by DeSloover (1964) in his study of the female gametophyte of *Encephalartos poggei.* During its free nuclear phase, the female gametophyte of cycads becomes surrounded by one or two layers of nucellar cells, which constitute a tapetumlike jacket, the cells of which convey nutrients from the nucellus to the enlarging female gametophyte.

The exact manner in which the young coenocytic female gametophyte becomes converted into a cellular structure, bearing archegonia, has been reviewed for the gymnosperms as a whole by Maheshwari and Singh (1967) and Singh (1978). At the end of the free nuclear period, each of the peripheral nuclei is connected to six adjacent nuclei by microtubules (Fig. 15-27, A). Anticlinal walls then begin to develop centripetally, resulting in the for-

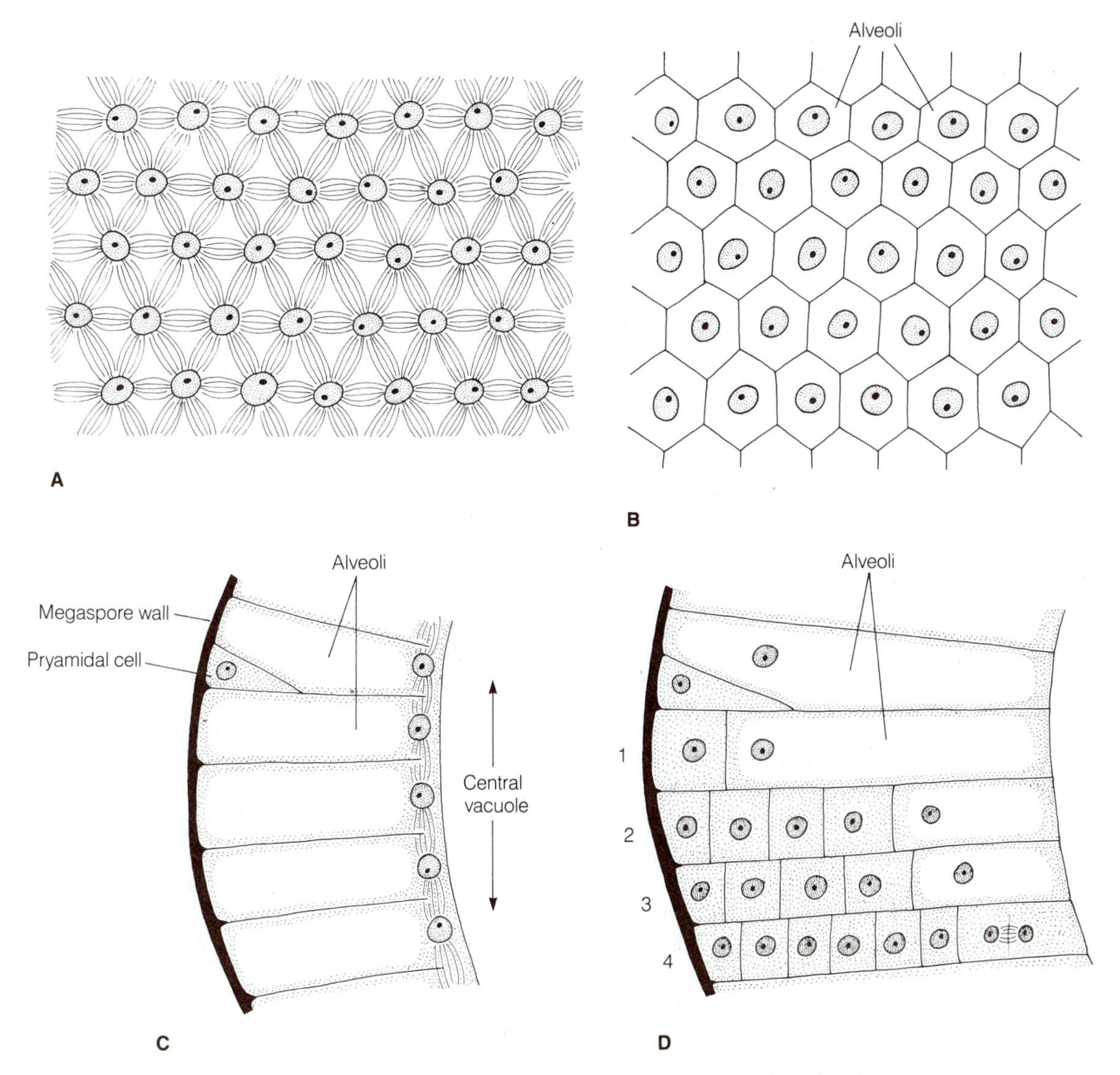

FIGURE 15-27 Schematic representation, stages in development of the female gametophyte of cycads and other gymnosperms. **A**, end of the multinucleate phase, nuclei equally spaced by microtubules (as seen in face view from the central vacuole); **B**, cell walls have formed between nuclei, forming alveoli; **C**, sectional view, showing alveoli in side view; precociously formed "pyramidal cell" in one alveolus; **D**, method of successive partitioning of alveoli, forming long files of cells (1–4). See text for details.

mation of long, six-sided tubular cells. These cells are termed *alveoli*. Each alveolus is hexagonal in transection, and its nucleus lies at the inner end, which is open to the cytoplasm next to the vacuolar membrane (Fig. 15-27, B, C). The female gametophyte has been likened at this stage to a honeycomb, each cavity representing a single alveolus. As the connected alveoli begin to extend toward the center of the gametophyte, some of them become closed precociously, resulting in the formation of "pyramidal cells" (Fig. 15-27, C). The other alveoli continue to develop centripetally. Before or after the alveoli have extended to the center of the gametophyte, each one becomes "closed" by the formation of an end wall (Fig. 15-27, D). The nucleus of each alveolus then moves back toward the periphery of the gametophyte. Through repeated periclinal divisions (with reference to the gametophytic surface) each alveolus is converted into a file of cells (Fig. 15-27, D). But, later, as a result of more divisions in irregular planes, this regular arrangement of partitioned alveoli may become lost. It is interesting to note that the processes just described, with some variations, also occur in the female gametophyte development of *Ginkgo* and living conifers. A detailed account of this development in the cycad *Encephalartos* can be found in the publication by DeSloover (1964).

After the female gametophyte has become cellular, certain of the surface cells at the micropylar end become differentiated as *archegonial initials,* the number fluctuating between different genera or even within the same species. In most cycads, the number of archegonia to reach maturity is usually two to six, according to Maheshwari and Singh (1967). *Microcycas* is apparently unique because large numbers of archegonia, arranged in groups, are initiated over the sides as well as the apex of the gametophyte. However, it is only the archegonia in the micropylar group that attain functional maturity.

The following account of the development of the archegonium in the cycads is based on Chamberlain's (1919, 1935) study on *Dioon edule.* The first step in ontogeny consists in the periclinal division of the archegonial initial into two unequal cells, a small outer *primary neck cell* and a much larger *central cell* (Fig. 15-25, F–G). Then the primary neck cell, by means of an anticlinal division, forms the two neck cells which, according to Chamberlain, are characteristic of the mature archegonia of all cycads. Following the formation of the neck, the central cell enlarges and becomes highly vacuolated. At this time the gametophytic cells that surround the central cell become defined as an "archegonial jacket," which possibly functions in the translocation of food materials to the archegonium (Fig. 15-25, H). A short while before fertilization—only a few days in *Dioon*—the nucleus of the central cell divides, forming the small *ventral canal-cell nucleus* and the much larger nucleus of the *egg cell.* No wall is formed between these nuclei, and the ventral canal-cell nucleus remains near the neck of the archegonium and soon degenerates, while the egg nucleus moves toward the center of the egg and becomes enormous in size. Chamberlain states that the egg nucleus in *Dioon* may reach a diameter of 500 micrometers, and Maheshwari and Singh (1967) remark that the egg in the cycads is perhaps the largest in the plant kingdom.

In concluding our description of the structure and ontogeny of the cycad archegonium, it should be noted that several recent studies fail to support Chamberlain's generalization that the neck of the archegonium consists of only two cells. Norstog and Overstreet (1965), for example, found that just prior to fertilization the archegonial neck in *Zamia integrifolia* consists of a tier of four cells as a result of divisions in the original two. According to Norstog (1972), increasing turgor pressure in each of the four cells results in their enlargement and partial separation, creating a channel between them. Norstog believes that distortion of the sperm must occur in fertilization since the neck opening is on the order of 50 to 70 micrometers in diameter, while the sperm is about 200 micrometers in diameter.

In addition to his intensive study on *Zamia,* Norstog found that the mature archegonial necks in *Encephalartos, Ceratozamia, Cycas,* and *Macrozamia* are likewise composed of four cells. On the basis of his studies, he reached the conclusion "that a four-celled neck apparatus is typical of cycads rather than the two-celled condition thought to be prevalent." (For a review on the archegonium in gymnosperms, see Maheshwari and Sanwal, 1963.)

FERTILIZATION. Although many aspects of fertilization in the cycads remain to be studied, the large multiflagellated sperm have attracted much attention ever since they were first observed in *Cycas revoluta* in 1896 by Ikeno. His discovery represented a milestone in the history of plant morphology because the motile sperm of cycads, like those of *Ginkgo,* are the only known authentic examples of flagellated sperm in the living seed plants. (See Ogura, 1967, for an interesting essay on the discovery of motile sperm in *Ginkgo* and *Cycas revoluta.*) Following the pioneering work of Ikeno, Webber (1901) and Chamberlain (1935) presented detailed accounts of the morphology and behavior of cycad sperm and the ultrastructure of the sperm of *Zamia* has been described by Norstog (1967, 1986).

The sperm of cycads are huge cells, those of *Zamia integrifolia* measuring about 180 micrometers in diameter while still in the pollen tube (Norstog, 1967); the sperm of *Dioon edule,* according to Chamberlain (1935), attain a diameter of 230 micrometers and a length of 300 micrometers after their release from the pollen tube. A fully mature sperm is top shaped, and the flagella are attached along the five to six turns of a lamellar band that occupies the distal half of the cell (Fig. 15-28). Norstog (1986) estimates that a sperm of *Zamia pumila* develops about 50,000 flagella!

Both Webber (1901) and Chamberlain (1935) observed that the two sperm move freely for some time within the intact pollen tube. Webber remarked that it is an interesting sight to see the two giant sperm moving around vigorously in the pollen tube, bumping against each other and the wall of the tube in their reckless haste. When the prothallial end of the pollen tube ruptures, the two sperm, together with some liquid, are discharged into the archegonial chamber. According to Swamy (1948), the entrance of the sperm into the archegonium appears to be aided by their pulsating and amoeboid movements. The exact physical mechanism, however, deserves study because the sperm are much larger than the opening created by the separation of the neck cells of the archegonium (Norstog, 1972). Once inside the cytoplasm of the egg, the flagellated band of the functional sperm becomes detached and remains plainly visible near the neck of the archegonium. The nucleus of the sperm comes into direct contact with the egg nucleus. This is fertilization. Although several sperm may enter an archegonium, only one male nucleus fuses with the nucleus of the egg; the other sperm remain near the top of the egg and finally disintegrate.

The male gametophytes of *Ginkgo biloba* and the cycad *Encephalartos altensteinii* have been grown successfully in artificial culture (Rohr, 1980; De Luca and Sabato, 1979). Only for *Encephalartos* did gametogenesis proceed to the final production of multiflagellate sperm, although the sperm were somewhat atypical in form.

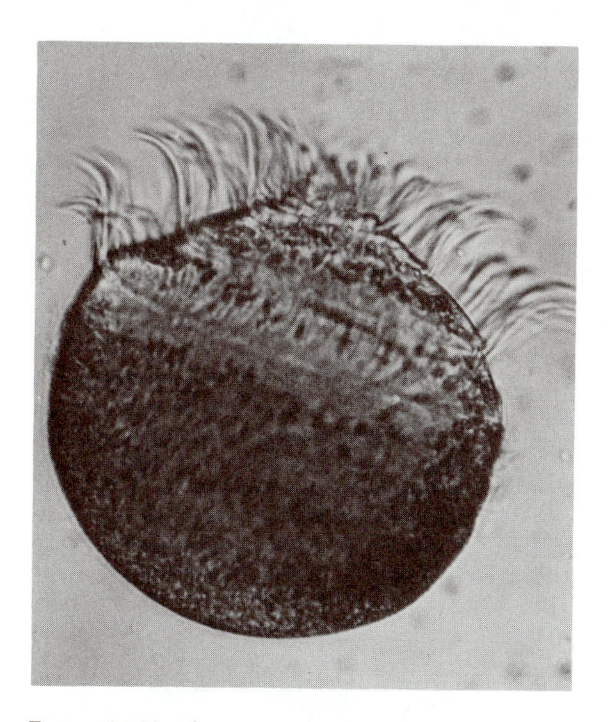

FIGURE 15-28 Photomicrograph of a living sperm of *Zamia integrifolia,* showing the flagellated helical band. (×300). [Courtesy of K. Norstog]

EMBRYOGENY AND SEED DEVELOPMENT. Embryogeny begins with a period of free nuclear divisions. The cycads as a whole are distinguished by the relatively large number of free nuclei produced before wall formation starts.

Following the division of the zygotic nucleus, successive divisions for some time are definitely synchronized. But after about eight successive divisions, which result in 256 nuclei, irregularities appear, some of the nuclei either failing to divide or at least not keeping pace with the rate of division of the remainder. Hence the ultimate number of free nuclei is likely to be less than the theoretical expectation. In *Dioon edule* Chamberlain found about 1,000 free nuclei before the initiation of cell walls. Other cycads have smaller numbers; the lowest recorded is sixty-four in the genus *Bowenia.*

Wall formation in the embryo begins at the lower end of the archegonium and progresses toward the neck end. In some genera this segmentation process extends completely throughout the entire mass of multinucleate cytoplasm. But in others only the inward-facing portion of the embryo develops cell walls — the remainder retains the

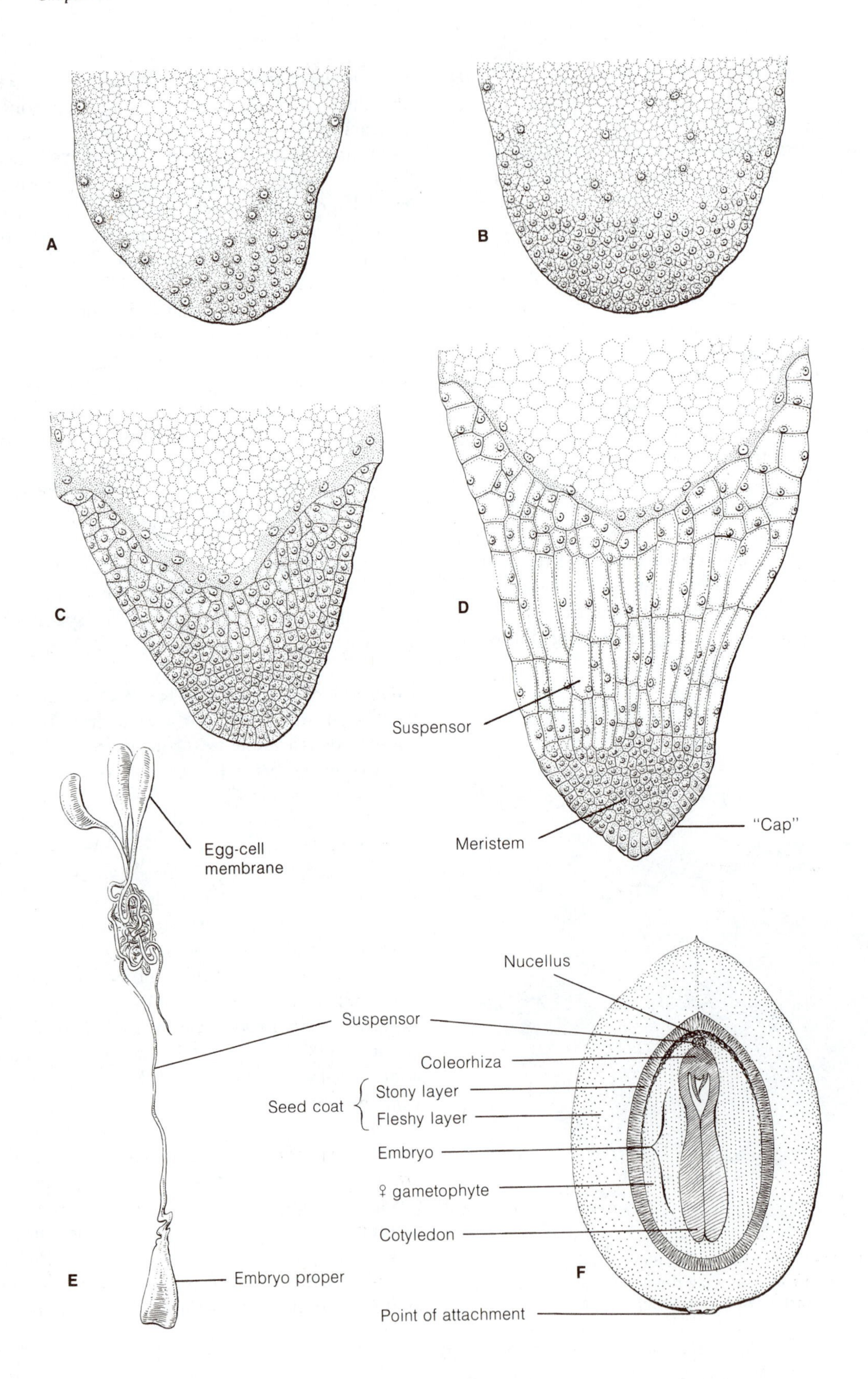
A
B
C
D
Suspensor
Meristem
"Cap"
Egg-cell membrane
E
Suspensor
Embryo proper
Nucellus
Coleorhiza
Seed coat
Stony layer
Fleshy layer
Embryo
♀ gametophyte
Cotyledon
F
Point of attachment

free nuclear condition and ultimately disintegrates. This is the situation in *Zamia,* and the salient features of its early embryogeny will now be outlined on the basis of the study by Bryan (1952).

At the final phase of the free nuclear period a large number of nuclei are aggregated at the inward-directed end of the embryo (Fig. 15-29, A). Two successive, simultaneous divisions of these nuclei occur, and during these divisions cell walls form progressively in an upward direction (Fig. 15-29, B, C). As a result, the embryo now consists of two well-defined regions: a tissue of walled cells, and a region of vacuolated cytoplasm with scattered nuclei. This latter portion of the embryo apparently serves a nutritive function for a time but finally disintegrates. Active cell division in the cellular part of the embryo results in the gradual differentiation of (1) a conspicuously meristematic zone, from which the main organs of the embryo ultimately arise, and (2) a posterior region of elongating cells arranged in vertical series, which mark the origin of the massive suspensor typical of cycad embryos (Fig. 15-29, D). The cells lying above the suspensor region are smaller and constitute a zone that possibly serves as a buffer, which may, to some degree, direct the downward extension of the suspensor. A distinctive feature of the young embryo of *Zamia* is the layer of discrete superficial cells that constitutes a cap over the meristematic zone (Fig. 15-29, D). Bryan found that the outer walls of the cap cells are extremely thick. Gradually the cap cells disintegrate, and the outermost cells of the meristematic zone become organized as a new surface layer. The functional significance, if any, of this "deciduous" cap in the embryo of *Zamia* is problematic. It may constitute a protective layer during the early period when the meristematic zone is pushed deeply into the gametophytic tissue by the elongating suspensor. Comparative embryological studies are needed in order to determine whether

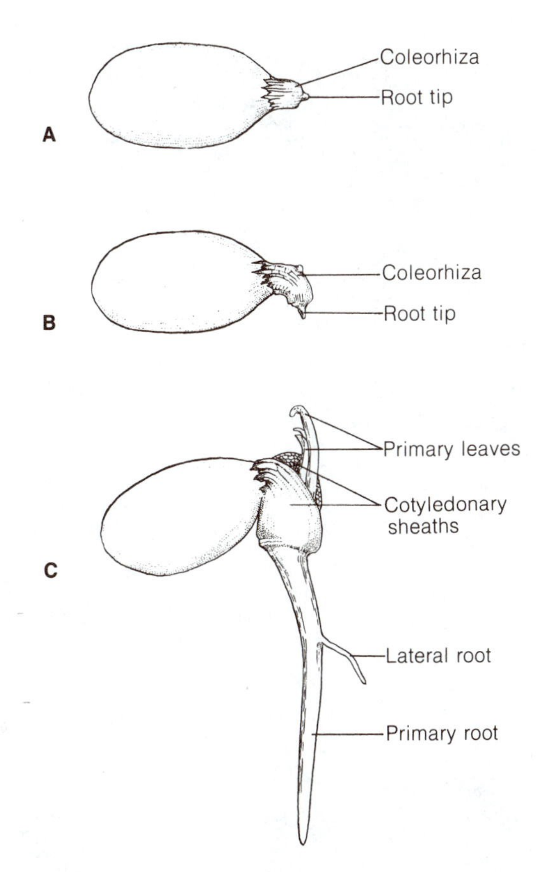

FIGURE 15-30 Seedling development in *Dioon edule.* **A, B,** early stages in seed germination, showing the emergence of the root tip from the coleorhiza; **C,** young seedling with several primary leaves enclosed by the sheaths of the two cotyledons. [Redrawn from *Gymnosperms. Structure and Evolution* by C. J. Chamberlain. The University of Chicago Press, Chicago. 1935.]

caps occur in the young embryos of other cycad genera.

In *Zamia,* as well as in other genera, the elongation of the suspensor forces the apex of the embryo through the wall of the original egg cell of the archegonium into contact with the nutritive tissue of

FIGURE 15-29 **A–D,** early embryogeny in *Zamia pumila.* **A,** free nuclear period; **B,** formation of walls between nuclei in lower end of embryo; **C,** beginning of cellular differentiation (note lower region of small actively dividing cells above which are the first elongating cells of the future suspensor); **D,** later stage showing clear differentiation of surface layer or "cap," meristem, suspensor, and "buffer" region (note that upper portion of embryo has remained in the free nuclear condition); **E,** young embryo of *Cycas* dissected from seed (note long, slender suspensor and, above, the persistent membranes of several egg cells); **F,** diagram of a longitudinal section of a cycad seed. [**A–D** redrawn from Bryan, *Amer. Jour. Bot.* 39:433, 1952; **E** redrawn from Swamy, *Amer. Jour. Bot.* 35:77, 1948.]

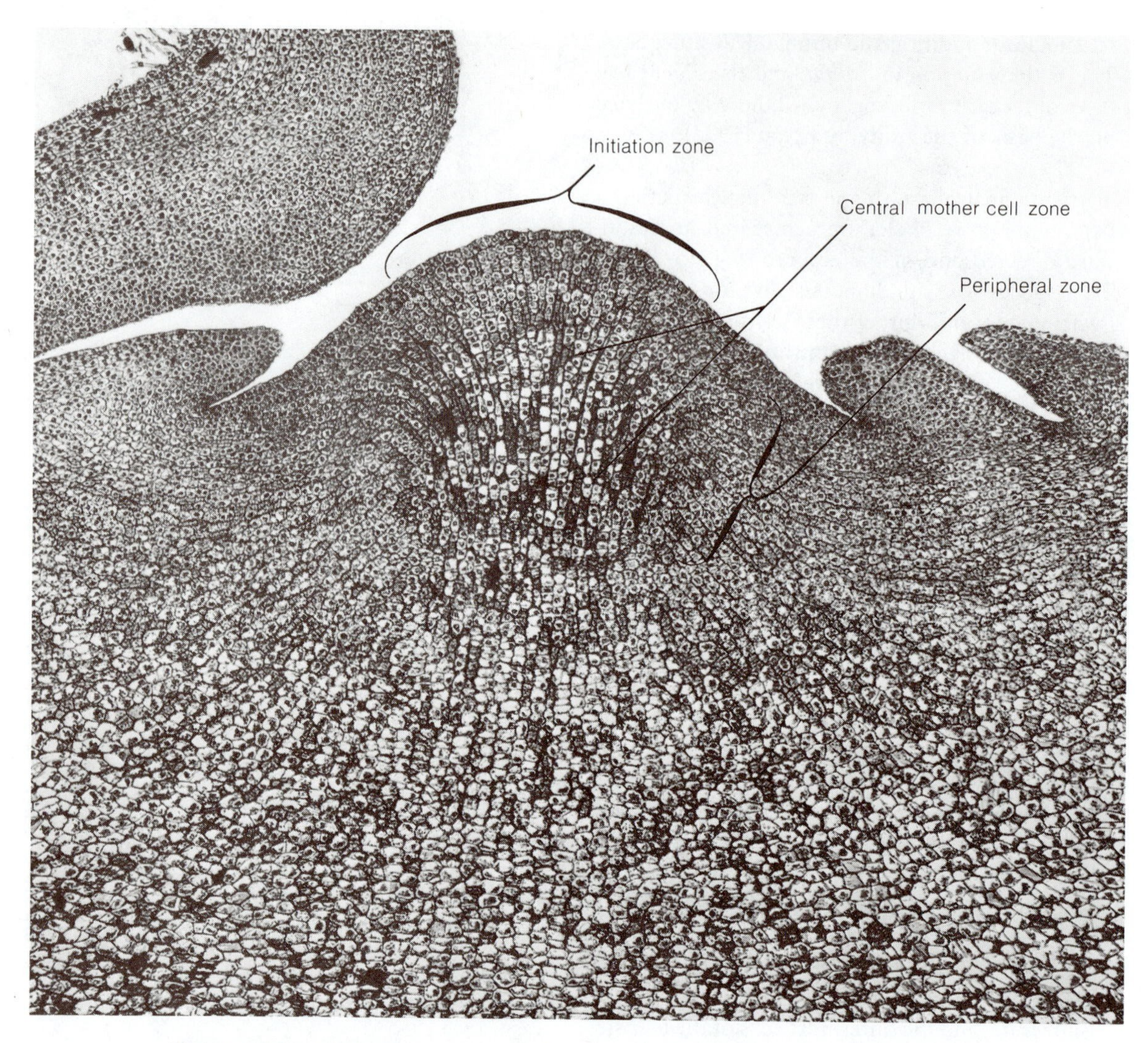

FIGURE 15-31 Median longisection of the shoot apex of *Microcycas calocoma.* Note particularly the fanlike arrangement of the tiers of cells that constitute the initiation zone, and the deeply situated zone of central mother cells, which gives rise to the shallow rib meristem zone and the inner portion of the peripheral zone. [From Foster, *Amer. Jour. Bot.* 30:56, 1943.]

the large gametophyte. During this invasion the suspensor usually becomes coiled or twisted and often reaches a considerable length. Simple polyembryony (i.e., the development of the zygotes of several archegonia in the same ovule) is extremely common in cycads. According to Chamberlain's investigations, the coiled suspensors of several young embryos may intertwine to form a compound structure (Fig. 15-29, E). Typically, only one embryo normally survives and is functional.

As stated above, the main organs of the embryo develop from the meristematic, inward-facing end of the embryo which is thus of the endoscopic type. (See Chapter 6 for a general discussion of endoscopic polarity in embryos.) Usually two cotyledons, which flank the shoot apex, are soon differentiated, although *Ceratozamia* is characterized by having a single cotyledon. The posterior end of the embryo in direct contact with the suspensor develops into the radicle. In some cycads the shoot apex forms several bud-scale primordia and possibly the primordium of the first foliage leaf before seed development is completed. The structure of a ripe cycad seed is shown in longisectional view in Fig. 15-29, F.

The diploid condition is not an absolute necessity for the formation of plantlets and/or embryos in cycads. An in vitro culture of the haploid female gametophyte of *Zamia integrifolia* resulted in the production of leaves and roots as well as a few organized embryos — termed embryoids — after 6 months in culture. Also, embryoids formed on cultured, diploid zygotic embryos. The two classes of embryoids of different ploidy closely resembled each other (Norstog, 1965).

The Seedling

There is no fixed period of seed dormancy in cycads, and germination may begin as soon as the seeds fall on the ground. The micropylar end of the stony layer of the seed coat is first ruptured by the elongation of the *coleorhiza*, a sheath of tissue that encloses the root tip. The root tip then protrudes through the coleorhiza and begins to grow down into the soil (Fig. 15-30, A, B). A distinctive feature of the subsequent development of the seedling is that the blades of the two cotyledons remain permanently within the seed, apparently functioning as haustorial organs, whereas the cotyledonary sheaths and the first primary leaves soon emerge from the ruptured seed coat (Fig. 15-30, C). In seedlings two or more years old, the seed may still remain attached by the withered cotyledons to the

FIGURE 15-32 Reconstructions of extinct cycads. **A,** *Leptocycas;* **B,** *Bijuvia simplex* (1) and *Palaeocycas integer* (2), a megasporophyll of *B. simplex*. [**A** from Delevoryas and Hope, *Postilla* No. 150: 1, 1971; **B** from *An Introduction to Paleobotany*, 1st edition, by C. A. Arnold. McGraw-Hill, New York. 1947.]

axis of the young plant. For several years after germination, the shoot apex of the seedling produces scale leaves *(cataphylls)* and, at intervals, *single* foliage leaves. As development of the young plant continues, the number of foliage leaves increases until well-defined crowns of foliage leaves, alternating with zones of scale leaves, become the established pattern of growth.

During the transition from seedling to adult sporophyte, the shoot apex increases greatly in size and in *Cycas revoluta* may attain a diameter of about 3.5 millimeters, a dimension greatly exceeding that typical of the apices of other vascular plants. The complex zonal structure of the large shoot apices of cycads is strikingly illustrated in *Microcycas* (Fig. 15-31). One of the most distinctive features of the apex of this cycad is the *convergence* of well-defined files of cells from the lateral part of the initiation zone toward the inner cells of the zone of central mother cells. This fanlike arrangement of cells, so strikingly displayed in *Microcycas,* presents a difficult problem in any effort to interpret the patterns of cell division and cell polarity in the shoot apex; perhaps additional examples of the *Microcycas* type of zonal structure will appear as comparative studies continue. (For further details on the shoot apex of *Microcycas* and other cycad genera, see Foster 1939, 1940, 1941, 1943; Johnson 1944, 1951; and Stevenson, 1980a).

Fossil Cycads

The cycads were distributed worldwide in the past. They reached their maximum distribution in the Mesozoic and then declined in numbers, being represented today by ten genera confined to restricted regions of the world as described previously (p. 358). The most complete and best preserved remains are from the Upper Triassic and Jurassic of the Mesozoic. Their existence is well known based upon fossil stems, leaves, sporophylls, and seeds. As reconstructed, the stem of *Leptocycas* had a slender trunk about 1.5 meters tall, with a crown of pinnate leaves (Fig. 15-32, A) that had haplocheilic stomata (see p. 414) similar to extant cycads (Delevoryas and Hope, 1971). Another reconstruction, commonly reproduced in textbooks, is *Bijuvia simplex*. The trunk is thought to have been stout and unbranched, reaching a height of about 3 meters. The leaves were simple, not pinnate, and a cluster of megasporophylls (form genus *Palaeocycas*) terminated the axis (Fig. 15-32, B). Exceedingly well-preserved stems *(Lyssoxylon)* were found in Arizona and New Mexico. The structure of the wood and the presence of persistent leaf bases and girdling leaf traces are indications of cycad affinities (Gould, 1971).

Without doubt the cycads also were present in the Lower Permian of the Paleozoic. In a series of publications Mamay (1976, and literature citations therein) has described a cycadean megasporophyll in which the lamina was undivided with ovules occurring in two rows at the base of the petiole (Fig.

FIGURE 15-33 Reconstruction of megasporophyll of *Phasmatocycas kansana* (Lower Permian), showing bilaterally arranged basal ovules. [From Mamay, U.S. Geological Survey, Professional Paper, No. 934. U.S. Government Printing Office, Washington, D.C. 1976.]

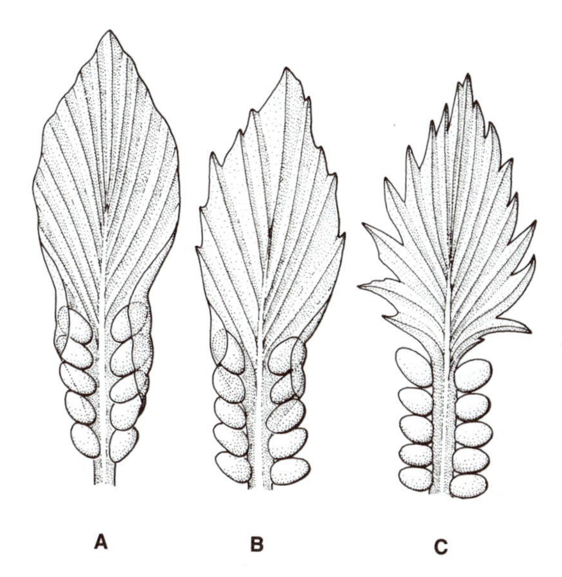

FIGURE 15-34 Hypothetical evolution of *Cycas revoluta*–like megasporophyll from *Archaeocycas* (**A**) by progressive reduction of the basal portion of the lamina and attachment of ovules to the axis of the sporophyll (**B, C**); **C**, distal lamina pinnatifid similar to *Cycas revoluta* (see Fig. 15-24, A). [From Mamay, U.S. Geological Survey, Professional Paper No. 934. U.S. Government Printing Office, Washington, D.C. 1976.]

15-33). In another form, *Archaeocycas,* the megasporophyll consisted of a sterile undivided distal blade and four to six pairs of basal ovules partially enclosed by laminal flaps. Beginning with this precursory type, Mamay proposed a hypothetical evolutionary sequence involving the elimination of the basal part of the lamina and the formation of a deeply pinnatifid distal lamina, resulting in a megasporophyll similar to that of the living cycad, *Cycas revoluta* (compare Figs. 15-24, A, and 15-34) considered to be primitive in *extant* cycads. Based upon his studies and those of others, Mamay suggested that the laminae of vegetative leaves, as well as megasporophylls, were originally simple and undivided in cycads. The pinnate type of leaf in modern cycads evolved by progressive dissection of an original simple leaf type. However, Delevoryas (1982) believes that there is no compelling evidence that the most primitive cycad leaf was a simple, entire leaf and favors the concept of the dissected leaf as the primitive condition. The occurrence of pinnately compound cycad or cycadlike leaves in the Triassic supports such a view.

REFERENCES

Abraham, A., and P. M. Mathew
1962. Cytological studies in the cycads: sex chromosomes in *Cycas. Ann. Bot.* n.s. 26:261–266.

Andrews, H. N., Jr.
1961. *Studies in Paleobotany.* Wiley, New York.
1963. Early seed plants. *Science* 142:925–931.

Arnott, H. J.
1959. Anastomoses in the venation of *Ginkgo biloba. Amer. Jour. Bot.* 46:405–411.

Baird, A. M.
1939. A contribution to the life history of *Macrozamia reidlei. Jour. Roy. Soc. West. Aust.* 25:153–175.

Brashier, C. K.
1968. Vascularization of cycad leaflets. *Phytomorphology* 18:35–43.

Bryan, G. S.
1952. The cellular proembryo of *Zamia* and its cap cells. *Amer. Jour. Bot.* 39:433–443.

Chamberlain, C. J.
1919. *The Living Cycads.* University of Chicago Press, Chicago.
1935. *Gymnosperms. Structure and Evolution.* University of Chicago Press, Chicago.

Crepet, W. L.
1972. Investigations of North American Cycadeoids: pollination mechanisms in *Cycadeoidea. Amer. Jour. Bot.* 59:1048–1056.
1974. Investigations of North American cycadeoids: the reproductive biology of *Cycadeoidea. Palaeontographica* 148(B):144–169.

Dehgan, B., and N. B. Dehgan
1988. Comparative pollen morphology and taxonomic affinities in Cycadales. *Amer. J. Bot.* 75:1501–1516.

Delevoryas, T.
1963. Investigations of North American cycadeoids: cones of *Cycadeoidea. Amer. Jour. Bot.* 50:45–52.
1968a. Investigations of North American cycadeoids: structure, ontogeny, and phylogenetic considerations of cones of *Cycadeoidea. Palaeontographica* 121(B):122–133.
1968b. Some aspects of cycadeoid evolution. *Jour. Linn. Soc. London Bot.* 61:137–146.
1982. Perspectives on the origin of cycads and cycadeoids. *Rev. Palaeobot. Palynol.* 37:115–132.

Delevoryas, T., and R. C. Hope
1971. A new Triassic cycad and its phyletic implications. *Postilla,* No. 150, pp. 1–21.

De Luca, P., and S. Sabato
1979. *In vitro* spermatogenesis of *Encephalartos* Lehm. *Caryologia* 32:241–245.

DeSloover, J.-L.
1961. Études sur les Cycadales. I. Méiose et mégasporogenèse chez *Encephalartos poggei* Asch. *Cellule* 62:105–116.
1964. Études sur les Cycadales. III. Nucelle, gamétophyte femelle et embryon chez *Encephalartos poggei* Asch. *Cellule* 64:149–200.

Downie, D. G.
1928. Male gametophyte of *Microcycas calocoma. Bot. Gaz.* 85:437–450.

Foster, A. S.
1939. Structure and growth of the shoot apex of *Cycas revoluta. Amer. Jour. Bot.* 26:372–385.
1940. Further studies on zonal structure and growth of the shoot apex of *Cycas revoluta. Amer. Jour. Bot.* 27:487–501.
1941. Zonal structure of the shoot apex of *Dioon edule. Amer. Jour. Bot.* 28:557–564.
1943. Zonal structure and growth of the shoot apex in *Microcycas calocoma.* (Mig.) A.D.C. *Amer. Jour. Bot.* 30:56–73.

Foster, A. S., and M. R. San Pedro
1942. Field studies on *Microcycas calocoma. Mem. Soc. Cubana Hist. Natur.* 16:105–121.

Gillespie, W. H., G. W. Rothwell, and S. E. Scheckler
1981. The earliest seeds. *Nature* 293:462–464.

Gould, R. E.
1971. *Lyssoxylon grigsbyi,* a cycad trunk from the Upper Triassic of Arizona and New Mexico. *Amer. J. Bot.* 58:239–248.

Gould, R. E., and T. Delevoryas
1977. The biology of *Glossopteris:* evidence from petrified seed-bearing and pollen-bearing organs. *Alcheringa* 1:387–399.

Harris, T. M.
1944. A revision of *Williamsoniella. Philos. Trans. Roy. Soc.* (London) 231B:313–328.

Hertrich, W.
1951. *Palms and Cycads.* Privately printed, San Marino, Calif.

Ikeno, S.
1896. Spermatozoiden von *Cycas revoluta. Bot. Mag.* (Tokyo) 10:367–368.

Jeffrey, E. C.
1917. *The Anatomy of Woody Plants.* University of Chicago Press, Chicago.

Johnson, L. A. S.
1959. The families of cycads and the Zamiaceae of Australia. *Proc. Linn. Soc. New South Wales* 84:64–117.

Johnson, M. A.
1944. On the shoot apex of the cycads. *Torreya* 44:52–58.
1951. The shoot apex in gymnosperms. *Phytomorphology* 1:188–204.

Kershaw, E. M.
1912. Structure and development of the ovule of *Bowenia spectabilis. Ann. Bot.* 26:625–646.

Lamont, B. B., and R. A. Ryan
1977. Formation of coralloid roots by cycads under sterile conditions. *Phytomorphology* 27:426–429.

Maheshwari, P., and M. Sanwal
1963. The archegonium in gymnosperms: a review. *Mem. Indian Bot. Soc.* No. 4:103–119.

Maheshwari, P., and H. Singh
1967. The female gametophyte of gymnosperms. *Biol. Rev.* 42:88–130.

Mamay, S. H.
1976. Paleozoic origin of the cycads. U.S. Geological Survey, Professional Paper No. 934. U.S. Government Printing Office, Washington, D.C.

Marchant, C. J.
1968. Chromosome patterns and nuclear phenomena in the cycad families Stangeriaceae and Zamiaceae. *Chromosoma* 24:100–134.

Millay, M. A., and D. A. Eggert
1974. Microgametophyte development in the Paleozoic seed fern family Callistophytaceae. *Amer. J. Bot.* 61:1067–1075.

Nathanielsz, C. P., and I. A. Staff
1975a. A mode of entry of blue-green algae into the apogeotropic roots of *Macrozamia communis. Amer. J. Bot.* 62:232–235.
1975b. On the occurrence of intracellular blue-green algae in cortical cells of the apogeotropic roots of *Macrozamia communis* L. Johnson. *Ann. Bot.* 39:363–368.

Norstog, K.
1965. Induction of apogamy in megagametophytes of *Zamia integrifolia. Amer. J. Bot.* 52:993–999.
1967. Fine structure of the spermatozoid of *Zamia* with special reference to the flagellar apparatus. *Amer. Jour. Bot.* 54:831–840.
1972. Role of archegonial neck cells of *Zamia* and other cycads. *Phytomorphology* 22:125–130.
1980. Chromosome numbers in *Zamia* (Cycadales). *Caryologia* 33:419–428.

1987. Cycads and the origin of insect pollination. *Amer. Scientist* 75:270–279.

Norstog, K., and R. Overstreet
1965. Some observations on the gametophytes of *Zamia integrifolia. Phytomorphology* 15:46–49.

Norstog, K., and D. W. Stevenson.
1980. Wind? or insects? The pollination of cycads. *Fairchild Tropical Garden Bull.* 35:28–30.

Norstog, K. J.
1986. The blepharoplast of *Zamia pumila* L. *Bot. Gaz.* 147:40–46.

Norstog, K. J., D. W. Stevenson, and K. J. Niklas
1986. The role of beetles in the pollination of *Zamia furfuracea* L. *fil.* (Zamiaceae). *Biotropica* 18:300–306.

Ogura, Y.
1967. History of discovery of spermatozoids in *Ginkgo biloba* and *Cycas revoluta. Phytomorphology* 17:109–114.

Oliver, F. W., and D. H. Scott
1905. On the structure of the Paleozoic seed *Lagenostoma lomaxi,* with a statement of the evidence upon which it is referred to *Lyginodendron. Phil. Trans. Roy. Soc. London* 197(B):193–247.

Pettitt, J. M., and C. B. Beck
1968. *Archaeosperma arnoldii*—a cupulate seed from the Upper Devonian of North America, *Contr. Mus. Paleontol. Univ. Mich.* 22:139–154.

Rattray, G.
1913. Notes on the pollination of some South African cycads. *Trans. Roy. Soc. S. Africa* 3:259–270.

Reynolds, L. G.
1924. Female gametophyte of *Microcycas. Bot. Gaz.* 77:391–403.

Rohr, R.
1980. Développement *in vitro* du pollen de *Ginkgo biloba* L. *Cytologia* 45:481–495.

Rothwell, G. W.
1977. Evidence for a pollination-drop mechanism in Paleozoic pteridosperms. *Science* 198:1251–1252.

Rothwell, G. W., and D. A. Eggert.
1986. A monograph of *Dolerotheca* Halle, and related complex permineralized medullosan pollen organs. *Trans. Roy. Soc. Edinburgh: Earth Sciences* 77:47–79.

Seward, A. C.
1917. *Fossil Plants,* Vol. III. Cambridge University Press, London.

Shapiro, S.
1951. Stomata on the ovules of *Zamia floridana. Amer. Jour. Bot.* 38:47–53.

Singh, H.
1978. *Embryology of Gymnosperms.* Handbuch der Pflanzenanatomie, Band X, Teil 2. Gebrüder Borntraeger, Berlin.

Smith, F. G.
1910. Development of the ovulate strobilus and young ovule of *Zamia floridana. Bot. Gaz.* 50:128–141.

Sterling, C.
1963. Structure of the male gametophyte in gymnosperms. *Biol. Rev.* 38:167–203.

Stevenson, D. W.
1980a. Radial growth in the Cycadales. *Amer. J. Bot.* 67:465–475.
1980b. Form follows function—in cycads. *Fairchild Tropical Gard. Bull.* 35:20–24.
1980c. Observations on root and stem contraction in cycads (Cycadales) with special reference to *Zamia pumila L. Bot. J. Linn. Soc.* 81:275–281.
1981. Observations on ptyxis, phenology, and trichomes in the Cycadales and their systematic implications. *Amer. J. Bot.* 68:1104–1114.
1988a. *Chigua,* a new genus in the *Zamiaceae. Adv. Cycad Research.* New York Botanical Garden, Bronx, NY (in press).
1988b. Strobilar ontogeny in the Cycadales. In P. Leins, S. C. Tucker, and P. K. Endress (eds.), *Aspects of Floral Development,* pp. 205–224. Gebrüder Borntraeger, Berlin.

Storey, W. B.
1968. Somatic reduction in cycads. *Science* 159:648–650.

Surange, K. R., and S. Chandra
1975. Morphology of the gymnospermous fructifications of the *Glossopteris* flora and their relationships. *Palaeontographica* 149B:153–180.

Swamy, B. G. L.
1948. Contributions to the life history of a *Cycas* from Mysore (India). *Amer. Jour. Bot.* 35:77–88.

Taylor, T. N., and M. A. Millay
1981. Morphologic variability of Pennsylvanian lyginopterid seed ferns. *Rev. Palaeobot. Palynol.* 32:27–62.

Thomas, H. H.
1925. The Caytoniales, a new group of angiospermous plants from the Jurassic rocks of

Yorkshire. *Phil. Trans. Roy. Soc.* (London) 213B:299–363.

Troll, W.
1938. *Vergleichende Morphologie der höheren Pflanzen.* Bd. I. Zweiter Teil, Lieferung I. Gebrüder Borntraeger, Berlin.

Webb, D. T.
1982. Effects of light on root growth, nodulation, and apogeotropism of *Zamia pumila* L. seedlings in sterile culture. *Amer. J. Bot.* 69:298–305.
1983a. Developmental anatomy of light-induced root nodulation by *Zamia pumila* L. seedlings in sterile culture. *Amer. J. Bot.* 70:1109–1117.
1983b. Nodulation in light- and dark-grown *Macrozamia communis* L. Johnson seedlings in sterile culture. *Ann. Bot.* 52:543–547.

Webber, H. J.
1901. *Spermatogenesis and Fecundation of Zamia.* U.S. Dep. Agr. Bur. Plant Ind. Bull. No. 2:1–100.

Wieland, G. R.
1906. *American Fossil Cycads.* (Publication 34, 1.) Carnegie Institution, Washington, D.C.
1916. *American Fossil Cycads. Taxonomy.* (Publication 34, 2.) Carnegie Institution, Washington, D.C.

CHAPTER 16

Ginkgophyta

KAEMPFER was the first European botanist to study the maidenhair tree. He observed it in cultivation in Japan in 1690 and later published a botanical description of the tree. Kaempfer proposed the name *Ginkgo,* and Linnaeus adopted this generic appellation in 1771, adding the descriptive specific epithet *biloba* as a reference to the deep notching that characterizes the lamina of the leaf on many specimens. Although there has been considerable difference of opinion regarding the etymology of the word "Ginkgo," there is, according to Li (1956), no justification for rejecting the name on the basis that it is a distortion of the word "Gink-yo." Li maintains that the maidenhair tree has been correctly designated as *Ginkgo* for nearly 250 years "and certainly will be known as such forever."

During the latter part of the nineteenth century, *Ginkgo biloba* was placed in the family Taxaceae under the order Coniferales because of the resemblance of its large fleshy seeds to the seeds of *Torreya* and *Cephalotaxus.* The discovery by Hirase (1896) of motile flagellated sperm in *Ginkgo* led to a complete reappraisal of the systematic position of *Ginkgo biloba,* which today is regarded as the only surviving representative of the Ginkgoales (Engler, 1954).

Like the cycads, *Ginkgo biloba* is a veritable "living fossil," and perhaps, as Arnold (1947) suggested, "may indeed be the oldest living genus of seed plants." Leaves very similar in form and venation to the leaves of the modern *Ginkgo* have been found as fossils in rocks deposited during the Mesozoic Era, when ginkgolike plants were apparently worldwide in distribution. For example, *Ginkgo digitata,* a Jurassic species, grew in such widely separated areas as western North America, Alaska, Australia, Japan, and England. Other fossil ginkgophytes are *Ginkgoites* and *Baiera.* The latter genus is particularly notable because its multilobed leaves resemble the deeply incised leaf blades often observed in seedlings and in coppice shoots of the living *Ginkgo.* According to Arnold (1947), many leaves of the form genus *Ginkgoites* so closely resemble *Ginkgo* as to be indistinguishable from those of the living species.

Following the Mesozoic, *Ginkgo* declined progressively in its distribution and, according to Seward (1938), "there can be no doubt that China was the last, if it is not the present, natural home of the maidenhair tree."

Whether *Ginkgo* still exists in the wild state in the more remote and poorly explored forests of China has been regarded an unsettled question by many writers. Some evidence has been produced in favor of the existence of native stands of *Ginkgo* trees, although many botanists contend that such

FIGURE 16-1 *Ginkgo biloba.* Young specimen illustrating dense foliage and characteristic excurrent habit of growth. This tree is vigorous and healthy despite its urban environment, which one might not expect to be favorable for its growth.

trees may represent the offspring of cultivated specimens. In a thorough discussion, however, Li (1956) presents evidence that *Ginkgo* still exists in the wild state in southeastern China, and that the last refuge of this living fossil is a mountainous area "along the northwestern border of Chekiang and southeastern Anhwei." From a broad evolutionary viewpoint, *Ginkgo* is to be regarded "as one of the wonders of the world; it has persisted with little change until the present through a long succession of ages when the earth was inhabited by animals and plants for the most part far removed, in kind as in time, from their living descendants. *Ginkgo* is one of a small company of living plants which illustrates continuity and exceptional power of endurance in a changing world" (Seward, 1938, p. 424).

Ginkgo biloba is widely cultivated as a park specimen or street tree in many temperate areas of the world (Figs. 16-1, 16-2). In accordance with the vitality that has enabled it to survive as a distinct organism for millions of years, the modern *Ginkgo* grows successfully amidst the smoke and gasoline fumes of modern cities. It is exceptionally resistant to the attacks of insects and parasitic fungi, except for the devastating fungus "Texas root-rot" or "Take-all" *(Phymatotrichum omniverum).* The outer fleshy coat of the seed emits an odor like rancid butter, and for this reason the male or microsporangiate trees are preferable to female trees for park or street planting. Several cities have learned this lesson the hard way.

Although there is no known way of separating male from female trees on the basis of external morphological characters prior to the formation of reproductive structures, the work of Lee (1954) and Pollock (1957) suggests that it is possible to differentiate between chromosomal complements in the two sexes. If a study of chromosome morphology proves practical in distinguishing the sex of young *Ginkgo* plants, it would then be possible to eliminate potential female trees from street plantings.

The kernel of the seed (i.e., the female gametophytic tissue and the embryo) of *Ginkgo* is highly nutritious and is used as food in China and Japan. For 3,000 years, or more, an extract of the *Ginkgo* leaf has been recommended in Chinese medicine as being "good for the heart and lungs." Extracts have shown promise for treatment of asthma, toxic shock, Alzheimer's disease, and various circulatory disorders. Recently the compound ginkgolide B has been synthesized, which eventually could lead to its widespread use in treating various disorders.

FIGURE 16-2 A large, microsporangiate specimen of *Ginkgo biloba* in leafless condition, on the campus of the University of California, Berkeley. Note very irregular pattern of branching.

Vegetative Organography and Anatomy

Habit

Young *Ginkgo* trees have a pronounced excurrent habit of growth, resembling that of many conifers (Fig. 16-1). With increasing age, the crown becomes broad and irregular and the pattern of branching variable (Fig. 16-2). Exceptionally robust trees, such as those found near certain temples and shrines in China and Japan, may attain a height of 30 meters.

Burls

In very old cultivated specimens of *Ginkgo,* formations resembling stalactites, called *burls*— known as "chichi" (nipples) to the Japanese—hang downward from the lower sides of many of the larger branches. These peculiar burls may occur either singly or in clusters and, according to Fujii (1895), may attain a length of 2.2 meters and a diameter of 30 centimeters. If one of these strange growths reaches the ground, it may take root and form leaves. Fujii's anatomical study revealed that a chichi, near its point of attachment to the parent branch, contains a central, deeply imbedded, spur shoot together with its associated buds. These buds keep pace with the secondary growth of the burl and appear as small protuberances on the outer surface of the thick cylinder of xylem. Although Fujii concluded that the chichi of *Ginkgo* represent a pathological formation, he provided no explanation of the causal factors responsible for their origin and unusual mode of development.

Shoot Dimorphism

During the development of a *Ginkgo* tree, a marked distinction becomes increasingly evident between two types of shoots: *long shoots,* distinguished by their widely separated nodes and leaves, and the more slowly growing short shoots, also called *spur shoots,* which are characterized by short, crowded internodes and the annual expansion of only a few leaves. Spur shoots begin their development as buds, which arise in the leaf axils of long shoots, and may continue their sluggish pattern of vegetative growth for many years (Fig. 16-5, A). Anatomical studies have shown that the zonal structure of the apical meristem is similar in the terminal buds of both long and spur shoots (Fig. 16-15); the histogenetic difference between the two shoot types is the result of the longer duration of cell division and cell elongation in the primary stem tissues derived from the terminal meristem of the long shoots (Foster, 1938).

It is interesting physiologically that the pattern of growth in the two types of shoots is reversible: a spur shoot may abruptly poliferate into a long shoot, and, conversely, the terminal growth of a long shoot may be greatly retarded for several seasons, thus simulating the growth pattern of a lateral spur shoot. The physiological basis for shoot dimorphism in *Ginkgo* has not been fully explained, although there is experimental evidence that auxin —produced in an elongated long shoot—inhibits the expansion of the axillary buds into long shoots but does not prevent the formation of spur shoots (Gunckel, Thimann, and Wetmore, 1949).

Leaves

One of the most distinctive morphological characters of *Ginkgo* is the foliage leaf, which consists of a petiole and a fan-shaped dichotomously veined lamina (Fig. 16-3). Although Linnaeus' specific epithet *biloba* correctly describes the form of the lamina of many *Ginkgo* leaves, there is an enormous range of variation, with respect to the *degree* of lobing and dissection, among the leaves of a single tree. Critchfield (1970) has studied "heterophylly" in *Ginkgo* and has shown that the form of the lamina correlates with the position of a given leaf or leaf series in the shoot system of the tree. On spur shoots and the basal region of long shoots, the leaf blades are either entire or divided by a distal notch into two lobes. Leaves of this type are present in a partly developed stage in the winter buds and complete their growth during bud expansion the following spring. In contrast, most of the leaves found on the upper part of long shoots are initiated and complete their development during the same season; the lamina of such leaves is always divided by a very deep sinus into two major lobes, each of which in turn is further dissected into segments. Critchfield suggested that auxin may control the extended period of growth of long shoots and the correlated formation of multilobed leaves.

The regular dichotomous pattern of venation in the lamina is one of the striking morphological characters of the leaf of *Ginkgo.* Two leaf traces, derived from separate sympodia of the stele, extend through the petiole and give rise to two systems of dichotomously branched veinlets which vascularize

FIGURE 16-3 Cleared leaf illustrating the dichotomous pattern of venation. The arrows indicate vein anastomoses. See text for further explanation. [From Arnott, *Amer. Jour. Bot.* 46:405, 1959.]

the two halves of the lamina (Fig. 16-3). This distinctive type of venation was also present in the leaves of many of the extinct members of the Ginkgoales (Florin, 1936; Arnold, 1947).

The literature on the morphology of *Ginkgo* describes its dichotomous venation as "open" in type —i.e., devoid of vein unions. Arnott's (1959) detailed survey of more than 1,000 leaves of *Ginkgo biloba* revealed that there are various types of anastomoses. Approximately 10 percent of the leaves that he studied showed one or more anastomoses; the largest number of vein unions observed in a single lamina was five (Fig. 16-3).

Ginkgo is deciduous, and before the leaves are shed in the autumn, they turn a beautiful golden yellow. During the leafless period (See Fig. 16-2), the dormant buds of the spur and long shoots are protected by a series of tightly imbricated bud scales.

Stem

Compared with cycads, the pith and cortex of young stems of *Ginkgo* are relatively small. The pith and cortical zones of spur shoots are larger than the corresponding tissue zones in the stems of long shoots. The stems in both types are eustelic, but vascular cambial activity is vigorous and sustained in long shoots and produces pycnoxylic secondary xylem with well-defined growth rings. Vascular cambial activity persists in spur shoots, but only a small amount of manoxylic secondary xylem is produced.

The Reproductive Cycle

Phenology

The series of events that culminates in the production of ripe seeds in *Ginkgo* takes approximately fourteen months, according to the detailed investigations of Favre-Duchartre (1956) in the area of Paris, France. As is shown diagrammatically in Fig. 16-4, pollination, the maturation of the male and female gametophytes, and fertilization occur in one season (April to September), and embryogeny is not completed until the spring of the following year. In Fig. 16-4, the upper part of the solid black spiral line represents the period during which the development of the female gametophyte (♀) takes place and the dotted line depicts the period of association between the developing male gametophyte (♂) and the megagametophyte within the ovule. The reader will find frequent reference to Fig. 16-4 helpful in connection with the detailed descriptions of sporogenesis, gametogenesis, and embryogeny that follow.

Microsporangiate and Megasporangiate Structures

Ginkgo is dioecious, and when the reproductive stage has been reached, the pollen-forming organs and the ovules are produced on separate trees. A distinctive feature is the restriction of the microsporangiate and ovuliferous structures to the spur shoots, where they are evident in the spring in the axils of the inner bud scales and the foliage leaves (Fig. 16-5).

THE MICROSPORANGIATE STROBILUS. A strobilus arises in the axil of a bud scale or foliage leaf on a spur shoot. It is a loose, pendulous, catkinlike structure, consisting of a main axis to which are attached numerous appendages, each of which bears two (sometimes three or four) pendant microsporangia at its tip (Fig. 16-5, A). Although it is not yet clear whether the microsporangia originate from superficial or hypodermal initials, the general plan of development is eusporangiate as in the cycads. A nearly mature microsporangium has a wall of several layers of cells (including the tapetum) that encloses a central group of microsporocytes. These microsporocytes, by meiotic division, produce numerous tetrads of haploid microspores.

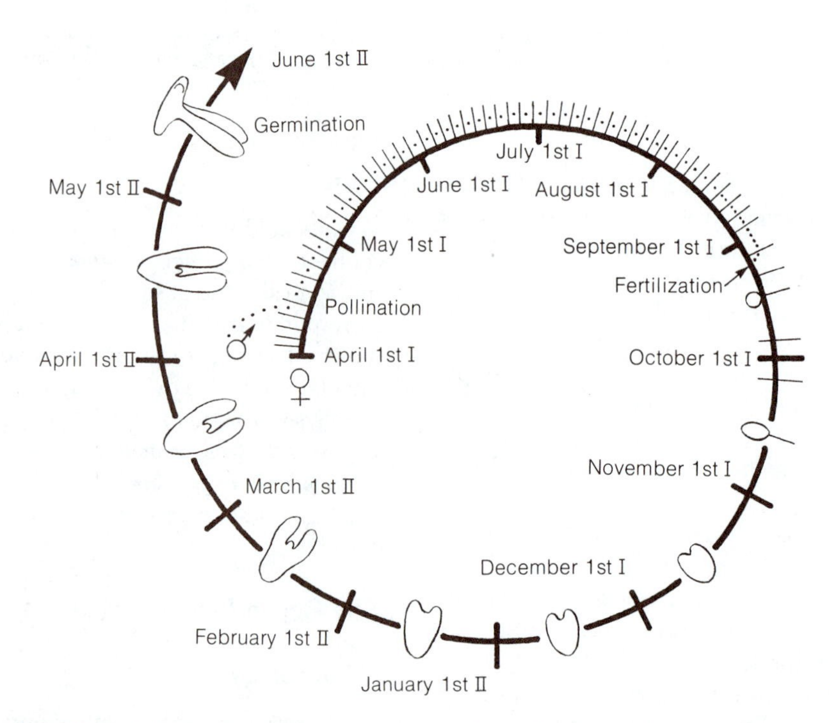

FIGURE 16-4 Diagrammatic representation of the cycle of reproduction in *Ginkgo biloba*. See text for explanation. [Adapted from Favre-Duchartre, *Phytomorphology* 8:377, 1958.]

A

B

FIGURE 16-5 *Ginkgo biloba.* **A**, spur shoot with expanding leaves and microsporangiate strobili; **B**, spur shoot with young leaves and pairs of ovules borne on slender stalks.

THE OVULIFEROUS STRUCTURE. In striking contrast to the megasporophylls of the cycads, each of the ovuliferous organs of *Ginkgo* arises in the axil of a leaf of the spur shoot and consists of a stalk or "peduncle" that bears at its tip two (occasionally three or more) erect ovules (Figs. 16-5, B; 16-13). Each ovule is subtended, below the point of divergence of the integument from the nucellus, by a rimlike outgrowth that was termed the "collar" by Chamberlain (1935). The so-called collar has been interpreted as a vestigial sporophyll and the entire ovuliferous structure as a strobilus. The histogenetic studies of Pankow and Sothmann (1967) provide no support for this speculative interpretation. They found that the collar originates as a rim of tissue *after* the integument of the ovule is well advanced in its ontogeny. This fact—coupled with the absence of any vascular tissue in the collar—led them to conclude that the ovules of *Ginkgo* are cauline and terminal on lateral axes, and that the assumed "foliar" nature of the collar is highly questionable, at least from an ontogenetic point of view (Figs. 16-6; 16-13).

The relatively extensive free portion of the nucellus of the ovule of *Ginkgo* is enclosed by a single integument which becomes anatomically differentiated into a thick outer fleshy layer, a more compact and thinner inner fleshy layer, and a hard stony middle layer; the latter constitutes the "shell" of the mature seed (Fig. 16-7). In contrast to the ovules of cycads, the vascular system of the ovule of *Ginkgo* is weakly developed and consists of two bundles which are restricted to the inner fleshy layer of the integument (Favre-Duchartre, 1956).

The Male Gametophyte (Pollen Grain)

The early ontogeny of the male gametophyte of *Ginkgo* resembles that typical for the cycads except that two prothallial cells—rather than one—are developed. After the second prothallial cell has been formed, the meristematic initial divides and produces the generative cell and the tube cell (see Fig. 17-31). At this four-celled stage, the young male gametophyte, enclosed within the wall of the

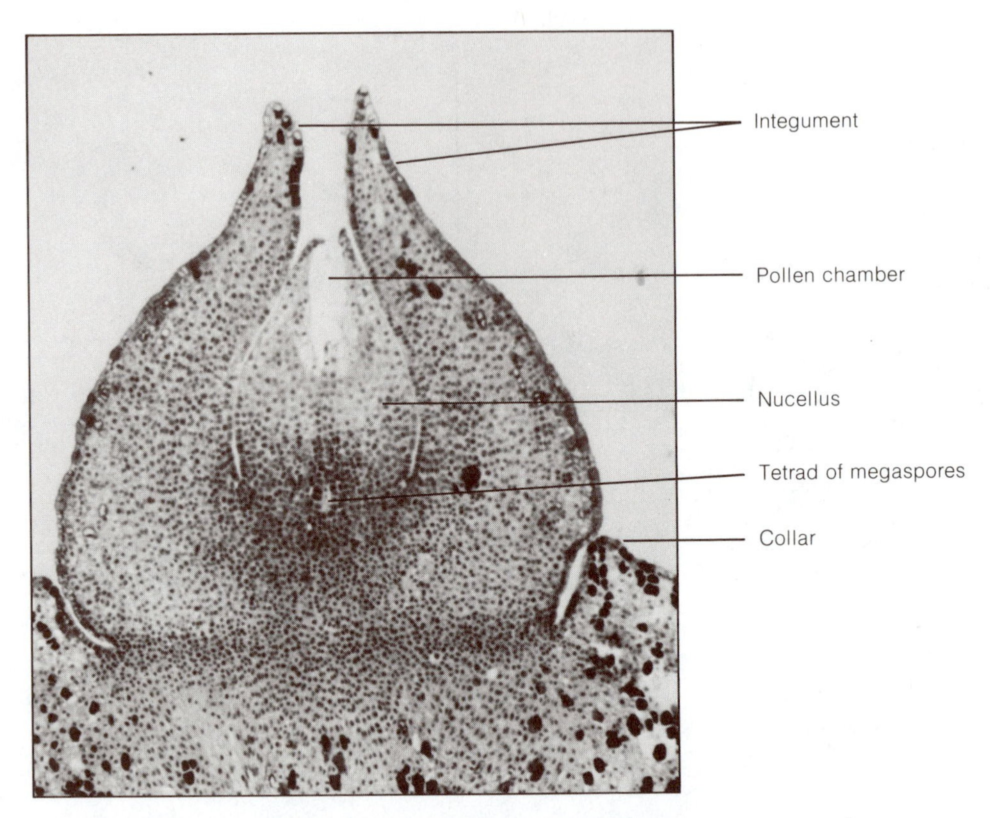

FIGURE 16-6 Median longisection of an ovule of *Ginkgo,* showing nucellus with conspicuous pollen chamber, enclosed by the single integument. Megasporogenesis has occurred and a linear tetrad of megaspores is present in the basal region of the nucellus. Note sectional view of the "collar" below the point of divergence of the ovular integument.

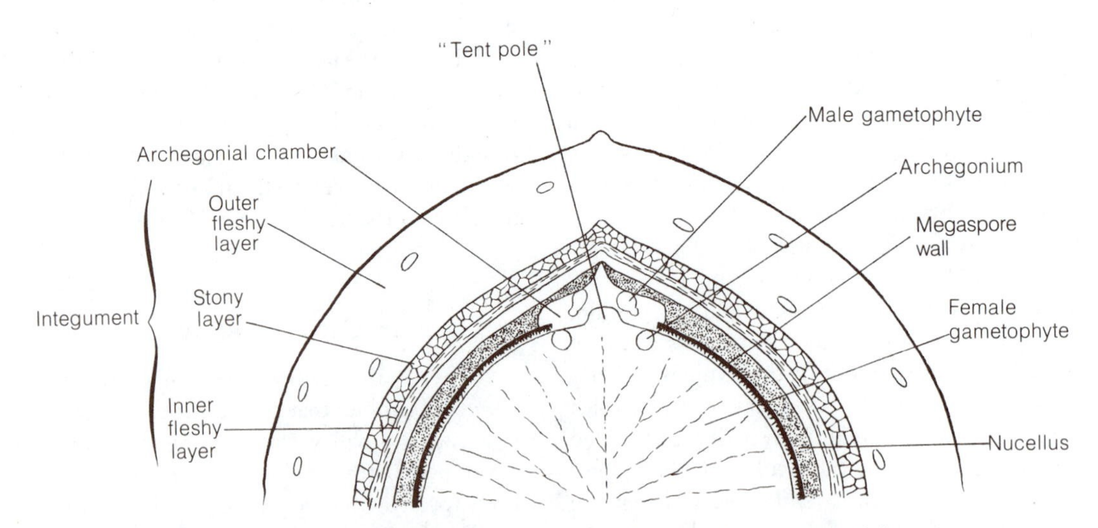

FIGURE 16-7 Longitudinal section (diagrammatic) showing the structure of the upper part of a mature ovule of *Ginkgo.* See text for further explanations. [Redrawn from Favre-Duchartre, *Rev. Cytol. Biol. Veg.* 17:1, 1956.]

pollen grain, is released into the air by the dehiscence of the microsporangium wall.

The pollen is carried to the megasporangiate tree by wind currents and adheres to the mucilaginous pollination droplet which exudes from the micropyles of the ovules. Retraction of this droplet brings the pollen into the pollen chamber where the formation of a much branched, rhizoidlike haustorial pollen tube (Fig. 16-8, A, B; Friedman, 1987) and the final stages in the development of the male gametophyte take place (see Fig. 16-4 for the approximate date of pollination).

Megasporogenesis and Pollen Chamber Development

According to Favre-Duchartre (1956), megasporogenesis in *Ginkgo* occurs toward the end of April. The single megasporocyte is a cell deeply situated in the ovule approximately at the level of separation of the integument from the nucellus (Fig. 16-6). Meiosis occurs, producing a row of four (or sometimes three) cells, the lowermost of which develops into the female gametophyte. In their study of the ultrastructure of the megaspore mother cell

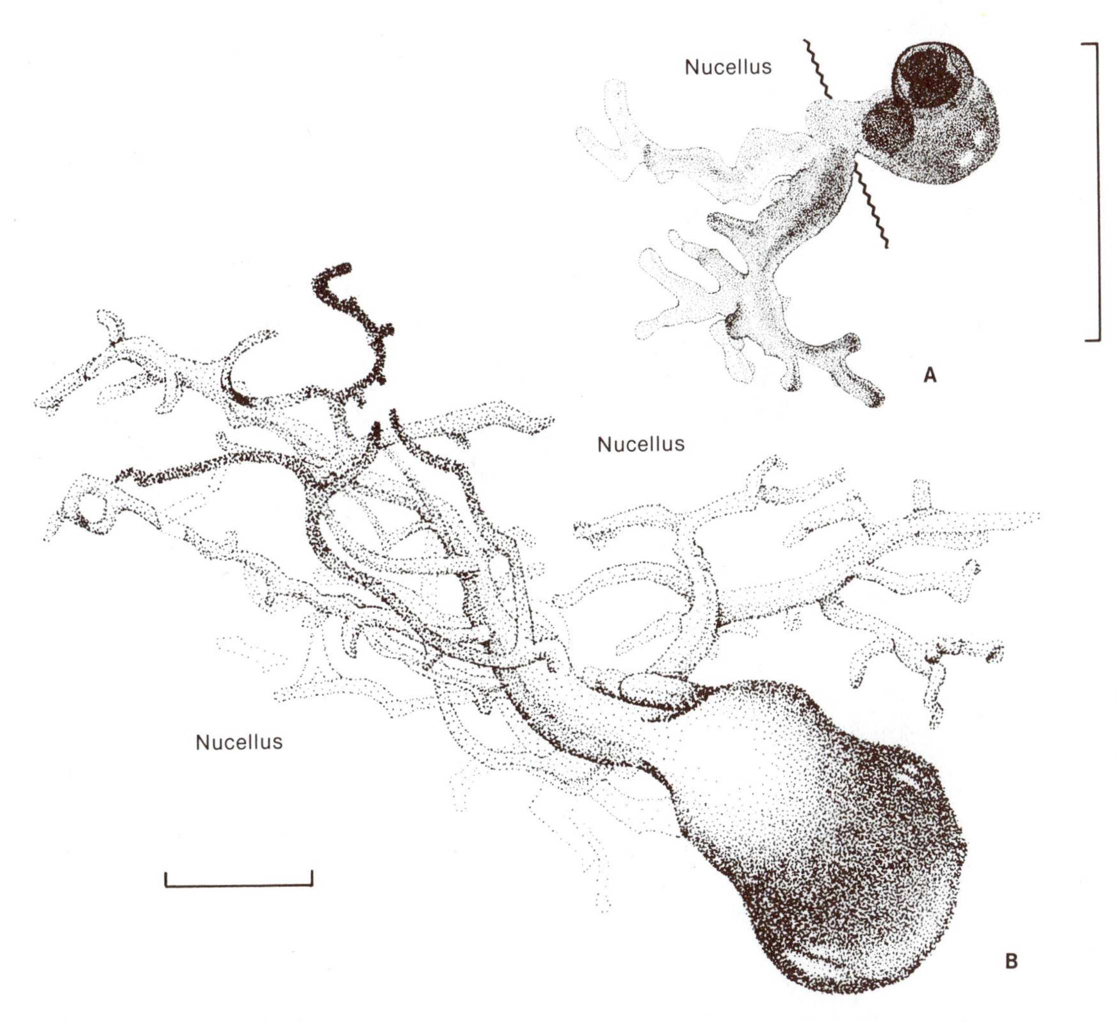

FIGURE 16-8 **A, B,** reconstructions of stereo computer images of the male gametophyte of *Ginkgo,* based on serial histological sections; note hyphal-like outgrowths of the pollen tube. Two ontogenetic stages are shown. Scale bars = 100 micrometers. [Courtesy of Dr. W. E. Friedman.]

in *Ginkgo,* Stewart and Gifford (1967) made the interesting observation that following meiosis I, *all* the plastids and mitochondria present at the chalazal end of the sporocyte become segregated in the lower dyad cell that is destined to produce the functional megaspore.

The investigations of Favre-Duchartre (1956) and DeSloover-Colinet (1963) have shown that megasporogenesis and the formation of the *pollen chamber* in the upper part of the nucellus are synchronized events. A group of internal cells, at the micropylar end of the nucellus, enlarge and then begin to degenerate, creating a cavity that is the beginning of the pollen chamber. The breakdown of nucellar tissue continues; finally the nucellar epidermis, which lies over the cavity, ruptures, and a large open pollen chamber is formed (Fig. 16-6).

The Female Gametophyte

As in the cycads, the development of the female gametophyte in *Ginkgo* begins with the *coenocytic* stage, characterized by extensive free nuclear divisions, and is followed by the *cellular stage,* during which the coenocyte becomes converted by wall formation into a cellular gametophyte bearing archegonia at its micropylar end. These two phases in the development of the megagametophyte may now be considered in more detail.

COENOCYTIC STAGE. This phase in development results from the enlargement of the functional megaspore accompanied by a succession of free nuclear divisions which occur in the peripheral cytoplasm situated between the megaspore wall and the large central vacuole. Favre-Duchartre (1958) reported that as the result of thirteen successive mitotic divisions, approximately 8,000 free nuclei are produced. He found that the free nuclear divisions are not synchronized but rather proceed from the chalazal to the micropylar end of the coenocyte. During the coenocytic phase, the megaspore wall becomes progressively thicker.

CELLULAR STAGE. Cellularization of the female gametophyte is similar to that of cycads. Chapter 15 (Fig. 15-27) presented the details of female gametophyte development in most gymnosperms. At the close of the period of free nuclear divisions, the nuclei are connected by microtubules (Fig. 15-27, A). Anticlinal walls begin to form between the nuclei, from the periphery of the coenocyte toward the central vacuole (Fig. 15-27, C). The six-sided *alveoli* that are produced are long tubular uninucleate cells that for some time are "open," i.e., devoid of internal walls. As seen in surface view, the developing gametophyte resembles a honeycomb of six-sided tubular cells (Fig. 15-27, B). Precociously formed "pyramidal cells" may be formed before many of the alveoli meet in the center of the gametophyte (Fig. 15-27, C). A wall then forms across their open ends, transforming them into complete cells. The closed alveoli divide by the formation of periclinally oriented walls, creating files of cells that are oriented radially with respect to the megaspore wall (Fig. 15-27, D). As the young cellular gametophyte increases in volume, anticlinal divisions also occur which disrupts the previous regular alignment of the files.

One would assume that all nuclei of the female gametophyte would have the haploid content of DNA. However, Avanzi and Cionini (1971) made the interesting observation that during the *early* phase of cellularization most of the nuclei are not haploid, but undergo an increase in DNA content by endoreduplication or amplification to the diploid level or higher. However, at gametophytic maturity the bulk of the cells have the haploid amount of DNA except for cells of the archegonial jacket, some of which undergo an increase in DNA content. This may be related to a high metabolic activity involved in the transfer of nutrients to the archegonia from the stored reserves in the surrounding gametophytic tissue (Cionini, 1971).

ARCHEGONIA. Although the number of archegonia varies from one to five, according to Favre-Duchartre (1956), the usual number is two (Fig. 16-9). Each archegonial initial is a superficial cell at the micropylar end of the gametophyte and, by means of periclinal division, forms a large *central cell* and a smaller *primary neck cell.* The latter soon divides anticlinally, forming a pair of neck cells. Coinciding with the enlargement of the archegonial central cell, a peculiar column of female gametophytic tissue—known as the "tent pole"—becomes elevated between the archegonia (Figs. 16-7; 16-9; 16-10). This column at first extends toward the part of the nu-

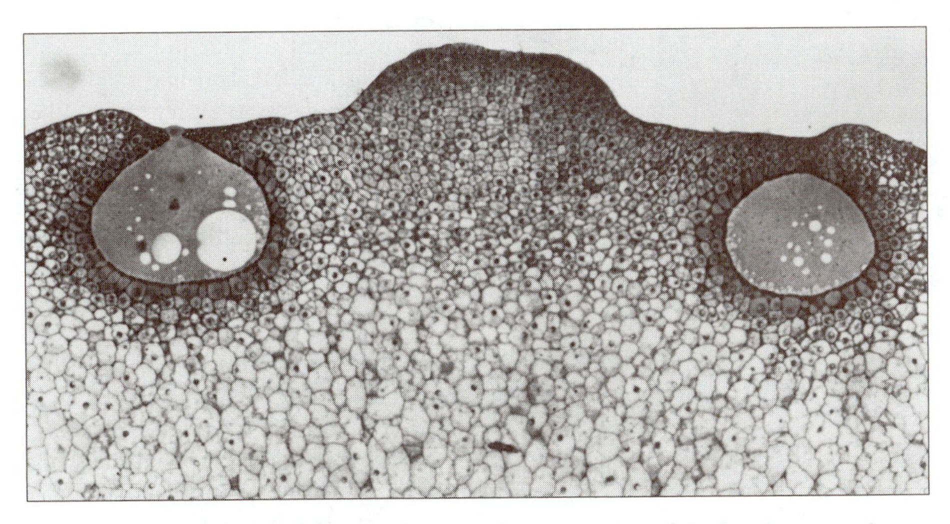

FIGURE 16-9 *Ginkgo biloba*. Longisection of upper portion of the female gametophyte showing "tent pole" and two archegonia. [Courtesy of Dr. W. E. Friedman.]

cellus lying below the pollen chamber. In late August, the nucellar tissue and the megaspore wall in this region become destroyed. This creates an *archegonial* or *fertilization chamber* surrounding the central tent pole (Fig. 16-7).

Fertilization

The final stages in development of the male gametophyte prior to fertilization, take place within the ovule. As mentioned previously, the pollen tube is haustorial, sending out delicate rhizoidlike processes between cells of the nucellus (Fig. 16-8, A, B). The basal end of the male gametophyte is freely suspended within the archegonial chamber (Fig. 16-10). The generative cell divides (commonly in July) giving rise to the sterile cell and the spermatogenous cell (Fig. 16-11) by a cell division similar to the first division of an antheridial initial in certain Filicales (Friedman and Gifford, 1988; see Chapter

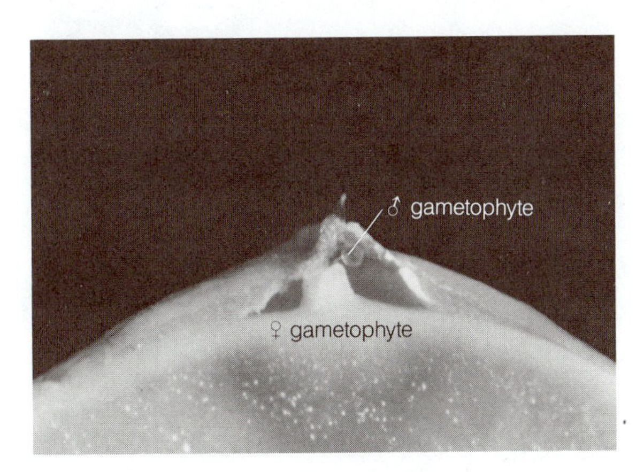

FIGURE 16-10 Integument removed from ovule of *Ginkgo* to reveal the suspended male gametophyte in the archegonial chamber and the "tent pole."

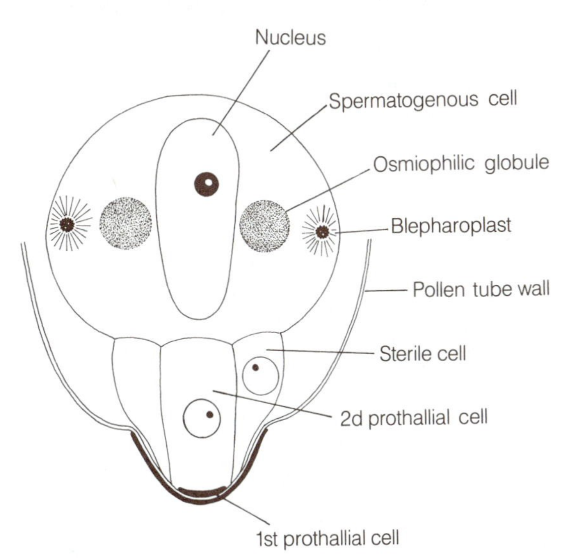

FIGURE 16-11 Schematic representation, basal portion of male gametophyte of *Ginkgo* prior to division of the spermatogenous cell.

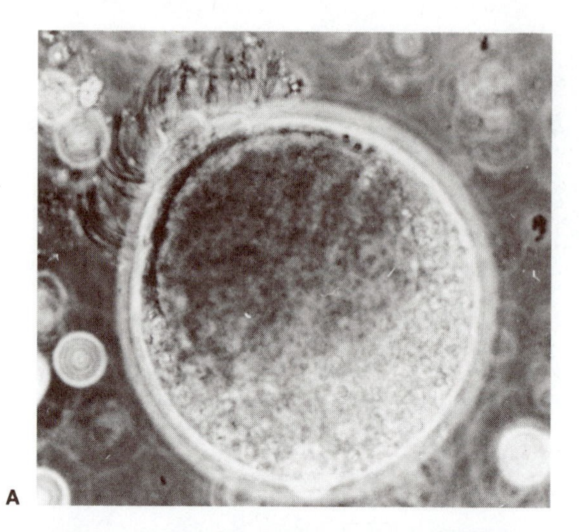

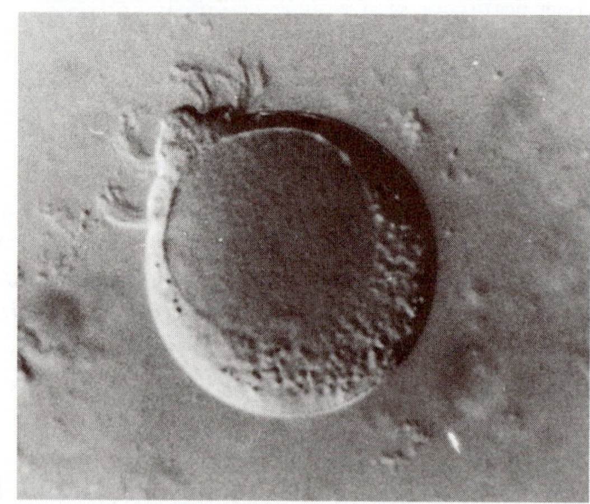

FIGURE 16-12 Sperm of *Ginkgo* seen with phase optics (**A**) and Nomarski optics (**B**). Note flagella at anterior end, and the globular nucleus (X ~ 500).

5, Fig. 5-5). The spermatogenous cell is of special interest. During August (in California) the spermatogenous cell consists of a large, discoid nucleus flanked by two "osmiophilic globules" (of unknown function) and two spherical organelles, the *blepharoplasts,* which arise *de novo* near the nuclear envelope (Fig. 16-11). Each blepharoplast appears to have been formed from a microtubule organizing center (Gifford and Lin, 1975; Gifford and Larson, 1980). Microtubules can be identified in the interior of the organelle and can also be seen extending into the cytoplasm. On the surface are about 1,000 centrioles or probasal bodies (Gifford and Lin. 1975). Concomitant with mitosis of the spermatogenous cell to form the two sperm, the blepharoplast breaks up, and the probasal bodies become aligned on a spiral, lamellar band becoming the basal bodies of flagella as in cycads. When the wall of the spermatogenous cell ruptures, the two sperm are released into the pollen tube. The basal end of the pollen tube then ruptures and the sperm are liberated, often forcefully, into the liquid of the archegonial chamber. The liquid would seem to come from the pollen tube and the breakdown of nucellar cells. The sperm swim in this liquid for some time before they enter an archegonium. The sperm aside from their smaller size, resemble those of cycads (compare Figs. 15-28 and 16-12).

During its passage between the swollen and reflexed cells of the archegonial neck, a sperm becomes greatly stretched, but the entire gamete, including the flagellated band, enters the cytoplasm of the upper part of the egg (Shimamura, 1937).

FIGURE 16-13 Portion of a branch of an ovulate tree of *Ginkgo biloba,* showing the attachment of the peduncles of the ripening ovules to the tips of the spur shoot.

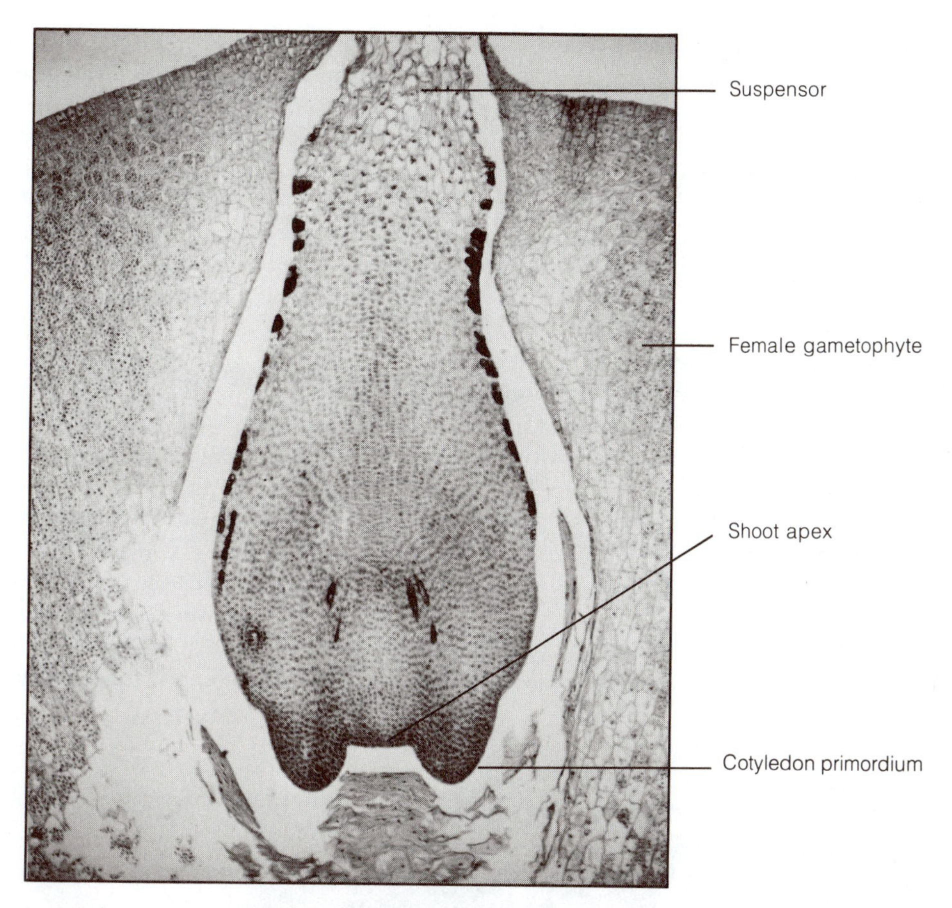

FIGURE 16-14 Longitudinal section of an advanced stage in the embryogeny of *Ginkgo biloba.* The future shoot apex region is flanked by a pair of developing cotyledon primordia. Note the poorly defined suspensor of the embryo and the adjacent food-storing tissue of the female gametophyte.

According to Lee (1955), usually only one of the two sperm produced by a pollen tube enters an archegonium; the other sperm, if it enters the same archegonium, soon degenerates and is eventually "absorbed." The nucleus of the functional sperm separates from the flagellated sheath and moves downward until it contacts the egg nucleus, forming a diploid zygote, which gives rise to the future embryo. Fertilization generally occurs in late August or early September.

Figure 16-13 illustrates the appearance of older ovules of *Ginkgo,* probably after fertilization had occurred. However, Eames (1955) confirmed earlier reports that fertilization and embryogenesis in *Ginkgo* may occur either on the tree or after the ovule has fallen to the ground. The latter possibility is of phylogenetic interest with reference to some of the Paleozoic gymnosperms (e.g., the Cordaitales) in which the embryo, like that of *Ginkgo,* may have only begun to develop after the ovules had been shed from the parent tree.

Embryogeny

The early phase of embryogeny in *Ginkgo* is characterized by numerous free nuclear divisions, as in the cycads. After a series of about eight successive divisions (256 nuclei), centripetal wall formation begins and the young embryo becomes cellular throughout. In contrast with cycad embryogeny, no well-defined suspensor is formed. The lower end of the embryo, by means of active cell divisions, becomes a meristem from which the shoot apex and cotyledons are developed (Fig. 16-14); the cells immediately behind this portion ultimately differentiate into the primary root or radicle (Ball, 1956a, b). Usually there are two cotyledons, but occasionally three are developed. In addition to the cotyledons, the embryo of ripe seeds commonly contains the primordia of several additional foliar structures which, together with the shoot apex, constitute the first terminal bud of the plant.

The germination of the seed closely resembles that typical of cycads. The primary shoot and root emerge by the rupture of the micropylar end of the seed, but the tips of the cotyledons remain within the nutritive tissue of the female gametophyte. The original seed may still cling to the base of a seedling a year or more in age.

The apical meristem of both the long and spur shoots of *Ginkgo* has a characteristic zonal structure. As shown in Fig. 16-15, the subsurface zone of the shoot apex consists of a conspicuous group of enlarged, highly vacuolated, central mother cells from which the more actively dividing and smaller cells of the peripheral and rib-meristem zones take their origin. The type of zonation in the apical meristem of *Ginkgo* has been very helpful in the interpretation of the structure and growth of the shoot apex in the cycads and in certain genera of the coniferales. (See Foster, 1938; Johnson, 1951; Esau, 1965; Gifford and Corson, 1971.)

Fossil Ginkgophytes

Fossils presumed to be ginkgophytes were present in the Lower Permian. Ginkgophytes increased in number in the Jurassic, and declined in the Upper Cretaceous; the only surviving representative is the "living fossil" *Ginkgo biloba*. Many species of *Ginkgoites* (fossil leaves resembling those of *Ginkgo biloba*) have been described as having been distributed worldwide during the Mesozoic. However, rather than representing many different

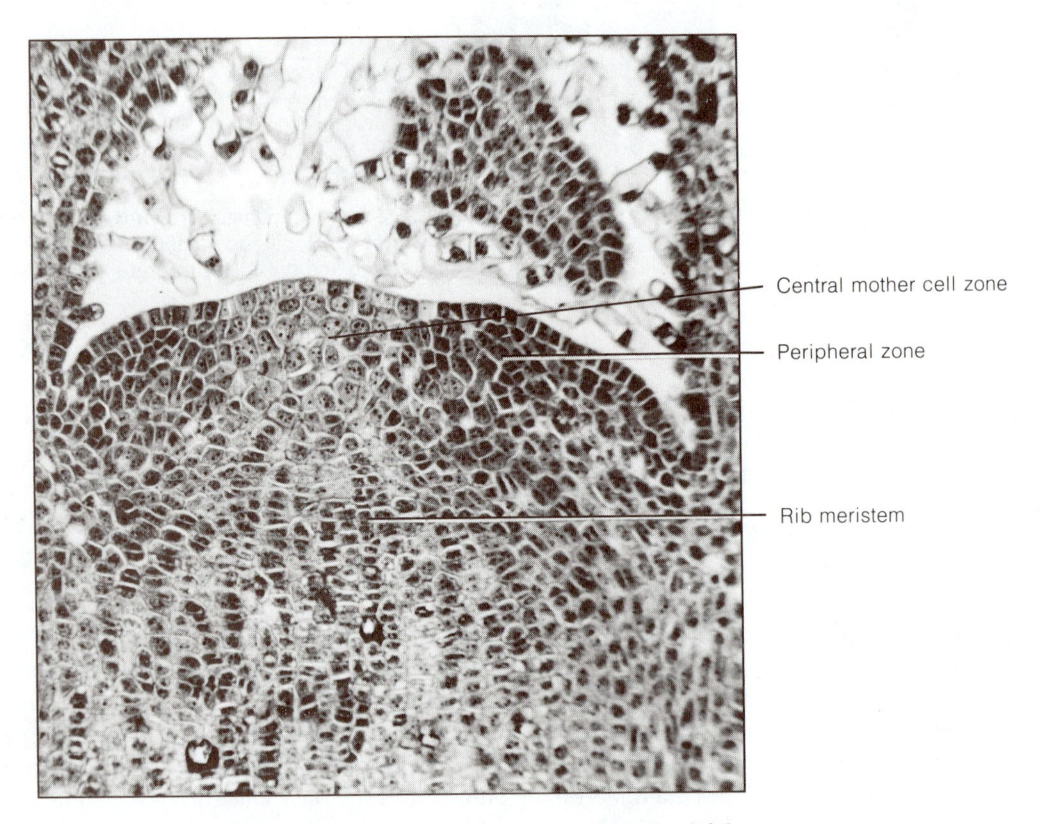

FIGURE 16-15 Median longisection of shoot apex of *Ginkgo biloba* showing typical zonal structure.

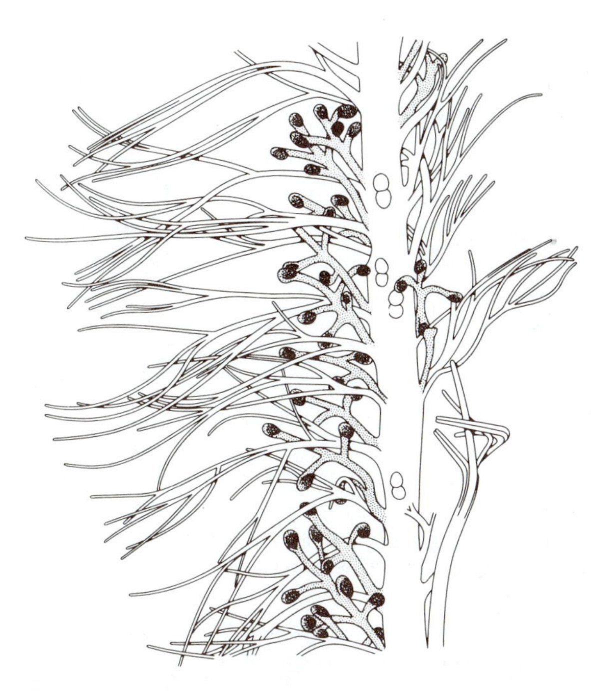

FIGURE 16-16 Reconstruction of *Trichopitys*, showing main axis and axillary, ovule-bearing shoots. [From *Studies in Paleobotany*, by H. N. Andrews, Jr., Wiley, New York. 1961.]

species, the fossils may instead represent only variations in leaf morphology similar to the variations found on the same tree of a living *Ginkgo biloba*. The fossil record of *Ginkgo* extends back to the Upper Triassic-Jurassic and is more extensive in the Lower Cretaceous. Convincing reproductive structures are known for ginkgophytes in the Jurassic-Cretaceous, some having a marked similarity to those of *Ginkgo biloba*. An older fossil, *Trichopitys*, from the Lower Permian indicates that the primitive type of ovuliferous structure in ginkgophytes probably consisted of a main axis with lateral branches bearing terminal ovules (Fig. 16-16). A fertile branch was axillary, similar to that of *Ginkgo biloba*. (For more information on the subject, see Stewart, 1983, and Taylor, 1981.)

REFERENCES

Arnold, C. A.
1947. *An Introduction to Paleobotany*. McGraw-Hill, New York.

Arnott, H. J.
1959. Anastomoses in the venation of *Ginkgo biloba*. *Amer. Jour. Bot.* 46:405–411.

Avanzi, S., and P. G. Cionini
1971. A DNA cytophotometric investigation on the development of the female gametophyte of *Ginkgo biloba*. *Caryologia* 24:105–116.

Ball, E.
1956a. Growth of the embryo of *Ginkgo biloba* under experimental conditions. I. Origin of the first root of the seedling *in vitro*. *Amer. Jour. Bot.* 43:488–495.
1956b. Growth of the embryo of *Ginkgo biloba* under experimental conditions. II. Effects of a longitudinal split in the tip of the hypocotyl. *Amer. Jour. Bot.* 43:802–810.

Chamberlain, C. J.
1935. *Gymnosperms. Structure and Evolution*. University of Chicago Press, Chicago.

Cionini, P. G.
1971. A DNA cytophotometric study on cell nuclei of the archegonial jacket in the female gametophyte of *Ginkgo biloba*. *Caryologia* 24:493–500.

Critchfield, W. B.
1970. Shoot growth and heterophylly in *Ginkgo biloba*. *Bot. Gaz.* 131:150–162.

De Sloover-Colinet, A.
1963. Chambre pollinique et gamétophyte mâle chez *Ginkgo biloba*. *Cellule* 64:129–145.

Eames, A. J.
1955. The seed and *Ginkgo*. *Jour. Arnold Arboretum* 36:165–170.

Engler, A.
1954. *Syllabus der Pflanzenfamilien*, 12th edition, Band 1. Gebrüder Bortraeger, Berlin.

Esau, K.
1965. *Plant Anatomy*, 2d edition. Wiley, New York.

Favre-Duchartre, M.
1956. Contribution à l'étude de la reproduction chez le *Ginkgo biloba*. *Rev. Cytol. Biol. Veg.* 17:1–218.
1958. *Ginkgo*, an oviparous plant. *Phytomorphology* 8:377–390.

Florin, R.
1936. Die fossilen Ginkgophyten von Franz-Joseph-Land nebst Erörterungen über vermeintliche Cordaitales mesozoischen Alters. *Palaeontographica* 81(B):71–173.

Foster, A.S.
1938. Structure and growth of the shoot apex in *Ginkgo biloba. Bull. Torrey Bot. Club* 65:531–556.

Friedman, W. E.
1987. Growth and development of the male gametophyte of *Ginkgo biloba* within the ovule (in vivo). *Amer. Jour. Bot.* 74:1797–1815.

Friedman, W. E., and E. M. Gifford
1988. Division of the generative cell and late development in the male gametophyte of *Ginkgo biloba. Amer. Jour. Bot.* 75:1430–1442.

Fujii, K.
1895. On the nature and origin of so-called "Chichi" (nipple) of *Ginkgo biloba* L. *Bot. Mag.* (Tokyo) 9:444–450.

Gifford, E. M., Jr., and G. E. Corson, Jr.
1971. The shoot apex in seed plants. *Bot. Rev.* 37:143–229.

Gifford, E. M., Jr., and S. Larson
1980. Developmental features of the spermatogenous cell in *Ginkgo biloba. Amer. J. Bot.* 67:119–124.

Gifford, E. M., Jr., and J. Lin
1975. Light microscope and ultrastructural studies of the male gametophyte in *Ginkgo biloba:* the spermatogenous cell. *Amer. J. Bot.* 62:974–981.

Gunckel, J. E., K. V. Thimann, and R. H. Wetmore
1949. Studies of development in long shoots and short shoots in *Ginkgo biloba* L. IV. Growth habit, shoot expression, and the mechanism of its control. *Amer. Jour. Bot.* 36:309–316.

Hirase, S.
1896. Spermatozoid of *Ginkgo biloba. Bot. Mag.* (Tokyo) 10:171 (in Japanese).

Johnson, M. A.
1951. The shoot apex in gymnosperms. *Phytomorphology* 1:188–204.

Lee, C. L.
1954. Sex chromosomes in *Ginkgo biloba. Amer. Jour. Bot.* 41:545–549.
1955. Fertilization in *Ginkgo biloba. Bot. Gaz.* 117:79–100.

Li, Hui-Lin
1956. A horticultural and botanical history of *Ginkgo. Bull. Morris Arboretum* 7:3–12.

Pankow, H., and E. Sothmann
1967. Histogenetische Untersuchungen an den weiblichen Blüten von *Ginkgo biloba* L. *Ber. Deutsch. Bot. Ges.* 80:265–272.

Pollock, E. G.
1957. The sex chromosomes of the Maindenhair tree. *Jour. Hered.* 48:290–294.

Seward, A. C.
1938. The story of the Maidenhair tree. *Sci. Progr.* (England) 32:420–440.

Shimamura, T.
1937. On the spermatozoid of *Ginkgo biloba. Fujii Jubilaei Volume,* pp. 416–423.

Stewart, K. D., and E. M. Gifford, Jr.
1967. Ultrastructure of the developing megaspore mother cell of *Ginkgo biloba. Amer. Jour. Bot.* 54:375–383.

Stewart, W. N.
1983. *Paleobotany and the Evolution of Plants.* Cambridge University Press, Cambridge.

Taylor, T. N.
1981. *Paleobotany: An Introduction to Fossil Plant Biology.* McGraw-Hill, New York.

CHAPTER 17

Coniferophyta

CONIFERALES, VOLTZIALES, AND CORDAITALES

The dominant and most conspicuous gymnosperms of the modern world belong to the order Coniferales. Included in this order are such familiar and widely cultivated trees as pine, spruce, fir, cedar, yew, and redwood (Fig. 17-1). Many conifer genera are of great economic importance as sources of lumber, wood pulp for the manufacture of paper, and turpentine. The evolutionary history of the Coniferophyta is at least as long as that of *Ginkgo* and the cycads, and extends from the Carboniferous to the present. For example, many of the extinct cordaites (Cordaitales) were tall trees and coexisted with the giant lycopods in the Upper Carboniferous (Fig. 17-44, A). The Voltziales, also an extinct group, were present in the late Upper Carboniferous and were more common in the Permian and Triassic of the Mesozoic. The Voltziales are considered to be the group that was ancestral to the Coniferales. The Cordaitales possessed coniferophyte characteristics and, as a group, predate the existence of the Coniferales and Voltziales in the fossil record.

Coniferales

According to Florin (1963), modern-day conifer families can be traced back to the Mesozoic. Families are recognizable by the Late Triassic or Early Jurassic whereas certain contemporary genera make their appearance as early as the Middle Jurassic (Miller, 1977). Students interested in the past distributions of the conifers and in a detailed analysis of their present geographical ranges should consult Florin's (1963) treatment of these complex subjects, as well as Miller's (1982) recent analysis of conifer phylogeny and Stockey's (1981) essay on the origin and evolution of conifers.

Unlike the living cycads, which consist of only a few relict genera, modern conifers are represented by fifty-one genera and approximately 550 species, and are widespread in both the Northern and Southern Hemispheres. Western North America and eastern Asia are regions of exceptional diversity of conifers, with some of the genera forming extensive forests. Parts of Australia and New Zealand are likewise notable for the abundance and diversity of their conifers. The conifers are plants of the more temperate regions of the world; in contrast to the cycads, only a comparatively small number of conifers are found in tropical areas.

Systematics and Geographical Distribution

Many efforts have been made to define genera and to group these genera into tribes and families. Prob-

FIGURE 17-1 **A,** *Pinus ponderosa,* from the Sierra Nevada mountains near Jackson, California; **B,** *Taxodium mucronatum* (bald cypress), Huntington Botanical Gardens, San Marino, California.

ably all of the proposed schemes of classification are artificial to some degree, because of the large amount of parallel evolution of the morphological structures that are employed as the basis for separating genera and families. As is true for other groups of vascular plants, an ideal classification of conifers should rest on the *totality* of morphological features, including the morphology and anatomy of the entire sporophyte as well as the gametophyte and embryo. This goal may ultimately be attained, but at present the systematic treatments emphasize characters of (1) the form, phyllotaxy, and anatomy of the leaves, (2) the organization of the strobili, especially the megasporangiate cones, and (3) to a lesser degree the mature pollen grains.

The various living genera of the Coniferales have been arranged by Pilger (1926) under seven families. This scheme of classification has found wide acceptance by many systematists and morphologists. Although the great majority of conifers produce their seeds in well-defined cones—as is illustrated by the familiar seed cones of pines, firs, and spruces—the family Taxaceae provides a notable exception to this generalization. In this family, each seed is terminal on a short lateral shoot and is partially *(Taxus)* or completely *(Torreya)* enclosed by an outer fleshy envelope known as the *aril* (Fig. 17-6). Because of the absence of a clearly defined seed cone, some botanists have removed the family Taxaceae from the Coniferales and placed it in the separate order Taxales (Sporne, 1965). Florin (1951), the

FIGURE 17-2 **A,** Branch of *Cedrus deodara* with two ripe pollen cones, each borne on a spur shoot; **B,** two ovulate cones of *Pseudotsuga* (Douglas fir) showing the prominent exserted tridentate bracts characteristic of this genus.

great student of the conifers, states that there is no evidence from the fossil record that the distinctive ovuliferous structure of the Taxaceae is the result of phylogenetic reduction. On the contrary, he favored the idea that the taxads have always had single, terminal ovules on short branches in the axils of foliage or scale leaves. This concept, however, remains controversial (see Miller, 1977).

In the following conspectus we present some of the distinguishing characters of each of the seven families of living conifers. The total number of genera in each family is shown by the number in parenthesis, but for the sake of brevity only a few of the representative genera of the larger families are listed.

PINACEAE (9): *Pinus, Pseudotsuga, Abies, Picea, Cedrus*. Monoecious trees with spirally arranged linear or needlelike leaves; each microsporophyll has two microsporangia; ovules borne in pairs on woody cone scales subtended by more-or-less free bracts; pollen grains usually have two wings (Fig. 17-2).

TAXODIACEAE (10): *Taxodium, Metasequoia, Sequoia, Sequoiadendron, Cryptomeria*. Monoecious trees with spirally arranged or opposite *(Metasequoia)* needlelike or linear leaves; microsporophylls with two to nine microsporangia; ovuliferous scale joined with its

FIGURE 17-3 Seed cones of two genera in the Taxodiaceae. **A**, branch of *Cryptomeria japonica* showing a group of seed cones; note that the ovuliferous scales terminate in four or five free tips, a distinctive character for this genus; **B**, branch of *Sequoia sempervirens* (California coast redwood) with five seed cones.

bract and bearing two to nine ovules; pollen grains devoid of wings (Fig. 17-3).

Cupressaceae (19): *Cupressus, Juniperus, Thuja, Calocedrus, Libocedrus.* Monoecious or dioecious trees or shrubs; leaves opposite or whorled, usually scalelike in form; microsporophylls with three to six (or more) microsporangia; ovulate cones woody or somewhat fleshy at maturity; ovuliferous scale united with its bract and bearing two to many ovules; pollen grains devoid of wings (Fig. 17-4).

Araucariaceae (2): *Araucaria, Agathis.* Dioecious or monoecious trees; leaves linear to broad, opposite or spirally arranged; microsporophylls with five to twenty pendant microsporangia; ovuliferous scale fused with its

Figure 17-4 Seed cones of two genera of the Cupressaceae. **A,** *Juniperus* sp. branch with tiny, scalelike leaves and compact seed cones, popularly known as "juniper berries;" **B,** *Calocedrus decurrens* (incense cedar); portion of flat shoot system with numerous small microsporangiate strobili terminating some of the distal branchlets and three open seed cones near base of specimen; two seed pairs, with united wings, were removed and are shown at lower right.

bract and bearing a single median ovule; pollen grains devoid of wings (Fig. 17-23, I–K).

PODOCARPACEAE (7): *Podocarpus, Dacrydium, Phyllocadus.* Trees or shrubs, most of which are dioecious with scalelike, linear or broad spirally arranged leaves; microsporophylls with two microsporangia; megasporangiate strobili conelike or greatly modified (in which case they consist of a few small scales, each with a single ovule, and a subtending bract); usually only one ovule matures and may be supported on a fleshy "receptacle"; pollen grains usually with two or more wings (Fig. 17-5).

CEPHALOTAXACEAE (1): *Cephalotaxus.* Trees or shrubs, most of which are dioecious; leaves needlelike and spirally arranged to two-ranked; microsporophylls with three to eight pollen sacs; ovulate cones with decussately arranged bracts, each bract subtending two ovules, only one of which generally develops into an olivelike seed; pollen grains devoid of wings.

TAXACEAE (4): *Taxus, Torreya.* Dioecious trees or shrubs; leaves linear or needlelike, spirally arranged; microsporophylls peltate with two to eight microsporangia; ovuliferous branch with a single terminal ovule partly or entirely enclosed at maturity by a fleshy aril; pollen grains devoid of wings. (Fig. 17-6; Fig. 17-23, L).

FIGURE 17-5 *Podocarpus* sp. Mature megasporangiate "cone," consisting of bracts (b), and a single seed (s) borne at the apex of a fleshy receptable (r). [Photo courtesy of Dr. T. E. Weier.]

Certain facts about the geographical distribution of the families and genera of conifers deserve brief mention. The majority of the members of the Pinaceae are found in the Northern Hemisphere. The largest genus is *Pinus,* which includes about ninety species. In contrast, the Araucariaceae and most of the Podocarpaceae are restricted to the Southern Hemisphere, the largest genus in these families being *Podocarpus,* comprising about 105 species. *Podocarpus* and *Araucaria* (e.g., the "monkey puzzle tree") are widely cultivated in the warmer parts of the world. The Cupressaceae and Taxodiaceae are large families that have representatives in both hemispheres. The family Taxodiaceae is of particular botanical and general interest for several reasons. First, it includes the coast redwood *(Sequoia sempervirens),* limited to a narrow coastal fog belt of northern California and southern Oregon, and the giant redwood, or "big tree" *(Sequoiadendron giganteum),* found only in a few groves on the western slopes of the Sierra Nevada in California. Both of the sequoias include individual trees—fortunately protected by the U.S. Park Service—which are among the largest trees in existence. The Taxodiaceae is also notable because this family now includes, as one of its most interesting species, the "dawn redwood" *(Metasequoia glyptostroboides).* Because *Metasequoia* had long been known only from its fossil record, uncovered in various parts of western North America and Asia, the discovery in 1944 by Chinese botanists of living specimens of the genus in Szechuan Province, China, was of exceptional scientific importance. The morphology and systematic affinities of the living species with other extant genera of the Taxodiaceae have been discussed in detail by Sterling

FIGURE 17-6 Ovulate structures in two genera of the Taxaceae. **A,** ripe seeds of *Taxus (yew),* each borne on a short axillary branch and partially enclosed by a cup-shaped fleshy aril; **B,** *Torreya californica* (California nutmeg), branch with a cluster of ripe seeds. Each seed is completely covered by a fleshy aril.

(1949). According to Chaney (1950), who investigated the paleobotanical aspects of fossil *Sequoia* and *Taxodium, Metasequoia* "was the most abundant and widely distributed genus of the Taxodiaceae in North America from Upper Cretaceous to Miocene time. There is no known record of its occurrence on this continent in rocks younger than Miocene; it survived into the Pliocene epoch in Japan, and a few hundred trees are still living in the remote interior of China." Thus, *Metasequoia,* somewhat like *Ginkgo biloba* and the living cycads, is truly a living fossil.

Habit and Longevity

Many of the conifers are trees that may attain an enormous size—e.g., *Sequoia, Sequoiadendron, Pseudotsuga,* certain species of *Pinus, Picea,* and *Abies,* and in the Southern Hemisphere, the genus *Agathis.* Very commonly the habit is prominently excurrent with a persistent central trunk and a tiered or whorled arrangement of branches (Fig. 17-1, A). Some species, however, may exhibit a more diffuse or deliquescent pattern of growth such as is true of the "digger pine" *(Pinus sabiniana),* bald cypress (Fig. 17-1, B), and the Monterey cypress *Cupressus macrocarpa).* In New Caledonia, *Parasitaxus ustus* has been reported to grow as a parasite on certain other conifers, a relationship which is apparently unique among all known gymnosperms (Laubenfels, 1959).

Except for a very few deciduous genera *(Larix, Pseudolarix, Taxodium,* and *Metasequoia),* the Coniferales are "evergreen"; i.e., the foliage leaves function for more than one season. In *Araucaria* the dead foliage leaves may remain attached for many years to the branches. Usually the leaves of conifers are individually shed, but in *Pinus, Sequoia sempervirens, Taxodium,* and *Metasequoia,* entire shoots are abscised from the older portions of the branch system.

It is of great interest to note the amazing longevity of certain genera in the Coniferales. Individual specimens of *Sequoiadendron giganteum* may reach an age of more than 3,000 years. During the past twenty years, much attention has been directed to the even greater longevity of the bristlecone pine *(Pinus longavea)*. The pioneering work on this species was done by Edmund Schulman, who concentrated most of his studies on the ancient trees growing—under the protection of the U.S. Forest Service—in the White Mountains of southeastern California. (For detailed information see Ferguson, 1968, and Lindsay, 1969.) By means of an increment borer, narrow cores of wood are removed from the trunks and careful determinations made of the number of growth layers or annual rings. Figure 17-7 shows the grotesque form of "Pine Alpha"—4,300 years old. Other trees in the same grove were determined to be older than 4,000 years. One of them—appropriately named "Methuselah" by Dr. Schulman—had reached an age of 4,600 years. How these trees have been able to survive through the millenia under the harsh conditions of their alpine environment is still a challenging problem in tree physiology. One possible factor responsible for their survival may be the retention of functional needles for twenty to thirty years. According to Ferguson (1968) "this insures a somewhat stable photosynthetic capacity that can carry a tree over several years of stress." As is shown in Fig. 17-7, "Pine Alpha" is largely a twisted, contorted "slab" of dead wood; only a narrow strip of living bark—and presumably some functional wood—connects its single leafy branch with the root system.

The oldest bristlecone pine thus far discovered grew in a grove of trees on Wheeler Peak in eastern Nevada (Currey, 1965). From a horizontal section, made near the base of the trunk, Currey determined that this specimen was approximately 4,900 years old. The life span of such organisms as the bristlecone pines can only be fully appreciated when viewed in comparison with events in man's own history. To quote from Lindsay's (1969) article, "Pine Alpha was growing when pyramids were being built in Egypt, and it was a very old tree when Christ was born." Hitch (1982) discusses in an interesting article the necessity of integrating radiocarbon (^{14}C) dating and the counting of growth rings to determine the age of exceedingly old trees such as *Pinus longavea.*

Vegetative Organography and Anatomy

Leaves

MORPHOLOGY AND VENATION. Although the foliage leaves of all *living* conifers are simple, they vary considerably in both size and form. In many conifers, such as spruce *(Picea)*, fir *(Abies)*, and pine *(Pinus)*, the leaves are commonly described as "needles" because of their narrow attenuated form. According to Chamberlain (1935), the slender needle-leaves of *Pinus palustris* may reach a length of 40 centimeters and are the longest leaves of any of the living coniferophytes. At the opposite extreme, in *Thuja* ("arbor vitae"), *Juniperus* (juniper), *Calocedrus* (incense cedar), and cypress *(Cupressus)*, the leaves are very small and scalelike, superficially resembling the microphylls of the shoots of *Lycopodium* and *Selaginella* (Fig. 17-4). From an anatomical standpoint, however, scale-leaves in the conifers are not strictly comparable with microphylls; the vascular supply of microphylls is derived from protosteles or siphonosteles that lack leaf gaps, whereas the traces of coniferous scale-leaves arise from the sympodial bundles of a eustele, as we have explained in Chapter 14, p. 342. Last, it must be noted that certain species of *Agathis, Araucaria,* and *Podocarpus,* native to the Southern Hemisphere, are distinctive because of their broad leaves, which present a striking contrast to the usual type of foliage in the conifers. Chamberlain (1935) states that the leaves of *Podocarpus wallichianus* reach a length of 12.5 centimeters and a width of 3.5 centimeters.

In most conifers the leaves develop on long shoots, and the phyllotaxy is spiral and alternate. A conspicuous exception is the Cupressaceae, all members of which have decussate or whorled phyllotaxy. The pattern of branching with reference to phyllotaxy is complex and variable. In such genera as *Picea, Abies,* and *Pseudotsuga* a large proportion of the needlelike leaves produced during a growing season are devoid of axillary buds; in these plants branching proceeds from a few axillary buds located in a pseudowhorl just below the terminal bud. The ramification of members of the Cupressaceae, by contrast, is often very profuse, and commonly results in the development of flattened spraylike branch systems (Fig. 17-4, B).

FIGURE 17-7 "Pine Alpha," a 4,300-year-old specimen of bristlecone pine *(Pinus longavea)* growing in Schulman Grove in the White Mountains of California. [Infrared photograph courtesy of Mr. Lloyd Ullberg, California Academy of Sciences.]

A few genera *(Larix, Cedrus, Pseudolarix)* produce their foliage leaves on spur as well as long shoots, thus recalling the marked dimorphism of shoots characteristic of *Ginkgo biloba* (see Chapter 16). With reference to spur shoots, the situation in *Pinus* is unique among all living conifers (Fig. 17-8). In this genus the photosynthetic needle-leaves of well-developed sporophytes are restricted to lateral spur shoots which are ultimately shed as units from the tree; following the seedling stage, the foliar appendages produced on the primary axes are exclusively nonphotosynthetic scale leaves. The spur shoots are often referred to as "leaf fascicles," a misleading term that tends to obscure the real morphology of these structures. A spur shoot in pine arises from a bud developed in the axil of a scale-leaf produced on the main axis. The young spur consists of a very short stem, a series of imbricated membranous bud scales and, depending on the species, from one to five foliage leaves. (See Sacher, 1955a, for a detailed study of bud scale ontogeny in *Pinus*.) The diminutive shoot apex is usually recognizable in a longisection of a mature spur, but the cells composing it are vacuolate and inactive in appearance, and the surface and subsurface cells may ultimately die, become desiccated, and collapse (Sacher, 1955b). However, for the extremely old bristlecone pine, Ewers (1983) has shown that a spur shoot may remain alive on a long shoot for thirty or more years and certain of them may proliferate into long shoots, depending upon their position in the yearly increment of spur shoots. For many years there have been attempts to propagate, with varying success, pine trees by rooting spur shoots. Ewers' work would indicate that it is essential to know, for each species, which of the annual increment of spur shoots retain a viable shoot apical meristem. The peculiar segregation of the foliage leaves to spur shoots can be followed in the progressive development of the pine seedling. When a pine seed germinates, the leaves that develop after the whorled cotyledons are green and needlelike, but are borne in a spiral series on the *main axis* of the seedling (Fig. 17-40). The first spurs arise as buds in the axils of certain of these juvenile leaves; all subsequently formed spurs, however, arise from buds subtended by nonphotosynthetic scale leaves. The deciduous spur shoots of *Pinus* not only are an interesting example of the vegetative specialization

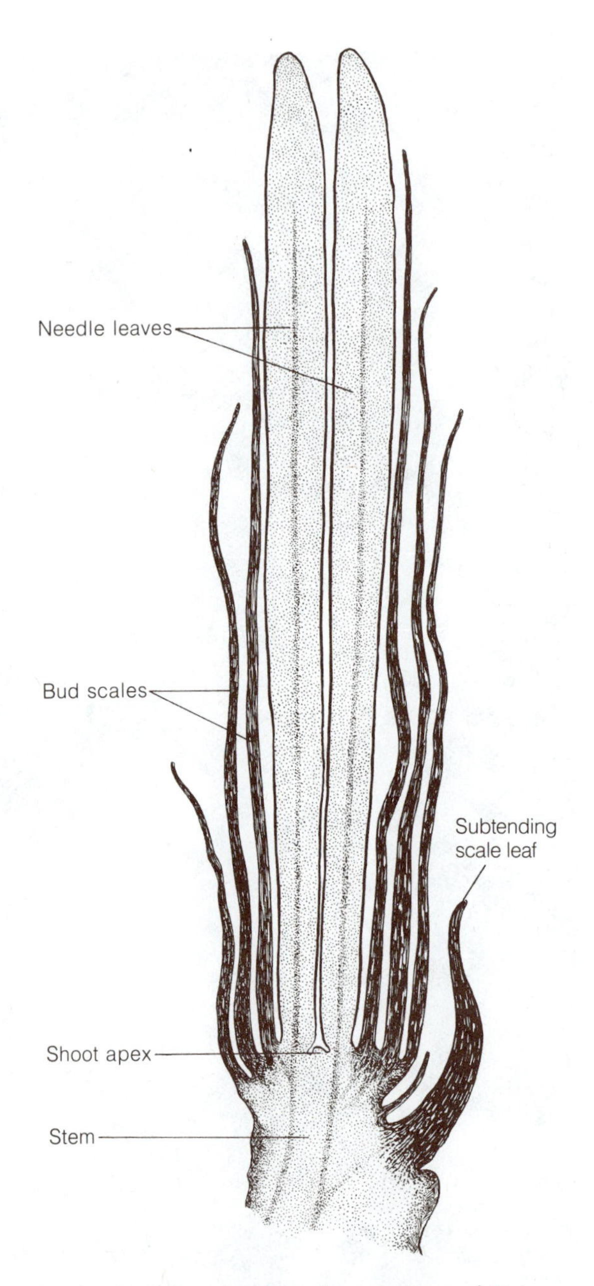

FIGURE 17-8 Longisection of a spur shoot of *Pinus nigra (laricio)* in axil of scale-leaf, showing the diminutive shoot apex between the bases of the two needle leaves. See text for further explanation. [Redrawn from *Gymnosperms; Structure and Evolution* by C. J. Chamberlain. The University of Chicago Press, Chicago. 1935.]

in the Coniferales but have figured prominently as an aid in the interpretation of the ovuliferous scales of the megasporangiate strobili in the conifers (Doak, 1935).

In a number of living conifers there is a marked difference in form between the seedling or juvenile foliage leaves and the leaves produced during the subsequent growth of the sporophyte. This heterophylly is particularly striking in certain genera of the Cupressaceae in which the juvenile leaves are needlelike and are ultimately followed by the appressed scalelike leaves characteristic of the adult plant. By means of cuttings it is possible to propagate the juvenile phase and to produce well-developed trees or shrubs bearing only (or largely) juvenile foliage. The various horticultural forms produced in this way are popular ornamentals.

Because the external form of the foliage leaf varies between genera or families and may even change during the ontogeny of the individual, it is difficult to trace the ultimate phylogenetic origin of the leaf of modern members of the Coniferales. Laubenfels (1953) has grouped the leaves of *living* conifers into the following major categories: *type 1* includes needlelike leaves that are univeined and tetragonal in transection. This type is regarded as the commonest form in fossil conifers, and it is widely distributed among the Pinaceae, Araucariaceae, Podocarpaceae, and Taxodiaceae; *type 2* comprises univeined leaves which are linear or lanceolate in contour and bifacially flattened. This type is regarded as the most common among living conifers, and members of all families "have type 2 in some genus at some period in ontogeny"; *type 3* comprises all scalelike forms of leaves, e.g., the adult foliage leaves of the Cupressaceae; *type 4* includes the broad, multiveined leaves of *Agathis*, and of species of *Araucaria* and *Podocarpus*.

What Is the Primitive Type of Conifer Leaf?

Our present knowledge of leaf morphology in the progymnosperms may provide an insight into the origin of conifer leaves. Several different types of leaves have been ascribed to the progymnosperms from the Upper Devonian. They ranged from simple, dichotomously branched segments to deeply dissected, filiform types, to laminate with many veins. The needlelike univeined leaves of many conifers could have evolved by the reduction of small, dichotomously branched segments. The larger, broader type of leaf, characteristic of certain species in *Araucaria*, could have evolved without much change from the *Archaeopteris macilenta* and *A. halliana* types (Figs. 14-3, 14-7). The same is true for the straplike leaves of *Cordaites* (Cordaitales, of the Upper Carboniferous to Permian). For an alternative view of the origin and evolution of the conifer leaf, see Rothwell (1982).

Florin (1950, 1951) has shown that by the Upper Carboniferous to Permian Periods two types of coniferophyte leaves (sometimes connected by intermediate forms on the same individual) were present in the order Voltziales. The leaves were needlelike to awl shaped and were helically arranged on the stem (Fig. 17-9, B). In the probably more primitive type, the leaf, although vascularized by a single trace, was dichotomously lobed or bifurcated and dichotomously veined. There were bifurcated leaves on the main axis and principal branches of *Lebachia* and *Ernestiodendron*, and they were followed on branches of the ultimate order by a second type which were univeined and not bifurcate. The occurrence of heterophylly in these very ancient conifers is of considerable interest, and led Florin to conclude that the simple leaf arose by reduction (Fig. 17-9, B). In certain genera, such as *Carpentieria*, even the leaves of small lateral shoots were forked (Fig. 17-9, A). This condition is interpreted by Florin as a slightly modified persistence of the juvenile type of foliage. From this brief discussion it may be concluded that (1) the needlelike type of leaf probably arose early in the phylogeny of the Coniferales from a dichotomously branched appendage, and (2) this simple leaf type, modified to varying degrees, has persisted to the present time as the characteristic foliar appendage of most living conifers.

Napp-Zinn (1966), in his comprehensive monograph on gymnosperm leaf anatomy, states that the leaves of most conifer genera are vascularized by a single median bundle. Examples of this common type of venation are found in the leaves of *Abies*, *Picea*, *Pseudotsuga*, all members of the Cupressaceae, and certain species of *Pinus* (Figs. 17-10; 17-12, B). As might be expected, the venation pattern developed in the broad leaves of *Araucaria*,

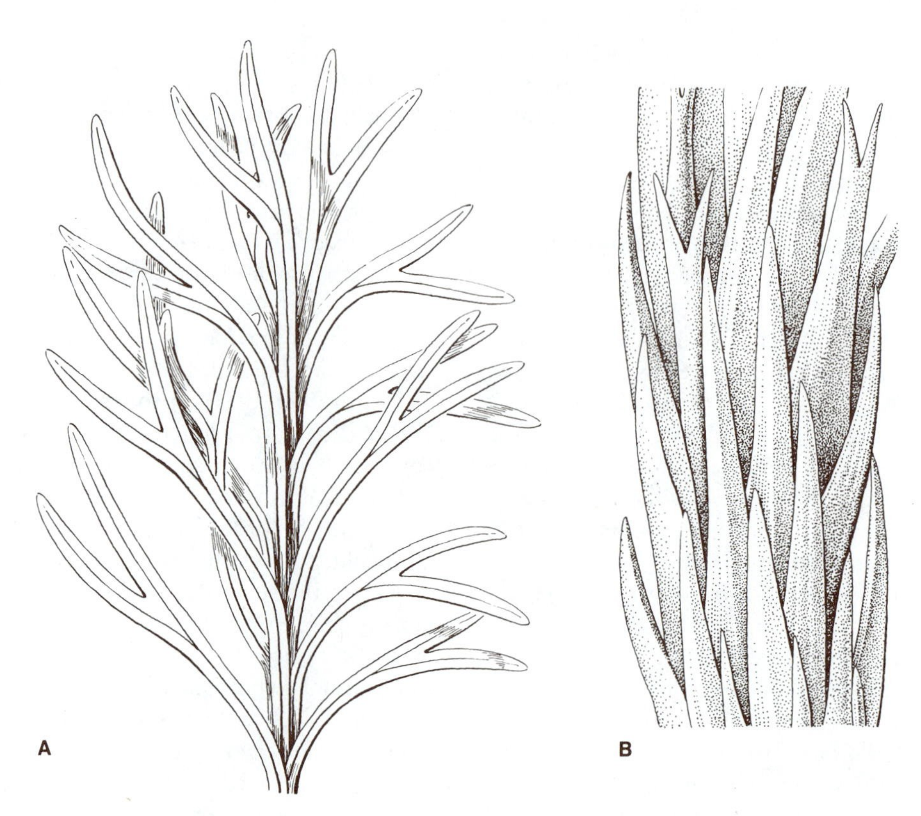

FIGURE 17-9 Foliage leaves of extinct conifers. **A,** dichotomously forked leaves of *Carpentieria frondosa;* **B,** simple and forked leaves at base of an ovuliferous cone of *Lebachia piniformis.* [Redrawn from Florin, *Palaeontographica* Abt. B, Bd. 85, 1944.]

Agathis, and *Podocarpus* is more complex. It consists of a longitudinal series of dichotomously branched veins, resembling in this respect the venation of the pinnae of the leaf in certain cycads (see Chapter 15, Fig. 15-21, A). An outstanding example of the systematic value of leaf venation is provided by the genus *Pinus.* In the "soft pines" (e.g., *Pinus strobus, Pinus monophylla,* and *Pinus lambertiana),* the vasculature of the mature needle consists of a single median vein; in the "hard pines" (e.g., *Pinus ponderosa* and *Pinus nigra),* the mature leaf has two more-or-less widely separated veins (Shaw, 1914) (Fig. 17-11). Many years ago, Koehne (1893) utilized this difference in vein number as a basis for dividing the genus *Pinus* into two subgenera, namely, *Haploxylon* (species with single-veined leaves) and *Diploxylon* (species with two-veined leaves). Although some morphologists in the past have interpreted the two-veined form as a "reversion" to an ancient pattern of dichotomous venation, this view is not supported by histogenetic studies. In pine leaves that at maturity are two-veined, the young developing needle *at first* is traversed by a single median vein with protophloem and protoxylem. As differentiation continues, a median raylike wedge of parenchyma is formed that alters the median portions of the protophloem and protoxylem and finally causes a division of the original bundle into two strands. In some instances, one or two additional rays are formed, dividing the original vein into three or four separated bundles. Thus, from an ontogenetic standpoint, the mature two-veined form seems to be a specialization of the basic univeined form that is so widespread in the leaves of conifers.

HISTOLOGY. Numerous comparative studies have shown that the histology of the simple leaves of conifers is extremely complex. The epidermal, fundamental, and fascicular tissue systems (see Chapter

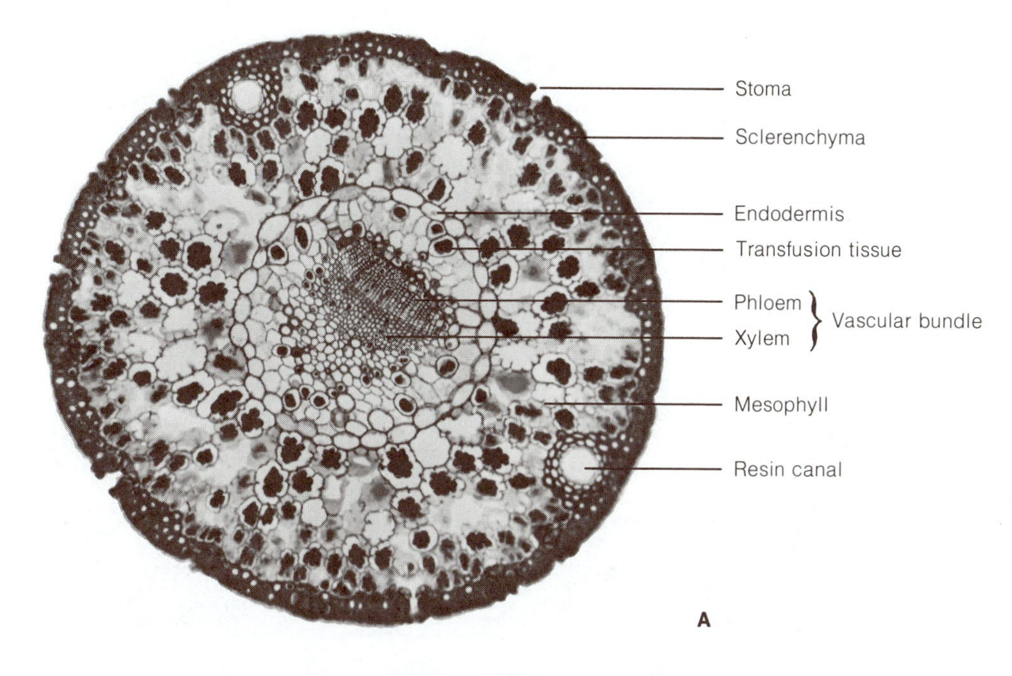

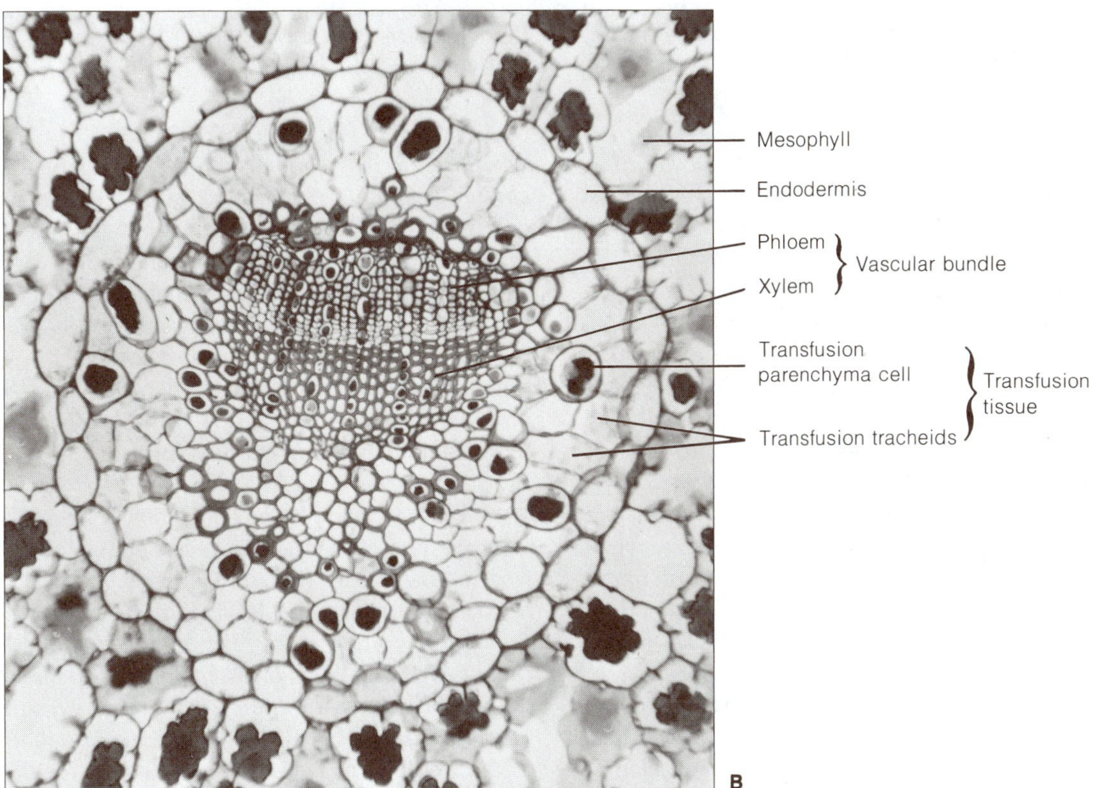

FIGURE 17-10 Histology of the leaf of *Pinus monophylla.* **A,** transection of leaf; **B,** details of endodermis, transfusion tissue, and vascular bundle.

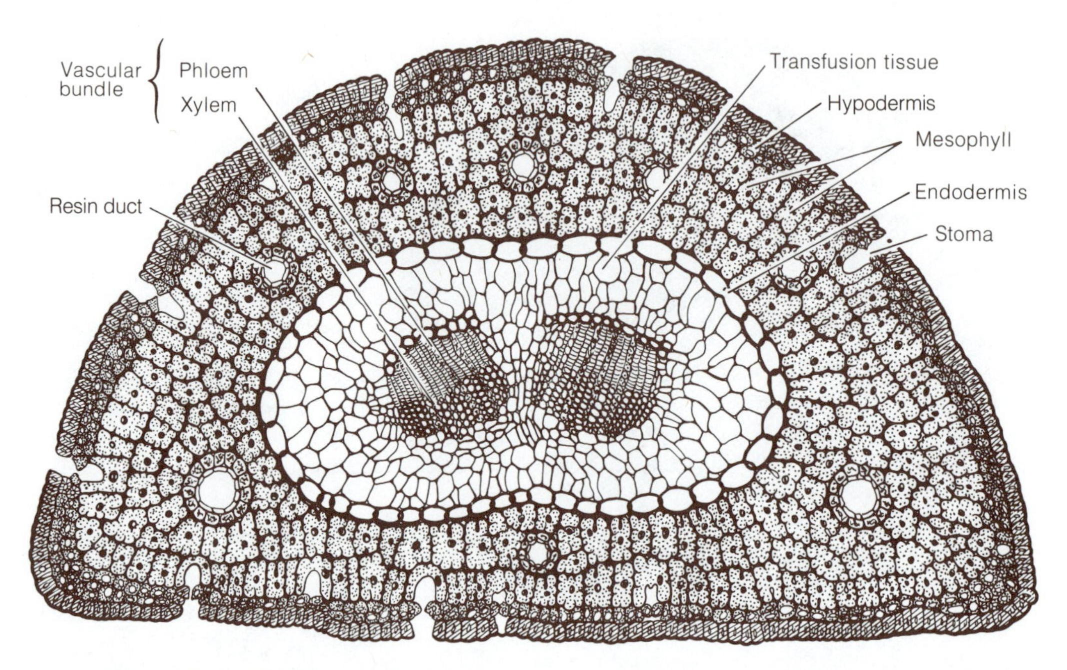

FIGURE 17-11 Transection of needle leaf of *Pinus nigra (laricio)*, showing the two vascular bundles characteristic of the leaves of the "hard pines." [Redrawn from *Gymnosperms. Structure and Evolution* by C. J. Chamberlain. The University of Chicago Press, Chicago. 1935.]

3) are characteristically well defined, but the cellular structure of these systems, particularly the last two, varies widely even among genera or species. The epidermal cells are generally thick walled with a cuticle of varying thickness. The stomata are generally sunken with overarching subsidiary cells. A fibrous hypodermis, composed of one or more layers, may occur beneath the epidermis except under the stomata. In some, a hypodermis is absent. The mesophyll may be differentiated into palisade and spongy parenchyma (e.g., *Abies, Taxus;* (Fig. 17-12). In others, palisade layers may occur on both sides of the leaf (e.g., *Araucaria).* The mesophyll of pine is unique in that it consists of a uniform tissue, the cell walls of which have ridgelike invaginations that protrude into the interior of the cell ("plicate mesophyll"). An endodermal sheath, the cells of which have casparian strips, often develop secondary walls with the suberin confined to the anticlinal walls (Figs. 17-10, 17-11). In others, the endodermis is represented only by a parenchymatous sheath. "Transfusion tissue" and a vascular bundle (or two, as in some pines) occur internal to the endodermal sheath. Over the years there have been reports of secondary growth in gymnosperm leaves. Recently, Ewers (1982) has shown conclusively for eleven conifer genera that the mature leaves produce secondary phloem each year but no secondary xylem. *Living* phloem remains constant in amount with advancing age, indicating yearly replacement of old by new phloem. Yearly increments are produced as long as the leaves remain on the tree — as long as thirty or more years for the bristlecone pine. The reader is referred to Figs. 17-10; 17-11; 17-12 for representative types of leaf histology.

STOMATA. In many of the living conifers, the stomata are on both surfaces of the leaf, but in certain taxa, stomata are restricted to either the abaxial or adaxial epidermis. The leaf of many species of *Pinus,* for example, is termed *amphistomatic* because there are stomata on all epidermal surfaces of the needle (Figs. 17-10, A; 17-11). A typical example of a *hypostomatic* leaf (having the stomata confined to the lower or abaxial epidermis) is provided by *Pseudotsuga.* Regardless of their topography, the stomata in the leaves of most genera are arranged in rather well-defined longitudinal rows.

Without question, the most comprehensive survey of the structure, ontogeny, and systematic value

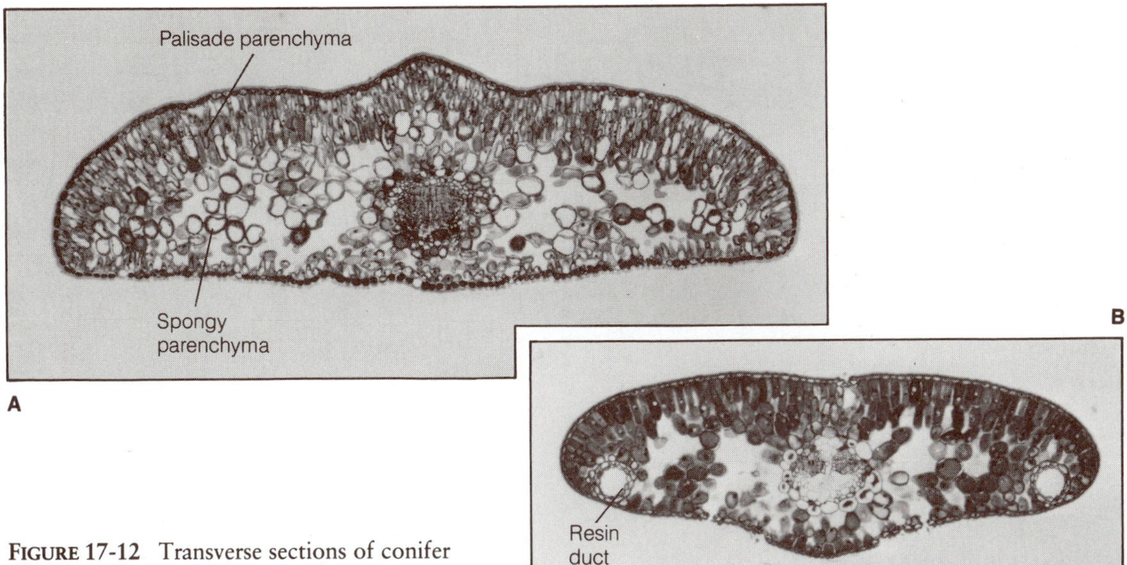

FIGURE 17-12 Transverse sections of conifer leaves. **A,** *Taxus,* yew; **B,** *Abies,* fir.

of stomata in gymnosperms was made by Florin (1931, 1951). The impetus for his detailed investigations was the extraordinarily well-preserved cuticular pattern of the epidermis of fossil gymnospermous leaves, which made it possible to compare the stomatal morphology of extinct and living taxa. In addition, he studied the ontogeny of the stomata in all of the major groups of living gymnosperms (i.e., cycads, *Ginkgo biloba,* Coniferales, and "gnetophytes"). On the basis of these investigations, he proposed a classification of gymnospermous stomata into two major types. In the *haplocheilic,* or "simple-lipped," type, the two guard cells of the stoma arise *directly* by the division of a protodermal mother cell (Fig. 17-13, A). In this type, the epidermal cells surrounding the stoma may function as subsidiary cells or each of them may divide into an inner subsidiary cell and one or more radially arranged "encircling cells". In the second, or *syndetocheilic,* type (compound lipped), a protodermal cell divides twice giving rise to two subsidiary cells. The initial remaining between the two subsidiary cells then divides, giving rise to the two guard cells. The distinctive feature of this type of stomatal development is the origin of the guard-cell initial and the two lateral subsidiary cells from the same "parent" cell (Fig. 17-13, B). One or both of the subsidiary cells may subsequently divide and produce an outer pair of "encircling cells" as is shown in Fig. 17-13, B, right).

It is particularly significant, from an evolutionary viewpoint, that both types of stomatal apparatus can be traced through the fossil record and that they serve to consistently define major gymnosperm taxa. The haplocheilic type is characteristic of (1) living and extinct conifers (e.g., *Lebachia, Ernestiodendron),* (2) the Pteridospermales, (3) Cordaitales, (4) Cycadales, (5) Ginkgoales, and (6) the genus *Ephedra* in the Ephedrales. The apparently advanced, syndetocheilic type is more limited in occurrence and is found in the extinct Cycadeoidales and in the living genera, *Welwitschia* and *Gnetum.* Some revision of the exclusive occurrence of the haplocheilic type in conifers may be in order. Johnson and Riding (1981) have shown that a polar subsidiary cell of a stomatal complex arises from the same protodermal cell as does the guard-cell initial in *Pinus strobus* and *Pinus banksiana.*

For more extensive descriptions of the ontogeny of stomata in gymnosperms, see the monographs of Florin (1931, 1951) and the accounts by Maheshwari and Vasil (1961), Pant and Mehra (1964), Martens (1971), and Kausik (1974).

TRANSFUSION TISSUE. The term transfusion tissue is applied to the tracheids and associated parenchyma cells, which are at the periphery of the leaf veins in living representatives of the gymnosperms (Worsdell, 1897; Abbema, 1934; Griffith, 1957, 1971).

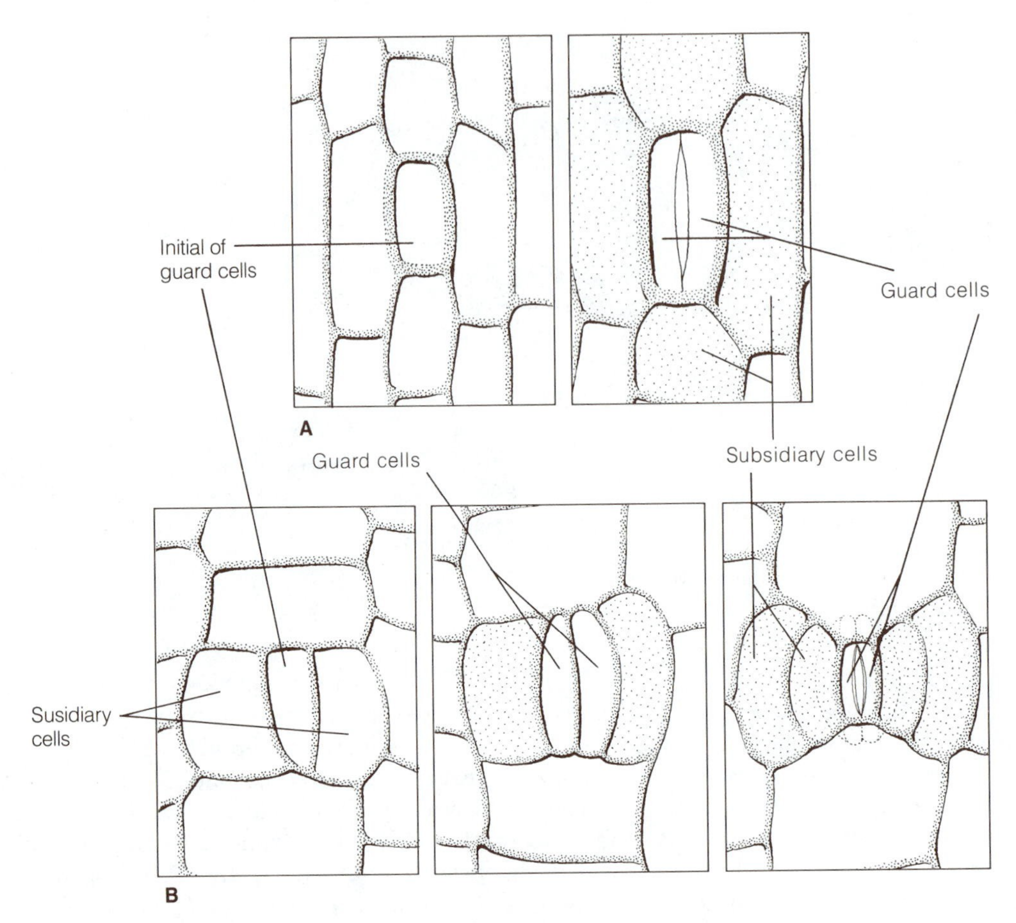

Figure 17-13 The two main types of stomata in gymnosperms. **A,** haplocheilic; **B,** syndetocheilic. See text for discussion. [Redrawn from Florin, *Acta Horti Bergiani* 15:285, 1951.]

As seen in transectional view, the transfusion tissue in a needle of *Pinus* consists of several layers of compactly arranged cells between the vascular bundle or bundles and the endodermis (Fig. 17-10). Sacher (1953), who has made a detailed ontogenetic study of the leaf of *Pinus lambertiana,* found that the transfusion tissue arises from the outer region of the procambium in the young needle, and that its maturation proceeds basipetally and centripetally. The transfusion tissue in *Pinus* consists of two principal cell types: *transfusion tracheids* with thick, lignified secondary walls and conspicuous circular bordered pits and *transfusion parenchyma* cells, which retain their cytoplasm and accumulate tanninlike substances (Fig. 17-10, B). These two cell types form a complex interconnected system and occur in direct contact with the cells of the endodermis. The physiological role of the transfusion tissue in *Pinus* and other gymnosperms is considered to be conduction of materials between the vascular bundle and the mesophyll.

The spatial arrangement of foliar transfusion tissue in certain other conifer genera is quite different from that of *Pinus.* In *Podocarpus macrophyllus,* for example, two winglike extensions of transfusion tissue occur at either side of the single vein (Fig. 17-14, A). Immediately adjacent to the xylem, the transfusion tissue consists only of short tracheids, whereas the outer parts of each wing—termed the accessory transfusion tissue by Griffith (1957)—consist of both parenchyma cells and tracheids, and extend between the palisade and spongy mesophyll

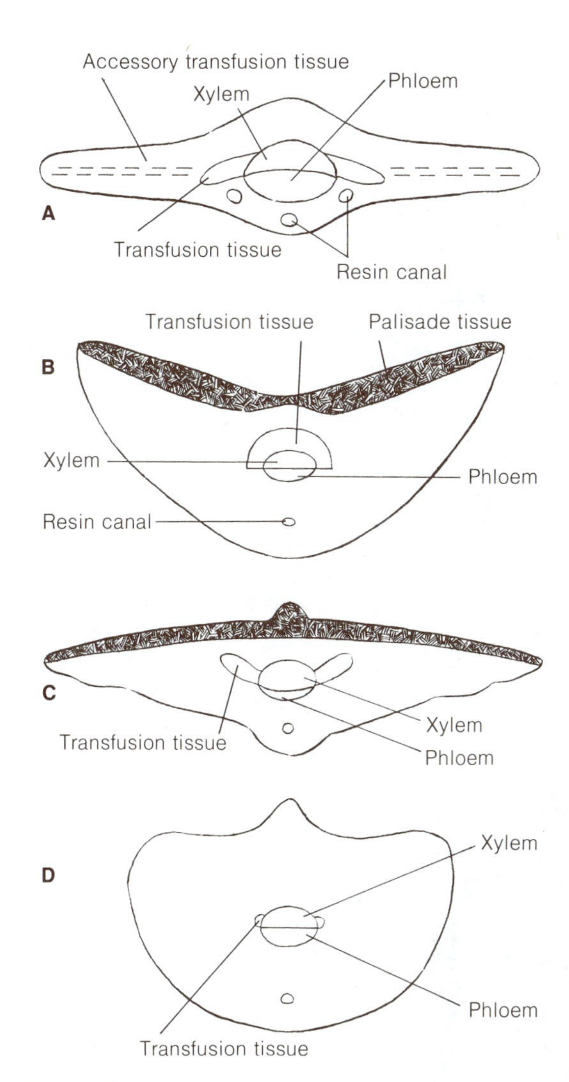

FIGURE 17-14 Diagrams showing arrangement of transfusion tissue in the leaves of conifers. **A,** *Podocarpus macrophyllus,* transection of leaf showing areas of transfusion tissue at each side of xylem of midvein and wings of accessory transfusion tissue extending to each margin of lamina; **B–D,** *Cephalotaxus harringtonia,* transections through terminal (**B**), middle (**C**), and basal (**D**) portions of same leaf. See text for further explanation. [**A** from Griffith, *Amer. Jour. Bot.* 44:705, 1957; **B–D** from Griffith, *Phytomorphology* 21:86, 1971.]

nearly to the leaf margin. In *Cephalotaxus,* according to Griffith (1971), the spatial relation of the transfusion tissue varies at different levels in the same leaf. Near the leaf apex, transfusion tissue forms an arc adjacent to the xylem (Fig. 17-14, B); in the middle region of the lamina, the xylem of the vein is bordered by two short wings of transfusion tissue (Fig. 17-14, C); and at the base of the leaf, only a few transfusion tracheids occur next to the xylem (Fig. 17-14, D). In both *Podocarpus* and *Cephalotaxus,* the transfusion tissue first appears in the apical region of the young leaf primordium and subsequently develops basipetally.

The association of transfusion tissue with the veins of the leaves in the Coniferales and other gymnosperms poses interesting but difficult morphological problems. According to one widely expressed view, transfusion tissue phylogenetically originated from the primitive, centripetal, primary xylem that is characteristic of the leaves of certain Paleozoic gymnosperms (Worsdell, 1897; Jeffrey, 1917). A contrary view holds that transfusion tissue represents a modification of the parenchyma surrounding or flanking the vascular bundle. Both of these interpretations appear conjectural, and a better understanding of the morphology and history of foliar transfusion tissue will depend on further studies. In this connection, it is interesting that in the leaf of such a living fossil as *Metasequoia* (dawn redwood), the vein is flanked on both sides by well-defined transfusion tracheids (Shobe and Lersten, 1967).

Stem Anatomy

The vascular cylinders of conifers are eustelic and the primary xylem is endarch in development. Conifers with helical phyllotaxy commonly have five, eight, thirteen, or twenty-one vascular sympodia. In a shoot with thirteen sympodia, for example, leaf fourteen and its trace are separated by thirteen nodes from leaf twenty-seven in the helical phyllotaxy (see Fig. 14-10). This pattern is referred to as an "open system." In some conifers with opposite phyllotaxy, four sympodia may be present and a leaf trace is derived from two adjacent sympodia, resulting in a "closed system."

Unlike stems of the cycads, the transection of a two-year-old stem of *Pinus* is characterized by a prominent secondary vascular system which begins early in stem development (Fig. 17-15, A). The system comprises secondary xylem and secondary phloem, both developing from the persistent type of vascular cambial activity characteristic of all conifers. Aside from resin ducts, the secondary xylem

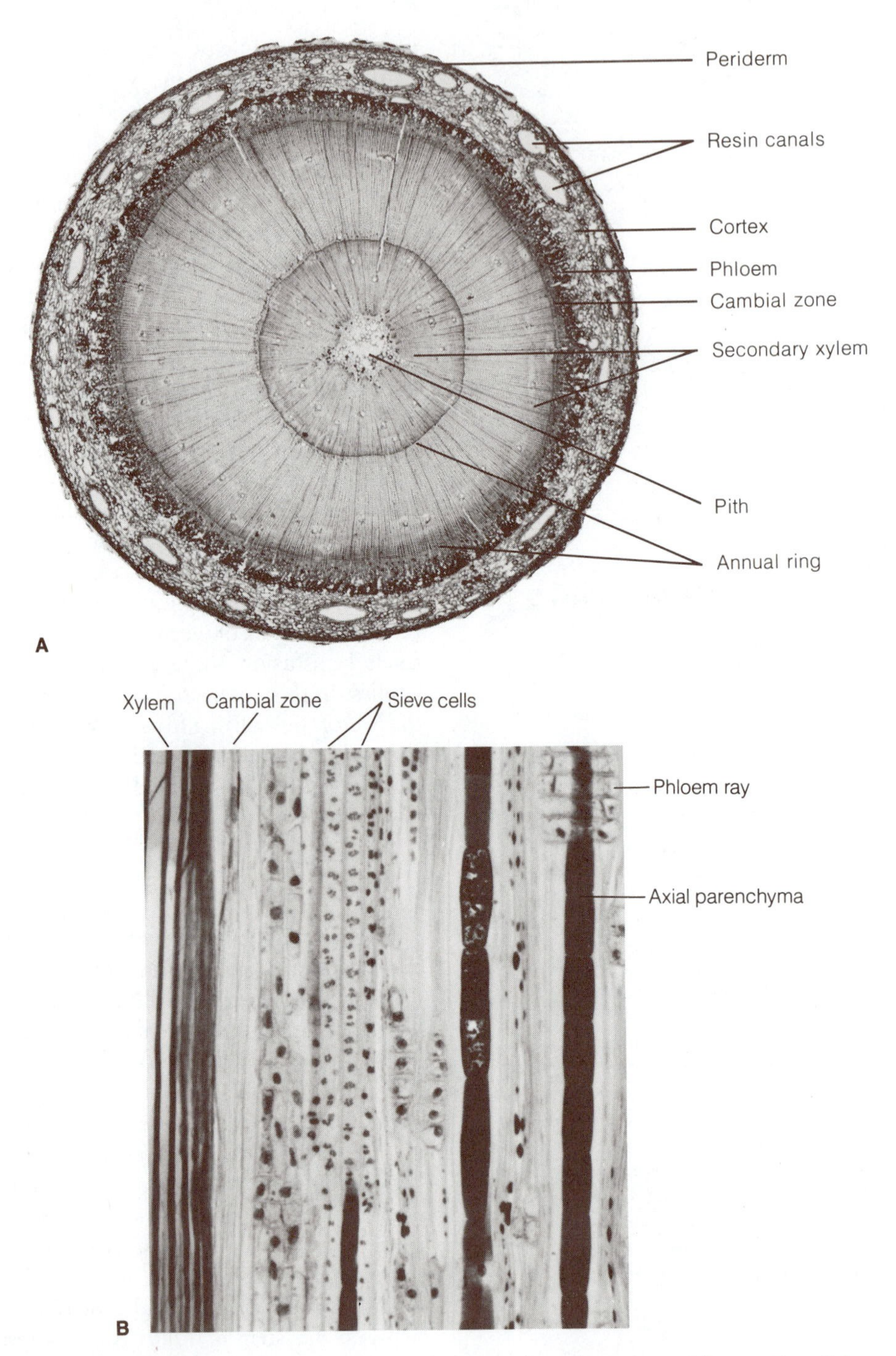

FIGURE 17-15 Stem anatomy of *Pinus*. **A,** transection of two-year-old stem; **B,** radial longisection of secondary vascular tissues and cambial zone. Dark areas on lateral walls of sieve cells are sieve areas.

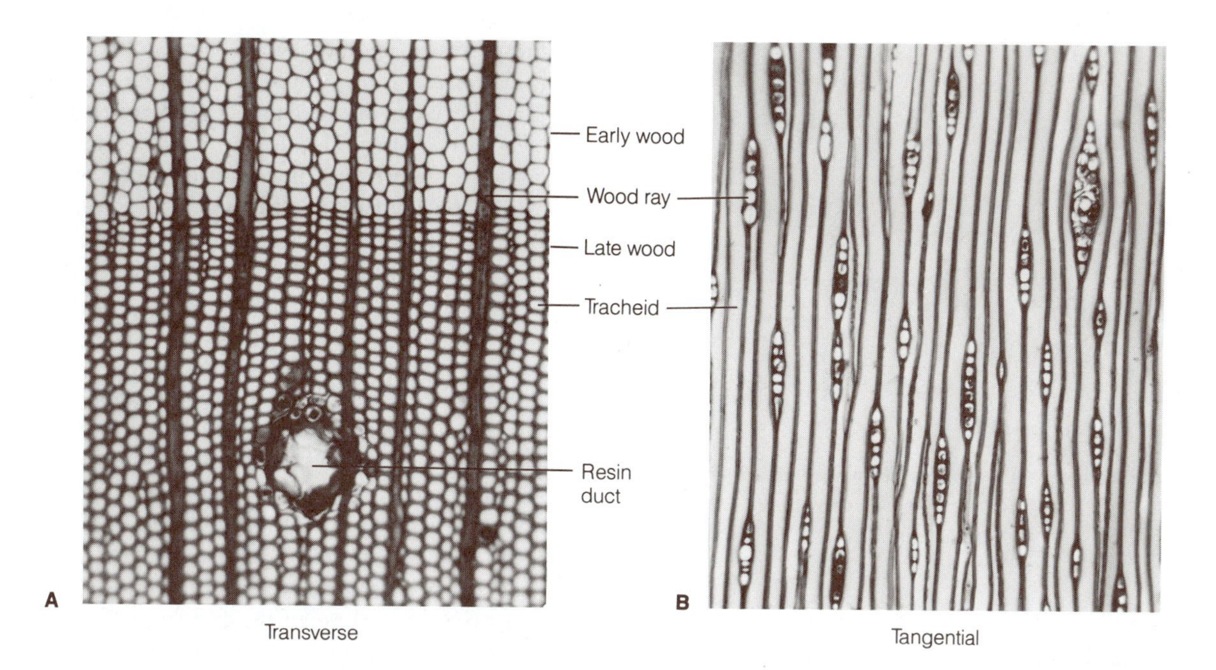

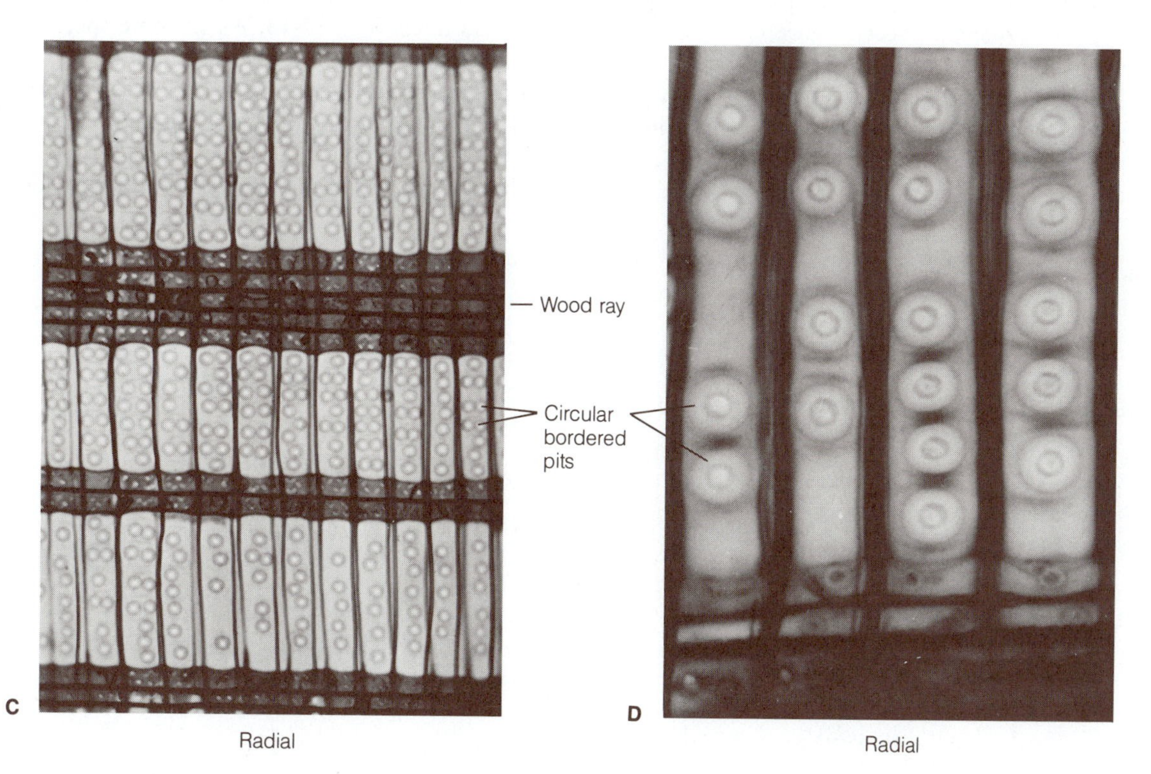

FIGURE 17-16 Representative examples of conifer wood seen in three planes of section. **A**, **B**, *Pinus,* as seen in transverse section (**A**) and tangential section (**B**); **C**, *Sequoia* (redwood), radial section, showing circular bordered pits; **D**, pine, radial section, showing circular bordered pits enlarged.

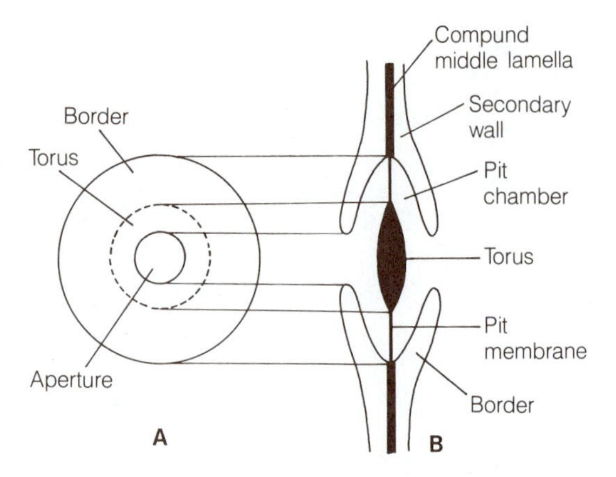

FIGURE 17-17 Diagram, circular bordered pit of the conifer type as seen in face view (**A**), and of a bordered pit-pair (of two adjacent tracheids) as seen in side view (**B**).

consists of elongated tracheids arranged in regular rows, and xylem rays of parenchyma cells arranged in radially directed sheets. Figure 17-16 shows the form and relationship of the tracheids and rays as seen in transverse, tangential, and radial sectional views. Large circular bordered pits occur primarily on the radial walls of tracheids (Figs. 17-16, C, D, and 17-17). The wood of *Pinus,* like that of all gymnosperms except members of the Gnetophyta, lacks vessels, and is more homogeneous and primitive than the xylem of the majority of angiosperms. The secondary phloem also displays a relatively simple structure consisting of elongated sieve cells, phloem, parenchyma, and rays (Fig. 17-15, B); in many conifers, fibers and sclereids also may develop in the secondary phloem. The pith and cortex of the pine stem consist largely of parenchyma with the cortex distinguished by the presence of large resin ducts. The epidermis has thickened outer walls overlaid by a thick cuticle, and later during the increase in stem circumference it is sloughed away by the development of a cylinder of periderm.

The number and course of resin ducts in the cortex varies in conifers. In the *long shoots* of *Pinus* there is a cylinder of ducts near the vascular cylinder, which may correspond to the number of vascular sympodia. At the level of the departure of a leaf trace to a *scale-leaf,* the trace is accompanied by two resin ducts derived from branches of two adjacent cortical ducts (Fig. 17-18, A). In some species of pine there is an additional peripheral series of ducts in the cortex that do not enter leaves. In other genera, e.g., *Araucaria,* the resin-duct system can be exceedingly complex. The ducts in leaves of still other genera are not connected to cortical ducts

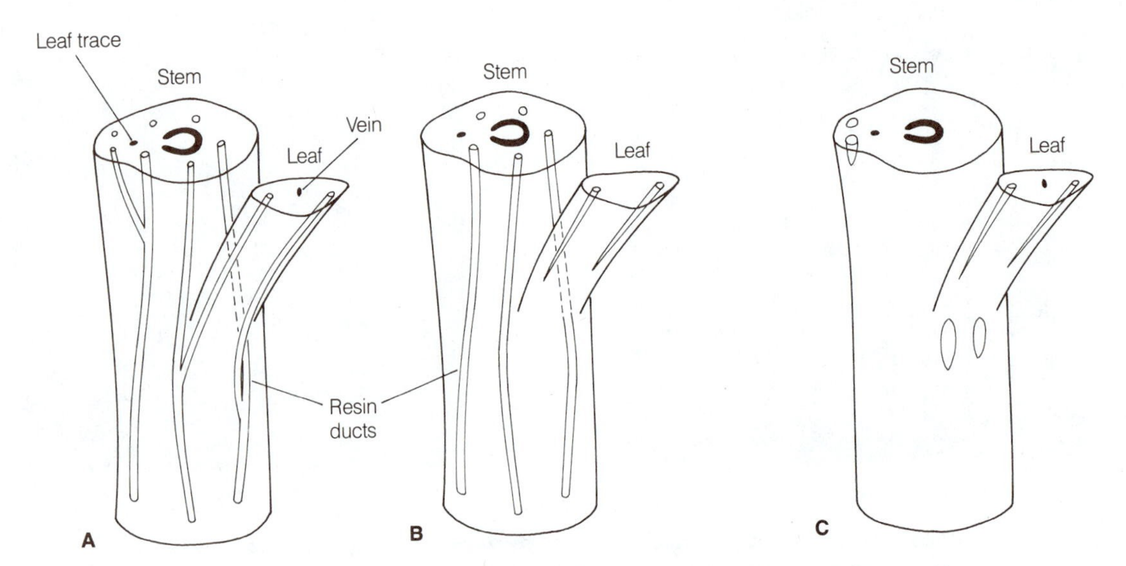

FIGURE 17-18 Schematic representation of the course of resin ducts in three conifers. See text for details. **A,** *Pinus* type; **B,** *Picea* type; **C,** *Larix* type. [Redrawn from Suzuki, *Bot. Mag.* (Tokyo) 92:333–353, 1979.]

(*Picea;* Fig. 17-18, B); or the ducts in the stem may be interrupted or may be discrete structures (Fig. 17-18, C) and are not connected to those of the leaf system (Suzuki, 1979a, b).

Because of the continued activity of the vascular cambium for hundreds or even thousands of years (e.g., *Sequoiadendron),* the stems of many conifers attain enormous diameters. Usually, as shown in Fig. 17-15, A, the so-called annual rings are clearly defined, each ring representing the increment of secondary xylem produced annually by the cambium. Not infrequently false annual rings are produced as the result of abnormal cambial activity; they, of course, must be considered when attempting to estimate the age of a given stem. The term "growth ring" rather than annual ring is probably a more acceptable term for the increments of secondary xylem.

The scientific interest in conifers and their importance as sources of lumber have been responsible for the numerous studies on cambial activity and wood histology in these plants. More detailed treatments of these topics and other aspects of stem anatomy are found in Esau (1965, 1977), and Brown, Panshin, and Forsaith (1949).

Strobili and Sporangia

One of the most uniform characteristics of the Coniferales is the monosporangiate nature of their cones or strobili. This means that in all normal instances two distinct types of strobili are formed: the microsporangiate, or pollen-bearing cone, and the megasporangiate, or seed cone (Figs. 17-19; 17-24). The latter is the larger, and is exemplified by the familiar cones of pines, firs, and spruces. Bisporangiate cones (structures with both microsporangia and ovules) have been observed in many genera of living conifers, but they clearly represent abnormalities and do not provide evidence about the phylogenetic history of the two cone types. According to Florin, no bisporangiate cones have been encountered in the fossil remains of either Paleozoic or

A

B

FIGURE 17-19 *Pinus.* **A,** terminal portion of branch showing clusters of microsporangiate strobili below the expanding spur shoots; **B,** tip of branch showing large terminal bud and, below it, three young megasporangiate strobili. [Courtesy of Dr. T. E. Weier.]

Mesozoic conifers — he concludes that monosporangiate cones represent a fundamental condition in the evolutionary history of conifers.

Living conifers are predominantly monoecious, both kinds of cones occurring on the same individual. This condition probably prevailed among the extinct conifers also. The dioecious condition, in which the two kinds of cones are produced by separate individuals of a species, is found in the Taxaceae, in a majority of the Araucariaceae, in *Podocarpus,* and in a number of genera in the Cupressaceae.

The Microsporangiate Strobilus (Pollen Cone)

Compared with those of cycads, microsporangiate strobili of most conifers are relatively small, commonly measuring only a few centimeters or less in length. The longest are those of *Araucaria bidwilli,* which may reach a length of 10 to 12 centimeters. The position of the microsporangiate strobili varies considerably within the Coniferales. In *Pinus* these strobili arise in the axils of scale leaves and are produced in subterminal clusters (Fig. 17-19, A), whereas in *Cedrus* solitary microsporangiate cones develop at the tips of certain of the spur shoots (Fig. 17-2, A). In other families, e.g., the Cupressaceae, the strobili develop terminally on certain specialized lateral shoots.

The microsporophylls vary in form from flattened leaflike appendages with expanded tips to peltate organs. In all species, the microsporangia develop on the lower surface of the sporophylls, but the number of sporangia varies throughout the order (Fig. 17-20). In the Pinaceae the number is consistently two, whereas in the other families there may be from two to seven sporangia on each sporophyll. Certain species of *Araucaria* and *Agathis* are notable because thirteen to fifteen sporangia may be produced by a single sporophyll.

It is reported that the microsporangium of conifers, although eusporangiate in its general pattern of development, *originates* from the periclinal division of a series of hypodermal cells (Chamberlain, 1935; Campbell, 1940; Haupt, 1953). In other words, in marked contrast with the *superficial* position of sporangial initials in the lower vascular plants, the initial cells of the microsporangia of conifers are asserted to lie *below* the surface or epidermal layer of the microsporophyll.

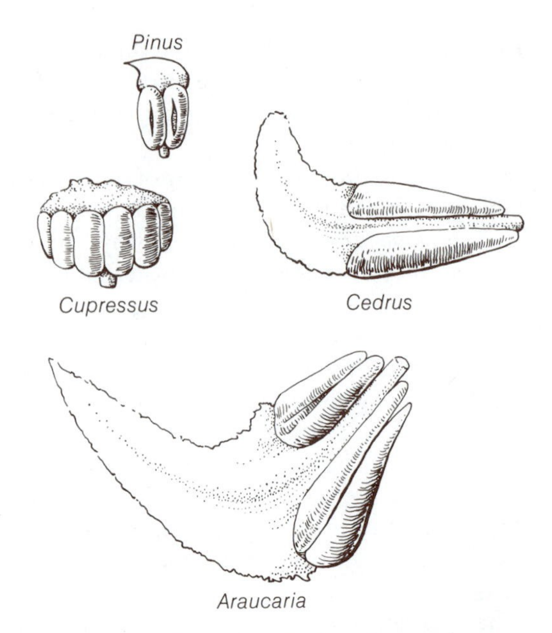

FIGURE 17-20 Microsporophylls in the conifers showing variations in their form and in the number of microsporangia. [Redrawn from *Gymnosperms. Structure and Evolution* by C. J. Chamberlain. The University of Chicago Press, Chicago. 1935.]

Several investigations reveal, on the contrary, that there are two different patterns of initiation. In two members of the family Pinaceae — *Pseudotsuga menziesii* (Allen, 1946b; Allen and Owens, 1972) and *Cedrus deodara* (Erspamer, 1952) — the sporangial initials are *superficial.* As in the lower vascular plants these initials divide periclinally, forming an outer series of from one to four cell layers that comprise the *sporangium wall* and, in most species, the tapetum and an inner group of *primary sporogenous cells* (Fig. 17-21, A, B). According to Fagerlind (1961), the microsporangia of *Pinus, Picea,* and *Larix* likewise originate from superficial initials. In all these plants, therefore, the superficial cells of the mature microsporangium are ontogenetically a part of the sporangium wall.

The mode of microsporangial initiation is different in *Cryptomeria japonica* (Erspamer, 1952; Singh and Chatterjee, 1963), *Chamaecyparis lawsoniana* (Erspamer, 1952), *Cupressus arizonica* (Owens and Pharis, 1967), *Taxus baccata* (Erspamer, 1952), and *Cephalotaxus drupacea* (Singh, 1961). In all these

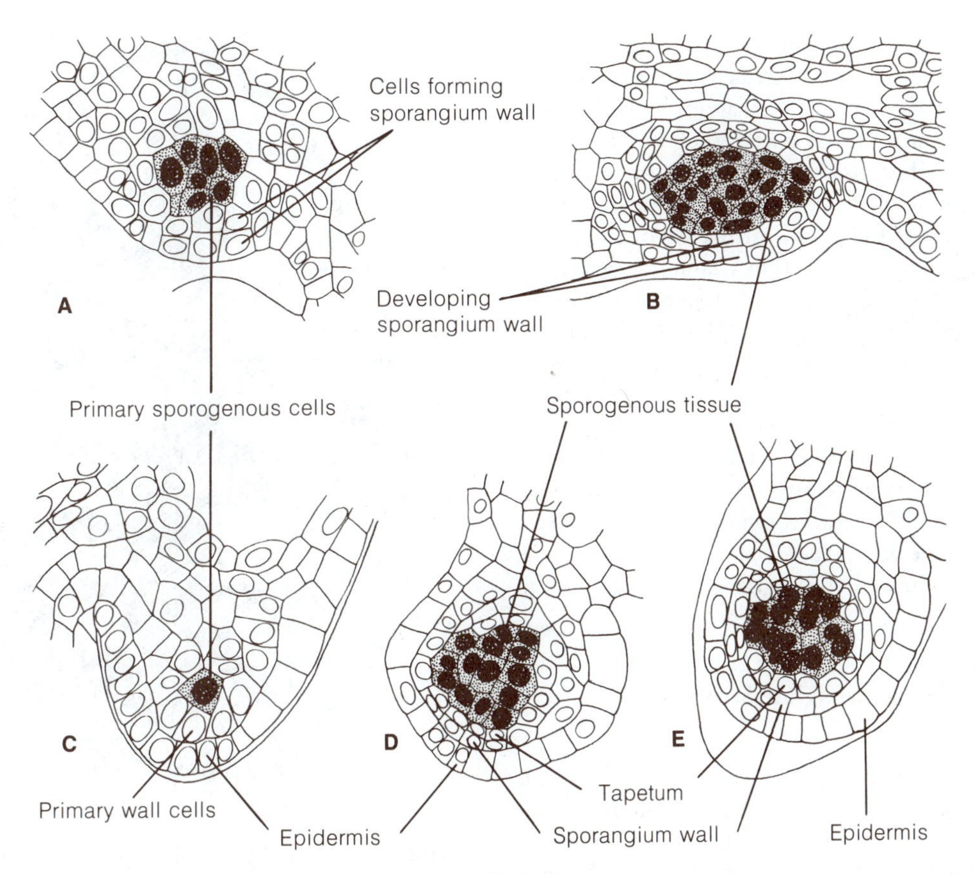

FIGURE 17-21 Initiation and early development of the microsporangium in conifers. **A, B,** *Cedrus deodara,* showing origin of microsporangium from superficial initials; **C–E,** *Chamaecyparis lawsoniana,* showing origin of microsporangium from hypodermal initials (note that epidermis is a distinct cell layer throughout development). [Redrawn from J. L. Erspamer, Ph.D. dissertation, University of California, Berkeley, 1952.]

conifers, the sporangial initials are strictly *hypodermal* and the mature eusporangiate microsporangium is externally bounded by an epidermal layer that had no role in the formation of the embedded sporangium (Fig. 17-21, C–E).

These studies emphasize the need for a comprehensive survey of the method of initiation and early ontogeny of the microsporangium throughout the Coniferales and other groups of the gymnosperms. Furthermore, all studies on sporangial development should consider (1) the structure and growth of the vegetative and reproductive apices and (2) the method of initiation of foliage leaves and microsporophylls. In members of the Pinaceae with superficial microsporangial initials, superficial cells of the apical meristem actively contribute to the initiation of leaf and sporophyll primordia. In contrast, hypodermal microsporangial initials in *Cryptomeria, Cupressus arizonica,* and *Taxus* is closely correlated with the presence in the shoot apex of a well-defined surface layer that does not contribute to the *inner tissue* of either the foliage leaf or the microsporophyll. Despite these apparent correlations, Fagerlind (1961) believes that "transitions" between the two patterns of microsporangial initiation may exist. He cites, in support of his opinion, an occasional periclinal division of epidermal cells in *Taxus,* contributing cells to the development of the underlying sporangium.

At maturity the wall of the microsporangium consists of one layer or, as in *Cedrus,* several layers of cells. Commonly many or all of the internal layers of the wall become crushed or obliterated. The outermost layer consists very often of cells with reticulate, helical, or annular thickenings, closely resembling the patterns of the secondary wall of tracheary elements. This surface layer is concerned with the mechanical rupturing of the sporangium at the period of release of the pollen grains. Although a tapetum characteristically develops, its origin is quite variable, since it may be the innermost layer of the sporangial wall *(Chamaecyparis)*, or a derivative of the sporogenous tissue *(Taxus)*. In most genera, particularly in the Pinaceae and Araucariaceae, a large number of microspores are produced in each sporangium (Fig. 17-22). A very characteristic feature of the pollen of most genera in the Pinaceae is the development of two air-filled lateral bladders or wings (Figs. 17-31, 17-32, A). Winged pollen grains are present also in members of the Podocarpaceae.

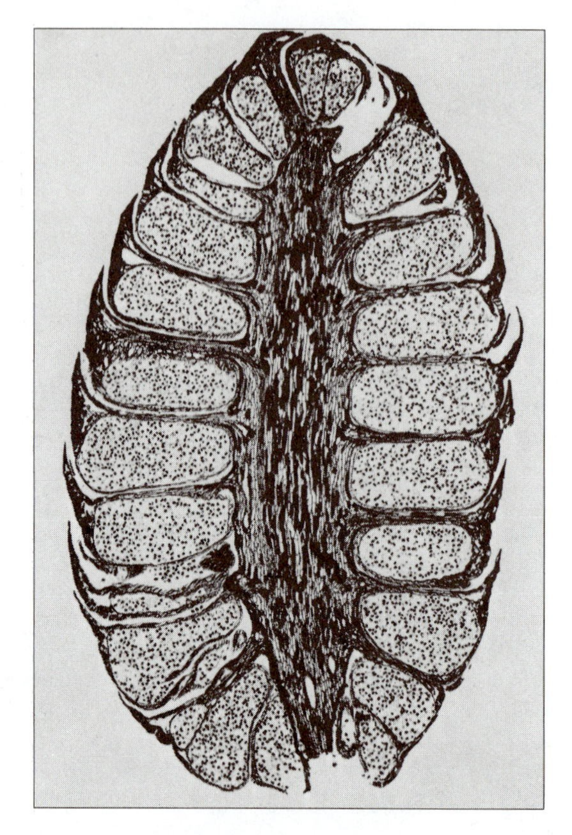

FIGURE 17-22 Longitudinal section of mature microsporangiate strobilus in *Pinus*.

The Megasporangiate Strobilus (Seed Cone)

The megasporangiate strobilus, or seed cone, is characteristic of the majority of the families of living conifers. The interpretation of the structure and evolution of the seed cone has been one of the most difficult and controversial problems in plant morphology. However, Rudolf Florin (1944, 1950, 1951) was able to make highly significant contributions to a better understanding of the structure and evolutionary history of the megasporangiate cone in the Coniferales.

The essence of the morphological problem posed by the seed cone of conifers is clearly illustrated in the family Pinaceae. Examination of a longisection of a young seed cone of *Pinus* reveals that each cone scale is associated with a small bract adnate to the abaxial basal region of the scale (Figs. 17-23, C, 17-24). Dissection of a young cone further reveals that each scale is ovuliferous in that it

FIGURE 17-23 *(facing page)* Morphology of ovuliferous structures in various conifers. **A,** *Pinus banksiana,* longisection of young seed cone showing relation of developing ovuliferous scales and their associated bracts; **B,** *Pinus banksiana,* longisection of vegetative bud showing the general structure and axillary position of spur-shoot primordia; **C,** *Pinus maritima,* longisection of immature seed cone showing several examples of the orientation of phloem (white) and xylem (black) in the vascular traces of the ovuliferous scale and its associated bract; **D,** *Pinus maritima,* longisection showing separate origin of bract trace and ovuliferous scale traces; **E,** *Pinus maritima,* transection showing vascular system of bract and ovuliferous scale (note inverted orientation of scale bundles); **F, G,** *Abies balsamea,* abaxial and adaxial views, respectively, of ovuliferous scale; **H,** *Chamaecyparis lawsoniana,* longitudinal section of seed cone showing union between ovuliferous scale and bract; **I, J,** *Araucaria bidwillii,* adaxial and end views, respectively, of ovuliferous scale joined with its bract; **K,** *Araucaria balansae,* longisection showing vascular supply of bract and scale (note single basal ovule); **L,** *Taxus canadensis,* median longitudinal section of vegetative shoot and of secondary lateral fertile shoot which terminates in a single ovule. [Redrawn from following sources: **A, B** from *Gymnosperms. Structure and Evolution* by C. J. Chamberlain. University of Chicago Press, Chicago. 1935; **C, D, E, H, K** from Aase, *Bot. Gaz.* 60:277, 1915; **F, G, I, J, L** from Florin, *Acta Horti Bergiani* 15:285, 1951.]

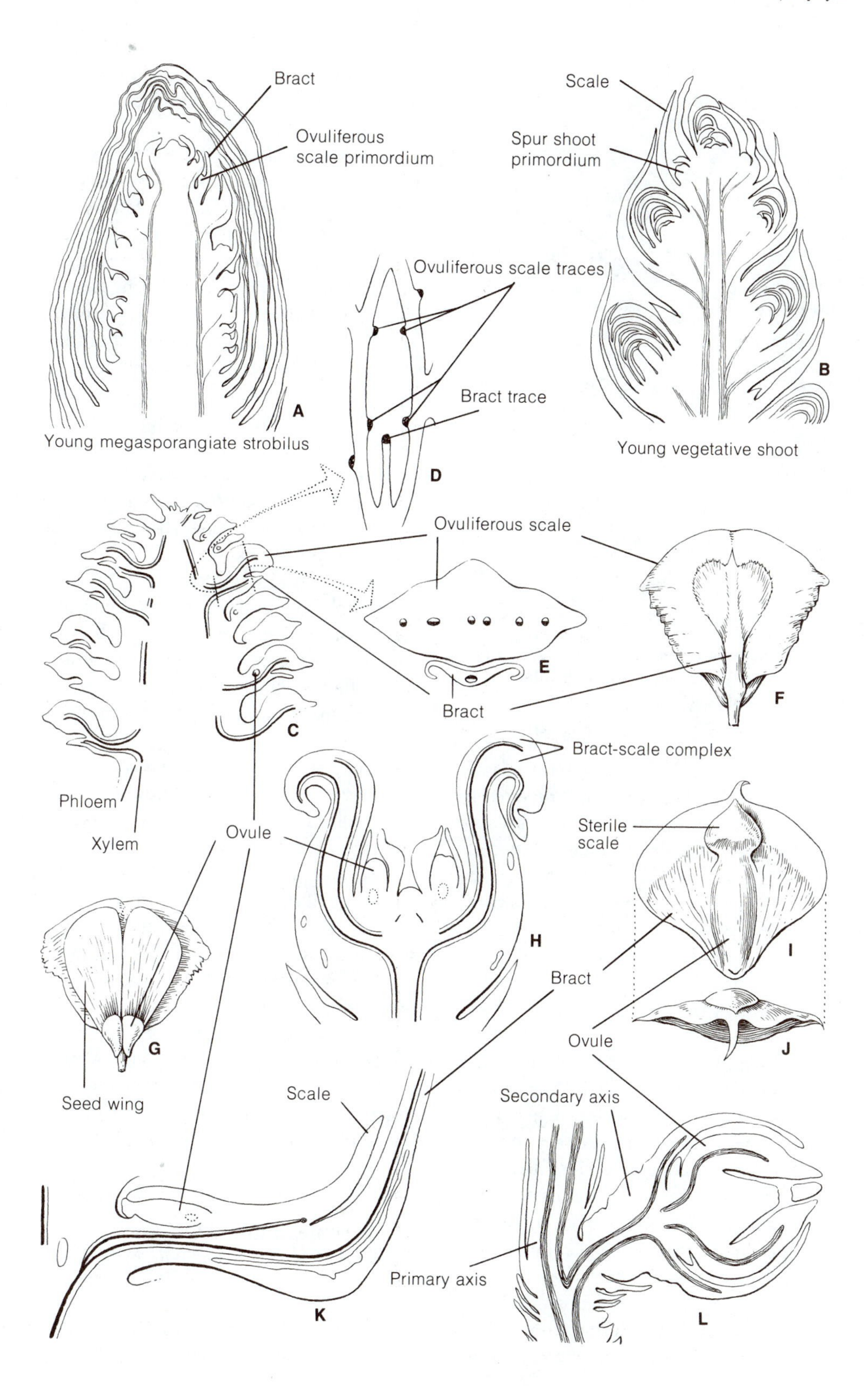
Bract
Ovuliferous
scale primordium
A
Young megasporangiate strobilus
Scale
Spur shoot
primordium
B
Young vegetative shoot
Ovuliferous scale traces
Bract trace
D
Ovuliferous scale
E
Bract
F
C
Phloem
Xylem
Ovule
Bract-scale complex
Sterile
scale
H
I
Bract
Ovule
J
G
Seed wing
Scale
Secondary axis
Primary axis
K
L

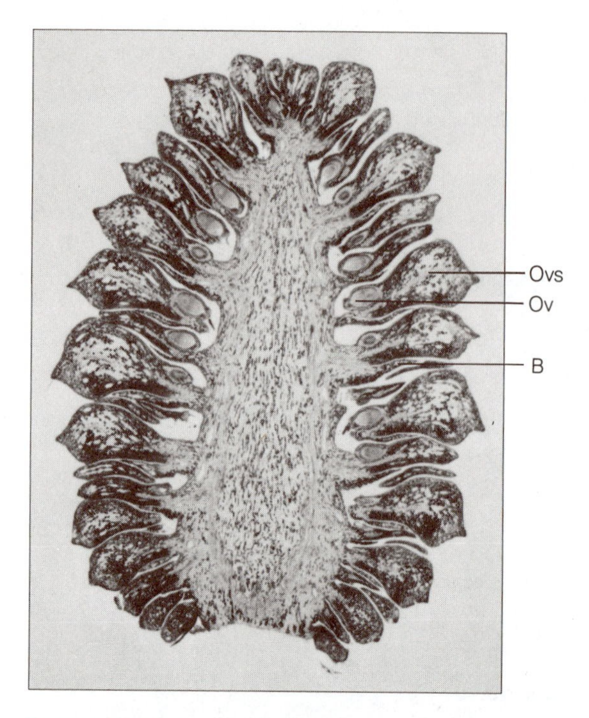

FIGURE 17-24 Longitudinal section, ovulate cone of *Pinus halepensis* shortly after pollination; B, bract; Ov, ovule; Ovs, ovuliferous scale.

bears a pair of ovules attached basally to the adaxial surface (Fig. 17-25, B). The ovules are *inverted* in orientation; that is, the micropyles are directed toward the base of the scale. As each ovuliferous scale is subtended by a bract, the seed cone is generally described as a *compound strobilus.* In contrast, the pollen cone represents a *simple strobilus* because the microsporangia are directly attached to the primary appendages or microsporophylls and not subtended by a bract (compare Figs. 17-22 and 17-23, C).

As shown in Figs. 17-23, C and 17-24, the bracts of the seed cone of *Pinus* are much shorter than their associated ovuliferous scales, whereas in *Pseudotsuga,* the tridentate bracts are very conspicuous in the mature seed cone (Fig. 17-2, B). Prominent exserted bracts are also characteristic of the seed cones of certain species of *Larix* and *Abies.*

Additional evidence that the seed cone in conifers is a compound strobilus is provided by its ontogeny (see review by Gifford and Corson, 1971). The type of structure and growth characteristic of a vegetative shoot apex to the pattern of growth typical of either the microsporangiate or megasporangiate strobili of conifers has received little attention.

According to Sacher (1954), the vegetative shoot apex of *Pinus ponderosa* is histologically demarcated into four tissue zones: an *apical initial zone,* a *peripheral zone* from which foliar appendages originate, a *central mother-cell zone,* and a *zone of rib meristem* which forms the pith of the shoot axis. Gifford and Corson (1971) use the term *apical zone* to include those cells at the summit of the apex and their immediate basal derivatives, i.e., the cells of the central mother-cell zone (Fig. 17-26). Gifford and Mirov (1960) found young undifferentiated megasporangiate cone primordia, each subtended by a scale, present by mid-September in the terminal buds of *Pinus ponderosa* trees growing at Placerville, California. The zonal structure of the apex of these cone primordia is similar to that of the vegetative shoot apex. In the latter part of November, the peripheral tissue zone of the cone apex appears as a mantle of small densely staining cells and the pith region of the cone axis increases in size. At this stage, the first *bract primordia* have been initiated from the mantle at the base of the young cone (Fig. 17-27, A). Gifford and Mirov's study revealed that the young seed cone continues to grow through the winter and early spring months in the vicinity of Placerville. By March of the year following cone initiation, the acropetal development of bracts is evident and procambial strands have differentiated into some of the lower bract primordia (Fig. 17-27, B). During April, the bracts attain nearly their full size, and the primordia of ovuliferous scales are developing in their axils. By May 1, well-defined primordia of the ovuliferous scales, each subtended by a maturing bract, are very conspicuous (Fig. 17-27, C). At this period of development, the young ovulate cone is structurally comparable to a young vegetative shoot with a bract subtending each of the developing spur shoots (compare Figs. 17-27, C, and 17-23, B).

Voltziales and the Origin of the Ovuliferous Scale

Although the evidence from organography, comparative anatomy, and ontogeny supports the

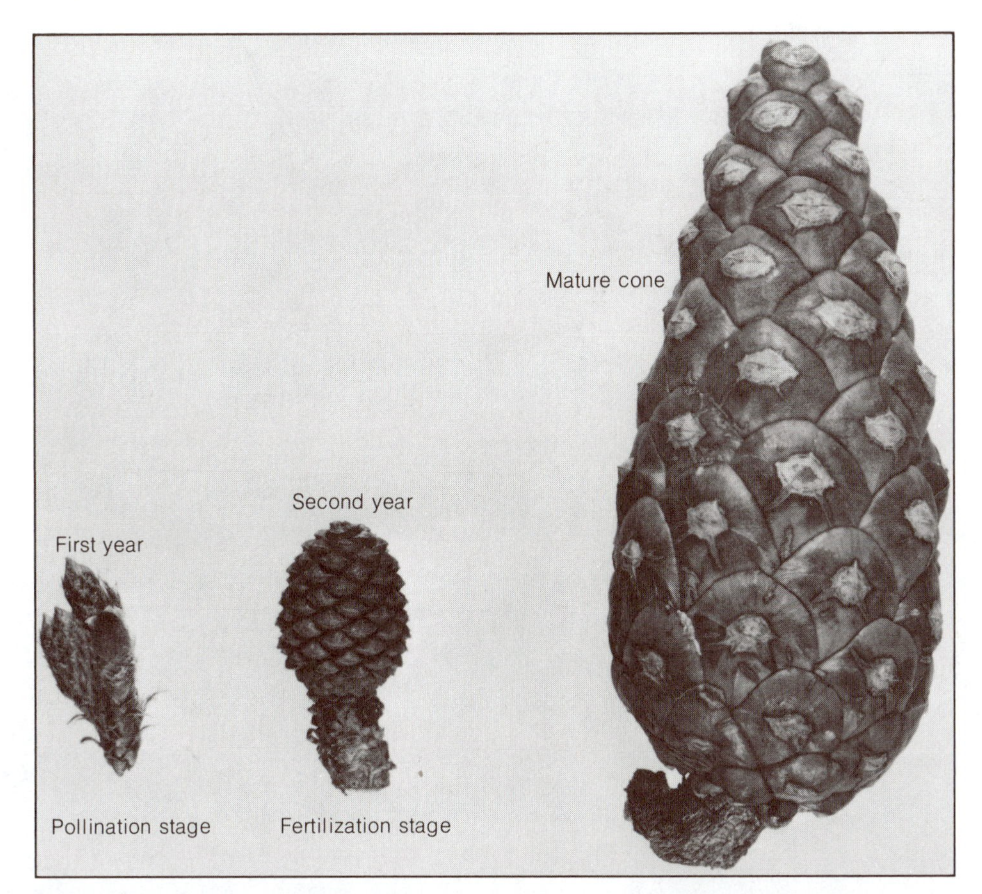

A

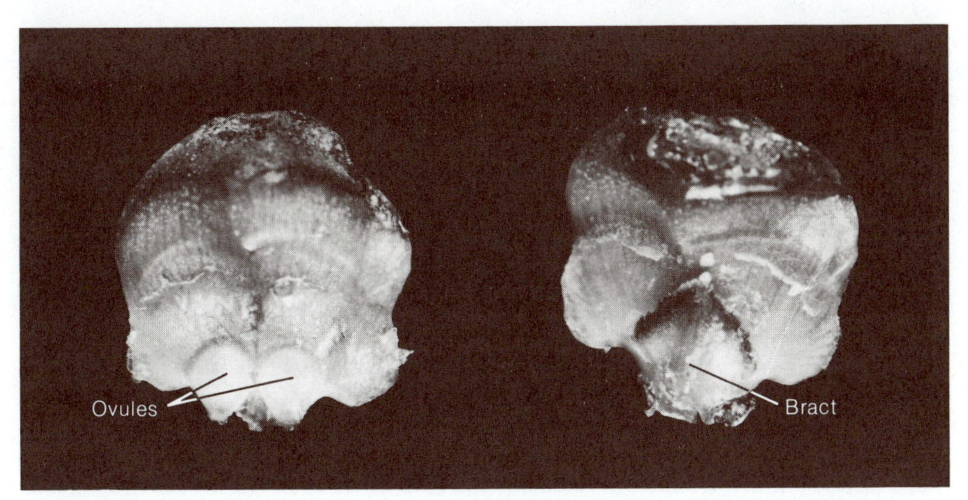

B

FIGURE 17-25 *Pinus halepensis.* **A,** developing seed cones showing numerous overlapping ovuliferous scales (cone scales); **B,** ovuliferous scale dissected from cone; adaxial view at left, showing the attachment of the two ovules; abaxial view at right, showing the short basally adnate bract.

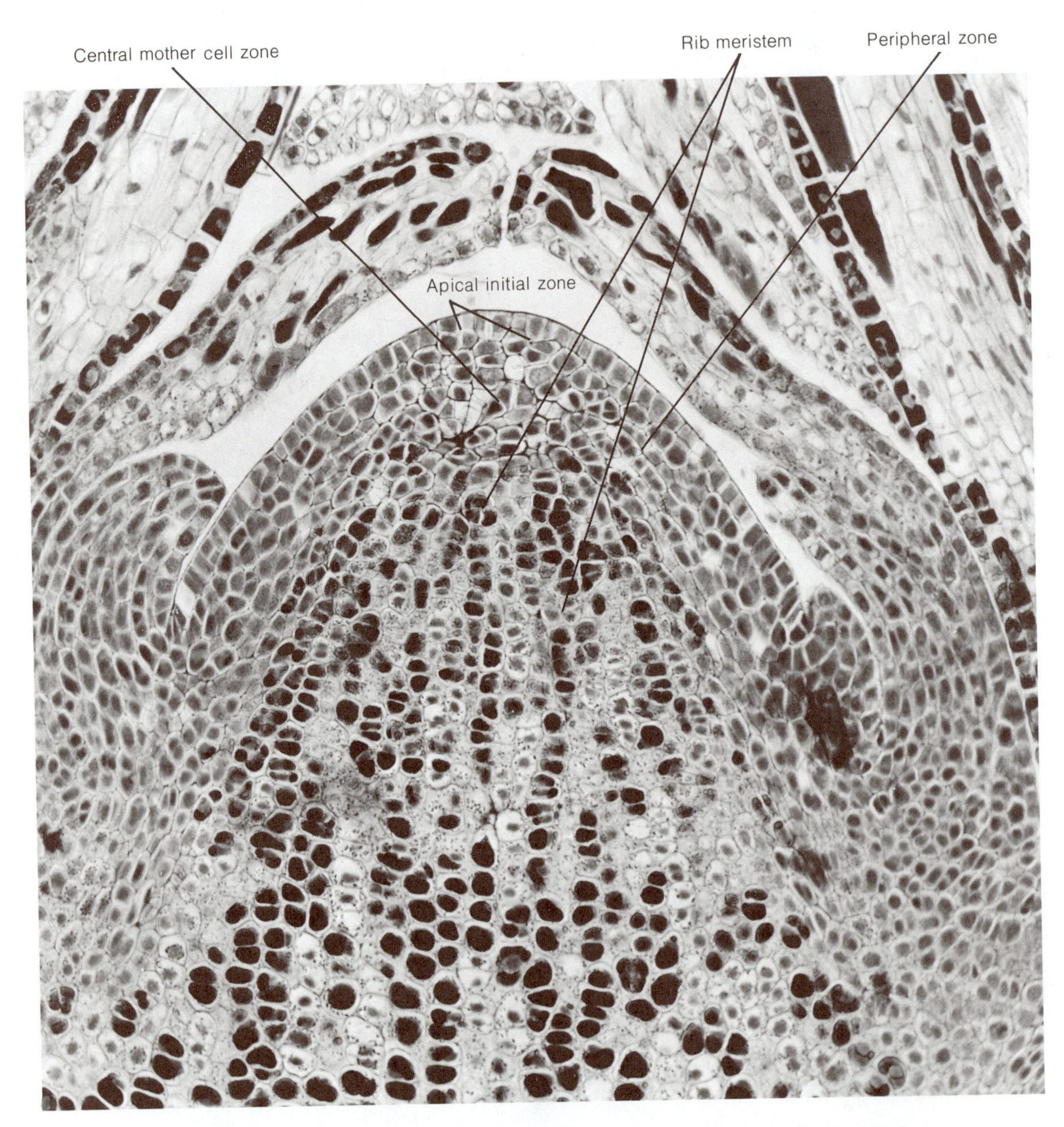

FIGURE 17-26 *Pinus ponderosa.* Median longisection showing zonal organization of the vegetative shoot apex. [From Sacher, *Amer. J. Bot.* 41:749, 1954.]

view that the ovulate cone in the conifers is a compound strobilus, the *phylogenetic* origin of the ovuliferous scale remains to be explained. The scale is similar to a sporophyll in that it bears ovules, but its axillary position with reference to a bract represents a puzzling situation from an evolutionary point of view. Of all the conflicting and involved theories that have been proposed during the past century, the most plausible and best-supported view was advanced by Florin: phylogenetically, the ovuliferous scale is a highly condensed and modified *fertile shoot* and hence is not a simple sporo-

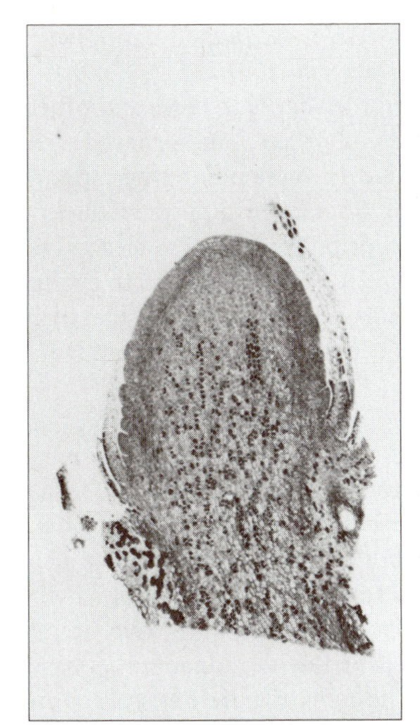

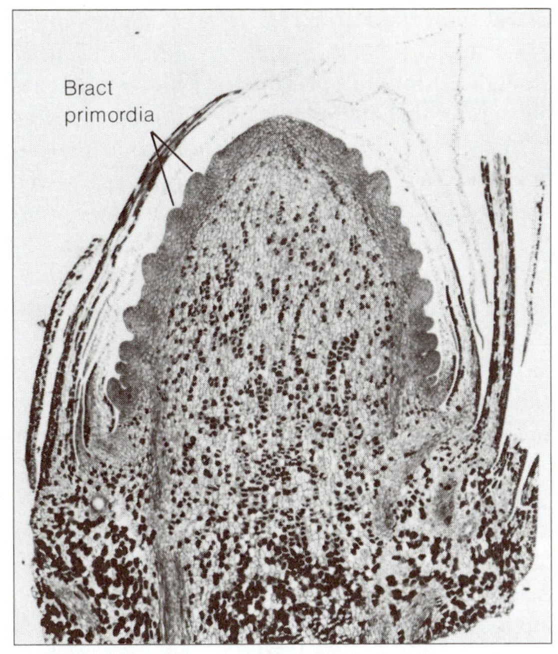

A B

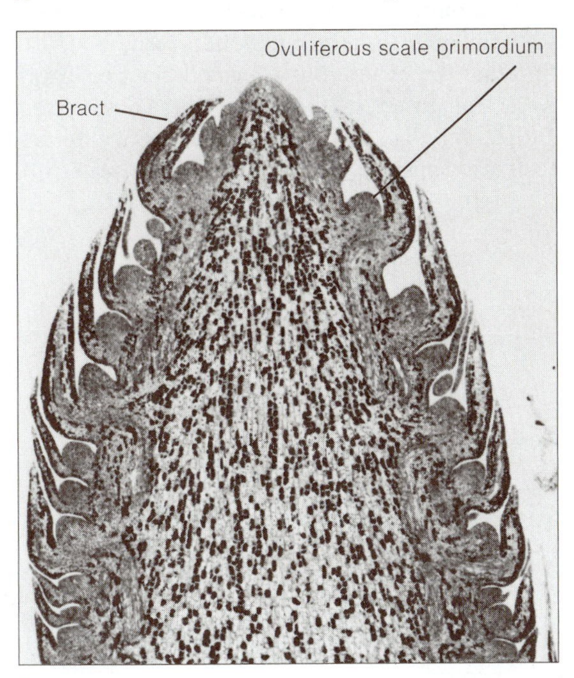

C

FIGURE 17-27 **A–C**, successively older stages in the development of the ovulate cone of *Pinus ponderosa*. **A**, young cone dissected from terminal bud on November 29, 1957, showing initiation of the first bract primordia; **B**, later stage (March 31, 1958), showing acropetal development of additional bract primordia; **C**, structure of cone on May 1, 1958, showing well-defined young ovuliferous-scale primordia in the axils of bracts (illustrations not at same magnification). [From Gifford and Mirov, *Forest Sci.* 6:19, 1960.]

phyll. In other words, the scale evolved from a leafy, ovule-bearing, dwarf shoot and its present "simple" appearance is the result of the fusion and specialization of both the sterile and fertile components of such an ancestral structure.

The earliest precursors of the modern *ovuliferous scale* occurred in *Lebachia* and *Ernestiodendron,* two genera of the Voltziales from the Upper Carboniferous-Permian. In *Lebachia piniformis,* a megasporangiate or ovulate cone consisted of a series of helically arranged, bifid bracts, in the axils of which developed short fertile shoots. Usually all but one of the scalelike leaves of the fertile shoot were sterile. The fertile scale (termed megasporophyll by some authors) consisted of a stalk and an erect terminal, bilaterally symmetrical ovule that faced the axis of the cone (Fig. 17-28, A). Most of the materials studied by Florin (1951) were impression-compression fossils. Recent studies of permineralized cones have supported, in general, Florin's concept of the early origin of conifer reproductive structures. A fertile shoot of *Lebachia lockardii* consisted of twenty-five to thirty sterile scales and one or two fertile scales at the base of the shoot (Fig. 17-28, B). There was one terminal inverted ovule per fertile scale (Mapes and Rothwell, 1984). Pollen grains have also been found in the pollen chamber. In contrast to *Lebachia,* all scales were fertile in *Ernestiodendron filiciforme* (Fig. 17-28, C). Florin termed the fertile shoots in these and other conifers "seed-scale complexes," and believed they support the view that the compound megasporangiate strobilus is the primary form in the Coniferales, with the exception of the Taxaceae. He regarded the Taxaceae as having a separate evolutionary history because they lack definable seed cones; this view has been questioned in recent years (Miller, 1977).

Genera from the Late Permian and early part of the Mesozoic, illustrate additional evolutionary steps toward the modern-day types of ovuliferous scales in the Coniferales. In *Pseudovoltzia liebeana,* for example, the fertile shoot consisted of five partially fused scales, forming a dorsiventral "ovuliferous scale." The middle lobe and the two lateral lobes each had a reflexed ovule, adnate to its base (Fig. 17-29, A, B). The entire ovuliferous scale complex was fused with the subtending bract to about its midpoint; vascular strands to the ovuliferous scale and bract, however, were not fused, an indication that the ovuliferous scale and bract probably were not fused in ancestral types (Schweitzer, 1963). In the Triassic genus *Voltzia,* the ovuliferous scale also consisted of five partially fused scales and

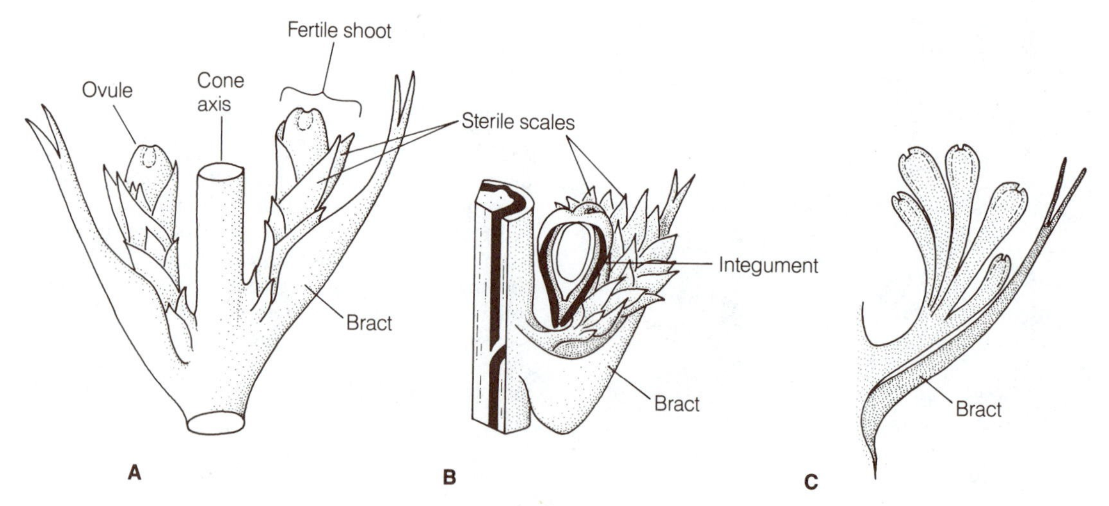

FIGURE 17-28 **A–C,** reconstructions of portions of the ovulate cones of the Voltziales from the Upper Carboniferous to Permian Periods. **A,** *Lebachia piniformis;* **B,** *Lebachia lockardii;* **C,** *Ernestiodendron filiciforme.* See text for explanations. [**A, C** redrawn from Schweitzer, *Palaeontographica* 113B: 1, 1963; **B** redrawn from Mapes and Rothwell, *Palaeontology* 27, Part 1:69–94, 1984.]

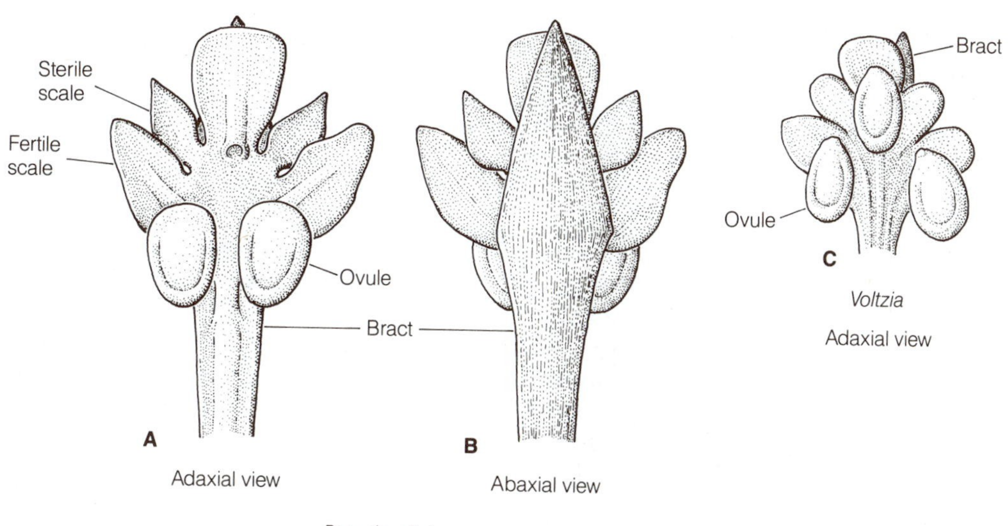

FIGURE 17-29 Bract, seed-scale complexes of the Voltziales from the Permian to Triassic Periods. **A, B,** *Pseudovoltzia liebeana;* ovule on middle lobe (scale) not shown; C, *Voltzia* sp. [**A–C**, redrawn from Schweitzer, *Palaeontographica* 113B:1, 1963.]

three inverted ovules whose stalks were adnate to the ovuliferous scale (Fig. 17-29, C).

In summary, the main trends in seed cone evolution have been toward (1) the elimination of all but a few sterile scales, which became fused into the so-called ovuliferous scale, (2) the recurvation of ovules and suppression of their stalks, and (3) their final incorporation with the lower adaxial side of the ovuliferous scale in derived conifers. On the basis of the ground plan-divergence method of cladistics, Miller (1982) has proposed a phylogeny of the Voltziales and modern-day families of the Coniferales based upon fourteen characters of the bract, seed-scale complex.

In modern conifers, evolutionary advancement in the ovulate cone is shown by the various degrees of fusion between the ovuliferous scale and its subtending bract. Throughout the Pinaceae, the bract is only basally adnate to the scale (Fig. 17-23, C), whereas these structures are more-or-less completely united in members of the Taxodiaceae, Cupressaceae, and Araucariaceae (Fig. 17-23, H–K). Comparative studies of the vascular anatomy of the ovulate cone in the Pinaceae reveal that the traces which enter the ovuliferous scale diverge from the stele in fundamentally the same manner as the branch traces of an axillary vegetative shoot (Fig. 17-23, D). In *Pinus maritima,* for example, Aase (1915) found that a single vascular strand, with the xylem oriented toward the adaxial surface, as in a vegetative leaf, extends as a separate trace into the bract (Fig. 17-23, D). In contrast, the branchlike vascular supply of the ovuliferous scale consists of three or four traces that diverge at a higher level from the stele; these strands branch dichotomously in the scale and form a series of veins with their xylem oriented toward the lower surface of the scale (Fig. 17-23, E). According to Florin (1951), the so-called "inversion" of the vascular bundles in the ovuliferous scale in the Pinaceae indicates its phylogenetically original radial symmetry. Each of the two ovules borne on the ovuliferous scale is vascularized by a single strand derived as a branch from an adjacent lateral bundle. (For a detailed account of the profuse dichotomous venation characteristic of the ovuliferous scales of *Abies, Pseudotsuga, Pinus,* and *Cedrus,* see Tison, 1913.)

Lemoine-Sebastian (1968, 1969) has studied the vasculature of the bracts and ovuliferous scales of the ovulate cones of various genera in the Taxodiaceae and Cupressaceae. The bract is vascularized by a single trace that only rarely branches in its course through the appendage. But the pattern of vasculature of the ovuliferous scale is usually very complex and varies according to the genus. In some genera, only a single system of inversely oriented strands is formed; in other genera, radial branching of the strands forms two vascular arcs: an adaxial series of strands with inverted orientation of the xylem and an abaxial system comprising normally oriented bundles (see Lemoine-Sebastian, 1968, 1969). These varied patterns of vasculature of the ovuliferous scales in the Taxodiaceae and Cupressaceae are of considerable morphological and systematic interest, but their phylogenetic significance is an open question; there are, however, recommendations that the two families should be merged, based upon analyses of numerous morphological characters (Eckenwalder, 1976; also see Miller, 1982).

The Reproductive Cycle in *Pinus*

Numerous comparative studies have demonstrated the wide variation in the details of sporogenesis, gametophytic development, and embryogeny within the living conifers. To choose one genus from the large number of investigated plants as being "typical" of the Coniferales as a whole is doubtless impossible. Most conifers, on the contrary, prove to have a blend of gametophytic and embryological characters, some advanced, some primitive, and still others shared by even the cycads and *Ginkgo*. For these reasons there is probably no living genus that can serve alone as the measure of phylogenetic trends within the order.

The selection of *Pinus* as the basis for the following discussion of reproduction in the conifers is admittedly arbitrary, yet not without some justification. Various species of pine are widely cultivated in many areas of North America and Europe and provide easily accessible material for study. Furthermore, prior to the publication of Allen and Owen's (1972) comprehensive monograph on the life history of the Douglas fir *(Pseudotsuga menziesii)*, the reproductive cycle of *Pinus* had been studied more comprehensively than that of any other conifer (Ferguson, 1904; Buchholz, 1918; Haupt, 1941; Johansen, 1950). As a result, it will be possible to present a rather complete and connected account of its cycle of reproduction. However, to ensure an understanding of the variations in the details of reproduction in the Coniferales as a whole, we also will compare *Pinus*, *Pseudotsuga*, and several other conifer genera.

Phenology

Many conifers, such as *Pseudotsuga* (Allen and Owens, 1972), are *both* pollinated and fertilized during the same season. The phenology of *Pinus*, however, is atypical among conifers in that a period of about twelve to fourteen months intervenes between the pollination of the ovules and the actual fertilization. If the period of initiation and early development of the strobili is included, the *complete* life cycle of *Pinus* extends over a period of three years, rather than the two-year period common to many other conifers.

The main events in the life cycle of pine are shown by the diagram in Fig. 17-30. To be explicit, we have depicted the cycle as beginning with cone initiation in 1972 and culminating with the maturation of the ripe seeds in 1974. We have deliberately avoided indicating specific months at which the various processes take place because there is considerable fluctuation and variation, depending on the latitude, altitude, and weather, of different parts of the geographical range of such a large genus as *Pinus*. This is illustrated clearly with respect to the exact period at which megasporangiate cones are initiated in different species of *Pinus*. Gifford and Mirov (1960) found that young ovulate cones could be identified within the terminal buds of *Pinus ponderosa* in mid-September in the vicinity of Placerville, California. In contrast, Konar and Ramchandani (1958), in their study of the Himalayan species *Pinus wallichiana*, reported that the ovulate cones are initiated in January. There are also interspecific differences in the exact timing of other events, which, of course, are not shown in the generalized life-cycle diagram illustrated in Fig. 17-30.

Sporogenesis

MICROSPOROGENESIS. The entire developmental history of the microsporangiate strobilus in certain species of pine extends over a period of approximately a year (Ferguson, 1904). The strobili are initiated in the axils of scale leaves in the spring or

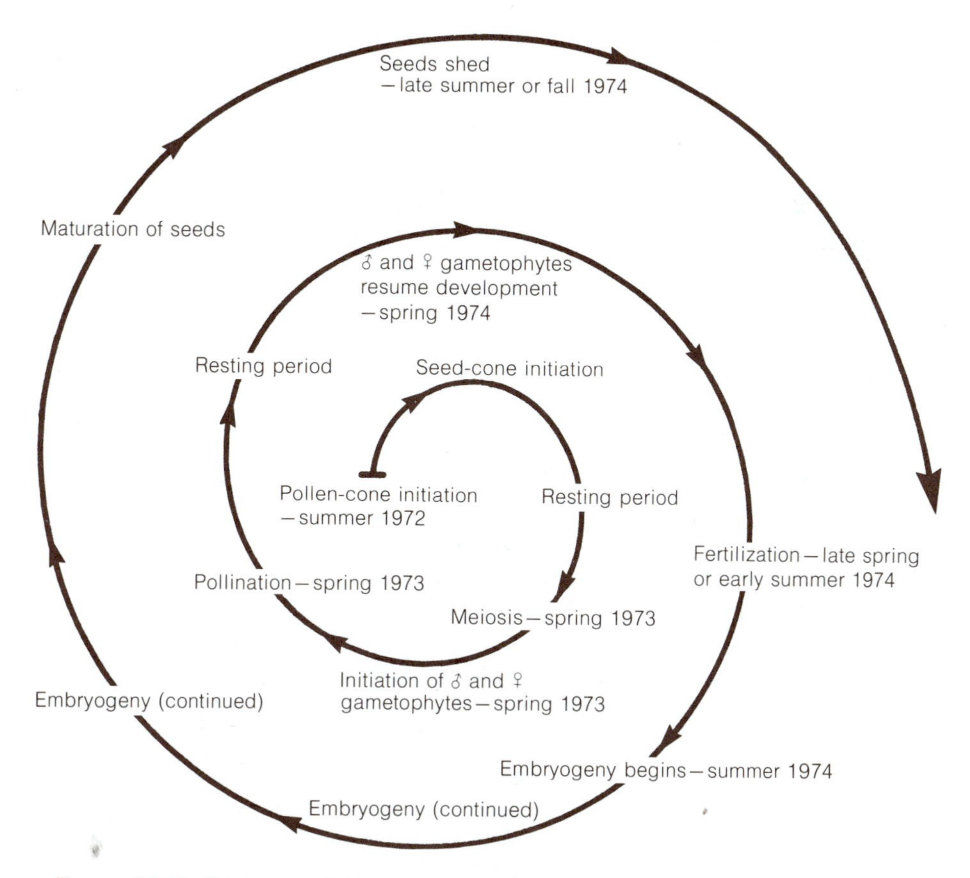

FIGURE 17-30 Diagrammatic representation of a complete reproductive cycle in *Pinus*. See text for detailed explanation.

early summer, and by winter the microsporangia contain well-defined sporogenous tissue. Meiosis and the formation of the characteristic winged pollen grains, however, do not occur until the following spring (see Fig. 17-30). Each functional microsporocyte gives rise to four haploid microspores, which remain enclosed within the original wall of the microsporocyte for some time (Fig. 17-31, A). According to Ferguson, the two bladders or wings of each spore are formed by the separation of the outer (exine) and inner (intine) layers of the microspore wall while the members of a spore tetrad are still surrounded by the common microsporocyte wall. The outer surface of the mature microspore, especially the wings, exhibits a reticulate sculpture (Fig. 17-32, A). The lower end of the spore wall, between the wings, is relatively thin and smooth; this region is the point at which the pollen tube later emerges (Fig. 17-31, B–F).

MEGASPOROGENESIS. Before describing the process of megasporogenesis, it is essential to review briefly the ontogeny of the ovuliferous scale in *Pinus*. Figure 17-23, A, represents somewhat diagrammatically a longisection of a young megasporangiate strobilus enclosed within a series of overlapping bud scales. In *Pinus ponderosa,* this stage in development is reached during the month of May (Gifford and Mirov, 1960). The young strobilus at this time consists of an axis bearing a series of bracts; in the axil of each is the primordium of an ovuliferous scale (Fig. 17-27, C). As in the microsporangiate strobilus, however, the differentiation of the young ovulate cone may be interrupted by the onset of winter. According to Ferguson (1904), the ovule development in certain pines does not occur until April or late May of the year following the initiation of the cone. Prior to megasporogenesis, the pine ovule consists of a nucellus and a

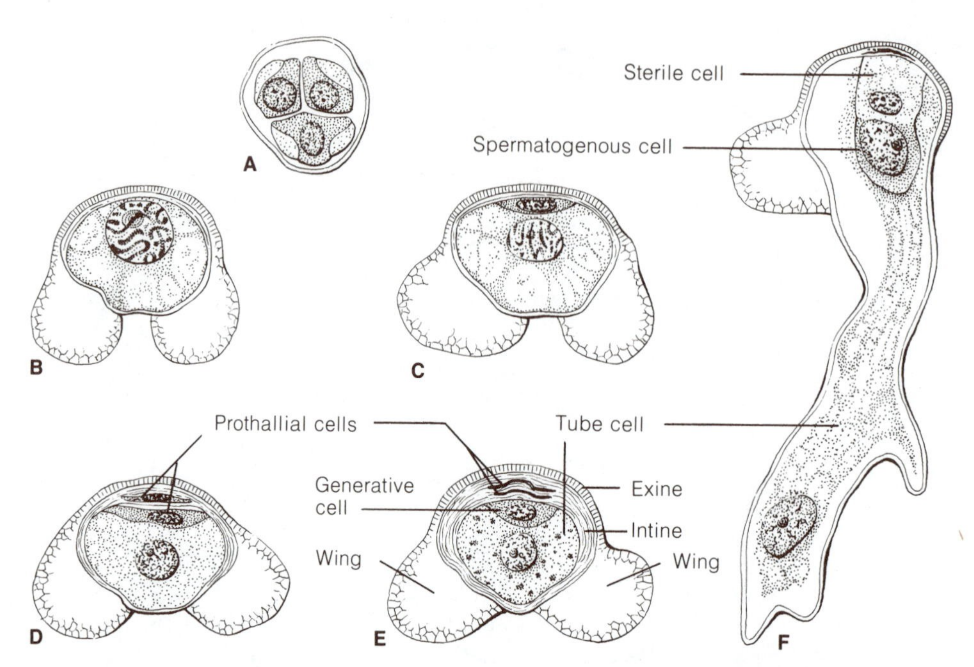

FIGURE 17-31 Development of the male gametophyte in *Pinus nigra (laricio)*. **A**, tetrad of young microspores enclosed within wall of microsporocyte (note developing wings of microspores); **B**, prophase of first division of microspore nucleus; **C**, the first prothallial cell has been formed and the microspore nucleus is preparing to divide again; **D**, a second prothallial cell has been formed; **E**, the microspore nucleus ("antheridial initial") has divided, producing the tube and generative cells; at this four-celled stage the male gametophyte is liberated from the microsporangium; **F**, young pollen tube as it would appear in the nucellus of ovule after pollination (note that generative cell has divided to form the sterile and spermatogenous cells). [Redrawn from *Gymnosperms. Structure and Evolution* by C. J. Chamberlain. University of Chicago Press, Chicago. 1935.]

single integument; the position of the ovule is inverted and the conspicuous micropyle points inward, toward the cone axis (Fig. 17-33).

In the species of *Pinus* that have been intensely studied, the single megasporocyte of each ovule just prior to meiosis is located deep within the nucellus (Fig. 17-33, A). According to Konar's (1960) study on *Pinus roxburghii*, a hypodermal initial cell, at the apex of the nucellus, divides periclinally into an outer *primary parietal cell* and the *megaspore mother cell.* As a result of the numerous divisions of the primary parietal cell, the megasporocyte ultimately becomes deeply embedded in the nucellus.

Theoretically the meiotic division of the megasporocyte should yield a tetrad of four haploid megaspores. Ferguson's (1904) detailed investigations, however, showed variation in the number of cells produced by the megasporocyte, not only among different species of pine but even within the same species. In *Pinus austriaca* a linear tetrad of four megaspores results from the division of *each* of the first two cells produced by the division of the sporocyte. In *Pinus strobus* and *Pinus rigida*, however, only three cells are formed. After meiosis I, the cell of the resulting dyad closest to the micropyle fails to divide. The cell furthest away from the micropyle does undergo meiosis II, resulting in a row of three cells. Regardless of these variations, the cell furthest from the micropyle becomes the functional haploid megaspore, and the two or three other cells soon degenerate.

The Male Gametophyte and Pollination

As in the cycads and *Ginkgo*, the early development of the endosporic male gametophyte takes

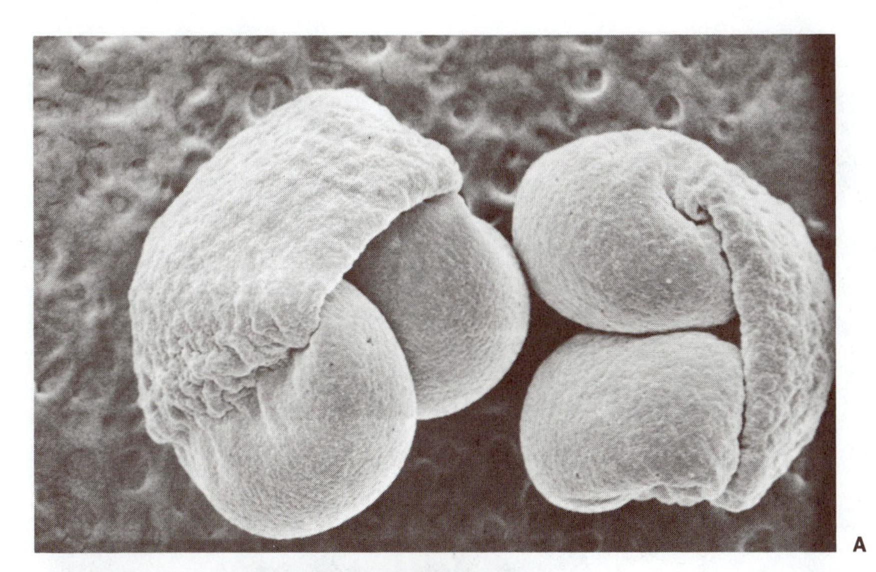

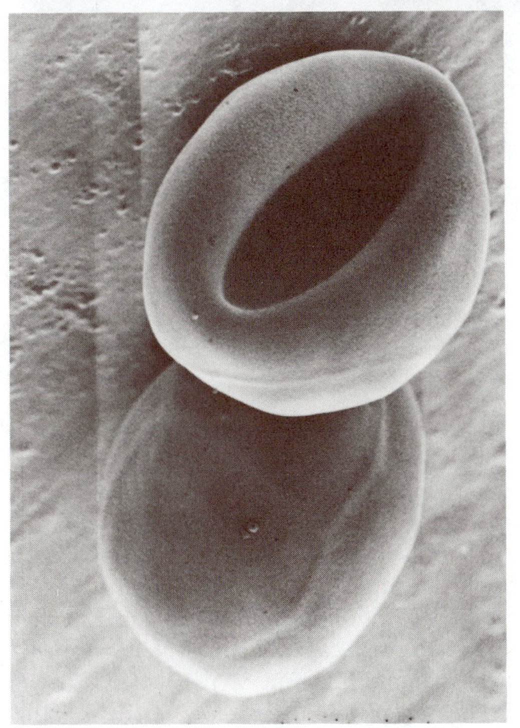

FIGURE 17-32 Scanning electron micrographs of conifer pollen. **A**, winged (vesiculate) pollen grain of *Pinus banksiana*, × 1,000; **B**, wingless pollen grains of *Pseudotsuga menziesii*, × 400. [Courtesy of Dr. G. Breckon.]

place before the dehiscence of the microsporangium. As a result of three successive nuclear divisions, the young male gametophyte consists of two prothallial cells (which soon become flattened and dead), a generative cell, and a tube cell. This is the usual stage in *Pinus* for the liberation of the pollen grains into the air (Fig. 17-31, E).

The ovules of *Pinus* are wind pollinated. During the late spring or early summer, the axis of the young megasporangiate cone elongates and the ovuliferous scales become separated. In a series of elegant experiments, Niklas (1982, 1984, 1985) has shown that scale-bract morphology and cone geometry play an important role in wind pollination.

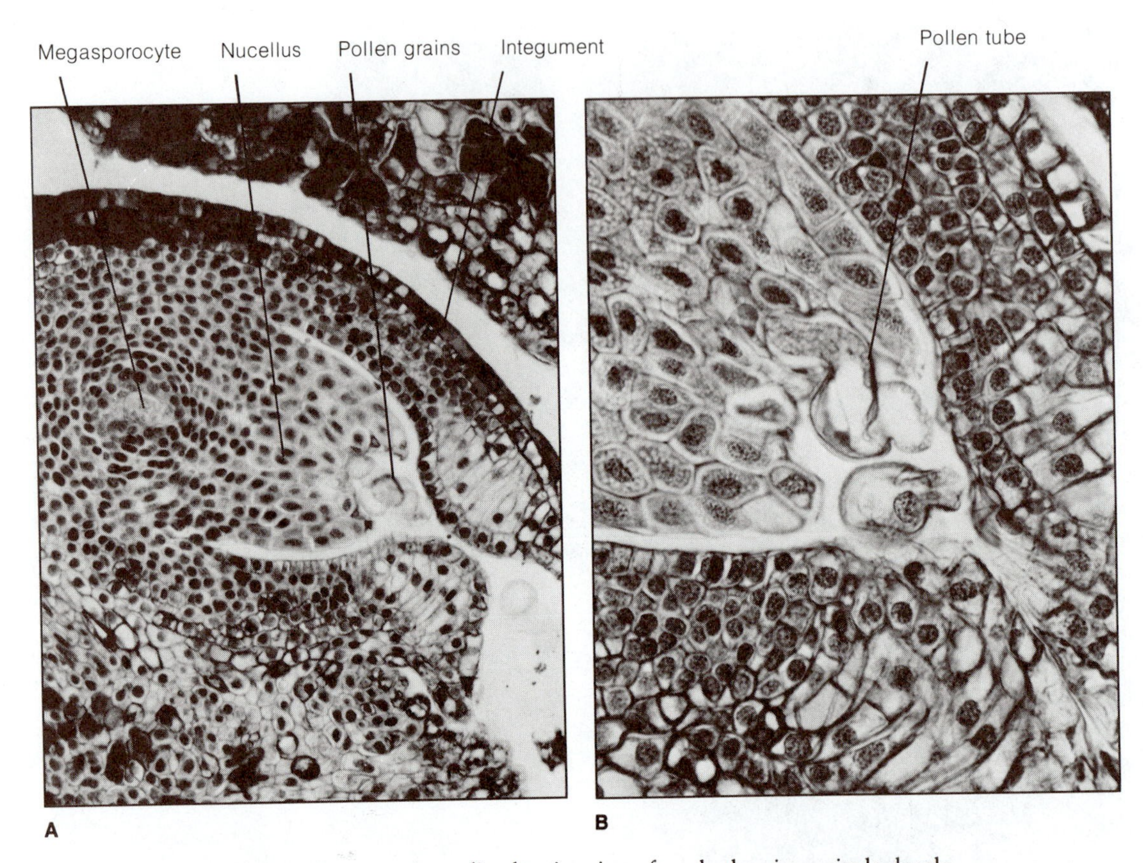

FIGURE 17-33 *Pinus* sp. **A**, median longisection of ovule showing a single deeply embedded megasporocyte (note large radially elongated cells of the middle layer of the integument and the pollen grains); **B**, longisection of ovule showing young pollen tube penetrating the nucellus.

The importance of these morphological features were determined by means of a bubble-streak method. In this technique, helium-filled bubbles were discharged into a wind tunnel toward models of cones. The motion of bubbles around and in the interior of the cone is seen as streaks when illuminated and photographed. The paths of wind-borne pollen grains around real ovulate cones of pine, fir, and larch correlate well with the bubble-streak method. In an air current, pollen grains roll along cone scales and then become redirected backward toward the ovules by the airflow pattern. Pollen grains passing around the cone are deflected back toward the cone where they either settle on cone scales or are deflected away by airflow passing under the cone. The orientation of a cone with respect to the direction of airflow also can affect pollination.

The pollen grains adhere to the pollination drops, which exude from the open ends of the *inverted* ovules. According to Doyle and O'Leary (1935), a pollen grain is very buoyant; at first it floats on the surface of the pollination drop and then begins to rise *upwardly* in the liquid, which fills the micropylar canal of the ovule. Doyle emphasizes that the function of the wings is to so orient the pollen grain that the end *away from* the prothallial cells is ultimately brought into direct contact with the exposed surface of the nucellus. In this position, the pollen grain germinates and forms a pollen tube.

Ferguson (1904) found that the free end of the nucellus is slightly depressed in *Pinus,* and that the pollen grains lie in this shallow cavity. According to her observations, the radial elongation of cells of the middle layer of the integument, a short distance above the apex of the nucellus, forms a rimlike outgrowth that seals off the outer part of the micropylar canal (Fig. 17-33, A). After pollination, the ovuliferous scales become drawn together and remain closely appressed until the seeds are released from the mature cones. In the so-called closed-cone pines (e.g., *Pinus attenuata),* heat produced by a forest fire is usually required to separate the cone scales and liberate the seeds.

A salient feature of the reproductive cycle in *Pinus* is the relatively long interval between pollination and fertilization. In many pines, this interval is about twelve months (Fig. 17-30). At the time of pollination, the male gametophyte consists of four cells and megasporogenesis has just begun in the ovule. Following pollination there is a slow period of development of the male gametophyte; the pollen tube emerges and the generative cell divides to form a sterile cell and a spermatogenous cell (see diagrams of male-gametophyte development shown in Figs. 17-34 and 17-31, F).

Concomitantly the single functional megaspore within the nucellus of the ovule enlarges, and there is a series of free nuclear divisions. According to Ferguson, about five successive nuclear divisions, yielding thirty-two nuclei, take place before the onset of dormancy. Thus, both the male and female gametophytes are in a comparatively early stage of development throughout the winter, and male gametes and archegonia do not form until the following spring.

The Female Gametophyte

The development of the female gametophyte of *Pinus* resembles in many respects that of the cycads and *Ginkgo.* At the close of the first period in its development, the young female gametophyte of *Pinus* is a more-or-less spherical sac, bounded by the megaspore wall and containing about thirty-two free nuclei embedded in a parietal layer of cytoplasm surrounding a large central vacuole. When development resumes in the spring, there is a very rapid formation of additional free nuclei. Ferguson estimated that about 2,000 free nuclei are present in the female gametophyte of *Pinus strobus* at the time when cell walls begin to form.

The method whereby the multinucleate female gametophyte becomes cellular is very similar to that occurring in cycads and *Ginkgo* (see Chapter 15, Fig. 15-27); that is, the centripetal development of tubular alveoli that become converted into files of cells. This regular arrangement may be obscured later by cell divisions in various planes.

The development of archegonia may be deferred until the female gametophyte is completely cellular, but not infrequently archegonia are initiated before wall formation has terminated. According to Ferguson, archegonia are detectable about two weeks before fertilization. The ontogeny of the archegonium is markedly like that of a cycad (Fig. 15-25, F–H). The initial, which is a superficial cell differentiated at the micropylar end of the gametophyte, divides periclinally into a small outer primary neck cell and a larger inner central cell. By means of two successive anticlinal divisions, the primary neck cell forms four neck cells, a condition apparently rather

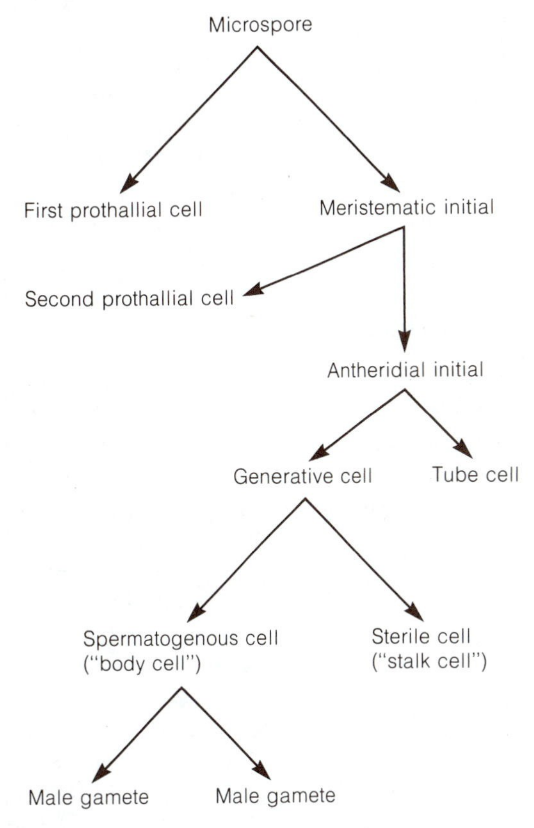

FIGURE 17-34 Diagram showing the successive stages in development of the male gametophyte in *Pinus.* Male gametophyte development is similar in *Ginkgo biloba.* [Adapted from Sterling, *Biol. Rev.* 38:167, 1963.]

common in *Pinus strobus.* In other species, however (e.g., *Pinus austriaca, Pinus rigida,* and *Pinus resinosa),* these four neck cells may divide periclinally, forming an eight-celled neck with the cells arranged in two tiers. Apparently the number of neck cells varies somewhat even in the same species. The central cell enlarges, but its nucleus remains close to the neck of the archegonium. Ultimately the nucleus of the central cell divides, forming a small ventral canal cell which is separated by a definite wall from the large egg cell. The ventral canal cell eventually degenerates, and the nucleus of the egg cell becomes enlarged and descends to a central position in the egg. When mature, the egg cell of the archegonium is jacketed by a distinct layer of cells, and contains many large and numerous small inclusions. The large inclusions are formed by the invagination of plastids, resulting in the enclosure of cytoplasm and organelles (Camefort, 1968). They exhibit a positive protein-staining reaction. The small inclusions are rich in phospholipids.

The number of archegonia produced by a single gametophyte varies considerably, depending on the species. According to Johansen (1950) there is usually a single archegonium in the Monterey pine *(Pinus radiata),* whereas in the majority of the other species investigated the number of archegonia ranges from two to six (Fig. 17-35).

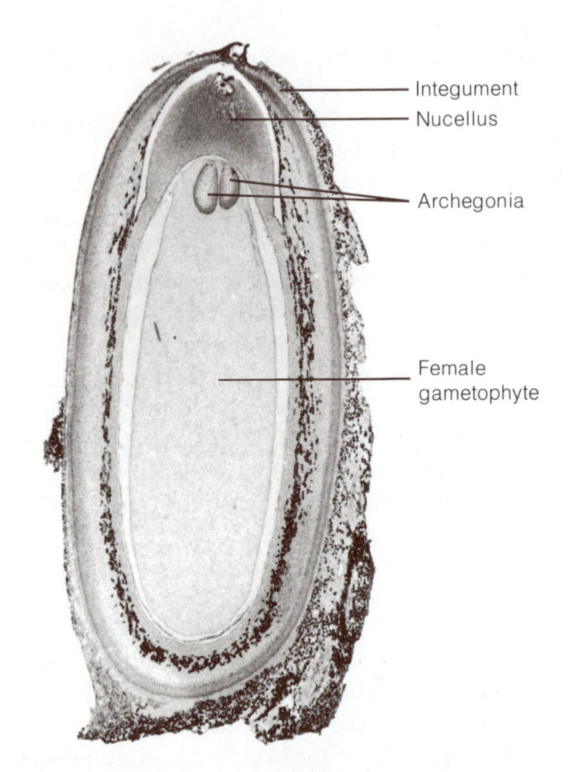

FIGURE 17-35 *Pinus* sp. Median longisection of ovule showing female gametophyte with two archegonia.

Fertilization

During the resumption of growth of the female gametophyte in the spring, the sterile cell and the spermatogenous cell of the male gametophyte move down toward the lower end of the pollen tube. The enzymes pectinase and cellulase are secreted from the pollen tube and dissolve the cell walls of the nucellus, facilitating the growth of the tube (Willemse and Linskens, 1969). Additional enzymes also have been identified, e.g., acid phosphatase, protease, and amylase (Pettitt, 1985). Approximately a week before fertilization, the spermatogenous cell divides to form two male gametes, which are reported to be unequal in size (Haupt, 1941; McWilliam and Mergen, 1958). In marked contrast with the large sperm found in *Ginkgo* and the cycads, the male gametes in *Pinus* and other conifers are small and devoid of flagella. The pollen tube continues its growth down through the nucellus (Fig. 17-36, A). Since several pollen grains may be present at the apex of the nucellus, a corresponding number of tubes may be formed; usually only two or three tubes, however, reach the female gametophyte. The tip of a pollen tube forces itself between the neck cells of an archegonium and then ruptures, discharging the two male gametes, the tube nucleus, and the sterile cell into the cytoplasm of the egg. Fertilization consists of the fusion of the nucleus of the larger of the two male gametes with the nucleus of the egg cell (Fig. 17-37, A). Usually the smaller male gamete, the tube nucleus, and the sterile cell can be seen at the top of the egg after sexual fusion has been accomplished. Eventually they degenerate.

Embryogeny and Seed Development

The early embryogeny of *Pinus,* like that of the majority of the Coniferales, is characterized by a short period of free nuclear divisions. The number of nuclei varies, according to the family, from eight

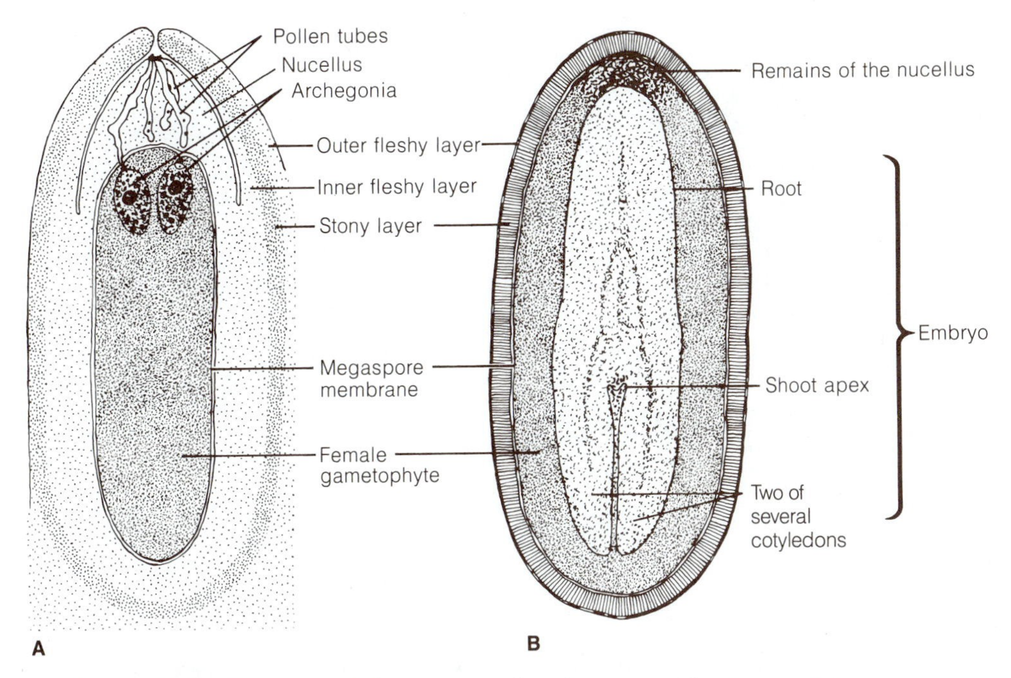

FIGURE 17-36 *Pinus nigra.* **A,** longisection of ovule at time of fertilization; **B,** longisection of mature seed. [Redrawn from *Gymnosperms. Structure and Evolution,* by C. J. Chamberlain. University of Chicago Press, Chicago. 1935.]

to sixty-four. Proembryogeny in *Pinus* differs from the more basic plan in conifers. However, to maintain the general theme of our discussion we will describe embryogeny in *Pinus*. For a description of the more basic plan in conifers, refer to p. 446. Following syngamy, the nucleus of the fertilized egg contains two distinct haploid groups of chromosomes, one paternal, the other maternal (Fig. 17-37, B). These two sets of chromosomes soon become arranged at the equatorial region of a common spindle, and at this time Ferguson (1904) and Haupt (1941) were able to count twenty-four chromosomes, apparently the typical diploid number for various species of *Pinus* (Fig. 17-37, C).

The division of the zygote nucleus yields two nuclei, each of which promptly divides, forming a total of four free nuclei located near the middle of the cytoplasm of the egg. These four nuclei represent the extent of the free nuclear period in *Pinus*. Soon after their formation, they move to the lower end of the archegonium where a third synchronized division accompanied by wall formation occurs (Fig. 17-37, D, E). The first wall formed is transverse to the long axis of the archegonium and separates the eight nuclei into two tiers, then vertical walls appear separating the nuclei in each tier; the proembryo now consists of a lower tier of four cells, completely bounded by walls and an upper tier devoid of walls adjacent to the cytoplasm of the egg (Fig. 17-37, E). The next synchronized division usually occurs in the cells of the upper tier and is followed by a similar transverse division in the lower tier. The *proembryo,* as it is termed, now consists of sixteen cells arranged in four superposed tiers (Fig. 17-37, F, G). This tiered arrangement of the cells of the proembryo is found in many genera of the Coniferales, and the precision with which it originates offers a marked contrast with the comparatively unstratified cell arrangement characteristic of the early embryogeny of the cycads and *Ginkgo* (see Chapter 16).

An interesting developmental event occurs at the first free nuclear division of the zygote nucleus. For several conifers a distinctive type of cytoplasm remains between and around the two resulting nuclei (see Singh, 1978, for a review). This ground substance has been termed neocytoplasm ("new" cytoplasm) and appears to represent nucleoplasm from

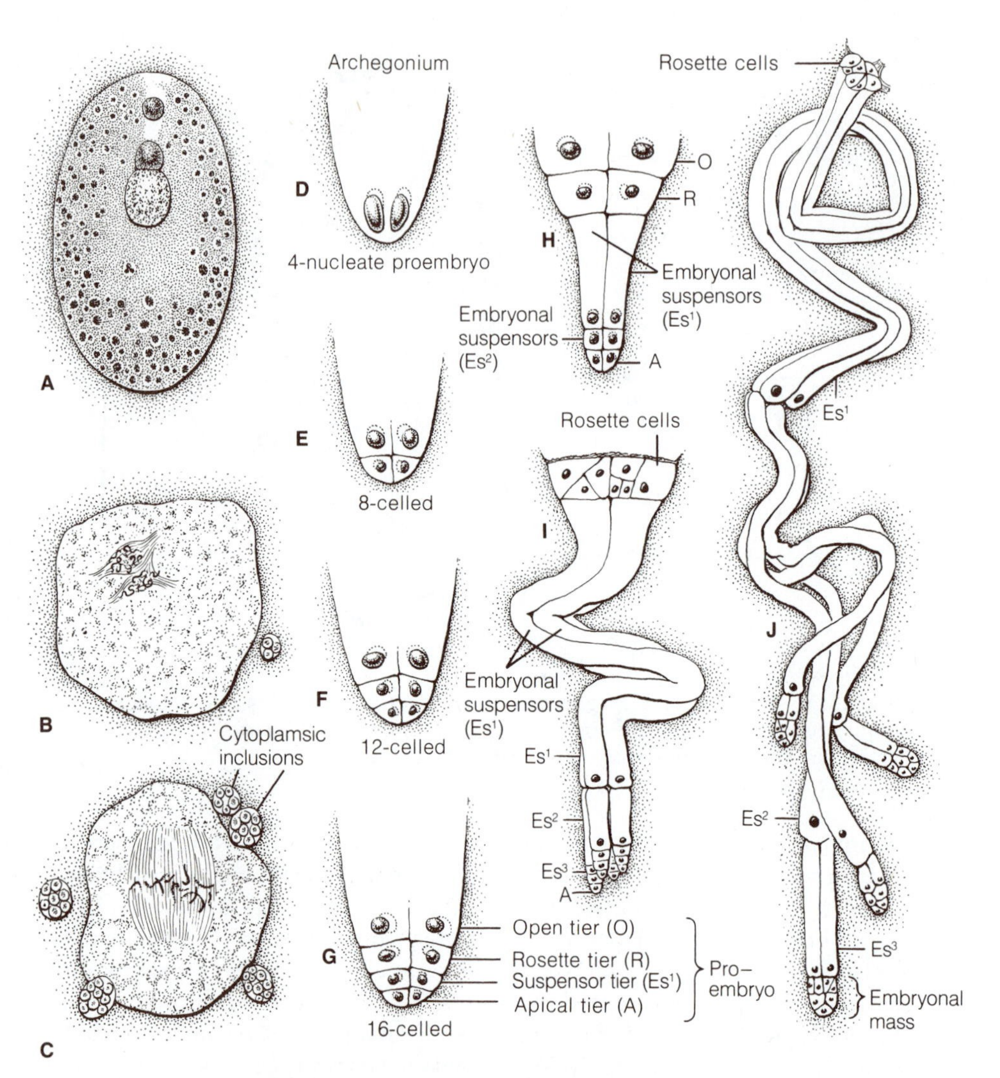

FIGURE 17-37 Fertilization and early embryogeny in *Pinus*. **A**, longisection showing the larger of the two male gametes in contact with egg nucleus (note abundant protein bodies in cytoplasm of egg cell); **B**, fertilized egg cell showing paternal and maternal groups of chromosomes within the egg nuclear membrane; **C**, chromosomes derived from paternal and maternal nuclei arranged at equatorial region of a common spindle; **D**, free nuclear phase in embryogeny with four nuclei; **E–G**, development of the four superposed cell tiers of the proembryo; **H**, elongation of the first embryonal suspensor cells (Es^1) and the formation of a second tier of embryonal suspensor cells (Es^2) that are derived from the apical tier (A); **I**, early separation of the proembryo into four potential embryos (cleavage polyembryony); note formation of additional embryonal suspensor cells (Es^3); **J**, series of four competing embryos formed by cleavage. [**A–C**, redrawn from Haupt, *Bot. Gaz.* 102:482, 1941; **D–J**, modified from Buchholz, *Trans. Illinois Acad. Sci.* 23:117, 1931.]

the male and egg nuclei. The neocytoplasm becomes progressively organized during the migration of the four nuclei to the base of the archegonium. Organelles become apparent in the neocytoplasm. In pine the mitochondria are apparently of maternal origin, and there is evidence that plastids are of paternal origin (Willemse, 1974); in *Larix* (larch), the plastids are definitely of paternal origin (Camefort, 1968, 1969). The neocytoplasm alone is the source of the proembryo cytoplasm; the remaining egg cell cytoplasm degenerates.

Polyembryony

One of the outstanding features of the later phases of embryogeny in *Pinus* (and in many other conifers) is the development, from the walled cells of the proembryo, of from four to eight separate competing embryos. This early formation of multiple embryos becomes particularly remarkable when it is recalled that embryogeny in pine — as in all vascular plants — begins with a single fertilized egg cell.

To understand clearly the process of *polyembryony* it is essential to begin with the further development of the cell tiers of the sixteen-celled proembryo shown in Fig. 17-37, G. The uppermost cell tier, sometimes termed the *open tier,* is in open communication with the egg cytoplasm and possibly serves for a short time to transmit reserve food materials to the lower portion of the proembryo; soon, however, the cells of this tier disintegrate. The next tier comprises the so-called *rosette* cells which may undergo several divisions but do not give rise to organized embryos, as was originally believed (Doyle, 1963).

Polyembryony in *Pinus* results from the ultimate separation of the lower cell tiers (apical cells) into four filamentous embryos (Fig. 17-37, H, I). This process is termed *cleavage polyembryony.* Very soon after the four tiers of the proembryo are formed, the cells next to the apical cells elongate markedly (Fig. 17-37, H). These are the first formed *embryonal suspensor* cells (Es^1), and their vigorous extension forces them and the apical tier of cells through the wall of the archegonium into the female gametophyte (Fig. 17-38). Starch-containing cells in the upper part of the female gametophyte break down, resulting in the formation of a "corrosion" cavity into which the growing system of embryos intrudes. Because of the limited confines of the corrosion cavity, the first formed embryonal suspensors (Es^1) soon become coiled and buckled (Fig. 17-37, I, J). During the early elongation of Es^1, cells of the apical tier give rise, by a series of transverse divisions, to several additional embryonal suspensors (Es^2 and Es^3, in Fig. 17-37, H–J). These cells quickly elongate like those of Es^1. Following the formation of the Es^2 embryonal suspensors, the lower end of the embryo system cleaves into four distinct series of cells. Each series consists of an apical cell and two or more embryonal suspensors, and represents an independently developing embryo (Fig. 17-37, J). Very soon derivatives of the apical cells undergo divisions in various planes to form a multicellular embryonal mass from which numerous additional suspensorlike cells, termed *embryonal tubes,* are formed.

Within the system of embryos derived by the cleavage of one proembryo, there is apparently intense competition. Only one of the four embryos continues to develop and becomes the differentiated embryo of the pine seed; the other embryos generally abort and cannot be detected in the mature seed.

Each proembryo in *Pinus* is theoretically capable of forming embryos from the cleavage process. If all six of the archegonia of a single female gametophyte are fertilized, an extraordinary number — as many as twenty-four — of separate embryos might begin development. These embryo systems grow in competition for variable periods of time, but only a single embryo among these normally reaches maturity in the ripe seed.

The later stages of development of the successful embryo are complex. According to Buchholz the terminal cell of the embryo soon takes on the character of a pyramidal apical cell which forms derivative cells or segments very much like the apical cell of *Equisetum* or a leptosporangiate fern (Fig. 17-37, J). Buchholz (1931) and Chamberlain (1935) regarded the presence of an apical cell in conifer embryogeny as a "primitive character." After the main body of the pine embryo consists of several hundred cells a definitive apical cell is no longer apparent, and the extreme apex of the embryo is occupied by a group of apical initials. The further

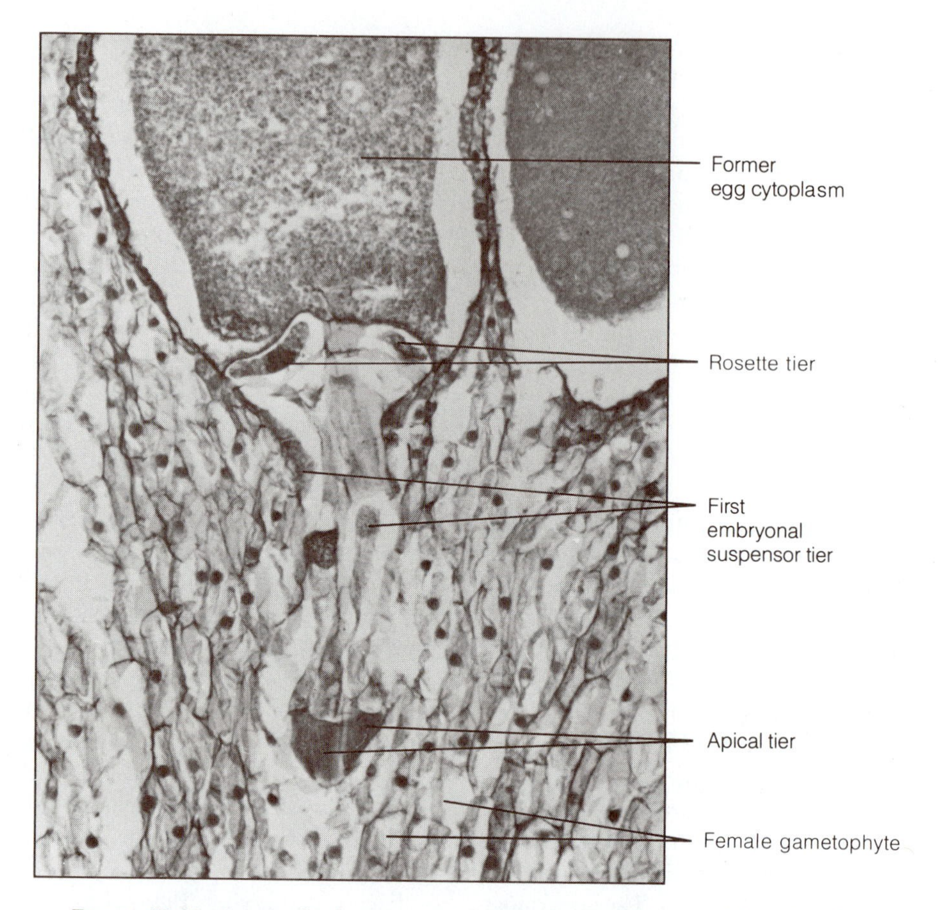

FIGURE 17-38 Longitudinal section of the proembryo of *Pinus* sp. Note that the elongation of the first embryonal suspensor cells has forced the apical tier deep into the tissue of the female gametophyte.

histogenesis of the embryo includes the differentiation, at the suspensor end, of the initial cells of the root apex, and the ultimate formation of a series of cotyledon primordia from the shoot apex region (Fig. 17-36, B). (For a detailed description of tissue and organ formation during the later stages of embryogeny refer to the work of Spurr, 1949).

The fully developed embryo of *Pinus* consists of a whorl of cotyledons (the average number in the genus, according to Butts and Buchholz, 1940, is 8.1) surrounding the shoot apex, a short hypocotyl, and a primary root or radicle. The embryo is embedded in the tissue of the massive female gametophyte which, in turn, is surrounded by the nucellus and seed coat (Fig. 17-36, B). The seed coat consists primarily of a hard outer coat, which is derived from the stony layer of the integument. The inner fleshy layer of the integument degenerates during seed development, and in the ripe seed is reduced to a thin papery membrane. Usually the remains of the nucellus can be seen at the micropylar end of the female gametophyte.

Although some kinds of pine (e.g., piñon pine) produce seeds devoid of wings (Fig. 17-39, A), in most species the seed, at the time it is shed from the cone, is attached to a thin membranous wing which apparently aids in its dispersal (Fig. 17-39, B). The seed wing, according to Sporne (1965), is formed by the separation of a portion of the adaxial surface of the cone scale adjacent to the ovule, and hence, morphologically, is not part of the seed itself.

When the pine seed germinates, the entire embryo emerges from the ruptured seed coat, which may adhere for a short time to the tips of the coty-

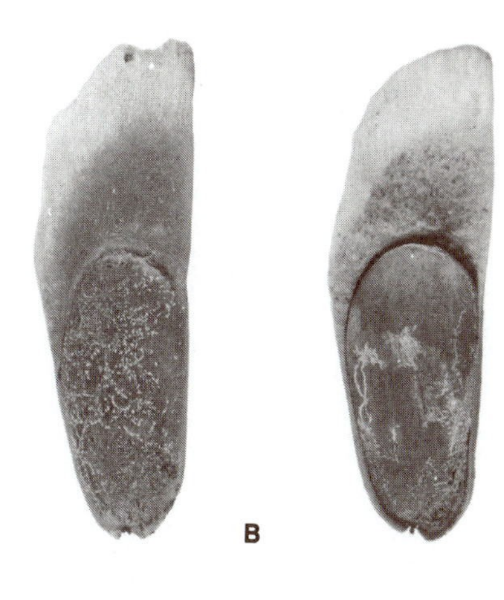

FIGURE 17-39 **A,** *Pinus monophylla* (piñon pine); portion of an open cone, showing pairs of wingless seeds attached to the adaxial surface of the reflexed ovuliferous scales; **B,** *Pinus sabiniana* (digger pine); seeds with prominent wings.

ledons (Fig. 17-40, A). The primary shoot formed by the terminal bud of the seedling at first bears only a spiral series of needlelike leaves (Fig. 17-40, B). Later the first spur shoots arise in the axils of some of the primary leaves.

Sacher (1954) has described in detail the zonal structure of the shoot apex in *Pinus lambertiana* and *Pinus ponderosa*. The structure of the apex of the latter species was considered in our account of the ontogeny of the ovuliferous cone (see p. 426 and Fig. 17-26).

Comparisons between *Pinus* and Other Conifers

In this book it is not possible to present a detailed comparison between the gametophytic and embryological features of *Pinus* and those of other conifers. For broad comparative treatments of the gametophytes and embryos of a very wide range of conifer genera, the reader is referred to Chamberlain's (1935) review of the older literature and to the more recent reviews of Sterling (1963), Maheshwari and Singh (1967), Doyle and Brennan (1972), and Singh (1978). The purpose of this account is to emphasize some of the most outstanding differences between *Pinus* and other conifers with reference to (1) the development and structure of the male gametophyte, (2) the types of pollination mechanisms, (3) the development and behavior of the proembryo, and (4) polyembryony.

The Male Gametophyte

Although the formation of two prothallial cells appears to be a consistent feature of the male gametophyte of *Pinus, Pseudotsuga* (Fig. 17-41, A), and other members of the Pinaceae, prothallial cells are absent from the pollen grains of the Taxodiaceae, Cupressaceae, Cephalotaxaceae, and Taxaceae. In these families, the first division of the microspore nucleus gives rise directly to the generative cell and the tube cell (Sterling, 1963). At the opposite extreme, the very numerous prothallial cells, typical of the Araucariaceae, have attracted study and aroused speculation about their phylogenetic significance (Burlingame, 1913; Eames, 1913; Chamberlain, 1935). In *Araucaria,* for example, the first two prothallial cells, instead of becoming senescent, as in members of the Pinaceae, undergo active divisions and form at first several tiers of cells. Then

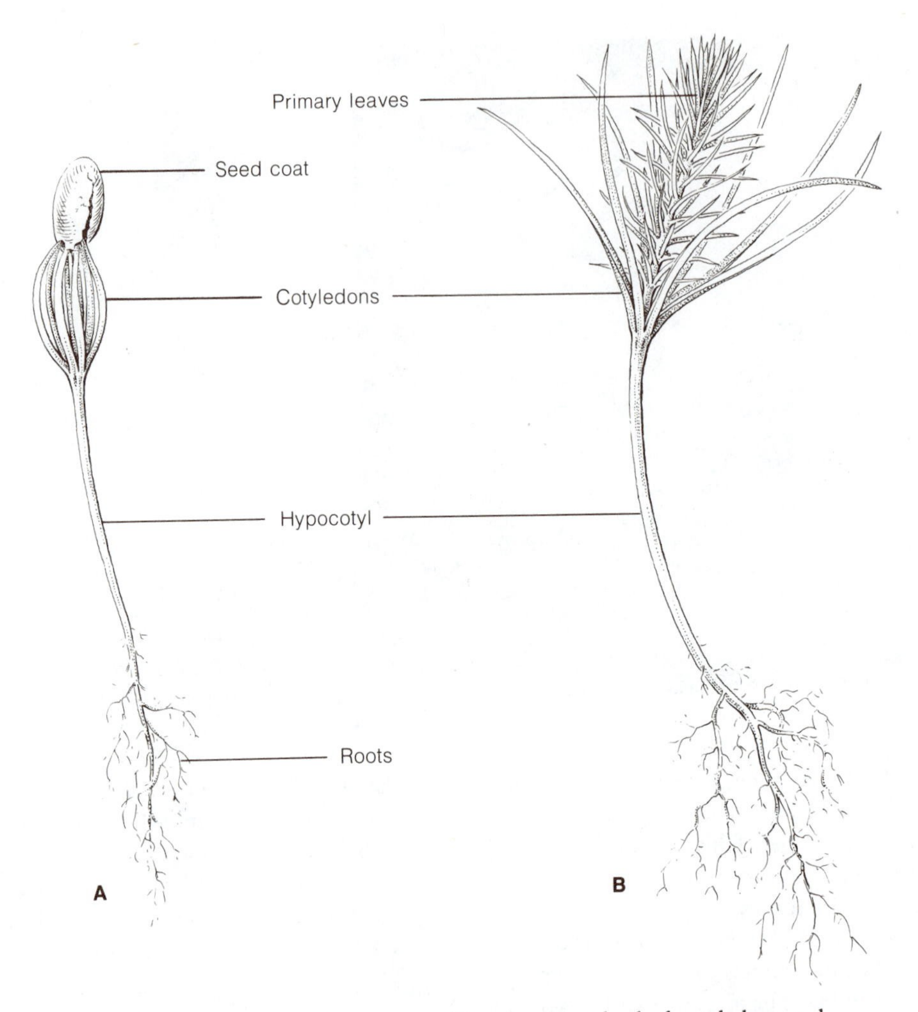

FIGURE 17-40 *Pinus edulis.* **A**, young seedling showing whorl of cotyledons and remains of seed coat; **B**, older seedling showing cotyledons and spirally arranged primary leaves. [Redrawn from *Gymnosperms. Structure and Evolution* by C. J. Chamberlain. University of Chicago Press, Chicago. 1935.]

the cell walls break down and the free prothallial nuclei are liberated into the general cytoplasm of the pollen grain (Fig. 17-41, B). There may be as many as forty prothallial cells (or nuclei) in the male gametophyte of *Agathis,* which is regarded by Zimmermann (1959) as the largest male gametophyte in living seed plants. The evolutionary significance of the large number of prothallial cells in the male gametophyte of the Araucariaceae is an unsettled question. Chamberlain (1935) regarded this as a primitive feature, whereas Eames (1913) believed that numerous prothallial cells in *Agathis* represent a phylogenetically derived condition.

In many conifers, the male gametes are represented as two nuclei (which may be associated with cytoplasm), as in *Pinus, Pseudotsuga,* and other members of the Pinaceae. But in certain genera of the Taxodiaceae and Cupressaceae, the male gametes have been described as well-defined cells after their formation in the spermatogenous cell (Singh, 1978). Aside from the absence of flagella, male gametes of this type resemble the young sperm of cycads and *Ginkgo.* As the pollen tubes of conifers function primarily to convey the male gametes directly to the archegonia, the apparently highly organized gametes of *Taxodium* and related plants

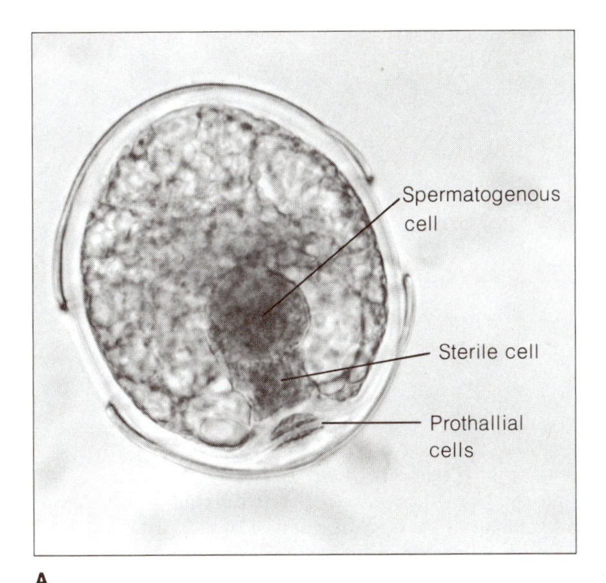

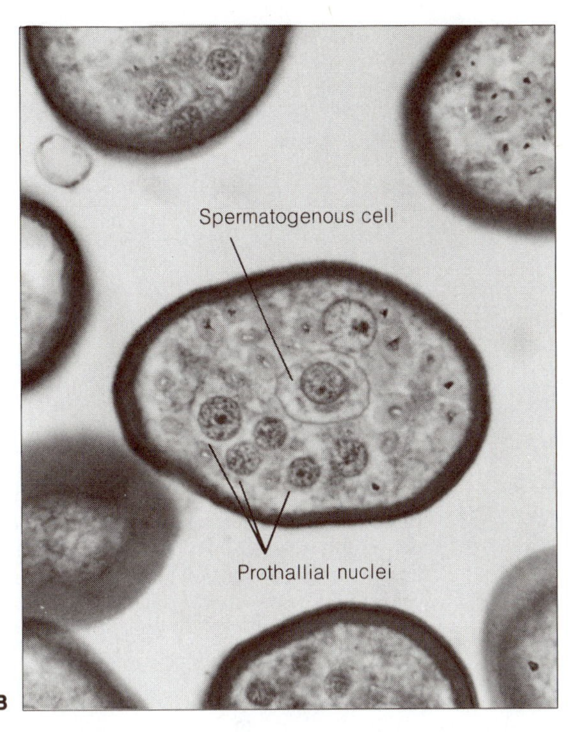

FIGURE 17-41 A, male gametophyte of *Pseudotsuga;* B, male gametophyte of *Araucaria.* See text for explanation.

may conceivably represent gametes that only recently—from a phylogenetic standpoint—have "lost" their locomotor apparatus. In this connection, it should be noted that Christiansen (1969) reported that he observed "ciliated" sperm in *Pseudotsuga* (Douglas fir) but this was not confirmed by Allen and Owens (1972).

Types of Pollination Mechanisms

The comprehensive investigations of Doyle (1945) revealed a surprising range in the types of pollination mechanisms in the living conifers. According to his interpretation, the exudation of pollination fluid, i.e., the pollination drop, and the formation of pollen grains with wings are salient and correlated aspects of the mechanism of pollination, not only in various members of the Pinaceae but in the conifers as a whole (Fig. 17-32, A). As we have previously noted, the function of the wings or bladders of the pollen grains of *Pinus* is to orient the grains as they rise through the fluid in the micropylar canal toward the nucellus of the inverted ovule.

According to Doyle, various "modifications" of the basal type, represented in *Pinus,* have evolved in the Coniferales as the result of (1) the loss of the pollination exudate, (2) the assumption of a "stigmatic" or receptive function by the free tips of the ovular integument, (3) the reduction and ultimate loss of the bladders or wings of the pollen grains, and (4) the germination of the pollen on the ovuliferous scale. *Pseudotsuga* may be used to illustrate one of the modified types of pollination mechanisms in the Pinaceae (see Fig. 17-42).

In *Pseudotsuga,* the pollen grains as shown in Fig. 17-32, B, are devoid of wings and there is no exudate of liquid from the ovule (Allen, 1963; Allen and Owens, 1972). The pollen grains, falling inwardly along the upper surface of the scales of the seed cone, land upon the sticky hairs of the unequal "lips" of the integument of the ovule (Fig. 17-42, B). At this time, a depression begins to form on the upper surface of the larger lip, apparently as the result of the collapse of surface and subsurface cells (Fig. 17-42, C–F). Some of the pollen grains sink into this shallow cavity and, as a result of the inward growth and approximation of the two lips, become enclosed within the upper end of the micropylar

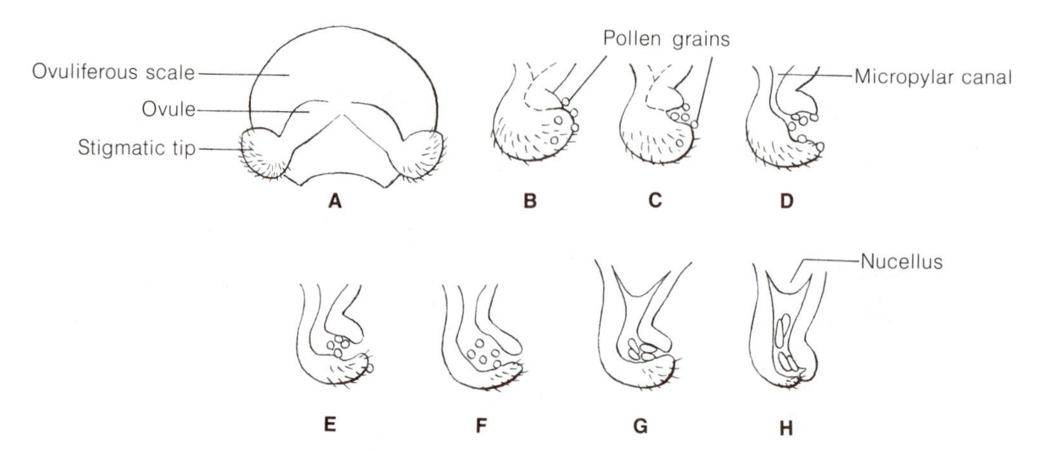

FIGURE 17-42 The pollination mechanism in Douglas fir *(Pseudotsuga menziesii)*. **A**, adaxial view of ovuliferous scale showing the stigmatic tips of the two ovules at the time of pollination; **B**, adherence of pollen to hairs on stigmatic tips; **C–F**, formation of depression in lower lip into which pollen grains sink; **G**, approximation of integumentary lips and enclosure of pollen grains in micropylar canal; **H**, elongation of pollen grains prior to the formation of pollen tubes. [Redrawn from *The Life History of Douglas Fir* by C. S. Allen and J. N. Owens. Information Canada, Ottawa, 1972.]

canal of the ovule (Fig. 17-42, G, H). In marked contrast to *Pinus*, the pollen grains of *Pseudotsuga* germinate within the micropylar canal rather than on the surface of the nucellus. A pollen tube does not form in *Pseudotsuga* until the young gametophyte has ruptured the exine of the spore wall and has become a much elongated structure. The pollen tube emerges only after the elongating gametophyte has come into contact with the nucellus (for further details and photographic illustrations, see Allen and Owens, 1972).

One of the most remarkable deviations from the *Pinus* or *Pseudotsuga* type of pollination mechanism occurs in the Araucariaceae. In this family, the pollen germinates on the ovuliferous scale—or in its axil—at a point far removed from the ovule. Eames (1913) describes in great detail how the pollen tubes of *Agathis* at first branch and grow into the tissue of the ovuliferous scale; some of the branches, in their haustorial growth, even penetrate the phloem and xylem tissues of the cone axis. Although it has been suggested that the germination of the pollen on the ovuliferous scale in araucarians may represent a primitive pattern in conifers, this idea is rejected by Doyle (1945), who maintains that the Carboniferous and Permian ancestors of the modern conifers "all had direct ovular reception of the pollen," and that hence the araucarian type of pollination mechanism should be regarded as a derived and advanced character.

The Development of the Proembryo

As compared with *Ginkgo* and the cycads, the period of free nuclear divisions of the proembryo in most conifers is brief and, as in *Pinus* and the majority of the Pinaceae, terminates after four nuclei have been formed. However, the pine type is not representative of the majority of conifers. In the more common type, the two nuclei resulting from division of the zygotic nucleus are enveloped by the "neocytoplasm" and migrate to the base of the archegonium. Additional free nuclei are then formed and lie in the dense neocytoplasm (Fig. 17-43, A, B). The number of nuclei varies from sixteen to thirty-two. Just before wall formation the nuclei take up a definite position in relation to the cell walls that arise later (Fig. 17-43, C). Following wall formation the proembryo consists of a group of *primary embryonal* cells and an upper group of cells organized in a single layer—the *primary upper tier* (Fig. 17-43, D). All of the cells then divide, which results in the doubling of the number of cells (Fig. 17-43, E). The proembryo now consists of a lower or distal

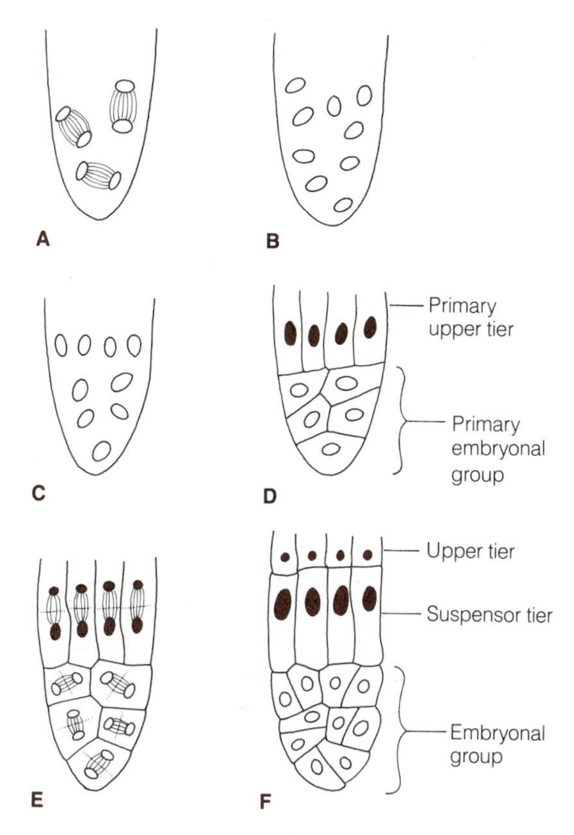

FIGURE 17-43 **A–F**, basic plan of conifer early embryogeny. **A–C**, free nuclear phase; **C**, **D**, cell walls have formed around nuclei; **E**, cell divisions leading to the establishment of three regions of the proembryo (**F**). See text for details. [Based on Doyle, *Proc. Roy. Irish Acad.* 62B:181, 1963.]

group of *embryonal* cells and an upper or proximal portion composed of an *upper tier* and a *suspensor tier* (Fig. 17-43, F). With continued growth, the suspensor cells elongate and additional embryonal tubes may be formed from the original embryonal group.

The early embryogeny of the Araucariaceae is rather unique. The free nuclear period in this family is relatively protracted, with thirty-two to sixty-four nuclei being produced before walls are formed. Unlike most conifers, the free nuclei do not migrate to the base of the archegonium but remain in the center where the proembryo organizes. A massive suspensor develops from the upper cells of the proembryo, whereas the outer cells of its lower portion form a protective cap, recalling the analogous structure formed during early embryogeny of *Zamia* (see Chapter 15, Fig. 15-29). The free end of the araucarian embryo, after the cap cells have degenerated, gives rise to the main body of the embryo (Johansen, 1950).

Another deviation from the basic plan is found in *Sequoia sempervirens* (coast redwood of northern California and southern Oregon). As far as is known, this taxon is the only conifer devoid of a free nuclear period in embryogeny. A transverse wall is formed at the division of the zygote nucleus, yielding a bicellular proembryo. However, in *Sequoiadendron giganteum* (Big Tree of the Sierra Nevada, California) embryogeny begins with the formation of eight free nuclei.

Polyembryony is widespread among members of the Coniferales and occurs in the majority of the genera that have been studied intensively. The occurrence of cleavage polyembryony, however, is variable even among genera of the same family. In the Pinaceae, for example, *Pinus, Cedrus,* and *Tsuga* exhibit cleavage polyembryony, whereas simple polyembryony seems to prevail in *Picea* and *Abies.*

Simple polyembryony arises when two or more fertilized eggs of one female gametophyte produce separate but undivided embryos. In other words, each proembryo produces a single embryo, and cells of the apical tier function as a unit without cleaving. All conifer genera show simple polyembryony, and in the great majority of them cleavage polyembryony also occurs regularly. *Pseudotsuga* exhibits a special type of simple polyembryony, sometimes termed *incipient cleavage.* The proembryo consists of three tiers of cells: an open tier in contact with the egg cytoplasm, a middle tier of suspensor cells, and an apical tier, for a total of twelve cells rather than sixteen as in pine (Fig. 17-37, G). Although the four cells of the apical tier may contribute more-or-less equally to the later development of the embryo, most commonly two of the cells of the apical tier become dominant and overtop the other two by their more active growth. Cells of the dominant pair undergo divisions and give rise to the main body of the embryo; there is no definite separation into two or more competing embryos (Allen, 1946a; Allen and Owens, 1972).

Although Buchholz (1926) regarded cleavage polyembryony as the primitive type of embryogeny in the conifers as a whole, this interpretation has been severely criticized by Doyle and Brennan (1972). They concluded that "the absence of cleavage was the primitive condition in conifers and taxads." From this standpoint, Doyle and Brennan maintain that cleavage polyembryony has arisen separately in many lines of conifers, and that the early embryogeny of *Pinus* "is not a prototype for conifers in general but shows a specialized derived condition." (For a detailed review on embryogeny of conifers see Chowdhury, 1962, and Singh, 1978.)

Embryogenesis and Biotechnology

There has been considerable interest in methods to propagate conifers vegetatively, especially trees that have desirable traits. *In vitro* propagation has been used with varying success to obtain clones of hybrids of controlled crosses. From callus tissue produced on excised seedling hypocotyls, cotyledons, young leaves, and stems, new plants can be produced which eventually can be placed in soil and later planted in the field. One difficulty in the procedure is that often shoots are formed but not roots. The shoots must be placed on a special rooting medium to complete organogenesis.

More recently, investigators have taken advantage of polyembryony to obtain numerous plantlets of loblolly and sugar pine. Embryos develop *in vitro* from a mucilaginous callus tissue derived from the suspensor cells of a zygotic embryo. This process is termed somatic polyembryogenesis. In general, the embryos pass through the usual phases of pine embryo development. Somatic embryos also have been produced by the hundreds from isolated protoplasts derived from proliferating embryonal-suspensor masses of loblolly pine (Gupta and Durzan, 1986, 1987a, b). For additional information on *in vitro* clonal propagation and embryogenesis see J. M. Bonga and D. J. Durzan, 1987.

Cordaitales

Members of the extinct Cordaitales occur earlier in the fossil record than the Voltziales. There is some evidence that the cordaites occurred in the Lower Carboniferous, and they formed large forests and were a dominant group in certain regions of the world during the Upper Carboniferous and the Permian. Some were shrubby, others were tall trees (some estimated to be 30 meters in height and 1 meter or more in diameter at the base). The trees were monopodial in growth habit; the leaves were strapshaped, helically arranged on the stem, and were up to 1 meter in length in some species (Fig. 17-44, A–C). The leaves had a parallel venation system, although dichotomies did occur.

The name *Cordaites* was originally used as a form genus for detached leaves, but the name is now generally used for the entire tree. The stems had a rather large pith and a eustelic primary vascular cylinder, similar to other coniferophytes. As was true of other coniferophytes, secondary growth occurred resulting in the production of a large amount of pycnoxylic secondary xylem.

The form genus *Cordaianthus* is used for microsporangiate (pollen) and megasporangiate (ovulate) reproductive structures. In both structures there was an elongate primary axis with bracts, in the axils of which were determinate fertile shoots (Fig. 17-44, B, D). In *Cordaianthus concinnus* a fertile shoot (or cone) consisted of many sterile scales of which the distal five to ten were fertile. Each fertile scale had six terminal microsporangia (pollen sacs) which were fused at the base (Fig. 17-44, D).

The ovulate fertile shoot also consisted of helically arranged scales; four to six of the distal scales were fertile, each bearing one or more bilaterally symmetrical ovules if the scale dichotomized (Fig. 17-45). Many examples of seed fossils have been found. The integument comprised three layers, the middle one consisting of thick-walled sclerified cells. Female gametophytes have been found in ovules with archegonia and a fleshy "tent pole" as in *Ginkgo* (Chapter 16, Fig. 16-9).

Conclusive evidence on the phylogeny of the Cordaitales is not available. Some paleobotanists favor the concept that the ancestors of the Cordaitales are to be found in the Upper Devonian progymnosperms because they, as well as the Voltziales and Coniferales, share several coniferophyte characteristics: simple leaves, pycnoxylic secondary xylem, the gymnosperm type of eustele, heterospory, and seeds (or "preovules" in the Archaeopteridales). The paleobotany textbook by Stewart (1983) provides details.

Florin (1951) favored the concept that the Cordaitales were ancestral to the Voltziales. When the

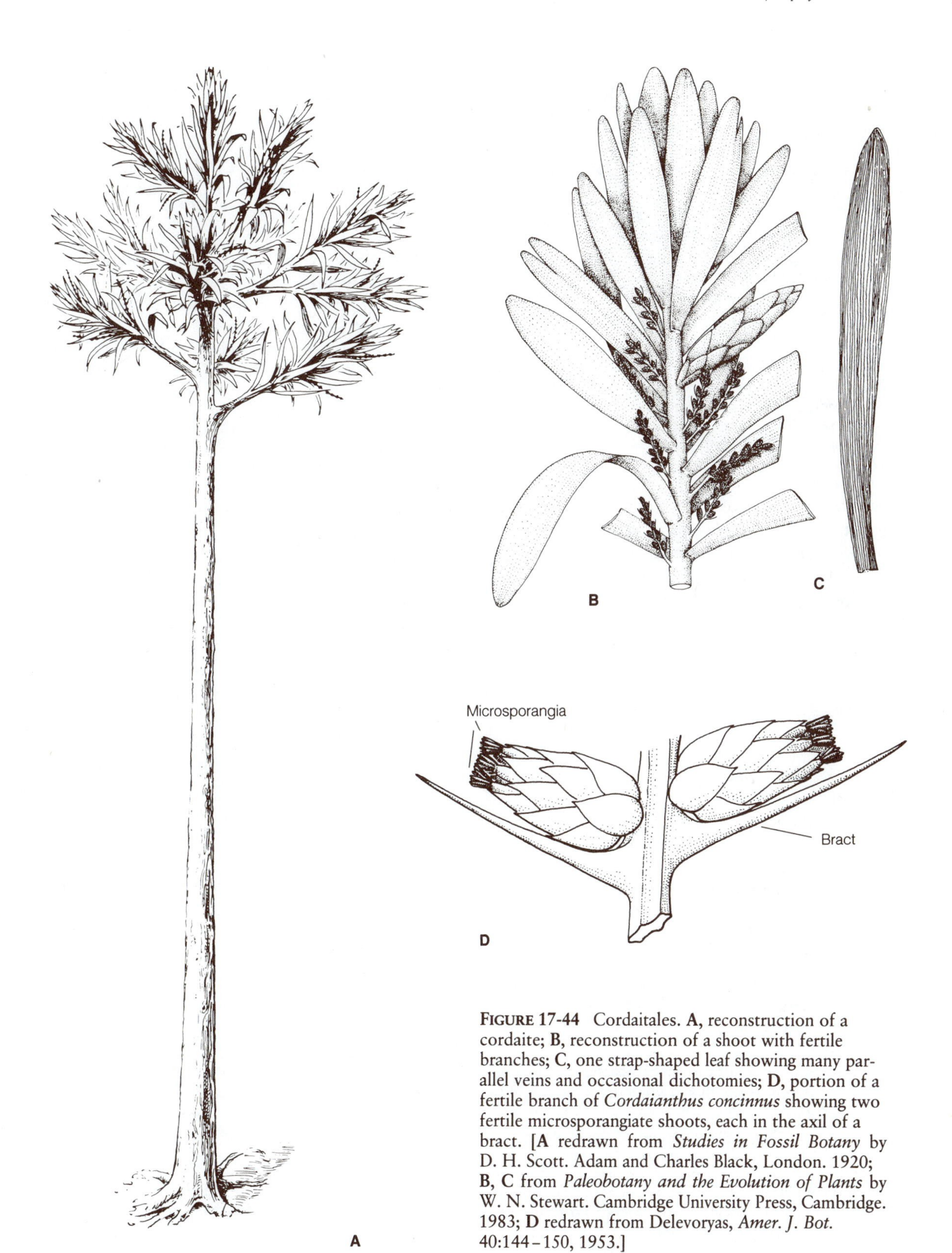

FIGURE 17-44 Cordaitales. **A,** reconstruction of a cordaite; **B,** reconstruction of a shoot with fertile branches; **C,** one strap-shaped leaf showing many parallel veins and occasional dichotomies; **D,** portion of a fertile branch of *Cordaianthus concinnus* showing two fertile microsporangiate shoots, each in the axil of a bract. [**A** redrawn from *Studies in Fossil Botany* by D. H. Scott. Adam and Charles Black, London. 1920; **B, C** from *Paleobotany and the Evolution of Plants* by W. N. Stewart. Cambridge University Press, Cambridge. 1983; **D** redrawn from Delevoryas, *Amer. J. Bot.* 40:144–150, 1953.]

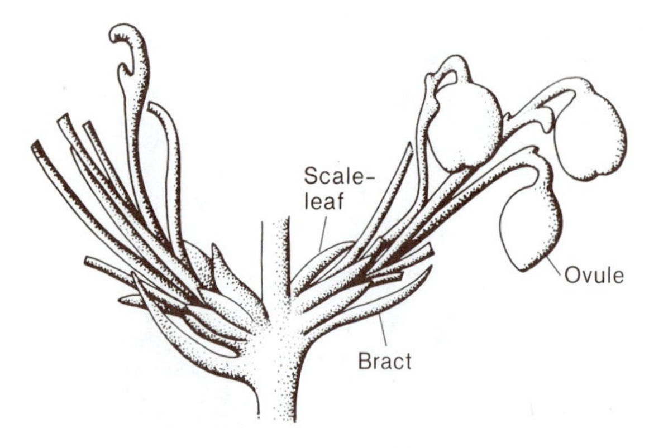

FIGURE 17-45 Portion of a fertile branch of *Cordaianthus* showing two fertile shoots, each in the axil of a bract, consisting of sterile scales and stalked ovules [From Taylor and Millay, *Rev. Paleobot. Palynol.* 27:329–355, 1979.]

ovulate reproductive structures are compared, one could visualize that the ovulate strobili of *Lebachia* and *Ernestiodendron,* for example, could have been evolved by the shortening of the primary axis of the cordaitalean reproductive structure, bringing the bracts and fertile shoots together into a definite cone. (Compare Figs. 17-28 and 17-45).

REFERENCES

Aase, H. C.
1915. Vascular anatomy of the megasporophylls of conifers. *Bot. Gaz.* 60:277–313.

Abbema, T. van
1934. Das Transfusiongewebe in den Blättern der Cycadinae, Ginkgoinae und Coniferen. *Trav. Bot. Neerl.* 31:310–390.

Allen, G. S.
1946a. Embryogeny and development of the apical meristems of *Pseudotsuga.* I. Fertilization and early embryogeny. *Amer. Jour. Bot.* 33:666–677.
1946b. The origin of the microsporangium of *Pseudotsuga. Bull. Torrey Bot. Club* 73:547–556.
1963. Origin and development of the ovule in Douglas fir. *Forest Sci.* 9:386–393.

Allen, G. S., and J. N. Owens
1972. *The Life History of Douglas Fir.* Information Canada, Ottawa, Canada.

Bonga, J. M., and D. J. Durzan (eds.).
1987. *Cell and Tissue Culture in Forestry.* Vols. 1, 2, 3. Martinus Nijhoff Publs., The Netherlands.

Brown, H. P., A. J. Panshin, and C. C. Forsaith
1949. *Textbook of Wood Technology,* Vol. I. *Structure, Identification, Defects, and Uses of the Commercial Woods of the United States.* McGraw-Hill, New York.

Buchholz, J. T.
1918. Suspensor and early embryo of *Pinus. Bot. Gaz.* 66:185–228.
1926. Origin of cleavage polyembryony in conifers. *Bot. Gaz.* 81:55–71.
1931. The pine embryo and the embryos of related genera. *Trans. Illinois Acad. Sci.* 23:117–125.

Burlingame, L. L.
1913. The morphology of *Araucaria brasiliensis. I.* The staminate cone and male gametophyte. *Bot. Gaz.* 55:97–114.

Butts, D. and J. T. Buchholz
1940. Cotyledon number in conifers. *Trans. Illinois Acad. Sci.* 33:58–62.

Camefort, H.
1968. Cytologie de la fécondation et de la proembryogénèse chez quelques gymnospermes. *Bull. Soc. Bot. France* 115:137–160.
1969. Fécondation et proembryogénèse chez les Abiétacées (notion de néocytoplasme). *Rev. Cytol. Biol. Vég.* 32:253–271.

Campbell, D. H.
1940. *The Evolution of the Land Plants.* Stanford University Press, Stanford, Calif.

Chamberlain, C. J.
1935. *Gymnosperms. Structure and Evolution.* University of Chicago Press, Chicago.

Chaney, R. W.
1950. A revision of fossil *Sequoia* and *Taxodium* in western North America based on the recent discovery of *Metasequoia. Trans. Amer. Phil. Soc.* 40:172–239.

Chowdhury, C. R.
1962. The embryogeny of conifers: a review. *Phytomorphology* 12:313–338.

Christiansen, H.
1969. On the pollen grain and the fertilization mechanism of *Pseudotsuga menziesii* (Mirbel) Franco. var. *viridis* Schwer. *Silvae Genet.* 18:97–104.

Currey, D. R.
1965. An ancient bristlecone pine stand in eastern Nevada. *Ecology* 46:564–566.

Doak, C. C.
1935. Evolution of foliar types, dwarf shoots, and

cone scales in *Pinus*. *Illinois Biol. Monogr.* 13:1–106.

Doyle, J.
1945. Developmental lines in pollination mechanisms in the Coniferales. *Sci. Proc. Roy. Dublin Soc.* 24:43–62.
1963. Proembryogeny in *Pinus* in relation to that in other conifers—a survey. *Proc. Roy. Irish Acad.* 62B:181–216.

Doyle, J., and M. Brennan, S. J.
1972. Cleavage polyembryony in conifers and taxads—a survey. II. Cupressaceae, Pinaceae, and conclusions. *Sci. Proc. Roy. Dublin Soc.* 4(A):137–158.

Doyle, J., and M. O'Leary
1935. Pollination in *Pinus*. *Sci. Proc. Roy. Dublin Soc.* 21:181–190.

Eames, A. J.
1913. The morphology of *Agathis australis*. *Ann. Bot.* 27:1–38.

Eckenwalder, J. E.
1976. Re-evaluation of Cupressaceae and Taxodiaceae: a proposed merger. *Madroño* 23:237–256.

Erspamer, J. L.
1952. Ontogeny and morphology of the microsporangium in certain genera of the Coniferales. Ph.D. Diss., University of California, Berkeley.

Esau, K.
1965. *Plant Anatomy*, 2d edition. Wiley, New York.
1977. *Anatomy of Seed Plants*, 2d edition. Wiley, New York.

Ewers, F. W.
1982. Secondary growth in needle leaves of *Pinus longavea* (bristlecone pine) and other conifers: quantitative data. *Amer. J. Bot.* 69:1552–1559.
1983. The determinate and indeterminate dwarf shoots of *Pinus longavea* (bristlecone pine). *Can. J. Bot.* 61:2280–2290.

Fagerlind, F.
1961. The initiation and early development of the sporangium in vascular plants. *Svensk Bot. Tidskr.* 55:299–312.

Ferguson, C.W.
1968. Bristlecone pine: science and esthetics. *Science* 159:839–846.

Ferguson, M. C.
1904. Contributions to the knowledge of the life history of *Pinus* with special reference to sporogenesis, the development of the gametophytes, and fertilization. *Proc. Wash. Acad. Sci.* 6:1–202.

Florin, R.
1931. Untersuchungen zur Stammesgeschichte der Coniferales und Cordaitales. *Svenska Vetensk. Akad. Handl.* Ser. 5. 10:1–588.
1944. Die Koniferen des Oberkarbons und des unteren Perms. *Paleontographica* 85(B):365–654.
1950. Upper Carboniferous and Lower Permian Conifers. *Bot. Rev.* 16:258–282.
1951. Evolution in Cordaites and Conifers. *Acta Horti Bergiani* 15:285–388.
1963. The distribution of conifer and taxad genera in time and space. *Acta Horti Bergiani* 20:121–312.

Gifford, E. M., Jr., and N. T. Mirov
1960. Initiation and ontogeny of the ovulate strobilus in Ponderosa Pine. *Forest Sci.* 6:19–25.

Gifford, E. M., Jr., and G. E. Corson, Jr.
1971. The shoot apex in seed plants. *Bot. Rev.* 37:143–229.

Griffith, M. M.
1957. Foliar ontogeny in *Podocarpus macrophyllus*, with special reference to transfusion tissue. *Amer. Jour. Bot.* 44:705–715.
1971. Transfusion tissue in leaves of *Cephalotaxus*. *Phytomorphology* 21:86–92.

Gupta, P. K., and D. J. Durzan
1986. Somatic polyembryogenesis from callus of mature sugar pine embryos. *Bio/Technology* 4:643–645.
1987a. Biotechnology of somatic polyembryogenesis and plantlet regeneration in loblolly pine. *Bio/Technology* 5:147–151.
1987b. Somatic embryos from protoplasts of loblolly pine proembryonal cells. *Bio/Technology* 5:710–712.

Haupt. A. W.
1941. Oogenesis and fertilization in *Pinus lambertiana* and *P. monophylla*. *Bot. Gaz.* 102:482–498.
1953. *Plant Morphology*. McGraw-Hill, New York.

Hitch, C. J.
1982. Dendrochronology and serendipity. *Amer. Scientist* 70:300–305.

Jeffrey, E. C.
1917. *The Anatomy of Woody Plants*. University of Chicago Press, Chicago.

Johansen, D. A.
1950. *Plant Embryology*. Chronica Botanica, Waltham, Mass.

Johnson, R. W., and R. T. Riding
1981. Structure and ontogeny of the stomatal

complex in *Pinus strobus* L. and *Pinus banksiana* Lamb. *Amer. J. Bot.* 68:260–268.

Kausik, S. B.
1974. Ontogeny of the stomata in *Gnetum ula* Brongn. *Bot. J. Linn. Soc.* 68:143–151.

Koehne, E.
1893. *Deutsche Dendrologie. Kurze Beschreibung der in Deutschland im Freien aushaltenden Nadel—und Laubholzgewächse.* F. Enke, Stuttgart.

Konar, R. N.
1960. The morphology and embryology of *Pinus roxburghii* Sar. with a comparison with *Pinus wallichiana* Jack. *Phytomorphology* 10:305–319.

Konar, R. N., and S. Ramchandani
1958. The morphology and embryology of *Pinus wallichiana. Phytomorphology* 8:328–346.

Laubenfels, D. J. de
1953. The external morphology of coniferous leaves. *Phytomorphology* 3:1–20.
1959. Parasitic conifer found in New Caledonia. *Science* 130:97.

Lemoine-Sebastian, C.
1968. La vascularisation du complexe bractée-écaille chez les Taxodiacées. *Trav. Lab. Foret. Toulouse* 7. Article 1.
1969. Vascularisation du complexe bractée-écaille dans le cone femelle des Cupressacées. *Bot. Rhedonica* Ser. 7:3–27.

Lindsay, G.
1969. The ancient bristlecone pines. *Pac. Discovery* 22:1–8.

Maheshwari, P., and H. Singh
1967. The female gametophyte of gymnosperms. *Biol. Rev.* 42:88–130.

Maheshwari, P., and V. Vasil
1961. The stomata of *Gnetum. Ann. Bot.* n.s. 25:313–319.

Mapes, G., and G. W. Rothwell
1984. Permineralized ovulate cones of *Lebachia* from Late Palaeozoic limestones of Kansas. *Palaeontology* 27, Part 1:69–94.

Martens, P.
1971. *Les Gnétophytes.* Handbuch der Pflanzenanatomie, Band 12, Teil 2. Gebrüder Borntraeger, Berlin-Nikolassee.

McWilliam, J. R., and F. Mergen
1958. Cytology of fertilization in *Pinus. Bot. Gaz.* 119:246–249.

Miller, C. N., Jr.
1977. Mesozoic conifers. *Bot. Rev.* 43:217–280.
1982. Current status of Paleozoic and Mesozoic conifers. *Rev. Palaeobot. Palynol.* 37:99–114.

Napp-Zinn, K.
1966. *Anatomie des Blattes. I. Blattanatomie der Gymnospermen.* Handbuch der Pflanzenanatomie, Band 8, Teil 1. Gebrüder Borntraeger, Berlin-Nikolassee.

Niklas, K. J.
1982. Simulated and empiric wind pollination patterns of conifer cones. *Proc. Nat. Acad. Sci.* (USA) 79:510–514.
1984. The motion of windborne pollen grains around conifer ovulate cones: implications on wind pollination. *Amer. J. Bot.* 71:356–374.
1985. The aerodynamics of wind pollination. *Bot. Rev.* 51:328–386.

Owens, J. N., and R. P. Pharis
1967. Initiation and ontogeny of the microsporangiate cone in *Cupressus arizonica* in response to gibberellin. *Amer. Jour. Bot.* 54:1260–1272.

Pant, D. D., and B. Mehra
1964. Development of stomata in leaves of three species of *Cycas* and *Ginkgo biloba. Jour. Linn. Soc. London Bot.* 58:491–497.

Pettitt, J. M.
1985. Pollen tube development and characteristics of protein emission in conifers. *Ann. Bot.* 56:379–397.

Pilger, R.
1926. Coniferae. In A. Engler and K. Prantl (eds.), *Die natürliche Pflanzenfamilien,* 2d edition, Vol. 13, pp. 164–166.

Rothwell, G. W.
1982. New interpretations of the earliest conifers. *Rev. Palaeobot. Palynol.* 37:7–28.

Sacher, J. A.
1953. Structure and histogenesis of the buds of *Pinus lambertiana.* Ph.D. Diss., University of California, Berkeley.
1954. Structure and seasonal activity of the shoot apices of *Pinus lambertiana* and *Pinus ponderosa. Amer. Jour. Bot.* 41:749–759.
1955a. Cataphyll ontogeny in *Pinus lambertiana. Amer. Jour. Bot.* 42:82–91.
1955b. Dwarf shoot ontogeny in *Pinus lambertiana. Amer. Jour. Bot.* 42:784–792.

Schweitzer, H. J.
1963. Der weibliche Zapfen von *Pseudovoltzia liebeana* und seine Bedeutung für die Phylo-

genie der Koniferen. *Palaeontographica* 113 B:1–29.

Shaw, G. R.
1914. *The Genus Pinus.* (Arnold Arboretum Publication No. 5.) Riverside Press, Cambridge, Mass.

Shobe, W. R., and N. R. Lersten
1967. A technique for clearing and staining gymnosperm leaves. *Bot. Gaz.* 128:150–152.

Singh, H.
1961. The life history and systematic position of *Cephalotaxus drupacea* Sieb. et Zucc. *Phytomorphology* 11:153–197.
1978. *Embryology of Gymnosperms.* Handbuch der Pflanzenanatomie, Band X, Teil 2. Gebrüder Borntraeger, Berlin.

Singh, H., and J. Chatterjee
1963. A contribution to the life history of *Cryptomeria japonica* D. Don. *Phytomorphology* 13:429–445.

Sporne, K. R.
1965. *The Morphology of Gymnosperms.* Hutchinson University Library, London.

Spurr, A. R.
1949. Histogenesis and organization of the embryo in *Pinus strobus. Amer. Jour. Bot.* 36:629–641.

Sterling, C.
1949. Some features in the morphology of *Metasequoia. Amer. Jour. Bot.* 36:461–471.
1963. Structure of the male gametophyte in gymnosperms. *Biol. Rev.* 38:167–203.

Stewart, W. N.
1983. *Paleobotany and the Evolution of Plants.* Cambridge University Press, Cambridge.

Stockey, R. A.
1981. Some comments on the origin and evolution of conifers. *Can. J. Bot.* 59:1932–1940.

Suzuki, M.
1979a. The course of resin canals in the shoots of conifers. II. Araucariaceae, Cupressaceae and Taxodiaceae. *Bot. Mag.* (Tokyo) 92:253–274.
1979b. The course of resin canals in the shoots of conifers. III. Pinaceae and summary analysis. *Bot. Mag.* (Tokyo) 92:333–353.

Tison, A.
1913. Sur la persistance de la nervation dichotomique chez les Conifères. *Bull. Soc. Linn. Normandie* 6ᵉ Ser. 4:31–46.

Willemse, M. T. M.
1974. Megagametogenesis and formation of neocytoplasm in *Pinus sylvestris* L. In H. F. Linskens (ed.), *Fertilization in Higher Plants,* pp. 97–102.

Willemse, M. Th. M., and H. F. Linskens
1969. Dévelopment du microgamétophyte chez le *Pinus sylvestris* entre la meiose et la fécondation. *Rev. Cytol. Biol. Vég.* 32:121–128.

Worsdell, W. C.
1897. On "transfusion tissue:" its origin and function in the leaves of gymnospermous plants. *Trans. Linn. Soc. London Bot.* Ser. II. 5:301–319.

Zimmermann, W.
1959. *Die Phylogenie der Pflanzen,* 2d edition. Gustav Fischer, Stuttgart.

CHAPTER 18

Gnetophyta

EPHEDRA, GNETUM, AND WELWITSCHIA

THE Gnetophyta is a small group of seed plants represented by three living genera, *Ephedra, Welwitschia,* and *Gnetum.* The distinctive organography and anatomy of the sporophyte and the many peculiar features of the reproductive cycle of these genera have attracted the attention of morphologists for more than a century and have resulted in a voluminous literature. Martens (1971) brought together in a single volume monographic treatments of each of the three genera of the gnetophytes and examined critically the many conflicting theories that have been advanced to explain their interrelationships and their affinities with other groups of living seed plants.

Throughout much of the literature on the gnetophytes, repeated efforts have been made to demonstrate that these plants form a "connecting link" between gymnosperms and angiosperms (Arber and Parkin, 1908). Among the "angiospermic" features usually mentioned in defense of this idea are (1) the compound nature of *both* the microsporangiate and megasporangiate strobili, which have been intepreted as "inflorescences" by some botanists, and (2) the presence of vessels in the xylem, a feature that has been regarded as a major divergence from the vessel-less wood characteristic of the cycads, *Ginkgo biloba,* and all members of the Coniferales. The presumed isolation of the gnetophytes from all living and extinct gymnosperms is clearly emphasized in Pulle's (1938) classification of the major groups of seed plants, in which *Ephedra, Welwitschia,* and *Gnetum* are segregated under the subdivision "Chlamydospermae," which is placed *between* the "Gymnospermae" and "Angiospermae." The word Chlamydospermae (from the Greek words meaning "seeds with an envelope or cloak") refers to the so-called outer integument of the ovule, which has been interpreted as a pair of fused appendages or "bracteoles" (see Fig. 18-13, B, outer envelope). In this connection, Martens (1971) astutely observed that an ovule provided with an accessory envelope is not "naked" in the sense employed to characterize the ovules of gymnospermous plants. Some authors have even compared the external envelope of the ovule of the gnetophytes to the "ovary" of the angiosperms.

For many years the existence of gnetophytes in the fossil record was based entirely upon the presence of pollen grains resembling those of *Ephedra* and *Welwitschia.* Recently, Crane and Upchurch (1987) discovered megafossils (leafy shoots) from the Lower Cretaceous that have gnetalean-type organization. The fossils are associated with masses of the *Welwitschia*-type pollen. In the past, *Ephedra,*

Welwitschia, and *Gnetum* were placed in the family Gnetaceae under the order *Gnetales*. The tendency now is to split this old order into three orders, namely, Ephedrales, Welwitschiales, and Gnetales, each consisting of a single family and a single genus (see Eames, 1952, and Martens, 1971). We have adopted this systematic treatment and have grouped the three orders, for convenience, under a single division, the Gnetophyta.

Geographical Distribution and Habit

The genus *Ephedra*, consisting of about thirty-five species, is confined to cool, usually arid regions in both the Eastern and Western Hemispheres (Fig. 18-1). In the New World, *Ephedra* is restricted to western North America, parts of Mexico, and a wide area in South America. According to Cutler's (1939) monograph, the sixteen species found in the United States occur in dry or desert areas of California, Nevada, Utah, Arizona, and New Mexico. Most species of *Ephedra* are profusely branched shrubs (Fig. 18-2), although a few are scandent, and one species, *Ephedra triandra*, native to Brazil, Uruguay, and Argentina, is a small tree.

In marked contrast, *Gnetum* inhabits tropical rain forests in parts of Asia, northern South America, Africa, and certain Pacific islands between Australia and Asia (Fig. 18-1). Most of the thirty or more described species of *Gnetum* are lianas that climb high into the crowns of various trees in the rain forest; one species, *Gnetum gnemon*, becomes a small tree (Maheshwari and Vasil, 1961).

The most bizarre and geographically restricted member of the Gnetophyta is the African genus *Welwitschia* consisting of the single species *Welwitschia mirabilis* (Fig. 18-3, A, B). The specific epithet "mirabilis" (Latin for "wonderful" or "marvelous") is quite appropriate because an adult plant, in habit and general organography, is unlike that of any known plant on earth. The exposed portion of an old plant consists of a short, woody, unbranched stem and a massive, woody, concave, "disc" or "crown" that bears only two huge strap-shaped leaves as the *permanent* photosynthetic organs of

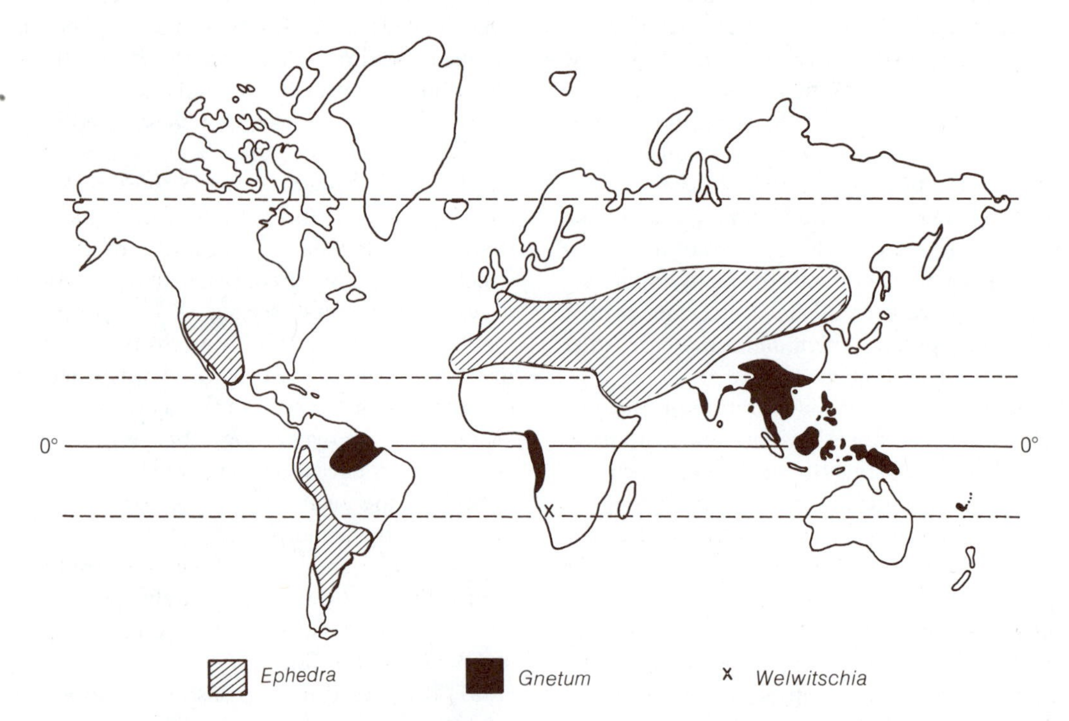

FIGURE 18-1 Present distribution of the three genera of the gnetophytes. See text for further explanation. [Based on Hutchinson, *Kew Bulletin of Miscellaneous Information*, No. 2, 1924. By permission of Her Majesty's Stationery Office.]

Figure 18-2 *Ephedra* sp. photographed near Monitor Pass, California. Note the large size of the basal stems of this shrubby species. [Courtesy of Dr. T. E. Weier.]

the plant (Fig. 18-7). These leaves become split and frayed in old plants, extending in a twisted and arching manner over the ground, and the tips become seared where they touch the ground. The leaves are perpetuated at the rate of 8 to 15 centimeters a year from a basal intercalary meristem. Bornman (1972) found that a leaf of one giant had an unbroken width of 1.8 meters and a length of 6.2 meters of which 3.7 meters were of living tissue. It is estimated that some plants may be 1,500 to 2,000 years old!

Welwitschia occurs in the Namib Desert near the coast of Angola and South-West Africa (Namibia) as well as inland to about 150 kilometers in a savannalike region. Rainfall of the Namib Desert varies from 0 to 100 millimeters per year whereas rainfall in the interior can exceed 200 millimeters. During the night for about nine months of the year, fog is blown inward from the coast to a distance of about 80 kilometers, but it dissipates by about 10:00 a.m. Moisture from the fog accumulates on the leaves. How does *Welwitschia* solve the problem of water economy and water uptake in a desert enviroment? *Welwitschia* has a large tap root that may extend downward for 1 to 1.5 meters before it divides into numerous thin roots. How far down the thin roots extend is uncertain. Some investigators believe that water from the fog enters the leaves, but the exact method, if it occurs, is unknown.

Many desert plants have a type of photosynthesis adapted for existence in a hot, dry desert environment. Stomata are open at night when the air is cool, and carbon dioxide is absorbed and fixed in organic acids. During the early daylight hours the carbon dioxide is released, not to the atmosphere because the stomata are now closed to prevent water loss by transpiration, but to chloroplasts of the mesophyll cells where it is transformed first into sugars and then into starch. *Welwitschia* has been described as having this type of photosynthesis—termed CAM (crassulacean acid metabolism) photosynthesis (Dittrich and Huber, 1975). In the more common type of photosynthesis, the carbon dioxide enters through stomata during the daylight hours and is fixed in the chloroplasts and then converted to sugars. The latter type of photosynthesis has been reported to occur in *Welwitschia* in the eastern part of its range where the climate is more moderate (Schulze and Schulze, 1976).

Recent physiological studies of plants from a variety of natural habitats support the belief that *Welwitschia* is not basically a CAM plant. Von Willert (1985) reported that carbon dioxide is taken up during the day through open stomata, leading to a tremendous loss of water by transpiration. A high transpiration rate prevents the occurrence of lethal leaf temperatures. These observations indicate that *Welwitschia* can tap water supplies in the soil not exploited by its associates. Von Willert (1985) sums up the enigma of *Welwitschia*: "If botanists should construct a plant best adapted to a desert environment, they would never come up with a monster like *Welwitschia mirabilis*."

For interesting and readable accounts of the morphology, anatomy, and physiology of *Welwitschia* see the publications of Bornman (1972, 1978).

Welwitschia is dioecious, and the microsporangiate and megasporangiate cones are borne terminally on ramified branch systems. As Martens (1971) emphasizes, these strobiliferous shoots represent the *only authentic branches* of this amazing plant (Figs. 18-22 A; 18-23). *Welwitschia* was discovered in 1860 in Angola by Dr. Frederic Welwitsch, in whose honor the plant was named by the

A

B

FIGURE 18-3 *Welwitschia mirabilis.* **A**, microsporangiate specimen growing in a desert area near Brandberg in South-West Africa (Namibia). Each of the two huge leaves has become split into a series of contorted and buckled "segments." Note the numerous clusters of strobili in the leaf axils at the periphery of the "woody" crown. **B**, extremely old plant growing on "*Welwitschia*-Flats," Namib Desert, South-West Africa; note fog in the background. [**A**, courtesy of Dr. R. J. Rodin; **B**, photograph by Dr. C. H. Bornman.]

British botanist, J. D. Hooker (1863). Dyer and Verdoorn (1972) claim that *Tumboa bainesii* is the valid name. However, the binominal *Welwitschia mirabilis* will undoubtedly continue to be used in descriptions of this bizarre plant; the name has been conserved in the International Code of Botanical Nomenclature, 1983.

Vegetative Organography and Anatomy

Despite the voluminous literature, which has been reviewed by Pearson (1929), Chamberlain (1935), and Martens (1971), much still remains to be done before an adequate treatment of the sporophytes of gnetalean plants as a whole can be accomplished. *Gnetum* and *Welwitschia* are both extremely complex anatomically and require further ontogenetic study.

The Leaf

The foliage leaves of *Ephedra, Gnetum,* and *Welwitschia* are strikingly different in form and venation and provide morphological characters that are definitive for each of the three genera.

In *Ephedra,* the phyllotaxis varies from *decussate* to *whorled,* and the leaves are basally joined by a membranous commissure to form a more-or-less conspicuous sheath at each node (Fig. 18-4, B). Throughout most of the literature on *Ephedra,* the leaves are characterized as "reduced" or "scalelike"

FIGURE 18-4 **A,** *Gnetum indicum.* Terminal portion of shoot showing seeds and three pairs of simple leaves with pinnate-reticulate venation (actual size); **B,** *Ephedra* sp. Tip of a vegetative shoot, illustrating the whorled arrangement of the reduced scalelike leaves (× 4).

in form. This generalization, however, needs to be qualified because in several species (e.g., *Ephedra foliata* and *Ephedra altissima)* the lamina is slender and needlelike, and, according to Stapf (1889), may attain a length of 3 centimeters and a width of 1 to 1.5 millimeters (Fig. 18-5, A). There also is considerable variation in the proportional development of the sheath and lamina in the successive leaves of a single shoot. In *Ephedra foliata,* for example, the leaves in the middle or upper part of a shoot may have well-developed laminae whereas the basal leaves of the same shoot tend to be much smaller, with laminae only 3 to 4 millimeters long (compare A and B in Fig. 18-5). In this and other species, there is obviously no sharp boundary between "well-developed" and "reduced" scalelike leaves. It is widely stated in the literature that the leaf in *Ephedra* is vascularized by a pair of traces that neither branch nor anastomose in their course through the sheath and lamina. The stem is eustelic, and each leaf is supplied by a trace from two adjacent sympodia (Marsden and Steeves, 1955). However, in *Ephedra chilensis* and *Ephedra fragilis* one of the leaves at a node may be three-veined. In a few instances, both members of a leaf pair are three-veined (Foster, 1972). Additional studies of the ontogeny and vascularization of three-veined leaves are needed.

The foliage leaf of *Gnetum* presents a complete morphological contrast to the leaf of *Ephedra.* In

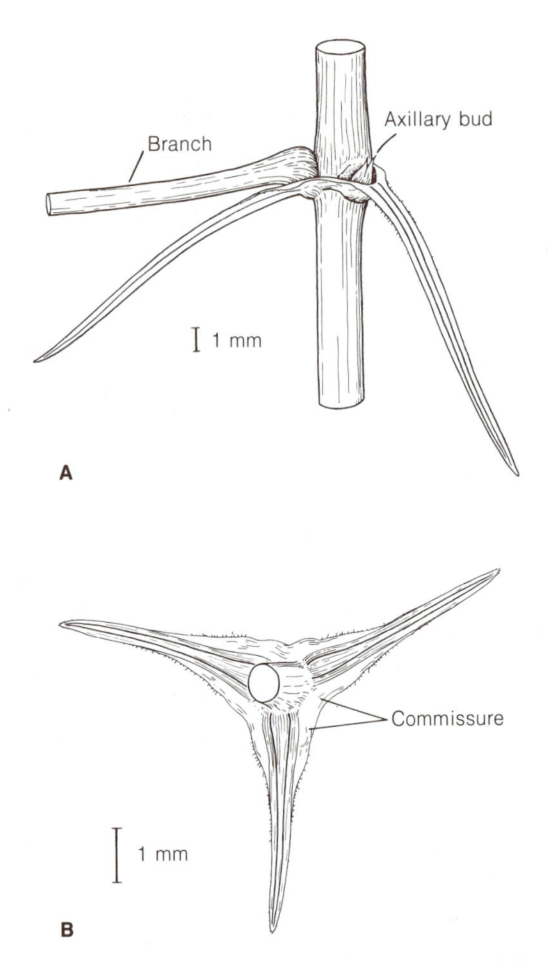

FIGURE 18-5 Variation in size and morphology of the leaves in *Ephedra foliata*. **A**, leaves with well-developed laminae and weakly connate sheaths; the leaf at the right is traversed by two freely terminating veins; **B**, whorl of three scale-leaves, detached from shoot and drawn from lower surface. Each leaf is vascularized by a pair of veins and is basally connected, by a prominent commissure, with adjacent leaves of the whorl. [From Foster, *Jour. Arnold Arboretum* 53:364, 1972.]

FIGURE 18-6 *Gnetum gnemon*, portion of branch system showing oppositely arranged leaves, and seeds. [General Biological Supply House, Inc., Chicago, Illinois.]

Gnetum the foliage leaves are arranged in pairs, and in their form and pinnate-reticulate venation bear a striking resemblance to the simple leaves of many dicotyledons (Figs. 18-4, A; 18-6). Very little comparative information is available regarding the number of leaf traces that vascularize the leaf in the various species of *Gnetum*. Rodin and Paliwal (1970) found that usually an *odd number of traces* (five to seven, or sometimes eight) extend into the leaf of *Gnetum ula*, but whether this type of nodal anatomy characterizes the genus as a whole is unknown. According to Rodin (1967), the five to seven leaf traces in *Gnetum gnemon* extend into the midrib of the young leaf as a series of longitudinal bundles that, at successive levels, dichotomize and form *pairs of secondary* veins that curve toward the margins of the lamina. In the submarginal region, each secondary vein bifurcates and the derivative branches unite, creating a series of coarse "meshes" at each side of the midrib. Within each mesh, a more delicate reticulum of veinlets is later differentiated. (For additional information and illustrations of venation patterns in several other species of *Gnetum*, see Rodin, 1966.)

The vegetative organography of *Welwitschia*, as we have already noted, is without parallel among all living vascular plants. Except for the two short-lived cotyledons, produced during the seedling phase, the *permanent* photosynthetic organs are represented by a single pair of enormous strap-shaped leaves, which continue to grow indefinitely in length through the activity of a meristem located at the base of each appendage (Fig. 18-7). According to Martens (1971), the young leaf of *Welwitschia* is at first vascularized by two and then four strands, and additional lateral bundles continue to differentiate as the leaf increases in width basally. A well-developed leaf in *Welwitschia* is traversed by numerous "parallel" longitudinal veins that become interconnected by smaller obliquely oriented veins. These may anastomose in various ways to form irregular areoles (i.e., meshes) or may fuse in pairs and terminate blindly in the mesophyll. This peculiar and highly distinctive type of venation also occurs in the blades of the cotyledons and is unique as compared with the venation of *Ephedra, Gnetum,* and other gymnosperms (see Rodin, 1953, 1958a, 1958b).

As described previously, stomata of the syndetocheilic type occur on both surfaces of a *Welwitschia* leaf and are somewhat sunken, although not as deeply as in most desert plants. Also, the cuticle is relatively thin in comparison with most xerophytes. Strands of thick-walled fibers of great tensile strength are present beneath the epidermis, and ramified sclereids occur in the mesophyll. The vascular bundles are collateral. The xylem consists of tracheids and vessels; the vessel members have large perforations in the end walls. The phloem consists of sieve cells and parenchyma; the sieve areas are much like those of conifers, and the nucleus of a sieve cell remains intact at maturity. The entire vascular bundle is surrounded by a layer of transfusion tissue (Evert et al., 1973a, b).

FIGURE 18-7 *Welwitschia mirabilis,* young sporophyte showing the two permanent foliage leaves. The specimen has been grown in a section of pipe to provide space for the development of the long tap root. [Courtesy of Dr. T. E. Weier.]

The Shoot Apex

Comparative studies have revealed a remarkable similarity in the *basic structure* of the shoot apex of the gnetophytes: in all three genera a more-or-less discrete surface layer, or *tunica,* is present, the con-

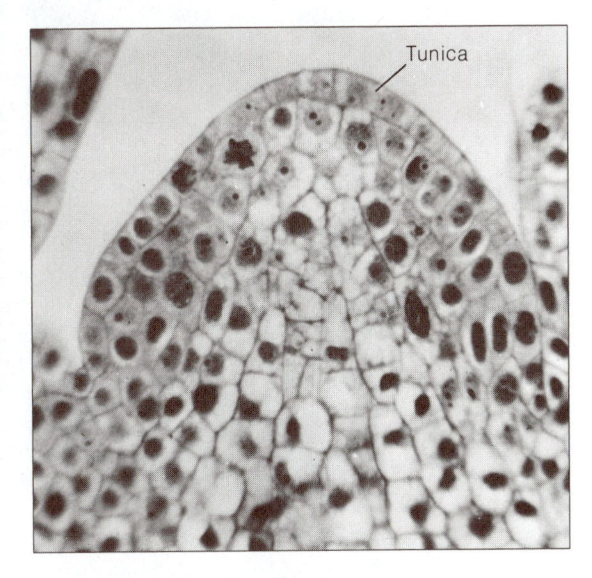

FIGURE 18-8 Median longisection of the shoot apex of *Ephedra altissima.* Note the single, clearly defined tunica.

tinuity of which is interrupted only by occasional periclinal divisions. The investigations of Gifford (1943) on *Ephedra altissima* and Seeliger's (1954) study on *Ephedra fragilis* var. *campylopoda* emphasized the infrequency of periclinal divisions in the surface layer of the apex and its tunicalike structure and growth (Fig. 18-8). Johnson (1950), in a study of eighty-five apices of *Gnetum gnemon,* found "no evidence of periclinal divisions in the tunica clothing the summit of the dome." The aberrant behavior of the shoot apex in *Welwitschia* presents a striking contrast to the "open system" of growth characteristic of *Ephedra* and *Gnetum*. According to the detailed study made by Martens and Waterkeyn (1963), the shoot apex of the seedling of *Welwitschia* first initiates a pair of primordia which

FIGURE 18-9 *Welwitschia mirabilis.* **A**, young plant showing the two permanent leaves and the two opposite scaly bodies at the bases of the leaves. **B**, older plant showing the two leaves and strobiliferous branches that arise from the basal or intercalary meristem; note that the scaly bodies have become much larger than in **A**, and the leaf to the right is split down to the basal meristem.

later develop into the two permanent leaves of the sporophyte. But contrary to the classical interpretation, an *additional pair* of appendages — the so-called scaly bodies — are formed by the apex in a plane at right angles to the plane of the foliage-leaf primordia (Fig. 18-9, A, B). Martens and Waterkeyn reject the prevalent view that the scaly bodies are cotyledonary buds, and interpret them instead as a *third pair of foliar appendages*. Through the activity of a basal meristem the scaly bodies increase in size, and become conspicuous structures on plants several years old or more (Fig. 18-9, B). Soon after the formation of scaly bodies, the shoot apex loses its meristematic character and finally degenerates. Meristematic activity is now shifted to quite another region — to that of the leaf bases of the two permanent photosynthetic leaves. This activity results in the bilobed condition of the crown which, in very old plants, may become a concave disc surmounted by a band of meristematic tissue that contributes new tissue to the two large foliage leaves and provides concentric crests of tissue in which reproductive structures are initiated. The precocious death of the terminal meristem of the *young sporophyte* of *Welwitschia* is unique among vascular plants and is an additional example of the bizarre morphology of this extraordinary plant. Martens (1977) considers *Welwitschia* to be a plant that has lost its head!

The phylogenetic significance of the angiospermic organization of the shoot apex of the gnetophytes is, of course, problematic and very possibly the result of evolutionary convergence. Martens (1971) believes that the structural similarity of the apices of the three genera of gnetophytes supports the evolutionary position of this group at a level intermediate between those of the conifers and angiosperms.

Vessels

The presence of vessels in the Gnetophyta distinguishes them from other living gymnosperms and has often been used as an argument for their presumed evolutionary relationship to the angiosperms. However, it is generally believed that the *method* of origin of the perforations in the vessel members of the gnetophytes differs from that of all other vascular plants including the angiosperms. In the angiosperms, as well as in certain species of *Pteridium* and *Selaginella*, the intial step in vessel evolution was the loss of the primary cell walls from the *transversely elongated* bordered pits situated at each sloping end of a tracheidlike cell. Further elaboration of these slitlike perforations led to the development of vessel members with well-defined scalariform perforation plates. The most advanced vessel members in angiosperms possess large circular or oval simple perforations which originated by the elimination (phylogenetically and ontogenetically) of the bars between the slitlike openings. In contrast, as Thompson (1918) has shown, the initial step in vessel evolution in the Gnetophyta began with the loss of primary cell walls from a series of *circular* bordered pits located near the ends of long tracheidlike cells. In *Ephedra*, there are transitional conditions between intact bordered pits and *bordered foraminate perforations*, that is, there are clear transition forms between typical tracheids and vessel members (Fig. 18-10). The vessel members of *Gnetum* commonly possess large circular or elliptical simple perforations and thus markedly resemble the specialized vessel members of many angiosperms. Thompson (1918) was of the opinion, however, that the *Gnetum* type of vessel perforation resulted from the further enlargement of a series of circular perforations and the elimination of the portions of the end wall between them.

Thus in the initial steps of their origin as well as in their subsequent specialization, the vessels of the gnetophytes are thought to have evolved differently and independently from those in the angiosperms. As Bailey (1953) remarked, "although the highly evolved vessels of *Gnetum* resemble those of comparably specialized vessels of angiosperms, the similarity cannot be used as an indication of close relationship, but provides a very significant illustration of convergent evolution in plants." However, on the basis of a detailed study of the xylem of two species of *Gnetum*, Muhammad and Sattler (1982) have suggested that tracheary element morphology does not rule out the possibility that *Gnetum* may have been close to the ancestral stock of at least some angiosperms. In the two species, scalariform as well as circular pits occur. Scalariform perforation plates are present as well as the large simple perforations, and these compare well with those of selected angiosperms.

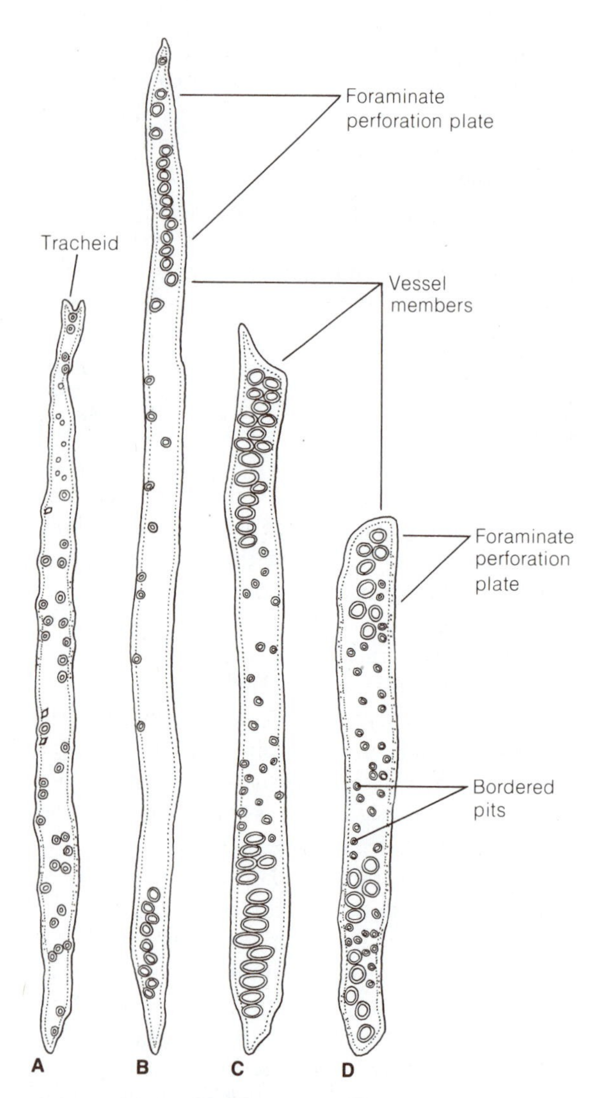

FIGURE 18-10 Tracheary elements from the secondary xylem of *Ephedra californica*. **A**, tracheid with numerous circular bordered pits; **B–D**, vessel members with foraminate perforation plates on end walls. [From *Plant Anatomy* by K. Esau. Wiley, New York. 1953.]

The Reproductive Cycle in *Ephedra*

By way of introduction we must comment briefly on the confusing "angiosperm-centered" terminology commonly used in describing the reproductive structures of the gnetophytes. To many writers the strobili of *Ephedra, Gnetum,* and *Welwitschia* are directly comparable with angiospermic inflorescences, and the parts of the gnetalean "flower" are very frequently designated by such terms as "perianth," "stamen," "anther," and "column." This kind of nomenclature is particularly confusing and misleading for *Ephedra,* which is notable for a morphology similar to conifers in its gametophytes and embryo; these structures are commonly described in the literature using "gymnosperm-centered" terminology. To eliminate the implication of homology between sporogenous structures of the Gnetophyta and those of angiosperms, the terms and general interpretations proposed by Eames (1952) will be adopted in the following resume of the life cycle of *Ephedra.*

The Strobili

Most species of *Ephedra* are strictly dioecious, and both the microsporangiate and ovulate cones are compound in structure; i.e., the cone axis bears pairs of bracts which subtend either microsporangiate or ovuliferous structures (Fig. 18-11). In certain monoecious species (e.g., *Ephedra campylopoda)* some of the strobili are *bisporangiate* with the microsporangiate structures developed in the axils of the lower bracts and the ovules located in the terminal part of the same cone.

The microsporangiate strobilus consists of a number of pairs of bracts; the lowest pairs are usually sterile, whereas each of the other pairs subtend microsporangiate shoots (Fig. 18-11; 18-12, A). The microsporangiate shoot consists of an axis bearing a pair of bracteoles, which enclose a stalked microsporophyll (or "microsporangiophore") with a number of sporangia (Fig. 18-12, B). The number of sporangia varies with the species. According to Eames (1952) the evidence from ontogeny and vascular anatomy indicates that the microsporangiophore of *Ephedra* is the result of the phylogenetic fusion of a pair of microsporophylls; in certain species, recognized as primitive by taxonomists, the two sporophylls are free and each bears a terminal cluster of four microsporangia.

The ovulate cone of *Ephedra* also consists of an axis bearing decussately arranged pairs of bracts. However, most of the bracts are sterile, and the cones of many species contain only two ovules, one

Figure 18-11 *Ephedra chilensis.* Microsporangiate (left) and megasporangiate (right) strobili. Note protrusion of tubular integument of the ovules of the megasporangiate strobili.

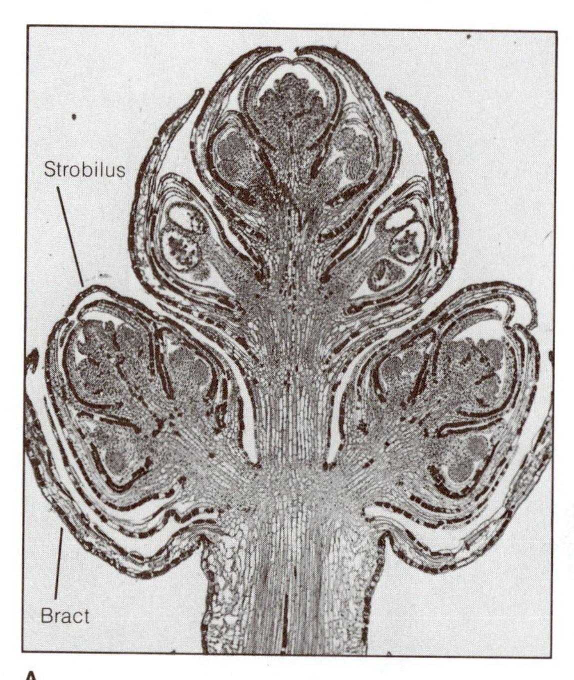

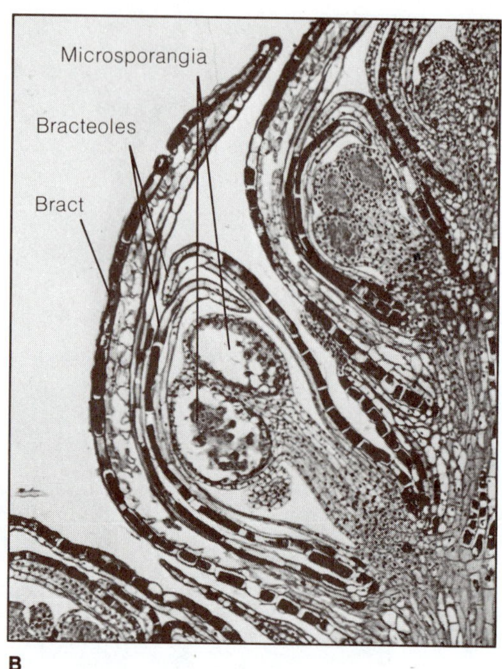

FIGURE 18-12 Structure of microsporangiate strobili in *Ephedra chilensis*. **A**, longisection of tip of reproductive shoot (each of the microsporangiate strobili is situated in the axil of a bract); **B**, enlargement of portion of **A** showing a microsporangiate shoot which consists of an axis with apical microsporangia enclosed by a pair of fused bracteoles.

in the axil of each of the upper bracts (Fig. 18-13, A). In some species the cones are *uniovulate,* and this condition is commonly the result of the abortion of one ovule and the crowding of the other into a false "terminal" position (see Eames, 1952).

The nucellus of the ovule of *Ephedra* is enclosed by two envelopes, *each* of which has been regarded as an ovular integument. Particularly notable is the marked elongation of the upper region of the integument, which protrudes from the tip of the ovule as a delicate open tube (Fig. 18-13, B). This *micropylar tube,* as it is frequently called, functions as a receptive organ for the pollen and represents one of the salient characters shared by the ovules of *all* gnetophytes. Description of the morphology of the outer envelope of the ovule of *Ephedra,* however, is highly controversial; Martens (1971) has summarized the conflicting opinions. We have adopted the view that the so-called outer integument of the ovule of *Ephedra* represents a pair of connate bracteoles, comparable to the pair of bracteoles of the microsporangiate shoot, whereas the "inner envelope" represents the only true integument.

Microsporogenesis and the Male Gametophyte

Microsporogenesis has been carefully investigated in several species of *Ephedra* (Land, 1904; Maheshwari, 1935; Singh and Maheshwari, 1962). The sporangial initials are hypodermal and divide periclinally, forming an outer layer of *primary wall cells.* The periclinal division of the layer of primary wall cells yields a single wall layer, which eventually becomes crushed, and the tapetum. During the meiotic division of the microsporocytes, the tapetal cells become multinucleate and finally degenerate. At maturity, the microsporangium contains tetrads of microspores enclosed within a thick-walled epidermis.

The early steps in the ontogeny of the endosporic male gametophyte of *Ephedra* closely paral-

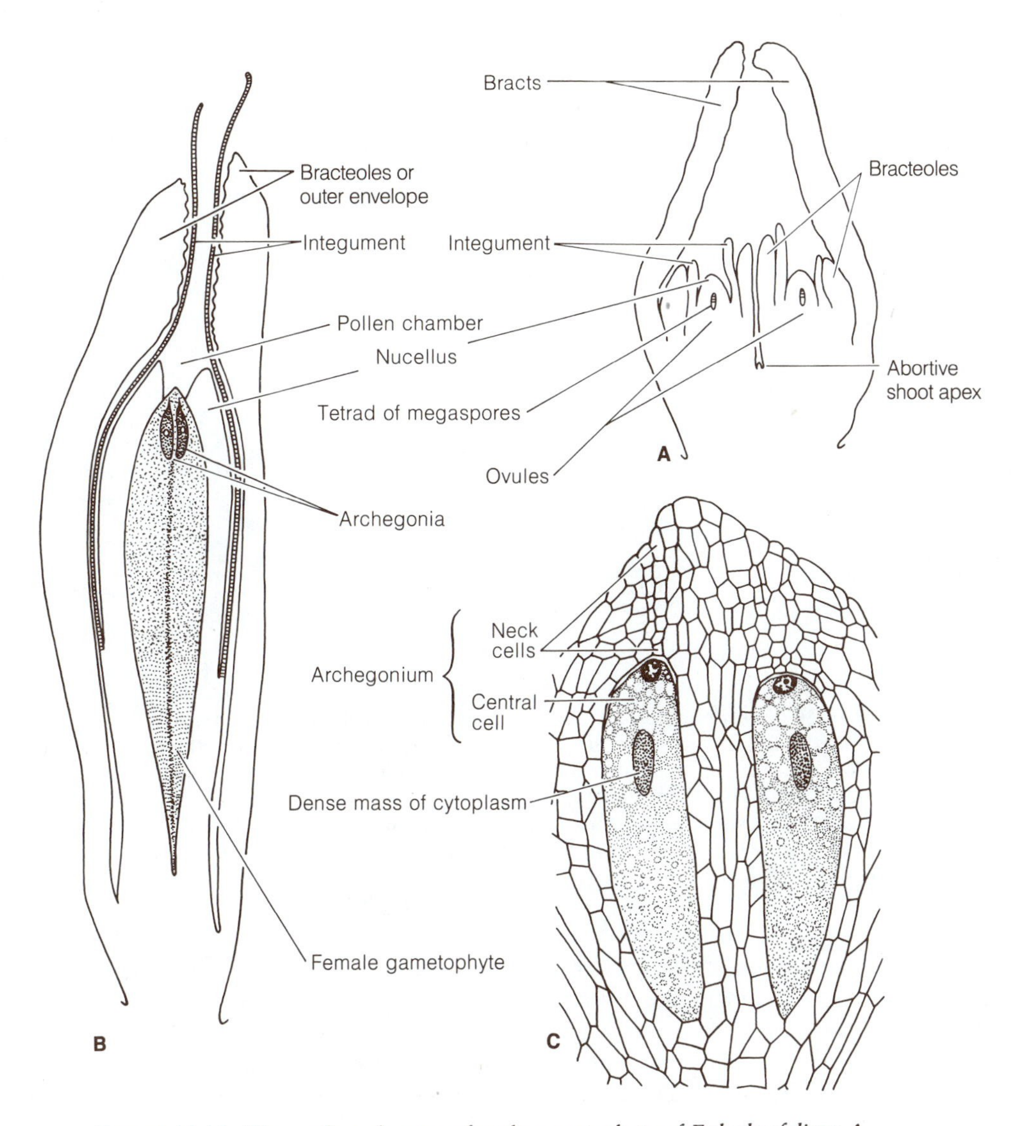

FIGURE 18-13 The ovule and mature female gametophyte of *Ephedra foliata*. **A**, longisection of megasporangiate strobilus showing two young ovules, each with a linear tetrad of megaspores; **B**, median longisection of an ovule showing the female gametophyte, the conspicuous pollen chamber, and elongated integument; **C**, details of micropylar region of female gametophyte showing structure of archegonia. [From Maheshwari, *Proc. Indian Acad. Sci.* 1:586, 1935.]

lel the mode of development of the male gametophyte of *Pinus* (see Chapter 17). The first two mitotic divisions yield two lens-shaped *prothallial cells* which begin to degenerate soon after their formation. Then the nucleus of the meristematic, or antheridial, initial divides again, forming the generative and tube cell (Fig. 18-14, A-E). The division of the generative cell produces a sterile cell and a spermatogenous cell (Fig. 18-14, E, F). In this five-celled stage in development, the pollen grain is shed from the microsporangium. There appears to be some species variation in the formation of persistent cellulose walls that delimit cells in the pollen grains (Singh and Maheshwari, 1962; Martens, 1971).

The mature pollen grains of *Ephedra* are ellipsoidal in form, and the exine is characteristically

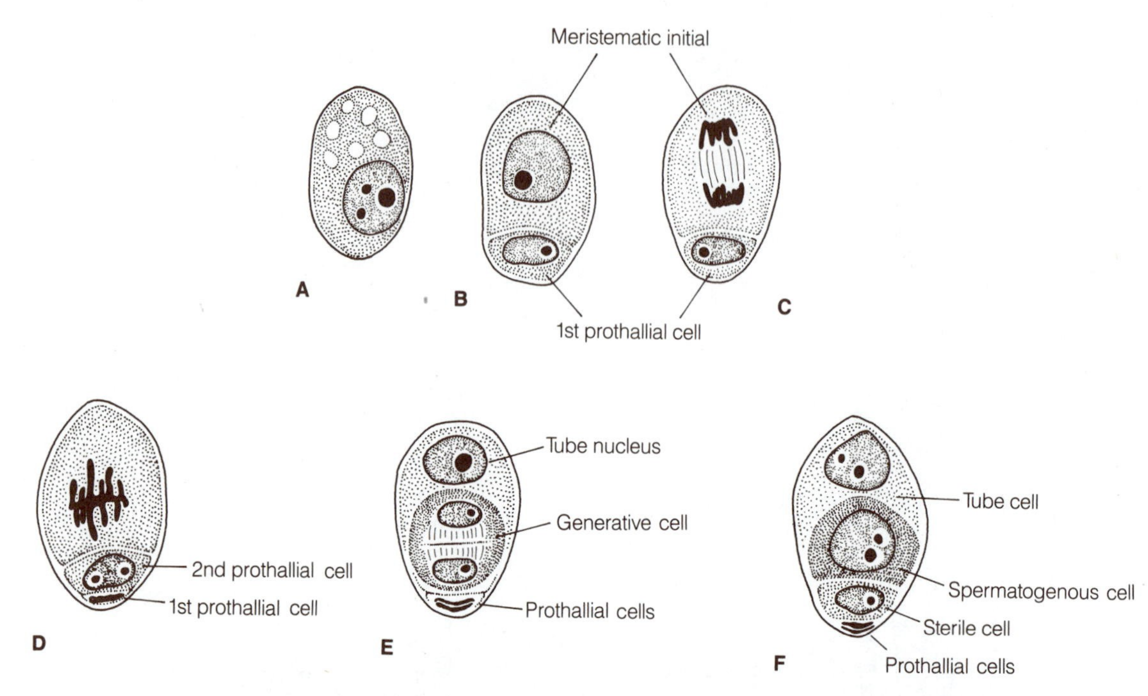

FIGURE 18-14 *Ephedra gerardiana*. Development of the male gametophyte. **A**, uninucleate microspore; **B**, microspore has divided resulting in the formation of the first prothallial cell and continuing meristematic initial; **C**, division leading to formation of second prothallial cell; **D**, metaphase, leading to formation of the generative cell and tube nucleus; **E**, division of generative cell to produce the spermatogenous cell and sterile cell as shown in **F**. [Redrawn from Singh and Maheshwari, *Phytomorphology* 12:361–372, 1962.]

sculptured into a series of ridges—extending from pole to pole—separated by longitudinal furrows (Fig. 18-15). Because of these definitive structural characters, it has been possible to identify *Ephedra* pollen in the fossil record (from the Cretaceous and Tertiary periods) and to draw structural comparisons between it and the pollen of living species (see the detailed study by Steeves and Barghoorn, 1959).

Megasporogenesis and the Female Gametophyte

The megasporocyte, prior to meiosis, is a large, conspicuous cell located rather deeply within the nucellus of the ovule (Fig. 18-13, A). According to some investigators (Maheshwari, 1935; Seeliger, 1954; Singh and Maheshwari, 1962), the sunken position of the megasporocyte is due (1) to the formation above it of several layers of parietal cells, derived from the early division of a hypodermal cell, and (2) the very active periclinal division of the cells of the nucellar epidermis. The meiotic divisions of the megasporocyte result in the formation of four megaspores (Fig. 18-13, A), although in *Ephedra distachya* wall formation between the megaspores is delayed which may give the illusion of a tetrasporic type of development found in angiosperms (Lehman-Baerts, 1967). The megaspore furthest from the micropyle enlarges, and a series of free-nuclear divisions occur in the peripheral cytoplasm that surrounds the large central vacuole. Land (1907) found 256 free nuclei in the coenocytic female gametophyte of *Ephedra trifurca* but higher numbers have been reported for other species—500 in *Ephedra foliata* and about 1,000 in *Ephedra distachya*.

Following the free nuclear phase, the female gametophyte passes through the usual *alveolation* stage as in most other gymnosperms. Anticlinal walls begin to form centripetally, forming long, tubular uninucleate alveoli which later become subdivided by periclinally oriented walls (Lehman-

Figure 18-15 Scanning electron micrograph of a group of pollen grains of *Ephedra* sp. Note the prominent longitudinal (meridional) ridges characteristic of the pollen of this genus (× 700).

Baerts, 1967). Subsequently, active cell divisions continue throughout the young gametophyte which soon becomes histologically differentiated into two regions: a lower zone (at the chalazal end) of small, compact, frequently dividing cells, and an upper zone (at the micropylar end) of longer, thinner-walled cells; certain of the superficial cells of the upper zone later function as *archegonial* initials. The number of archegonia formed varies from two to eleven, according to Martens (1971).

The archegonial initial divides periclinally into an outer *primary neck cell* and an inner *central cell.* The most distinctive feature of archegonial development in *Ephedra* is the formation of a massive neck which is produced by the repeated periclinal and anticlinal divisions of the derivatives of the primary neck cell. At first the divisions are so regular that three to five tiers of cells are produced, each tier composed of a quartet of cells. Subsequent divisions are less regular, and the cells of the archegonial neck merge with the adjacent cells of the gametophyte (Fig. 18-13, C). The mature archegonial neck may consist of about thirty to forty cells; Land (1904) remarks: "of all gymnosperms, *Ephedra* has the longest-necked archegonium." The enlargement of the central cell is followed by the division of its nucleus to form the *ventral canal nucleus* and the *egg nucleus*. Sometimes the ventral canal nucleus appears to degenerate soon after its formation, but in some species it persists and remains intact near the upper part of the archegonium (Lehman-Baerts, 1967).

Following the first division of the archegonial initial, the contiguous cells of the gametophyte divide, producing a jacket around the entire developing archegonium.

Pollination

Prior to pollination, the nucellar tissue lying directly above the archegonia of the gametophyte begins to break down, creating a very conspicuous funnel-shaped *pollen chamber* (Fig. 18-13, B). During pollen-chamber formation, the disintegrating cells, at the summit of the nucellus, produce a liquid, rich in sugar, which fills the micropylar canal of the ovule and, at the time of pollination, exudes as a "pollination drop" from the open end of the exserted integument (Fig. 18-16).

Figure 18-16 *Ephedra* sp. Basal cluster of three megasporangiate strobili. Note conspicuous "pollination drop" at tip of two of the strobili. See text for further discussion. [Courtesy of Dr. E. G. Cutter.]

Wind pollination is undoubtedly the important process in *Ephedra*. However, earlier accounts of insect pollination have been substantiated (Bino, Dafni, and Meeuse, 1984; Bino, Devente, and Meeuse, 1984). The pollination drops on ovules as well as drops of nectar on other parts of both pollen and ovulate cones contain about 10 percent sucrose; in contrast, pollination drops in pine contain only about 1.25 percent. The flying insects visit the drops of nectar and in so doing transfer pollen grains from their bodies to the ovules. By whichever pollination method, the pollen grains adhere to and float on the sticky surface of the pollination drop. As the water begins to evaporate, the column of liquid in the micropylar canal shortens and the pollen grains are pulled inward into the pollen chamber.

Fertilization

According to Land (1907), the interval between pollination and fertilization in *Ephedra trifurca* may be as short as ten hours, which is a remarkable contrast with the more extended interval typical of *Pinus* and a number of other conifers. A unique morphological feature of *Ephedra* is the fact that at the time of pollination, the archegonial end of the female gametophyte is freely exposed at the base of the deep pollen chamber (Fig. 18-13, B). As a result, when the pollen grain germinates, the pollen tube penetrates *gametophytic tissue,* i.e., the tissue of the archegonial neck. This represents a striking contrast to the growth of the pollen tube through the nucellar — i.e., *sporophytic* — tissue of the ovule, characteristic of the conifers.

Following the emergence of the pollen tube, the exine of the pollen grain is shed and the spermatogenous cell divides, forming two male gametes. After reaching the egg, the tip of the pollen tube ruptures and the two gametes, together with the sterile cell and the tube nucleus, are discharged into the egg cytoplasm.

Although one of the male gametes unites with the egg nucleus and forms a diploid zygote, the behavior of the other male gamete is extraordinary in certain species. Khan (1943), working on *Ephedra foliata,* reported that the second male gamete may fuse with the ventral canal nucleus. Although an embryo does not result from this fusion, Khan observed in one specimen, two nuclei that he considered to have been formed by the division of the fertilized ventral canal nucleus. Likewise, Moussel (1978) reported one instance in *Ephedra distachya* in which the ventral canal nucleus, which otherwise degenerates, probably fused with the second male gamete. No nutritive tissue was formed from this union as in the case of "double fertilization" in angiosperms (see Chapter 20). Moussel speculated that additional proembryonic nuclei may be formed from divisions of the second diploid nucleus.

Embryogeny

In *Ephedra* there is a process of free nuclear division, beginning with the first mitosis of the zygotic nucleus. Each of the eight (or more) free diploid nuclei that are produced becomes surrounded by a densely staining sheath of cytoplasm; later a cellulosic wall is formed around each nucleus (Fig. 18-17, A). As each of these proembryonic cells may develop *independently* into an embryo (Fig. 18-17, A–C), *Ephedra* exhibits a distinctive and precocious type of cleavage polyembryony. From an ontogenetic standpoint, polyembryony in *Ephedra* has been "pushed back" to the free nulear stage in embryogeny.

Each proembryonal cell puts out a tubular outgrowth. The nucleus may divide before the formation of the outgrowth or it may move into the tube and divide. A transverse wall is then formed, giving rise to a terminal embryonal cell and an embryonal suspensor which elongates (Fig. 18-17, C, D). As an embryo grows through the archegonium and into the female gametophyte, the terminal embryonal group of cells increases in number; some of the proximal cells elongate, forming embryonal tubes (Fig. 18-17, E), a feature characteristic of all other gymnosperms (Chapter 17). With further development, two cotyledons and the shoot apex are formed at the distal end of the embryo, and a root apical meristem becomes organized at the proximal end (Fig. 18-17, F).

Of the several embryos which competitively develop in a single ovule, only one normally reaches a fully developed stage in the seed. Kahn (1943) observed in one ovule eighteen to nineteen separate developing embryos. He interpreted this large num-

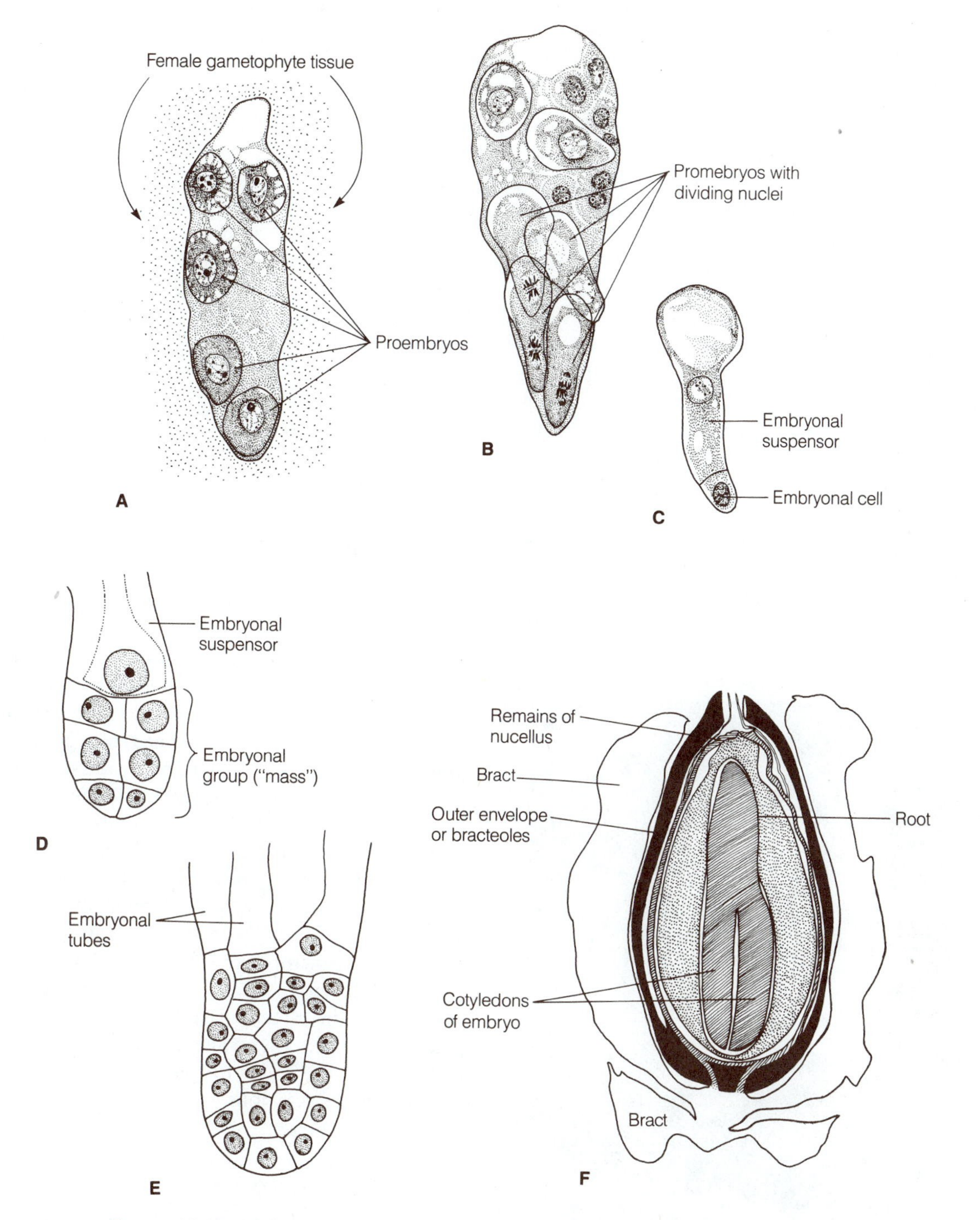

FIGURE 18-17 Embryogeny and seed structure in *Ephedra*. **A**, longitudinal section of zygotic cell showing five proembryos; **B**, reconstruction from several sections showing six proembryos in various stages of development (note mitoses and origin of embryonal suspensors in the lower proembryos); **C**, embryo consisting of embryonal cell and suspensor cell; **D**, group of embryonal cells derived from derivatives of embryonal cell in C; **E**, older developing embryo with embryonal tubes derived from the embryonal group ("mass"); **F**, longisection of mature seed. [**A–C, F** redrawn from Khan, *Proc. Nat. Acad. Sci. India* 13:357–375, 1943; **D, E** redrawn from Lehmann-Baerts, *Cellule* 67:51–87, 1967.]

ber to be the result of the combination of simple and cleavage polyembryony.

The general structure of a ripe seed in *Ephedra foliata* is shown in Fig. 18-17, F. The conspicuous embryo, with its two large cotyledons, is embedded within the tissue of the female gametophyte, and the remains of the nucellus are evident as a disorganized sheath of cells. At the micropylar end of the seed the remains of the "true" integument are evident, and the entire seed is externally enclosed by the fused bracteoles (= outer envelope), many cells of which develop thick hard walls. As the seed matures in *Ephedra foliata,* as well as many other species, the adjacent subtending bracts of the ovulate strobilus become thick and fleshy, forming an additional investment (Fig. 18-18). The bracts may become ivory, red, or orange in color. According to Land's (1907) study on *Ephedra trifurca,* there appears to be no resting, or "dormant," period for the seed, which may even begin to germinate within the parent strobilus.

FIGURE 18-18 *Ephedra chilensis.* Mature ovulate strobili, showing thick fleshy bracts.

In addition to the formation of diploid embryos through the normal processes of fertilization and embryogeny, Konar and Singh (1979) demonstrated that plantlets, consisting of shoots and roots, can be formed from the haploid female gametophyte of *Ephedra.* Callus was initiated in vitro on a basal nutrient medium plus 2 ppm 2,4-D (2,4-dichlorophenoxyacetic acid). The callus when subcultured on a medium containing 2 ppm kinetin led to the formation of plantlets.

Morphological Comparisons Between *Ephedra, Gnetum* and *Welwitschia*

The three genera of gnetophytes share the following morphological characters: compound microsporangiate and megasporangiate strobili, an extended micropylar tube formed by the integument of the ovule, vessels in the secondary xylem, shoot apices having a well-defined surface layer (tunica), decussate phyllotaxis, and embryos with two cotyledons. The phylogenetic and taxonomic significance of these points of resemblance, however, must be judged in the light of equally impressive *differences* among the genera with reference to (1) the organization of strobili, (2) the development and structure of both the male and female gametophytes, (3) the methods of fertilization, and (4) the types of embryogenesis. The details of reproduction in *Ephedra* already have been described and will not be repeated here. The reproductive features of *Gnetum* and *Welwitschia* will be emphasized in the following discussions.

The Reproductive Cycle in *Gnetum*

The strobili in *Gnetum* are compact or elongate axes with conspicuous nodes and internodes. In a microsporangiate strobilus there are two fused bracts at a node forming a cupulelike structure that partially surrounds numerous fertile shoots (often referred to as "flowers"). Each fertile shoot consists of two fused bracteoles, enclosing a microsporophyll (or microsporangiophore). Generally there are two separate microsporangia at the tip of the sporophyll. In *Gnetum gnemon,* the top whorl consists of abortive ovules (Fig. 18-19, A, B).

At each node of a megasporangiate strobilus, the "cupule" or "collar" subtends a whorl of eight to ten or fewer ovules (Fig. 18-20, A). An ovule is surrounded by three concentric insheathing structures—the outer and inner "envelopes" and the integument (Fig. 18-20, B). Many authors do not consider the outer and inner envelopes as true integuments, but Rodin and Kapil (1969) have shown that the inner envelope can become sclerified, and may become fused with the outer envelope to form a hard seed coat.

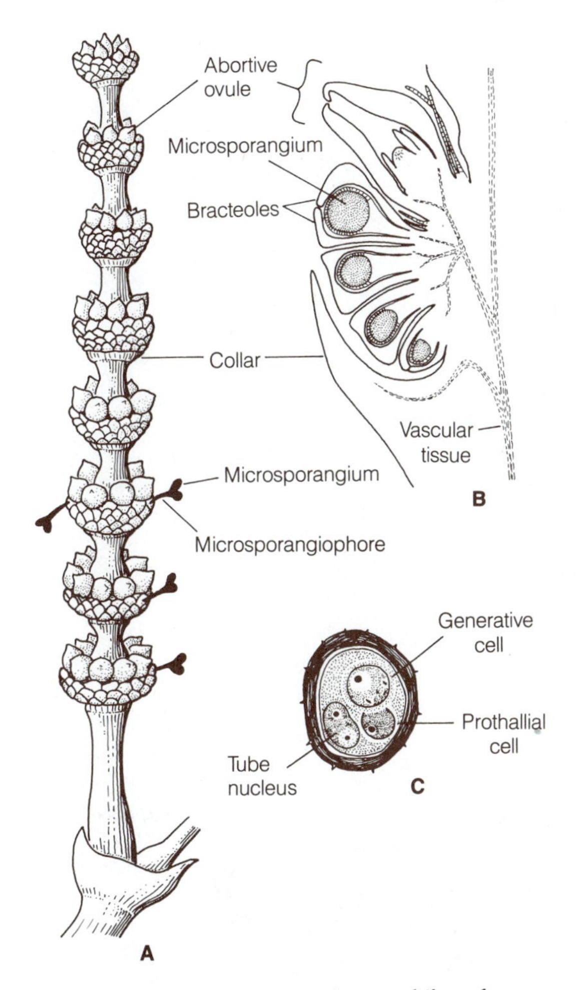

FIGURE 18-19 **A,** microsporangiate strobilus of *Gnetum gnemon;* **B,** Longitudinal section through a node, showing four developing microsporangiate fertile shoots and one abortive ovule; **C,** *Gnetum ula:* pollen grain at time of shedding from microsporangium. [**A, B** redrawn from Sanwal, *Phytomorphology* 12:243–264, 1962; **C** redrawn from Vasil, *Phytomorphology* 9:167–215, 1959.]

There are several accounts of the reproductive cycle in *Gnetum.* However, investigators differ in their descriptions, even for the same species. The following discussion is more or less representative for the genus. Terminology is that of Singh (1978).

A pollen grain at the time of shedding consists of a tube nucleus, a prothallial cell, and a generative cell (Fig. 18-19, C). When the pollen grain germinates in *Gnetum,* the generative cell in the pollen tube produces two male gametes just prior to fertilization.

In *Gnetum,* walls do not form between the four megaspore nuclei; the result is a *tetrasporic* type of female gametophyte development found in certain angiosperms (Chapter 20). The four nuclei undergo a period of free nuclear divisions resulting in 250 to 1,000 nuclei. Wall formation begins at the lower (chalazal) end of the female gametophyte, and may be initiated before or after fertilization. At the upper (micropylar) end single nuclei or groups of two or more nuclei become surrounded by membranes. These are the egg cells (Fig. 18-21, A). One or more pollen tubes push into the female gametophyte and come to lie next to the egg cells (Fig. 18-21, A). Only one male gamete in a pollen tube may fertilize an egg, but there are reports that both male gametes may be functional and two eggs could become fertilized. Following fertilization the micropylar end of the female gametophyte becomes cellular. A zygote undergoes cell division, resulting in two cells that branch repeatedly, each branch having one nucleus at its tip (Fig. 18-21, B, C). The cells elongate as they grow deep into the cellular female gametophyte; the greatly elongated cells are termed suspensor tubes (Fig. 18-21, C, D). A nucleus at the tip of a tube undergoes mitosis, followed by wall formation. By repeated cell divisions of the embryonal cell, an embryonal group or mass is formed. Some of the proximal cells elongate, forming embryonal tubes as in other gymnosperms (Fig. 18-21, E–L). Generally only one embryo develops to maturity in the seed. This type of embryogeny is another example of cleavage polyembryony. The embryo becomes dicotyledonous and has a conspicuous structure, the "feeder," that arises from the hypocotyl region and develops into

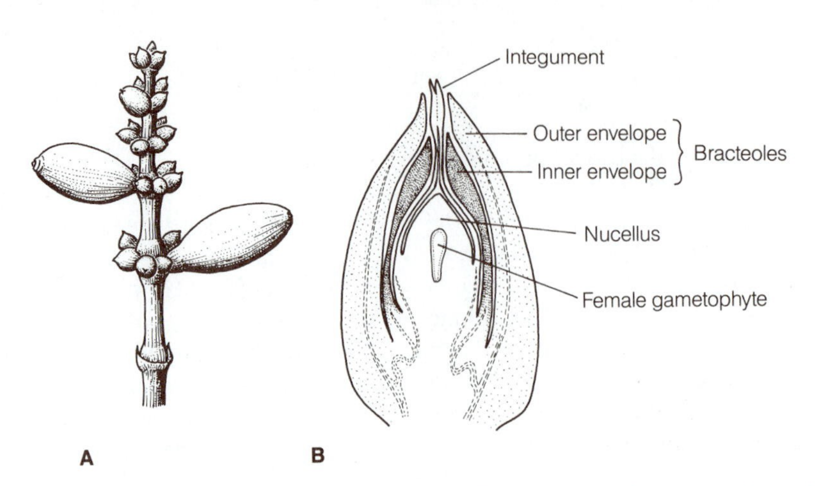

FIGURE 18-20 *Gnetum gnemon.* **A,** megasporangiate strobilus, showing ovules and partially developed seeds; **B,** longitudinal section of young ovule. [Redrawn from Sanwal, *Phytomorphology* 12:243–264, 1962.]

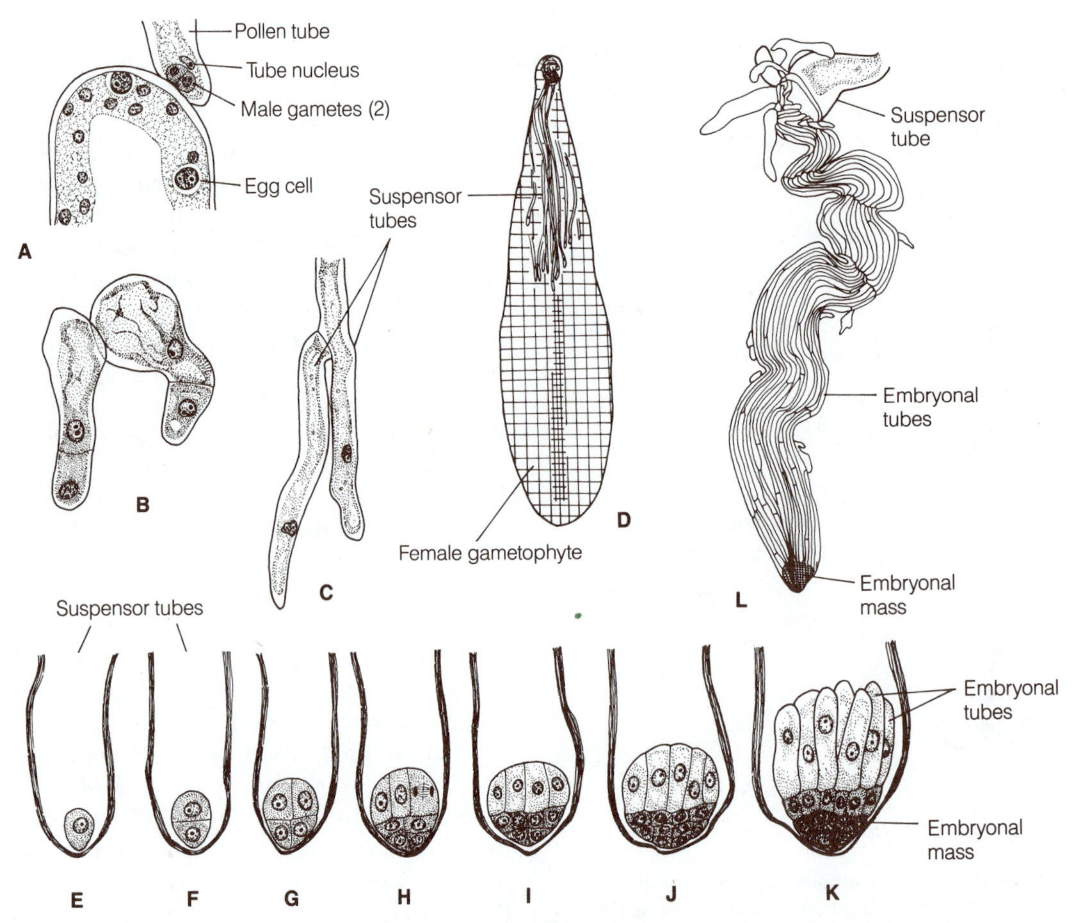

FIGURE 18-21 Early embryogeny in *Gnetum.* **A,** pollen tube appressed to female gametophyte with two differentiated egg cells; **B,** zygotes have divided forming suspensor tubes; **C,** branching suspensor tubes; **D,** longitudinal section of cellular female gametophyte with numerous suspensor tubes; **E–K,** tips of suspensor tubes showing embryonal cell and its derivatives; **L,** older developing embryo. [A redrawn from Thompson, *Amer. J. Bot.* 3:135–184, 1916; **B–D, L** redrawn from Vasil, *Phytomorphology* 9:167–215, 1959; **E–K** redrawn from Swamy, *Phytomorphology* 23:176–182, 1973.]

a large absorptive structure. Seeds are reported to be shed during early embryogeny (suspensor-tube stage); further development occurs on the ground, as in *Ginkgo*. [For additional reading on the subject, see the following: Waterkeyn (1954, 1959), Vasil (1959), Sanwal (1962), Martens (1971), Swamy (1973), and Singh (1978).]

The Reproductive Cycle in *Welwitschia*

The branched, reproductive shoot systems arise as buds from tissues of the "crests" at the bases of the two permanent leaves. Both microsporangiate and megasporangiate cones are compound in organization (Figs. 18-22, A; 18-23, A, B).

A

B

FIGURE 18-22 *Welwitschia mirabilis*. **A**, portion of a microsporangiate shoot system showing strobili, bracts, and protruding microsporangiophores; **B**, pollen grains; note median longitudinal furrow (sulcus) (× 700). [**A** courtesy of Dr. D. W. Stevenson.]

A

B

Figure 18-23 *Welwitschia mirabilis.* **A**, plant from Namib Desert showing megasporangiate (ovulate) branch systems at the base of a leaf; scars of previous branches can be seen at left; **B**, ovulate strobili showing bracts and exserted integuments with pollination droplets; from greenhouse grown plant. [**A** from Lindsay, *Pacific Discovery.* California Academy of Sciences, Vol. 35, No. 5, 1982; photograph courtesy of Mr. Edward S. Ross]

A microsporangiate strobilus (pollen cone) consists of four rows of bracts (decussate phyllotaxy). In the axil of a bract is a fertile shoot consisting of (1) two lateral unfused bracteoles, (2) two fused bracteoles forming an envelope, (3) six microsporangiophores, fused at their bases into a sheath or cup, and (4) an abortive (sterile) ovule that terminates the axis of the fertile shoot. The integument of the ovule forms a flange of tissue at the apex (Fig. 18-24). At the tip of each microsporangiophore are three fused sporangia with radial lines of dehiscence across the top. The external morphology of the pollen grain is similar to that of *Ephedra* (compare Figs. 18-15 and 18-22, B). At the time of dehiscence the pollen grain of *Welwitschia* consists of an ephemeral prothallial cell, a generative cell, and a tube cell (Martens and Waterkeyn, 1974). As with *Gnetum* some authors apply angiosperm terminology to a fertile shoot. Hence, the bracteoles constitute the perianth, the microsporangiophores are the stamens (androecium), and the abortive ovule occupies the position of the gynoecium. The entire fertile shoot is referred to as a flower. However, gymnosperm terminology will be used in this book (or, at least, noncommital terms) because the phylogeny of the gnetophytes is still to be resolved.

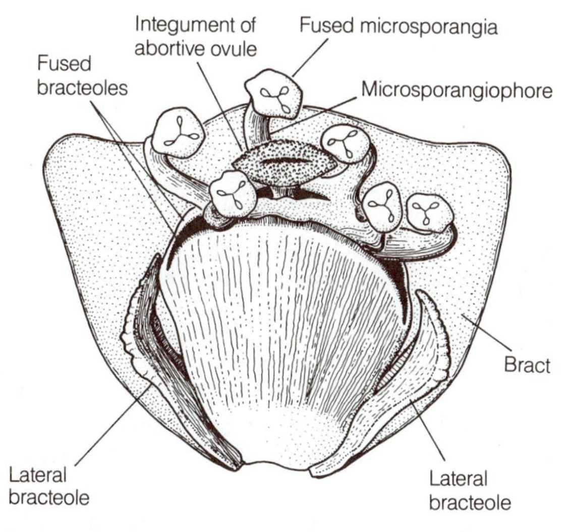

FIGURE 18-24 Fertile shoot and subtending bract from microsporangiate strobilus of *Welwitschia.* [Based on Les Gnétophytes by P. Martens. In *Handbuch der Pflanzenanatomie,* Band XII, Teil 2. Gebrüder Borntraeger, Berlin. 1971.]

The organography of a fertile shoot in a megasporangiate strobilis (ovulate cone) is very similar to that of a microsporangiate fertile shoot. There are two pairs of oppositely arranged bracteoles. Those of the lower pair are relatively small and unfused. Bracteoles of the second pair are elongate in the plane of the bract and are fused, forming an envelope around the ovule that has an exserted integument (Fig. 18-25). In the mature seed, the fused bracteoles surrounding the ovule produce winglike extensions that aid in seed dispersal.

Megasporogenesis in *Welwitschia* is similar to *Gnetum,* and in constrast to *Ephedra,* since walls are not formed between the four nuclei resulting from meiosis. The four haploid nuclei then enter a period of free nuclear divisions. The formation of alveoli does not occur as in the majority of gymnosperms. Rather, wall formation results in the enclosure of two to eight nuclei per cell in the micropylar end and six to twelve nuclei in cells at the basal or chalazal end of the female gametophyte. At the basal end the nuclei fuse to form polyploid cells that undergo cell divisions. Subsequent development is without parallel in other gymnosperms or angiosperms. Figure 18-26 should be consulted in following the sequence of developmental events. The multinucleate cells at the micropylar end of the female gametophyte grow up into the nucellus, forming female gametophytic tubes (FGTs) into which the nuclei migrate. Pollen tubes growing downward in the nucellus meet and fuse with the upward growing multinucleate FGTs. Fertilization

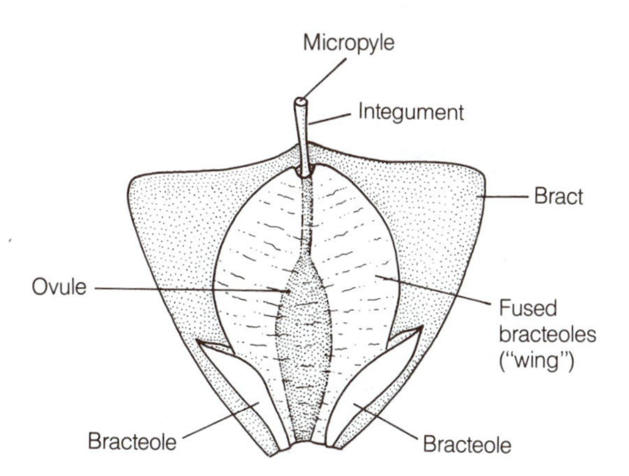

FIGURE 18-25 Fertile shoot and subtending bract from megasporangiate strobilus of *Welwitschia.*

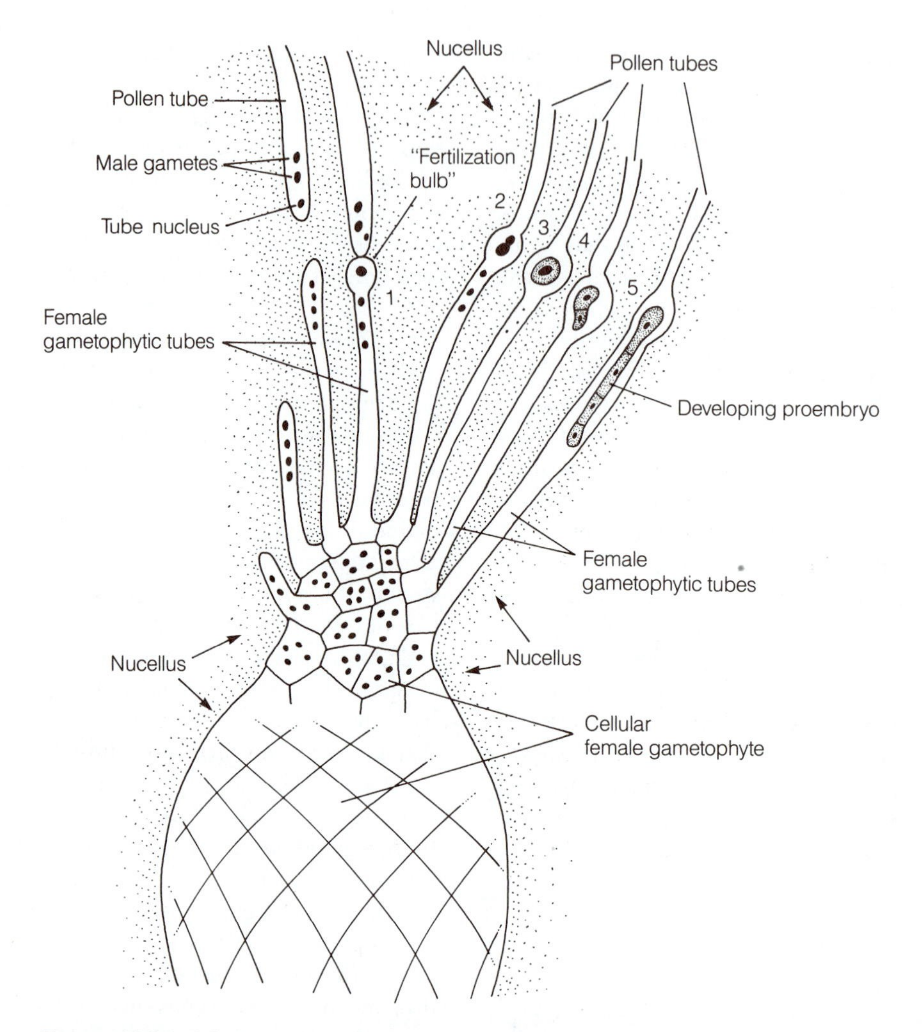

FIGURE 18-26 Schematic representation of the developmental events leading up to fertilization and proembryo development in *Welwitschia*. Starting from the left, certain multinucleate female gametophyte cells form tubes that grow up into the nucellus. Pollen tubes growing down in the nucellus meet the upward growing tubes of the female gametophyte (**1**); wall dissolution occurs and fertilization occurs in the "fertilization bulb" (**2**) to form a zygote (**3**); zygote divides, forming a two-celled proembryo (**4**); the proembryo grows down inside the female gametophytic tube toward the nutritive tissue below (**5**).

of one of the nuclei by a male gamete occurs in a bulbous tip ("fertilization bulb") of a FGT. A zygote is formed which, upon division, gives rise to an embryonal suspensor cell and an embryonal cell. The latter cell divides repeatedly, giving rise to additional embryonal suspensor cells that elongate, and the entire proembryo grows toward the female gametophyte *inside* the FGT. During the growth of the embryo, a terminal group of embryonal cells (embryonal "mass") is produced from which numerous embryonal tubes are formed as they are in other gymnosperms (Fig. 18-27). Many embryos begin development but normally only one completes its development.

Later stages of development occur within the female gametophyte from which the developing

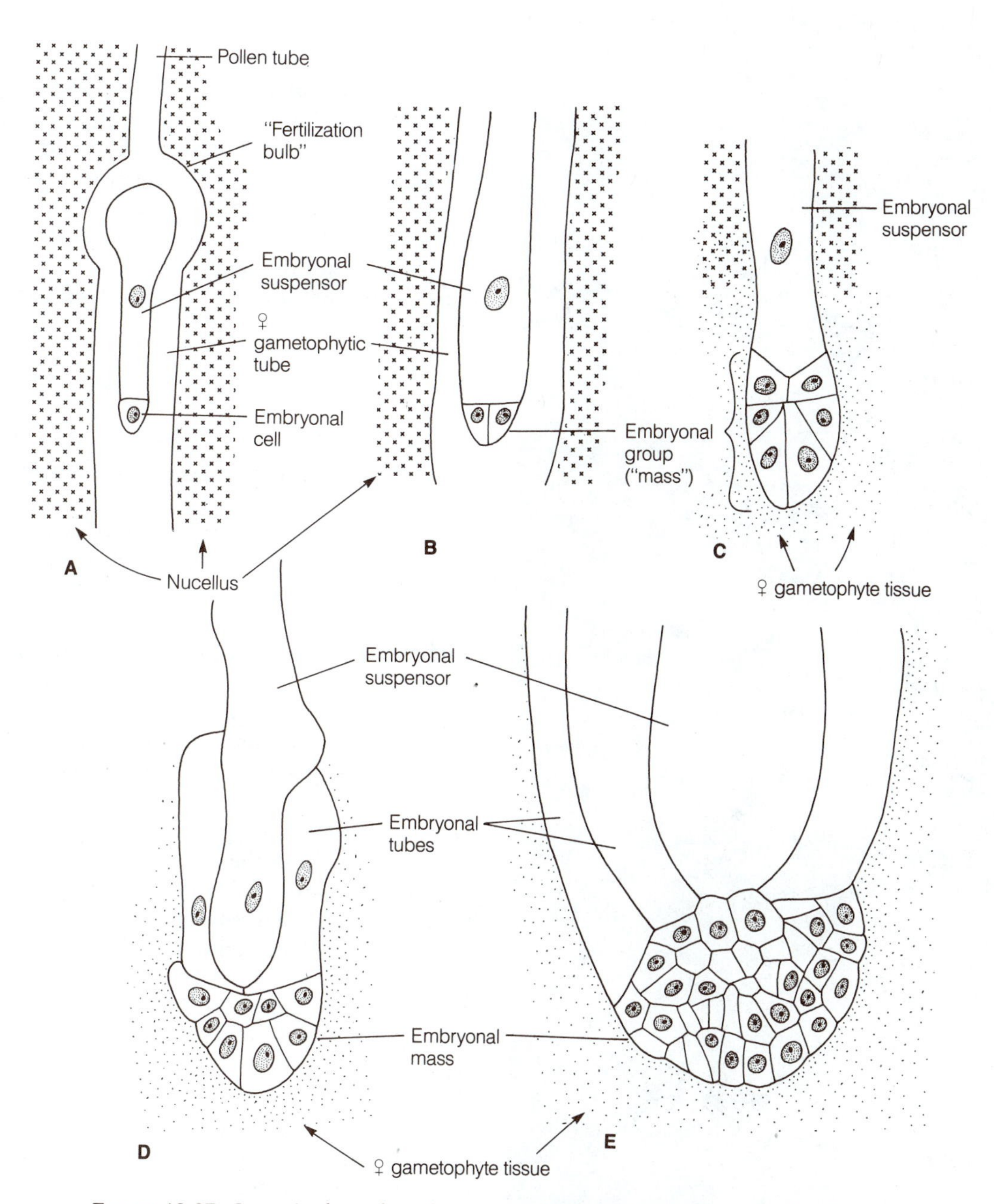

FIGURE 18-27 Stages in the early embryogeny of *Welwitschia*. **A**, the zygote has divided forming a two-celled proembryo growing inside the female gametophytic tube; **B**, embryonal cell has divided to form two embryonal cells; **C**, developing young embryo has grown into female gametophyte tissue; **D**, **E**, later stages in which elongate embryonal tubes have been produced by cells of the embryonal group (mass). [Based in part on Martens and Waterkeyn, *Cellule* 70:163–258, 1974.]

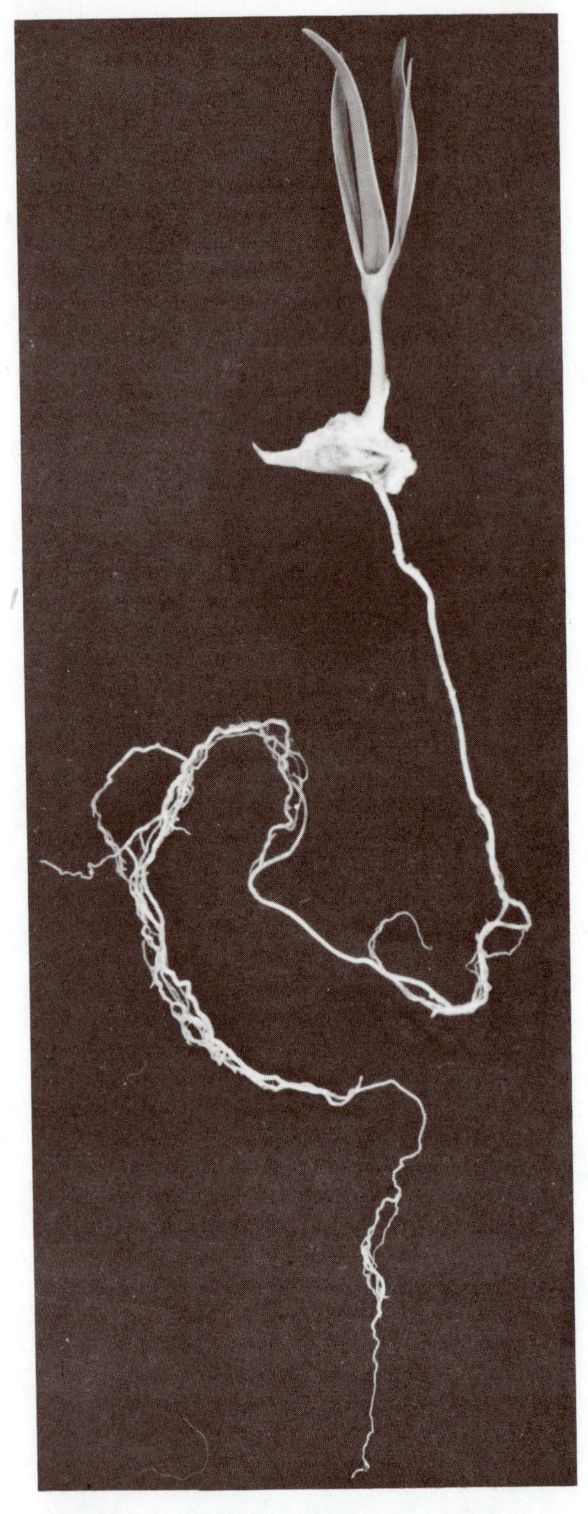

FIGURE 18-28 Seedling of *Welwitschia* showing two cotyledons and the two young permanent leaves; note that a prominent tap root is formed. [Courtesy of Dr. B. Dehgan.]

embryo receives its nutrition. In a ripe seed the embryo consists of a shoot apex, two cotyledons, hypocotyl, primary root (radicle), and a lateral bulge that develops into a "feeder" when the seed germinates (Fig. 18-28). Cells of the female gametophyte contain numerous lipid and protein bodies. When the seed germinates, enzymes are released that break down the food reserves, making them available to the growing embryo (Butler et al., 1979).

Gnetophytes, a Heterogeneous Group

We must emphasize that the gnetophytes represent, morphologically and phylogenetically, a paradoxical group of seed plants. On the one had, as Chamberlain (1935) maintained, the *combination* of certain characters — such as the compound nature of strobili and the presence of vessels in the xylem — is sufficient to keep the three genera together. On the other hand, Eames (1952) considered *Ephedra* to be closer to the conifers than to *Gnetum* or *Welwitschia* based upon differences in the organization of strobili. Details of the reproductive cycles also tend to support this belief (Table 18-1).

As can be seen in Table 18-1, *Gnetum* and *Welwitschia* share certain reproductive characters, but the methods of fertilization and embryo development are quite different and specialized in the two. The differences among all three genera suggest that the gnetophytes are a very heterogeneous assemblage of plants. This was the view of Martens (1971), who remarked that he was more struck by the contrasts among the three genera than by the traits they have in common. *Gnetum*, at various times, has been proposed as a possible progenitor of angiosperms, but the gnetophytes as well as all extant gymnosperms lack "pollenkitt" (a sticky lipoidal substance on the surface of pollen grains). Pollenkitt is apparently restricted to angiosperms, leading one scientist to rule out the gnetophytes as progenitors of angiosperms (Hesse, 1984). However, the gnetophytes for many years have been implicated in angiosperm phylogeny; more will be said on the subject in later chapters.

Table 18-1 Comparison of certain reproductive characteristics in *Ephedra, Gnetum,* and *Welwitschia*

	Ephedra	Gnetum	Welwitschia
Sterile (stalk) cell in the pollen grain	Present	Absent	Absent
Female gametophyte monosporic in origin	Present	Absent (Tetrasporic)	Absent (Tetrasporic)
Alveolation in development of female gametophyte	Present	Absent	Absent
Archegonia	Present	Absent	Absent
A free nuclear period in embryogenesis	Present	Absent	Absent

REFERENCES

Arber, E. A., and J. Parkin
1908. Studies on the evolution of the angiosperms. The relationship of the angiosperms to the Gnetales. *Ann. Bot.* 22: 489–515.

Bailey, I. W.
1953. Evolution of the tracheary tissue of land plants. *Amer. Jour. Bot.* 40:4–8.

Bino, R. J., A. Dafni, and A. D. J. Meeuse
1984. Entomophily in the dioecious gymnosperm *Ephedra aphylla* Forsk. (=*E. alte* C. A. Mey.), with some notes on *E. campylopoda* C. A. Mey. I. Aspects of the entomophilous syndrome. *Proc. Kon. Ned. Akad. Wetensch.* (Amsterdam), Ser. C, 87:1–13.

Bino, R. J., N. Devente, and A. D. J. Meeuse
1984. Entomophily in the dioecious gymnosperm *Ephedra aphylla* Forsk. (=*E. alte* C. A. Mey.), with some notes on *E. campylopoda* C. A. Mey. II. Pollination droplets, nectaries, and nectarial secretion in *Ephedra. Proc. Kon. Ned. Akad. Wetensch.* (Amsterdam), Ser. C, 87:15–24.

Bornman, C. H.
1972. *Welwitschia mirabilis:* Paradox of the Namib Desert. *Endeavor* 31(113):95–99.
1978. *Welwitschia: Paradox of a Parched Paradise.* C. Struik Publishers, Cape Town.

Butler, V., C. H. Bornman, and W. A. Jensen
1979. *Welwitschia mirabilis:* Fine structure of the germinating seed. II. Cytochemistry of gametophyte and feeder, *Z. Pflanzenphysiol.* 91:197–210.

Chamberlain, C. J.
1935. *Gymnosperms. Structure and Evolution.* University of Chicago Press, Chicago.

Crane, P. R., and G. R. Upchurch, Jr.
1987. *Drewria potomacensis* gen. et sp. nov., an early Cretaceous member of Gnetales from the Potomic group of Virginia. *Amer. J. Bot.* 74:1722–1736.

Cutler, H. C.
1939. Monograph of the North American species of the genus *Ephedra. Ann. Missouri Bot. Gard.* 26:373–428.

Dittrich, P., and W. Huber
1975. Carbon dioxide metabolism in members of the Chlamydospermae. In M. Avron (ed.), *Proc. Third Intl. Congress on Photosynthesis,* pp. 1573–1578. Elsevier, Amsterdam.

Dyer, R. A., and I. C. Verdoorn
1972. Science or sentiment: the *Welwitschia* problem. *Taxon* 21:485–489.

Eames, A. J.
1952. Relationships of the Ephedrales. *Phytomorphology* 2:79–100.

Evert, R. F., C. H. Bornman, V. Butler, and M. G. Gilliland
1973a. Structure and development of the sieve-cell protoplast in leaf veins of *Welwitschia. Protoplasma* 76:1–21.
1973b. Structure and development of sieve areas in leaf veins of *Welwitschia. Protoplasma* 76:23–34.

Foster, A. S.
1972. Venation patterns in the leaves of *Ephedra. Jour. Arnold Arboretum* 53:364–378.

Gifford, E. M., Jr.
1943. The structure and development of the shoot apex of *Ephedra altissima* Desf. *Bull. Torrey Bot. Club* 70:15–25.

Hesse, M.
1984. Pollenkitt is lacking in Gnetatae: *Ephedra* and

Welwitschia; further proof for its restriction to the angiosperms. *Pl. Syst. Evol.* 144:9–16.

Hooker, J. D.
1863. On *Welwitschia,* a new genus of Gnetaceae. *Trans. Linn. Soc. London* 24:1–48.

Johnson, M. A.
1950. Growth and development of the shoot of *Gnetum gnemon* L. I. The shoot apex and pith. *Bull. Torrey Bot. Club* 77:354–367.

Khan, R.
1943. Contributions to the morphology of *Ephedra foliata* Boiss. II. Fertilization and embryogeny. *Proc. Nat. Acad. Sci. India* 13:357–375.

Konar, R. N., and M. N. Singh
1979. Production of plantlets from the female gametophytes of *Ephedra foliata* Boiss. Z. Pflanzenphysiol. 95:87–90.

Land, W. J. G.
1904. Spermatogenesis and öogenesis in *Ephedra trifurca. Bot. Gaz.* 38:1–18.
1907. Fertilization and embryogeny in *Ephedra trifurca. Bot. Gaz.* 44:273–292.

Lehmann-Baerts, M.
1967. Études sur les Gnétales. XII. Ovule, gamètophyte femelle et embryogenèse chez *Ephedra distachya* L. *Cellule* 67:51–87.

Maheshwari, K.
1935. Contributions to the morphology of *Ephedra foliata* Boiss. I. The development of the male and female gametophytes. *Proc. Indian Acad. Sci.* 1:586–606.

Maheshwari, P., and V. Vasil
1961. *Gnetum.* Council of Scientific and Industrial Research, New Delhi.

Marsden, M. P. F., and T. A. Steeves
1955. On the primary vascular system and the nodal anatomy of *Ephedra. Jour. Arnold Arboretum* 36:241–258.

Martens, P.
1971. *Les Gnétophytes. (Handbuch der Pflanzenanatomie,* Band 12, Teil 2.) Gerbrüder Borntraeger, Berlin.
1977. *Welwitschia mirabilis* and neoteny. *Amer. J. Bot.* 64:916–920.

Martens, P., and L. Waterkeyn
1963. The shoot apical meristem of *Welwitschia mirabilis* Hooker. *Phytomorphology* 13:359–363.
1974. Études sur les Gnétales. XIII. Recherches sur *Welwitschia mirabilis.* V. Évolution ovulaire et embryogenèse. *Cellule* 70:163–258.

Moussel, B.
1978. Double fertilization in the genus *Ephedra Phytomorphology* 28:336–345.

Muhammad, A. F., and R. Sattler
1982. Vessel structure of *Gnetum* and the origin of angiosperms. *Amer. J. Bot.* 69:1004–1021.

Pearson, H. H. W.
1929. *Gnetales.* Cambridge University Press.

Pulle, A.
1938. The classification of the spermatophytes. *Chron. Bot.* 4:109–113.

Rodin, R. J.
1953. Seedling morphology of *Welwitschia Amer. Jour. Bot.* 40:371–378.
1958a. Leaf anatomy of *Welwitschia.* I. Early development of the leaf. *Amer. Jour. Bot.* 45:90–95.
1958b. Leaf anatomy of *Welwitschia.* II. A study of mature leaves. *Amer. Jour. Bot.* 45:96–103.
1966. Leaf structure and evolution in American species of *Gnetum. Phytomorphology* 16:56–68.
1967. Ontogeny of foliage leaves in *Gnetum. Phytomorphology* 17:118–128.

Rodin, R. J., and R. N. Kapil
1969. Comparative anatomy of the seed coats of *Gnetum* and their probable evolution. *Amer. J. Bot.* 56:420–431.

Rodin, R. J., and G. S. Paliwal
1970. Nodal anatomy of *Gnetum ula. Phytomorphology* 20:103–111.

Sanwal, M.
1962. Morphology and embryology of *Gnetum gnemon* L. *Phytomorphology* 12:243–264.

Schulze, E.-D., and I. Schulze
1976. Distribution and control of photosynthetic pathways in plants growing in the Namib Desert, with special regard to *Welwitschia mirabilis* Hook. *fil. Madoqua* 9(3):5–13.

Seeliger, I.
1954. Studien am Sprossvegetationskegel von *Ephedra fragilis var. campylopoda* (C. A. Mey.) Stapf. *Flora* 141:114–162.

Singh, H.
1978. *Embryology of Gymnosperms. (Handbuch der Pflanzenanatomie,* Band X, Teil 2.) Gebrüder Borntraeger, Berlin.

Singh, H., and P. Maheshwari
1962. A contribution to the embryology of *Ephedra gerardiana. Phytomorphology* 12:361–372.

Stapf, I.
1889. Die Arten der Gattung *Ephedra*. *Denkschr. Akad. Wien. Math.-Naturw.* Cl. II. 56:1–112.

Steeves, M. W., and E. S. Barghoorn
1959. The pollen of *Ephedra*. *Jour. Arnold Arboretum* 40:221–255.

Swamy, B. G. L.
1973. Contributions to the monograph on *Gnetum*. I. Fertilization and proembryo. *Phytomorphology* 23:176–182.

Thompson, W. P.
1918. Independent evolution of vessels in Gnetales and angiosperms. *Bot. Gaz.* 65:83–90.

Vasil, V.
1959. Morphology and embryology of *Gnetum ula* Brongn. *Phytomorphology* 9:167–215.

von Willert, D. J.
1985. *Welwitschia mirabilis*—new aspects in the biology of an old plant. *Adv. Bot. Res.* 11:157–191.

Waterkeyn, L.
1954. Études sur les Gnétales. I. Le strobile femelle, l'ovule et la graine de *Gnetum africanum* Welw. *Cellule* 56:103–146.
1959. Études sur les Gnétales. II. Le strobile mâle, la microsporogenèse et le gamétophyte mâle de *Gnetum africanum* Welw. *Cellule* 60:5–78.

CHAPTER 19

Magnoliophyta (Angiosperms)

GENERAL MORPHOLOGY AND EVOLUTION

THE angiosperms, or flowering plants, constitute the dominant and most ubiquitous vascular plants of modern floras on earth. The term angiosperm (literally, a vessel seed) designates one of the most definitive characteristics of flowering plants, namely the enclosure of the ovules or potential seeds within a hollow ovary. In this respect angiosperms are considered to be advanced, as compared with the naked seeded gymnosperms.

Angiosperms far exceed all other major groups of living plants in number and diversity of form and structure; more than 200,000 different species have been named and classified. Flowering plants occupy a very wide range of ecological habitats (including both salt and fresh water), and extend far toward the polar extremities of the earth. Although scientific study of the classification, morphology, and geographical distribution of angiosperms has been pursued for more than 200 years, we still do not possess even a reasonably complete census of living flowering plants. Modern scientific botany had its origin in the studies of the north temperate angiosperms of England and Europe, and our knowledge of the richly diversified floras of tropical areas is still far from complete. Such regions of the earth as New Guinea, New Caledonia, tropical Asia and Africa, and the vast Amazon rain forest still await comprehensive botanical exploration. A fuller knowledge of the angiosperms in these regions may ultimately modify many of our present concepts.

Aside from bacteria and pathogenic fungi, angiosperms are the plants that most obviously affect the existence of man on earth. The basic food supply of the world is derived from the seeds and fruits of angiosperms (rice, wheat, corn, are outstanding examples), and fibers, wood, drugs, and other products of great economic value come from them.

Thus far the fossil record of the angiosperms has failed to provide any reliable information of a specific group of plants ancestral to the angiosperms. Leaf form, venation, and pollen indicate that the angiosperms were present in the Lower Cretaceous (Hickey and Doyle, 1977), and by mid-Cretaceous they had reached a high degree of morphological specialization. Even if allowance is made for errors in identifying genera from leaf impressions, it is clear that by the latter part of the Mesozoic many modern families were clearly defined. How do we interpret this apparently sudden appearance of angiosperms in the Cretaceous? Does it indicate that angiosperms evolved at a faster rate than did gymnosperms and lower vascular plants, or does this abrupt appearance of angiosperms in the fossil record indicate a long pre-Cretaceous period in their evolutionary development of which we have no definitive proof?

It may not be possible to answer these questions in any satisfactory manner. Some paleobotanists believe that the angiosperms probably evolved over an extremely long period of time. Their so-called abrupt rise to dominance may be only an illusion. Axelrod (1952) has discussed the paleobotanical evidence suggesting the existence of angiosperms during the Triassic and Jurassic; he postulates that they may even have been in existence at the end of the Paleozoic.

There is the possibility that angiosperms developed initially in upland areas subject to erosion and thus most unlikely areas for the preservation of plant parts (Axelrod, 1952, 1961, 1970); others (e.g., Hughes, 1974) take exception to the "upland origin hypothesis." An interesting, frequently cited example of an early angiosperm in the Late Triassic is provided by impressions of striated laminae of a monocot-type leaf named *Sanmiguelia lewisii* (Brown, 1956; Tidwell et al., 1977). More recently, better preserved specimens have been found from the Late Triassic in Texas (Cornet, 1986). Fossils include stems with roots attached, leaves, and reproductive structures that are angiospermlike in many ways.

Because the fossil record is sparse and equivocal, theories regarding the ancestral stock or stocks of the angiosperms are extremely speculative and contradictory. Nearly every group of vascular plants—including nonseed plants—has been suggested at one time or another as a possible "precursor" of the angiosperms. At present one of the most widely held views is that the "ancestors" of the angiosperms should be sought amongst the gymnosperms. At various times many gymnospermous orders have been suggested as the progenitors, e.g., Pteridospermales, Caytoniales, Glossopteridales, Cycadales, Cycadeoidales, and the gnetophytes. As pointed out by Dilcher (1979) all of these gymnosperms do exhibit certain angiospermlike tendencies: the occurrence of one or more ovules in cupules (Pteridospermales, Caytoniales) and the aggregation of reproductive structures into strobiloid structures protected by bracts during ontogeny (Cycadeoidales). In the living gnetophytes there are vessels in the xylem, and they possess the so-called perianth parts (bracteoles) surrounding the "stamens" (microsporangiophores) and ovules. For more than seventy years the most generally accepted concept of the *most primitive* flower has been the magnolia-type—that is monoclinous (stamens and carpels in one flower) with numerous spirally arranged perianth parts, numerous stamens and carpels, and with a flower that was entomophilous (insect pollinated). According to this concept all other angiosperm flowers have been derived from the "prototype" by reduction in floral parts, fusion of similar parts, adnation, and modification to a diclinous condition (stamens and carpels in separate flowers). The latter condition is commonly correlated with wind pollination (anemophily). However, as Dilcher (1979) has pointed out, the fossil record does not support the tenet that the magnolia-type flower is *the primitive* condition in angiosperms, but may have been only one of the prototypes. By the mid-Cretaceous there were other types, such as catkins (elongate axes), bearing either pollen-producing florets or carpellate florets. Of course the "abominable mystery" of the *origin* of angiosperms, as stated by Darwin, still remains to be solved.

Classification

The studies of John Ray in the seventeenth century on the structure of seeds eventually led to the recognition of two major groups within living angiosperms: Dicotyledoneae and Monocotyledoneae (see Engler, 1926, 1964). According to Cronquist's (1968) rough estimate, approximately 165,000 species of dicotyledons and about 55,000 species of monocotyledons have been described. In the second edition of this book, these two groups were recognized as two subclasses of the class Angiospermopsida in the division Tracheophyta. As discussed in Chapter 1 we have given divisional rank to major groups of vascular plants. We have adopted the classification of Cronquist et al. (1966) whereby the angiosperms are designated by the term Magnoliophyta (based upon *Magnolia* as the type genus) and further divided into two classes (Cronquist,

Table 19-1 Comparison of the systems of angiosperm classification used in the second edition of this book and the present one. Only the major taxa are shown. See Cronquist (1981) for a complete outline of classification

Second Edition	Present Edition
Tracheophyta (division)	abandoned
Angiospermopsida (class)	Magnoliophyta (division; angiosperms)
Dicotyledoneae (subclass)	Magnoliopsida (class; dicotyledons)
Monocotyledoneae (subclass)	Liliopsida (class; monocotyledons)

1981), the Magnoliopsida and Liliopsida, corresponding to the dicotyledons and monocotyledons of other systems of classification (Table 19-1). For purposes of discussion the two latter, more common terms will be used rather than the technical terms Magnoliopsida and Liliopsida.

Some of the general morphological characteristics of dicotyledons and monocotyledons are shown in Table 19-2. The difference in the number of cotyledons present in the embryo — one in the monocotyledons and two in the dicotyledons — provides the most familiar distinction between the two classes of angiosperms, although as discussed in Chapter 20, this classical distinction is by no means without exceptions. As with cotyledon number, the additional distinctions between dicotyledons and monocotyledons intergrade or have exceptions, as will be pointed out at appropriate places in the presentation of the vegetative and reproductive morphology of the angiosperms (Chapters 19 and 20).

Organography and Anatomy

The extensive descriptions and interpretations of organography given by Goebel (1905) and Troll (1935 – 1969), and the comprehensive accounts of comparative anatomy found in the treatises by Solereder (1908), Metcalfe and Chalk (1950), Bailey (1954), Esau (1965), and Fahn (1974), ably demonstrate the structural complexity and diversity of living angiosperms. But it should be emphasized that many of our past concepts and generalizations, es-

Table 19-2 Comparison of certain morphological characters between dicotyledons and monocotyledons.

Magnoliophyta (division)	
Magnoliopsida (class) (dicotyledons)	Liliopsida (class) (monocotyledons)
1 Two cotyledons in embryo (exceptions)	1 Single cotyledon
2 Vascular bundles of stem arranged in a cylinder	2 Vascular bundles commonly irregularly disposed in stem, or in one or more rings as seen in transverse section
3 Herbaceous and woody forms; woody species with vascular cambium giving rise to secondary xylem and phloem	3 Essentially herbaceous; some woody forms with specialized cambium giving rise to secondary vascular bundles
4 Leaves with reticulate venation; ultimate veins often ending freely in mesophyll	4 Leaves frequently have "parallel" veins but which are fused near the tip of the leaf; also have small interconnecting veins
5 Flower parts commonly in fours or fives or multiples thereof, or in indefinite numbers; certain floral parts greatly reduced in numbers, or absent in many species	5 Flower parts basically in threes or multiples of three

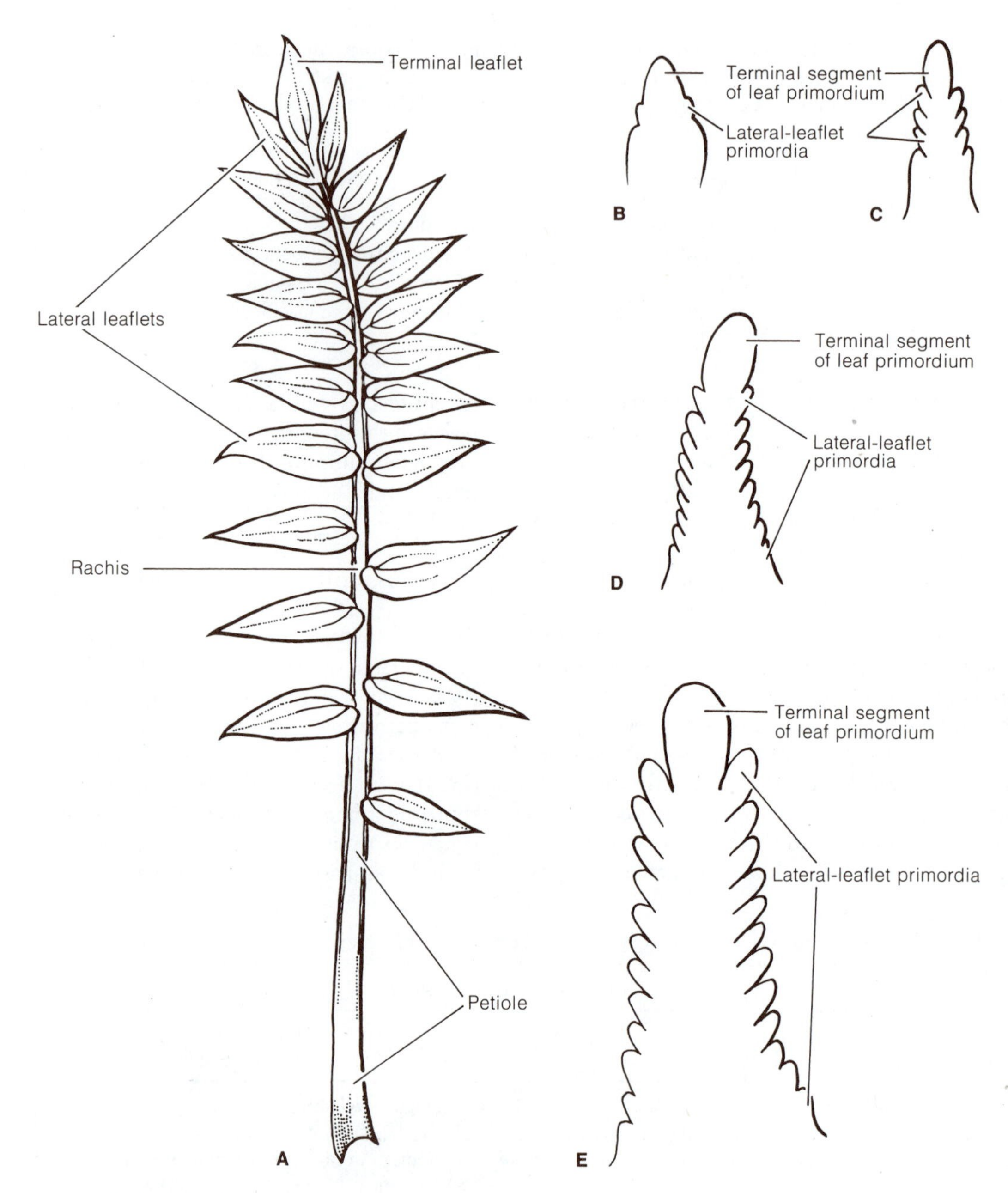

FIGURE 19-1 Morphology and development of the leaf of *Polemonium caeruleum*. **A**, adult leaf; **B–E**, successive stages in the basipetal formation of lateral leaflet primordia. [Redrawn from Troll, *Nova Acta Leopoldina* n.f. 2:315, 1935.]

pecially of comparative anatomy, were based to a very large extent on the study of the dicotyledons rather than on a comprehensive and balanced knowledge of *both* of the classes of the angiosperms. Fortunately, broad surveys of the systematic anatomy of the vegetative organs of the monocotyledons were initiated at the Jodrell Laboratory, Royal Botanic Gardens, Kew, England. The result of these and later studies have been published, under the able editorship of Dr. C. R. Metcalfe, as a series of volumes bearing the general title *Anatomy of the Monocotyledons* (see Metcalfe and

Chalk, 1960, 1971; Tomlinson, 1961, 1969; Cutler, 1969; Ayensu, 1972). These monographs are not only important contributions to the comparative anatomy and systematics of the monocotyledons, but they are reminders that great caution must be observed when generalizations are attempted regarding the trends of anatomical specialization in this group of flowering plants.

General Morphology of Foliage Leaves

The truly enormous range in form and organization of the foliage leaves of angiosperms precludes a comprehensive presentation. Our account, therefore, will be necessarily brief and limited to the general morphology of leaves and the unsolved problems of their evolutionary history. Without question, Wilhelm Troll's (1935a, 1935b, 1937, 1938, 1939a, 1939b) comprehensive analyses of leaf morphology are outstanding in quality, and his publications are valuable for detailed information and for references to the voluminous literature.

Although the foliage leaf in some angiosperms appears to consist only of a leaf blade — and for this reason is described as *sessile* — more commonly the leaf consists of three parts that are more-or-less well defined: (1) the *leaf base*, developed, in many species, as a sheath or provided with a pair of stipules; (2) the *petiole*, and (3) the leaf blade, or *lamina*. The form, proportions, and structure of each of these parts, but particularly of the lamina, vary widely, not only between the leaves of different taxa but even among the succession of leaf types produced during the seedling and post-seedling phases of development of a single species. In "simple leaves," the undivided lamina is highly variable in form and may have entire, dentate, serrate, or crenate margins. Leaves in which the lamina is more-or-less conspicuously divided into pinnately or palmately arranged *lobes* are also common, especially in various genera or species of dicotyledons (Fig. 19-9, H).

The most complex type of lamina organization occurs in "compound leaves," characterized by the formation of separate pinnae or leaflets attached in various ways to the portion of the leaf axis known as the *rachis*. In palmately compound leaves, the individual leaflets radiate in a digitate pattern from a very short rachis; in pedately compound leaves, such as those of *Kingdonia*, the lamina is composed of a series of lobed "segments" attached to a transversely expanded rachis (Fig. 19-12). In the most familiar and widespread type of compound leaf, the rachis is elongated and bears two rows of simple or divided leaflets; the leaflets may be arranged alternately or in pairs along the rachis (Fig. 19-1). There is a superficial resemblance between a shoot with simple leaves and a pinnately compound leaf, but the latter usually subtends an axillary bud and ontogenetically, unlike a shoot, it is a strictly "determinate" organ — i.e., a structure in which all apical growth ceases at an early stage in development.

One of the most interesting features of pinnately compound leaves in the dicotyledons is the variable patterns in which lateral leaflets are formed during the early phases of leaf ontogeny. Leaves that at maturity appear very similar in general morphology may prove strikingly different with respect to the order of formation of lateral-leaflet primordia. Three general developmental sequences of leaflet initiation are recognized, namely *basipetal, acropetal,* and *divergent* (Fig. 19-2).

The basipetal order of formation of lateral leaflets is the most common pattern, according to Troll, and a clear example is provided by leaf development in *Polemonium caeruleum* (Fig. 19-1). At a very early stage, the young lamina consists of a prominent terminal segment — destined to develop into the terminal leaflet — and a pair of lateral-leaflet primordia (Fig. 19-1, B). As development continues, two rows of lateral-leaflet primordia develop basipetally (i.e., downwardly) toward the base of the leaf primordium (Fig. 19-1, C–E). Lateral-leaflet initiation is accompanied by the intercalary elongation of the rachis, and its continued extension finally results in the separation of the leaflets along the mature rachis. As shown in Fig. 19-1, A, the lowest leaflets may become alternately arranged whereas the upper leaflets are more crowded and are disposed in well-defined pairs.

The acropetal order of leaflet initiation represents the reverse of the pattern illustrated in *Polemonium* in that the two rows of lateral leaflet primordia develop upwardly, from the base towards its apex (Fig. 19-3, C). According to Troll, the acropetal formation of leaflets occurs only in leaves that exhibit pronounced apical growth for a relatively long period in leaf ontogeny. In this connection it is interesting to note that in *bipinnate leaves,* the first order of lateral-leaflet primordia

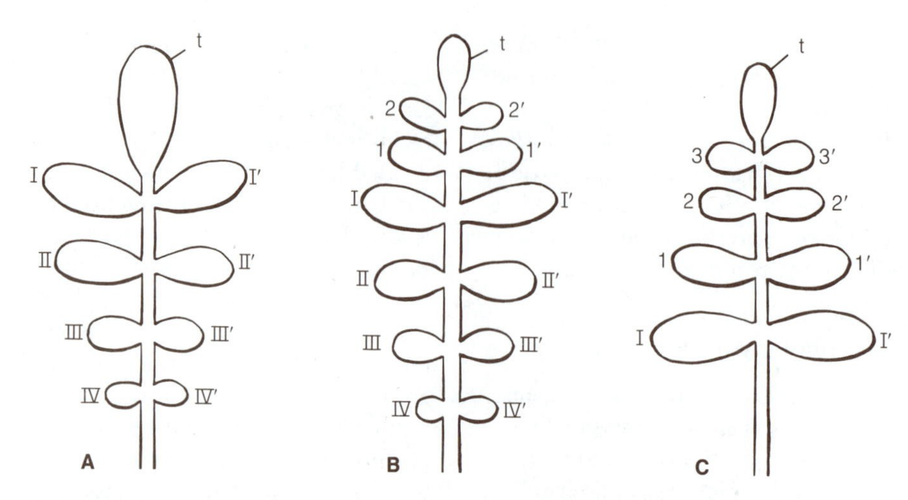

FIGURE 19-2 Diagrams showing the basipetal (A), divergent (B), and acropetal (C) orders of lateral-leaflet formation in pinnately compound leaves. In each diagram, the basipetal series is indicated by Roman numerals (I–I′, II–II′, etc.), the acropetal series by Arabic numerals (1–1′, 2–2′, etc.). In each figure, t designates the terminal leaf segment. [Redrawn from Troll, *Nova Acta Leopoldina* n.f. 2:315, 1935.]

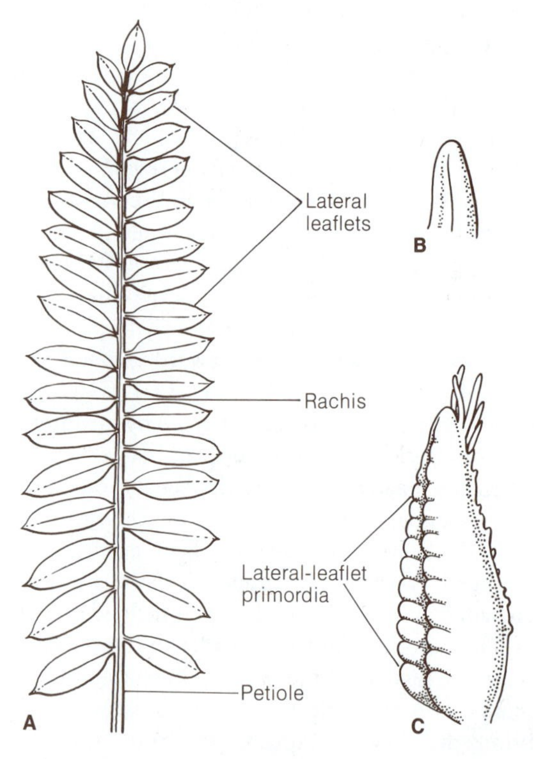

FIGURE 19-3 Morphology and development of the leaf of *Astragalus cicer.* **A,** adult leaf; **B,** adaxial view of leaf primordium showing the two marginal ridges from which the lateral-leaflet primordia will develop; **C,** later stage in leaf development, showing acropetal order of leaflet formation. [Redrawn from Troll, *Nova Acta Leopoldina* n.f. 2:315, 1935.]

commonly develops basipetally and the second order of leaflets, or "pinnules," arising from them form in an acropetal sequence.

In the divergent type of leaflet development, lateral-leaflet initiation begins near the middle of the young primordium and proceeds *both* basipetally and acropetally (Fig. 19-4). This type, therefore, combines the basipetal and acropetal orders of leaflet initiation. In extreme examples, illustrated diagrammatically in Fig. 19-2, B, all but the lowest pair of leaflets belong to the acropetally developed series; in *Achillea,* there is a marked emphasis on the formation of a larger number of basipetally formed pinnae (see Fig. 19-4).

Despite the considerable morphogenetic interest of Troll's studies, they do not resolve the difficult question of the phylogenetic relationship between "simple" and "compound" leaves in the dicotyledons. Which is the primitive type? In members of certain families, such as the Quiinaceae, the juvenile leaves are pinnately compound or pinnately lobed and "merge" gradually into the simple leaves characteristic of the adult tree (Foster, 1950, 1951). This type of *heterophylly* might be used to support the idea that a pinnately compound leaf is the ancestral type *in this family.* More commonly, however, the seedling or juvenile leaves in angiosperms tend to be simpler than the adult type of foliage, and this encourages the speculation that simple leaves, rather

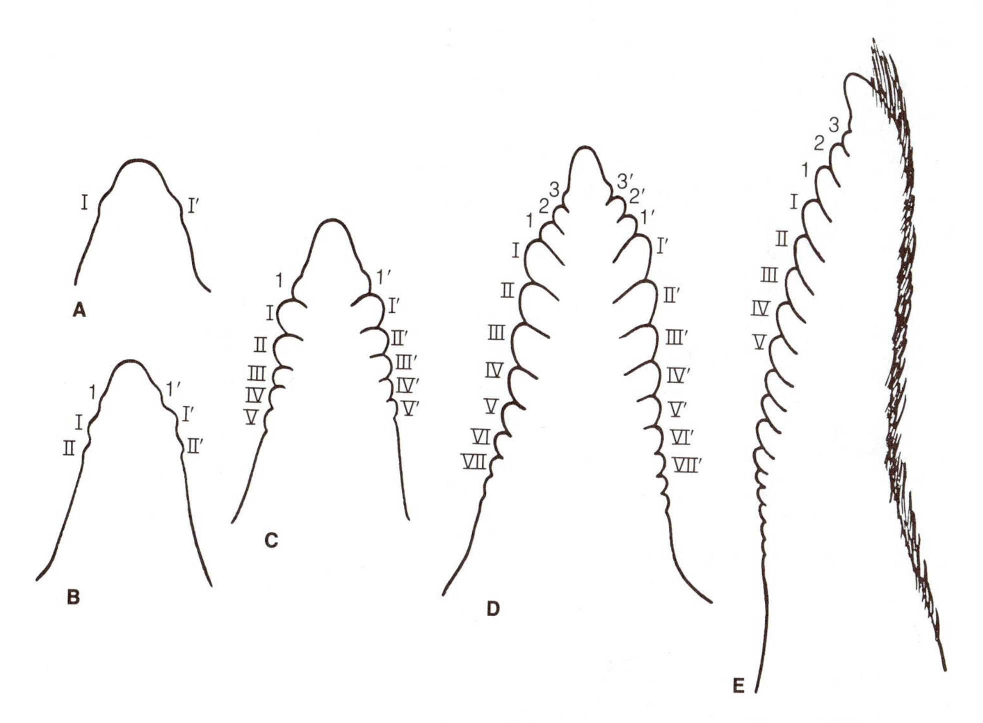

FIGURE 19-4 The divergent pattern of lateral-leaflet formation in *Achillea millefolium*. **A–D,** adaxial views of successive stages in development, showing the basipetal series (I–I′ to VII–VII′) and the acropetal series (1–1′ to 3–3′) of lateral-leaflet primordia; **E,** side view of later stage in leaf development. [Redrawn from *Vergleichende Morphologie der höheren Pflanzen* by W. Troll. Gebrüder Borntraeger, Berlin. 1937.]

than compound ones, are the primitive form. Fossil evidence is now available in support of this speculation. An intensive study of angiosperm leaves of the Lower Cretaceous by Hickey and Doyle (1977) revealed that both simple and compound leaves were present at that time, although the simple leaf appears to be the more primitive, confirming previous suggestions (Eames, 1961; Cronquist, 1968). More will be said on this subject later in the chapter.

In general, the leaves of monocotyledons develop a simple lamina and compound leaves are extremely uncommon. The most striking examples of compound leaves in monocotyledons are found in palms, in certain genera of which the leaf reaches a length of 9 to 15 meters and is the largest among seed plants. However, the leaflets or subdivisions of the lamina in both fan and pinnate-leaf types of palms do not arise from separate leaflet primordia as do the compound leaves of dicotyledons. Some investigators have reported that the leaflets of a palm leaf result from tissue separation (splitting) early in the development of the lamina (Eames, 1953; Padmanabhan, 1963, 1967). On the contrary, others (Kaplan et al., 1982a, b; Dengler et al., 1982) have demonstrated that the young lamina of a palm leaf first develops a series of compressed "folds" (plications) through differential growth. A pattern of alternating adaxial and abaxial ridges separated by furrows is established. Only later in development is there a complete separation of the units (leaflets) in pinnate leaves or incomplete separation in some fanlike leaves. A palm, *Palmoxylon,* was described from the Jurassic (Tidwell et al., 1970a, b), but subsequently it was shown that the deposits were actually of Tertiary age (Scott et al., 1972). There is good evidence for the presence of monocotyledons in the Lower Cretaceous. The leaves were small, simple, and the venation was on the plan of modern-day monocotyledons. Palms appear to have originated relatively late in the history of monocotyledons (Doyle, 1973) and probably represent a specialized group of monocotyledons.

Another problem in leaf morphology among monocotyledons is the interpretation of the so-

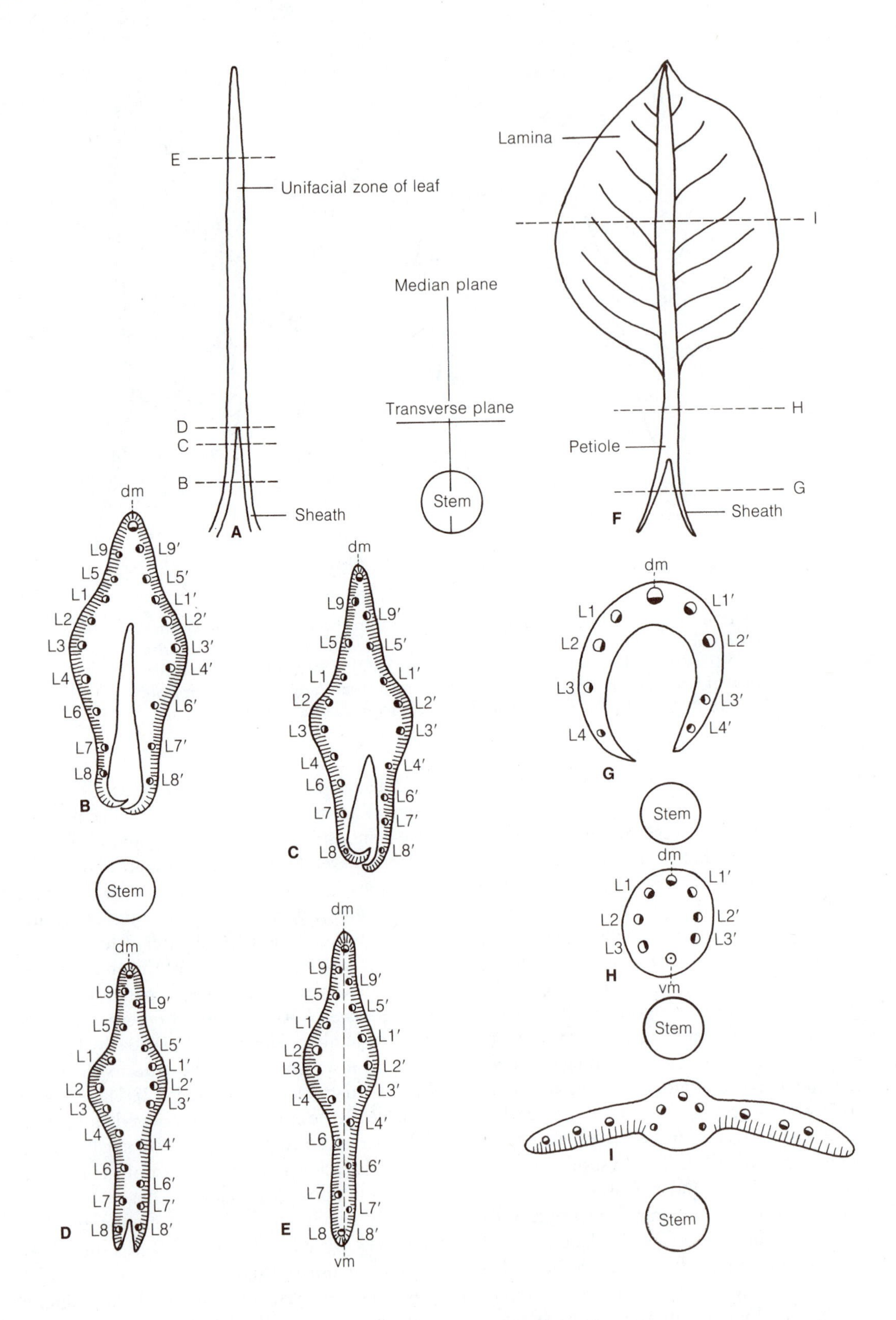
E
Unifacial zone of leaf
D
C
B
Sheath
A
Median plane
Transverse plane
Stem
Lamina
I
H
Petiole
G
Sheath
F
dm
L9 L9′
L5 L5′
L1 L1′
L2 L2′
L3 L3′
L4 L4′
L6 L6′
L7 L7′
L8 L8′
B
C
Stem
D
E
vm
G
H
I
Stem

called unifacial type of leaf of certain members of the Liliaceae, Araceae, Iridaceae, Amaryllidaceae, and Juncaceae. A unifacial leaf differs from a "typical" bifacial monocotyledonous leaf in the following ways: (1) although it generally possesses a sheathing dorsiventral leaf base, the remainder of the leaf is not differentiated into petiole and lamina; (2) it is flattened in a median rather than in a transverse plane, and (3) the collateral vascular bundles are arranged radially rather than in the crescentic pattern characteristic of a bifacial lamina. These features are diagrammatically represented in Fig. 19-5, which is reproduced from Kaplan's (1970) study on leaf histogenesis in *Acorus calamus* (Araceae). His investigation revealed that in this species apical growth ceases at an early stage and that marginal growth in the upper part of the developing leaf is suppressed in favor of *radial development,* which occurs by means of an active meristem situated on the *adaxial side* of the leaf primordium. A very similar pattern of histogenesis was described earlier by Boke (1940) in his study of the *Acacia* phyllode, a foliar structure which he interpreted as equivalent to the petiole-rachis regions of a pinnate foliage leaf. However, Kaplan (1980) concluded from a detailed ontogenetic study of several species of *Acacia* that the phyllode "leaf" of this dicotyledon is the positional homologue of the entire lamina-petiole region of the pinnate-type leaf in *Acacia* rather than just of petiolar origin. The reader is referred to the publication of Kaplan (1975) in which he compares, from an ontogenetic viewpoint, the problems of development and structural relationships between the unifacial foliar appendages of monocots and the more conventional dorsiventral leaves in dicotyledons.

Stipules

An enigmatic character of the foliage leaf of many dicotyledons is the presence of a pair of small appendages, known as stipules, near the base of the leaf (Fig. 19-6). The expression "free lateral stipules" is often used to describe such stipules that *appear* to be "independent" organs. The impression that they are independent is strengthened by the early abscission of lateral stipules during leaf expansion in many trees and shrubs; in such plants (e.g., *Quercus, Fagus, Populus, Ulmus*) the deciduous stipules leave separate "stipule scars" on the stem at each side of the leaf. But in many other plants stipules are apparently fused with the leaf base, and these are termed "adnate stipules" (Fig. 19-7, A). In plants with decussate phyllotaxis, *interpetiolar stipules* may develop *between* the two leaves at each node. Interpetiolar stipules are dual structures, i.e., they represent the partial or complete union of the adjacent stipules at each side of the node. In *Magnolia,* the margins of the two stipules of a leaf are loosely joined to form a conical

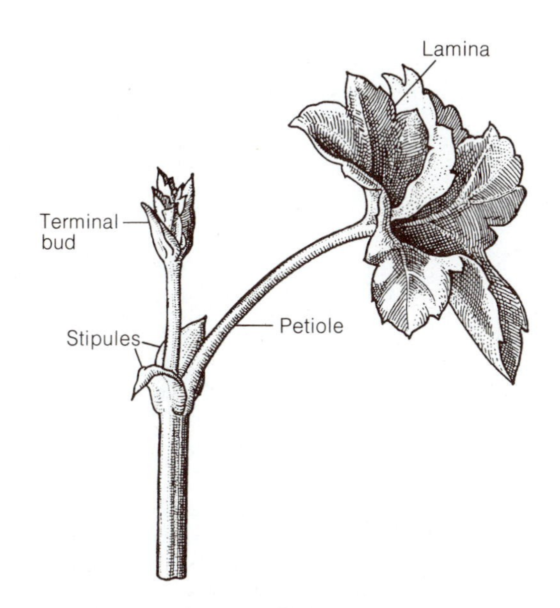

FIGURE 19-6 Shoot tip of *Pelargonium peltatum,* showing the pair of free lateral stipules associated with a recently expanded leaf. [Redrawn from *Vergleichende Morphologie der höheren Pflanzen* by W. Troll. Gebrüder Borntraeger, Berlin. 1939.]

FIGURE 19-5 Diagrams contrasting the morphology and vascular anatomy of the unifacial leaf of *Acorus calamus* with the type of foliage leaf characteristic of most Araceae. **A,** adaxial view of leaf of *Acorus;* **B–E,** transections at the levels indicated by broken lines in **A. F,** adaxial view of leaf typical of other members of the Araceae; **G–I,** transections at the levels indicated by the broken lines in **F.** Phloem of vascular bundles represented in white, xylem in black. Photosynthetic tissue indicated by hatching. dm, dorsal median bundle; L1–L1′, L2–L2′, etc., lateral vascular strands; vm, ventral median bundle. [Redrawn from Kaplan, *Amer. Jour. Bot.* 57:331, 1970.]

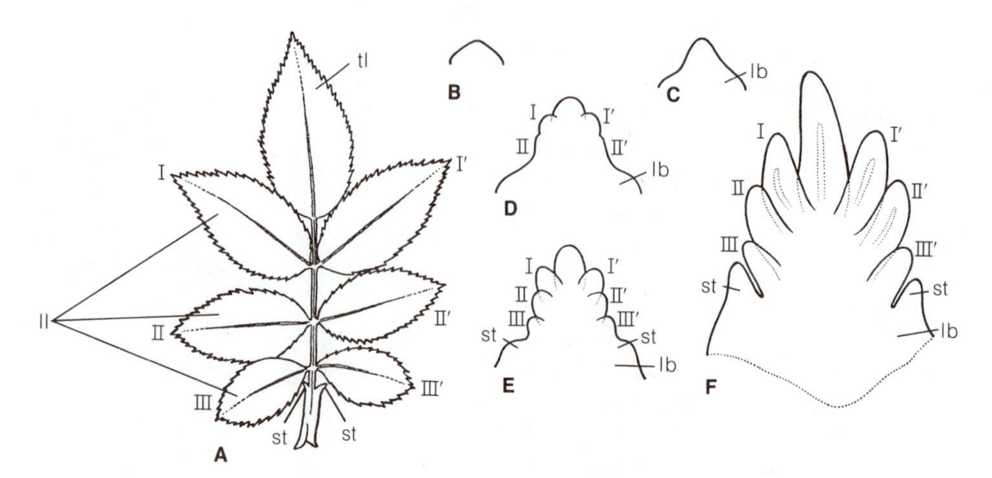

FIGURE 19-7 Leaf development in *Rosa.* **A,** adult leaf with adnate stipules (st) and pinnately compound lamina, composed of terminal leaflet (tl) and three pairs of lateral leaflets (ll). **B–F,** successively later stages in leaf development, showing origin of stipules (st) from leaf base (lb) and basipetal order of formation of lateral-leaflet primordia (I–I′ to III–III′). [Redrawn from *Vergleichende Morphologie der höheren Pflanzen* by W. Troll. Gebrüder Borntraeger, Berlin. 1938.]

"stipular sheath" which encloses the younger leaves of the terminal bud.

Regardless of the great variability in size, form, and methods of attachment of stipules, they are integral parts of a leaf and not "accessory" or "independent" appendages of the shoot. It can be shown ontogenetically that stipules develop from the leaf-base region of a foliage leaf (Fig. 19-7). Also, there are examples of the intimate relationship between stipules and the leaf where the vascular supply of each member of a pair of stipules is derived from the corresponding lateral leaf trace departing from the stem into the petiole (Fig. 19-8). The question of whether the presence or the absence of stipules is primitive has not been resolved conclusively. Decisive fossil evidence is lacking, but the presence of stipules in both monocotyledons and dicotyledons and their common association with the more primitive dicotyledonous families suggest that stipules are primitive (Hickey and Wolfe, 1975). In taxa in which stipules occur they are generally associated with either multilacunar or trilacunar nodes (see p. 508).

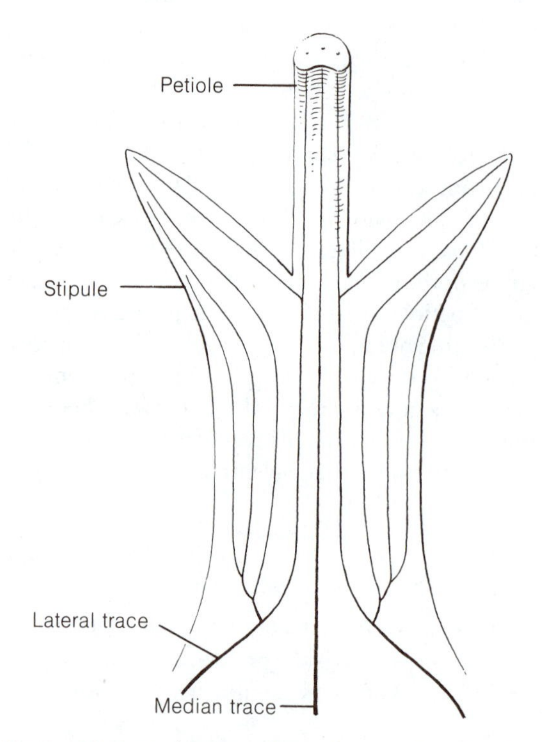

FIGURE 19-8 *Trifolium medium,* diagram showing that the veins entering the paired adnate stipules represent branches of the corresponding lateral leaf traces departing from the vascular cylinder of the stem. [Redrawn from *Vergleichende Morphologie der höheren Pflanzen* by W. Troll. Gebrüder Borntraeger, Berlin. 1939.]

Cataphylls

The development of the shoot of many angiosperms is accompanied by the periodic formation of scale leaves, or cataphylls. The term "cataphylls" (derived from the German word *Niederblätter,* meaning "lower leaves") collectively designates the scales of the rhizomes of monocotyledons and dicotyledons as well as the protective scales of the winter buds of shrubs and trees. Cataphylls, like foliage leaves, originate from primordia formed by the shoot apex, and for this reason the two types of appendages may be regarded as *serially homologous,* despite their adult divergence in form, anatomy, and function (see Chapter 1 for a discussion of serial homology). In many dicotyledons, the homology between the bud scale and the foliage leaf is clearly demonstrated by the presence of a *rudimentary lamina* at the apex of each of the sheathlike scales (Fig. 19-9, A–D), and by the occurrence of "transitional forms" between typical scales and foliage leaves (Fig. 19-9, E, F). This type of bud-scale morphology is found in many taxa and has been studied ontogenetically in *Aesculus* (Foster, 1929)

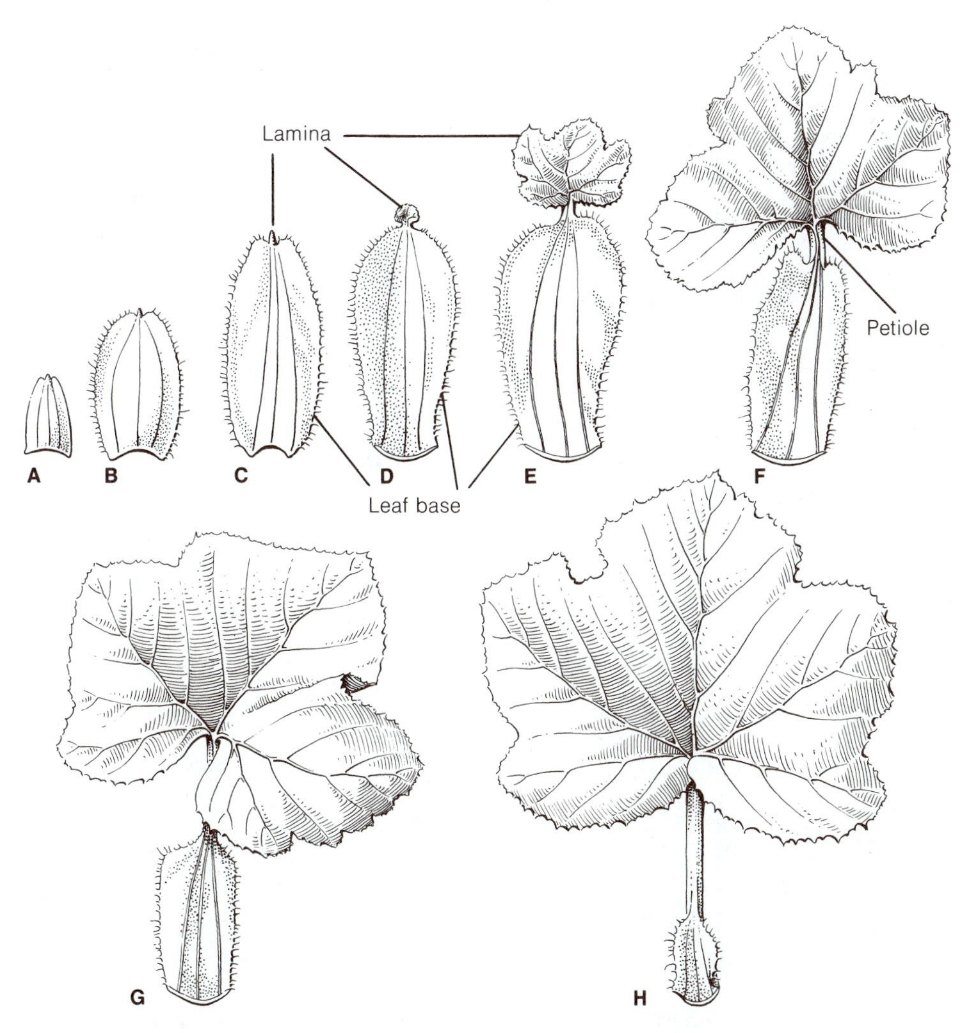

FIGURE 19-9 *Ribes sanguineum.* **A–D,** outer scales of a bud showing enlarged leaf base and rudimentary lamina; **E, F,** transitional forms between bud scales and foliage leaves (note enlarged lamina and short petiole of the appendages); **G, H,** foliage leaves each with a well-developed palmately lobed lamina, an elongated petiole, and a sheathing leaf base. [Redrawn from *Vergleichende. Morphologie der höheren Pflanzen* by W. Troll. Gebrüder Borntraeger, Berlin. 1939.]

and *Paeonia* (Foster and Barkley, 1933). In other genera, however, such as *Carya,* the *morphological divergence* between bud scale and foliage leaf takes place at such an early stage in ontogeny that a *direct comparison* between the mature unsegmented scale and corresponding parts of the pinnately compound foliage leaf cannot be drawn (see Foster, 1931, 1932, 1935). The bud scales of some woody dicotyledons (e.g., *Betula, Alnus, Fagus*) represent modified stipules, and each pair of scales is associated with a rudimentary lamina. Stipular bud scales in some genera (e.g., *Malus, Prunus*), are tridentate in form, with the median pointlet representing the rudimentary lamina and the two lateral pointlets corresponding to a pair of adnate stipules.

Venation

A striking morphological character of modern angiosperms is the complex and diversified venation patterns of their leaves. Even casual inspection of the gross venation of the foliar lamina of the

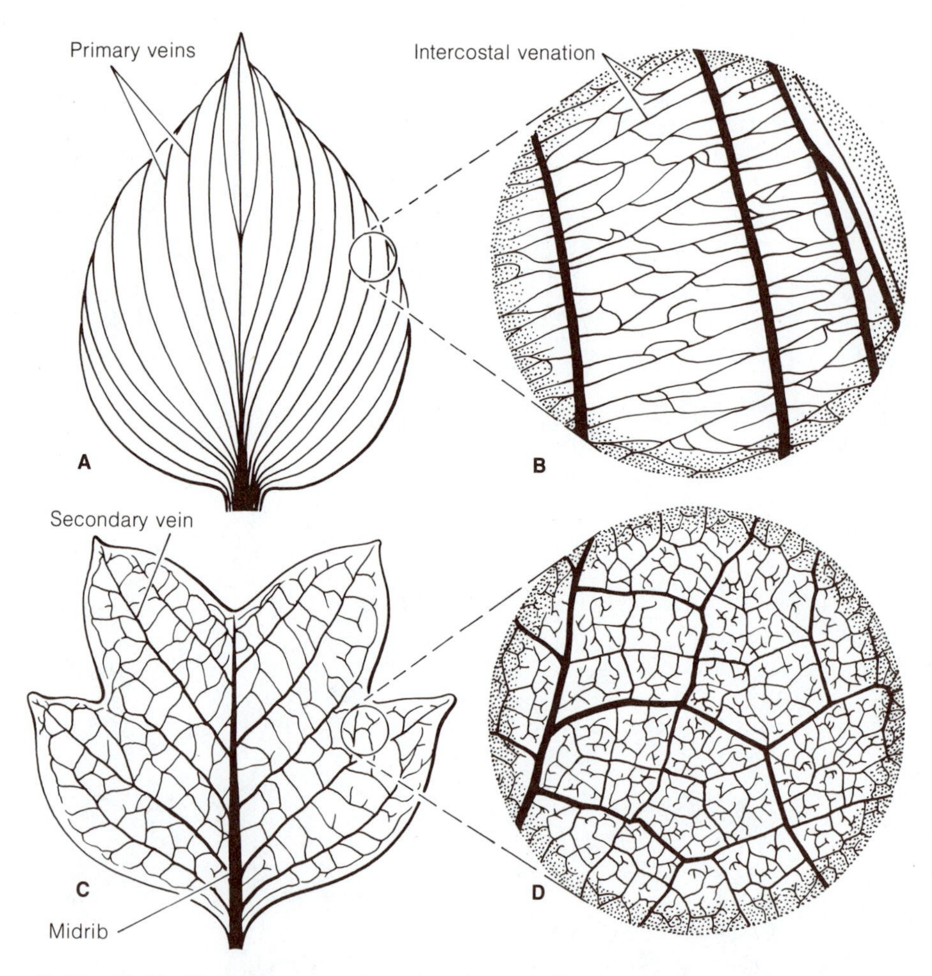

FIGURE 19-10 Venation patterns in angiosperms. **A,** the campylodromous venation of *Hosta caerulea* (note the arcuate course and the progressive union, from base of lamina upwards, of the primary veins); **B,** details of network of veinlets from area indicated in the circle of **A** (note the freely terminating vein endings); **C,** major venation in the leaf of *Liriodendron tulipifera;* **D,** details of area indicated in the circle of **C,** showing complex reticulum of veinlets and numerous branched vein endings. [**A, B** redrawn from Pray, *Amer. Jour. Bot.* 42:611, 1955; **C, D** redrawn from Pray, *Amer. Jour. Bot.* 41:663, 1954.]

majority of dicotyledons reveals a more or less prominent midvein that terminates at the apex of the lamina, and a series of pinnately arranged secondary veins (Fig. 19-10, C). The "course," i.e., the direction and mode of termination, of the major secondary veins in pinnate venation fluctuates widely between genera (or even among species) and was used more than a century ago by Ettinghausen (1861) to establish a series of "venation types." However, his system was largely overlooked or ignored by the majority of plant morphologists. Some of his terms for describing the course of the major veins have now proved useful in the revival of interest in the phylogeny of angiosperms. More will be said later of his classification of venation patterns.

The vasculature of the intercostal areas (regions between the major veins) consists of a complex network of anastomosed "minor" veins and veinlets, resulting in what is termed *reticulate* venation. Reticulate venation is more pronounced in dicotyledons. When the minor venation is carefully studied at high magnification, it becomes evident that the ultimate (smallest) meshes, or areoles, contain solitary or often branched vein endings that terminate freely in the mesophyll (Fig. 19-10, D). An intercostal venation system also occurs to varying degrees in the leaves of monocotyledons (Fig. 19-10, A, B). For the past twenty years, increased attention has been given to the ontogenetic and histological aspects of minor venation in the leaves of both monocotyledons and dicotyledons (Pray, 1963; Lersten, 1965; Esau, 1967). More recently, extensive and intensive comparative studies have been made of the total vasculature (i.e., both the major and minor venation), the results of which can be used in determining taxonomic and phylogenetic relationships in angiosperms (Hickey, 1973; Hickey and Wolfe, 1975).

Venation Patterns in Dicotyledons and Monocotyledons

A detailed description and comparison of the extremely varied types of angiosperm venation patterns cannot be undertaken in the present book. Only representative types will be discussed and only the major venation patterns will be emphasized. We will use the terminology of Ettinghausen (1861) as modified by Hickey (1973). A common type in dicotyledons has a prominent midvein (primary vein) and a system of pinnately arranged secondary (lateral) veins ending at the margin of the leaf (*craspedodromous* type, from the Greek, *kraspedon,* meaning "edge, border"; Fig. 19-11, A). If the secondary veins do not end at the leaf margin but rather join together in a series of prominent arches, this type is termed *camptodromous* or, more specifically, *brochidodromous* (Greek, *brochos,* "a noose"; Fig. 19-11, B). In the second general type, the major venation system is basically palmate with three or more primary veins diverging radially from a single point at the base or near the base of the lamina (*actinodromous,* from the Greek, *aktinos,* "radiating arms"; Fig. 19-11, C). If two or more primary or strongly developed secondary veins run in convergent arches toward the apex, this type is designated *acrodromous* (Greek, *akros,* "toward the tip"; Fig. 19-11, D). In the *campylodromous* (Greek, *kampylos,* "a bend or curve") type, several

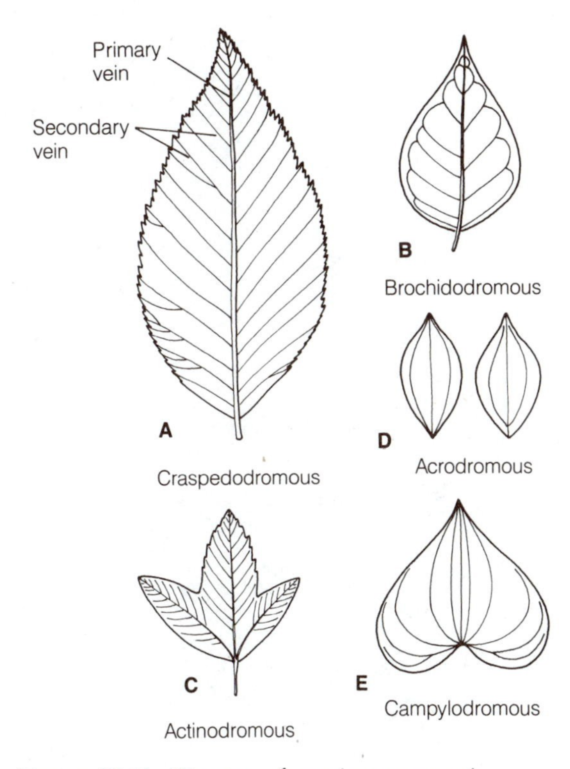

FIGURE 19-11 Diagrams of certain representative major venation patterns in dicotyledons. See text for details. [Redrawn from Hickey, *Amer. J. Bot.* 60:17–33, 1973.]

primary veins or their branches originate at a single point and run in strongly developed recurved arches before converging toward the leaf apex (Fig. 19-11, E). Space does not permit a discussion of the numerous other aspects of "leaf architecture," such as intercostal venation patterns. The reader is referred to the publications of Hickey (1973) and Hickey and Wolfe (1975).

Before leaving this discussion of venation in dicotyledons, one rather unique type should be considered — that of open dichotomous venation. This pattern is known only for two genera of small, herbaceous plants that have limited distribution in the high montane areas of Asia. Cronquist (1981) placed the genera in one family (Circaeasteraceae) in the order Ranunculales. In *Circaeaster agrestis* the leaf is simple and dentate at its distal end. The isotomous, open dichotomous venation is generally very regular with the ultimate dichotomies terminating in corresponding marginal teeth. In some instances fusions occur between adjacent branches of two vein dichotomies (Foster, 1963, 1966). The leaf of *Kingdonia uniflora* (Fig. 19-12) is pedately lobed, consisting of five segments, each of which has an elegant system of open dichotomous venation with only occasional vein fusions (Foster, 1959, 1961a, b; Foster and Arnott, 1960). What is the phylogenetic significance of dichotomous ven-

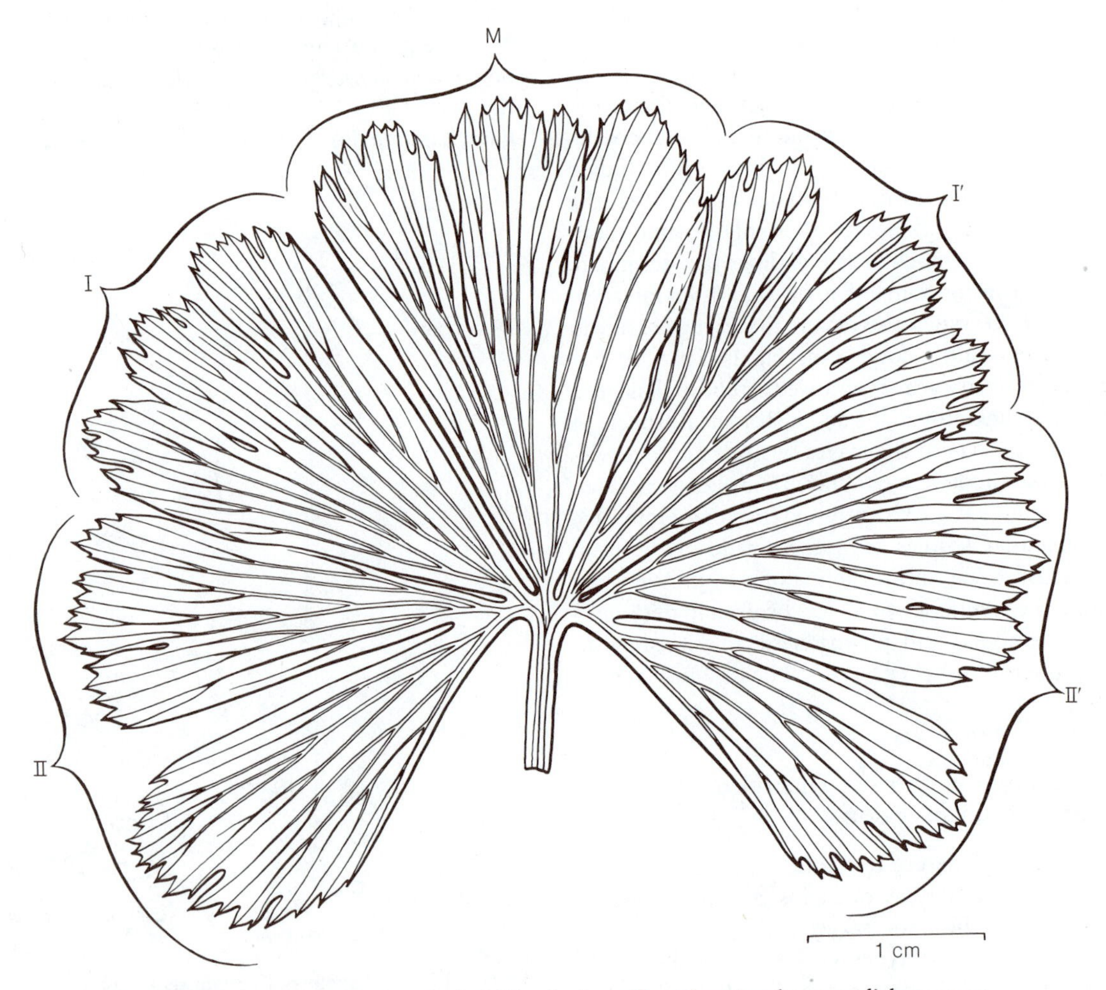

FIGURE 19-12 Mature foliage leaf of *Kingdonia uniflora* showing the open dichotomous venation of the five major segments (M, I – I′, II – II′) of the pedately divided lamina. [Redrawn from Foster and Arnott, *Amer. Jour. Bot.* 47:684, 1960.]

ation in these plants? Do they illustrate the *persistence* of a very ancient (ancestral?) form of venation as found in *Ginkgo,* cycads, and some conifers, or should the dichotomous venation be interpreted as the result of "reversion" to an ancient type of foliar vasculature? Partial answers to these questions will be discussed later.

It is commonly stated in elementary textbooks that monocotyledons are characterized by "parallel venation" in contrast to the reticulate pattern of venation of dicotyledons. Although it is true that reticulate venation is highly developed in dicotyledons, the concept of so-called parallel venation should be rejected because it is both inaccurate and misleading. In grasses, for example, the leaf superficially appears to be traversed by a closely spaced series of independent, longitudinal, parallel veins. But careful examination, using cleared and stained* leaves, reveals that the main veins do not extend equidistant throughout their course but, on the contrary, *converge* and progressively anastomose toward the apex of the lamina. These features were clearly analyzed by Troll (1938), who termed this type of closed venation *striate* rather than parallel (Fig. 19-13). It should be further emphasized that the longitudinally striate type of venation found in grasses is by no means representative of the monocotyledons as a whole. In certain families (Liliaceae, Araceae), the diverging lateral veins form an *arcuate* pattern in which many of the lateral veins do not reach the apex of the lamina but join the adjacent inner veins as they converge in the upper region of the leaf (Pray, 1955). This distinctive venation pattern is well illustrated in the liliaceous genus *Hosta* (Fig. 19-10, A).

Accompanying the continued inaccurate use of the term parallel venation there has been a conspicuous neglect of the minor venation usually present between the main longitudinal veins in the leaves of monocotyledons. In grass leaves, the minor venation is relatively simple and consists of delicate transverse or oblique *commissural veinlets* which interconnect the primary veins (Blackman, 1971). Commissural veinlets are also well defined in the leaf of *Maranta* (Troll, 1938) and are so regularly spaced in the leaf of *Bambusa* (bamboo) that a reticulum that is mosaic like is produced. More complex types of intercostal minor venation are found in members of the Araceae and Orchidaceae and also are present in the leaf of *Hosta* (Fig. 19-10, B). One

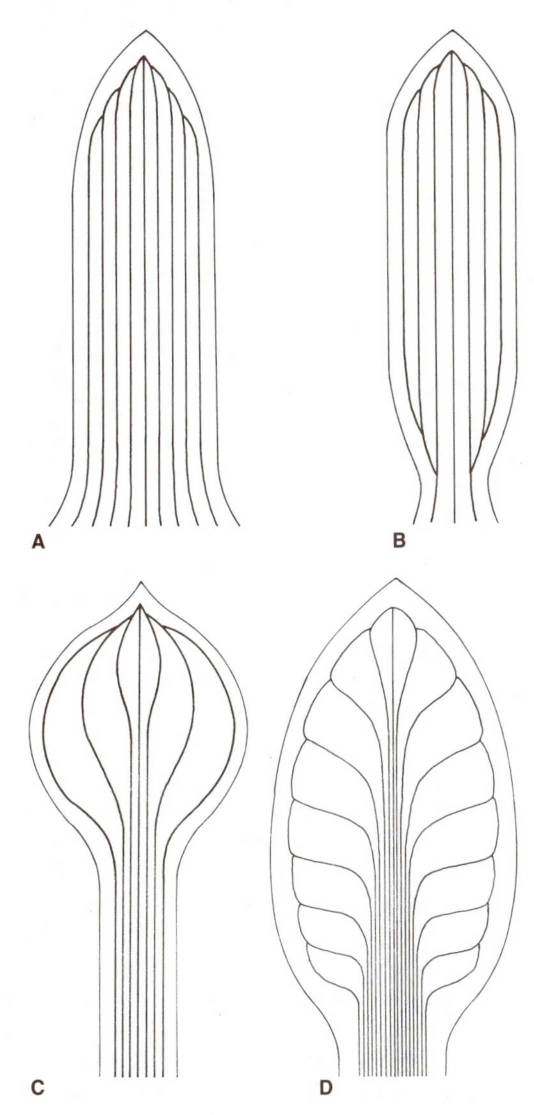

FIGURE 19-13 Diagrams showing various types of striate venation in the leaves of monocotyledons. **A, B,** longitudinally striate venation typical of the leaves of the Gramineae; note apical convergence and fusion between successive primary veins; **C,** arcuate-striate venation pattern; **D,** pinnate-striate venation pattern. [Adapted from *Vergleichende Morphologie der höheren Pflanzen* by W. Troll. Gebrüder Borntraeger, Berlin. 1938.]

*For information on the techniques used to clear and stain leaves, see Foster (1953, 1955) Shobe and Lersten (1967), and Kurth (1978).

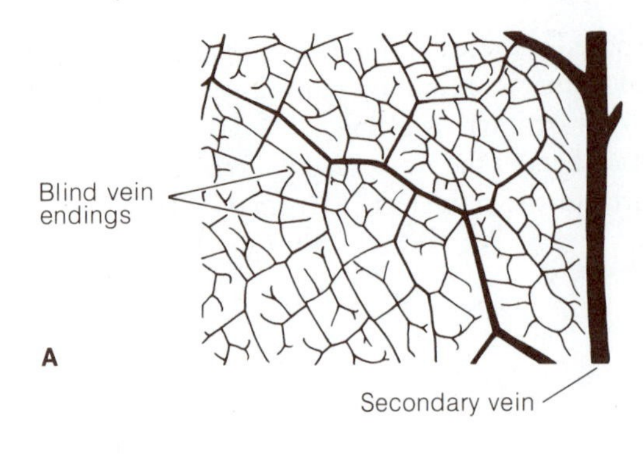

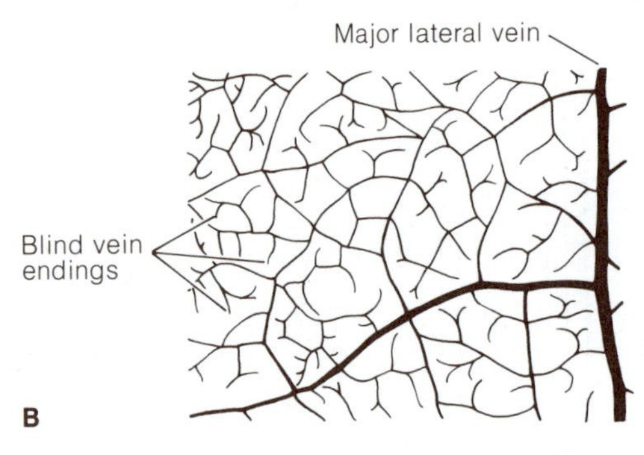

FIGURE 19-14 Illustrations of reticulate venation patterns in angiosperm leaves. **A,** in *Liriodendron,* a dicotyledon; **B,** in *Smilax,* a monocotyledon. Note remarkable similarity in overall venation pattern including the presence of vein endings. [Redrawn from Pray, *Phytomorphology* 13:60, 1963.]

of the most interesting types of intercostal minor venation may be observed in the leaf of *Smilax.* According to Pray (1963) the minor venation "is composed of a reticulum of polygonal areoles . . . and is remarkably similar in pattern to that of *Liriodendron* and many other dicotyledons" (Fig. 19-14). The example of *Smilax* clearly demonstrates that with our present knowledge, there is no single type of foliar venation which distinguishes the leaves of monocotyledons from those of dicotyledons. This fact should be carefully considered in efforts to reconstruct the phylogenetic origin and systematic relationship of the monocotyledons.

Phylogeny of Venation Patterns in Angiosperms

Many years ago Sinnott and Bailey (1915) postulated that the primitive angiospermic leaf probably was three-lobed and palmately veined. In the intervening years this view has been supplanted by the idea that primitive angiosperm leaves were simple with entire leaf margins, and had pinnate-reticulate venation. One major observation supporting this hypothesis is the occurrence of simple pinnately veined leaves in many living members of presumably primitive families (e.g., Magnoliaceae, Degeneriaceae, Winteraceae). Leaves with this type of venation are also found in a number of families that have vessel-less xylem (e.g., Winteraceae, Amborellaceae). If the idea that pinnate venation is primitive is accepted, it follows, according to Takhtajan (1959), that palmate venation and the so-called parallel venation are derivative types from a phylogenetic standpoint. According to Takhtajan, the most primitive form of pinnate venation is the *craspedodromous* type with all the main secondary veins terminating at the lamina margin (Fig. 19-11, A).

What does the actual fossil record tell us of the earliest known angiosperm leaves? Some answers to this question have come from intensive studies of Lower Cretaceous leaves from deposits in eastern United States extending from Delaware to Virginia; the fossil beds are termed the Potomac Group (Doyle and Hickey, 1976; Hickey and Doyle, 1977). Leaves with dicotyledonous or monocotyledonous characteristics are present in the Lower Cretaceous. A dicotyledon-type leaf was generally simple with an entire margin. Representative leaves exhibit the *brochidodromous* type of pinnate venation—that is, the secondary veins do not extend to the leaf margin (Fig. 19-15) and the intercostal venation was irregularly and poorly differentiated. This leaf type characterizes leaves of living dicotyledons considered to be primitive (Hickey and Wolfe, 1975). Palmate and compound pinnate leaves were also present in the Lower Cretaceous but occurred later and appeared to be derived forms. Small, simple leaves with sheathing leaf bases also coexisted with dicotyledonous-type leaves. The venation system was essentially *acrodromous,* with the secondary veins converging toward the leaf apex. Some cross veins were present between the secondary veins. Leaves with venation patterns similar to those of modern-day monocotyledons also are known from the Upper Cretaceous. If the leaves from the Lower Cretaceous are those of bona fide monocotyledons, the two major groups of angiosperms were separate phyletic lines at least by the Lower Cretaceous.

Histology of Leaves

In Chapter 3 the general application of Sachs' scheme of tissue systems to the description of leaf anatomy was briefly discussed. An examination of leaf surfaces reveals a great variety of trichomes (hairs, glands, scales), and they often provide valuable diagnostic characters in the taxonomic definition of species and genera. Stomata are ubiquitous structures in the epidermal system and are often restricted to the abaxial (lower) epidermis but may be present on both leaf surfaces. Stomata are the pores for the exchange of gases and for transpira-

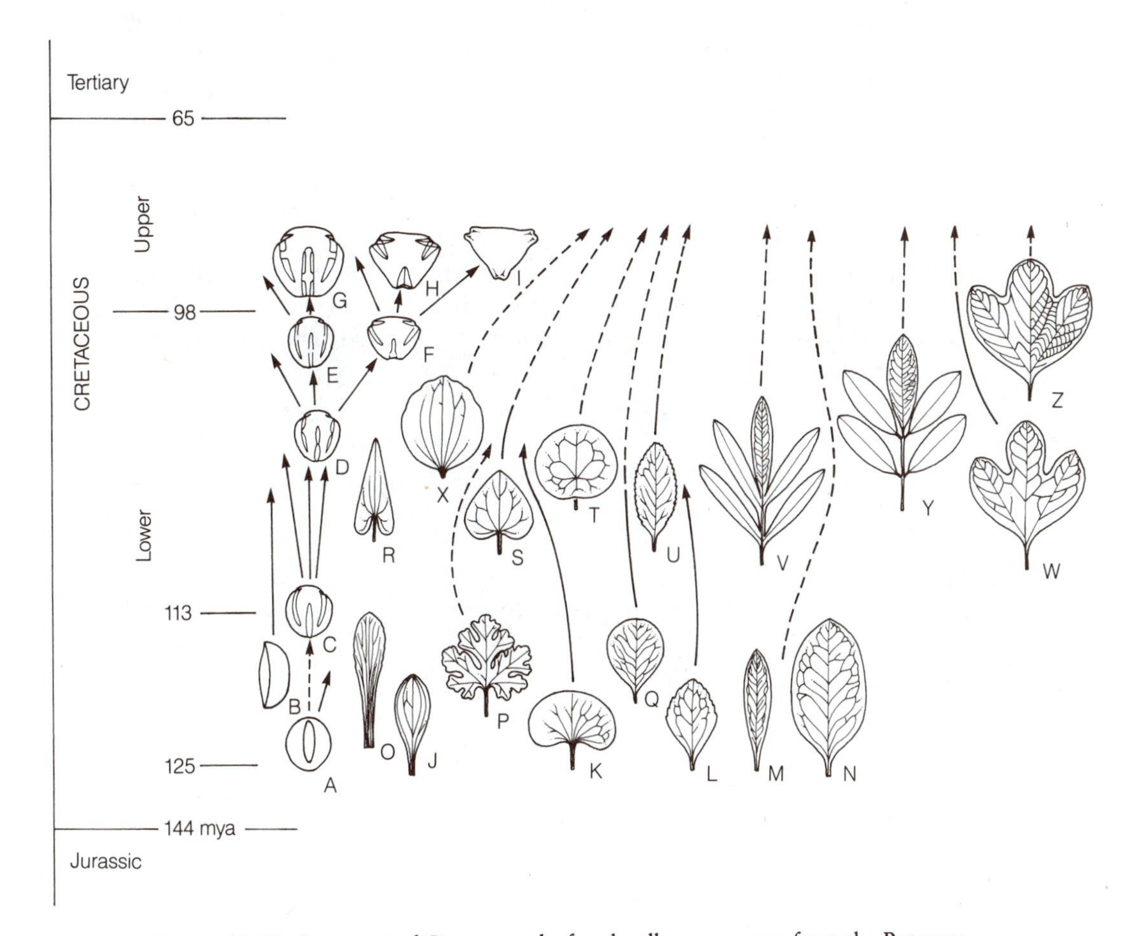

FIGURE 19-15 Summary of Cretaceous leaf and pollen sequences from the Potomac Group. *Leaves.* Note that the earliest leaf types in the Lower Cretaceous were simple; **J**, acrodromous, monocotyledonous type; **L–N**, pinnately veined, brochidodromous, dicotyledonous types. By the late Lower Cretaceous there were palmately lobed (**W**) and pinnately lobed leaves (**V, Y**), all showing brochidodromous venation. *Pollen.* The earliest pollen was monosulcate (**A**); later forms were tricolpate (**C**), tricolporate (**D**), and triporate (**I**).

For leaves, solid arrows indicate upward extensions of morphological complexes within the Potomac sequence; dashed arrows indicate range extensions inferred from other areas. For pollen, arrows indicate both evolutionary transformations and stratigraphic range extensions. The dashed arrow (**A** to **C**) indicates a transformation for which there is only indirect evidence. See p. 546 for a discussion of a possible pollen evolutionary sequence. [Redrawn from Hickey and Doyle, *Bot. Rev.* 43:3–104, 1977.]

tion and thus are important to the normal functioning of a leaf. The term *stoma* (singular) is generally used to designate the entire unit, the pore and the two guard cells. Stomata may be flanked or surrounded by cells that differ in morphology from the neighboring epidermal cells and are termed *subsidiary cells.* In these cases the entire unit is called a stomatal complex. The diversity of stomatal types offers one of the most important epidermal characters for comparative purposes.

As in gymnosperms (Chapter 14) there are variations in stomatal development. Guard cells and subsidiary cells may or may not arise from the same initial. For our purposes only the fully mature stomatal complexes will be described. (See Esau, 1977, and Mauseth, 1988, for a more complete discussion of development.) Seven or more types can be recognized in angiosperms, of which only five will be described in detail (Fig. 19-16): (1) *anomocytic type* — stoma surrounded by cells indistinguishable from other epidermal cells; (2) *anisocytic type* — stoma surrounded by three subsidiary cells, one of which is smaller than the other two; (3) *diacytic type* — stoma enclosed by a pair of subsidiary cells whose common wall is at right angles to the guard cells; (4) *paracytic type* — stoma accompanied on either side by one or more subsidiary cells parallel to the long axis of the pore and guard cells; (5) *tetracytic type* — stoma surrounded by four subsidiary cells, two lateral and two polar ones. For more complete discussions of stomatal types, consult Van Cotthem (1970) and Dickison (1975).

The importance of using stomatal types in determining phylogenetic trends in angiosperms is at present unclear. Some families are characterized by one type; in others, one or more types may be present. One might think that the simple type — anomocytic — might be primitive in dicotyledons. However, in a survey of certain families in the Magnoliales (considered to be primitive on other bases),

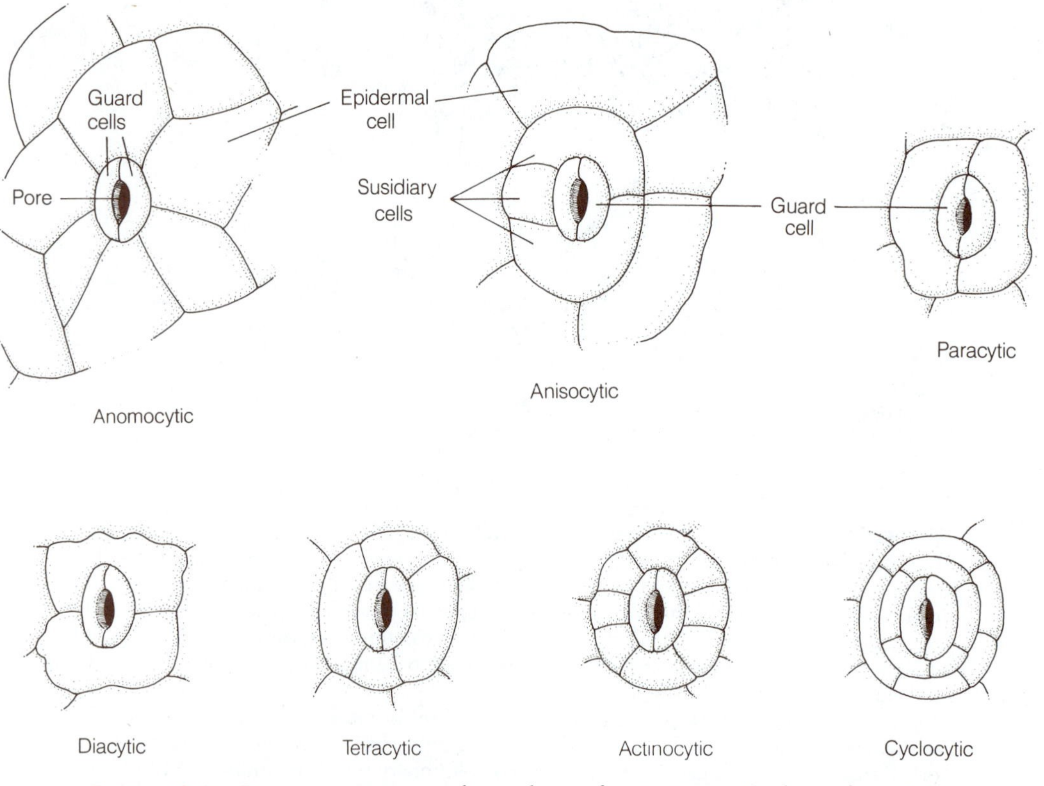

FIGURE 19-16 Representative stomatal complexes of angiosperms. [Redrawn from A. E. Radford, W. C. Dickison, J. R. Massey, and C. R. Bell, *Vascular Plant Systematics,* Harper and Row, New York. 1974]

Baranova (1972) has shown that the paracytic type is the prevailing condition, but it should be noted that the paracytic type is also found in a large number of advanced families. The tetracytic type is common in the monocotyledons but Tomlinson (1974) reports that there are no major groups of monocotyledonous families characterized by a specific type of stomatal development.

Figure 19-17 shows transections of portions of the leaf blades of *Pyrus* (a dicotyledon) and *Lilium*

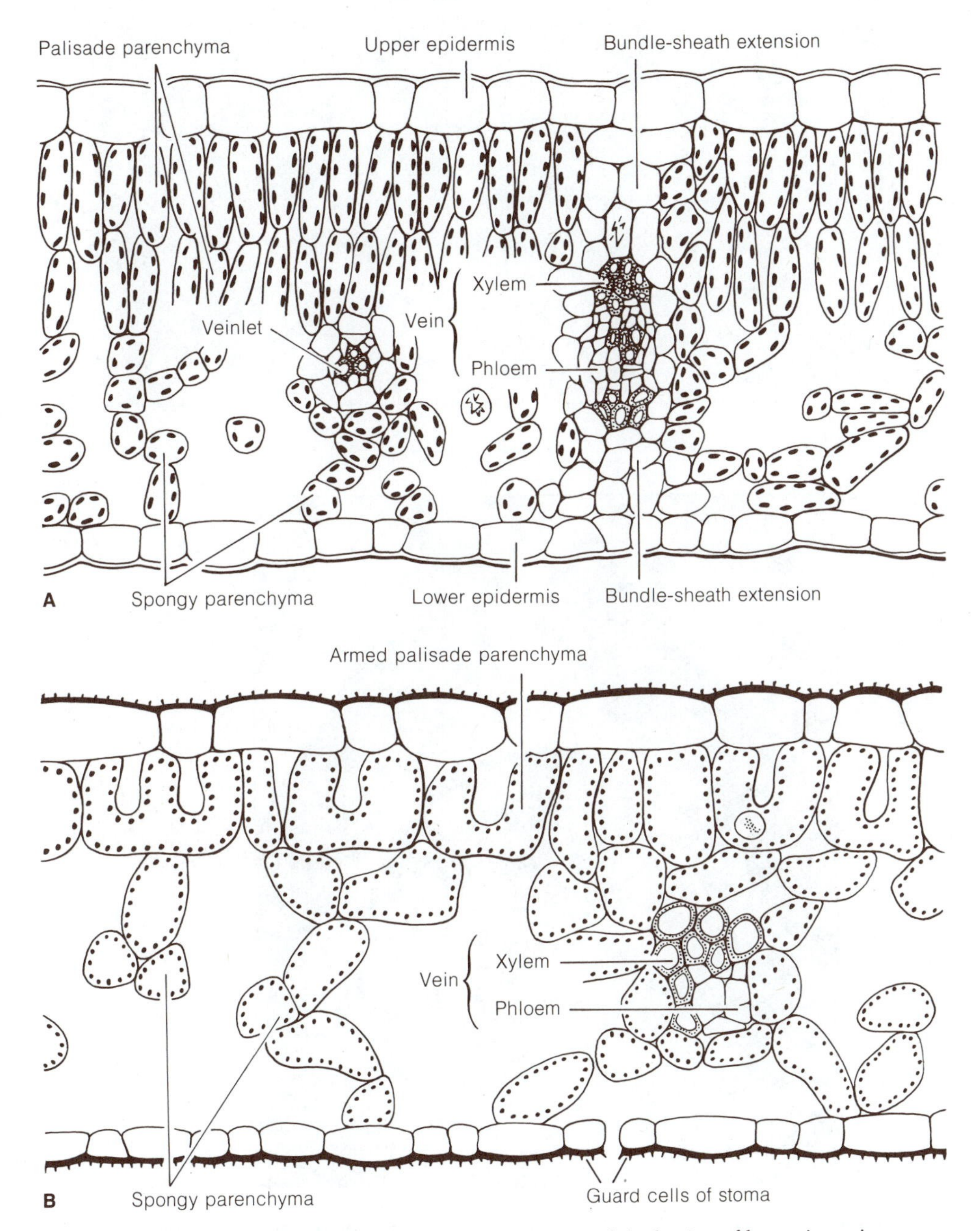

FIGURE 19-17 Transections illustrating the histology of the lamina of leaves in angiosperms. **A**, *Pyrus;* **B**, *Lilium* (see text for discussion of this figure). [Redrawn from *Plant Anatomy* by K. Esau. Wiley, New York. 1953.]

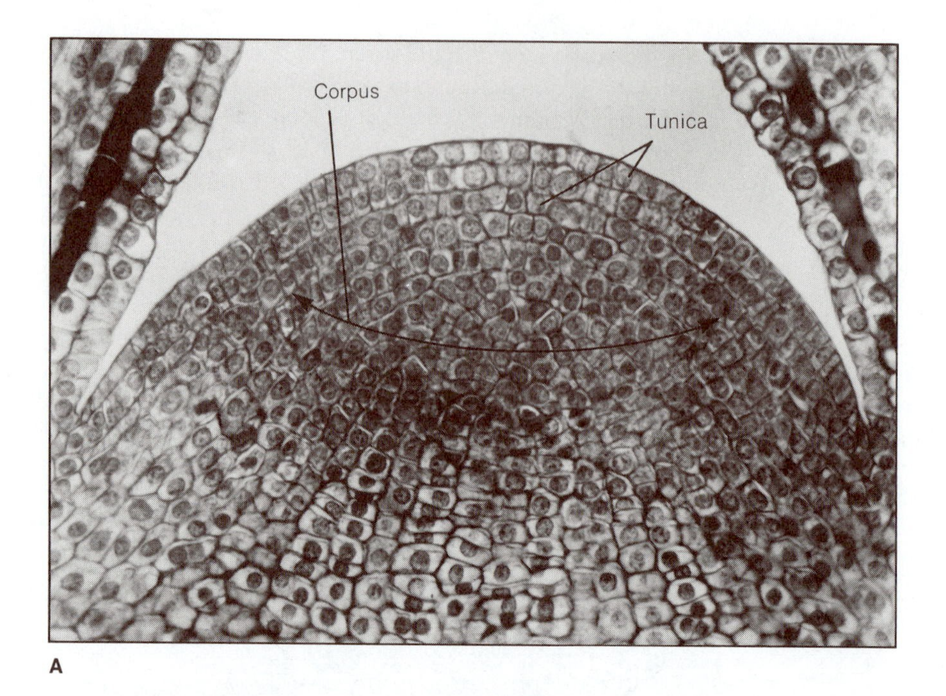

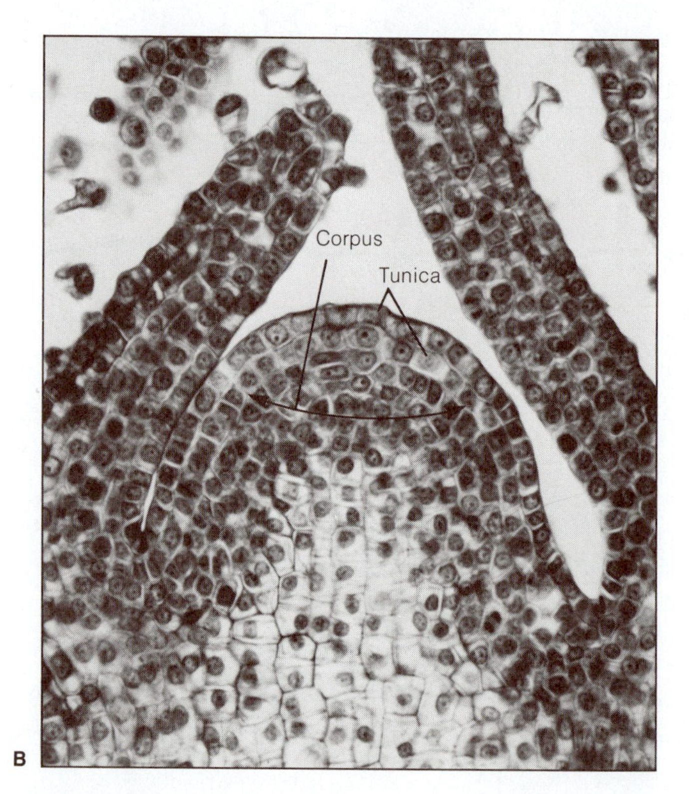

FIGURE 19-18 Median longitudinal sections of vegetative shoot apices, showing tunica-corpus organization. **A,** *Trochodendron aralioides;* **B,** *Chenopodium album.* (See also diagrammatic representation in Fig. 19-19.)

(a monocotyledon), and will be used to illustrate the internal anatomy. The internal tissue systems of the lamina are the mesophyll, or *fundamantal system,* and the veins and veinlets, which comprise the fascicular or *conducting* system. The mesophyll is composed of living, thin-walled cells rich in chloroplasts, and in many dicotyledons it is clearly differentiated into one or more relatively compact layers of palisade parenchyma and a region of more loosely arranged spongy parenchyma. Very commonly the palisade parenchyma is situated directly below the upper epidermis (dorsiventral type of leaf), but is often found below both epidermal layers (isolateral type of anatomy). Palisade cells are usually columnar in form, but in some angiosperms these cells are armed or lobed (Fig. 19-17, B). The mesophyll functionally represents the chief photosynthetic tissue of the plant.

The histology of veins and veinlets in the lamina depends on their degree of development, and also varies widely according to the species or genus. Each of the larger veins and veinlets consists of a collateral strand of phloem and xylem, or less frequently a bicollateral strand, and is enclosed within a bundle sheath. Bundle sheaths may be parenchymatous or sclerenchymatous. Bundle sheaths extend to the ends of the veins. In many woody dicotyledons the bundle sheaths are connected with one or both epidermal layers by cells known as bundle-sheath extensions (Fig. 19-17, A). These structures are believed to conduct liquids between the veins and the epidermis. Minor veins are highly specialized by having densely cytoplasmic cells, often with wall ingrowths, that transfer photosynthates from the mesophyll cells to the sieve elements. The vascular tissue of the smaller veins and veinlet ends is derived from the procambium and thus is exclusively primary. Cambial activity, if it occurs in a given leaf, is restricted to the coarser veins in the lamina and to the vascular system of the petiole. Generally speaking, there is relatively little secondary xylem and phloem in the vascular system of the leaf.

This account of leaf histology reflects very little of the wide variation in leaf structure. Readers interested in more extensive discussions of the structure of angiospermous leaves should consult Metcalfe and Chalk (1950), Esau (1965, 1977), Fahn (1974), and the monographs on the comparative anatomy of monocotyledons cited on p. 488 of this chapter.

The Shoot Apex

A prominent feature of the shoot apices in angiosperms is the presence of a *tunica-corpus* type of cellular zonation. (Figs. 19-18, 19-19). This kind of zonal structure is not restricted to angiosperms, but is also found in the shoot apices of such conifers as *Araucaria* and *Agathis,* as well as in the gnetophytes *Ephedra* and *Gnetum* (Chapter 18). The *tunica* is a layer or layers of cells in which the anticlinal plane of cell division predominates except at the sites of leaf initiation, where cells of the second and deeper tunica layers divide periclinally. In contrast to the prevailingly anticlinal plane of division in cells of the tunica, the *corpus* is a zone in which cells divide in varied planes. The boundary between tunica and corpus zones may fluctuate even in the same species because (1) at certain phases of growth, the outer cells of the corpus may be arranged in a tunicalike layer, and (2) periclinal division may occur, in certain monocotyledons, in the summital cells of the outermost tunica layer.

In addition to the topographical delimitation of tunica and corpus, most angiosperm shoot apices also exhibit a *cytohistological* pattern of zonation (Fig. 19-19). A *central zone* is apparent which includes some cells of both the tunica and corpus. Cells of the central zone are frequently larger, more vacuolate, and often exhibit a lower rate of mitotic division than the cells in the other regions of the shoot apex. In contrast, the *peripheral zone,* located lateral to the central zone, consists of cells that are relatively smaller and may have a higher mitotic rate than those of the central zone. Cells of the *pith rib meristem* zone found at the base of the central zone divide predominantly in a transverse plane (with reference to the longitudinal axis of the stem) and form more-or-less well-defined files of cells which differentiate into the pith.

Additional information on the structure of shoot apices, leaf initiation, and differentiation of primary tissues in angiosperms can be found in the review articles by Wardlaw (1965), Cutter (1965), and Gifford and Corson (1971), as well as in the more recent article by Niklas and Mauseth (1980) on computer simulations and ontogenetic patterns

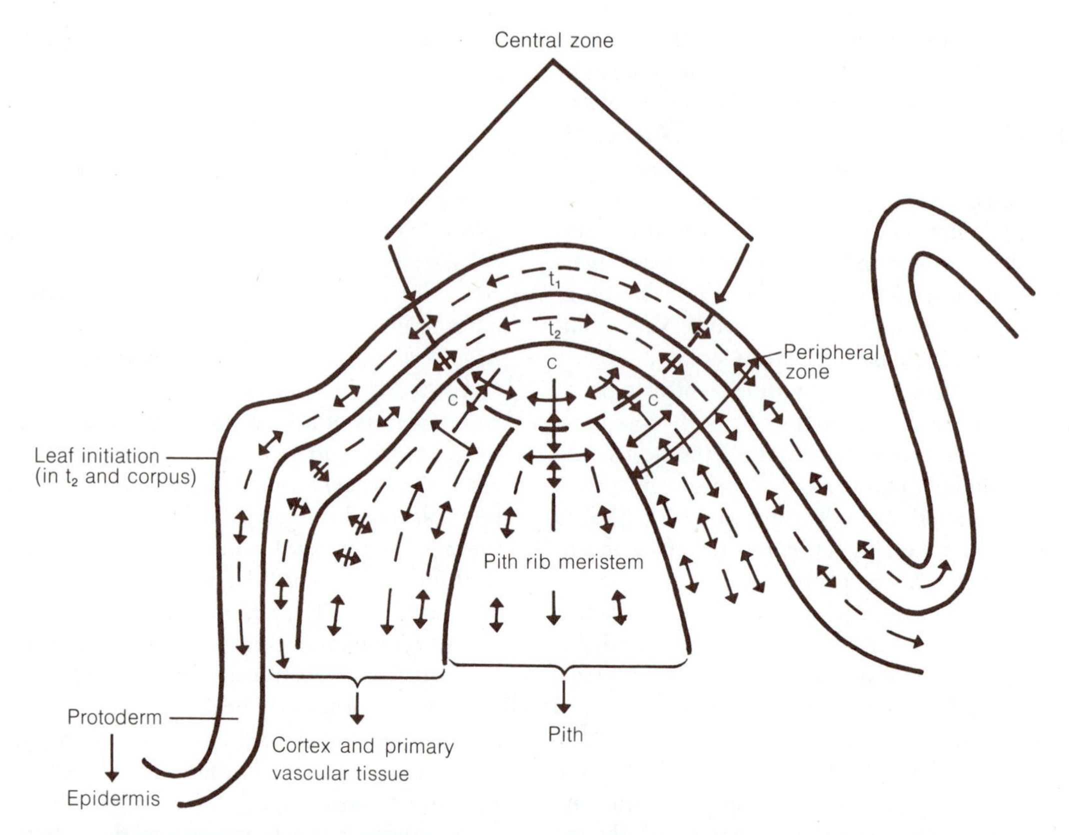

FIGURE 19-19 Diagrammatic representation of zonation and growth in the vegetative shoot apex of a hypothetical dicotyledon possessing a two-layered tunica. The distance between a pair of arrowheads indicates the degree of mitotic activity; the shorter the distance, the greater the mitotic activity. Contributions of various zones of the shoot apex to primary tissue layers and regions of the stem are indicated. t_1, first tunica layer; t_2, second tunica layer; c, corpus.

in shoot apices, and on the ultrastructure of the shoot apices in cacti (Mauseth, 1984).

Stems: General Structure

The comparative anatomy of the stem in angiosperms is a subject of great complexity and can be treated only in very brief outline here. In addition to the wide variation in the structure of the primary tissue systems, the stems of all woody and of many herbaceous dicotyledons develop secondary vascular tissues from a cambium and usually also form cork and phelloderm tissues from the phellogen or cork cambium. Extensive secondary growth results in the crushing and ultimate elimination of the epidermis, cortex, and primary phloem of a stem. It is generally held that herbaceous angiosperms phylogenetically have arisen from woody types—from this point of view, the narrow vascular cylinder or the ring of vascular bundles commonly developed in herbaceous dicotyledons represent highly reduced and specialized types of vascular systems (Fig. 19-20, A). In the stems of many monocotyledons the vascular bundles are situated throughout the fundamental tissue system, and there is no definable boundary between cortex and pith (Fig. 19-20, B). Monocotyledons lack the type of cambial activity commonly present in dicotyledons. Many species have what is termed a cylindrical primary thickening meristem that arises beneath the leaf bases near the shoot apex and is responsible for the initial thickening of the stem (DeMason, 1983). In a few genera (e.g., *Yucca, Dracaena, Cordyline*) the primary thickening meristem is continuous lon-

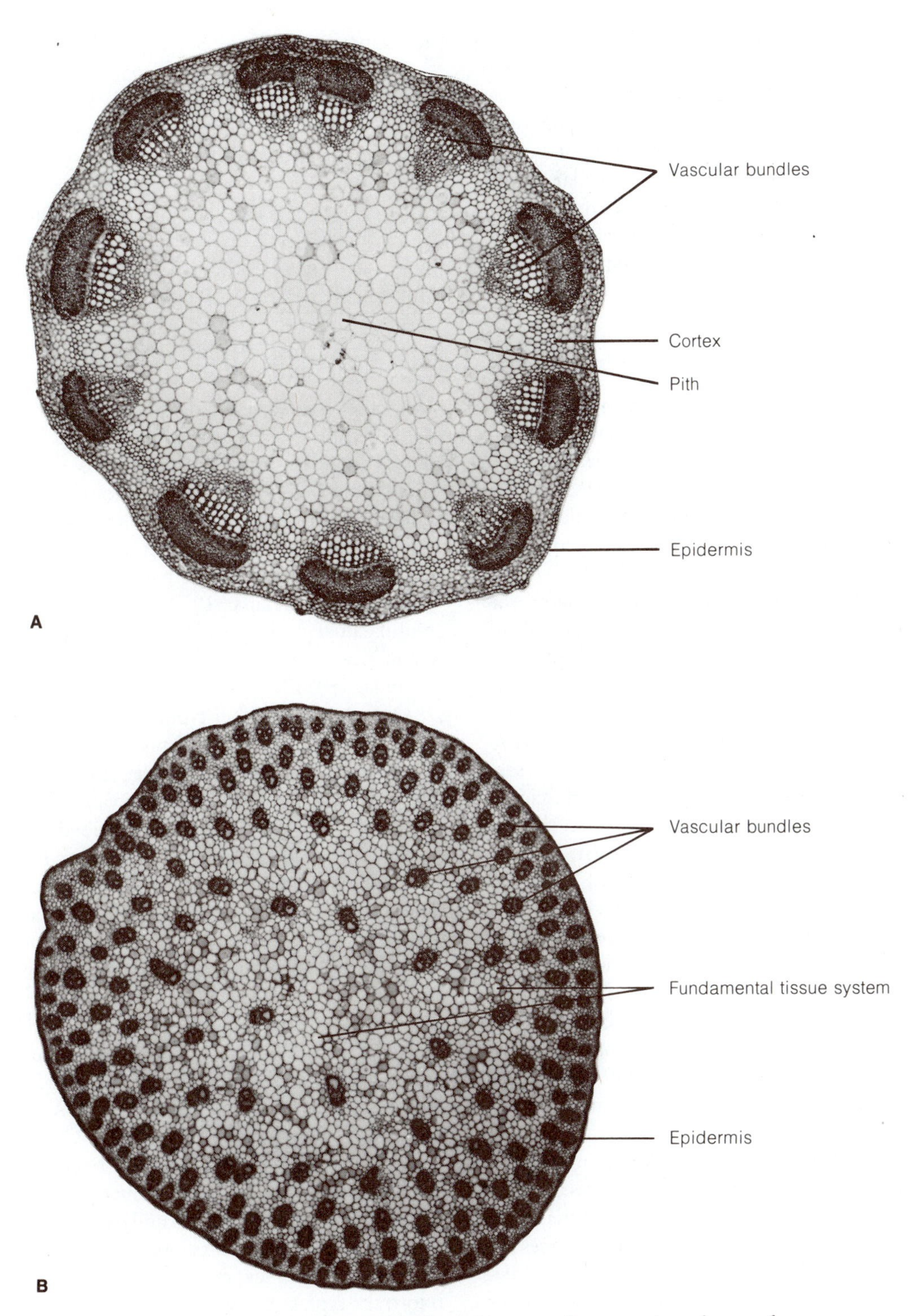

Figure 19-20 Types of stem anatomy in angiosperms. **A,** transection of stem of *Trifolium* (a dicotyledon); **B,** transection of stem of *Zea mays* (a monocotyledon) (note scattered arrangement of vascular bundles and the absence of definable cortex and pith).

gitudinally with a secondary thickening meristem that gives rise centripetally to ground tissue (parenchyma) and individual secondary vascular bundles. The bundles in *Yucca* consist of phloem surrounded by xylem — termed amphivasal bundles (Diggle and DeMason, 1983a, b). Vascular bundles of the primary body have been shown by cinematographic techniques to be continuous with secondary bundles in *Dracaena* (Zimmermann and Tomlinson, 1970).

Nodal Anatomy

A salient feature of stem anatomy that has received a renewed interest is the vascular anatomy of the nodes. In dicotyledons, for example, the vascular supply of a foliar appendage consists of one or more leaf traces, which diverge at the node into the leaf base and are associated with corresponding parenchymatous regions in the vascular cylinder. These parenchymatous regions have for many years been termed "leaf gaps" in a eustele. The difficulties which arise in attempting to define precisely what is meant by a leaf gap with reference to the primary vascular system was discussed in Chapter 14, p. 342. Briefly stated, the parenchymatous regions in the primary vascular cylinders of seed plants probably should be defined as interfascicular regions, and they are not comparable to the leaf gaps of fern siphonosteles. Nevertheless, in stems of dicotyledons that have experienced some secondary growth, serial transections of a node reveal well-defined parenchymatous regions — i.e., "gaps" in the secondary vascular cylinder — each related to the divergence of one or more leaf traces. The term leaf gap, used in a purely descriptive sense, is a useful term in describing the nodal anatomy of dicotyledons, and will be used interchangeably with the term *lacuna* (from Latin, meaning "space or break").

The significant investigations of Sinnott (1914) revealed that there are three primary forms of nodal anatomy in the dicotyledons as a whole: the *unilacunar node,* in which one, three, or more traces are related to a single gap (Fig. 19-21, A, C); the *trilacunar node,* characterized by the association of three gaps with the diverging leaf traces (Fig. 19-21, D); and the *multilacunar node,* in which five or more leaf traces are related to a corresponding number of leaf gaps (Fig. 19-21, E). Subsequently, another type was added to the list — the *unilacunar, two-trace type* (Marsden and Bailey, 1955; Fig. 19-21, B). More recently another nodal type has been recognized. In certain species that have opposite phyllotaxy, for example, two lateral traces (associated with two gaps) "split" soon after their departure from the stele. The divided traces then become the marginal veins of each leaf of the pair. This condition is referred to as a node with "split-lateral" traces (Howard, 1970).

One might ask which type of nodal anatomy is primitive in dicotyledons. Most families of dicotyledons tend to have a uniform nodal anatomy, but some families have variable nodal structure. The trilacunar node occurs in the majority of dicotyledons, which led Sinnott (1914) to conclude that this condition is primitive and that other types are derived. In a survey of five families in the Magnoliales, considered to have retained some other primitive characters, Sugiyama (1979) concluded that the multilacunar node is the basic type and the trilacunar type is the derived form in the order (see also Ozenda, 1949).

Over the years, other types have been suggested to be the primitive condition (see Dickison, 1975, for a discussion of the subject). Space does not permit a complete analysis of the various theories. However, one concept will be described. In an extensive survey of the range of variability in *cotyledonary* nodal anatomy, Bailey (1956) determined that 77 percent of them have an *even* number of traces, and 60 percent of them have *two* independent traces that are related to a single gap. This is in marked contrast to the foliage-leaf nodes where the large majority of dicotyledons have an *odd* number of traces. The double-trace unilacunar node associated with foliage leaves has been reported in only a limited number of families, including Austrobaileyaceae, Lactoridaceae, Lauraceae, Chloranthaceae. However, based upon its occurrence in families considered to have certain other primitive features, and the fact that the unilacunar, two-trace system occurs in certain gymnosperms (e.g., *Ephedra, Ginkgo*), Bailey considered the unilacunar, two-trace type to be primitive and the other types derived. Benzing (1967b) rejects Bailey's conclusions on the evolution of nodal types and maintains that "the anatomy of cotyledonary nodes does not

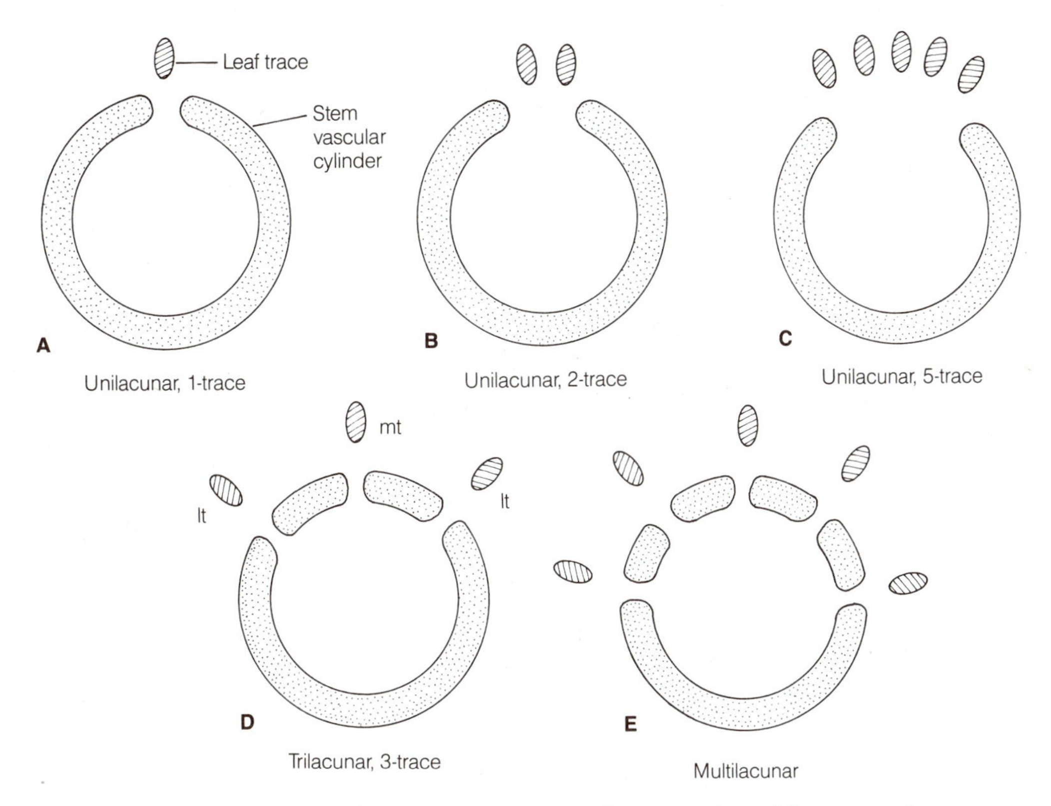

FIGURE 19-21 **A–E,** schematic representation of variations in nodal structure of dicotyledons. mt, median trace; lt, lateral trace. Consult text for details.

necessarily reflect ancestral conditions in the mature stem," an opinion shared by Conde and Stone (1970).

There is as yet no definitive answer as to what constitutes the most primitive type of nodal anatomy in angiosperms. Future paleobotanical discoveries of early Cretaceous angiosperms may provide an answer. Our present state of knowledge is based primarily upon comparative studies of extant plants. As pointed out by Beck et al. (1982), consideration also should be given to the possible adaptive value of different types of nodal anatomy. In plants with large leaves, the trilacunar and multilacunar nodes would be functionally more adaptive than unilacunar nodes. Among angiosperms, the trilacunar condition may have been derived in some groups from the unilacunar condition, but in other groups the trilacunar condition may have evolved early and been retained because it was highly adaptive. Of course nodal anatomy cannot be divorced from a consideration of the primary vascular patterns in shoots, as explained in the following section.

Primary Vascular Systems in Dicotyledons

In recent years there has been a renewed interest in the primary vasculature of angiosperms. The usual practice is to show the vascular cylinder as though it were split open and laid out in one plane. There are generally two basic types of vasculature —the "open" type and the "closed" type. Among dicotyledons with helical phyllotaxy, the *open* type is the most common. In the open type, leaf traces diverge laterally from axial bundles or sympodia. (A sympodium consists of an axial vascular bundle and its branches.) There are no fusions between leaf traces and/or axial bundles. Commonly there are five sympodia. In its simplest form, one leaf trace departs to a leaf at each node (Fig. 19-22, A), resulting in a unilacunar node that becomes more promi-

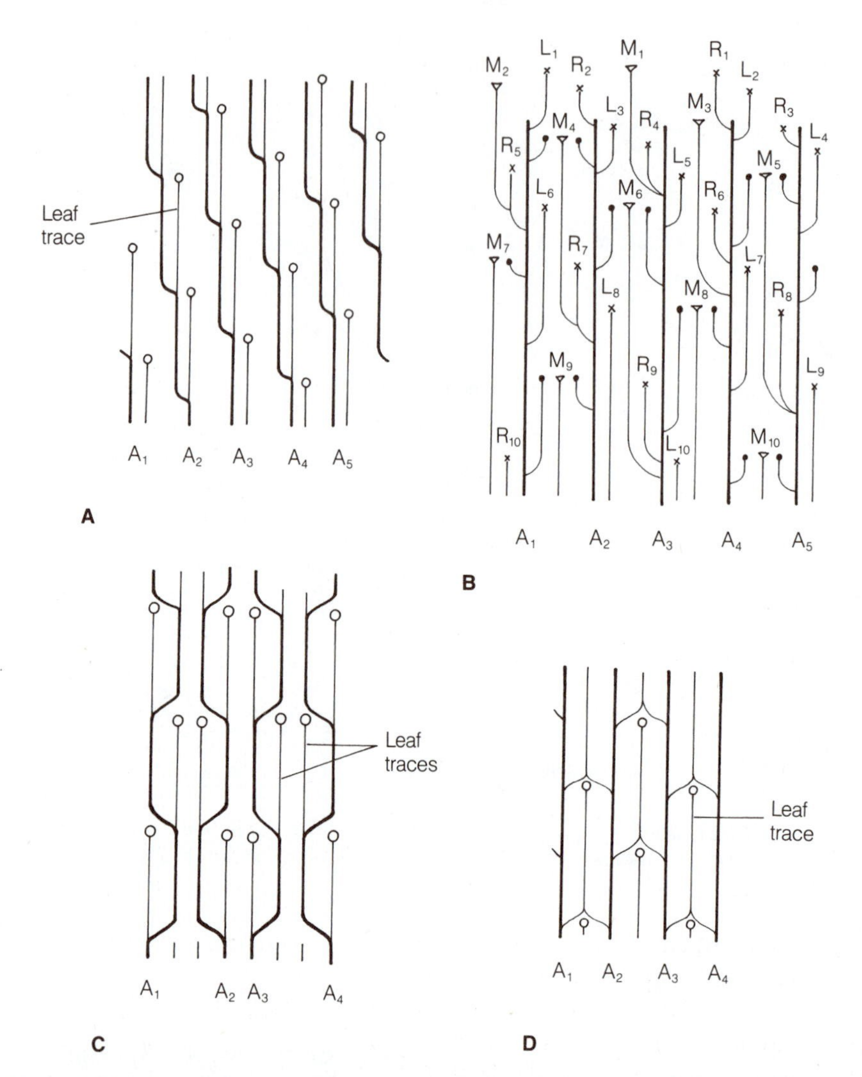

FIGURE 19-22 Longitudinal diagrammatic representations of stem primary vascular patterns. The vasculature is shown as if the stem were split along one side and laid out flat. **A–C**, open systems. **A,** *Godetia:* five sympodia, each node is unilacunar with one leaf trace; **B,** *Potentilla:* five sympodia, nodes trilacunar; M, L, and R, median, left lateral, and right lateral leaf traces, respectively; branch-trace bundles represented by small black circles; **C,** *Ascarina:* four sympodia, nodes unilacunar with two leaf traces; **D,** *Kalanchoe:* closed system; four sympodia, unilacunar; a single leaf trace originates from the fusion of strands from two adjacent axial bundles. In all diagrams, A_1, A_2, etc. refers to the axial bundles of the sympodia. [**A–C** based on Beck et al., *Bot. Rev.* 48:691, 1982; **D** based on Jensen, *Amer. J. Bot.* 55:553, 1968.]

nent after some secondary growth has occurred, as shown diagrammatically in Fig. 19-21, A. In *Potentilla fruticans,* the median and lateral traces diverge from different sympodia and are separated by two sympodia and leaf traces to other leaves; hence, a trilacunar type of nodal anatomy (compare Figs. 19-22, B and 19-21, D). The open system also can be found in plants with decussate (opposite) phyllotaxy. In *Ascarina lucida* there are four sympodia. A leaf trace is formed from each of two sympodia, resulting in a unilacunar node with two leaf traces (Fig. 19-22, C). This type is not common in dicotyledons. Most dicotyledons with decussate phyllotaxy have the *closed*-type system, in that the leaf trace is the product of the fusion of traces from each of two sympodia, resulting, for example, in a unilacunar, one-trace system (Fig. 19-22, D).

In a review of phyllotaxy and the vasculature of *seed plants,* Beck et al. (1982) have concluded that helical phyllotaxy is primitive, other types are derived; steles that consist of five sympodia are probably primitive, those consisting of more or fewer are derived; open primary vascular systems are primitive, closed systems are derived. Unilacunar nodes are primitive, multilacunar nodes are derived except among dicotyledons, in which unilacunar and trilacunar nodes may be equally primitive. [For additional information on the subject, the reader is referred to the following selected references and the literature cited therein: Benzing (1967a, b), Jensen (1968), Devadas and Beck (1972), and Dormer (1972).]

Wood Anatomy

One of the most productive approaches of modern comparative investigations on the systematics and phylogeny of the angiosperms is the study of the structure and development of secondary xylem ("wood") that constitutes the major *permanent* tissue of the stems and roots of woody plants (see Metcalfe and Chalk, 1950; and Cheadle, 1956). From a histological point of view, secondary xylem is a highly complex tissue which may be resolved, in its basic organization, into two integrated systems. These are clearly distinguished by the type and orientation of the component cells. In the *axial system,* which is composed of tracheary elements (tracheids, vessel members), wood fibers, and axial strands of living parenchyma, the cells are arranged with their long axes parallel to the longitudinal axis of the stem and root. The *radial system,* in contrast, comprises the xylem rays, which are sheets (one or more cells in thickness) of parenchyma cells, oriented radially with reference to the cells of the axial system. The cells of the two systems, interconnected by means of pit pairs, originate from the vascular cambium. The *fusiform initials* of the cambium, by means of tangential divisions, add new cells to the axial system; the *ray initials,* by similar divisions, contribute additional cells to the existing rays (Fig. 19-23). As the xylem cylinder continues to grow in diameter, new fusiform and ray initials originate within the cambium.

To understand the complex structure of any wood, it is essential to examine the tissue in transverse, radial, and tangential planes of section. By way of illustration, a small portion of the secondary xylem of *Liriodendron* (Magnoliaceae) is shown in Fig. 19-23 in the form of a three-dimensional "block diagram." This type of representation is particularly helpful for the visualization of the structure and spatial extent of the wood rays. In transverse section, the wood rays appear as radially oriented series of living parenchyma cells which extend from the vascular cambium for varying distances into the wood. Radial sections reveal that each xylem ray consists of several tiers of "procumbent cells" flanked marginally by a series of "upright cells" that are conspicuously elongated in the vertical direction (Fig. 19-23). A ray of this type is termed *heterocellular,* in contrast to the *homocellular* rays found in the woods of certain other dicotyledons that consist of only procumbent or only upright cells. Tangential sections reveal the rays in transverse view; in this plane of section, each ray is characteristically lenticular or elliptical (Fig. 19-23). The combined use of transverse, radial, and tangential sections is also essential to an understanding of the axial system of the wood, particularly with reference to (1) the distribution and structure of vessel members and other cell types, (2) the location and types of perforation plates in the vessel members, and (3) the nature of the pit pairs on the lateral walls of contiguous cells. As shown in Fig. 19-23, for example, scalariform perforation plates, characteristic of vessel members of *Liriodendron,* are displayed in face and sectional views, respectively, in the radial and tangential sections of the wood.

Increasing attention has been focused on the problems of the origin, structure, and systematic distribution of *vessels* in angiosperm woods. Vessels seem to provide some of the most reliable evidence regarding the general phylogeny of the xylem, not only in the angiosperms but in vascular plants as a whole (Bailey, 1944, 1953). The salient feature of a vessel is that it consists of a series of cells interconnected by *perforations*—i.e., open areas located most commonly on the end walls of the cells (Fig. 19-24, F–I). The cells of a vessel are

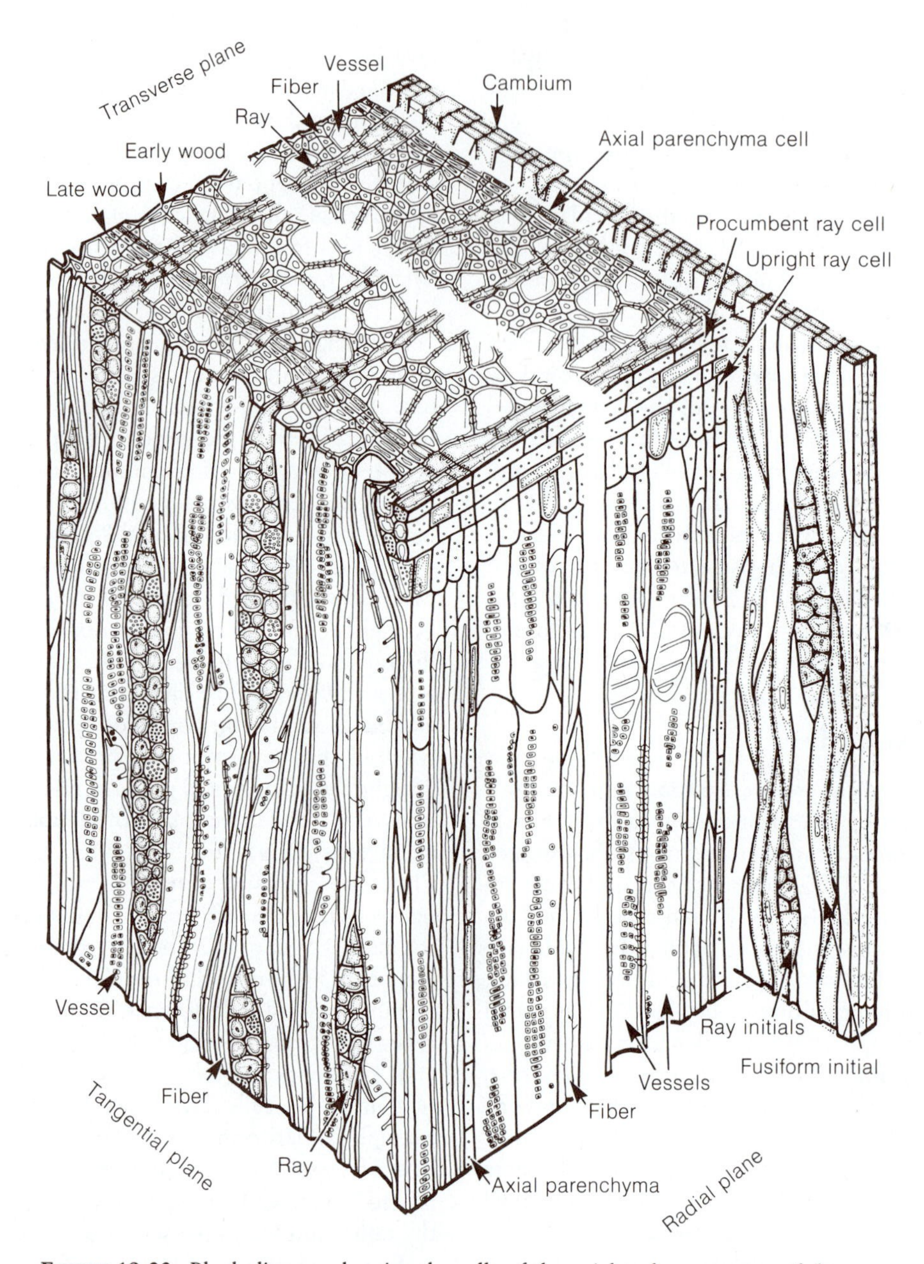

FIGURE 19-23 Block diagram showing the cells of the axial and ray systems of the secondary xylem of *Liriodendron tulipifera* in transverse, radial, and tangential planes of section. A small portion of the vascular cambium (separated from the wood for the sake of clarity) is shown at the right of the diagram. See text for further explanation. [From *Plant Anatomy,* 2d edition, by K. Esau. Wiley, New York. 1965.]

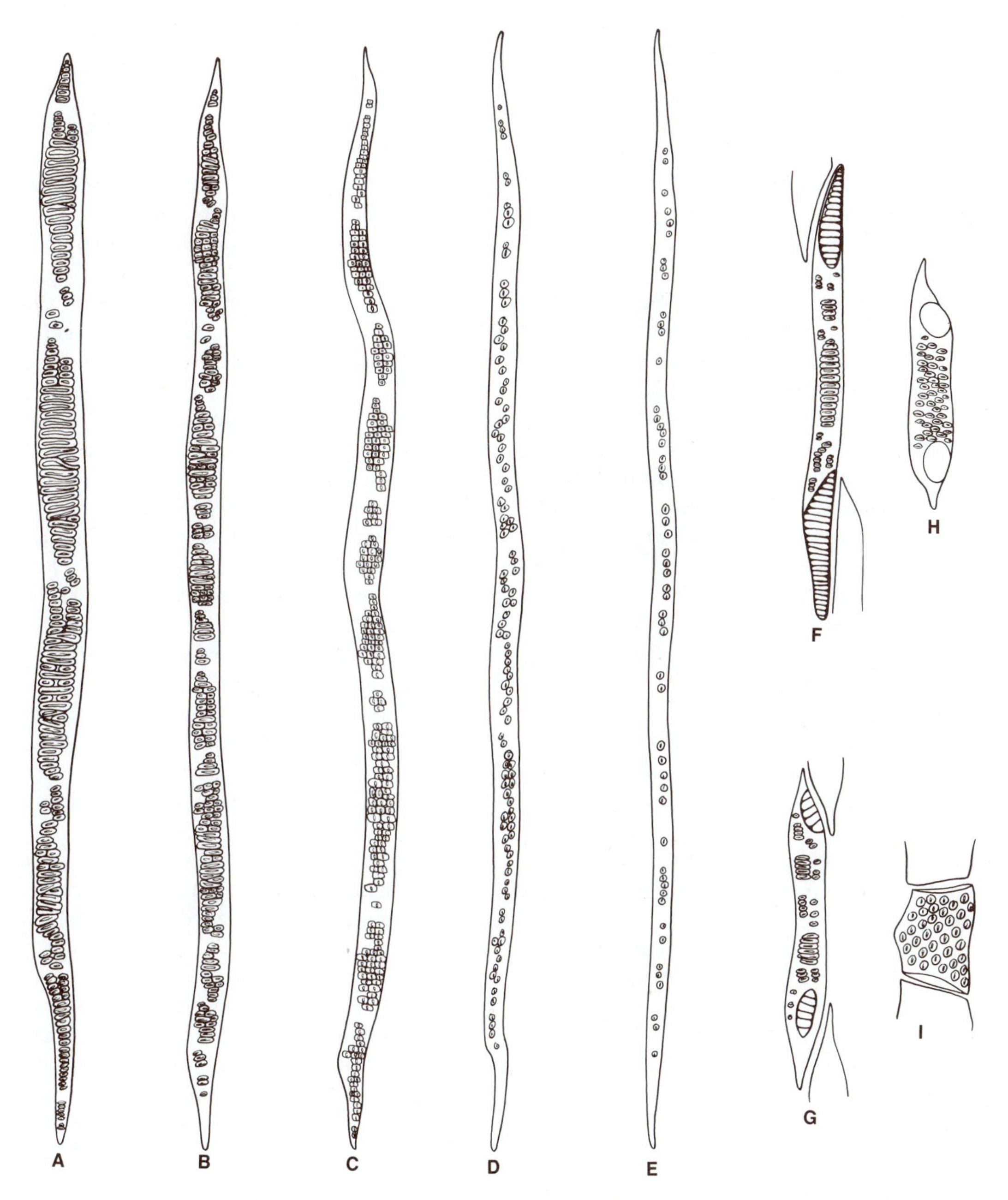

FIGURE 19-24 Various types of tracheary elements of the secondary xylem of dicotyledons. **A–E,** tracheids characteristic of the primitively vessel-less wood of *Tetracentron, Trochodendron,* and *Drimys,* showing scalariform bordered pits (**A**), circular bordered pits (**D, E**) and transitions between scalariform and opposite pitting (**B, C**). **F–I,** various types of vessel members formed in more specialized woods: **F, G,** vessel members with scalariform perforation plates and lateral walls with scalariform-opposite pitting; **H, I,** vessel members with simple perforations and lateral walls with alternately arranged bordered pits. In **F, G,** and **I,** portions of the vessel members lying above and below the cell shown in detail are represented in outline only. See text for further discussion. [Redrawn from Bailey and Thompson, *Ann. Bot.* 32:503, 1918.]

termed *vessel members.* In contrast to vessel members, *tracheids* are imperforate cells interconnected by bordered pit pairs and not forming well-defined longitudinal cell series.

Comprehensive surveys on the comparative anatomy of the xylem of extinct and living vascular plants have revealed that in all probability vessel members originated phylogenetically from tracheids, as the result of the dissolution of the primary cell walls of bordered pit pairs located at the terminal points of contact of the members of a vertical series of cells. In the Gnetophyta, as we have already pointed out (Chapter 18), the *initial steps* in the formation of vessel elements resulted from the loss of the primary cell walls during ontogeny between groups of circular bordered pit pairs. But in *Selaginella,* in certain ferns *(Pteridium),* and in the angiosperms, the initial step in vessel evolution was the loss of the primary cell walls of a series of transversely elongated pit pairs located near the ends of long tracheidlike cells. These facts all point to a very important general conclusion: vessels originated *independently* in a number of the major taxa of vascular plants, and hence the presence of vessels is not by itself a valid proof of systematic affinity, at least for the major groups of vascular plants (Bailey, 1944, 1953; Cheadle, 1953; Cheadle and Tucker, 1961).

Within the angiosperms — both monocotyledons and dicotyledons — the conclusion seems inescapable that the most primitive type of vessel member closely resembles a long scalariformly pitted tracheid but differs by having a series of transverse slitlike openings or scalariform perforation plates at each end of the cell (Fig. 19-24, F). The prototype of such a tracheid, from which such a primitive vessel member originated, fortunately is preserved in the woods of a series of living vessel-less plants (Bailey and Thompson, 1918).

At the beginning of the present century, vessel-less wood had been found in *Trochodendron, Tetracentron,* and various genera in the family Winteraceae. Tieghem (1900), the celebrated French anatomist, was so impressed with this unique type of xylem structure that he grouped all these vessel-less dicotyledons into a special order, the "Homoxylées." This taxonomic segregation was, however, rejected by later anatomists as it became increasingly evident that, although *Trochodendron* and *Tetracentron* show some degree of relationship, they clearly have no direct affinity with the Winteraceae when morphological characters other than wood structure are given due consideration (Bailey and Nast, 1945a, b). Moreover, two additional genera of vessel-less dicotyledons have been discovered and studied: *Amborella,* a monotypic genus endemic to New Caledonia (Bailey and Swamy, 1948; Bailey, 1957), and *Sarcandra,* an Asiatic genus in the Chloranthaceae (Swamy and Bailey, 1950). However, Carlquist (1987) has reported the occurrence of vessels in the root secondary xylem of *Sarcandra,* but not in the stem. It is evident nevertheless that the living, primitively vessel-less dicotyledons can no longer be regarded as a small "anomalous" group of angiosperms. On the contrary, the known genera are morphologically diversified and in habit range from shrubs *(Amborella)* to large, long-lived trees (*Trochodendron, Tetracentron,* and certain representatives of the Winteraceae). It seems likely that, as comparative anatomical studies on the angiosperms continue, other examples of vessel-less dicotyledons will be discovered.

Because of the phylogenetic importance of vessel-less angiosperm wood, we will describe, as an example, the wood of *Tetracentron sinense,* a large tree native to the montane forests of central and western China. As seen in transverse section, the wood of *Tetracentron* consists of well-defined growth layers (i.e., "annual rings") in which the tracheids of both the early and late wood are arranged in very regular radial rows, much like the

FIGURE 19-25 Structure of the primitively vessel-less wood of the stem of *Tetracentron sinense.* **A,** transection, showing part of a growth layer ("annual ring") composed of tracheids arranged in very regular radial rows and the characteristic uniseriate and multiseriate rays. Note the sharp demarcation between the early, or spring, wood (large, thin-walled tracheids) and the late, or summer, wood (smaller, thicker-walled tracheids) (×92); **B,** tangential section showing the characteristic elongated tracheids with overlapping ends, and the structure of the uni- and multiseriate rays (×124); **C,** radial section passing through the early wood, showing the conspicuous scalariform pitting on the lateral walls of the tracheids. See text for further comments (×124). [Photomicrographs made by A. A. Blaker, from sections prepared by Charles Quibell.]

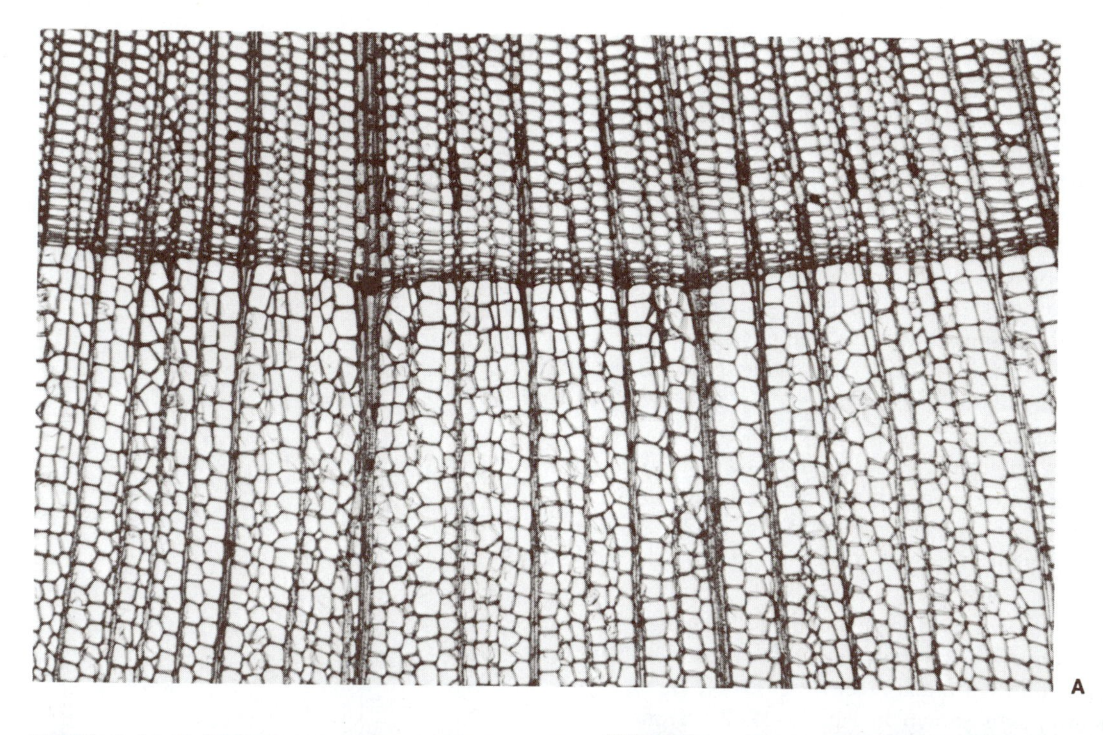
A

B

C

orderly series of tracheids in the wood of a gymnosperm (Fig. 19-25, A). The rays, as seen in a transection of the wood, are relatively numerous and include both uniseriate and multiseriate types. Tangential sections (Fig. 19-25, B) emphasize the structural difference between the two kinds of rays and also reveal the very long tracheids with their characteristically overlapping ends. One of the most significant aspects of tracheid structure in *Tetracentron* is the character of the pitting on the lateral walls. As seen in radial section (Fig. 19-25, C) the walls of the tracheids of the early wood have typical scalariform pitting; i.e., the closely spaced, transversely elongated pits are arranged like the rungs of a ladder, hence the term "scalariform." (See also the isolated cell shown in Fig. 19-24, A, and the diagram in Fig. 19-26, A). In contrast, the thicker-walled tracheids of the late wood of *Tetracentron*—as in the late wood of *Trochodendron* and *Drimys*—are provided with circular bordered pits on their radial facets (Fig. 19-24, D, E). Frequently, as is shown in Fig. 19-24, B, certain parts of the wall of a tracheid may show transitions between typical scalariform and opposite pitting (i.e., the formation of several oval pits in a transverse row) as well as transitions between opposite and alternately arranged pits (see Fig. 19-26 for diagrammatic representations of the various patterns of bordered pits of tracheary elements).

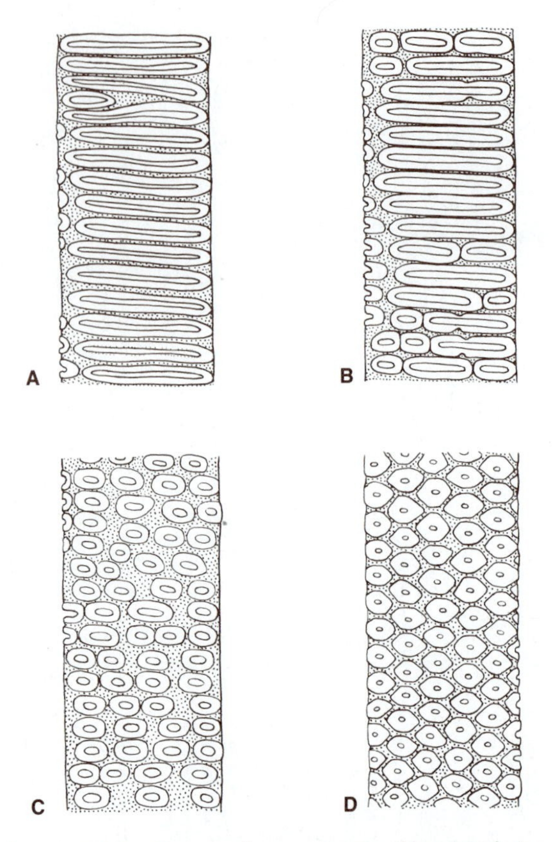

FIGURE 19-26 Form and arrangement of bordered pits on the lateral walls of tracheary elements. **A,** transversely elongated pits arranged in a scalariform series; **B,** scalariform opposite pitting; **C,** pits arranged in horizontal rows (opposite pitting); **D,** alternately arranged, crowded, bordered pits. [Redrawn from *Plant Anatomy* by A. Fahn. Pergamon, Oxford. 1967.]

It is significant that in dicotyledonous woods with vessels, scalariform pitting is frequently, although by no means invariably, associated with vessel members that develop long scalariform perforation plates (Fig. 19-24, F). Such primitive vessel members are thus similar to the longer scalariformly pitted tracheids of *Tetracentron, Trochodendron,* and other vessel-less dicotyledons. The various steps in the further evolutionary development of vessel members from tracheidlike cells are extensively preserved in the secondary xylem of more advanced dicotyledons and have been analyzed in detail by Bailey (1944). The major trends of specialization include (1) the marked *decrease* in the length of the vessel members, a change which reflects a corresponding decrease in length of the fusiform cambial initials, (2) an *increase* in the lateral expansion of vessel members during their ontogeny, culminating, in the most specialized examples, in short drum-shaped elements with truncated ends, (3) a gradual transition, by a reduction in the number of perforations, from scalariform to simple perforations, and (4) the replacement of scalariform intervessel pitting by circular bordered pits, arranged first in opposite and finally in alternate seriation. These four major trends in the phylogeny of vessel members are represented diagrammatically in Fig. 19-24, F–I.

Vessels may well have originated in several phylads of dicotyledons from primitive angiosperms, followed by vessel specialization. There is evidence that primitive angiosperms were restricted to mesic habitats. Vessel origin, followed by specialization, would have permitted radiation to new habitats leaving a few relict groups of vessel-less dicotyledons in the mesic pockets where we find them today (Carlquist, 1987).

Other views have been expressed with regard to whether angiosperms are primitively vessel-less. Young (1981) agrees with the tenet that a vessel member with scalariform perforation plates and scalariform lateral pitting is the primitive condition in angiosperms. However, from a cladistic analysis of selected families, he suggests that the absence of vessels in the living genera of "primitive" woody dicotyledons represents a secondary loss. A similar view was expressed by Jeffrey and Cole in 1916. Young also suggests that if the angiosperms originated in dry areas, as several investigators have suggested, vessels in the earliest angiosperms would have been highly adaptive. Vessels were lost in the ancestors of living vessel-less taxa during their secondary radiation into mesic habitats, along with modifications of other morphological characters.

Thus far our discussion of vessel evolution in the angiosperms has been restricted to the *dicotyledons,* in which the evidence is clear that vessels first appeared in the *secondary xylem* and subsequently "worked backward" in their ontogenetic appearance through the primary xylem (Bailey, 1944). A different and *independent mode of origin* of vessels, however, was proposed for the monocotyledons. According to the voluminous data assembled through the research of Cheadle (1953, 1955), vessels in the monocotyledons first appeared in the *metaxylem* (i.e., the last formed part of the primary xylem) *of the roots.* Indeed, in many monocotyledonous taxa, vessels are entirely restricted to the roots and the aerial parts of such plants are vessel-less! Because of this remarkable fact, Cheadle, in his detailed studies, has been able to trace the progressive development of vessels from an organographic standpoint (see also Cheadle and Tucker, 1961). A low level of specialization, for example, would be represented by those taxa that have vessels only in the roots, an intermediate level by taxa that have vessels in both roots and stems, and the highest level of specialization by those that have vessels throughout all parts (stems, inflorescence axes, leaves) of the plant. It is highly significant that Cheadle's investigations also show that the *structural evolution* of vessel members in monocotyledons closely parallels that characteristic of the dicotyledons. Vessel members in monocotyledons were derived from scalariformly pitted tracheids; subsequent specialization, as in dicotyledons, involved a decrease in length and an increase in diameter of vessel members accompanied by a transition from scalariform to simple perforations.

In conclusion, the broad systematic implications of the independent origin and parallel trends of specialization of vessels in dicotyledons and monocotyledons need consideration. Bailey (1944) remarked that "if the angiosperms are monophyletic, the monocotyledons must have diverged from the dicotyledons before the acquisition of vessels by their common ancestors. This renders untenable all suggestions for deriving monocotyledons from vessel-bearing dicotyledons or *vice versa.*" In Cheadle's (1953) view, however, the most likely hypothesis is "that the monocotyledons were derived (by reduction of vascular cambium activity, or by modification of this activity, as illustrated in the Agavales and elsewhere) from undiscovered fossil or living woody dicotyledons that lacked vessels throughout the plant."

Cronquist (1981) has taken quite a different position on the origin of monocotyledons. He believes that the monocotyledons were derived from aquatic members of "premonocotyledonous dicots" quite similar to members of the modern day aquatic, herbaceous Nymphaeales. The order is generally placed in the dicotyledons but members of the order do have certain monocotyledonous features. Members of the Nymphaeales are vessel-less except for the roots in species of *Nelumbo* (Kosakai et al., 1970). Cronquist points to the presence of leaves much like those of the Nymphaeales in the Lower Cretaceous along with the more conventional dicotyledonous types (Fig. 19-15). He considers that the monocotyledons have had an "aquatic ancestry, that these aquatic ancestors either lost or did not have a functional . . . vessel system to begin with . . . ." When some members returned to dry land, their further evolutionary progress was hindered by the absence of cambial activity and the absence of vessels. Some of the derivative monocots did "manage" to acquire a vessel system (Cronquist, 1969).

Earlier in this chapter, reference was made to the fossil *Sanmiguelia lewisii* from the Late Triassic (ca. 200 million years ago). This plant is considered to have been semiaquatic and, according to the description by Cornet (1986), to have had monocotyledonouslike leaves and reproductive structures, but had secondary xylem of the dicotyledonous type. Vessels were present in the secondary xylem of roots but not in the stem. Some paleobotanists believe that *Sanmiguelia* can be more accurately described as a gnetophyte. Additional fossils from time periods earlier than the Cretaceous may help to resolve the phylogeny of vessel numbers in angiosperms.

Mauseth (1988) has taken a different approach to the evolution of vessels. He agrees that vessel members did have their origin from tracheids with features considered to be primitive as we described previously, but the shape and anatomical features of a vessel member depends primarily upon its adaptive value in relation to its position in the plant and to the external environment. This concept emphasizes the importance of the structural-functional relationship rather than the phylogeny of vessel members.

Phloem

Phloem constitutes the tissue involved in the translocation of organic substances, especially photosynthates. A discussion of phloem histology was presented in Chapter 3 (p. 41). In review, the axial system of the phloem in angiosperms consists of sieve tubes and parenchyma. Phloem fibers also are frequently encountered. In secondary phloem the parenchymatous cells of the rays (radial system) arise from the same ray initials of the vascular cambium that give rise to ray cells in the secondary xylem. A sieve tube is a long conduit consisting of cells termed *sieve-tube members.* The latter are living but usually lack nuclei at functional maturity, and are characteristically associated with nucleated *companion cells* in angiosperms, especially in the metaphloem and secondary phloem. The end walls of sieve-tube members in angiosperms are oblique to transverse and have sieve areas or one large sieve area on each end wall. An end wall with a few to numerous sieve areas is termed a *compound sieve plate* (Fig. 3-18, D). A *simple sieve plate* consists of a single more highly differentiated sieve area on an end wall (Fig. 3-18, E–G). Sieve areas have *pores,* lined with cylinders of callose ((1 $\rightarrow$ 3)-β-D-glucan), through which the cytoplasm of contiguous sieve-tube members is interconnected. In general, the pores of simple sieve plates are larger than those of sieve areas on lateral walls and serve to enhance the longitudinal conduction of organic substances. Sieve tubes and vessels are analogous in the sense that they are made up of cells — sieve-tube members and vessel members — that have perforations and are therefore considered to be more efficient in conduction. Of course, a vessel member is dead at maturity.

PHYLOGENY. The evolutionary trends in specialization of sieve-tube members in monocotyledons are more clearly evident than for dicotyledons because of the comprehensive studies of Cheadle and Whitford (1941) and Cheadle (1948). The general trends have been (1) the change from sieve-tube members having very oblique end walls with compound sieve plates to having slightly oblique or transverse end walls with simple sieve plates, and (2) the sieve areas on the lateral walls of adjacent sieve-tube members becoming progressively less conspicuous. In monocotyledons the sieve tube first developed in the aerial portions of the plant and specialization then spread to the roots. This progression is opposite to that of the specialization of vessel members in monocotyledons. The reasons for these differences are probably understandable when we consider the functions of shoots and roots.

Possible phylogenetic trends in the phloem of dicotyledons, especially of the sieve-tube members, have been based primarily upon studies of secondary phloem. In general it is believed that the major trends in the morphology of sieve-tube members have been (1) a change in the orientation of the end walls from very oblique to transverse, (2) a change from compound to simple sieve plates with larger pores, (3) a progressive decrease in the number of sieve areas on the lateral walls, and (4) a decrease in length of the sieve-tube member, although this character is somewhat obscured by the fact that long potential sieve-tube members derived from the vascular cambium may, in some species, become subdivided obliquely or transversely before the file of cells differentiate into sieve-tube members. Therefore, the phylogenetic shortening of sieve-tube members is not entirely comparable to the phylogenetic decrease in length of vessel members in dicotyledons. Long sieve-tube members with oblique sieve plates tend to occur in the Magnoliales-Laurales, orders considered to be primitive for other reasons, and may be correlated with the occurrence of only tracheids or vessel members in the secondary xylem that have scalariform perforation plates. However, in many taxa there is no strict correlation between the degree of primitiveness or advancement in the morphology of the conducting cells of the phloem and xylem. Of course it is well known that tissues have evolved independently and at different rates. This may be related to the growth habit of various groups and the adaptive value in the past of both systems in specialized environments, the significance of which is not completely understood at present. For additional information see the following publications: Zahur (1959), Esau (1969, 1977, 1979), Dickison (1975), Carlquist (1975) and Mauseth (1988).

Roots

Although roots perform very important functions, such as anchorage, absorption and conduction of water and mineral solutes, and food storage, these essential organs of vascular land plants have never been studied morphologically or anatomically as thoroughly as leaves and stems. Some of the reasons for the neglect of broad comparative studies on roots are (1) the technical problems of dealing satisfactorily with the growth and structure of subterranean organs, (2) the wide divergence in anatomy between "typical" subterranean roots and the highly modified roots of tropical epiphytes, mangroves, and climbing plants, and (3) the absence of convincing paleobotanical evidence regarding the phylogenetic origin of the root and the trends in its evolutionary specialization.

Among the most comprehensive of modern publications on roots are the exhaustive monographic work by Troll (1967) and the detailed histological treatments of the primary structure of the root in gymnosperms and angiosperms provided by Guttenberg (1941, 1968). Since it is obviously impossible in the present book to consider in any detail the development and structure of roots, the following account is simply a brief resume of the highlights of a very complex and inadequately explored subject.

GENERAL MORPHOLOGY. In Chapter 6 we explained briefly that the terms *homorhizic* and *allorhizic* were introduced by Goebel (1930) to designate the two main patterns of origin of the first root in the embryo of vascular plants. In allorhizic plants (living gymnosperms and angiosperms) the embryo is bipolar—i.e., root and shoot apices lie at opposite ends or poles of the embryonic axis (Chapter 20). In homorhizic plants (all lower vascular plants) the first root is lateral with reference to the embryonic axis (Chapter 6, Fig. 6-2).

Allorhizic plants, or "allorhizophytes" as Troll (1949) designates them, may form a well-defined root system consisting of the main or primary root (derived from the radicle of the embryo) and its lateral branches (Fig. 19-27, A). On the other hand, in typical "homorhizophytes," such as *Lycopodium* and ferns, aside from the first root in the embryo, all subsequent roots arise from the stem of the sporophyte (Chapter 3, Fig. 3-6). According to Troll's (1967) view, plants characterized by "primary homorhizy" (i.e., the lower vascular plants) do not have a definable root system because *all* roots are shoot borne. An analogous kind of organization, termed "secondary homorhizy" by Troll (1949), is characteristic of many angiosperms and is virtually the predominant type of root organization throughout monocotyledons. In angiosperms that exhibit secondary homorhizy, the primary root of the seedling is short lived or else weakly developed, and lateral roots form mainly from various regions of the stem (Fig. 19-27, B).

Despite the often consistent pattern of origin of shoot-borne roots from specific nodal or internodal regions of the stem, such roots are very commonly termed "adventitious." We believe this term is inappropriate, because the shoot-borne "prop roots" of *Zea*, and the aerial roots of English Ivy, to cite striking examples, are as much a part of the "normal" morphology of these plants as their leaves or flowers! It therefore seems more in keeping with the facts to restrict the expression "adventitious roots" to those roots which arise *de novo* as a result of regenerative processes in callus tissue of stem cuttings or from the "dedifferentiation" of cells in detached leaves of such plants as *Begonia* and African violet *(Saintpaulia ionantha)*. In these instances, as well as in laboratory cultures of callus tissue, the formation of root primordia is truly an adventive

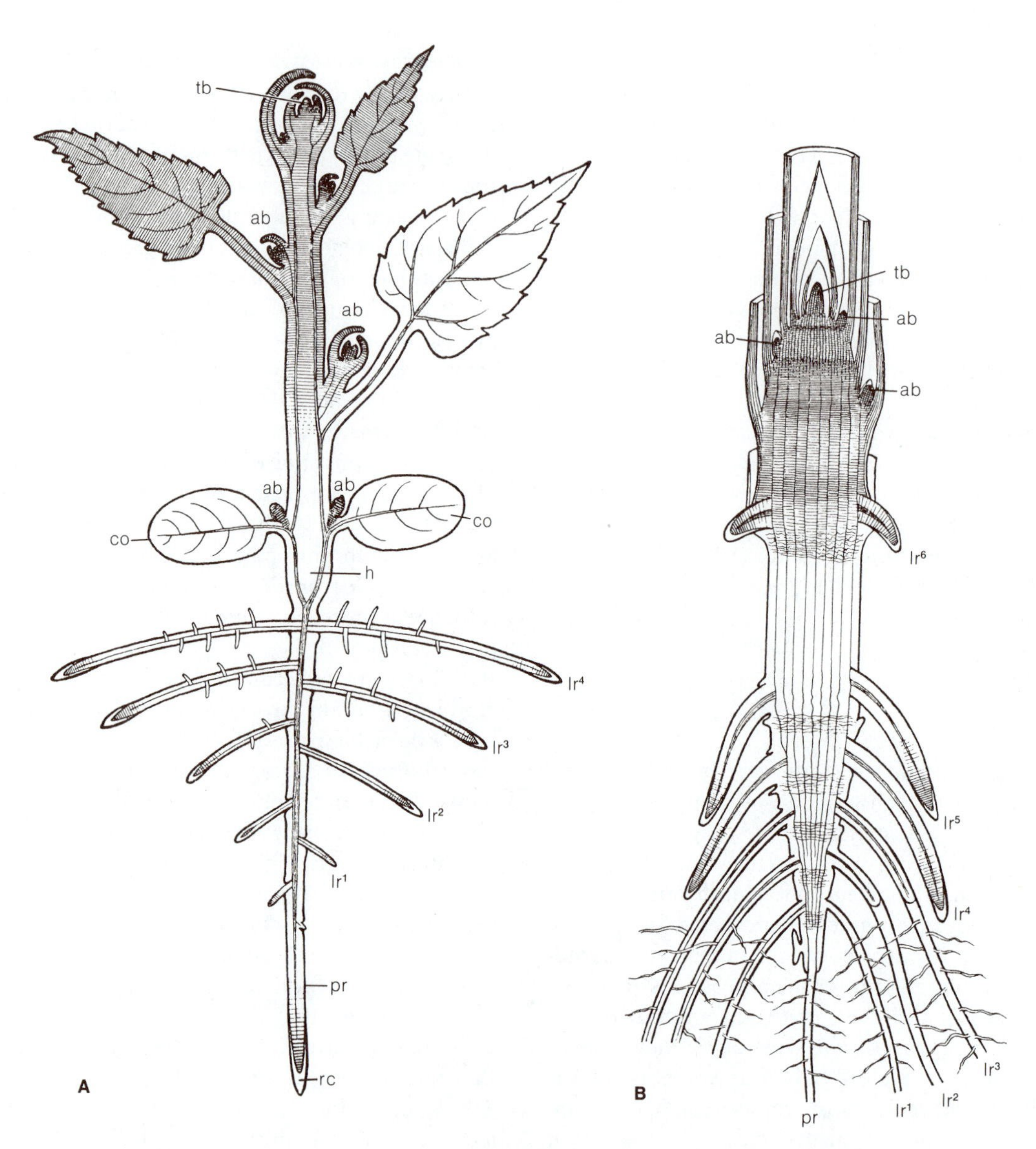

FIGURE 19-27 **A,** diagrammatic representation of a typical allorhizic dicotyledonous plant, as seen in longisectional view. Note acropetal sequence in formation of the endogenous lateral roots from primary root; **B,** diagram of a young homorhizic monocotyledonous plant *(Zea mays),* as seen in longitudinal section, to illustrate development of lateral roots (lr^1 – lr^6) from lower internodes of stem. Note weak development of primary root and its branches. ab, axillary bud; co, cotyledon; h, hypocotyl; lr, lateral root; pr, primary root; rc, root cap; tb, terminal bud. [Redrawn from *Vergleichende Morphologie der höheren Pflanzen* by W. Troll. Gebrüder Borntraeger, Berlin. 1935.]

phenomenon and not part of the normal developmental morphology of the plant as a whole.

THE ROOT APEX. The apices of shoots and roots in vascular plants differ strikingly in their growth and structure. In the shoot, the apical meristem is *superficial* and is overarched by the exogenous foliar organs that it has produced, whereas the corresponding meristem in the root tip is, strictly speaking, *subterminal* because it is covered by a protective root cap.

Expressed in ontogenetic terms, cell differentiation in the root tip is bidirectional, in that the outwardly produced cells become part of the root cap whereas the inwardly derived cells are added to the body of the root. The precision of these opposed patterns of cell formation is vividly illustrated in certain lower plants (e.g., *Pteris* and other leptosporangiate ferns, and *Equisetum*) in which all tissues of the root are *ultimately* derived from the segments of a single conspicuous four-sided apical cell (Fig. 19-28). Divisions parallel to the *outer* or *distal face* of the apical cell contribute exclusively to the root cap; divisions parallel to its three inner (lateral) faces yield additions to the root body (Gifford, 1983). An apical cell is also present in the *shoot apex* of the same plants, but there are no divisions parallel to the outer face of the apical cell, and thus a cap is not produced. In angiosperms the root cap originates in various ways; in many mono-

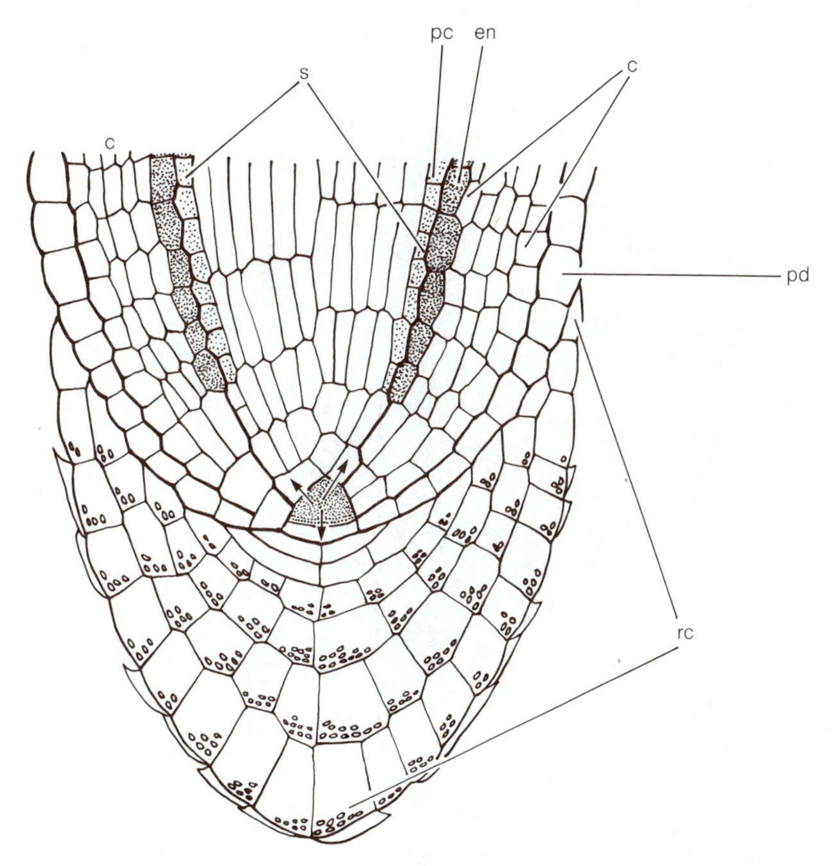

FIGURE 19-28 Diagrammatic representation of median longisection through the root tip of *Equisetum arvense,* showing conspicuous apical cell (demarcated by stippling). The arrows indicate that cell lineages of root cap as well as root body originate from segments (merophytes) derived from the apical cell. c, cortex; en, endodermis; pc, pericycle, pd, protoderm; rc, root cap; s, stele. [Redrawn from *Grundzüge der Pflanzenanatomie* by B. Huber. Springer-Verlag, Berlin. 1961.]

cotyledons the root cap arises from a separate multicellular layer ("histogen") termed the *calyptrogen.* In a wide range of dicotyledons both the root cap and the protoderm layer of the root body are derived from a common meristem, referred to as a "dermatocalyptrogen."

There is no single conspicuous apical cell in the roots of seed plants. Rather, cell lineages of the various zones of the root can be traced back either to a common meristem or to tiers of cells termed histogens. For many years these cells were considered to be the origin of all tissues of the root. During the past thirty years it has been shown conclusively that the rate of divisions in the histogens and a variable number of surrounding cells is low; the cells are relatively inactive mitotically and constitute the so-called *quiescent center* (Clowes, 1961). The quiescent center may be hemispherical in shape but does not include the initials of the root cap located distally to the quiescent center. Meristematic activity resulting in contributions to the main body of the root is located around the proximal edge of the quiescent center. This meristematic region is termed the promeristem or *proximal meristem* (Fig. 19-29). For a complete discussion of root types or experimentation designed to determine the significance of the quiescent center, the reader is referred to the following publications and the literature cited therein: Clowes (1961, 1984), Torrey and Feldman (1977), and Feldman (1984).

The Flower

Because angiosperms are commonly designated the flowering plants, one might assume that there is

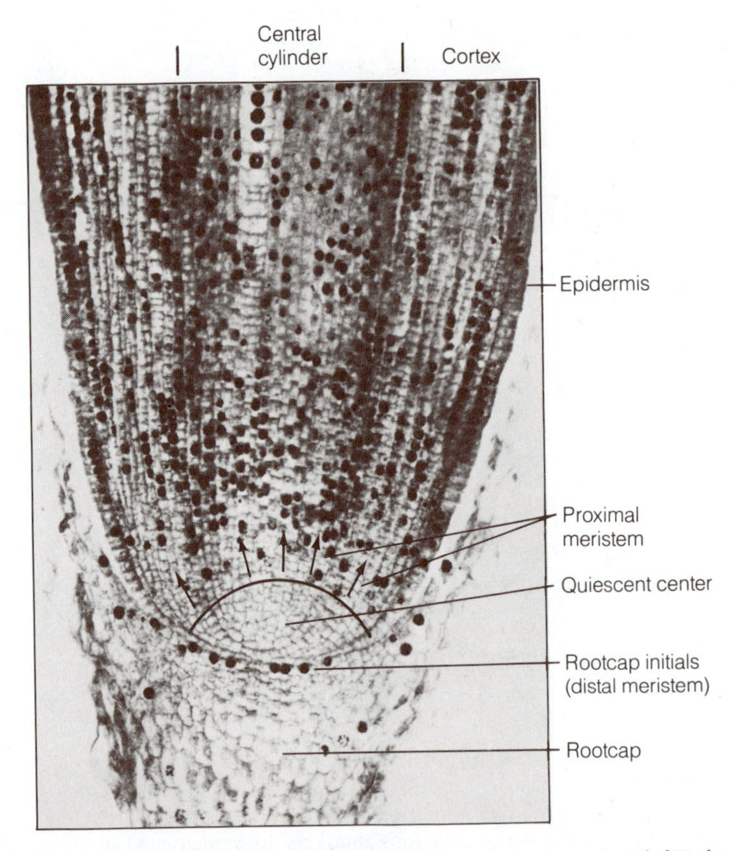

FIGURE 19-29 Longitudinal section of *Zea mays* (corn) root treated with ^{3}H-thymidine to reveal sites of nuclear DNA synthesis in preparation for mitosis and cell division. Note the absence of labeled nuclei in the quiescent center. Cells of the proximal meristem serve as initials for the longitudinal files of cells in the main axis of the root. [Courtesy of Dr. L. J. Feldman.]

general agreement about the scientific concept of a flower. The term flower, however, has been used to describe the cones of conifers, for example, and reproductive structures of the Gnetophyta. In this book we will restrict our use of the term to the angiosperms regardless of whether the flower is elaborate or greatly reduced in the number of floral parts.

One of the most definitive organs of the flower is the carpel, which resembles a megasporophyll in general *function* but is morphologically distinct because the ovules (or megasporangia) are typically enclosed within its hollow basal portion designated the *ovary*. Furthermore, in contrast to the megasporophylls of gymnosperms, carpels terminate in a *stigma*, which serves as a receptive structure for the pollen.

What is the primitive type of flower in angiosperms? Many morphologists favor the so-called classical theory which describes the *Magnolia-* or *Liriodendron*-type flower (Fig. 19-41) as a retained primitive condition and deviation from it as a specialization and thus a derived state. These genera and others also have retained certain vegetative characters considered to be primitive. What does the paleobotanical record tell us? Flowers have been described from the Lower and Upper Cretaceous that exhibited radial symmetry and had several to many parts for each category (sepals, petals, stamens, carpels). Coexisting with these flowers were simple or greatly reduced types, often composed of either stamens or carpels (Fig. 19-52). A more complete account of the types of early angiosperm flowers is presented later in this chapter.

The following sections will describe briefly a few types of flowers in living angiosperms. We will also present some evidence from vascular anatomy and ontogeny upon which the classical theory of the flower depends. The following references provide background on the theory: Goethe (1790), Arber (1937, 1946), and Eyde (1975a).

Flower Types

There is endless variation in floral types, and a complete description of even a few of them is beyond the scope of this book. Figure 19-30 illustrates the basic parts of a flower. The terminal portion of the floral axis is the receptacle to which the floral parts are attached. The basal segments of the generalized flower are the *sepals,* collectively termed the *calyx*. Above the calyx is the *corolla,* which consists of a variable number of *petals.* A *stamen* consists of an *anther,* in which pollen is produced, and a *filament.* The stamens collectively are referred to as the *androecium*. The *gynoecium* may consist of one or more separate *carpels* or the carpels may be fused into one unit called a *pistil.* A single carpel or the fused carpels are composed of three regions—the *stigma, style,* and *ovary*. The latter contains the enclosed ovules which will become the seeds in the mature ovary (fruit). Actual living examples of the type of flower just described are *Liriodendron* and *Sedum* (Figs. 19-31, A; 19-41).

Flowers such as *Liriodendron* and *Sedum* are often cited in support of the classical theory, which states that floral parts are helically arranged and represent, in an evolutionary sense, modifications of a basic type of foliar structure. The stamens of most angiosperms do not resemble leaves, but there are genera in which the stamens are flat, leaflike structures. Any modification in the organization of

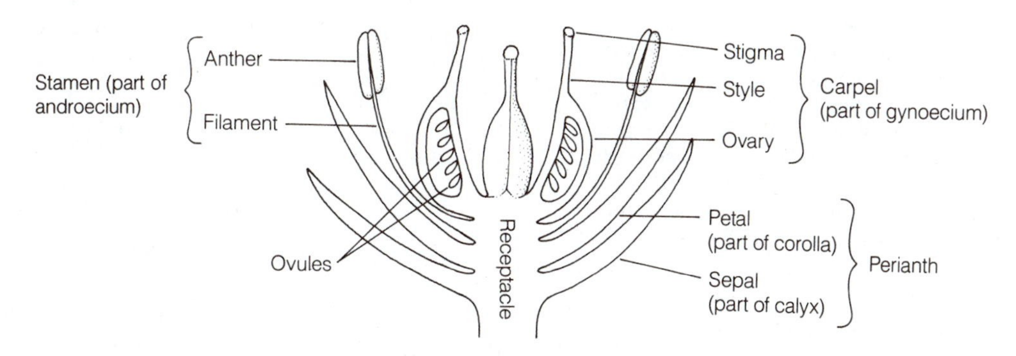

FIGURE 19-30 Schematic representation of a flower showing the basic floral parts and their relationship to each other.

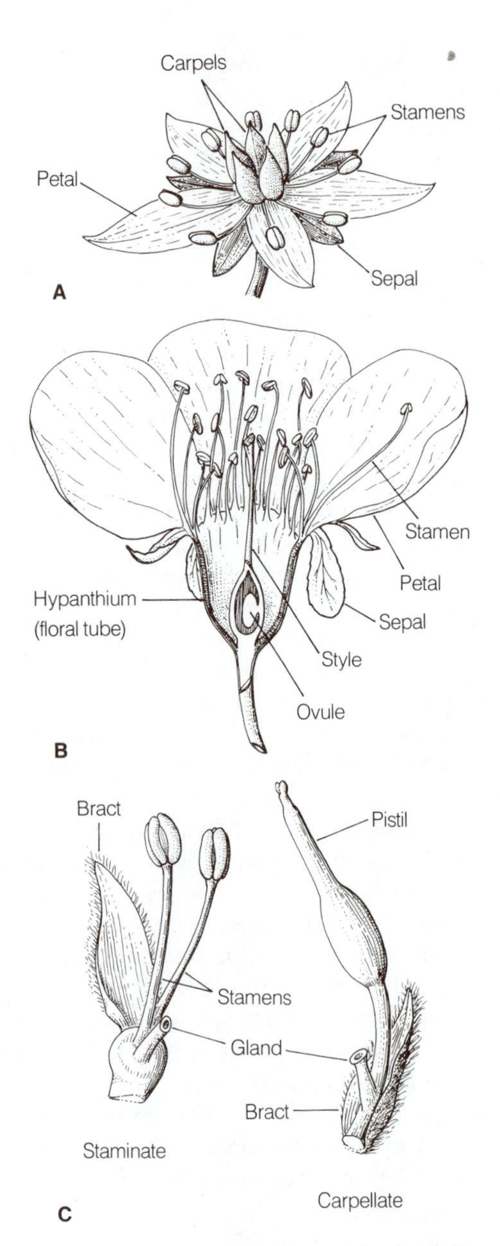

FIGURE 19-31 Examples of flower types. **A,** *Sedum* (hypogynous flower); **B,** longitudinal section of the perigynous flower of cherry; **C,** staminate and carpellate flowers of *Salix* (willow). [**A, C,** redrawn from *Plant Taxonomy* by E. L. Core. Prentice-Hall, Englewood Cliffs, N.J. 1955; **B** redrawn from *Textbook of General Botany,* 4th edition, by R. M. Holman and W. W. Robbins. Wiley, New York. 1938.]

the basic flower is interpreted as an evolutionary advancement or specialization—such as a change in the helical arrangement of parts to a whorled condition, fusion between similar parts *(connation),* fusion between different parts *(adnation),* and the change from radial symmetry to bilateral symmetry. The terms applied to fusion (connation) among similar floral parts—sepals, petals, stamens, and carpels—are *synsepaly, sympetaly, synandry,* and *syncarpy,* respectively. An example of adnation would be the fusion of stamen filaments to petals of the corolla (Fig. 19-31, B). In the course of evolution, according to the theory, certain parts became reduced not only in size but in number (Fig. 19-31, C). The flowers of *Salix* (willow) represent an extreme in reduced form. The flowers are either staminate or carpellate (pistillate). The staminate flower consists of a subtending bract and two partially fused stamens. The carpellate flower is composed of one stalked gynoecium (or pistil) consisting of two fused carpels (Fig. 19-31, C).

One of the many morphological characters taken into account in the grouping of families into orders is the position of the ovary. In our examples of flowers that presumably show certain primitive features (Figs. 19-31, A; 19-32, A), the *ovary* is attached to the receptacle above the other floral parts and is referred to as being *superior* in position. These are examples of *hypogynous* flowers. If the basal portions of the sepals, petals, and stamens are fused into a cup (floral tube), the flower is termed *perigynous* and the ovary is still superior (Fig. 19-32, B). In others, the floral tube may be partially fused with the ovary (Fig. 19-32, C), or completely fused, and the *free* portions of the calyx, corolla, and androecium are at the summit of the ovary (Fig. 19-32, D). Such a flower is termed *epigynous,* and the ovary is said to be *inferior* in position. According to the classical theory, any departure from the basic type—the hypogynous flower—constitutes a specialization and is considered an advanced character.

Another significant aspect of floral anatomy is the position of ovules in the ovary. In certain examples where the carpels are free from each other (the gynoecium is said to be *apocarpous*), the ovules are generally in two longitudinal rows attached to two strips of specialized tissue termed the *placentae* (Fig. 19-33, A). When the carpels are fused (syncarpy) and the placentae are at the center, the placentation

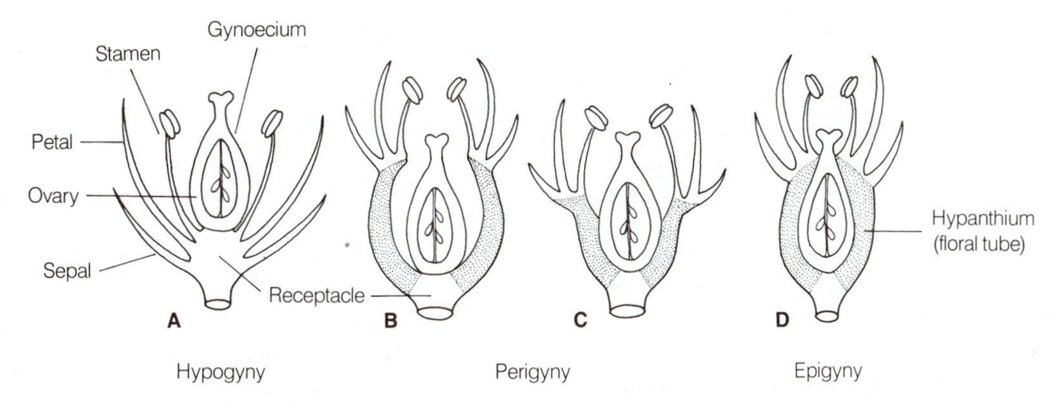

FIGURE 19-32 **A–D,** diagrams showing the organization of hypogynous, perigynous, and epigynous flowers, and the position of the ovary with respect to other floral parts.

is referred to as *axile* (Fig. 19-33, C). If the partitions (septa) between the carpels of the gynoecium are absent, resulting in a one-chambered ovary, the placentation is *parietal* (Fig. 19-33, B). The placentation is *free central* if there are no septa between the carpels and the placentae are on a central stalk that arises from the floor of the ovary (Fig. 19-33, D).

Vascular Anatomy

Since the classic investigations of Tieghem (1875) on the anatomy of the pistil and fruit, many studies have been devoted to the vasculation of the flower (Moseley, 1967; Carlquist, 1969; Eyde, 1975b; Schmid, 1972).

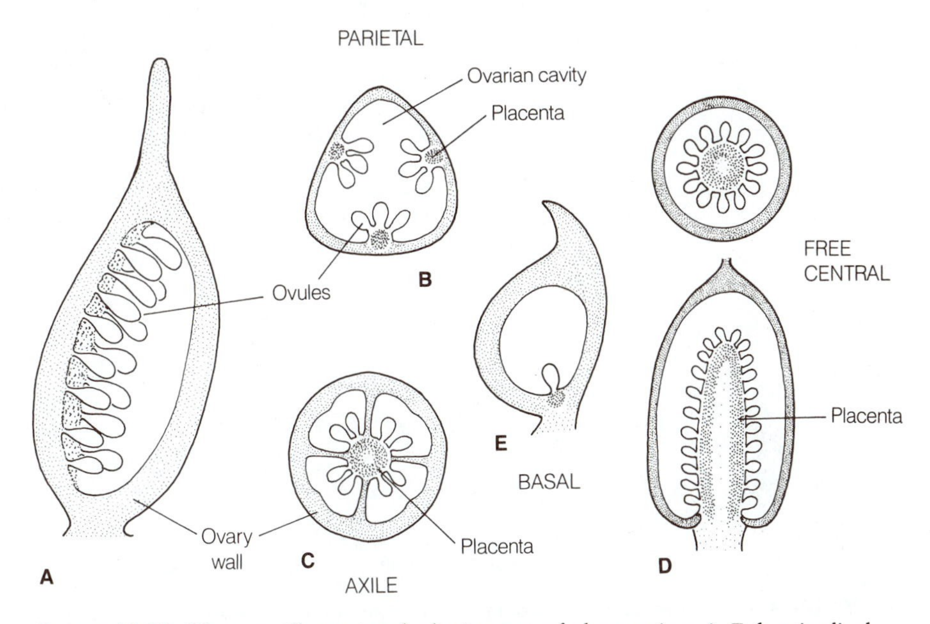

FIGURE 19-33 Diagrams illustrating the basic types of placentation. **A, E,** longitudinal sections of individual carpels in which there are two longitudinal rows of ovules **(A),** or one ovule attached basally **(E); B, C,** transverse sections of gynoecia with fused carpels; **D,** free central type placentation seen in transverse and longitudinal section. [**B–E** redrawn from *Plant Taxonomy* by E. L. Core. Prentice Hall, Englewood Cliffs, N.J. 1955.]

In addition to providing data of importance in the morphological interpretation of fusion, adnation, and the inferior ovary, the anatomical method has, in general, furnished strong support for the classical theory of the flower. A. J. Eames (1931), an outstanding leader in the study of floral anatomy, maintained that "flowers, in their vascular skeletons, differ in no essential way from leafy stems." He emphasized, however, that because of the determinate pattern of growth of a flower and the crowding and fusion of the numerous appendages on the short floral axis, anatomical interpretation is often difficult, and, to be successful, requires broad comparative knowledge of the vasculature of both flowers and vegetative shoots. Puri (1951), who reviewed the extensive literature on floral anatomy, concluded that evidence derived from the anatomical method has made a significant contribution to a better understanding of the angiosperm flower. But he emphasized the necessity of regarding vasculation as only *one* of the important sources of morphological ideas—evidence from organography and floral ontogeny also needs full consideration.

Numerous comparative studies on the vasculation of angiosperm flowers make it possible to select relatively simple types of flowers for brief consideration. In many respects, the flower of *Aquilegia*, a genus in the Ranunculaceae, is ideal because the successive floral appendages are free from one another and are borne on a well-defined stemlike axis or receptacle. According to the detailed investigations of Tepfer (1953) the flower of *Aquilegia formosa* var. *truncata* commonly consists of an axis bearing seventy appendages as follows: a calyx of five sepals, a corolla of five petals, an androecium of forty-five stamens and ten staminodia (petaloid organs devoid of anthers), and a gynoecium of five carpels, free at first but becoming basally concrescent later. Serial transections of the flower of *Aquilegia* reveal that the vascular system

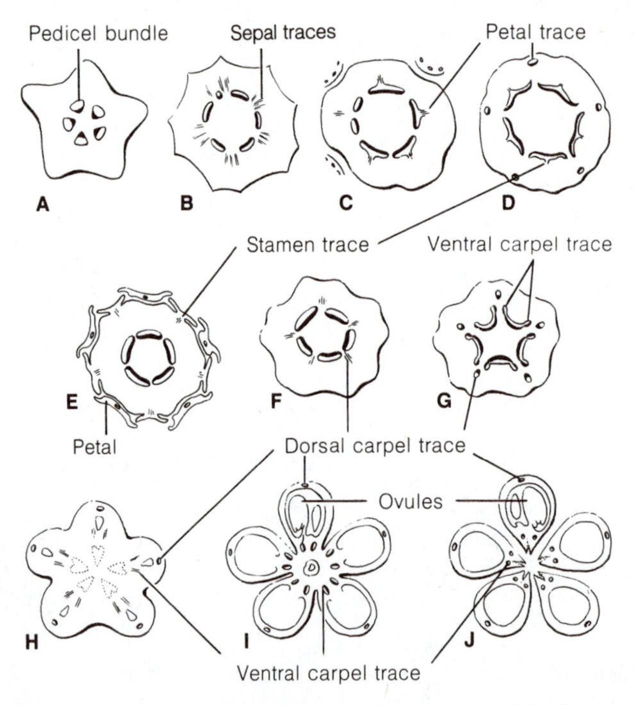

FIGURE 19-34 **A–J,** transections illustrating vascular anatomy of the flower of *Aquilegia*. **A,** transection of pedicel; **B, C,** departure of sepal and petal traces; **D, E,** departure of stamen traces; **F,** departure of dorsal traces of each carpel; **G,** departure of ventral carpel traces; **H,** transection of basal region of united carpels; **I,** the five carpels nearly free; **J,** the carpels at higher level, each with a dorsal and two ventral bundles. [Redrawn from Eames, *Amer. Jour. Bot.* 18:147, 1931.]

of the pedicel and receptacle is stemlike, consisting of a dissected cylinder of phloem and xylem from which are derived the traces to the successive floral appendages (Fig. 19-34, A–J). These floral traces, like leaf traces, are associated with definable "gaps" *after* secondary growth has occurred (Fig. 19-35).

As is true of a great many flower types which have been studied anatomically, there are remarkable differences in *Aquilegia* in the number of traces and the venation characteristic of the successive sets of floral appendages. Figure 19-35 illustrates the basic plan of vasculation in the flower, and Fig. 19-36 depicts the venation patterns typical of the sepal, petal, stamen, staminode, and carpel. At the nodal level of the calyx, each of the five sepals receives three traces which typically are associated with a single gap (Fig. 19-35). The three traces become more-or-less joined as a single bundle in the base of the sepal, and then separate into a palmate series at the base of the lamina, the venation of which consists of a series of irregularly dichotomizing and interconnected bundles (Fig. 19-36, A). The five petals, and each of the numerous stamens and staminodia, in contrast, are vasculated by single traces, each trace related to a distinct and separate gap (Fig. 19-34, C–E). Like the sepals, petals* of the flower exhibit a complex pattern of venation that differs markedly from the simple unbranched vascular strand that traverses each of the stamens and the staminodia (Fig. 19-36, B–D). Each of the five carpels receives three traces (associated with a single gap)—a dorsal trace which extends up the abaxial side of the carpel into the terminal region of the style, and two ventral traces which traverse the fused adaxial margins of the carpel and end in the style a short distance below the dorsal bundle (Fig. 19-34, F–J). As shown in Fig. 19-36, E, F, these ventral carpellary bundles give rise to lateral dichotomizing veins, one series of which enters the stalks of the ovules, the other series constituting the venation of the two walls of the ovary. From this description it is evident that the basic vasculation of

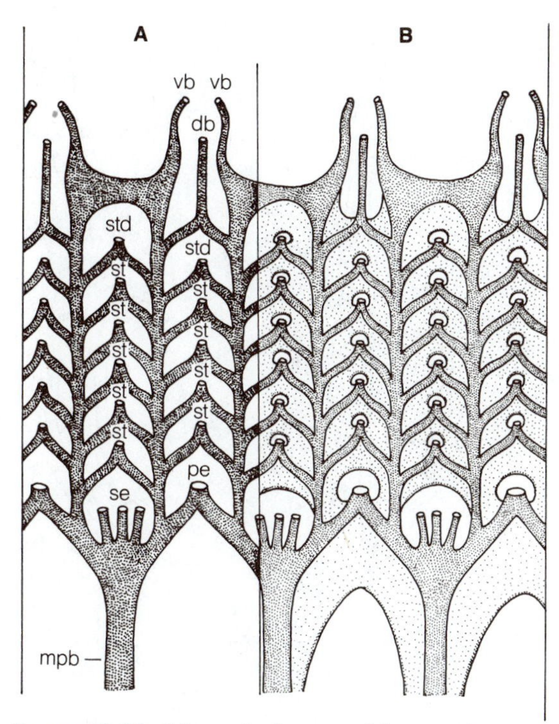

FIGURE 19-35 Schematic diagram of the vascular cylinder of the receptacle of *Aquilegia* as seen from the side, split longitudinally, and spread out in one plane. **A** represents the pattern of primary vascular tissue of a flower at anthesis. **B** represents the vascular cylinder after the production of some secondary vascular tissue (stippled). Floral parts served by the traces are labeled in **A.** se, sepal traces; pe, petal traces; st, stamen traces; std, staminodium traces; db, dorsal (median) trace of the carpel; vb, ventral (lateral) traces of the carpel; mpb, major pedicel (flower stalk) bundles. [Redrawn from Tepfer, *Univ. Calif. Publ. Bot.* 25:513, 1953.]

the flower of *Aquilegia* closely resembles that typical of a vegetative shoot. In both the shoot and the flower the successive appendages are vascularized by traces associated with gaps in the secondary vascular cylinder of the axis (i.e., stem or receptacle). Indeed, the vascular plan of the flower in *Aquilegia* appears to lend strong support to the classical interpretation of the flower, and serves as an example of a simple and possibly primitive type of floral vasculature in the angiosperms.

Many flowers, however, are more complex because of the reduction in size or phylogenetic loss of certain organs, or because of the very common tendency toward the fusion between floral organs. Fusion may consist in partial or complete lateral

*The venation patterns of petals in other taxa in the Ranunculaceae range from open dichotomous in *Adonis* (Hiepko, 1965) to dichotomous-reticulate in *Ranunculus* (Arnott and Tucker, 1963, 1964). For further information on the vasculature of sepals and petals in other angiosperms see Glück (1919), Chrtek (1962, 1963) and Hiepko (1965).

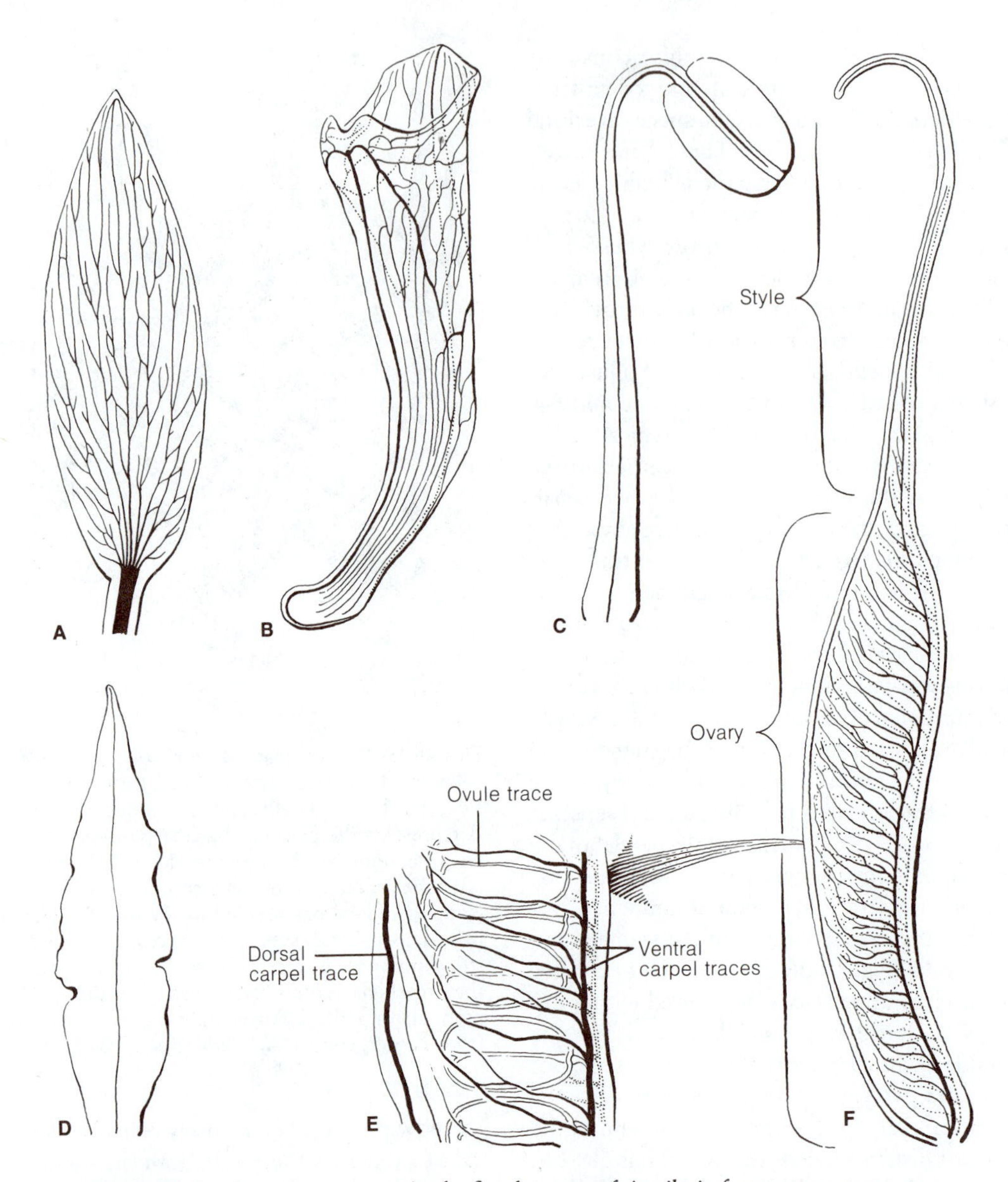

FIGURE 19-36 Venation patterns in the floral organs of *Aquilegia formosa* var. *truncata*. **A,** sepal; **B,** petal; **C,** stamen; **D,** staminodium; **E,** details of vasculature of portion of ovary (note the lateral, dichotomizing branches of the ventral traces and the derivation of the ovule traces); **F,** total venation of a carpel (the ovules, ovule traces, and wall venation of lower side are shown by dotted lines). [Redrawn from Tepfer, *Univ. Calif. Publ. Bot.* 25:513, 1953.]

union or connation between the adjacent members of the calyx, corolla, androecium, or gynoecium of a flower (Fig. 19-37, A, K–M); or it may consist in adnation between members of successive floral whorls, for example, the union of stamens to petals or to the tube of a sympetalous corolla (Fig. 19-37, B–F). These organographic specializations are, in turn, reflected in corresponding modifications of the vasculature of complex types of flowers.

It will not be possible in this book to discuss fully the application of the anatomical method to the interpretation of the diversified levels of specializa-

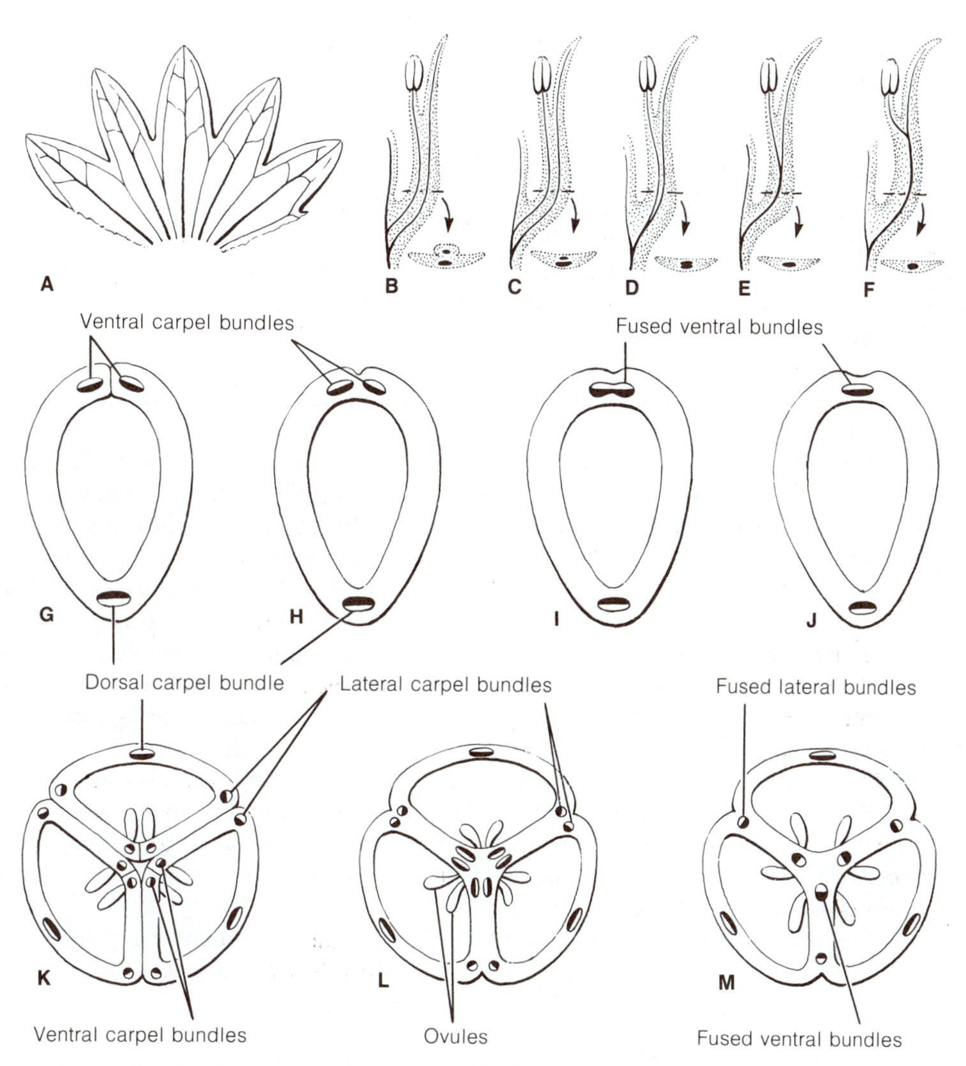

FIGURE 19-37 The effects of connation and adnation on the vasculature of floral organs. **A**, the calyx of *Ajuga reptans*, showing the five basally fused sepals (the adjacent lateral veins of each sepal are united below the sinus); **B–F**, corresponding longitudinal and transectional diagrams showing the results of successive degrees of adnation between stamen and petal upon the pattern of vasculature; **B, C**, weak adnation, with traces of stamen and petal separate; **D–F**, progressive steps in the radial fusion between stamen and petal bundles; **G–J**, transectional diagrams showing effect of union of carpel margins on ventral bundles; **G, H**, the ventral bundles are distinct; **I**, a double ventral bundle; **J**, fusion of ventrals to a single bundle; **K–M**, transections showing effects of fusion between three carpels; **K**, carpels in close contact but with distinct margins and paired ventral bundles; **L**, carpels fused, the ventral and lateral bundles of adjacent carpels arranged in pairs; **M**, the pairs of ventral and lateral bundles united. [Adapted from *An Introduction to Plant Anatomy*, 2d edition, by A. J. Eames and L. H. MacDaniels. McGraw-Hill, New York. 1947.]

tion in angiosperm flowers, but a few examples, selected from anatomical studies on the gynoecium, will serve at least to illustrate the effects of connation. In a free carpel the margins may become closely appressed without fusion, or, as in *Aquilegia,* may fuse during early phases in carpel ontogeny (Tepfer, 1953). In either case, the two ventral bundles are separate and distinct throughout the carpel (Figs. 19-36, F; 19-37, G). In instances in which the carpellary margins fuse earlier in ontogeny, the two ventral bundles may be joined as a double or an apparently single bundle at various levels in the carpel (Fig. 19-37, I, J). Syncarpous gynoecia reflect in their vasculature the varied degrees of lateral connation between adjacent carpels. In the simplest examples, the paired ventral bundles remain distinct and constitute the vascular system of the axile placenta (Fig. 19-37, K). When the carpels are closely united to form a tri- or multiloculate ovary, the placental region is often vascularized by one-half the expected number of bundles. This is interpreted as the result of fusion either between the ventrals belonging to each carpel or between the ventral bundles of laterally adjacent carpels (Fig. 19-37, L, M). Readers interested in the further complexity in the vasculature of gynoecia, stamens, sympetalous corollas, and synsepalous calyces of specialized flowers should consult Eames (1931), Puri (1951), Eames and MacDaniels (1947), and Eyde (1975b).

Ontogeny

In recent years there has been a revival of interest in the subject of apical ontogeny of vegetative shoots, inflorescences, and flowers. In addition to numerous studies on the structure of the vegetative shoot apex and the development of leaves and stem tissues, much attention has been given to the ontogeny of the reproductive apex. In general, whether a single flower or an inflorescence is formed, there are often changes in the size of the reproductive apex. Commonly the apex increases in height and width, associated with an increase in mitotic activity. During transition to flowering the organization characteristic of the vegetative apex may persist, or may become obscured by the formation of a rather uniform meristematic mantle of cells overlying initials of the developing receptacle of the flower. The floral parts are initiated in the mantle. Histochemical and autoradiographic studies have shown that there may be increases in concentrations of RNA and total proteins during transition to flowering. Likewise, there are changes in the amount and distribution of certain enzymes and growth substances. See the following references for additional information: Gifford (1963), Nougarède et al. (1965), Gifford and Corson (1971), Bernier et al. (1981), and Orr (1984).

In general, developmental studies appear to strengthen the classical interpretation and to com-

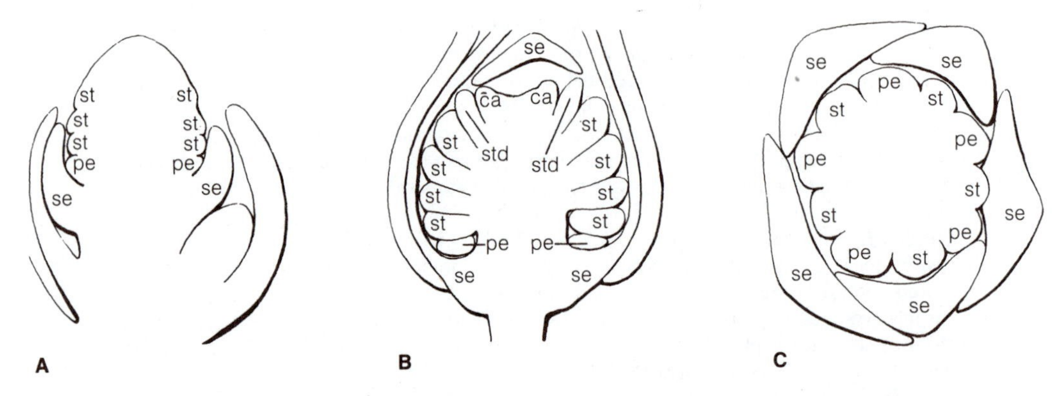

FIGURE 19-38 Organogeny of the flower of *Aquilegia formosa* var. *truncata.* **A, B,** outlines of longisections of the floral apex, illustrating acropetal development of sepals (se), petals (pe), stamens (st), staminodes (std), and carpels (ca); **C,** outline of a transection of a developing flower showing the five sepals and the primordia of petals and stamens. [Redrawn from Tepfer, *Univ. Calif. Publ. Bot.* 25:513, 1953.]

plement the conclusions reached through the study of the vasculature of flowers. An example is Tepfer's (1953) study of floral ontogeny in *Aquilegia* and *Ranunculus:* In both of these genera the sepals, petals, stamens, and carpels are produced in an acropetal sequence from the meristematic floral apex (Fig. 19-38). During the period of initiation of the appendages, however, there are rather striking changes in the dimensions and cellular structure of the floral apex. These changes involve a transition from a relatively small apex of uniformly and densely staining cells to a broad dome-shaped apical meristem with a conspicuously vacuolated parenchymatous core overlaid by several layers of meristematic cells. The latter structure is particularly characteristic of the apex of *Ranunculus* during the formation of the carpel primordia. In other angiosperms, for example, *Vinca, Umbellularia, Laurus,* and *Frasera,* the structure and growth of the floral apex during the formation of appendages resemble closely the organization of a vegetative shoot apex (McCoy, 1940; Boke, 1947; Kasapligil, 1951).

Tepfer's investigations further demonstrate that the method of initiation and the early phases of cellular differentiation of sepal, petal, stamen, and carpel primordia are fundamentally similar and closely agree with the early ontogeny of bracts and foliage leaves. The striking resemblances include the method of development of the procambium, which is produced acropetally in *both* vegetative and floral apices. The early ontogeny of the carpels in *Aquilegia* and *Helleborus* is of particular morphological interest. Both show a similar mode of initiation and development of carpels. Soon after the initiation of a carpel on the flanks of the floral apex, the carpel becomes somewhat cup shaped, resulting in a median cleft or crease (Fig. 19-39). Two longitudinal ridges or flanges are formed that become extended and increase in length as the carpel grows in height. The carpel now resembles a pitcher, but is still open on the adaxial side (Fig. 19-39, C). At this

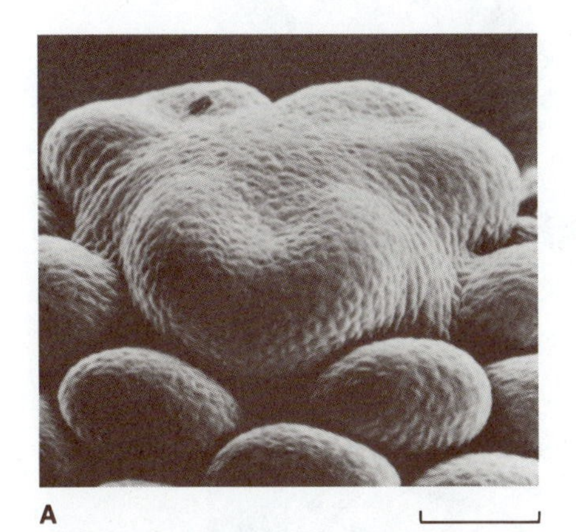

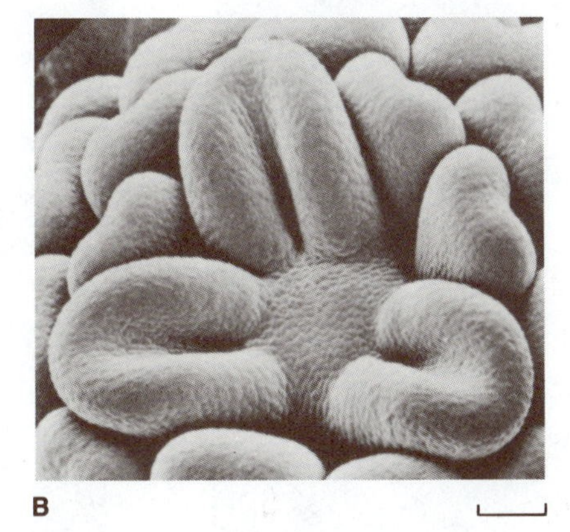

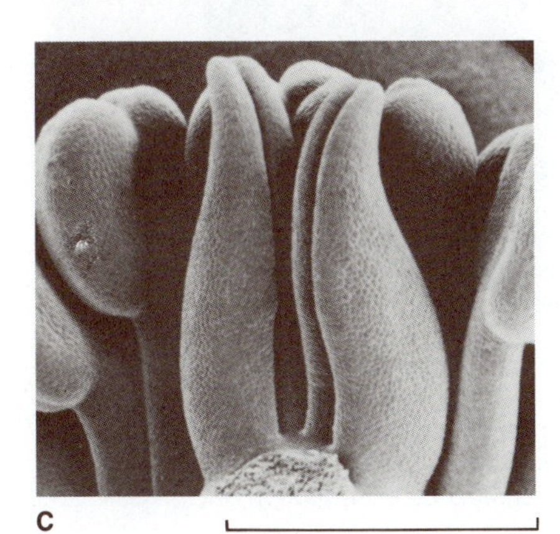

Figure 19-39 A–C, carpel development in *Helleborus foetidus* (Ranunculaceae); scale bar = 0.1 millimeter. See text for details. [From van Heel, *Blumea* 27:499–522, 1981.]

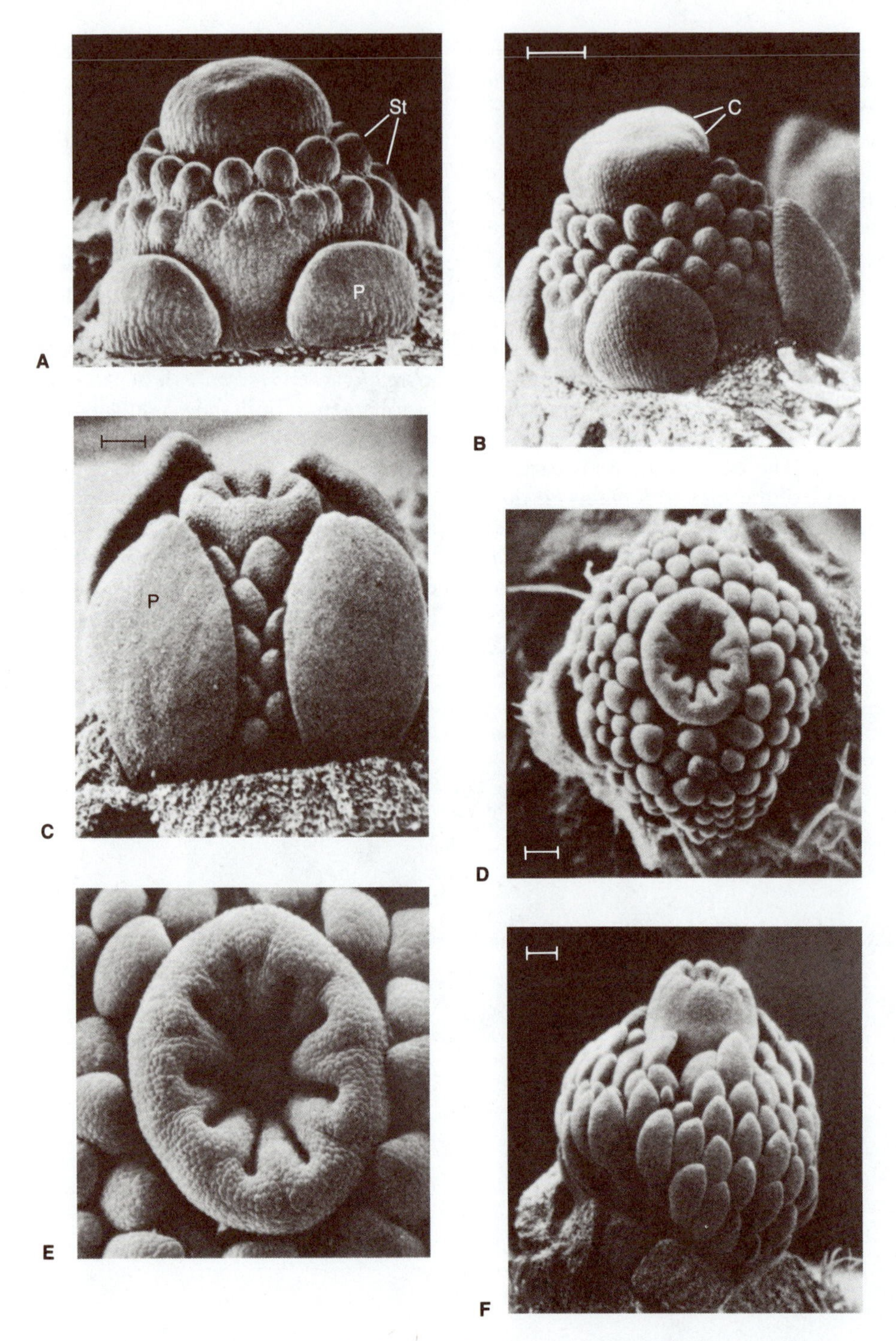

FIGURE 19-40 Ontogeny of the flower bud and syncarpous gynoecium of *Capparis spinosa* var. *inermis*. **A,** petal (P) and stamen primordia (St), and rounded receptacle; **B,** simultaneous origin of carpels (C) that are fused laterally, from their time of inception (**C**); **D–F,** later stages in the development of the gynoecium and stamens. Scale bars = 100 micrometers. [From Leins and Metzenauer, *Bot. Jahrb. Syst.* 100:542–554, 1979.]

stage in growth the carpel is horseshoe shaped as seen in transverse section. Ultimately, the marginal portions of the ridges become fused ontogenetically in the formation of the ovarian cavity. The line of fusion is generally evident in a mature carpel and is referred to as the *suture;* it is along the suture that dehiscence occurs in the mature fruit. Ovules are initiated from hypodermal cells situated about midway between the fused edges of the carpel and the inner surface of the ovarian cavity. During the early phases of gynoecial development in *Aquilegia* the initially free carpels become fused at their bases by connation of their meristematic epidermal layers. Thus in *Aquilegia* the union of the carpel margins and the fusion between the bases of adjacent carpels are truly ontogenetic processes. There are many examples of such postgenital carpel fusions in angiosperms. Of course, there are also numerous examples in which the carpels are essentially united at the time of their initiation and the gynoecium grows as a fused entity, resulting in an ovary with two or more locules, each of which morphologically represents a carpel (Fig. 19-40).

It is beyond the scope of this book to present the many well-documented descriptions of *transition from vegetative growth to flowering.* We will, however, describe the ontogeny of the flower in *Liriodendron tulipifera* (tulip tree), which provides some support for the classical theory of the angiosperm flower. The flower of the tulip tree embodies several presumably primitive features of organization: (1) the presence of numerous and helically arranged parts that are not fused and (2) the determinate nature of the floral apex is expressed only after an extended period of morphogenesis. Morphologically, the flower consists of three sepals, six petals, thirty to fifty stamens, and about 100 carpels on an elongate receptacle (Fig. 19-41).

From a topographical standpoint the vegetative shoot apex is a low dome (Fig. 19-42, A) and possesses a tunica and corpus (see p. 505 for discussion of tunica-corpus organization). Cells of the tunica divide anticlinally except during the initiation of leaf primordia, where cells of the second tunica layer divide periclinally. Cells of the corpus divide in various planes. During August or September (depending upon the latitude) the growth of some buds on a tree changes from vegetative to reproductive. There is little structural change in the vegetative apex as the meristem enters the period of floral development. A biseriate tunica remains as distinct as it was in the vegetative apex. The receptacle begins to elongate at the time of sepal initiation, and there is an increase in mitotic activity in all regions of the apex. Petals are initiated, and then the floral meristem enlarges greatly during stamen initiation (Fig. 19-42, C). Carpels are initiated acropetally, and throughout stamen and carpel development a rib meristem is active in the formation of the greatly elongated receptacle (Fig. 19-42, B – D). This account is based upon a study by Vertrees (1962). [See Sattler (1973) for photographs of three dimensional floral organogenesis in many species.]

Phylogeny of Stamens and Carpels

The subject of floral morphology may be appropriately concluded by a consideration of certain modern views on the evolution of the stamen and carpel in the angiosperms. Usually these sporangium-bearing organs are quite different from foliage leaves in their general form and appearance, and there are still conflicting views about their morphology and evolution. In the light of the evidence already presented in this chapter, stamens and carpels seem to be leaflike appendages. But how should we attempt to visualize a primitive stamen or a primitive carpel?

A common type of modern-day stamen consists of a filament that bears at its tip the anther, commonly consisting of four embedded microsporangia (pollen sacs). Some representative types are shown in Fig. 19-43 which illustrate the form and method of dehiscence of anthers. Are these types of stamens highly specialized or relatively conservative (primitive) in character? As for the carpel, the classical view regards this structure fundamentally as foliar with fused involute margins serving as the point of attachment of the enclosed ovules. Does this concept of an involutely folded, leaflike structure provide a reliable idea of the nature and method of origin of the carpel in ancient flowering plants? Provisional answers to these questions have been attempted, and they merit careful consideration.

On the basis of a series of investigations on the vasculation of stamens, Wilson (1937, 1942, 1953) proposed that these organs be interpreted with the aid of the telome theory. This theory, discussed

Figure 19-41 Flower of *Liriodendron tulipifera* showing external perianth, numerous stamens with abaxially embedded sporangia, and central group of closely crowded carpels. [From *Botany. An Introduction to Plant Science,* 2d edition, by W. W. Robbins, T. E. Weier, and C. R. Stocking. Wiley, New York. 1957.]

earlier in Chapter 3, attempts to derive vegetative megaphylls as well as sporogenous organs from primitive, dichotomously branched axes. In Wilson's view the modern angiosperm stamen, with its slender filament and compact anther, represents the end product of reduction and fusion of a dichotomous branch system with terminal sporangia. According to his idea the number of sporangia ultimately was reduced to four, and the modern anther morphologically is a synangium consisting of four

Figure 19-42 *Liriodendron tulipifera.* **A,** vegetative shoot tip. **B–D,** stages in development of the flower from a vegetative shoot apex. Note the great enlargement of the young floral apex (**B, C**) and the elongate form of the receptacle (**D**) near the end of the phase of carpel initiation. See text for more complete explanation.

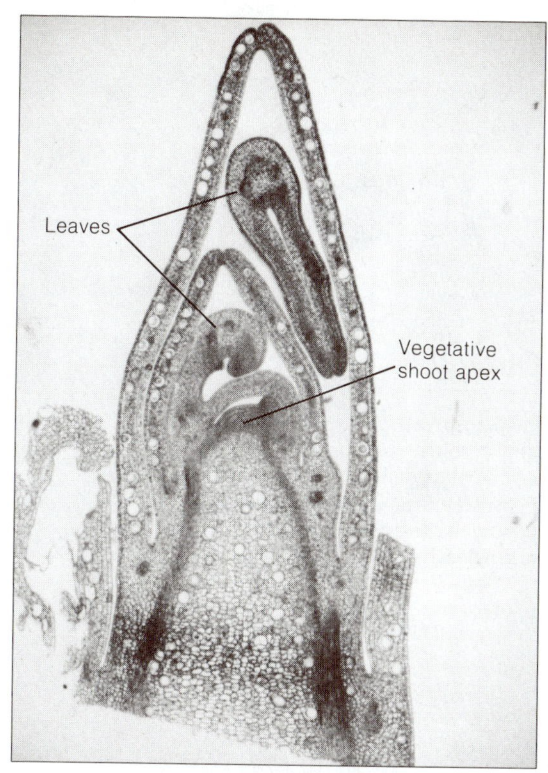
Leaves
Vegetative
shoot apex

A

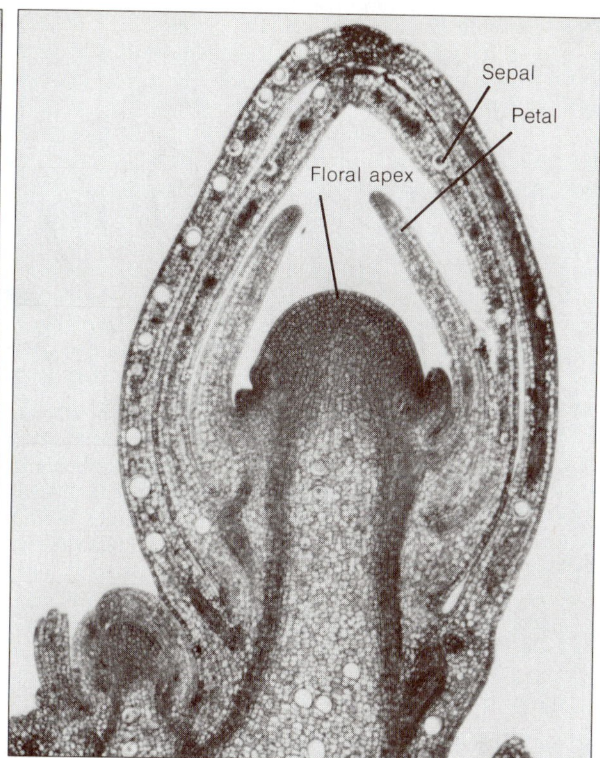
Sepal
Petal
Floral apex

B

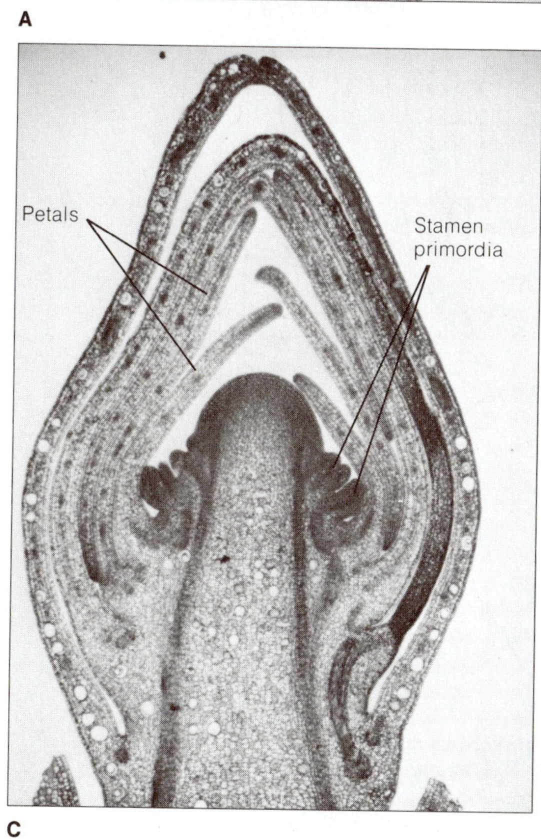
Petals
Stamen
primordia

C

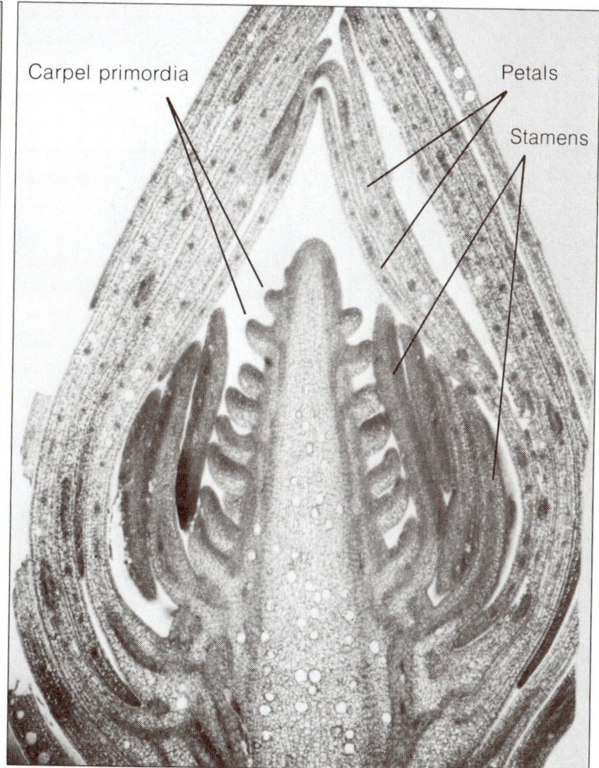
Carpel primordia
Petals
Stamens

D

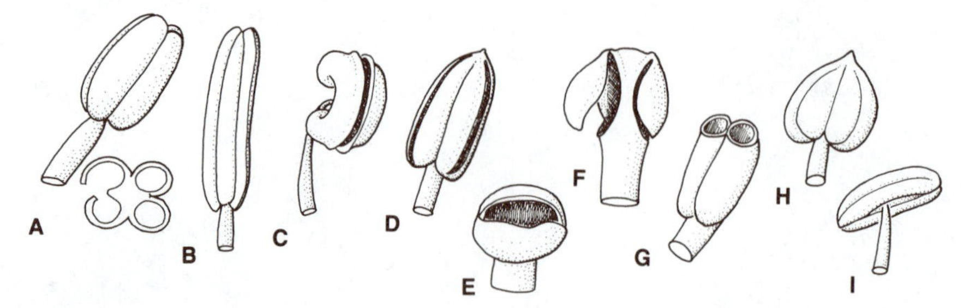

FIGURE 19-43 Representative types of stamens in angiosperms, and methods of dehiscence. **A,** anther consists of four microsporangia (pollen sacs); as shown in transverse section, each pair of pollen sacs becomes confluent at dehiscence; **D,** longitudinal dehiscence; **E,** transverse dehiscence; **F,** valvular flaps; **G,** apical pores; **I,** filament attached at middle of anther (versatile). [Redrawn from *An Introduction to Plant Taxonomy* by G. H. M. Lawrence. Macmillan, New York. 1955.]

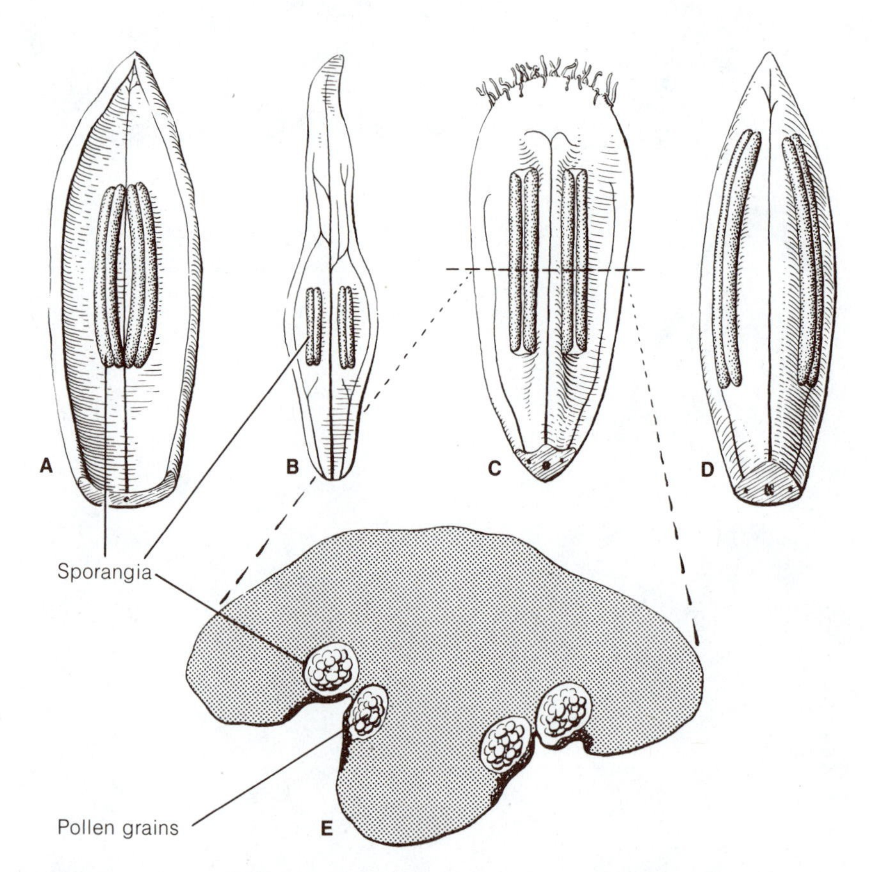

FIGURE 19-44 Primitive types of stamens. **A,** *Austrobaileya maculata,* adaxial surface showing paired sporangia at each side of midvein; **B,** *Himantandra baccata,* abaxial surface showing sporangia and relatively complex venation; **C,** *Degeneria vitiensis,* abaxial view showing pairs of sporangia between lateral veins and midvein; **D,** *Magnolia maingayi,* adaxial view showing venation and paired sporangia; **E,** diagrammatic view of transection of stamen of *Degeneria* showing the four sporangia embedded in abaxial surface. [Redrawn from Canright, *Amer. Jour. Bot.* 39:484, 1952.]

microsporangia. Wilson concluded also that the carpel originated from a series of fertile telomes which became webbed to form ultimately a leaflike structure bearing marginal ovules. Infolding of the margins of such a hypothetical structure was the final step in producing the modern ovary that encloses the ovules.

If the telome interpretation of the stamen and carpel were accepted, it would appear somewhat futile to draw any *direct* morphological comparisons between vegetative and floral organs in present-day angiosperms. On the contrary, the comprehensive surveys made by I. W. Bailey and his associates on the comparative morphology of many primitive families produced a new and hopeful line of attack on the problem of stamen and carpel evolution (Bailey, 1954). These surveys indicate that within *living* woody members of certain families there have persisted not only primitive trends of wood specialization — including primitively vesselless xylem — but also types of stamens and carpels that appear relatively primitive and unspecialized in character.

Degeneria, an investigated genus native to Fiji, provides a striking illustration of primitive stamens and carpels. The stamen of *Degeneria* is *not* differentiated into filament, anther, and connective but is a broad, foliaceous, three-veined sporophyll which develops four slender, elongated microsporangia deeply embedded in its abaxial surface (Fig. 19-44, C, E). It should be noted that the paired sporangia are laminal rather than marginal in position, and that they lie between the lateral veins and the midvein of the stamen. There are closely similar types of broad microsporophylls in other genera (Figs. 19-44, 19-41) such as *Austrobaileya, Himantandra,* and certain members of the Magnoliaceae (Canright, 1952; Ozenda, 1952). The stamen of *Degeneria* is considered by Canright (1952) to be "the closest of all known types to a primitive angiosperm stamen." The general trends of specialization in the angiosperm stamen would appear to be (1) the narrowing of the lamina, (2) restriction of the microsporangia (pollen sacs) to the margin of the anther, (3) the development of a definite elongate filament, and (4) a reduction in the number of veins from three to one. As pointed out by Carlquist (1969) in a critique of stamen evolution as just presented, the one-trace nature of most stamens is related to their filiform nature and their brief longevity at anthesis. Not all stamens of presumably more advanced taxa have only one vascular bundle. Stamens with more numerous or branched traces have a greater longevity, which is probably related to particular modes of pollination (Carlquist, 1969).

The carpel of *Degeneria* differs in many fundamental ways from the typical angiosperm carpel which is differentiated into a closed ovary, style, and stigma. In external form the carpel of *Degeneria* consists of a stigmatic surface (or stigmatic crest) and a rather broad base that is attached to the receptacle of the flower. When viewed in transverse section the carpel appears to be folded lengthwise (conduplicately folded) rather than involute (Fig. 19-45). The term "conduplicately folded" is only a convenient descriptive term; the carpel does not actually develop initially as a flat laminar structure which then becomes folded. At its inception the carpel resembles a shallow cup (Fig. 19-46, A). Cell divisions and subsequent growth continue to take place along the rim of the cup, increasing the depth of the ovarian cavity. The rim continues to grow out in the form of flanges, whose margins flare apart (Fig. 19-46, B, C). A narrow cleft remains between

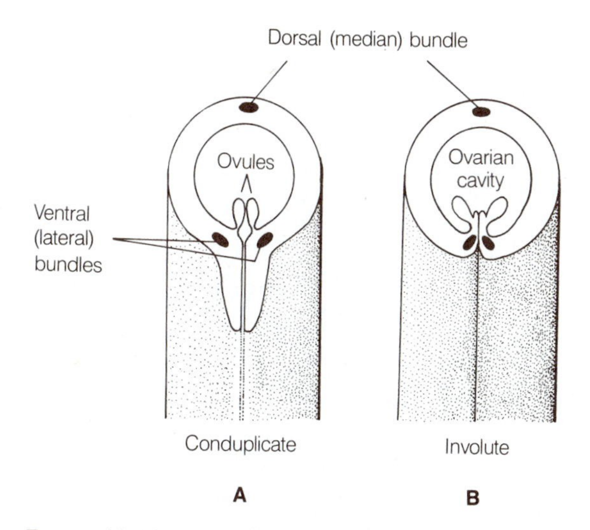

FIGURE 19-45 **A, B,** schematic diagrams showing conduplicate and involute carpels. In the conduplicate carpel, the inner (adaxial) epidermal layer is enclosed; the margins are flangelike and more-or-less completely united. The involute carpel is folded so that the carpel is closed along the abaxial surface for a short distance.

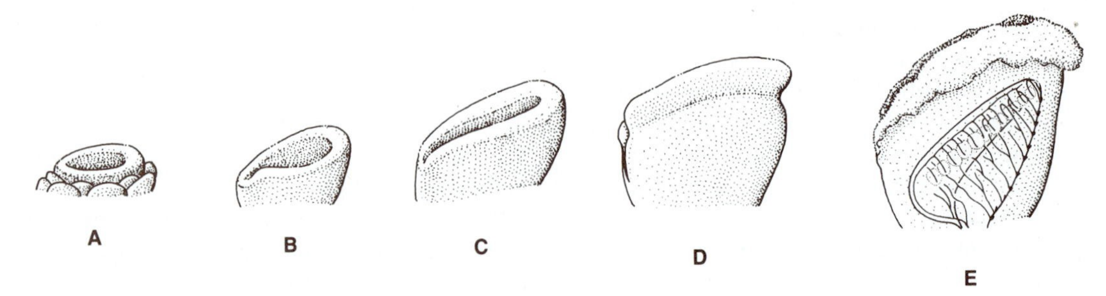

FIGURE 19-46 **A–E,** ontogeny of the carpel in *Degeneria.* See text for description. [**A–D** redrawn from Swamy, *J. Arnold Arboretum* 30:10, 1949; **E** redrawn from Bailey and Swamy, *Amer. J. Bot.* 38:373, 1951.]

the flanges, extending from the exterior to the inner cavity or locule.

Swamy's (1949) investigation shows that, beginning with the flared carpellary margins, an extensive development of epidermal hairs proceeds inwardly toward the locule (Fig. 19-47). The space between the closely adjacent carpellary margins becomes filled with interlocking hairs, whereas the trichomes on the inner surface of the carpel are short and papillate. Collectively, all these hairs represent a stigmatic surface. Pollination requires the deposition of pollen grains on the adaxial hairy divergent margins of the open carpel. Swamy observed that the developing pollen tubes grow into the cavity of the carpel between the hairs and along the papillate surface—in no instance do the tubes penetrate the

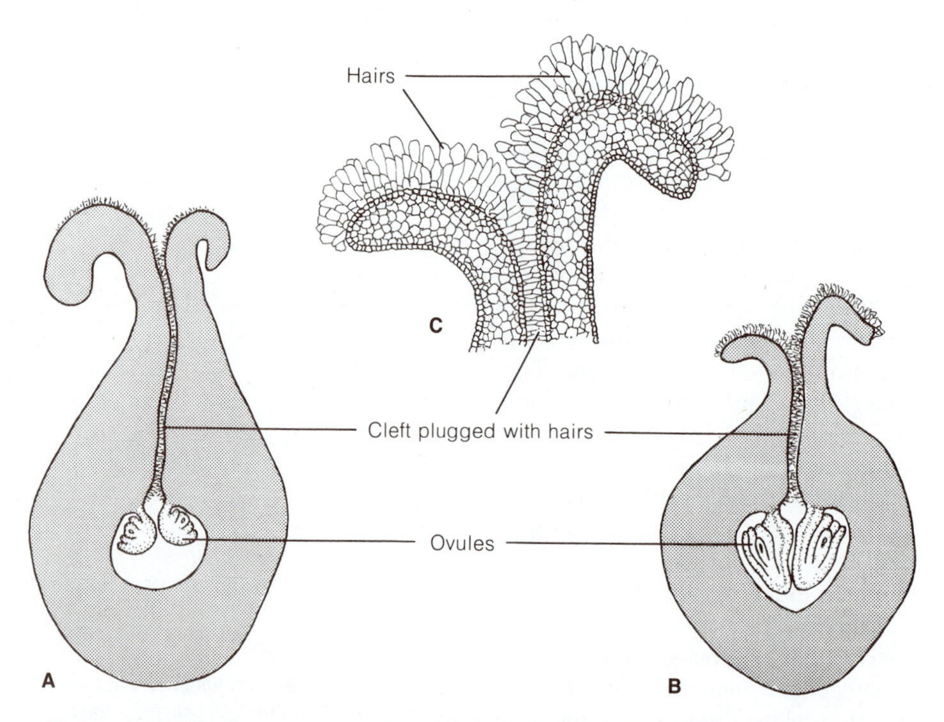

FIGURE 19-47 **A, B,** transections of conduplicate carpel of *Degeneria* showing position of ovules and progressive occlusion of cleft between margins by development of hairs; **C,** detailed structure of dense mat of hairs within the cleft and on the adaxial flared ventral edges of the carpel. [Redrawn from Swamy, *Jour. Arnold Arboretum* 30:10, 1949.]

tissue of the carpel. The position and source of vascular supply of the ovules in *Degeneria* are also remarkable. The two rows of ovules are remote during initiation and development from the true margins of the carpel, and some are vascularized from branches of the ventral bundles, some from branches of the dorsal bundle, and still others by strands derived from both the main ventral and dorsal bundles. Following pollination and the fertilization of the ovules, the adjacent adaxial surfaces

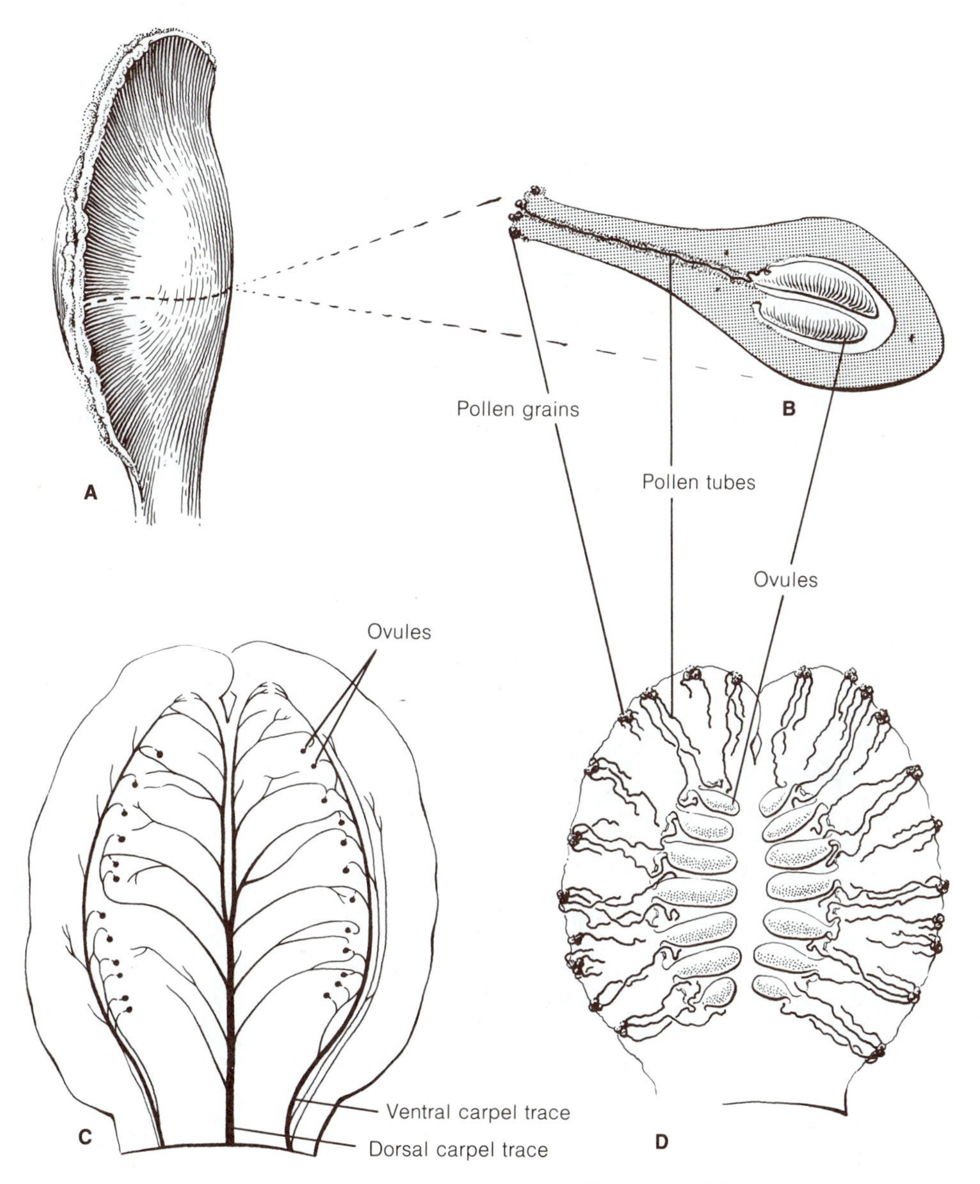

FIGURE 19-48 The primitive conduplicate carpel of *Tasmannia (Drimys) piperita.* **A,** side view showing the paired stigmatic crests; **B,** transection showing attachment of ovules and growth of a pollen tube through the mass of hairs which lies between the ventral surfaces of the carpel; **C,** cleared unfolded carpel showing vasculature and variations in derivation of ovule traces; **D,** unfolded carpel showing the course of the pollen tubes and the laminal placentation of the ovules. [Redrawn from Bailey and Swamy, *Amer. Jour. Bot.* 38:373, 1951.]

of the carpel become concrescent, and the recurved portions of the margins persist as parallel corky ridges on the mature fruit.

Additional examples of conduplicate carpels with extensive marginal stigmatic surfaces are found in the Winteraceae, another very interesting family exhibiting many primitive morphological characters including the total absence of vessels in the xylem. According to Bailey and Swamy (1951) "the least modified form of surviving carpel" occurs in *Tasmannia piperita (Drimys piperita)* and allied species of this genus. The young carpel is

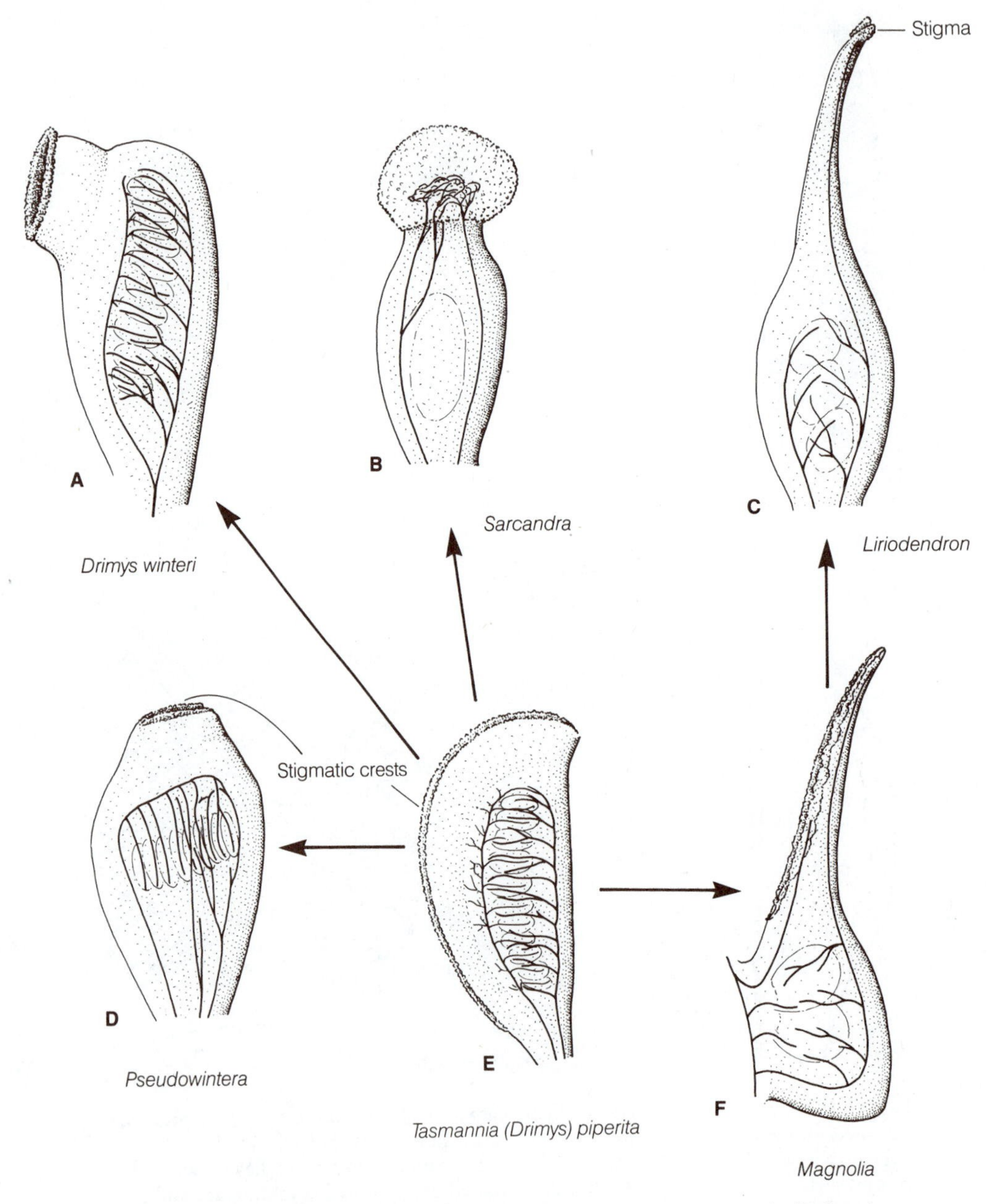

FIGURE 19-49 Initial trends in modification of the primitive *Tasmannia (Drimys) piperita*-type carpel; all figures are oriented with the adaxial (ventral) side of the carpel to the left. [Redrawn from Bailey and Swamy, *Amer. J. Bot.* 38:379, 1951.]

stalked, but, like the carpel of *Degeneria,* consists of a "conduplicately folded" lamina enclosing a series of ovules attached to the adaxial inner surface, distal to the carpellary margins (Fig. 19-48, B, D). This carpel, during the period of anthesis, can readily be unfolded and cleared, revealing that the ovules are attached to areas between the dorsal and two lateral veins (Fig. 19-48, C). As in *Degeneria,* the ovules of *Tasmannia piperita* are vascularized by extensions from either the dorsal or ventral carpellary bundles or by bundles that arise from strands derived from both the median and lateral systems. Pollen grains adhere to the external stigmatic marginal hairs, and, as in *Degeneria,* the pollen tubes reach the ovules by growing through the mat of hairs which extend inwardly to the surface of the locule (Fig. 19-48, B, D). Initial trends in the modification of the *Tasmannia piperita* type of carpel leading to other types are shown in Fig. 19-49. Examples of carpel ontogeny in the Winteraceae can be found in the publications by Tucker (1959) and Tucker and Gifford (1966 a, b).

The *Tasmannia piperita*-type carpel has, as yet, not been discovered in the fossil record. However, an important discovery of a magnoliid type of reproductive axis from the mid-Cretaceous supports the general concept that the conduplicate carpel was a basic type in at least one line of ancient angiosperms (Dilcher and Crane, 1984b). The flowers had an elongate receptacle bearing helically arranged conduplicate seed-bearing carpels (follicles) with two longitudinal adaxial crests, resembling stigmatic crests, that were covered with long trichomes (Fig. 19-50).

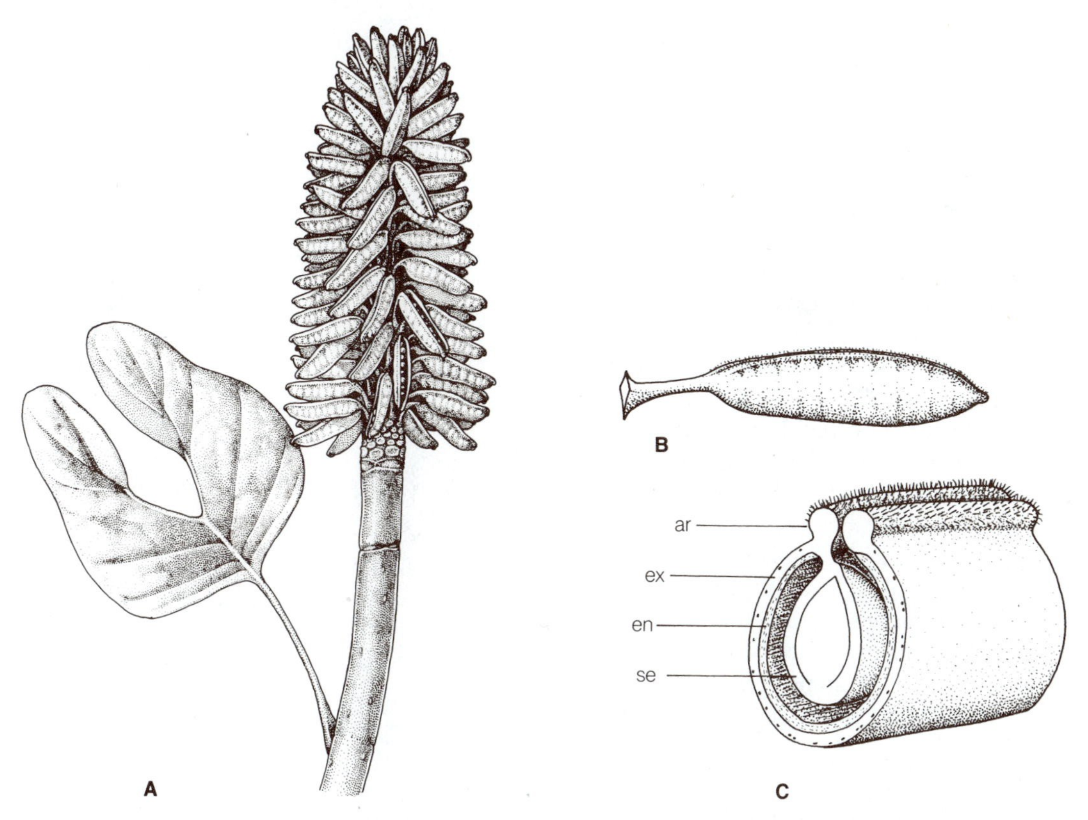

FIGURE 19-50 *Archaeanthus linnenbergeri* (mid-Cretaceous). **A,** reconstruction of leafy twig bearing numerous conduplicate carpels (the mature fruit is a follicle); **B,** reconstruction of a single follicle; **C,** section through a follicle; ar, adaxial ridge; en, endocarp; ex, exocarp containing resin bodies; se, seed. [From Dilcher and Crane, *Ann. Missouri Bot. Gard.* 71:351–383, 1985.]

FIGURE 19-51 Reconstruction of a flower from the mid-Cretaceous, showing pentamerous whorls of sepals, petals, stamens, and connate carpels. [From Basinger and Dilcher, *Science* 224:511, 1984.]

What is the Primitive Flower?

There are, of course, many ideas concerning the origin of the flower and possible ancestors of angiosperms. Over the years the paleobotanical evidence of the earliest angiosperm flowers has been rather sketchy. Dilcher (1979) reevaluated what information is available on angiosperms of the Lower to Upper Cretaceous. Most parts of flowers have been discovered, although a particular flower may not have all of its component parts. Some parts were probably not preserved. More recently, some important discoveries have been made of mid-Cretaceous flowers. Flowers have been described that exhibited radial, pentamerous symmetry and had both stamens and carpels in the same flower (Fig. 19-51; Basinger and Dilcher, 1984; Dilcher and Crane, 1984a). In others there were catkinlike axes to which either staminate or carpellate florets were attached (Fig. 19-52). To some botanists this type of information suggests a polyphyletic origin of angio-

FIGURE 19-52 Reconstructions of *Caloda delevoryana* (mid-Cretaceous). **A,** fruiting axis consisting of clusters of fruit-bearing secondary axes; **B,** enlargement of a secondary axis showing stalked conduplicate carpels. [From Dilcher and Kovach, *Amer. J. Bot.* 73:1230–1237, 1986; original drawings by Sally Wolfe.]

FIGURE 19-53 Reconstruction of a leafy shoot and terminal flower of *Archaeanthus linnenbergeri* from the mid-Cretaceous. [From Dilcher and Crane, *Ann. Missouri Bot. Garden* 71:351, 1985; original art work by Megan Rohn.]

sperms rather than a derivation from one specific ancestral group (monophyletic origin). In a more recent report, Cornet (1986) has described the reproductive axes of the plant of the Late Triassic, *Sanmiguelia lewisii,* in which numerous antherlike structures occurred on branched axes. On other shoots, clusters of carpel-like structures terminated the axes and were surrounded by a structure like a perianth composed of bracts. Some paleobotanists may consider *Sanmiguelia* to be a gnetophyte, but Cornet (1986) concludes that the "microsporangiate and megasporangiate organs of *Sanmiguelia* can no more easily be compared to those of the gnetales than the carpels and anthers of the angiosperms." If *Sanmiguelia* is a bona fide flowering plant it would be the earliest record thus far known for angiosperm macrofossils. Recent phylogenetic analyses (cladistics) have confirmed earlier suggestions that there are similarities between the gnetophytes (especially *Gnetum* and *Welwitschia*) and angiosperms, but each has its own specializations with neither group directly ancestral to the other (Crane, 1985; Doyle and Donoghue, 1986).

For many years the large, monoclinous flower (with both stamens and carpels) with many free, helically arranged parts was considered to be the primitive type in angiosperms. Pollination was probably achieved by beetles, one of the oldest groups of insects. The present-day magnolia or tulip-tree flower (Fig. 19-41) would more-or-less typify the *primitive* flower according to this concept. There are examples today of beetle pollination in a number of dicotyledons that presumably have retained a number of primitive characteristics. Originally this concept of the primitive flower was based upon rather scanty fossil evidence. There is now good evidence to believe that such a flower did exist in the mid-Cretaceous (Dilcher and Crane, 1984b), as illustrated in Fig. 19-53. However, as Dilcher (1979) has shown, both monoclinous and small diclinous flowers (having either stamens or carpels), organized into catkins, also were present in the early and mid-Cretaceous (Fig. 19-52). The diclinous flowers probably were wind pollinated. From all available evidence, Dilcher believes that floral organization as typified by the fossil *Archaeanthus* (Fig. 19-53) or the magnolia-type flower should no longer epitomize the primeval flower. The Magnoliales may well represent *one* of the early lines of angiosperm evolution that led to the coevolution of various types of specialized flowers and their insect pollinators. Dilcher (1979) concludes that independent lineages of some groups with anemophilous (wind pollinated) flowers developed early and perhaps separately from entomophilous groups from a common ancestral stock. Whether the angiosperms are monophyletic or polyphyletic, the two types of pollination methods undoubtedly go far back in the history of flowering plants. For additional reading on pollination biology, see Fae-

FIGURE 19-54 Scanning electron micrograph of a tetrad of pollen grains of *Drimys winteri* var. *chilensis*. This is the shedding condition of pollen in *Drimys* and most other members of the Winteraceae. Note the large pore evident in the distal face of one of the grains in the tetrad and the coarsely reticulate sculpture of the exine. [Courtesy of Dr. R. H. Falk.]

gri and van der Pijl (1979), Baker and Hurd (1968), Proctor and Yeo (1972), Stebbins (1974), and Crepet (1979).

Pollen

A relatively new branch of modern botany is the field of palynology, which deals with the varied and complex aspects of spores and pollen grains. The outer layer of these structures contains *sporopollenin* which consists of polymers of carotenoids and carotenoid esters. Sporopollenin is resistant to various chemicals, high temperatures, and decay organisms (bacteria, fungi). All of these features are responsible for the well-preserved condition of spores and pollen grains in the fossil record. Palynology has many contacts with other scientific disciplines, including paleobotany, taxonomy, economic geology, and medicine (such as the study of pollen as a cause of certain types of allergies, such as "hay fever").

There is no field in comparative plant morphology in which there is such a rich and changing terminology as that associated with the descriptions of pollen-grain structure. Heslop-Harrison (1968) refers to the "intimidating terminology" that has developed with reference to the wall structure of pollen grains. Pollen morphology, aside from its own inherent attractions, has increasingly become a "tool" in systematic and phylogenetic studies and for this reason deserves some attention in this book. As far as possible, every effort has been made to define, as simply and clearly as possible, those basic terms that are indispensable in describing the general morphology of angiosperm pollen.

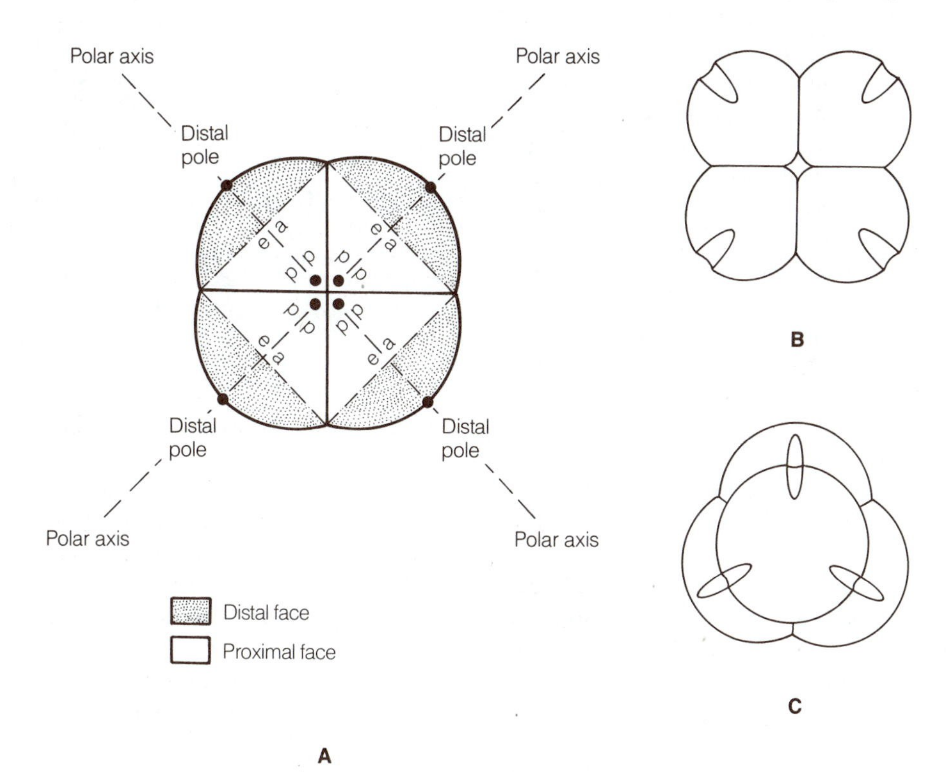

FIGURE 19-55 Spatial relationships of pollen grains in pollen tetrads. **A,** diagram of a tetragonal pollen tetrad (all pollen grains in same plane); ea, equatorial axis, pp, proximal pole; **B,** diagram, tetragonal pollen tetrad composed of four monosulcate (anasulcate) pollen grains; **C,** tetrahedral pollen tetrad composed of four tricolpate pollen grains; the top pollen grain is shown in distal-polar view. [Redrawn from *Origin and Early Evolution of Angiosperms* by J. W. Walker. pp. 241–299. Columbia University Press, New York. 1976.]

The following discussion will be restricted to a presentation of two of the most salient features of pollen grains: (1) the morphology and positions of *apertures*—the thin areas in the wall that are normally the points of exit for the emerging pollen tubes—and (2) the organization of the outer wall or *exine* of the grain. Before doing this, however, we must discuss the polarity of a pollen grain.

The process of microsporogenesis yields tetrads of haploid microspores, each of which, prior to the dehiscence of the anther, gives rise to a two- or three-celled endosporic male gametophyte (pollen grain; see Chapter 20). Usually the pollen grains become separated while still in the ripe anther and are shed as solitary grains known as *monads.* In this event, the position of the aperture or apertures with reference to the two "poles" of the grain is difficult to ascertain in the absence of developmental studies. But in some angiosperms the pollen is shed in the form of coherent tetrads; pollen units of this type permit an accurate analysis of pollen-grain polarity (Fig. 19-54). The pollen grains are so arranged in a tetrad that each grain is bisected by an imaginary "polar axis" which passes from the *distal* (outer) face to the *proximal* (inner) face of the pollen grain. An *equatorial* plane perpendicularly bisects the polar axis and forms the boundary (equator) between the outer distal face and the inner proximal face (Fig. 19-55). We are now able to discuss the relevance of these terms with reference to the position and form of the apertures.

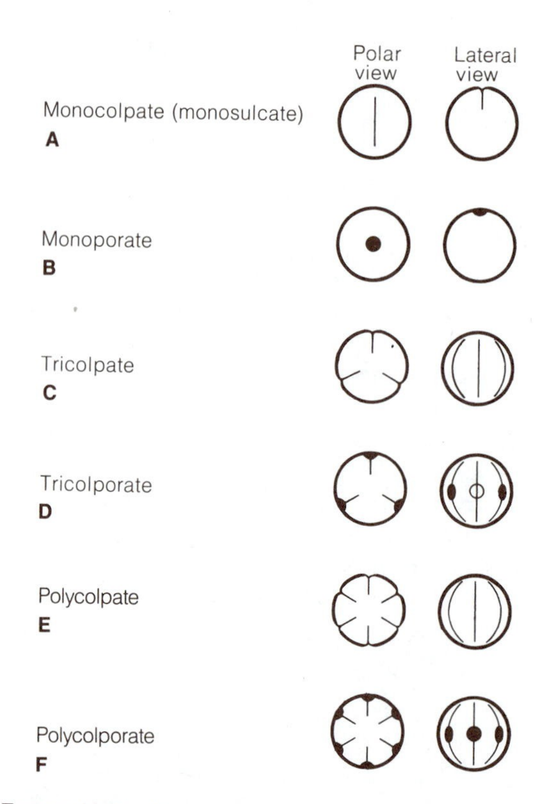

FIGURE 19-56 Diagrammatic representations of various types of apertures found in the pollen grains of angiosperms. [Redrawn from *Textbook of Modern Pollen Analysis* by K. Faegri and J. Iversen. Ejnar Munksgaard, Copenhagen. 1950.]

APERTURES. An aperture is a specific area on the pollen grain through which a pollen tube generally emerges after pollination. With respect to their *position,* three basic types of apertures are recognized: (1) *polar apertures,* located at or toward a pole, (2) *equatorial apertures,* located at the equator, and (3) *global apertures,* more or less uniformly scattered over the surface of the pollen grain (Walker and Doyle, 1975).

Three general types of apertures, distinguished by their *form,* are generally recognized: (1) elongate, furrowlike apertures (*colpate* pollen), (2) round, porelike apertures (*porate* pollen), and (3) encircling, ring-shaped or band-shaped apertures (*zonate* pollen). Furrows may have a pore at their center, a condition termed *colporate* (Fig. 19-57, B). Representative pollen types are shown in Figs. 19-56, 19-57, and 19-58).

According to Walker and Doyle (1975), the most primitive form of aperture in angiosperm pollen is the *sulcus,* a groove or furrow that is located at the distal pole of a pollen grain (Figs. 19-56, A; 19-57, A). The grain is *monosulcate,* specifically *anasulcate.* From a broad phylogenetic standpoint, the monosulcate pollen type is found in the majority of the primitive Magnoliales, and it is the only type characteristic of twelve genera and approximately twenty-three species of the family Magnoliaceae. The monocotyledons are basically a monosulcate group. It is interesting to note that the monosulcate pollen grain is correlated with the occurrence of primitive angiosperm leaf types from the Lower Cretaceous as described by Hickey and Doyle (1977) (see Fig. 19-15).

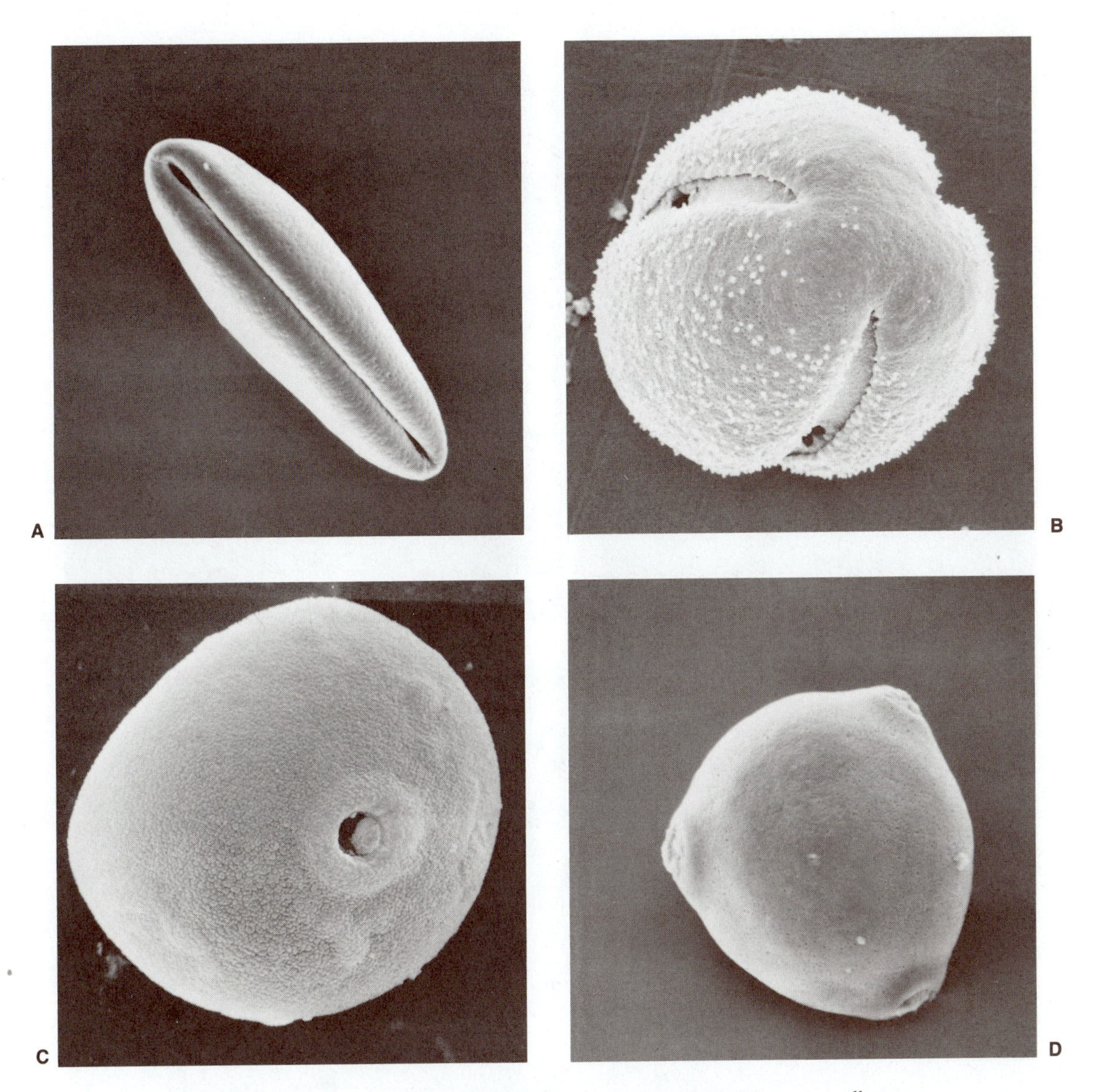

FIGURE 19-57 Scanning electron micrographs of representative angiosperm pollen-types. **A,** *Magnolia grandiflora:* monosulcate (×700); **B,** *Scaevola glabra:* tricolporate; the operculum (cap) over a pore detached during acetolysis (×1500); **C,** *Oryza sativa:* monoporate (×2,000); **D,** *Cucumis sativus:* triporate (×1,000). [Courtesy of Dr. J. Ward and Dr. L. Sunell.]

Tricolpate and tricolpate-derived pollen types are restricted to the majority of the advanced dicotyledonous taxa. The origin of the tricolpate pollen type is open to speculation. On the basis of an intensive study of dicotyledonous pollen, Walker (1974) and Walker and Doyle (1975) have postulated that from the primary monosulcate type there were a number of evolutionary endpoints in pollen aperture types in the more primitive dicotyledons along with the common occurrence of the inaperturate (without apertures) type. From the inaperturate type, colpate and porate types were secondarily derived in certain other primitive dicotyledons as well as in advanced dicotyledonous

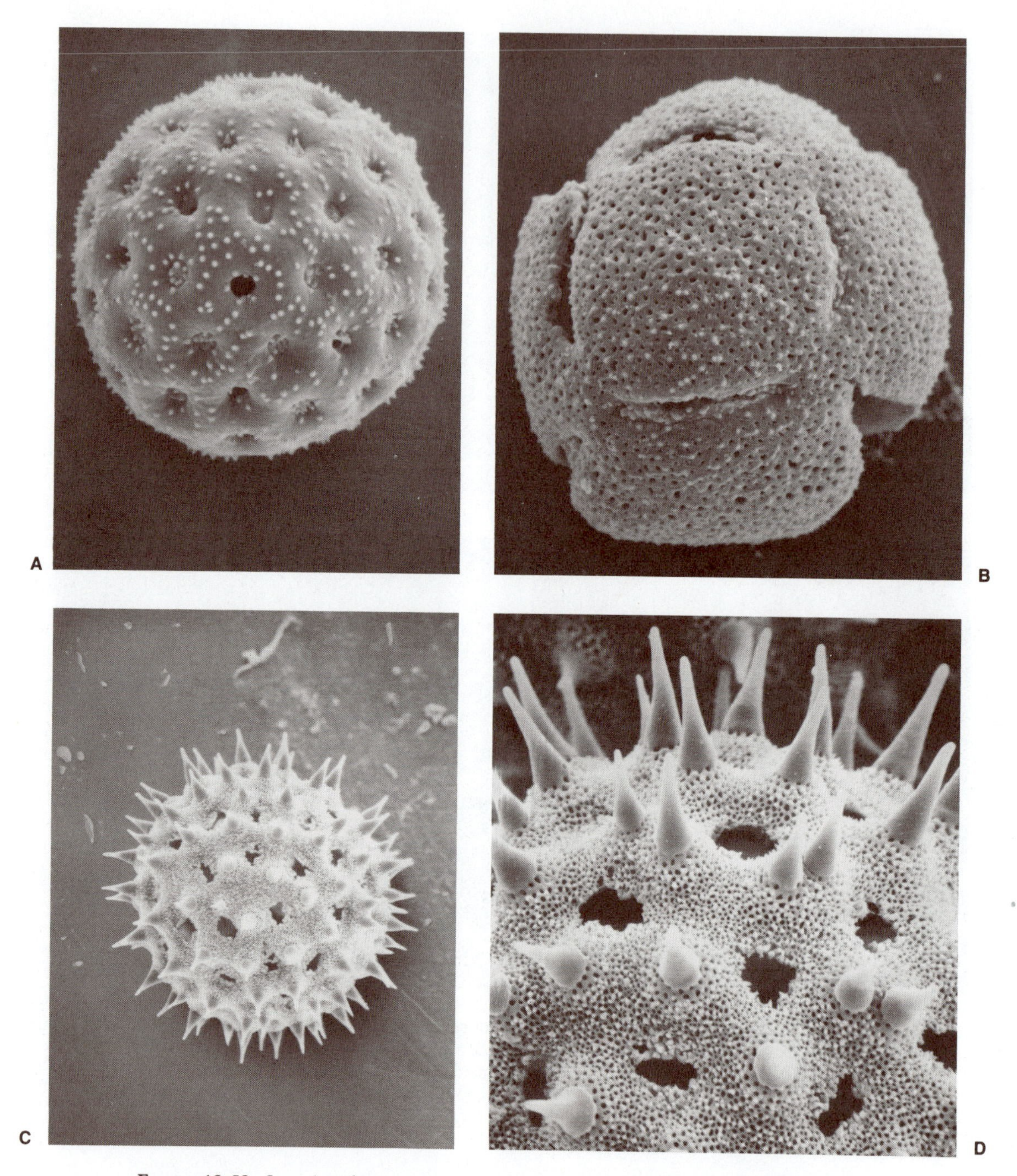

FIGURE 19-58 Scanning electron micrographs of representative angiosperm pollen-types. **A,** *Chenopodium oahuense:* polyporate (×4,000); **B,** *Pereskia grandifolia:* polyrugate (twelve slitlike apertures) (×1,700). **C,** *Ipomoea wolcottiana:* polyporate (×800); **D,** *Ipomoea,* enlarged view, surface of exine showing spines, germinal pores, and perforations in the tectum (×2,200). [Courtesy of Dr. J. Ward.]

taxa. Other theories regarding the phylogeny of colpate pollen and derivative types have been proposed, and a discussion of these concepts can be found in Walker (1974) and Takhtajan (1980).

SPORODERM. The term sporoderm is sometimes used to designate collectively the two major wall layers of a mature pollen grain. The inner layer is the *intine* that is produced by the male gametophyte during development of the pollen grain. The outer layer is the *exine* that is derived from tapetal substances during sporogenesis and from the gametophyte to some extent. The exine alone survives the drastic chemical treatment, known as "acetolysis," used to prepare pollen for microscopical study; for this reason, emphasis has centered on the structure and external sculpturing of the exine. When the surface of the exine is studied at high magnifications, particularly with the scanning electron microscope, patterns of sculpturing in the form of warts, reticula, spines, and striations can be observed (Figs. 19-57, 19-58). Surface features of pollen grains, together with various types and arrangements of apertures, provide a rich source of information which can be applied in the elucidation of taxonomic and phylogenetic problems in the angiosperms (Walker, 1976).

From the standpoint of its internal structure, the *exine* can usually be resolved into two principal layers. The terminology of Erdtman (1969) is now generally in use, whereby the outer layer is termed the *sexine* and the inner layer is the *nexine*. Frequently the nexine can be resolved into two layers. The layer of the nexine toward the outside of the grain is commonly called the *foot layer* to which *columellae* are attached. The heads of the columellae are fused laterally to form the *tectum* which may be imperforate or perforate (Figs. 19-59; 19-60). The majority of the angiosperms have the columellate type of pollen wall structure. It is in the spaces or interstices where compounds (such as proteins), derived from the tapetum, are found. (See p. 590 for the importance of stored tapetal substances in pollen-stigma interaction.) If the columellae (or bacula) remain free, the exine is *intectate;* the columellae may be rod-shaped or have swollen heads (pilate exine).

The exine of certain primitive living angiosperm families (Magnoliaceae, Degeneriaceae, Annonaceae) is *atectate* (without a tectum) and possesses no trace of either a columellate or granular structure. Walker (1976) considers this to be the most primitive type of exine structure in flowering plants. From this type of organization the granular and columellate types were derived during the course of angiosperm evolution. However, the earliest fossil pollen from the Lower Cretaceous already had a columellate type of organization, which means that the pollen grains were already too specialized to reveal much about the earliest stages of pollen evolution in angiosperms if the theory is correct (Walker and Walker, 1984). Living primitive taxa, as mentioned previously, may have retained the earliest type of exine structure, and the validity of the concept may be confirmed in future studies of angiosperm fossil pollen.

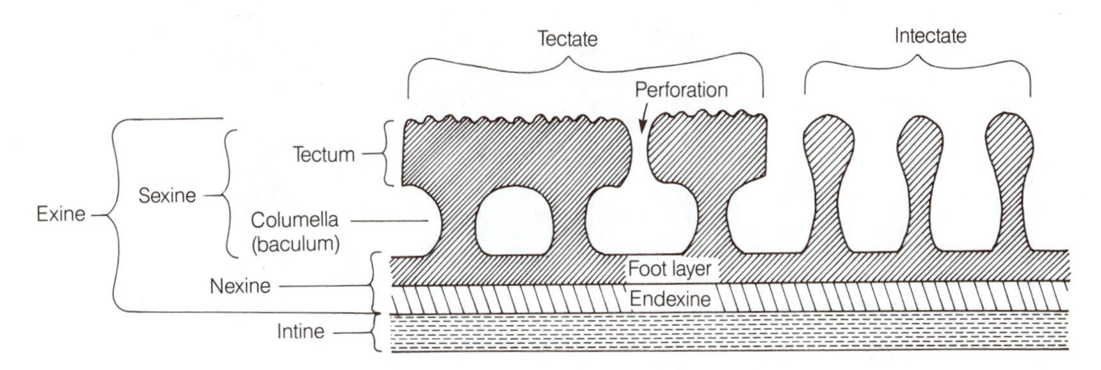

FIGURE 19-59 Composite diagram showing representative pollen-wall structure of some angiosperms, as seen in sectional view. See text for details.

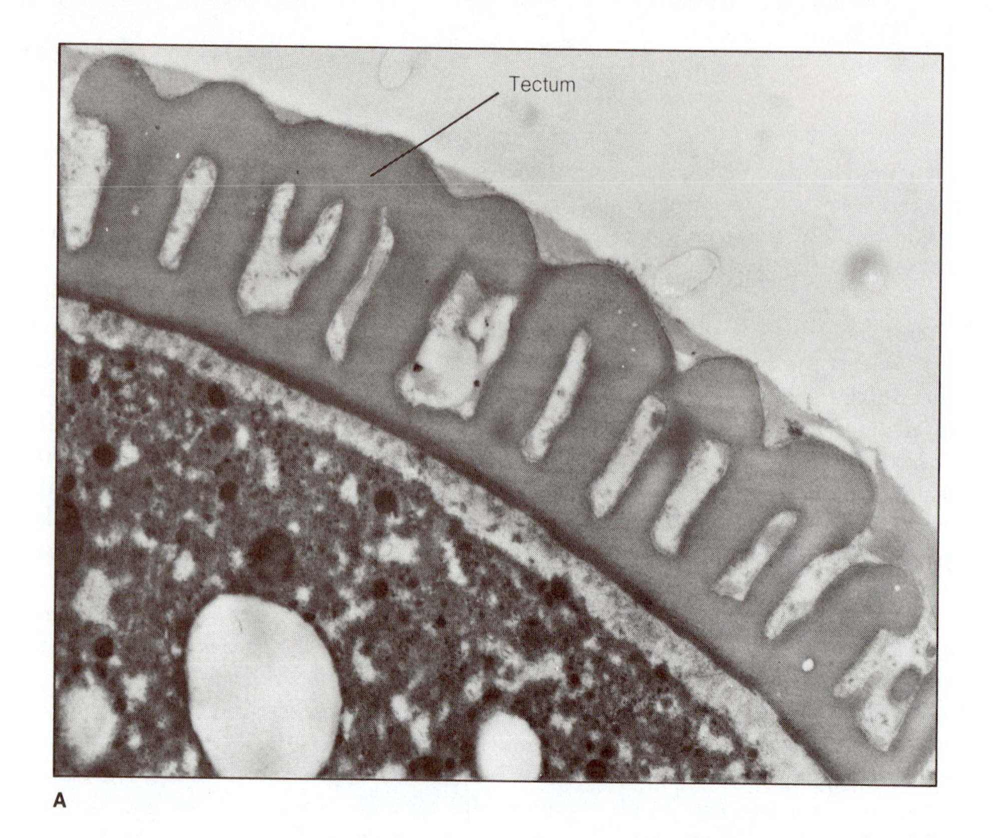

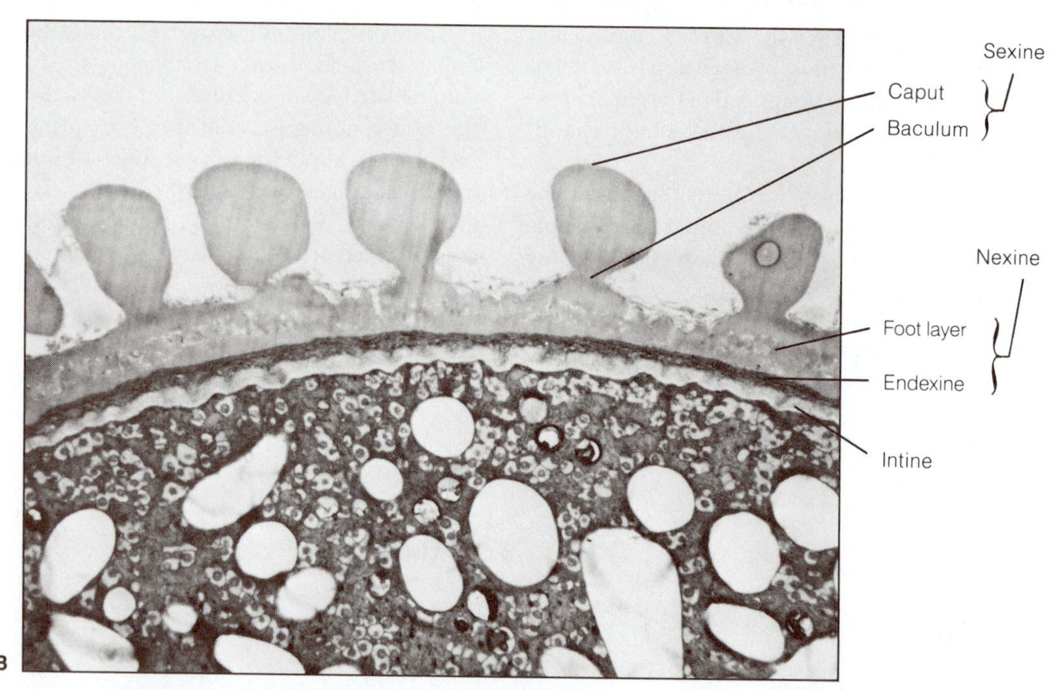

FIGURE 19-60 Electron micrographs of small portions of the sporoderm of pollen grains. **A,** sporoderm of *Euphorbia esculenta* in which the fusion of the bacula of the sexine forms a tectum. **B,** sporoderm of *Cnidoscolus rotundifolia,* illustrating an intectate type of sporoderm. [Courtesy of James Tanno.]

Inflorescences

In some angiosperm taxa, the flowers are solitary and are borne in a terminal position on the axis. Although this position has been regarded by certain botanists as primitive in the angiosperms, the phylogenetic significance of large terminal flowers in such genera as *Paeonia, Magnolia, Liriodendron, Calycanthis,* and *Eupomatia* is a somewhat open question because the *original* angiosperms are undoubtedly extinct and not represented in modern-day taxa. We do know that by mid-Cretaceous times there were angiosperms with large terminal flowers as well as those with catkinlike inflorescences (see p. 542). Among living angiosperms, most solitary terminal flowers probably represent the surviving members of either terminal or lateral clusters of flowers (Eames, 1961). In the Winteraceae, for example—a family distinguished by several primitive features—there are numerous transitions between complex branched inflorescences and solitary axillary or terminal flowers. In one genus of this family, *Zygogynum,* the solitary terminal flower represents "the end of a reduction series" (Bailey and Nast, 1945a). According to Troll's (1969) comprehensive survey, the inflorescence has been reduced to a solitary terminal flower in a great many dicotyledonous genera, some of which are uniflorus, whereas in others only certain of the species have solitary terminal flowers. Stebbins (1974) agrees in general with this reduction series and has developed arguments for the adaptive value of certain inflorescence types in the early evolution of angiosperms.

Most commonly the flowers of angiosperms develop in more-or-less well-demarcated clusters known as inflorescences. During the long history of taxonomy, many efforts have been made to characterize and classify the bewildering "types" of angiosperm inflorescences. As might be expected, a correspondingly rich terminology has inevitably arisen, and we are indebted to Rickett (1944, 1955) for his scholarly review of the subject and his effort to redefine the confusing terms that are still used to classify inflorescences. Without question, the most comprehensive modern account of inflorescences has been given by Troll (1964, 1969) in his treatises on inflorescence morphology.

Organography of Inflorescences

From a broad organographic standpoint, an inflorescence "is a flower-bearing branch or system of branches" (Rickett, 1944). When inflorescences are characterized in this way, no sharp boundary can be drawn between terminal bracteate clusters of flowers and floriferous leafy shoots in which the lateral flowers arise in the axils of foliage leaves. Diagrammatic examples of well-demarcated inflorescences are shown in Fig. 19-61, A, B. Each flower may be subtended by a minute—and often ephemeral—bract, or a progressive reduction in the size of the bracts may culminate in the formation of bractless flowers in the upper region of the inflorescence. According to Troll (1964), many floral bracts are merely small, greatly simplified leaf blades or else correspond morphologically to the basal sheath region of foliage leaves. In the type of organization which Troll designates as a "frondose inflorescence" (Fig. 19-61, D), each flower is subtended by a slightly reduced foliage leaf. As is shown in Fig. 19-61, C, there are transitional forms between typical "bracteate" and "frondose" types of inflorescences.

Ontogeny of Inflorescences

Inflorescences have long been classified, on a developmental basis, into two major categories: "closed" or determinate and "open" or indeterminate types (Fig. 19-62). Troll (1964) has strongly emphasized this traditional method of treatment, particularly with reference to his typological classification of inflorescences.

In closed inflorescence development (Fig. 19-62, A), the meristematic apex of the developing inflorescence ceases to produce bracts and gives rise to a terminal flower. The initiation of the terminal flower is thus precocious, meaning it occurs when the uppermost lateral flowers are still in a very early stage in their ontogeny. Correlated with this fact, the terminal flower, as a rule, opens before the expansion of the lateral flowers. From a comparative standpoint, it is interesting that lateral flowers, in a closed type of inflorescence, may open acropetally, basipetally, or in a divergent pattern with respect to the terminal flower. In the divergent sequence, ex-

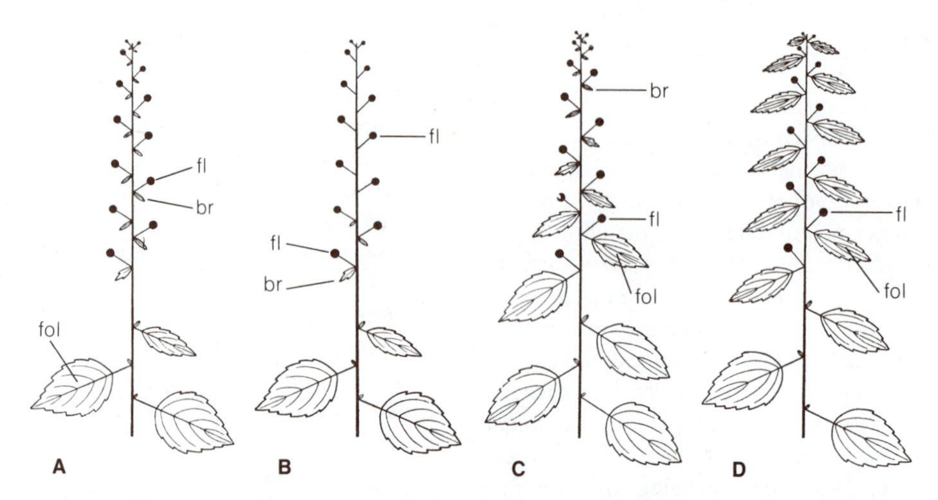

FIGURE 19-61 Diagrams showing variations in the types of foliar organs that subtend the lateral flowers (fl) of inflorescences. **A,** well-defined terminal raceme with each flower subtended by a diminutive bract (br); **B,** similar type of inflorescence illustrating progressive reduction and ultimate suppression of bracts; **C,** transitional form, the lower flowers arising in the axils of small foliage leaves (fol); **D,** leafy or "frondose" inflorescence with all flowers subtended by small petiolate foliage leaves. [From *Die Infloreszenzen* by W. Troll. Gustav Fischer, Stuttgart. 1964.]

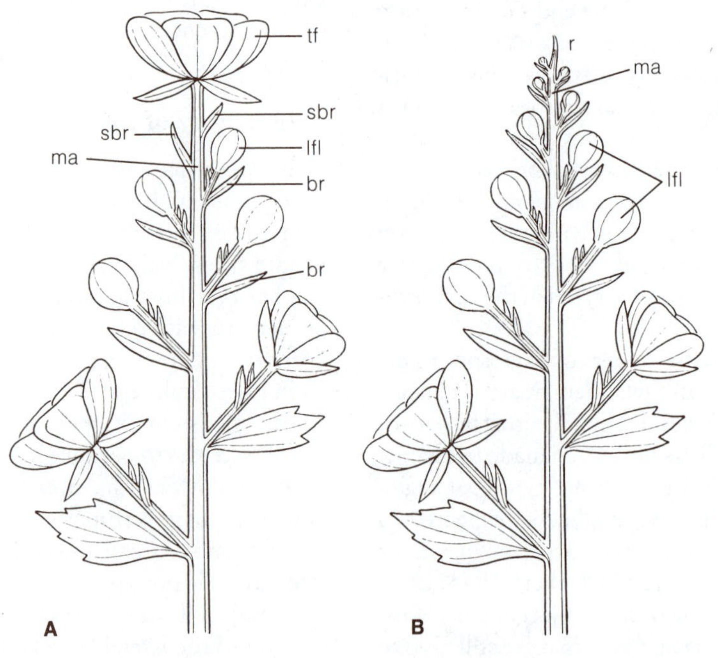

FIGURE 19-62 Diagrams contrasting the morphology of "closed" and "open" types of inflorescences. **A,** closed type, the main axis (ma) of the inflorescence ending in a terminal flower (tf). Note the two sterile bracts (sbr) intervening between the terminal and uppermost lateral flower (lfl); **B,** open type, the main axis (ma) ending blindly in the form of a small terminal "rudimentary" tip (r); br, bract. See text for further explanations. [From *Die Infloreszenzen* by W. Troll. Gustav Fischer, Stuttgart. 1964.]

pansion of the lateral flowers begins in the middle region of the inflorescence and continues both acropetally and basipetally (Troll, 1964).

In the open inflorescence, the apex of the young inflorescence, in place of initiating a terminal flower, continues in an "indefinite" manner to form bracts (with lateral flowers) but ultimately ceases growth and ends in the form of a rudimentary stub (Fig. 19-62, B, r). Troll describes in detail the considerable variation in morphology of the abortive tip in open inflorescences. In some instances, *all* the lateral flowers complete their development, and the true apex of the inflorescence may be minute or even unrecognizable. In contrast, the entire terminal region of the inflorescence may consist of the apical meristem together with a few undeveloped bracts and lateral flower primordia.

Classification of Inflorescences

Many of the terms that are still used to characterize the common types of inflorescences, such as those presented together with diagrams in Fig. 19-63, were used as early as 1751 by Linnaeus in his *Philosophia Botanica.* During the long history of morphology and taxonomy, the meaning of some of these terms changed very little or not at all, while other terms became more-or-less drastically modified. In addition, new terms were introduced during the nineteenth century. As a result, there now exists in botanical literature a confused labyrinth of terminologies which has been analyzed in great detail in Rickett's (1944, 1955) papers.

Several major factors appear to have contributed to the present rather chaotic state of inflorescence classification. In the first place, we are still uninformed about the evolutionary history of angiosperms, i.e., what constitutes the basic or "primitive" type of inflorescence? Does such an apparently "simple" type as the *raceme,* which consists of an axis bearing pedicillate lateral flowers subtended by bracts, represent a "prototype" of the other forms of inflorescences, such as the *spike, panicle, corymb,* and *umbel?* (See Fig. 19-63 for illustrations of these types.) Or should we begin our phylogenetic series with complex branched inflorescences—such as the panicle—and postulate that from them were derived, for example, the *umbel* and *head* in which the flowers are congested at the end of the inflorescence apex? (See Fig. 19-63 for illustrations of these types.) Theoretical trends of evolution of this sort, however, are of doubtful value because it seems evident that *parallel evolution* has played a major role in the formation of angiosperm inflorescences. In other words, comparative studies clearly suggest that not all racemes, umbels, corymbs, or heads are homologous; types which appear similar may have had different methods of origin during the evolution of the various taxa of the angiosperms.

An outstanding example of confusion in terminology is provided by the traditional recognition of two major groups of inflorescences: "racemose" and "cymose" types. The group of racemose inflorescences is commonly taken to include the raceme, panicle, and corymb in which, as represented in Fig. 19-63, there is presumed to be a more-or-less indefinite formation of flowers by the apex of the inflorescence. In contrast, the main axis of so-called cymose inflorescences ("cymes")—and the axes of the lateral flowering branches—are said to end in terminal flowers (Fig. 19-63). If defined in this way, the terms racemose and cymose are imprecise and inaccurate. Many simple racemes end in a functional terminal flower (Fig. 19-62, A); conversely some inflorescences, classified as cymose, are devoid of a terminal flower (Rickett, 1955). In this connection, Troll (1964) states the problem clearly by saying that although *cymose branching* may be found in various types of inflorescences, a "cymose inflorescence" does not exist as a separate morphological type.

When the various forms of inflorescences are precisely defined, as Rickett (1955) has attempted to do, important "characters" are provided for the descriptive needs of systematic botany. But the continued use of the often ambiguous morphological concepts of "racemose" and "cymose" to designate major categories of inflorescences, is highly questionable from either an ontogenetic or comparative point of view (see Rickett, 1944).

Unfortunately, there has been a conspicuous tendency in the past to treat inflorescences as static structures rather than as "modes of flowering." As Rickett (1944) critically remarks, "the names (of inflorescences) have outlived the theories, to our ultimate confusion." The reader is referred to Troll (1964, 1969) who took a new and broader view of

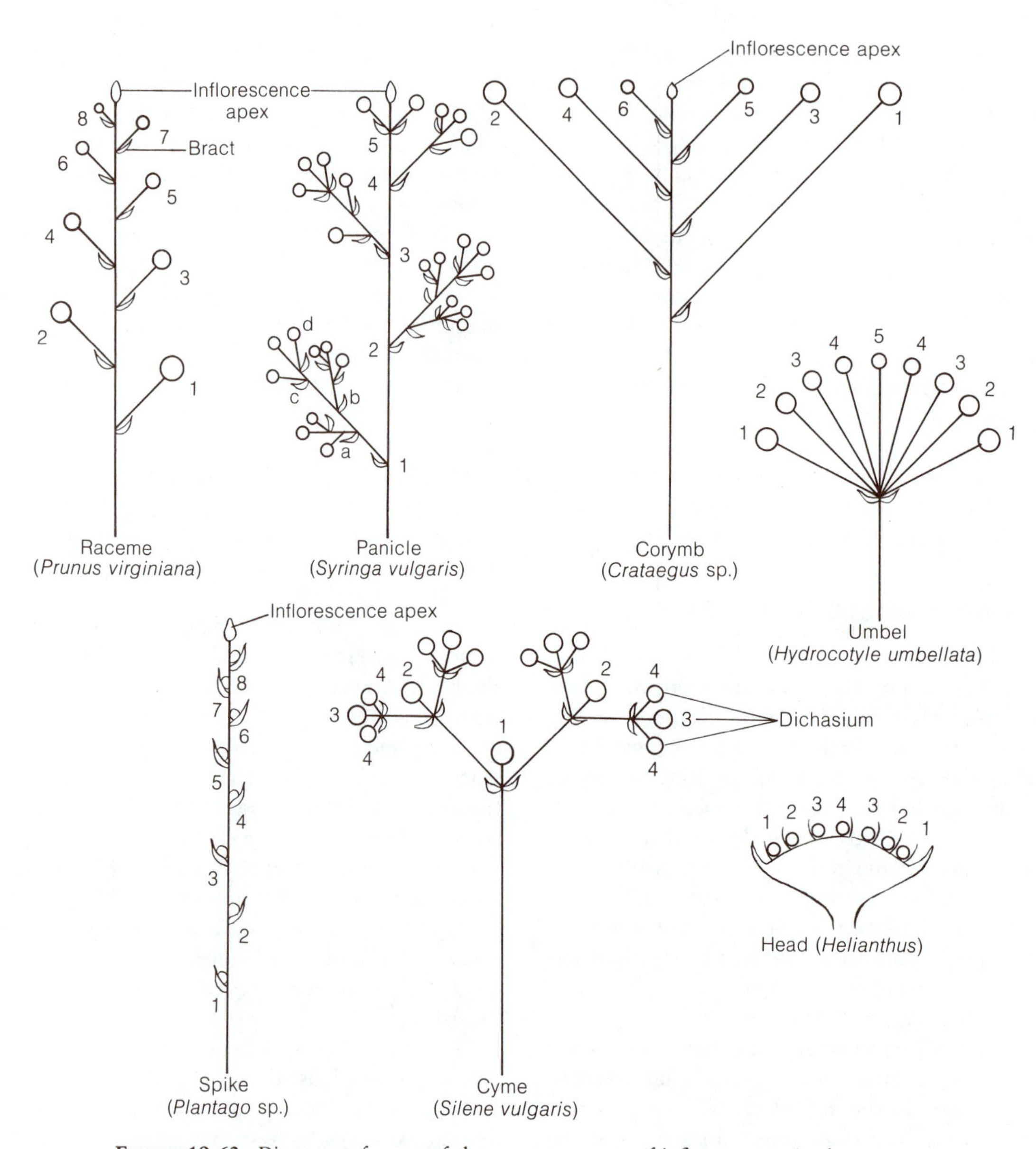

FIGURE 19-63 Diagrams of some of the common types of inflorescences in the angiosperms. Flowers are represented by circles and their order of development indicated in numerical sequence; in each inflorescence, flower 1 is the oldest in the group. The letters a–c, in the diagram of the panicle, depict the sequence of flower development in one of the lowest branches. See text for further explanation. [From *A Textbook of General Botany*, 2d edition, by R. M. Holman and W. W. Robbins. Wiley, New York. 1928.]

the problem of inflorescence morphology and typological classification. (Also, see Weberling, 1965, 1981; Stebbins, 1974; and Tucker, 1987.)

REFERENCES

Arber, A.
1937. The interpretation of the flower: a study of some aspects of morphological thought. *Biol. Rev.* 12:157–184.
1946. Goethe's botany. *Chron. Bot.* 10:67–124.

Arnott, H. J., and S. C. Tucker
1963. Analysis of petal venation in *Ranunculus*. I. Anastomeses in R. *repens v. pleniflorus*. *Amer. Jour. Bot.* 50:821–830.
1964. Analysis of petal venation in *Ranunculus*. II. Number and position of dichotomies in *R. repens v. pleniflorus*. *Bot. Gaz.* 125:13–26.

Axelrod, D. I.
1952. A theory of angiosperm evolution. *Evolution* 6:29–60.
1961. How old are the angiosperms? *Amer. Jour. Sci.* 259:447–459.
1970. Mesozoic paleogeography and early angiosperm history. *Bot. Rev.* 36:277–319.

Ayensu, E. S.
1972. *Anatomy of the Monocotyledons.* VI. *Dioscoreales.* Clarendon Press, Oxford.

Bailey, I. W.
1944. The development of vessels in angiosperms and its significance in morphological research. *Amer. Jour. Bot.* 31:421–428.
1953. Evolution of the tracheary tissue of land plants. *Amer. Jour. Bot.* 40:4–8.
1954. *Contributions to Plant Anatomy.* Chronica Botanica, Waltham, Mass.
1956. Nodal anatomy in retrospect. *Jour. Arnold Arboretum* 37:269–287.
1957. Additional notes on the vesselless dicotyledon, *Amborella tricopoda* Baill. *Jour. Arnold Arboretum* 38:374–378.

Bailey, I. W., and C. G. Nast
1945a. The comparative morphology of the Winteraceae. VII. Summary and conclusions. *Jour. Arnold Arboretum* 26:37–47.
1945b. Morphology and relationships of *Trochodendron* and *Tetracentron*. I. Stem, root, and leaf. *Jour. Arnold Arboretum* 26:143–154.

Bailey, I. W., and B. G. L. Swamy
1948. *Amborella trichopoda* Baill., a new morphological type of vesselless dicotyledon. *Jour. Arnold Arboretum* 29:245–254.
1951. The conduplicate carpel of dicotyledons and its initial trends of specialization. *Amer. Jour. Bot.* 38:373–379.

Bailey, I. W., and W. P. Thompson
1918. Additional notes upon the angiosperms *Tetracentron, Trochodendron,* and *Drimys,* in which vessels are absent from the wood. *Ann. Bot.* 32:503–512.

Baker, H. G., and P. D. Hurd, Jr.
1968. Intrafloral Ecology. *Ann. Rev. Entomol.* 13:385–414.

Baranova, M.
1972. Systematic anatomy of the leaf epidermis in the Magnoliaceae and some related families. *Taxon* 21:447–469.

Basinger, J. F., and D. L. Dilcher
1984. Ancient bisexual flowers. *Science* 224:511–513.

Beck, C. B., R. Schmid, and G. W. Rothwell
1982. Stelar morphology and the primary vascular system of seed plants. *Bot. Rev.* 48:691–815.

Benzing, D. H.
1967a. Developmental patterns in stem primary xylem of woody Ranales. I. Species with unilacunar nodes. *Amer. J. Bot.* 54:805–813.
1967b. Developmental patterns in stem primary xylem of woody Ranales. II. Species with trilacunar and multilacunar nodes. *Amer. Jour. Bot.* 54:813–820.

Bernier, G., J.-M. Kinet, and R. M. Sachs
1981. *The Physiology of Flowering.* Vol. II. *Transition to Reproductive Growth.* CRC Press, Boca Raton, Fl.

Blackman, E.
1971. The morphology and development of cross veins in the leaves of Bread Wheat (*Triticum aestivum* L.). *Ann. Bot. n.s.* 35:653–665.

Boke, N. H.
1940. Histogenesis and morphology of the phyllode in certain species of *Acacia. Amer. Jour. Bot.* 27:73–90.
1947. Development of the adult shoot apex and floral initiation in *Vinca rosea* L. *Amer. Jour. Bot.* 34:433–439.

Brown, R. W.
1956. *Palmlike Plants from the Dolores Formation (Triassic) in Southwestern Colorado.* U.S. Geological Survey, Professional Paper 274–H, pp. 205–209.

Canright, J. E.
1952. The comparative morphology and relationships of the Magnoliaceae. I. Trends of specialization in the stamens. *Amer. Jour. Bot.* 39:484–497.

Carlquist, S.
1969. Toward acceptable evolutionary interpretations of floral anatomy. *Phytomorphology* 19:332–362.
1975. *Ecological Strategies of Xylem Evolution.* University of California Press, Berkeley.
1987. Presence of vessels in wood of *Sarcandra* (Chloranthaceae): comments on vessel origins in angiosperms. *Amer. J. Bot.* 74:1765–1771.

Cheadle, V. I.
1948. Observations on the phloem in the Monocotyledoneae. II. Additional data on the occurrence and phylogenetic specialization in structure of the sieve tubes in the metaphloem. *Amer. J. Bot.* 35:129–131.
1953. Independent origin of vessels in the monocotyledons and dicotyledons. *Phytomorphology* 3:23–44.
1955. The taxonomic use of specialization of vessels in the metaxylem of Gramineae, Cyperaceae, Juncaceae, and Restionaceae. *Jour. Arnold Arboretum* 36:141–157.
1956. Research on xylem and phloem. Progress in fifty years. *Amer. Jour. Bot.* 43:719–731.

Cheadle, V. I., and J. M. Tucker
1961. Vessels and phylogeny of Monocotyledoneae. *Rec. Adv. Bot.* 1:161–165.

Cheadle, V. I., and N. B. Whitford
1941. Observations on the phloem in the Monocotyledoneae. I. The occurrence and phylogenetic specialization in structure of the sieve tubes in the metaphloem. *Amer. J. Bot.* 28:623–627.

Chrtek, J.
1962. Der Verlauf der Nervatur in den Kronblättern. bzw. Kronen der Dikotyledonen, pp. 3–10. *Novitates Bot. Horti. Bot. Univ. Carolineae Pragensis.*
1963. Die Nervatur der Kronblätter bei den Vertretern der Ordnung Rosales s. l. *Acta Horti. Bot. Pragensis.* pp. 13–29.

Clowes, F. A. L.
1961. *Apical Meristems.* Blackwell, Oxford.
1984. Size and activity of quiescent centres of roots. *New Phytol.* 96:13–21.

Conde, L. F., and D. E. Stone
1970. Seedling morphology in Juglandaceae, the cotyledonary node. *Jour. Arnold Arboretum* 51:463–477.

Cornet, B.
1986. The leaf venation and reproductive structures of a Late Triassic angiosperm, *Sanmiguelia lewisii. Evolutionary Theory* 7:231–309.

Crane, P. R.
1985. Phylogenetic analysis of seed plants and the origin of angiosperms. *Ann. Missouri Bot. Gard.* 72:716–793.

Crepet, W. L.
1979. Some aspects of the pollination biology of Middle Eocene angiosperms. *Rev. Palaeobot. Palynol.* 27:213–238.

Cronquist, A.
1968. *The Evolution and Classification of Flowering Plants.* Houghton Mifflin, Boston.
1969. Broad features of the system of angiosperms. *Taxon* 18:188–193.
1981. *An Integrated System of Classification of Flowering Plants.* Columbia University Press, New York.

Cronquist, A., A. Takhtajan, and W. Zimmermann
1966. On the higher taxa of Embryobionta. *Taxon* 15:129–134.

Cutler, D. F.
1969. *Anatomy of the Monocotyledons.* IV. *Juncales.* Clarendon Press, Oxford.

Cutter, E. G.
1965. Recent experimental studies of the shoot apex and shoot morphogenesis. *Bot. Rev.* 31:7–113.

DeMason, D. A.
1983. The primary thickening meristem: definition and function in monocotyledons. *Amer. J. Bot.* 70:955–962.

Dengler, N. G., R. E. Dengler, and D. R. Kaplan
1982. The mechanism of plication inception in palm leaves: histogenic observations on the pinnate leaf of *Chrysalidocarpus lutescens. Can. J. Bot.* 60:2976–2998.

Devadas, C., and C. B. Beck
1972. Comparative morphology of the primary vascular systems in some species of Rosaceae and Leguminosae. *Amer. J. Bot.* 59:557–567.

Dickison, W. C.
1975. The basis of angiosperm phylogeny: vegetative anatomy. *Ann. Missouri Bot. Gard.* 62:590–620.

Diggle, P. K., and D. A. DeMason
1983a. The relationship between the primary thickening meristem and the secondary

thickening meristem in *Yucca whipplei* Torr. I. Histology of the mature vegetative stem. *Amer. J. Bot.* 70:1195–1204.

1983b. The relationship between the primary thickening meristem and the secondary thickening meristem in *Yucca whipplei* Torr. II. Ontogenetic relationship within the vegetative stem. *Amer. J. Bot.* 70:1205–1216.

Dilcher, D. L.

1979. Early angiosperm reproduction: an introductory report. *Rev. Palaeobot. Palynol.* 27:291–328.

Dilcher, D. L., and P. R. Crane

1984a. In pursuit of the first flower. *Natur. Hist.* 93(3):56–61.

1984b. *Archaeanthus:* an early angiosperm from the Cenomanian of the western interior of North America. *Ann. Missouri Bot. Gard.* 71:351–383.

Dormer, K. J.

1972. *Shoot Organization in Vascular Plants.* Chapman and Hall, London.

Doyle, J. A.

1973. Fossil evidence on early evolution of the monocotyledons. *Quart. Rev. Biol.* 48:399–413.

Doyle, J. A., and M. J. Donoghue

1986. Seed plant phylogeny and the origin of angiosperms: an experimental cladistic approach. *Bot. Rev.* 52:321–431.

Doyle, J. A., and L. J. Hickey

1976. Pollen and leaves from the mid-Cretaceous Potomac Group and their bearing on early angiosperm evolution. In C. B. Beck (ed.), *Origin and Early Evolution of Angiosperms,* pp. 139–206. Columbia University Press, New York.

Eames, A. J.

1931. The vascular anatomy of the flower with refutation of the theory of carpel polymorphism. *Amer. Jour. Bot.* 18:147–188.

1953. Neglected morphology of the palm leaf. *Phytomorphology* 3:172–189.

1961. *Morphology of the Angiosperms.* McGraw-Hill, New York.

Eames, A. J., and L. H. MacDaniels

1947. *An Introduction to Plant Anatomy,* 2d edition. McGraw-Hill, New York.

Engler, A.

1926. Angiospermae. In A. Engler and K. Prantl (eds.), *Die natürlichen Pflanzenfamilien,* Bd. 14a.

1964. *Syllabus der Pflanzenfamilien,* 12th edition, Vol 2. Gebrüder Borntraeger, Berlin.

Erdtman, G.

1969. *Handbook of Palynology. An Introduction to the Study of Pollen Grains and Spores.* Hafner, New York.

Esau, K.

1965. *Plant Anatomy,* 2d edition. Wiley, New York.

1967. Minor veins in *Beta* leaves: structure related to function. *Proc. Amer. Phil. Soc.* 111:219–233.

1969. The Phloem. In W. Zimmermann and P. Ozenda (eds.), *Encyclopedia of Plant Anatomy,* Vol. 5, Part 2. Gebrüder Borntraeger, Berlin.

1977. *Anatomy of Seed Plants,* 2d edition. Wiley, New York.

1979. Phloem. In C. R. Metcalfe and L. Chalk (eds.), *Anatomy of the Dicotyledons,* 2d edition, Vol. 1. *Systematic Anatomy of Leaf and Stem, with a Brief History of the Subject,* pp. 181–189. Clarendon Press, Oxford.

Ettinghausen, C. R. von

1861. *Die Blatt-Skelette der Dicotyledonen mit besonderer Rücksicht auf die Untersuchung und Bestimmung der fossilen Pflanzenreste.* Kais. Kön. Hof und Staatsdruckerei, Wien.

Eyde, R. H.

1975a. The foliar theory of the flower. *Amer. Scientist* 63:430–437.

1975b. The bases of angiosperm phylogeny: floral anatomy. *Ann. Missouri Bot. Gard.* 62:521–537.

Faegri, K., and L. van der Pijl

1979. *The Principles of Pollination Ecology,* 3d edition. Pergamon Press, Oxford.

Fahn, A.

1974. *Plant Anatomy,* 2d edition. Pergamon Press, Oxford.

Feldman, L. J.

1984. The development and dynamics of the root apical meristem. *Amer. J. Bot.* 71:1308–1314.

Foster, A. S.

1929. Investigations on the morphology and comparative history of development of foliar organs. I. The foliage leaves and cataphyllary structures in the horsechestnut (*Aesculus hippocastanum* L.). *Amer. Jour. Bot.* 16:441–501.

1931. Investigations on the morphology and comparative history of development of foliar organs. II. Cataphyll and foliage leaf form and organization in the black hickory (*Carya*

Buckleyi var. *arkansana*). *Amer. Jour. Bot.* 18:864–887.

1932. Investigations on the morphology and comparative history of development of foliar organs. III. Cataphyll and foliage leaf ontogeny in the black hickory (*Carya Buckleyi* var *arkansana*). *Amer. Jour. Bot.* 19:75–99.

1935. A histogenetic study of foliar determination in *Carya Buckleyi* var. *arkansana*. *Amer. Jour. Bot.* 22:88–147.

1950. Morphology and venation of the leaf in *Quiina acutangula* Ducke. *Amer. Jour. Bot.* 37:159–171.

1951. Heterophylly and foliar venation in *Lacunaria*. *Bull. Torrey Bot. Club* 78:382–400.

1953. Venation patterns in the leaves of angiosperms, with special reference to the Quiinaceae. *Proc. Seventh Int. Bot. Congr.* (Stockholm, 1950) p. 380.

1955. Comparative morphology of the foliar sclereids in *Boronella* Baill. *Jour. Arnold Arboretum* 36:189–198.

1959. The morphological and taxonomic significance of dichotomous venation in *Kingdonia uniflora* Balfour f. et W. W. Smith. *Notes Bot. Gard. Edinb.* 23:1–12.

1961a. The floral morphology and relationships of *Kingdonia uniflora*. *Jour. Arnold Arboretum* 42:397–410.

1961b. The phylogenetic significance of dichotomous venation in angiosperms. *Rec. Adv. Bot.* 2:971–975.

1963. The morphology and relationships of *Circaeaster*. *Jour. Arnold Arboretum* 44:299–321.

1966. Morphology of anastomoses in the dichotomous venation of *Circaeaster*. *Amer. Jour. Bot.* 53:588–599.

Foster, A. S., and H. J. Arnott
1960. Morphology and dichotomous vasculature of the leaf of *Kingdonia uniflora*. *Amer. Jour. Bot.* 47:684–698.

Foster, A. S., and F. A. Barkley
1933. Organization and development of foliar organs in *Paeonia officinalis*. *Amer. Jour. Bot.* 20:365–385.

Gifford, E. M., Jr.
1963. Developmental studies of vegetative and floral meristems. *Brookhaven Symposia in Biology* No. 16:126–137.
1983. Concept of apical cells in bryophytes and pteridophytes. *Ann. Rev. Plant Physiol.* 34:419–440.

Gifford, E. M., Jr., and G. E. Corson, Jr.
1971. The shoot apex in seed plants. *Bot. Rev.* 37:143–229.

Glück, H.
1919. *Blatt-und blütenmorphologische Studien.* Gustav Fischer, Jena.

Goebel, K.
1905. *Organography of Plants*, Part II. English edition by I. B. Balfour. Clarendon Press, Oxford.
1930. *Organographie der Pflanzen*, 2d edition, Part 2, p. 1145. G. Fischer, Jena.

Goethe, J. W. von
1790. *Versuch die Metamorphose der Pflanzen zu erklären.* Gotha.

Guttenberg, H. von
1941. *Der primäre Bau der Gymnospermenwurzel.* (*Handbuch der Pflanzenanatomie*, Band 8, VIII. Lief. 41.) Gebrüder Borntraeger, Berlin.
1968. *Die primäre Bau der Angiospermenwurzel.* (*Handbuch der Pflanzenanatomie*, 2d edition, Band 8, Teil 5.) Gebrüder Borntraeger, Berlin.

Heslop-Harrison, J.
1968. Pollen wall development. *Science* 161:230–237.

Hickey, L. J.
1973. Classification of the architecture of dicotyledonous leaves. *Amer. Jour. Bot.* 60:17–33.

Hickey, L. J., and J. A. Doyle
1977. Early Cretaceous fossil evidence for angiosperm evolution. *Bot. Rev.* 43:3–104.

Hickey, L. J., and J. A. Wolfe
1975. The bases of angiosperm phylogeny: vegetative morphology. *Ann. Missouri Bot. Gard.* 62:538–589.

Hiepko, P.
1965. Vergleichend-morphologische und entwicklungsgeschichtliche Untersuchungen über das Perianth bei den Polycarpicae. Teil I. *Bot. Jahrb.* 84:359–426.

Howard, R. A.
1970. Some observations on the nodes of woody plants with special reference to the problem of the 'split-lateral' versus the 'common gap.' *Bot. Jour. Linn. Soc.* (Suppl. 1) 63:195–214.

Hughes, N. F.
1974. Angiosperm evolution and the superfluous upland origin hypotheses. *Birbal Sahni Inst. Palaeobot.* (Lucknow) 1:25–29.

Jensen, L. C. W.
1968. Primary stem vascular patterns in three

subfamilies of the Crassulaceae. *Amer. J. Bot.* 55:553–563.

Kaplan, D. R.
1970. Comparative foliar histogenesis in *Acorus calamus* and its bearing on the phyllode theory of monocotyledonous leaves. *Amer. Jour. Bot.* 57:331–361.
1975. Comparative developmental evaluation of the morphology of unifacial leaves in the monocotyledons. *Bot. Jahrb. Syst.* 95:1–105.
1980. Heteroblastic leaf development in *Acacia:* morphological and morphogenetic implications. *Cellule* 73:135–203.

Kaplan, D. R., N. G. Dengler, and R. E. Dengler
1982a. The mechanism of plication inception in palm leaves: problem and developmental morphology. *Can. J. Bot.* 60:2939–2975.
1982b. The mechanism of plication inception in palm leaves: histogenetic observations on the palmate leaf of *Rhapis excelsa. Can. J. Bot.* 60:2999–3016.

Kasapligil, B.
1951. Morphological and ontogenetic studies of *Umbellularia californica* Nutt. and *Laurus nobilis* L. *Univ. Calif. Publ. Bot.* 25:115–240.

Kosakai, H., M. F. Moseley, Jr., and V. I. Cheadle
1970. Morphological studies of the Nymphaeaceae. V. Does *Nelumbo* have vessels? *Amer. Jour. Bot.* 57:487–494.

Kurth, E.
1978. A modified method for clearing leaves. *Stain Techn.* 53:291–293.

Lersten, N.
1965. Histogenesis of leaf venation in *Trifolium wormskioldii* (Leguminosae). *Amer. Jour. Bot.* 52:767–774.

Marsden, M. P. F., and I. W. Bailey
1955. A fourth type of nodal anatomy in dicotyledons, illustrated by *Clerodendron trichotomum* Thunb. *Jour. Arnold Arboretum* 36:1–50.

Mauseth, J. D.
1984. Effect of growth rate, morphogenic activity, and phylogeny on shoot apical ultrastructure in *Opuntia polyacantha* (Cactaceae). *Amer. J. Bot.* 71:1283–1292.
1988. *Plant Anatomy.* Benjamin/Cummings, Menlo Park, Calif.

McCoy, R. W.
1940. Floral organogenesis in *Frasera carolinensis. Amer. Jour. Bot.* 27:600–609.

Metcalfe, C. R., and L. Chalk
1950. *Anatomy of the Dicotyledons.* 2 vols. Clarendon Press, Oxford.
1960. *Anatomy of the Monocotyledons.* I. *Gramineae.* Clarendon Press, Oxford.
1971. *Anatomy of the Monocotyledons.* V. *Cyperaceae.* Clarendon Press, Oxford.

Moseley, M. F., Jr.
1967. The value of the vascular system in the study of the flower. *Phytomorphology* 17:159–164.

Niklas, K. J., and J. D. Mauseth
1980. Simulations of cell dimensions in shoot apical meristems: implications concerning zonate apices. *Amer. J. Bot.* 67:715–732.

Nougarède, A., E. M. Gifford, Jr., and P. Rondet
1965. Cytohistological studies of the apical meristem of *Amaranthus retroflexus* under various photoperiodic regimes. *Bot. Gaz.* 126:281–298.

Orr, A. R.
1984. Histochemical study of enzyme activity in the shoot apical meristem of *Brassica campestris* L. during transition to flowering. II. Cytochrome oxidase. *Bot. Gaz.* 145:308–311.

Ozenda, P.
1949. Recherches sur les dicotylédones apocarpiques, contribution a l'étude des angiospermes dites primitives. *Publ. Lab. de l'École Norm. Sup., Sér. Bio., Fasc.* II. Paris.
1952. Remarques sur quelques interprétations de l'étamine. *Phytomorphology* 2:225–231.

Padmanabhan, D.
1963. Leaf development in palms. *Curr. Sci.* 32:537–539.
1967. Direct evidence for schizogenous splitting in palm-leaf lamina. *Curr. Sci.* 36:467–468.

Pray, T. R.
1955. Foliar venation of angiosperms. III. Pattern and histology of the venation of *Hosta. Amer. Jour. Bot.* 42:611-618.
1963. Origin of vein endings in angiosperm leaves. *Phytomorphology* 13:60–81.

Proctor, M., and P. Yeo
1972. *The Pollination of Flowers.* Taplinger, New York.

Puri, V.
1951. The rôle of floral anatomy in the solution of morphological problems. *Bot. Rev.* 17:471–553.

Rickett, H. W.
1944. The classification of inflorescences. *Bot. Rev.* 10:187–231.

1955. Materials for a dictionary of botanical terms. III. Inflorescences. *Bull. Torrey Bot. Club* 82:419–445.

Sattler, R.
1973. *Organogenesis of Flowers: A Photographic Text-Atlas.* Univ. of Toronto Press, Toronto, Canada.

Schmid, R.
1972. Floral bundle fusion and vascular conservatism. *Taxon* 21:429–446.

Scott, R. A., P. L. Williams, L. C. Craig, E. S. Barghoorn, L. J. Hickey, and H. D. MacGinitie
1972. "Pre-Cretaceous" angiosperms from Utah: evidence for Tertiary age of the palm woods and roots. *Amer. J. Bot.* 59:886–896.

Shobe, W. R., and N. R. Lersten
1967. A technique for clearing and staining gymnosperm leaves. *Bot. Gaz.* 128:150–152.

Sinnott, E. W.
1914. Investigations on the angiosperms. I. The anatomy of the node as an aid in the classification of angiosperms. *Amer. Jour. Bot.* 1:303–322.

Sinnott, E. W., and I. W. Bailey
1915. Investigations on the phylogeny of the angiosperms. V. Foliar evidence as to the ancestry and early climatic environment of the angiosperms. *Amer. Jour. Bot.* 2:1–22.

Solereder, H.
1908. *Systematic Anatomy of the Dicotyledons.* 2 vols. Clarendon Press, Oxford.

Stebbins, G. L.
1974. *Flowering Plants: Evolution above the Species Level.* Harvard University Press, Cambridge.

Sugiyama, M.
1979. A comparative study of nodal anatomy in the Magnoliales based on the vascular system in the node-leaf continuum. *Jour. Fac. Sci., Univ. of Tokyo,* Sec. III, Vol. XII, pp. 199–279.

Swamy, B. G. L.
1949. Further contributions to the morphology of the Degeneriaceae. *Jour. Arnold Arboretum* 30:10–38.

Swamy, B. G. L., and I. W. Bailey
1950. *Sarcandra,* a vesselless genus of the Chloranthaceae. *Jour. Arnold Arboretum* 31:117–129.

Takhtajan, A.
1959. *Die Evolution der Angiospermen.* Gustav Fischer, Jena.
1980. Outline of the classification of flowering plants (Magnoliophyta). *Bot. Rev.* 46:225–359.

Tepfer, S. S.
1953. Floral anatomy and ontogeny in *Aquilegia formosa* var. *truncata* and *Ranunculus repens. Univ. Calif. Publ. Bot.* 25:513–648.

Tidwell, W. D., S. R. Rushforth, J. L. Reveal, and H. Behunin
1970a. *Palmoxylon simperi* and *Palmoxylon pristina:* two pre-Cretaceous angiosperms from Utah. *Science* 168:835–840.

Tidwell, W. D., S. R. Rushforth, and A. D. Simper
1970b. Pre-Cretaceous flowering plants: further evidence from Utah. *Science* 170:547–548.

Tidwell, W. D., A. D. Simper, and G. F. Thayn
1977. Additional information concerning the controversial Triassic plant: *Sanmiguelia. Palaeontographica* 163 B:143–151.

Tieghem, Ph. van
1875. Recherches sur la structure du pistil et sur l'anatomie comparée de la fleur. *Mem. Acad. Sci. Inst. Imp.* France 21:1–262.
1900. Sur les dicotylédones du groupe homoxylées. *Jour. Bot.* 14:259–297; 330–361.

Tomlinson, P. B.
1961. *Anatomy of the Monocotyledons.* II. *Palmae.* Clarendon Press, Oxford.
1969. *Anatomy of the Monocotyledons.* III. *Commelinales—Zingiberales.* Clarendon Press, Oxford.
1974. Development of the stomatal complex as a taxonomic character in the monocotyledons. *Taxon* 23:109–128.

Torrey, J. G., and L. J. Feldman
1977. The organization and function of the root apex. *Amer. Sci.* 65:334–344.

Troll, W.
1935a. Vergleichende Morphologie der Fiederblätter. *Nova Acta Leopoldina* n. f. 2:311–455.
1935b. *Vergleichende Morphologie der höheren Pflanzen,* Erster Bd. Lieferung I. Gebrüder Borntraeger, Berlin.
1937. *Vergleichende Morphologie der höheren Pflanzen,* Bd. I. Erster Teil, Lieferung 3. Gebrüder Borntraeger, Berlin.
1938. *Vergleichende Morphologie der höheren Pflanzen,* Bd. I. Zweiter Teil, Lieferung 1. Gebrüder Borntraeger, Berlin.
1939a. *Vergleichende Morphologie der höheren Pflanzen,* Bd. I. Zweiter Teil, Lieferung 2. Gebrüder Borntraeger, Berlin.
1939b. *Vergleichende Morphologie der höheren Pflanzen,* Bd. I. Zweiter Teil, Lieferung 4. Gebrüder Borntraeger, Berlin.

1949. Über die Grundbegriffe der Wurzelmorphologie. *Osterr. Bot. Zeit.* 96:444–452.

1964. *Die Infloreszenzen.* Bd. 1. Gustav Fischer, Stuttgart.

1967. *Vergleichende Morphologie der höheren Pflanzen.* Erster Band, Dritter Teil. (Authorized reprinting of original published in 1943.) Gebrüder Borntraeger, Berlin.

1969. *Die Infloreszenzen,* Bd. 2, Teil 1. Gustav Fischer, Stuttgart.

Tucker, S. C.

1959. Ontogeny of the inflorescence and the flower in *Drimys winteri* var. *chilensis Univ. Calif. Publ. Bot.* 30:257–336.

1987. Floral initiation and development in legumes. In C. H. Stirton (ed.), *Advances in Legume Systematics,* Part 3, pp. 183–239. Royal Bot. Gardens, Kew.

Tucker, S. C., and E. M. Gifford, Jr.

1966a. Organogenesis in the carpellate flower of *Drimys lanceolata. Amer. J. Bot.* 53:433–442.

1966b. Carpel development in *Drimys lanceolata. Amer. J. Bot.* 53:671–678.

Van Cotthem, W. R. J.

1970. A classification of stomatal types. *Bot. J. Linn. Soc.* 63:235–246.

Vertrees, G. L.

1962. Plastochronic changes in the vegetative apex and the ontogeny of the floral apex of *Liriodendron tulipifera* L. M. S. thesis, University of California, Davis.

Walker, J. W.

1974. Aperture evolution in the pollen of primitive angiosperms. *Amer. J. Bot.* 61:1112–1136.

1976. Evolutionary significance of the exine in the pollen of primitive angiosperms. In I. K. Ferguson and J. Muller (eds.), *The Evolutionary Significance of the Exine,* pp. 251–308. Linnean Society Symposium, Series No. 1.

Walker, J. W., and J. A. Doyle

1975. The bases of angiosperm phylogeny: palynology. *Ann. Missouri Bot. Gard.* 62:664–723.

Walker, J. W., and A. G. Walker

1984. Ultrastructure of Lower Cretaceous angiosperm pollen and the origin and early evolution of flowering plants. *Ann. Missouri Bot. Gard.* 71:464–521.

Wardlaw, C. W.

1965. The organization of the shoot apex. Pp. 966–1076 in *Handbuch der Pflanzenphysiologie.* Vol. 15, Pt. 1. Springer-Verlag, Berlin.

Weberling, F.

1965. Typology of inflorescences. *Jour. Linn. Soc. London Bot.* 59:215–221.

1981. *Morphologie der Blüten und der Blütenstände.* Eugen Ulmer, Stuttgart.

Wilson, C. L.

1937. The phylogeny of the stamen. *Amer. Jour. Bot.* 24:686–699.

1942. The telome theory and the origin of the stamen. *Amer. Jour. Bot.* 29:759–764.

1953. The telome theory. *Bot. Rev.* 19:417–437.

Young, D. A.

1981. Are the angiosperms primitively vesselless? *Syst. Bot.* 6:313–330.

Zahur, M. S.

1959. Comparative study of secondary phloem of 423 species of woody dicotyledons belonging to 85 families. *Cornell Univ. Agric. Exp. Sta., Mem.* 358:1–160.

Zimmermann, M. H., and P. B. Tomlinson

1970. The vascular systems in the axis of *Dracaena fragrans.* 2. Distribution and development of secondary vascular tissue. *Jour. Arnold Arboretum* 51:478–491.

CHAPTER 20

The Reproductive Cycle in Angiosperms

In the great majority of lower vascular plants the gametophyte generation, although comparatively small in size, is completely independent of the sporophyte and is a free-living plant. In marked contrast, the gametophyte generation of gymnosperms is not only reduced in size, but the female gametophyte is permanently enclosed within the ovule and thus entirely dependent for its nutrition on the sporophyte. The general evolutionary trend toward simplification and dependence reaches its culmination in the small, few-celled gametophytes of the angiosperms.

Our present ontogenetic concepts of angiospermous gametophytes and our interpretation of the striking, novel aspects of sexual reproduction and seed formation in this group go back to the inspired and careful investigations that were made in Europe during the latter part of the nineteenth century. [See Maheshwari (1950) and Johri and Ambegaokar (1984) for excellent historical sketches of these early pioneering studies.) Modern studies on the details of the reproductive cycle in angiosperms are both voluminous and diversified, and they include systematic-phylogenetic surveys as well as the initiation of experimental work on the gametophytes and the developing seed. Some of these studies have resulted in clarification of the proper systematic disposition of certain genera or families. Others have greatly enriched our knowledge of the wide range in patterns of development of the endosperm tissue and the embryo. We hope that the continuous accumulation of knowledge in this area will eventually throw new and important light on the evolutionary relationships of flowering plants.

Recently, a very valuable book edited by B. M. Johri (1984) has provided morphologists with up-to-date resumes of the salient aspects of sporogenesis, gametophyte development, fertilization, endosperm formation, and embryogeny in flowering plants.

One of the characteristic features of sexual reproduction in angiosperms is the marked telescoping of the processes of sporogenesis and gametophyte development. Microsporogenesis, for example, is directly followed by one or two cell divisions which yield a pollen grain consisting only of two or three cells at the time of anther dehiscence. Megasporogenesis may involve, as in gymnosperms, the formation of a linear tetrad of morphologically discrete megaspores. But in many angiosperms, wall formation does not occur between the haploid megaspore nuclei, all four of

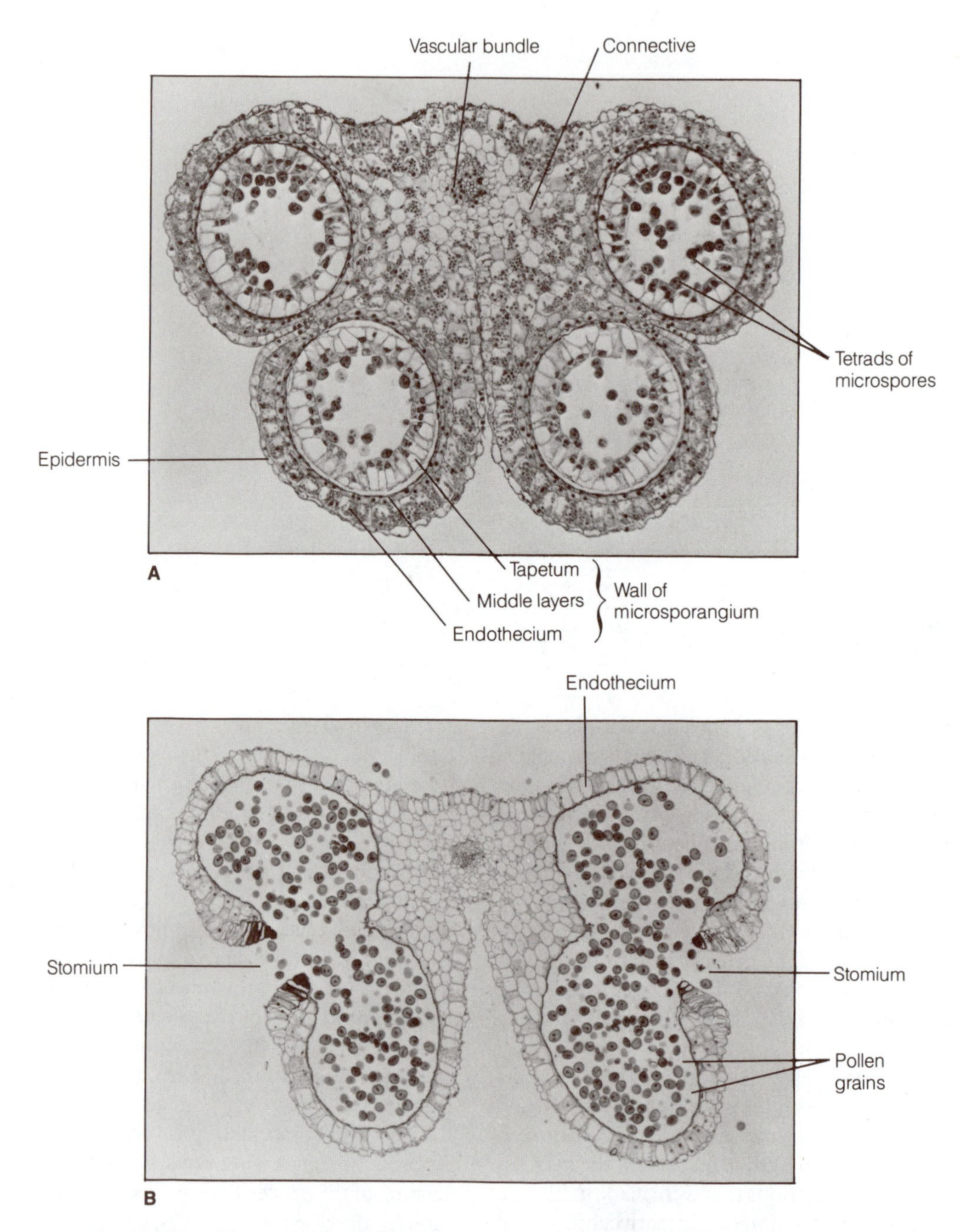

FIGURE 20-1 Transections of *Lilium* anthers. **A,** anther with two pairs of microsporangia, each sporangium containing tetrads of microspores; tapetum is still intact. **B,** anther at time of dehiscence; the tissue between the members of each pair of sporangia has broken down, forming two pollen-containing sacs. The stomium marks the region of longitudinal dehiscence of the anther. Note that the tapetum and inner wall layers have collapsed and are no longer evident.

which, by further divisions, contribute to the development of the female gametophyte. In short, the demarcation between the spore and the early phase in ontogeny of the gametophyte is frequently not sharp, and correct interpretation depends on a full appraisal of cytological details, especially the point at which meiosis occurs and the subsequent behavior and fate of the haploid nuclei.

In Chapter 19 the angiosperm flower was discussed from a broad morphological point of view, with particular emphasis on the ontogeny and vasculation of floral organs. In this chapter we are concerned with the role of the flower in reproduction, and an attempt will be made to present a coherent account of sporogenesis, the development of the gametophytes, pollination, fertilization, and the formation of endosperm and embryo in the seed. Because of the rapid accumulation in recent years of information on the life cycles of a wide range of angiosperms, it is now possible to present a synthetic description of the *basic features* common to all reproductive cycles in the angiosperms. This is done with full realization of the numerous—and possibly significant—departures in detail from the general situation as illustrated, for example, by the various types of embryo-sac development that are now recognized. Important deviations from what appears to be the the typical condition will be reserved for brief discussion at the end of each topic.

Microsporangia

Structure and Development

Chapter 19 provided a description of reputedly primitive stamens in which two pairs of linear sporangia occur on either side of the midvein of a laminarlike structure (Fig. 19-44). In the majority of angiosperms, however, a stamen consists of a delicate filament and an anther consisting of four microsporangia arranged in pairs. Members of each pair are separated from each other by a plate of sterile tissue and the central tissue or *connective*, and the anther is traversed by a strand of vascular tissue (Fig. 20-1, A, B). In certain families there are only two sporangia. [See Bhandari (1984) for a discussion of additional types.]

In lower vascular plants such as *Lycopodium* and *Selaginella* (Chapter 9) and in gymnosperms, including the majority of conifers, the origin of a microsporangium is in the surface layer of cells (Chapters 15 through 17). In angiosperms, microsporangia originate in cells of the *hypodermal* layer.

Comparative studies on a wide range of angiosperm taxa have revealed considerable variation in the number and structure of the various layers of the microsporangium wall (Davis, 1966). Before we describe specific examples that illustrate this variability, it is essential to indicate the general pattern of histogenesis that appears to predominate in angiosperms as a whole. This can be most easily accomplished by means of the schematic representation shown in Fig. 20-2.

In a typical tetrasporangiate stamen, each of the four microsporangia originates by the periclinal division of a group of hypodermal initials situated at the corners of the young anther. These periclinal divisions form a subepidermal *primary parietal layer* and an inner *primary sporogenous layer* (Fig. 20-2). The cells of the primary sporogenous layer may divide further prior to microsporogenesis, or else give rise directly to the diploid *microsporocytes*. When meiosis occurs in the sporocytes, tetrads of haploid microspores are formed, which, in turn, differentiate into pollen grains (Fig. 20-2). The cells of the primary parietal layer, by means of further periclinal and anticlinal divisions, form a variable number of concentrically arranged layers from which, by special histological differentiation, the various layers of the mature microsporangium wall take their origin. Frequently the epidermis is considered to be part of the sporangial wall even though it does not participate in the origin of parietal cells and the sporogenous tissue.

As is shown diagrammatically in Fig. 20-2, the outermost layer of the wall is commonly termed the *endothecium*. Because of the characteristic thickenings, in the form of bars or networks, which develop on certain of the cell walls, the endothecium is termed by some morphologists the "fibrous layer." The endothecium appears to be a highly specialized wall layer concerned with the dehiscence of the anther (Fig. 20-1). Below the endothecium there is usually a layer—or several layers—of tabular thin-walled cells which collectively represent the *middle layers* of the sporangium wall.

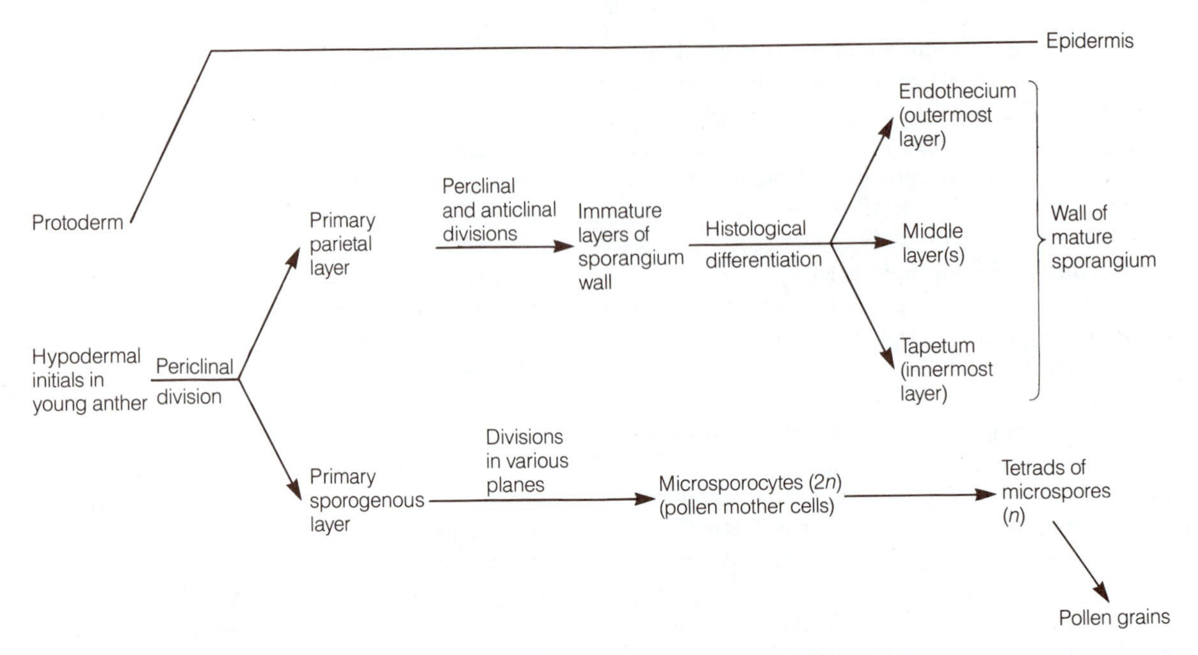

FIGURE 20-2 Diagram summarizing the general pattern of development of the microsporangium in the angiosperms.

These middle layers become compressed or completely destroyed during the formation of the microspores (Fig. 20-1). The innermost layer of the sporangium wall is the tapetum, a glandular layer which produces the enzymes, amino acids, and other nutritive materials used during microsporogenesis. Many studies have shown that there is a delicate interaction between the tapetum and the developing sporogenous tissue. Any ontogenetic disturbance in the structure and/or function of the tapetum leads to degeneration of pollen, so characteristic of the so-called cytoplasmic male sterile angiosperms (Laser and Lersten, 1972; Bhandari, 1984). Additional information on (1) the formation of tapetal cells at the *inner side* of the sporogenous tissue and (2) the morphological types of tapeta will be given later.

As already mentioned, Fig. 20-2 is intended only as a generalized histogenetic diagram which makes no attempt to show the wide variation in the *sequence* of periclinal divisions that may occur in the *derivatives* of the primary parietal layer. However, Davis (1966) made a comprehensive systematic survey of the patterns of sporangium wall development, which she maintains are consistent enough at the family level to justify classifying the methods of wall formation into four major types. She bases her recognition of each of the types on the behavior of the two "secondary parietal layers." These layers, according to her interpretation, are produced by the *first* periclinal division of the primary parietal layer.

A particularly clear example of microsporangium wall development which does not readily fit into any of the types proposed by Davis is provided by Boke's (1949) study on *Vinca rosea* (Apocynaceae) (Fig. 20-3). In *Vinca,* the embryonic anther, as seen in transection, consists of a mass of ground meristem enclosed by a discrete protoderm (Fig. 20-3, A). The protoderm develops into the anther-epidermis, and groups of 2 to 4 hypodermal cells, at the four corners of the differentiating anther, function as the initials of the microsporangia. Each initial group divides periclinally into an outer series of *primary parietal cells* and an inner layer of *primary sporogenous cells* (Fig. 20-3, B). The latter, with or without preliminary divisions, give rise to the microspore mother cells. The primary parietal cells at first may divide *either* anticlinally or periclinally (Fig. 20-3, C). After periclinal divisions have appeared, the inner cells become part of the future tapetum while the outer parietal cells divide periclinally once or twice and the derivative cells may then divide anticlinally. During this critical stage of

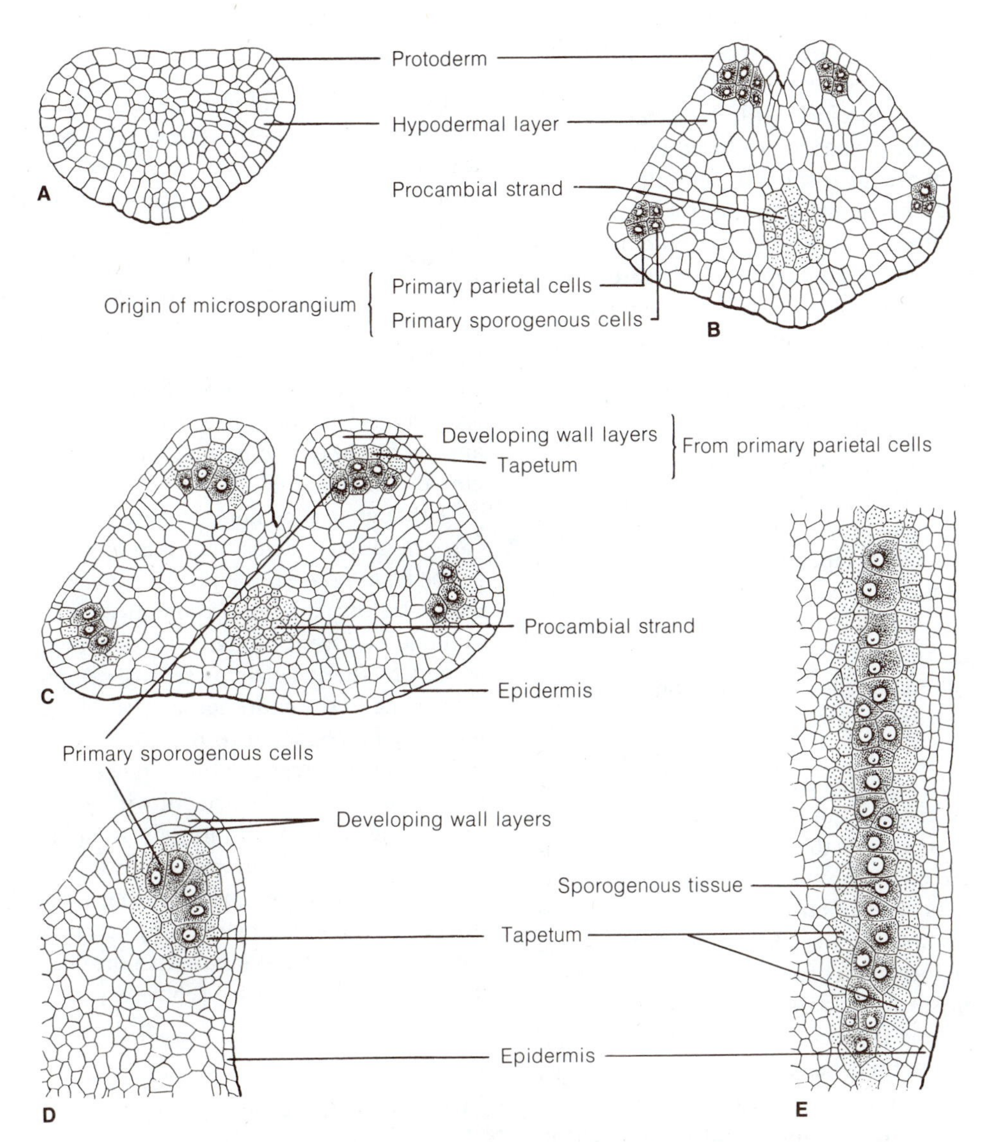

FIGURE 20-3 Origin and development of microsporangia in the stamen of *Vinca rosea*. **A–D,** transections of successively older stages in development of the anther. **A,** young anther with clearly defined hypodermal layer; **B,** later stage showing origin of microsporangia by periclinal divisions in four separate groups of hypodermal cells; **C,** stage illustrating the origin of a portion of the tapetum from the primary parietal cells; **D,** details of a later stage in development of a single microsporangium showing the developing wall layers and the group of sporogenous cells surrounded by the tapetum; **E,** longisection of portion of anther showing origin of inner portion of tapetum from cells adjacent to the sporogenous tissue (note the continued development of the wall layers below the epidermis). [Redrawn from Boke, *Amer. Jour. Bot.* 36:535, 1949.]

wall development in *Vinca* there is apparently no fixed sequence in the plane of cell division, and hence no evidence of the presence of definable "secondary parietal layers," in the sense of Davis' concept of the early ontogeny of either the basic or dicotyledonous types of sporangium wall formation (see Fig. 20-3, D, E).

As the microsporangia of *Vinca* approach maturity, the inner wall cells (including the tapetum) collapse so that only two wall layers are seen at the period of maturation of the pollen grains (Boke, 1949). It is not evident from Boke's account whether an endothecium, with characteristic secondary wall thickenings, is produced by the outermost layer of the microsporangium wall.

In *Vinca,* the innermost layer of the sporangium wall constitutes a considerable portion of the tapetum; the remainder of the tapetum originates, independently of the sporangium wall, from connective cells of the anther which lie adjacent to the inner boundary of the sporogenous tissue (Fig. 20-3, C, D). According to Boke's study, the inner part of the tapetum is frequently several layers in thickness in contrast to the predominantly one-layered structure of the outer tapetum which is a part of the sporangium wall proper (Fig. 20-3, E). A dual method of origin of tapetal cells is not confined to *Vinca* but has been observed in a number of other genera in the angiosperms (see Periasamy and Parameswaran, 1965).

During the later stages of development of the microsporangium in angiosperms, the cells of the tapetum enlarge and, in many cases, become bi- or multinucleate after failure of cell-wall formation following successive mitotic divisions. Recently it has also been shown that tapetal nuclei may increase in size and in chromosome number as a result of the process of *endomitosis*—that is, the division of chromosomes within the nuclear membrane without the formation of a spindle. By this process, the conspicuously enlarged nuclei become polyploid to varying degrees. [See Carniel (1963), and Bhandari (1984) for detailed reviews of the complex changes in nuclear and chromosome number which may accompany the maturation of tapetal cells in the angiosperms.]

From a structural as well as a physiological standpoint, it is interesting that, as in lower vascular plants, two major types of tapeta, distinguished on the basis of cell behavior during microsporogenesis, are recognized within the angiosperms: the *glandular* or *secretory tapetum,* the cells of which remain in their original position but finally become disorganized and obliterated (Fig. 20-1, A, B), and the *periplasmodial tapetum,* characterized by the preliminary breakdown of the cell walls and the fusion of the tapetal protoplasts to form a multinucleate periplasmodium which intrudes between the pollen mother cells.

Dehiscence

In a comparatively few families (e.g., Ericaceae, Epacridaceae, Melastomaceae), the pollen is released from a small opening situated at one end (usually the distal end) of the anther. This is termed poricidal dehiscence. More commonly, the anther dehisces *longitudinally* along the *stomium,* or furrow, situated at each side of the anther between the members of a pair of sporangia (Fig. 20-1, A, B). In the common type of anther represented in *Lilium,* the epidermal cells lining the base of each stomium are very small and easily ruptured, causing the anther to split at two points, thus releasing the enclosed pollen grains.

Although it is commonly stated that the actual force resulting in dehiscence is in some way related to the *endothecium,* the precise "mechanism" still awaits full explanation. Many investigators maintain that dehiscence is a hygroscopic phenomenon related to the peculiar pattern of secondary wall thickening of the endothecial cells. These thickenings may be restricted to the inner tangential and radial walls whereas the outer tangential walls remain thin. In an endothecium with this structure, the delicate outer walls are said to collapse more than the inner walls upon losing water, and hence the two edges of the anther lobe curl, or reflex, in dry air. But it is also believed by certain investigators that *cohesion mechanisms* in the endothecium may play a role in anther dehiscence.

Microsporogenesis

During the formation of the sporangium wall, the sporogenous cells may divide in various planes, the cells finally separating from one another and functioning as microsporocytes. The microsporocytes, or pollen mother cells, of angiosperms provide readily accessible material for the study of meiosis. Indeed, it is possible, under favorable or controlled conditions, to predict with considerable accuracy when various phases of the meiotic process will be most abundantly displayed. The laborious and time-consuming method of sectioning fixed anthers and subsequently staining the serial sections has been largely replaced by the process of "squashing" the anthers and then staining the masses of dividing pollen mother cells by the aceto-carmine, Feulgen, or crystal-violet techniques. [See Jensen (1962) and Berlyn & Miksche (1976) for detailed descriptions of the various procedures.] The use of these techniques also greatly facilitates the determination of the diploid and haploid number of chromosomes and the study of their form and structure.

Each microsporocyte, by means of meiosis and cytokinesis, gives rise to a tetrad of haploid microspores (Fig. 20-1, A). As is shown diagrammatically in Fig. 20-4, the arrangement of the four microspores in a tetrad varies widely and depends to some extent on the taxon in question. According to Maheshwari (1950) the *tetrahedral* and *isobilateral* types of microspore arrangement are very common, but in some genera several kinds of arrangements may occur, even in the same species. The *linear* arrangement — paralleled by the common arrangement of megaspores in angiosperm ovules — occurs in the family Asclepiadaceae but apparently from Davis' (1966) survey is a comparatively rare type.

When meiosis begins, a callose ((1→3)-*β*-D-glucan) wall is secreted around each of the microsporocytes, and later a wall of similar composition delimits each of the microspores in the tetrad. There are variations in the time and mode of callose secretion (Bhandari, 1984). The development of callose at the microsporocyte stage, and later between the individual microspores, and its degradation after the completion of meiosis suggest that the callose layers play an important function. It has been suggested, and supported experimentally, that the callose layers establish a partial seal between genetically different haploid cells (microspores), which must pass through their developmental stages unexposed to the influence of other spores and somatic tissue (Heslop-Harrison, 1966; Bhandari, 1984).

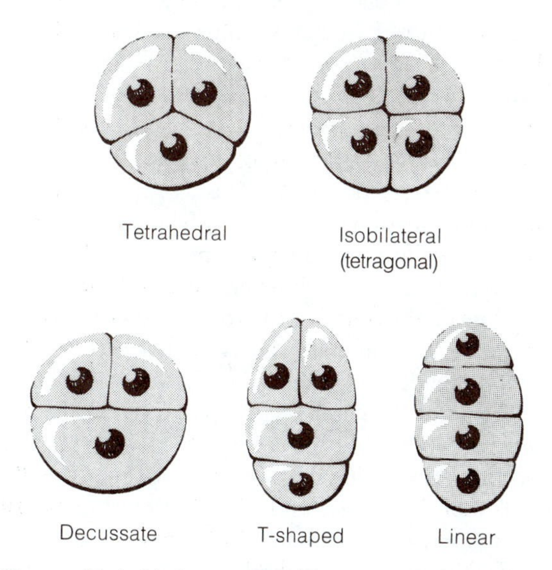

FIGURE 20-4 Various types of arrangement of microspores in a tetrad. [Adapted from *An Introduction to the Embryology of Angiosperms* by P. Maheshwari. McGraw-Hill, New York. 1950.]

The process of cytokinesis that takes place during microspore formation has been intensively studied and has led to the recognition of two basic methods of wall formation.

In the so-called successive type illustrated by *Zea mays*, a *centrifugally* developing cell plate is formed at the end of meiosis I, dividing the microsporocyte into two cells (Fig. 20-5, A–C). Then the second meiotic division takes place in each of these two cells followed again by the centrifugal formation of cell plates (Fig. 20-5, D, E). In this way, an *isobilateral* type of microspore tetrad is produced in *Zea*. This method of cytokinesis — the successive formation of two cell plates — is prevalent in the monocotyledons and has also been reported in some dicotyledonous families.

The distinguishing character of the so-called simultaneous type of cytokinesis, illustrated in Fig. 20-5, F–I, consists in the centripetal development of "constriction furrows," which usually first appear after the completion of meiosis. The furrows originate at the surface of the original wall of the microsporocyte and develop inwardly until they join in the center, resulting in the accompanying formation of walls that partition the sporocyte into four microspores. The simultaneous type of cytokinesis is found in members of both classes of the angiosperms; it is regarded by Maheshwari (1950) as the prevailing type within the dicotyledons.

Studies that utilize indirect immunofluorescence have shown that centripetal wall deposition at the periphery of the sporocyte proceeds along planes marked by interaction of opposing arrays of nuclear-based microtubules (Brown and Lemmon, 1989).

Although it is evident from the extensive literature that microspore tetrads may be formed either

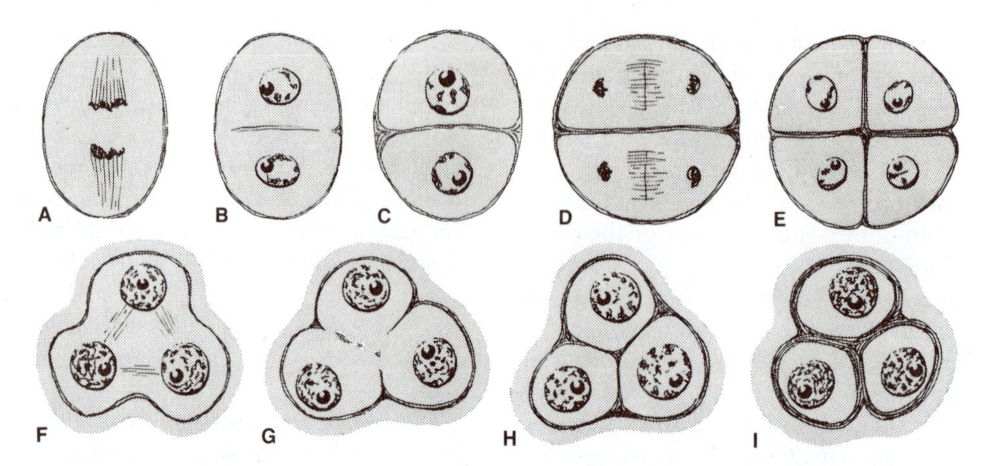

FIGURE 20-5 Types of cytokinesis in microsporocytes. **A–E,** the successive type in *Zea mays:* **F–I,** the simultaneous type in *Melilotus alba* (note centripetal direction of furrows separating the microspore protoplasts as shown in **G, H**). [Adapted from *An Introduction to the Embryology of Angiosperms* by P. Maheshwari. McGraw-Hill, New York. 1950.]

by centrifugal cell-plate development or by centripetal furrowing, the widespread use of the terms "successive" and "simultaneous" to designate respectively these types of cytokinesis has been criticized by Sampson (1963, 1969). His review of the problem reveals that these two terms are, to some degree, misleading. In the genus *Schisandra* and in *Pseudowintera traversii,* for example, wall formation is delayed until the completion of meiosis II when a simultaneous, rather than successive formation of centrifugal cell plates takes place (Kapil and Jalan, 1964; Sampson, 1970). Conversely, centripetal furrowing is not always simultaneous but, as in *Zygogynum, Annona,* and *Asimina,* may take place in two steps, beginning at the end of meiosis I (Swamy, 1952; Periasamy and Swamy, 1959). Such variations indicate that in drawing a distinction between the two types of cytokinesis, the major emphasis should be placed on the *contrasted processes* of cell-plate formation versus centripetal furrowing, rather than upon the times in the meiotic cycle when these processes take place.

To decide which of the two basic types is primitive in angiosperms is difficult. The dicotyledons are characterized mainly by the simultaneous type, while the monocotyledons primarily have the successive type. The families Winteraceae and Magnoliaceae, regarded to have retained other primitive characters, possess intermediate types of cytokinesis in microsporocytes. Davis (1966) considers the successive type with centrifugal wall formation to be the primitive process. For additional information see Sampson (1969, 1970) and Bhandari, 1984.

Remarkable patterns of microsporogenesis, characterized by the *absence of cytokinesis* during the meiotic divisions of the microsporocyte, have been found in members of two angiosperm families and deserve brief consideration at this point. In the family Epacridaceae, an essentially Australian group of heaths, an extraordinary diversity of pollen types has been described and classified (Smith-White, 1959; Venkata Rao, 1961). Besides the formation of "permanent" tetrads, each consisting of four normal functioning microspores, deviant types of tetrads occur, composed of three, two, or only one functional microspore. The most extreme pattern is illustrated in the genus *Styphelia,* as is shown in Fig. 20-6. Following meiosis in the microsporocyte, the four free microspore nuclei become more-or-less uniformly spaced, either in a tetrahedral or quadrant arrangement (Fig. 20-6, A). Before partition of the parent cell, three of the nuclei migrate to one end of the cell and the fourth nucleus becomes situated at the opposite end (Fig. 20-6, B). Cell walls then arise and separate the three small microspores, which eventually become compressed and obliterated, from the large functional microspore (Fig. 20-6, C–E). Then the nucleus of the surviving microspore divides mitotically, forming the generative and vegetative cells of the young

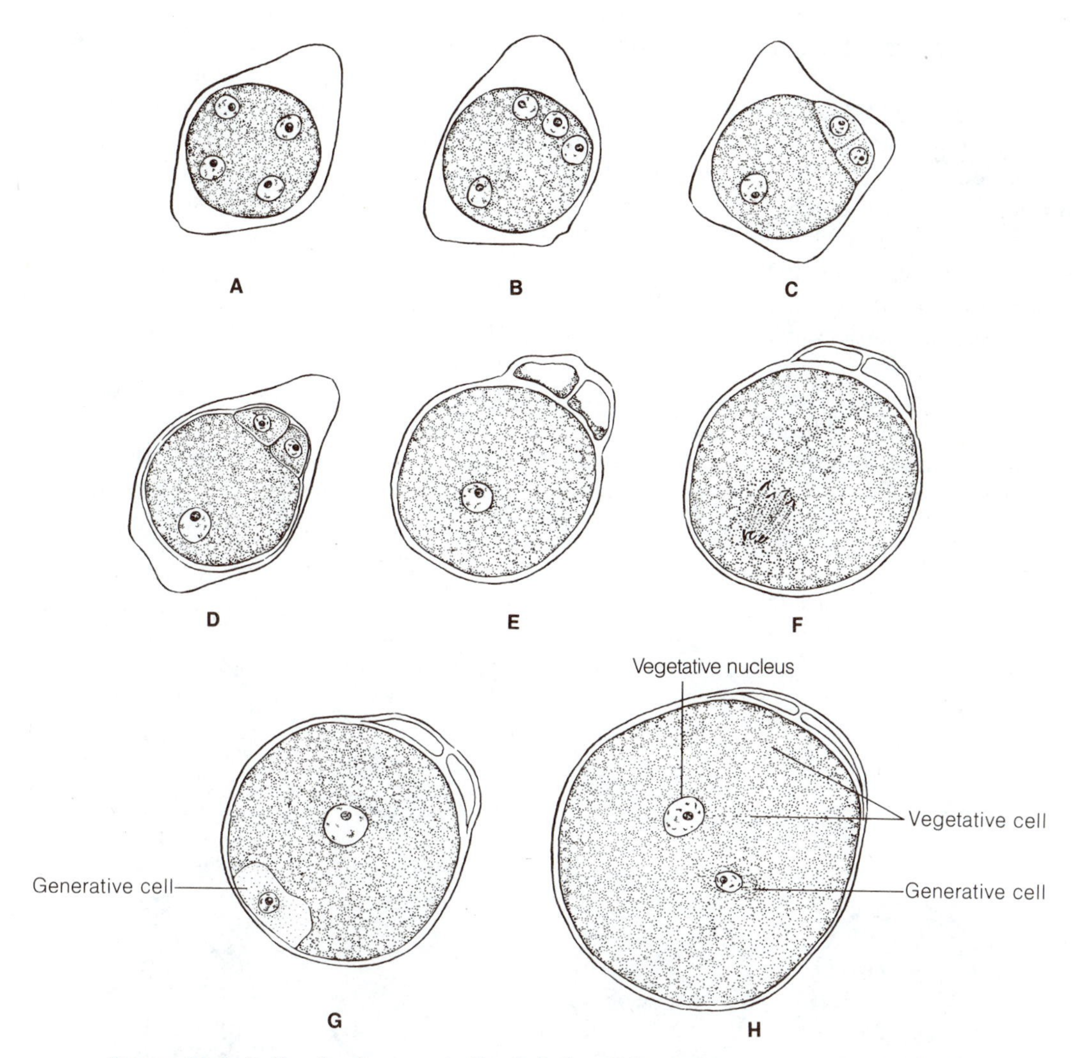

FIGURE 20-6 Pollen development in *Styphelia longifolia.* **A,** microsporocyte, immediately following meiosis, with four free microspore nuclei; **B,** migration of three microspore nuclei to one end of mother cell; **C,** formation of walls separating a large functional microspore from a cluster of three smaller microspores; **D–H,** death and progressive collapse of the three nonfunctional microspores; **F, G,** mitotic division of the nucleus of the functional microspore and formation of generative and vegetative cells of male gametophyte. [Redrawn from Smith-White, *Proc. Linn. Soc.* (New South Wales) 84:8, 1959.]

male gametophyte (Fig. 20-6, F–H). Thus the mature "tetrad" of *Styphelia,* which is termed a "monad" by Smith-White (1959), consists of one normal pollen grain and the degenerated remains of three microspores. For a detailed analysis of the morphogenetic, cytological, and genetic implications of the highly unusual pattern of microsporogenesis in *Styphelia* and other genera in the Epacridaceae, the reader is referred to Smith-White's paper. Venkata Rao (1961) regards the "monad" type of pollen development of *Styphelia* as morphologically comparable to the monosporic type of embryo-sac development in angiosperms, in which three of the potential megaspores are eliminated and only a single megaspore is functional (see Fig. 20-13, E).

The Male Gametophyte (Pollen Grain)

After a resting period, which may vary according to the taxon from a few days to several months in arctic species, each functional microspore develops into a pollen grain that consists of the young male gametophyte enclosed within the pollen wall. Unlike the more complex and variable structure of the male gametophytes in gymnosperms, the male gametophyte in the angiosperms is relatively simple in organization.

Figure 20-7 presents schematically the main steps in the ontogeny of the male gametophyte in the angiosperms. Prior to the first mitotic division, the nucleus of the microspore becomes displaced from its original central position to a peripheral position where it divides to form two unequal cells: a smaller *generative cell* lying next to the spore wall and a much larger *vegetative cell* (Fig. 20-7, A–D). This first step in gametogenesis determines the polarity of the generative cell but its exact position, following mitosis, can only be determined accurately in taxa in which the pollen grains remain united in tetrads, in the absence of developmental stages. In such taxa, the generative cell may form either at the proximal (inner) side of each pollen grain, as in *Pseudowintera* and *Zygogynum* (Fig. 20-8, G, H) or next to the distal (outer) face of each grain, as in *Degeneria, Cananga,* and certain members of the Orchidaceae. These two locations ap-

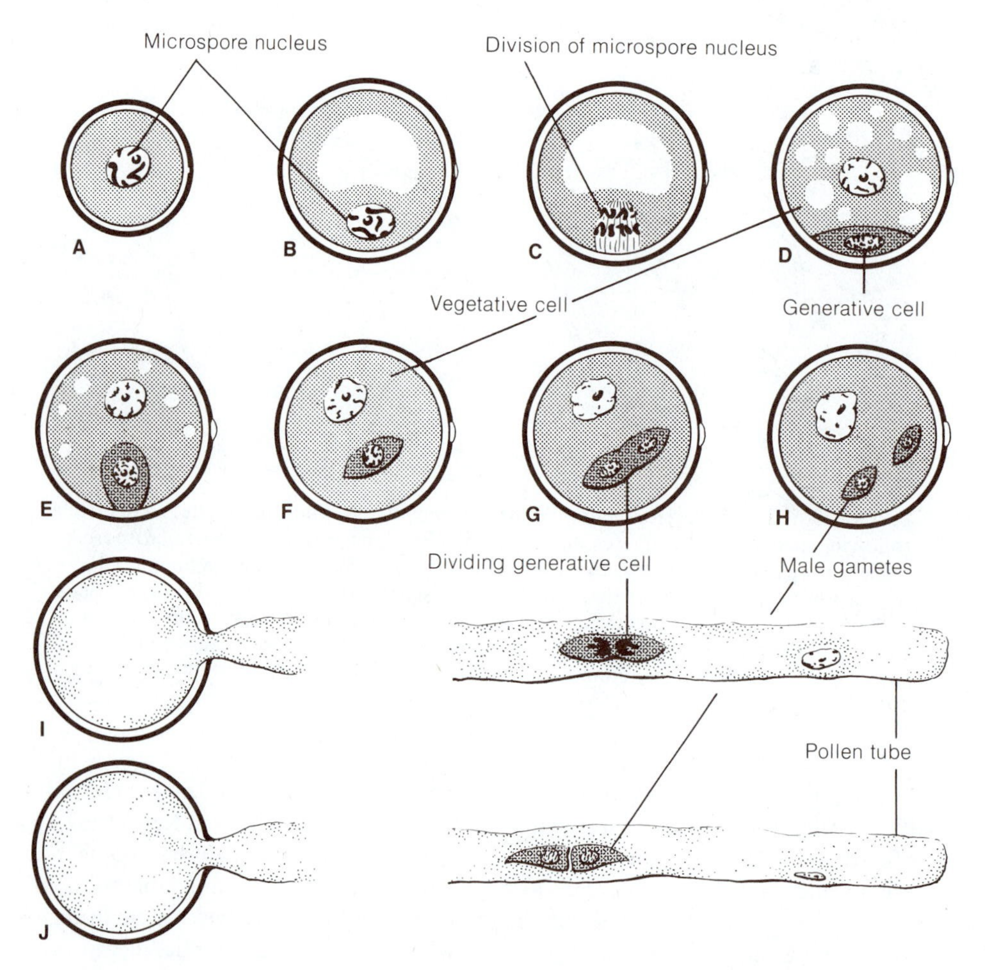

FIGURE 20-7 Development of the male gametophyte in angiosperms. See text for full discussion of this diagram. [Redrawn from Maheshwari, *Bot. Rev.* 15:1, 1949.]

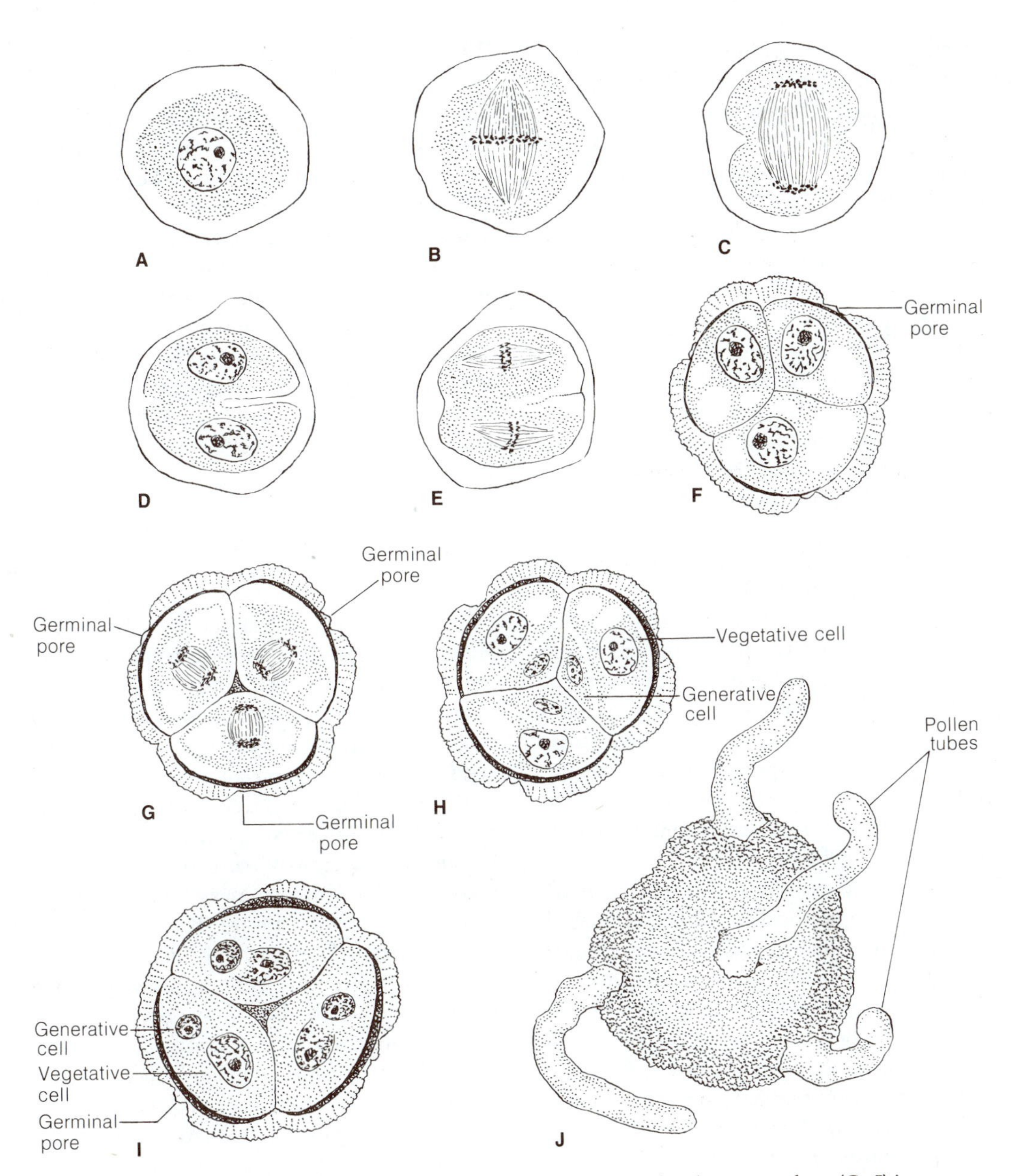

FIGURE 20-8 Microsporogenesis (**A–F**) and development of male gametophyte (**G–I**) in *Zygogynum bailloni* (Winteraceae). **A,** microsporocyte; **B,** metaphase of meiosis I; **C,** late anaphase of meiosis I, showing preliminary equatorial constriction of cytoplasm; **D,** further progress in centripetal furrowing; **E,** meiosis II; **F,** tetrad of microspores; **G,** mitotic division of nucleus in each microspore; **H,** formation of generative cell next to proximal side of each pollen grain; **I,** tetrad of two-celled pollen grains at time of shedding; **J,** germination of each of the pollen grains of a tetrad removed from surface of the stigma. [Redrawn from Swamy, *Proc. Nat. Inst. Sci. India* B 18:399, 1952.]

pear sufficiently constant within taxa that they may be of systematic significance (Swamy, 1949b, 1952).

The generative cell, soon after its formation, moves away from its peripheral position and appears as an ellipsoidal, lenticular, or horseshoe-shaped cell that lies free in the cytoplasm of the vegetative cell, but is surrounded by a definite cell wall (Fig. 20-7, E, F). The vegetative cell contains a lobed central nucleus and is packed with organelles, starch grains, lipids, and proteins (see Knox, 1984, and the literature cited therein). The generative cell contains only a limited number of organelles. Apparently in the majority of investigated angiosperms, the male gametophyte is in the two-celled stage of development when the pollen grains are shed from the anther, and the division of the generative cell to form the two male gametes takes place in the pollen tube (Fig. 20-7, F, I, J). But in certain taxa, in both the monocotyledons and dicotyledons, the generative cell divides prior to the dehiscence of the anther and the pollen is shed at the three-celled stage in development of the gametophyte (Fig. 20-7, G, H). From a phylogenetic standpoint, the morphologically simple male gametophyte of modern angiosperms seems to be the end product of a profound evolutionary reduction, possibly from some hypothetical organization that is similar to that in gymnosperms.

The systematic distribution and phylogenetic significance of the two-celled and three-celled types of angiosperm pollen grains have been studied in detail by Brewbaker (1967). In approximately 70 percent of the species included in his survey, the pollen is released from the anther at the two-celled stage in development of the gametophyte. This type of pollen appears to predominate in primitive families and is regarded by Brewbaker as the primitive form in the angiosperms as a whole. The three-celled type of pollen grain is thus a derivative type which, in his opinion, has originated independently many times during the evolution of the angiosperms.

The assumption that the male gametes and the vegetative nucleus (tube nucleus) travel independently in the pollen tube is not true for some species. In *Plumbago zeylanica,* for example, the male gametes and the vegetative nucleus are linked and remain essentially joined until the contents of the pollen tube are discharged into the embryo sac (Russell and Cass, 1981, 1983).

The Ovule

Since the ovule is the structure within which meiosis and megaspore formation take place, it corresponds *functionally* to a megasporangium and thus is the "promise" of a future seed. But, as in the gymnosperms, the angiosperm ovule, from a morphological standpoint, is not simply a spore case or sporangium since it usually consists of a nucellus invested by one or two integuments and a stalk or *funiculus* attaching it to the placenta of the ovary. Moreover, angiosperm ovules may develop a vascular system within the outer integument or, in a few taxa, there may be tracheids even in the lower part of the nucellus.

Placentation

As described earlier in Chapter 19 (p. 525), ovules in most taxa are attached to the adaxial surface of the carpel. The position of the ovule-bearing regions or *placentae* vary widely in angiosperms. In many types of free, simple carpels, the placentation is submarginal, meaning that the ovule primordia arise very near the approximated or fused margins of the carpel (Figs. 19-33, A; 19-45, B). However, in reputedly primitive taxa such as *Degeneria* and *Tasmannia piperita,* the two rows of ovules are situated a considerable distance away from the carpellary margins (Figs. 19-47, 19-48). This type of placentation is termed "laminar" by Eames (1961), who maintains that it represents the primitive ovule position from which the submarginal position has been derived by the restriction of ovules to the near-marginal position. In many syncarpous gynoecia, for example *Lilium* and *Fritillaria,* the two rows of ovules in each locule of the ovary are attached to placental regions formed by the fusion of the ventral margins of the carpels. This pattern is designated axile placentation (Fig. 19-33, C). See Chapter 19 (p. 525) for other types of placentation.

Types of Ovules

The ovules of angiosperms have long been classified into various types, depending upon whether the axis of the ovule is erect, with the micropyle facing away from the point of attachment of the funiculus, or curved, with the micropyle facing the placenta or the base of the funiculus. We will describe briefly some of the principal types of ovules classified according to shape and orientation.

The *orthotropous ovule,* which is found in members of such families as Polygonaceae, Juglandaceae, and Najadaceae, is erect and devoid of curvature, with the micropyle distal and directed away from the placenta (Fig. 20-9, A). In sharp contrast, the *anatropous ovule*—a very common type throughout the angiosperms—is inverted in its orientation as a result of the approximately 180° curvature of the funiculus (Bersier, 1960). The longitudinal axis of the nucellus, in this type, is parallel to the funicular axis, and the micropyle thus faces down toward the placenta (Fig. 20-9, C). Very commonly, orthotropous and anatropous ovules are regarded as "basic types," but there is no convincing proof which one represents the ancestral form in the angiosperms as a whole. Eames (1961), for example, on the basis of the prevalence of anatropous ovules in presumably primitive monocotyledons and dicotyledons, regards the orthotropous ovule as a derivative type. On the other hand, Bocquet and Bersier (1960) suggest that the orthotropous ovule may be more ancient. Support for this interpretation is in part derived from a study of ovule development, which shows that in many angiosperms the primordium of an anatropous ovule is "orthotropous" prior to the initiation of the integuments; subsequent development is asymmetric and results in a curvature (of the upper region of the funiculus) and the final inverted orientation of the micropyle. A very clear example is provided by the development of the ovule in *Degeneria* in which, according to Swamy (1949a), "the apex of the ovule undergoes a curvature of 90° . . . and by the time of sporogenesis, becomes bent on itself, thereby assuming a completely anatropous position" (Fig. 20-9, C). It should be noted further that in the *hemitropous* ovule the degree of curvature is intermediate between the typical orthotropous and anatropous ovules (Fig. 20-9, B).

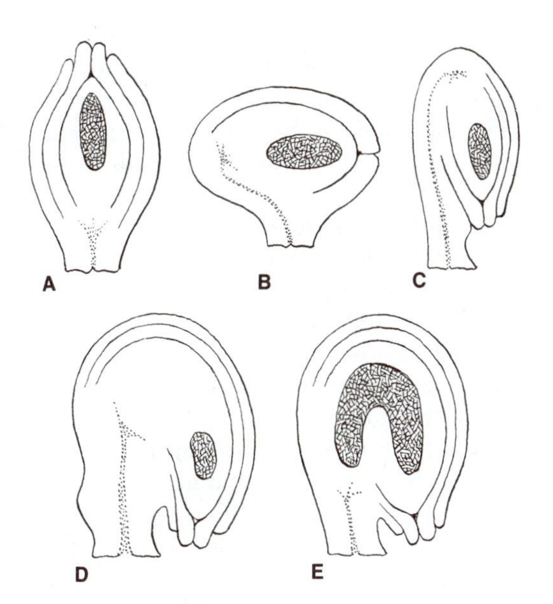

FIGURE 20-9 Diagrammatic longisectional views of the main types of ovules. The position of the embryo sac in each figure is shown by hatching and the course of the vascular strand by stippling. **A,** orthotropous; **B,** hemitropous; **C,** anatropous; **D,** campylotropous; **E,** amphitropous. [Redrawn from *Syllabus der Pflanzenfamilien,* 12th edition, Vol. 2. by A. Engler. Gebrüder Borntraeger, Berlin. 1964.]

Two additional types of ovules remain to be considered: the *campylotropous* and the *amphitropous* (Fig. 20-9, D, E). According to the detailed analysis given by Bocquet and Bersier (1960), these are subordinate types which may appear *successively* as ontogenetic modifications of *either* the orthotropous or anatropous types of ovule. Figure 20-10 presents diagrammatically the orthotropous and anatropous series. In each series, the campylotropous ovule develops a micropyle which is directed toward the base of the funiculus because of the ontogenetic curvature, or distortion, of the nucellus. A more exaggerated curvature in the development of the nucellus leads to the amphitropous type of ovule, the embryo sac of which becomes kidney shaped. This excessive bending of the contour of nucellus (and embryo sac) is produced by the intrusion of a chalazal pad of specialized nucellar tissue—known as the *hypostase,* or basal body—into the main body of the ovule. A clear example of an amphitropous ovule is found in *Capsella* (Fig. 20-30); campylotropous ovules are characteristic of members of such families as Caryophyllaceae, Capparidaceae, and Leguminosae.

Bocquet's diagram (Fig. 20-10) shows that his basis for a decision about which of the two series a given ovule type should be assigned to depends in part on the course of the ovule trace. In the orthotropous series, regardless of the degree of curvature of the campylotropous and amphitropous derivatives, the ovule trace extends vertically upward

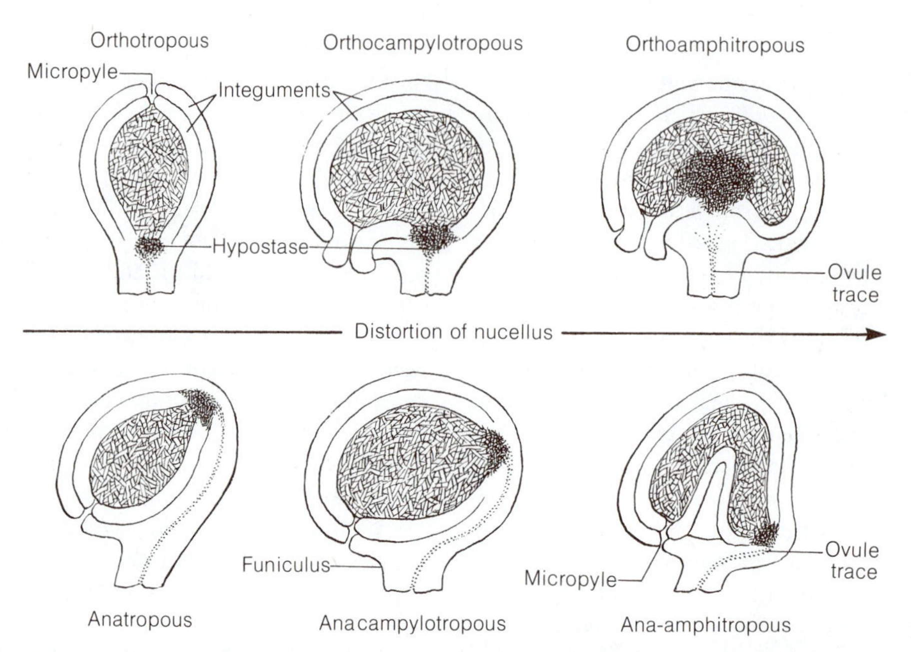

FIGURE 20-10 Diagram showing that both campylotropous and amphitropous ovules may arise by the successive ontogenetic modification of *either* an orthotropous or an anatropous ovule. Top row, the orthotropous series; lower row, the anatropous series. The hatched area in each ovule represents the embryo sac. [Adapted from Bocquet, *Phytomorphology* 9:222, 1959.]

through the funiculus and terminates below the weakly developed hypostase. On the other hand, in the anatropous series, because of the fundamental curvature of the prototype, the ovule trace in the derivative forms is likewise curved in its course before terminating at the base of the hypostase.

From the standpoint of the diagnostic value of ovule morphology as a taxonomic character, the observations and conclusions of Bocquet and Bersier merit full consideration. If indeed a parallel, independent development of campylotropous and amphitropous ovules has taken place during the evolution of the angiosperms, this possibility must be fully recognized whenever the form, orientation, and vasculature of the ovule are used in an effort to determine the systematic relationships between genera or families.

Integuments

In the angiosperms, the ovule develops one or two integuments which completely enclose the nucellus except for the terminal *micropyle* (Fig. 20-13, J). The micropyle is a small opening in the apical portion of the ovule and is usually the point of entrance of the pollen tube into the interior of the ovule. In a few families of dicotyledons the ovule lacks an integumentary system.

There may be considerable variation in the number of integuments characteristic of the ovules of members of a genus, family, or order, although, for example, *unitegmic* ovules (those with a single integument) appear to be almost universal in advanced families of dicotyledons and *bitegmic* ovules (those with two integuments predominate in the monocotyledons and in many families of dicotyledons (Fig. 20-11). Bitegmy and unitegmy occur in species of many dicotyledonous families. Bitegmy is considered the *primitive* condition in angiosperms and is the prevailing type in the Magnoliales. Unitegmy has probably occurred several times in the evolution of angiosperms through modified ontogenies leading to congenital fusion of two integumentary primordia or by the phylogenetic loss of one of the two integuments. [See Bouman (1984) for a review of the subject.]

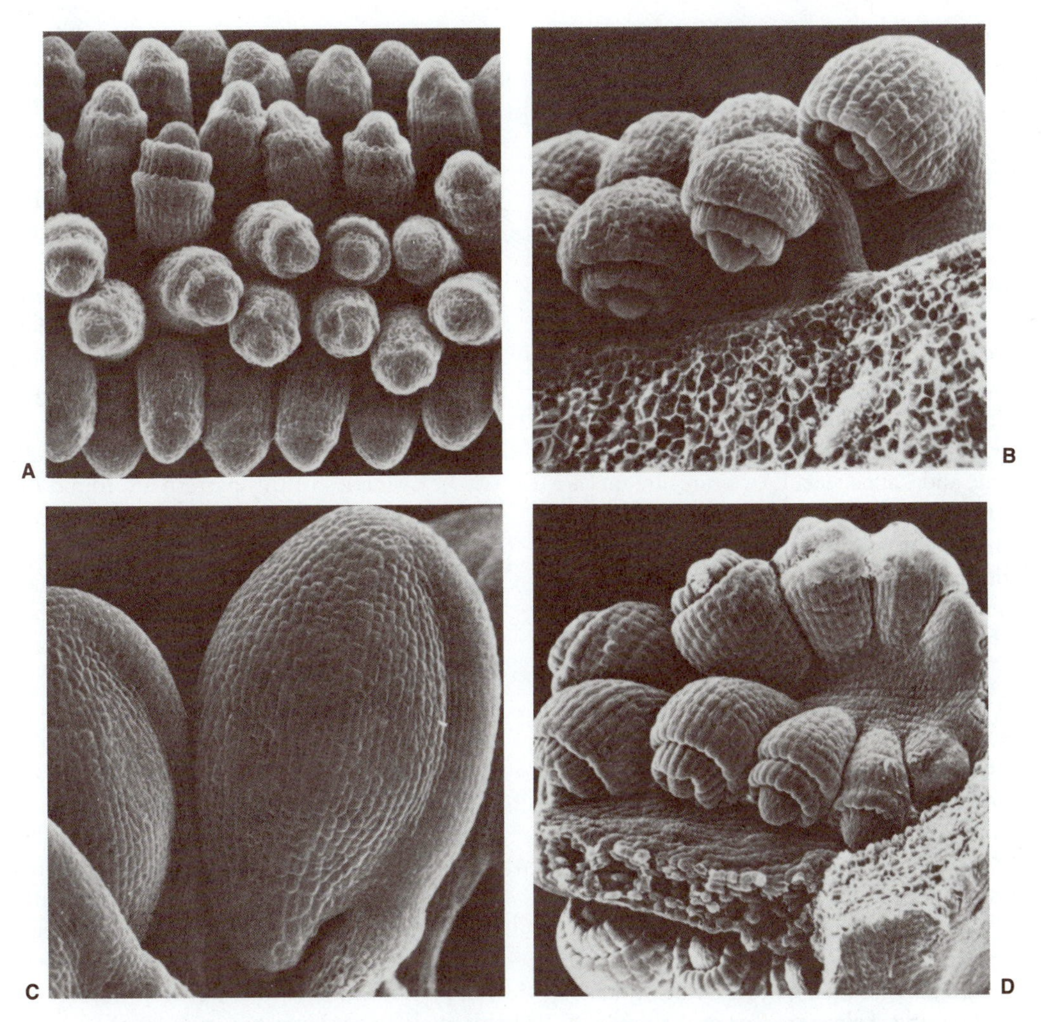

FIGURE 20-11 Scanning electron micrographs of developing ovules. **A,** *Passiflora racemosa,* ovule primordia showing initiation of inner and outer integuments and the beginning of anatropous curvature; **B, C,** *Passiflora verspitilio,* showing anatropous curvature of ovules and growth of nucellus beyond the integuments (**B**), and mature (**C**) where nucellus and inner integument are no longer visible externally; **D,** *Agrostemma gracile,* showing ovule initiation and development of integuments. [From Bouman, in *Embryology of Angiosperms* (B. M. Johri, ed.). Springer-Verlag, Berlin. 1984.]

According to Bouman (1984) the initiation and early development of integuments have not been studied adequately. In general, the inner integument is usually initiated before or sometimes simultaneously with the outer integument (Roth, 1957). The inner integument is almost always of dermal (protodermal) origin and becomes two to three cells thick. The outer integument is initiated in either the dermal or subdermal layer. A survey of the literature indicates that the outer integument is subdermal in several dicotyledonous families, including the Magnoliaceae and the Ranunculaceae, while a dermal origin is found in most monocotyledons and in several orders of dicotyledons.

Vasculature of Ovules

Most commonly, ovules develop a comparatively simple type of vasculature: the ovule trace, usually a single bundle, is derived as a branch of one of the carpel veins (commonly a ventral carpel bundle) and extends through the funiculus, ending abruptly at the base of the chalaza. This simple pattern of vasculature is found—but not exclusively—in all the main types of ovules, as is shown in Fig. 20-9. The classical study of seed venation by LeMonnier (1872) and the later comprehensive survey made by Kühn (1928) have revealed that vascular strands, producing a wide range of venation patterns, may develop in the integuments of ovules. According to Kühn, vascularized integuments are found in more than 100 angiospermous genera distributed among thirty families, including such monocotyledonous families as Palmae, Amaryllidaceae, and Cannaceae. If two integuments are formed, the vasculature is usually restricted to the outer integument, but a few examples of ovules in which both of the integuments are vascularized have been discovered, particularly in members of the Euphorbiaceae (e.g., *Ricinus, Jatropha, Dalechampia,* and *Aleurites*).

Kühn (1928) classified the diversified patterns of integumentary venation into a series of types that she conveniently arranged under the three main categories of ovules—i.e., orthotropous, anatropous, and campylotropous. Figure 20-12, based on Kühn's figures and interpretations, depicts some of the extraordinary range in complexity of integumentary vascular systems in the angiosperms.

In the comparatively infrequent type of orthotropous ovule, the vascular bundle of the funiculus, upon reaching the chalazal region of the ovule, divides into numerous veins which extend, with or without branching, through the integument, ending blindly near the micropylar end (Fig. 20-12, A). Examples of this pattern are found in *Juglans, Myrica,* and *Sterculia.*

The patterns of integumentary venation in anatropous ovules are extremely diversified. Before considering them, however, it is necessary to define the term *raphe.* As a result of the approximately 180° curvature which takes place during the ontogeny of an anatropous ovule, fusion may occur between the funiculus and the immediately adjacent portion of the outer integument (Bersier, 1960). This produces externally a ridge, termed the raphe, through which the upward continuation of the ovule trace extends in its course to the chalaza. The special term "raphe bundle" designates this vein and its position is indicated by r in Fig. 20-12, B–F.

In the simplest type of integumentary venation found in anatropous ovules, the raphe bundle extends as a curved and unbranched vein nearly to the micropylar region of the integument (Fig. 20-12, B). Kühn observed this type of vasculature in more than thirty widely scattered genera of angiosperms (e.g., *Anemone, Himantandra, Symphoricarpos,* and *Lonicera*). More complex, open venation patterns occur when the raphe bundle subdivides, near the chalazal end of the ovule, into a system of dichotomously branched veins which extend toward the micropylar end of the integument (Fig. 20-12, C, D). Examples of this pattern are found in *Thea* (Theaceae), *Castanea* (Fagaceae), and *Prunus* (Rosaceae).

Finally, a surprisingly complex reticulate type of venation is characteristic of the integuments of the ovule in several genera of the Cucurbitaceae (e.g., *Momordica, Cyclanthera,* and *Trichosanthes*). In this type, the raphe bundle continues beyond the chalaza almost entirely around the edge of the anatropous ovule and encloses an intricate *anastomosed* system of veinlets (Fig. 20-12, E).

According to Kühn, integumentary vasculature is not commonly developed in campylotropous ovules. Figure 20-12, F, shows the pattern of dichotomous-reticulate venation in the integument of *Lupinus luteus* (Leguminosae). A similar venation pattern was also observed in the integument of the ovule of the common cultivated legume, *Phaseolus vulgaris.*

What phylogenetic or taxonomic conclusions can be made on the basis of the data provided by Kühn's survey of integumentary vasculature? From a phylogenetic viewpoint, it has frequently been maintained that when vasculature is present in angiosperm ovules, it is vestigial in character and represents the results of reduction of the type of integumentary vascular system that characterized the seeds of pteridosperms, and which has persisted in the ovules of such surviving relics as the living cycads and *Ginkgo biloba* (Chapters 15 and 16). Kühn (1928) rejected this concept on the basis of the sporadic occurrence of vasculated angiosperm ovules and the differences in vascular patterns between living and extinct gymnosperms and the angiosperms. In her view, integumentary patterns of

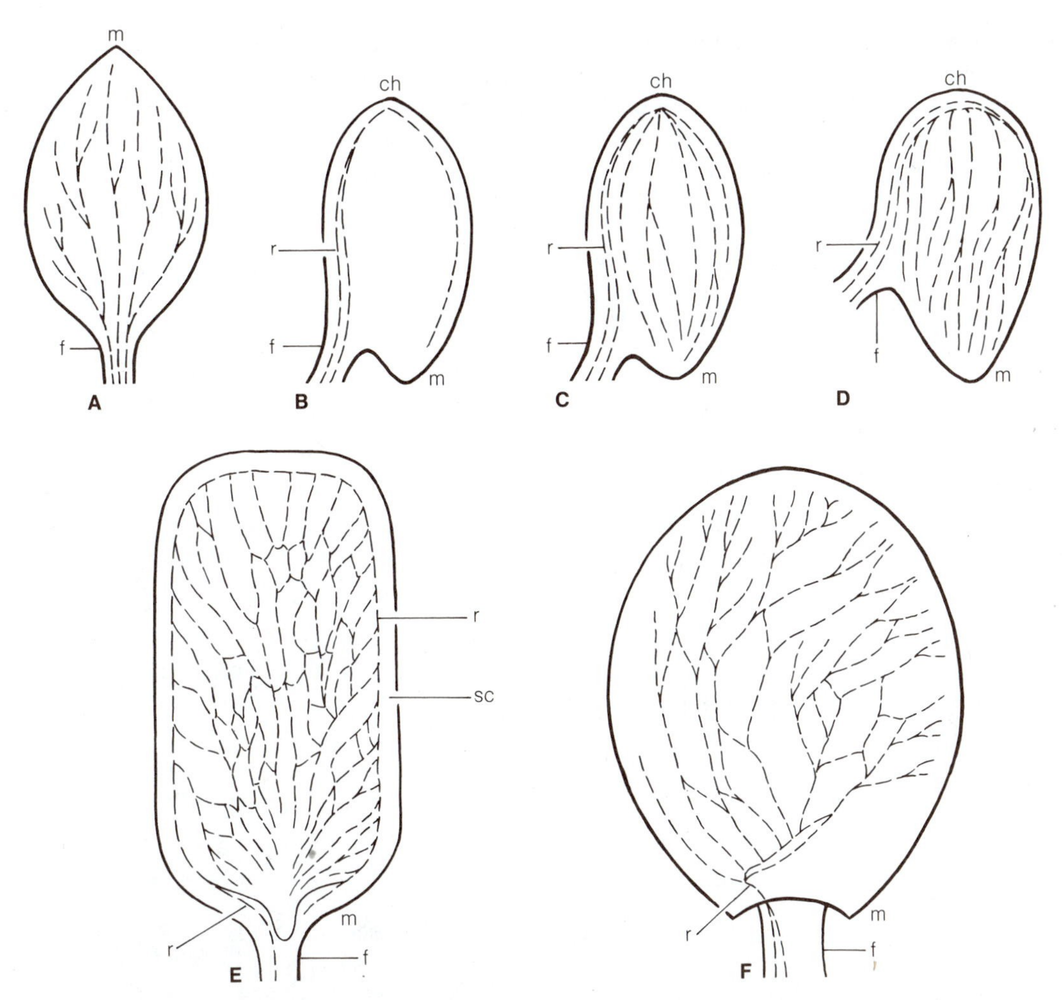

FIGURE 20-12 Types of venation patterns in the integument of the ovules of various angiosperms. **A,** tangential view (diagrammatic) of palmate-dichotomous venation in the orthotropous type of ovule; **B–E,** diagrams of venation patterns in anatropous ovules: **B,** simple venation, consisting of single unbranched vein, as seen in longisectional view of ovule; **C, D,** tangential views of integument, showing variations in point of origin of dichotomized lateral veins; **E,** tangential view of integument of *Momordica* showing complex reticulate pattern of venation; **F,** tangential view of dichotomous-reticulate venation in integument of campylotropous ovule of *Lupinus luteus*. Abbreviations for all figures: f, funiculus; ch, chalaza; m, micropyle; r, raphe bundle; sc, hard part of seed coat. [Redrawn from Kühn, *Bot. Jahrb.* 61:325, 1928.]

vasculature in angiosperms should be regarded as "new formations" which have developed independently of one another in both primitive and advanced taxa.

In conclusion, it must be admitted that neither the method of initiation of integuments nor the presence of integumentary venation throw much light on the vexing question of the morphology of ovular integuments. While the ontogenetic origin of the integuments and the presence of integumentary vascular bundles may invite the conclusion that ovular envelopes are modified "foliar structures," this idea is conjectural in view of our ignorance of the evolutionary origin of the angiosperms and the nature of the ovule in proangiosperms (see Maheshwari, 1960). If Zimmermann's telome theory is applied to the elucidation of the problem, one might conclude that the integument represents a "syntelome," which is the result of the fusion of a ring of vascularized sterile telomes to form a cupulelike envelope surrounding a terminal sporangium. [See Boesewinkel and Bouman (1967), van Heel and Bouman (1972), Stebbins (1974), and Stewart (1983) for discussions of various phylogenetic theories about the origin and nature of the integuments in angiosperm ovules.]

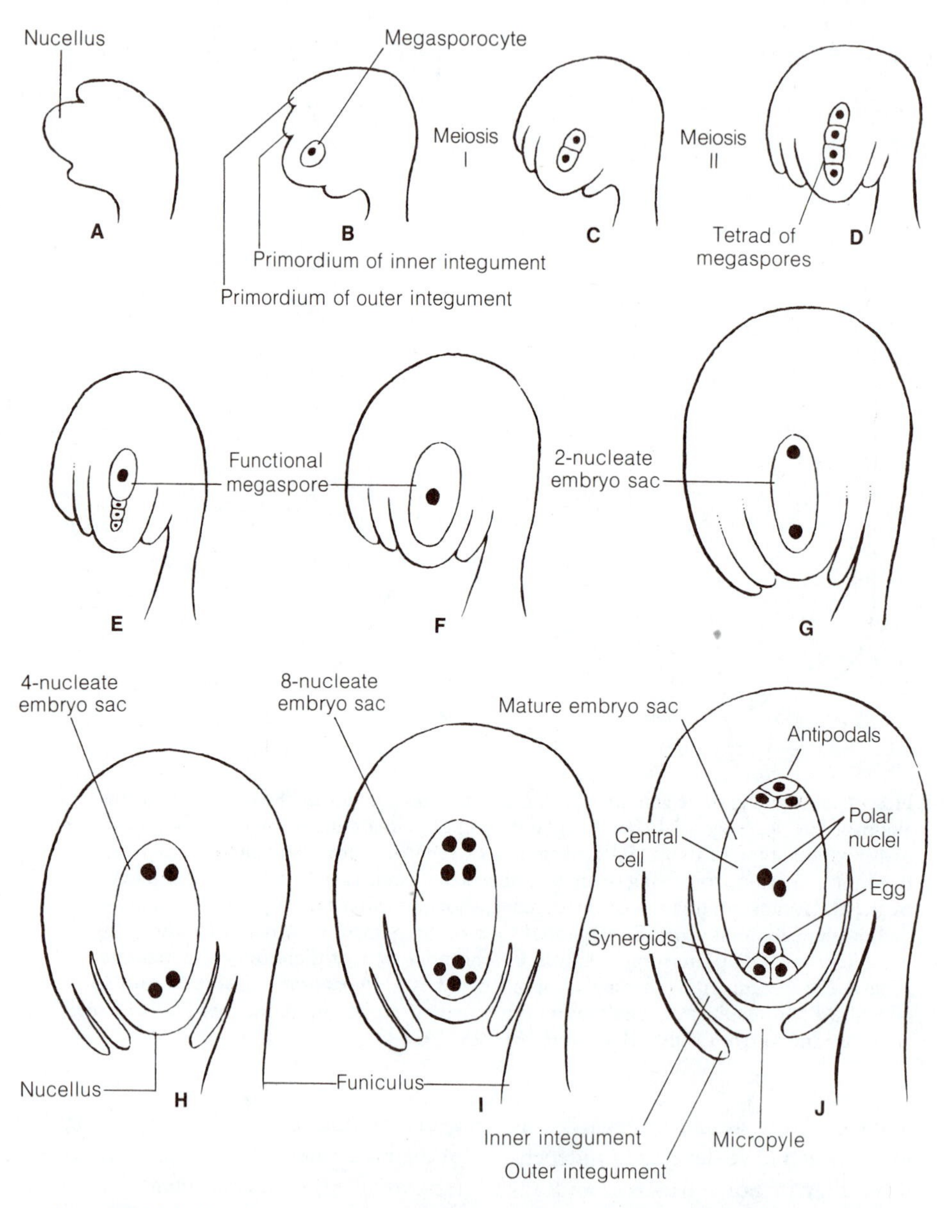

FIGURE 20-13 Development of an angiosperm ovule, beginning with the formation of the integuments and the single megasporocyte (**A, B**), continuing through the formation of megaspores (**C–E**), and concluding with the successive stages in development of the embryo sac (**F–J**). See text for detailed discussion of this figure. [Redrawn from *A Textbook of General Botany,* 4th edition, by R. M. Holman and W. W. Robbins. Wiley, New York. 1951.]

Structure of Nucellus and Origin of Megasporocyte

Early in the ontogeny of the ovule, one or several *internal* cells near the apex of the nucellus enlarge and differentiate. These distinctive cells are "potential" megasporocytes, as each of them, by meiotic divisions, can produce a tetrad of haploid megaspores. Most frequently only a single megasporocyte is formed in the developing nucellus (Fig. 20-13, B). But in the ovules of some angiosperms, a group of potential megasporocytes may arise, each of which theoretically can give rise to a tetrad of megaspores. A particularly striking example of the formation of "multiple megasporocytes" has been observed during the development of the massive nucellus of the ovule of *Paeonia californica*. In this species Walters (1962) found that thirty to forty megasporocytes differentiate and many of them complete meiosis and form linear tetrads of megaspores. Even more remarkable is that at the time of fertilization, several of the chalazal megaspores, from different tetrads, may have formed female gametophytes. During later ontogenetic stages, however, all but one of the megasporocytes and megagametophytes begin to degenerate and the mature seed contains only one embryo.

Because of the relatively infrequent—and apparently erratic—occurrence of multiple megasporocytes in the angiosperms as a whole, it is not possible to reach a conclusion about the phylogenetic significance of this condition. Perhaps, as Walters suggests, the development of multiple megasporocytes has occurred "independently numerous times during the evolution of angiosperm families as well as gymnosperms."

More typically, in both monocotyledons and dicotyledons, only a single functional megasporocyte differentiates in the ovule, and its *ultimate position* is usually correlated with the size and cellular organization of the nucellus. The terms *crassinucellate* and *tenuinucellate*, originally introduced by van Tieghem (1898), are widely used in current literature to designate two contrasted types of nucellar organization. In a "typical" tenuinucellate ovule, the megasporocyte originates *directly* from a hypodermal cell located at the apical region of the relatively delicate nucellus consisting of an epidermal layer and comparatively few nonsporogenous internal cells (Fig. 20-14, A). On the other hand, in crassinucellate ovules, a more-or-less prominent hypodermal initial *first* divides periclinally into an outer *parietal cell* and an inner *sporogenous cell*. The latter functions as the megasporocyte while the parietal cell by means of further periclinal divisions forms one or many layers of cells. Thus, by this *indirect method of origin*, the megasporocyte ultimately becomes deeply embedded within a relatively massive type of nucellus. In some species, the division of parietal cells above the megasporocyte may be accompanied by periclinal divisions in the nucellar epidermis (Fig. 20-14, B). In such species, the deeply sunken position of the megasporocyte is very obvious, and nucelli with this kind of epidermal proliferation tend to be particularly massive in structure.

In conclusion, it must be noted that the ovules of certain angiospermous genera cannot be rigidly classified as either crassinucellate or tenuinucellate if the presence or absence of parietal cells is used as a fundamental defining character. Dahlgren (1927), in his comprehensive and critical treatise on the nucellus, has emphasized that parietal cells may be developed in ovules, which on the basis of their small size and weakly developed nucellus would be classified as "tenuinucellate." Furthermore, there may even be variation between the ovules of the same species, some forming parietal cells and others being devoid of such cells. Conversely, not all "massive" ovules, which at maturity might be classified as crassinucellate, develop parietal cells during the differentiation of the megasporocyte. For example, in two species of *Calycanthus* investigated by Peter (1920), the nucellar epidermis produces a massive "nucellar cap" but the multiple megasporocytes originate *directly* from hypodermal cells without the formation of parietal cells. Davis (1966) proposed the new term "pseudocrassinucellar" for ovules with this type of nucellar organization. It should thus be evident that although it is generally useful to be able to classify ovules into two major types—crassinucellate and tenuinucellate—there are intergradations and deviations, and these must be fully recognized whenever the organization of the nucellus is used as a taxonomic character.

During the development of the embryo sac, endosperm, and embryo in the ovule, most or all of the nucellar tissue becomes crushed and ultimately

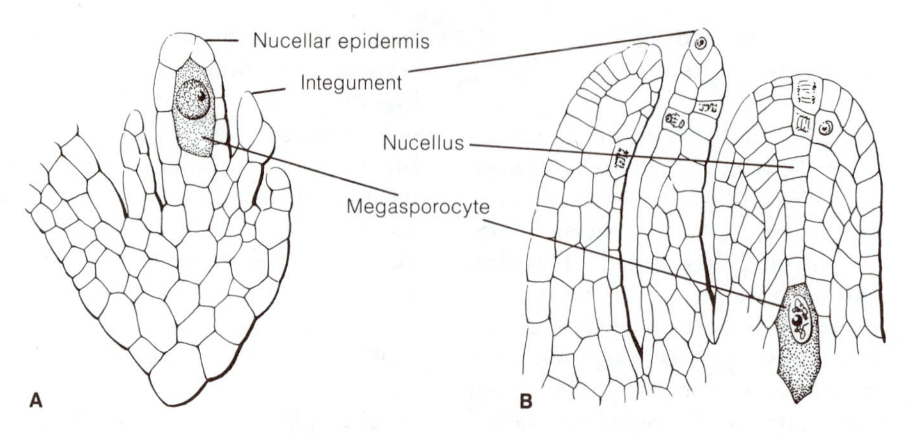

FIGURE 20-14 **A,** tenuinucellate type of ovule in *Orchis maculatus,* in which the megasporocyte is directly below the nucellar epidermis; **B,** crassinucellate type of ovule in *Quisqualis indica* (note deeply embedded position of megasporocyte and the active periclinal divisions in the nucellar epidermis). [Adapted from *An Introduction to the Embryology of Angiosperms* by P. Maheshwari. McGraw-Hill, New York. 1950.]

destroyed. In a few plants, however, such as various species of *Yucca,* a portion of the nucellus survives and forms a nutritive tissue, known as *perisperm,* in the mature seed (see Arnott, 1962, and Kapil and Vasil, 1963, for details).

Megasporogenesis

In lower heterosporous vascular plants and in most gymnosperms, the female gametophyte arises from the growth and division of a *single* haploid megaspore. There is a comparable method of origin for the megagametophyte, or embryo sac, according to Maheshwari (1950) "in at least 70 percent of the angiosperms now known." The salient features of this widespread type of megasporogenesis are illustrated in Fig. 20-13. Each of the two divisions (meiosis I and II) is accompanied by wall formation resulting in a tetrad of megaspores. The first division is always in the transverse plane and produces two dyad cells (Fig. 20-13, C). Usually the second division is followed by the formation of a transverse wall in each dyad cell. As a result, a file, or linear tetrad, of megaspores is produced (Fig. 20-13, D). In some cases, however, the plane of division in the micropylar dyad cell is longitudinal, resulting in a ⊥-shaped arrangement of the megaspores. If the division plane in the lower dyad cell is vertical, a T-shaped tetrad of spores is formed. Linear and T-shaped megaspore tetrads may, according to Maheshwari, arise in different ovules of the same ovary, but the factors controlling this variability are at present unknown.

In the very common *Polygonum* type of embryo-sac origin, three megaspores of the linear tetrad degenerate while the chalazal megaspore enlarges and, by means of three successive mitotic divisions, gives rise to an eight-nucleate embryo sac (Fig. 20-13, E–I). Thus, beginning with the megasporocyte, it takes a total of *five* synchronized nuclear divisions, of which two are meiotic, to produce the haploid female gametophyte (Fig. 20-13, J).

However, striking deviations from this common pattern of megasporogenesis and embryo-sac development have been discovered as the result of comparative studies on a wide range of angiosperm genera. One of the most remarkable deviations is the participation of more than one megaspore nucleus in the formation of the embryo sac. This very peculiar—and probably specialized—condition is the result of the partial or complete failure of wall formation during meiosis. In the most extreme expression, the completion of meiosis I and II results in four megaspore nuclei, which lie free in a cell that is termed the *coenomegaspore* (Maheshwari, 1950). Embryo sacs that arise from a coenomegaspore are thus *tetrasporic* and represent a type of female gametophyte origin that is without parallel in vascular plants, with the exception of two genera of gymnosperms, *Gnetum* and *Welwitschia* (see Waterkeyn, 1954, and Martens, 1962).

Among the noteworthy aspects of megasporogenesis are the reports that callose is formed in the walls of the megasporocyte and megaspores. Just as in microsporogenesis (see p. 569), the callose is believed to temporarily seal off the megaspores, enabling them to embark upon an independent course of development. Callose formation has been reported for certain species with mono- and bisporic types of embryo-sac development. However, to date callose has not been found in species with tetrasporic development (see Bouman, 1984). For a review of the ultrastructural changes that occur during megasporogenesis and development of the embryo sac, see Willemse and van Went (1984) and Cass et al. (1985, 1986).

Types of Embryo-sac Development

Numerous studies have revealed a surprising diversity in the details of megasporogenesis and embryo-sac development within the angiosperms, and various typological classifications have been proposed (see Maheshwari, 1950, and Johri, 1963). The different types of embryo-sac development are distinguished on the basis of the following characters: (1) the number of megaspores or megaspore nuclei which participate in the formation of the embryo sac, (2) the total number of mitotic nuclear divisions during megagametogenesis, (3) the presence or absence of nuclear fusions, and (4) the number, arrangement, and chromosome number of the cells and free nuclei present in the fully mature embryo sac. The chart represented in Fig. 20-15 shows the principal types of megasporogenesis and embryo-sac development discussed in the review by Johri (1963). Although a detailed account of the eleven types represented in this diagram will not be given in this book, selected examples, under each of the three major categories of embryo-sac origin that are generally recognized, will be described. These major categories are *monosporic development* (from a single megaspore), *bisporic development* (from two megaspore nuclei), and *tetrasporic development* (from four megaspore nuclei).

Monosporic Development

Under this category are grouped the *Polygonum* and *Oenothera* types of embryo sacs. Both types begin development from a single megaspore, and there are three successive mitotic divisions during megagametogenesis in the *Polygonum* type but only two in the *Oenothera* type (Fig. 20-15).

In the *Polygonum* type — so named because of its discovery in this genus by Strasburger (1879) — the megaspore farthest from the micropyle is functional, and its first division yields two nuclei which move to the poles of the embryo sac (Fig. 20-16, D–G). Each of these nuclei then divides and a final division of the four nuclei produces a total of eight nuclei, arranged in quartets, at the micropylar and chalazal ends of the embryo sac. Three of the nuclei at the micropylar pole become differentiated as cells and constitute the *egg apparatus* consisting of the *egg* and flanked by the two *synergids*. At the opposite end of the embryo sac, three of the four nuclei differentiate as the *antipodal cells*. The two remaining nuclei — termed *polar nuclei* — migrate from the opposite ends of the *central cell* (the central region of the embryo sac) (Fig. 20-13, J). The polar nuclei may remain separate until the discharge of the male gametes into the embryo sac, or they may fuse, prior to fertilization, to form a diploid *secondary nucleus* (Figs. 20-16, I; 20-22). The pattern of development that we have just described thus produces a seven-celled embryo sac which consists of a three-celled egg apparatus, three antipodal cells, and a central cell with either one or two nuclei.

The development of the *Oenothera* type of embryo sac, now recognized as characteristic of the family Onagraceae, presents a very interesting variant of the monosporic type of development. Usually the micropylar, rather than the chalazal spore of the tetrad, is functional, and a total of two (rather than three) nuclear divisions result in a very distinctive type of embryo sac consisting of a three-celled egg apparatus and a single polar nucleus (Fig. 20-15). Because of the smaller number of mitotic divisions during megagametogenesis, an additional polar nucleus and the antipodal cells are not formed. The *Oenothera* type of embryo sac is of particular morphological interest because, following double fertilization, the primary endosperm nu-

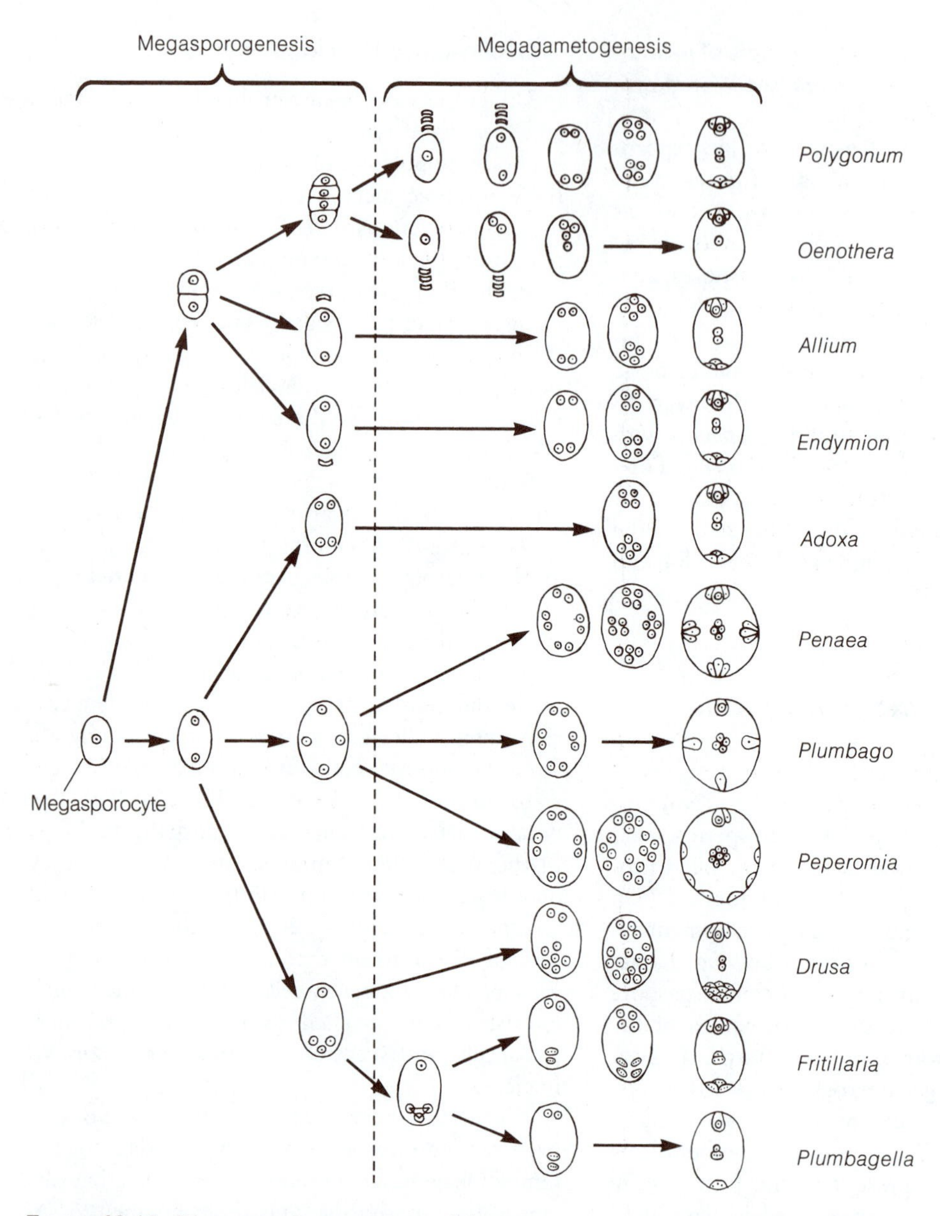

FIGURE 20-15 Diagrammatic comparison of the main types of megasporogenesis and megagametogenesis in the angiosperms. See text for detailed descriptions of selected examples of monosporic, bisporic, and tetrasporic patterns of development. Micropyle is toward the top of the page. [Redrawn and modified from Johri, in *Recent Advances in the Embryology of Angiosperms* (P. Maheshwari, ed.), p. 69. University of Delhi, Delhi. 1963.]

FIGURE 20-16 Megasporogenesis and development of the monosporic type of embryo sac in *Anemone patens.* **A,** ovule primordium with large hypodermal megasporocyte; **B, C,** meiosis I; **D,** linear tetrad of megaspores following meiosis II; **E, F,** enlargement and mitotic division of chalazal megaspore; **G,** binucleate embryo sac; **H,** four-nucleate embryo sac; **I,** mature embryo sac. [Redrawn from *Plant Morphology* by A. W. Haupt. McGraw-Hill, New York. 1953.]

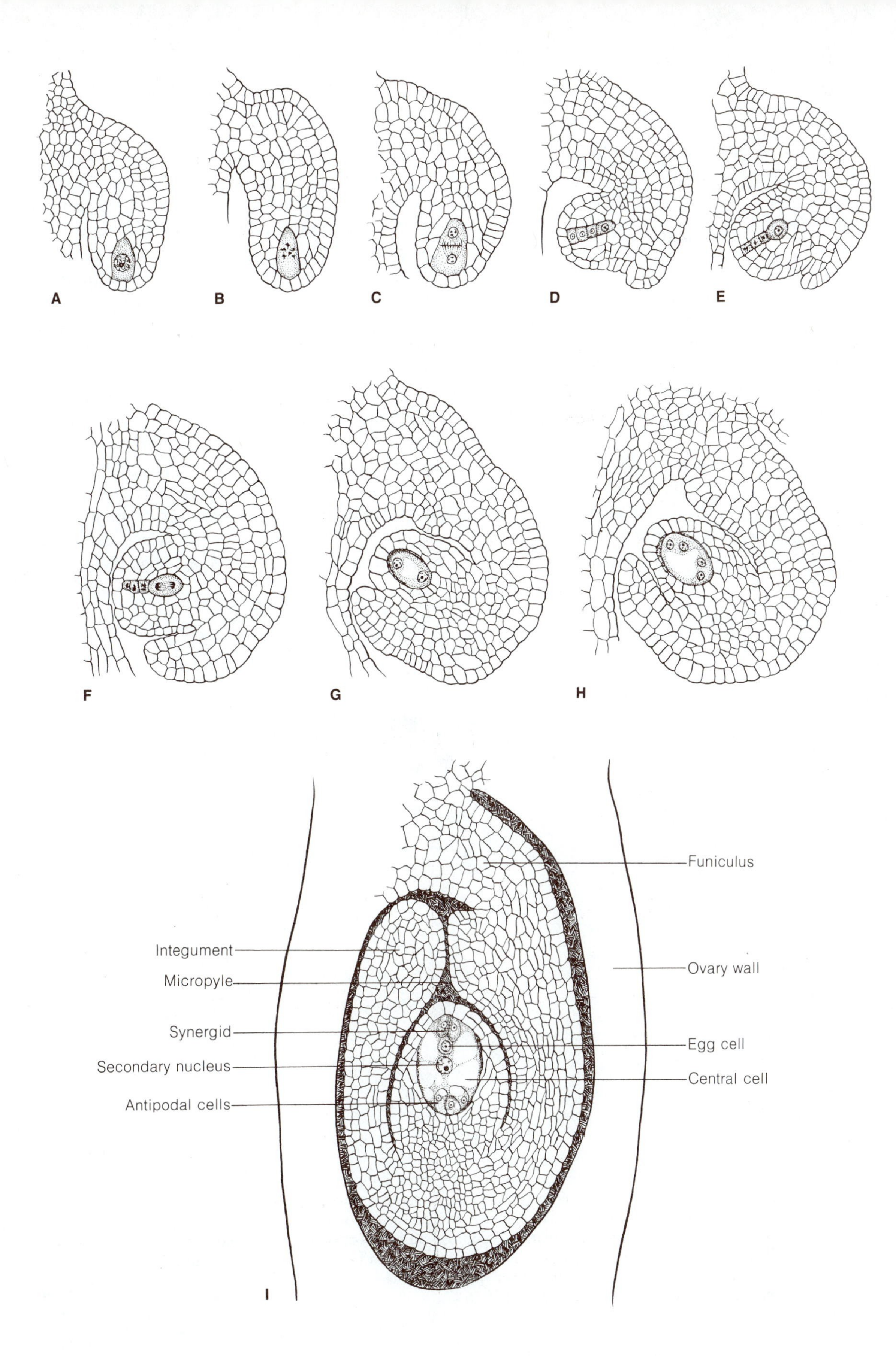
A
B
C
D
E
F
G
H
I
Funiculus
Integument
Micropyle
Ovary wall
Synergid
Egg cell
Secondary nucleus
Central cell
Antipodal cells

cleus is diploid, rather than triploid as is the very widespread *Polygonum* type (Fig. 20-23).

Bisporic Development

In the bisporic type of embryo-sac development, one of the haploid binucleate dyad cells produced during meiosis gives rise to the mature female gametophyte (see Maheshwari, 1955). As the bisporic condition was first described for *Allium fistulosum* by Strasburger (1879), this method of embryo-sac origin is usually designated as the *Allium* type (Fig. 20-15).

The definitive features of the *Allium* type are illustrated by Jones and Emsweller's (1936) description of embryo-sac development in *Allium cepa* (Fig. 20-17). Following the meiotic I division in the megasporocyte (Fig. 20-17, D), two dyad cells, separated by a thin transverse wall, are produced (Fig. 20-17, E). The micropylar dyad cell soon aborts, and meiosis II, without the accompanying formation of a wall, produces two free haploid megaspore nuclei in the chalazal dyad cell (Fig. 20-17, F, G). Two successive mitotic divisions then take place in the functional dyad cell, yielding an eight-nucleate embryo sac which ultimately becomes organized into an egg apparatus, a group of antipodal cells, and two polar nuclei. (Contrast the *Allium* and *Polygonum* types of embryo-sac development shown diagrammatically in Fig. 20-15.)

An additional type of bisporic embryo-sac development was proposed by Battaglia (1958) and termed by him the *Endymion* type. In *Endymion*, a genus allied to *Scilla* (Liliaceae), the embryo sac originates from the micropylar binucleate dyad cell, rather than from the chalazal one (Fig. 20-15). Ac-

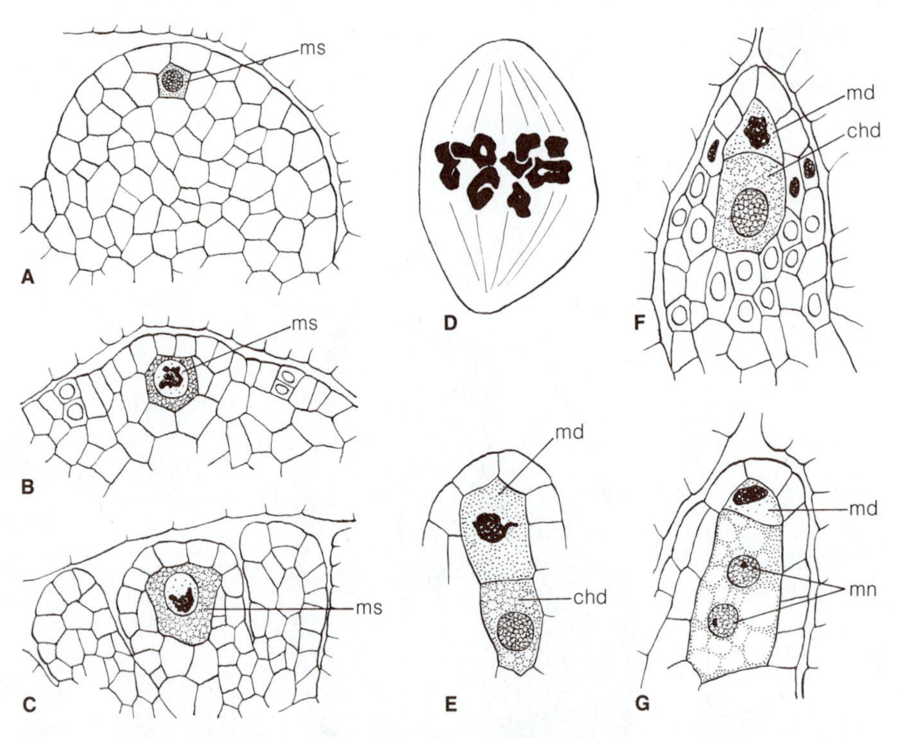

FIGURE 20-17 Origin and early development of bisporic type of embryo sac in *Allium cepa*. **A,** longisection of nucellus showing hypodermal megasporocyte; **B, C,** prophase of meiosis **I; D,** meiosis I division of megasporocyte; **E,** formation of the two dyad cells; **F,** chalazal dyad cell beginning to enlarge, accompanied by abortion of micropylar dyad cell; **G,** formation of two megaspore nuclei in chalazal dyad cell. chd, chalazal dyad cell; md, micropylar dyad cell; mn, megaspore nuclei; ms, megasporocyte. [Redrawn from Jones and Emsweller, *Hilgardia* 10:415, 1936.]

cording to Battaglia, the chalazal dyad cell, by means of two mitotic divisions, may reach a four-nucleate phase before it collapses.

Tetrasporic Development

This category includes a series of complex and puzzling types of embryo-sac development, all of which, however, show two basic characteristics: (1) there is no wall formation during megasporogenesis and (2) all four megaspore nuclei participate in various ways in the formation of the mature embryo sac. Reference to Fig. 20-15 will show that, following meiosis, the megaspore nuclei exhibit three main types of polarity, or arrangement, in the coenomegaspore. In the *Adoxa* type, the megaspore nuclei show a 2 + 2 arrangement—i.e., one pair is situated at the micropylar end and the other pair at the chalazal end of the coenomegaspore. A tetrapolar, or 1 + 1 + 1 + 1, arrangement of megaspore nuclei follows meiosis in the *Penaea, Plumbago,* and *Peperomia* types of embryo-sac development. In the 1 + 3 arrangement a single megaspore nucleus is located at the micropylar end of the sac and the other three nuclei are at the chalazal pole. This peculiar kind of polarity characterizes the *Drusa, Fritillaria,* and *Plumbagella* types of embryo-sac development, and, in the two latter genera, the three chalazal megaspore nuclei fuse later in development. The *Fritillaria* type will now be discussed in some detail, followed by a brief discussion of two other types of tetrasporic development.

FRITILLARIA TYPE. The type of tetrasporic embryo-sac ontogeny illustrated in the genus *Fritillaria* is of particular interest because it occurs in a rather wide variety of genera including *Lilium,* which had long been used in botany courses to illustrate "typical" embryo-sac development and structure in angiosperms. It was only after the careful investigations of Bambacioni (1928a, 1928b) and Cooper (1935) that the true sequence of events in the embryo-sac development of *Lilium* and *Fritillaria* was fully revealed (Fig. 20-15).

In *Lilium* and *Fritillaria,* the four megaspore nuclei resulting from meiosis behave in a very distinctive manner. Three of the nuclei migrate to the chalazal end of the coenomegaspore while the remaining nucleus becomes situated at the micropylar pole. This 3 + 1 arrangement of the megaspore nuclei represents the *first, four-nucleate* stage in the development of the embryo sac of both *Fritillaria* and *Lilium.* These four nuclei next undergo mitosis but the events which occur at the two poles of the coenomegaspore are strikingly dissimilar. The division of the micropylar nucleus is normal and yields two haploid nuclei. In contrast, the spindles formed by the divisions of the three chalazal nuclei first unite to form a multipolar spindle. Then *all* the three sets of chromosomes become arranged on a bipolar spindle, and the two nuclei which are reconstituted are therefore $3n$ in chromosome number. As a result of all of these events a *second, four-nucleate* stage is produced which consists of two haploid micropylar nuclei and two larger triploid chalazal nuclei. A final mitotic division produces a micropylar quartet of haploid nuclei and a chalazal quartet of triploid nuclei. The organization of the mature embryo sac, with reference to the chromosome number of its cells and the two polar nuclei, is strikingly different from the *Polygonum* type. In *Fritillaria* (and *Lilium*), the egg apparatus is haploid, the antipodal cells are triploid, and one polar nucleus (derived from the micropylar quartet) is haploid while the other polar nucleus (derived from the chalazal quartet) is triploid. When double fertilization occurs, a normal diploid zygote is produced, but the primary endosperm nucleus is $5n$ in chromosome number as the result of the union between a male gamete and the two polar nuclei (Fig. 20-23). In the Adoxa type, the two meiotic divisions yield four megaspore nuclei arranged in pairs at opposite poles of the coenomegaspore. Megagametogenesis entails a single mitosis of each megaspore nucleus, producing two quartets of nuclei which then become differentiated to form a *Polygonum* type of embryo-sac organization (Fig. 20-15). The Penaea type presents features of exceptional interest. In this type the four megaspore nuclei are arranged in a tetrapolar—rather than a bipolar—pattern in the coenomegaspore. Each megaspore nucleus divides mitotically resulting in the formation of eight nuclei. Each of these eight nuclei undergoes mitosis, and the embryo sac now consists of sixteen nuclei. Four of the nuclei, one from each quartet, function as polar nuclei and fuse in the center of the sac to form a large, lobed $4n$ secondary nucleus while the other three nuclei in each group become organized as an egg apparatus (Fig. 20-15). According to Kapil (1960), only the egg cell of the micropylar group enlarges and becomes functional, the other three groups ultimately degenerate. Despite this fact, the embryo sac appears to be organized as though in fact it were a "compound" structure made up of four completely "fused" megagametophytes, each consisting of an egg apparatus and a single polar nucleus.

Pollination

In angiosperms the male and female gametophytes are remote and isolated from each other because of the enclosure of the ovules within the ovarian cavity of the gynoecium. Unlike gymnosperms, in which the ovule itself is directly pollinated, pollination in angiosperms involves the transferral of pollen grains from the opened anther of the stamen to the *receptive stigma* of the carpel.

The agents or vectors responsible for pollen transfer in angiosperms have been grouped into two main categories: *abiotic agents*—inanimate forces in nature such as wind currents, gravity, and water—and *biotic agents*—various animal pollinators, with pollination by insects termed *entomophily*, by birds *ornithophily*, by bats *chiropterophily*, and by certain rodents, lemurs, and marsupials (Sussman and Raven, 1978).

What is the primitive mechanism of pollination—entomophily or anemophily, (wind pollination)? The current view is that the ancient angiosperms were entomophilous and that their pollinators very probably were beetles (Coleoptera), known to be one of the oldest groups of insects. Hymenoptera and Lepidoptera, so important in pollination today, appeared later. Among living angiosperms, pollination by beetles is characteristic in a number of primitive dicotyledonous genera such as *Calycanthus*, *Eupomatia*, *Magnolia*, and in some species of Ranunculaceae. Anemophily occurs in such specialized taxa as the Gramineae, Cyperaceae, and in many members of the "Amentiferae" (such as some species of *Quercus*, Betulaceae, Juglandaceae). It should be emphasized, however, that some angiosperms with flowers that were probably wind pollinated occurred in the mid-Cretaceous along with angiosperms that were insect pollinated (Dilcher, 1979). Stebbins (1974) also believes that the earliest angiosperms were probably insect pollinated, and that wind pollination originated several times in relatively primitive groups. Adaptive radiation toward wind pollination was favored in groups of angiosperms inhabiting dry or cold climates, where pollinators were scarce and the winds were strong. Secondary reversion to insect pollination probably took place in some derivatives of wind-pollinated angiosperms that became readapted to moister, more equable climates. [For additional information on pollination biology, see Grant (1950), Faegri and van der Pijl (1979), Baker and Hurd (1968), Proctor and Yeo (1972), Carlquist (1969), and Crepet (1979).]

Among present-day angiosperms, various types of "attractants" seem primarily responsible for the establishment of direct relationships between flowers and their animal pollinators. One of the most evident attractants—and perhaps the oldest in a phylogenetic sense—is pollen, which constitutes the food either of an adult insect visitor or else is used by the insect to feed its brood of larvae. The outstanding examples of this relationship are provided by bees. Nectar, secreted from special glandular organs known as *nectaries*, is also an important (or even primary) source of food for certain groups of insects and birds and doubtless plays a significant role in the syndrome of pollination in many angiosperms. [For descriptions of the varied structure and position of floral nectaries, see Eames (1961), Esau (1965), and Fahn (1982).]

The study of the fascinating and complex interrelationships between inflorescence and flower structure and animal pollinators, especially insects, constitutes an important aspect of modern plant biology. In some angiosperms, there appears to have developed an *obligatory* relationship in which a particular type of plant is dependent upon a particular type of insect for pollination. Classic examples are provided by the relationship between *Yucca* and its moth pollinator *Tegeticula* (= *Pronuba*), the orchid *Ophrys speculum* and its hymenopteran pollinator, and *Ficus* and its wasp pollinator *Blastophaga*. Other functional associations between specific pollinators and flower types have also been widely explored. Readers interested in this rapidly developing field of research should consult the extensive treatment of the intricate

FIGURE 20-18 Pollen on stigmatic heads. **A,** *Hibiscus* sp.: "dry" stigma; note pollen attached to unicellular trichomes (× 30). **B,** *Dendromecon* sp.: "dry" stigma with papillate surface (× 600). **C,** *Mahonia* sp.: "wet" stigma (× 70). **D,** *Mahonia* sp., pollen adhering to secretion on surface of stigma (× 300).

problems given by Faegri and van der Pijl (1979) and the review of intrafloral ecology by Baker and Hurd (1968).

It should be emphasized that a distinction must be drawn on the one hand between the *functional role* that animal pollinators (preeminently the insects) have played in floral evolution and diversification and, on the other hand, the unknown factors in the remote past which were responsible for the *phylogenetic origin* of the flower itself. "Adaptation" of flower structure to specific pollinators is a phenomenon that can be determined by direct field observations and experiments on modern angiosperms, but adaptation per se leaves unsolved the abiding mystery of the origin and early development of the angiosperm carpel. Perhaps the most significant biological role of the carpel in present-day flowering plants is that of the primordium of the *fruit*, a structure of considerable importance in the protection and dispersal of the seeds.

Pollen-Stigma Interaction and Growth of the Pollen Tube

In the strict sense, the term pollination designates the actual transfer of pollen to some type of receptive surface of the carpel. Fertilization, on the other hand, is a process of gametic fusion that takes place within the confines of the embryo sac. But as the ovules with their enclosed female gametophytes are situated within the ovary, there is inevitably an "intermediate phase" in the process of sexual reproduction which includes (1) the germination of the pollen grains on the stigma, and (2) the pollen tube's traversing a relatively long path through the style before finally penetrating the tissue of the ovule and entering the embryo sac.

Pollination triggers a series of intricate interactions between pollen and the stigma. The pollen exine is composed of sporopollenin that is resistant to biodegradation. The cavities in the exine are filled with proteins (mainly enzymes), glycoproteins, lipids, and certain allergens derived from the tapetum of the pollen sacs late in the maturation of pollen grains. The proteins carry sporophytic recognition factors because of their origin from the tapetum. In contrast, the proteins and polysaccharides of the pollen intine are products of the haploid spore and hence are of gametophytic origin. The proteins include a variety of enzymes.

Some angiosperms have "wet" and others have "dry" stigmas (Heslop-Harrison and Shivanna, 1977). In the wet type there is often a copious amount of a free-flowing secretion on the surface, generally consisting of proteins, amino acids, and lipids. In the dry type there is a hydrated layer on the surface of the stigmatic cuticle consisting of proteins, carbohydrates, and a small amount of lipid. These substances are of sporophytic origin from cells at the surface of the stigma. Pollen grains, after becoming attached to the stigma (Fig. 20-18), hydrate and swell within a few minutes and release the wall-held proteins and the other components. By this process, "recognition" factors from the pollen and stigma come into contact with each other, determining whether a pollen grain will germinate (by formation of a pollen tube) or not. About two-thirds of the families of angiosperms exhibit self-incompatibility whereby germination and tube growth are inhibited as a mechanism to favor outbreeding. A pollen grain may germinate but pollen-tube growth is arrested at the stigmatic surface or in some instances within the style. Recognition in the style is controlled by the interaction of substances produced by the male gametophyte and the stylar tissue (Fig. 20-19).

This overview by no means covers the large volume of literature on the subject or the advances being made to determine the genetic basis for pollen-stigma-style interactions. The reader is referred to the following references for additional information: Heslop-Harrison (1975), Knox (1979, 1984), and Went and Willemse (1984).

Subsequent to mutual recognition and acceptance of pollen grains, the pollen tubes grow between cells of the stigma and into a column of tissue in the style termed the *transmitting tissue*, the cells of which have many of the cytological characteristics of the stigmatic cells. The pollen tubes grow between cells of the transmitting tissue and are nourished by secretions from the transmitting tissue. In many monocotyledons and certain groups of dicotyledons the style is hollow and the transmitting tissue is represented by the glandular epidermis that lines the stylar canal. In such "open" styles, growth of a pollen tube may be along the epidermis or the canal may be filled with a secretion through which the pollen tubes grow on their way to the ovary.

An interesting part of the physiology of pollen tubes is the synthesis, within the growing tube, of *callose*. In germinating pollen grains and in elongating pollen tubes, callose occurs not only in the innermost lamella of the intine but is locally deposited in the form of *callose* plugs. These latter remarkable structures have been observed in a wide variety of angiosperms, both in tubes growing natu-

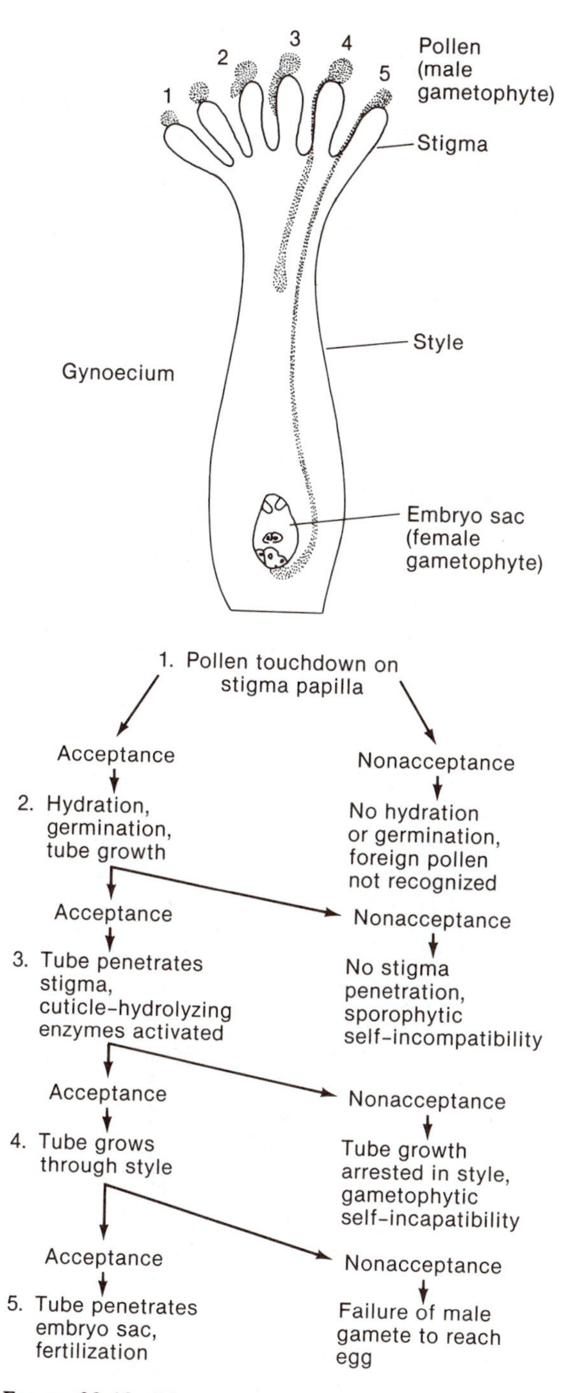

FIGURE 20-19 Diagrammatic representation of the behavior of pollen in terms of events leading to a compatible pollination and the various incompatibility options. [Redrawn from R. B. Knox, *McGraw-Hill Yearbook of Sci. and Technol.*, 1979.]

rally within the transmitting tissue of the style and in pollen tubes grown on artificial media. A good example of the abundant formation of callose is provided by Borthwick's (1931) study of carrot, *Daucus carota* (Fig. 20-20). He found that plugs were most conspicuous in those portions of the tubes growing in styles and were 3 to 10 micrometers in length and spaced at intervals of approximately 25 micrometers. Portions of the tubes that

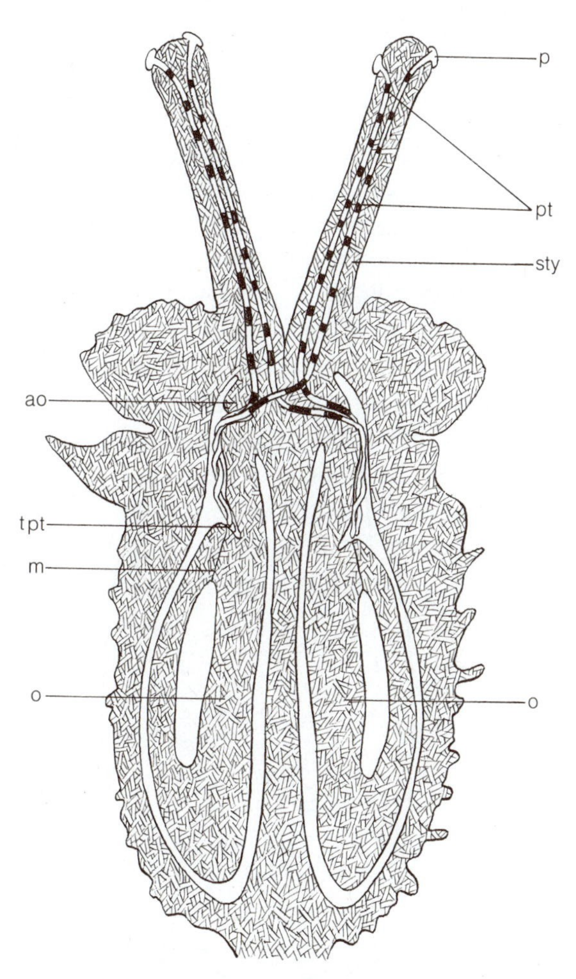

FIGURE 20-20 Diagram of median longisection of gynoecium of *Daucus carota* showing the course of the pollen tubes. The series of black areas in each of the pollen tubes represent callose plugs. ao, abortive ovule; m, micropyle; o, functional ovule; p, pollen grain on stigma; pt, pollen tubes with callose plugs (black areas); sty, style of bicarpellate gynoecium; tpt, tip of pollen tube. [Redrawn from Borthwick, *Bot. Gaz.* 92:23, 1931, University of Chicago Press.]

had reached the ovarian cavity or the micropyle of the ovule showed few or no callose plugs, yet at this region the entire wall of the tube is frequently thickened with callose. The pollen tube grows at the tip and is devoid of callose. The vegetative nucleus and generative cell, or the male gametes (if the generative cell has divided) remain near the tip and are not isolated between callose plugs.

The physiological significance of callose formation—particularly the formation of plugs—is problematic. It is evident that the formation of plugs is not limited to tubes grown in vitro, since callose plugs also form within tubes situated in tissue of the style (Fig. 20-20). One theory advanced as to the significance of callose plugs is that the formation of a succession of plugs serves to isolate the terminal gamete-containing portion of the pollen tube, and thus serves to maintain its integrity and restrict the requirements for growth to only the tip portion of the tube.

Entry of the Pollen Tube into the Ovule

After reaching the ovary, the pollen tube may enter an ovule by several possible routes. Most commonly the tip of the pollen tube enters the micropyle and pushes through the nucellar tissue to the egg-apparatus end of the embryo sac. This type of ovule penetration is designated as *porogamy*. But some taxa (e.g., *Casuarina* and several members of the "Amentiferae") are said to be *chalazogamous* because the tip of the pollen tube penetrates the chalazal end of the ovule, and then continues its growth along the surface of the embryo sac before reaching the egg apparatus. Chalazogamy was formerly considered the primitive condition in angiosperms but at present the exact mode of ovule penetration in a given taxon is believed to have physiological rather than phylogenetic significance. A very peculiar—and apparently rare—mode of

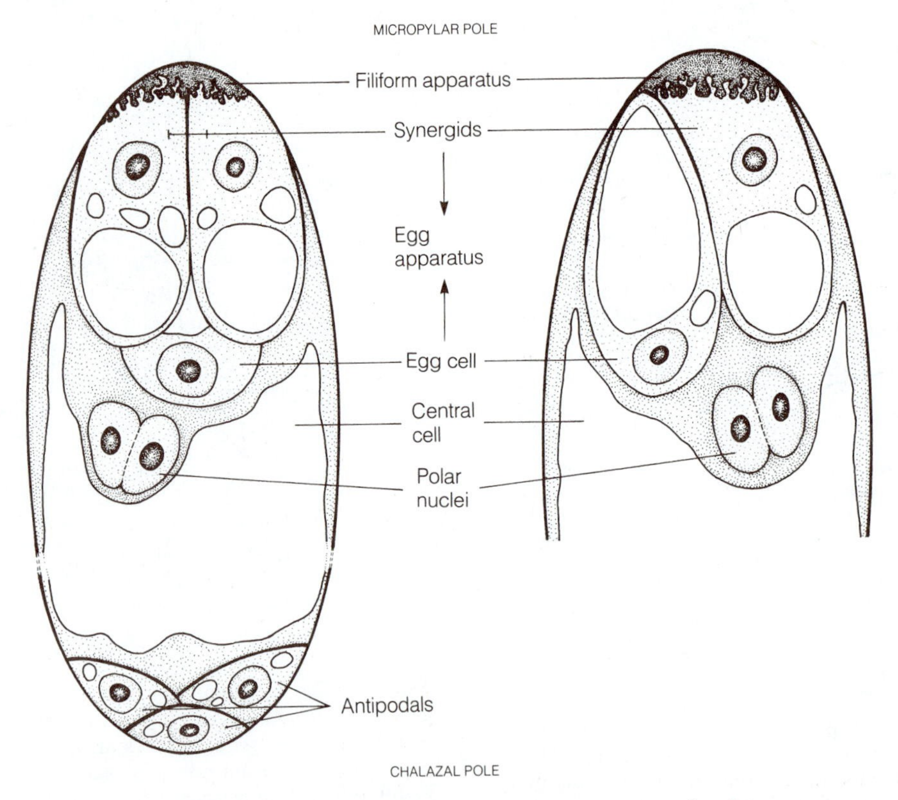

Figure 20-21 Diagrammatic representation of the common type of embryo sac at maturity. If a cell wall is distinctly present, the line is thickened and includes the plasma membrane. [Redrawn from Willemse and van Went, in *Embryology of Angiosperms* (B. M. Johri, ed.). Springer-Verlag. Berlin. 1984.]

entry of the pollen tube is termed *mesogamy* and is illustrated in the monotypic *Circaeaster agrestis* (Junell, 1931). In this plant, the pollen tube penetrates laterally the single integument of the ovule before reaching the egg apparatus of the embryo sac.

The final phase in the growth of the pollen tube consists of its entry into the micropylar end of the embryo sac. There are exceptions to micropylar penetration as noted previously. In the prevailing type of embryo sac, the egg apparatus consists of two synergid cells and the egg cell. Electron microscopy has shown that the synergids contain a large number of organelles and numerous ribosomes which are indicative of high metabolic activity. One additional characteristic is the presence of a *filiform apparatus* at the extreme micropylar end of each synergid. The apparatus consists of a convoluted and thickened cell wall which results in an increased plasma membrane surface area (Fig. 20-21). It is assumed, with some supporting evidence, that the synergids are the origin of a chemotropic secretion that results in directional growth of the pollen tube into the micropyle and nucellar tissue (Fig. 20-22, B). The pollen tube grows through the filiform apparatus and into one synergid (Fig. 20-22, C). Degeneration then begins in the penetrated synergid. In some species one of the synergids may begin to show signs of degeneration before the pollen tube arrives. After penetration of the synergid, most of the contents of the tube, including the two male gametes and the vegetative nucleus are released into the synergid. Subsequent to discharge of the pollen tube contents, two darkly stained bodies have been observed that originally were designated X-bodies because of the uncertainty of their origin (Fig. 20-22, D). Subsequently, it was shown that they contain DNA and were identified as degenerating nuclei of the synergid and of the vegetative nucleus of the pollen tube (Fisher and Jensen, 1969).

Fertilization

One of the outstanding characteristics of angiosperms is the participation of *both* male gametes in an act of fusion: one unites with the egg to form a diploid zygote, from which the *embryo* originates, while the other gamete fuses with one or several polar nuclei (or with the secondary nucleus) to form the *primary endosperm* nucleus, from which the *endosperm* tissue originates (Fig. 20-22). These events constitute the process of *double fertilization.* The mechanisms are not well known by which one male gamete moves to the egg-cell membrane and enters the egg cytoplasm prior to fusion with the egg nucleus nor by which the other one enters the central cell prior to fusion with the polar nuclei (or with the secondary nucleus) in the formation of the primary endosperm nucleus (see Went and Willemse, 1984). There is some evidence that male gametes are deposited in the embryo sac as complete cells (Cass and Jensen, 1970; Russell and Cass, 1983) which could have a bearing on plastid inheritance. [For additional information, see Jensen (1972, 1973) and Russell (1982).]

Endosperm

The term "endosperm" designates the tissue, formed during the development of an angiosperm seed, that provides essential nutrients utilized in the growth of the embryo and, in many cases, of the young seedling during seed germination. Aside from a few angiosperms in which endosperm fails to develop or else degenerates early in ontogeny, the presence of endosperm is a salient and consistent feature of seed ontogeny. In some angiosperms (such as beans, peas, and vetch) the endosperm is completely digested during embryogenesis. In such *exalbuminous* seeds, the nutrients are stored in the fleshy cotyledons. Other angiosperms (such as wheat, corn, onions, and palms) have *albuminous* seeds, in which copious amounts of endosperm tissue are present at the time of seed germination. The histological structure of endosperm and the types of reserve nutrients in its cells vary widely. The cells may store starch, proteins, lipids, and growth regulators in varying proportions, depending upon the species (Chopra and Sachar, 1963; Jacobsen, 1984). In certain palms (e.g., the so-called ivory-nut palm) the endosperm cells develop very thick walls which are composed of hemicellulose, an important reserve-food material during germination (Corner, 1966).

The endosperm of the palm, *Cocos nucifera,* is interesting in that it consists of both a cellular component and "coconut milk" which contains free

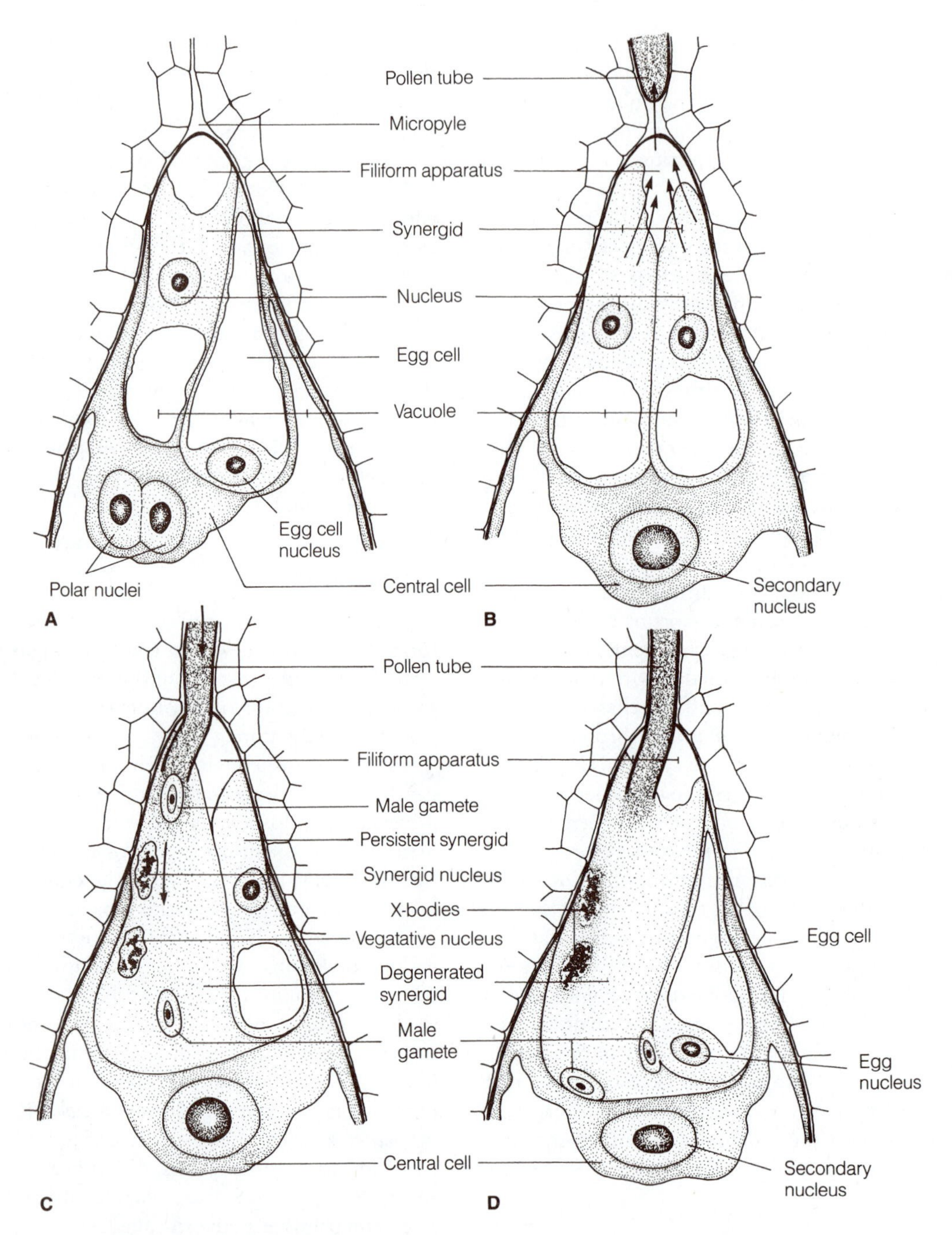

FIGURE 20-22 Schematic representation, pollen tube entrance into the embryo sac, tube discharge, and male gametic transfer. **A,** micropylar portion of mature embryo sac; note fusion of polar nuclei to form secondary nucleus; **B,** proposed chemotropic activity of synergids, resulting in directed growth of pollen tube; **C,** discharge, contents of pollen tube; **D,** fusion of a male gamete with egg membrane, leading to fertilization of egg nucleus by one of the male gametes; after passage of second male gamete through synergid membrane, it will fuse with the secondary nucleus resulting in the formation of the primary endosperm nucleus. See text for details. [Modified from van Went and Willemse, *Embryology of Angiosperms* (B. M. Johri, ed.), p. 299. Springer-Verlag, Berlin. 1984.]

nuclei, protein granules, oil droplets, and growth hormones. Coconut milk has proved beneficial—or essential—as part of the media used for in vitro cultures of somatic embryos and excised plant tissue (Steward, 1968). The activity of coconut milk at the biochemical level, however, is still an enigma. Today, most investigators use well-defined media for in vitro cultures.

The initiation of endosperm normally requires an act of fusion between a male gamete and one or more of the polar nuclei in the embryo sac (Fig. 20-23). This process is one of the most definitive features of the reproductive cycle of angiosperms and indeed forms a most striking point of comparison with the type of seed ontogeny characteristic of the gymnosperms. In the latter, the nutritive tissue of the seed is already present *before fertilization* as the massive cellular female gametophyte. In other words, the nutritive tissues in gymnosperm and angiosperm seeds are *analogous* rather than *homologous,* from a morphological point of view. It seems necessary to emphasize this point because even in certain modern texts and research articles, the term endosperm is still used in a loose sense for the reserve-food-containing gametophytic tissue of gymnosperm seeds.

Extensive comparative studies have revealed considerable variation in the mode of development of endosperm, and the following resume presents the salient features of the three main ontogenetic types that are now recognized.

Types of Endosperm Development

NUCLEAR TYPE. In the nuclear type, the mitotic division of the primary endosperm nucleus is followed by a variable number of subsequent free nuclear divisions without wall formation; in some plants several hundred nuclei may be produced. During the preliminary phase of development the center of the embryo sac (central cell) is frequently occupied by a large vacuole, and the nuclei lie peripherally in the cytoplasm between the embryo-sac cell wall and the vacuolar membrane of the central cell (Fig. 20-29, B, C). Although the entire development of nuclear endosperm in certain species is limited to the production of free nuclei and the

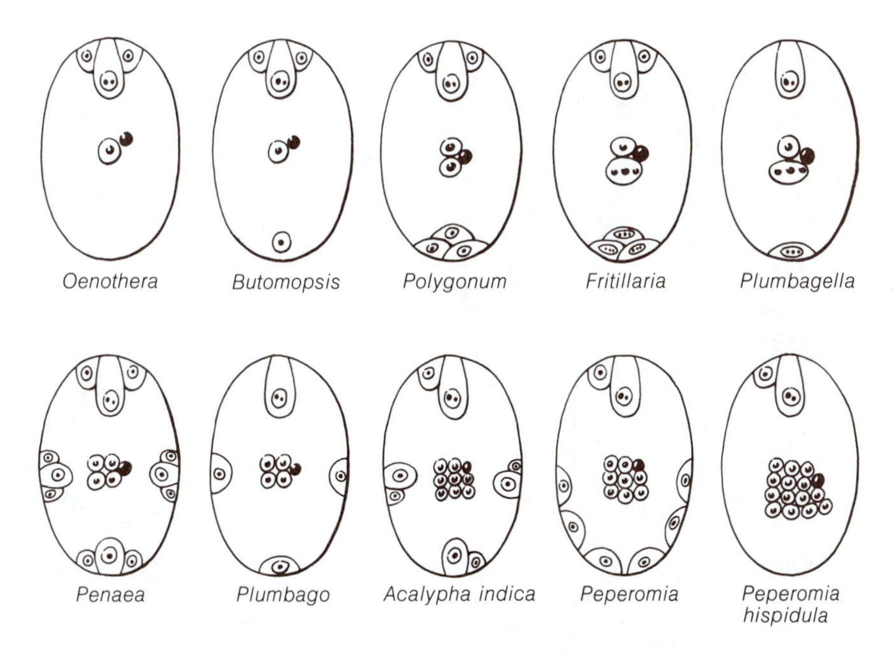

FIGURE 20-23 Embryo sacs. Note the wide variation in the number of polar nuclei that join with the male gamete (shown in black) in the initiation of endosperm. [Adapted from *An Introduction to the Embryology of Angiosperms* by P. Maheshwari. McGraw-Hill, New York. 1950.]

resultant formation of a multinucleate mass of protoplasm, in most species there is a second phase in development. This consists of the centripetal formation of cell walls much like the development of the female gametophyte in cycads and *Ginkgo* (Chapters 14, 15). In such cases all or most of the endosperm is converted into a cellular tissue (Figs. 20-29, D; 20-30, B). The endosperm cells can increase markedly in size and DNA content through, for example, endomitosis (without spindle formation) resulting in endopolyploidy, or by endoreduplication of DNA. According to the species, nuclei can range from 192*n* to 384*n* or even 3,072*n* (D'Amato, 1984). Increases in DNA are undoubt-

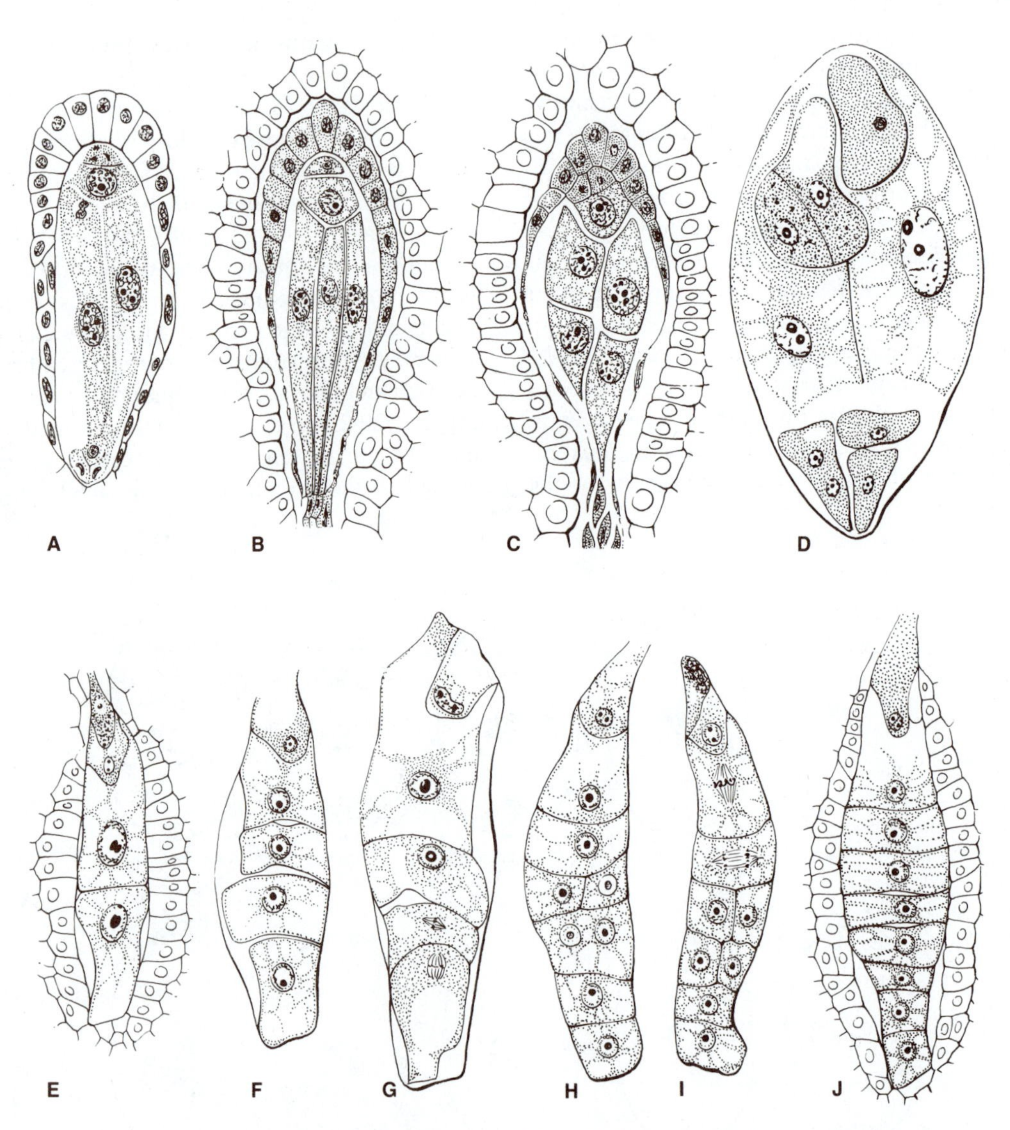

FIGURE 20-24 Early stages in development of the cellular type of endosperm in various angiosperms. In **A** and **D,** the plane of the first wall is longitudinal; in **E,** the first division wall is transverse. **A–C,** successive stages in *Adoxa moschatellina;* **D,** first longitudinal wall in *Centranthus macrosiphon;* **E–J,** successive stages in *Villarsia reniformis.* [Adapted from *An Introduction to the Embryology of Angiosperms* by P. Maheshwari. McGraw-Hill, New York. 1950.]

edly designed to enhance macromolecule synthesis, especially proteins and carbohydrates required for growth of the endosperm and also later for growth of the embryo.

CELLULAR TYPE. The cellular type is well demarcated from the nuclear type because division of the primary endosperm nucleus is followed by the formation of either a longitudinal or transversely oriented cell wall in the central cell (Fig. 20-24, A, E). When the plane of the first division is transverse, the plane of the second and sometimes of the third division may also be transverse, resulting in a row of cells. In many species the early sequence of transverse and longitudinal divisions do not conform to a set pattern, and the arrangement of cells is variable and irregular.

An interesting feature of the cellular type (although not confined to this type) is the formation of *haustoria* that penetrate various parts of the ovule, such as the integuments and nucellus. The haustorial processes may be extensions of cells or multicellular. The nuclei of the haustoria may undergo several cycles of DNA synthesis and are rich in cytoplasmic RNA and proteins. It is assumed, with some evidence, that the haustoria have an absorptive function and transfer nutrients from surrounding tissues to the endosperm proper rather than to the embryo during early embryogeny (see Vijayaraghavan and Prabhakar, 1984).

HELOBIAL TYPE. The helobial type of endosperm development was first designated as such by Schnarf (1929), probably because of its prevalence among those monocotyledons classified under the order Helobiae (Helobiales), a taxon recognized by plant systematists at the time. The helobial type is distributed throughout the monocotyledons, and according to Swamy and Parameswaran (1963) no dicotyledon duplicates the basic ontogenetic events found in monocotyledons. Thus, they consider the helobial type to be a distinctive feature of monocotyledons.

The mode of initiation and the early method of formation of helobial endosperm is a remarkably uniform process. Swamy and Parameswaran (1963) recognize a basic type, or "norm," and five principal variants or deviations (Fig. 20-25). In the norm

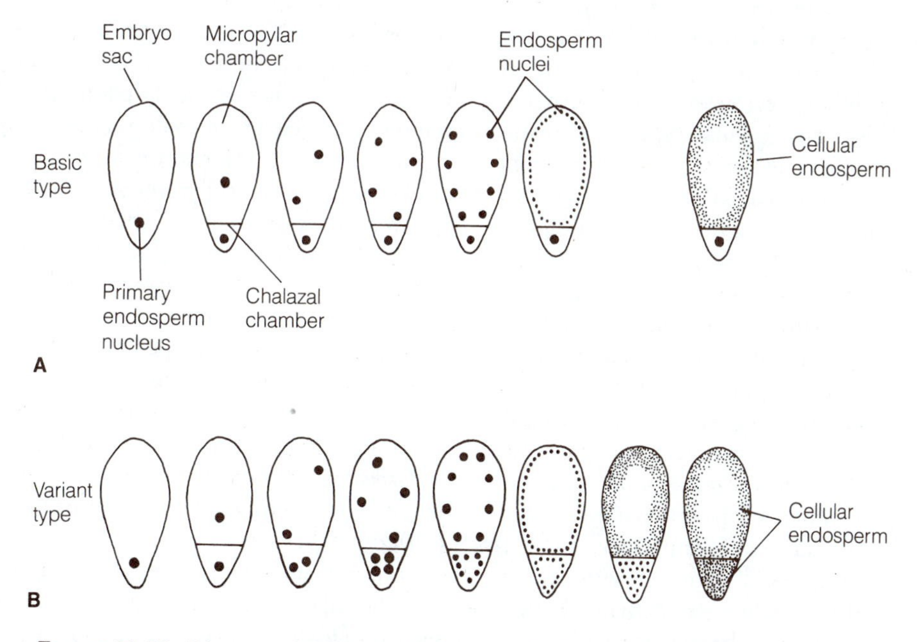

FIGURE 20-25 Diagrams showing the basic type of helobial endosperm development (**A**) and one variation (**B**). In each developmental series, the formation of a transverse wall divides the embryo sac into a small lower (i.e., chalazal) chamber and a larger micropylar chamber. [Adapted from Swamy and Parameswaran, *Biol. Rev.* 38: 1, 1963.]

as well as in all deviations from it, the primary endosperm nucleus is always found at the chalazal end of the embryo sac. When the primary endosperm nucleus divides, a transverse wall is produced, dividing the sac into a small chalazal cell and a much larger micropylar cell. In the norm, the nucleus in the chalazal cell does not divide further, and the cell appears to function as a haustorium in later stages of development. In contrast, there are numerous free nuclear divisions in the larger micropylar cell. The free nuclei may degenerate or, more commonly, wall formation may take place (Fig. 20-25, A).

The deviations from the norm fundamentally represent a series of increasing numbers of free nuclear divisions in the chalazal cell, although they are never as numerous as in the micropylar cell. Generally the chalazal chamber does not become cellular, or may do so in one of the variants (Fig. 20-25, B).

Phylogenetic Relationships Between the Types of Endosperm

The phylogenetic interpretation of nuclear, cellular, and helobial types of endosperm represents a most difficult and controversial problem in angiosperm morphology. Maheshwari (1950), for example, considered that the helobial type is "intermediate" in its developmental pattern between the nuclear and cellular types, but maintained that "whether the series is to be read from the nuclear toward the cellular type or vice versa is not clear." Other authors have taken a more positive stand. Sporne (1980), using computerized statistical methods on many families, found that the nuclear-type endosperm is significantly correlated with a group of reputedly primitive characters in dicotyledons, including woody habit, presence of stipules, free petals, numerous stamens, ovule with two integuments, vascular tissue in integuments, and crassinucellate ovules. On the other hand, nuclear endosperm is involved in certain negative correlations of high statistical significance, such as heterogeneous rays, vessel elements with scalariform pitting, and apotracheal parenchyma. Earlier, Sporne (1954) maintained that nuclear endosperm is a primitive character, but now considers it to be an enigma phylogenetically, and remains more so now than twenty years ago (Sporne, 1967, 1980).

On the basis of a detailed review of the reported occurrence of helobial endosperm in dicotyledons, Swamy and Parameswaran (1963) maintain that true helobial endosperm is restricted to the monocotyledons. The frequent reports, many very old, of its presence in certain dicotyledons are due to inadequate and often incorrect interpretations. The authors concluded that the helobial-type endosperm is not intermediate in pattern of development, but rather is a distinctive feature of monocotyledons. The same conclusions were reached by Swamy and Krishnamurthy (1973).

Obviously additional survey work is needed on the typology of endosperm as well as on the morphogenetic role of various factors within the ovule that may shape the pattern of its development. There is also a need for additional experimental studies on the biochemical factors that are necessary for the expression of the varied potentialities of this remarkable tissue when grown in vitro. [See Johri and Rao (1984) for a review of experimental studies on endosperm.]

RUMINATE ENDOSPERM. Among the other types of endosperm morphology that merit brief consideration is the peculiar one known as ruminate endosperm, which has been observed in members of at least thirty families of dicotyledons as well as in a number of genera of palms (Periasamy, 1962; Corner, 1966). In the mature seeds that form this type of endosperm, the outer surface of the endosperm tissue is irregular or furrowed to varying degrees. Ontogenetic studies have shown that the irregular contour of ruminate endosperm is the result of either the formation of outgrowths from the endosperm surface or the penetration, i.e., the "invasion," of the endosperm by portions of the integument of the ovule. A good example of the latter method of origin is found during seed development in *Degeneria*. According to Swamy (1949a) the ruminate endosperm of this plant is produced by localized patches of cells of the innermost layers of the outer integument that grow as wedges of tissues into the adjacent endosperm (Fig. 20-26, A). As the seed of *Degeneria* matures, the inner part of the outer integument and its protruding wedges differentiate into a hard sclerenchymatous tissue, which forms the stony inner seed coat (Fig. 20-26, B).

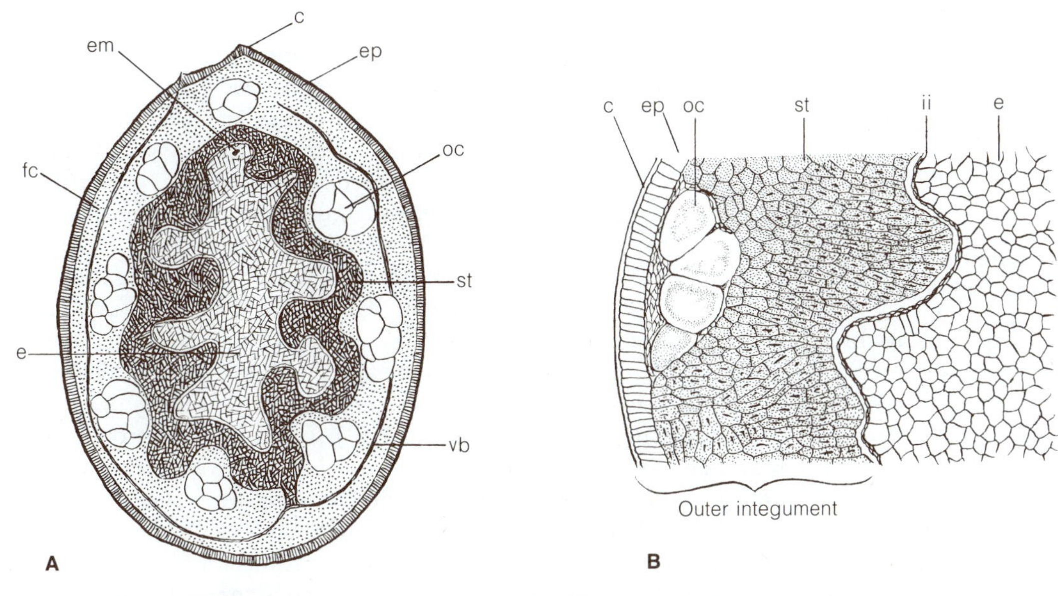

FIGURE 20-26 Structure of the mature seed of *Degeneria*. **A,** diagrammatic longisection showing minute embryo and ruminate endosperm; **B,** portion of mature seed, showing intrusion of stony layer of seed coat into the adjacent cellular endosperm. c, cuticle; e, endosperm; em, embryo; ep, epidermis of seed coat; fc, fleshy layer of seed coat; ii, inner integument; oc, oil-bearing cells; st, stony layer of seed coat; vb, vascular bundle. [Redrawn from Swamy, *Jour. Arnold Arboretum* 30:10, 1949.]

Interpretations of Endosperm

Beginning with the classic studies of the past century and extending to the present, the significance of the virtually universal formation of endosperm in angiosperms has been the subject of controversy and the most varied interpretations. Strasburger (1900) regarded the process of fusion of one of the two male gametes with two polar bodies as a necessary stimulus for the rapid development of the nutritive endosperm. He interpreted the endosperm phylogenetically as equivalent to the tissue of the female gametophyte that has been delayed in its development. A similar interpretation was later adopted by Coulter and Chamberlain (1912). On the other hand, if the fusion of a male gamete with two polar nuclei is regarded as an act of true fertilization, or syngamy, the endosperm might be regarded as a malformed or "unorganized" second embryo, the normal development of which was prevented by the participation of the chalazal polar nucleus in the formation of the triploid primary endosperm nucleus. This ingenious theory, proposed by Sargant (1900), obviously is not valid as an explanation of endosperm development in *Oenothera* and *Butomopsis,* in which the male gamete fuses with a single polar nucleus and hence the primary endosperm nucleus, like the zygotic nucleus, is diploid. In these remarkable instances it would be interesting, with the use of in vitro culture techniques, to determine whether the diploid primary endosperm nucleus could experimentally be induced to form a normal embryo, comparable in development to the embryo formed by the zygote.

The "fertilization" of polar nuclei has been interpreted from a genetic-physiologic point of view by Brink and Cooper (1947). According to their hypothesis (1) the fusion of polar nuclei and a male gamete is essential for the inception and rapid development of endosperm and (2) as endosperm nuclei contain both maternal and paternal chromosomes, endosperm possesses *hybrid vigor* and hence is a physiologically aggressive tissue, undergoing early development before cell division begins in the zygote. But as Maheshwari (1950) has pointed out, the theory of Brink and Cooper fails to explain why endosperm is sooner or later "digested" by the diploid embryo. It has been suggested that the advantage of endosperm becoming cellular could be a way of more efficiently storing nutrients for use later by the embryo.

Embryogeny

Origin and Development of the Proembryo

In its earliest stage of development the embryo of angiosperms is usually designated as the *proembryo.* The application of this term is, however, largely a matter of descriptive convenience because only an arbitrary boundary can be drawn between the "proembryo" and the subsequent "embryo proper." Johansen (1950) defines the proembryo of angiosperms as "a more or less filamentous row of cells constituting the earliest phase of embryogenesis. It ends with the fourth cell generation when it consists of approximately sixteen cells." In contrast, the French embryologist Souèges, according to Crété (1963), extended the concept of the proembryo to include those early stages of embryogeny prior to the appearance of the cotyledonary primordia. In view of the great variation in the number, arrangement, and future "destiny" of the cells of very young embryos, it would appear that Souèges' concept is more appropriate and flexible in the light of our present knowledge of comparative embryogenesis.

The extensive literature on embryogenesis in angiosperms indicates that the development of the endoscopic embryo begins usually with a transverse division of the zygote (Figs. 20-28, 20-32, and 20-34). Exceptions to the prevailing mode of embryo initiation have been observed and are usually classified as *irregular types.* In *Juglans regia,* for example, the plane of the first division of the zygote may be transverse *or* more-or-less conspicuously oblique (Nast, 1941). Additional examples of irregular types are provided by the vertical or oblique plane of the first division (or divisions) in the zygote of *Scabiosa* and members of the family Piperaceae (see Maheshwari, 1950; Johansen, 1950; and Crété, 1963). The morphogenetic or phylogenetic significance of vertical or oblique division planes in the initiation of an embryo is obscure at present, although Crété considers the peculiar embryogenesis of *Scabiosa* to be "primitive."

It was briefly emphasized in our general account of embryogeny in vascular plants (see Chapter 6) that the genus *Paeonia* exhibits a type of embryonic development remarkably different from that of the other angiosperms which have been investigated. The unique feature of *Paeonia* is a preliminary phase of *free nuclear divisions* during the early development of the proembryo (Fig. 20-27, A–E). The number of nuclei formed in the coenocytic proembryo varies, not only between different species but also within a single species. According to Cave et al. (1961), six or seven synchronous mitoses yield a proembryo of 64 to 128 free diploid nuclei. Carniel's (1967) investigations revealed that occasionally as many as 256 free nuclei may develop. Eventually, by means of wall formation, the multinucleate proembryo becomes differentiated into a cellular proembryo which, however, appears to lack the well-defined endoscopic polarity characteristic of the proembryo of other angiosperms and of those gymnosperms that show a free nuclear phase in their early embryogeny (Fig. 20-27, F). In *Paeonia,* numerous *embryo primordia*—as many as twenty-five in the species studied by Yakovlev (1967)—begin to differentiate at the periphery of the cellular proembryo (Fig. 20-27, G). Usually only one of these embryo primordia survives and gives rise to the single dicotyledonous type of embryo found in the maturing seed of *Paeonia* (see Cave et al., 1961, and Carniel, 1967, for details). Yakovlev (1967) lists *Paeonia* as an outstanding example among angiosperms of "proembryonal polyembryony."

The original discovery of the free nuclear phase in the development of the proembryo of *Paeonia* was made by Yakovlev and Yoffe (1957). Their observations have been fully confirmed by a series of independent investigations (see Cave et al., 1961; Walters, 1962; Matthiessen, 1962; Wunderlich, 1966; and Carniel, 1967). The only disagreement in the interpretation of the facts was voiced by Murgai (1959, 1962) who maintained that (1) the first division of the zygote is accompanied by the formation of a transverse wall and (2) free nuclear divisions are restricted to the basal cell, termed the "suspensor haustorium," from which the cellular proembryo

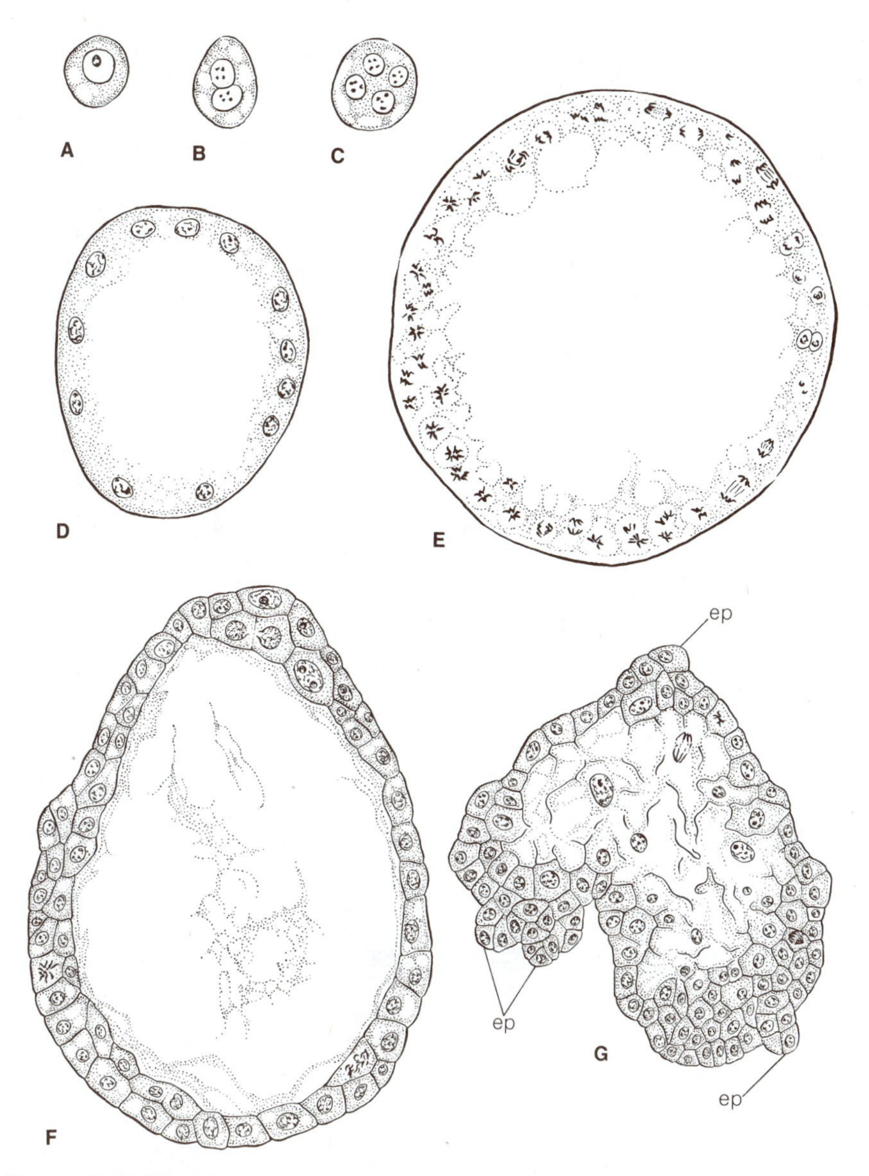

FIGURE 20-27 Development of the proembryo of *Paeonia.* **A,** zygote; **B–E,** successive stages in free-nuclear divisions in young proembryos; **F,** early stage in cell formation; **G,** later stage in development of cellular proembryo, showing several peripheral embryo primordia (ep). [Redrawn from Wunderlich, *Osterr. Bot. Zeit.* 113:395, 1966.]

later takes its origin. However, the work of Cave et al. strongly indicates that a persistent synergid, appressed to the lower side of the zygote, may have been incorrectly interpreted by Murgai as the small apical cell of a two-cell proembryo.

Although Yakovlev and Yoffe (1957) believed that the coenocytic proembryo of *Paeonia* may provide a connecting link between the usually divergent embryogenesis of gymnosperms and angiosperms, this interpretation has been criticized by several other investigators. Cave et al. (1961), for example, regard the multinucleate phase of the proembryo of gymnosperms and *Paeonia* as an example of convergent evolution; they conclude that the peculiar free nuclear phase in the development of the proembryo of *Paeonia* is a derived rather than a primitive character. A similar type of conclusion was also reached by Matthiessen (1962) and Wunderlich (1966). However, Stebbins (1974) is of the opinion that the initially coenocytic proembryo is vestigial and not a derived condition because *Paeonia* displays many other primitive characteristics and its low chromosome number renders unlikely its descent from any modern family.

Comparative Embryogeny of Dicotyledons

During the present century much attention has been paid to the early phases of development of the proembryo. Elaborate and complex embryonic classifications have been devised, based on the origin, position, and histogenic role of specific cells, or of cell tiers. Special emphasis has been placed on the plane in which the *terminal cell* divides during the

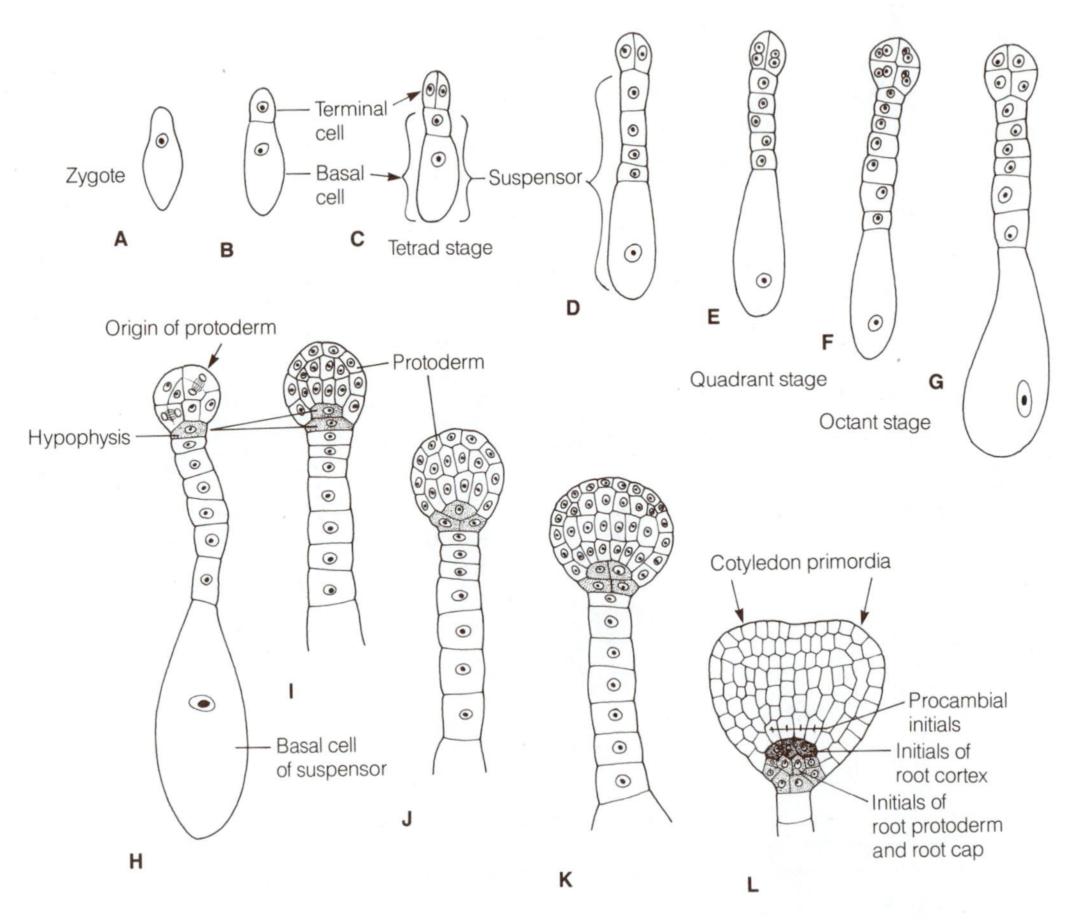

FIGURE 20-28 Early embryogeny in *Capsella bursa-pastoris*. **A, B,** division of zygote into terminal and basal cell; **C,** tetrad stage, the result of a longitudinal division in the terminal cell and a transverse division of the basal cell, resulting in the establishment of the suspensor; **D–F,** establishment of the octant stage; two of the nuclei in **E** and two nuclei in each tier in **F** are nuclei of cells in which the cell walls are in the plane of the page; **G,** octant stage as seen in sectional view; **H,** origin of protoderm and differentiation of hypophysis; **I–L,** results of division of the hypophysis in the establishment of root initials and the origin of the cotyledons. [Based on Souèges, *Ann. Sci. Nat. Bot. Ser.* 10:1–28, 1919.]

formation of the four-celled proembryo, or "tetrad," as it is sometimes called. Most commonly, the terminal cell of the young proembryo divides, forming either a vertical wall, as in *Capsella* (Fig. 20-28, C), or a transverse wall, as in *Nicotiana* (Fig. 20-32, B, C). However, embryological research also has revealed examples of dicotyledonous proembryos in which the plane of division of the terminal cell is clearly oblique. On the basis of these three possible planes of division of the terminal cell, Crété (1963) classified the four-celled stages of proembryos into three major series which form a significant feature of his embryonic classification.

In the classification of embryonal types adopted by Johansen (1950) and Maheshwari (1950), considerable use is made, not only of the plane of division of the terminal cell but also of the degree to which the basal cell of the proembryo contributes to the formation of the embryo proper. In these schemes of classification, the name of each of the major types is derived from that family in which examples of that type are found. Thus, six principal types of embryogeny are recognized: the piperad, onagrad, asterad, caryophyllad, solanad, and chenopodiad types. A further critical review of the various—and often extremely complicated—classifications of embryonal types is beyond the scope and intention of the present book. Although the successive steps in *early* embryogeny of many species which have been studied in detail seem remarkably precise, there is obviously considerable overlapping among the proposed types.

In order to illustrate specifically some of the salient features of embryogenesis typical of certain dicotyledons, we have selected *Capsella bursa-pastoris* (an example of the onagrad type) and *Nicotiana* (an example of the solanad type).

CAPSELLA BURSA-PASTORIS. The embryology of *Capsella* has long served as an instructional model for the discussion of the dicotyledonous embryo. The first division of the zygote yields a *terminal* and *basal* cell (Fig. 20-28, A, B). The basal cell then divides transversely, and a *longitudinal division* follows in the terminal cell. As a result, a four-celled proembryo is developed—the *tetrad* stage (Fig. 20-28, C). It will be necessary at first to trace separately the further division of the upper and lower pairs of cells. Each of the two terminal cells divides longitudinally, resulting in a *quadrant* stage (Fig. 20-28, E). There are then transverse divisions in each of these four cells, yielding the *octant* stage of the embryo (Figs. 20-28, F, G; 20-29, A). A critical histogenetic event then occurs: each of the eight cells divides periclinally into an outer protodermal cell and an inner cell. As a consequence the young embryo proper now consists of eight external protodermal cells (only four cells appear in sectional view), destined to produce by additional anticlinal divisions the embryonic surface layer or epidermis (Fig. 20-28, H, I). The eight internal cells will give rise eventually to the ground meristem and procambial system of the hypocotyl and cotyledons (Fig. 20-28, I, J). The *suspensor* develops from the lower two cells of the tetrad stage (Fig. 20-28, C). The lowermost cell toward the micropyle may fail to divide, but instead progressively enlarges to form a vesicular cell termed the *basal cell* (Figs. 20-28, C–H; 20-31, A–C). A variable number of transverse divisions of the upper cell (next to the embryo proper) and its derivatives produce additional suspensor cells. The suspensor cell in contact with the base of the globular proembryo represents the *hypophysis* (Fig. 20-28, H). This cell plays an important role in the formation of the root apical meristem. By transverse and longitudinal divisions three tiers of cells are produced (Figs. 20-28, H–L; 20-29, C, D). Cells of the upper tier function as cortical initials of the embryonic root, or radicle (Figs. 20-28, L; 20-29, C, D), while cells of the middle tier are common initials of the root protoderm–root cap (Figs. 20-28, L; 20-29, C, D). Central cells at the root apex have been referred to as "initials," but it should be pointed out that they may actually remain somewhat mitotically quiescent, constituting the quiescent center as discussed in Chapter 19, p. 522. Cells of the quiescent center may become reactivated during seed germination but then return to a quiescent condition.

To recapitulate, the young embryo of *Capsella,* as a result of a rather well-coordinated sequence of cell divisions, consists of a spherical group of cells attached to a filamentous suspensor (Fig. 20-28, K). The paired cotyledons arise as two ridges of tissue derived from the distal portion of the embryo (Fig. 20-28, L), and a few cells situated between the bases

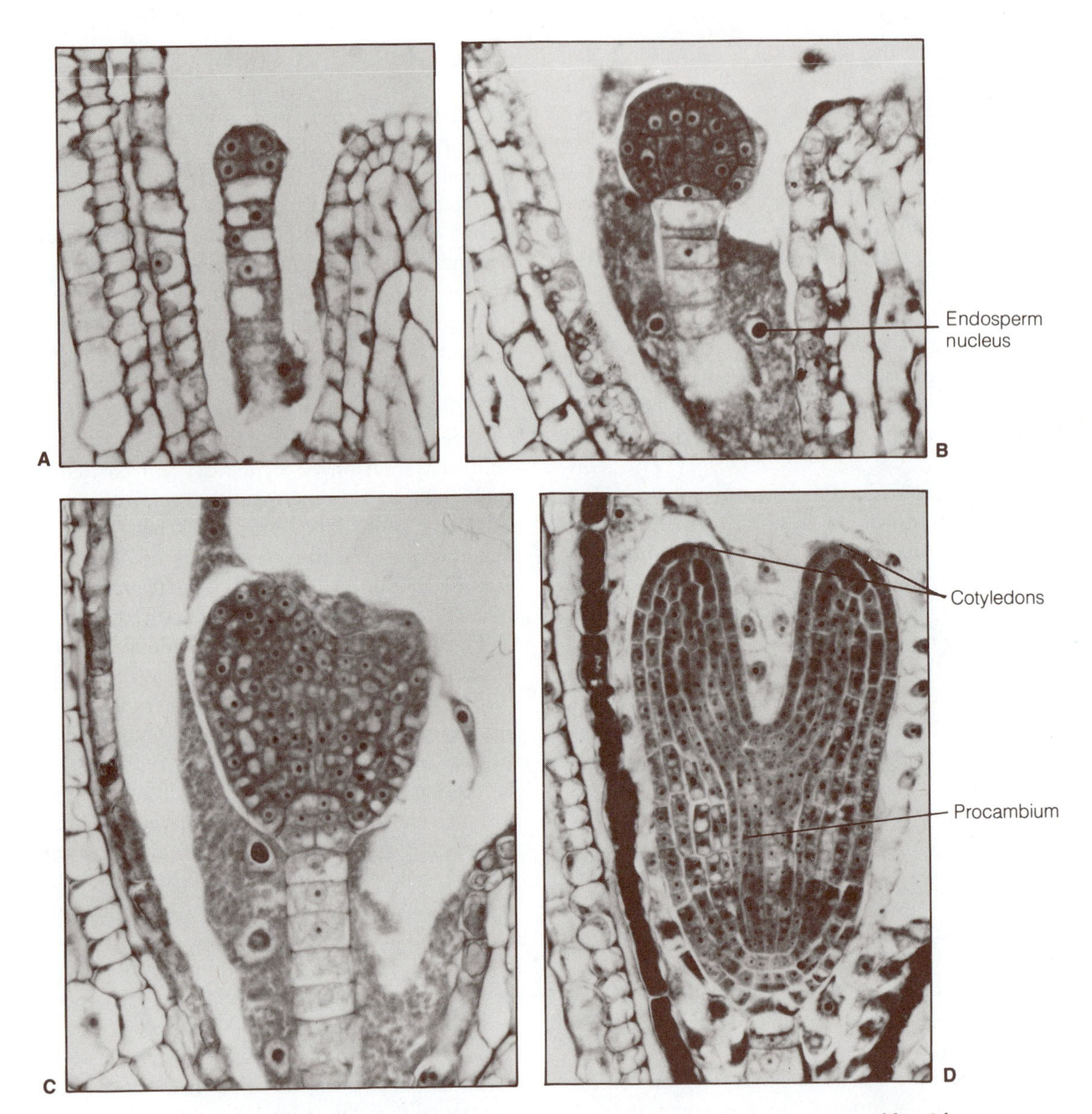

FIGURE 20-29 Embryogeny in *Capsella*. **A,** octant stage of proembryo, comparable with Fig. 20-28, **G; B,** later stage of proembryo with protoderm layer comparable with Fig. 20-28, **I** (note lightly stained hypophysis cell at base of globular embryo); **C,** later stage, showing cell tiers derived from hypophysis and initiation of cotyledons; **D,** embryo with well-developed cotyledon primordia and procambium. Free nuclear stage in endosperm development shown in **B, C;** beginning of cellular endosperm tissue in **D.**

of the cotyledon primordia remain undifferentiated and constitute the future shoot apex (Figs. 20-29, D; 20-30, B). During cotyledon development, the division and differentiation of cells in the lower part of the embryo gradually produce the young axis or hypocotyl of the embryo. At this general stage the embryo is somewhat heart-shaped, as seen in longitudinal section. Continued enlargement of the hypocotyl and cotyledons results in a pronounced curvature of the cotyledons which lie parallel to the

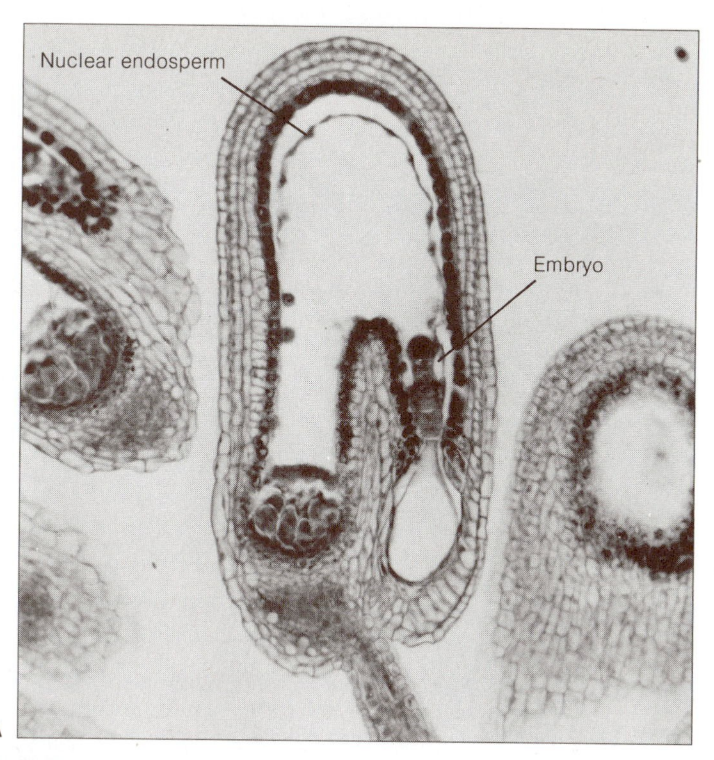

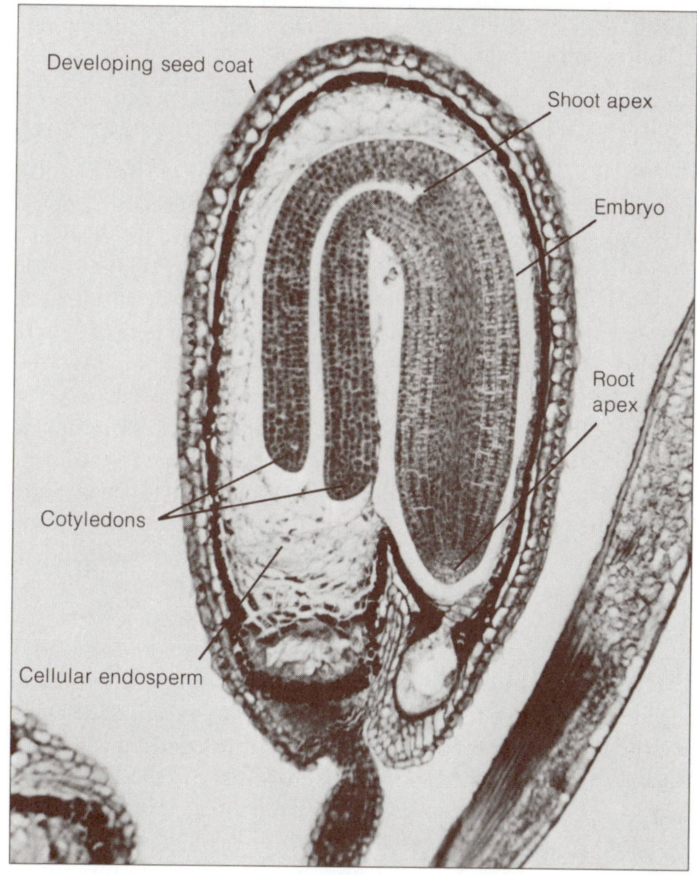

FIGURE 20-30 Longisections of developing seeds in *Capsella*, showing early embryogeny and free nuclear endosperm in **A** and nearly mature embryo and cellular endosperm in **B.**

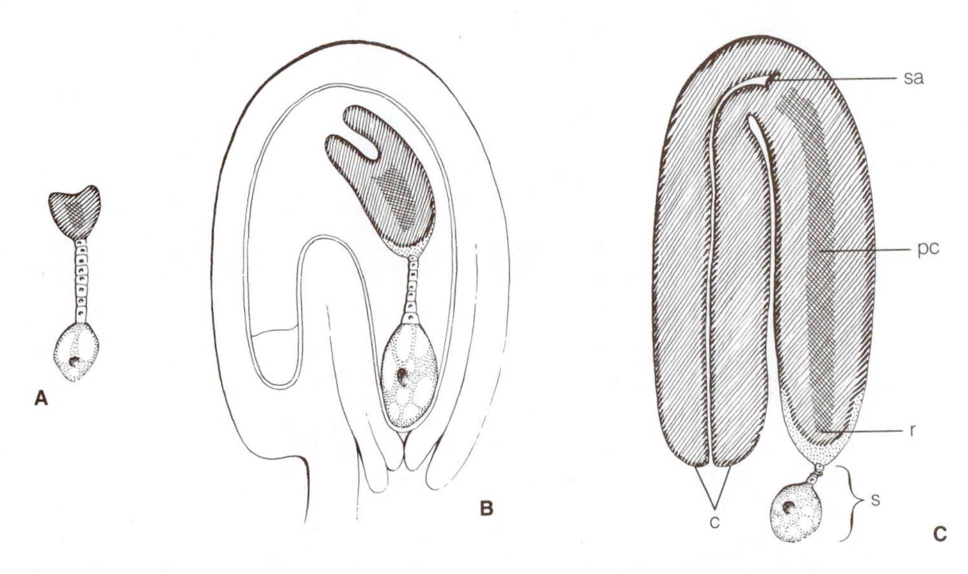

FIGURE 20-31 **A, B,** late stages in embryogeny of *Capsella.* **A,** the cordate form of a longisection of an embryo at stage of initiation of cotyledons; **B,** longisection of developing seed showing orientation and general structure of an embryo with two cotyledons; **C,** longisection of an embryo from a mature seed. c, cotyledons; pc, procambium; r, tip of root; s, suspensor, sa, shoot apex. [**A** and **C** redrawn from Schaffner, *Ohio Nat.* 7:1, 1906; **B** after Bergen and Caldwell and redrawn from *A Textbook of General Botany,* 4th edition, by R. M. Holman and W. W. Robbins. Wiley, New York. 1951.]

axis of the embryo in the mature seed (Fig. 20-31, A–C). [For information on the ultrastructure of the developing embryo, see Schulz and Jensen (1968a, b, c) and Natesh and Rau (1984).]

NICOTIANA. We have already noted that in the current schemes for classifying the varied types of embryogeny considerable importance is attached to the plane of cell division of the terminal cell of the proembryo. In *Capsella* the terminal cell divides longitudinally, and it is primarily on this basis that the embryo is classified by Johansen (1950) as the onagrad type. A contrasting pattern of proembryo development is illustrated by *Nicotiana* because the terminal and basal cell *both* divide *transversely* and thereby produce a linear four-celled (tetrad) proembryo (Fig. 20-32, C); on this basis, Johansen classifies *Nicotiana* as the solanad type. The two upper cells (t^1 and t^2) by successive longitudinal divisions, give rise first to the *quadrant* stage and then to the *octant* stage (Fig. 20-32, D, E). The suspensor becomes four cells in length and the cell next to the embryo proper is the *hypophysis* (Fig. 20-32, E, F). Periclinal division in each of the octant cells separates the protoderm from the internal cells of the proembryo (Fig. 20-32, F). Then the lower internal cells of the hypocotyledonary region, by means of transverse divisions, form the cortical initials of the future root (Fig. 20-32, F–H). Concomitantly, the hypophysis cell, by longitudinal and transverse divisions, produces two tiers of cells that complete the organization of the root apical meristem (Fig. 20–32, F–I). It will be noted, in contrast to *Capsella,* that the hypophysis gives rise only to a set of initials for the root protoderm–root cap.

Two aspects of embryogeny that deserve attention are the morphology and function of suspensors. Suspensors vary in form from a single file of cells with an enlarged basal cell *(Capsella),* to a biseriate column of short cells, to a massive elongate structure 2 to 6 cells wide. In some species the suspensor is haustorial, penetrating the surrounding tissues (e.g., integuments, nucellus). Research has revealed several functions of suspensors. In addition to orienting the embryo, the cell walls of certain investigated species have invaginations that would permit an increase in the absorption of nutrients from surrounding tissues. It has been shown for a number of species that suspensor nuclei undergo endoreduplication of DNA which may be

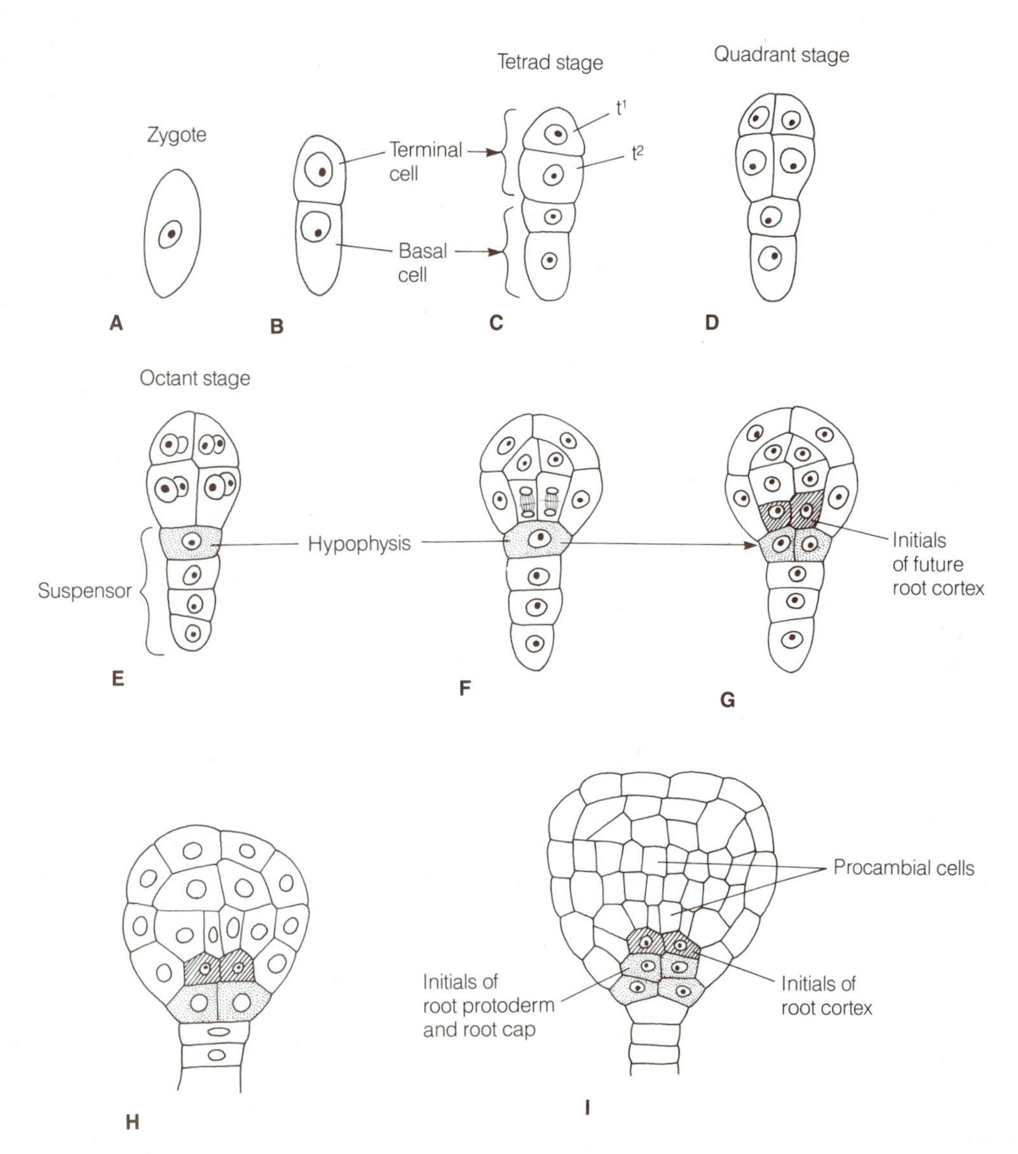

FIGURE 20-32 **A–I,** early development of the embryo of *Nicotiana*. See text for details. [Redrawn from Souèges, *C. R. Acad. Sci.* (Paris) 170:1125, 1920.]

related to the synthesis of gene products related to growth. Also, there is evidence for the production of hormones in the suspensor and the possible transfer to the growing embryo (Raghavan, 1986).

Comparative Embryogeny of Monocotyledons

The differences between the organography of the "mature" embryos of monocotyledons and dicotyledons are striking and have long been used to demarcate these two taxa in the angiosperms. With relatively few exceptions (e.g., three cotyledons, one cotyledon) the embryo typical of dicotyledons develops a pair of lateral cotyledons between the bases of which is situated the terminal shoot apex (Fig. 20-33). In contrast, the embryo of monocotyledons typically produces a single cotyledon and the shoot apex appears to be lateral in position in a well-developed embryo (Fig. 20-33), but see later

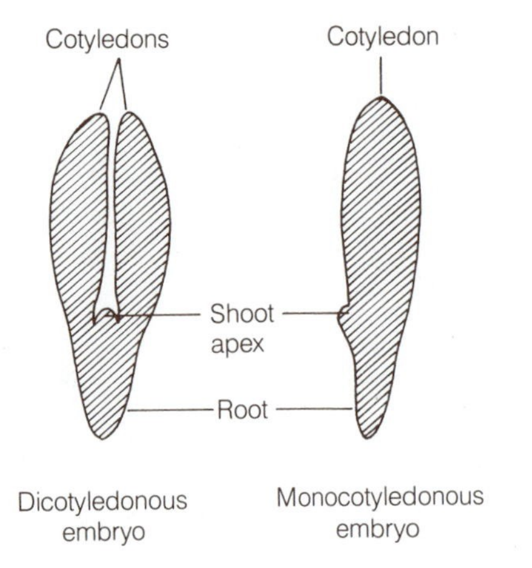

Figure 20-33 Differences between the organography of the "mature" embryo in dicotyledons and monocotyledons.

discussions in this section with regard to the origin of the shoot apex.

One of the most interesting conclusions reached through comparative studies is that the early phases of embryogenesis are very similar in both dicotyledons and monocotyledons, and certain genera have been classified under the same embryonic type (see Crété, 1963).

For many years the embryogenesis of *Sagittaria sagittaefolia* was regarded as "typical" of monocotyledons as a whole. The one cotyledon was interpreted as being terminal in position, and the future shoot apex took its origin laterally from a different, subjacent tier of initials (Souèges, 1931). On the basis of a re-investigation of the same species, Swamy (1980) has shown that the cotyledon and future shoot apex share a common origin from derivatives of the terminal cell. As shown in Fig. 20-34, A, the zygote divides transversely in the formation of a terminal cell and a basal cell. This division separates the future cell lineages leading to the formation of the cotyledon and shoot apex from those producing the hypocotyl-root axis. The cotyledon grows rapidly and overtops the future shoot apex, the cells of which initially do not divide very often (Fig. 20-34, H–J). Procambium is formed initially in the hypocotyl region, and future root initials become evident at the lower end of the hypocotyl.

The long-held dogma that the single cotyledon in monocotyledons is strictly terminal in embryogenesis has been challenged over the years by several investigators. As early as 1952, Haccius (1952, a) showed that both the cotyledon and the shoot apex are differentiated side by side from the terminal portion of the proembryo in *Ottelia alismoides* (Fig. 20-35, A). In other words, the shoot apex is terminal in origin like that of a dicotyledonous embryo, and its subsequent "displacement" to one side results from the early aggressive growth of the single cotyledon (Fig. 20-35, B). Another example of the lateral origin of the cotyledon is that of *Halophila ovata* (Swamy and Lakshmanan, 1962; Lakshmanan, 1972). The cotyledon and shoot apex both arise from derivatives of the terminal tier by division of the *original* terminal cell during embryogenesis (Fig. 20-36). The cotyledon and the future shoot apex arise from groups of adjacent cells (Fig. 20-36, H). The cotyledon, by active growth, soon overtops the shoot apex, the cells of which initially remain somewhat mitotically quiescent (Fig. 20-36, I).

In other monocotyledons such as *Potamogeton indicus* (Potamogetonaceae) and *Pistia stratiotes* (Araceae), the single cotyledon and the shoot apex also jointly arise from the terminal cell tier of the proembryo, but in *Lemna gibba* (Lemnaceae) the cotyledon is reported to be strictly terminal in method of origin (Swamy and Parameswaran, 1962; Haccius and Lakshmanan, 1966).

It is interesting to note that in a number of so-called "monocotyledonous dicotyledons" the single cotyledon arises in a more or less obviously "lateral" position and there is no evidence of a "rudimentary" or "abortive" second cotyledon. Haccius (1954) found complete agreement between the relative position of the shoot apex and the single cotyledon in the young embryos of *Clatonia virginica* (a member of the dicotyledonous family Portulacaceae) and *Ottelia* (a monocotyledon). Additional examples of monocotyledonous dicotyledons are provided by *Pinguicula* (Lentibulariaceae), *Anemone apennina* (Ranunculaceae), certain geophilous members of the Umbelliferae, and *Cyclamen* in the Primulaceae (Haccius and Hartl-Baude, 1957; Haccius and Fischer, 1959; Haccius, 1952b; Haccius and Lakshmanan, 1967).

In summary, recent studies of a number of monocotyledons have shown that the proembryos of dicotyledons and monocotyledons are very simi-

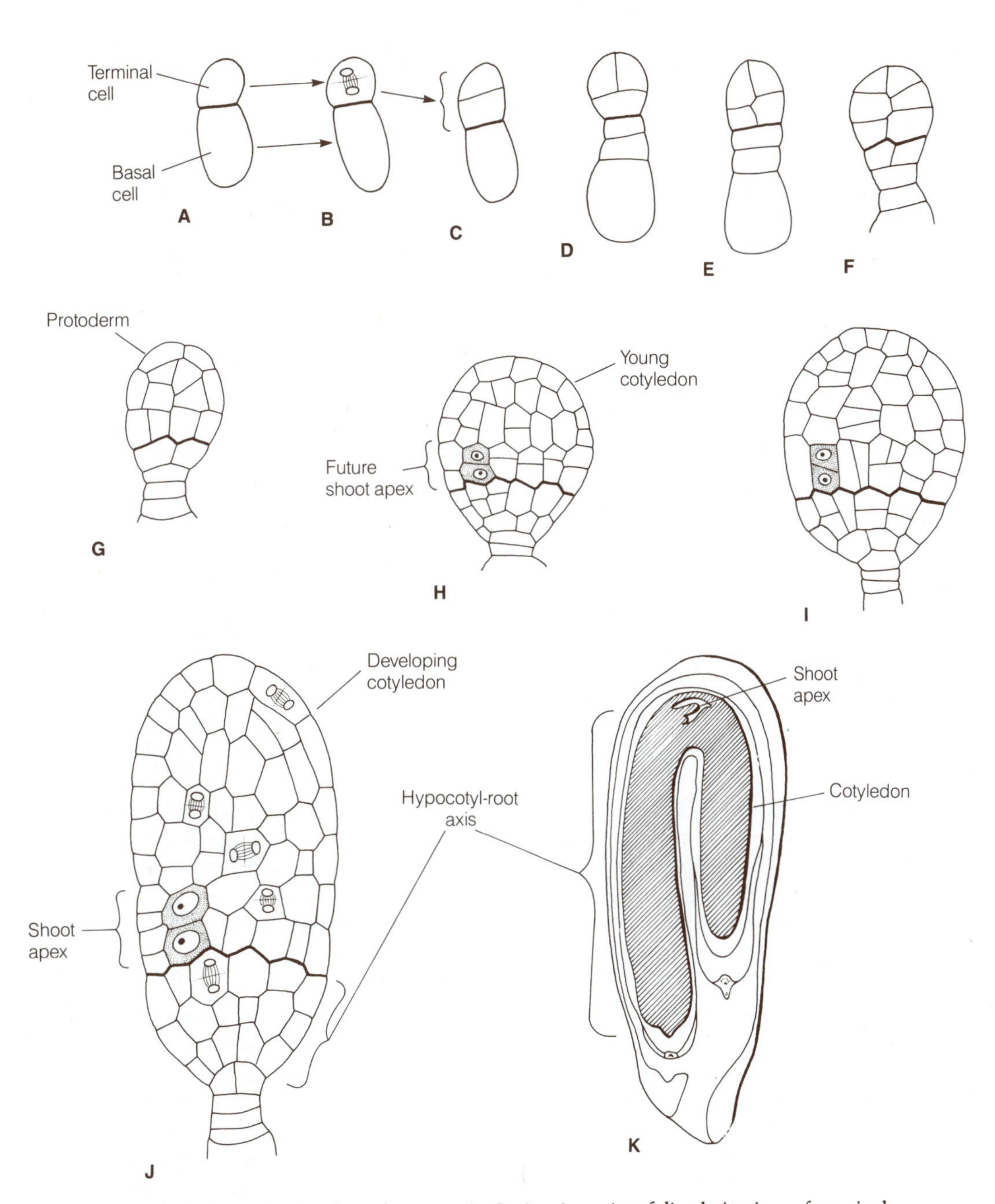

FIGURE 20-34 **A–J,** early embryogeny in *Sagittaria sagittaefolia;* derivatives of terminal cell and basal cell separated by heavy line; **K,** nearly mature embryo in seed. Note that the shoot apex and single cotyledon are derived from derivatives of the original terminal cell. [**A–J,** redrawn from Swamy, *Phytomorphology* 30:204–212, 1980; **K,** redrawn from Schaffner, *Bot. Gaz.* 23:252, 1897.]

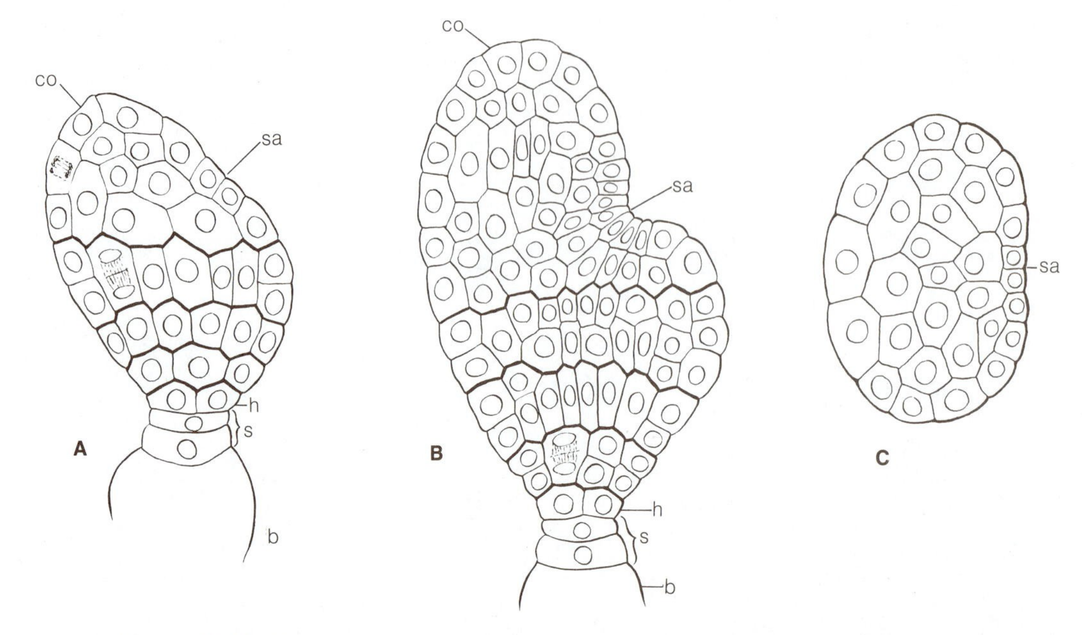

FIGURE 20-35 Young embryos of *Ottelia alismoides.* **A,** common origin of cotyledon primordium and shoot apex from terminal region (demarcated by heavy upper line) of proembryo; **B,** later stage, showing lateral displacement of shoot apex by growth of cotyledon; **C,** transverse section at level of shoot apex of stage represented in **A.** b, basal cell of proembryo; co, primordium of cotyledon; h, hypophysis; s, suspensor; sa, shoot apex. [Redrawn from Haccius, *Planta* 40:443, 1952.]

lar in development up to the globular (sphere) stage. In dicotyledons the median axial cells constitute the future shoot apex and are more-or-less initially quiescent in contrast to the rapid growth of the two lateral cotyledons. In monocotyledons one-half or less of the terminal-cell derivatives have retarded growth, but will function eventually as the shoot apex, whereas the remainder of the cells show a rapid rate of cell division and growth in the formation of the one cotyledon. The seemingly lateral position of the shoot apex in later stages of organogenesis is the result of the rapid growth of the one cotyledon (Natesh and Rau, 1984).

In addition to the problems posed by the position of the cotyledon in the embryos of monocotyledons, there still remains the question of the morphology and phylogeny of the cotyledon. Is this structure the homologue of a single foliar organ or is it in reality a "double" structure, equivalent to the fusion of a pair of cotyledons? It is generally believed that monocotyledons have been derived from primitive dicotyledons. This belief is probably true, but Stebbins (1974) believes that all attempts to derive the monocotyledons from any living order of dicotyledons have failed. Nevertheless, there are several theories with regard to the origin of "monocotyledony." Considerable effort was made by Sargant (1902, 1903) to demonstrate that the solitary cotyledon in monocotyledons represents two congenitally united ancestral cotyledons. A similar view has been expressed by Cronquist (1968). According to the latter's concept, two ancestral cotyledons became fused by their margins toward the base, forming a bilobed, basally tubular, compound cotyledon. In the course of evolution one of the lobes was reduced and lost, still leaving a single cotyledon with a sheathing tubular base. In this connection, Stebbins (1974) has emphasized the adaptive value of a sheathing tubular cotyledon. The cotyledonary tube would provide protection for the shoot apex in geophilous species during seasonal drought or cold in those species that undergo a prolonged period of dormancy soon after early stages of seed germination. A more widely held

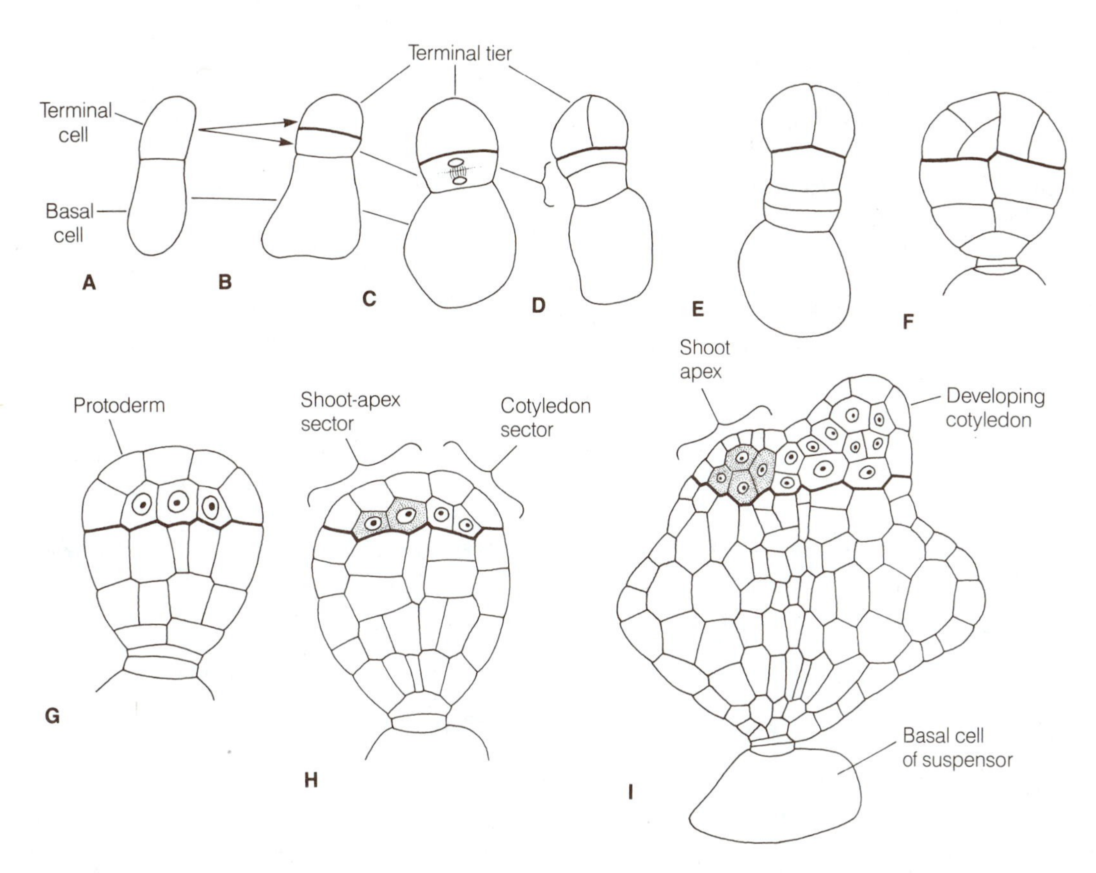

FIGURE 20-36 **A–I,** embryogeny of *Halophila ovata.* Derivatives of the terminal tier and subjacent tiers separated by a heavy line. Note that the shoot apex and cotyledon arise from adjacent groups of cells derived from the terminal tier of the proembryo. See text for details. [Redrawn from Swamy and Lakshmanan, *Ann. Bot.* 26:243, 1962.]

view regards the solitary cotyledon of monocotyledons as a surviving member of an ancestral pair (Eames, 1961; Takhtajan, 1969). The results of all recent ontogenetic studies support this concept. There is no well-documented evidence of "dicotyledony" in living monocotyledons. Of course, ontogenetic studies of living species do not provide conclusive evidence of the phylogeny of an organ in the absence of a fossil record. However, through countless modified ontogenies a second cotyledon may have been lost, and the remaining tubular or sheathing cotyledon assumed an important protective function in monocotyledons, as described previously. This progression may have evolved repeatedly in the history of several lines of monocotyledons.

Pollen Embryogenesis

Pollen grains would seem to be unlikely initials of embryos. However, the formation of embryoids from pollen is now known in more than 170 species distributed in 68 genera and 28 families of angiosperms. The original discovery was made when excised anthers of *Datura innoxia* at the pollen grain stage were cultured in vitro on a mineral salt medium supplemented with certain growth substances. Embryo-like structures appeared on the sides of the anthers. Careful examination revealed that the embryoids did arise from the haploid pollen grains and not diploid tissue of the anther. There are two modes of sporophytic type growth in angiosperms. In the first, the pollen grain directly forms an embryoid. In the second, a callus is formed from which eventually plantlets are formed by organogenesis (production of shoots and roots) or by somatic embryogenesis. Pollen embryogenesis may begin by divisions in the vegetative cell or in other cases the generative cell is the progenitor of an embryoid. There are other variations of embryogenesis. [See Raghavan (1986) for an interesting and informative account of pollen embryogenesis.]

REFERENCES

Arnott, H. J.
1962. The seed, germination, and seedling of *Yucca. Univ. Calif. Publ. Bot.* 35:1–164.

Baker, H. G., and P. D. Hurd, Jr.
1968. Intrafloral ecology. *Ann. Rev. Entomol.* 13:385–414.

Bambacioni, V.
1928a. Ricerche sulla ecologia e sulla embriologia di *Fritillaria persica* L. *Ann. Bot.* 18:7–37.
1928b. Contributo alla embriologia di *Lilium candidum* L. *Rend. Accad. Naz. Lincei* 8:612–618.

Battaglia, E.
1958. L'abolizione del tipo embriologico *Scilla* e la creazione dei nuovi tipi *Endymion* ed *Allium. Caryologia* 11:247–252.

Berlyn, G. P., and J. P. Miksche
1976. *Botanical Microtechnique and Cytochemistry.* The Iowa State University Press, Ames.

Bersier, J. D.
1960. L'ovule anatrope: Ranunculaceae. *Bull. Soc. Bot. Suisse* 70:171–176.

Bhandari, N. N.
1984. The Microsporangium. In B. M. Johri (ed.), *Embryology of Angiosperms,* pp. 53–121. Springer-Verlag, Berlin.

Bocquet, G., and J. D. Bersier
1960. La valeur systématique de l'ovule: développements tératologiques. *Arch. Sci.* (Geneve) 13:475–496.

Boesewinkel, F. D., and F. Bouman
1967. Integument initiation in *Juglans* and *Pterocarya. Acta Bot. Neerl.* 16:86–101.

Boke, N. H.
1949. Development of the stamens and carpels in *Vinca rosea* L. *Amer. Jour. Bot.* 36:535–547.

Borthwick, H. A.
1931. Development of the macrogametophyte and embryo of *Daucus carota. Bot. Gaz.* 92:23–44.

Bouman, F.
1984. The ovule. In B. M. Johri, (ed.), *Embryology of Angiosperms,* pp. 123–157. Springer-Verlag, Berlin.

Brewbaker, J. L.
1967. The distribution and phylogenetic significance of binucleate and trinucleate pollen grains in the angiosperms. *Amer. Jour. Bot.* 54:1069–1083.

Brink, R. A., and D. C. Cooper
1947. The endosperm in seed development. *Bot. Rev.* 13:423–541.

Brown, R. C., and B. E. Lemmon
1989. Microtubules associated with simultaneous cytokinesis of coenocytic microsporocytes. *Amer. J. Bot.* (in press).

Carniel, K.
1963. Das Antherentapetum. Ein kritischer Überblick. *Osterr. Bot. Zeit.* 110:145–176.
1967. Über die Embryobildung in der Gattung *Paeonia. Osterr. Bot. Zeit.* 114:4–19.

Carlquist, S.
1969. Toward acceptable evolutionary interpretations of floral anatomy. *Phytomorphology* 19:332–362.

Cass, D. D., and W. A. Jensen
1970. Fertilization in barley. *Amer. Jour. Bot.* 57:62–70.

Cass, D. D., D. J. Peteya, and B. L. Robertson
1985. Megagametophyte development in *Hordeum vulgare.* 1. Early megagametogenesis and the nature of cell wall formation. *Can. J. Bot.* 63:2164–2171.
1986. Megagametophyte development in *Hordeum vulgare.* 2. Later stages of wall development and morphological aspects of megagametophyte cell differentiation. *Can. J. Bot.* 64:2327–2336.

Cave, M. S., H. J. Arnott, and S. A. Cook
1961. Embryogeny in the California peonies with reference to their taxonomic position. *Amer. Jour. Bot.* 48:397–404.

Chopra, R. N., and R. C. Sachar
1963. Endosperm. In P. Maheshwari (ed.), *Recent Advances in the Embryology of Angiosperms,* pp. 135–170. University of Delhi, Delhi.

Cooper, D. C.
1935. Macrosporogenesis and development of the

embryo sac of *Lilium henryi. Bot. Gaz.* 97:346–355.

Corner, E. J. H.
1966. *The Natural History of Palms.* University of California Press, Berkeley and Los Angeles.

Coulter, J. M., and C. J. Chamberlain
1912. *Morphology of Angiosperms.* Appleton, New York.

Crepet, W. L.
1979. Some aspects of the pollination biology of Middle Eocene angiosperms. *Rev. Palaeobot. Palynol.* 27:213–238.

Crété, P.
1963. Embryo. In Maheshwari, P. (ed.), *Recent Advances in the Embryology of Angiosperms,* pp. 171–220. University of Delhi, Delhi.

Cronquist, A.
1968. *The Evolution and Classification of Flowering Plants.* Houghton Mifflin, Boston.

D'Amato, F.
1984. Role of polyploidy in reproductive organs and tissues. In B. M. Johri (ed.), *Embryology of Angiosperms,* pp. 519–566. Springer-Verlag, Berlin.

Dahlgren, K. V. O.
1927. Die Morphologie des Nuzellus mit besonderer Berücksichitigung der deckzellosen Typen. *Jahr. Wiss. Bot.* 67:347–426.

Davis, G. L.
1966. *Systematic Embryology of the Angiosperms.* Wiley, New York.

Dilcher, D. L.
1979. Early angiosperm reproduction: an introductory report. *Rev. Palaeobot. Palynol.* 27:291–328.

Eames, A. J.
1961. *Morphology of the Angiosperms.* McGraw-Hill, New York.

Esau, K.
1965. *Plant Anatomy.* Wiley, New York.
1969. *The Phloem.* (*Handbuch der Pflanzenanatomie,* Band 5, Teil 2.) Gebrüder Borntraeger, Berlin.

Faegri, K., and L. van der Pijl
1979. *The Principles of Pollination Ecology,* 3d edition. Pergamon, London.

Fahn, A.
1982. *Plant Anatomy,* 3d edition. Pergamon, Oxford.

Fisher, D. B., and W. A. Jensen
1969. Cotton embryogenesis: the identification, as nuclei, of the X-bodies in the degenerated synergid. *Planta* 84:122–133.

Grant, V.
1950. The protection of the ovules in flowering plants. *Evolution* 4:179–201.

Haccius, B.
1952a. Die Embryoentwicklung bei *Ottelia alismoides* und das Problem des terminalen Monokotylen—Keimblatts. *Planta* 40:443–460.
1952b. Verbreitung und Ausbildung der Einkeimblättrigkeit bei den Umbelliferen. *Osterr. Bot. Zeit.* 99:483–505.
1954. Embryologische und histogenetische Studien an "monokotylen Dikotylen." I. *Claytonia virginica. Osterr. Bot. Zeit.* 101:285–303.

Haccius, B., and E. Fischer
1959. Embryologische und histogenetische Studien an "monokotylen Dikotylen." III. *Anemone apennina* L. *Osterr. Bot. Zeit.* 106:373–389.

Haccius, B., and E. Hartl-Baude
1957. Embryologische und histogenetische Studien an "monokotylen Dikotylen." II. *Pinguicula vulgaris* L. und *P. alpina* L. *Osterr. Bot. Zeit.* 103:567–587.

Haccius, B., and K. K. Lakshmanan
1966. Vergleichende Untersuchung der Entwicklung von Kotyledon und Spross-scheitel bei *Pistia stratiotes* und *Lemna gibba,* ein Beitrag zum Problem der sogenannten terminalen Blattorgane. *Beitr. Biol. Pflanzen* 42:425–443.
1967. Experimental studies on monocotyledonous dicotyledons: phenylboric acid-induced "dicotyledonous" embryos in *Cyclamen persicum. Phytomorphology* 17:488–494.

Heslop-Harrison, J.
1966. Cytoplasmic continuities during spore formation in flowering plants. *Endeavour* 25:65–72.
1975. Incompatibility and the pollen-stigma interaction. *Ann. Rev. Plant Physiol.* 26:403–425.

Heslop-Harrison, Y., and K. R. Shivanna
1977. The receptive surface of the angiosperm stigma. *Ann. Bot.* 41:1233–1258.

Jacobsen, J. V.
1984. The seed: germination. In B. M. Johri (ed.), *Embryology of Angiosperms,* pp. 611–646. Springer-Verlag, Berlin.

Jensen, W. A.
1962. *Botanical Histochemistry. Principles and Practice.* W. H. Freeman and Company, San Francisco.

1972. The embryo sac and fertilization in angiosperms. *University of Hawaii, Harold L. Lyon Arboretum Lecture,* No. 3.
1973. Fertilization in flowering plants. *Bio-Science* 23:21–27.

Johansen, D. A.
1950. *Plant Embryology.* Chronica Botanica, Waltham, Mass.

Johri, B. M.
1963. Female gametophyte. In P. Maheshwari (ed.), *Recent Advances in the Embryology of Angiosperms,* pp. 69–103. University of Delhi, Delhi.
1984. *Embryology of Angiosperms.* Springer-Verlag, Berlin.

Johri, B. M., and K. B. Ambegaokar
1984. Embryology: then and now. In B. M. Johri (ed.), *Embryology of Angiosperms,* pp. 1–52, Springer-Verlag, Berlin.

Johri, B. M., and P. S. Rao
1984. Experimental embryogeny. In B. M. Johri (ed.), *Embryology of Angiosperms,* pp. 735–802. Springer-Verlag, Berlin.

Jones, H. A., and S. L. Emsweller
1936. Development of the flower and macrogametophyte of *Allium cepa. Hilgardia* 10:415–423.

Kapil, R. N.
1960. Embryology of *Acalypha* Linn. *Phytomorphology* 10:174–184.

Kapil, R. N., and S. Jalan
1964. *Schisandra michaux*—its embryology and systematic position. *Bot. Notis.* 117:285–306.

Kapil, R. N., and I. K. Vasil
1963. Ovule. In P. Maheshwari (ed.), *Recent Advances in the Embryology of Angiosperms.* University of Delhi, Delhi.

Knox, R. B.
1979. Flower. In *McGraw-Hill Yearbook of Science and Technology,* pp. 198–200. McGraw-Hill, New York.
1984. Pollen-pistil interactions. In H. F. Linskens, and J. Heslop-Harrison (eds.), *Encyclopedia of Plant Physiology,* Vol. 17, *Cellular Interactions,* pp. 508–608. Springer-Verlag, Berlin.

Kühn, G.
1928. Beiträge zur Kenntnis der intraseminalen Leitbündel bei den Angiospermen. *Bot. Jahrb.* 61:325–379.

Lakshmanan, K. K.
1972. Monocot embryo. In T. M. Varghese and R. K. Grover (eds.), *Vistas in Plant Sciences,* Vol. 2, pp. 61–110. Searchmates Publications, India.

Laser, K. D., and N. R. Lersten
1972. Anatomy and cytology of microsporogenesis in cytoplasmic male sterile angiosperms. *Bot. Rev.* 38:425–454.

LeMonnier, G.
1872. Recherches sur la nervation de la graine. *Ann. Sci. Nat. Bot.* Ser. 5. 16:233–305.

Maheshwari, P.
1950. *An Introduction to the Embryology of Angiosperms.* McGraw-Hill, New York.
1960. Evolution of the ovule. In *A. C. Seward Memorial Lectures,* Ser. 7, pp. 1–13. Birbal Sahni Institute of Palaeobotany, Lucknow.

Maheshwari, S. C.
1955. The occurrence of bisporic embryo sacs in angiosperms—a critical review. *Phytomorphology* 5:67–99.

Martens, P.
1962. Études sur les Gnétales. VI. Recherches sur *Welwitschia mirabilis.* III. L'ovule et le sac embryonnaire. *Cellule* 63:309–329.

Matthiessen, A.
1962. A contribution to the embryogeny of *Paeonia. Acta Horti Bergiani* 20:57–61.

Murgai, P.
1959. The development of the embryo in *Paeonia. Phytomorphology* 9:275–277.
1962. Embryology of *Paeonia* together with a discussion on its systematic position. In *Plant Embryology: A Symposium,* pp. 215–223. Council of Scientific and Industrial Research, New Delhi.

Nast, C. G.
1941. The embryogeny and seedling morphology of *Juglans regia* L. *Lilloa* 6:163–205.

Natesh, S., and M. A. Rau
1984. The embryo. In B. M. Johri (ed.), *Embryology of Angiosperms,* pp. 377–443. Springer-Verlag, Berlin.

Periasamy, K.
1962. The ruminate endosperm. Development and types of rumination. In *Plant Embryology: A Symposium,* pp. 62–74. Council of Scientific and Industrial Research, New Delhi.

Periasamy, K., and N. Parameswaran
1965. A contribution to the floral morphology and embryology of *Tarenna asiatica. Beitr. Biol. Pflanzen* 41:123–138.

Periasamy, K., and B. G. L. Swamy
1959. Studies in the Annonaceae. I. Microsporogenesis in *Cananga odorata* and *Miliusa wightiana*. *Phytomorphology* 9:251–263.

Peter, J.
1920. Zur Entwicklungsgeschichte einiger Calycanthaceen. *Beitr. Biol. Pflanzen* 14:59–86.

Proctor, M., and P. Yeo
1972. *The Pollination of Flowers*. Taplinger, New York.

Raghavan, V.
1986. *Embryogenesis in Angiosperms: A Developmental and experimental study*. Cambridge University Press, Cambridge.

Roth, I.
1957. Die Histogenese der Integumente von *Capsella bursa-pastoris* und ihre morphologische Deutung. *Flora* 145:212–235.

Russell, S. D.
1982. Fertilization in *Plumbago zeylanica:* entry and discharge of the pollen tube in the embryo sac. *Can. J. Bot.* 60:2219–2230.

Russell, S. D., and D. D. Cass
1981. Ultrastructure of the sperms of *Plumbago zeylanica*. 1. Cytology and association with the vegetative nucleus. *Protoplasma* 107:85–107.
1983. Unequal distribution of plastids and mitochondria during sperm cell formation in *Plumbago zeylanica*. In D. L. Mulcahy and E. Ottaviano (eds.), *Pollen: Biology and Implications for Plant Breeders,* pp. 135–140. Elsevier Biomedical, New York.

Sampson, F. B.
1963. The floral morphology of *Pseudowintera,* the New Zealand member of the vesselless Winteraceae. *Phytomorphology* 13:403–423.
1969. Cytokinesis in pollen mother cells of angiosperms, with emphasis on *Laurelia novae-zelandiae* (Monimiaceae). *Cytologia* 34:627–634.
1970. Unusual features of cytokinesis in meiosis of pollen mother cells of *Pseudowintera traversii* (Buchan.) Dandy (Winteraceae). *Beitr. Biol. Pflanzen* 47:71–77.

Sargant, E.
1900. Recent work on the results of fertilization in angiosperms. *Ann. Bot.* 14:689–712.
1902. The origin of the seed-leaf in monocotyledons. *New Phytol.* 1:107–113.
1903. A theory of the origin of monocotyledons, founded on the structure of their seedlings. *Ann. Bot.* 17:1–92.

Schnarf, K.
1929. *Embryologie der Angiospermen. (Handbuch der Pflanzenanatomie,* Band 10, Teil 2.) Gebrüder Borntraeger, Berlin.

Schulz, Sister Richardis, and W. A. Jensen
1968a. *Capsella* embryogenesis: the synergids before and after fertilization. *Amer. Jour. Bot.* 55:541–552.
1968b. *Capsella* embryogenesis: the egg, zygote, and young embryo. *Amer. Jour. Bot.* 55:807–819.
1968c. *Capsella* embryogenesis: the early embryo. *Jour. Ultrastruct. Res.* 22:376–392.

Smith-White, S.
1959. Pollen development patterns in the Epacridaceae. A problem in cytoplasm-nucleus interaction. *Proc. Linn. Soc. New South Wales* 84:8–35.

Souèges, R.
1931. L'embryon chez le *Sagittaria sagittaefolia* L. Le cone végétatif de la tige et l'extrémité radiculaire chez les monocotylédones. *Ann. Sci. Nat. Bot.*, Ser. 10, 13:353–402.

Sporne, K. R.
1954. A note on nuclear endosperm as a primitive character among dicotyledons. *Phytomorphology* 4:275–278.
1967. Nuclear endosperm: an enigma. *Phytomorphology* 17:248–251.
1980. A re-investigation of character correlations among dicotyledons. *New Phytol.* 85:419–449.

Stebbins, G. L.
1974. *Flowering Plants: Evolution above the Species Level.* Harvard University Press, Cambridge.

Steward, F. C.
1968. *Growth and Organization in Plants.* Addison-Wesley, Reading, Mass.

Stewart, W. N.
1983. *Paleobotany and the Evolution of Plants.* Cambridge University Press, Cambridge.

Strasburger, E.
1879. *Die Angiospermen und die Gymnospermen.* Jena.
1900. Einige Bemerkungen zur Frage nach der "doppelten Befruchtung" bei Angiospermen. *Bot. Zeit.* II. 58:293–316.

Sussman, R. W., and P. H. Raven
1978. Pollination by lemurs and marsupials: an archaic coevolutionary system. *Science* 200:731–736.

Swamy, B. G. L.
1949a. Further contributions to the morphology of

the Degeneriaceae. *Jour. Arnold Arboretum* 30:10–38.

1949b. Embryological studies in the Orchidaceae. *Amer. Midland Natur.* 41:184–201.

1952. Some aspects in the embryology of *Zygogynum Bailloni* V. Tiegh. *Proc. Nat. Inst. Sci. India* 18:399–406.

1980. Embryogenesis in *Sagittaria sagittaefolia. Phytomorphology* 30:204–212.

Swamy, B. G. L., and K. V. Krishnamurthy
1973. The helobial endosperm: a decennial review. *Phytomorphology* 23:74–79.

Swamy, B. G. L., and K. K. Lakshmanan
1962. The origin of epicotylary meristem and cotyledon in *Halophila ovata* Gaudich. *Ann. Bot.* n.s. 26:243–249.

Swamy, B. G. L., and N. Parameswaran
1962. On the origin of cotyledon and epicotyl in *Potamogeton indicus. Osterr. Bot. Zeit.* 109:344–349.
1963. The helobial endosperm. *Biol. Rev.* 38:1–50.

Takhtajan, A.
1969. *Flowering Plants: Origin and Dispersal.* Translated from the Russian by C. Jeffrey. Oliver and Boyd, Edinburgh.

Van Heel, W. A., and F. Bouman
1972. Note on the early development of the integument in some Juglandaceae together with some general questions on the structure of angiosperm ovules. *Blumea* 20:155–159.

Van Tieghem, P.
1898. Structure de quelques ovule et parti qu'on peut tirer pour améliorer la classification. *Jour. Bot.* 12:197–220.

Venkata Rao, C.
1961. Pollen types in the Epacridaceae. *Jour. Indian Bot. Soc.* 40:409–423.

Vijayaraghavan, M. R., and K. Prabhakar
1984. The endosperm. In B. M. Johri (ed.), *Embryology of Angiosperms,* pp. 319–376. Springer-Verlag, Berlin.

Walters, J. L.
1962. Megasporogenesis and gametophyte selection in *Paeonia californica. Amer. Jour. Bot.* 49:787–794.

Waterkeyn, L.
1954. Études sur les Gnétales. I. Le strobile femelle, l'ovule et la graine de *Gnetum africanum. Cellule* 56:105–146.

Went, J. L. van, and M. T. M. Willemse
1984. Fertilization. In B. M. Johri (ed.), *Embryology of Angiosperms,* pp. 273–317. Springer-Verlag, Berlin.

Willemse, M. T. M., and J. L. van Went
1984. The female gametophyte. In B. M. Johri (ed.), *Embryology of Angiosperms,* pp. 159–196. Springer-Verlag, Berlin.

Wunderlich, R.
1966. Zur Deutung der eigenartigen Embryoentwicklung von *Paeonia. Osterr. Bot. Zeit.* 113:395–407.

Yakovlev, M. S.
1967. Polyembryony in higher plants and principles of its classification. *Phytomorphology* 17:278–282.

Yakovlev, M. S., and M. D. Yoffe
1957. On some peculiar features in the embryogeny of *Paeonia. Phytomorphology* 7:74–82.

Index